WITHDRAWN

W9-CQZ-408

PATHOPHYSIOLOGY

Altered Regulatory Mechanisms
In Disease

Contributors

François M. Abboud
J. Wesley Alexander
Frederic C. Bartter
Michael J. Brody
Thomas A. Bruce
T. E. Bynum
William A. Cain
Reuben M. Cherniack
James Christensen
William J. Crowley, Jr.
Horace W. Davenport
Leo V. DiCara
Harriet P. Dustan
Leonard P. Eliel
William W. Faloon
Joanne Finstad
Edward D. Frohlich
Ann E. Gabrielsen
Robert A. Good
Clarence A. Guenter
James W. Hampton
Lerner B. Hinshaw
Alan F. Hofmann
Eugene D. Jacobson
Edward J. Lennon
Robert D. Lindeman
Mortimer B. Lipsett

Robert B. Livingston
Ove Lundgren
John F. Maher
Farahe Maloof
Victor J. Marder
Donald B. Martin
H. Page Mauck, Jr.
Robert A. Mitchell
Gordon K. Moe
Don H. Nelson
Karl D. Nolph
Arthur R. Page
Solomon Papper
Amadeo J. Pesce
Richard J. Pickering
Victor E. Pollak
Oscar D. Ratnoff
Eugene D. Robin
Sami I. Said
Sol Sherry
Lawrence M. Simon
William O. Smith
Konrad H. Soergel
Robert C. Tarazi
Carlos A. Vaamonde
Walter H. Whitcomb
Stewart G. Wolf

PATHOPHYSIOLOGY
Altered Regulatory Mechanisms In Disease

Edited by

Edward D. Frohlich, M.D.

*Professor of Medicine and of Physiology and
Biophysics and Director, Division of Hypertension,
Department of Medicine, The University of Oklahoma
Medical Center, Oklahoma City, Oklahoma*

J. B. Lippincott Company
Philadelphia · *Toronto*

Copyright © 1972, by J. B. Lippincott Company

This book is fully protected by copyright and, with the
exception of brief excerpts for review, no part of it may be
reproduced in any form, by print, photoprint, microfilm,
or by any other means, without the written permission of
the publishers.

Distributed in Great Britain by
Blackwell Scientific Publications
Oxford and Edinburgh

ISBN-0-397-50285-O

Library of Congress Catalog Card Number 75-157909

Printed in the United States of America

1 3 5 7 6 4 2

Dedication

The clinical investigator should not be considered the young physician of the "shining ivory tower" who is unconcerned with reality and practicality. To the contrary, he is any physician, deeply concerned with mechanisms of disease, the vastly complicated inter-relationships of function in the total human being, and a refocusing of knowledge to the specific management of his patient. This concern for the "why's" and "how's" of disease is as old as Medicine itself; and indeed, this is its life blood and vitality.

It is for these reasons that we dedicate this volume to continued basic and clinical knowledge—to our esteemed clinical investigators of the Past and to those of the Future.

Contributors

François M. Abboud, M.D.

Professor of Medicine and Director of Cardiovascular Division, Department of Internal Medicine, College of Medicine, University of Iowa, Iowa City, Iowa

J. Wesley Alexander, M.D., Sc.D.

Associate Professor of Surgery and Director of the Transplantation Division, Department of Surgery, University of Cincinnati Medical Center, Cincinnati, Ohio

Frederic C. Bartter, M.D.

Chief, Endocrinology Branch, National Heart and Lung Institute, National Institutes of Health, Bethesda, Maryland

Michael J. Brody, Ph.D.

Professor of Pharmacology, Department of Pharmacology, College of Medicine, University of Iowa, Iowa City, Iowa

Thomas A. Bruce, M.D.

Professor of Medicine and Chief of Cardiovascular Section, Department of Medicine, The University of Oklahoma Medical Center, Oklahoma City, Oklahoma

T. E. Bynum, M.D.

Assistant Professor of Medicine and of Physiology and Biophysics, The University of Oklahoma Medical Center, Oklahoma City, Oklahoma

William A. Cain, Ph.D.

Assistant Professor of Microbiology and Immunology and Associate Professor of Research Medicine, The University of Oklahoma Medical Center, Oklahoma City, Oklahoma

Reuben M. Cherniack, M.D.

Professor of Medicine, University of Manitoba; Medical Director, D. A. Stewart Center for the Study and Treatment of Respiratory Disease; Director, Joint Respiratory Programme, University of Manitoba and Sanitorium Board of Manitoba; and Director, Respiratory Division, Clinical Investigation Unit, Winnipeg General Hospital, Winnipeg, Manitoba, Canada

James Christensen, M.D.

Associate Professor of Medicine, Department of Internal Medicine, The University of Iowa School of Medicine, Iowa City, Iowa

William J. Crowley, Jr., M.D.

Associate Professor of Neurology, The University of Oklahoma Medical Center, Oklahoma City, Oklahoma

Horace Davenport, D.Sc.

Professor and Chairman, Department of Physiology, The University of Michigan, Ann Arbor, Michigan

Leo V. DiCara, Ph.D.

Professor of Psychiatry and Psychology, University of Michigan Medical Center, Ann Arbor, Michigan

Harriet P. Dustan, M.D.

Staff Member, Research Division, Cleveland Clinic, Cleveland, Ohio

Leonard P. Eliel, M.D.

Professor of Medicine and Executive Vice-President, The University of Oklahoma Medical Center, Oklahoma City, Oklahoma

William W. Faloon, M.D.

Chief of Medicine, Highland Hospital of Rochester, and Professor of Medicine, University of Rochester School of Medicine and Dentistry, Rochester, New York

Joanne Finstad, M.S.

Pediatric Research Associate, Variety Club Heart Hospital, University of Minnesota School of Medicine, Minneapolis, Minnesota.

Edward D. Frohlich, M.D.

Professor of Medicine and of Physiology and Biophysics and Director, Division of Hypertension, Department of Medicine, The University of Oklahoma Medical Center, Oklahoma City, Oklahoma

Ann E. Gabrielsen, Ph.D.

Principal Research Scientist, N. Y. State Kidney Disease Institute and Department of Pediatrics, Albany Medical College, Albany, New York

Robert A. Good, Ph.D., M.D.

Professor of Pathology, Pediatrics, and Microbiology and Chairman, Department of Pathology, University of Minnesota School of Medicine, Minneapolis, Minnesota

Clarence A. Guenter, M.D.

Professor of Medicine and of Physiology and Biophysics, and Chief of Pulmonary Section, Department of Medicine, The University of Oklahoma Medical Center, Oklahoma City, Oklahoma

James W. Hampton, M.D.

Professor of Medicine and Chief of Hematology Section, Department of Medicine, The University of Oklahoma Medical Center, Oklahoma City, Oklahoma

Lerner B. Hinshaw, Ph.D.

Professor of Physiology and Biophysics, Research Professor of Surgery, The University of Oklahoma Medical Center, and Physiologist, Veterans Administration Hospital, Oklahoma City, Oklahoma

Alan F. Hofmann, M.D.

Associate Director, Gastroenterology Unit, Professor of Physiology and Medicine, Mayo Graduate School of the University of Minnesota, Rochester, Minnesota

Eugene D. Jacobson, M.D.

Professor and Chairman, Department of Physiology and Biophysics, and Associate Professor of Medicine, The University of Oklahoma Medical Center, Oklahoma City, Oklahoma

Edward J. Lennon, M.D.

Professor of Medicine, Department of Medicine, and Associate Dean for Clinical Affairs, Medical College of Wisconsin, Milwaukee, Wisconsin

Robert D. Lindeman, M.D.

Professor of Medicine and of Physiology and Biophysics, Chief of Renal Section, Department of Medicine, The University of Oklahoma Medical Center, and Associate Director for Professional Services, Veterans Administration Hospital, Oklahoma City, Oklahoma

Mortimer B. Lipsett, M.D.

Associate Scientific Director, Reproductive Biology, National Institute of Child Health and Human Development, Bethesda, Maryland

Robert B. Livingston, M.D.

Professor of Neurosciences, The University of California at San Diego, La Jolla, California

Ove Lundgren, M.D.

Department of Physiology, University of Göteborg, Göteborg, Sweden

John F. Maher, M.D.

Professor of Medicine and Director of Division of Nephrology, Department of Medicine, University of Missouri Medical Center, Columbia, Missouri

Farahe Maloof, M.D.

Chief, Thyroid Clinic, Massachusetts General Hospital; Assistant Professor of Medicine, Harvard Medical School, and Adjunct Associate Professor of Biochemistry, Brandeis University, Boston, Massachusetts

Victor J. Marder, M.D.

Associate Professor of Medicine, Department of Medicine, Temple University, School of Health Sciences, Philadelphia, Pennsylvania

Donald B. Martin, M.D.

Chief, Diabetes Unit, Massachusetts General Hospital and Associate Professor of Medicine, Harvard Medical School, Boston, Massachusetts

H. Page Mauck, Jr., M.D.

Associate Professor of Medicine and Pediatrics, Medical College of Virginia, Health Sciences Center, Virginia Commonwealth University, Richmond, Virginia

Robert A. Mitchell, M.D.

Associate Professor of Physiology and Medicine (Anesthesia) and Staff Member of the Cardiovascular Research Institute, University of California, San Francisco, California

Gordon K. Moe, M.D., Ph.D.

Director, Masonic Medical Research Laboratory, Utica, New York, and Professor of Physiology, State University of New York, Syracuse, New York

Don H. Nelson, M.D.

Chief of Medicine, Latter-Day Saints Hospital, and Professor of Medicine, The University of Utah College of Medicine, Salt Lake City, Utah

Karl D. Nolph, M.D.

Assistant Professor of Medicine, University of Missouri School of Medicine, Columbia, Missouri

Arthur R. Page, M.D.

Associate Professor of Pediatrics, Department of Pediatrics, University of Minnesota School of Medicine, Minneapolis, Minnesota

Solomon Papper, M.D.

Professor of Medicine, University of Colorado School of Medicine and Chairman, Department of Medicine, General Rose Memorial Hospital, Denver, Colorado

Amadeo J. Pesce, Ph.D.

Director, Biochemical Division, Michael Reese Hospital and Medical Center, Adjunct Associate Professor, Illinois Institute of Technology and Established Investigator, American Heart Association, Chicago, Illinois

Richard J. Pickering, M.D.

Research Physician, New York State Kidney Disease Institute and Associate Professor of Pediatrics, Albany Medical College, Albany, New York

Victor E. Pollak, M.D.

Director, Renal Division, Michael Reese Hospital and Medical Center and Professor of Medicine, Pritzker School of Medicine, University of Chicago, Chicago, Illinois

Oscar D. Ratnoff, M.D.

Professor of Medicine, Case Western Reserve University School of Medicine and University Hospitals of Cleveland and Career Investigator, American Heart Association, Cleveland, Ohio

Eugene D. Robin, M.D.

Professor of Medicine and Physiology, Stanford University School of Medicine, Stanford, California

Sami I. Said, M.D.

Professor of Medicine, Medical College of Virginia, Health Sciences Center, Virginia Commonwealth University, Richmond, Virginia

Sol Sherry, M.D.

Professor and Chairman, Department of Medicine, Temple University School of Health Sciences, Philadelphia, Pennsylvania

Lawrence M. Simon, M.D.

Assistant Professor of Medicine and Physiology, Stanford University School of Medicine, Stanford, California

William O. Smith, M.D.

Professor of Medicine and Vice Chairman, Department of Medicine, the University of Oklahoma Medical Center, Oklahoma City, Oklahoma

Konrad H. Soergel, M.D.

Professor of Medicine and Chief, Section of Gastroenterology, Department of Medicine, Medical College of Wisconsin, Milwaukee, Wisconsin

Robert C. Tarazi, M.D.

Staff Member, Research Division, Cleveland Clinic, Cleveland, Ohio

Carlos A. Vaamonde, M.D.

Associate Professor of Medicine, University of Miami School of Medicine and Chief, Nephrology Section, Veterans Administration Hospital, Miami, Florida

Walter H. Whitcomb, M.D.

Associate Professor of Medicine, The University of Oklahoma Medical Center and Chief of Staff, Veterans Administration Hospital, Oklahoma City, Oklahoma

Stewart G. Wolf, M.D.

Director, Marine Biomedical Institute, University of Texas, Galveston, Texas and Professor of Medicine and of Physiology and Biophysics, The University of Oklahoma Medical Center, Oklahoma City, Oklahoma

Foreword

Countless biomedical scientists—physiologists, biochemists, pharmacologists and physicians—have attempted to breathe enthusiasm into teaching clinicians basic science. With what degree of success no one knows, except a few clinicians. I say "a few" because only a few know how to measure success.

In my view, when each patient's problem conjures in the physician's mind logical basic mechanisms which may describe how the patient "got that way" and how he may get out of it, success has been achieved. Medicine is then no longer a vocation, an application of cook-book directions, an uninspired routine of diagnosing and treating instead of a true profession in which the questions why and how may be answered. The extraordinary experiments that both nature and man perform otherwise go unnoticed, the chance to add to medical knowledge is lost, and the first two years of medical school are a waste.

Basic and applied sciences are exciting, albeit demanding. Clinical medicine is equally rewarding but each is incomplete without the other.

Much of the failure of the two sides of this coin to appreciate the other lies in the attitudes of clinicians and scientists. The former tend to disdain what they think to be the needless complexities of science. They imagine themselves much too busy curing people to be bothered with science. And the scientists imagine the clinician too dull and unappreciative to be taught. It seems so much more fun to overwhelm him with erudition, and not bother with the difficult problem of simplicity, clarity and screening of the relevant from the irrelevant.

Over the years there has been much discussion of how to order biomedical knowledge and to assign sensible priorities. Few have agreed on any single system.

I have long opposed the plan whereby learning is compartmentalized. Rather, I much prefer to think and teach according to functional systems of the body (Are You Listening and Reading? Modern Medicine, August 28, 1967, pp. 22-24). They provide a framework on which my mind can hang as few, or many details as I care to remember. Each system can be thought of as a vignette which encompasses and unifies all the classified subdivisions of medicine ranging from anatomy to bedside care. By providing structure and continuity, priorities are much more easily ordered.

Medicine today is much like organic chemistry was a century ago. Until chemical structure was introduced into the body of knowledge, the growing number of compounds being discovered was creating a jungle, the many components of which no one could remember. But the ring systems, valence, and stereochemistry brought order, aiding in associative memory and logical progression of thought. Facts are not remembered out of context but fall into a system with meaning. The same may well happen in medicine. Knowledge will be remembered through use of basic principles and building from them a logical structure which is relevant to the functions of the body. It was the wholeness of biomedical knowledge that suggested, for example, the use of a mosaic to describe the varied mechanisms of hypertension.

Fortunately times are changing and, with them, attitudes. I see much greater willingness, even eagerness of younger physicians to close the gap between chaotic empiricism and logical synthesis to make themselves more complete physicians.

I hope this book will help. It is written

by some very capable young authors and its editor, Dr. Frohlich, has had experience helping to launch the series called "Physiology for Physicians" published on behalf of the American Physiological Society by the New England Journal of Medicine.

I suspect one of the greater lacks today is not the availability of good pedagogic material but adequate time for the physician to read selectively and learn. He must be careful to protect himself from being overwhelmed by the sheer volume of printed and spoken words and my only advice is learn what you learn well enough so that the patient's problems conjure a logical image of causal mechanisms in your mind. Science should be an integral part of diagnosis and treatment. The patient and you will be the better for it.

Irvine H. Page, M.D.

Preface

Most diseases encountered by physicians are without known cause. Thus, whether it be related to cardiology (arteriosclerosis, myocardiopathies, or hypertension), endocrinology (diabetes mellitus, obesity, thyrotoxicosis), respirology (emphysema, chronic bronchitis, bronchiectasis), gastroenterology (peptic ulcer, colitis, cholecystitis), or the broad areas of oncology, we are unable to ascribe a specific etiology. Nevertheless, with modern diagnostic tools and therapy the physician is usually quite effective in his treatment and help of patients suffering from these problems. Often life has been remarkably prolonged and the patients' well-being considerably improved. Usually the treatment involves specific correction of abnormal mechanisms involved in the disease process. Thus, in hypertension arterial pressure is reduced to normotensive levels; in atherosclerosis and myocardiopathies ventricular failure is reversed and the myocardium recompensated; in diabetes the carbohydrate and lipid metabolism is controlled; and in thyrotoxicosis the increased catabolism is rendered normal. Often the mechanistic approach used by the physician is so "natural" that it verges on empiricism which might seem to be reflex or intuitive. However, the rationale for such therapeutic maneuvers has been well-thought-out and was at one time or another painstakingly documented in the great variety of journals of clinical investigation. Most important to students and practicing physicians is the continuing need to understand the mechanisms underlying the diseases and therapeutic programs and their rationale should be applied generally in the day-to-day practice of medicine. Such thinking leads to good medicine, improved health care, and less complications of treatment. The purpose then of this book is to present a mechanistic view of

pathology which hopefully will provide the student and physician with a means of conceptualization of disease.

This book is not intended to be another textbook of medicine, of pathology, physiology, or biochemistry. There are already many well-written textbooks available satisfying the need for an all-encompassing tour-de-force of disease or physiology. Nor is this book intended to be used for an explanation for each described illness, sign, or symptom; such texts are also already available. The book, then, is intended to present a way of thinking of disease. The book has been divided into eight sections of convenience concerned with the physicians' orientation to systems or disciplines each edited by a well-respected and knowledgeable Section-Editor with broad clinical and academic background. Rather than subdivide each section into the component organs and their respective diseases, well-defined mechanisms of normal function have been selected about which the respective chapter authors describe the role of the mechanism in achieving homeostasis and how, when disease ensues, the regulation of these mechanisms goes awry. Each mechanism is described by an author of broad clinical competence in his respective area who also has been acknowledged for his fine ability as a teacher and writer. Just as the book is not intended to be all-encompassing with respect to disease, it is not all-inclusive with respect to mechanisms. Other mechanisms or areas might well have been covered and still may be discussed in the future; treatment of the concerned mechanisms was chosen to provide the reader with the broadest approach to clinical thinking without unduly entering into those areas which still remain speculative or controversial and with deliberate attempt to keep the book within a reasonable length. The individual chapters dealing with mechanisms are not intended to enter into a lengthy discourse of all diseases concerned with that mechanism under discussion. The authors present the underlying physiological concept, pertinent diseases which exemplify best the disarray of the controlling mechanism, and establish a way of thinking which will enable the reader to conceptualize other clinical problems not covered in the subject material.

It is hoped that this new approach to physiology, pathology, and internal medicine can be used on a multiplicity of levels in medical education. Early in medical school training new courses are appearing in medical curricula designed to present Pathophysiology or Correlative Medicine; this volume is intended to be a companion text for such programs of instruction. As the student continues further with his medical training, either in school or at the postgraduate level, it is hoped this text will provide the matrix upon which his broader clinical experience will develop. Ultimately, we hope that this approach to disease will provide the learning reinforcement to the practicing physician in his continuing experience and thinking about clinical medicine.

Since the framework of the book is a mechanistic approach to the regulation of function, we begin our text with the mechanisms involved in the major systems. The discussions of the final two sections, concerned with the neuromuscular and immunological mechanisms, may at first thought seem to duplicate some earlier sections. Thus, in the neuromuscular section we present discussions on the neural control of the cardiovascular and gastrointestinal sytems. This was constructed intentionally to provide the reader with an alternative approach to the complexities of the integrative role of the entire body. By discussing these areas separately, an alternative manner of conceptualization is provided which may be more comfortable to some readers. Similar areas may exist in the pulmonary, endocrine-metabolic or renal sections; but hopefully the reader will profit from another way of thinking. A few chapters have been included which, perhaps, may have less pertinence in today's practice of medicine; however, we believe these areas (learning and conditioning, autoallergy, complement, transplantation, metabolic function of the lung) are likely to be of increasing importance in development of our clinical concepts of the future.

Each chapter concludes with a selection of annotated references chosen by the author to amplify the points made within the chapter. Then, at the conclusion of each section the authors, Section-Editors and Editor have chosen specific "classical" references for one of two appendices. The other appendix is designed for those readers who are interested in being referred to specific clinical tests for each of the mechanisms under discussion. Normal values and brief discussions concerning interpretation of the normal function may also be included to provide the reader with further opportunity to apply the information presented in the section.

EDWARD D. FROHLICH, M.D.

Acknowledgment

The Editor expresses the deep appreciation of each of the contributors to their associates who have permitted them to collate their experiences, thoughts, and work into the written word. No single published contribution could be possible without these individuals, but to list each would be almost impossible.

Similarly I should like to express my appreciation to my associates not only for the above reasons, but also for permitting me to take time from my daily obligations to organize and collate this volume. I particularly want to express my sincere gratitude to Doctor James F. Hammarsten for providing me with this extra time, for his moral support, and for the assistance of his Department. Additional expression of thanks is indeed due to my secretary, Mrs. Irene Smith, for the many hours of work added to an already burdened schedule; to my research technicians Miss Kay Cheadle and Mrs. Janice Pfeffer for their most competent help, and to my students, fellows and colleagues for acting as "sounding boards" for many of these chapters. I also thank Messrs. George F. Stickley and Lewis Reines of J. B. Lippincott Company for their fine cooperation in making publication of this book possible.

The vast accumulation of clinical knowledge and the rapid surge of medical progress during the past 25 years result in no small way from the material encouragement of fundamental and applied research by the United States Public Health Service's National Institutes of Health and the many non-Federal lay organizations (including the American Heart Association, National Tuberculosis and Respiratory Disease Association). Certainly without these agencies the contributors to this volume would not have received much of their support for their work which has permitted them to share their experiences and knowledge. However, rather than citing each separately with each chapter we collectively express our deep appreciation with the fervent hope that this close cooperation will continue to benefit all mankind.

Finally, one personal note of appreciation is allowed by usurping whatever editorial prerogatives are afforded me. I can attest to the long hours taken from time which would otherwise be spent with one's family. Not only do I lovingly appreciate this gift from my wife, Sherry, and children, Margie, Bruce, and Lara, but I can never repay my wife for her constant patience and encouragement.

Contents

SECTION ONE: CARDIOVASCULAR MECHANISMS............................ 1
Edward D. Frohlich, M.D., Section Editor

Introduction 3

1 **Cardiac Adaptation** 5
Thomas A. Bruce, M.D.

Morphological Features as Related to Function 5
Measurements of Performance 9
Myocardial Adaptability.. 11
Cardiac Dilatation and Hypertrophy 13
Cardiac Failure 14
Mechanisms of Myocardial Failure.................................. 16
The Multiple Levels of Nutrient Failure 18

2 **Tissue Perfusion** 23
Michael J. Brody, Ph.D., and François M. Abboud, M.D.

Introduction 23
Changes in Vascular Smooth Muscle Contraction 24
Structural Factors Altering Resistance to Blood Flow 33
Rheologic and Mechanical Factors Altering Resistance to Blood Flow 35
Intraorgan Blood Flow Distribution 36

3 **Pressor Mechanisms** 41
*Harriet P. Dustan, M.D., Edward D. Frohlich, M.D.,
and Robert C. Tarazi, M.D.*

Introduction 41
Hemodynamics... 42
Volume Mechanisms.. 48
Renal Pressor Mechanisms... 55
Catecholamine and Neural Mechanisms............................. 60

4 **Depressor Mechanisms** 67
Lerner B. Hinshaw, Ph.D.

Introduction 67
Cardiovascular Factors.. 67
Nervous Factors 70
Normal and Stressful Environmental Stimuli 76
Pathophysiological States 77

5 **Cardiac Rhythmicity** 83
   Gordon K. Moe, M.D., Ph.D.

   Impulse Generation 83
   Impulse Propagation 85
   Block and Re-entry 87
   Electrical Conversion of Cardiac Arrhythmias 97
   Conclusions 98

   Appendix A 100

   Appendix B 102

SECTION TWO: PULMONARY MECHANISMS 107
   Clarence A. Guenter, M.D., Section Editor

   Introduction 109

6 **Control of Respiration** 111
   Robert A. Mitchell, M.D.

   Introduction 111
   Automatic Respiratory Control Systems 112
   Reflex Control of Breathing 119
   Effects of Central Nervous System Disease on Breathing in Man 120

7 **Ventilation, Perfusion and Gas Exchange** 127
   Reuben M. Cherniack, M.D.

   Ventilatory Function 127
   Exchange of Gases in the Lungs 131
   Respiratory Insufficiency 134
   The Assessment of Pulmonary Function 136
   Gas Exchange 140
   Acid-Base Balance 141
   Cardiopulmonary Response to Exercise 141

8 **Oxygen Transport and Cellular Respiration** 145
   Eugene D. Robin, M.D., and Lawrence M. Simon, M.D.

   Introduction 145
   Pulmonary Oxygen Uptake 145
   Cellular Oxygen Delivery 146
   Movement of Oxygen from Plasma to Intracellular Sites of Utilization 152
   Mixed Venous-Metabolizing Cell Oxygen Gradient (V-c Oxygen Difference) 153
   Tissue Oxygen Stores 155
   Oxygen Utilization 155
   Anaerobic Glycolysis 159

Monitoring Abnormalities of Mitochondrial Oxygen Utilization 162
Oxygen and Nonenergy Providing Processes 163
Changes in Organ Function with Hypoxia 164
Therapy of Hypoxia 166

9 Metabolic Events in the Lung 167
Sami I. Said, M.D.

Introduction 167
Cellular Sites of Metabolism 167
Biosynthesis of Lipids and Phospholipids 172
The Maintenance of Alveolar Stability 173
Protein Synthesis and Secretion 173
Carbohydrate Metabolism 174
The Lung and Hematological Mechanisms 175
Relation to the Kallikrein-Kinin System 176
The Uptake, Alteration, Storage and Release of Vasoactive Substances 176
Defense Against Infection 178
Immunological Responses 179
The Lung in Allergic Reactions 180
The Potential for Hormone Secretion: Endocrine Syndromes in Lung Disease . 180
Conclusion 181

Appendix A: Normal Values for Respiratory Function 185

Appendix B: Major References in Respirology 186

SECTION THREE: RENAL MECHANISMS 187
Robert D. Lindeman, M.D., and William O. Smith, M.D., Section Co-Editors

Introduction 189

10 Maintenance of Body Protein Homeostasis 195
Victor E. Pollak, M.D., and Amadeo J. Pesce, Ph.D.

The Glomerulus as a Molecular Sieve 195
Tubular Absorption and Secretion of Proteins 199
Proteinuria in Disease 204
The Metabolism of Albumin and Other Plasma Proteins 207
The Consequences to the Body of the Loss of Large Amounts of Albumin
and Other Proteins in the Urine 211

11 Maintenance of Fluid and Electrolyte Homeostasis 215
John F. Maher, M.D., and Karl D. Nolph, M.D.

Normal Sodium and Water Metabolism 215
Altered Sodium and Water Metabolism 220
Potassium in Health and Disease 227

12 **Maintenance of Body Tonicity** 233
    Solomon Papper, M.D., and Carlos A. Vaamonde, M.D.

    Introduction 233
    Terminology 233
    Pathophysiology of Urine Concentration 235
    Pathophysiology of Urine Dilution 241

13 **Body Buffering Mechanisms** 249
    Edward J. Lennon, M.D.

    Introduction 249
    Definitions 249
    Buffers 250
    Normal Mechanisms 251
    Endogenous Acid Production 253
    Hypercapnia and Hypocapnia 258
    Metabolic Acidosis and Alkalosis 259
    Clinical Evaluation of Acid-Base Disturbances 263

    Appendix A: Renal Functional Tests 269

    Appendix B: Supplemental Textbooks and Monographs 281

SECTION FOUR: ENDOCRINE-METABOLIC MECHANISMS 283
    Leonard P. Eliel, M.D., Section Editor

    Introduction 285

14 **Regulatory Mechanisms of the Pituitary and Pituitary-Adrenal Axis** 287
    Don H. Nelson, M.D.

    Pituitary-Adrenal Axis 287
    Hypothalamic Control of Secretion 287
    Chemical Nature of ACTH 290
    Abnormalities of ACTH Secretion 290
    Adrenocortical Secretion 291
    Regulation of Growth Hormone Secretion 297
    Panhypopituitarism 298

15 **Regulatory Mechnisms of the Pituitary-Thyroid Axis** 301
    Farahe Maloof, M.D.

    Introduction 301
    Hypothalamic Control of TSH 301
    Pituitary Secretion of TSH 302
    Biosynthesis of Thyroid Hormones 304
    Circulating Thyroid Hormones 308
    Kinetics of Thyroxine and Triiodothyronine Metabolism 308
    Excess Synthesis and Secretion of T_4 and T_3 308

Decreased Thyroxine and Triiodothyronine Synthesis and Secretion 312
Alteration in Plasma Binding Proteins 315
Metabolic Derangements Associated with Thyroid Carcinoma 316

16 **Endocrine Control Mechanisms of Electrolyte and Water Metabolism** 319
Frederic C. Bartter, M.D.

Body Water 319
Body Sodium 324
Body Potassium 325
Body Hydrogen Ion 326
Calcium 330

17 **Metabolism and Energy Mechanisms** 335
Donald B. Martin, M.D.

General Features of Metabolism 335
Organ Physiology 337
Altered States and Diseases 344

18 **Endocrine Mechanisms of Reproduction** 355
Mortimer B. Lipsett, M.D.

Introduction 355
The Hormones 355
Mechanism of Action 359
Testis 360
The Ovary 364
Amenorrhea 366
The Feto-Placental Unit 369
Contraceptives 371

Appendix A 375

Appendix B: Tests of Thyroid Function 375

SECTION FIVE: GASTROINTESTINAL MECHANISMS 383
Eugene D. Jacobson, M.D., Section Editor

Introduction 385

19 **Motility** 389
James Christensen, M.D.

The Anatomy of Gastrointestinal Muscle 389
The Nerves of the Gut 390
Integration of Contraction in Smooth Muscle 392
The Pharynx and Esophagus 394
Disorders of Esophageal Motor Function 396
The Stomach 398

Disorders of Gastric Motor Function   400
The Gallbladder and Bile Ducts   401
Motor Disorders of the Biliary System   401
The Small Intestine   402
Disorders of Small Bowel Function   403
The Colon   403
Disorders of Colonic Motor Function   404

20   Mechanisms of Gastric and Pancreatic Secretion   407
Horace W. Davenport, D.Sc.

The Secretions   407
Control of Gastric and Pancreatic Secretions   408

21   Absorption   423
Konrad H. Soergel, M.D., and Alan F. Hofmann, M.D.

General Considerations   423
Mechanisms of Intestinal Membrane Transport   426
Absorption of Water and Water-Soluble Solutes   431
Absorption of Lipids   439
Malabsorption and Diarrhea   448

22   Hepatic Mechanisms   455
William W. Faloon, M.D.

Bilirubin Metabolism and Jaundice   455
Portal Hypertension   459
Pathological Physiology of Ascites Formation   461
Pathological Physiology of Renal Failure in Severe Liver Disease   464
Pathological Physiology of Hepatic Encephalopathy   465
Pathological Physiology of Cholesterol and Bile Acid Abnormalities in Liver
    Disease   467
Disturbances in Metabolism of Carbohydrate, Protein, Fat and Other Nutri-
    ents   468

Appendix A: Classical References   475

Appendix B: Laboratory Tests for Disturbed Gastrointestinal Functions   476

SECTION SIX: HEMATOLOGICAL MECHANISMS   485
James W. Hampton, M.D., Section Editor

Introduction   487

23   Erythropoiesis   489
Walter H. Whitcomb, M.D.

The Erythron   489
Regulation of Erythropoiesis   492

Cellular Proliferation 495
The Mature Erythrocyte 506

24 **Leukopoiesis** 515
James W. Hampton, M.D.

    Introduction 515
    "Lifespan" of Erythrocytes 515
    Stem Cell Compartment 516
    Control Mechanisms 516
    Circulating Leukocytes 517
    Bone Marrow Leukocytes 520
    Leukopoiesis in Disease 522

25 **Hemostasis** 527
Oscar D. Ratnoff, M.D.

    Physiology and Pathology of the Blood Clotting Mechanism 527
    Disorders of Blood Platelets 537
    Disordered Hemostasis Due to Vascular Pathology 543

26 **Fibrinolysis** 547
Victor J. Marder, M.D., and Sol Sherry, M.D.

    Introduction 547
    Components of the Fibrinolytic System 547
    Physiological Mechanism for Fibrinolysis 551
    Abnormal States of Fibrinolysis 554

    Appendix A 559

    Appendix B 563

SECTION SEVEN: NEUROMUSCULAR MECHANISMS 565
Stewart G. Wolf, M.D., Section Editor

    Introduction 567

27 **Neural Integration** 569
Robert B. Livingston, M.D.

    Introduction 569
    Some General Features of Brain Organization Relating to Regulation 569
    Pathophysiological Mechanisms Affecting Nervous System Regulation 582

28 **Autonomic Control of Cardiovascular Function** 599
H. Page Mauck, Jr., M.D.

    Introduction 599
    Integrated Neural Control 599
    Higher Central Neural Control 602

Reflex Control 603
Neuroeffector Function 607
Central Nervous System Cardiac Control 608

29 **Autonomic Control of Gastrointestinal Function** 615
Ove Lundgren, M.D.

Introduction 615
Anatomical Considerations 615
Motility 616
Blood Flow 625

30 **Neural Control of Skeletal Muscle** 631
William J. Crowley, Jr., M.D.

Mechanisms of Development and Maintenance of Normal Muscle Bulk ... 631
Mechanisms of Development and Maintenance of Normal Muscle Tone ... 632
Mechanism for Development and Maintenance of Muscular Strength 637
Mechanisms of Movement 642

31 **Learning Mechanisms** 653
Leo V. DiCara, Ph.D.

Conditioning and Learned Responses in Visceral Control 653
Autonomic Nervous System 653
Learning and Conditioning and Visceral Pathology 656

Appendix 663

SECTION EIGHT: IMMUNOLOGICAL MECHANISMS 665
William A. Cain, Ph.D., Section Editor

Introduction 667

32 **Inflammatory Mechanisms and Fever** 671
Arthur R. Page, M.D.

Introduction 671
The Local Inflammatory Response 671
Increased Vascular Permeability 671
Systemic Manifestations of Inflammation 677
Interferons 679
Anti-inflammatory Drugs 679
Diseases Associated with Abnormalities of Inflammation 680

33 **Adaptive Immunity** 683
Robert A. Good, M.D., and Joanne Finstad, M.S.

Introduction 683
Host-Parasite Relationships 683

Humoral Immunity.. 686
The Role of the Thymus... 691
Immunological Deficiency Diseases of Man............................. 694
Phylogenetic Development of Immunity.................................. 702
Relationship of Malignant Adaptation and Immunity.................. 707
Immunity and Aging... 708

34 Mechanisms Involving the Complement System......................... 711
Richard J. Pickering, M.D., and Ann E. Gabrielsen, Ph.D.

Introduction... 711
Nomenclature... 711
Characteristics of Complement Proteins................................ 712
Functions of the Complement System.................................... 712
Complement Abnormalities in Man....................................... 719

35 Autoallergic Mechanisms and Transplantation......................... 731
J. Wesley Alexander, M.D., Sc.D.

Mechanisms of Immunological Injury..................................... 731
Transplantation.. 733
Diseases of Autoallergy.. 738
Immunosuppressive Therapy... 742

Appendix A.. 747

Appendix B.. 749

Index.. 757

Introduction

ARTHUR C. GUYTON, M.D.

Department of Physiology and Biophysics
University of Mississippi School of Medicine
Jackson, Mississippi

During all my professional life I have been devoted to the proposition that the practice of good medicine is a science. Therefore, I feel rewarded personally with each significant new contribution to the synthesis of medical science. When I first read the list of authors who had agreed to contribute to this book and then further learned the topics to be covered, I awaited with anticipation until I should actually read the manuscripts. As these began to assemble it soon became clear that I would not be disappointed, because the topics and the detailed subject matter were chosen well to fill voids between the individual disciplines of physiology and clinical medicine. They were chosen to teach, to illustrate the good sense that prevails when a clinician treats a patient with full knowledge of what is happening in the diseased body, knowledge that allows a multiplicity of approaches rather than the practice of stereotyped routines. For these reasons, even I, though engaged most of my life in interfacing basic physiology with clinical medicine, find a host of new ideas and principles that will carry over into my future teaching.

Shortly, I wish to speak of the new horizons in clinical physiology, but before doing so I think it worthwhile to tell the readers something about the person who has spent so much time and intelligence organizing, directing, and producing this book, Dr. Edward Frohlich. For those who do not already know, this is not Dr. Frohlich's first major contribution to medicine

nor to basic physiology. It was evident as long as a decade ago that Dr. Frohlich, who is even now still young, would be a major force in teaching basic physiological concepts in the practice of clinical medicine both to students and to professionals in the fields of physiology and clinical medicine. As one can tell from his chapter with Drs. Dustan and Tarazi, Dr. Frohlich is an accomplished clinician. He is also thoroughly versed in basic animal physiological experimentation; he has achieved the art of clinical research measurement; and he has an introspective mind that delves into the deeper aspects of basic physiological concepts while searching constantly for ways to make this knowledge fruitful to the patient. It is this rare combination of talents that is necessary for the successful undertaking of a book of this type. Such a book appears about once every five years, and its impact is immeasurable when its editor's talents fill the measure.

Now, let us look at the borderland that lies between basic medical science and the practice of clinical medicine. For those readers who are clinicians, how often have you had to treat a patient with a disease that still falls within one of our scientific voids, and you have said to yourself, if only I could understand what I am doing? For those of you who are students and are studying the basis of medicine and by now know the molecular configurations of DNA, how does this fit into your life goal of understanding and treating disease? And for those of you who are working on the forefronts of detailed basic physiological concepts, how often have you been driven to your experiments because of your desire to explain and to contribute in your way to the final solution of human problems? The answers to all of these questions can best be given in the form of examples.

One of the most exciting fields for the fertile physiological and clinical mind during the past fifty years has been that of hypertension. In the nineteenth century and in the early part of the twentieth century there was a strong general feeling that hypertension was in some way related to fluids and electrolytes of the body, though most of the reasons for these beliefs were circumstantial. Then, in the early nineteen thirties, most of the momentum shifted to humoral mechanisms as the cause of hypertension, mainly because of the very successful animal preparation made by Goldblatt to cause well-controlled hypertension in animals and because of the subsequent isolation, characterization, and physiological study of the renin-angiotensin system. Later, with still more knowledge, the role of aldosterone in hypertension and the welding of the renin-angiotensin system with aldosterone as well as with salt and water mechanisms came to the forefront. Through all of these periods there also ran an undercurrent of belief among many physicians and experimentalists that hypertension, especially essential hypertension, has a neurogenic basis. Finally, in recent years the systems physiologists, the biomedical engineers and a host of clinicians have come to believe strongly that hypertension can be a manifestation of any one of many different abnormalities affecting one or more specific parts of an overall and complex arterial pressure regulatory system. Furthermore, our knowledge of these regulatory systems, even of the mathematics of most of their aspects, is beginning to emerge, so much so, indeed, that many types of hypertension can be simulated almost in detail by mathematical models on computers. Much of the story of the science of hypertension will be told in future chapters of this book, and references from these chapters will lead one into one of the most rewarding of stories of the scientific approach to medical problems, a story that interweaves the deepest of our basic physiological concepts with one of the most important of clinical problems.

Another field that captures the imagination, not only of the medical scientist but even of the public, is the pathophysiology of immunological mechanisms, especially their relation to autoimmunity, transplantation, and even survival of the human race in the face of rising population density with its attendant exponential growth of epidemiologic potential. It was hardly ten years ago that we wrote in our textbooks that blood lymphocytes might be nothing more than cast-off products with probably little function once they had been "excreted" by the lymph nodes into the blood, merely waiting their arrival in some tissue where they would be cleansed from the blood by phagocytosis. Yet, at the same time we spoke in terms of phenomena such as tissue immunity or other vague types of immunity that could not be explained on the basis of circulating protein antibodies in the plasma. Then, suddenly, the vista of the lymphatic system opened wide to entwine together all these segments of half-knowledge, displaying the beauty of our body's defense against disease and at the same time beginning to explain the abnormalities that lead to autoimmunity, allergic phenomena, and so forth. And simultaneously came the saga of man's attempts to obviate this defense system so that he can practice his cherished dream of reassembling human beings from organs no longer needed by others. The story of our defense system is still unfolding; a major segment of this scientific forefront will be told in the appropriate section of this book.

I could continue indefinitely to detail the excitement, the fun, and the intellectual reward that characterize what is happening today in the area between basic medical science and clinical medicine. I could tell how basic mathematicians have helped to solve problems of respiratory patients such as asthmatics, emphysematous patients, and patients with capillary-alveolar block. I could tell how the mechanisms for regulation of our body fluids, their electrolyte compositions, their acid-base balance, their pressures, and their volumes, are all beginning to fall into place. Indeed, modern practice of renology and daily control of body fluids in sick patients are both highly dependent on this new and exacting knowledge.

The basic chemistry and genetic control of both normal and abnormal hormones secreted in the body are emerging as a new science and are becoming basic to the practice of endocrinology. And presiding over almost all systems of the body—over the cardiovascular system, the gastrointestinal system, the kidneys, and the lungs—is always the nervous system. Its primary role is to control our rapid bodily functions such as our quick motor acts, motility of the gut, degree of constriction of the blood vessels, and much of secretion by the endocrine glands. But in the background are literally thousands of other control systems that act more slowly, sometimes subservient to the nervous system, sometimes not; sometimes opposing the nervous system, sometimes supporting it; sometimes weak in comparison with the nervous controls, sometimes very potent. These other controls include even the genes themselves, each of which is a basic block in a much larger system of control.

It is the body's control systems that makes it possible for the human being to live, indeed, that demand that he live, that drive him to live, so much so that he has to perform a positive act to make himself die. And it is abnormalities in these control systems or in their target organs that are the bases of disease. The modern doctor is inheriting a new and formidable task to understand these systems so that he might practice a better type of medicine, but the compensations are commensurately great, for a doctor's pleasure in his work is directly related to the extent that he understands what he is doing.

Section One

Cardiovascular Mechanisms

Introduction

Much information in regard to the cardiovascular system has been amassed in the past three decades, greatly augmenting knowledge gained over the preceding three hundred years. Nevertheless, the four most common cardiovascular diseases (those resulting from hypertension, shock, arteriosclerosis, and myocardiopathies) remain, for the most part, of unknown cause. True, much knowledge has been gained concerning pressor mechanisms, and with the introduction of antihypertensive drugs great inroads have been made into morbidity and mortality from the hypertensive diseases; but, in the final analysis, the bulk of patients with hypertension have chronic illnesses of unknown cause. The picture of arteriosclerosis is still more primitive. Not only are we ignorant of the causes of the various diseases of atherosclerosis, arteriosclerosis, and the associated anomalies of lipid and protein metabolism, but we have only begun to be provided with inklings of the mechanisms underlying this group of afflictions. Likewise, only in the past few years, with the development of more refined biochemical methods, diagnostic techniques and pharmacological tools, have we come to think of the various possibilities underlying the ischemic heart diseases and the manifold possible types of myocardiopathies.

Through association with engineers and biophysics-oriented physiologists, as well as investigators with a strong interest in biochemistry, we have come to think in terms of total integration of the cardiovascular system. We no longer consider the heart as the central organ "calling the plays," as it were, for the entire systemic circulation and its associated organs. We are at least enough sophisticated clinically now to appreciate such concepts as: local regulation of flow; local tissue demands regulating the output of blood from the heart; the important interdependency of the autonomic nervous, endocrine, and renal systems with the cardiovascular system; and the role of the various endocrine activities of the cardiovascular system as it pertains to the fluxes of vasoactive agents liberated from the various organs.

The lessons learned from the various diseases comprising what we so glibly diagnose as "hypertension" are beginning to "pay off." We now realize that what we have lumped collectively as "essential hypertension" is a host of unhomogeneous problems manifested by varying degrees of disregulation of the many controlling pressor mechanisms. From this physiological "sorting-out" process will come descriptions of other discrete diseases; hence, by detecting operable pressor mechanisms in a particular patient we are better able to understand his particular problem and, perhaps, apply more intelligently specific, life-prolonging therapy.

As our knowledge of atherogenesis grows, and we are able better to apply this newfound information to the differing aspects of the arteriosclerosis problem (as *it* unfolds), we might be able to devote a special area of cardiovascular mechanisms to this problem. Perhaps, as our knowledge of biochemical and hereditary defects increases, a discussion concerned especially with those mechanisms will be required.

Thus, at present, after considerable thought we were only able to outline five particular mechanisms as they remain uniquely under the purview of the physician-physiologist concerned with the cardiovascular system. The first of the following five chapters presents a discussion of the contractile mechanisms of the heart and their pertinence to *myocardial performance*. This discussion will be concerned not only with concepts of muscle mechanics, relationships of pressure, flow, and volume, and dilatation and hypertrophy but also— in biochemical terms, wherever possible — as they are related to the diseases of cardiac "inflow," cardiac "outflow," and intrinsic "pump failure." A natural follow-through discussion is concerned with the local aspects of *tissue perfusion*, the mechanisms of local regulation of flow, and the relationships of systemic vascular, endocrine, metabolic, and biochemical altera-ations to the concepts of tissue nutrition, and how they pertain to disease. Not only should we be concerned with the problems of mechanical obstructions to the flow of blood, we are now concerned with the role of circulating chemical and vasoactive agents, and the interrelationships of the cardiovascular with the nervous and endocrine systems and systemic metabolism in various pathological states. The following two chapters, on *pressor* and *depressor mechanisms*, present the various physiological mechanisms which serve to maintain arterial pressure as normotensive levels. They describe how, when these delicately interrelated mechanisms are disrupted and go awry, various diseases develop, manifested either by increases or by decreases in arterial pressure. Finally, no cardiovascular discussion would be complete without

elucidation of the peculiar and intrinsic property of cardiac muscle to contract rhythmically. Rather than a discussion of specific arrhythmias, a concept of *rhythmicity* and the nature of the various electrophysiological mechanisms whereby rhythm disorders may develop—impulse propagation, block, escape, reentry, ectopic pacemakers—are presented so application to specific rhythm disturbances can be drawn therefrom.

Classically, clinical cardiology and cardiovascular physiology have been presented in terms of physical diagnostic techniques and various radiographic and electronic diagnostic tools. Following such an introduction, the specific congenital, rheumatic, hypertensive, endocardial, myocardial, and pericardial diseases are discussed in terms of the specific pathophysiological criteria of each disease. Presentations of this type are abundantly accessible. It is therefore genuinely hoped that, by presenting this material with respect to the concepts concerned, the physiological problems of clinical management and the various possibilities for a mechanistic approach to therapy will further stimulate the student and physician to practice with the goal of overcoming our present ignorance of disease causes by understanding *clinical* mechanisms.

Edward D. Frohlich, M.D.

1

Cardiac Adaptation

Thomas A. Bruce, M.D.

The heart is a carefully balanced neuromuscular organ capable of adaptation to a wide range of total systemic needs. Its performance is regulated both intrinsically and extrinsically, with the degree of change determined by the hemodynamic and metabolic needs of the various organ systems of the body. As the circulatory pump it is the source for considerable heat production. Defects in its performance are manifest in myriad ways, often as failure of other distant organs or tissues.

MORPHOLOGICAL FEATURES AS RELATED TO FUNCTION

There is perhaps no more splendid demonstration in the body of the integration of anatomical structure, biochemical activity and physiological performance than the myocardium.

Macrostructure. At the gross anatomical level, the muscle bundles all arise from the tendinous structures at the base of the atrioventricular valves (Fig. 1-1). Those bundles that constitute the ventricular myocardium converge in curving courses toward the apex and then turn spirally upward to be inserted on the opposite side of the same tendinous material. Muscle bundles are arranged in layers, each spread in a different direction so that the deep and the superficial fibers lie approximately at right angles. Each layer is incorporated into the wall of both ventricles and into the interventricular septum. There is little collagenous tissue separating the bundles and, from a functional standpoint, the muscles act as a single syncytium. There is therefore no such entity as a uniquely left ventricular or right ventricular muscle bundle or group. Consequently, when the muscular wall in one ventricle is caused to hypertrophy by a pressure or volume load, some degree of hypertrophy is usually seen in the nonloaded ventricle also.

Thin-walled atria arise from the same tendinous base as do the ventricular muscles and are suited for distensibility and a primary reservoir function. Numerous fig-

BASE   APEX

FIG. 1-1. Spiral course of muscle fibers of the heart. (*Left*) Base of ventricles (atria removed). (*Right*) Apex of heart. Note especially that all fibers form a continuous muscle "bundle," arising from the valve rings at the base, encircling both ventricular walls and reinserting close by their point of origin.

ure-of-eight fibers also arise from the base, forming the ring of the semilunar valves in the ventricular outflow tracts. All valves are of delicate but extremely resilient material which supports a lifelong durability under prolonged phasic stress.

Lymphatics. Surprisingly little is known about the cardiac lymphatics. A rich plexus can be found in the subepicardial zone, collecting into right and left "current" channels that follow the pathways of the coronary veins to the base of the heart and end in lymph nodes near the tracheal bifurcation or in the anterior mediastinum. The thick myocardium is not known to have lymphatic vessels (unless the minimal amount of intercellular connective tissue acts as a "sponge" in some obscure way). A subendocardial network is present but its drainage characteristics are poorly understood; whether the valves, chordae tendinae and the atria are a part of this subendocardial plexus is not known. Congenital or acquired obstruction of this subendocardial lymph system seems to be related to the presence of fibroelastosis; in addition, the role of vasohumoral substances such as serotonin and polypeptides of the kallikrein-kinin group as possible factors in induction of lymphatic obstruction needs to be explored. Abnormalities in lymphatic flow have been proposed as a mechanism of some of the myocardial diseases of unknown origin; without data this must remain conjectural.

Integration. The function of the entire organ is coordinated by a specialized "nervous" conduction pathway (Fig. 1-2) in which electrical depolarization of each chamber is precisely timed, thereby providing a unified, synchronous contraction. The pacemaker function is usually centered in the sino-atrial (S-A) node, induced by the automaticity and the low excitability threshold of its P cells. The impulse propagates through two routes, radially through the atrial wall and directly through the specialized internodal and interatrial conducting pathways. At the atrioventricular (A-V)

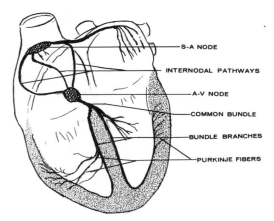

FIG. 1-2. Intracardiac conduction pathways. The intrinsic pacemaker function results from spontaneous P cell depolarization in the sino-atrial (S-A) node; the impulse is then carried through specialized muscle pathways to the terminal Purkinje network throughout both ventricles, and thence to the myocardial working cells.

node there is some slowing of the impulse. Transmission is then routed through the common atrioventricular bundle (bundle of His), the bundle branches and the peripheral Purkinje fibers to depolarize the myocardial contractile fibers.

[**Note:** There is some confusion about terminology when these events are followed electrocardiographically. Since the A-V node is located in the central fibrous body separating the atria and the ventricles, the ventricular conducting pathways are sometimes misinterpreted as being part of the QRS depolarization complex on the electrocardiogram. Actually, the His-Purkinje pathways form the terminal portion of the P-Q (P-R) "atrial" interval, and the QRS represents actual ventricular muscle activation.]

The anatomy of the ventricular conducting pathways provides the explanation for a number of pathological problems. The common His bundle descends to the junction of the membranous and muscular portions of the interventricular septum, where it divides into a distinct right bundle branch and a highly variable left bundle branch which is usually comprised of a network of smaller branches. The left bundle network separates into an anterior group (supplying the anterior wall of the left ventricle) and a posterior group (supplying the apical and posterior portions of the left ventricle).

The Purkinje system of the right bundle branch supplies the right ventricular wall and the lower portion of the right septum. The Purkinje system of the left bundle branches terminates in the anterior and posterior papillary muscles and their adjacent myocardium, plus the midportion of the left septum. Very few Purkinje fibers are distributed to the base of the interventricular septum or to the posterobasal portion of the left ventricle, resulting in late depolarization of these areas.

Thus, it can be seen that disease or infarction of the anteroseptal wall frequently disrupts the main right bundle branch and the anterior ramus of the left bundle. Conversely, apical or posterior wall disease often compromises the posterior ramus of the left bundle. When multifocal damage is present, complete block of the conducting system may result. In such an event, ventricular contraction and survival of the patient depend on the development of an idioventricular pacemaker. Fortunately, most of the conducting pathways have the capability for automaticity and new pacemaker function; their rate, however, tends to be slow and is initially rather unreliable, accounting for the frequent instances of Stokes-Adams syncope on development of complete atrioventricular block. It is also obvious that patients with the congenital anomalies which so frequently involve the membranous interventricular septum and the "A-V canal," or who undergo the surgical procedures designed to correct defects in this area, have a very high incidence of atrioventricular conduction disturbances.

Ultrastructure. Four basic types of cells are seen microscopically: P cells, transitional cells, Purkinje cells, and myocardial cells. *The P cells* (pale, primitive, pacemaker) have a simple internal organization; they are located predominantly in the sinus

node and, to a lesser extent, in the A-V node. Numerous nerve endings lie nearby but do not touch the specialized cell membrane which seems to be involved in the characteristic automaticity of these cells. Electrical impulses are passed from the P cells to *transitional cells*, also located in the sinus and A-V nodes. Their internal structure is intermediate in complexity between P cells and myocardial cells; myofibrillar contractile components are not numerous nor are they arranged for very effective contraction. *Purkinje cells*, located in the internodal and ventricular conduction pathways, receive impulses from the transitional cells and transmit them to the working myocardial cells. The Purkinje cells are short and broad, with an end-to-end alignment which facilitates their impulse-conducting properties. *Myocardial cells* are large and complex, having many myofibrils in parallel bundles and a full complement of supporting mitochondrial and sarcotubular support elements (see below). These cells are arranged in longitudinal "fibers," similar to skeletal muscle cells, but with the difference that there are many end-to-side cross connections.

Cell Membranes. Cellular function is regulated by the several types of membranes present. The doubled cytoplasmic membrane, or sarcolemma, contains an external basement membrane and an inner plasma membrane. Mitochondria also have double membranes, the inner portion of which is folded into numerous cristae which are the sites of fuel substrate transformation into energy stores. Membranes also delineate the voluminous, convoluted sarcoplasmic reticulum (sarcotubular system), the cell nucleus, and the lysosomes. All membranes are composed of complex lipoprotein molecules, with a core of lipid (predominant phospholipid) surrounded by protein. Each membrane is unique, however, in the percentage of specific phospholipid components contained. Cholesterol is present in significant amounts only in the plasma membranes, thereby controlling cellular fluidity. The composition of intracellular membranes is so specific that an expert can easily guess the membrane origin by knowing the type and amounts of phospholipid present. Although the mechanisms remain unknown, membrane transport is undoubtedly related to lipoprotein structure.

Intercalated Discs. Myocardial cells are joined by a wavy intercalated disc (Fig. 1-3, *left*). Here the cell membrane (sarcolemma) divides, with the external basement membrane portion continuing on to the next cell and the plasma membrane turning inward to pair with the plasma membrane of the adjoining cell and apparently fusing at certain spots. The discs are perforated for entrance of the sarcotubular system. Each disc functions prominently as a point of low resistance for electrical impulse propagation.

Myofibrils. Inside the cell, the contractile elements are present as long chains of myofibrils, each divided into tiny functional units called sarcomeres (Fig. 1-3 *center*). Each sarcomere is delineated by a dark line known as the Z band, and is composed of sliding actin filaments (light color) and myosin filaments (dark color, heavy). A single myocardial cell contains hundreds of the parallel myofibrils.

Mitochondria. Sandwiched between the myofibrillar chains are numerous large mitochondria. Myocardial cells contain greater numbers of mitochondria than does any other cell in the body, evidence of the tremendous energy requirements of contraction. Both the external and the internal membranes of the mitochondrion have enzymatic functions related to their protein "signatures." Specific enzymes of the respiratory-cytochrome system are located at specific sites on the laminae of the internal cristae. This is the only site for intracellular oxidative metabolism and, therefore, the major site of energy release and storage (predominantly as ATP).

Sarcotubular System. The intricate sar-

FIG. 1-3. Characteristic features of myocardial cells. Three views of the cardiac cell at increasing magnification focus attention on the functional anatomy. (*Left*) Anastomosing fibers as typically seen in the specialized conduction pathways. Note the wavy intercalated discs which mark the ends of each cell, and the central position of the nucleus. (× 130) (*Center*). A portion of one cell, with rows of large mitochondria sandwiched between the myofibrils which are divided by Z lines into sarcomeres. The sarcomere, the basic functional unit of the cell, is ultimately responsible for shortening by coupling of its thin filament, (actin) and thick filaments (myosin). (× 5000) (*Right*) Two sarcomeres, illustrating the transverse tubular system along the Z line axis and the interconnecting longitudinal tubular system, site of the activator calcium storage. (× 15,000)

coplasmic reticulum (Fig. 1-3 *right*), composed of transverse and longtitudinal sarcotubules, assumes the vital role of intracellular communication and coordination. The sarcolemmal cell membrane invaginates at the end of each sarcomere (Z band), forming the transverse sarcotubular system or T system. Each myofibril is surrounded by a portion of the longtitudinal network, and the two systems meet at the level of the Z band.

Mechanism of Coupling. The electrical depolarization impulse is received from an adjoining cell at the intercalated disc via the sarcolemmal membrane and is distributed throughout the cell via the transverse sarcotubular system. In some unknown way this stimulates ATPase activity in the longtitudinal sarcotubules, where calcium is stored. Release of ATP-stored energy activates the "calcium pump," and ionized calcium is released directly into the region of the actin and myosin contractile units. Actomyosin shortening occurs when the ionized calcium binds a "restraining" protein, tropinin, present in trace amounts among the contractile elements. Relaxation occurs as calcium is rebound within the longtitudinal sarcotubules. The energy cost of contraction can therefore be seen to reside in the binding and release of Ca^{++} from the sarcotubules as much as in the shortening of the contractile proteins themselves.

Digitalis and Coupling. The purified glycosides have greater use and value today than any other group of cardiovascular drugs. Although beneficial effects in increasing the force of cardiac contraction and in regulating cardiac rate and rhythm are well known, the mechanism of digitalis action is still elusive. A key to its great efficacy in congestive heart failure is the improved cardiac efficiency which results— i.e., greater effective work produced per unit of energy expended. The mechanism seems to reside in the improved coupling and uncoupling of the contractile proteins rather than in the availability of more energy in the form of ATP or creatine phosphate. Potassium ions compete with the digitalis molecule for a site on the cell membrane and therefore regulate indirectly the degree of pharmacological activity, but potassium does not seem to be the ultimate factor in actomyosin contraction. Regulation of Ca^{++} binding and release seems much more likely, but the exact mechanism of digitalis effect remains unknown. It has been suggested that an increase in sodium influx may be the trigger mechanism for increasing calcium release from the sarcoplasmic reticulum. Digitalis also stimulates ATPase in low doses and inhibits it in larger (or possibly toxic) doses; since ion transport across the cell membrane is regulated by the ATP-dependent "sodium pump," this ATPase activity may be a critical step in digitalis action.

Cell nuclei are located in the center of the other intracellular elements and are surrounded by a double membrane, site of the nucleic acid and protein synthesis which is so necessary to maintain contractile elements, enzymes and structural membranes in working order. Nearby also are the *lysosomes*, containing catabolic enzymes for cellular "cleanup and repair" purposes and a variety of other membrane-enclosed inclusion bodies of unknown function.

MEASUREMENTS OF PERFORMANCE

Because of its critical role in determining pressure and flow relationships throughout the vascular system, the performance of the cardiac pump is usually discussed clinically in terms of two measurements, cardiac output and cardiac work. A third measure, cardiac energy, although not as commonly used, is also important to understand.

Cardiac output (expressed as volume of ventricular outflow per minute) is carefully regulated even in the presence of severe cardiac disease; a fall to critically low levels occurs only after the considerable variety of reserve mechanisms have been exhausted. The volume of output is ultimately determined by two variables—the amount of blood ejected with each heart beat (stroke volume), and the number of beats per minute. Stroke volume is regulated by several factors, which are both intrinsic and extrinsic to the ventricular pump per se. Most basic are the volume of blood available for ejection (which is dependent upon the quantity of venous blood returned to the ventricle), the size of the ventricular chamber, and the amount of residual blood left in the ventricle from the previous stroke. Stroke volume also depends on the rate and force of contraction generated by the ventricular muscle and the net mechanical resistance developed in the outflow pathways. Cardiac rate, on the other hand, is almost entirely regulated by extracardiac factors, most notably the autonomic nervous system; it is this variable of cardiac output which is most readily available for rapid increases, such as that engendered during the beginning of exercise or by walking into an overheated room. Although small phasic differences may be noted, cardiac output is ultimately the same from the right (pulmonary) ventricle and from the left (systemic) ventricle, provided that no anatomical shunts are present to alter the distribution of blood flow.

In practice, the cardiac output is calculated either by the Fick method (total body oxygen consumption divided by the difference in oxygen content between arterial and mixed venous blood) or by one of the indicator-dilution techniques which, incidentally, also utilize the Fick principle (exact quantity of dye injected divided by the concentration of the mixed dye in the circulation). In the latter, an easily identified indicator dye is assumed to be completely and rapidly mixed with venous blood by the time the bolus passes through the pulmonary circulation and then out of the left ventricle. The rate of appearance of the indicator in the arterial blood is inversely proportional to the rate of flow. Stated in simpler terms, sluggish blood flow causes a marked prolongation of the concentration-time dye appearance curve, whereas rapid

flow results in a much smaller curve. The cardiac output is an extremely useful clinical tool in quantitating the effects of therapy or in determining the myocardial resting and reserve function by contrasting the circulatory effects of various activity states (e.g., sleep, exercise).

Cardiac work is another expression of ventricular performance which considers both the effects of pressure generation and blood flow (as determined above). Two types of work are commonly calculated: *potential* work is the product of force and distance and, with respect to cardiac function, is expressed in terms of the generated ventricular pressure and the volume of blood moved. Determination of ventricular work is usually measured on a beat-to-beat basis, using a pressure-volume diagram (Fig. 1-4). In fact, the mechanical work achieved (pressure-volume product in the ascending aorta) is always somewhat less than the ventricular potential work because of the resistance created by the ventricular outflow tract, the viscosity of the blood, etc. Measurement of *kinetic work* attempts to quantitate this difference by determining the velocity of flow through the aortic orifice

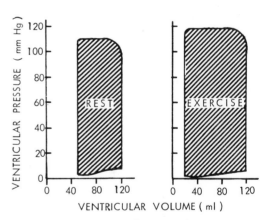

FIG. 1-4. Pressure-volume work loops. These diagrams demonstrate graphically the contrast between the left ventricular potential work per stroke, at rest and during exercise. Note that the increase in exercising stroke volume is primarily the result of a decrease in end-systolic volume.

and calculating its work equivalent. It is but a small fraction of total ventricular work at resting conditions; however, it can amount to sizable fractions during high levels of cardiac performance and in disease states.

The calculation of cardiac work is especially useful clinically in estimating the cost of circulatory performance to the heart itself. Thus, it makes a great deal of difference to the patient whether he is pumping five liters of blood per minute at normal arterial pressure (120/80 mm. Hg) or at hypertensive levels (200/100 mm. Hg); whether there is aortic valvular stenosis or a polycythemia-induced change in viscosity.

Cardiac Energy. If one measures the amount of fuel used by the heart to accomplish its work load, it becomes apparent that the efficiency of ventricular work is remarkably low, in the range of 5 to 10 percent. The bulk of the energy utilized is liberated as heat, similar to most of the artificial pumps used in industry. Because temperature changes are conducted so rapidly to surrounding tissues, it is hard to measure cardiac heat production with accuracy in the ordinary hospital laboratory. Nonetheless, the concepts of *energy* and *efficiency* are critical to understanding the true performance of the heart. Efficiency may be expressed by the following equations:

$$\text{Efficiency} = \frac{\text{Mechanical work}}{\text{Total energy consumed}} =$$

$$\frac{\text{Ventricular potential work} - \text{Kinetic work}}{\text{Potential work} + \text{Heat production}}$$

If the coronary arterial blood supply is compromised because of atherosclerotic occlusion, the amount of cardiac energy available is fixed at a level which may be lower than that necessary to sustain normal pump function. The physician must therefore know how to improve cardiac efficiency if he hopes to keep the patient alive. Often correction of acid-base imbalance, ablation

of arrhythmias, or the strengthening of cardiac contraction with digitalis will significantly improve mechanical work, thereby decreasing the proportion of energy converted into heat.

MYOCARDIAL ADAPTABILITY

One of the limitations of the concepts of cardiac output, work and energy is that they are indices of the pump functions of the heart but do not measure the true performance potential of the ventricular myocardium. Frequently, as in the presence of valvular deformity, we need further information about the state of the cardiac muscle per se. As a result, careful physicians have returned to concepts of basic muscle behavior to understand ventricular function.

Two basic properties of cardiac muscle are involved in adaptation: change in diastolic fiber length and change in contractility. The former, commonly referred to as the Frank-Starling phenomenon, can be demonstrated by the stretch of an isolated muscle fiber. The greater the stretch of the fiber (up to a maximal capacity), the greater the degree of fiber shortening during contraction. Starling demonstrated this by varying the volume of venous return to the heart in an experimental heart-lung preparation; the larger the return volume, the greater the stretching of the ventricular walls and the volume of blood ejected during the next cardiac beat. This finding has been confirmed more recently in intact animals with cardiac denervation, and there is little doubt now that such an entity as intrinsic myogenic regulation does exist.

The role of diastolic fiber length in controlling cardiac performance can be seen in a common clinical experience: The patient with complete heart block is unable to develop a ventricular rate faster than 40 beats per minute but, nonetheless, is able to maintain a normal cardiac output. This is made possible by a prolonged diastolic filling period and an increased end-diastolic volume; stroke ejection volume is enhanced greatly enough that resting cardiac output is not embarrassed. In congestive heart failure the diastolic fiber strength is regularly called into play to aid in regaining compensation. The response of the failing heart muscle to increased stretch is apparently comparable to response of the nonfailing muscle.

The contractile force generated by myocardial wall tension has been the subject of intensive investigation during the past ten years. The level of myocardial tension has been difficult to quantitate accurately, and a variety of indices have been used, most notably the tension-time index (the product of integrated ventricular pressure and pulse rate) and dP/dt, or rate of change of ventricular pressure rise prior to ejection. Both of these indices have correlated reasonably well, but not perfectly, with measurements of myocardial oxygen consumption. Again, it has been helpful to return to basic muscle physiology for guidance in understanding better this contractile force element.

The most fundamental property of muscle is its force-velocity relation, in which the velocity of shortening is inversely proportional to the magnitude of the load. If there is no load at all, there is maximum velocity of shortening (V_{max}), whereas, if the load is so heavy that the muscle cannot shorten, velocity is zero and contraction is said to be *isometric*. V_{max} is never achieved in a clinical situation, of course, and can only be calculated mathematically. Isometric contractions, on the other hand, are frequently seen in premature cardiac beats, where the force of contraction is inadequate to open the aortic valve and no ejection occurs from the ventricle.

When the length of an isolated strip of cardiac muscle is progressively increased, and the results plotted, a series of force-velocity curves can be derived. All curves extrapolate to the same point at zero load

(V_{max}), suggesting that the *contractile* state of the muscle strip is functioning in a predictable and measurable fashion. When this concept is extended to the intact heart, similar conclusions may be drawn. In particular, useful insight into the contractility of the myocardium can be gained by plotting the velocity of shortening as a function of ventricular pressure, extrapolating this relation to zero pressure as an estimate of V_{max}.

Although contractility, as defined by V_{max}, is constant under any given set of circumstances, it can be changed by a variety of neural, humoral and pharmacological interventions, so that new pressure-velocity relationships become apparent. It is uncommon in clinical usage for one isolated factor (such as the contractile state) to change while others remain constant. As an example, epinephrine increases both V_{max} and the maximum pressure of contraction. The concept of V_{max} is therefore of greatest usefulness in the experimental laboratory where mechanisms of action can be carefully controlled. Nonetheless, the estimation of V_{max} can provide clinically valuable information about the inotropic state of the heart in any patient in whose case there is a question about relative therapeutic goals. For instance, a patient with congestive heart failure who is having frequent ventricular extrasystoles may become much worse when given many of the standard antiarrhythmic drugs. A fall in V_{max} would be a highly sensitive test of these drugs prior to long-term utilization in the individual patient concerned, since it would reflect a dangerous drop in the capacity of the myocardium to adapt long before changes in pressure and flow became apparent.

In summary, two distinct and basic properties of cardiac muscle are involved in adaptation: changes in end-diastolic fiber length and changes in contractility. These are measured in actual practice as changes in ventricular end-diastolic volume and ejection velocity. When to these are added the absolute values for stroke volume and pressure, cardiac rate and heat production, all the variables of myocardial performance are subject to analysis. The precise measurement of these few determinants of ventricular function is the goal of the excellent cardiologist.

CARDIAC DILATATION AND HYPERTROPHY

Although beat-to-beat variation in end-diastolic fiber length is a manifestation of continuing cardiac autoregulation, the overall size of the ventricular chamber is kept surprisingly constant to within a few cubic centimeters at any particular level of cardiac function, such as rest or exercise. Dilatation of the ventricle as a clinically recognizable entity is an entirely different mechanism of adaptation. In this situation the primary factor is a marked increase in the residual volume of the ventricle due to a rearrangement of the muscle fibers. There is no increase in fiber stretch, or end-diastolic fiber length, over that seen in ventricles of smaller size. This muscular rearrangement can be demonstrated by the effects of the dilatation process on a sheet of myocardium one inch wide and 10 cell layers thick: If the muscle fibers spread apart and interdigitate so that the sheet is only 5 cell layers thick, the width will have increased 100 percent, from one to two inches. Such dilatation remains easily reversible unless other pathological processes, such as inflammation or fibrosis, supervene. Dilatation occurs early in all volume-overload syndromes, such as that induced by valvular insufficiency or intracardiac shunts.

There is definite improvement in the ability of the heart to eject a volume-load from the dilated ventricle. If each cardiac fiber shortens as much as before, an increase in surface area should bring about a proportionate increase in the volume of blood ejected with each stroke, regardless of the amount of residual blood in the ventricle at the end of the ejection cycle. The energy cost of such work is disproportionately high, however, and may be devastating when the cause of the dilation is related to coronary arterial occlusion and myocardial ischemia. As dilatation and concomitant increase in oxygen demand progress, the heart may develop acute manifestations of myocardial insufficiency, with congestive heart failure or angina pectoris as the clinical consequences.

Prolonged hyperactivity of the heart in any capacity (e.g., increased contractility, increased end-diastolic fiber length, dilatation) inevitably results in *hypertrophy* of the cardiac fibers. Indeed, the very processes that activate the intracellular energy control mechanisms (see Mechanisms of Cardiac Failure, p. 16) also stimulate nucleic acid and protein synthesis. Although there is no increase in the total number of myocardial cells, a remarkable increase in myocardial mass can develop owing to an increased number of the contractile elements (myofibrils) in the cell. This is accompanied by an increase both in the supportive cellular elements, such as the number and size of mitochondria and the amount of sarcoplasmic reticulum, and in the number of capillaries supplying the cell. Ultimately a new level of compensation is reached in which the work load is distributed so that, per unit of myocardial mass, there is normal contractile function, energy production and protein synthesis. Aortic valvular stenosis, for example, imposes a pressure load on the left ventricle and thereby stimulates generalized myocardial hypertrophy. If the stenosis is not progressive, at some point sufficient increase in muscle mass will have accumulated so that the distribution of cardiac work *per cell* again reaches normal levels; at that time, hypertrophy ceases and ventricular function reaches a new plateau of stability.

CARDIAC FAILURE

Exhaustion of cardiac reserve and myocardial adaptive capacity can occur at multiple levels. Circulatory events entirely independent of adequate function of the cardiac pump sometimes appear which are incompatible with life. Thus, hemorrhage can produce a state of hypovolemia so severe that no pump force, however great, can maintain adequate cardiac output. The converse is equally true—that is, marked hypervolemia (caused, for example, by uncontrolled administration of intravenous fluids) can exceed the capacity of the pump and its vascular bed, resulting in loss of fluid into perivascular tissues (classically demonstrated by the pulmonary edema syndrome). Severe hypermetabolism induced by extreme heat, or hypoxia from poisoning of the oxidative pathways (e.g., with sodium cyanide) can induce tissue circulatory demands beyond the capacity of a normal pump.

More common hypermetabolic disorders such as fever or anemia produce high output demands which the normal heart usually meets adequately. Table 1-1 lists several causes of high output syndromes and their mechanisms. These hyperkinetic states, however, are beyond the capacity of a diseased heart which has already called into play its adaptive mechanisms to maintain circulatory homeostasis. For example, it is common to see patients with compensated rheumatic heart disease develop overt congestive failure during an acute febrile illness.

A variety of purely mechanical problems can interfere with cardiac function. These are most important to recognize, since many are correctable. Into this category fall the "inlet" and the "outlet" disorders, the "shunt" syndromes and various rhythm disturbances.

Inlet disorders are characterized by an obstruction to cardiac inflow and result in markedly increased venous blood proximal to the obstruction (often with increased venous pressure) and low output arterial syndromes distally. Constrictive pericarditis produces a textbook picture of this type entity; engorged veins of the head, neck and arms, hepatomegaly with ascites, dependent leg edema, arterial hypotension, and effort dyspnea and fatigue. Other common causes are vena cava obstruction from extrinsic masses or fibrosis, pericardial tamponade with blood or serous effusion, myxomatous tumors in the right atrium, and congenital tricuspid stenosis or deformity (Ebstein's disease). The left ventricle has its own kinds of inlet disorders, including obstruction of the pulmonary arteries from

TABLE 1-1. Common High-Output Circulatory States

Syndrome	Mechanism
Normal exercise	*Increased tissue substrate demands*
Fright; anxiety	*Release of circulating epinephrine*
Thyrotoxicosis	*Thyroid hormone stimulation of cyclic AMP*
Fever	*Generalized accelerated enzyme reactions*
Severe anemia	*Decreased oxygen transport to tissues*
Arteriovenous fistula	*Vascular shunt with localized distal ischemia*
Pregnancy	*Intrauterine shunting; accelerated vasoregulatory hormones*
Paget's disease	*Hyperostosis with vascular shunting*
Hyperkinetic states	*Increased responsiveness of beta-adrenergic receptors*
Malignancy	*(?) Circulating pyrogen, oxygen-dependent tumor, arteriovenous shunting, metabolic hormones*

thromboembolism or parenchymal pulmonary disease, pulmonary venous constriction (e.g., in heroin intoxication), left atrial thrombi or tumors, mitral stenosis, and the like.

The *outlet disorders* produce the same types of arterial underperfusion but have their most damaging effects on the ventricles instead of the proximal veins. These also should be classified into right ventricular and left ventricular groups. Pulmonary valvular stenosis is the outstanding example of the former, with marked right ventricular dilatation and hypertrophy. A similar syndrome can be caused by tricuspid valvular insufficiency, infundibular muscular hypertrophy, pulmonary arterial strictures and any of the causes of pulmonary hypertension (including the left heart failure syndromes, primary obliterative lung disease, multiple pulmonary emboli, severe hypoxia from high altitudes, pulmonary vasculitis, etc.) Analogous mechanical interference with left ventricular outflow produces hypertrophy, dilatation and failure of the left heart, emanating from mitral insufficiency, aortic stenosis or insufficiency, coarctation of the aorta and the systemic hypertensive diseases.

Shunt disorders may be present between the cardiac chambers or in the great vessels and are either congenital or acquired in origin. Their manifestations are protean, depending on the precise site and size, and the stage of development of the cardiovascular system. Blood flow is generally shunted from the higher pressure left-sided chambers into the lower pressure right chambers, producing intracardiac and pulmonary hypervolemia. If the shunt is large enough, right and left pressures become equalized and the shunting of blood may be bidirectional. Not infrequently, severe pulmonary hypertension develops to the degree that right-sided pressures are higher than those on the left; blood then flows from right to left and the patient becomes cyanotic. The most frequently found

causes of congenital shunts are atrial septal defect, ventricular septal defect, patent ductus arteriosus and anomalous insertion of pulmonary veins into the venae cavae or right atrium. Acquired shunts are often the results of a ruptured ventricular septum in myocardial infarction or are of traumatic origin.

Finally, extreme variations of heart rate and a variety of arrhythmias are common "mechanical" causes of poor cardiac performance. Marked bradycardia may result in increases in stroke volume such that the heart stays compensated only under basal conditions. Severe tachycardia may induce so much shortening of ventricular filling time and so much increase in cardiac energy requirements that cardiac insufficiency may occur even in the absence of intrinsic heart disease. Several times each year medical students are seen in the emergency room, in a state of circulatory collapse with paroxysmal atrial tachycardia or fibrillation, induced by fatigue, anxiety or drug stimulants. The skin is cool and clammy; there is marked weakness with mental confusion and restlessness, and the radial pulse is feeble or absent. The response to conversion of the arrhythmia is dramatic and immediate! Evidence of underlying cardiac disease is rarely found.

The arrhythmias that most frequently produce decreased cardiac performance are the atrial and ventricular extrasystoles. When these appear soon after the preceding beat, there is inadequate filling of the ventricle, inept ventricular contraction and pressure development and, often, total failure of ventricular emptying. Ventricular fibrillation is an extreme manifestation of this same mechanism, in which the totally incoordinated contractions of the ventricular fibers produce no filling or emptying of the cavity, and death occurs within a very few minutes.

In summary, the state of health of the cardiac muscle is not the primary factor in determining competent performance of the heart in the high output and mechanical disorders; sometimes, nevertheless, congestive failure may be the secondary or end result. In its true sense, and in contrast to the situations just described, cardiac failure means myocardial failure, with intrinsic abnormality of the heart muscle itself. Hyperactivity states and mechanical problems can certainly precipitate reactive changes in the myocardium if they remain present for a sufficient period of time. More commonly there are other etiological changes, such as a decrease in nutrient blood flow, inflammation, cellular enzymatic blockade or infiltrative and degenerative diseases present.

We shall next focus on the mechanisms of response of the muscular part of the pump.

MECHANISMS OF MYOCARDIAL FAILURE

The syndrome of congestive heart failure seems to present no conceptual difficulty when the mechanism is progressively inadequate coronary arterial perfusion: One expects to find progressive myocardial atrophy, with focal or diffuse fibrosis· and perhaps spotty areas of compensatory hypertrophy in areas of improved collateral blood flow. This indeed is found in a certain number of patients; frequently, however, a microscopic picture of reasonably normal or generally hypertrophied myocardium is seen, with minimal fibrosis or inflammatory cell infiltration. Since cardiac failure occurs so frequently in the absence of severe myocardial atrophy, emphasis in pathogenesis has shifted from a critical loss of muscle cells to a functional change in the viable cells. To be more specific, an understanding of alteration in the physiological chemistry of the myocardium is fundamental to an understanding of the failing heart.

Intracellular metabolic events of potential interest in explaining cardiac failure may be grouped into four general areas: (1) avail-

ability and usage of "raw materials" to produce energy; (2) adequate storage of the energy produced in high energy bond compounds (predominantly ATP and, to a lesser extent, creatine phosphate); (3) harnessing of the high energy compounds to produce satisfactory myofibrillar contraction, and (4) synthesis of nucleic acids and protein to maintain or increase cellular function as needed.

The essential raw materials for liberation of energy are oxygen, carbon, hydrogen and, possibly, nitrogen. The primary fuel substrates are glucose and fatty acids, with minor amounts of amino acids being utilized. Very careful studies have reconfirmed that glucose, fatty acid and amino acid utilization for energy does not significantly differ in normal and chronically failing hearts, in vitro or in vivo, when results are expressed as amount utilized per unit of myocardial muscle. (Acute heart failure is an exception, since marked biochemical and physiological hyperfunction occurs without an increase in tissue mass.)

Severe hypoxia can undoubtedly lead to improper function of the myocardial cell, since the amount of ATP produced by anaerobic metabolism is only a fraction of that produced aerobically (Fig. 1-5). That hypoxia is the underlying mechanism for chronic cardiac failure, however, is difficult to prove. Accumulation of lactate seems to be reliable evidence for cardiac hypoxia, as shown in Figure 1-5, and this has been a reasonably constant observation during angina pectoris; it is occasionally present, but uncommon, in chronic heart failure. Moreover, numerous studies have found no measurable decrease in coronary blood flow at resting or exercising states in the majority of patients with chronic failure. Relative myocardial hypoxia may exist, of course, but it has been difficult to demonstrate by comparison of the volume of flow with the mass of functioning myocardium.

Production of high energy phosphate compounds in the myocardium has been a difficult and elusive subject in man. Based

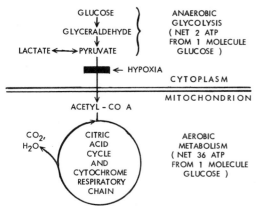

GLUCOSE
↓
GLYCERALDEHYDE
↓
LACTATE ←→ PYRUVATE

ANAEROBIC GLYCOLYSIS (NET 2 ATP FROM 1 MOLECULE GLUCOSE)

← HYPOXIA

CYTOPLASM

MITOCHONDRION

ACETYL - CO A

CO_2, H_2O ←

CITRIC ACID CYCLE AND CYTOCHROME RESPIRATORY CHAIN

AEROBIC METABOLISM (NET 36 ATP FROM 1 MOLECULE GLUCOSE)

Fig. 1-5. Schematic of intracellular metabolic pathways. Note the enormous difference in efficiency of the aerobic degradation cycle, compared with the anaerobic degradation cycle, in the production of high-energy compounds. The constantly contracting fibers of the myocardium require vast supplies of high-energy stores; thus, it is easily understandable that the myocardium should be much more oxygen-dependent than other tissues.

Hypoxia results in arterial lactate accumulation—a contrast to the usual finding of lactate utilization as fuel.

on experimental work in animals the following conclusions have been reached: During acute ventricular failure resulting from pressure or volume overload, an actual increase of ATP production occurs per unit of muscle. Because of the extremely high energy demands, however, tissue stores of creatine phosphate are very low and those of ATP are borderline. During chronic failure, both ATP production and tissue concentration have a tendency to be low, a finding in harmony with the observation in man that total mitochondrial mass is decreased and mitochondrial catabolism is increased during chronic heart failure. However, it seems unlikely that depressed ATP production can be the essential factor initiating cardiac failure, since additional ATP does not reverse the syndrome.

Utilization of High-Energy Stores. The ability of the heart to utilize the high energy stores in effective cardiac contraction is of great importance. A variety of attempts have been made to demonstrate changes in the physical properties of the

contractile proteins themselves. No consistent results have been found, and such "end organ" abnormalities must remain conjectural at this time. The main reason for suspecting such an actomyosin abnormality is a decrease in the capacity of the heart to hydrolyze ATP during chronic cardiac failure. This ATPase, or energy-relating, activity is for the most part associated with myosin function. Experiments have shown that, in spite of depressed ATPase activity in heart failure, the conversion of chemical energy to mechanical work is normal. Such observations reduce the likelihood of serious defects in energy production and, consequently, emphasize the situations that limit the rate of energy utilization.

Hormonal control of the energy utilization process is of considerable interest. The final common pathway for these compounds is now known to reside in the transformation of ATP into cyclic 3′,5′ AMP via the enzyme adenyl cyclase, located in the subcellular membranes (Fig. 1-6). Cyclic AMP thus produced is responsible for transformation of phosphorylase to a metabolically active form, initiating the entire chain of glycogen release and further energy supply. Moreover, cyclic AMP seems to regulate the calcium-pump system in the sarcotubules, which controls the coupling and uncoupling of the contractile proteins. It appears that ionized calcium achieves this goal by inactivating tropinin, one of the trace proteins that inhibits actin and myosin coupling. The removal of calcium and consequent release of tropinin to the effector sites uncouples the reaction, initiating muscle relaxation.

Only two hormone groups appear capable of activating the adenyl cyclase system—catecholamines and glucagon, each working via a different intermediary receptor. Little is known about endogenous glucagon metabolism in cardiac failure, but cardiac norepinephrine stores have been shown repeatedly to be seriously depleted.

Still other end-products of ATP metabolism—nucleic acid and protein synthesis—

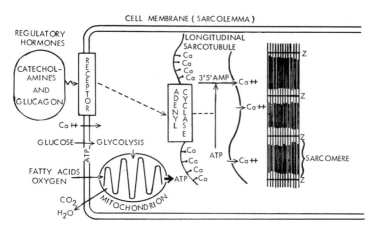

CELL MEMBRANE (SARCOLEMMA)

REGULATORY HORMONES

CATECHOL-AMINES AND GLUCAGON

RECEPTOR

LONGITUDINAL SARCOTUBULE

ADENYL CYCLASE

3'5'AMP

GLUCOSE → GLYCOLYSIS

FATTY ACIDS OXYGEN

MITOCHONDRION

CO_2
H_2O

SARCOMERE

Fig. 1-6. Regulation of coupling-uncoupling. The neurohumoral hormones appear to set the level of contractile force of the myofibril via the adenyl cyclase enzyme system located in the cell membranes. Adenyl cyclase determines the rate of cyclic AMP formation from precursor ATP and appears to be critical to proper cell function. Ca^{++} is released from the longitudinal sarcotubular system by cyclic AMP, initiating actomyosin by "salting out" the inhibitor protein, tropinin.

are defective in cardiac failure. These may be the most elemental problems of all, since the depressed regenerative capacity is responsible both for the loss of mitochondria and for the decrease in myofibrillar ATPase activity. A vicious circle is thus initiated, in which impaired protein synthesis is responsible for defective ATP production and utilization, and impaired energy metabolism further interferes with new protein generation—all resulting in a progressive decrease in cardiac contractile force.

In summary, there is evidence in varying degrees for metabolic inadequacy in all the essential stages of cellular performance during congestive heart failure, ranging from oxygen unavailability to inadequate ATP production and utilization and, finally, to impaired structural repair with failure to generate new contractile elements.

THE MULTIPLE LEVELS OF NUTRIENT FAILURE

There is a very common tendency among clinicians to think that all ischemic heart disease is related to occlusive coronary arteriosclerosis. Few would question that arteriosclerosis is the "black plague" of the twentieth century, nor would the author. There is increasing evidence, however, that "coronary insufficiency," "angina pectoris,"

"coronary artery disease," "arteriosclerotic heart disease" and "myocardial ischemia" are not synonymous terms. If the definition of the ischemic syndrome focuses on the end organ, the cardiac muscle, it becomes evident that many of the affected patients do not have plaques occluding the major coronary vessels. Figure 1-7 depicts several sites where myocardial ischemia may be produced.

The clinical manifestations of the various types of ischemic heart disease are likely to be the same: angina pectoris, acute infarction of the myocardium, congestive heart failure, and a variety of intracardiac conduction disorders often terminating as arrhythmias. Angina is the hallmark syndrome; in its classical form it is characterized by precordial constricting pain induced by emotional or physical stress and relieved promptly by sublingual nitroglycerin. A great variety of atypical pain syndromes—in the epigastrium, chest, back, neck or arms—may also be angina equivalents, and temporary dyspnea and palpitations may be associated symptoms—or, indeed, the only complaints present. Angina is a referred pain, of course, since there are no pain fibers per se in the ischemic myocardium; the reflex is carried by afferent sympathetic fibers to ganglia at the base of the neck and, thence, to pain structures in the chest wall and the shoulder girdle.

Intercostal nerve block is an effective way to relieve intractable anginal pain. The exact substance in the ischemic myocardium that triggers the sympathetic reflex arc is unknown. Nitroglycerin induces its dramatic antianginal effect by reducing cardiac work through its peripheral vasodilator properties; possibly an increment in collateral flow to the ischemic myocardium may be a result of its coronary vasodilator action, but this seems unlikely in the presence of major vascular obstruction from arteriosclerosis.

Atherosclerotic occlusion of the main coronary conduits comprises perhaps 50 percent of the entire group (Fig. 1-7A). Under other circumstances there may be circulatory impairment due to widespread small vessel (arteriolar) disease, as in diabetes or polyarteritis nodosa (Fig. 1-7B). The hemoglobin affinity for oxygen may be abnormal, as has been noted in 2,3-diphosphoglyceride deficiency states (Fig. 1-7C). Abnormalities at the capillary-cell membrane level (Fig. 1-7D) have not yet been described clinically, but may be analogous to the alveolar-capillary block syndrome in the lung. Other changes at this level may

be inadequacy in the number of capillaries supplying a hypertrophic muscle fiber, or insufficient oxygen availability in the center of a hypertrophic cell. The latter two situations have been demonstrated in a number of experimental situations which are thought to be relevant to the hypertrophied myocardium of man.

Intracellularly, other problems may arise which might ultimately result in ischemic heart disease. The transfer of energy from oxygen to ATP depends upon mitochondrial integrity, including the enzymes of the intermediary citric acid cycle, the respiratory oxidative chain, and the cytochrome C system (Fig. 1-7E). Even when ATP is produced by anaerobic glycolysis, at least 16 enzymes have been shown to be involved.

Beriberi is a good example of the production of ischemic heart disease at this level. Beriberi results from thiamine deficiency, and, since thiamine is an essential component of the decarboxylation enzymes, the oxidative cycle is severely maimed. The resultant ATP deficiency is noted throughout the body, especially in the highly ATP-dependent cardiac muscle.

Still other clinical problems may be re-

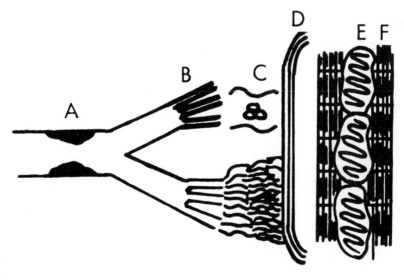

FIG. 1-7. Sites of etiological factors in myocardial ischemia. (A) Coronary main stem atheroma. (B) Small vessel disease. (C) Abnormal oxy-hemoglobin dissociation. (D) Capillary-membrane permeability changes. (E) Changes in ATP production. (F) Inept ATP utilization or contractile protein abnormalities.

lated to improper storage or utilization of the high energy phosphate compounds (see Mechanisms of Myocardial Failure, above). In thyrotoxicosis, for example, there is much waste in the use of ATP in muscle contraction, as a result of thyroid hormone and norepinephrine excess, producing the so-called uncoupling of oxidative phosphorylation.

Finally, one may assume that certain ischemic disease states ultimately will be expressed as defective end-organ responsiveness, or abnormalities in the myofibrils per se (Fig. 1-7F). Some conditions classified loosely as "primary cardiomyopathy" undoubtedly fall into such a category. Biochemical definition of these fairly common clinical states has not yet been achieved.

ANNOTATED REFERENCES

SELECTED REVIEW ARTICLES AND SYMPOSIUM REPORTS

Badeer, H. S.: Metabolic basis of cardiac hypertrophy. Progr. Cardiovasc. Dis., *11*:53, 1968.

Bajusz, E. (ed.): Symposium on experimental "metabolic" cardiopathies and their relationship to human heart diseases. Ann. N. Y. Acad. Sci., *156*:1, 1969 (The 42 papers presented at this 1967 conference express the basis of our knowledge on the cardiomyopathies.)

Bing, R. J. (ed.): Symposium on myocardial metabolism. Am. J. Cardiol. 22:297, 1969. (Eight papers are included, primarily concerned with oxidative metabolism. The editor is in many ways the "father" of myocardial metabolism and the most innovative and productive investigator in this field.)

Braunwald, E., and Pool, P. E.: Mechanism of action of digitalis glycosides. Mod. Concepts Cardiovasc. Dis., 37:129, 1968. (A good summary of the enigmas of this, the earliest known cardioactive drug)

Brest, A. N., and Moyer, J. H. (eds.): Symposium on congestive heart failure. Am. J. Cardiol., 22:1-48, 151-194, 1968. (Several papers explore the etiologic factors, pathophysiology and management of this important entity.)

Bruce, T. A., and Bing, R. J.: Coronary sinus catheterization, including a review of coronary blood flow and myocardial metabolism. *In:* Zimmerman's Intravascular Catheterization, Ed. 2, p. 888. Springfield, Illinois, Charles C Thomas, 1966.

Bruce, T. A., Chapman, C. B., Baker, O., and Fisher, J. N.: The role of autonomic and myocardial factors in cardiac control. J. Clin. Invest., *42*:721, 1963. (Progressive and total denervation studies indicate the capacity of the heart to increase its output and work performance during exercise at all experimental levels.)

Burch, G. E., and Giles, T. D.: The burden of a hot and humid environment on the heart. Mod. Concepts Cardiovasc. Dis., *39*: 115, 1970.

Chapman, C. B. (ed.): Physiology of muscular exercise. Circulation Res., *20*:Suppl. I:1, 1967. (This carefully structured symposium serves as an excellent summary of present knowledge in the field of the physiology of physical exercise.)

Cohen, L.: Contributions of serum enzymes and isoenzymes to the diagonsis of myocardial injury. Mod. Concepts Cardiovasc. Dis., *36*:43, 1967.

Davila, J. C. (ed.): Symposium on measurement of left ventricular volume. Am. J. Cardiol., *18*:1, 208, 1966.

Ebashi, S., Kadama, A., and Ebashi, F.: Tropinin. I. Preparation and physiological function. J. Biochem., *64*:465, 1968. (Much of our information about the "inhibitor" proteins comes from these investigators, and they have provided stimulating and refreshing insight into the area of actin-myosin contraction.)

Elliott, W. C., and Gorlin, R.: The coronary circulation, myocardial ischemia, and angina pectoris. Mod. Concepts Cardiovasc. Dis., 35:111, 1966. (This review set the stage for many of our present concepts about the interrelationships of the physiology and pathology of the coronary vascular bed and the myocardium.)

Entman, M. L., Levey, G. S., and Epstein, S. E.: Mechanism of action of epinephrine and glucagon on the canine heart: Evidence for increase in sarcotubular calicum stores mediated by cyclic 3',5' AMP. Circulation Res., 25:429, 1969. (The authors explore the significance of the enzyme adenyl cyclase, and its effect on cardiac performance. The work delineates one of the giant steps of the past decade in our understanding of the physiological chemistry of the myocardium.)

Evans, J. R. (ed.): Symposium on structure and function of heart muscle. Circ. Res., *14* (Suppl. 2):1, 1964. (A balanced monograph containing the proceedings of a symposium held in Toronto in 1964. There are several papers in each of four fields: mechanism of muscular contraction, metabolism and myocardial function, regulation of cardiac performance and rhythmicity-conduction.)

Fishman, A. P. (ed.): Symposium on the myocardium—Its biochemistry and biophysics. Circulation, *24*:324, 1961. (For a decade, the outstanding monograph of its kind, with the most beautiful electron photomicrographs yet published on cardiac ultrastructure and an in-depth look at the physiology and metabolism of heart muscle and its contractile proteins)

Grant, R. P.: Architectonics of the heart. Am. Heart J., *46*:405, 1953. (A scholarly showpiece embracing a special "hobby" interest of this great humanitarian, teacher and scientist)

Hoffman, B. F., Cranefield, P. F., and Wallace, A. G.: Physiological basis of cardiac arrhythmias. Mod. Conc. Cardiovasc. Dis., *35*:103, 1966.

Hultgren, H. N., and Flamm, M. D.: Pulmonary edema. Mod. Concepts Cardiovasc. Dis., *38*:1, 1969.

James, T. N., and Sherf, L.: Ultrastructure of myocardial cells. Am. J. Cardiol., *22*:389, 1968. (A definitive report on cellular classification and function in heart muscle)

Kaplan, M. H. (ed.): Symposium on immunity and the heart. Am. J. Cardiol. *24*:457, 1969. (A vital "new" field of interest to the student of cardiovascular medicine and surgery. The symposium explores cardiac antigens, autoimmunity, anaphylaxis and problems in cardiac transplantation.)

Kumar, S., and Spodick, D. H.: Study of the mechanical events of the left ventricle by atraumatic techniques: Comparison of methods of measurement of their significance. Am. Heart J., *80*:401, 1970.

Leonard, J. J., and de Groot, W. J.: The thyroid state and the cardiovascular system. Mod. Concepts Cardiovasc. Dis., *38*:23, 1969.

Mason, D. T., Spann, J. F., Zelis, R., and Amsterdam, E. A.: Alterations of hemodynamics and myocardial mechanics in patients with congestive heart failure: Pathophysiologic mechanisms and assessment of cardiac function and ventricular contractility. Prog. Cardiovasc. Dis., *12*:507, 1970.

Meerson, F. Z.: The myocardium in hyperfunction, hypertrophy and heart failure. Circulation Res., *25* (Suppl. II):1, 1969. (The foremost Russian scientist in his field summarizes the contributions to our understanding of myocardial metabolism, physiology and pathophysiology.)

Neal, R. W., Nair, K. G., and Hecht, H. H.: A pathophysiological classification of cor pulmonale: With general remarks on therapy. Mod. Concepts Cardiovasc. Dis., *37*:107, 1968.

Shabetai, R. (ed.): Symposium on pericardial disease. Am. J. Cardiol. *26*:445, 1970.

Spann, J. F., Mason, D. T., and Zelis, R. F.: Recent advances in the understanding of congestive heart failure. Mod. Concepts Cardiovasc. Dis., *39*:73, 1970.

Stock, T. B., Wendt, V. E., Hayden, R. O., Bruce, T. A., Gubdjarnason, S., and Bing, R. J.: New concepts of angina pectoris. Med. Clin. N. Am., *46*:1497, 1962.

Sutherland, E. W.: On the biological role of cyclic AMP. JAMA, *214*:1281, 1970. (An up-to-date summary of the multiple levels of biochemical control by the acknowledged master in the field)

Wendt, V. E., Stock, T. B., Hayden, R. O., Bruce, T. A., Gudbjarnason, S., and Bing, R. J.: The hemodynamics and cardiac metabolism in cardiomyopathies. Med. Clin. N. Am. *46*:1445, 1962.

Wood, E. H. (ed.): Symposium on use of indicator-dilution techniques in the study of the circulation. Circulation Res., *10*:377, 1962.

2

Tissue Perfusion

Michael J. Brody, PH.D., *and François M. Abboud,* M.D.

INTRODUCTION

Cellular functions are limited by and dependent upon a system for delivery of chemical substances required for metabolic activity and for removal of the waste products of that activity. In the complex multicellular arrangement comprising the human body, the delivery and removal system is provided by the cardiovascular system. The adequacy with which the cardiovascular system performs its functions is determined ultimately by the magnitude of tissue perfusion. If the tissues are perfused with a sufficient amount of normally constituted blood to meet metabolic requirements and to provide for disposal of waste materials, the purpose of the cardiovascular system is satisfied.

Tissue perfusion is controlled by two major factors: arterial blood pressure, which provides the energy for driving blood through the vascular channels; and the resistance to blood flow provided by the physical characteristics of the blood vessels and of the blood itself.

Blood Pressure. The control of arterial blood pressure is considered in detail in other chapters. Very briefly, arterial pressure is the result of cardiac output and the resistance to flow provided by the vessels. In the cardiovascular system pressure is developed by the active contraction of the heart. This contraction produces the cardiac output which is distributed throughout the system in a manner dependent upon the resistance to flow provided by individual segments of the vascular tree. Two

elementary considerations should be emphasized: First, without adequate myocardial contractile activity, perfusion is severely restricted. Secondly, although blood pressure is determined by both cardiac output and the resistance to blood flow, in vascular beds with low potential for producing active changes in vascular resistance (e.g., when neurogenic influences to the vessel are low and the vessels themselves respond poorly to vasomotor influences) blood pressure becomes the major determinant of perfusion. As will be seen later on, in certain circumstances adequate tissue perfusion may be directly related to the magnitude of arterial pressure.

Vascular Resistance. Emphasis will be placed in this chapter upon those factors which play prominent roles in determining the adequacy of tissue perfusion by alterations in the resistance to blood flow. Vascular resistance may be broadly defined as the resistance to blood flow provided by the geometry of the vascular tree and by the physical properties of the fluid circulating through the vascular compartment. According to the Poiseuille relationship, the length and the radius of a tube are the geometric components providing resistance to flow through the tube. Although this relationship does not hold strictly for the distensible vascular compartment, it can be concluded that radius (which is an exponential function) is the major determinant of resistance to flow. Radius is contributed to and altered by passive elastic components of the vessel wall, muscular elements of the vessel wall that contract actively, and structural alterations in the vessel wall that may provide obstruction to flow.

The resistance to blood flow provided by the total cardiovascular system can be estimated by dividing the difference between arterial pressure and right atrial pressure by the cardiac output. This value, referred to as total peripheral resistance, can be a useful index of the average or overall radius of the vascular compartment. Total peripheral resistance does not, however, provide any estimate of the distribution of blood flow between organs or of the adequacy of perfusion of any or all organs. In addition, total peripheral resistance does not provide any direct measure of the contribution of active or passive tension of vascular smooth muscle to the level of radius in the vessel wall. For example, the administration of a ganglionic blocking agent, by interfering with neurogenic vasoconstrictor influence on the vessel wall, may be expected to reduce active tension in the vessel wall and thus reduce vascular resistance. However, total peripheral resistance is unchanged following administration of the ganglionic blocker because cardiac output and blood pressure are reduced in approximately the same proportion. The average caliber of the vascular tree remains unchanged because of the reduction in arterial (i.e., distending) pressure which allows the vessels to adjust passively to unchanged caliber. This example is used to illustrate the point that active tension may change when there is no change in computed total peripheral resistance if at the same time arterial pressure is altered.

Resistance to blood flow provided by individual organs can also be determined with degrees of difficulty that vary, depending upon the organ under consideration. These measurements of vascular resistance depend upon some accurate determination of blood flow to the organ. Flow to the forearm (primarily muscle) and skin (e.g., a digit) may be determined with relative ease using plethysmographic techniques.

Renal blood flow can be estimated with the clearance of PAH, but any accurate measurement of blood flow to the kidney requires that the extraction of PAH be determined, since virtually all interventions that alter renal blood flow change intrarenal distribution of that flow, and thus clearance measurements per se can be misleading. Dye dilution has also been used clinically for renal blood flow measurements, but the error associated with this measurement may be large. Flow measurements for other individual organs such as the brain, heart, liver, intestine, etc. are much more difficult and less routinely performed. In terms of the estimate of the adequacy of tissue perfusion, virtually all methods presently available for the determination of vascular resistance in individual organs are inadequate. Ultimately the determination that proves to be most meaningful is the estimation of the intraorgan distribution of flow which provides estimates of how effectively the available flow reaches that portion of the circulation where exchange between blood and tissues may occur.

CHANGES IN VASCULAR SMOOTH MUSCLE CONTRACTION

The contractile state of vascular smooth muscle is perhaps the major physiological determinant of tissue perfusion. Three factors regulate the contractile activity of vascular smooth muscle. For purposes of this discussion these are referred to as *humoral, neurogenic* and *intrinsic* factors.

Humoral Factors Regulating Smooth Muscle Contraction

A large number of substances with vasoactive properties are carried in the blood and exert important effects on vascular smooth muscle tension and, thus, on tissue perfusion. In any portion of the vascular tree the level of active tension in smooth muscle affected by humoral factors is the algebraic sum of stimulant and relaxant influences exerted upon the muscle.

Adrenergic. Catecholamines found in the blood arise from several sources. The major source of epinephrine is the adrenal medulla, another source being extraneous chromaffin tissue. Norepinephrine is derived primarily from its secretion from adrenergic nerve terminals. Catecholamines may produce either vasodilation or vasoconstriction, depending on the affected organ. Thus, epinephrine is a powerful human skeletal muscle vasodilator. Both epinephrine and norepinephrine have the potential to constrict coronary blood vessels; but their net effect on the coronary tree is relaxation produced secondarily to their stimulant effects on myocardial contractility and rate. This coronary dilatation is perhaps mediated by adenine nucleotides. Prominent vasoconstrictor effects of the catecholamines are observed in the cutaneous, renal, and mesenteric circulation, whereas the cerebral vascular bed is much less reactive to adrenergic stimulation. Ordinarily, the levels of catecholamines circulating in the blood probably exert minimal effects on tissue perfusion. However, under certain conditions increased blood levels of catecholamines may produce important effects on distribution of blood flow and on tissue perfusion. Prominent examples are pheochromocytoma, hypotensive shock, anxiety state, exercise, hypertension, and cardiac failure.

In *pheochromocytoma* arterial pressure may be increased by vascular effects or cardiac effects or a combination of the two. Vasoconstriction alone may raise arterial pressure, and this increased pressure may be reduced dramatically by the administration of an alpha-adrenergic receptor blocking agent such as phentolamine. Intense stimulation of the heart by high levels of circulating catecholamines may also lead to hypertension. In such a situation the administration of a beta-adrenergic receptor blocking agent such as propranolol would be more likely to lower arterial pressure.

In *hypotensive shock,* cardiovascular reflexes attempt to compensate for reduced arterial pressure by increasing sympathetic discharge. If these reflexes are at all intact, the secretion of catecholamines from adrenergic nerves and from the adrenal medulla is enhanced and the circulating blood levels of these substances will be increased. Poor tissue perfusion in the shock state may be the result primarily of two contributing factors. Reduced arterial pressure alone will limit the amount of flow available for adequate perfusion of the tissues; if in vital organs this is combined with vasoconstriction promoted by high circulating levels of catecholamines, perfusion is further compromised. In hypotensive shock the level of arterial pressure is not an informative indicator of the adequacy of tissue perfusion, since intense vasoconstriction may be supporting arterial pressure at the expense of adequate perfusion. Treatment of hypotensive shock should always be directed at improving tissue perfusion. Ordinarily, this requires that the output of the heart be improved in addition to whatever measures are taken to relieve vasoconstriction produced by exaggerated sympathetic activity. Alpha-adrenergic receptor blocking agents such as phentolamine and phenoxybenzamine have been used in conjunction with plasma expanders to improve tissue perfusion by relieving vasoconstriction. Although catecholamines probably play a role in producing vasoconstriction in hypotensive shock, they may be used with considerable efficacy in the treatment of shock if special attention is paid to their cardiostimulant properties.

Anxiety states may be associated with increased blood levels of catecholamines. Several cardiovascular signs observed in anxiety are probably attributable to the catecholamines—for example, cold clammy palms, tachycardia, dilated pupils and hypertension.

In *exercise* an extraordinary demand for increased perfusion is made by skeletal muscles and heart. The distribution of blood flow in exercise is appropriate to the demands for flow required by skeletal muscle. Flow to muscle is increased dramatically while flow to other organs such as the kidneys, skin and splanchnic circulations is reduced or remains unchanged. In any event, the *fraction* of cardiac output delivered to an organ such as the kidney is reduced considerably. Although the blood levels of catecholamines are increased during exercise, it is not really known whether these substances play any role in providing for the redistribution of blood flow to skeletal muscle, since infusions of catecholamines do not mimic precisely the cardiovascular response to exercise. It is possible that the vasoconstrictor effects of circulating catecholamines on vascular beds other than skeletal muscle contribute to skeletal muscle hyperemia observed in exercise.

Hypertension. Except for its demonstrated role in pheochromocytoma, an adrenergic humoral mechanism has not been shown to play a significant role in the maintenance of hypertension. Using improved and more precise methods for the detection of catecholamines in blood it has been shown recently that there is a small increase in the level of circulating catecholamines in the blood of essential hypertensives over normotensives. The significance of this observation is not known, although it could reflect some change in the turnover rate of catecholamines in the tissues or in the metabolism of the substances.

Cardiac Failure. Tissue perfusion is impaired in congestive heart failure. Catecholamines in the blood are elevated in cardiac failure and this elevation is also reflected in increased excretion of these amines and their metabolites in urine. It is not known whether the increased levels of catecholamines through their vasoconstrictor effects contribute to the poor tissue perfusion found in congestive heart failure. It

is likely, however, that the adrenergic drive to the decompensated myocardium is derived from a humoral rather than a neurogenic source, since the heart in failure is depleted of catecholamines.

Renin-Angiotensin System. The pathophysiological role of the renin-angiotensin system in hypertension is considered in detail elsewhere in this book. With specific reference to tissue perfusion it seems likely that this system plays an important role only in malignant stages of hypertensive disease. Here the high circulating blood levels of renin and the vasoconstrictor polypeptide angiotensin undoubtedly contribute significantly to the high vascular resistance. It is also likely that the vasoconstrictor properties of angiotensin are prominent in renal vascular hypertension and in eclampsia. This is not to say that the renopressor system does not contribute importantly to normal homeostasis. Clearly, it is a determinant of aldosterone excretion and must also function in a more subtle fashion in control of pressure.

Kinins. A variety of potent vasodilator polypeptides are formed by the actions of certain proteolytic enzymes on plasma protein precursors. The substance bradykinin serves as the prototype for this class of endogenous substances. Their major physiological role has been proposed to be in the local regulation of blood flow and function of such organs as the salivary glands and pancreas. The increased blood flow needed for metabolic activity of such glands is thought to derive at least in part from the vasodilator action of kinins formed within these glands by the secretion of proteolytic enzymes. On the pathophysiological side, kinins are believed to play a part in hyperemia associated with inflammation and as vasodilators in hypotension produced by anaphylactic reactions. Their most prominent pathophysiological role appears to be in the carcinoid syndrome, where the vasodilator peptide bradykinin may be responsible for the characteristic cutaneous flushing.

In these patients, administration of epinephrine causes release of the enzyme kallikrein
from metastatic lesions in the liver. Kallikrein liberates a decapeptide, lysylbrady-
kinin, which is converted to vasoactive
bradykinin. Serotonin and histamine may
also be involved in the hypertensive flushing attacks of carcinoid syndrome.

Prostaglandins. The prostaglandins are a
family of endogenous acidic, lipid-soluble
materials with a wide distribution in the
body and with diverse cardiovascular effects. A given member of the family may
possess the ability to stimulate the myocardium, increase coronary blood flow, constrict venous smooth muscle, relax arterial
smooth muscle, or facilitate adrenergic
transmission to vascular smooth muscle. The
prostaglandins received a great deal of attention in reference to hypertension. Several members of the prostaglandin family,
classified by structure as *E* and *F,* have
been found in renal medullary tissue (perhaps in renal interstitial cells) and have
been shown to be released into renal venous
blood following acute renal ischemia. There
have been suggestions that the prostaglandins may play an antihypertensive role
through their vasodilator properties. Recently prostaglandins have been found to be
elevated in renal venous blood in patients
with proven renovascular disease. Several
of the prostaglandins behave like angiotensin in their ability to facilitate adrenergic
transmission to blood vessels. An unanswered question is whether they influence
the level of blood pressure in the normotensive or hypertensive states.

Vasopressin (Antidiuretic Hormone).
Vasopressin is an endogenous octapeptide
liberated from the posterior pituitary gland,
the prominent physiological role of which
is in control of water balance. However, in
addition to its effects on permeability of
renal tubules, vasopressin possesses vasoconstrictor properties which may come into
play if the blood levels of the material are
elevated sufficiently. Conditions in which
these levels are attained are general anesthesia, cardiopulmonary bypass and hypotension.

Release of vasopressin is inhibited by
stretch of left atrial receptors which are
sensitive to changes in circulating blood
volume. A similar inhibition is achieved by
activation of the arterial mechanoreceptors,
but these play a secondary role. When reflex
inhibition of vasopressin is reduced or lost,
for example when the left atrium collapses
during cardiopulmonary bypass, high circulating levels of vasopressin ensue. The coronary vessels are quite sensitive to the vasoconstrictor action of vasopressin. In animals
high levels of vasopressin reduce coronary
blood flow and, secondarily, cardiac output.
A depressant effect of vasopressin on the
heart remains to be demonstrated in clinical
situations where the levels of the polypeptide are high.

Neurogenic Factors Regulating Smooth Muscle Contraction

The sympathetic nervous system is responsible for rapid circulatory adjustments
in response to a variety of stressful physiological and pathological conditions such as
diving, exercise, temperature changes, heart
failure, myocardial infarction, hemorrhagic
shock, hypoxia and emotional stress. A sympathetic discharge causes an adjustment in
the peripheral circulation through the regulation of vascular resistance in each organ
which, in turn, governs the fraction of cardiac output supplied to that organ, or, in
other words, the perfusion of that organ.
There are several components to the sympathetic nervous system. The adrenergic
component, which has norepinephrine as a
neurotransmitter, is the major pathway, and
it moderates the level of vasoconstriction in
each vascular bed. Other sympathetic components mediate vasodilatation through the
release of either acetylcholine or histamine.

Adrenergic Component. The resistance
of each vascular bed is determined by the

density of sympathetic adrenergic innervation, the responsiveness of resistance vessels to the released neurotransmitter norepinephrine, and the frequency of efferent sympathetic discharge to that bed. Each of these factors varies significantly in different organs. For example, stimulation of the sympathetic nerve supply to the kidney, to the forelimb, and to the heart of dogs causes vasoconstriction in the renal vessels and in vessels of the forelimb (greater in the former than in the latter) and, in contrast, vasodilatation of the coronary vessels. Thus, if cardiac output is constant, a sympathetic discharge would cause a reduction in blood flow to the limbs and kidneys and an increase in flow to the coronary vessels. It appears also that the responsiveness of cerebral vessels to norepinephrine is negligible, so that, when a generalized sympathetic discharge occurs, it would favor the distribution of blood flow to the cerebral and the coronary circulations. Since the heart and brain depend on aerobic metabolism for their function, the peripheral circulatory adjustment is essential to maintain their perfusion in clinical situations in which cardiac output is limited or in which the conservation of oxygen is necessary.

Activation of sympathetic efferent vasoconstrictor pathways is not uniform. Activation of an afferent input, for example through stimulation of chemoreceptors, may result in a selective activation of efferent sympathetic fibers such that vasoconstriction is noted in some beds (i.e., muscle) and vasodilatation in others (i.e., coronary circulation). Such a differential response is certain to modify tissue perfusion in each organ even if cardiac output remained constant.

Inhibition of neurogenic vasoconstriction such as is observed with stimulation of arterial mechanoreceptors (baroreceptors) during a rise in pressure or stimulation of myocardial receptor during an acute increase in cardiac size tends to reflexly decrease arterial pressure. Depending upon the degree of inhibition of vasoconstrictor tone in different circulations, such reflexes would cause a redistribution of flow to various organs. In the following paragraphs the role of neurogenic factors in producing maldistribution of blood flow and defective perfusion of certain organs in clinical situations will be considered.

Orthostatic Hypotension and Decreased Cerebral Perfusion. Upright tilt decreases cardiac filling pressure and cardiac output. Afferent impulses originating in low- as well as high-pressure baroreceptor areas trigger the peripheral compensatory adjustment, consisting in tachycardia and vasoconstriction predominantly in skin, muscles, splanchnic beds and kidneys. This reflex maintains arterial blood pressure and cerebral blood flow.

In the presence of an intact adrenergic system, orthostatic hypotension may result from severe hypovolemia. Thus, under this circumstance compensatory reflexes are stimulated but are insufficient to oppose the marked reduction in cardiac output. Reflex tachycardia, very low central venous pressure, and increased blood catecholamines in a patient with orthostatic hypotension all suggest that the hypotension is caused by hypovolemia and should be treated with volume replacement.

In the absence of reflex tachycardia and peripheral vasoconstriction during upright tilt, the percentage of cardiac output supplying each organ remains unchanged but the fall in cardiac output will be reflected uniformly in all organs. The resulting reduction in cerebral blood flow may cause dizziness or syncope. Such a defective adrenergic reflex may be seen, for example, after prolonged recumbency or in diabetic patients with peripheral neuropathy. The presence of anhydrosis, normal or high venous pressure, and the absence of arterial pressure overshoot immediately after termination of the Valsalva maneuver all suggest that the cause of the orthostatic hypotension is autonomic neuropathy rather than hypovolemia.

A defect in adrenergic transmission may also result from a metabolic fault in synthesis of norepinephrine rather than structural damage to autonomic nerves. Such a defect may be found in patients with familial dysautonomia, and may be demonstrated by reduced levels of urinary VMA and manifestations of sympathetic paralysis. In some patients, the orthostatic hypotension may result from a failure of the integration of the baroreceptor reflex in the central nervous system such as may be seen in the Shy-Drager anomaly. This syndrome is characterized by upper motor neuron paralysis, cerebellar damage, impotence, urinary incontinence, constipation, incoordination, and severe orthostatic hypotension.

In some patients vasodilation has been observed during orthostatic hypotension. This dilatation is analogous to that seen in vasodepressor syncope where severe bradycardia and a marked increase in forearm blood flow precede the fainting. Thus, not only may the defective adjustment in orthostatic hypotension result from an absence of vasoconstrictor activity in nonvital organs, but it may be aggravated by an active vasodilatation in vessels of skeletal muscle, which tends to reduce still further cerebral perfusion by diverting an already reduced cardiac output away from the cerebral circulation. The mediator of this vasodilatation is not known but it could very well be acetylcholine. This dilatation may also represent activation of vascular beta adrenergic receptors.

Hypertension and Defective Renal Perfusion. Considerable controversy exists concerning the role of neuronal catecholamine synthesis, storage and turnover in hypertension. Several observations suggest that a restricted pool size and a reduction in binding or inactivation of norepinephrine might account for elevated arterial pressure by making larger concentrations of circulating catecholamines available for receptor sites. If this phenomenon were predominant in the kidney it might cause a reduction in renal perfusion with the release of renin, production of angiotensin and augmented neurogenic vasoconstriction. There is evidence that renal vascular resistance is indeed elevated in essential hypertension, and it is also evident that increased circulating angiotensin augments the vasoconstrictor response to sympathetic nerve stimulation. Although these interactions have not been identified clearly, they may have important implications relevant to the pathogenesis of essential and renal hypertension.

Ischemic Vasospastic Disease in the Extremities (Raynaud's phenomenon). It has been proposed that excessive adrenergic stimulation directed selectively to the vessels of the upper limb is responsible for the intermittent vasospasm and associated pallor and cutaneous ulcerations seen in the hands of young patients with Raynaud's phenomenon. More recent evidence, however, favors a different mechanism such as a defect in basal heat production limiting the ability of such patients to dilate their cutaneous vessels. The unusual sensitivity to cold that these patients exhibit and their tendency to shiver readily support this contention. The therapeutic regimen in which a combination of triiodothyronine and reserpine is administered to raise basal heat production and eliminate adrenergic influence has been proposed and in some cases methyl androstenediol has been added to the regimen.

Reserpine administered into the brachial artery of patients suffering from Raynaud's phenomenon (in doses of 0.5 mg.) causes a sustained vasodilatation without interruption of sympathetic transmission. These results have been particularly satisfactory in the young patients with clear-cut vasospastic manifestations. The relief of pain and the rapid healing of ulcers have been the most obvious manifestations of clinical improvement.

Cholinergic Component. Cholinergic innervation in skeletal muscle and skin has been identified in both man and animal but it does not appear to be involved in baroreceptor reflexes. Activation of this system is seen in responses to emotional stimuli, under stressful situations, during vasovagal syncope and possibly in anticipation of exercise. Activation of this cholinergic pathway increases blood flow primarily in skeletal muscle and, in the absence of increased cardiac output (in vasovagal syncope in contrast to exercise), compromises cerebral perfusion. There is evidence that a vagal cholinergic vasodilator system supplies the coronary vessels. This system may be involved in mediating coronary vasodilatation during stimulation of chemoreceptors, since there does not appear to be any sympathetic cholinergic innervation of the coronary vessels.

Other Vasodilator Systems. A histaminergic pathway supplying the limbs of experimental animals is activated by stimulation of carotid baroreceptors. This system may have pathophysiological significance, since the vasodilator response to its activation by stimulation of carotid baroreceptors is reduced in animals with renal hypertension. One might postulate that a defective vasodilator mechanism may contribute to the increased peripheral vascular resistance found in hypertension.

Another sympathetic vasodilator pathway has been identified by direct nerve stimulation in cutaneous vessels of experimental animals. The pathway is noncholinergic and nonhistaminergic and appears to be activated during stimulation of chemoreceptors. Conceivably, a defect in this cutaneous vasodilator system may contribute to the exaggerated vasoconstrictor responsiveness in patients with ischemic vascular disease of the limbs.

Intrinsic Tissue Factors Regulating Smooth Muscle Contraction

Autoregulation. Blood vessels possess the intrinsic ability to regulate the level of active tension of their smooth muscle. The ability to regulate tissue perfusion independently of neurogenic and humoral influences has been called autoregulation of blood flow. Different vascular beds vary with respect to their ability to regulate blood flow over a wide range of perfusion pressures. For example, skeletal muscle exhibits relatively poor autoregulatory capacity as compared to the kidney, which can autoregulate blood flow almost perfectly over a range of pressures between 80 and 200 mm. Hg. By this ability, the kidney can, by increasing its vascular resistance, keep blood flow and glomerular filtration rate constant when arterial pressure is raised.

The mechanism of this intrinsic response of vascular smooth muscle remains obscure. Some earlier explanations were based on physical factors such as changes in viscosity brought about by plasma skimming, or changes in tissue pressure brought about by increased filtration of fluid at elevated hydrostatic pressures. It is generally agreed, however, that autoregulation involves active responses of vascular smooth muscle wall. The rate of oxygen delivery to arterioles and precapillary sphincters could regulate blood flow by altering the rate of contraction of these small vessels. Accumulation of metabolites, when flow is reduced, may cause vasodilatation and allow flow to be restored. Finally, it has been suggested that the activity of smooth muscle pacemaker cells in vessel walls may be increased by stretch, causing greater smooth muscle contraction.

Autoregulation of renal blood flow has been implicated in the pathogenesis of hypertension. Several authors have suggested that the first phase of hypertension involves increased cardiac output without any change in peripheral vascular resistance. The hypothesis suggests that renal vessels especially increase their resistance in an effort to maintain blood flow constant in the face of increased arterial pressure. Local renal vasoconstriction might lead to the elaboration of renin and further accelerate the development of hypertension. Although this hypothesis is intriguing, there is at present little experimental evidence to support the concept that autoregulatory responses are involved in either initiating or sustaining high arterial pressure.

Metabolic Factors. Such factors may be released in response to absolute or relative tissue ischemia and cause vasodilatation to restore optimal blood flow, adequate tissue perfusion, and oxygenation. The mediators of this metabolic vasodilatation are not known but it is likely that the combination of changes in oxygen and carbon dioxide tension, hydrogen ion and other cation concentrations, changes in osmolarity, and the amount of adenosine compounds released, and accumulated Krebs intermediate metabolites in the immediate environment of the blood vessels contribute to adjustments in vascular tone. The cerebral and coronary vessels are particularly sensitive to changes in their metabolic environment. Excess carbon dioxide is a potent cerebral vasodilator, whereas hypoxia causes significant coronary vasodilatation.

Deficiency of tissue oxygenation may result either from the failure of supply of oxygen to tissues when their oxygen demand is normal or from an increased oxygen consumption in the hypermetabolic state, creating a situation in which there is relative oxygen deficiency.

Failure of delivery of oxygen to the tissues has many causes. It may result from pulmonary arteriovenous fistulae. A right-to-left shunt which may be large enough to cause significant depression of arterial oxygen saturation to levels of 70 to 75 percent produces cyanosis, polycythemia and digital clubbing.

High Altitude Exposure. The hypoxia resulting from high altitude exposure could result in a net peripheral vascular response which is the resultant of at least three effects. One is the direct vasodilator effect of hypoxia seen mostly in skeletal muscle vessels, but which would also be expected to occur in coronary vessels; the second is a sympathetic discharge causing peripheral vasoconstriction predominantly in skin and, possibly, in mesenteric and renal vessels; and, third, a direct inhibitory effect of hypoxia on adrenergic responses is possible whereby the vasoconstrictor action of sympathoadrenal stimulation, possibly at a peripheral site, is inhibited or blocked. The degree to which these various effects influence the different vascular beds is presently unknown, but is of critical importance if we are to understand the mechanisms whereby peripheral blood flow is distributed to various organs in hypoxic states.

Thiamine deficiency and other diseases in which coenzymes necessary for oxidative decarboxylation of pyruvic acid are lacking are associated with failure of tissue oxygen delivery. Deficiency of this and other respiratory enzymes resulting from severe chronic vitamin deficiency may trigger the release of vasodilator metabolites and produce clinical manifestations of a high cardiac output state as seen in beriberi. It should be remembered that excessive and too rapid replacement of thiamine reverses the vasodilation and constricts the peripheral circulation, and may precipitate cardiac decompensation.

Laennec's Cirrhosis. The hyperkinesis seen in Laennec's cirrhosis may be caused in part by a deficiency in respiratory enzymes analogous to that found in beriberi. There may also be small portal and pulmonary arteriovenous communications which may contribute to the low peripheral resistance and high cardiac output.

Anemia. All forms of anemia result in reduced ability of the blood to deliver adequate amounts of oxygen to the tissues. Because of the large oxygen reserve in the venous effluent of most organs, tissue oxygenation may be partially maintained (despite the presence of significant anemia) through the process of more complete extraction of oxygen. However, venous oxygen content is low in the coronary circulation, and in the presence of severe anemia an increased myocardial oxygen requirement must be met by an increased coronary blood flow, achieved by either dilatation or increased perfusion pressure, or both. However, anemia must be rather severe, with a hemoglobin level of less than 7 grams per 100 ml. of blood, before significant cardiovascular changes are evident.

An abnormal hemoglobin also may reduce oxygen-carrying capacity of the blood significantly. Such situations arise in methemoglobinemia, carboxyhemoglobinemia and sulfhemoglobinemia. Tissue hypoxia, as estimated by the plasma lactic acid: pyruvic acid ratio, does not occur in resting patients who have levels of hemoglobin above 6 grams per 100 ml. In such patients, compensatory mechanisms must take place to deliver more oxygen to tissues. These may include vasodilatation, increased cardiac output, and a rise in 2,3-diphosphoglycerate which is known to facilitate oxygen dissociation from hemoglobin.

Increased oxygen consumption may be the result of increased metabolism which may or may not be associated with the formation of high energy phosphate bonds in adenosine triphosphate (ATP). During exercise, pregnancy, and anabolic processes following acute illness a hyperkinetic state provides for delivery of greater than normal amounts of oxygen to tissues, which is then utilized for the formation of high energy phosphate bonds. This represents an efficient oxidative phosphorylation. However, in certain states hypermetabolism may occur without the formation of ATP. This may take place in non-shivering thermogenesis, pheochromocytoma, adrenergic calorigenesis, diabetic ketosis, idiopathic hyperkinetic heart syndrome and possibly in hyperthyroidism. The increased oxygen utilization may represent an activation of metabolic pathways that are non-ATP-dependent. This respiration without phosphorylation appears to be wasteful but may be essential for normal cellular processes.

In both idiopathic hyperkinetic heart syndrome and hyperthyroidism there is a high cardiac output and stroke volume, a bounding pulse, and vasodilatation. Idiopathic hyperkinesis may represent a state of neurogenic vasodilatation accompanied by excessive neurogenic cardiac stimulation. It may also reflect increased sensitivity of beta-adrenergic receptors to adrenergic stimulation or an overactivity of such receptors. On the other hand, hyperkinetic states associated with anemia or hyperthyroidism do not appear to be influenced by beta-adrenergic receptor blocking drugs and, therefore, may not be dependent on activation of sympathoadrenal pathways involving such receptors. In fact, one might use the responsiveness to beta-adrenergic blockade as an index of the contribution of the sympathoadrenal system to the hyperkinesis.

There is little information concerning the relative degrees of vasodilatation of various vascular beds in the hyperkinetic states beyond the fact that cutaneous blood flow must be increased for the dissipation of heat. One might expect that the organs consuming the largest amounts of oxygen in these states would be the ones having the greatest degree of vasodilatation. Since cardiac output seems to parallel oxygen consumption in the hyperkinetic states, it may be assumed that the peripheral vasodilatation does not represent autonomic or physiological shunting, and the increase in cardiac output occurs because of increased tissue demand for blood flow.

STRUCTURAL FACTORS ALTERING RESISTANCE TO BLOOD FLOW

Arterial insufficiency of vital organs is the major cause of death in the United States. The largest percentage of such insufficiency results from arteriosclerotic changes in vessel walls and atherosclerotic lesions in the intima, particularly of the coronary vessels, leading to ischemic heart disease. Other diseases (autoimmune, inflammatory) associated with structural changes of the vessel wall are of lesser importance because of their lower incidence. Diabetes also may cause a specific degenerative change in the capillaries which results in decreased tissue perfusion and ischemic changes.

Arteriosclerosis

Metabolic studies of arterial wall demonstrate that the movements of nutrients, substrates, and catabolic products are directed through the luminal as well as the adventitial sides of the vessel wall. The vasa vasora in the adventitia do not supply the wall beyond a point halfway in the media and do not reach the intima. The integrity of the endothelium is important for transfer of nutrients to the intimal layer. Inflammatory intimal lesions lead to fibrosis and hyalinoid changes deeper in the vessel wall. These lesions are independent of cholesterol and lipids and should be considered as distinct entities separate from atherosclerotic lesions or from Mönckeberg's sclerosis, which is essentially a focal calcification of the media. The initial damage to the endothelial layer may also promote thrombus formation. Organization of such thrombi and formation of new capillaries within them might favor intramural hemorrhage and promote further inflammatory reaction in the wall.

Atherosclerosis

This is basically an involvement of the intimal surface of the arteries with fatty plaques. The first lesions appear as fatty streaks. As the lesions progress and enlarge, they become vascular, but the blood supply may be insufficient so that necrosis occurs in the core of the atheroma. The aorta and the left coronary artery are most commonly involved, and the aortic involvement is frequently at its terminal bifurcation into the iliac arteries. Involvement of the coronary vessels is sometimes out of proportion to atherosclerosis elsewhere, and the epicardial segments of the coronary vessels appear to be more heavily involved than the transmural segments. In the cerebral vessels distribution of atherosclerosis is patchy. It is difficult to recognize clinically the development of atherosclerosis except possibly by angiographic visualization of the narrowed lumen. Otherwise, one has to depend upon the clinical manifestation of ischemia in the various organs. The manifestations are often catastrophic: cardiac arrest, following myocardial infarction; paralysis, from occlusion of a cerebral vessel; gangrene, following occlusion of peripheral limb vessels; hypertension, after occlusion of renal vessels; and intestinal gangrene, from occlusion of the mesenteric arteries. Of the less alarming, but nonetheless serious, manifestations of advanced vascular disease are symptoms of intermittent claudication in the legs, Raynaud's phenomenon in the hands, and angina pectoris with electrocardiographic changes compatible with ischemic myocardial disease.

Risk Factors. The concepts concerning the prevalence of atherosclerosis and its association with biological and social variables have evolved primarily from epidemiological studies in which a high correlation was found between ischemic heart disease and certain risk factors. In descending order the major factors appear to be hyperlipidemia, cigarette smoking, hypertension, obesity, and diabetes.

HYPERLIPIDEMIA. Plasma lipids of clinical importance are cholesterol, triglycerides and free fatty acids, all circulating bound to protein. Free fatty acids circulate with albumin, and cholesterol, and triglycerides are bound to lipoproteins.

Types of Hyperlipoproteinemia. Hyperlipoproteinemia may be conveniently classified into five types according to the scheme proposed by Fredrickson, Levy and Lee. However, most Americans who develop atherosclerotic coronary heart disease have mild elevation of serum cholesterol levels and perhaps only slight elevation of serum triglyceride concentration and, therefore, might not necessarily be classified as having one of the distinctive types of hyperlipidemia.

Type I is a very rare genetic recessive disease associated with deficiency of lipoprotein lipase, and appears in early childhood, manifested primarily by marked elevation in chylomicrons. After an overnight fast, the creamy layer on top of a clear plasma will indicate the presence of these fat globules. Clinical manifestations are primarily those of lipemia retinalis, eruptive xanthomas, hepatosplenogemaly, and abdominal pain. Type V hyperlipoproteinemia is a similar disorder, also associated with chylomicronemia and similar clinical manifestations, except that it appears in early adulthood and may often be associated with pancreatitis and insulin-dependent diabetes mellitus. The treatment of Types I and V hyperlipidemia consists of an extremely low animal fat diet.

Type II hyperlipoproteinemia is common and is associated with elevated serum cholesterol and normal triglycerides. The appearance of the plasma is clear and the clinical manifestations include xanthelasma, tendon xanthomas, and juvenile corneal arcus. Type III is relatively uncommon. It is associated with elevation of both cholesterol and triglycerides, giving a cloudy plasma; its clinical characteristics include xanthomas on palmar creases and tendon xanthomas and is usually seen in adulthood. Type IV disease is common and is associated with elevation of triglycerides and normal cholesterol; the plasma is turbid. Types II, III, and IV are associated with accelerated vascular disease. Types II and IV are associated primarily with coronary vascular disease, Type III with a high incidence of both coronary and peripheral vascular atherosclerotic lesions. The treatment of Types III and IV, which have high levels of triglycerides, consists of restriction of dietetic carbohydrates which cause the elevated serum triglycerides. A diet low in cholesterol, high in fats, and low in carbohydrates is often beneficial. Clofibrate, which lowers serum triglycerides, should be used in the treatment of Types III and IV.

Type II patients, who have hyperlipidemia associated with hypercholesterolemia, may have reduction in lipids on a diet low in cholesterol and moderately low in fat. In some instances endogenous cholesterol may contribute to the hypercholesterolemia. The use of cholestyramine which increases secretion of cholesterol in the form of bile acids in the feces may often be effective and beneficial.

Vascular Anomalies

Congenital vascular anomalies, such as are seen in lungs of patients with hereditary hemorrhagic telangiectasia, have been reported as causes of a hyperkinetic circulatory state. The hyperkinetic circulation in such individuals may be related to the deficiency in tissue oxygenation associated with severe anemia rather than to the vascular dysplasia. Anemia in such patients is caused by frequent attacks of bleeding.

Acquired Vascular Anomalies. *Paget's Disease.* Acquired arteriovenous communication may be seen in Paget's disease of bone, but it is not known whether the increased vascularity is the primary defect in the involved bone. Approximately 30 percent of the bones are involved in those patients who show the hyperkinetic syndrome associated with Paget's disease.

In pregnancy the excessive blood flow

through the elongated and tortuous uterine arteries into the large sinuses surrounding the placental villi and back into the uterine vein may be associated with 30 to 40 percent increases in cardiac output. The increased blood volume observed in the third trimester may also contribute to the high output. The ease with which the heart meets these temporary increases in demand, often despite severe valvular deformities, testifies to the tremendous reserve of the heart muscle.

Another group of acquired vascular abnormalities, which may have significant hemodynamic effects by reducing renal blood flow, are the fibrosing lesions of the renal artery that result in renal hypertension. This problem is discussed in detail in Chapter 3.

Collagen Diseases

These diseases involve the medium-sized and small arteries and arterioles in various parts of the body with inflammatory fibrinoid changes. The vascular involvement may be the result of an autoimmune disease which also affects other tissues.

Polyarteritis nodosa is a collagen disease that manifests itself primarily in younger men, with fever, abdominal pain, hypertension, polyarteritis, and eosinophilia. Several organs may be involved but the most frequently damaged is the kidney, resulting in severe hypertension.

Systemic lupus erythematosus is a collagen disease of unknown cause which may involve all organs; its lesions are most often seen in the walls of small arteries and arterioles. Here, also, the presence of focal or diffuse glomerulonephritis may be the terminal event in this unrelenting disease.

In *scleroderma,* the involvement of the skin or subcutaneous tissue may be the most visible manifestation; however, the more serious complication involves pulmonary fibrosis and pulmonary vascular obstruction, resulting in emphysema, atelectasis, bron-chiectasis, and pulmonary hypertension with cor pulmonale. Scleroderma also involves the gastrointestinal tract and particularly the lower esophagus; in addition, obliterative vascular lesions in the gastrointestinal tract may result in ulceration, perforation, infarction and hemorrhage. In this, as in the other collagen diseases, involvement of the muscular, skeletal and cardiovascular systems is common.

RHEOLOGIC AND MECHANICAL FACTORS ALTERING RESISTANCE TO BLOOD FLOW

Altered Viscosity. The physical characteristics of the blood can, under certain circumstances, be important determinants of resistance to blood flow. According to the Poiseuille relationship, vascular resistance is directly related to blood viscosity. Viscosity varies relatively little in the range of hematocrits between 0 and 40 percent, but it increases rather steeply when hematocrits increase above the normal level. If all other factors remain unchanged, an increase in hematocrit from approximately 40 to 70 percent will, by doubling the relative viscosity, double the resistance to blood flow. Noncellular constituents may have predictably similar effects on viscosity. For example, chylomicrons present in large quantities in hyperlipidemia or macroglobulins in Waldenström's macroglobulinemia could increase blood viscosity substantially. The high hematocrits found in polycythemia undoubtedly contribute to increased resistance to blood flow and, thus, to the increase in blood pressure seen in this condition. There is very little effect on tissue perfusion in anemic states with low hematocrits, because the relationship between hematocrit and viscosity is altered very little at low hematocrit levels.

Red Cell Wall Rigidity. Normally, the red cell is easily deformed and undergoes considerable change in shape as it passes through the capillary. If the wall should change in respect to this physical characteristic so that it resisted deformation, the ease with which cells could pass the capillary would be reduced; thus tissue perfusion could be altered by this physical change in the nature of the cell wall alone. Some recent evidence suggests that the red cell wall becomes more rigid in hypoxia. Further investigation is necessary, however, to determine whether red cell wall rigidity is an important regulator of tissue perfusion.

Clotting, Thrombosis and Platelet Aggregation. Physical obstruction of vascular channels can be produced by clotting or thrombosis. Obstruction of major vessels, such as pulmonary, coronary, or cerebral arteries, may lead to catastrophic consequences if the obstruction severely compromises perfusion of the organ.

The role of altered blood coagulation in shock has received attention. At the irreversible stage of shock there may be damage to the endothelial lining of small vessels and capillaries, with subsequent fibrin deposition, accumulation of microthrombi, and intravascular coagulation. This, in addition to diffuse to extensive vasospasm, may severely restrict tissue perfusion and cause cellular death. The major example of a tissue lesion that follows disseminated intravascular coagulation is renal cortical necrosis, associated with gram-negative septicemia. Other organs may also be involved in this generalized reaction which mimics the Shwartzman reaction.

INTRAORGAN BLOOD FLOW DISTRIBUTION

From a functional point of view circulation within organs can be divided into two compartments; the first of these a series of channels in which diffusion between blood and tissues may occur, and the second a series of channels in which there is relatively little surface area available for diffusion to occur. Thus, any given level of

blood flow to an organ can be divided into its *nutritional* and *non-nutritional* components. Adequate tissue perfusion depends upon the existence of nutritional blood flow and is not necessarily related to the total blood flow delivered to an organ. It is convenient to think of nonnutritional flow as passing through an arteriovenous shunt, although there may be little anatomical evidence for the existence of such shunts.

Techniques have been developed for studying shifts in the distribution of blood flow within organs. Changes in the extraction of oxygen provide a convenient index for determining whether the amount of blood flow reaching sites where diffusion may occur is altered. More sophisticated techniques involve determination of the clearance of a freely diffusible isotope such as rubidium. Coronary, skeletal muscle and skin circulation can all be shown to possess significant potential for functional shunting of blood away from nutritional vascular channels.

The pathophysiological role of altered nutritional blood flow is not well understood and can only be speculated about.

Angina pectoris is an excellent example of a condition in which nutritional circulation may be inadequate despite the existence of coronary blood flow in the normal range. In fact the most effective antianginal agents (e.g., nitroglycerin) can relieve anginal pain effectively without altering total coronary blood flow. Their efficacy may well be related to their ability to divert flow to nutritional channels in ischemic areas and to a reduction in myocardial oxygen demands.

The kidney represents an organ where shunting of blood flow can also cause significant loss of function. Blood supply to functional nephrons is concentrated in the outer portion of the renal cortex. A variety of studies have suggested that interventions that alter total renal blood flow can produce significant redistribution of flow within the kidney. For example, renal vasoconstriction following adrenergic discharge tends to shunt blood away from outer cortex, whereas renal vasodilators such as furosemide and ethacrynic acid tend to shunt blood toward outer cortical nephrons. In experimental congestive heart failure, the percentage of renal blood flow supplying outer cortical nephrons is reduced. This observation suggests that intrarenal blood flow distribution may well be important in the retention of sodium and the formation of edema associated with failure. In several forms of renal disease the fraction of blood flow going to the outer cortex may be reduced.

ANNOTATED REFERENCES

Barger, A. C.: Renal hemodynamic factors in congestive heart failure. *In:* The Physiology of Diuretic Agents. Ann. N. Y. Acad. Sci., *139* (part 2):276, 1966. (Changes in intraorgan distribution of blood flow can have profound effects on the function of the organ; e.g., excretion of sodium by the kidney could be depressed if blood were shunted away from primary areas of filtration and reabsorption. This may well be the case in congestive heart failure.)

Bayliss, L. E.: The rheology of blood. *In:* Hamilton, W. F. (ed.): Handbook of Physiology. Section 2, Circulation. Vol. 1, p. 137. Washington, D. C., American Physiological Society, 1962. (Physical characteristics of blood are often ignored as causes of altered tissue perfusion. Altered viscosity is perhaps the most important of these.)

Berne, R. M.: Regulation of coronary blood flow. Physiol. Rev., *44*:1, 1964. (The effects of catecholamines on the coronary vessels are complex, depending upon the sum of direct vasoconstrictor effects, which are relatively weak, and indirect potent vasodilator actions which result secondarily from metabolic changes associated with increased myocardial contractility. Adenine nucleotides have been implicated as possible metabolic mediators of coronary vasodilatation produced by catecholamines.)

Bergström, S., Carlson, L. A., and Weeks, J. R.: The prostaglandins: a family of biologically active lipids. Pharmacol. Rev., *20*:1, 1968. (Prostaglandins have diverse actions upon the cardiovascular system. Small modifications in structure produce qualitative changes in activity of members of the class, e.g., from depressor to pressor. Their extreme potency make them potential candidates as chemical mediators of a wide spectrum of cardiovascular phenomena.)

Blumenthal, H. T. (ed.): Cowdry's Arteriosclerosis. A Survey of the Problem. Springfield, Illinois, Charles C Thomas, 1967. (Structural changes in the blood vessel lumen alter tissue perfusion A diverse spectrum of pathophysiological events relating to altered tissue perfusion result from arteriosclerotic and atherosclerotic lesions.)

Brigden, W., Howarth, S., and Sharpey-Shafer, E. N.: Postural changes in the peripheral blood-flow of normal subjects with observations on vasovagal fainting reactions as a result of tilting, the lordotic posture, pregnancy and spinal anesthesia. Clin. Sci., *9*: 78, 1950. (The hemodynamic contributions to orthostatic hyoptension are unmasked when normal reflex adjustments are diminished or absent. These include a fall-in cardiac output, a lack of peripheral vasoconstriction, or even active vasodilatation in skeletal muscle.)

Brod, J., Fencl, V., Hejl, Z., Jirka, J., and Ulrych, M.: General and regional hemodynamic pattern underlying essential hypertension. Clin. Sci., *23*:339, 1962. (The kidney contributes to the elevated total vascular resistance seen in hypertension. The cause of the increased renal vascular resistance is not known, nor has it been determined if the kidney plays any role in the pathogenesis of essential hypertension.)

Calvelo, M. G., Abboud, F. M., Ballard, D. R., and Abdel-Sayed, W.: Reflex vascular responses to stimulation of chemoreceptors with nicotine and cyanide: activation of adrenergic constriction in muscle and noncholinergic dilatation in dog's paw. Circulaton Res., *27*:259, 1970. (It is now becoming apparent that cardiovascular reflexes, once thought to produce uniform qualitative effects on blood flow distribution, can differentially effect different vascular beds, constricting some and dilating others.)

Chidsey, C. A., Braunwald, E., and Morrow, A. G.: Catecholamine excretion and cardiac stores of norepinephrine in congestive heart failure. Am. J. Med., *39*:442, 1965. (Blood catecholamines are elevated in congestive heart failure. Neither the cause of this change nor its role in the circulatory status in failure is known.)

Chien, S.: Role of the sympathetic nervous system in hemorrhage. Physiol. Rev., 47: 214, 1967. (In hypotensive shock, generalized sympathetic discharge plays a prominent role in altering tissue perfusion, through both neurogenic and humoral mechanisms.)

Engelman, K., Portnoy, B., and Sjoerdsma, A.: Plasma catecholamine concentrations in patients with hypertension. Circulation Res., 27 (Suppl. 1):141, 1970. (There is slightly more catecholamine in the blood of essential hypertensives compared with normotensives. The difference is unlikely to account solely for the hypertension but may reflect some disturbance in catecholamine metabolism.)

Frederickson, D.: Atherosclerosis and other forms of arteriosclerosis. *In:* Wintrobe, M. M., Thorn, G. W., Adams, R. D., Bennett, I. L. Jr., Braunwald, E., Isselbacher, K. J., and Petersdorf, R. G. (eds.): Harrison's Principles of Internal Medicine, ed. 6, p. 1239. New York, McGraw-Hill, 1970.

Landis, E. M., and Pappenheimer, J. R.: Exchange of substances through the capillary wall. *In:* Hamilton, W. F. (ed.): Handbook of Physiology. Section 2, Vol. 2., p. 961. Washington, D. C., American Physiological Society, 1965. (Circulation in an organ is of two types, one devoted to the distribution of flow to sites of diffusion, the other a series of channels that largely bypass sites of diffusion. Pathophysiological changes in either or both of these may exist in certain diseases, but these are still to be adequately documented.)

McRaven, D. R., Mark, A. L., Abboud, F. M., and Mayer, H. E.: Responses of coronary vessels to adrenergic stimuli. J. Clin. Invest. 50:773, 1971. (The coronary vessels have beta receptors which mediate the dilator action of isoproterenol. The dilator action of norepinephrine or of sympathetic nerve stimulation is indirectly related to metabolic myocardial factors. There is a paucity of alpha receptors in coronary vessels.)

Margetten, W.: Local tissue damage in disseminated intravascular clotting. Am. J. Cardiol., 20:185, 1967. (Shwartzman reaction type of phenomena can profoundly alter tissue perfusion by physical obstruction of both small and large vessels.)

Mellander, S.: Comparative studies on the adrenergic neurohormonal control of resistance and capacitance blood vessels in the cat. Acta physiol. scand., 50 (Suppl. 176): 1-86, 1960. (The responsiveness of various vessels to catecholamines is considerably different and is an important determinant of the overall effects of these substances on tissue perfusion. Thus, the distribution of blood flow to several organs in proportion to the degree of vasoconstriction produced in each.)

Nickerson, M., and Gourzis, J. T.: Blockade of sympathetic vasoconstriction in the treatment of shock. J. Trauma., 2:399, 1962. (Interruption of neurogenically and humorally mediated adrenergic vasocontriction can improve tissue perfusion in certain hypotensive shock states. This can be achieved with alpha-adrenergic blockade and should be accompanied by volume expansion to take up the "increase in circulatory capacity" produced by venular and arteriolar dilation.)

Oates, J. A., Pettinger, W. A., and Doctor, R. B.: Evidence for the release of bradykinin in carcinoid syndrome. J. Clin. Invest., 45: 173, 1966. (Under special conditions such as carcinoid syndrome, bradykinin is liberated in sufficient quantities to have vasoactive effects. The skin flushing of carcinoid may be an example of the activity of an endogenous vasodilator.)

Share, L.: Vasopressin, its bioassay and the physiological control of its release. Am. J. Med., 42:701, 1967. (Although the precise role of vasopressin in altering tissue perfusion remains to be determined, large quantities, with considerable potential to influence cardiac performance and blood flow distribution, are liberated in certain states.)

Shepherd, J. T.: Behavior of resistance and capacity vessels in human limbs during exercise. Circulation Res., 20:(Suppl. 1):70, 1967. (Exercising muscle, through metabolic need, shows enormously increased blood flow, probably at the relative expense of other vascular beds.)

Symposium: Active neurogenic vasodilatation. Fed. Proc., 25:1583, 1966. (Histamine appears to be the chemical mediator of baroreceptor reflex vasodilatation. The precise nature of the storage in, and release from, sympathetic nerves of the substance is under active investigation. The histamine store involved in reflex vasodilatation appears to be distinct from mast cell stores.)

Symposium: Autoregulation of blood flow. Circulation Res., *14* (Suppl. 1): 1-291, 1964. (Blood vessels respond to changes in distending pressure with active changes in vessel wall tension.)

Uvnäs, B.: Cholinergic vasodilator innervation to skeletal muscles. Circulation Res., *20* (Suppl. 1):83, 1967. (Reflex vasodilator responses from baroreceptor activation are not altered by cholinergic blockade. The sympathetic cholinergic pathway, in fact, appears to be anatomically distinct from other vasomotor pathways which originate in medullary reticular formation.)

Whelan, R. F.: Vasodilatation in human skeletal muscle during adrenalin infusion. J. Physiol., *118*:575, 1952. (In the human being epinephrine is a profound vasodilator of skeletal muscle vessels. This action is essential to the pattern of blood flow distribution produced by this catecholamine.)

Wilson, W. R., and Abboud, F. M.: Hyperkinetic circulatory states. *In:* Gordon, B. L., Carleton, R. A., and Faser, L. P. (eds.): Clinical Cardiopulmonary Physiology. ed. 3, p. 219. New York, Grune & Stratton, 1969. (Oxygen demand is an important determinant of tissue perfusion. A large variety of circulatory disorders result from situations when oxygen requirements are different from blood flow.)

Zimmerman, B. G., Brody, M. J., and Beck, L.: Mechanism of cardiac output reduction by hexamethonium. Am. J. Physiol., *199:*319, 1960. (Interruption of neurogenic influence on vasculature does not reduce vascular resistance if arterial pressure falls. Passive vasoconstriction can, therefore, mask loss of active tension in vessel wall.)

3

Pressor Mechanisms

Harriet P. Dustan, M.D., Edward D. Frohlich, M.D., and Robert C. Tarazi, M.D.

INTRODUCTION

The heart supplies the energy for the circulation of blood. With each contraction, it ejects a small volume (about 75 ml.) into a container of limited capacity, the arterial system, and by this action displaces blood out of this system into capillaries. The ease with which this displacement is accomplished depends on the diameter of the arterioles; these supply outflow resistance. Thus, pressures within the arterial tree are: that generated by cardiac contraction, the systolic pressure; and that exerted by outflow resistance, the diastolic pressure.

Hypertension indicates elevated arterial pressure, either systolic or diastolic or, as is often the case, both. This discussion will deal with what is called diastolic hypertension, although systolic pressure is elevated as well. It results from functional and/or structural abnormalities of the arterioles. It represents an important medical problem because it affects large numbers of people and can result in premature disability and death.

Since diastolic hypertension occurs in several unrelated diseases (Table 3-1), there must be a variety of ways to produce the arteriolar abnormalities that cause elevated arterial pressure. These are called pressor mechanisms. In a sense, this term is a misnomer because, with the exception of pheochromocytoma, the physiological abnormalities associated with hypertension have not been shown to be causal. Although these abnormalities represent distinct aberrations of a number of cardiovascular con-

TABLE 3-1. Most Common Types of Hypertension

TYPE	CAUSE
Hypertension of unknown cause— usually called essential	
Renal	*Renovascular*
	Renal parenchymal disease
Cardiovascular	*Thoracic aortic coarctation*
Adrenal Medullary	*Pheochromocytoma*
Adrenal Cortical	*Aldosteronoma*
	Idiopathic adrenal hyperplasia
	Cushing Syndrome
	Enzymatic deficiencies of steroid biosynthesis

trol systems, the degrees to which they participate in arterial pressure elevation are not known.

In a very real sense, hypertension represents a unique cardiovascular disease because the entire system is involved from the beginning. This is in contrast to valvular or myocardial heart diseases, in which the initial abnormalities are cardiac only, or to atherosclerosis which is symptomless until it becomes occlusive enough to jeopardize blood flow to an organ. Accordingly, to understand hypertension one should consider the cardiovascular system as a whole and not just arterial pressure, which is only one hemodynamic function. These considerations should include cardiac output and peripheral resistance and the ways in which they can be modified by sympathetic neural activity and hormonal factors. Important also is the volume of blood

41

that distends the vascular system, because there is evidence to suggest that arterial pressure, in part, reflects a relationship between the capacity of the vasculature and the intravascular volume. This leads, then, to consideration of neural and hormonal factors that determine vascular capacity, as well as to mechanisms that control body fluid volumes.

HEMODYNAMICS

Basic Considerations

Alone among the disorders of the cardiovascular system, arterial hypertension is defined by a quantitative alteration of a biophysical measurement. It is therefore essential to consider some laws of hydrodynamics and their translation into clinical terms.

The rate of flow of any fluid along a tube is related to a pressure gradient along that tube and to the resistance it meets. Resistance (R) cannot be measured directly and, therefore, is calculated as the ratio of the pressure gradient (Δ P) to the rate of flow (F):

$$R = \Delta P / F \quad (1).$$

If the flow of a fluid within a cylindrical vessel is laminar, its rate can be predicted, depending on its viscosity (v), from the dimension of the tube and the associated pressure gradient, according to the Poiseuille formula:

$$F = P_1 - P_2 \left(\frac{r^4}{l.v} \right) k \quad (2)$$

where r is radius, l, the length of the tube, and k a constant ($\pi/8$) factor arising from calculus integrations. From operations (1) and (2), it is obvious that

$$R = \frac{v.l}{r^4} \times \frac{8}{\pi}$$

Within the usual physiological limits of blood viscosity and assuming an unchanging vascular length in the same individual,

variations in resistance result usually from active, passive or structural changes of vessel diameter. Since the radius is magnified to the fourth power in the equation, flow and pressure may be markedly affected by relatively small changes in vessel diameter.

Translation of these mathematical symbols into clinical terms results in the basic equation expressing the relationship of arterial pressure, cardiac output and "total peripheral resistance" (TPR):

$$MAP = CO \times TPR$$

where CO (cardiac output) is the equivalent of F; and MAP (mean arterial pressure), the equivalent of ΔP, is the integrated arterial pressure over one cardiac cycle. This integration can be obtained either by electronically integrating pulsatile pressure, by planimetry of the pressure wave, or by calculating it as diastolic pressure $+$ 1/3 pulse pressure. The marked difference between mean arterial pressure and central venous pressure, as well as the relatively small fluctuations of the latter, allow its disregard in calculations of TPR.

The total peripheral resistance is the composite of the vascular resistance for the entire systemic circulation. Resistance to flow obeys the same laws as do series and parallel arrangements of electrical resistances. It is clear, therefore, that a change in TPR does not necessarily indicate that similar quantitative or even similar directional changes are occurring in all individual vascular territories. Further, a word of caution is needed against unqualified translation of TPR into an index of peripheral arteriolar vasoconstriction, because this simplistic approach fails to recognize the important role that large and small arteriovenous shunts, precapillary sphincters, passive arterial variations, and structural vascular changes, as well as collateral vessels, may sometimes play in that respect.

Despite both of these restrictions and the well-recognized limitations in the application of Poiseuille's equation to the intact organism, these calculations are very useful in assessing the relative parts played by cardiac output and resistance in changes of arterial pressure. Cardiac output is frequently expressed in relation to body surface area or cardiac index (L./min./M.^{2}). Total peripheral resistance can be calculated from either cardiac output or cardiac index in arbitrary units (mm. Hg/L./min., mm. Hg/L./min./M.^{2}, or arbitrary PRU) or in fundamental units of force (dynes-sec./cm.^{5}). To convert to the latter expression, pressure (in mm. Hg) must be converted to dynes/cm.^{2} and flow must be expressed as cm.^{3}/sec. To accomplish this, 1 mm. Hg is equivalent to 1,333 dynes/cm.^{2}; and 1 L./min. is equivalent to 1,000/60 cm.^{3}/sec. The fundamental resistance unit is expressed as dynes-sec./cm.^{5} (or, more simply, as PRU $\times$ 80).

Aorta and Large Vessels. As the left ventricle ejects blood intermittently into a partially filled arterial tree, arterial pressure varies between a maximum at the peak of ejection and a minimum at the end of diastole. Whereas mean arterial pressure depends on the relation between flow and resistance defined above, actual systolic and diastolic levels and their difference (pulse pressure) are influenced by additional factors which may not necessarily affect mean pressure. Wiggers has repeatedly emphasized the importance of large arteries in that respect. The systolic pressure rise in the aorta depends in part on its distensibility: the less its wall stretches before the ejected blood the higher the pressure will go. Obviously, other variables such as speed of ejection, magnitude of runoff during systole, presence of localized constrictions (coarctation) or ectasias (aneurysms) also have important effects. Conversely, the rate and level of pressure decline during diastole depend not only on blood flow out of the arteries but also on the extent of aortic elastic recoil on the diminishing

blood content. This role of the aorta has been variously described as "aortic compression chamber" or the "windkessel (windbreaker) effect."

Variations in heart rate, stroke volume, and peripheral resistance thus affect not only mean pressure but also pulse pressure. The latter changes can be deduced from the following equation defining aortic distensibility (K):

$$K = dp/dv \ V$$

dp being the increment in pressure per increment in volume (dv) and V the aortic end diastolic volume. Experimentally, with aortic distensibility constant, increasing peripheral resistance will be associated with declining pulse pressure. The increased pulse pressure found in many hypertensives may therefore reflect a secondary loss of large vessel distensibility. Variations in distensibility not only may be due to structural changes from longstanding hypertension but also may be induced acutely by vasoconstrictor agents.

Resting Hemodynamics in Hypertensive Diseases

As can be readily deduced, hypertension in the final analysis depends on a disproportion between cardiac output and total peripheral resistance. Practically all early studies of hypertensive patients showed a normal cardiac output in the absence of left ventricular failure, and it was generally assumed that the hemodynamic "fault" in hypertensive diseases lay entirely in the peripheral resistance. However, recent work has challenged this conclusion, and, with the advent of more refined techniques and, especially, with more specific differentiation of various types of hypertension, a more complex hemodynamic picture has emerged (Table 3-2 and Fig. 3-1).

Established Essential Hypertension. As a group, patients with well-established essential hypertension (hypertension of un-

TABLE 3-2. Hemodynamic Characteristics of Different Types of Hypertension°

GROUP	HR (beats/m.)	CI (L/m./M.²)	SI (ml./beat/M.²)	TPR (units/M.²)
Normotensive	68	3.05	45	30
Hypertensives				
Labile	75	3.36	46	31
Essential (normal sized hearts)	76	2.91	39	45
Essential (cardiac enlargement)	74	2.67	36	53
Renovascular	80	3.48	44	37
Renal parenchymal disease	76	3.23	43	41

°Modified from Frohlich, E. D., Tarazi, R. C., and Dustan, H. P.: Am. J. Med. Sci., 257:9, 1969.

Increase of cardiac index in labile and renovascular hypertensives is statistically significant, as is its reduction in essential hypertensives with cardiac enlargement. Note the higher heart rate in all types of hypertension as compared with normotensive subjects.

HR, heart rate; CI, cardiac index; SI, stroke index; TPR, total peripheral resistance.

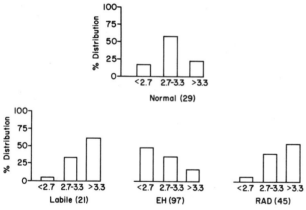

FIG. 3-1. Cardiac output characteristics in normal persons and patients with labile, essential (EH) and renal arterial disease (RAD) hypertensions. The ordinates represent percent distribution of the various levels of output (abscissae) in the four groups.

known cause) are characterized by a normal cardiac output and elevated peripheral resistance. Within this general pattern, however, a certain gradation has been described wherein a progressive reduction of resting cardiac output occurred with development of progressive cardiac involvement in the course of the disease. Thus, even before cardiac decompensation occurred, cardiac output and rate of ventricular ejection were lower (and total peripheral resistance further increased) in hypertensive patients with definite left ventricular hypertrophy. Patients without cardiac involvement had normal output and ventricular ejection rate despite obvious increase in cardiac work due to hypertension.

Studies of regional circulations have shown that the elevated resistance in established hypertension seems to be uniformly distributed in practically all vascular territories—except in the kidney, where it may be more intense, and in the skeletal muscular system, where it is slightly less marked. This pattern was likened by Brod to a constant preparedness for exercise or response to unspecified stress.

"Borderline" (or "Labile") Hypertension. An increasing number of studies from many clinical laboratories in the United States and Europe have established that a large proportion of patients with essential hypertension, especially those with mildly or intermittently elevated arterial pressure, have increased cardiac output and normal, near normal or subnormal values of total peripheral resistance. The data are impressive; the pattern of "increased output—normal resistance" is real and reproducible, but the implications and significance are not yet clear. Two basic questions remain unanswered—one in regard to the prognosis of this hemodynamic pattern, and the other, to its genesis.

From information available, it is not yet evident whether this hemodynamic pattern of increased cardiac output represents an early stage in the development of hypertension or a qualitatively different hemodynamic type of the disease. Comparison of group averages and analogies with the time-course of experimental renovascular hypertension suggest the former hypothesis, but these cannot replace longitudinal clinical studies, which are as yet unavailable.

The finding of increased cardiac output in borderline hypertension reawakened great interest in the old suggestion that an "augmented force" of the heart beat could play a role in the genesis of hypertension. Stimulation of hypothalamic centers or of left stellate ganglion in dogs is known to produce striking increases in both systolic and diastolic arterial pressure even when a constant heart rate is maintained with an artificial pacemaker. Recent description of increased cardiac response to isoproterenol in some hypertensive patients suggested a possible role for increased beta-adrenergic receptor site responsiveness in certain predisposed individuals. In the majority of patients, no cause for the increased cardiac

output is evident; oxygen consumption is normal for that level of output, but basal heart rate may be elevated. The shift of emphasis toward the role of the heart, although clearly justifiable, must not cloud the fact that some abnormality in peripheral circulatory adjustment is demonstrable both at rest and on exercise, even in those patients with elevated outputs. Though their total peripheral resistance is correctly said to be "within normal range," it most likely is abnormally high for their level of output; this becomes evident by comparison with normal subjects whose cardiac output is increased to equivalent levels. Finally, as discussed in a later section, variations in output may represent only a normal response of the heart to a redistribution of blood to the central circulation, rather than primary myocardial effects.

Renal Hypertensions

One of the important results of recent studies has been the differentiation of hypertensive states associated with renal disease into at least two types; one is related (at least initially) to activation of the renopressor mechanism, the other, to loss of renal parenchyma and, possibly, of an antipressor effect. The first is exemplified by renal arterial stenosis and the second by the anephric state (Fig. 3-2). In a schematic form that admits of many exceptions, the former is characterized in man by elevated plasma renin activity (especially in severe hypertension), low plasma volume, and indices of increased neurogenic activity. The latter type is marked by a direct relationship between blood volume and arterial pressure, very low or absent plasma renin activity, and neurogenic activity which fluctuates inversely with the degree of volemia. Patients with parenchymal kidney disease present varying mixtures of these two extremes with either the "renal" or the "renoprival" element predominating according to the type or stage of the lesion. (The reader is referred to more

Renal and Renoprival Factors in Renal Hypertension

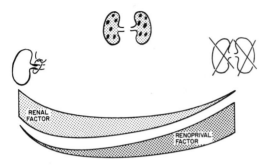

FIG. 3-2. A schematic representation of the relative participations of renal pressor and renoprival factors in renal hypertensions. The renal pressor factor seems to be the main influence in hypertension associated with renal arterial stenosis or experimental renal arterial narrowing. With renal disease both factors may participate whereas, in the absence of kidneys or, often, in terminal renal failure the renoprival, or volume dependent, factor predominates.

detailed discussions of these specific mechanisms, below).

Renovascular Hypertension. A large body of knowledge has accumulated concerning this type of hypertension since renovascular hypertension can be produced experimentally by a variety of ways in different animals. Essential to proper evaluation of experimental studies is the realization of the extent of differences that may result from variations in techniques, the timing of observations, and the variety of animal species used. Early studies in man have shown that both cardiac output and total peripheral resistance were elevated in patients with renovascular hypertension. Continued observations confirmed the increase of output in the majority of patients (but not in all) and demonstrated further that this increase was not alone responsible for human renovascular hypertension. Total peripheral resistance was increased in patients with normal as well as those with increased cardiac output; following successful surgical repair of the arterial lesion or nephrectomy, reduction in resistance was observed more often than was reduction in output. In a study of nine renovascular hypertensive patients, Brod et al. found an uneven distribution of this increased resistance, with higher values in the kidney and near normal values in the forearm muscles.

This common participation of varying degrees of increased output and resistance in the maintenance of renovascular hypertension in man corresponds to findings in recent experimental studies in rats and dogs. Though the early rise of arterial pressure following clipping of renal arteries or cellophane wrapping of the kidney is due primarily to increased cardiac output, later stages are characterized by a delayed rise in peripheral resistance with a return of output toward normal.

Two hypotheses, in the main, have been advanced to account for the increased cardiac output noted both at the onset of experimental hypertension and in most renovascular hypertensive patients. Transient expansion of extracellular fluid and plasma volume was noted in rats and, though there was no correlation between the increased blood pressure and the volume expansion, the implication was that hypertension was initiated by fluid retention. However, no change in blood volume was noted in the initial stages of cellophane perinephritic hypertension in dogs when cardiac output was rising; indeed, in both man and dogs with chronic renovascular hypertension plasma volume was reduced. The combination of lower intravascular volume and increased cardiac output suggested an increased tone of capacitance vessels leading to intravascular redistribution of blood from the periphery to the central cardiopulmonary circulation, since a highly positive correlation was found between central blood volume and cardiac output. This postulate was later supported by the finding in dogs of increased mean circulatory pressure in the face of normal or reduced plasma volume. Whether a similar mechanism is also responsible for the increased output in borderline or labile hypertension remains pure conjecture. More recently angiotensin was found to influence specifically some vasomotor center in the area postrema in the medulla. Subpressor doses infused into the vertebral artery led to arterial pressure increase for the duration of the infusion; this rise was usually due to an increase in peripheral resistance and occasionally to increase in cardiac output. From various studies on the effect of angiotensin on nerve function, catecholamine uptake and central vasomotor mechanisms, there is emerging an interesting concept (not yet fully proven) suggesting that antigiotensin modulates neural function by affecting neurohumoral transmission.

The changing relationship between cardiac output and peripheral resistance during development of hypertension in experimental animals has not yet been fully explained. Autoregulation, or adjustment of blood flow to the need of the tissues, is a term often used in this context, but, unfortunately, it appears to mean different things to different authors. This area of study should ultimately provide much information concerning the hemodynamic events in the development of fixed diastolic hypertension.

Hypertension and Renal Parenchymal Disease. This type of hypertension is so often complicated by the features attendant on diminished kidney function (anemia, overhydration, acidosis, etc.) that acceptable studies of its mechanisms in man have been very difficult to obtain. Further, the time course of renal decompensation can be compressed into a few days or extended over several years, so that hypertensive mechanisms may be quite different from one patient to another.

The hypertension of acute glomerulone-phritis has been related to hypervolemia with consequent circulatory congestion, high ventricular filling pressure and increased cardiac output, the total peripheral resistance remaining inappropriately normal in the face of increased blood flow.

In patients with end-stage renal disease, conflicting findings have been reported, probably reflecting the varying proportions in each particular instance of the renal and the renoprival elements outlined above. Various investigators have commented upon the differences between two varieties of hypertension in uremic patients. The first (and more common) is characterized by satisfactory reduction of arterial pressure with dehydration, and pressure levels are apparently related to the magnitude of total blood and extracellular fluid volumes without respect to peripheral plasma renin activity. The second group of patients apparently do not respond well to dehydration or vasodepressor agents; their hypertension correlates better with peripheral plasma renin activity and is markedly ameliorated by nephrectomy.

The hypertension of renal parenchymal disease and the variations of arterial pressure with hemodialysis have been ascribed to changes either in cardiac output or in total peripheral resistance. The wide spectrum of changes is probably related to individual differences in volemia, myocardial status or degree of anemia. However, even among patients without severe anemia or renal decompensation, resting cardiac index varied so widely that difference from normotensive or other hypertensive controls was not obvious. Total peripheral resistance, however, was usually increased, as in essential hypertension, but its distribution among the various vasculatures was not similar in the two conditions. Brod et al. interpreted this as a definite evidence distinguishing these two types of hypertension pathophysiologically.

Coarctation of the Aorta. The hypertension associated with coarctation of the aorta is an experiment of nature of considerable hemodynamic interest. It has demonstrated that the body can adjust peripheral resistance differently in organs above and below the coarctation so as to provide all of them with a normal blood flow. Unfortunately, the mechanism of this precise adjustment has not yet been determined, although it offers the best possible example for autoregulation of blood flow. Cardiac output is usually increased; and the ejection of a large stroke volume into an aorta with decreased capacity and relatively limited runoff accounts for the large pressure above the aortic constriction. As output and rate of ejection are increased with exercise, so is the systolic blood pressure—often to alarming levels—even in patients who have near-normal pressure at rest. By diminishing this increase of cardiac output with propranolol, arterial pressure rise with exercise is attenuated despite the significant increase in total peripheral resistance induced by the drug.

Hemodynamic Responses to Stress and Exercise in Hypertensive Patients

Studies at rest under basal conditions may not reveal the full scope of possible hemodynamic abnormalities in hypertension. Various stressful stimuli such as application of cold, production of ischemic pain, disturbing demands of interviews, exercise, and pharmacologic agents have therefore been used to detect hypothetical "hyperresponsiveness" of hypertensive patients. By and large, few consistent differences between the normotensive individual and hypertensive patient have been demonstrated with regard to magnitude of pressure rise with stress.

However, compared with normotensive controls, the duration of the rise seemed longer, and, perhaps more important, its hemodynamic pattern appeared different in hypertensive patients.

Increase of arterial pressure during a stressful interview was due in most normal subjects to increased cardiac output, whereas most hypertensive patients responded by an increase in total peripheral resistance. In the absence of cardiac failure, response to dynamic exercise in hypertensive patients is much the same as in normal subjects. Cardiac output increases proportionately to oxygen uptake, and peripheral resistance falls. This fall is never to normal levels, so that blood pressure remains high —indeed, the extent of reduction in peripheral resistance may not be commensurate with the rise in output, especially in young subjects. These observations, as well as the greater tendency to respond to stress by raising vascular resistance, point to an early and persistent anomaly in peripheral reactions in hypertension.

VOLUME MECHANISMS

Extracellular Fluid—Distribution and Measurement

Total body water accounts for about 60 percent of body weight and is contained in two compartments—the intracellular and the extracellular. The latter is the subject of this discussion because of abnormalities that often accompany hypertension. The suggestion has been made that intracellular fluid may also play a role in regulating arterial pressure by determining the volume of arteriolar cells. Thus excess fluid should increase cell size, thicken the arteriolar wall, and narrow the lumen. This would increase resistance to blood flow and could raise arterial pressure. Although this is an attractive hypothesis, methods for measuring intracellular fluid are not yet precise enough to test its validity.

Extracellular fluid (ECF) has two components—the plasma and the interstitial fluid volumes—which are in dynamic equilibrium at the capillary level. The plasma volume (PV) and red cell mass comprise the intravascular volume. Considering these basic features, it should not be surprising to find that the ECF and the vascular system are intimately related because the intravascular volume distends the vascular bed and the interstitial fluid (IF) surrounds it.

The extracellular fluid volume is about 20 percent of body weight; roughly speaking, interstitial fluid comprises about 80 percent of this and plasma about 20 percent. Actually, the PV/IF ratio is normally maintained around 0.233. Measurements of these volumes are readily available and are based on the dilution principle. The substances used for these purposes have been chosen because they distribute homogeneously within the space to be measured. They are injected in precisely determined amounts, and the degree to which they are diluted allows a measure of the diluting volume. For measuring plasma volume it is necessary to use a substance that stays within the vascular system for a sufficiently long period to allow adequate mixing. This requirement is supplied by radioiodinated human serum albumin (using ^{125}I or ^{131}I) and by Evans blue dye that binds to plasma proteins. Values obtained with these two substances are practically identical. In contrast, many of the substances used for measuring total extracellular fluid yield differing results. This is because some (e.g., inulin, sucrose, and thiosulphate) diffuse into the interstitial fluid of tendon and other supporting structures so slowly that they give falsely low values, whereas others (e.g., radiobromine, chloride and sodium) get into cells in small amounts, giving larger than normal values. However, each compound has its own distribution characteristics and gives results within a reasonably narrow range. Thus, in any study of ECF, the same substance must be used throughout.

Under normal circumstances, volumes of these fluid compartments depend upon the size of the individual. Also, the percentage of body weight they represent is inversely related to amount of body fat. Thus, men have larger volumes than women and fat people relatively less than thin people. The latter feature creates difficulties in expressing results so that individuals and groups can be compared, regardless of how obese they may be. Relating these volumes to lean body mass provides the best expression; but, since that measurement is difficult, other means of expression must be used. The most usual one employed, even with its recognized limitations, is the relationship to body weight. Body height is a better reference, because it is not influenced by the degree of obesity and because it remains fairly constant over many years. Body surface area has not been used extensively as a reference index; it has the advantage of providing some consideration of the amount of water contained in fatty tissue, which use of height does not.

Intravascular Volume and the Vascular System

Broadly speaking, the vascular system has three components, and each of these has its own characteristics. The arterial segment has limited distensibility, and is maintained at high pressure and low volume; it is thought to contain about 20 percent of the total blood volume. Although the capillary bed is of considerable length, it contains only 5 percent of the total blood volume. Capillary pressure is determined by the amount of constriction of precapillary arterioles and postcapillary venules. The venous side of the circulation is a low-pressure, highly distensible compartment which contains about 75 percent of the intravascular volume.

Both the arterial and venous components are importantly affected by sympathetic vasomotor outflow, but, characteristically, these effects are different. Since the volume of the arterial side is relatively small compared to the venous, its capacity is insignificantly changed by neural influences, although in this vascular segment resistance to blood flow and perfusion pressure can be greatly modified. In contrast, sympathetic vasomotor activity plays a large role in determining the capacity of the venous side, whereas, at the same time, it has little effect on pressure.

The heart, of course, supplies the energy for tissue perfusion but the amount of blood that it pumps in the absence of cardiac failure depends on the amount returned to the left atrium. This, in turn, is a function of the total intravascular volume and the central blood volume. The latter, the volume of blood in the heart and lungs, is determined in large measure by the capacity of the systemic (peripheral) venous compartment. Thus it is possible on the one hand to have a large blood volume, venous pooling, low central blood volume and low cardiac output while, on the other, a small blood volume, diminished venous capacity, a disproportionately high central blood volume and a normal or slightly increased cardiac output.

Control of Extracellular Fluid Volumes

Although highly compartmentalized, plasma and interstitial fluid comprise one system, with the locus of contiguity at the capillary level. Their electrolyte compositions are practically identical, with the minor concentration differences reflecting the protein content of plasma. Since these two fluids are in dynamic equilibrium, any change in electrolyte composition is quickly transmitted across the capillary endothelium. The volume relationships of plasma and interstitial fluid are controlled by capillary (hydrostatic) filtration pressure, the oncotic pressure of plasma proteins, the tissue pressure, and venular pressure. The precapillary arterioles determine the amount of arteriolar pressure that reaches the capillary bed. The difference between this outward force and the sum of tissue pressure and plasma protein oncotic pressure is the filtration pressure. On the venular side of the capillary bed reabsorption of interstitial fluid is determined by oncotic

pressure, which facilitates it, and by venular pressure which reflects the relationship between the capacity of the systemic venous compartment and the volume of blood it contains. Broadly speaking, interstitial fluid is controlled by the volume and the composition of the plasma, and this, in turn, is controlled by external water and electrolyte exchanges which are primarily a renal function. A subsidiary system, the lymphatics, removes the small amounts of protein filtered and any excess of filtered fluid over that reabsorbed.

Although fluid losses occur through skin, lungs, and bowel, these are normally of minor importance in homeostasis compared to the control exerted by the kidneys because of the large volumes of ECF that are constantly being processed. In brief, water and electrolyte balances are achieved through the amount of glomerular filtrate formed and the amounts reabsorbed. According to Pitts, 67 to 87 percent of glomerular filtrate is reabsorbed in the proximal tubule. With a normal GFR of 150 liters per day, as much as 50 or as little as 20 liters leave the proximal tubule to be processed at more distal sites. If the 24-hour urine volume is 1.5 liters, this represents excretion of only 1 percent of glomerular filtrate. With these amounts in mind, it is easy to comprehend how changes in GFR and in proximal or distal tubular reabsorption can modify the amount of plasma water excreted and, thus, the volume and composition of the ECF.

Glomerular filtration rate is determined by renal blood flow and the filtration pressure within glomerular capillaries. When the intrarenal vascular system is intact—which is often not the case in hypertension—intrarenal mechanisms for control of blood flow and filtration pressure are the afferent and efferent arterioles. Thus, afferent arteriolar constriction can decrease renal blood flow and glomerular capillary pressure, but the latter can be normalized by a narrowing of the efferent arteriole. Obviously, when arterial pressure is elevated, the task of regulating renal pressure and flow is a large one. This mechanism of readjusting local blood over a wide range of arterial pressure is called autoregulation and is developed to a high degree in the kidney.

Reabsorption of filtrate in the proximal tubule is isosmotic. Mechanisms for it, although not completely understood, are probably a combination of active transport processes of the tubular cells and hydrostatic forces within the peritubular capillaries. In the distal tubule and collecting ducts, sodium and water are influenced by aldosterone and antidiuretic hormone respectively. (For further details, see Chap. 16.)

With extracellular fluid volume expansion natriuresis and diuresis normally occur, which cannot be accounted for completely by changes in GFR or aldo-

sterone. This has suggested the possibility of a natriuretic hormone, popularly called third factor (i.e., a factor in addition to GFR and aldosterone). As yet such a substance has not been identified.

Plasma and Interstitial Fluid Volumes in Hypertensive Patients

The foregoing discussion of the distribution and control of extracellular fluid is a necessary background for consideration of the abnormalities of plasma volume that accompany hypertension.

Plasma volume decreases have been reported in hypertensive patients repeatedly since 1948. At first it was not possible to separate patients into various diagnostic groups because techniques for this were not available. More recently, however, plasma volume characteristics have been described for essential and renovascular hypertension and for pheochromocytoma, primary aldo-

TABLE 3-3. Plasma Volume (PV) and Red Cell Mass (RCM)
in Normal People and in Hypertensive Patients

| | PV (ml./per cm. of height) | | RCM (ml./per cm. of height) | |
	Mean	S.E.M.	Mean	S.E.M.
MEN				
Normal	18.4	0.33	12.0	0.22
Essential hypertension				
diastolic BP<105 mm. Hg	18.3	0.54	13.0	0.52
diastolic BP>105 mm. Hg	16.0	0.37	11.9	0.44
Renal hypertension				
arterial stenosis	17.6	0.82	12.2	0.46
parenchymal disease	18.2	0.56	13.0	0.90
Primary aldosteronism*	19.8		11.8	
WOMEN				
Normal	15.3	0.25	8.5	0.16
Essential hypertension	14.1	0.19	8.3	0.34
Renal hypertension				
arterial stenosis	14.3	0.39	7.9	0.33
parenchymal disease	14.7	0.29	8.2	0.64
Primary aldosteronism*	15.3		9.1	

* Numbers too small for calculation of S.E.M.
S.E.M., Standard Error of Mean.

steronism, and renal parenchymal disease; in some groups plasma volume is decreased, in others it is increased; while in yet others, it is inappropriately normal (Table 3-3).

In both men and women with essential hypertension, plasma volume is lower than normal and, in the men, the higher the arterial pressure the lower the volume. Similar decreases have been found in patients with stenosing lesions of the renal artery and with pheochromocytoma. Primary aldosteronism is often accompanied by a modest expansion of plasma volume, as is acute glomerulonephritis also. In each of these instances, since red cell mass is usually normal, plasma volume is a faithful reflection of the total intravascular volume; if it is diminished so is total blood volume, and vice versa. In hypertension associated with renal parenchymal disease— and most obviously in chronic renal failure —fluid retention is a common feature, and, although plasma volume is often much increased, this has occurred, in part, as compensation for the reduction of red cell mass. Even at that, increased total blood volume is often present.

The relationship of plasma to interstitial fluid volumes in these various hypertensions has not been thoroughly explored, and this limits understanding of the mechanisms of plasma volume changes. In one study of essential hypertensive men, extracellular fluid was found to be normal; but, since plasma volume was decreased, the ratio of plasma to interstitial fluid volume was decreased from a normal value of 0.223 to 0.194. This information suggests that, in this particular group of hypertensive patients, factors controlling the distribution of fluids in the ECF compartment were abnormal whereas those controlling total volume were not. One interpretation would be that capillary hydrostatic pressure is slightly elevated, thereby causing a small portion of plasma water (normally kept in the intravascular compartment) to be translocated to the interstitial space. In support of this interpretation is the common experience that plasma volume expansion is a frequent accompaniment of antihypertensive drug treatment (diuretics and propranolol excepted), which can occur without weight gain and suggests a transfer of interstitial fluid to the intravascular compartment.

Physiological Relationships of Plasma Volume in Hypertension

To consider plasma volume alone provides no insight into its causative mechanisms and/or functions in patients with elevated arterial pressure. But to consider it in relationship to other physiological variables such as arterial pressure, cardiac output, circulating renin activity, and neurogenic vasomotor tone allows a glimpse of its contributions to circulatory dynamics. Before proceeding to a discussion of these relationships, two points should be re-emphasized. One is that plasma volume is really a reference to total intravascular volume, since the red cell mass is usually normal in hypertension. The other is that since the bulk of intravascular volume is contained in the venous limb, distribution of blood between the systemic and the central venous compartments is an important consideration.

Diastolic arterial pressure is inversely related to plasma volume in men with essential and renovascular hypertension and patients with pheochromocytoma (Fig. 3-3). Although plasma volume is reduced in women with essential and renovascular hypertension, it does not seem to be correlated with diastolic pressure (perhaps because of variability due to the menstrual cycle). In contrast to this inverse relationship, pressure and volume are directly correlated in patients with chronic renal parenchymal disease. In acute glomerulonephritis, although hypervolemia is charac-

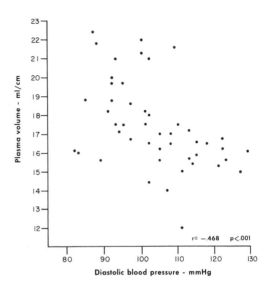

FIG. 3-3. Relationship between plasma volume and supine diastolic pressure calculated as weekly average from four daily measurements of 47 essential hypertensive men (r = — 0.468, P < 0.001). (Tarazi, R. C., Dustan, H. P., Frohlich, E. D., *et al.:* Plasma volume and chronic hypertension. Relationship to arterial pressure levels in different hypertensive diseases. Arch. Intern. Med., *125:836,* 1970)

teristic, it apparently is not causally related to arterial pressure.

Cardiac Output. In essential and renovascular hypertension, cardiac output was found to be well-maintained—and, sometimes, slightly elevated—even in the face of considerable reduction of plasma volume. Since the central blood volume is normal and total blood volume is reduced, a greater proportion must be redistributed to the central circulation from the periphery. Although cardiac output bears no direct relationship to plasma or total blood volume, it is directly correlated with the ratio of central blood volume to total intravascular volume. Since the capacity of the systemic venous compartment seems limited, in these hypertensive states it is tempting to ascribe this to increased sympathetic activity; however, the evidence is strictly inferential. In contrast, in acute glomerulonephritis cardiac output is directly associated with total blood volume, suggesting that, in this congested circulatory state, the distribution of blood between the systemic and the central venous compartments is normal. This suggestion is compatible with the postulate that this type of hypertension results from failure of peripheral vasodilatation in the presence of a hyperdynamic circulation.

Neural Mechanisms. Since the capacity of the vascular bed is determined by the amount of vasoconstriction in both arterial and systemic venous compartments, and since this can be considered a neurogenic function, it would be anticipated that intravascular volume and sympathetic vasomotor tone would be inversely related. Thus, the smaller the blood volume the greater would be neurogenic activity. As logical as this association may seem, it is difficult to test clinically, because, although measurement of intravascular volume is relatively easy, sympathetic control of the vasculature can be only inferred. Reduction of arterial pressure by sympathetic inhibition in patients with renal insufficiency is achieved with greater ease when plasma volume is contracted. When arterial pressure is similarly reduced (using the ganglion blocking drug trimethaphan) in patients with essential or renovascular hypertension, reduction of arterial pressure was negatively correlated to intravascular volume only in those with essential hypertension. Although there is no explanation for the differences between the renovascular and the essential hypertensives, these results provide a basis for understanding some earlier observations. In the late 1940's and early 1950's, when there was much investigation of the antihypertensive effects of low sodium diets, it was found that dietary sodium restriction, which reduces plasma volume, enhanced the depressor effects of the ganglion blocking agent tetraethylammonium chloride (TEAC) and that sodium repletion diminished this response.

Plasma renin activity is also related to intravascular volume. Thus, plasma volume and plasma renin activity have been found to be inversely related in normal men, men with essential hypertension and patients (men and women) with renovascular hypertension (Fig. 3-4). Although renin release is known to be stimulated by hemorrhage, the mechanism whereby intravascular volume and plasma renin activity are related in normal and hypertensive individuals has

yet to be elucidated. Thus, renin release may be stimulated either by adrenergic factors induced by contracted intravascular volume, or by less stretch of the juxtaglomerular apparatus in the kidney, other factors, or a combination of any or all.

Control of ECF in Hypertension-Exaggerated Natriuresis. Hypertensive patients excrete an intravenously administered salt load faster than do normotensive individuals. This phenomenon is called *exaggerated natriuresis* and it is directly correlated with height of arterial pressure. Since normotensives achieve the same rates of urinary sodium excretion as hypertensives after prolonged saline infusion, exaggerated natriuresis can be viewed as a mechanism for protecting the hypertensive from ECF expansion. It would be tempting to relate this to the contracted plasma volume so often found in hypertensives, except that exaggerated natriuresis can also be demon-

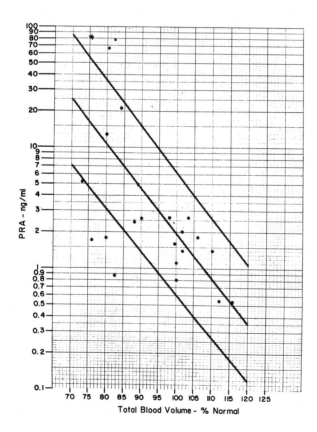

Fɪɢ. 3-4. Regression line, ($\pm$) 2 standard deviations for the association of plasma renin activity (PRA) with intravascular volume in 21 patients with renal arterial hypertension. Total blood volume, expressed as percentile deviation from normal because of naturally occurring sex differences, is plotted against the logarithm of the plasma renin value because the relationship was found to be exponential. Correlation coefficient for the entire group was -0.584 (P < 0.02). However, when the four values that fell below 2 standard deviations were excluded from the calculation, the correlation coefficient was -0.820 (P < 0.001). (Dustan, H. P., Tarazi, R. C., and Frohlich, E. D.: Functional correlates of plasma renin activity in hypertensive patients. Circulation, *41*: 560, 1970. By permission of The American Heart Association, Inc.)

strated in patients with primary aldosteronism who often have expanded plasma volume.

This facilitated salt excretion cannot be explained by increases in glomerular filtration rate; and, since it occurs in the presence of increased aldosterone production, cannot be related to changes in that hormone. Although a natriuretic hormone, "third factor," has been postulated the available evidence suggests that exaggerated natriuresis has a renal hemodynamic mechanism that is probably the renal expression of a systemic hemodynamic response to volume expansion. Thus, it has been shown that hypertensive patients have an elevated renal vein wedge pressure that rises yet higher with saline administration; this is taken to indicate a higher than normal pressure in peritubular capillaries which would serve to limit sodium reabsorption. In addition, such patients respond to volume expansion with a rise in cardiac output, which may well reflect the hypertensives' tendency to maintain a greater proportion of intravascular volume in the central circulation because of restricted capacity of the systemic venous compartment.

RENAL PRESSOR MECHANISMS

The Renal Pressor System—Description and Measurement. The importance of this system in regard to hypertension is that its end product, angiotensin II, is the most potent pressor substance known. The system, itself, is really two enzyme systems in series (Fig. 3-5). In the first reaction, a proteolytic enzyme of renal origin, called *renin*, reacts with a circulating globulin (renin substrate, or angiotensinogen) to release the decapeptide angiotensin I. This serves as substrate for converting enzyme which, by releasing two amino acids, produces the octapeptide, angiotensin II. This is rendered inactive by plasma and tissue angiotensinases. Although renin comes from the kidney, it is active in circulating blood and is stored in arterial walls. The converting enzyme is present in plasma and tissues, and there is evidence to suggest that in some species the major conversion of angiotensin I to II occurs in the lung.

Angiotensin I has no effect on arterial pressure, whereas only nanogram amounts of angiotensin II are sufficient to produce substantial elevations. Information currently available suggests that the plasma concentration of angiotensin II is usually less than 100 picograms per ml. Of course, this substance is important in understanding the relationships of the renal pressor system to hypertension; but current information about it is fragmentary because a radioimmunoassay method for the measurement of angiotensin II has only recently been developed.

Current information concerning the renal pressor system comes from estimation of plasma renin activity (PRA). Under

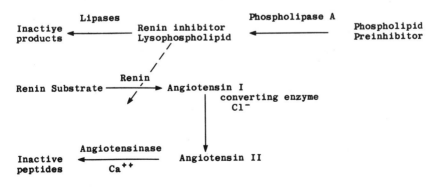

FIG. 3-5. A schema of the renal pressor system.

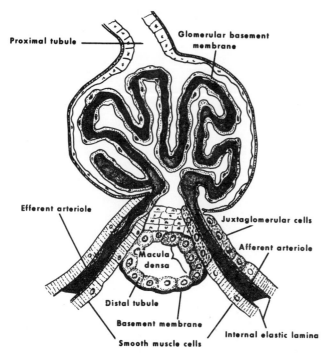

Fig. 3-6. A schematic representation of the juxtaglomerular apparatus, showing the relationships of juxtaglomerular cells of the afferent arteriole to macula densa cells of the distal tubule. These granulated juxtaglomerular and macula densa cells form the juxtaglomerular apparatus.

proper conditions of pH and temperature the endogenous plasma renin is allowed to act on endogenous substrate. By inhibiting plasma angiotensinases enough angiotensin is formed to allow its quantification by bioassay. Renin substrate can also be measured by incubating plasma with enough added human renin to completely exhaust the protein and then using the bioassay to measure the angiotensin formed.

It should be emphasized that the methods widely used provide an estimate of the activity of renin but do not measure it directly.

Source of Renin—the Juxtaglomerular Apparatus. Renin comes from the kidney, hence the term renal pressor system. It is formed and stored in the juxtaglomerular (JG) apparatus, or complex, at the vascular pole of the glomerulus (Fig. 3-6). At this point the macula densa portion of the distal tubule is in close proximity to the afferent and efferent arterioles. The JG complex is composed of granular cells in the afferent arterioles, the macula densa and the polkissen, a group of cells in the triangle formed by the afferent and efferent arterioles and the distal tubule. The complex is richly innervated by sympathetic nerve fibers.

The granular cells of the afferent arterioles seem to be the primary location of renin. Experimentally, there is a close correlation between the granularity of these cells and renal renin content. The role of the macula densa has not been defined but there is evidence to suggest that the amount of sodium reabsorbed at that site is an important determinant of renin release.

Considering the facts that the granular cells are in the afferent arterioles and that the macula densa is in close anatomical proximity to them it should not be surprising to find that both arterial pressure and sodium excretion are among the factors known to affect renin production and/or release.

Renin Release. There are a number of stimuli that strongly influence renin release and, therefore, circulating renin. These stimuli include height of arterial pressure, magnitude of intravascular volume, sympathetic vasomotor outflow and the amount of sodium thought to be present at the macula densa.

Practically all information concerning renin release in man has come from the clinical application of animal studies. Thus, it has been found that rapid arterial pressure reduction with intravenous administration of sodium nitroprusside or diazoxide increases PRA. This reduces renal perfusion pressure; renal arterial stenosis also reduces renal perfusion pressure, a situation that often, but not always, results in increased levels of circulating renin and, if the lesion is unilateral, an increase in PRA in renal venous blood from the affected side.

Rapid reduction of intravascular volume by controlled bleeding also increases PRA. A slower reduction is produced by the negative sodium balance that accompanies a low sodium diet or diuretic drug treatment and this also elevates PRA in most individuals. The converse is also true, since plasma volume expansion turns off the stimulus for renin release and reduces PRA. So predictable are these responses that dietary sodium restriction and intravenous sodium chloride infusion are used as standard tests of the renal pressor system.

The sympathetic nervous system also plays a role in renin release. Normal cardiovascular adjustments during upright posture depend on augmented sympathetic vasomotor outflow, and one of the effects of this is a rise in PRA. This rise is often absent in patients with idiopathic orthostatic hypotension. In addition, both alpha- and beta-adrenergic blocking drugs have been shown to diminish PRA. Upright posture, as well as manipulation of sodium intake, is frequently used as a clinical test of renin release.

Serum sodium concentration in some way influences the amount of renin produced or released into the circulation. This has been shown clearly in animal experiments utilizing renal perfusion techniques with which it is possible to vary sodium concentration of the perfusate without varying that of the whole body. Hyponatremia increased renin release, and hypernatremia diminished it. Whether this reflects a direct effect on the afferent arteriolar component of the JGA or on the amount of sodium available for macula densa reabsorption is not known. At any rate, there is no information to suggest that serum sodium concentration has any ongoing control over PRA under normal conditions.

It is impossible to measure the amount of sodium present in tubular fluid at the level of the macula densa. Thus, the macula densa theory of renin control cannot be tested in man. Certainly, dietary sodium restriction results in diminished sodium excretion and increased PRA, whereas sodium chloride infusions have the opposite effects. However, the former is associated with decreased intravascular volume which increases sympathetic vasomotor outflow, and this, in itself, could be responsible for augmenting renin release. It has been suggested that diuretic drugs increase PRA not only by decreasing plasma volume but also by paralyzing the sodium reabsorptive capacity of the macula densa. Unfortunately, there is no way to study this possibility in man.

Potassium, as well as sodium, may play some role in renin release; it has been shown that potassium loading can diminish PRA and depletion can elevate it, independent of change in aldosterone secretion. How important potassium concentration is in the normokalemic individual is yet to be determined.

Plasma Renin Activity in Hypertensive Patients. Measurement of plasma renin activity was expected to permit definition of a renal component in any hypertension. Generally speaking, at the present time, plasma renin activity has not provided a precise measurement of any mechanism but only indications of one. This is not surprising on two counts. In the first place, knowledge of angiotensin II levels must be available before conclusions concerning renal pressor factor can be reached. Also, angiotensin II has a number of physiological effects, in addition to its direct vasoconstrictor action that could modify arterial pressure. These include both central and peripheral augmentation of sympathetic nervous activity, a role in catecholamine release from the adrenal medulla, and a major regulatory function in aldosterone production.

That there is more than the renal pressor system to any hypertension—even the most fulminant renal type—is shown by the marked increases in PRA that accompany normotensive states such as hepatic cirrhosis and the nephrotic syndrome. There are, however, hypertensive states in which PRA is usually elevated, usually normal, or usually decreased. It is not now possible to determine its contribution to the hypertension in any of these.

Peripheral plasma renin activity may be elevated in patients with renal arterial stenosis and seems to bear a direct relationship to the height of diastolic arterial pressure (Fig. 3-7). This is important to remember, since these patients in a hospital setting often have mild labile hypertension and the finding of a normal PRA does not mean that the hypertension is nonrenal. Because peripheral PRA is inconsistently elevated in renovascular hypertensives, measurements in renal venous blood have been advocated. This is based on the likelihood that in unilateral renal arterial stenosis the affected kidney produces more renin than the unaffected kidney will produce, and that bilateral stenoses, being unequally severe, will have a similar effect. This approach seems likely to be helpful in diagnosis, but the extent of its value has not yet been established. Moreover, the results of surgery in patients without elevation in renal venous renin activity must also be known.

At the other end of the spectrum of renal arterial stenosis is the patient with malignant hypertension, very high PRA and secondary aldosteronism as indicated by hypokalemia and normal or slightly depressed serum sodium concentration. These are some of the characteristics of malignant hypertension of any type (except, perhaps, that accompanying adrenal hyperplasia), but this does not eliminate the possibility of renal arterial stenosis or renal paren-

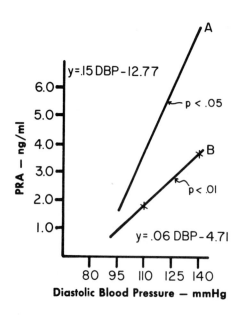

FIG. 3-7. Regression lines and formulas for the relationship of plasma renin activity (PRA) to diastolic arterial pressure in patients wth renal arterial stenosis. Line A indicates the relationship for all 21 patients of the group; line B for 19 patients, excluding the two with the highest values for renin activity. (Dustan, H. P., Tarazi, R. C., and Frohlich, E. D.: Circulation, 41:563, 1970. By permission of the The American Heart Association, Inc.)

chymal disease as a cause of hypertension.

In terminally uremic patients maintained by chronic dialysis, measurement of PRA can indicate a renal pressor component. In these patients two types of hypertension exist. One is volume dependent and can be controlled by hemodialysis, since it occurs only with salt and water excesses; in these patients PRA is not elevated. The other type seems to depend on continued activation of the renal pressor system. These patients have elevated levels of PRA and their hypertension cannot be controlled by dialysis. This pressor component is not found if bilateral nephrectomy is performed; in such cases, hemodialysis will provide excellent control of arterial pressure by providing adequate control of the volume component. (See Fig. 3-2).

There is no indication of renal participation in most hypertensives but there is a group in which abnormally low PRA has diagnostic significance. This group is comprised of the few patients with primary aldosteronism and a larger number whose hypertension has not been explained. In both types of hypertensives, the usual stimuli used to increase circulating renin are ineffective. Thus, low sodium diets, upright posture, and rapid arterial pressure reduction fail to elevate PRA. In both primary aldosteronism and the "low renin" hypertension of undetermined cause there is no suggestion that the renal pressor system is a factor in the hypertension. In regard to primary aldosteronism, it seems likely that depressed PRA reflects the modest positive salt and water balance that results from aldosterone excess.

The wide use of oral contraceptive agents has brought attention to another type of hypertension that may depend on the renal pressor system. The estrogen component of these medications increases renin substrate and, although renin itself is not necessarily increased, more angiotensin is formed. Since there is no way to block the enzymes of the renal pressor system, there is no way of knowing the exact nature of the hypertension that occurs in a few of the women so treated. However, when the medication is discontinued, the components of the renal pressor system return to normal levels and the hypertension disappears.

Physiological Relationships of Renin in Hypertensive Patients. It seems reasonable to expect that PRA could be related to the factors that have been shown to influence renin production and release. Although the information is fragmentary, a review of what has been learned suggests this to be the case.

The importance of intravascular volume in determining renin release is expressed by the inverse correlation between plasma or total blood volume and PRA in normal men, men with essential hypertension and patients with renovascular hypertension.

In regard to the possible influence of plasma sodium concentration on circulating renin, PRA has been found to be inversely related to it in a large study of many types of hypertension. In another study this inverse correlation was found only in patients with renovascular hypertension but not those with essential hypertension or renal parenchymal disease.

As indicated above, there is one report of a direct relationship of PRA and diastolic arterial pressure in patients with renal arterial stenosis but no correlation of these two variables in essential hypertensives or patients with renal parenchymal disease. The fact that this has not been previously reported may reflect a failure to consider the wide range of arterial pressure found in groups of patients with renovascular hypertension.

In animal experiments sympathetic neural activity strongly influences renin release, but in man this influence is difficult to study quantitatively. For instance, although increased sympathetic vasomotor outflow seems a major factor in the PRA increases with upright posture, the amount of neural activity cannot be measured. Change in plasma norepinephrine with standing would be the function to study in relationship to PRA, but that correlation is not available.

Cardiac output and left ventricular ejection rate are in part determined by neural activity and may be taken as indices of neural cardiovascular control. These have been found to be directly correlated with PRA in essential, renovascular and renal parenchymal disease hypertensions.

CATECHOLAMINE AND NEURAL MECHANISMS

Nervous Control of the Circulation. The autonomic nervous system plays a major role in control of the circulation. It does this by influencing cardiac rate and contractility, arteriolar resistance, and distensibility of the venous compartment. One of its chief responsibilities is to adjust circulatory hemodynamics in response to a variety of stimuli so that tissue perfusion may proceed optimally.

These neural circulatory controls are carried out through an afferent sensing system, a vasomotor center and an efferent effector system. The afferent nerve endings are pressure sensors (mechanoreceptors, or baroreceptors). They occur primarily in the arterial system above the cardiac level. Chief among these are those in the carotid sinus and the aortic arch. Fibers from the carotid sinus travel cephalad in the glossopharyngeal nerve; those from the aortic arch travel in the vagus. Of less importance for control of arterial pressure are those receptors in the atria, the left ventricle and lungs, the fibers from which are vagal in location. Impulses traveling centrally from these receptors influence the medullary vasomotor centers to increase or decrease control of the heart and the peripheral circulation. Impulses emanating from these centers travel along the two components of the efferent system—the sympathetic and the parasympathetic. Of these, the sympathetic is of greater im-

portance by far because parasympathetic efferent fibers go only to the S-A and A-V nodes and both atria, whereas the sympathetic are distributed to these areas as well as to both ventricles and the peripheral vasculature. There are two types of receptors for sympathetic neurons—alpha and beta. The former influence the peripheral circulation primarily and determine the degree of arteriolar and venous tone; the latter primarily affect cardiac rate and contractility.

Simply stated, under normal circumstances, this integrated system acts to raise arterial pressure when it is decreased and to decrease it when it rises. Perhaps the best way to visualize this control is to consider what happens to the circulation with a change from the supine to the upright posture, either by standing up or by being tilted up on a tilt table. The purpose of these circulatory adjustments is to counteract the effect of gravity on the distribution of blood within the vascular system. Thus, when a person stands up, there is, sequentially, a pooling of blood in the highly distensible venous system below cardiac level, a decrease in central volume—and, as a result, a fall in cardiac output and systolic pressure. If this sequence should continue unchecked, fainting would occur because of inadequate cerebral blood flow. However, an intact autonomic nervous system prevents this. The fall in systolic pressure is immediately sensed by the mechanoreceptors, and the information is relayed to the vasomotor centers, resulting in an almost instantaneous increase in sympathetic vasomotor outflow that affects both venous and arterial systems as follows: It limits venous distention, thereby stabilizing cardiac output and systolic pressure. It constricts arterioles and, thus, raises diastolic pressure. Pulse pressure is narrowed a little and mean arterial pressure, in normal persons, is changed by ±10 mm. Hg. As part of this sympathetic stimulation, heart rate increases slightly.

When a tilt table is used, it is possible to study readjustments that occur with return to the supine position. As the table is tilted back, there is a transient rise in arterial pressure—the tilt-back overshoot. This occurs when the pooled blood is returned to the central circulation, suddenly increasing cardiac output into the now-constricted arterial system. This is normalized quickly and within seconds arterial pressure is returned to the pretilt levels. That the tilt-back overshoot reflects increased sympathetic vasomotor tone is established by its disappearance following sympathectomy or use of drugs that suppress sympathetic vasomotor activity.

Another good example of neural circulatory control is the response to a sudden increase in intrathoracic pressure, as with the Valsalva maneuver. This sharply diminishes venous return and decreases cardiac output and systolic pressure, which, in turn, stimulates an increased vasomotor outflow. Suddenly, straining is stopped, blood rushes into the thorax and the resurgent cardiac output is thrust into an arterial system whose outflow resistance has been increased. This results in a brief overshoot of arterial pressure. As the pressure rises, the opposite reflex readjustment is seen in a slowing of heart rate and decline of pressure to pre-strain levels.

Thus, the sympathetic nervous system works constantly to maintain proper pressure-flow relationships in the various vascular beds and this knowledge immediately raises the question concerning its inability to prevent hypertension. The answer is not known but considering the complexities of the system and the demands it must meet, it is not surprising that abnormalities are found in hypertensive patients, and there must be many more than have, by now, been described.

Adrenergic Neuroeffector Mechanisms: Catecholamines. Since it is the sympathetic component of the autonomic nervous system that plays the major role in neural control of the circulation, this discussion will be restricted to adrenergic neuroeffector mechanisms. Like the renal pressor system, these are made up of enzyme systems in series (Table 3-4).

The amino acid, tyrosine, is taken up by brain, peripheral neurons and chromaffin cells, most importantly those of the adrenal medulla, and is hydroxylated by tyrosine hydroxylase to l-dopa in mitochondria. The l-dopa migrates into the cytoplasm, where it is decarboxylated to dopamine. Dopamine, in turn, moves into a specialized, cellular subcompartment, (a vesicle) where it is acted upon by dopamine β-oxidase and becomes norepinephrine. Heart, adrenal medulla and uterus are the only organs known to have the capacity to methylate norepinephrine (NE) to form epinephrine (E) because they alone contain the enzyme phenylethanolamine-N-methyl transferase.

Our concern is primarily with NE because it is the major catecholamine released from adrenergic nerve endings both upon stimulation and as a continuous "leaking" process. It is the neuroeffector agent exerting its physiological effects by attaching to receptor sites that are in close proximity to the nerve terminals.

Typical of Nature's profligacy, more NE is released than is needed; so there are a number of ways in which it is handled. First, some escapes unchanged into circulating blood and is rendered inactive by an enzyme, catechol-O-methyl transferase (COMT), which is present in large quantities in liver and kidney. Probably, however, the bulk of physiological inactivation occurs through reuptake of NE by nerve terminals where it is again stored in granulated vesicles.

Through the action of COMT, normetanephrine is formed and is excreted in the urine as such. Deamination, another process of chemical inactivation, occurs within the nerves themselves through action of the enzyme monoamine oxidase. The compounds thus produced are physiologically inactive and on release into the circulation are methylated in tissues containing COMT to form vanillylmandelic

TABLE 3-4. Catecholamine Synthesis and Metabolism

Catecholamine Synthesis

SUBSTRATE	SITE OF REACTION IN SYMPATHETIC NERVES	ENZYME	PRODUCT
Tyrosine	mitochondria	tyrosine hydroxylase	dihydroxyphenylalanine (DOPA)
DOPA	cytoplasm	aromatic L-amino acid decarboxylase	dopamine
Dopamine	granulated vesicles	dopamine β-oxidase	norepinephrine
Norepinephrine	cytoplasm	phenylethanolamine N-methyl transferase*	epinephrine

* Present in adrenal medulla and nerve endings of heart and uterus.

Fate of Norepinephrine Released from Nerve Endings

INITIAL STEP	SITE OF INACTIVATION	INACTIVATING ENZYME	COMPOUND FORMED	FURTHER METABOLISM	URINARY COMPOUND
Reuptake	intraneuronal	monoamine oxidase	dihydroxymandelic acid	(COMT in liver and kidneys)	VMA
Release into circulation	none				free norepinephrine
Release into circulation	liver, kidney	catechol-O-methyl transferase (COMT)	normetanephrine	none	normetanephrine

acid (VMA) and methoxyhydroxyphenyl-glycol (MHPG).

Catecholamines and their metabolites are excreted in the urine. The amounts of free NE and E are very small because the greater part of the NE that escapes reuptake and is released into the circulation is transformed into normetanephrine and metanephrine, and the NE metabolized by nerve endings appears mostly as VMA.

Catecholamines in Hypertensive Patients. It would seem easy to estimate neural control of circulation by measuring catecholamines and metabolites in the urine. Such, however, is not the case because of the variety of ways in which catecholamines are handled before excretion, because there is no way of knowing the magnitude of nerve terminal reuptake, and because of the various physiological factors that determine the different types of this inactivation. There have been a number of studies attempting to circumvent these difficulties by use of radio-labeled catecholamines, either NE or its precursors. The results, however, have not proved conclusively that there is any abnormality of catecholamine metabolism in patients with essential or renovascular hypertension. The exception to this, of course, is pheochromocytoma. These tumors either continuously or sporadically secrete quantities of catecholamines in sufficient amounts to produce measurable increases in urinary excretion rates of metabolites and often of free NE and E. The ability to measure metanephrines and VMA has facilitated the diagnosis of pheochromocytoma. Previously, only measurements of free NE and E were available, and these measurements were relatively unreliable, presumably because most of the catecholamine released by the tumor was metabolized. To allow greater diagnostic precision, measurement of urinary excretion rates of free catecholamines was always augmented by use of histamine to stimulate release from the tumor or phentolamine (Regitine) to block the pressor effects of circulating catecholamines. Now, with the ability to measure urinary metabolites, such pharmacological tests are practically never necessary.

At present, study of plasma concentrations of NE and E may be the best way to assess participation of the adrenergic nervous system in maintenance of hypertension. This, of course, would hold true only if the pathways of inactivating circulating catecholamines were not efficient enough to maintain plasma levels within a narrow range. Until now, most methods for measuring plasma NE and E have not been sensitive enough for such a study. Recently, a double isotope derivative method has been developed, and, with its use, slightly increased amounts of circulating catecholamines have been found in patients with essential hypertension. Such results must be interpreted with caution, because most of such patients have reduced plasma volume and the increased levels reported may not reflect an increased quantity of circulating catecholamine but, rather, a normal amount in a reduced intravascular volume.

Neural Reflexes in Hypertensive Patients. Although studies of catecholamine plasma concentrations and urinary excretion rates have failed to suggest abnormalities of the sympathetic nervous system in hypertensive patients, the striking effectiveness of antihypertensive drugs that suppress adrenergic function indicates that neural factors operate in some way to maintain hypertension. Evidence that this is so can be shown by studying cardiovascular reflex responses in hypertensive patients.

As indicated previously, the sympathetic nervous system plays a major role in hemodynamic adjustments during upright posture, and arterial pressure varies normally ±10 mm. Hg (mean arterial pressure) during 5 minutes of 50° head-up tilt. Most hypertensive patients respond normally to this stimulus; but there are some in whom orthostatic hypotension develops whereas in others pressure rises abnormally (orthostatic hypertension). In one study of these responses, the patients with orthostatic hypertension had a much greater increase in total peripheral resistance during upright tilt than the others, strongly suggesting that the exaggerated rise in arterial pressure resulted from either increased sympathetic vasomotor outflow or an exaggerated vasomotor response to normal outflow.

Further evidence for increased neural circulatory control or responsiveness in these patients with orthostatic hypertension was a greater overshoot of diastolic pressure following the Valsalva maneuver. Although this study is a qualitative one, it indicates that neural reflex responses are not identical in all hypertensive patients. Further, in the study just described, patients in the three groups had different clinical characteristics. Thus, patients with orthostatic hypertension had the mildest hypertension, the least severe vascular disease, and the best responses to the antihypertensive drugs used most of which suppressed sympathetic adrenergic activity. In contrast, those with orthostatic hypotension had highest diastolic pressure, most severe vascular disease, and poorest responses to prolonged antihypertensive treatment. The patients with normal pressure responses to tilt had intermediate characteristics. At least in these patients, it would seem that exaggerated neural circulatory control was a factor in those with the mildest hypertension, whereas in those with the severest vascular disease other factors were operating.

As far as centrally mediated stimuli are concerned, it has been found that patients with hypertension respond to mental arithmetic and other stressful experiences with exaggerated rises in arterial pressure. Again, however, it should be pointed out that these experiments do not differentiate between increased sympathetic outflow from central vasomotor centers and increased vascular responsiveness to normal outflow.

Physiological Relationships of Sympathetic Vasomotor Activity in Hypertensive

Patients. Considering the extent of neural control of metabolic functions and the vascular system, it is not surprising to find quantitative relationships in hypertensive patients between estimations of sympathetic vasomotor activity and other systems that influence arterial pressure.

Since pressure within the vascular system can be looked upon as being partly determined by the capacity of the system and the volume of its contents and since the capacity can, in large measure, be controlled by sympathetic activity it might be expected that intravascular volume would be inversely related to sympathetic activity. Using the fall in arterial pressure produced by the ganglion blocking agent trimethaphan as an index of vasomotor tone, it has been found in hypertensive patients that, generally speaking, the smaller the plasma volume the greater the depressor effect of trimethaphan. At present, there is no way of knowing whether intravascular volume is reduced because increased sympathetic tone has diminished vascular capacity or whether increased sympathetic vasomotor outflow is a compensatory response to a reduced plasma volume.

Renin release is influenced by neural stimulation, and presumably this is the reason for increased plasma renin activity (PRA) during upright posture. The usual procedure of four hours of standing and walking is certainly not a standard stimulus: individual variations are common, and, in any case, the degree of sympathetic stimulation so produced is impossible to quantify. However, cardiac contractility is partly determined by neurogenic activity and certain hemodynamic functions can be used for a study of neural and renal pressor relationships. In patients with essential, renal parenchymal disease and renovascular hypertension, correlation of PRA with left ventricular ejection rate and cardiac index was found to be highly significant, suggesting an ongoing neural influence on circulating renin.

There is a large body of evidence from

animal experiments that angiotensin increases sympathetic vasomotor activity by affecting central vasomotor outflow directly and catecholamine handling by peripheral adrenergic nerves. The clinical study referred to above could just as well be interpreted as providing evidence for these effects rather than for a common sympathetic stimulus affecting both heart and kidney. However attractive this hypothetical relationship may be, the common clinical experience that renovascular hypertension can be effectively treated with drugs such as guanethidine or alpha methyldopa, which suppress adrenergic activity, strongly suggests that these renal pressor–sympathetic nervous relationships exist in at least one clinical form of hypertension.

ANNOTATED REFERENCES

Abbrecht,, P. H., and Vander, A. J.: Effects of chronic potassium deficiency on plasma renin activity. J. Clin. Invest., *49*:1510, 1970.

and

Brunner, H. R., Baer, L., Sealey, J. E., *et al.*: Influence of potassium balance and potassium deprivation on plasma renin activity in normal and hypertensive patients. J. Clin. Invest., *49*:2128, 1970. (These papers present the first systematic evidence that potassium balance, itself, plays a role in renin release.)

Baldwin, D. S., Biggs, A. W., and Goldring, W., *et al.*: Exaggerated natriuresis in essential hypertension. Am. J. Med., *24*:893, 1958. (A study showing that excretion of an intravenous salt load in both hypertensives and normotensives is inversely related to dietary sodium intake and that the exaggerated natriuresis that characterizes hypertension can be produced in normals by continuing the infusion beyond the usual test time.)

Brust, A. N., and Ferris, E. B.: Varying patterns of blood pressure response to autonomic blockade: implications concerning the interplay of neurogenic and humoral factors in control of vascular tone. Hypertension, *4*:41, 1956. (Shows, among many other things, how low sodium diets enhance responses to ganglioplegic drugs.)

Burton, A. C.: Physiology and Biophysics of the Circulation. Year Book Medical Publishers, Chicago, 1965. (Sections 2 and 3 are a clear discussion of energetics and laws of the circulation; Section 5 is a simplified introduction to overall regulation of the circulatory system.)

Conn, J. W., Cohen, E. L., and Rovner, D. R.: Suppression of plasma renin activity in primary aldosteronism. J.A.M.A., *190*:213, 1964. (One of the first reports of the hyporeninemia of primary aldosteronism)

Cottier, P. T.: Renal hemodynamics, water and electrolyte excretion in essential hypertension. *In:* Bock, K. D., and Cottier, P. T. (eds.): Essential Hypertension: an International Symposium. Berlin, Springer-Verlag, 1960. (A good review and bibliography of exaggerated natriuresis up to 1960)

DeFazio, V., Christiansen, R. C., Regan, T. J., *et al.*: Circulatory changes in acute glomerulonephritis. Circulation *20*:190, 1959; *and* Fleisher, D. S., Voci, G., Garfunkel, J., *et al.*: Hemodynamic findings in acute glomerulonephritis. J. Pediatrics, *69*:1054, 1966. (Both papers describe the congested circulation of acute glomerulonephritis. The former relates cardiac output to blood volume and the latter describes hypervolemia and hydremia.)

Dustan, H. P., and Page, I. H.: Some factors in renal and renoprival hypertension. J. Lab. Clin. Med., *64*:948, 1964. (Shows the dependence of hypertension on overhydration in 3 terminally uremic patients maintained by chronic dialysis.)

Dustan, H. P., Tarazi, R. C., and Frohlich, E. D.: Functional correlates of plasma renin activity in hypertensive patients. Circulation, *41*:555, 1970. (Indicates that intravascular volume and hemodynamic factors may be important determinants of plasma renin activity. Also shows a direct relation of plasma renin to arterial pressure in patients with renovascular hypertension.)

Early, L. E., and Daugharty, T. M.: Sodium metabolism. New Eng. J. Med., *281*:72, 1969. (An excellent presentation of the evidence that intrarenal hemodynamic factors may play a major role in excretion of salt and water)

Elkinton, J. R., and Danowski, T. S.: The Body Fluids. Baltimore, Williams and Wilkins, 1955. (An excellent text, still of value particularly in regard to descriptions, volume and composition of body fluids and methods for measurement)

Frohlich, E. D., Tarazi, R. C., and Dustan, H. P.: Re-examination of hemodynamics of hypertension. Am. J. Med. Sci., 257:9, 1969. (A report of hemodynamic studies in 117 untreated patients with different types of hypertension, with a concise discussion of characteristics of each type and fairly complete bibliography of the subject)

José, A., Crout, J. R., and Kaplan, N. M.: Suppressed plasma renin activity in essential hypertension. Ann. Int. Med., 72:9, 1970. (One of the many descriptions of a subgroup of hypertensive patients with plasma renin activity hyporesponsive to low sodium diet and upright posture)

Laragh, J. H., Cannon, P. J., and Ames, R. P.: Interaction between aldosterone secretion, sodium and potassium balance and angiotensin activity in man: studies in hypertension and cirrhosis. Canad. M. A. J., 90:248, 1964. (One of a number of articles in this special issue [the journal] devoted to renin-aldosterone relationships in health and disease)

Lowenstein, J., Beranbaum, E. R., Chasis, H., et al.: Intrarenal pressure and exaggerated natriuresis in essential hypertension. Clin. Sci., 38:359, 1970. (Relates exaggerated natriuresis to elevated intrarenal venous pressure, providing a hemodynamic explanation for this phenomenon.)

Lund-Johanssen, P.: Hemodynamics in early essential hypertension. Acta med. Scand. (Suppl. 482), 1967; *and* Sannerstedt, R.: Hemodynamic response to exercise in patients with arterial hypertension. Acta med. Scand. (Suppl. 28), 1966. (Both reports discuss in great detail hemodynamic findings at rest and during exercise in patients with essential hypertension. Both come to the same conclusion—of a graded inverse variation in cardiac output and total peripheral resistance in different stages of the disease and an early involvement of both heart pump and peripheral vessels from the first observable stage.)

Newton, M. A., Sealey, J. E., Ledingham, J. C. G., et al.: High blood pressure and oral contraceptives: studies of renin, angiotensinogen and aldosterone. Hypertension, 16:51, 1968. (Relates the hypertension produced in oral contraceptive treatment to estrogen-induced increases in angiotensinogen.)

Pitts, R. F.: Chapter 7 *in* Physiology of the Kidney and Body Fluids. Ed. 2. Chicago, Year Book Medical Publishers, 1968. (A lucid, thought-provoking presentation of the magnitude and mechanisms of renal tubular salt and water transport)

Tarazi, R. C., Dustan, H. P., and Frohlich, E. D.: Relation of plasma to interstitial fluid volume in essential hypertension. Circulation, 40:357, 1969. (Discusses the distribution of the ECF volume between intravascular and interstitial compartments in normals and hypertensives.)

Tarazi, R. C., Dustan, H. P., Frohlich, E. D., et al.: Plasma volume and chronic hypertension. Relationship to arterial pressure levels in different hypertensive diseases. Arch. Intern. Med., 125:835, 1970. (Presents information concerning plasma volume abnormalities in essential hypertension, pheochromocytoma, primary aldosteronism and renovascular and renal parenchymal disease hypertensions; also shows the association between these abnormalities and height of arterial pressure.)

Tobian, L., and Binion, J.: Artery wall electrolytes in renal and DCA hypertension. J. Clin. Invest., 33:1407, 1954. (Describes intra- and extracellular water and electrolyte changes in various experimental hypertensions.)

Ulrych, M., Frohlich, E. D., Tarazi, R. C., et al.: Cardiac output and distribution of blood volume in central and peripheral circulations in hypertensive and normotensive man. Brit. Heart J., 31:570, 1969. (Shows the relationship between central blood volume (CBV) and cardiac output and that in hypertensives with reduced plasma and total blood volumes (TBV), cardiac output is maintained normal by a disproportionately high CBV/TBV.)

Vertes, V., Canziano, J. L., Berman, L. B., et al.: Hypertension in end-stage renal disease. New Eng. J. Med., 280:978, 1969. (Describes two types of hypertension accompanying terminal renal parenchymal disease: one, "volume dependent" and controlled by proper dialysis; the other "renin dependent" and controlled only by bilateral nephrectomy.)

Weil, J. V., and Chidsey, C. A.: Plasma volume expansion resulting from interference with adrenergic function in normal man. Circulation 37:54, 1968. (One of the first papers showing that an antihypertensive drug can increase intravascular volume)

Wurtman, R. J.: Catecholamines. Boston, Little, Brown and Co., 1966. (An excellent discussion of biosynthesis, storage and metabolism of catecholamines)

4

Depressor Mechanisms

Lerner B. Hinshaw, PH.D.

INTRODUCTION

Since Stephen Hales, in the eighteenth century, first measured arterial pressure in a restrained mammal, it has been obtained in most warm-blooded animals. It varies only moderately between species, there being no relationship between the weight of the animal and the height of its arterial pressure. Among the many factors influencing arterial pressure are such variables as age, sex, physical build, digestion, emotion, exercise, posture, and time of day.

Arterial pressure is ordinarily maintained at a higher level than required to ensure blood flow to peripheral tissues. Thus, safety margin is provided so that tissue metabolic needs are adequately met. Local vasodilation may intervene, however, in order to satisfy the body requirements by adjusting tissue blood flow sufficiently to maintain the organism in a condition of homeostasis (see Chap. 2). Mean arterial pressure may also be elevated in certain emergency states in order to provide adequate tissue flow with the increased pressure necessary to propel blood through the systemic vascular circuit. Finally, arterial pressure may rise or fall transiently, depending on the presence of normal and abnormal conditions; however, its steady state value is ordinarily maintained within fairly narrow limits. A certain critical value of arterial pressure is required in vital organs such as the brain, heart, liver, and kidneys in order to maintain normal function.

Mean arterial pressure is the product of total peripheral resistance and cardiac output, assuming a constant and low mean right atrial pressure. This relationship may be utilized in evaluating changes in the entire organism as well as in a given organ such as the kidney. A discussion of depressor factors should begin with a consideration of the determinants of arterial pressure (i.e., flow and resistance). Mean arterial pressure is the time integrated mean of the arterial pressure curve, and as such is not simply the average of diastolic and systolic pressures. In clinical practice, it may be estimated by adding to the diastolic pressure one third of the difference between the systolic and the diastolic pressures. Mean arterial pressure is essentially the same in all large arterial segments, and only where the vessels narrow significantly is there an appreciable decrement in mean pressure.

CARDIOVASCULAR FACTORS

Cardiac Output (Fig. 4-1)

Since cardiac output is one of the main determinants of arterial pressure, it is apparent that pressure may be decreased if cardiac output falls. There are two normal mechanisms by which cardiac output may be decreased. First, a decrease in venous return to the right atrium (since the heart ordinarily ejects the blood that comes to it) decreases cardiac output without any impairment of cardiac integrity. For example, if the blood is temporarily trapped peripherally (i.e., upright position, lower extremity tourniquets, etc.) venous return is reduced and cardiac output falls. Second, direct reduction of cardiac pumping action —by neurohumoral stimuli, among others— reduces the amount of blood ejected from the heart. This may be achieved through a reduction of either stroke volume or heart rate or both, since cardiac output is the product of heart rate and stroke volume.

Total Peripheral Resistance (Fig. 4-1)

Peripheral resistance, the second major determinant of arterial pressure, is itself determined by vascular geometry and blood viscosity. Peripheral resistance is dependent primarily on the caliber of small vessels—arterioles mainly and, to a lesser extent, the capillaries—and on the viscosity of blood. Under conditions of both marked vasodilatation and decreased blood viscosity, a very great fall in arterial pressure can occur.

Vasodilation is a very potent mechanism for reducing arterial pressure and may be caused by direct action, either by neurohumoral stimuli or by locally released vasodilator agents, unless compensatory actions take place. For example, when the splanchnic blood vessels alone are fully dilated, they are capable of accommodating almost the entire blood volume; in this instance arterial pressure would fall nearly to zero. Approximately fifty percent of the total resistance to blood flow occurs in the arterioles which can be dilated tremendously either by inhibition of sympathetic nerve impulses or by humoral factors. Since the resistance in any given arteriole may vary enormously, one of the most important influences in reducing arterial pressure is the opening up of high-resistance vasculatures by arteriolar dilation.

An increase in blood osmolality will decrease vascular resistance, presumably through an action on the vascular wall which increases net vessel diameter.

Viscosity is the frictional resistance which is developed between the parts of the liquid itself. If viscosity decreases, less pressure is required to force blood through the vessels; the internal friction is reduced and the coefficient of viscosity is therefore lower. Water is approximately one fifth as viscous as blood; therefore, if the hematocrit decreases, particularly in its higher range (50 to 60%), viscosity also decreases, and the result is a decline in peripheral resistance; mean arterial pressure likewise falls unless compensatory mechanisms intervene. An increase in temperature lowers viscosity, thereby decreasing both resistance and pressure, assuming other conditions remain constant.

Other Factors

The capillaries serve an important function in maintaining a normal arterial pressure primarily by virtue of their influence on circulating blood volume and cardiac output. Under normal conditions, a state of equilibrium exists at the capillary membrane in which the volume of fluid leaving the active circulation equals the quantity of fluid returned to the circulation at the venous ends of the capillaries and through the lymphatics. If this balance is disturbed —for example, by increasing capillary hydrostatic pressure, increasing capillary permeability, or decreasing plasma colloid osmotic pressure—circulating blood volume will decrease, and, other factors remaining constant, arterial pressure will fall. Other factors influencing transfer of fluid at the capillary level are tissue pressure and osmotic pressure exerted in the tissue spaces.

The capillaries should not be regarded as inert channels; it has been shown that they undergo wide changes in internal diameter that are independent of intraluminal pressure produced by alteration and size of the arterioles and venules. Whether capillary diameter is influenced by back pressure

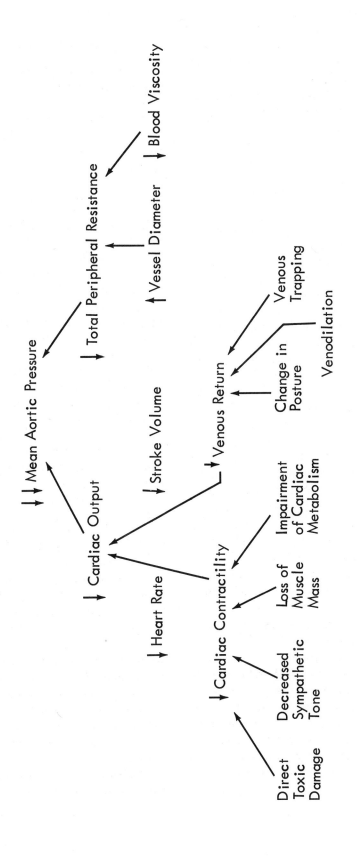

Fig. 4-1. Factors decreasing blood pressure.

from venous constriction, or arteriolar dilatation, or increased flow, all such effects would alter capillary segmental resistance.

Intravascular volume and **vascular elasticity** are two other factors affecting pressure. A decrease in the total blood volume diminishes venous return and cardiac output, tending to reduce arterial pressure. A decrease in the elasticity of the aorta and large arterial walls may increase vascular resistance.

The veins have a peculiarly important role in the regulation of arterial pressure. They are not simply a passive system of draining tubes. The venous system appears to be as reactive and well controlled as any of the other vascular segments. Veins possess at least two major dynamic functions, *resistance* and *capacitance*, that are influenced directly by passive and active (neurohumoral) factors and indirectly by other factors. Any factor that alters venous capacitance or resistance may be expected to alter cardiac output and arterial pressure. If venous return is decreased because of venous dilation and pooling, cardiac output and arterial pressure will fall. Even though the magnitude of venous resistance changes is ordinarily very small when compared to the total contribution of precapillary vascular segments, its physiological significance is virtually without limit. Hence, if venous pressure increases in one particular vascular bed, a rise in organ blood volume may take place. Therefore, with only a minimal increase in venous resistance, capillary pressure rises, resulting initially in pooling upstream from the constricted veins, then in a net loss of fluid into the extravascular space and, finally, in a decrease in venous return and cardiac output. This, of course, assumes that no other compensatory changes intervene in response to the increase in venous pressure which could conceivably precipitate a variety of other actions.

NERVOUS FACTORS

A variety of peripheral and central mechanisms are involved in decreasing mean arterial pressure. Among these are: the carotid sinus mechanism; axon, cardiac, spinal vasomotor, and other reflexes; and the action of the medullary depressor center (Fig. 4-2).

Mechanoreceptors (baroreceptors)

The main systemic arterial receptors are found in the walls of the aortic arch and carotid sinus and have been termed mechanoreceptors. Similar receptors are found in the right subclavian carotid angle, certain segments of the common carotid arteries, the pulmonary artery and its branches, the subendocardial layers of the atria and ventricles, and the vena cavae and pulmonary veins. Mechanical stimulation (by stretch) of the walls of the mechanoreceptors causes a reflex systemic vasodilatation and bradycardia. Therefore, a rise in arterial pressure in the vascularly isolated but innervated and perfused carotid sinus (for example) causes systemic hypotension. These receptors are altogether different from the so-called chemoreceptors which are mainly restricted to the carotid and aortic bodies. It is important to remember that the *mechanoreceptors are not stimulated by a fall in arterial pressure.* Therefore, hypotension exerts a positive reflex effect, because it *decreases* afferent neural discharges that are continually causing reflex vasodilatation and cardiac inhibition. The fall in arterial pressure following stimulation of these stretch receptors is due largely to a reflex diminution in adrenergically mediated peripheral vascular resistance through arteriolar dilatation. The blood vessels participating in this response are found primarily in splanchnic tissue, skin, and skeletal muscle.

The carotid sinus mechanoreceptor, located within the wall of the carotid artery,

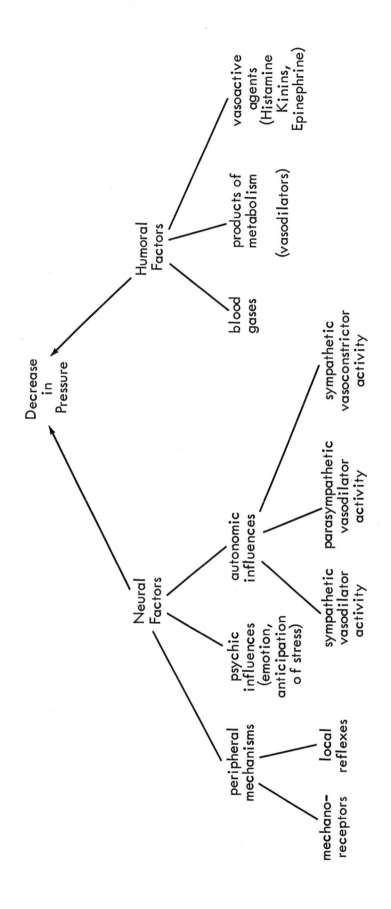

Fig. 4-2. Neurohumoral and central factors resulting in lowered arterial pressure.

consists of nerve endings distributed widely within the arterial wall which are stimulated by deformation (stretch)—i.e., expansion during systole. Sensory nerve fibers pass from the sinus to the medulla via the vagus and glossopharyngeal nerves. Any abnormal form of stretch will stimulate these nerve endings. Thus, tugging on vessels near this region during surgery or applying external pressure to the overlying skin may evoke nervous discharges that can produce reflex systemic hypotension. The hypotension probably results from reduction of arteriolar tone owing to inhibition of vasoconstrictor nerve activity. Electrical stimulation of the carotid sinus nerve carried out during neck operations, or distention of the carotid sinuses by applying a negative pressure to the outside of the neck in conscious man will decrease arterial pressure, presumably owing, in part, to reduction in peripheral vascular resistance.

Pathological manifestations typically may occur if the mechanoreceptors are functioning improperly. A sudden change from a lying to a standing position may cause a person whose pressor receptor system is inactive to faint because of inadequate reflex peripheral vasoconstriction, decreased venous return, and a consequent decreased cerebral perfusion pressure. External pressure to the carotid sinus (by a cervical rib or anomalous scalenus muscle insertion) may activate the pressor receptor reflex and the peripheral vessels may dilate as though to correct an increased arterial pressure. In this instance, the arterial pressure is abruptly and inappropriately reduced, and the individual may experience lightheadedness or even become unconscious. The sensitivity of the carotid sinus mechanism may be so greatly increased that mild external compression at the carotid sinus region may evoke syncope associated with an abrupt fall in mean arterial pressure (presumably due to bradycardia or sinus node arrest). Circulatory collapse accompanied by a precipitous fall in arterial pressure has been elicited in erect normal subjects by mechanical stimulation of the brachial arterial wall.

Arterial pressure is most apt to fall as a result of a sudden transition from lying down to arising, especially in patients bedridden for prolonged periods or older patients with small vessel disease of the brain in whom posture-suppressed sympathetic reflex vasoconstrictor tone is further inhibited. A most common example is found in patients receiving any of a variety of drugs that inhibit sympathetic reflex vasoconstriction.

Patients lacking normally functioning carotid sinus reflex have been studied. Some, with idiopathic orthostatic hypotension (an unusual condition), show very little alteration in vasomotor and cardiac activity when changing posture, others show an exaggerated reflex response that fails to maintain a normal level of arterial pressure following vasodilator drugs and exaggerated pressor responses following norepinephrine.

Deeply anesthetized patients may lack a normally functioning pressor-receptor reflex, and, as a result, changes in posture produce very marked changes in arterial pressure. Therefore, operations under anesthesia of the usual type are ordinarily carried out with the patient in a nearly horizontal position and with particular care to minimize blood loss or right atrial return (i.e., compression of the inferior vena cava).

Mechanoreceptors have been identified in other vascular areas: systemic hypotension has been reported to occur as a reflex response to elevated pressure in the pulmonary artery, but it is not known whether the hypotension in this instance is due mainly to reduction of cardiac output or to arteriolar-venular dilatation. Experimental findings suggest the presence of pressure-sensitive receptors in the cranial cavity. Reflex responses can also be obtained in the innervated perfused canine hindleg when the descending aorta between the left subclavian artery and the diaphragm is perfused at varying pulsatile pressures. Other reflexes arise from intrathoracic vessels, including the cardiac chambers. Included among these are receptors responding to changes in pressures in the walls of the atria and great veins, reflecting changes in blood volume and, hence, stretch of the vessel wall. An increase in right atrial pressure produces systemic hypotension; increased left ventricular pressure may result in systemic hypotension and peripheral vasodilatation. It has been proposed that the function of ventricular mechanoreceptors is proprioceptive, presumably signalling information of cardiac overloading.

Sensory Factors in Dilatation and Local Depressor Responses

Electrical stimulation of the cut central end of a sensory nerve from the skin often produces a local cutaneous vasodilation in the area supplied by the nerve, by means of impulses that travel antidromically down other sensory nerves to this specific area. This response (the so-called axon reflex) apparently does not require participation of spinal vasomotor neurons or higher centers. For instance, dilatation of hindlimb blood vessels can be produced by stimulating the dorsal roots of nerves going to the limb, far below the point of origin from the cord of the sympathetic fibers. The stimulus in this type of reflex is applied to one afferent nerve branch, setting up an impulse that travels centrally to the point of division where it is reflected down the other branch to an effector organ. This response is of significance only in tissues with a relatively rich distribution of pain fibers, such as the skin and mucous membranes. Local cutaneous stimulation by mechanical means or by irritants cause afferent impulses to pass up the sensory nerves and down their collateral branches, evoking local arteriolar dilatation. In certain inflammatory conditions involving the dorsal root ganglia, there is reddening of

the skin in the area supplied by the respective roots. Several substances have been indicated as possible transmitters, among them a histaminelike substance, acetylcholine, and adenosine triphosphate. The local cutaneous axon reflex is induced by any factor that causes damage to exposed tissues—e.g., trauma, heating, or frostbite. Thus, this local mechanism contributes to local defense and repair in surface tissues by creating an increase of local blood flow in response to harmful stimuli.

Primary Central Factors

Modifications of active vasoconstrictor tone dominate in the control of the caliber of resistance vessels; that is, vasodilatation may be considered to be a result of inhibition in existing constrictor tone. Tonic neural control of the resistance and capacitance vessels as exercised by sympathetic vasoconstrictor fibers is influenced mainly by discharges from the brain. Circulatory homeostasis usually does not involve participation of specific vasodilator fibers; rather, vasodilatation is ordinarily achieved by inhibition of vasoconstrictor discharge. There is a balance between the effects of central vasoconstrictor discharge and locally produced chemical vasodilator substances. Transection of the spinal cord at any cervical level causes an immediate and profound fall in arterial pressure, owing to elimination of the tonic activity of the bulbar vasoconstrictor mechanism, and to depression of pressure reflexes.

Sympathetic vasodilator fibers are active when fainting occurs spontaneously or when caused by passive tilting of the human subject. Forearm cutaneous vasodilatation in response to body heating has been ascribed to activation of cholinergic sympathetic vasodilator fibers. Vasodilator impulses that are antidromic emerge from the central nervous system by the thoracolumbar sympathetic outflow, the parasympathetic cranial outflow, the sacral outflow of the pelvic nerve, and the posterior spinal nerve pathway.

There is evidence for efferent dilator nerves distributed specifically to skeletal muscle. This pathway appears to be associated with the mental anticipation of skeletal muscle activity so that impulses originating from the cerebral cortex would result in local dilatation. This vasodilator skeletal muscle outflow is probably activated in alarm reactions when blood flow is redistributed from the viscera to muscle.

Organs supplied by parasympathetic vasodilator nerve fibers include certain cerebral vessels, the tongue, salivary glands, external genitalia, and perhaps the bladder and rectum. Vasodilator fibers are believed to be activated not only in the fainting reaction but also during emotional stress.

The roles of sympathetic and parasympathetic vasodilator fibers in man are still somewhat obscure, and much more information is required to make a general application of their use in the body.

Emotional and Volitional Factors

Mechanisms such as volitional activity, with its accompanying emotional components, involve excitation of numerous central and sympathetic spinal pathways where, in certain circumstances, the vasodilator fibers are excited to elicit a specific discharge pattern. Other types of emotional changes such as resentment, anger, and fear can also cause profound changes in the circulation via cortical hypothalamic pathways, causing many different hemodynamic patterns of response. Emotions such as hostility and resentment can produce marked vasodilatation of the skin or the gastrointestinal mucosa or, often, both. Embarrassment provokes cutaneous vasodilatation which is usually, but not always, confined to the face and shoulders. Blushing was one of the earliest of phenomena recognized as demonstrating the influence of nerves on circulation. Whether it results

from sympathetic vasodilation or inhibition of chronic vasoconstrictor fiber discharges is unknown. There are other and equally well established emotional influences that exert a depressant effect on cardiovascular activity—for example, in emotional fainting and the yoga trance in man and in the "playing dead" observed in some animals. Depressor responses can be originated from certain cortical autonomic areas. Profound, generalized inhibition of sympathetic vasoconstrictor tone may result, the hemodynamic consequence being a decreased peripheral resistance, venous relaxation, and hypotension. Generalized sympathetic inhibition apparently causes greater vasodilation in skeletal muscle than in other tissue. Thus, muscle blood flow increases at the onset of a faint in man despite the fall in arterial pressure, suggesting excitation of the sympathetic vasodilator fibers.

Unquestionably, local vasodilation is rapidly produced in active muscles, and muscle contraction assists venous return to the heart. Evidence now exists that the initiation of nervous adjustments required during exercise emanates from higher nervous structures *prior* to anticipated exercise and not only as a consequence of exertion. It is important to realize that the initiating central drive, insofar as it anticipates muscle activity, provides a means of developing cardiovascular performance before the metabolic needs arise rather than adjusting it in response to altered tissue requirements.

Direct Vascular Tissue Influences

Often, activation of efferent nervous pathways may act in concert with the release of local substances to produce regional vasodilation and a reduction in arterial pressure. Hypercarbia and local hypoxia cause peripheral vasodilation, but the relative vascular effects of variations in blood gases and pH, within the physiological range, are not clearly known. For some regional circulations, however, one can distinguish between local vascular actions of intravascular chemical changes and central nervous effects. Thus, small increases in carbon dioxide have a direct vasodilating action on cerebral and skin vessels in warm-blooded mammals. The hindlimb of an animal perfused with blood from a heart-lung preparation shows vasodilation when the lungs are ventilated by air containing a small amount of carbon dioxide or with reduced oxygen content. These reactions probably result from a local chemical stimulation of vascular smooth muscle. There is some evidence that marked hypoxia does not affect local overall vessel resistance in man, and it has not been known until recently whether the effects of circulatory dilatation which are observed with respiratory or metabolic acidosis result from the elevated partial pressure of carbon dioxide or the fall in pH, or a combination of both. It now appears that the cardiovascular effects arise from the decreased pH, and the increased hydrogen ion concentration during hypercapnia seems to be the determinant factor.

Oxygen lack and excess carbon dioxide may cause capillary dilatation by a direct action. The role of the capillaries must be taken into account in consideration of their segmental resistance contributions in the total vascular circuit which would influence blood pressure in a local area or systemically. Therefore, prominent capillary dilatation might result in a notable depressor response, in which mean arterial pressure would decline.

The endogenous release of certain naturally occurring vasoactive biochemical agents may have profound effects on vascular resistance, causing a fall in arterial pressure. Among these agents is bradykinin, a powerful vasodilator. Histamine, another potent vasodilator, affects arterioles primarily, and may act upon capillaries directly or indirectly. Epinephrine, in low concentrations, is a dilator of skeletal muscle, hepatic and coronary vessels.

There is no doubt that a great many vasodilator agents as yet unidentified are released in normal and pathological states and greatly influence blood pressure. Recent studies have defined the cardiovascular action of the prostaglandins; others indicate the possible role of lysosomes in the release of agents that are potently vasoactive and extensive research will be required to evaluate their actions.

Reactive Hyperemia and Autoregulation

In man, hypoxia causes forearm vasodilatation, and a number of miscellaneous metabolites, released into the circulation during increased activity, may be capable of producing vasodilatation in specific vascular regions. These metabolic substances are being produced constantly, though in small amounts, by resting tissues, and when the blood supply is temporarily arrested, they accumulate. When blood flow is resumed, the vessels dilate widely—a condition termed *reactive hyperemia*. The exact nature of the role of metabolites in this response is uncertain. Among the possible agents are acetylcholine or some related substance, epinephrine, adenosine and its related phosphates, and the Krebs intermediary metabolites. Indeed, the vasodilatation during physical activity (*active hyperemia*) is greater than can be accounted for by lactic acid and carbon dioxide accumulation.

The question arises as to why arterioles are affected by substances that enter the bloodstream downstream from their anatomical location. It is possible that a nervous local reflex may cause arteriolar dilatation in such a situation, or the diffusion of agents upstream through interstitial or other pathways may occur. It may also be that the tone of the resistance vessels increases when the transmural pressure (difference between the intravascular and extravascular tissue pressures) rises, and decreases when it falls. During circulatory arrest the arterial pressure (and presumably the transmural pressure of the resistance vessels) falls and tone decreases. The continuation of this decreased tone after restoration of the blood flow accounts in part for the reactive hyperemic phenomenon. Thus, reactive hyperemia demonstrates a type of autoregulation; that is, as arterial pressure falls, arterioles dilate. The mechanism of tissue autoregulation has been debated through the years and there are possibilities of both physical and metabolic mechanisms to explain the dilator response in the face of a lowered arterial pressure. (The reader is referred to Chapter 2 for further discussion of mechanisms of Tissue Perfusion.)

NORMAL AND STRESSFUL ENVIRONMENTAL STIMULI

Exercise. During exertion the arterial pressure is generally higher than at rest; however, since muscle blood flow is exceedingly high, primarily because of the increase in locally produced metabolites, there is extensive local vasodilatation, and the resistance to flow is greatly lowered. Muscle blood flow is increased about fifteen times in strenuous exercise, implying that smooth muscle of resistance blood vessels in the resting state has considerable tone. The onset of muscular exercise brings into operation some important variables: the vasodilator metabolites released by the exercising muscle, which tend to lower arterial pressure by virtue of arteriolar dilatation; and muscle contractions which increase right atrial venous return by mechanical action, thereby increasing the cardiac output. In considering a depressor response in exercise, attention must be directed to regional vascular beds. Thus, the muscle bed may be maximally dilated, and other vascular beds, such as the renal, hepatic and splanchnic, may be vasoconstricted, which would result in a significant

redistribution of the visceral blood flow to active muscular tissue.

Although the extreme vasodilator mechanisms induce additional blood flow through muscle, it has been found that the clearance of radioactive materials from muscle is somewhat retarded and oxygen consumption is decreased. These observations have been interpreted as evidence that the sympathetic vasodilator system tends to open arteriovenous shunts, reducing flow through true capillaries in the muscle tissues and, thereby, producing a significant metabolic stress.

Posture. A subject, previously normal, may show marked impairment of postural circulatory reflex adjustments following a short period of illness in bed. It is not known whether the tendency to faint on arising from bed after illness is due to loss of venous tone adjustment, changes in response of venous muscle to stretch, suppressed sympathetic discharge, or lack of reponsiveness of central vasomotor responses in the baroreceptor reflex input. Assumption of upright posture, by standing or by tilting, causes a fall in hydrostatic pressure at the level of the carotid sinuses. Simultaneously, venous distention due to the influence of gravity on infracardiac veins causes a peripheral pooling of blood. The resultant reduction in venous return lowers the filling pressure of the right atrium, resulting in a decrease in cardiac output. However, if reflex peripheral vasoconstriction and cardioacceleration are normally operative, the end result is a maintenance of mean arterial pressure.

When a normal person arises from the recumbent to the erect position, there is a redistribution of regional blood flows. Blood flow through the hand promptly diminishes and then rises to reach levels somewhat lower than those recorded initially, while blood flow through the leg decreases significantly. If the mechanisms stabilizing blood pressure are inoperative, hypotension and syncope occur. Diminished arterial pressure, accompanied by dizziness, sweating, visual disturbances and loss of consciousness, may be observed following sudden assumption of the erect posture by someone who has been seated or lying down, especially if there is pathological or pharmacological suppression of the adrenergic nervous system. The principal effects of arising from recumbent position result from the hydrostatic pressures of vertical columns of blood. Thus, arterial and venous pressures increase in the dependent regions of the body, and, in regions above the heart, veins collapse and arterial pressure diminishes by an amount equivalent to the height of the column of blood above the cardiac level. A fall in arterial pressure on standing ordinarily signifies either a reduction in cardiac output without vasoconstriction or a drop in peripheral resistance without a corresponding increase in cardiac output.

Temperature. Moderate degrees of exercise by man in a hot environment may precipitate syncope as a result of a marked fall in venous return and cardiac output associated with a lowered resistance to blood flow. Prolonged heat exposure may cause extreme sweating, resulting in a seriously depleted blood volume which may reduce the cardiac output and arterial pressure.

Heat loss is mainly effected in man by cutaneous vasodilatation and increased sweating. Both mechanisms are called into play when hypothalamic circuits are activated by increase in afferent impulses arising from cutaneous thermal receptors and by the increased temperature of the blood perfusing the hypothalamus. The chain of events on heat exposure would be dilatation of the skin vessels and increased muscle blood flow. The skeletal muscle blood vessels are believed to be dilated by local metabolic agents. The overall demand for blood flow may result in a four-fold increase in cardiac output. If the increased output cannot match the enormous reduction in peripheral resistance in muscle and cutaneous regions, arterial pressure will fall and remain low, and, since reflex vasoconstrictor discharge is not as effective during heat dissipation, *heat stroke* occurs.

PATHOPHYSIOLOGICAL STATES

Sympathectomy and Spinal Shock

If postural hypotension and syncope result from an inadequate sympathetic constrictor response, elimination of the sympathetic outflow should produce a similar reaction which is more severe. Sectioning of sympathetic fibers running to various organs releases vasoconstrictor tone and increases resting blood flow through most tissues except brain, heart and kidney. Since the sympathetic vasoconstrictor mechanism is a major factor in the maintenance of arterial pressure, its abolition produces a prompt and severe reduction in arterial pressure. The increased blood flow produced by the diminished peripheral resistance does not persist, even if the entire length of both sympathetic chains is removed bilaterally. Immediately after sympathectomy, patients suffer from hypotension on arising. This so-called orthostatic hypotension persists in very severe forms for days or weeks but is gradually reduced in most instances, as a result of nerve regeneration or altered neurohumoral activity. In time, function may become relatively normal.

Transection of the spinal cord at any cervical level causes an immediate and profound fall in arterial pressure. This is the result of elimination of the tonic activity of the bulbar vasoconstrictor mechanism; moreover, when the spinal nerves are isolated from the medullary centers, the important coordination between the state of the blood vessels and subsidiary centers in the spinal cord is lacking. In the intact body, these spinal centers are regarded as subordinated to the higher vasoconstrictor center in the medulla.

Shock and Systemic Hypotension

Shock, an abnormal state ordinarily associated with poor tissue perfusion, is usually characterized by systemic hypotension. The decrease in pressure may be elicited by different mechanisms brought into play by a variety of insults. A few forms of shock will be discussed and the mechanisms whereby systemic hypotension is produced will be outlined.

Basically there are two discrete ways that hypotension can result: a decrease in total peripheral resistance or a decrease in cardiac output. These two hemodynamic indices not only change with the particular form of shock, but also may be species dependent. Hemorrhagic shock and the resultant hypotension in the dog are considered to be due primarily to a decrease in cardiac output. In both endotoxin shock and hemorrhagic shock in the dog, although total peripheral resistance remains elevated, the reduction in cardiac output is at times so severe that mean arterial pressure falls below control values. In contrast, in the monkey and baboon, both cardiac output and total peripheral resistance fall, explaining the very marked reduction in arterial pressure. During a portion of the postendotoxin period in the primate, it is found that cardiac output may be near normal; however, the fall in peripheral resistance is so profound and persistent that arterial pressure is significantly lowered. The primary mechanism for the depression in peripheral resistance is a decrease in arteriolar tone; the main cause of the fall in cardiac output in the primate species is a diminished venous return. This decrease in venous return has been traced to both peripheral pooling in the systemic circuit and pooling in the pulmonary intra- and extravascular compartments.

Several theories have been advanced to explain the causes of systemic hypotension in various forms of shock. The "cardiac theory" postulates that shock is due to a low cardiac output resulting primarily from a failing heart. The "dilator-agent theory" suggests that histamine, bradykinin, or some unknown vasodilating agent elaborated during shock, causes dilatation of arterioles and a sustained reduction in arterial pressure. It may be, however, that death in shock results from multiple mechanisms, all of which are effective in producing irreversible systemic hypotension.

Several forms of shock are of great clinical significance; these may be classified as hemorrhagic, septic, myocardial, traumatic and anaphylactic.

Hemorrhage decreases mean arterial pressure by virtue of a marked reduction in venous return. The effects of blood loss depend on the amount, the rate, the site of bleeding and the time elapsed after hemorrhage, and, therefore, all degrees of shock severity may result. Knowledge of mechanisms of shock due to hemorrhage has been derived largely from animal studies, and marked species differences reportedly occur, particularly between the dog and the monkey. However, of the various forms of clinical shock, hemorrhagic shock is probably the best studied and most consistent with its experimentally produced counterpart.

Septic shock in man has remained a persistent clinical problem, mortality being little changed during the past several decades. It has been studied largely in animal models, mainly the dog. Recent studies in subhuman primates such as the monkey or baboon, which utilized intravenous injections of endotoxin or live bacteria, have elicited findings quite similar to those in man in clinical septic shock.

Anaphylactic shock results from an antigen-antibody reaction and a drastic fall in cardiac output and arterial pressure occurs. The kinds of antigenic offenders producing shock in man are numerous and commonly include penicillin and insect venom. Allergic shock is particularly serious because of its unpredictable nature and rapidity of onset.

Trauma to the body, which may involve hemorrhage as well as tissue damage is one of the most common causes of shock in man. Toxic factors and CNS dysfunction may aggravate the degree of insult. An increase in vascular capacitance, together with hemorrhage, may severely reduce venous return and arterial pressure.

Cardiogenic shock results when the heart is the primary organ to fail. The heart has a diminished ability to pump blood and may assume a variety of degrees of decompensation. A common cause of cardiogenic shock is acute myocardial infarction. In man, the primary hemodynamic defect is a reduction in cardiac output and an increase in total peripheral resistance. In the experimental animal model, the circulatory effects of myocardial infarction include systemic hypotension, decreased coronary flow and lowered coronary vascular resistance.

Role of Veins

Because pressure can be transmitted upstream from veins to capillaries, changes in postcapillary resistance may exercise profound influences on the capillary bed and the general circulation. A normal function of veins is their ability to store blood (capacitance function), up to as much as 80 percent of the total blood volume. The role of the veins is complex; although it is true that a mild venous constriction should result in an increased venous return on the basis of a decrease in venous capacitance, constriction of the veins may also act as a dam in a stream, causing pooling of blood upstream from the constricted site. This is true particularly where the venous constriction is intense and prolonged. For example, conservation of postcapillary response beyond precapillary response in hemorrhagic shock has been reported to result in an outward movement of

capillary fluid from the vascular bed as well as intravascular sequestration of blood proximal to constricted veins.

In primate endotoxin shock, significant venous pooling occurs and may be so generalized and intense as to substantially reduce venous return and cardiac output, accounting for the notable decrease in arterial pressure. In dogs, gram-negative endotoxin shock has been associated with early profound hepatic venoconstriction, presumably in the post-sinusoidal region. Thus, after intravenous injection of endotoxin there is an immediate and precipitous decline in arterial pressure, a simultaneous elevation of portal venous pressure, and a marked decrease in venous return and cardiac output. This reaction has been related primarily to hepatic trapping of blood, which is prevented by hepatectomy or exclusion of the liver from circulation. Intestinal veins and venules also participate in this response, and microscopic findings have shown that the splanchnic veins and venules are increased significantly in diameter shortly after endotoxin injection during the initial phase of increased portal vein pressure. The veins, however, become significantly smaller within one hour. The mechanism for the pooling of blood during the latter stages of endotoxin shock presumably involves the development of active tension in the intestinal small veins (or constriction of posthepatic venous sphincters) in the canine species. These findings suggest the very prominent part performed by peripheral small veins in endotoxin shock. Similar responses of veins may also occur during some phase of canine hemorrhagic shock.

Responsiveness of small veins to injected vasoconstrictor agents increases after endotoxin, and the progressive development of splanchnic pooling in shock may be due primarily to active constriction of small veins on the basis of an increased responsiveness to circulating catecholamines.

Species differences are extensive in shock.

The relatively gradual development of hypotension that usually occurs in the monkey administered endotoxin or retransfusion of blood in late hemorrhagic shock does not appear to be explainable on the basis of hepatic venous constriction, and hepatic venous responses in man are totally unknown in any form of shock.

Transcapillary Exchange

Decreases in pressure can be produced by intravascular pooling of blood within the capillaries or transudation of fluid across capillary membranes into the interstitial space. Effects at the capillary level can be produced by direct or indirect actions. Several forms of shock associated with potent precapillary vasoconstriction (canine hemorrhagic and endotoxin shock) can result in poorly perfused capillary regions, resulting ultimately in failure of the capillary membrane to function in a normal capacity because of the reduced blood flow. It is also possible that potent vasodilatation of the arterioles may engorge the capillaries with blood at a high pressure, thus leading to extravasation of fluid into extracapillary regions. This may be observed in "low resistance" shock as found in septic shock in the primate, including man. Venoconstriction may impede right atrial venous return, causing blood to be pooled within capillaries, and, if the capillary pressure is sufficiently elevated, transudation will occur. Lastly, direct action on the capillary membrane by hypoxia, circulating toxins or vasoactive agents may increase capillary permeability, thereby permitting transudation of fluid. The common denominator in all of these situations is decreased venous return due to peripheral pooling, with reduced cardiac output and, if continued without other intervening compensatory mechanisms, severe hypotension.

With sustained and severe reduction of blood flow to critical regions in shock, functional deterioration of the microvasculature as well as alterations in tissue metabolism undoubtedly occur. The decrease in oxidative metabolism of the microvasculature may affect the functional integrity of the capillary endothelium and thus promote capillary stasis, increased permeability and the resultant hemodynamic effects of reduced intravascular volume and venous return. In addition, several endogenously released substances, including histamine and carbon dioxide, act upon the capillary to promote fluid loss. Other agents, including lactic acid, bradykinin, proteins, and polypeptides may participate as a result of a systemic inflammatory response. Because of loss of plasma, hemoconcentration results at the venous end of the capillary, bringing to a virtual standstill movement of red blood cells and resulting in the formation of an obstruction proximal to the venules. Plasma continues to enter the arterial end of the capillary, but at a much slower rate than before because of the blockage in the downstream end of the capillary. Eventually, flow ceases entirely in the capillary, and, if the injury is severe, the capillary stasis is irreversible.

If the capillary wall is regarded as a membrane with fixed pores, an increase in permeability should indicate that in some way these pores have become enlarged by the agents mentioned above. A more nearly complete view of capillary permeability takes into account properties of the intercellular ground substance and basement membrane colloids whose behavior also controls the extravascular distribution of water and solutes. Experimental studies suggest that the ground substance of the capillary membrane becomes altered in various forms of stress, allowing the passage of materials into the extravascular compartment.

Blood Volume

A severe loss of blood, approximating 40 percent of the total blood volume, may result in irreversible hypotension and death. The effects of blood loss on mean arterial pressure depend upon the amount, the rate and the site of bleeding as well as the amount of time elapsed after hemorrhage. If a large artery is severed, blood loss may be so rapid that nearly complete exsanguination occurs, and death may result in a few minutes. In contrast, when only small vessels are severed, bleeding may occur relatively slowly, with little effect on arterial pressure.

In man, total peripheral resistance does not rise significantly in the early phase after hemorrhage, thus preventing a compensatory increase in arterial pressure on the basis of vasoconstriction. With severe hemorrhage there is only a narrow range between the degree of blood loss that can be compensated for by the organism, resulting in recovery, and the extent of loss causing acute circulatory failure and rapid death. When the insult is large, the effects of hemorrhagic shock are readily observable. If the blood volume is reduced to 60 percent of normal, signs of shock are clearly evident. If arterial pressure is maintained at a fixed low level in this situation, ultimately a decompensatory state intervenes. Prolonged and severe hypoxia in such a state may cause depression of vital brain centers, thus reducing sympathetic efferent impulses to the cardiovascular system, lowering cardiac output and further depressing arterial pressure. A decrease in blood pH and the development of other adverse metabolic alterations may result in a marked reduction in cardiovascular reactivity to catecholamines, thereby further reducing arterial pressure. Another adverse effect, occurring at the capillary region during prolonged hemorrhagic shock, is a withdrawal of sympathetic vasoconstriction which may decline more rapidly in the precapillary than in the postcapillary vascular segment. This may be expected to result in pooling of blood in the microcirculation and further reduction in pressure, because of a progressively decreased venous return. Progressive reduction in flow in such a situation not only causes deterioration of blood vessels but also contributes to derangements in the metabolism of critical organs and certain tissues. Prolonged shock may lead to the release of a variety of agents into the bloodstream, which in turn may cause a further deterioration of the cardiovascular system. Cardiac failure ultimately intervenes, resulting in a further reduction of cardiac output. Elevated central venous pressure upstream from the failing heart brings about a further reduction in cardiac output on the basis of diminished venous return. During this period, simple volume replacement procedures are ineffective in elevating cardiac output. Eventually, after the various insults are successively brought to bear on the organism, survival is no longer possible and death intervenes. Some of the vicious feedback systems described above may be observed in other forms of shock also, during certain time periods after the primary insult. It is possible that to a variable extent several common final pathways may ultimately occur in the various forms of shock, assuring the development of irreversibility and death.

ANNOTATED REFERENCES

Bard, Philip: Regulation of the Systemic Circulation. *In:* Medical Physiology. ed. 12. Vol. 1. Chap. 11. St. Louis, V. B. Mountcastle, 1968. (Excellent discussion of the role of mechanoreceptors in regulating the systemic, circulation, antidromic vasodilator impulses; characteristics of spinal shock)

Best, C. H., and Taylor, N. B.: Regulation of Pressure and Flow in the Systemic and Pulmonary Circulation. *In:* The Physiological Basis of Medical Practice. ed. 7. Chap. 21, Baltimore, Williams & Wilkins, 1961. (Discussion of peripheral resistance)

Burton, A. C.: General View of Homeostatic Control Mechanisms and of Overriding Controls. *In:* Physiology and Biophysics of the Circulation. Chap. 19. Chicago, Yearbook Medical Publishers, 1966. (Discussion of pressoreceptors, active vasodilation)

Catchpole, H. R.: The Capillaries, Veins and Lymphatics. *In:* Physiology and Biophysics.

ed. 19. Chap. 32. Philadelphia, W. B. Saunders, 1965.

Chien, S., and Gregersen, M. I.: Hemorrhage and Shock. *In:* Medical Physiology. ed. 4. Vol. 1. Chap. 15. St. Louis, C. V. Mosby, 1968.

Davson, H., and Eggleton, M. G.: *In:* Principles of Human Physiology. ed. 14. Chap. 12, Philadelphia, Lea and Febiger, 1968. (Discussion of autonomic vasoconstrictor and vasodilator influences; vasodilator substances upon vascular smooth muscle)

Folkow, B., Heymans, C., and Neil, E.: Integrated Aspects of Cardiovascular Regulation. *In:* Handbook of Physiology. Vol. 3. Chap. 49. Washington, American Physiological Society, 1965. (Discussion of mechanoreceptors) emotional influences on circulation; effects of posture, temperature)

Guyton, A. C.: Regulation of Mean Arterial Pressure and Hypertension. *In:* Textbook of Medical Physiology. ed. 3. Chap. 25, Philadelphia, W. B. Saunders, 1966.

Landis, E. M., and Pappenheimer, J. R.: Exchange of Substances through the Capillary Walls. *In:* Handbook of Physiology, Vol. 2. Section 2. Chap. 29, 1963. (Factors influencing capillary filtration and hemodynamics)

Rushmer, R. F.: Effects of Posture. *In:* Cardiovascular Dynamics, ed. 2. Chap. 7, Philadelphia, W. B. Saunders, 1965. (Effects of posture and exercise on circulation are discussed very clearly.)

Scher, A. M.: Control of Arterial Blood Pressure: Measurement of Pressure and Flow. *In:* Physiology and Biophysics, Chap. 34. Philadelphia, W. B. Saunders, 1965.

5

Cardiac Rhythmicity

Gordon K. Moe, M.D., PH.D.

All disorders of cardiac rhythm are the result of abnormal impulse generation, abnormal impulse conduction, or both. Much of what is known about arrhythmias (more properly, dysrhythmias, because some abnormalities are quite precisely rhythmic) has been derived from astute observation and analysis of ectopic rhythms in the clinic, but models of most of these disorders can be reproduced in the laboratory, many of them in healthy tissue. The relationship between pathology and physiology, between clinic and laboratory, is thus closer than in many other disease states. The recent surge of interest in cardiac dysrhythmias, largely the result of the development of intensive care units, has created an area in which the physiologist, the pharmacologist, and the clinician have come together in forums and conferences of great mutual benefit. The development of cellular cardiac physiology, made possible by the invention of the Ling-Gerard microelectrode, has greatly influenced this marriage; clinical cardiologists talk about membrane potentials, ion fluxes, and dV/dt; while physiologists have discovered the WPW syndrome, parasystole, and PAT with block.

IMPULSE GENERATION

The Normal Site. Spontaneous rhythmicity, or automaticity, is a property of embryonic heart cells that develops before cardiac innervation is established. Normally restricted to the sinus node in the adult heart, pacemaker activity can also appear in certain cells of the atria (probably in the specialized internodal tracts and in Bachmann's bundle, the interatrial band which is presumably the "preferred" or faster route of communication between the S-A node and the left atrium), in cells at the upper and lower margins of the A-V node, and in cells of the His-Purkinje system. Except in grossly abnormal circumstances (e.g., absence of sodium), automaticity is said not to occur in *myocardial* cells of atria or ventricles; ectopic impulses that are not coupled at a fixed interval to a preceding discharge are probably the result of true pacemaker activity in the *specialized conduction tissue* of either the atria or the ventricles. Many instances of premature atrial or ventricular contractions can be ascribed to enhanced automaticity in such sites.

Pacemaker activity is characterized by slow diastolic (phase 4) depolarization (Fig. 5-1, B). Immediately following the phase 3 repolarization of an automatic discharge, the intracellular potential begins to diminish—i.e., it becomes less negative with respect to the external medium. This diastolic depolarization is probably the result of a progressive decrease in the permeability of the membrane to potassium ions, resulting in a drift away from the potassium equilibrium potential (i.e., the potential difference that may be expected from the difference between K^+ concentration on the inside and the outside of the cell). When the membrane potential reaches a threshold value (TP, Fig. 5-1, B), the cell discharges. It is undoubtedly important that agencies

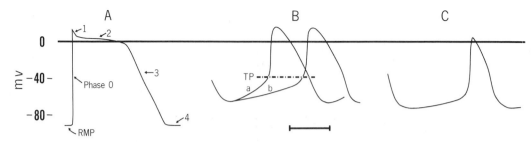

FIG. 5-1 (A) Schematic transmembrane action potential of a ventricular muscle cell, illustrating the various phases as numbered by Hoffman and Cranefield. RMP, the resting membrane potential, is primarily due to the electrochemical potential established by the difference in intracellular versus extracellular potassium concentrations. It depends, of course, on the permeability of the cell membrane to potassium ions. No potential difference could be recorded across a completely impermeable membrane. The horizontal line represents zero potential difference, recorded when the inside and the outside of the cell are at the same voltage. During electrical diastole (phase 4), the inside of the cell is approximately 90 mV negative with respect to the outside. When this value approaches 60 mV, the cell "fires," an event accompanied by an explosive increase in the permeability of the membrane to sodium ions. This "depolarization" (which is really a reversal of potential) is called phase 0. The initial brief abortive repolarization is phase 1, the "plateau" is designated as phase 2, and the more or less rapid phase of repolarization which restores the resting membrane potential is phase 3. Except in pacemaker cells, the resting membrane potential (phase 4) remains constant until the cell is again activated. The duration and configuration of action potentials vary in the several cardiac tissues. The refractory period, in general, lasts from phase 0 until repolarization is about 2/3 complete. In cells of the A-V node, refractoriness lasts beyond full repolarization.

(B) Schematic representation of S-A nodal action potentials. TP, threshold potential; i.e., that membrane voltage at which the cell fires. Note that phase 4 is not stable, that phase 0 (depolarization) is much slower than in muscle cells, and that the membrane potential immediately following phase 3 is less negative. The action potentials *a* and *b* indicate the change in cycle length which would result from a change in the slope of "phase 4" depolarization. (Redrawn from Hoffman, B. E., and Cranefield, P. F.: Electrophysiology of the Heart. New York, McGraw-Hill, 1960.)

(C) Action potential recorded from a spontaneously active Purkinje fiber excised from a dog's heart. Note similarity in shape to S-A nodal potentials. In the absence of pacemaker activity, this fiber would exhibit action potentials much more like the schematic one in Figure 5-1A.

that tend to induce partial depolarization (for example, reducing the RMP from –90 to perhaps –50 mV or less) also commonly favor the development of slow oscillations of the membrane potential, which may reach threshold and induce active pacemaker responses. Such agencies are ischemia and digitalis. Increased serum K+ concentration, on the other hand, which also reduces the RMP, tends to suppress ectopic pacemakers, probably because increased extracellular K+ *increases* the permeability of the membrane to potassium.

The frequency with which a pacemaker fires depends on the slope of phase 4 depolarization and on the threshold membrane voltage. Modulation of the frequency of the S-A node through either cholinergic influences or adrenergic influences is characterized in the former case, by a decrease and, in the latter, an increase in the slope of phase 4 depolarization, without a significant change in the threshold at which the spontaneous discharge is triggered. Although both the slope and threshold may be modified by a "direct" action of drugs, many of the cardioactive and antiarrhythmic drugs act, at least in part, through facilitation or inhibition of the effects of the autonomic neurohumors. Digitalis, for example, may enhance vagal effects on the S-A node and, also, antagonize adrenergic effects; the latter effect of digitalis is commonly referred to as its direct or nonvagal influence. Quinidine, procainamide or lidocaine, on the other hand, may increase heart rate by antagonizing cholinergic effects on phase 4 depolarization. It should be emphasized that a drug that interferes with cholinergic effects cannot cause an acceleration of the heart unless "vagotonic" effects are present, nor can an antiadrenergic agent (e.g., propranolol) cause a decrease of heart rate in the absence of adrenergic influences.

Ectopic Sites. An abnormal site of pacemaker activity can be exposed by failure of the normal rhythmicity of the S-A node, by failure of penetration of the normal impulse to a subordinate pacemaker ("entrance block"), or by enhanced activity of an ectopic focus. All three of these mechanisms can be produced in experimental models; no major impairment of normal cardiac physiology is required.

Failure of normal dominance can occur if S-A nodal rhythmicity is severely depressed, as by an episode of enhanced vagotonia. The resultant rhythm, commonly referred to as a vagal escape, may be due to impulse generation in the junction between atrium and A-V node, in the nodal-His junction, or in the His bundle itself. For want of direct evidence on the precise site of origin, such escape rhythms are referred to as junctional rhythms. There is, however, no compelling evidence that the A-V node in man may not assume a pacemaker role.

The simplest example of ectopic pacemaker function occurring as a result of entrance block is the junctional or His bundle rhythm which occurs in the presence of A-V block. Clearly, without the emergence of a subordinate pacemaker, complete A-V block would be invariably fatal.

Escape rhythms in a "normal" heart are slow—in the range of 30 to 50 beats per minute. However, junctional or idioventricular rhythms may emerge at much faster rates when subordinate pacemaker activity is enhanced. If the ectopic discharge frequency exceeds that of the S-A node, an ectopic tachycardia of atrial, junctional, or ventricular origin may result. Agencies commonly responsible for such rhythms are digitalis, which can enhance subordinate pacemaker activity while depressing normal impulse formation in the S-A node, and adrenergic discharge, which, at least in some conditions, may locally enhance pacemaker activity.

When a subordinate pacemaker is "protected" by entrance block from discharges propagated from the sinus node, it may only occasionally produce an extrasystole recognizable in the ECG. A parasystolic rhythm results.

It is of considerable interest that when a tissue (for example, a Purkinje fiber) begins to discharge spontaneously, the contour of its action potential changes remarkably. A pacemaker site in the Purkinje system yields action potentials that closely resemble those recorded in the normal sinus node (Fig. 5-1, C). In spite of this similarity, the behavior of ectopic pacemakers is, at least quantitatively, quite different from that of the S-A node. An idioventricular pacemaker, or even an atrial ectopic focus, is relatively immune to vagal stimulation and is only moderately influenced by adrenergic discharge. Thus, the ventricular rate in complete A-V block is only slightly increased by atropine or by adrenergic drugs.

IMPULSE PROPAGATION

Normal Mechanisms. To say that the heart is a syncytium is not to say that the heart is homogeneous; cells specialized for impulse generation connect with cells specialized for conduction which deliver excitation to cells specialized for contraction. The specialized conduction pathways of the ventricle — the His-Purkinje system — have long been recognized (Purkinje died more than 100 years ago). But only recently has it been functionally demonstrated that specialized conduction pathways, conducting more rapidly than "ordinary" atrial myocardium, connect the S-A and A-V nodes, and the right and left atria. Because these special pathways, anatomically characterized by Thomas James, are more resistant than atrial muscle to the depolarizing effects of hyperkalemia and of digitalis, it is possible for the normal impulse of sinus origin to be transmitted to the A-V node and the ventricles without discernible P waves (sinoventricular rhythm). It does not follow that transmission from sinus node to ventricles would be impossible in the absence of such special internodal tracts; the appropriate model, however, has not been described.

Conduction in the atria has an apparent velocity of about 80 cm. per sec. On arrival at the A-V node, abrupt deceleration occurs. In isolated preparations of the rabbit A-V node, conduction appears to be continuous, but at a very slow velocity; approximately 50 msec. are required for transmission across a distance of only about 2 mm. (i.e., an average velocity of 4 cm. per sec.).

The speed of conduction in excitable tissue is a function of the stimulating efficacy of the action potential and the excitability of the tissue. The stimulating efficacy of the action potential is, in turn, a function of its amplitude (i.e., the total voltage swing from resting potential to peak voltage) and its rate of rise (dV/dt). Weidmann showed that the rate of rise is a function of the membrane potential from which the action potential develops. In tissue in which excitation takes place from a high resting potential ("higher" meaning more negative inside) the rate of rise is rapid (approaching 1,000 V/sec. in Purkinje fibers). In a tissue in which the membrane potential, the rate of rise of the action potential, and the excitability are low, the con-

duction velocity is correspondingly slow. Such is the situation in the A-V node. Conduction velocity through the A-V node, like pacemaker frequency in the S-A node, is enhanced by adrenergic influences and depressed, even to the point of complete block, by cholinergic influences. Electrocardiographically, these effects are expressed respectively as an abbreviation and as a prolongation of the P-R interval. The effective refractory period (RP) of the A-V node is abbreviated by adrenergic and prolonged by cholinergic influences. Here again, cardioactive drugs may influence A-V conduction, either directly or by altering the effects of the autonomic neurohumors. Digitalis enhances vagal effects and antagonizes adrenergic effects, and, in high enough doses, incomplete or complete A-V block may result. Quinidine, procainamide, atropine, or lidocaine may facilitate A-V transmission and alleviate incomplete block by opposing cholinergic action. It is for this reason that these agents may lead to an undesirable increase in the ventricular rate in cases of supraventricular tachyarrhythmias. Propranolol, by antagonizing adrenergic effects on the A-V node, may enhance the degree of A-V block in atrial tachycardias—an action that can be used therapeutically to reduce the ventricular rate in atrial flutter or fibrillation.

Slowing of propagation within the A-V node may be the result of cell-to-cell depression of conduction velocity, but it may also represent actual but temporary arrest at critical junctions. A slowly rising action potential proximal to a junction may lead, in distal elements, to a "local response" which only slowly reaches the firing threshold. With sufficient depression of conductivity, complete A-V block may occur. Although the terminology may not be quite accurate, depression of this magnitude is commonly referred to as "decremental" conduction.

Upon reaching the His bundle, the normal cardiac impulse accelerates to velocities which may exceed 2 m./sec., only to decelerate again when the ventricular myocardium is reached. Of the total conduction time represented by the P-R interval (say 160 msec.), approximately 40 msec. are accountable to intra-atrial conduction time, about 80 msec. to transnodal conduction, and about 40 msec. to conduction within the His-Purkinje system of the ventricle. In the human heart, estimation of these subdivisions of the P-R interval has recently been made possible with the advent of His-bundle branch recordings through the medium of transvenous intracardiac electrode probes, first demonstrated by Giraud and co-workers; a considerable improvement in the interpretation of complex arrhythmias has already resulted from this approach.

Subnormal and Supernormal Conduction. Normal impulse propagation takes place, once the sinus node has discharged, in tissue that has fully recovered from its refractory period. Conduction velocity, like excitability, is also a function of the degree of recovery of the tissue. The situation is, however, complicated. Tissue that has not yet fully repolarized remains closer to its threshold potential (i.e., its excitability is "supernormal") but its rate of depolarization is limited. Whether conduction is depressed or supernormal during this stage depends on which of these determinants of conduction velocity is predominant. Purkinje fibers normally go through a phase of supernormal excitability before repolarization is complete; supernormal conduction may accompany this phase. During phase 4 depolarization similar considerations may apply. As the transmembrane potential diminishes, membrane excitability and resistance increase. Accordingly, conduction velocity may be enhanced as a result of incipient pacemaker activity. When depolarization in phase 4 becomes excessive, the reverse may occur: the decreased rate of rise of the action potential may now become the predominant factor. This pos-

sibility has, in fact, been emphasized as a possible explanation of entrance block in parasystolic foci by Singer and co-workers. Supernormal conduction has been demonstrated only in the specialized conduction system of atria and ventricles; examples of so-called supernormal A-V conduction have been explained by other mechanisms by Moe, Childers, and Merideth.

In general, a response initiated very early in the recovery phase (phase 3) of the action potential is propagated at subnormal speed. Impulses that encounter tissue in its relatively refractory period will be delayed in transit, and may, if they encounter tissue sufficiently refractory, be altogether suppressed. The phenomena of propagation in relatively refractory tissue have been extensively studied, and numerous models of the generation and maintenance of cardiac arrhythmias have been developed through such studies. Refractoriness can, for example, be utilized to demonstrate the phenomenon of *unidirectional block* which may be shown by the following example. If a region, *A,* has a refractory period intrinsically shorter than an adjacent region, *B,* then propagation of an early impulse from *A* may be blocked at the junction; the earliest impulse that could be generated in B would, however, successfully cross the junction in the opposite direction.

BLOCK AND RE-ENTRY

Many clinical examples of ectopic impulse generation are undoubtedly the result of "true" pacemaker activity, but the possibility of re-entry, or circus movement, excitation has been postulated for many years, and is easily demonstrable in the laboratory. Re-entry demands the following conditions: (1) unidirectional block at some junctional site; (2) slow propagation over a parallel route in the cardiac syncytium; therefore (3) delayed excitation of the tissue beyond the blocked junction; and (4) re-excitation of the tissue proximal to the block. It is implicit in these conditions that the time for conduction over the parallel route and for excitation of the distal elements must exceed the refractory period (RP) of the tissue proximal to the block.

The primary event, the sine qua non of re-entry, is the existence of block. An impulse that sweeps in an orderly fashion through the entire cardiac syncytium cannot, by definition, re-enter. Furthermore the long RP of cardiac cells operates to prevent re-entry. Nevertheless, even normal cardiac tissue can be tricked into a situation that permits re-entry. We may proceed to describe some experimentally induced re-entrant rhythms which almost certainly have their clinical counterparts.

Reciprocal Rhythm

Re-entry has been proposed as the mechanism of closely coupled premature beats, or of beats with fixed coupling, for many years. Since their first description it has seemed probable that ectopic beats which are attached at a fixed interval to a previous event must somehow be dependent upon that event; one way in which such dependence could be established is through the mechanism of re-entry. Within the atrium or within the ventricles, such a re-entrant circuit is difficult to demonstrate (although we shall describe a suitable model below), but many clinical and experimental studies have established re-entry as the mechanism for reciprocal beats, whether isolated or repetitive.

The earliest observations of reciprocal beats, reported by White in 1915 and 1921, suggested that an impulse originating in the A-V node (a junctional beat, in modern parlance) could return to the atria over one portion of the node, cause retrograde excitation of the atria ("negative" or "inverted" P-wave), and return via an alternate pathway through the node to the ventricles. This combination of events, namely, two QRS complexes of supraventricular con-

figuration bracketing an inverted P-wave, has been frequently described. As in many such situations, clinical observation and analysis preceded laboratory investigation. Scherf and Shookhoff demonstrated a similar event experimentally in 1926, and their observations were later confirmed and extended by Moe and co-workers in 1956, and by Rosenblueth in 1958. The sequence of events—from White's observation in 1915 to experimental demonstration in 1926 and and then again in 1956—epitomizes the kind of regrettable lack of communication that existed between the clinician and the physiologist during that period. It was not until recently when, through the use of intracardiac stimulation and recording, confirmation of the animal experiments was achieved by Schuilenberg and Durrer in human subjects that the mechanism of reciprocal beats was accepted by cardiologists in general.

Basically, reciprocal beats ("echoes" to the physiologists) are an illustration of the fundamental conditions for re-entry that have long been accepted by clinician and physiologist alike. These conditions are illustrated in Figure 5-2. An impulse initiated in the ventricle (or His bundle) enters the A-V node at a time when one of the two available pathways in the upper node is still refractory; the impulse is blocked at the refractory junction (condition *1*). Because the alternate or parallel pathway is still partially refractory, retrograde passage through the node is slow (condition *2*). Activation of the atria results in delayed antegrade activation of the tissue above the site of block (condition *3*). Because the total transit time exceeds the RP of the tissue below the site of block, re-excitation of that tissue (final common pathway, or FCP in the figure) occurs, yielding a ventricular echo (condition *4*). This schema does not differ in any important detail from the verbal description in White's case reports; he emphasized that a long R-P interval was necessary to permit recovery

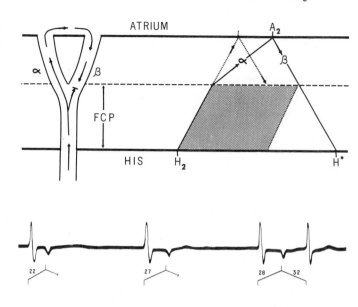

Fig. 5-2 *(Above)* Schematic representation of longitudinal dissociation in the A-V node. In the sketch at the left it is assumed that the upper portion of the node is functionally dissociated. At the time of arrival of a "junctional" beat, the pathway at the left (α) can support retrograde transmission, but that on the right (β) is blocked. The impulse returns to the atrium, enters the β pathway from above, and returns to the site of block. If the time taken for the passage through α, atrium, and β exceeds the refractory period of the lower node and His bundle (final common pathway, FCP) a reciprocal beat, or "echo," will discharge the ventricles.

In the diagram at the right, the time course of this event is illustrated. A premature beat initiated in the His bundle (H_2) traverses the FCP, leaving it refractory for the time indicated by the stippled area. If propagation back to the atrium over α and down to the FCP over β occurred at full speed (dotted lines), no echo would appear. Slow transmission over one or both of the upper nodal pathways (solid lines) permits the reciprocal response to clear the RP of the lower node, leading to an echo (H^*). (Reproduced from Moe, G. K., and Mendez, C.: The physiologic basis of reciprocal rhythm. Prog. Cardiov. Dis., 8:461, 1966.)

The diagram illustrates unidirectional block exposed by a premature beat which arrives in the node before complete recovery from a prior response (i.e., while the node is relatively refractory). It is evident that any agency which causes unequal depression of two pathways in the node could also lead to reciprocal beats.

(Below) An example of reciprocation in a case of junctional rhythm. The first three QRS complexes are followed by retrograde P waves, conducted at increasing R-P intervals (retrograde Wenckebach phenomenon). When the R-P interval reaches 0.28 sec., a reciprocal beat occurs. (Redrawn from Katz, L. N., and Pick, A.: Clinical Electrocardiography. I. Arrhythmias. Philadelphia, Lea and Febiger, 1956.)

from refractoriness of what we have chosen to call the final common pathway. The only difference, and it is not an important one, is that his reciprocal beats occurred "spontaneously"—i.e., unidirectional block or longitudinal dissociation in the node was present as the result of pathology or of medication. In the experimental situation, longitudinal dissociation and unidirectional block occurred as a result of relative refractoriness. An early premature response enters the node during a stage of nonuniform recovery; some elements can conduct while others fail to fire.

If dissociation can occur in the upper node in response to premature beats initiated during the relatively refractory period, it can also occur when nodal conductivity is depressed by other agencies; e.g., digitalis, hypoxia, or ischemic damage. The difference between the clinical situations and the experimental conditions is only a technical one. The important point is that cardiac tissue is not physically homogeneous; depression of function will not be uniformly distributed.

If dissociation can occur in the retrograde direction it should also be demonstrable in the normal direction. Fewer examples of atrial than of ventricular echoes have been described in the clinical literature, but atrial reciprocation has been clearly demonstrated both in the animal and, recently, in the clinical laboratory.

Reciprocal Tachycardia. If the events in Figure 5-2 can occur once, they can, given the proper time relations between conduction time and RP, occur repetitively. This would result in a paroxysmal supraventricular tachycardia—a mechanism proposed long ago by Barker and others and described at least once experimentally by Moe and co-workers. Bigger and Goldreyer reported recently quite compelling evidence that a number of cases of "junctional" tachycardia are indeed the result of repetitive reciprocation.

How could such a tachycardia develop, and how should it respond to intervention? It could develop, as in the laboratory, from a single premature beat, either atrial or ventricular. It could develop as a result of nonuniform depression of nodal conductivity, as in White's description. It could also develop as a result of an anatomic abnormality. Suppose that an aberrant communication exists between atria and ventricles. Such a connection need not conduct with equal facility in both A-V and V-A directions. If it regularly conducts in the A-V direction, the Wolff-Parkinson-White (WPW) configuration of the QRS complex will be manifest. If it fails in the A-V sense, no abnormality will be apparent; retrograde conduction from ventricle to atrium would occur, under "normal" circumstances, at a time when the atria are still refractory. Suppose, now, that A-V conduction is delayed as in Wenckebach periods. If the P-R interval is sufficiently prolonged, retrograde excitation of the atria may now be possible (i.e., the atria will have recovered from refractoriness before activation of the aberrant communication is achieved). A retrograde P-wave will result, and re-entry of the ventricles can now

occur over the A-V nodal pathway. This circuit, too, can sustain itself, and much evidence suggests that the patient with intermittent preexcitation is prone to episodes of paroxysmal tachycardia. Thus, in the original report establishing the Wolff-Parkinson-White syndrome as a clinical entity, reference was made to the common occurrence of supraventricular tachycardias.

The response of a reciprocal rhythm to intervention depends in large measure upon the basic mechanism. If the circuit is completed through the A-V node itself (i.e., down one longitudinally dissociated pathway, up the other) one should expect that vagal stimulation (reflexly induced by carotid sinus stimulation, digitalis, or cholinergic drugs) or any other agency that depresses A-V nodal conductivity should terminate the attack. If one limb of the circuit is through an aberrant A-V communication, vagal intervention would succeed only if complete block of A-V conductivity were achieved. Whatever the communication between atria and ventricles, the introduction of one or more stimuli to the atrium or ventricle (e.g., through a transvenous electrode) should interrupt the circuit and restore normal rhythm. This maneuver, predictable and shown by Moe and co-workers in 1963, has also been effective clinically. The mode of termination, in fact, supports the diagnosis of the mechanism (see section on electrical conversion of cardiac arrhythmias, below).

Sinoatrial Reciprocation. Barker and others suggested that some instances of paroxysmal supraventricular tachycardia were due to a self-sustained circuit involving atrial muscle and S-A node. The sinus node, like the A-V node, propagates impulses slowly, and its conductivity can be depressed by vagal stimulation. It is, however, difficult to demonstrate conclusively that reciprocation can occur between atrium and sinus node. Recent experiments by Han and others strongly suggest this mechanism; if it does in fact occur in man, then the susceptibility to vagal stimulation can be readily explained.

Atrial Flutter: Re-entry or Focus?

We have chosen reciprocal rhythm as the prototype of re-entry because it is easy to produce experimentally and because it has been amply confirmed clinically. It is equally easy to produce a circus movement flutter but not so easy to demonstrate that this is indeed a clinical entity.

An atrial tachycardia in the frequency range of flutter (5 or 6 impulses per second) could certainly result if an intra-atrial ectopic pacemaker elected to discharge at such a rate. That tachycardias in that frequency range can be produced by local application of alkaloids such as aconitine provides no evidence that ectopic pacemakers are indeed the cause of the dysrhythmia; nor does the existence of "continuous" atrial activity in the ECG constitute proof that the mechanism is a circus movement. A more realistic approach is to recognize that both mechanisms may exist and to attempt to define, in each individual case, which of the two (and there can hardly be more than two) applies.

Production of Experimental Flutter. Flutter that is clearly the result of a circus movement about an obstacle can be produced in the dog heart, provided that the obstacle is large enough. (The reader is referred to Rosenblueth and Garcia Ramos' classical demonstration of experimental circus movement flutter in the dog heart.) As in the experimental model of reciprocal rhythm, the round trip conduction time must exceed the RP of the tissue in the circuit. This in itself poses no major problem. Thus, if the conduction velocity in atrial muscle is 50 cm./sec. (assuming relative refractoriness), and if the RP of the atrium at the flutter frequency is 150 msec., then the obstacle need be no more than 7.5 cm. in circumference; the resulting frequency will be about 6.6 impulses per second. The experimental problem is: how

can one-way conduction be established around a natural or artificial obstacle of such dimensions?

Practically, this problem is "finessed" by applying stimuli at a frequency grossly exceeding the ability of the atria to follow —namely, in the range of 20 to 50 impulses per second. At such stimulus frequencies, atrial fibrillation occurs. Upon termination of the stimuli, chance dictates whether or not flutter will ensue; in the experimental situation, if an obstacle of sufficient size exists naturally, or has been created by crushing atrial tissue, the probability is in the range of 50 per cent for any given period of stimulation. What determines the percentage?

Initiation of One-Way Conduction. For a flutter to be established as a circus movement, the primary condition, the sine qua non, is unidirectional block. Suppose, as in the original experiments of Mines, we have a slender ring of excitable tissue. It is not difficult to imagine that a single early premature stimulus applied to such a ring would find the tissue on the left refractory, and the tissue on the right excitable. One-way conduction would ensue. If, on the other hand, we cut a hole in the center of an infinite sheet of conducting tissue, it is extremely unlikely that one-way conduction could ever be established. An impulse, initially blocked at the left by refractory tissue, would proceed to the right. Soon after the origin of the impulse, recovery would occur at the initially blocked site. As the impulse progressed it would invade the blocked site and proceed to engage the tissue around the obstacle in both directions; no circuit would be possible. Suppose now that the sheet of excitable tissue is of finite dimensions, and that a relatively narrow isthmus separates one obstacle from another (e.g., the inferior vena cava from the tricuspid orifice). A *single* premature beat might fail to encounter unidirectional block at the isthmus, but the chance is good that the chaotic excita-

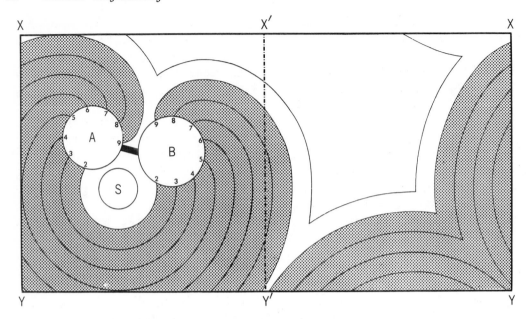

FIG. 5-3. Diagrams illustrating how a circus movement could be established. The tissue is assumed to be a closed surface in the form of a square envelope. For visual display of both front and rear surfaces, the envelope has been slit along the edges XX', XY, and YY', and unfolded; the dotted line X'Y' represents one intact seam. Accordingly, the segments XX', YY', and XY on the left are to be regarded as contiguous with the corresponding segments on the right.

In the upper diagram it is assumed that a premature stimulus is applied at S at a time when a band of tissue between the obstacles (indicated by the black bar) has not yet recovered from a prior excitation; this establishes the condition for one-way conduction, which proceeds to envelop both obstacles. At the time indicated by isochron 9, the wave front has nearly completed the circuit around the smaller obstacle, A, and by this time the initially refractory band would have long since recovered. As indicated by the "rear view" of the envelope, the remotest parts of the tissue would not yet have been invaded when re-entry of the isthmus occurs.

The lower diagram illustrates the first re-entrant circuit (isochrons 10 to 18). The rear view is not shown because it would not influence the pattern of the circuit. If we assume the isochrons to be 20 msec apart, then the circuit time is approximately 180 msec and the frequency is 333 per minute. If we assume a conduction velocity of 50 cm./sec., the circumference of obstacle A is 9 cm. The shaded area in each diagram represents the duration of the "absolutely" RP, or 140 msec.

tion characteristic of a brief period of atrial fibrillation may result in temporary occlusion of the isthmus. Once unidirectional block is established, and provided that the obstacle is of sufficient size, flutter is the inevitable result (Fig. 5-3). If the isthmus is narrow enough, and if the RP of the tissue within it is long enough, one-way conduction could be established by a single premature beat.

Determinants of Flutter Frequency. The frequency of a circus movement flutter is determined by the mean conduction velocity of the circulating wave front and the circumference of the obstacle. The conduction velocity, in turn, will depend, as in any conducting tissue, upon the stimulating efficacy of the action potential (amplitude and rate of rise) and upon the excitability of the tissue. If the circumference of the obstacle is small, the advancing wave-front will encroach upon the relatively refractory wake of the preceding wave, and the conduction velocity will diminish. If the obstacle is large, conduction will proceed at full speed in fully recovered tissue; i.e., about 80 cm./sec. To support a flutter at a frequency of 300 per min., full-speed propagation would require an obstacle with a perimeter of 16 cm.—a rather large hole for even a dilated human atrium. It is likely, therefore, that the impulse travels at less than full speed through relatively refractory tissue surrounding a smaller orifice.

Influence of Increased Refractory Period. How will alterations of the RP influence a circus movement flutter? An increase in the duration of the RP must reduce the velocity (and, therefore, the frequency) or terminate the rhythm. The larger the obstacle, the more stable the rhythm, and the less vulnerable it will be to prolongation of the refractory period. In a hypothetical case in which the obstacle is so large that a considerable length of fully recovered tissue lies between the relatively refractory tail of one impulse and the advancing wave

front of the next, an increase in the duration of the RP will have no operational effect; the impulse will continue to travel in fully excitable tissue. If, however, the impulse is travelling in relatively refractory tissue (i.e., if the pathway is a "tight fit"), any prolongation of the RP must reduce the circuit speed. Consider now what must happen if we permit conduction velocity to be continuously graded from 80 cm./sec. to zero. This is, of course, unrealistic; but if it were true, an infinite prolongation of the RP would never arrest the arrhythmia; it would merely reduce the frequency to a value approaching zero. There is, however, a lower limit to the conduction velocity that cardiac tissue can sustain repetitively without failure of transmission. The lower limit is, unfortunately, not known; however, a figure of 50 percent of normal is probably not unrealistic. Accordingly, if we now consider some reasonable parameters for a circus movement, we can estimate whether or not a moderate prolongation of the RP will break the circuit. If the flutter rotates in a "tight fit" loop, any specific prolongation of RP must slow the wave front, leading to an increase in the cycle length and, in turn, to a further increase of the RP. In a mathematical model of this situation, it turns out that the larger the initial obstacle, the greater the prolongation of the RP necessary to terminate the arrhythmia. It is possible that some cases of flutter are resistant to quinidine and similar antiarrhythmic agents, not because they are not due to a circus movement, but because the obstacle size is large.

Thus far we have not considered specific effects on conduction velocity. Clearly, any agent that prolongs the RP but also depresses conduction velocity will be less effective than an agent that does not depress conduction. This has been admirably demonstrated by Mendez and others in a comparative study of various antihistaminic compounds that have antiarrhythmic properties. One of these, clemizole, prolongs atrial refractory period without specifically depressing conduction; the "wave length" of an experimental circus movement is therefore prolonged, and the circuit is promptly interrupted. In a number of clinical cases of flutter, the drug was also effective. As the authors suggest, the agent may provide a tool for distinguishing a circus movement flutter from an ectopic focus.

Influence of RP Abbreviation. The response of a circus movement flutter to prolongation of the RP is predictable, given logical assumptions, but what should be expected of an agency that abbreviates the atrial RP? A reduction of RP should increase the flutter frequency, provided that the wave front was coursing through relatively refractory tissue. Increased vagal discharge, reflexly or pharmacologically induced, will diminish the *mean* RP of the atria, but it will not do so uniformly. Alessi and co-workers demonstrated that some areas are profoundly affected, whereas neighboring areas exhibit little or no effect. If the circus movement is initially a "tight fit," it will invade some areas of atrial tissue that are just barely able to respond. If the flutter frequency is increased even moderately by increased vagal discharge, these areas will respond irregularly or will be forced to respond 2:1; fractionation of the wave front will occur, and conversion of the flutter to fibrillation will result. This is the text-book picture of the response of flutter to digitalis administration. What happens when digitalis is withdrawn? The situation is closely related to the experimental preparation in which a circus movement flutter can be induced by a brief period of fibrillation. When the digitalis-induced enhancement of vagal tone wanes, the atria may no longer be able to support fibrillation (i.e., the "stimulator" is turned off); chance then dictates, as in the experimental animal, whether flutter resumes, or sinus rhythm supervenes.

Atrial Fibrillation

It is clear from the above remarks that fibrillation may be converted to flutter and flutter to fibrillation, merely by manipulation of the atrial refractory period. It is not surprising that spontaneous drift from one to the other might occur (flutter-fibrillation—or, more colorfully "flitter"[*]) with spontaneous variations of vagal tone. Does it follow, then, that these dysrhythmias are manifestations of the same disorder?

There have been conflicting views of this question for nearly half a century. One view holds that flutter and fibrillation are both due to repetitive discharge from an ectopic pacemaker. When the ectopic focus fires at a frequency that permits regular activation of all areas of the atria, the picture is flutter. If the frequency is so fast that some areas fail to respond to every impulse, fractionation and irregularity of successive wave fronts occur, and the picture will be the chaotic electrical pattern of fibrillation. This mechanism, which can be duplicated in the laboratory by local application of aconitine, is certainly possible. How such an ectopic focus is generated, and how it can be stable enough to sustain an arrhythmia for years, are unanswered questions. The alternate view, that flutter and fibrillation are both examples of circus movement activity, differing only with respect to frequency, is probably not tenable. Flutter can lead to fibrillation, and fibrillation can convert to flutter, but the mechanisms by which the dysrhythmias are sustained are probably different.

Production of Fibrillation

Let us first consider in some detail how fibrillation can be produced. Suppose rapidly repetitive stimuli are applied to a site in a sheet of tissue which is perfectly uniform with respect to refractory period

---

[*] With apologies to Dr. J. A. Abildskov.

duration, excitability, and conduction velocity. Tissue under the stimulating electrodes will respond with a new action potential as soon as excitability has recovered from the RP of a prior response. The new wave front will spread in a smoothly concentric pattern over the whole sheet; the temporal interval between successive responses will be fixed by the minimal RP (determined by the limiting frequency), and the spatial interval (the "wave length") will be determined by the limiting conduction velocity and the frequency. The important point is that an electrical record of such activity would be absolutely regular: fibrillation would not be possible, and activity would cease as soon as the stimulator was turned off.

Behavior of a Nonuniform Matrix. Consider now what must happen if the tissue, like any biological system, is *not* absolutely uniform from point to point and from moment to moment. We need consider only variations in the duration of the refractory period, recognizing that nonhomogeneity in architecture, excitability, conduction velocity, ionic environment, and nerve supply must add to the complexity of the conducting matrix. A stimulus applied at a time when this nonhomogeneous matrix is fully excitable (late in electrical diastole) will be propagated in a roughly concentric fashion from the stimulated site. However, the first of a series of premature stimuli, applied at the earliest possible moment after the basic response, will encounter an irregularly excitable field. Some immediately surrounding areas will still be refractory and incapable of responding. Other elements, still partially refractory, will conduct the new wave front slowly, and others, more completely recovered, will conduct more rapidly. The premature wave front, in other words, will not be even roughly concentric; it will be irregularly serrated to conform with the retreating edge of refractoriness of the

prior response. As the wave front progresses, the initial delay imposed by refractory tissue will permit time for recovery of more distant areas; the wave front will accelerate. Furthermore, areas initially refractory will have recovered, and may be invaded from their distal margins. The possibility of re-entry thus exists. To the initial variability (the intrinsic variation in refractory period duration in the nonhomogeneous matrix) will be added an additional variation: those elements that responded promptly to the first premature stimulus will, because of the brief preceding cycle, be subjected to a still further abbreviation of the RP; those that, being refractory, were excited only after a delay will not undergo an equivalent shortening. The initial temporal dispersion will thus be amplified: the system will be, so to speak, divergent. Turbulence will develop near the stimulated site after one or more repetitions of this event, and will spread to include the whole matrix. These concepts have been extensively studied by Moe and co-wokers in a computer model; the behavior of the model conforms with the known behavior of fibrillation in dog atria in all respects in which suitable tests have been applied.

Self-Sustaining Turbulence. Once turbulence is established in the model, it sustains itself indefinitely. It does not continue because an ectopic focus develops (no ectopic impulse generator was incorporated in the program); it does not persist because a circus movement revolves about an obstacle (no obstacles were built into the matrix). The arrhythmia sustains itself because multiple small wave fronts course irregularly through the "tissue." These wavelets will accelerate in units that have more completely recovered from the refractory state, decelerate in relatively refractory tissue, block at the margins of completely refractory areas, and divide and reunite around refractory islets. The wave fronts will change continuously in number, breadth, velocity, and direction. The process will be sustained indefinitely unless, through intervention or by chance, all wandering wavelets coalesce.

The conditions that determine whether fibrillation will persist or not are: the total mass (or area) and the geometry of the tissue; the mean duration of the refractory period and its range of variation; and the conduction velocity. A large area can support more individual wavelets than a small one; chance coalescence is thus less likely. An unobstructed area will permit greater turbulence than one with many perforations. It can be shown in the model that numerous obstacles separated by less than one wavelength will, if the obstacles are large enough, permit rhythmic activity (flutter) but not fibrillation. Prolongation of the RP will force an increase of the wavelength, thereby reducing the number of wavelets; and if variation of RP duration is eliminated, the initially turbulent activity becomes organized and rhythmic, or may cease altogether.

Comparison of Model with Clinical Fibrillation. A table of numbers in a mathematical model is not identical with the dilated atria of mitral stenosis, but the similarities are nevertheless striking. The atria in cardiac insufficiency are dilated (i.e., the surface area is increased); this is, as in a model, an important factor in the maintenance of the arrhythmia. Prolongation of the RP, by quinidine in man, or by manipulation of an equation in the model, results in termination of the disorder. Abbreviation of the RP, especially if irregularly distributed, increases the likelihood of a selfsustained arrhythmia—in man (digitalis), as in the model. Fibrillation, in the presence of suitable obstacles, may convert to flutter, and vice versa—in man, and in the model. None of these observations proves the mechanisms of fibrillation in man; they do provide an explanation that cannot be excluded.

Drugs that prolong the RP of atrial tissue may be expected to terminate atrial fibrillation, but they need not do so invariably. If the atrial surface area is very large, as in severe mitral stenosis and atrial dilatation of long standing, the chance of conversion is minimized. Larger doses of quinidine, which specifically depress conduction velocity, will not further enhance the probability. Drugs that, by depressing phase 4 depolarization ("true" pacemaker activity) without prolonging the RP (e.g., propranolol, diphenylhydantoin, and lidocaine) should not be expected to terminate either atrial flutter or atrial fibrillation, unless these are the result of ectopic pacemaker activity. The case for re-entrant activity in both of these arrhythmias is supported by the lack of efficacy, in general, of these drugs in supraventricular arrhythmias.

Ventricular Premature Contractions

Premature beats of the ventricle, unless closely coupled, are commonly regarded as benign manifestations of pacemaker activity. Benignity is not, however, synonymous with simplicity. If these events are in truth the result of pacemaker activity, it is necessary to explain why they escape discharge and resetting by the normal cardiac impulse. If, on the other hand, they result from re-entry, it is necessary to consider how re-entry may occur within the ventricle.

We have already indicated that four conditions must be met before re-entry can occur. The prime condition is unidirectional block. The other conditions can be readily accounted for in reciprocal rhythm, but it is difficult to imagine that a sufficiently long loop could account for even a closely coupled premature beat in the ventricles. Recent observations of Sasyniuk and Mendez have removed the major stumbling block. In their experiments, unidirectional block in an isolated Purkinje-muscle preparation was achieved by introducing premature stimuli during the relatively refractory period. Early premature responses initiated in the Purkinje fibers were blocked at some, but not at all, Purkinje-muscle junctions. The important feature of this event is that the action potential of tissue just proximal to a site of block is remarkably abbreviated. This abbreviation, which is the result of an electrotonic current flow between the last elements to fire (terminal Purkinje fibers) and the first elements that fail (underlying muscle), is accompanied by an equivalent abbreviation of the RP. Slow propagation over parallel junctions, and slow propagation through the muscular syncytium, can then result in re-excitation (re-entry of the terminal Purkinje fiber). With suitable time relations, re-entry can occur as a closely coupled premature beat. Because of the extraordinarily brief action potential in the blocked fiber, the re-entrant circuit need be no more than a few millimeters in length. It is also possible for a "silent" re-entry to occur (silent in the sense that it does not appear in the body surface ECG); repetition of the circuit can then lead to a complete re-entry which may or may not be closely coupled. It is important to recognize that the coupling interval does not necessarily differentiate between a re-entrant and a pacemaker interval. It is also important to recognize that, whether re-entrant or pacemaker, unidirectional block provides the essential background.

The use of premature beats to expose "spontaneous" re-entry may appear to be artificial. It is not. *Any* agency that results in unidirectional block facilitates re-entry. Local or general increase in extracellular potassium can result in localized block; the localized ischemia of small arterial occlusion can also be responsible. In any isolated clinical situation it may be impossible to define the mechanism of a premature beat, but the laboratory demonstration of re-entry can hardly be a phenomenon unique to the experimental preparation, never duplicated in nature.

ELECTRICAL CONVERSION OF CARDIAC ARRHYTHMIAS

When high voltage stimuli are applied to the chest wall to terminate ventricular tachycardia or fibrillation, or to arrest atrial fibrillation or flutter, it is probable that the current density through the heart is high enough to depolarize every excitable cell (including, of course, intrathoracic and intracardiac nerves). All cells will therefore be thrown into the refractory state simultaneously, and any wave front or fronts in existence at that instant will be extinguished.

The astonishingly high success rate of this technique, recently reviewed by Zipes, does not provide proof that the mechanism of the susceptible arrhythmias is re-entry, but it is suggestive that many cases of atrial flutter, atrial fibrillation and ventricular tachycardia and fibrillation are in fact the result of self-maintaining circuits—single and regular in flutter, multiple and irregularly wandering in fibrillation. Perhaps the one model of tachycardia that both physiologist and clinician would ascribe to ectopic pacemaker activity is the ventricular tachycardia induced by digitalis, and it is in precisely this situation that "cardioversion" fails.

More suggestive than the results of massive electrical stimuli are the effects of brief locally applied stimuli at little more than threshold intensity. Such stimuli can be delivered through an intra-atrial or intraventricular probe at any desired moment in the cardiac cycle, and may initiate a propagated response without widespread simultaneous depolarization of cardiac and neural tissue.

In an experimental case of "paroxysmal supraventricular tachycardia," reported by Moe, Cohen, and Vick, the mechanism appeared to be a reciprocal rhythm traversing the A-V node. Properly timed atrial or ventricular beats (initiated, in this case, by a pair of stimuli) could terminate the episodes by occluding the circuit pathway. The mechanism by which premature evoked responses could accomplish this objective was illustrated in a later publication by Moe and Mendez. This observation has been repeatedly confirmed in a number of clinical cases. In a recent study by Bigger and Goldreyer, strategically timed responses to discrete local stimuli could either initiate or terminate a reciprocal rhythm. It is difficult to see how an ectopic focus could be switched on or off, more or less permanently, by stimuli of little more than threshold amplitude; one may conclude that these cases of reciprocal rhythm must in fact be due to an A-V nodal circuit or, in some cases, a circuit that similarly dissociates the S-A node.

Some clinical cases of flutter can also be "captured" (as they invariably can in the laboratory) by stimulating the atria at a frequency slightly higher than the spontaneous rate, as shown by Haft and others in 1967. Failure of conversion does not exclude a circus movement mechanism; success strongly supports it.

CONCLUSIONS

Models of many disorders of rhythm can be constructed in "normal" cardiac tissue. In some cases these are undoubtedly the result of ectopic pacemaker activity.

The characteristics of some experimental arrhythmias strongly point to re-entry; the results of drug therapy, discrete intracardiac stimulation, and cardioversion support this interpretation in a variety of clinical situations. The initiation of a tachysystolic rhythm may be due to a single premature impulse; the maintenance of the rhythm need not be due to continuing discharge of the ectopic focus. Accordingly, drugs or other agencies that prevent recurrent attacks may have a mechanism of action different from that of those agencies that terminate the attacks.

ANNOTATED REFFERENCES

Alessi, R., Nusynowitz, M., Abildskov, J. A., and Moe, G. K.: Nonuniform distribution of vagal effects on the atrial refractory period. Am. J. Physiol., *194*:406, 1958.

Barker, P. S., Wilson, F. N., and Johnston, F. D.: The mechanism of auricular paroxysmal tachycardia. Am. Heart J., *26*:435, 1943.

Bigger, J. T., and Goldreyer, B. N.: The mechanism of supraventricular tachycardia. Circulation, *42*:673, 1970.

Giraud, G., Puech, P., and Latour, H.: Acad. Nat. Med., *363*, 1960. (Published the first records of His bundle activity in man. This technique has been widely applied since the routine use of catheter recordings was reported by Scherlag, *et al.*: Circulation, *39*: 13, 1969.)

Han, J., Malozzi, A. M., and Moe, G. K.: Sino-atrial reciprocation in the isolated rabbit heart. Circulation Res., *22*:355, 1968.

Haft, J. I., Kosowsky, B. D., Lau, S. H., Stein, E., and Damato, A. N.: Termination of atrial flutter by rapid electrical pacing of the atrium. Am. J. Cardiol., *20*:239, 1967.

Hoffman, B. F., and Singer, D. H.: Effects of digitalis on electrical activity of cardiac fibers. Prog. Cardiovasc. Dis., 7:226, 1964. (This review includes an excellent summary of action potential characteristics, including the behavior of normal and ectopic pacemakers.)

James, T. N.: Am. Heart J., *66*:498, 1963. (Describes the anatomy of the internodal tracts. Special physiological characteristics of these fibers are detailed by Vassalle and Hoffman. Circulation Res., *17*:285, 1965 and by Wagner *et al.*: Circulation Res., *18*:502, 1966.)

Mendez, R., Kabela, E., Pastelin, G., Martinez Lopez, M., and Sanchez Perez, S.: Antiarrhythmic actions of Clemizole as pharmacologic evidence for a circus movement in atrial flutter. Arch. Exp. Path. Pharm., *262*: 325, 1969.

Moe, G. K., Childers, R. W., and Merideth, J.: An appraisal of "supernormal" A-V conduction. Circulation, *38*:5, 1968.

Moe, G. K.: Cohen, W., and Vick, R. L.: Experimentally induced paroxysmal A-V nodal tachycardia in the dog heart. Am. Heart J., 65:87, 1963.

Moe, G. K., and Mendez, C.: The physiologic basis of reciprocal rhythm. Prog. Cardiovasc. Dis., 8:461, 1966.

Moe, G. K., Preston, J. B., and Burlington, H.: Physiologic evidence for a dual A-V transmission system. Circulation Res., 4:357, 1956.

Moe, G. K., Rheinboldt, W. C., and Abildskov, J. A.: A computer model of atrial fibrillation. Am. Heart J., 67:200, 1964.

Rosenblueth, A.: Ventricular "echoes." Am. J. Physiol., 195:53, 1958.

Rosenblueth, A., and Garcia Ramos, J.: Studies on flutter and fibrillation. Am. Heart J., 33:677, 1947. (The classical demonstration of experimental circus movement flutter in the dog heart. A mathematical treatment of impulse transmission in a sheet of excitable tissue with obstacles was also published by Rosenblueth and the late Norbert Wiener. It is on this treatment that figure 3 is based.)

Sasyniuk, B., and Mendez, C.: A mechanism for re-entry in ventricular tissue. Circulation Res., 28:3, 1971.

Scherf, D., and Shookhoff, C.: Arch. Int. Med. 67:372, 1941. A later version of the original observations published in Vienna in 1926.

Schuilenberg, R. M., and Durrer, D.: Ventricular echo beats in the human heart elicited by induced ventricular premature beats. Circulation, 40:337, 1969.

Singer, D., Lazzara, R., and Hoffman, B. F.: Interrelationship between automaticity and conduction in Purkinje fibers. Circulation Res., 21:537, 1967. (It has been suggested that conduction velocity is depressed in the vicinity of pacemaker sites, and that the "entrance block" which "protects" a parasystolic focus may be due to local phase 4 depolarization.)

Weidmann, S.: J. Physiol., 129:568, 1955. (Discovered the important relationship between membrane potential and rate of rise of the action potential. His studies, together with related observations, are discussed in the somewhat out-of-date but still excellent book by Hoffman and Cranefield: Electrophysiology of the Heart, New York, McGraw-Hill, 1960.)

White, P.: Arch. Int. Med. 16:571, 1915; 28:213, 1921. (Two interesting reports on reciprocal rhythm)

Wolff, L., Parkinson, J., and White, P.: Bundle branch block with short P-R interval in healthy young people prone to paroxysmal tachycardia. Am. Heart J., 5:685, 1930. (In the original study which established the Wolff-Parkinson-White Syndrome as a clinical entity, reference is made to the common occurrence of supraventricular tachycardias. This has been amply confirmed in many later studies, although the mechanism of the conduction anomaly originally proposed is no longer a valid explanation. See, for example, Wolferth and Wood, Am. Heart J., 8:297, 1933.)

Zipes, D.: Clinical applications of cardioversion. Cardiovasc. Clin., 2:240, 1970. (The techniques, and the name *cardioversion*, were popularized by Dr. B. Lown; Dr. Zipes reviews the status of electrical conversion of cardiac arrhythmias as of 1970.)

Appendix A

The references given below will be of great help to the reader who wishes to obtain a deeper understanding of the subject material that has been presented in this section. In Part 1 he will find those textbooks and monographs that present up-to-date and comprehensive discussions of a variety of cardiovascular subjects and mechanisms. In Part 2, we have selected major contributions to the cardiovascular literature that have become outstanding and frequently cited references; however, we deliberately set a cut-off date after the introduction of cardiac catheterization in deference to our many colleagues actively working to provide a better understanding of the mechanisms of cardiovascular pathophysiology.

PART 1:

TEXTBOOKS OF EXCELLENCE

Cardiovascular Physiology

Altman, P. L.: Handbook of Circulation. Philadelphia, W. B. Saunders, 1959. (This book is a compilation of biological data referable to cardiovascular function, normal values, abnormal findings, and literature references.)

Berne, R. M., and Levy, M. N.: Cardiovascular Physiology. St. Louis, C. V. Mosby, 1967. (Well written description of cardiac function)

Braunwald, E., Ross, J. Jr., and Sonnenblick, E. H.: Mechanisms of Contraction of the Normal and Failing Heart. Boston, Little Brown and Company, 1967.

Brecher, G. A.: Venous Return. New York, Grune & Stratton, 1956.

Burton, A. C.: Physiology and Biophysics of the Circulation. Chicago, Year Book Medical Publishers, 1965. (A well written and easily understood text, biophysically oriented)

Guyton, A. C.: Circulatory Physiology Cardiac Output and Its Regulation. Philadelphia, W. B. Saunders, 1963. (Well written discussion of theory, physiology, control, and clinical disease related to control of cardiac output)

Hamilton, W. F., and Dow, P. (eds.): Handbook of Physiology. Section 2, Circulation. Vols. 1, 2, and 3. Washington, D. C., American Physiological Society, 1965. (This is a three volume set of detailed monographs on an extensive range of cardiovascular physiological subjects by the world's outstanding cardiovascular investigators. The bibliographies are in fact comprehensive.)

Krogh, A.: The Anatomy and Physiology of Capillaries. New York, Hafner, 1959. (A classic book on capillary physiology)

Rushmer, R. F.: Cardiovascular Dynamics. Ed. 3. Philadelphia, W. B. Saunders, 1970. (The newest edition of a well presented and practical approach to cardiovascular physiology and pathology)

Shepherd, J. T.: Physiology of the Circulation in Human Limbs in Health and Disease. Philadelphia, W. B. Saunders, 1963.

Clinical Cardiology

Fowler, N. O.: Physical Diagnosis of Heart Disease. New York, Macmillan, 1963. (A broad range of discussions of cardiovascular disease and its associated physical findings)

Friedberg, C. K.: Diseases of the Heart. Ed. 3. Philadelphia, W. B. Saunders, 1966. (One of the most widely used cardiology textbooks)

Hurst, J. H., and Logue, R. B.: The Heart Ed. 2. New York, McGraw-Hill, 1970. (An excellent cardiology textbook written by a number of authors, each contributing a section in the field in which his expertise has been acknowledged)

New York Heart Association: Diseases of the Heart and Blood Vessels, Nomenclature and Criteria for Diagonsis. Ed. 6. Boston, Little Brown, 1964. (The basic guide for classification of cardiovascular disease)

Taussig, H. B.: Congenital Malformations of the Heart (2 volumes). Boston, Harvard University Press, 1960. (A comprehensive encyclopedia of congenital cardiovascular disease)

Wood, P. H.: Diseases of the Heart and Circulation. Ed. 3. Philadelphia, J. B. Lippincott, 1968. (An excellent clinical description of cardiovascular disease, with physical diagnosis an outstanding aspect)

Cardiovascular Pathology

Edwards, J. E.: An Atlas of Acquired Diseases of the Heart and Great Vessels. 3 Volumes. Philadelphia, W. B. Saunders, 1961.

Gould, S. E.: Pathology of the Heart and Blood Vessels. Ed. 3. Springfield, Ill., Charles C Thomas, 1968.

Netter, F. H.: The Ciba Collection of Medical Illustrations, Vol. 5, Heart. Summit, N. J. Ciba, Inc. 1969.

Specialized Cardiovascular Topics

Allen, E. V., Barker, N. W., and Hines, E. A., Jr.: Peripheral Vascular Diseases. Ed. 3. Philadelphia, W. B. Saunders, 1962.

Bayley, R. H.: Biophysical Principles of Electrocardiography (Vol. 1), and Clinical Applications of Electrocardiography (Vol. 2). New York, Paul B. Hoeber, 1958.

Dickinson, C. J.: Neurogenic Hypertension. Oxford, Blackwell Scientific Publications, 1965.

Heymans, C., and Neil, E.: Reflexogenic Areas of the Cardiovascular System. Boston, Little Brown and Company, 1958.

Levine, S. A., and Harvey, P. W.: Clinical Auscultation of the Heart. Philadelphia, W. B. Saunders, 1959.

Moore, F. D.: The Body Cell Mass and Its Supporting Environment. Philadelphia, W. B. Saunders, 1963.

Page, I. H., and McCubbin, J. W. (eds.): Renal Hypertension. Chicago, Year Book Medical Publishers, 1968.

Pickering, K. K.: High Blood Pressure. Ed. 2. New York, Grune & Stratton, 1968.

Sodi-Pollares, D.: New Bases of Electrocardiography. St. Louis, C. V. Mosby, 1956.

Tavel, M. E.: Clinical Phonocardiography and External Pulse Recording. Chicago, Year Book Medical Publishers, 1967.

Wurtman, R. J.: Catecholamines. Boston, Little Brown and Company, 1966.

Zimmerman, H. A. (ed.): Intravascular Catheterization. Ed. 2. Springfield, Ill., Charles C Thomas, 1966.

PART 2: HALLMARKS OF CARDIOVASCULAR PROGRESS

Braun-Menendez, E., Fasciolo, J. C., Leloir, L. E., and Munoz, J. M.: The substance causing renal hypertension. J. Physiol., 98:283, 1940.

Bright, R.: Reports of medical cases selected with a view of illustrating symptoms and cure of diseases by a reference to morbid anatomy. Volume I. London, Longman, Rees, Orme, Brown, and Greene, 1827.

Cournand, A.: Measurements of cardiac output in man using right heart catheterization; description of technique, discussion of validity and of place in study of circulation. Fed. Proc., 4:207, 1945.

Fick, A.: Über die Messung des Blutquandems in den Herzventrikeln. Sitz. der Physik, Med. ges. Würzburg, p. 16, 1870.

Forssmann, W.: Die Sondierung des rechten Herzens. Klin. Wschr., 8:2085, 1929.

Frank, O.: On the dynamics of cardiac muscle (translated by Chapman, C. B., and Wasserman, E.). Am. Heart J., 58:282, 467, 1959.

Goldblatt, H., Lynch, J., Hanzal, R. F., and Summerville, W. W.: Studies on experimental hypertension. I. The production of persistent elevation of systolic blood pressure by means of renal ischemia. J. Exp. Med., 59:347, 1934.

Hales, S.: Statical essay: containing hemostatics; or, an account of some hydraulic and hydrostatical experiments made on the blood and blood vessels of animals. Ed. 3. London, W. Innys, 1769.

Hamilton, W. F., Moore, J. W., Kinsman, J. M., and Spurling, A. G.: Studies on the circulation; IV. Further analysis of the injection method, and of changes in hemodynamics under physiological and pathological conditions. Am. J. Physiol., 99:534, 1932.

Havery, W.: Exercitatio anatomica de motu cordis et sanguinis in animalibus. London, 1628, Guilielmi Fitzeri. (Translated by R. Willis. Barnes, Surrey, England, 1847.)

Hill, A. V.: The heat of shortening and the dynamic constants of muscle. Proc. Roy. Soc. Lond. [Sero. B.], 126:136, 1938.

Landis, E. M.: Capillary pressure and capillary permeability. Physiol. Rev., 14:404, 1934.

Page, I. H., and Helmer, O. M.: A crystalline pressor substance (angiotensin) resulting from the reaction between renin and renin activator. J. Exp. Med. 71:29, 1940.

Pappenheimer, J. R.: Passage of molecules through capillary walls. Physiol. Rev., 33:387, 1953.

Starling, E. H.: On the absorption of fluids from the connective tissue spaces. J. Physiol., 19:312, 1896.

Starling's Linacre Lecture: On the Law of the Heart. New York, Longmans Greene and Co., 1918.

Stewart, G. W.: Researches on the circulation time and on the influences which affect it. IV. The output of the heart. J. Physiol., 22:159, 1897.

Symposium: The regulation of the performance of the heart. Physiol. Rev., 35:90, 1955.

Withering, W.: An account of the foxglove and some of its medical uses: with practical remarks on dropsy, and other diseases. Birmingham, M. Swimmey, 1785.

Appendix B

This appendix is essentially a "glossary of tools" that are clinically useful in understanding cardiovascular function, and hence, mechanisms of cardiovascular disease. Guidelines for the use of these tools may be found in the textbooks tabulated in Appendix A. Also included are outlines and reference tables for normal electocardiographic intervals and axis positions, and for normal hemodynamic measurements.

We emphasize not only that normal values must be defined with respect to a "normal population" but also that they must be interpreted within the frame of reference appropriate for that individual patient. Thus, in a patient who is febrile from acute lobar pneumonia the cardiovascular system may be functioning normally and appropriately, and the heart rate of 124 beats per minute therefore does not reflect an abnormality of the heart.

Glossary of Tools Clinically Useful in Measuring Cardiovascular Performance

Medical History and Physical Examination: A comprehensive review of the life structure of each patient, noting significant deviations from normal in past and present, provides the foundation for all further evaluation. Without this groundwork, all other diagnostic aids become aimless and without meaning. Circulatory performance cannot be measured without knowledge of the status of all organ systems of the body.

Radiography: Standard six foot posteroanterior and lateral chest x-rays have become routine procedures. When evaluating specific cardiac chambers and great vessels, right and left oblique projections with barium-filled esophagus should be obtained. For special problems other procedures (e.g., fluoroscopy and laminography) can be utilized.

Electrocardiography (ECG): Recording the body-surface electrical potential generated by cardiac depolarization and repolarization is routine. The ECG is of particular value in assaying cardiac rate, rhythm and chamber enlargement. It may reflect also a spectrum of myocardial electrolyte and other metabolic disorders. The method has limitations of definition in each of these areas, some of which have been circumvented by the study of differential intracardiac tracings, high frequency electrocardiograms, intracellular electrograms, and the like.

Vectorcardiography (VCG): This technique incorporates three planar views (horizontal, frontal, sagittal) of the body-surface electrical potential during one cardiac cycle. It can supplement information obtained from the electrocardiogram but is limited in evaluating a continuous series of cardiac cycles (rhythm). It is of particular value in evaluating chamber configuration and conduction.

Phonocardiography (PCG): The record of cardiac sounds and murmurs has its greatest use in timing the various cardiac events. It also documents changes over a period of time; with very few exceptions, however, it is inferior to the human ear in discerning murmurs.

Pulse Wave Recordings: The peripheral *venous* (i.e., jugular) wave is an indirect manifestation of right atrial pressure unless there is some type of intervening obstruction. An analysis of the wave form provides information about central venous pressure, atrial transport and rhythm, and tricuspid valve function. The peripheral *arterial* (i.e., carotid) wave is a slightly modified or dampened version of the pressure wave at the aortic root. It can be recorded directly or by indirect methods using a pressure-sensitive device placed over the artery, in which case the wave form is further altered by the intervening skin and subcutaneous tissue. It provides information on left ventricular contraction, aortic valve function and the various factors that determine peripheral resistance.

Apex Cardiography (ACG): This procedure records the sequence of cardiac contraction from the chest wall at the point of maximal impulse. It is, in reality, a very low frequency phonocardiogram, and is primarily of value in timing cardiac events. It may be of special interest in documenting poor left ventricular contraction.

Sonocardiography: The ultra sound wave has particularly good potential for the evaluation of structural cardiac changes. It has been used primarily to evaluate pericardial effusions and valvular dysfunction.

Thermography: Mapping of the cutaneous temperature changes over the chest wall is of considerable interest in ischemic heart disease because of the frequent sympathetic outflow pathways referred from the myocardium to the thoracic cage. This technique is in the embryonic stage of its development and practicability is still to be determined.

Cardiac Output: Determination of left ventricular outflow is a useful guide to systemic metabolic rate and to the effect of a variety of pharmacological or physiological interventions on cardiac performance.

Exercise Testing: The response of the individual to a standard exercise stress test (treadmill, bicycle, two step, etc.) can be measured in various ways—electrocardiographically, and by changes in various ventricular hemodynamic and metabolic functions, oxygen consumption, coronary blod flow and the like. Obviously, such stress tests reveal more information relative to cardiovascular function than do studies at basal, resting conditions alone.

Stress Testing Other Than Exercise: A variety of techniques have been divided into sub-groups: autonomic responses from procedures such as the Valsalva maneuver, tilting, pain, and startle; change in heart rate, such as can be produced by artificial atrial pacing techniques; changes in pressure load, such as that induced by isometric hand grip; and changes in volume load, such as that produced by rapid changes in cardiac venous return.

Clinical Pharmacological Testing: Changes in cardiovascular performance induced by drugs are more difficult to standardize than are some of the physical stress tests outlined above; however, they may be equally valuable in providing information about circulatory responses. Examples of drugs that can be used are digitalis, nitroglycerin, and those causing selective autonomic stimulation or blockade.

Cardiac Catheterization and Selective Vascular Angiography: This procedure involves selective catheter placement in the individual cardiac chambers and great vessels to determine a variety of pressure, flow and metabolic parameters. It has achieved widespread acceptance as a routine preoperative and postoperative tool to evaluate the results of cardiac surgery, and has a variety of other diagnostic uses. It frequently is used in conjunction with selective visualization of cardiac chambers and arterial and venous circulations.

Coronary Blood Flow: There is no ideal way to measure total coronary flow, since there are at least two arterial inlets and a variety of venous outlets. Flow can be measured in each coronary artery by a catheter-tip flow transducer, although this is not generally available for man at this time. More important, it can be measured as "nutrient" or "tissue" flow, i.e., the amount of uptake or release of some easily identifiable indicator dye such as nitrous oxide or radioisotope-labeled potassium or rubidium. Coronary flow studies are important in determining the rate of myocardial metabolic activity.

Myocardial Balance Studies: Selective catheterization of the great cardiac vein (coronary sinus) is necessary in order to measure coronary arteriovenous differences of constituent substances. If coronary flow is determined, the arteriovenous difference (extraction) can be calculated quantitatively as total myocardial consumption or production. Elements commonly studied are the energy substrates—oxygen, glucose, fatty acids, pyruvate, and lactate—but any component of blood could as easily be evaluated. Myocardial oxygen consumption is a highly reliable index of cardiac function, since the heart is almost entirely oxygen-dependent for its energy production. Coronary lactate and pyruvate studies reflect the tissue balance of NAD-NADH and, therefore, have been commonly used as an index of the oxidation-reduction state of the myocardium.

Differential Enzymology: Serum enzymes have been used to follow the activity of a variety of intracellular processes, and these are most commonly used as an index of cellular catabolism or destruction. Certain enzymes (iso-enzymes) seem to be organ-specific and are therefore of particular value in clinical detection of organ damage. In myocardial balance studies, however, enzymes are more useful as an index of cardiac function. Selected enzymes are studied from glycolytic, citric acid, pentose phosphate shunt, fatty acid synthesis or other metabolic pathways; changes in the level of their myocardial production reflect indirectly changes that are occurring inside the myocardium.

Myocardial Biopsy: Study of heart muscle obtained by biopsy techniques has been slow to gain widespread acceptance because of the hazardous nature of the procedure itself and the relatively sparse amount of therapeutically useful information that can be obtained by microscopic evaluation. As improved methods decrease the risk involved (catheter needle biopsy of the ventricular septum seems particularly promising), and as our diagnostic capability increases with electron microscopy and differential enzyme staining techniques, etc., this procedure undoubtedly will prove to be of value.

"Noninvasive" Hemodynamics Testing: It is frequently desirable to have information on cardiovascular function in patients or normal subjects in whom cardiac catheterization and other "invasive" studies seem unwarranted. Indirect studies involve simultaneous use of electrocardiogram, apexcardiogram, phonocardiogram and carotid arterial pulse tracing. The measurements that have provided the greatest amount of information are the isovolumic contraction and pre-ejection periods and the left ventricular ejection time.

Lymphangiography: Special contrast studies of the lymphatic circulation can be obtained in certain areas of the body by injecting radiocontrast material into subcutaneous tissues which is then picked up in local lymphatics and transported to the thoracic lymph duct. The procedure is especially useful in evaluating the presence of deep abdominal and thoracic lymphadenopathy.

Pericardiocentesis: Percutaneous withdrawal of pericardial fluid has been employed primarily in the treatment or prevention of cardiac tamponade. However, it has been very useful in regard to analysis of the constituents of the fluid for differential cell count, total protein, glucose and amylase, as well as the performance of microbiological and pathological cell block studies. It is helpful to inject 50 to 100 ml. of room air into the pericardial sac after fluid has been removed in order to obtain repeat cardiac x-rays in both right and left recumbent positions. This provides information on the thickness of the parietal pericardial membrane and on any loculation of fluid or possible filling defects within the pericardial cavity.

Measurement of Body Fluid Compartments: Blood volume determination has become a simple procedure with isotopic dilution techniques, and provides diagnostic and therapeutic information in patients with potentially contracted or expanded intravascular compartments, such as the surgical patient and the patient with cardiac failure, hypertension, polycythemia, etc.

NORMAL CARDIOVASCULAR INDICES

ELECTROCARDIOGRAPHIC INTERVALS AND AXIS POSITIONS

A. Heart Rate (beats per minute)
  1. In adults, 50-60 to 100
  2. In children, wide variation with age, crying, etc.

B. Electrocardiographic Intervals
  1. PR interval
    a. Normally 0.10 to 0.21 sec., varies with rate and age.
    b. PR interval greater than normal indicates a partial A-V block
    c. PR interval less than 0.12 sec. should lead to consideration of ventricularization of pre-excitation syndromes (Wolff-Parkinson-White (WPW) syndrome)
  2. QRS interval
    a. Normally 0.06 sec. in infants to 0.10 sec. in adults.
  3. QT interval
    a. Normally 0.30 to 0.40 sec., but it varies with rate. If an extremely long QT exists, look for U waves and metabolic disturbances.
    b. Corrected QT (QTc): subtract all over 0.09 sec. of QRS duration from measured QT.

  c. Short QT: consider digitalis effect or increased ionized Ca.

C. Axes
  1. Mean QRS axis varies between $-30°$ and $+90°$ (except in infants: may be 100°) and varies about 30° in space.
  2. Mean P axis tends to follow the anatomical axis of heart.
  3. Mean T axis shorter than QRS and has less excursion. It is posterior in infants and children and is in frontal plane or forward in adults.
  4. Ventricular gradients (the sum of the QRS and T axis) is ⅓ longer and lies within 15° of a normal QRS and close to the frontal plane.
  5. First 0.04 vector points away from missing forces and is normally base apex in direction and in the frontal plane.
  6. Last 0.04 vector (block) points toward last part of the ventricular muscle activated or leads with no terminal negative. Normally the higher posterior wall.
  7. ST segment vector points toward the injured region and is opposite in direction to the T vector.

TABLE I-1A.   Upper Limits of the Normal PR Interval

	HEART RATE				
	Below 70	71 to 90	91 to 110	111 to 130	Above 130
Large adults	0.21	0.20	0.19	0.18	0.17
Small adults	0.20	0.19	0.18	0.17	0.16
Children, ages 14 to 17 yrs	0.19	0.18	0.17	0.16	0.15
Children, ages 7 to 13 yrs	0.18	0.17	0.16	0.15	0.14
Children, ages 1½ to 6 yrs	0.17	0.165	0.155	0.145	0.135
Children, ages 0 to 1½ yrs	0.16	0.15	0.145	0.135	0.125

TABLE I-1B.   Upper Limits of the Normal QT Interval

Heart Rate	40	50	60	70	80	90	100	110	120	130	150
QT Interval	0.50	0.46	0.43	0.40	0.38	0.36	0.34	0.33	0.31	0.30	0.28

TABLE I-2. Normal Values of Pressure and Oxygen Saturation*

LOCATION	PRESSURES (MM. HG)		OXYGEN SATURATION (%)	
	Mean	Range	Mean	Range
Aorta			96	94-100
Systolic	120	90-140		
Diastolic	70	60- 90		
Left Ventricle			96	94-100
Systolic	120	90-140		
End-Diastolic	7	4- 12		
Left Atrium	7	4- 12	96	94-100
Pulmonary Artery, main			78	73- 83
Systolic	24	15- 28		
Diastolic	10	5- 16		
Mean	16	10- 22		
Right Ventricle			79	71- 87
Systolic	24	15- 28		
End-Diastolic	4	0- 8		
Right Atrium, mean	4	−1- +8	80	74- 86
Venae Cavae, mean	6	1- 10		
Superior			77	67- 87
Inferior			83	77- 89

* Modified from Hurst and Logue

(*Note*: For normal values and ranges of cardiac index, heart rate, stroke volume, and total peripheral resistance, the reader is referred to Chapter 3. Caution is suggested in attributing abnormality of any index to a given individual unless the physician considers the patient's age, sex, body habitus, possible drugs the patient is receiving at the time of study, time of day, etc.)

Section Two

Pulmonary Mechanisms

Introduction

Human respiration may be considered to include all the processes whereby oxygen is removed from the environment and delivered to metabolizing cells and carbon dioxide from the cells is delivered to the environment. Few body functions are as urgently life-sustaining, and, at the same time, as clearly understood, readily evaluated and reasonably manipulated therapeutically. Abnormal ventilation due to malfunction of respiratory control centers, neuromuscular disease, or excess work of breathing may be easily documented and therapeutically modified. Problems of gaseous exchange in the lungs are easily assessed, and appropriate oxygenation or ventilation may be controlled so exquisitely that acceptable levels of arterial gases can be achieved in all but the most severe disease. The laboratory can quantify the blood hemoglobin and its effectiveness as an oxygen carrier. Measurement of total blood flow or cardiac output may permit calculation of the total oxygen delivery to the tissues. Only when he attempts to assess regional tissue blood flow, and the even more elusive factors involved in gas exchange between the systemic capillary and intracellular organelles, is the clinician limited in his assessment of the total process of oxygenation. In contrast to the frequently limited oxygen delivery system, the reserve mechanisms for carbon dioxide transport, from metabolizing cells to the lungs, generally exceed demands and therefore rarely require evaluation. Furthermore, development of sustaining environments for travel through oxygen-free outer space and the extreme hyperbaric conditions (more than 30 atmospheres of pressure) encountered in man's deep sea explorations testify to the pragmatic role of this science.

A major deficit in our understanding of the pathophysiology of respiration lies in the problem of defining normalcy or disease. Thus, when the statistically defined ranges of normal are established for a population, they have limited application to the individual; e.g., if a normal range of vital capacity includes 20 percent above and below a predicted mean, a given individual may have initially scored at the upper limit and subsequently deteriorated by one third to the lower limit, but still be classified as normal. Such limitations cause the investigator to base his understanding of altered function in disease on analysis of the unequivocally or severely abnormal patient. Thus, patients with severe chronic bronchitis or emphysema typically have extreme alterations in blood gas exchange as manifest by wasted ventilation (increased dead space) high alveolar-arterial oxygen gradients, low arterial oxygen tensions and, frequently, carbon dioxide retention. They uniformly have decreased maximum expiratory flow rates. Important though this knowledge is, the clinician is more frequently called upon to interpret symptoms of early stages of the disease when the definable physiological abnormalities are minimal. Such patients may be symptomatic only during stress (e.g., exercise) or during associated illness, or they may be so sensitive that they are aware of altered function at milder levels of abnormality than generally pertains. The development of principles for evaluation and treatment of altered function in such individuals presents a cardinal clinical challenge.

Physicians through the ages have dealt with functional manifestations long before they recognized the cause of a disease. It comes as no surprise, then, that the diseases most commonly associated with abnormal function as outlined in this section do not have well defined etiologies. Although manipulation of these abnormal functions may improve the patient's status and relieve his symptoms in a way gratifying to both patient and physician or may even be life saving, the underlying disease process is frequently unaltered—e.g., when bronchodilators temporarily relax bronchial smooth muscles effectively, both physician and asthmatic patient are relieved, but the underlying disease remains, only to become manifest at another time. Great progress has been made in understanding the etiology of some diseases and, consequently, the physiological accompaniments of pulmonary infections for example, are of only temporary significance during the successful institution of specific antimicrobial therapy.

Although our understanding of the offending organism far exceeds our knowledge of the host, the delineation of pulmonary defense mechanisms is being pursued vigorously and will doubtless permit an earlier approach even to therapy of infectious diseases. Similarly, the pulmonary effects of diseases involving pulmonary arteries (e.g., pulmonary embolism) and pulmonary veins (e.g., left ventricular failure) have been substantially clarified. Other major diseases such as hyaline membrane disease of the newborn, cystic fibrosis, bronchial asthma and emphysema (particularly the variety associated with serum alpha$_1$ antitrypsin deficiency) have demonstrable biochemical or immunological manifestations which may soon be linked to fundamental etiological factors.

Small wonder that the physiologist so keenly welcomes the advances in biochemistry, immunology, pharmacology and pathology which will help to elucidate mechanisms of altered lung function, structure, injury and repair. This excitement is exemplified in the chapter on metabolic functions of the lung, a frequently speculative treatise that comprehensively explores many new vistas in the broad interests of the student of respirology.

These chapters relate scientific knowledge, currently considered to be well established, with the priorities identified by their authors. They introduce the reader to controversial areas and speculation in regard to phenomena not well understood. Large areas of fundamental knowledge regarding environmental, stressful and humoral effects on respiratory function are merely touched on. We invite the student to read, learn, and apply this knowledge and, then, to join the never-ending quest for answers to questions he discovers.

Clarence A. Guenter, M.D.

6

Control of Respiration

Robert A. Mitchell, M.D.

INTRODUCTION

Breathing is functionally under the control of two systems, a metabolic or automatic homeostatic system, which has been the primary concern of physiologists, and a voluntary or behavioral system, which has been studied primarily by neurologists. The metabolic respiratory control system serves a fundamental need of all cells—the maintaining of an adequate supply of oxygen for the whole body at rest—and rapidly adjusts to meet the oxygen demands during increased activity. Simultaneously, breathing eliminates carbon dioxide, thus providing a mechanism for rapid adjustments in whole body acid-base balance. The neural structures mediating this metabolic control of breathing are the classical respiratory centers located in the pons, medulla and upper cervical cord. These are influenced primarily by chemoreceptors, both peripheral and central, and by mechanoreceptors in the lungs and chest wall. The behavioral or voluntary control of breathing is manifest during willed apnea or hyperpnea, which represent gross respiratory control. In more complex acts such as speech and singing the respiratory muscles must respond with great precision to produce the appropriate volume and tonal qualities of the voice. The neurological structures mediating the voluntary or behavioral control of breathing are located largely in the forebrain.

Under normal circumstances the metabolic and the voluntary control systems appear to be integrated functionally at many levels. The events that occur during breath holding demonstrate both the anatomical separation of pathways and functional integration. If one holds his breath in an inspiratory position with a closed glottis, the inspiratory muscles (external intercostals and diaphragm) become tonically active and rhythmic breathing movements cease. However, as breath holding continues, there is a resumption of rhythmic movements in the external intercostals. The rate and amplitude of these respiratory efforts increase progressively and rates may approach 60 to 80 per minute. If, just prior to the tolerance limit for breath holding, one makes a maximum inspiratory and expiratory effort against a closed glottis, the rate of the rhythmic respiratory effort falls to 20 to 30 per minute and the tolerance of breath holding is prolonged. Electromyographic studies show that rhythmic activity occurs in the abdominal and intercostal muscles, whereas the diaphragm shows a tonic discharge which is needed to support the negative intrathoracic pressure. Observations made in this simple maneuver support the following conclusions: the motor pathways for a voluntary action do not act by way of motor neurons of the metabolic respiratory center; the input to various groups of respiratory motor neurons may be partially (intercostal and abdominal muscles) or completely (diaphragmatic muscles) suppressed by a willed action; and there is some direct cortical control of the metabolic respiratory center, as indicated by the slowing of rhythmic

111

respiratory efforts following a maximal inspiratory and expiratory effort against a closed glottis during the breath hold.

AUTOMATIC RESPIRATORY CONTROL SYSTEMS

Neural Respiratory Control

Pontine and Medullary Respiratory Center. Three experimental approaches have been employed to determine the location of the neural structures involved in rhythmic breathing: transection of brain stem or localized lesions, focal electrical stimulation, and recording of neuronal activity with microelectrodes. From the results of studies utilizing these techniques the classical concept of a group of respiratory centers in the pons and medulla which control breathing has emerged.

The most rostral of these centers is the pneumotaxic center, a term first used to describe the respiratory function of the rostral pons (Fig. 6-1). Midpontine transection or focal lesions in the lateral tegmentum of the rostral pons result in a slowing of breathing with an increase in tidal volume. If, in addition, the vagus nerves are cut, the normal breathing pattern ceases and is replaced by a deep sustained inspiration called apneusis. Thus, it was concluded that normal respiratory rhythm was caused by two competing feedback loops (the pneumotaxic center and the vagus nerves) which set the medullary respiratory centers at a faster rate than their spontaneous pace. If the brain stem is again transected at the pontomedullary junction, apneusis ceases and rhythmic breathing is restored, although it is usually of a gasping character. To explain apneusis and its release, a tonically active apneustic center which was periodically inhibited by both the pneumotaxic center and the vagus nerves was postulated. Also, to explain the return of rhythmic breathing, inherent

rhythmicity of the medullary respiratory centers was postulated. The location of the medullary centers was first described as a pair of overlapping inspiratory and expiratory centers in the medial medulla at the level of the obex. Electrical stimulation of these areas produced either an inspiratory or an expiratory shift in breathing. In addition, stimulation of these areas produced changes in arterial pressure and tendon reflexes, suggesting that this region is a nonspecific respiratory center.

Recent investigations, in which the activity of respiratory neurons was recorded by way of microelectrodes driven into the brain stem, has increased our knowledge of the location and function of the respiratory centers. It has been demonstrated that the nucleus parabrachialis medialis (NPBM) is probably the site of the pneumotaxic center and functions as the normal rhythm generator as well as the site of integration of somesthetic and sensory information.

Respiratory cells in the medulla are located primarily in two paired nuclei, the nucleus ambiguus and the nucleus retroambiguus (Fig. 6-1). The former is a motor nucleus of cranial nerves (IX, X), and the axons of these respiratory cells directly innervate the striated muscles of the larynx and pharynx as well as the bronchoconstrictor smooth muscle in the lung. The discharge pattern of these cells is primarily all inspiratory. In contrast, the respiratory cells in the nucleus retroambiguus are upper motor neurons whose axons descend through the spinal cord, carrying information to the spinal respiratory motor neurons. Within the nucleus the inspiratory cells are concentrated rostrally and the expiratory cells caudally. Since a "medullary" experimental animal preparation is capable of rhythmic respiration, it has been suggested that the medullary respiratory cells must be able to generate a respiratory rhythm. However, the response of these cells to hypercapnia, hy-

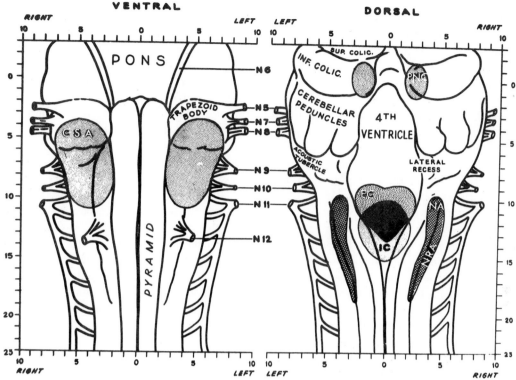

Fig. 6-1.   Ventral and dorsal views of the cat medulla oblongata and pons. The chemosensitive area (CSA) represents the superficial regions of sensitivity to high H+ and CO_2 tension. Areas EC and IC represent the classical expiratory and inspiratory centers in the reticular substance of the medulla. Area PNC represents the pneumotaxic centers. Areas NA and NRA represent the nucleus ambiguus and nucleus retroambiguus. N6 to N12 indicate the cranial nerves.

poxia and electrical stimulation of sensory nerves that stimulate respiration suggests that these cells do not generate the normal respiratory rhythm.

The precise mechanism by which respiratory rhythm is generated is unknown. The hypothesis most frequently advanced is that of a bistable oscillatory system made up of groups of inspiratory and expiratory neurons in self re-excitatory chains which can raise their activity by positive feedback. Mutual inhibition of these inspiratory and expiratory groups is postulated to enforce reciprocal activity. Whether or not this is the exact mechanism that is operative in the pneumotaxic center, it is important to remember that the respiratory rhythm generator (whatever its

mechanism) does not function without the presence of a tonic input into the respiratory system.

Respiratory Control at the Spinal Level. Three anatomically separate and functionally different inputs to the spinal motor neurons arise from higher levels in the central nervous system. The corticospinal tract, which carries information from the voluntary or behavioral centers, crosses in the medulla and descends in the lateral funiculus to the motor cells in the ventral horn. This probably also sends collaterals to the respiratory center, as suggested by the effect of breath holding on the rate of rhythmic breathing efforts. The axons from the respiratory cells in the nucleus retroambiguus descend in a discrete tract in the ventral column, with a majority of the axons crossing to descend on the contralateral side of the cord. Most of the axons from the medial nuclei of the medulla are uncrossed and descend in the ventral and ventrolateral columns. The input from these three systems, one rhythmic and two nonrhythmic, partially determine the membrane potential of the spinal motor neurons and the discharge pattern evoked by each cycle from the rhythmic input from the medullary respiratory neurons. In addition to these inputs a fourth important input to the spinal motor neuron pool arises segmentally from muscle spindles in the intercostal muscles. Section of the dorsal root causes a marked decrease in the impulse activity of the motor nerves to the intercostals associated with that segment. Thus, it appears likely that the respiratory muscles, in common with other skeletal muscles, have their tension controlled by a feedback system from the muscle spindles.

Two types of neurons are present in the spinal motor pool in the anterior horn: the larger of the cells are the alpha motor neurons which pass through the ventral roots and directly innervate the muscle fibers; the smaller cells, the gamma motor neurons, innervate the intrafusal fibers of the muscle spindles which are in parallel with the main muscle mass. Contraction of the muscle spindle causes an increased tension in the nuclear bag located in the central part of the muscle spindle. As a result, impulses are generated in the annulospiral endings surrounding the nuclear bag and transmitted directly to the alpha motor neurons, thus making a monosynaptic reflex arc. (For further discussion of the function of muscle spindles see Chapter 30.) Fundamentally, the muscle spindle serves as a detector of length change in the extrafusal fibers of its parent muscle through the reflex pathway described above. This reflex may drive the intercostal muscles in response to a demand for a certain tidal volume from the respiratory centers. During

breathing, the descending signals activate both alpha and gamma motor units, and the intra- and extrafusal fibers contract together and the tension in the nuclear bag is unchanged if the movement is unopposed. However, if the shortening of the muscles is opposed, the intrafusal fibers will shorten more than the extrafusal, increasing the tension in the nuclear bag and generating impulses in the annulospiral endings. These impulses raise the excitatory state in the alpha motor neurons and cause the tension in the main muscle mass to increase and the muscle to shorten further. With this further shortening of the extrafusal fibers, the tension in the nuclear bag in the spindle is reduced and impulses generated in the annulospiral endings are decreased. Alternatively, the same result may be obtained if the volume demand from the respiratory center is conveyed to the gamma motor neurons, causing the intrafusal fibers of the muscle spindle to contract and generate afferent signals. These signals, in turn, stimulate the alpha motor neurons, causing the main muscles to contract until they match the shortening of the muscle spindles. In this situation the motor neurons of the main muscles operate on an error signal in such a way that any difference in length between the muscle spindles and the main muscle mass is minimized and the effects of variation in loads tend to be compensated. Thus, when the command for a given tidal volume is initiated by the respiratory center, it is interpreted in the spinal cord to mean that the respiratory muscles should contract with whatever force is necessary to effect that tidal volume and not simply to contract with a certain force. This servomechanism produces an immediate effect, to compensate for changes in load on the respiratory muscles —changes that occur long before the effects of alterations in blood gas tensions could affect the output from the respiratory centers through the central and peripheral chemoreceptors.

Chemical Respiratory Control

One can only say that it is "the wisdom of the body" that sets the arterial oxygen tension at 100 mm. Hg, carbon dioxide tension at 40 mm. Hg, and pH at 7.40 in man at sea level. However, it is logical to conclude that chemosensitive cells, strategically located in various parts of the body, maintain these values by sensing changes in blood gas tensions and pH and initiating homeostatic mechanisms to restore them to normal values. Thus, they resemble the temperature-sensitive thermostat that controls the heat source in the house. Just as the temperature-sensitive element is the simplest sensor necessary to regulate temperature, it is logical to suppose that sensors responding to changes in oxygen, pH and carbon dioxide would provide a simple system for observed homeostasis in blood. Carbon dioxide has acid-forming properties when dissolved in water:

$$CO_2 + H_2O \leftrightarrows H_2CO_3 \leftrightarrows H^+ + HCO_3^-$$

Thus, the regulation of arterial carbon dioxide (CO_2) tension also may be accomplished adequately by a pH-sensitive sensor. Arterial pH, CO_2 tension, and bicarbonate (HCO_3^-) are related as defined by the Henderson-Hasselbalch equation:

$$pH = pK_1 + (HCO_3^-/S_{P_{CO_2}})$$

where pK_1 is the first dissociation constant for carbonic acid and S is the solubility coefficient for CO_2. Hence, arterial pH may be altered by two interrelated homeostatic mechanisms: ventilation, which regulates arterial CO_2 tension; and renal excretion, which regulates the arterial HCO_3^- concentration. If specific regions of the body are to have more precise regulation of hydrogen ion (H^+) at a pH different from that of blood, a second set of homeostatic mechanisms with H^+ sensitive sensors in the environment of the region are needed. The constancy of cerebrospinal fluid (CSF) pH at a normal value of about 7.32 in subjects with chronically abnormal acid-base balance suggests the presence of such a regional mechanism.

Carotid and Aortic Chemoreceptors (Peripheral Chemoreceptors). It seems logical to assume that the chemoreceptors that regulate the overall oxygen tension and assist in overall regulation of H^+ should be exposed to arterial blood. The carotid bodies are well-suited for this purpose. They are located near the bifurcation of the common carotid arteries. Their arterial blood supply comes from one or more branches of the carotid artery. Sympathetic nerve fibers coming fom the superior cervical ganglia supply the arterioles with vasoconstrictor fibers that may influence blood flow to the chemoreceptor cells. The carotid bodies ordinarily have a very high blood flow in proportion to their metabolism, so that little change in blood pH, CO_2 and O_2 tension occurs in transit through these structures and they therefore respond as arterial receptors.

Impulses originating in the chemosensitive cells or in the nerve endings of carotid bodies are carried to the central nervous system by the carotid branch of the glossopharyngeal nerve.

The physiological stimuli to the chemoreceptive elements in the carotid body are: (a) a decrease in arterial O_2 tension rather than a decrease in O_2 content; (b) a decrease in arterial pH or an increase in arterial CO_2 tension; (c) a decrease in blood flow to the chemoreceptor cells, relative to their metabolic needs, which may be caused by hypotension or local vasoconstriction, and (d) increase in blood temperature–warm blood reflexly increases breathing, cold blood decreases breathing.

The aortic chemoreceptors are comprised of scattered groups of cells, histologically similar to those in the carotid body, located above the aortic arch between the right subclavian and right common carotid and between the left subclavian and left common carotid arteries, and below the aortic arch between the aortic arch and pulmonary artery. In the adult, the blood supply to these structures is from the systemic circulation. Afferent impulses from the aortic bodies reach the central nervous system (CNS) by way of the vagus nerve. The physiological stimuli to the chemosensitive cells of the aortic bodies are similar to those for the carotid bodies.

The physiological response to stimulation of the various peripheral chemoreceptors is, in most instances, qualitatively similar but quantitatively different. Stimulation of the carotid body produces strong respiratory changes, whereas stimulation of the aortic body produces predominantly cardiovascular changes. Local stimulation of the *carotid bodies* produces: (a) increased rate, depth and minute volume of ventilation; (b) vasoconstriction of limb vessels; (c) bradycardia; (d) hypertension; (e) increased bronchiolar tone; (f) increased pulmonary vascular resistance, and (g) increased adrenal medullary and cortical secretions.

Responses to stimulation of the *aortic bodies* include (a) increased rate, depth, and minute volume of ventilation; (b) systemic vasoconstriction; (c) tachycardia, and (d) hypertension.

Physiological Importance of the Peripheral Chemoreceptors. Prior to the discovery of the peripheral chemoreceptors, it was generally assumed that hypoxemia, hypercapnia, and metabolic acidosis directly stimulated the medullary respiratory centers. Subsequent studies in dogs, deprived of their peripheral chemoreceptors, have demonstrated that acute hypoxia depresses ventilation. However, with sustained hypoxia there is a delayed increase in minute ventilation owing to an increased rate of breathing but a diminished tidal volume, which persists for some time after the hypoxemia is eliminated. This increased dead space ventilation serves no useful purpose. It is probably fair to assume that in the intact animal the useful hyperpnea of hypoxemia

originates reflexly from the carotid and aortic bodies.

The respiratory effects of metabolic acidosis on the aortic and carotid bodies appear to be similar to the effect of low O_2 tension. Denervation of the aortic and carotid bodies eliminates the respiratory stimulation otherwise produced by metabolic acidosis or the depression produced by metabolic alkalosis between arterial blood pH values of 7.3 and 7.5. However, at pH below 7.3 and possibly above 7.5, H^+ leaks across the barrier separating blood from CSF, which normally regulates CSF H^+, and directly affects the intracranial chemoreceptors.

The effect of denervation on the response to hypercapnia is different. The hyperpnea of CO_2 inhalation is depressed only slightly (up to 20%) by denervation of the aortic and carotid bodies. The remainder of the hyperpnea originates from much more sensitive receptors located intracranially. However, when increased P_{CO_2} or H^+ acts simultaneously with the effect of decreased oxygen tension in the peripheral chemoreceptors, the respiratory sensitivity to changes in both pH or P_{CO_2} may be increased and may provide a significant portion of the total response to inhaled CO_2.

The peripheral receptors are therefore of primary importance in the response to disturbances of arterial oxygen homeostasis and in the respiratory compensation for metabolic acidosis. They contribute little to the response to inhaled CO_2 except possibly when combined with low O_2 tension.

Intracranial Chemoreceptors. That portion of the respiratory response to inhaled CO_2 that cannot be accounted for by the peripheral chemoreceptors results from stimuli originating within the central nervous system. For many years it was assumed that the integrating and coordinating neurons of the respiratory centers in the reticular formation were responding directly to changes in their own environment. However, there is good evidence that the central chemoreceptors are anatomically separate from the classical medullary respiratory centers. This concept was developed following the demonstration that respiration was stimulated when CSF with a high CO_2 tension and low pH was perfused through the cerebral ventricles and subarachnoid space about the medulla. In the cat, chemosensitive cells or their sensory nerve endings, which account for the hyperpnea, are located, not in the classic medullary respiratory center, but 5 to 6 mm. away, on the ventrolateral surface of the medulla near the roots of cranial nerves IX and X (see Fig. 6-1). However, there has been no histological identification of the receptors, cell bodies or nerve fibers running to the respiratory centers. The specific stimulus appears to be H^+ in the CSF or extracellular environment of the receptors. At present, an independent action of CO_2, aside from its acid-forming properties, cannot be excluded. Local application of acetylcholine and nicotine to the chemosensitive areas causes hyperpnea; cyanide causes depression of ventilation; procaine, in dilute concentrations, causes apnea even at elevated arterial CO_2 tension, suggesting the importance of these structures in breathing.

In some situations the stimulus arising from the chemoreceptors is essential for breathing. A reduction of arterial P_{CO_2} in anesthetized man and animals causes apnea. Also, respiratory efforts are temporarily suspended in awake, trained dogs when arterial CO_2 is lowered by artificial ventilation. In awake man the results are variable; however, apnea is frequently observed in subjects immediately following hyperventilation with a gas containing elevated concentrations of O_2. The stimulus for continued breathing in subjects who fail to manifest apnea presumably arises from other sensory and neural inputs to the respiratory center. These stimuli come by way of the reticular activating system and maintain the respiratory center's activity above the threshold necessary for rhythmic activity.

The CSF environment of the medullary H^+ receptors is ideal for a receptor that senses changes in CO_2 tension as a result of its acid-forming properties. Unlike blood, CSF has no effective cation buffer, so that a change in CO_2 tension produces a maximum change in H^+.

The location of the medullary H^+ receptor (about 200 μ below the surface of the brain) places it functionally between brain tissue, which is well perfused with blood, and unperfused CSF and is probably important in determining the time required to achieve a new steady state level of breathing after an abrupt change in arterial CO_2 tension. The highly perfused peripheral chemoreceptors equilibrate rapidly with arterial blood and respond maximally, within a few seconds, to changes in arterial CO_2 tension. However, their contribution to the total ventilation is small. In contrast, CSF must receive its CO_2 indirectly from the surrounding tissues, which are perfused with blood. Therefore, following an abrupt change in arterial CO_2 tension, CSF slowly reaches a new level of CO_2 tension and pH and, thus, delays the maximal stimulation of the superficial receptors. This may contribute to the 6 to 10 minute delay in achieving a new steady state of respiration when CO_2 is inhaled.

Regulation of Cerebrospinal Fluid Hydrogen Ion. Recent studies of the acid-base characteristics of CSF that suggest regulation of CSF H^+ by the electrical potential between CSF and blood and by active transport clarify many of the remaining problems surrounding the regulation of ventilation.

The normal composition of CSF is determined by a balance between three processes: (a) secretion in bulk by the choroid plexus; (b) passive diffusion through the barrier between blood and CSF along the electrochemical gradient; and (c) specific active transport mechanisms. Cerebrospinal fluid obtained from the ventricles, cisterna magna, or lumbar sac differs significantly from a dialysate of plasma. The concentration of Na$^+$ and H$^+$ is greater in CSF than in dialysate of plasma, whereas K$^+$ and HCO$_3^-$ are present in lower concentrations. The electrical potential between blood and CSF which should influence the distribution of ions between blood and CSF is normally 4 mv, CSF positive. This potential difference may be a major factor in regulation of CSF pH, since it varies inversely in a nonlinear fashion with plasma pH. Thus, this potential would act as a negative feedback system, apparently regulated by arterial pH which would oppose movement of H$^+$ into the CSF in metabolic acidosis and out of CSF during metabolic alkalosis. A third mechanism, increased brain lactic acid formation, also appears to operate to stabilize the CSF pH during the respiratory alkalosis associated with hypoxic hyperpnea. By these mechanisms and by active transport, acute changes in CSF Pco$_2$, independent of their cause (i.e., hypoventilation, hyperventilation), are compensated for by proportional changes in the HCO$_3^-$ of CSF that restore the CSF pH and the medullary H$^+$ receptor activity to normal.

Assuming the level of breathing to be primarily the result of the stimuli originating from the medullary H$^+$ and carotid and aortic receptors, we may now describe the sequence of events in altitude acclimatization, adaptation to elevated arterial CO$_2$ tension, and metabolic acid-base disturbances.

Unified Concept of the Chemical Regulation of Respiration. When man ascends to high altitude, the arterial oxygen tension falls, stimulating the carotid and aortic chemoreceptors. This causes hyperventilation and a decrease in arterial and CSF CO$_2$ tension. The rise in arterial and CSF pH above normal values (line A-B, Fig. 6-2, A), shifts CSF pH above 7.32 and diminishes the normal stimulus originating

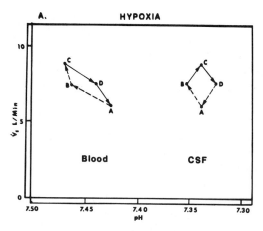

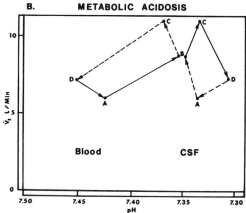

Fig. 6-2. Sequence of events relating ventilation to changes in arterial and CSF pH in (A) ascent to and descent from high altitude, and (B) development of and recovery from metabolic acidosis. Solid line represents an acid shift in pH which should stimulate the appropriate receptors, dashed line an alkaline shift which would depress the receptors.

(A) Adaptation to high altitude: AB, effect of acute hypoxia; BC, adaptation to hypoxia over the next few days; CD, sudden return to sea level; DA, gradual return of ventilation to normal over the next few days to weeks. (B) Metabolic acidosis: AB, development of acute acidosis; BC, development of chronic acidosis over the next 1 or 2 days; CD, acute correction of acidosis; DA, return to normal ventilation over the next few days or weeks.

from the medullary H$^+$ receptors, thus, partially offsetting the effect of impulses originating from peripheral chemoreceptors.

After a few days at high altitude, the CSF pH is restored to 7.32 (line B-C), restoring the medullary H⁺ activity to normal. Ventilation, which is now the resultant of an increased peripheral chemoreceptor activity and a normal medullary H⁺ receptor activity, is increased. This explains why chronic hypoxemia is a more potent stimulus to breathing than is acute hypoxemia. When fully adapted to altitude, both blood and CSF pH are normal, and the hyperventilation originates solely from the effect of low O_2 unopposed by either blood or CSF alkalosis. Once adapted, a further change in ventilation and CO_2 tension is actively opposed by medullary chemoreceptors. Termination of hypoxia by returning to sea level now fails to restore breathing immediately to normal. As ventilation decreases when oxygen is increased, CO_2 rises and causes an acid shift in CSF below pH 7.32 (line C-D, Fig. 6-2, A). This stimulates the medullary H⁺ receptors and prevents ventilation from falling to normal. As HCO_3^- rises in CSF over a period of days (presumably by the mechanism described earlier), the pH of CSF is restored to normal and ventilation returns to normal. The chemoreceptors are reset to maintain an arterial P_{CO_2} of 40 mm. Hg at an arterial pH of 7.40 and CSF pH of 7.32.

The sequence of events and the homeostatic mechanisms involved in adaptation to high CO_2 are probably similar, but opposite in direction, to those described above. With prolonged exposure, the chemoreceptors are reset to maintain elevated CO_2 and resist further changes in ventilation and arterial CSF P_{CO_2} by acute hyperventilation.

Clinically, adaptation to abnormal CO_2 levels is seen in conditions demonstrating hyperventilation or hypoventilation. Patients who have been hyperventilated by mechanical respiration continuously for prolonged periods continue to hyperventilate when removed from the respirator. This presumably results from a decrease in CSF HCO_3^-, which has restored the CSF pH to normal and reset the medullary chemoreceptor to discharge normally at a low CO_2 tension.

Like the return from high altitude, any fall in ventilation causes an acid shift in CSF pH, preventing a return of ventilation to normal until readjustments in CSF HCO_3^- have restored CSF pH to normal. Similarly, it seems logical that the restoration of arterial CO_2 tension and ventilation to normal following prolonged CO_2 retention, as in emphysema, would require a prolonged period of assisted ventilation to allow the CSF HCO_3^- to fall to normal values.

Adaptations to metabolic acidosis are similar. Acute acidosis stimulates the aortic and carotid chemoreceptors, producing increased ventilation and decreased arterial and CSF CO_2 tension. The CSF pH shifts in an alkaline direction from 7.32, diminishing the full effect of the peripheral chemoreceptors (line A-B, Fig. 6-2, B). After about 24 hours, the pH of CSF is restored to normal by a proportionate reduction in CSF HCO_3^-, restoring the contribution of the medullary H⁺ receptors to normal and increasing breathing (line B-C, Fig. 6-2, B). Acute correction of the metabolic acidosis in the blood fails to restore ventilation to normal because any fall in ventilation is associated with a rise in arterial and CSF P_{CO_2} and an acid shift in CSF pH from 7.32 (line C-D, Fig. 6-2, B). The return to normal breathing again depends upon restoration of CSF pH to normal by the CSF H⁺ regulatory mechanisms (line D-A, Fig. 6-2, B). This sequence of events would explain the observation of continued hyperventilation after acute correction of diabetic or renal acidosis.

REFLEX CONTROL OF BREATHING

We have already discussed two reflexes, the chemoreceptor and muscle spindle reflexes. A number of other reflexes have effects on breathing; however, only four additional reflexes arising from the lung will be considered. These reflexes are of particular importance either because they play a role in the regulation of breathing in the normal individual or because they have been implicated in reflex hyperpnea, bronchoconstriction and dyspnea associated with a number of pulmonary conditions. The afferent limb of these reflexes is in the vagus nerve, as indicated by the loss of reflex response when the vagus nerves are transected.

Inflation Reflex. The inflation reflex (Hering-Breuer inhibito-inspiratory reflex) is mediated by slowly adapting pulmonary stretch receptors located in the smooth muscle of the airways. Inflation of the lung stretches the receptors and generates impulses that inhibit inspiration and cause bronchodilatation. The function of this reflex seems to be regulation of the breathing pattern so that it is most economical in terms of work and of the inspiratory force of breathing.

Cough Reflex. The cough reflex is initiated by receptors in the trachea and large bronchi, primarily those airways outside the lung parenchyma. Mechanical stimulation of these receptors initiates the complex act of coughing whose function is to clear the upper airways of the irritant.

Type J Receptor Reflex. The type J receptors were originally called deflation receptors. However, because they responded poorly to deflation of the lung or atelectasis but were stimulated by lung congestion and microemboli, the original name was dropped. They are thought to be located in the juxtacapillary region and, therefore, have been renamed *type J receptors*. Stimulation of these receptors causes bradycardia, hypotension and apnea or rapid shallow breathing, but their function in normal breathing is not known.

Irritant Receptor Reflex. The lung irritant receptors have been proposed as possibly playing a major role mediating dyspnea (see below). These receptors are located between the epithelial cells of the bronchi and bronchioles. They are stimulated by irritants, constriction of bronchial smooth muscles, sudden distention or col-

lapse of the bronchial walls and atelectasis, as well as by conditions that increase lung compliance. Stimulation of these receptors cause vagal reflex hyperventilation and bronchoconstriction. The function of these receptors during normal breathing in healthy man is unknown.

EFFECTS OF CENTRAL NERVOUS SYSTEM DISEASE ON BREATHING IN MAN

We have discussed the theories concerning the control of breathing which have been derived from animal experiments; to extend these observations to man would be difficult if it were not for a few well-controlled clinical studies. From these studies several distinct patterns of breathing associated with diseases of the CNS have been described (Table 6-1). In disease these CNS lesions are rarely the result of a single discrete lesion of the type produced experimentally, so that the clinical picture often is associated with other neurological abnormalities.

Forebrain Lesions. The mechanisms controlling voluntary (or behavioral) breathing are in the forebrain, and knowledge of these comes almost entirely from neurological descriptions of clinical problems, since they cannot be studied adequately in the anesthetized experimental animal.

Loss of ability to breathe deeply or to hold one's breath has been termed respiratory apraxia. It is frequently associated with the loss of ability to initiate voluntarily the act of swallowing. Patients exhibiting this phenomenon are elderly and show evidence of mild to moderate arteriosclerotic cerebrovascular disease; but they are neither demented nor unable to follow other commands requiring voluntary motor acts. As yet there have been no postmortem studies on these patients; however, the clinical syndrome suggests that their defects could be explained best by a diffuse, bifrontal dysfunction of the extrapyramidal system.

TABLE 6-1   Effect of Various Lesions on Breathing

CLINICAL DEFECT	SITE OF LESION
Lesions affecting voluntary (behavioral) central systems	
Aphasia	Left hemisphere language area
Loss of ability for deep breathing or breath holding	Probably diffuse bifrontal lesions
Loss of voluntary control of breath	Any site in the corticospinal tract from the internal capsule to the pons
Forced pseudo-bulbar laughing or crying	Limbic system and subcortical connections
Epileptic respiratory arrest	Medial temporal lobe—limbic system
Increased responsiveness to CO_2 often associated with Cheyne-Stokes respiration	Bilateral pyramidal motor system
Quadriplegia with loss of voluntary control: rhythm breathing intact	Bilateral destruction of pyramidal tracts at any level
Lesions affecting automatic (metabolic) control systems	
Apneustic breathing	Lateral pontine tegmentum
Central hyperventilation	Medial portion of rostral pons
Ondine's curse: Loss of automatic breathing	Damage to medullary respiratory neurons in the medulla or bilateral lesions in descending respiratory motor pathways in the ventrolateral spinal cord
Abnormalities affecting the chemoreceptor drive to ventilation	
Loss of response to inhaled CO_2	Unknown; presumably injury to central chemoreceptor
Insensitivity to hypoxia	Unknown; presumably due to insensitivity of the carotid body to hypoxia

Cheyne-Stokes respiration, a cyclic waxing and waning of tidal volume, has been observed in patients both with CNS lesions and with cardiovascular disease. When associated with CNS lesions the patients hyperventilate, even during the phase of diminished tidal volume, and have a rather marked increase in their sensitivity to CO_2. Either an increased sensitivity to CO_2 or a prolongation of the time required for the blood-borne CO_2 to reach and equilibrate with the central chemoreceptors may cause this abnormal pattern of breathing. The mechanisms involved might be compared to the erratic control of a new driver. When a new driver with normal reaction time deviates to the right of the road he frequently over-reacts and, instead of returning to the center, he deviates to the left and then continues to oscillate from one side of the road to the other. In this analogy the reaction time is comparable to the time required for the CO_2 in the pulmonary capillary blood to reach the central chemoreceptors, and the over-response of the driver is analogous to the increased sensitivity of the respiratory control mechanisms. The increased CO_2 sensitivity that triggers Cheyne-Stokes respiration may result from bilateral damage of descending motor pathways that have an inhibitory or damping effect on ventilation. The exact site of these pathways is unknown; however, all patients exhibiting this phenomenon had bilateral cerebral infarctions or traumatic lesions of the hemispheres.

Cheyne-Stokes respiration associated with circulatory disorders are not the result of a neurological disorder. In this situation the respiratory arrhythmia is caused by an increased circulatory delay from the lungs to the brain. The mechanism involved may be considered comparable to an intoxicated driver with a delayed reaction time. In this situation when the car deviates to the right the driver's response is delayed, and when he finally makes a proper correction to return the car to the center of the road he overshoots his mark because of the prolonged reaction time and ends up on the left side and thus continues to weave down the road. Both the respiratory control system and the analogy presented involve a closed system with negative feedback, which operates on an error that initiates the correction. Such a system operating on an error signal is inherently unstable and tends to oscillate, and oscillations are frequently seen in normal individuals, especially after periods of voluntary hyperventilation. However, the tendency toward oscillation is damped by the short lung-to-brain circulation time and the normal sensitivity of the respiratory centers to CO_2. It becomes operative in disease when the respiratory centers are inappropriately sensitive to CO_2 or the lung-to-brain circulation time is prolonged.

Pontine and Medullary Lesions. A second form of central hyperventilation, different from Cheyne-Stokes breathing, is observed in patients with thrombosis or embolism of the midpontine portion of the basilar artery. This lesion, which is associated with necrosis of the medial tegmentum of the midbrain and rostral pons, causes very severe regular hyperventilation of the type seen in severe acidosis. Also, in contrast to neurogenic Cheyne-Stokes respiration, the sensitivity to CO_2 is not increased. In experimental animals an analogous syndrome produced by midbrain-pontine compression has been accompanied by a fall in Pa_{O_2}. This has not been observed in man, although it has been claimed that the Pa_{O_2} is relatively low for the degree of hyperventilation and hypocapnia. Hypoxia does not reflexly cause

this hyperventilation, since the administration of O_2 does not significantly slow the breathing. Neither is it believed that this phenomenon is caused by release from inhibition arising from higher centers; rather, it has been suggested that the hyperpnea results from pulmonary receptors.

It has long been known that patients with head injuries, subarachnoid hemorrhage or increased intracranial pressure frequently have pulmonary edema and focal pulmonary hemorrhages. The following mechanism has been suggested to account for the pulmonary pathological changes and hyperventilation following head trauma (as well as midbrain-pontine compression or infarction). Head injury initiates an intense sympathetic discharge to the lung, causing pulmonary vasoconstriction, pulmonary hypertension and edema. The edema results in a decreased lung compliance, but both decreased compliance and pulmonary edema stimulate the irritant receptors which could reflexly cause hyperpnea and bronchoconstriction. Thus, labored hyperventilation may be produced. In partial support of this concept is the report that the pulmonary lesion produced by head trauma in experimental animals can be prevented by cutting the sympathetics to the lung.

Lesions in the lateral tegmentum have the opposite effect of medial pontine tegmentum lesions. Bilateral lesions in man, in the region defined as the pneumotaxic region in animals, produce "ataxic" breathing. This pattern consists of totally irregular rate and depth of breathing, with periods of tonic inspiratory cramps alternating with expiratory cramps. In spite of the marked irregularity, the mean level of breathing is sufficient to sustain life for days, and blood gas tensions are normal.

Because of the close proximity of the descending motor pathway from the forebrain to the respiratory centers in the medulla, traumatic or vascular lesions in the medulla usually cause cessation of breathing, both voluntary and automatic. However, under some circumstances, where there is compression of the medulla, the nuclear structures are affected before the descending motor pathways. The clinical syndrome that results has been named "Ondine's Curse," from its description in German legend: Ondine, a water nymph, took a mortal lover who subsequently became unfaithful. Because of this action a curse was cast upon him which took away all of the less automatic functions. Only by remaining awake could he sustain life. Finally, from sheer exhaustion, he fell asleep and respiration ceased. Patients with this syndrome are capable of breathing during wakefulness by voluntary efforts; on falling asleep, respiration ceases.

This syndrome has been observed in medullary compression, the early stages of bulbar poliomyelitis and after bilateral cordotomy. However, the underlying mechanism is probably the same in all. In medullary compression the syndrome results from selective suppression of the automatic control of breathing while voluntary control is maintained. In bulbar polio, presumably the respiratory motor neurons in the nucleus ambiguus and nucleus retroambiguus are initially impaired, so that the automatic

control of breathing fails. After bilateral ventrolateral cervical cordotomy for pain, at surgery, on completion of the section, respiration ceases. As the patient begins to recover from the depressant effects of anesthesia, he can often be made to make respiratory efforts upon command. When fully awake, these patients are capable of maintaining normal blood gas tensions by voluntary efforts which cannot be readily distinguished from eupnea; they are not dyspneic. However, some are aware that respiration will stop if they fall asleep. Pulmonary function studies are usually normal except that the respiratory response to inhaled CO_2 is greatly reduced. The production of Ondine's curse by bilateral cordotomy results from the section of the descending motor pathways from both the medial medullary reticular nuclei (Pitts' medullary respiratory region) and the nucleus retroambiguus (Fig. 6-3).

Thus, the three clinically separate conditions in which this syndrome has been reported have a common link: the suppression or interruption of metabolic or automatic respiratory centers or motor pathways while pathways for voluntary breathing are unaffected. These clinical observations are strong evidence that voluntary and meta-

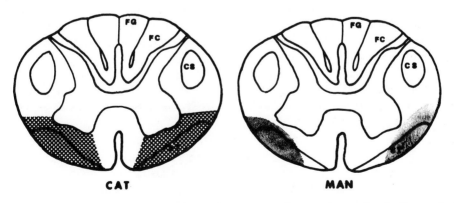

CAT   MAN

Fig. 6-3. Location of descending respiratory pathways (cross hatched areas) in the cat, and the location of spinal cord incisions which produced Ondine's curse in man (stippled areas). F. G., Funiculus gracilis. F. C., Funiculus cuneatus. C. S., Corticospinal tract. S. T., Spinothalamic tract.

bolic regulation of breathing travel by different pathways through medulla and spinal cord.

Abnormal Chemoreceptor Function. Alterations in the level of breathing may be initiated by an increased stimulus to the chemoreceptors. If the central chemoreceptors are anatomically separated from the metabolic respiratory centers, one would expect to occasionally see patterns of breathing that result from decreased function of the central chemoreceptors. There are a few clinical reports of such an abnormality; these patients, without intrinsic pulmonary disease, usually hyperventilate, and they have little or no response to inhaled CO_2 as the only significant abnormality in pulmonary function. Careful post-mortem studies have not been performed in these patients so that the nature and location of the lesion is unknown.

Of perhaps greater clinical interest is the effect of loss of function of the peripheral chemoreceptors. As indicated earlier the carotid bodies are of primary importance in arterial O_2 homeostasis and in man appear to contribute 10 to 20 percent of the normal drive to ventilation. Bilateral carotid endarterectomy, which frequently denervates the carotid body, results in hypoventilation and a rise in arterial P_{CO_2} of 5 to 6 mm. Hg. This degree of CO_2 retention is well-tolerated by patients with normal lungs. However, in patients with intrinsic lung disease, severe CO_2 retention and hypoxemia may occur.

A second form of insensitivity to hypoxia is observed in individuals who were subjected to chronic hypoxia at birth and/or during early childhood. It has long been known that natives of high altitude have a higher P_{CO_2} and lower ventilation than acclimatized newcomers. It has now been shown that the relative hypoventilation in altitude natives results from decreased sensitivity to hypoxia. The decreased sensitivity to hypoxia is not genetic, since it also occurs in sea level natives who have been cyanotic since birth. This defect appears to be irreversible, since the insensitivity to hypoxia persists after an altitude dweller returns to sea level or after the cyanosis resulting from congenital heart disease is alleviated by corrective cardiac surgery (Fig. 6-4). Thus, an individual deprived of his carotid body depends entirely upon his central chemoreceptors to drive ventilation. Total reliance on these central mechanisms has one inherent danger: hypoxia directly depresses them.

Dyspnea has been defined as difficult, labored, uncomfortable breathing; it is an unpleasant type of breathing although not painful in the usual sense of the word. It is not hyperventilation, hyperpnea, polypnea or tachypnea, although it may be associated with these patterns of breathing. Also, since dyspnea is a subjective sensation, it cannot be studied in experimental animals. Furthermore, the quality of dyspnea varies. A patient with bronchial asthma is able to distinguish the dyspnea resulting from that disease from that of hypoxia. Likewise the dyspnea resulting from a pneumothorax has a different quality from that arising from severe muscular exercise. These clinical observations suggest that multiple factors are involved in the mechanism of dyspnea. In recent years two mechanisms that may contribute to dyspnea have received most attention: muscle tension inappropriateness, and vagal afferents.

As discussed above, the muscle spindles are detectors of length change in the intercostal muscles. When the degree of shortening following a command from the respiratory center is inappropriate, there is increased afferent discharge from the muscle spindles. According to the hypothesis, not only do these impulses act monosynaptically on the alpha motor neurons but also they stimulate ascending sensory pathways. This increased sensory activity is perceived

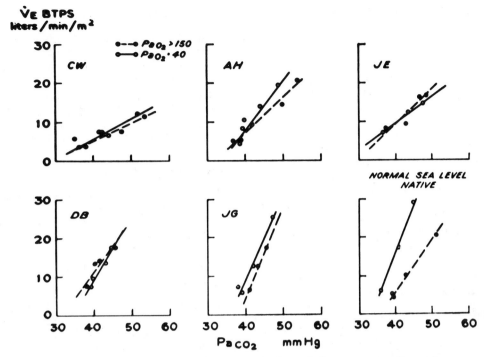

FIG. 6-4. Ventilatory response to CO_2 at high and low Pa_{O_2} in five subjects before and after correction of cyanotic congenital heart disease (tetralogy of Fallot). The CO_2 response curves of a normal subject are shown for comparison (Sørensen, S.C., and Severinghaus, J. W.: J. Appl. Physiol., *25*:221, 1968. Reproduced by permission of J. Appl. Physiol.)

centrally as dyspnea. If this were the only cause of dyspnea, complete motor paralysis should alleviate dyspnea. Such a procedure does prolong breathholding time, but severe hypoxia, hypercapnia and pulmonary congestion still cause distress in paralyzed patients.

Bilateral blockade of the vagus nerve also increases breath holding time and, in addition, relieves the dyspnea in patients with certain lung diseases. It has been suggested, therefore, that vagal afferents play an important role in dyspnea; however, they are not the only cause, since subjects with bilateral vagal block still experience severe dyspnea if they are forced to breathe through a high resistance. Vagal afferents from the irritant receptors may be involved in the production of dyspnea, since all the conditions known to stimulate these recep-

tors also cause dyspnea in man, whereas no such correlation exists between dyspnea and stimulation of the other receptors.

ANNOTATED REFERENCES

General Reviews of Regulation of Respiration

Kellogg, R. H.: Central chemical regulation of respiration. *In:* Handbook of Physiology. Section 3. Respiration. Vol. 1, p. 507. Baltimore, Waverly Press, 1964.

Oberholzer, R. J. H., and Tofani, W. O.: The neural control of respiration. *In:* Handbook of Physiology. Section 1. Neurophysiology. Vol. 2, p. 1111. Baltimore, Waverly Press, 1960.

Sorensen, S. C.: Acta physiol. scand. (Suppl.), *361*:9, 1971.

Wang, S. C., and Ngai, S. H.: General organization of central respiratory mechanisms. *In:* Handbook of Physiology. Section 3. Respiration. Vol. I, p. 487. Baltimore, Waverly Press, 1964.

Neural Respiratory Centers

Bertrand, F., and Hugelin, A.: Respiratory synchronizing function of nucleus parabrachialis medialis: Pneumotaxic mechanisms. J. Neurophysiol., 34:189, 1971.

Merrill, E. G.: The lateral respiratory neurons of the medulla: Their associations with nucleus ambiguus, nucleus retroambigualis, the spinal accessory nucleus and the spinal cord. Brain Research, 24:11, 1970.

Salmoiraghi, G. C., and Burns, B. D.: Localization and patterns of discharge of respiratory neurones in brain-stem of cat. J. Neurophysiol., 23:2, 1960.

Function of Peripheral and Central Chemoreceptors

Biscoe, T. F.: Carotid body: Structure and Function. *In:* Physiol. Rev. (In press).

Brooks, C. M., Kao, F. F., and Lloyd, B. B.: Cerebrospinal Fluid and the Regulation of Ventilation. p. 421. Philadelphia, F. A. Davis, 1965.

Comroe, J. H., Jr.: The peripheral chemoreceptors. *In:* Handbook of Physiology. Section 3. Respiration Vol. 1, p. 557. Baltimore, Waverly Press, 1964.

Fencl, V., Vale, J. R., and Brock, J. A.: Respiration and cerebral blood flow in metabolic acidosis and alkalosis in humans. J. Appl. Physiol., 27:67, 1961.

Mitchell, R. A.: Cerebrospinal fluid and the regulation of respiration. *In:* Advances in Respiratory Physiology. p. 1. London, Arnold, 1966.

Torrance, R. W.: Arterial Chemoreceptors. Oxford, Blackwell, 1968.

Mechanisms of Dyspnea

Plum, F.: Neurological Integration of Behavioral and Metabolic Control of Breathing. *In:* Breathing. p. 159. London, J. and A. Churchill, 1970.

Plum, F.: Breathlessness in Neurological Disease: The Effects of Neurological Disease on the Act of Breathing. *In:* Breathlessness. p. 203. Oxford, Blackwell, 1966.

Respiratory Reflexes

Widdicombe, J. G.: Respiratory Reflexes. *In:* Handbook of Physiology. Section 3. Respiration. Vol. 1, p. 585. Baltimore, Waverly Press, 1964.

Abnormal Respiratory Control Mechanisms

Belmusto, L., Brown, E., and Owens, G.: Clinical observations on respiratory and vasomotor disturbance as related to cervical cordotomies. J. Neurosurg., 20:225, 1963.

Guyton, A. C., Crowell, J. W., and Moore, J. W.: Basic oscillating mechanism of Cheyne-Stokes breathing. Am. J. Physiol., 187:395, 1956.

Mitchell, R. A.: Respiration. Ann. Rev. Physiol., 32:415, 1970.

Wade, J. G., Larson, C. P., Hickey, R. F., Ehrenfeld, W. K., and Severinghaus, J. W.: Effect of carotid endarterectomy on carotid chemoreceptor and baroreceptor function in man. New Eng. J. Med., 282:823, 1970.

Severinghaus, J. W., and Mitchell, R. A.: Ondine's curse—failure of respiratory center automaticity while awake. Clin. Res., 10:122, 1962.

7

Ventilation, Perfusion and Gas Exchange

Reuben M. Cherniack, M.D.

The major functions of the lung are to provide oxygen to the blood perfusing it so that it may be carried to the tissues, and to remove carbon dioxide which has been produced in the tissues. This exchange of gases is accomplished by the convective movement of air into the lungs by the breathing movements (ventilation) and the convective movement of blood through the lungs by the pumping action of the heart (perfusion).

VENTILATORY FUNCTION

Lung Volumes. The breathing movements take place within the framework of the total lung capacity (T.L.C.), which can be divided into several components or subdivisions. The absolute values of these subdivisions are dependent upon the age, sex, and size of the person, although the proportion of the total lung capacity that each occupies is remarkably similar. Normally the resting level or functional residual capacity (i.e. the end-expiratory position) is about 40 percent of the T.L.C. The lungs cannot be entirely emptied voluntarily, and the gas that remains at the end of a maximum expiration is the residual volume. For many years, an increased residual volume, particularly if it was greater than 30 percent of the total lung capacity, was believed to be indicative of emphysema. However, an increase in residual volume indicates only that the lungs are hyperinflated, and this can occur as a result of obstruction of the airways, as in bronchial

asthma. In addition, the residual volume increases with age, and may be as much as 50 percent of the total lung capacity in an elderly healthy person.

In order to carry out the breathing movements, the respiratory muscles must overcome the elastic and nonelastic resistances of the lungs and chest wall. The elastic resistance is related to the amount of distention of the respiratory system or the tidal volume; the nonelastic resistance is related to the rate of change in tidal volume or air flow. Definitive information about these mechanical resistances can be gained by simultaneous measurement of the volume changes and the rate of air flow and of the variations in pressure that are involved.

Elastic Resistance. The elastic properties, or the compliance, of the respiratory system are determined by measurement of the number of liters of distention induced by a change in pressure of one centimeter of water. The pressure change is measured under static conditions when there is no air flowing. The compliance of the lungs is determined by relating the change in volume to the change in transpulmonary pressure from end-expiration to end-inspiration. Similarly, the compliance of the chest wall is determined by relating the change in lung volume to the difference between the intrathoracic pressure and the pressure exterior to the chest wall.

Examples of the elastic behavior, or the compliance, of three different types of lung are shown in Figure 7-1. In the normal lung

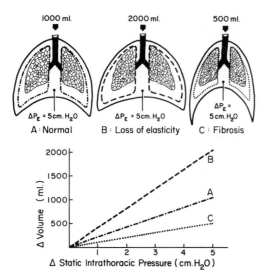

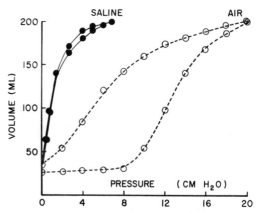

FIG. 7-1. The distention produced by a change in intrathoracic pressure of 5 cm. H_2O in a normal lung (A), a lung which has lost elasticity (B), and a lung which has become fibrosed (C). (Cherniack, R. M., Cherniack, L., and Naimark, A.: Respiration in Health and Disease. Philadelphia, W. B. Saunders, 1961.)

FIG. 7-2. Static pressure volume relationships when the lungs are inflated with air and with saline. (Cherniack, R. M., Cherniack, L., and Naimark, A.: Respiration in Health and Disease. Philadelphia, W. B. Saunders, 1961.)

(A), a change in the intrathoracic pressure of 5 cm. H_2O results in an inspiration of one liter of air, so that its compliance is 0.200 L./cm. H_2O. When the lung loses its elasticity (B), as in chronic obstructive emphysema, the same change in intrathoracic pressure leads to an inspiration of 2 liters of air so that the compliance is 0.400 L./cm. H_2O. If the lungs are stiff (C), as in pulmonary fibrosis or congestion, the same change in the intrathoracic pressure results in an inspiration of only 0.5 liters, so that the compliance is only 0.100 L./cm. H_2O.

Because one cannot empty the lungs completely, it is impossible to describe the pressure volume behavior of the lungs completely in vivo. However, much can be learned about the elasticity of lungs from study of excised lungs. The pressure-volume characteristics of the excised lung filled with air are different from those found filled with saline (Fig. 7-2); thus, at any given volume transpulmonary pressures are much greater when the lung is filled with air. This is because the surface tension at the air-liquid interface during air filling is much greater than the negligible amount at the liquid interface during saline filling. More than one half of the elastic recoil of the lungs is due to the surface-active forces in the lung. Despite these surface forces, the lungs normally do not collapse completely, even after a maximal expiration. A surface-active material (surfactant) that lines the surface of the terminal lung units (alveoli, alveolar ducts, respiratory bronchioles) allows the lung to remain inflated at low pressures and also allows alveoli of different size to coexist at the same pressure. A deficiency of this material may render the lung unstable and promote collapse (atelectasis). Impaired surface activity has been demonstrated in lungs of children dying from hyaline membrane disease, in patients exposed to high concentrations of oxygen, as well as after open heart surgery and in a wide variety of experimental conditions. In addition, in severe hypoxemia, acidemia, pulmonary embolism, drowning, aspiration or shock, a state of diminished pulmonary capillary perfusion (pulmonary hypoperfusion syndrome, or "shocked lung") may develop. These conditions may be associated with altered alveolar cell function, with a deficiency in available surfactant, which results in alveolar instability and a tendency to develop focal areas of atelectasis which, in turn, aggravates any gas exchange abnormality.

Nonelastic Resistance. In addition to inertia, which must be overcome during acceleration and deceleration, airway resistance and tissue viscance must be overcome during breathing. Because the resistance to movement of air in the airways accounts for about 85 percent or more of the resistance, it is the airway resistance or the total nonelastic resistance that is usually assessed in patients.

In the normal subject, the total nonelastic resistance of the lungs is approximately 1.8 cm. H_2O/L./sec. of airflow, the greater part of this resistance being due to laminar flow, and only about one tenth of it to turbulence. In patients suffering from obstruction of the airway, the total nonelastic resistance is increased, owing predominantly to turbulence, and may be greater than 5 cm. H_2O/L./sec. of airflow. When the ventilation is increased above normal, such as during exertion, this turbulent resistance becomes exceedingly high.

Interaction Between Nonelastic Resistance and Elastic Resistance. In addition to knowledge of the overall resistance to air flow and the compliance of the lungs, the relationship between the resistance to air flow and the compliance in individual portions of the lungs is important, for this influences the distribution of the inspired gas. When two or more parallel units of lung are subjected to the same inflation or deflation pressure, each fills or empties at a rate determined by its time constant (i.e., the product of its resistance and compliance). If the time constants of the units are equal, they will fill or empty uniformly; conversely, if the time constants are unequal filling or emptying will be nonuniform. When there is no localized disease, a tidal volume is equally distributed to different

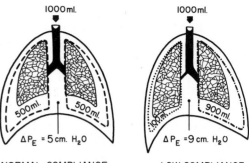

FIG. 7-3. The effect of a local bronchial obstruction on the pressure-volume relationship of the lungs when 1000 ml. of air is inhaled. (Cherniack, R. M., Cherniack, L., and Naimark, A.: Respiration in Health and Disease. Philadelphia, W. B. Saunders, 1961.)

areas of lung (Fig. 7-3, *left*), and in normal individuals this appears to be what happens, indicating that the time constants are relatively uniform throughout the lung. However, when localized disease such as an airway obstruction is present, the air tends to move into the areas of the lung that offer the least resistance (Fig. 7-3, *right*). For the same volume of air to be inspired, a greater intrathoracic presure must develop, and this results in a fall of the calculated compliance. When the time constants are unequally distributed in the lung, the inspired air is distributed unequally. This disturbance of distribution increases concomitantly with a fall in compliance with increasing respiratory frequency. Alterations in compliance related to increasing respiratory rate have been termed frequency-dependent compliance.

Work of Breathing. The mechanical work necessary to overcome the resistances offered by the lung and the chest wall during breathing is performed by the respiratory muscles. During quiet breathing, almost all of the muscular work is carried out during inspiration, for the elastic recoil of the lungs is sufficient to overcome the nonelastic resistance of both the air and the tissues during expiration. If the expiratory resistance is high, however, expiratory muscular work may be required. In normal subjects, the total mechanical work performed on the lungs has been estimated to be approximately 0.3 to 0.7 kg./m./min. In a patient suffering from diffuse bronchial obstruction (such as bronchial asthma or chronic obstructive emphysema), the mechanical work necessary to overcome the nonelastic resistance is increased considerably; in pulmonary fibrosis, much more work must be performed in order to overcome the high elastic resistance of these "stiff lungs."

In order to perform the mechanical work, the respiratory muscles require oxygen. Figure 7-4 illustrates the change in oxygen consumption associated with increasing ventilation in normal individuals, in patients suffering from emphysema, congestive heart failure, or obesity. In the normal subject, the oxygen cost of breathing at rest varies from 0.3 to 1.0 ml./L. of ventilation (about 2 percent of the total oxygen consumption), and the minute ventilation can be increased considerably with little alteration in the total oxygen consumption. After a certain level, the oxygen consumption rises disproportionately with further increases in

ventilation. In the patients in whom the mechanical resistances are great, the oxygen consumption rises considerably even with small increases in ventilation, and may amount to as much as 50 percent of the total oxygen consumption. The oxygen cost of breathing has been found to be between 3.0 and 18.0 ml./L. of ventilation in patients suffering from emphysema, 1.0 to 8.0 ml./L. of ventilation in obese individuals, and 1.5 to 4.5 ml./L. of ventilation in patients with congestive heart failure. When the mechanical resistance to breathing is high, the disproportionate increase in oxygen consumption with increasing ventilation occurs at a much lower level of ventilation than it does in normal individuals. The high oxygen requirements of the respiratory apparatus means that the proportion of the oxygen consumption available for the other muscles of the body is reduced; this is extremely important clinically, particularly during exertion. Another important aspect of the oxygen cost of breathing, particularly in respiratory disease, is that, when ventilation is increased, the tendency to lower the alveolar carbon dioxide tension may be offset by the increased metabolic production of carbon dioxide by the respiratory muscles. Even though the ventilation could be increased still further, it would serve no useful purpose, because the greater ventilation would merely increase the tendency toward carbon dioxide retention. This has been estimated to occur in normal individuals at a ventilation of about 140 L./min.; but in patients with an increased oxygen cost of breathing, the level of ventilation that is maximally effective in lowering the carbon dioxide tension may be as low as 15 to 20 L./min.

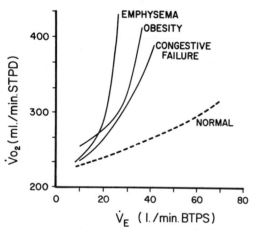

Fig. 7-4. The changes in oxygen consumption ($\dot{V}O_2$) associated with increasing ventilation in a normal subject and patients with congestive heart failure, obesity and emphysema. (Cherniack, R. M., Cherniack, L., and Naimark, A.: Respiration in Health and Disease. Philadelphia, W. B. Saunders, 1961.)

In addition, there is apparently a relationship between the mechanical work of breathing and the rate at which an individual breathes. For any given alveolar ventilation, both normal subjects and patients with respiratory disease appear to breathe at a rate and depth at which the work of breathing is minimal. When the elastic resistance is increased, as in pulmonary fibrosis or kyphoscoliosis, the respirations tend to become rapid and shallow, probably because of the increased work required to overcome the elastic resistance with even small increases in tidal volume. In contrast, when the nonelastic resistance is increased, as in diffuse bronchial obstruction, the respirations tend to become slower and deeper—because a faster respiratory rate leads to an increase in the resistance to airflow.

EXCHANGE OF GASES IN THE LUNGS

The air that enters the alveoli takes part in gas exchange with mixed venous blood in the pulmonary capillaries, so that the gas tensions of the pulmonary capillary blood and the alveolar air come almost into equilibrium. This transfer of gases at the alveolar level is entirely a result of diffusion and thus is determined by the partial pressures of the gases on both sides of the alveolo-capillary membrane. Normally, the partial pressures of oxygen and of carbon dioxide of the blood coming to the pulmonary capillaries are 40 and 46 mm. Hg respectively, while those of the blood leaving the alveoli and entering the pulmonary veins are 100 and 40 mm. Hg, respectively. If the amount of air taking part in gas exchange is reduced, or if there is failure of the gas tensions in the alveolar and pulmonary capillary blood to come into equilibrium, the pulmonary venous or arterial blood gas tensions will be abnormal. Reduction in the amount of air taking part in gas exchange relative to the metabolic production of carbon dioxide is called alveolar hypoventilation. Failure of gas tensions in the alveoli and the pulmonary capillary blood to equilibrate may be the result of uneven ventilation/perfusion ratios in different areas of the lungs, true venous admixture, or impaired diffusion.

Alveolar Ventilation. In a healthy person only about 70 to 80 percent of each tidal volume reaches the alveoli and supplies oxygen to, and removes carbon dioxide from, the pulmonary capillary blood (alveolar component), the remainder being wasted and not taking part in gas exchange (dead space component). In many patients suffering from pulmonary disease, the physiological dead space is increased, so that the proportion of the inspired air that takes part in gas exchange is reduced. Normally, an increase in the tidal volume could compensate for the effect of an enlarged dead space, but the patient with chronic respiratory disease is frequently unable to increase his ventilation sufficiently to provide an adequate alveolar ventilation because of the mechanical disturbances in the lungs.

Whenever the level of the alveolar ventilation is inadequate to cope with the metabolic production of carbon dioxide, arterial carbon dioxide tension will be elevated. The hypercapnia is always associated with hypoxia unless the individual is inhaling an oxygen-enriched gas mixture. This situation is encountered in conditions in which the total ventilation is decreased (e.g., barbiturate poisoning or muscular paralysis), the respiratory pattern is rapid and shallow (e.g., obesity or kyphoscoliosis), or the dead-space ventilation is increased without a concomitant increase in minute ventilation (e.g., emphysema).

Even though the total ventilation may appear to be normal, hypoxia and hypercapnia could be present if the CO_2 production is high for that particular alveolar ventilation. If Figure 7-4 is considered to be representative of the relationship between alveolar ventilation and carbon dioxide production rather than oxygen consumption, it will be seen that any alveolar ventilation will be associated with a greater than normal CO_2 production in patients suffering from cardiopulmonary insufficiency. Clearly then, the high cost of breathing found in respiratory disease has important implications in the development of respiratory insufficiency, and any exer-

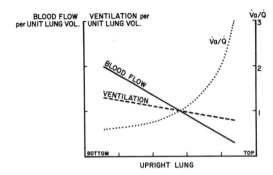

FIG. 7-5. The distribution of ventilation ($\dot{V}_A$), blood flow ($\dot{Q}$) and ventilation/perfusion relationships in the upright lung. (Cherniack, R. M., Cherniack, L., and Naimark, A.: Respiration in Health and Disease. Philadelphia, W. B. Saunders, 1961.)

tion or respiratory insult that requires an increase in ventilation may be associated with aggravation of hypoxia and hypercapnia.

Distribution of Ventilation. Even in young healthy persons—and, particularly, in older persons—the inspired air is not distributed uniformly throughout the lungs (Fig. 7-5). Because of the effect of gravity on transpulmonary pressure, the air spaces at the top of the lung are expanded more than those at the bottom. The proportion of inspired air going to different areas of lung varies with the lung volume at which one is breathing. At the normal resting level, less air goes to the air spaces at the top of the lung than to those at the bottom of the lung during inspiration. In contrast, if anyone should breathe near his residual volume, the air spaces at the top of the lung would be ventilated more than those at the bottom.

In addition to the gravity effect and regional alterations in the mechanical resistances offered by the lung, the airways and the extrapulmonary structures affect the distribution of the inspired gas. Normally, the inspired gas enters the two areas of lung almost synchronously and equally, and expiration takes place in the same fashion. This is illustrated in Figure 7-6, A, which presents the continuous analysis of nitrogen concentration in the expired air following a single inspiration of pure oxygen. There is a definite end expiratory plateau in the nitrogen concentration curve. However, when there is localized airway disease, the inspired oxygen moves into the areas of lung that offer the least resistance before it enters the others, and moves out of the unobstructed areas before it leaves the obstructed areas during expiration. The asynchronous delivery of air from the two lungs results in a rising alveolar nitrogen concentration (Fig. 7-6, B).

Distribution of Perfusion. The distribution of blood flow is also affected by gravity. The pressures in the arteries and the veins near the top of the lung are lower than those at the bottom, so that blood flow is less at the apex than at the base of the upright lung (see Fig. 7-5). The apical-basal blood flow differences are largely abolished in the supine position, since the two parts of the lung are nearly at the

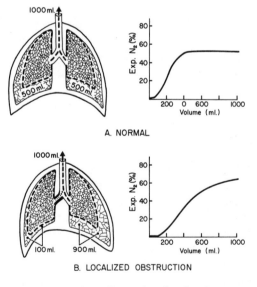

FIG. 7-6. The effect of a localized airway obstruction on the distribution of air. (Cherniack, R. M., Cherniack, L., and Naimark, A.: Respiration in Health and Disease. Philadelphia, W. B. Saunders, 1961.)

same hydrostatic level. If the subject lies on his side, the dependent lung is better perfused, in comparison to the contralateral lung. During exercise as a result of increased pulmonary artery pressure, the total blood flow increases at the apex relatively more than at the base, so that there may be a slightly more nearly uniform distribution of blood.

Ventilation/Perfusion Relationships. Even though the total alveolar ventilation and the total pulmonary blood flow may be normal, hypoxia develops if the distribution of the ventilation in relation to the perfusion of lung units is not uniform throughout the lungs. Figure 7-5 demonstrates that even in the normal upright lung ventilation/perfusion ratios are high at the apex (*top*) and low at the base (*bottom*). Clearly the gas concentrations in the alveoli must also differ normally from region to region, but these regional differences are not large enough to interfere materially with gas exchange, so that the mixed arterial blood normally has nearly the same composition as the mixed alveolar gas. However, in patients with cardiorespiratory disease there may be gross variations in ventilation/perfusion relationships throughout the lungs, so that the difference between the alveolar and the arterial partial pressures of oxygen (the A-a Po_2 gradient) increases.

When there is perfusion of inadequately ventilated alveoli or, in the extreme case, perfusion of nonventilated alveoli (i.e., low ventilation/perfusion areas), poorly aerated venous blood leaves these pulmonary capillaries and mixes with fully "arterialized" blood coming from the other pulmonary capillaries, causing hypoxemia and slight carbon dioxide retention in the arterial blood. This is called venous-admixture-like perfusion. Carbon dioxide retention may not develop if there is sufficient hyperventilation of the remaining well-perfused alveoli; but, because of the shape of the oxyhemoglobin dissociation curve, hyper-

ventilation does not significantly hyperoxygenate the blood leaving these regions and by itself does not correct the arterial hypoxia to any significant degree.

Where the ventilation of alveoli is maintained but the blood perfusion is limited, or, in the extreme case, when there is no perfusion (i.e., high ventilation/perfusion areas) the gas leaving such alveoli tends to have the same composition as the gas in the tracheo-bronchial tree and, thus, contributes to the physiologic dead space. This is called dead-space-like ventilation. The blood that perfuses these alveoli becomes fully oxygenated and, probably, excessively depleted of carbon dioxide, so that the arterial gas tensions will be normal or low as long as normally perfused alveoli are well ventilated (i.e., there is no associated venous-admixture-like perfusion).

Recently, by use of techniques employing the inhalation and intravenous injection of radioactive tracer materials such as xenon 133, much has been gained in the understanding of factors that normally influence the distribution of ventilation and perfusion of the lungs (e.g., posture, as well as the impact of respiratory disease). However, it is important to understand that the perfusion calculated using these techniques is only that perfusing ventilated alveoli. If there is perfusion of nonventilated regions of lung, or of lung units whose conducting airways are obstructed (i.e., venous-admixture-like perfusion) this will not be recognized by the gaseous isotope techniques.

True Venous Admixture. Mixed venous blood that does not come into contact with alveoli but mixes with blood in the pulmonary or the systemic circulation (true venous admixture) may also cause a lower than normal arterial oxygen tension. Even in normal persons, this amounts to approximately 2 to 4 percent of the total pulmonary blood flow. An increased amount of true venous admixture may be intra-

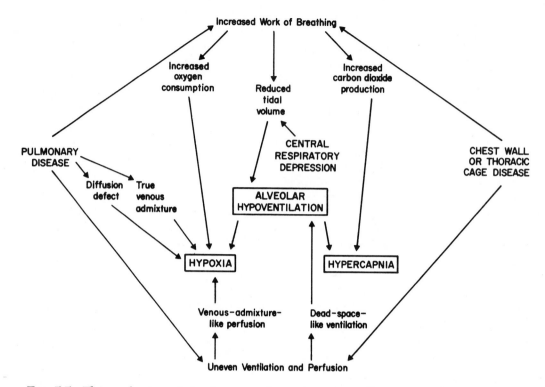

FIG. 7-7. The mechanism of development of respiratory insufficiency. (Cherniack, R. M., Cherniack, L., and Naimark, A.: Respiration in Health and Disease. Philadelphia, W. B. Saunders, 1961.)

cardiac (as in some congenital heart diseases) or from the pulmonary artery to the pulmonary vein (as in pulmonary arteriovenous fistulae). A picture resembling that of true venous admixture can also be encountered if areas of lung are nonventilated but still perfused. Although considerable hypoxia may be present, carbon dioxide retention does not occur in true venous admixture, because the lungs are usually healthy and capable of compensatory hyperventilation with resultant hypocapnia.

Diffusion. A diffusion defect may develop in disease when the alveolo-capillary membrane is markedly thickened (as in diffuse fibrosis) or the capillary bed is considerably reduced (as in thromboembolic obstruction) or if a large volume of lung is destroyed by disease or removed by surgery. Such a diffusion defect probably only contributes to a high A-a Po_2 gradient when such patients breathe a low concentration of oxygen during heavy exercise. Since diffusion of carbon dioxide takes place twenty times slower than oxygen diffusion, impaired diffusion probably never significantly limits its exchange in the lungs.

RESPIRATORY INSUFFICIENCY

Respiratory insufficiency may develop in a wide variety of patients who are suffering from a multitude of conditions. Basically however, these patients may be divided into three major groups: those with bronchopulmonary disease, those with chest wall or thoracic cage disease and those with central nervous system depression (Fig. 7-7).

The findings in the arterial blood in acute and chronic alveolar hypoventilation are

TABLE. 7-1. Arterial Blood Gas Findings in
Alveolar Hypoventilation

ARTERIAL BLOOD	ACUTE	CHRONIC
Po_2	↓	↓
Pco_2	↑	↑
HCO_3	↔	↑
CO_2 content	↔	↑
pH	↓	↔

* Unless patient is receiving oxygen
↑, Increased; ↓, Decreased; ↔, Unchanged

presented in Table 7-1. In acute respiratory failure, the bicarbonate and total carbon dioxide are little elevated. In chronic respiratory failure there has been compensation for the elevated partial pressure of carbon dioxide through elimination of chloride and retention of base and bicarbonate, so that the carbon dioxide content is elevated and the serum chloride is low.

Respiratory insufficiency frequently develops in two phases. Initially there is a stage in which there is compensation for abnormal gas exchange by an increase in ventilation and/or perfusion, so that near-normal blood gases are maintained at the expense of increased respiratory or cardiac work (patients with obstructive pulmonary disease who do this have been labeled "pink puffers"). Later, because of the marked increase in the resistance to breathing, this compensation may become inadequate and alveolar hypoventilation with severe hypoxia and carbon dioxide retention may develop (patients with obstructive pulmonary disease in whom this develops with associated right sided heart failure have been labeled "blue bloaters"). The tempo of progression from the first to the second stage is highly variable, and indeed, the order in which they appear may vary. In addition, there seems to be no correlation between the degree of compensation and the severity of the patient's symptoms. In fact, many patients who are maintaining relatively normal blood gas tensions, although at the cost of a marked increase in respiratory work, are much more disabled than others who have chronic hypoxia and hypercapnia. Nevertheless, progressive decompensation is almost invariably associated with an increasing severity of symptoms and is attended by secondary effects of hypoxia and acidosis and, in some instances, hypercapnia, which further compromise respiratory and cardiac function.

Whereas, in normal individuals, mild to moderate hypoxia usually is a mild respiratory stimulant, hypoxia may replace carbon dioxide as the major stimulus to respiration in those with chronic CO_2 retention. When high oxygen concentrations are administered to such persons, the hypoxic stimulus is removed and the respiratory failure may be exaggerated because ventilation falls.

Both hypoxia and acidosis lead to vasoconstriction of the pulmonary vasculature, thereby increasing the pulmonary vascular resistance. The pulmonary vascular resistance may be elevated further by the increased blood viscosity due to secondary polycythemia, or by obliteration of the pulmonary capillaries. The high pulmonary vascular resistance leads to pulmonary hypertension, with subsequent right ventricular hypertrophy and eventually, right ventricular failure—one of the most serious manifestations of chronic respiratory insufficiency.

It has been suggested that chronic elevation of the arterial carbon dioxide tension may be associated with fluid retention through other, yet unexplained, mechanisms. This might explain why tissue edema is encountered most frequently in patients who have hypoxia and hypercapnia ("blue bloaters") and is rarely seen in "pink puffers." In addition, particularly in elderly patients with respiratory disease, there may be an element of left ventricular failure, perhaps because of hypoxic depression of myocardial function, the added load of a high circulating blood volume, or excessive bronchopulmonary anastomoses. The presence of pulmonary congestion would increase the work of breathing further and, also, aggravate the respiratory insufficiency.

THE ASSESSMENT OF PULMONARY FUNCTION

There are a variety of tests that may be used to detect disturbed pulmonary function, and the judicious use of such tests is essential to the evaluation of the patient with respiratory complaints. The tests vary widely in degree of sophistication and complexity, but, in general, they fall into two groups: those relating to the ventilatory function of the lungs and chest wall, and those relating to gas exchange.

Ventilatory Function

Ventilatory function is assessed by determination of static lung volumes, which are predominantly a reflection of the elastic resistance or distensibility of the lungs and thorax, and dynamic lung volumes, which are predominantly a reflection of the nonelastic resistance.

Static Lung Volume. The distensibility of the respiratory system can be assessed indirectly by determining the vital capacity (V.C.), which varies according to age, body size, and sex. Vital capacity is related to the compliance of the lung and thorax, decreasing as the total compliance decreases. Thus, it is reduced in any condition in which the compliance of the lungs is decreased, such as pulmonary fibrosis or congestion, or in which the compliance of the chest wall is decreased, such as obesity or kyphoscoliosis. However, a low vital capacity is not necessarily indicative of respiratory disease. Since a great deal depends on the cooperation of the patient, it is essential that the test be repeated several times in order to be certain that the maximal value has been obtained.

If a reliable value has been obtained, the vital capacity should be interpreted in the

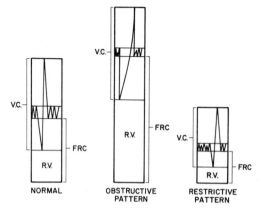

FIG. 7-8. Total lung capacity and its subdivisions normally, in the obstructive pattern and in the restrictive pattern. V.C. = Vital Capacity, FRC = Functional Residual Capacity, R.V. = Residual Volume. (Cherniack, R. M., Cherniack, L., and Naimark, A.: Respiration in Health and Disease. Philadelphia, W. B. Saunders, 1961.)

light of any associated changes in lung volume. Figure 7-8 illustrates that the vital capacity may be low in both restrictive disease and obstructive lung disease. In restrictive disease this is because all components of the lung volumes are reduced. In obstructive disease, the vital capacity is low because the residual volume (R.V.) and functional residual capacity (F.R.C.) are increased, (i.e., the individual is breathing at an increased resting level).

Dynamic Lung Volumes. Considerable information about the nonelastic resistance can be obtained by determination of the rates at which both an inspiratory and an expiratory vital capacity take place. In particular, it is the forced expired vital capacity (F.V.C.) which is usually used. The assessment of the expiratory nonelastic resistance from the F.V.C. can be carried out in several ways, but there are cogent reasons why certain assessments are more informative than others.

The forces tending to narrow the airways (peribronchial pressure and the force exerted by contraction of bronchial smooth muscle) are balanced by those tending to widen them (intraluminal pressure and the tethering action of the surrounding connective tissue). During forced expiratory efforts the pleural and the alveolar pressure become greater than atmospheric pressure, while the intrabronchial pressure decreases from values near alveolar pressure in the peripheral airways to atmospheric pressure near the airway opening. At some point in the airway, intrabronchial pressure may be equal to or less than pleural pressure so that there is a net force tending to narrow the airways at that point, thereby limiting air flow. In patients with diseased airways, which are more collapsible, flow limitation occurs at even lower levels of pleural pressure. In this situation the obstruction to airflow may be due to loss of tissue retraction or elastic recoil. This is in contrast with patients who have narrowing of the airway lumen by muscle spasm, mucosal thickening, or secretions, with resultant increased resistance to airflow. Obviously, both loss of tissue retraction (as in emphysema) and increased resistance within the bronchi (as in asthma or bronchitis) may be present in a single patient.

As is demonstrated in Figure 7-9, the maximal inspiratory flow rate at every lung inflation depends primarily on the force developed. In contrast, the maximal flow rate achieved during expiration depends on the degree of lung inflation. At lung volumes near the total lung capacity, expiratory flow increases as the pressure increases. It is apparent that an assessment of expiratory flow resistance, which is based predominantly on analysis of the flow rate at high degrees of lung inflation, may be fraught with inconsistency. This is because the value obtained may be related more to patient cooperation and effort than to alterations in intrapulmonary mechanics. At lesser lung inflations, expiratory flow increases as pressure increases—up to a certain point, beyond which more effort fails to elicit any further increase, and may even result in a slight decrease in flow. At lesser degrees of lung inflation, then, the influence

of variation in patient cooperation is minimized because maximum flow does not require maximal effort. Thus the maximum mid-expiratory flow rate (M.M.F. or F.E.F. 25-75 percent), which ignores the first 25 percent of the F.V.C., is to be recommended over a measurement such as that of peak expiratory flow for assessment of expiratory flow resistance. Many laboratories also use the forced vital capacity in the first second of maximum expiration (F.E.V.$_{-1.0}$) as a measure of expiratory flow resistance. In patients with airway obstruction, the F.E.V.$_{-1.0}$ is less than 80 percent of the total vital capacity and the M.M.F. is low. On the other hand, in patients with a reduced compliance but no associated

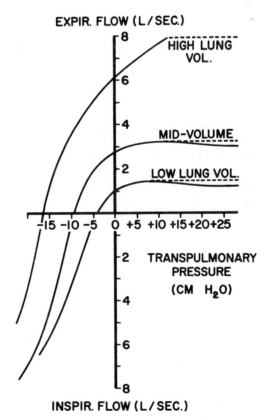

EXPIR. FLOW (L / SEC.)

HIGH LUNG VOL.

MID-VOLUME

LOW LUNG VOL.

TRANSPULMONARY PRESSURE (CM H$_2$O)

INSPIR. FLOW (L / SEC.)

FIG. 7-9. The relationship between pressure and maximal flow at high lung volumes, mid volume, and low lung volume. (Cherniack, R. M., Cheniack, L., and Naimark, A.: Respiration in Health and Disease. Philadelphia, W. B. Saunders, 1961.)

airway obstruction, as in diffuse pulmonary fibrosis, these indices of air flow are usually normal.

Since the very small airways ($<$ 2 to 3 mm. in diameter) contribute less than 20 percent of the airway resistance, even a marked increase in peripheral airway resistance will not be reflected in tests used as indices of airway obstruction such as the F.E.V.$_{.1.0}$ or M.M.F., for these tests are relatively insensitive to changes in the peripheral portions of the airways and are mainly a reflection of the large airways. In diseases of small airways, measurement of the A-a Po$_2$ gradient and dynamic lung compliance at increased respiratory frequencies may be the only way to detect an abnormality.

Inspiratory obstruction to air flow also may be considerable in patients with bronchial obstruction. Although flow rates during an inspiratory vital capacity may be analyzed in the same manner as the expiratory vital capacity, Figure 7-9 demonstrates that the air flow achieved during a forced inspiration is related more to the degree of patient effort than to changes in intrapulmonary mechanical properties.

The overall effects of alterations in the mechanical properties of the lungs and chest wall can be assessed by determining the maximal breathing capacity (M.B.C.),

also called the maximum voluntary ventilation (M.V.V.). Because maximal ventilation can be attained only by voluntary effort, the subject must be encouraged constantly during the performance of the test, and one must be convinced that the patient has made a maximal effort before a low value is interpreted as being abnormal. Since the rate of air flow is markedly increased during the performance of the M.B.C., this test is particularly affected by alterations in the nonelastic resistance and, to a much lesser extent, by changes in the elastic resistance.

Assessment of the simultaneous volume and flow changes during the forced vital capacity also yields information about the overall mechanical resistances to breathing. It has been demonstrated that calculation of the ratio of the change in volume to the change in flow over the second or the third quarter of the expired forced vital capacity correlates well with the overall mechanical time constant of the lung calculated from direct measurement of lung compliance and total nonelastic resistance. Clearly the M.M.F. which is measured over the same volume change is a rough index of this calculation from flow/volume diagrams. Normal values of various indices of ventilatory function calculated from a forced vital capacity are shown in Table 7-2.

TABLE 7-2. Normal Values of Ventilatory Function

Elastic Resistance	
Vital Capacity	size, sex
Air Flow Resistance	
Peak Expiratory Flow	8.2-10.4 L./sec.
Timed Volumes	
F.E.V. 0.5	2.8-3.2 L. (58-66% F.V.C.)
F.E.V. 1.0	3.0-4.0 L. (80-85% F.V.C.)
Air Flow/Volume Relationships	
F.E.F. 200-1200 ml	5.8-9.2 L./sec.
F.E.F. 25-75% V.C.	2.7-4.5 L./sec.
Absolute Flow At Specific Volumes	
at 0.25 V.C.	4.8-8.4 L./sec.
at 0.50 V.C.	2.5-5.3 L./sec.
at 0.75 V.C.	0.9-2.1 L./sec.

In addition, assessment of gas distribution or mixing in the lung provides information about the distribution of altered mechanical properties in the lungs. The severity of impairment of distribution or mixing of gas in the lung is indicative of the degree of inequality of time constants in lung units.

Patterns of Impaired Ventilatory Function

Figure 7-10 presents the types of records of ventilatory function obtained by spirometry from a normal subject and two patients suffering from pulmonary disease; and Table 7-3 contrasts the pattern of abnormality found in patients with obstructive pulmonary disease and patients with restrictive pulmonary disease.

Obstructive Pattern. In the patient suffering from chronic obstructive lung disease the one second forced expiratory volume ($F.E.V._{.1.0}$), the MMF and the maximal breathing capacity (M.B.C.) are markedly reduced, whereas the vital capacity is often only slightly impaired. Vital capacity may be considerably reduced even though the total lung capacity is-

TABLE 7-3. Disturbances of Ventilatory Function in Obstructive and Restrictive Disease

Test	Obstructive Disease	Restrictive Disease
V.C.	←→ ↓	↓
$F.E.V._{.1.0}$	↓	←→ ↓
M.M.F.	↓	←→ ↓
M.B.C.	↓	←→ ↓
R.V.	↑	↓
F.R.C.	↑	↓
T.L.C.	↑	↓

↑, Increased; ↓, Decreased; ←→ Unchanged

greater than normal if the residual volume is increased and the patient is breathing at a more inspiratory position than normal. In addition, when performing an M.B.C., patients with obstructive disease do so at a higher lung volume because the expiratory resistance is easier to overcome at this degree of lung inflation where the bronchi are wider and the elastic recoil is greater. From these tests, one may infer that there is little increase in elastic resistance but that there is a marked increase in the resistance to air flow, which is usually distributed nonuniformly, so that the distribution of the inspired air is impaired.

It is essential that diffuse obstruction to air flow be recognized both by clinical and spirometric means, and it is useful to determine whether air flow can be improved by the administration of a nebulized bronchodilating agent. An increase in M.M.F. and M.B.C. of more than 10 percent indicates that the nonelastic resistance has fallen after inhalation of the bronchodilator, and that the bronchial obstruction is at least partially reversible. The vital capacity often also increases after bronchodilation. This suggests that the elastic resistance has changed—perhaps because completely obstructed airways have been opened, or as is more likely, because the residual volume has decreased as resistance to air flow diminished.

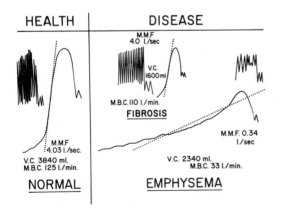

HEALTH | DISEASE

M.M.F 4.0 l./sec
V.C. 1600ml
M.B.C. 110 l./min.
FIBROSIS

M.M.F 4.03 l./sec.
V.C. 3840 ml.
M.B.C. 125 l./min.
NORMAL

M.M.F. 0.34 l./sec
V.C. 2340 ml.
M.B.C. 33 l./min.
EMPHYSEMA

FIG. 7-10. Ventilatory function in a normal subject and patients with pulmonary fibrosis and emphysema. V.C. = Vital Capacity, M.M.F. = Maximum Mid-Expiratory Flow Rate, M.B.C. = Maximum Breathing Capacity. (Cherniack, R. M., Cherniack, L., and Naimark, A.: Respiration in Health and Disease. Philadelphia, W. B. Saunders, 1961.)

Restrictive Pattern. In the patient suffering from restrictive disease such as pulmonary fibrosis, the vital capacity and total lung capacity are reduced, while the F.E.V.$_{1.0}$, M.M.F. and M.B.C. are very well maintained. From these measurements one may infer that the distensibility or compliance of the lung is reduced and that the airway resistance is essentially normal. If the increase in elastic resistance is relatively uniformly distributed, the inspired air will be distributed uniformly.

This pattern of disturbance in ventilatory function is also found in patients who have had surgical removal of lung tissue, or have an increased elastic resistance of the chest cage, as in obesity or kyphoscoliosis, or whose respiratory muscles are unable to perform normally because of neuromuscular disease or paralysis.

GAS EXCHANGE

Major alterations of gas exchange in the lungs are readily assessed by analysis of the partial pressures of oxygen and carbon dioxide in the blood. More precise assessment is obtained by analysis of simultaneously collected samples of arterial blood and expired air during a "steady state" (when the metabolic respiratory quotient is > 0.07, < 1.0), at rest, during exercise and while breathing 100 percent oxygen. From these samples, arterial gas tensions and pH, minute ventilation, oxygen consumption, carbon dioxide production, respiratory quotient, physiological dead-space, and A-a Po$_2$ gradient can be calculated. It is therefore possible to delineate the physiological disturbance (or disturbances) responsible if hypoxia is demonstrated. If there is hypercapnia, alveolar hypoventilation is present. If this is unaccompanied by other disturbances, the alveolar Po$_2$ will be low, but the alveolar-arterial oxygen (A-a Po$_2$) gradient will be within normal limits. An A-a gradient greater than 10 mm. Hg is indicative of a defect in blood-

gas equilibrium. Although this may result from a diffusion defect under special circumstances, by far the most common cause of a failure of blood and gas to equilibrate is a mismatching of the distribution of ventilation and perfusion in the lungs.

The amount of true venous admixture is assessed by having the subject breathe 100 percent oxygen. Under normal circumstances the arterial Po$_2$ will rise to about 500 mm. Hg when 100 percent oxygen is being inspired, and a value of less than 500 mm. Hg suggests that there is an abnormal amount of true venous admixture. Thus, the finding of an elevated A-a Po$_2$ gradient while breathing ambient air and of a Po$_2$ greater than 500 mm. Hg while breathing oxygen indicates that nonuniform distribution of ventilation perfusion relationships is responsible for the hypoxemia.

Clearly then, the finding of a high A-a gradient is usually due to nonuniformity of ventilation/perfusion relationships in the lung. The finding of abnormal differences between alveolar and arterial oxygen tensions in the presence of relatively normal spirometric tests may be particularly significant, for this may be indicative of peripheral airway disease. Small airway disease may result in inequalities of time constants throughout the lung, which, as pointed out previously, may be reflected by the frequency dependence of lung compliance, or by unequal distribution of inspired air. If these poorly ventilated regions are still well perfused, the A-a gradient will be increased.

Carbon monoxide is usually used to determine the diffusing capacity (Dco). The single-breath method is the easiest to perform, but values obtained by the steady state technique are more likely to be relevant to the actual conditions of gas exchange than are those obtained under artificial conditions of breath holding. In contrast, the steady-state techniques, particularly that utilizing end-tidal sampling to estimate the alveolar partial pressure of carbon monoxide, can be fraught with error because gas from the dead space may be included in the end-tidal sample. In any case, the tests used to calculate diffusion are affected by uneven ventilation/perfusion ratios and the ability to measure the alveolar-capillary gas tension gradient accurately. Thus, the findings of a low diffusing capacity should be interpreted with caution. It has been suggested that the interpretation of the Dco can be aided by measurement of fractional carbon monoxide removal, and, if both measurements are low, either the abnormality in ventilation/perfusion distribution is fairly extreme or there is a reduction of the internal surface of the lung.

ACID-BASE BALANCE

Since alterations in gas exchange can affect the arterial pH, with subsequent renal compensatory measures, and metabolic disturbances lead to compensatory measures by the respiratory system, estimation of the acid-base balance is an essential component of the assessment of respiratory disease.

The current availability of relatively simple equipment for measurement of the arterial pH and Pco_2 has expedited the study of acid-base balance to the extent that it should be as much a part of the assessment of a patient with metabolic or respiratory disturbances as is the hemoglobin and leukocyte count. From knowledge of any two of the variables of the Henderson-Hasselbalch equation, the acid-base status can be established and proper therapy instituted. (For further details concerning acid-base relationships the reader is referred to Chapter 13.)

CARDIOPULMONARY RESPONSE TO EXERCISE

The assessment of gas exchange during exercise is probably the most important and informative of all tests of respiratory func-

TABLE 7-4. Gas Exchange Abnormalities During Exercise

PARAMETER	VENTILATORY IMPAIRMENT	V/Q IMBALANCE	CARDIOVASCULAR IMPAIRMENT OR PHYSICAL UNFITNESS
V_E	↑	**⬆**	**⬆**
R	↓ ↔	↔	**⬆**
V_D/V_T (%)	↓ ↔	↕	↓ ↔
A-a Po_2	↔	**⬆**	↔ ↑
Pa_{O_2}	**⬇**	**⬇**	↔ ↓
Pa_{CO_2}	**⬆**	↔ ↓	↔ ↓
pH	**⬇**	↔ ↓	**⬇**

V_E, Ventilation

R, Respiratory quotient

V_D/V_T(%) Ratio of dead space to tidal volume

A-a Po_2, Alveolar-arterial oxygen tension gradient

Pa_{O_2}, Arterial partial pressure of oxygen

Pa_{CO_2}, Arterial partial pressure of carbon dioxide

↑, Increased; ↓, Decreased; ↔, Unchanged

The most characteristic abnormalities are indicated by the bold arrows.

tion. Measurement of the arterial partial pressures of oxygen and carbon dioxide and of pH, ventilation, oxygen consumption, dead-space/tidal air ratio and A-a gradients during exercise may demonstrate an impairment when all other pulmonary function measurements are normal, since most patients experience disability particularly on exertion. Thus, gas exchange during exercise should be assessed in all patients complaining of dyspnea.

A reduction in exercise tolerance may result from a variety of mechanisms (Table 7-4). If the work of breathing is excessive, the ventilatory response to exercise may be limited; in this instance, even though the ventilation is greater than at rest, the high O_2 consumption and CO_2 production will result in a decreased alveolar and arterial Po_2 and an increased alveolar and arterial Pco_2. If there is mismatching of blood and gas distribution or if the diffusing capacity for oxygen fails to increase in proportion to the oxygen consumption, the A-a gradient will rise and arterial Po_2 will fall further. If the cardiovascular system is unable to satisfy the increased tissue demands, or if the individual is not physically fit, the distribution of systemic blood flow to the exercising muscles may be inadequate; this will result in tissue hypoxia, with excessive lactate production, acidemia, and increased dissociation of carbonic acid to produce CO_2 and water. This will result in increased release of CO_2 from the body; thus, the finding of a respiratory quotient greater than 1.0 and a fall in pH during the exercise period suggests that there is excessive lactate production, possibly as a result of anaerobic metabolism. When patients with cardiorespiratory insufficiency undertake a program of exercise training, considerable improvement in exercise tolerance may be achieved. This training effect is often due not to improvement of ventilatory function but rather to improved circulatory function, i.e., in the cardiac response or in the distribution of peripheral blood flow to the exercising muscles.

ANNOTATED REFERENCES

General references regarding lung volumes, mechanics, gas exchange, and alterations in disease:

Bates, D. V., and Christie, R. V.: Respiratory Function in Disease. Philadelphia, W. B. Saunders, 1964.

Cherniack, R. M., and Cherniack, L.: Respiration in Health and Disease. Philadelphia, W. B. Saunders, 1961.

Comroe, J. H., Jr., Forster, R. E., DuBois, A. B., Briscoe, W. A., and Carlsen, E., Jr.: The Lung. Chicago, Year Book Medical Publications, 1962.

Fenn, W. O., and Rahn, H.: Handbook of Physiology. Section 3: Respiration. Vols. 1 and 2. Baltimore, Waverly Press, 1965.

West, J. B.: Ventilation/Blood Flow and Gas Exchange. Oxford, Blackwell Scientific Publications, 1965.

Lung volumes and mechanical properties of lungs:

DuBois, A. B., Botelho, S. Y., Bedell, G. N., Marshall, R., and Comroe, J. H., Jr.: A rapid plethysmographic method for measuring thoracic gas volume. J. Clin Invest., 35: 322, 1956.

Kory, R. C., Callahan, R., Boren, H. G., and Syner, J. C.: The Veterans Administration-Army Cooperative Study of Pulmonary Function. I. Clinical spirometry in normal men. Am. J. Med., 30:243, 1961.

Mead, J.: Mechanical properties of lungs. Physiol. Rev., 41:281, 1961.

Mead, J., Turner, J. M., Macklem, P. T., and Little, J. B.: Significance of the relationship between lung recoil and maximum expiratory flow. J. Appl. Physiol., 22:95, 1967.

Otis, A. B., McKerrow, C. B., Bartlett, R. A., Mead, J., McIlroy, M. B., Selverstone, N. J., and Radford, E. P. Jr.: Mechanical factors in distribution of pulmonary ventilation. J. Appl. Physiol., 8:427, 1956.

Distribution of ventilation and perfusion in the lungs, and ventilation/perfusion relationships:

Anthonisen, N. R., and Milic-Emili, J.: Distribution of pulmonary perfusion in erect man. J. Appl. Physiol., 21:760, 1966.

Ball, W. C., Stewart, P. B., Newsham, L. G. S., and Bates, D. V.: Regional pulmonary function studied with xenon[133]. J. Clin. Invest., *41*:519, 1962.

Cumming, G.: Gas mixing efficiency in the human lung. Resp. Physiol., *2*:213, 1967.

West, J. B.: Ventilation/Blood Flow and Gas Exchange. Oxford, Blackwell Scientific Publications, 1965.

Some physiological consequences of airway obstruction:

Cherniack, R. M., and Snidal, D. P.: The effect of obstruction to breathing on the ventilatory response to CO_2. J. Clin. Invest., *35*:1286, 1956.

Cohen, J. J., and Schwartz, W. B.: Evaluation of acid-base equilibrium in pulmonary insufficiency. Am. J. Med., *41*:163, 1966.

Levison, H., and Cherniack, R. M.: Ventilatory cost of exercise in chronic obstructive pulmonary disease. J. Appl. Physiol., *25*:21, 1968.

Woolcock, A. J., Vincent, N. J., and Macklem, P. T.: Frequency dependence of compliance as a test for obstruction in the small airways. J. Clin. Invest., *48*:1097, 1969.

Spectrum of diseases associated with altered respiratory function:

American Thoracic Society: Definitions and classification of noninfectious reactions of the lung. A statement of the Committee on Diagnostic Standards in Respiratory Diseases. Am. Rev. Resp. Dis., *93*:965, 1966.

Scadding, J. G.: Patterns of respiratory insufficiency. Lancet., *1*:701, 1966.

8

Oxygen Transport and Cellular Respiration

Eugene D. Robin, M.D., *and Lawrence M. Simon*, M.D.

INTRODUCTION

The lung is only one unit of a complex functional and structural arrangement that provides molecular oxygen for critical oxygen-consuming reactions occurring in cells. This chapter will describe the steps involved in oxygen transport from external environment to intracellular sites of utilization; the nature of cellular metabolic processes that require oxygen; the methods available for monitoring abnormalities of oxygen transport and metabolism (Table 8-1); the consequences of oxygen depletion; and the therapy for abnormalities of oxygen transport and metabolism.

Oxygen transport and metabolism may be classified conveniently into pulmonary oxygen uptake, cellular oxygen delivery, and cellular oxygen utilization.

PULMONARY OXYGEN UPTAKE

Pulmonary oxygen uptake has been described in Chapter 7. As outlined, abnormalities of pulmonary structure and function frequently result in abnormal gas exchange within the lung, leading to a reduction of arterial oxygen tensions (Pa_{O_2}). In practice, measurements of Pa_{O_2} provide a convenient and accurate estimate of the adequacy of pulmonary oxygen uptake. A normal value of Pa_{O_2} (90 ± 10 mm. Hg while breathing room air) establishes normal pulmonary oxygen uptake. In the normal subject Pa_{O_2} tends to decrease with advancing age (being perhaps 70 mm. Hg

TABLE 8-1. Measurement Available for Monitoring Oxygen Transport and Metabolism

PROCESS	MEASURED VARIABLE
Pulmonary Oxygen Uptake	Pa_{O_2} (arterial oxygen tension)
Cellular Oxygen Delivery	
Oxygen Supply	Arterial O_2 content Hb Concentration
	Sa_{O_2} (arterial oxygen saturation)
	Cardiac output
	$P\bar{v}_{O_2}$ (mixed venous oxygen tension)
State of binding	$P_{50}O_2$
Cellular Oxygen Utilization	Reflected cytoplasmic NAD^+/NADH (plasma lactate/pyruvate)
	Free cytoplasmic NAD^+/NADH (cytoplasmic lactate/pyruvate)
	Free mitochondrial NAD^+/NADH (mitochondrial β-hydroxybutyrate / acetoacetate)
	Reflected mitochondrial NAD^+/NADH (appropriate plasma redox pair)?
	Erythropoietin

at ages greater than 60) and is several mm. Hg lower during sleep than during the waking state.

The term *hypoxemia* refers to values of Pa_{O_2} that are abnormally low and is frequently associated with *hypoxia,* a term that describes abnormal tissue utilization

145

of oxygen. However, mild hypoxemia may be present without evidence of hypoxia, and hypoxia can be present in the absence of hypoxemia (because of abnormalities of cellular oxygen delivery or oxygen utilization). The precise Pa_{O_2} at which impairment of tissue oxygen utilization occurs is variable. It depends, among other factors, on the rate of development of hypoxemia. All things being equal, the more rapid the onset of hypoxemia, the more extensive the tissue abnormalities. In general, Pa_{O_2} values persistently less than 50 mm. Hg are associated with abnormal tissue oxygen metabolism. However, patients with chronic values as low as 30 mm. Hg have been reported, and such values are occasionally compatible with long-term survival.

CELLULAR OXYGEN DELIVERY

Oxygen delivery to the tissues is determined by oxygen supply, which, in turn, depends on both arterial oxygen content and cardiac output; state of oxygen binding to hemoglobin; diffusion of oxygen from capillary plasma to intracellular sites of utilization; relationship of regional perfusion to diffusion at the tissue level; and intracellular oxygen binding.

Oxygen Supply

Arterial blood is the "inspired" fluid of cells, just as ambient air is the inspired fluid of the lung. In this sense, the amount of oxygen supplied to the tissues depends upon the product of arterial oxygen content and cardiac output.*

Arterial oxygen content is determined by the value of Pa_{O_2} and the hemoglobin concentration. As previously noted, Pa_{O_2} depends on pulmonary gas exchange and, in turn, determines the amount of oxygen that will be taken up by hemoglobin.

---

* Pulmonary oxygen supply is equal to the volume of ventilation $\times$ concentration of oxygen in inspired air. Cellular oxygen supply is equal to volume of arterial blood flow (cardiac output) $\times$ concentration of oxygen in arterial blood.

Each gram of hemoglobin can combine with 1.34 ml. of oxygen. In contrast, the absolute quantity of oxygen represented by dissolved oxygen (P_{O_2}) is small because of the relatively poor solubility in plasma (0.003 ml. of oxygen per 100 ml. of plasma per mm. Hg at 37° C.). Thus, a P_{O_2} of 450 mm. Hg would be required to provide the same oxygen content in 100 ml. of arterial blood as would be achieved by a full saturation of hemoglobin at a concentration of 1 g./100 ml.! It is obvious that an adequate tissue supply of oxygen requires adequate hemoglobin concentration. Anemia (reduced red cell mass) is, therefore, an important cause of abnormal oxygen supply, and measurements of hemoglobin concentration are important in monitoring adequacy of oxygen supply.

Cardiac Output. An adequate cardiac output is required for normal oxygen delivery. Not only must total cardiac output be quantitatively sufficient, but the regional distribution of output must match the variable oxygen requirements of the different tissues. (Specific details of normal circulatory dynamics and the alterations of tissue oxygen delivery produced by disease are described in Chapter 2.)

Interruption of circulation interferes more drastically with tissue oxygen utilization than does cessation of ventilation. In the former instance the oxygen supply of any organ is limited to that contained in its residual blood volume, whereas, in the latter case, the oxygen supply of the entire circulating blood volume is available (see below).

With inadequate cardiac output, there are mechanisms that provide for redistribution of blood flow so that more critical regions (i.e., brain) receive a major portion of the total output at the expense of less critical areas (i.e., skin, gut, liver).

Quantitative estimates of the adequacy of cardiac output may be obtained by direct measurement; however, this is not entirely satisfactory, since the precise level of cardiac output appropriate for a given metabolic state is difficult to estimate. Moreover, the heterogeneous distribution of cardiac output to various organs makes it difficult to infer regional deficiencies in oxygen supply from measurements of total cardiac output. The adequacy of cardiac output in terms of oxygen supply may also be monitored by measurements of mixed venous oxygen tension ($P\dot{v}_{O_2}$). Normal resting $P\dot{v}_{O_2}$ is approximately 40 mm. Hg; with an inadequate cardiac output this value tends to fall as a result of increased tissue extraction. This measurement also fails to evaluate the heterogeneous distribution of blood flow and regional variations in tissue oxygen utilization. Moreover, patients inhaling high oxygen concentrations may have $P\dot{v}_{O_2}$ values somewhat higher than 40 mm. Hg, even with an inadequate cardiac output. In such patients a falling $P\dot{v}_{O_2}$, independent of its absolute value, may indicate circulatory insufficiency.

Oxygen Binding to Hemoglobin (Oxyhemoglobin Dissociation Curve)

Tissue oxygen supply (as well as pulmonary oxygen uptake) depends critically on the affinity of hemoglobin for oxygen. Affinity is described by the relationship between P_{O_2} and hemoglobin saturation and is quantitatively defined by the oxyhemoglobin dissociation curve (Fig. 8-1). The curve is sigmoid shaped and the physiological range of oxygen tensions encompasses values of approximately 10 to 100 mm. Hg. At the relatively high oxygen tensions pres-

ent in alveolar air, oxygen combines with hemoglobin, leading to a high oxygen saturation and a high oxygen content. At the relatively low oxygen tension of tissues, oxygen is released from hemoglobin; the unbound oxygen (quantitatively reflected by P_{O_2}) then diffuses from the red cell through the plasma, capillary wall and interstitial space to enter intracellular water. Shifts of the curve to the right result in a smaller quantity of oxygen bound to hemoglobin at any oxygen tension (decreased affinity). All things being equal, a shift to the right will impair pulmonary uptake of oxygen but facilitate unloading of oxygen to the tissues. It should be noted that the upper part of the curve (P_{O_2} greater than 60 mm. Hg) is relatively flat. Thus, with a normal end-pulmonary capillary P_{O_2}, a shift to the right produces only a moderate decrease in arterial oxygen saturation. However, patients with abnormal gas exchange resulting in low arterial P_{O_2} have values that fall on the steep part of the curve. A shift to the right in these patients produces a substantial additional decrease of oxygen uptake by hemoglobin in the lung.

Conversely, shifts of the curve to the left (increased affinity) will increase the amount of oxygen bound to hemoglobin at any oxygen tension. A shift to the left tends to increase oxygen uptake in the lung but impairs the release of oxygen to the tissues. In patients with a sharp reduction in Pa_{O_2} a shift to the left could produce considerable improvement in pulmonary

FIG. 8-1. Oxyhemoglobin dissociation curve: the effects of changes in pH and P_{CO_2} are indicated by the arrows. Note that the P_{50} (P_{O_2} at 50% saturation) equals 26 mm. Hg.

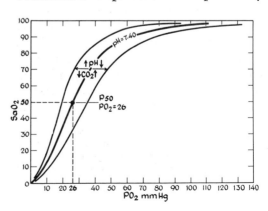

oxygen uptake. At a normal Pa_{O_2} the effect on pulmonary loading is of little importance; however, the effect on tissue unloading is considerable.

Affinity of hemoglobin for oxygen is regulated by a number of normal control mechanisms which may be modified by disease. In addition, there are a number of diseases unrelated to these normal regulatory processes which are characterized by alterations of oxyhemoglobin affinity. The physiological and pathophysiological factors modifying the oxyhemoglobin dissociation curve are summarized in Table 8-2.

Red Cell pH, P_{CO_2}, Temperature. Decreases in pH (acidosis) decrease hemoglobin affinity for oxygen and increases in pH (alkalosis) increase hemoglobin affinity. This effect of pH is known as the Bohr effect.* At the tissue level, at which plasma and red cell pH are relatively low, the Bohr effect serves to facilitate oxygen dissociation from hemoglobin. Conversely, in the lung, where plasma and red cell pH are relatively high, the combination of oxygen with reduced hemoglobin is facilitated. Normally, the time required for the operation of the Bohr effect is exceedingly brief and its contribution to pulmonary and tissue oxygen exchange is of decisive importance. As a result of acid-base disturbances, the Bohr effect may operate to impair oxygen delivery. Severe alkalosis will reduce the release of oxygen from hemoglobin and may compromise oxygen delivery to the tissues, whereas severe acidosis will decrease the uptake of oxygen by hemoglobin and may impair pulmonary oxygen uptake, particularly in patients with reduced Pa_{O_2}.

Carbon dioxide influences the affinity of hemoglobin for oxygen both by affecting red cell pH and by a direct effect as carbon dioxide, which combines with α-amino

* Christian Bohr (1835-1911) was a Danish physiologist who made the study of hemoglobin, blood gases and gas transport his life work. He was the first to describe the sigmoid shape of the O_2 dissociation curve and the shift of the curve to the right with increasing P_{CO_2}.

TABLE 8-2. Conditions Associated with Alterations of O_2-Hb Affinity

O_2-Hb Dissociation Curve	
Shift to Left	**Shift to Right**
Increased pH	Decreased pH
Decreased P_{CO_2}	Increased P_{CO_2}
Decreased temperature	Increased temperature
Decreased 2,3-DPG	Increased 2,3-DPG
1. decreased pH	1. increased pH
2. stored blood	2. hypoxemia
3. increased ADP	3. anemia
4. phosphate depletion	4. phosphate retention
5. red cell pyruvate kinase excess	5. red cell pyruvate kinase deficiency
6. red cell hexokinase deficiency	
7. chemical inhibition of glycolysis (e.g., mono-iodoacetate)	
Decreased 2,3-DPG binding to Hb (fetal hemoglobin)	
Abnormal Hemoglobins	Abnormal Hemoglobins
Hereditary	Hereditary
Hb Ranier	Hb Kansas
Hb Barts	Hb Seattle
Hb H	Hb S
Hb Yakima	
Hb J Capetown	
Hb Chesapeake	
Hb Kempsey	
Hb Hiroshima	
Hb Little Rock	
Acquired	
Carboxyhemoglobin	
Methemoglobin	

groups of hemoglobin to form carbamate. Hemoglobin with carbamino groups binds oxygen less avidly. As a result, increased P_{CO_2} leads to decreased affinity (shift to the right), and decreased P_{CO_2} to increased affinity.

Increased temperature decreases the affinity of hemoglobin for oxygen and decreased temperature has the opposite effect. In the basal state this effect is of little consequence; however, during severe exercise, metabolically active nonpulmonary tissue may be substantially warmer than the lung and, consequently, the unloading of oxygen to the tissues may be facilitated. In febrile disorders, especially those occurring in patients with a decreased Pa_{O_2}, pulmonary uptake of oxygen may become impaired because of this mechanism. Conversely in hypothermic states, unloading of oxygen from hemoglobin at the tissue level may be impaired.

Organic Phosphates. A variety of anions, especially organic phosphates, have profound effects on the affinity of hemoglobin for oxygen and are involved in the physiological regulation of oxygen transport as well as its pathophysiological aberrations. Dilute hemoglobin solutions containing no phosphates show marked affinity for oxygen. Addition of organic phosphates to the solution markedly reduces this affinity.

From the quantitative standpoint the major organic phosphate in human red cells is the compound 2,3-diphosphoglycerate (2,3-DPG). Although both inorganic phosphate and other organic phosphates such as ATP have similar effects, 2,3-DPG is the major anion involved in the regulation of oxygen transport in man. Increasing concentrations of 2,3-DPG leads to progressive decreases in affinity of hemoglobin for oxygen. This effect is mediated by the ability of 2,3-DPG to combine reversibly with reduced, but not oxygenated, hemoglobin. The combination of 2,3-DPG with reduced hemoglobin results in increased stability of the deoxyhemoglobin, so that a higher Po_2 is required to achieve a given degree of combination between oxygen and hemoglobin. Thus, increased 2,3-DPG concentrations lead to a rightward shift of the oxyhemoglobin dissociation curve and decreased values to a leftward shift.

Red cell organic phosphates, including 2,3-DPG, are generated metabolically by anaerobic glycolysis, the major energy-providing metabolic pathway in mature

mammalian erythrocytes. 2,3-Diphosphoglycerate is found in a uniquely high concentration only in the red cells of most mammals, its basal concentration in man amounting to 5 mM per liter of red cells. The formation of 2,3-DPG represents a deviation from the classical glycolytic pathway (see below). The glycolytic intermediate 1,3-diphosphoglycerate is converted to 2,3-DPG by the action of the enzyme 2,3-diphosphoglycerate mutase with 3-phosphoglyceric acid serving as a cofactor. It is degraded to 3-phosphoglyceric acid by the enzyme 2,3-diphosphoglycerate phosphatase, which splits the phosphate from the 2- position, leaving 3-phosphoglycerate, an intermediary on the main pathway of glycolysis. The concentration of 2,3-DPG in the red cell reflects the balance between its rate of formation and its rate of breakdown. The precise mechanisms that influence rate of 2,3-DPG synthesis versus rate of breakdown are not known, but factors that influence the overall rate of glycolysis in the erythrocyte affect its concentration.

Glycolytic rate in the red cell is increased by alkalosis and decreased by acidosis. These changes are at least partially related to the increased activity of phosphofructokinase (a rate-limiting step in glycolysis) that results from increased pH. Changes in glycolytic rate induced by pH changes lead to corresponding changes in 2,3-DPG concentration.

Concentrations of ADP, ATP and inorganic phosphate likewise play important roles in the regulation of glycolysis and thus influence the metabolism of 2,3-DPG. For example, low levels of ADP lead to the conversion of 1,3-DPG to 2,3-DPG rather than to the phosphorylation of ADP to ATP (see Fig. 8-5).

Hypoxemia produces alterations of diphosphoglyceromutase activity as a result of an inhibition of the enzyme by free 2,3-DPG. With hypoxemia there is increased binding of 2,3-DPG, which leads to increased diphosphoglyceromutase activity that results in augmented 2,3-DPG synthesis.

An interesting mechanism involving 2,3-DPG has been observed in placental oxygen exchange between mother and fetus. There are comparable concentrations of 2,3-DPG in the red cells of mother and fetus. However, reduced fetal hemoglobin binds less avidly to 2,3-DPG than does adult hemoglobin. As a result, the unloading of oxygen from maternal blood is facilitated and its association with fetal blood increased.

These fundamental mechanisms operate in a number of different clinical states which are characterized by changes in red cell 2,3-DPG concentrations and corresponding shifts in the oxyhemoglobin dissociation curve. Metabolic alkalosis increases and metabolic acidosis decreases red cell 2,3-DPG concentrations. In turn, this produces shifts in the oxyhemoglobin dissociation curve which are opposite to those produced by the direct effect of pH. Stored red cells are acidotic and have a decreased glycolytic rate. Levels of 2,3-DPG are reduced and there is increased affinity of hemoglobin for oxygen. Therefore, patients transfused with large amounts of stored blood may have impairment of tissue oxygen supply because of the leftward shift of the oxyhemoglobin dissociation curve.

Genetic abnormalities involving red cell glycolytic enzymes may produce changes in 2,3-DPG concentration and the oxyhemoglobin dissociation curve. Erythrocyte pyruvate kinase deficiency leads to increased 2,3-DPG concentrations. These patients may show sharp shifts of the dissociation curve to the right. Patients have been described with increased red cell pyruvate kinase activity who have decreased 2,3-DPG concentrations. Some are polycythemic, and this increase in red cell mass presumably reflects decreased tissue oxygen availability resulting from a leftward shift of the dissociation curve. Decreased 2,3-DPG

concentrations have also been reported in patients with deficient red cell hexokinase activity, and they too have a leftward shift of the dissociation curve.

As serum inorganic phosphate levels change, there are corresponding changes in red cell concentrations of 2,3-DPG. Phosphate retention, as in renal failure, is associated with increased 2,3-DPG concentrations and phosphate depletion is associated with reduced concentrations. Acute and chronic hypoxemia are associated with increased 2,3-DPG concentrations (presumably because of increased diphosphoglyceromutase activity) and a shift of the dissociation curve to the right. This has been demonstrated most clearly in individuals exposed to high altitudes and it is suggested that the rightward shift is an adaptive mechanism improving tissue oxygen delivery. Some forms of anemia (a type of hypoxia) are associated with shifts to the right related to increased 2,3-DPG concentrations.

Acquired Abnormal Hemoglobins. High concentrations of carboxyhemoglobin (resulting from the inhalation of carbon monoxide) and high concentrations of methemoglobin (resulting from increased oxidation of the ferrous moiety of hemoglobin to the ferric form) produce increased oxyhemoglobin affinity, with a leftward shift of the dissociation curve. Consequently, these patients have two abnormalities of oxygen delivery. The fraction of hemoglobin in the abnormal form (carboxy or methemoglobin) does not combine reversibly with oxygen, and thus the amount of hemoglobin available for oxygen transport is reduced. Like anemia, this results in a reduction in effective red cell mass. In addition, the leftward shift of the remaining normal hemoglobin results in deficient unloading of oxygen to the tissues.

Heritable Abnormalities of Hemoglobin. A number of heritable disorders of hemoglobin structure may be associated with either increased or decreased oxygen affinity. Such changes may occur as a result of single amino acid substitutions in one or more of the hemoglobin chains, or replacement of an entire chain by an abnormal chain; also, the chains making up the hemoglobin may all be of one type or there may be structural modifications of the heme group (Chapter 23). These hemoglobinopathies not only are of great intrinsic interest but serve to demonstrate the potential pathological effects of shifts in the dissociation curve.

Shifts to the Right. Heterozygotes with hemoglobin (HB) Seattle have a modest shift of the oxyhemoglobin dissociation curve to the right. At the normal arterial Po_2 (approximately 90 mm. Hg) hemoglobin saturation averages approximately 85 to 90 percent. At the usual mixed venous blood Po_2 (40 mm. Hg) saturation is only 45 percent. This leads to a wide difference between arterial and venous oxygen content at normal oxygen tensions. Red cell mass is reduced in some of these patients, and it has been suggested that this results from decreased erythropoietin activity related to supranormal oxygen delivery to the tissues.

Heterozygotes with hemoglobin Kansas show a marked decrease in oxygen affinity. At normal arterial Po_2 values, saturation is approximately 60 percent and at normal mixed venous Po_2 hemoglobin is only 30 percent saturated. These individuals show gross cyanosis but appear to be normal in other respects. In particular, red cell mass is not increased, suggesting no tissue hypoxia.

The hemoglobin associated with sickle cell disease (Hb S) shows a moderate decrease in oxygen affinity. However, this abnormality appears not to play an important role in the manifestations of sickle cell disease.

The practical importance of *decreased hemoglobin affinity for oxygen* can be summarized as follows: With normal arterial Po_2, these patients appear to have no striking abnormality of oxygen utilization. Pulmonary and cardiac function and the general clinical status are normal. The development of lung disease with reduced Pa_{O_2} may result in a further substantial decrease in pulmonary oxygen uptake, leading to severe hypoxemia. There is a possibility that increased tissue oxygen delivery results in decreased erythropoietin activity and may account for the reduced red cell mass seen in some of these patients. Screening of patients with hypoxemia and cyanosis* should include studies of hemoglobin affinity for oxygen. The possibility of a rightward shift is particularly pertinent in patients who are asymptomatic, with normal pulmonary and cardiac function and rather than being polycythemic have a normal or reduced red cell mass.

Shifts to the Left. Hemoglobins Rainier, Yakima, J Capetown, Chesapeake, Kempsey, Hiroshima, San Francisco and Little Rock are associated with increases in oxygen affinity. These shifts to the left are of sufficient degree in the heterozygote to interfere with oxygen delivery to the tissues, as judged by the development of polycythemia and as judged by increased erythropoietin activity in some of the above types.

Hemoglobin H and hemoglobin Barts have profound shifts to the left; in the former the shift is so extreme that the dissociation curve resembles that of myoglobin (see below), and, therefore, this hemoglobin is useless in oxygen transport. Both of these hemoglobins are found in patients with α-thalassemia and hemoglobin Barts is associated with death of the affected fetus.

In summary, *increased hemoglobin affinity for oxygen* may produce abnormal tissue oxygen utilization in spite of a normal Pa_{O_2}, normal arterial oxygen saturation and, fre-

---

* Cyanosis may also be caused by an excess concentration of methemoglobin and sulfhemoglobin. In addition to the acquired forms previously described, there are several congenital types of methemoglobin (Hemoglobin M Boston, Iwate, Milwaukee, Saskatoon, and Hyde Park).

quently, increased erythropoietin activity and increased hemoglobin content.

MOVEMENT OF OXYGEN FROM PLASMA TO INTRACELLULAR SITES OF UTILIZATION

Oxygen moves from capillary plasma to intracellular sites of oxygen utilization by the process of diffusion. The amount of oxygen that diffuses per unit time depends on: driving pressure; surface area available for diffusion; diffusion distance; character of the membranes across which diffusion occurs; and the time available for diffusion.

Driving Pressure

This refers to the difference between the Po_2 of capillary plasma and the Po_2 at various intracellular sites where oxygen is utilized. This pressure differential is generated by the various oxygen-dependent metabolic processes. Thus, the greater the metabolic rate, the lower the Po_2 at a given site and the greater the driving pressure. This arrangement provides for the highest availability of oxygen at the intracellular sites where oxygen need is maximal. For example, during muscular exercise, mitochondrial oxygen utilization increases, the capillary-mitochondrial driving pressure is increased, and oxygen delivery is augmented. Since there are variable rates of oxygen consumption at different intracellular sites, there may be regional variations of Po_2 within the cell. Measurements of a single intracellular Po_2, even if technically feasible, cannot be interpreted simply. Although exact driving pressure along the entire length of capillaries is difficult to estimate, a maximal figure can be derived. The Po_2 at the arterial end of the capillary is 90 mm. Hg. Estimates of mitochondrial Po_2 suggest values of less than 1 mm. Hg. Thus, normal maximal driving pressure has been estimated to be in excess of 89 mm. Hg.

Decreases in driving pressure can occur as a result of reduction in Pa_{O_2} or intracellu-

lar oxygen utilization. The latter has been demonstrated in dramatic fashion in experimental cyanide poisoning. Cyanide inhibits mitochondrial oxygen utilization so that intracellular Po_2 and Pa_{O_2} become equal, the driving pressure is zero and transfer of oxygen ceases.

In considering cellular oxygen uptake, it is important to distinguish between oxygen content and partial pressure. The major fraction of the oxygen content of blood is combined with hemoglobin. Therefore, content may be considered as a reservoir serving the function of providing an adequate overall supply, whereas the partial pressure serves as a driving pressure to permit adequate diffusion of oxygen into the cell. Increasing of Po_2 values above 100 mm. Hg provides only a small increase in blood oxygen content; however, marked increases in Po_2 proportionately increase the driving pressure for diffusion of oxygen into the cell. Under normal circumstances, a driving pressure of 90 mm. Hg is adequate for ensuring an adequate oxygen supply, and there is little advantage to increasing Po_2 further. In special circumstances, such as interstitial or intracellular edema, a normal driving pressure might be inadequate, and there could be a theoretical advantage to increasing arterial Po_2 to supranormal values, thereby increasing driving pressure.

Surface Area

A major determinant of the surface area available for oxygen transport is the number of capillaries per unit of tissue mass. Under normal circumstances, there may be an excess number of capillaries in most tissues. This excess provides an important reserve, and it is probable that not all capillaries are perfused during the resting state. With increased flow, additional (unused) capillaries may be recruited, and the surface area available for diffusion of oxygen is thereby increased. During chronic hypoxia and with exercise training, there is suggestive evidence that new capillaries may form in tissues such as skeletal muscle. This increased capillarity serves the function of maintaining oxygen diffusion in the face of a reduced driving pressure or increased oxygen requirement. In some diseases, the ratio of capillaries to unit tissue mass may be decreased. For example, hypertrophy of myocardial fibers is not generally accompanied by a corresponding increase in the number of capillaries. As a result, the hypertrophied fiber may be relatively ischemic in spite of the fact that the absolute number of capillaries has not decreased.

Diffusion Distance and Pathway

The distances and pathways involved in the movement of oxygen from capillary to mitochondria have not been quantitated precisely. Estimates depend on assumptions in regard to the geometry of the transport pathway. Qualitatively, it appears reasonable that the development of interstitial edema, interstitial fibrosis, intracellular edema, or changes in the physical character of cell membranes may increase the length of the diffusion pathway or decrease the rate of diffusion and thus interfere with oxygen delivery. Theoretically, inhalation of oxygen-rich mixtures to produce supranormal Pa_{O_2} values and increase the driving pressure may be helpful in these conditions.

Physicochemical Properties of the Membranes

The membranes that offer resistance to oxygen flow include capillary endothelium, with its basement membrane; the plasma membranes of the various cell types throughout the body, with their basement membranes; the double mitochondrial membrane; and the limiting membranes of other subcellular structures in which oxygen-consuming reactions take place. Although no precise characterization of the diffusibility of oxygen through these membranes is available, it is reasonable to assume that structural alterations by disease could serve to impair oxygen diffusion.

Time Available for Diffusion

Although the rate constants of tissue oxygen diffusion have not been determined, a marked decrease in the capillary transit time might be expected to decrease capillary to tissue transport.

MIXED VENOUS-METABOLIZING CELL OXYGEN GRADIENT (V-c OXYGEN DIFFERENCE)

In the lung, the difference between alveolar Po_2 and arterial Po_2 (A-a oxygen difference) is an important variable. Analysis of the factors that operate to produce the A-a oxygen difference (diffusion limitation, physiological shunt, anatomical shunt) serves to clarify the mechanism of pulmonary gas exchange (see Chapter 7).

There is likewise a mixed venous-cellular Po_2 difference. Normal mixed venous Po_2 is approximately 40 mm. Hg; mean cellular oxygen tensions are difficult to measure and interpret, but this value for most cell types is 20 mm. Hg or less. Thus, from the standpoint of whole body tissue gas exchange, a V-c Po_2 difference of 20 mm. Hg or greater is present. As with the A-a oxygen difference, the V-c difference arises from diffusion limitation, physiological shunting and/or anatomical shunting. Oxygen exchange in the lung is compared to oxygen exchange in the tissues in Figure 8-2.

Diffusion Limitation

This would operate to produce an oxygen gradient between systemic end capillary blood and the mean cellular Po_2. A useful integrated approach to diffusion is to consider that the time available for diffusion is equivalent to the time required for a unit volume of blood to traverse the capillary (transit time). Diffusion limitation, in this sense, could be defined as a period of time insufficient for equilibration to occur.

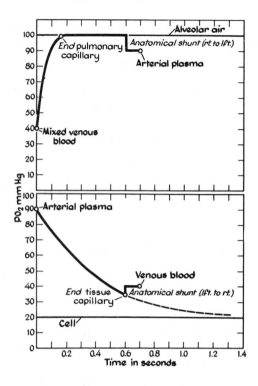

FIG. 8-2. A hypothetical comparison of pulmonary vs. tissue oxygen exchange: note diffusion equilibrium in the lung and lack of diffusion equilibrium in the tissues. Anatomical shunting produces a decrease in arterial plasma Po_2 but an increase in venous plasma Po_2.

The time spent in the capillary increases with decreasing cardiac output and is affected by both proximal (arteriolar) and distal (venular) vascular resistance. It is well known that decreases in cardiac output (increasing the time available for diffusion) result in a decreased mixed venous Po_2. This presumably means that the V-c gradient is decreased and strongly suggests that diffusion limitation is an important factor in the generation of the V-c oxygen difference.

Physiological Shunting

Oxygen exchange in the tissues involves liquid-to-liquid interfaces. In other forms of liquid-to-liquid gas exchange (gill, placenta) the dynamics of flow operate so that a portion of the flow does not participate in gas exchange. This may be considered as resulting from unevenness of flow distribution with respect to the gas exchange surface. It is likely that this is also true of capillary flow. This process, then, results in some degree of physiological shunting, with resultant inequality between end-capillary and cellular oxygen tensions.

Anatomical Shunting

In some organs (e.g., skin) there are arteriovenous shunts which bypass capillaries entirely. Flow through such shunts does not participate in cellular gas exchange. In some forms of shock there also may be capillary to venous shunting, with failure to venulize the blood which is shunted. Such anatomical shunts would be expected to contribute to the inequality between mixed venous and cellular oxygen tensions.

The same general mechanisms that operate in the lung to produce A-a oxygen differences operate in tissues to produce V-c oxygen differences. Little quantitative information is available; but it appears that diffusion limitation is more important in the tissues than the lung. A major difference between the two areas is that diffusion limitation, physiological shunting and anatomical shunting produce higher Po_2 values and oxygen contents in the efferent blood of the tissues rather than the lower values seen in the lung (Fig. 8-2).

TISSUE OXYGEN STORES (MYOGLOBIN)

Myoglobin, an intracellular heme protein, is present in various tissues, including skeletal and cardiac muscle. Like hemoglobin, it is capable of a reversible combination with oxygen, but, unlike hemoglobin, it contains only one heme prosthetic group. As a result, the oxymyoglobin dissociation curve is hyperbolic rather than sigmoid (Fig. 8-3). Myoglobin is fully saturated with oxygen at P_{O_2} values greater than 5 mm. Hg. Although its precise physiological role is unknown, it has been suggested that it serves as an emergency storage form of oxygen. At relatively high cellular P_{O_2} (resting), oxygen is bound and represents a potential reserve. When cell P_{O_2} drops below 5 mm. Hg (during severe muscular contraction), oxygen is released from myoglobin. It is of some interest that, under conditions of chronic hypoxia (and in some animals capable of withstanding severe oxygen depletion), myoglobin concentrations may be increased.

OXYGEN UTILIZATION

The utilization of molecular oxygen by cells subserves two general biochemical functions: Oxygen is the terminal electron acceptor in substrate oxidation, a process by which energy is made available for various endergonic processes, and oxygen is an obligatory oxidant in a wide variety of biosynthetic processes. It is also involved in a number of oxidative processes that do not provide energy directly.

Oxygen and Energy Provision

In most forms of life the major source of energy is provided by the oxidation of substrate by molecular oxygen, with energy being made available in the form of ATP.

Mitochondrial Structure. The reactions involved in this pathway are localized in the mitochondrion, which may be regarded as a biological energy transducer. This pathway accounts for well over 75 percent of total cellular oxygen consumption.

Mitochondrial structure closely subserves mitochondrial function. The chemical reactions involved in substrate oxidation are facilitated by the spatial arrangements. This cell organelle has as its limiting structure two continuous membranes; the outer one is smooth, and the inner membrane is thrown into a series of folds called cristae. Within the inner membrane is a semiliquid substance called the matrix. The enzymes of the Krebs cycle are located in the matrix, whereas the enzymes directly involved in respiration are located on the inner membrane in the same sequence in which the chemical reactions occur. Abnormalities of mitochondrial structure have been reported

FIG. 8-3. Myoglobin dissociation curve: note the hyperbolic shape and the essentially complete saturation at oxygen tensions greater than 10 mm. Hg.

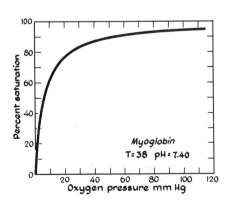

in a rare variety of skeletal muscle disease (giant mitochondria) and in alveolar epithelial cells following exposure to high oxygen concentrations and various noxious gases.

The diffusion pathway of oxygen from cytoplasm to the active site of oxygen utilization (cytochrome oxidase) includes cytoplasmic water, the outer and inner membranes of the mitochondrion, and the active site on cytochrome oxidase. With intracellular edema or structural abnormalities of the mitochondrial membranes, intracellular oxygen diffusion might be impaired.

Chemical Reactions in the Mitochondria. Two general processes are involved in the reactions culminating in mitochondrial oxygen utilization. By means of the Krebs tricarboxylic acid (TCA) cycle (Fig. 8-4),

the three classes of foodstuff (proteins, fats and carbohydrates) are oxidized to carbon dioxide and water. In this cycle, one molecule of acetic acid, in a special activated form, acetyl CoA, is oxidized in 8 steps to carbon dioxide, with the net provision of 4 pairs of electrons. The TCA cycle may be regarded as a common pathway for the ultimate use of all types of foodstuffs as a source of energy. These transformations are not involved primarily in energy provision but may be considered as providing electrons which are then used in energy transformations.

In each "revolution" of the Krebs cycle there are four steps in which electrons are provided. In three of these oxidation steps, the molecule nicotinamide adenosine dinucleotide (NAD^+) serves as the electron

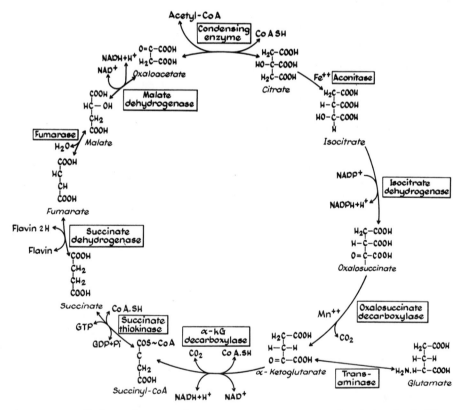

Fig. 8-4. Krebs cycle: note that no net energy conservation is involved. The Krebs cycle is linked to the electron transport chain via the $NAD^+/NADH$ steps and the flavin/flavin 2H steps.

acceptor. Nicotinamide adenosine dinucleo-tide, which is found in all cells, serves as an electron carrier in a wide variety of reactions. When oxidized, NAD^+ accepts electrons and is converted to the reduced form, NADH. In the fourth oxidation step a pair of electrons is removed from succinic acid, with flavine adenine dinucleotide (FAD) serving as the electron acceptor.

The direct provision of energy in a utilizable form (ATP) occurs in the electron transport chain. Electrons enter the chain from the Krebs cycle at a fairly high energy level which decreases progressively as electrons flow down the chain. The energy released is conserved in the form of ATP. The electron transport chain consists of a series of cytochromes, heme-containing enzymes which can participate reversibly in oxidation-reduction reactions. The iron atom of a given cytochrome molecule alternates between the oxidized ferric state (Fe^{+++}) and the reduced ferrous state (Fe^{++}). In the oxidized Fe^{+++} state electrons are accepted, with reduction to the reduced Fe^{++} state. The reduced form, in turn, donates electrons to the next cytochrome, and ultimately the final cytochrome (cytochrome oxidase) donates electrons to molecular oxygen. The sequence of reactions from substrate to molecular oxygen may be summarized as follows:

oxide and water, amounts to approximately 686,000 calories. This results in the generation of 38 moles of ATP from ADP and inorganic phosphate. The theoretical energy yield from 1 mole of ATP is approximately 7,000 cals and thus the minimal energy provided by the oxidation of 1 mole of glucose to carbon dioxide and water is equal to $7,000 \times 38$ cals per mole or approximately 266,000 calories per mole. The overall reaction may be summarized as follows:

$$1 \text{ Glucose} + 38 \, \substack{\text{Inorganic} \\ \text{phosphate}} + 38 \text{ ADP} + 6 \text{ O}_2 \rightarrow$$

$$6 \text{ CO}_2 + 44 \text{ H}_2\text{O} + 38 \text{ ATP}.$$

Metabolic Control. The various intracellular reactions involved in provision of energy must operate in a regulated and integrated fashion. This is necessary so that widespread fluctuations in activity can be met by appropriate changes in energy availability. For example, total oxygen consumption may vary from approximately 250 ml. per minute at rest to 3,500 ml. per minute during maximal exercise. Moreover, changes in oxygen consumption vary widely from one organ to another. A marked increase in sodium reabsorption by renal tubular cells may require increased renal oxygen consumption, although other organs may have no change in oxygen requirements. The mechanisms by which energy

Krebs Cycle	Respiratory Chain

Substrate→ NADH→ FAD→ FAD→ Cytochrome b→ Cytochrome c→ Cytochrome oxidase (a and a_3)→ molecular O_2
derived
electrons

The flow of electrons is associated with a progressive decrease in energy. The energy level involved in the successive oxidation-reduction steps may be quantitated as a series of voltages expressed as oxidation-reduction potentials. In overall terms, the total energy change involved in the passage of electrons from one mole of glucose to oxygen, with the production of carbon di-

availability is geared to energy requirement are collectively termed metabolic control.

The processes involved in metabolic control are complex and, in many cases, poorly understood. It is likely that changes in metabolic rate can occur either as a result of an actual change in the concentration of a given enzyme or as a result of the alteration of activity of the enzyme independent of a change in concentration. Modification of some of these mechanisms may occur in response to hypoxia as adaptive changes or in relation to disease; however, specific changes are poorly defined at present.

Regulatory mechanisms for mitochondrial oxygen utilization may involve either short-term or long-term control factors.

Short-Term Regulation. Short-term control factors are concerned with the second-to-second regulation of energy output. In general, these mechanisms involve changing concentrations of low molecular weight molecules such as ADP, ATP and inorganic phosphate. In the case of mitochondrial oxygen utilization, emphasis has been placed on the ratio of ADP to ATP. During times of rapid energy expenditure (e.g., during muscular exercise), ATP is utilized rapidly and this ratio increases, resulting in increased mitochondrial oxygen consumption. During more basal activities, ATP utilization is low and the ADP/ATP ratio is decreased. As a result, mitochondrial oxygen uptake is decreased. Absolute levels of ADP and ATP depend not on the rate of utilization solely, but also on the rate of formation and rate of breakdown. A large number of factors other than energy utilization may influence the rate of formation or rate of breakdown.

Long-term Regulation. Although short-term control mechanisms are important, there are a number of situations in which the rate of oxygen consumption does not appear to depend on differences in the concentration of low molecular weight species. With exercise training, skeletal muscle acquires an increased mitochondrial oxygen uptake capacity. This is associated with a general increase in mitochondrial protein and specific increases in the various cytochromes, including cytochrome oxidase. Under conditions of severe hypoxemia, there may be reduction of mitochondrial mass and reduction in cytochrome oxidase activity. Therefore, the biogenesis of mitochondria and the biosynthesis of key electron transport chain enzymes may be geared to variable requirements for oxygen consumption. Recent work suggests that the biosynthesis of some of the respiratory enzymes and mitochondrial biogenesis are regulated by a distinct type of DNA present in mitochondria (mitochondrial DNA). Mitochondrial DNA is physically and chemically distinct from classical nuclear DNA and is regulated by different controls. It is possible that oxygen availability or oxygen utilization may play an important role in modifying the rate of synthesis of electron transport chain enzymes by affecting nuclear and, particularly, mitochondrial DNA metabolism.

Disorders Affecting Mitochondrial Oxygen Utilization. Abnormalities of pulmonary oxygen uptake and oxygen supply are the commonest type of disorder affecting mitochondrial oxygen utilization. In addition, various components of the enzymatic pathways can be inhibited by one or another chemical species, some of which are of clinical importance. In appropriate concentrations, cyanide ion inhibits cytochrome oxidase, which prevents passage of electrons to molecular oxygen, thereby stopping mitochondrial oxygen utilization. This inhibition is the basis of the extreme toxicity of cyanide (and explains its popularity as an agent for suicide, homicide, and execution). Carbon monoxide also inhibits cytochrome oxidase, but the concentrations required are so high that patients may well die of the effects of this gas on hemoglobin before the effects on cytochrome oxidase become important. A chronic form of carbon monoxide intoxication characterized by nonspecific central nervous system abnormalities may be related to low level exposure, with resultant partial inhibition of cytochrome oxidase.

British antilewisite (BAL), a therapeutic agent used in heavy metal intoxication, inhibits cytochrome b, although it is not certain that this effect is of clinical importance. Barbiturates inhibit the transfer of electrons from NADH to cytochrome b and inhibit flavoprotein. This may be the mode of action of this group of drugs and could explain the toxic effects seen in barbiturate overdosage. Chloramphenicol is a relatively specific inhibitor of mitochondrial protein synthesis and, in pharmacologic doses, inhibits the biosynthesis of mitochondrial enzymes (i.e., cytochrome oxidase). This inhibition may be the basis of the aplastic anemia which develops in some patients treated with this drug.

Thyroid hormone plays a critical role in the regulation of cellular oxygen consumption. The precise mechanism involved is not clear but the hormone is probably involved in the coupling of oxygen consumption to the generation of ATP. With excessive amounts of hormone (hyperthyroidism), partial uncoupling may occur and excess energy may be dissipated as heat rather than being available in a useful form. With inadequate concentrations of thyroid hormone (hypothyroidism), oxygen consumption becomes inappropriately low, leading to an inadequate energy supply.

Luft's syndrome is a rare disorder in which oxygen consumption is markedly increased (hypermetabolism) without evidence of thyroid abnormality. The site of the hypermetabolism is in skeletal muscle. Profuse sweating and extreme heat intolerance are important clinical features. There is excess heat production; cardiac output is high, and there is hyperventilation. The disorder is characterized by a large overgrowth of bizarre mitochondria of skeletal muscles, which micrographically show mitochondrial hyperplasia and hypertrophy with abnormal mitochondrial cristae. Biochemical examination of these mitochondria indicates some uncoupling of oxidation from phosphorylation. Chloramphenicol, which depresses mitochondrial biogenesis, produced a partial remission in one patient.

Pellagra is a disorder characterized by dementia, dermatitis and gastrointestinal symptoms. It results from inadequate availability of the vitamin nicotinamide, usually on the basis of dietary deficiency. Nicotinamide is an important constituent of NAD. It is reasonable to conclude, therefore, that pellagra results from deficient NAD supplies, leading to abnormalities in mitochondrial electron transfer.

In addition to these more or less well defined disorders, it should be expected that other specific disorders involving the Krebs cycle and the electron transport chain will be uncovered. Given the large number of different enzymes and complex control factors, the probability of heritable disorders or environmental alterations is high.

ANAEROBIC GLYCOLYSIS

When energy supplied by mitochondrial oxygen utilization is inadequate, supplemental energy is provided by glycolysis. During glycolysis, glucose is oxidized in a series of 11 reactions to pyruvate and lactate. This pathway has several general functions: It provides a supplemental source of energy under normal circumstances when the delivery of oxygen is insufficient to meet metabolic needs and during pathologic hypoxic states; in some cell types (e.g., the mature mammalian erythrocyte) it represents the sole source of cellular energy; and it serves as an important source of pyruvate for the Krebs cycle and ultimate oxidation of substrate to carbon dioxide and water. (The last mentioned function is normally subserved by aerobic glycolysis—that is, glycolysis occurring in cells, with no limitation of oxygen supply or abnormality of oxygen utilization. Under these conditions the rate of glycolysis is slow and pyruvate is generated at a rate equal to that at which it enters the Krebs cycle as acetyl coA.) One of the products of glycolysis, lactic acid, subserves important metabolic functions in some cell types in the absence of oxygen.

Structural Basis

The reactions involved in glycolysis are largely localized in the cytoplasm. The glycolytic enzymes are distributed more or less randomly in the cytoplasmic brei, and, in contrast to provision of oxygen dependent energy, no precise anatomical arrangement is required for the reactions to proceed. Exposure of substrate to the mixture of enzymes is adequate to produce the sequence of chemical reactions.

Chemical Reactions

The sequential steps of glycolysis (shown in Fig. 8-5) involve the conversion of glucose to pyruvate. Pyruvate may be converted to lactate by the enzyme(s) lactate dehydrogenase. The overall reaction may be summarized as follows (see also Chapter 17):

glucose + 2 ATP

+ 2 phosphate + 2 ADP →

2 lactate + 2 ADP + 4 ATP

In terms of energy supply, the oxidation of 1 mole of glucose requires 2 moles of ATP but results in the production of 4

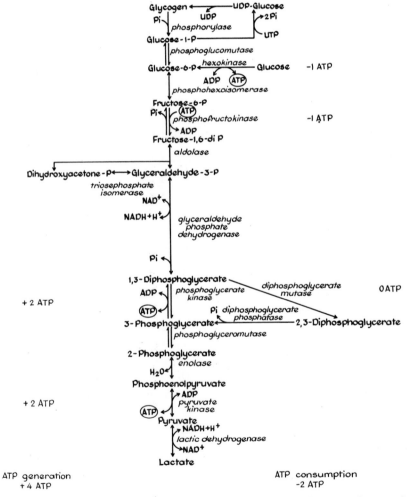

FIG. 8-5. Anaerobic glycolysis including the 2,3-DPG bypass: the usual pathway results in the net generation of 2 ATPs whereas the 2,3-DPG pathway results in the net generation of 0 ATPs.

moles of ATP. Thus, there is a net gain of 2 moles of ATP (or approximately 14,000 calories).*

Metabolic Control

Metabolic control of glycolysis is exceedingly complex and the potential mechanisms for regulation of mitochondrial oxygen utilization reviewed above are pertinent to anaerobic glycolysis. It is useful to consider both short-term and long-term control factors.

Short-Term Regulation. The various enzymes in the glycolytic sequence can be stimulated or inhibited by the various products of the reactions. Hexokinase activity is inhibited by its product glucose-6-phosphate. Phosphofructokinase activity is stimulated by ADP and inorganic phosphate and is inhibited by ATP. The activity of this enzyme is unusually sensitive to H^+ concentrations, increasing with alkalosis and decreasing with acidosis. The activity of glyceraldehyde phosphate dehydrogenase depends on NAD^+ and inorganic PO_4 concentrations. Concentrations of ADP decisively influence the activity of phosphoglycerate kinase and pyruvate kinase. Lactate dehydrogenase is strongly influenced by cytoplasmic NADH concentrations. In overall terms, the short-term regulation of glycolytic rate depends heavily on intracellular ATP concentrations. This, in turn, depends on the balance between ADP synthesis and ADP removal.

Long-Term Regulation. A number of the glycolytic enzymes, including hexokinase, phosphofructokinase and pyruvate kinase, are rate-limiting under one condition or another. Thus, the mechanisms that regu-

---

* It may be noted that two of the ATPs are generated by the conversion of 1,3-diphosphoglycerate to 3-phosphoglycerate. If 1,3-DPG were completely converted to 2,3-DPG, these two ATPs would not be formed. The oxidation of 1 mole of glucose to pyruvate would produce 2 moles of ATP, and, as 2 moles of ATP are utilized, no net energy would be provided. It has therefore been suggested that the 2,3-DPG bypass serves as an energy clutch providing for dissipation of excess energy in the erythrocyte.

late the concentrations of these enzymes must play a decisive role in overall metabolic control. For example, there is a precise relationship between the maximal glycolytic rate in various organs and the pyruvate kinase activity of these tissues. This suggests that the biosynthesis of this enzyme must be closely geared to the requirements of the individual organ for glycolysis. Furthermore, in the chronically exercised heart, there is an increase in pyruvate kinase activity, which presumably increases myocardial glycolytic capacity. These changes in enzyme activity and glycolytic capacity are independent of the short-term regulating factors described above. Cytoplasmic enzyme biosynthesis depends on the nuclear DNA–microsomal RNA system. It appears that long-term variations in the rate of glycolysis depend on modifications of this system

Interrelationship Between Mitochondrial Oxygen Utilization and Anaerobic Glycolysis (The Pasteur Effect)

When cells are exposed to anaerobic environment, there is a brisk increase in the rate of glycolysis. This is known as the Pasteur effect. It may be evoked not only by anaerobiosis but also by inhibition of mitochondrial oxygen utilization by agents such as cyanide. The mechanism of the Pasteur effect probably depends on the availability of ADP in the cytoplasm versus the mitochondrion. Under conditions of normal rate of oxidative phosphorylation, there is greater mitochondrial than cytoplasmic affinity for ADP. Thus, intracytoplasmic ADP concentrations are low and the glycolytic rate is minimal. With decreased mitochondrial oxygen consumption, less ADP is used in the mitochondrion, cytoplasmic concentrations rise and anaerobic glycolysis is progressively accelerated. The precise quantitative relationship between decreases in mitochondrial oxygen utilization and increased glycolysis is not known. As a result, the precise threshold of the Pasteur effect is unclear and it is possible that substantial abnormalities of oxygen utilization may occur before a measurable increase in glycolysis is present.

MONITORING ABNORMALITIES OF MITOCHONDRIAL OXYGEN UTILIZATION

The ability to monitor abnormalities of mitochondrial oxygen utilization in patients would be of great importance. No entirely satisfactory approach is available. Variables such as PaO_2, Pv_{O_2}, etc. (see Table 8-1) reflect processes which are proximal to the energy-providing reactions. With only moderate abnormalities of these proximal processes, or in the face of adaptive changes, these measurements provide a poor quantitative estimate of the degree of abnormality of mitochondrial oxygen utilization.

Lactate

One approach to the monitoring of abnormalities of oxygen utilization has been based on measurements of lactate concentrations in blood or plasma. The chemical basis for this approach is the Pasteur effect. However, lactate concentration in plasma depends not only on its rate of intracellular generation but also on other factors: the rate at which lactate is delivered from cells to blood, which, in turn, is dependent on the permeability of cell membranes to lactate and on tissue perfusion; and the rate at which lactate is cleared from the blood, which is dependent on the rate at which lactate is metabolized (chiefly by the liver) and the rate at which it is excreted by the kidney. Moreover, lactate may be generated by processes other than the gly-

colytic pathway. When hepatic blood flow and liver function are normal, the capacity of the liver to metabolize lactate is extensive. As a result, in pathological conditions, increases in blood lactate are usually not found unless there is circulatory insufficiency (decreased liver blood flow) or severe functional impairment of liver function, or both. Thus, measurements of blood lactate concentration are most useful as an indicator of abnormal oxygen utilization in conditions associated with inadequate circulation such as shock and severe forward heart failure.

$NAD^+/NADH$ (Lactate/Pyruvate)

The intracellular oxidation-reduction potential (redox state) is a fundamental property of biological systems, influencing the chemical behavior of all oxidizable or reducible components of the system. The importance of $NAD^+–NADH$ as an electron carrier in energy conservation reactions has already been described. Because of this key position, attempts have been made to quantitate the ratio of NAD^+ to NADH as a reflection of the redox state of individual organs. Direct chemical measurements of $NAD^+/NADH$ are not appropriate, since these molecules exist in bound form intracellularly and it is only the free or unbound form that would reflect the state of intracellular oxidation-reduction potential. In addition, measurements in blood could not be expected to reflect events within the intracellular compartment.

To circumvent these difficulties, measurement of the ratio of various substrates that participate in $NAD^+–NADH$ linked reactions has been proposed. The theoretical basis of this approach is as follows:

$$\text{pyruvate} + \text{NADH} \leftrightharpoons \text{lactate} + NAD^+ + H^+$$

by the mass action equation:

$$\frac{(\text{Lactate})\,(NAD^+)\,(H^+)}{(\text{Pyruvate})\,(\text{NADH})} = K \text{ equilibrium constant of the reaction}$$

$$\text{or:} \quad \frac{(NAD^+)}{(\text{NADH})} = \frac{(\text{Pyruvate})\,(H^+)}{(\text{Lactate})\,(K' \text{ equil.})}$$

Since pyruvate and lactate concentrations are relatively easy to measure, and the value of the equilibrium constant is known, the use of lactate/pyruvate ratio measurements has been proposed in various ways. A semiquantitative approach for measurement does not detect early changes.

Mitochondrial NAD^+/NADH ratios have been calculated in alveolar macrophages by measuring the substrates involved in a reaction localized in the mitochondrial matrix:

$$\text{Acetoacetate} + \text{NADH} \leftrightarrows \beta\text{-Hydroxybutyrate} + NAD^+ + H^+,$$

so that

$$\text{mitochondrial } \frac{NAD^+}{NADH} = \frac{(\text{Acetoacetate})\ (H^+)}{(\beta\text{-Hydroxybutyrate}) K_{eq}}$$

ments in blood based on so-called "excess lactate" has been studied extensively. Excess lactate is defined as the difference in lactate concentration from that which would be expected from a simple rise in pyruvate concentration with no change in NAD^+/NADH. This approach has considerable theoretical limitations. Empirically, the determination of so-called excess lactate does not appear to be more helpful in assessing abnormalities of tissue oxygen utilization than do simple measurements of blood lactate concentrations.

Measurements of lactate/pyruvate ratios have been used in the calculation of NAD^+/NADH in liver cells and in alveolar macrophages. As discussed above, the conversion of pyruvate to lactate is largely cytoplasmic in location. These tissue studies have established the existence of separate cytoplasmic and mitochondrial NAD^+ and NADH pools. In the case of the macrophage, it has been shown that abnormalities of mitochondrial oxygen utilization may occur before there are measurable changes in cytoplasmic NAD^+/NADH (lactate/pyruvate), and the mitochondrial changes persist longer than cytoplasmic changes. The absolute value of lactate/pyruvate ratios in arterial plasma does not appear to differ from that found in liver or macrophage extracts. This suggests that measurements of lactate/pyruvate ratios in plasma may be useful in detecting far-advanced changes in mitochondrial oxygen utilization; unfortunately, however, this

This measurement, made on tissue samples, appears to enable detection of early mitochondrial hypoxia. Similar determinations in arterial plasma may prove useful in monitoring early abnormalities of mitochondrial oxygen utilization.

Erythropoietin

As described in Chapter 23, tissue hypoxia is frequently associated with increased levels of the humoral agent erythropoietin, which stimulates hemoglobin synthesis. Apparently, there is a fairly rapid increase in circulating erythropoietin with abnormalities of mitochondrial oxygen utilization. Assays of erythropoietin activity in blood or urine have been used to assess the existence of abnormal tissue O_2 exchange, notably in patients with abnormal hemoglobins.

OXYGEN AND NONENERGY PROVIDING PROCESSES

Molecular oxygen is involved in a series of intracellular chemical reactions that are not directly involved in energy provision. Some of these reactions involve the biosynthesis of a number of biologically important molecules, including pigments, steroids and fatty acids. Oxygen is also involved in the oxidative transformation of a variety of different compounds, including steroids, drugs, polycyclic hydrocarbons, insecticides and fatty acids, as well as the degradation of heme to biliverdin, with the release of carbon monoxide. These diverse functions require a special heme-containing enzyme, cytochrome P-450, which is generally associated with the endoplasmic reticulum of tissues such as liver, kidney and intestine. Its concentrations in these organs may be much greater than the concentration of respiratory chain cytochromes in mitochondria.

Oxygen is involved in the intracellular generation of hydrogen peroxide in certain special cells such as macrophages. The liberated peroxide may play an important role in the intracellular killing of various microorganisms (Chap. 32).

The precise intracellular localization of many of these reactions is not known, nor have the exact kinetics of these types of oxygen-consuming reactions been established. In general, it may be assumed that hypoxia results in altered mechanisms of biosynthesis, detoxification and defense, by modification of this group of important chemical reactions. No methods are currently available for the monitoring of abnormalities of these effects of oxygen deficiency.

CHANGES IN ORGAN FUNCTION WITH HYPOXIA

Abnormalities of oxygen-dependent reactions are manifest clinically as abnormal organ function. It is useful to describe the effects of hypoxia on several key organs.

Central Nervous System

The brain is critically dependent on an adequate oxygen supply. Brain oxygen consumption averages approximately 3 ml. of oxygen per 100 g. per minute. In an adult with a brain weighing 1,500 g., this amounts to 45 ml. per minute (or approximately 20% of basal oxygen consumption). Profound oxygen depletion rapidly produces abnormal cerebral function; in fact, an abnormal electroencephalogram may be seen within seconds after oxygen depriva-

tion. The effects of profound cerebral oxygen depletion may be illustrated by considering cardiac arrest. With cardiac arrest, the cerebral oxygen supply is limited to the oxygen contained in the residual capillary volume of blood in the brain. The total blood volume of the brain is approximately 75 ml.; assuming that capillary volume is 30 ml. and oxygen content is 20 ml. per 100 ml., this would amount to only 6 ml. of oxygen, an amount that would be exhausted in 10 to 15 seconds. It is not surprising, therefore, that consciousness is lost seconds after circulatory arrest. Cessation of ventilation, with maintenance of circulation, provides a somewhat longer period before irreversible parenchymal changes occur, because, theoretically, total body blood oxygen stores are available. An important factor in the development of nervous system abnormalities produced by oxygen depletion is the functional change in brain capillaries, which become more permeable to water and solutes, with consequent cerebral edema. Chronic hypoxia of moderate degree may produce impairment of judgment, psychological abnormalities, and increased neuromuscular irritability.

Heart

The basal cardiac oxygen consumption is approximately 10 ml. per 100 g. per minute. With a heart of average weight of 350 g., total myocardial oxygen consumption is about 35 ml. per minute or approximately 15 percent of total oxygen consumption. Approximately two thirds of myocardial oxygen consumption is involved in contractility, the remaining one third subserving noncontractile energy consuming processes. For example, conducting tissue has an unusually high oxygen requirement, and this may account for the increased irritability of the hypoxic heart. This oxygen is made available by an unusually complete extraction of oxygen from arterial blood, as indicated by extremely low coronary venous Po_2 and oxygen content. Even under basal circumstances, these values are lower than in any other vascular bed. Since extraction is virtually complete, increased oxygen needs of the heart can be met only by increased flow. In contrast to most systemic capillary beds, diffusion is not limiting and there is little physiological or anatomical shunting of coronary blood flow.

Chronic myocardial hypoxia may produce severe structural alterations, including dilatation, hypertrophy and fibrosis, which, in turn, may further limit oxygen supply.

Acute decreases in oxygen supply produce a decrease in myocardial ATP. There is a substantial increase in glycolytic rate through the Pasteur effect, as indicated by substantial increases in glucose utilization.

Pulmonary Vascular Bed

Hypoxia produces vasoconstriction of small pulmonary precapillary vessels, resulting in increased pulmonary vascular resistance. This effect is evoked not only by decreases in alveolar oxygen tension but by agents such as dinitrophenol, which interfere with cellular respiration. This suggests that the basis of hypoxic vasoconstriction is an interference with mitochondrial oxygen utilization in the smooth muscle cells of affected vessels. Hypoxic vasoconstriction is of particular importance in chronic hypoxia due to lung disease as well as in normal high altitude dwellers.

Kidney

Renal oxygen consumption averages 6 ml. per 100 g. per minute. With a renal weight of approximately 300 grams, renal oxygen consumption amounts to approximately 18 ml. per minute, or 8 percent of basal oxygen consumption. Renal blood flow is high and oxygen extraction low, so that renal venous blood has unusually high Po_2 and oxygen content. A major fraction of the energy provided by renal oxygen consumption is used in the active transport of sodium, and there is a close correlation between renal oxygen consumption and net

tubular reabsorption. There is little specific knowledge in regard to the effects of oxygen depletion on renal function; however, acute renal ischemia produces the structural and functional abnormalities collectively termed acute renal failure.

Liver

The liver possesses a dual blood supply which is important in determining the effects of hypoxia. Much of the hepatic parenchyma derives its blood supply from the portal vein, with its relatively low Po_2. In addition, the anatomical arrangement is such that peripheral cells in the hepatic lobule receive blood before cells in the center of the lobule. As a result, centrilobular cells receive a marginal oxygen supply even under normal conditions. Therefore, under pathological conditions, particularly circulatory insufficiency, these cells are quite vulnerable to hypoxia, and necrosis and centrilobular fibrosis are common consequences of decreased hepatic oxygen supply.

Skeletal Muscle

Skeletal muscle possesses a reasonably high capacity for oxygen utilization. However, vigorous muscular activity generally requires a quick source of additional energy because the rate at which oxygen can be supplied under these circumstances is limited. It is not surprising that the glycolytic capacity of skeletal muscle is higher than that of other tissues. There are a number of muscle diseases in which abnormalities of glycolysis lead to muscle dysfunction during exercise. The model for these diseases is the contracting frog muscle treated with the metabolic inhibitor monoiodoacetate (MIA). This compound inhibits glycolysis at the 3-glyceraldehydephosphate step. Exercised frog muscle treated with MIA develops a prolonged state of contracture. The clinical equivalents of this experimental model include muscle phosphorylase deficiency (McArdle's syndrome) and muscle phosphofructase kinase deficiency. Patients with these disorders develop painful, prolonged contractures of exercised muscle because of inability to obtain adequate energy ATP from glycolysis during periods of inadequate oxygen supply. In similar fashion, rigor mortis has been attributed to ATP deficiency following muscle death and loss of metabolic activity.

THERAPY OF HYPOXIA

The aim of therapy of hypoxia is to improve tissue oxygen utilization. The potential benefit of oxygen-enriched mixtures is limited to those situations in which the arterial Po_2 is decreased. With normal Pa_{O_2}, normal hemoglobin becomes fully saturated, and any further increase in oxygen content depends upon physically dissolved oxygen. This amounts to a small increment under normal atmospheric conditions and does not appreciably augment tissue oxygen supply. Abnormalities in the tissue oxygen diffusion pathway may be one exception to this, since increased driving pressure resulting from supranormal Pa_{O_2}'s may be of some benefit. Excess oxygen administration often not only is of negligible benefit but also may produce significant pathological abnormalities. Prolonged high oxygen exposure produces alterations of pulmonary structure and function. Changes in red cell membranes and other cell membranes may occur as a result of peroxidation of important lipid constituents of these membranes. Furthermore, hyperbaric oxygen interferes with normal carbon dioxide exchange and may produce severe central nervous system abnormalities on a metabolic basis.

An awareness of the multiple determinants of tissue oxygen supply is important in the treatment of hypoxia. Normal Pa_{O_2}'s indicate normal pulmonary oxygen uptake but give little information as to the adequacy of cellular oxygen delivery. Determination of arterial oxygen content as well as circulatory dynamics must be carried out. The augmentation of cellular oxygen delivery by increasing the arterial oxygen content depends on an adequate cardiac output, and therapy intended to achieve a decrease in tissue hypoxia must be based on consideration of both variables. For example, although continuous positive pressure ventilation—a recent advance in the treatment of acute respiratory failure—may be beneficial in that it increases arterial oxygen content, it may be detrimental when it decreases cardiac output.

New techniques designed to monitor mitochondrial oxygen utilization directly should prove helpful in determining the adequacy of therapy at the cellular level.

ANNOTATED REFERENCES

Chemistry of Energy Metabolism

Mahler, H. R., and Cordes, E. H.: Biological Chemistry. New York, Harper and Row, 1966.

Lehninger, A. L.: Bioenergetics: The Molecular Basis of Biological Energy Transformations. New York, W. A. Benjamin, 1965.

Oxygen Metabolism and Cellular Function

Jobsis, F. F.: Basic Processes in Cellular Respiration. *In:* Fenn, W. O., and Rahn, H., (eds.): Handbook of Physiology. Section 3. Respiration. Vol. 1, p. 63, Baltimore, Waverly Press, 1964.

Oxygen Transport in Blood

Bromberg, P. A.: Editorial: Cellular cyanosis and the shifting sigmoid: the blood oxygen dissociation curve. Am. J. Med. Sci., *260:* 1, 1970.

Conley, C. L., and Charache, S.. Inherited hemoglobinopathies. Hosp. Prac., *4:*35, 1969.

Stamatoyannopoulos, G., Bellingham, A. J., Lenfant, C., and Finch, C. A.: Abnormal hemoglobins with high and low oxygen affinity. Ann. Rev. Med., *22:*221, 1971.

Oxidation-Reduction in the Cytoplasm and Mitochondria of Alveolar Macrophages

Mintz, S., and Robin, E. D.: Redox state of free nicotinamide-adenine nucleotides in the cytoplasm and mitochondria of alveolar macrophages. J. Clin. Invest., *50:*1181, 1971.

9

Metabolic Events in the Lung

Sami I. Said, M.D.

INTRODUCTION

The fundamental importance of the lung in providing oxygen and eliminating carbon dioxide is well known, and the effects of failure of this function are widely appreciated. The lung appears, however, to have another critical role. As the site of numerous and important metabolic processes, the lung can regulate and modify the functions of many other organs. Some of these metabolic activities are essential to the normal performance of pulmonary gas exchange. Impairment of pulmonary metabolic activities could, therefore, have far-reaching repercussions on many organ systems. Several examples of systemic disturbances related to lung disease are already recognized, and more will likely become apparent as we probe more deeply into this new area of pulmonary physiology and pathophysiology.

CELLULAR SITES OF METABOLISM

Along with our increasing awareness of the importance of lung metabolism, and perhaps contributing to this awareness, has been a rapid growth in our knowledge of pulmonary structure and ultrastructure. Wider use of electron microscopy and application of histochemical and radioautographic techniques have been of great use in elucidating the fine structure of the lung and in relating structure to function.

If a piece of lung is removed from a living animal and quick-frozen, sectioned and incubated in appropriate medium, the presence of oxidative enzymes can be demonstrated by a special indicator (tetrazolium) which forms a colored salt (formazan) with the completion of the enzymatic reaction (Fig. 9-1). The color not only localizes the reaction but also provides an index of its intensity. By the use of different specific substrates and coenzymes, several specific oxidative enzymes have now been shown to occur in the lung (Fig. 9-2). Figure 9-3 shows the enzymes of oxidative and biosynthetic pathways to be concentrated in bronchial epithelium and in certain alveolar cells, including the large alveolar cell and the alveolar macrophage.

It is appropriate to review the main features of the various types of alveolar and bronchiolar cells and their probable or possible metabolic functions.

Large (or Great) Alveolar Cell (granular, type II pneumonocyte, corner cell, Fig. 9-4). This cuboidal cell has a cytoplasm that is rich in mitochondria, a granular endoplasmic reticulum, an extensive Golgi apparatus and multivesicular inclusion bodies, and is filled with lipids and phospholipids. The nucleus is large and there are numerous microvilli directed toward the alveolar surface. This cell is active in the biosynthesis of lipids and phospholipids and probably secretes the surface-active agent lining the alveoli.

Small (Flat, Squamous) Alveolar Epithelial Cell (type I). This cell has an attenuated

167

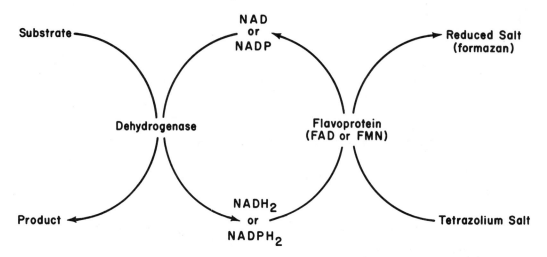

FIG. 9-1. Principle of tetrazolium reaction used for the histochemical demonstration of oxidative enzymes.

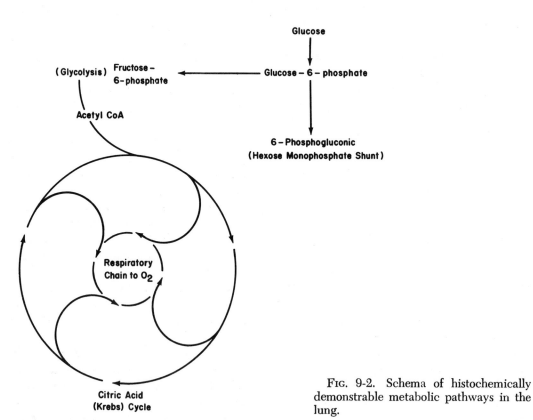

FIG. 9-2. Schema of histochemically demonstrable metabolic pathways in the lung.

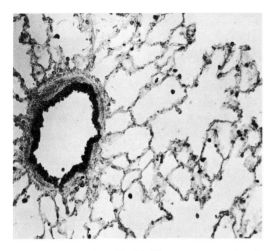

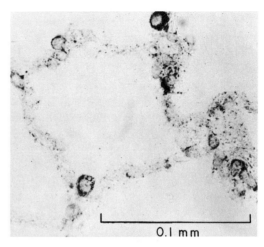

FIG. 9-3. *(Left)* Photomicrograph of quick-frozen lung section, showing reaction for NADP-diaphorase. Strongly reacting cells appear darker than others. (Dog lung, unstained, × 500) *(Right)* Higher magnification of similar section, showing reaction for NAD-diaphorase.

FIG. 9-4. Profile of a great alveolar cell which lies in a pocket of the alveolus (A) and adjoins the flat epithelial cells. In addition to the nucleus (N), the great alveolar cell contains lamellar inclusion bodies (arrows) and mitochondria (M). (Rat lung, × 15,000) (Leary, W. P., and Smith, U.: Life Sciences. Pergamon Press. In press.) (Courtesy Dr. Una Smith. With permission.)

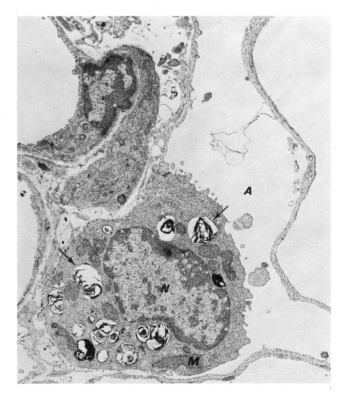

cytoplasm that is relatively deficient in organelles; it probably possesses little metabolic capacity.

Alveolar Brush Cell (type III pneumonocyte). Also cuboidal or columnar, this cell possesses large, square microvilli, numerous vacuoles and vesicles, and is rich in glycogen. The function of this newly described cell is uncertain; a similar cell occurs in the trachea.

Alveolar Macrophage (Fig. 9-5). This cell is morphologically similar to the great alveolar cell but is distinguishable by its location and, cytochemically, by being richer in hydrolytic enzymes (e.g., acid phosphatase) and poorer in certain oxidative enzymes (e.g., glucose-6-phosphate dehydrogenase) and lipids. The alveolar macrophage is unique among phagocytic cells in its strong dependence on oxidative phosphorylation.

Mast Cell. A component of mesenchymal tissue throughout the body, this cell is plentiful in the lung (Fig. 9-6), particularly around smaller blood vessels but also in alveolar and bronchial walls. Mast cells

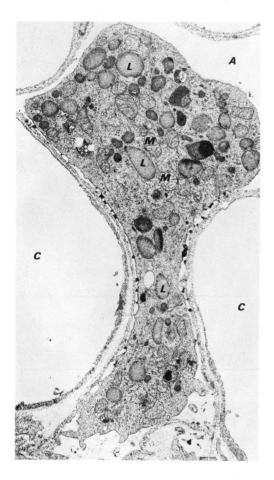

Fig. 9-5. Detail of an alveolar macrophage lying within the alveolar space (A). The wall between the capillary lumen (C) and the alveolus is, as usual, composed of endothelium, basement membrane and epithelium. The macrophage contains a variety of cellular components, including abundant lysosomes (L) and mitochondria (M). (Rat lung, × 21,000) (Leary, W. P., and Smith, U.: Life Sciences. Pergamon Press. In press.) (Courtesy Dr. Una Smith. With permission.)

Fig. 9-6. Quick-frozen section of lung, stained with toluidine blue to show distribution of mast cells. Cells are identified by their basophilic, metachromatic granules, which appear in figure as darker spots. (Dog lung, × 220)

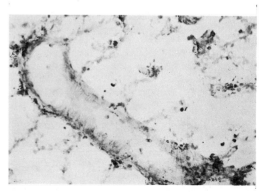

are packed with basophilic, electron-dense granules that are rich in histamine, heparin, and other biologically active substances.

Endothelial Cell (Fig. 9-7). Although not commonly thought of in such terms, this cell might well play an important role in metabolism, especially in the uptake of certain vasoactive materials.

Clara Cell (Fig. 9-8). Occurring among ciliated cells in terminal bronchioles, this cell is heavily endowed with large, rounded mitochondria and granular endoplasmic reticulum and is thus a good candidate for

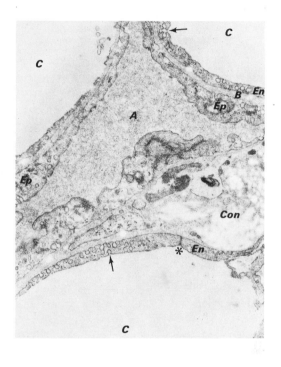

Fig. 9-7. Micrograph showing the components of the interalveolar septum, which is composed mainly of blood capillaries (C). The luminal surfaces of the capillaries are lined with a single layer of endothelial cells (En) which characteristically show large numbers of pinocytotic vesicles (arrows). An intercellular junction between endothelial cells is indicated by an asterisk. The alveolar spaces (A) (unusually small in this field) are bounded by epithelial cells (Ep); occasionally, great alveolar cells are interposed (compare with Fig. 9-4). Both the endothelial cells and the epithelial cells secrete their own basement membrane (basal lamina) but these may appear as a single fused layer (B). Connective tissue (Con), composed chiefly of collagen and elastin, is also present. In this figure, as in Figures 9-4 and 9-5, the lungs had been perfused, causing the capillaries to be empty. (Rat lung, × 43,000) (Smith, U., and Ryan, J. W.: Advances in Experimental Medicine and Biology, 8:249, 1970.) (Courtesy Dr. Una Smith. With permission.)

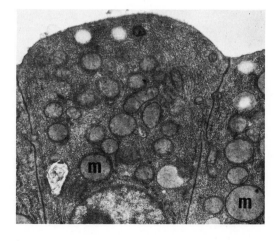

Fig. 9-8. Electron photomicrograph showing detail of nonciliated bronchiolar (Clara) cell. M, mitochondria. (Mouse lung) Sorokin, S. P.: The cells of the lung. *In:* Proceedings of the Conference on Morphology of Experimental Respiratory Carcinogenesis. AEC Symposium Series. U. S. Atomic Energy Commission, Oak Ridge, Tenn., 1969. Washington, D. C., U. S. Govt. Printing Office. In press.) (Courtesy Dr. Sergei P. Sorokin.)

some active synthetic or secretory role. Its precise function, however, is yet to be determined.

Argyrophil (argentaffin, enterochromaffin, Feyrter, Kultschitzky) Cells. Although little is known about the prevalence and distribution of these cells in the lung, they may be of considerable physiological and pathological interest. Probably derived from neural ectoderm, one member of this group of cells, which occurs in several organs, may be responsible for the secretion of polypeptide hormones. Another name for these cells, "APUD," is derived from certain cytochemical characteristics they have in common: a high level of *a*mines (catecholamines, serotonin), amine *p*recursor *u*ptake and amino acid *d*ecarboxylases (which form the amines from their precursor amino acids). The potential ability of these cells to secrete peptide hormones may express itself in the many endocrine syndromes which sometimes complicate certain tumors and other lesions of the lung.

Of the cells named above, the great and flat alveolar cells, the macrophage and the Clara cell are unique to the lung. The brush cell, mast, endothelial and APUD cells are known to occur in other organs. All bronchiolar and alveolar cells receive their blood supply from the pulmonary arterial system.

BIOSYNTHESIS OF LIPIDS AND PHOSPHOLIPIDS

Mammalian lung is active in the biosynthesis of lipids and phospholipids. One aspect of this activity which has been the subject of much investigation is the synthesis of the phospholipid dipalmitoyl phosphatidyl choline (dipalmitoyl lecithin). The particular biological importance of this compound stems from the fact that it is the chief component of the surface-active alveolar lining (surfactant).

The fatty acids needed for lecithin synthesis (Fig. 9-9) either are extracted from circulating blood or are themselves synthesized in the lung. From blood they may be derived directly from "free" fatty acids, or from the hydrolysis of circulating triglycerides by the action of lipoprotein lipase which is abundant in lung tissue. Pulmonary synthesis of fatty acids can be by one of two main pathways—de novo, or chain elongation. Both mechanisms require NADPH as an energy source, so that they are dependent on glucose metabolism via the hexose monophosphate shunt.

The incorporation of fatty acids into

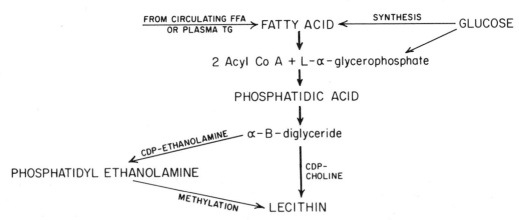

Fig. 9-9. Schema of lecithin synthesis in the lung. Modified from Harlan, W. R., and Said, S. I.: Selected aspects of lung metabolism. *In:* Bittar, E. E., and Bittar, N. (eds.): The Biological Basis of Medicine. Vol. 6, p. 357. New York, Academic Press, 1969.

lecithin can be direct—from the diglyceride by addition of cytidine diphosphate choline, by methylation of phosphatidyl ethanolamine, or by addition of a fatty acyl CoA to lysolecithin. The direct route appears to be the most important in the adult lung.

THE MAINTENANCE OF ALVEOLAR STABILITY

ALVEOLAR SURFACTANT

The alveoli of human and other mammalial lungs are lined with a thin layer of surface-active material (surfactant), which regulates surface tension at the air-liquid interface. Investigation into the composition, biosynthesis, cellular origin, secretion and metabolism of surfactant has been a major stimulus to the study of pulmonary metabolism as a whole.

The main component of alveolar surfactant is dipalmitoyl lecithin, probably complexed to a protein. Surfactant is probably synthesized and secreted by the great alveolar cell. The primary function of surfactant is to stabilize the alveoli by preventing excessive increase or unevenness in alveolar surface forces. Reduction of surface tension by surfactant reduces the amount the pressure required to fill the alveoli during inspiration and helps to maintain alveolar patency at a given pressure during expiration. It is also a factor in guarding against transudation of fluid into the alveoli.

The formation and maintenance of normal surfactant depends upon several factors, including the maturity of the great alveolar cells and biosynthetic enzyme systems, the adequacy of blood flow (normally from pulmonary arterial circulation) to the alveolar walls, a normal rate of turnover, and the absence of inhibitors. There are other influences that are still of uncertain physiological importance. Glucocorticoids can accelerate the differentiation of the great alveolar cells and secretion of surfactant in fetal animals. The possible effect of

innervation on function of alveolar cells is unknown.

Numerous clinical and experimental situations are associated with inadequacy of surfactant. This inadequacy could be due in large measure to insufficient formation (e.g., in prematurity), to inactivation, perhaps by constituents of serum and certain lipids (e.g., in pulmonary edema and alveolar proteinosis), or to both factors (e.g., after pulmonary arterial occlusion). Excessively rapid depletion and incomplete regeneration of surfactant may complicate breathing at large tidal volumes, as in extended respiratory therapy. The relative importance of surfactant deficiency in these and similar states is difficult to ascertain. Even when it may not be the sole or primary lesion, defective surfactant as a secondary phenomenon could seriously compromise lung function.

PROTEIN SYNTHESIS AND SECRETION

The lung is capable of rapid protein synthesis. Some of this protein is actively secreted in an energy-dependent fashion. Synthesis of glycoprotein by the lung has also been demonstrated. The full determinants of protein synthesis and secretion by the lung remain largely unexplored.

Among the numerous biologically important proteins in the lung are secretory antibodies, interferon, and various proteolytic and fibrinolytic enzymes and activators.

Collagen and elastin are two proteins responsible for the structural integrity and elasticity of the lung. Alterations in the chemical structure and internal arrangement of these molecules result in altered mechanical properties of the lung.

Proteolysis in Relation to Lung Disease. The possible role of proteolytic enzymes in causing pulmonary disease is receiving considerable clinical investigative attention. The first indication of a possible relation-

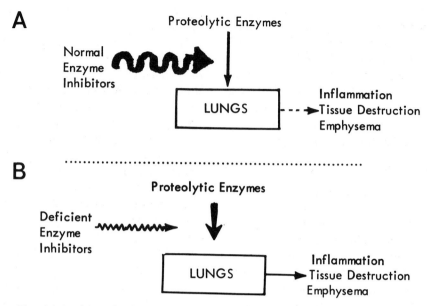

FIG. 9-10. (*A* and *B*): Schema of the possible role of proteolytic enzymes in the pathogenesis of lung disease (see text).

ship came with the realization that deficiency of a serum protease inhibitor, α_1-antitrypsin, predisposes to pulmonary emphysema. Then came the discovery that, in rats, intratracheal instillation of papain, a potent proteolytic enzyme, produces lesions resembling human emphysema. Actually, the papain animals often show evidence of pulmonary inflammation and hemorrhage before developing the structural changes of emphysema. The question naturally arose whether proteolytic enzymes could be formed or released within the lung, and if they could be responsible for lung damage similar to that which follows papain.

There are several possible sources of proteases in the lungs; leukocytes (from the bloodstream), alveolar macrophages, bacteria, and activation of pulmonary kallikrein (see later). Of these, the lysosomal enzymes of leukocytes and alveolar macrophages are possibly most important. The discharge of these enzymes is subject to the interplay of numerous factors which tend to decrease the stability of lysosomal membranes (e.g., some bacterial toxins, O_2

excess, ultra-violet and x-rays, and complement), and a few factors which enhance the stability of these membranes (e.g., cyclic AMP, glucorcorticoids and prostaglandin E_1). To relate these experimental observations to the clinical association of emphysema with α_1-antitrypsin deficiency, the hypothesis may be presented (Fig. 9-10, *A* and *B*) that the lung is normally protected from the potential damaging effects of proteolytic enzymes by the presence of adequate inhibitors; absence (or lack) of such inhibitors renders the lung vulnerable to acute and chronic tissue destruction.

CARBOHYDRATE METABOLISM

As noted previously, the oxidation of glucose provides energy (in the form of reduced NADP) and acetate, required for the synthesis of fatty acids and lecithin. The glycolytic path of sugar metabolism supplies L-α-glycerophosphate, a precursor for the synthesis of glycerol. Glucose also enhances protein synthesis in the lung. In the developing lung, glycogen is an important metabolite for alveolar epithelium;

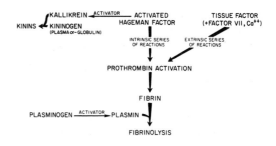

FIG. 9-11. Simplified diagram of clotting and fibrinolytic mechanisms and interactions with kallikrein-kinin system.

its importance decreases toward maturity.

Mucopolysaccharides are normal constituents of bronchial mucus. Under certain pathological influences, such as sulfur dioxide, the glycoprotein of normal mucus (sialomucin) is transformed into sulfomucin, which is more viscous and therefore more difficult to move up the bronchial tree. Patients with cystic fibrosis of the pancreas (mucoviscidosis) have abnormal glycoproteins in bronchial and other secretions. The accumulation of these secretions leads to obstruction and secondary complications including infection and respiratory failure.

Carbohydrates in the lung may have other important roles that are yet incompletely appreciated. For example, mucopolysaccharides are a component of an alveolar lining layer that is demonstrable histochemically, but their physiological significance in that location is uncertain; also, the significance of heparin in the pulmonary mast cells is not established.

THE LUNG AND HEMATOLOGICAL MECHANISMS

In the normal and abnormal formation of blood clots and in the dissolution of these clots several important steps are strongly influenced through interaction with factors in the lung.

Thus, the extrinsic path of activation of prothrombin, eventually leading to fibrin deposition, can be initiated by "tissue factors" (thromboplastin), present in higher concentration in the lung than in any other organ. Once a clot has formed, its lysis depends on the action of plasmin which must be generated by the activation of plasminogen. Again, the lung is unusually rich in plasminogen activator and, hence, in fibrinolytic activity (Fig. 9-11). Several other interrelated reactions are known to occur; for example, thrombin can accelerate some of the earlier reactions in the extrinsic pathway, leading to further formation of thrombin. The generation of thrombin leads not only to the conversion of fibrinogen to fibrin, but also to platelet aggregation.

The megakaryocytes, parent cells of the platelets which are all-important in the intrinsic sequence of clotting, are concentrated in the pulmonary vascular bed. Trapping and aggregation of platelets in the pulmonary circulation occur in experimental pulmonary embolism. In hemorrhagic shock there is also sequestration of leukocytes and sludging of blood. These changes could lead to release of vasoactive materials and proteolytic enzymes, which could contribute to pulmonary injury.

Heparin, a widely used anticoagulant, is a major constituent of mast cells which are prevalent throughout the lung. A functional significance for lung heparin, however, remains to be determined. When it is clear that a number of hematological mechanisms are closely related to pulmonary metabolism, it is possible to begin understanding the greater susceptibility of patients with a variety of pulmonary disorders to certain clotting and fibrinolytic anomalies. A higher incidence of thromboembolism in association with pulmonary malignancy; increased fibrinolytic activity following lung surgery; disseminated intravascular coagulation (consumption coagulopathy) complicating viral pneumonitis and other pulmonary lesions—all these exemplify the possible influence of lung factors.

RELATION TO THE KALLIKREIN-KININ SYSTEM

A series of chemical reactions culminating in liberation of the potent kinins can be triggered by the same principle that initiates the clotting sequence in a test tube—the Hageman factor. The reactions (Fig. 9-11) start with the activation of kallikrein, a proteolytic enzyme occurring in the pancreas, salivary glands, and other organs including the lung. Activated kallikrein then acts on a plasma α_2-globulin (kininogen), releasing the biologically active polypeptide bradykinin and other kinins which dilate systemic blood vessels, contract nonvascular smooth muscle organs, increase vascular permeability and induce pain. The kallikrein-kinin chain of reactions is potentially important in the mediation of inflammatory

responses and in other physiological and pathological situations. Some of these conditions include septicemia, shock and the syndrome of disseminated intravascular coagulation.

THE UPTAKE, ALTERATION, STORAGE AND RELEASE OF VASOACTIVE SUBSTANCES

The lung has the only capillary bed through which passes the entire blood flow, making the pulmonary capillary circulation uniquely suited for a controlling influence on circulating vasoactive hormones. These hormones have many and diverse actions on all smooth-muscle organs such as the blood vessels, the bronchi and alveolar ducts, the gastrointestinal system and the uterus. Certain vasoactive substances (e.g., prostaglandins, catecholamines) also affect various metabolic functions throughout the body, including the metabolism of lipids, carbohydrates and cyclic AMP.

Some vasoactive compounds are normal constituents of blood; others are usually

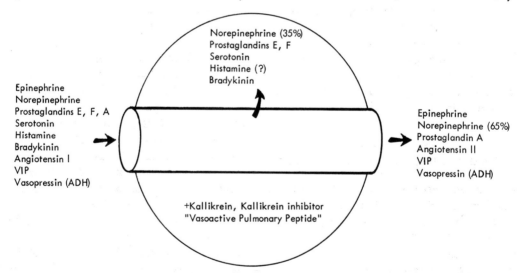

Fig. 9-12. Schema of handling of vasoactive substances by lung. Many compounds entering pulmonary capillaries are taken up, stored or modified, but some pass through the lung relatively unchanged. Kallikrein and vasoactive peptide also occur in lung tissue. The (?) after histamine indicates lack of quantitative data on extent of histamine uptake by lung. (Said, S. I.: Pulmonary responses to inhaled particles: Role of vasoactive substances. Arch. Int. Med., *126*:475, 1970.) (Modified with permission.)

present only in abnormal circumstances, on release from injured or inflamed tissue. In either case, pulmonary handling of the particular active agent can modulate its physiological or pharmacological effects. A summary of the metabolic alterations of some vasoactive agents by lung is given in Figure 9-12. The conversion of angiotensin I to angiotensin II is the only known example of biological activation by passage through the pulmonary circulation. Angiotensin II, up to 50 times more active than its precursor, is unaffected by passage through the lung. The angiotensin-converting activity of lung is many times greater than that of plasma. Many vasoactive materials are partially or completely inactivated by the lung. Among those which are almost completely ($> 80\%$) removed or inactivated are serotonin (5-hydroxytryptamine), bradykinin, adenosine-5^1-triphosphate (ATP), and prostaglandins E_1, E_2 and F_2. Norepinephrine and histamine are taken up to a smaller degree.

Vasoactive hormones that pass through the lung without significant loss or gain in activity include epinephrine, prostaglandins A_1 and A_2, some bradykininlike peptides such as eledoisin and polisteskinin, angiotensin II, vasopressin (ADH), and a newly isolated vasoactive intestinal peptide (VIP) which may circulate in the presence of hepatic failure. (The reader is referred to Chapter 22 for possible significance of VIP in the ascites associated with portal hypertension.)

Much of the information concerning this facet of lung metabolism has been gathered by an ingenious technique in which systemic arterial and mixed-venous blood are assayed simultaneously for their content of the active agent under study, after that agent has been injected intravenously or intraarterially. The assay is based on the response of selected smooth muscle preparations which are "superfused" by arterial or venous blood.

An interesting feature of the pulmonary metabolism of vasoactive hormones is its high selectivity. One member of a given group of substances (e.g., catecholamines, prostaglandins, kinins) may be removed in one passage, while another member of the same group is permitted to go through without change. In the case of compounds such as serotonin and norepinephrine, loss of activity in passage across the lung is mainly due to uptake and storage. In other cases the inactivation is by enzymic action, e.g., by a specific dehydrogenase acting on prostaglandins E and F, and by a peptidase breaking down bradykinin.

Although the cellular sites of these metabolic alterations are still undetermined, the endothelial cell, in intimate contact with blood, is a reasonable possibility. There is evidence that the pinocytotic vesicles of this cell, many of which communicate directly with the capillary lumen, take up ATP and other adenine nucleotides. Other cells, including the macrophage, the Kultschitzky and the mast cells, may participate in the metabolism of vasoactive hormones.

Vasoactive substances that are normally kept in storage within the lung may be discharged into the circulation under certain pathological influences. In anaphylaxis, for example, the lung releases histamine, "slow-reacting substances," bradykinin, prostaglandins and other pharmacologically active substances. Certain other pathological or experimental conditions, including bronchial asthma, pulmonary embolism, and mechanical distortion of the lung, lead to the release of potent chemicals which may then contribute to the pathogenesis of complications such as systemic hypotension and bronchial and alveolar-duct constriction (Table 9-1).

Other vasoactive agents which are known to occur in lung tissue, and could therefore be released or activated following tissue damage, include kallikrein, a newly extracted systemic vasodilator peptide, and other incompletely identified but active substances.

TABLE 9-1. Some Experimental and Clinical Conditions Associated with Release of Vasoactive Substances from the Lung

SUBSTANCE	CONDITION
Histamine	Anaphylaxis, extrinsic asthma, pulmonary embolism, byssinosis, (?) hypoxia, tumor
Prostaglandins	Anaphylaxis, pulmonary embolism, mechanical distortion
Serotonin	Pulmonary embolism, tumor
Kallikrein-kinins	Anaphylaxis, (?) inflammation
Other (incompletely identified)	Anaphylaxis

Aside from the acute release of vasoactive hormones, alterations in the pulmonary metabolism of these hormones may have important implications in human disease. In the carcinoid syndrome, the usual absence of left-sided cardiac lesions may be explained by the efficient pulmonary removal of the two main humoral mediators of the disease, serotonin and bradykinin. At present there are sufficient experimental data to permit the speculation that either inactivation failure or repetitive release of these potent substances could, indeed, be an important factor in the pathogenesis of some pulmonary diseases. Periodic release of histamine has been demonstrated in patients with byssinosis, and kinin formation has been invoked as a factor in the inflammatory response to silicosis.

DEFENSE AGAINST INFECTION

Phagocytosis. Because the pulmonary alveolar surface is constantly exposed to the external environment, it is constantly threatened by the onslaught of contaminating bacteria, dusts, chemical irritants and other particulate material. The lung, however, proves to be well prepared for defense against this challenge, and the alveolar macrophages are critically important in this defense.

As the only cells in the body which are normally present on the epithelial surface, the alveolar macrophages can be described as effective policemen of this extensive surface. Thanks to them in large measure, lung tissue is normally kept sterile. These specialized cells originate both from hematopoietic elements and from fixed alveolar cells, notably the great alveolar cell. Biochemically, the alveolar macrophages have extremely high metabolic activity, and they differ from other phagocytes (e.g., the monocyte) in their dependence on oxidative phosphorylation as a source of energy for phagocytosis. They are capable of protein and lipid synthesis, and are especially well endowed with hydrolytic enzymes which permit them to digest material they engulf. After completing their phagocytic function these cells are carried up in the mucociliary stream, to be coughed up or swallowed.

The phagocytic ability of alveolar macrophages may be depressed by ethanol, cigarette smoke, hypoxia, hyperoxia, starvation, corticocosteroids, nitrogen dioxide and ozone. Their antibacterial activity is decreased in the presence of metabolic or respiratory acidosis, azotemia and acute viral infections.

Mucociliary Action. The other major mechanism of clearance is the highly coordinated beating of the cilia, by which mucus is transported up the bronchial tree, ultimately to be eliminated. Energy for ciliary motion is derived mainly from ATP. As in the case of alveolar macrophages, ciliary function can be depressed by a number of agents including alcohol, sulfur dioxide, nitrogen dioxide and cigarette smoke. Mucus itself is said to have bactericidal activity, but abnormalities of mucus could slow down its clearance even though the cilia themselves may be normal. Mucus is abnormal if its viscosity is increased (by dehydration, e.g., from insufficient humidification or silica and other hydrophilic dusts, or by chemical alteration) or if it is produced excessively, as in chronic bronchitis.

The role of impaired alveolar and bronchial clearance in the production of pulmonary disease has not been established. Failure of pulmonary bactericidal mechanisms is probably a factor in the pathogenesis of some chronic inflammatory, granulomatous and fibrotic lesions, though its contribution is difficult to quantify. Alveolar macrophages that are overwhelmed by certain toxic agents might release their lysosomal enzymes which, in turn, could induce or aggravate inflammation and fibrosis. Finally, by allowing prolonged exposure to inhaled carcinogens, impairment of mucus transport (regardless of mechanism) may be a factor in the pathogenesis of bronchogenic carcinoma.

IMMUNOLOGICAL RESPONSES

Besides contributing the highly specialized alveolar macrophages which have major responsibility in the defense against microorganisms, the lung produces certain proteins with antibody and antimicrobial activity.

IgA Immunoglobulins. The existence of a separate system of antibodies in external secretions, distinct from and independent of the circulating (systemic) antibodies, has been recently established. The *secretory (mucosal) antibodies* are found in bronchial and nasal secretions, the gastrointestinal tract, tears and milk. Chemically, the secretory molecule is composed of a dimer of 7S immunoglobulin IgA (γA), synthesized locally by plasma cells, which is complexed to a stabilizing transport or secretory "piece" — a glycoprotein that is found in epithelial cells.

Much has been learned about the secretory immune system, but more remains to be known about the factors controlling the synthesis and secretion of these immune globulins, their precise function, and their possible role in human disease. At present, there is sufficient evidence to suggest that these antibodies are important in immunity against certain viral infections of the respiratory tract such as rhinovirus, Type 1 parainfluenza virus, and measles. A direct correlation between resistance to these infections and the presence of the antibodies has been shown. Many patients who lack respiratory tract IgA (e.g., in ataxia telangiectasia) suffer from frequent recurrences of pulmonary infections. Further, the greater efficacy of vaccines made from live or attenuated virus, compared to those made from inactivated virus, is attributable to the stimulation of secretory IgA by the former vaccines. These antibodies may also be important in the pathogenesis of some mucosal allergic reactions, such as hay fever.

IgE Immunoglobulins. A small fraction of respiratory mucosal antibodies is made up of a distinct class of immunoglobulins, IgE or γE. The special significance of this type of globulin is that it is the reaginic antibody involved in atopic reactions such as anaphylaxis. Fluorescent antibody studies show a relative concentration of γE-producing plasma cells in respiratory and gastrointestinal mucosa. Upon sensitization by γE, lung tissue releases histamine and other pharmacologically active substances.

Interferon. Another contribution of the lung to immunological responses is in the formation of *interferon,* a protein that appears early during viral infection and is capable of inhibiting viral multiplication within the affected cells. Interferon, therefore, may serve to limit the spread of viral infections and promote recovery. Experimental evidence indicates that the lung is one of the most active sites of interferon production. Though the responsible cells in the lung are unknown, they may be elements of the lymphoid or reticulo-endothelial system, e.g., macrophages. (For a further discussion of these and other immune mechanisms, the reader is referred to the Immunology Section.)

THE LUNG IN ALLERGIC REACTIONS

The lung is often involved in allergic reactions that represent departures from the normal responses to antigenic substances and can be manifested in serious and even dramatic clinical states. Allergic reactions occurring in the lung have been conveniently classified, on the basis of the underlying mechanism, into four main types:

Type I (Anaphylactic) Response. This response is initiated by the reaction of an antigen with tissue cells passively sensitized by reaginic antibody (IgE). The reaction leads to a release of biologically active mediators including histamine, "slow-reacting substance of anaphylaxis (SRS-A)," prostaglandins and, possibly, other substances. Type I response is exemplified by anaphylaxis, extrinsic (atopic) asthma, allergic rhinitis and Ascaris pneumonitis.

Type II (Cytotoxic) Response. This response involves a reaction between antibody and either an antigenic component of tissue or an antigen (or hapten) attached to the tissue. Transfusion reactions are included in this category. The Goodpasture syndrome, in which patients develop diffuse pulmonary capillary bleeding and focal glomerulonephritis, is a probably example of pulmonary reaction of this type. In man, alveolar and renal glomerular basement membranes are antigenically similar; injections of heterologous antilung antibodies can produce an experimental model resembling the human syndrome.

Type III (Arthus type) Response. The reaction betwen antigen and antibody, in antigen excess, forming complexes that are locally toxic to tissues is termed the Arthus response. The classical example of this reaction is serum sickness. The lung exhibits the same sort of reaction in response to a wide variety of antigens. Clinical examples include an assortment of reactions to organic dusts, such as Farmers' lung, bronchopulmonary aspergillosis, bagassosis, mushroom-picker's lung, bird-fancier's lung, and related conditions. In all of these examples the reactions are to inhaled allergens, but similar pathological responses may result from ingested or injected antigens. Possibly belonging in this latter group are polyarteritis nodosa, Wegener's granulomatosis and Löffler's syndrome.

Type IV (Delayed or Cell-Mediated) Response. This response is characterized by the presence of tubercle-like lesions containing epithelioid cells and lymphocytes. *Histoplasma capsulatum* (and other fungi), silica and beryllium in addition to *Mycobacterium tuberculosis* may cause this type of allergic response. A decreased delayed sensitivity is often seen in association with sarcoidosis.

THE POTENTIAL FOR HORMONE SECRETION: ENDOCRINE SYNDROMES IN LUNG DISEASE

Malignant tumors can secrete hormones, causing ectopic endocrine syndromes. Tumors of the lung, notably bronchogenic carcinoma, have shown the capacity to produce virtually all the polypeptide hormones. The commoner examples of these hormonal syndromes are listed in Table 9-2. The most frequent offender is the oat-cell bronchogenic carcinoma, although certain syndromes (hyperparathyroidism and osteoarthropathy) are more common with squamous-cell carcinomas.

The association between endocrine syndromes and certain malignant tumors is of obvious importance in the recognition of these tumors and their management. Aside from this, the retention by the neoplastic cell of the ability to produce normal peptide hormones is of considerable interest to the endocrinologist, the oncologist and the geneticist. To the student of the lung, the greater incidence of these endocrine syn-

TABLE 9-2. Some Endocrine Syndromes Associated with Lung Tumors

SYNDROME	HORMONE
Hyponatremia (Inappropriate Antidiuresis)	ADH (arginine vasopressin)
Cushing's syndrome	ACTH
Hypercalcemia	Parathormone
Gynecomastia	Gonadotropins
Hyperthyroidism	Thyrotropin
Hypoglycemia	(?) Insulinlike substance
Carcinoid syndrome	Serotonin (+ kinins)
Osteoarthropathy and acromegaly	Growth hormone
Eton-Lambert (myasthenia-like) syndrome	
Multiple syndromes	ACTH + MSH, etc.

dromes found with tumors of the lung than is found with tumors of any other organ is of added interest.

Although hormonal secretion by the lung is largely a manifestation of malignant disease, endocrine activity may be associated with nonmalignant lesions. For example, ACTH production has been reported in bronchial adenoma, and ADH secretion in pulmonary tuberculosis, abscess, and other conditions.

Normal lung may also have endocrine functions. An antidiuretic and sodium-retaining factor from the lung has been described, though its role in physiological regulation is undetermined. The importance of the lung in the activation, inactivation, storage and release of vasoactive substances has been outlined earlier.

CONCLUSION

It is clear that the lung serves the body in many ways besides the exchange of oxygen and carbon dioxide. It is equally apparent that impairment of these functions can be manifested in a wide variety of ways, many of which are not obviously connected with the process of respiration. No specific means are now available for testing how well the lung is performing a particular metabolic function in patients. Perhaps such tests will soon be introduced. It seems certain, however, that the exploration of metabolic and endocrine events in the lung is an area of investigation that touches on many organ systems and promises to grow actively in the coming years.

ANNOTATED REFERENCES

The many metabolic activities of the lung are reviewed in separate treatments of different aspects of the subject:

Harlan, W. R., and Said, S. I.: Selected aspects of lung metabolism. *In:* Bittar, E. E., and Bittar, N. (eds.): The Biological Basis of Medicine. Vol. 6. p. 357. New York, Academic Press, 1969.

Heinemann, H. O., and Fishman, A. P.: Nonrespiratory function of mammalian lung. Physiol. Rev., 49:1, 1969.

Massaro, D.: Biochemistry of lung. *In:* Sulavik, S. B., and Carrington, C. B. (eds.): Diseases of the Lung. Yale University Press (in press).

Said, S. I.: The lung as a metabolic organ. New Eng. J. Med., 279:1330, 1968.

Morphologic, developmental and functional features of cellular elements in the lung:

Sorokin, S. P.: The cells of the lung. *In:* Proceedings of the Conference on Morphology of Experimental Respiratory Carcinogenesis. AEC Symposium Series, U. S. Atomic Energy Commission, Oak Ridge, Tenn., 1969. Washington, U. S. Govt. Printing Office (in press).

Rhodin, J. A. G.: Ultrastructure and function of the human tracheal mucosa. Ann. Rev. Resp. Dis., 93 (Suppl.):1, 1966.

Cytochemical characteristics and possible functional significance of the argyrophil cells:

Pearse, A. G. E.: Common cytochemical and ultrastructural characteristics of cells producing polypeptide hormones (the APUD series) and their relevance to thyroid and ultimobranchial C cells and calcitonin. Proc. Roy. Soc. (Ser. B), *170*:71, 1968.

Weichert, R. F., III: The neural ectodermal origin of the peptide-secreting endocrine glands. Am. J. Med., *49*:232, 1970.

Physiological and clinical importance of surfactant:

Clements, J. A.: Pulmonary surfactant (Editorial). Ann. Rev. Resp. Dis., *101*:984, 1970.

Scarpelli, E. M.: The Surfactant System of the Lung. Philadelphia, Lea and Febiger, 1968.

The relationship between hereditary antiprotease deficiency and emphysema:

Eriksson, S.: Studies in alpha₁-antitrypsin deficiency. Acta med. scand., *177* (Suppl. 175): 1, 1965.

Falk, G. A., and Briscoe, W. A.: Chronic obstructive pulmonary disease and heterozygous alpha₁-antitrypsin deficiency. Ann. Intern. Med., *72*:595, 1970.

Gross, P., Babyak, M., Tolker, E., and Kaschak, M.: Enzymatically produced pulmonary emphysema: A preliminary report. J. Occup. Med., *6*:481, 1964.

Guenter, C. A., Welch, M. H., and Hammarsten, J. F.: Alpha₁-antitrypsin deficiency and pulmonary emphysema. Ann. Rev. Med., *22*:283, 1971.

The role of the lung in metabolizing vasoactive hormones:

Vane, J. R.: The release and fate of vaso-active hormones in the circulation. Brit. J. Pharmacol., *35*:209, 1969.

Role of the lung in relation to blood clotting and fibrinolysis:

Astrup, T.: Tissue activators of plasminogen. Fed. Proc., *25*:42, 1966.

Chu, J., Clements, J. A., Cotton, E. K., Klaus, M. H., Sweet, A. Y., and Tooley, W. H.: Neonatal pulmonary ischemia: Clinical and physiological studies. Pediatrics, *40*:709, 1967.

Deykin, D.: Clinical challenge of disseminated intravascular coagulation. New Eng. J. Med., *283*:636, 1970.

Possible physiological and pathological roles of the kallikrein-kinin system in man:

Kellermeyer, R. W., and Graham, R. C., Jr.: Kinins—possible physiologic and pathologic roles in man. New Eng. J. Med., *279*:754, 802, 859, 1969.

Mechanisms of clearance of airways:

Green, G. M.: The J. Burns Amberson lecture —In defense of the lung. Ann. Rev. Resp. Dis., *102*:691, 1970.

Kilburn, K. H., and Salzano, J., Eds.: Symposium on the structure, function and measurement of respiratory cilia. Ann. Rev. Resp. Dis., *92* (Suppl.:1, 1966).

Reid, L.: Evaluation of model systems for study of airway epithelium, cilia, and mucus. Arch. Intern. Med., *126*:428, 1970.

Rylander, R.: Studies of lung defense to infections in inhalation toxicology. Arch. Intern. Med., *126*:496, 1970.

Local immune mechanisms in the lung:

Editorial: Viral infections and local IGA. New Eng. J. Med., *280*:666, 1969.

Ishizaka, K.: Human reaginic antibodies. Ann. Rev. Med., *21*:187, 1970.

Tomasi, T. B., Jr.: Structure and function of mucosal antibodies. Ann. Rev. Med., *21*:281, 1970.

Tomasi, T. B., Jr., and DeCoteau, E.: Mucosal antibodies in respiratory and gastrointestinal disease. *In:* Stollerman, G. H. (ed.): Advances in Internal Medicine. Vol. 16. p. 401. Chicago, Year Book Medical Publishers, 1970.

Current concepts regarding interferon:

DeClercq, E., and Merigan, T. C.: Current concepts of interferon and interferon induction. Ann. Rev. Med., *21*:17, 1970.

Basic mechanisms of allergic responses of the lung:

Parish, W. E., and Pepys, J.: The lung in allergic disease. *In:* Gell, P. G. H., and Coombs, R. R. A., (eds.): Clinical aspects of Immunology. ed. 2. Oxford, Blackwell, 1968.

Willoughby, W. F., and Dixon, F. J.: Experimental hemorrhagic pneumonitis produced by heterologous antilung antibody. J. Immunol., *104*:28, 1970.

Vasquez, J. J.: Immunopathologic aspects of lung disease. Arch. Intern. Med., *126*:471, 1970.

Endocrine syndromes arising from non-endocrine tumors:

Gellhorn, A.: Ectopic hormone production in cancer and its implication for basic research on abnormal growth. *In:* Stollerman, G. H. (ed.): Advances in Internal Medicine. Vol. 15. p. 299. Chicago, Year Book Medical Publishers, 1969.

Lipsett, M. D.: Hormonal syndromes associated with neoplasia. *In:* Levine, R., and Luft, R. (eds.): Advances in Metabolic Disorders. Vol. 3. p. 111. New York, Academic Press, 1968.

Omenn, G. S.: Ectopic polypeptide hormone production by tumors (Editorial). Ann. Intern. Med., *72*:136, 1970.

Appendix A

NORMAL VALUES FOR
RESPIRATORY FUNCTION

Ideal values for a given individual with respect to specific physiological parameters can be estimated only by knowledge of these values in other presumed ideal individuals. Any estimate of average values in a population may deviate from the ideal and may even deviate from the normal. Several parameters of respiratory function vary predictably in relation to age, sex, body size, environment (e.g., altitude) and metabolic activity. Predicted values based on numerous studies in various populations of "normal" subjects are summarized in the textbook by Bates and Christie (*see* Appendix B). The following table of normal values for a 70-kilogram man illustrates some quantitative aspects of ventilatory and gas exchange data.

Lung Volumes

	(ml.)
Vital Capacity (VC)	4,600
Inspiratory Capacity	3,500
Expiratory Reserve Volume	1,100
Functional Residual Capacity (FRC)	2,300
Residual Volume (RV)	1,200
Total Lung Capacity (TLC)	5,800
RV/TLC	20%-25%

Measures of Air Flow and Resistance

Forced Vital Capacity (FVC)	Same as vital capacity
Forced Expiratory Volume (FEV) 0.5 sec.	50% of FVC
Forced Expiratory Volume (FEV) 1.0 sec.	80% of FVC
Maximum Mid-Expiratory Flow (FEF 25-75%)	greater than 2 L. per min.
Maximum Voluntary Ventilation (MVV)	120-180 L. per min.
Airway Resistance (cm. H_2O per L. per sec.)	1.6

Ventilation and Blood Gas Exchange in the Lungs

Minute Ventilation ($\dot{V}_E$) ml. per min.	6-8,000
Frequency (f) breaths/min.	12-18
Tidal Volume (V_T)	500-700 ml.
Respiratory Dead Space (V_D)	140-170 ml.
Oxygen Consumption ($\dot{V}_{O_2}$)	250 ml. per min.
Carbon Dioxide Production (V_{CO_2})	200 ml. per min.
Respiratory Exchange Ratio (R)	0.7-1.0
Arterial Blood Gases pH	7.38-7.42
Carbon Dioxide Tension (Pa_{CO_2})	36-44 mm. Hg
Oxygen Tension (Pa_{O_2})	80-100 mm. Hg
Oxygen Saturation	97%
Alveolar-Arterial Oxygen Tension Gradient (A-a P_{O_2})	10 mm. Hg
Diffusing Capacity for Carbon Monoxide	Varies with above factors and specific techniques

Oxygen Transport to the Tissues

Cardiac output	5.0 L. per min.
Hemoglobin	15.0 g. per 100 ml.
Oxygen Content Arterial	20 ml. per 100 ml.
Venous	15 ml. per 100 ml.
Oxygen Tension Arterial (Pa_{O_2})	80-100 mm. Hg
Mixed Venous ($P_{V_{O_2}}$)	40 mm. Hg
Oxyhemoglobin Dissociation (Expressed as Oxygen tension at 50% oxyhemoglobin saturation (P_{50})	26-27 mm. Hg

* The above values relate only to the resting state. Normal values for various degrees of exercise and the responses to altered inspired air (e.g., inhalation of supplemental oxygen) are available in references listed (Appendix B).

Appendix B

MAJOR REFERENCES IN
RESPIROLOGY

Pulmonary Physiology

Altman, P. L., and Dittmer, D. S.: Respiration and Circulation. Biological Handbooks. Federation of American Societies for Experimental Biology, 1971.

Aviado, D. M.: The Lung Circulation, Volumes I and II. Elmsford, N. Y., Pergamon Press, 1965.

Caro, C. G.: Advances in Respiratory Physiology. London, Edward Arnold Publishers, 1966.

Comroe, J. H.: Physiology of Respiration. Chicago, Year Book Medical Publishers, 1965.

Comroe, J. H., Jr., Forster, R. E., II, DuBois, A. B., Briscoe, W. A., and Carlsen, E.: The Lung. Chicago, Year Book Medical Publishers, 1962.

Dejours, P.: Respiration. Translated by L. E. Farhi. New York, Oxford University Press, 1966.

Fenn, W. O., and Rahn, H.: Handbook of Physiology. Section 3: Respiration. Vols. 1 and 2. American Physiological Society. Baltimore, The Williams & Wilkins Company, 1965.

Fishman, A. P., and Hecht, H. H.: The Pulmonary Circulation and Interstitial Space Chicago, The University of Chicago Press, 1969.

Slonim, N. B., Bell, B. P., and Christensen, S. E.: Cardiopulmonary Laboratory, Basic Methods and Calculations. Springfield, Ill., Charles C Thomas, 1967.

Winters, R. W., Engel, K., and Dell, R. B.: Acid Base Physiology in Medicine; A Self Instruction Program. London Company of Cleveland, 1967.

Pathophysiology

Bates, D. V., Macklem, P. T., and Christie, R. V.: Respiratory Function in Disease. Philadelphia, W. B. Saunders, 1971.

Bendixen, H. H., Egbert, L. D., Hedley-White, J. Laver, M. B., and Pontoppidan, H.: Respiratory Care. St. Louis, The C. V. Mosby Company, 1965.

Cherniack, R. M., and Cherniack, L.: Respiration in Health and Disease. Philadelphia, W. B. Saunders, 1961.

Cherniack, R. M., Cherniack, L., and Naimark, A.: Respiration in Health and Disease. Ed. 2. Philadelphia, W. B. Saunders (In Press)

Filley, G. F.: Pulmonary Insufficiency and Respiratory Failure. Philadelphia, Lea and Febiger, 1967.

Howell, J. B. L., and Campbell, E. J. M.: Breathlessness; Proceedings of an International Symposium. London, Blackwell Scientific Publications, 1966.

Sykes, M. K., McNicol, M. W., and Campbell, E. J. M.: Respiratory Failure. London, Blackwell Scientific Publications, 1969.

Pulmonary Disease

Baum, G. L.: Textbook of Pulmonary Disease. Boston, Little, Brown and Company, 1971.

Crofton, J., and Douglas, A.: Respiratory Diseases. London, Blackwell Scientific Publications, 1969.

Fraser, R. G., and Paré, J. A. P.: Diagnosis of Diseases of the Chest. Volumes 1 and 2. Philadelphia, W. B. Saunders, 1970.

Hinshaw, H. C.: Diseases of the Chest. Philadelphia, W. B. Saunders, 1969.

Pulmonary Pathology

Liebow, A. A., and Smith, D. E.: The Lung. Baltimore, The Williams & Wilkins Company, 1968.

Reid, L.: The Pathology of Emphysema. Chicago, Year Book Medical Publishers, 1967.

Spencer, H.: Pathology of the Lung. Elmsford, N. Y., Pergamon Press, 1968.

Section Three

Renal Mechanisms

Introduction

The kidney is responsible for elimination from the body of most of the nonvolatile waste products of metabolism. Equally important is its role in maintenance of a constant internal environment of electrolyte concentrations and fluid volume. By selective reabsorption and secretion of electrolytes, the normal kidney possesses an enormous capacity to maintain precisely fluid and electrolyte balance. Most fluid and electrolyte imbalances result not from loss of renal function or reserve, but rather from extrarenal pathology that initiates transmission of inappropriate or faulty information to renal regulatory mechanisms.

The kidney consists of approximately two million functional units or nephrons. The elaboration of urine by these nephrons can be divided, for our purposes, into two components. The first is the formation of an ultrafiltrate in the glomerular capillary bed; the second consists of the selective active and passive reabsorption and secretion of the wide spectrum of filtered solutes (and of water osmotically obligated to these solutes), as the ultrafiltrate proceeds through the tubule.

The *glomerular ultrafiltrate* is similar in composition to plasma, except that high molecular weight substances, primarily proteins, are largely excluded. The rate at which this filtrate is formed can be quantified by determining the clearance of any substance that is freely filterable (molecular weight less than 10,000), is not bound to protein and is neither secreted nor reabsorbed in passage through the tubule. Inulin and mannitol are the best examples of such substances.

The *filtered load* of any solute can be determined by multiplying the plasma concentration of the solute (P_{conc}) by the glomerular filtration rate (GFR). If this filtered solute passes unaltered through the tubular system, the amount excreted in the urine (measured by multiplying urine concentration (U_{conc}) by urine volume per unit of time (V) must equal filtered load. This is expressed by the equation:

$$P_{conc} \times GFR = U_{conc} \times V$$

Since all components other than the GFR can be measured directly, one can calculate the GFR from the following equation:

$$GFR = \frac{U_{conc} V}{P_{conc}} \cdot$$

The rate at which glomerular filtrate is formed is dependent primarily upon the rate of plasma flow through the kidney. Normally about 20 percent of the plasma circulating through the kidney is filtered, but this may vary substantially under various physiological and pathological conditions. Glomerular filtration rate is dependent upon glomerular capillary hydrostatic pressure minus intracapsular pressure and plasma oncotic pressure. Hydrostatic pressure is dependent upon the balance between afferent and efferent glomerular arteriolar constriction or resistance and renal arterial and venous blood pressures. Changes in glomerular structure, such as those seen in acute and chronic glomerulonephritis, also contribute to the rate of formation of glomerular filtrate.

The urinary clearances of some endogenous metabolic waste products such as urea and creatinine approximate those of inulin and mannitol. The quantity of these metabolites formed daily remains fairly constant under normal conditions. Plasma concentrations of these metabolites remain constant only as long as daily endogenous production is matched by urinary excretion. Since the urinary excretion of these solutes is dependent primarily upon filtered load (determined by multiplying GFR by plasma concentration of the solute), it is apparent that a reciprocal relationship must exist between the GFR and plasma concentration. When plasma concentrations of these and other unidentified metabolites become sufficiently elevated in response to a fall in GFR, the symptom complex of *uremia* develops. Blood urea nitrogen and serum creatinine concentrations can be utilized to estimate the extent of renal dysfunction; however, abnormal concentrations are not responsible for the array of symptoms seen in renal failure.

Other urinary solutes have a clearance much below or above the glomerular filtration rate so that total urinary excretion is regulated not only by the amount filtered but also by the rate of tubular reabsorption or secretion. The clearance of any urinary solute can be quantified using the same technique. By comparing this clearance value with the inulin clearance, *net* reabsorption or secretion of any solute can be measured. For example, if the clearance of a specific solute was calculated to be 10 ml. per minute in a person with a simultaneous inulin clearance of 100 ml. per minute, net reabsorption of this solute would be 90 percent of the filtered load. Tubular reabsorptive and secretory activity are governed by local structural and metabolic conditions and a whole host of hormonal effects.

In 1960, the "intact nephron" hypothesis was proposed by Bricker, Morrin and Kime. They suggested that in chronic renal disease, regardless of origin, surviving nephrons either functioned normally or did not contribute significantly to final urine formation. This report focused the attention of many investigators on the functional consequences of chronic renal disease. Their studies demonstrated that as glomerular function declined, there was a simultaneous and proportional decline in a variety of tubular functions. For example, the reduction in the tubular maximums for glucose absorption and para-aminohippurate secretion closely paralleled the decrease in glomeru-

lar filtration rate in the unilaterally diseased kidney compared to a normal control kidney.

This view was challenged by Biber and associates, whose micropuncture and microdissection studies in animals with chemically induced acute renal damage indicated that damaged nephrons apparently do contribute to urine formation. Thus, with progressive loss of renal function, the remaining nephrons must retain an ability to adapt or adjust appropriately so that a normal internal environment can be maintained. If this did not occur and significant glomerulotubular imbalances resulted from selective tubular disruptions, fluid and electrolyte disturbances incompatible with life would invariably develop with time.

More recently Bricker, in updating his original hypothesis, emphasized that changes in the excretion rates for each solute by residual nephrons follow an orderly, predictable and appropriate pattern for the maintenance of homeostasis. Such balanced and regulated changes in glomerulotubular balance for sodium, potassium, hydrogen and a whole host of other solutes do not exclude a contribution to function by diseased nephrons or nephron segments. Rather, it emphasizes that, *despite* impairment in glomerular and tubular functions in individual nephrons or nephron segments due to structural damage, the remaining intact or normal nephrons or nephron segments adapt or compensate appropriately to maintain the constant internal environment necessary for the individual to survive.

A variety of finely tuned regulatory mechanisms maintain fluid, electrolyte and buffer balances within a narrow, normal range. These mechanisms are discussed in detail in the four subsequent chapters. Pathological conditions are used to illustrate how normal control mechanisms become overwhelmed or ineffective and imbalances result.

We initiate presentation of major renal mechanisms by discussing the maintenance of serum protein—more specifically, albumin—concentrations. The serum proteins are the only endogenous intravascular solutes that are not readily exchangeable with the interstitial fluids and, hence, provide the colloid osmotic pressure necessary to maintain intravascular volume. One may no longer be able to maintain normal serum albumin concentrations when the liver is unable to synthesize new albumin adequately, when there are substantial losses of albumin via the gastrointestinal tract or kidney or when there is accelerated endogenous albumin catabolism. When serum albumin concentration decreases, fluid may shift from the intravascular spaces to the interstitial spaces, and edema develops. Considerable controversy has been generated over a period of years over the relative roles of glomerular versus tubular pathology in the development of proteinuria in different forms of renal dysfunction. Although our understanding of the mechanisms of proteinuria still remains limited, this chapter provides a comprehensive review of the current consensus.

Discussion of renal mechanisms is continued by a presentation of the means whereby the kidney regulates total body sodium and potassium stores. Maintenance of extracellular fluid volume is primarily dependent upon the osmotic effects exerted by extracellular sodium. An intricate network of renal regulatory mechanisms exists to maintain extracellular sodium within a narrow, normal range despite wide variations in oral intake and extrarenal loss. When gastrointestinal or renal salt losses exceed intake, extracellular fluid volume decreases. Volume depletion stimulates antidiuretic hormone release in an attempt to maintain volume. Continued salt loss results in development of hyponatremia.

Any decrease in intravascular volume promotes renal retention of salt and water. In a variety of edematous conditions, such as congestive heart failure and cirrhosis, local arterial and intracardiac receptors responsible for regulating intravascular volume sense a volume depletion when, in fact, total volume may be expanded but abnormally distributed. Further salt and water retention under these conditions leads to development of edema. When "apparent" intravascular volume is decreased, antidiuretic hormone release also occurs. If fluid intake continues, water in excess of sodium is retained and a dilutional hyponatremia develops.

Potassium is the primary intracellular solute providing the osmotic force necessary to maintain intracellular volume. Only a small portion of total body potassium is contained in the extracellular fluid compartment. Therefore, the serum concentration of potassium may fail to reflect accurately total body potassium. A potassium flux into cells occurs with cell growth, intracellular nitrogen and glycogen deposition, and increases in extracellular pH; potassium leaves the cells with cell destruction, glycogen utilization and decreases in extracellular pH. In interpreting the significance of any given serum potassium concentration, consideration must be given to these factors, which affect the ratio of intracellular to extracellular concentrations.

The role of the renal concentrating and diluting mechanisms is discussed in terms of maintenance of body fluid volumes. Isotonic polyuria results either from structural or functional defects in the countercurrent concentrating mechanisms (e.g., hypokalemia, hypercalcemia and sickle cell disease) or from an osmotic diuresis. When the distal nephron remains impermeable to water, owing to either a deficit of or tubular unresponsiveness to antidiuretic hormone, a more striking polyuria develops, with elaboration of a hypotonic urine. Persistent polyuria leads to significant dehydration or volume depletion only when thirst mechanisms are impaired or when free access to water is denied by physical handicap or other such limitation.

Dilutional hyponatremia resulting from

an inability to excrete water not obligated to solute most commonly occurs in persons with apparently inappropriate ADH release —i.e., ADH levels remain high in spite of decreased serum osmolality and increased volume. In some instances, such as congestive heart failure, nephrotic syndrome and cirrhosis, volume receptor sites in the cardiovascular system locally react as though intravascular volume was decreased, whereas, in reality, it is increased. In other conditions, such as pulmonary tumors and central nervous system disorders, the continued secretion of ADH remains unexplained. A second mechanism also is operative in patients with salt retention. A greatly increased proximal tubular reabsorption decreases delivery of sodium to the distal diluting segments of the nephron. This greatly limits the renal capacity to generate water not obligated to solute ("free" water), and, if water intake is sufficient, hyponatremia develops. The capacity of the kidney to excrete a dilute urine under other conditions is so great that dilutional hyponatremia develops only when fluid intake is high *and* impairment of renal function is severe.

The section concludes with considerations of the extracellular and intracellular buffer systems and the pulmonary and renal regulatory mechanisms that are coordinated to maintain a constant serum hydrogen ion concentration or pH. Increased production of fixed acid or retention of hydrogen ions leads to development of a metabolic acidosis; retention of carbon dioxide, which is hydrated to the weak acid, carbonic acid, leads to a respiratory acidosis. Conversely, loss of hydrogen ion produces a metabolic alkalosis; loss of CO_2 (hyperventilation) produces a respiratory alkalosis. The most common of the acid-base disturbances encountered clinically is metabolic acidosis. A healthy person on a normal diet must excrete about 35 mEq. of fixed acid daily by the kidneys to eliminate the fixed acid formed by normal metabolic processes. The

capacity to maintain plasma pH at 7.4 may be exceeded following administration of an exogenous acid load, increased endogenous acid production, and/or decreased renal acid excretion resulting from either a decrease in functioning nephrons or an intrinsic tubular defect (renal tubular acidosis). Although hyperventilation, with loss of the weak acid CO_2, raises the plasma pH toward normal, this partial compensation is ineffective for prolonged maintenance of pH. Permanent correction of acidemia can be achieved only by administration of a base or by renal or gastrointestinal loss of fixed acid.

Although these four chapters cover four of the most important regulatory functions of the kidney, numerous other examples may be cited. For example, mainfestations of gout may result from an elevation of serum uric acid levels. This elevation may result from excessive endogenous production of uric acid or from an increased *net* tubular reabsorption of uric acid. Since uric acid is both secreted and reabsorbed at different levels in the tubule, increased net reabsorption could mean either increased reabsorption or decreased secretion of uric acid, or both.

Another example is stone formation in the urinary tract of the patient with cystinuria. Although there is a common defect resulting in decreased net tubular reabsorption of ornithine, arginine, lysine and cystine, only cystine, which has the lowest solubility coefficient, precipitates and produces disease. Other proximal tubular defects in phosphate, amino acid, glucose and bicarbonate reabsorption may be present as isolated abnormalities or in combination, resulting in a whole spectrum of disease.

The role of the kidney in maintenance of blood pressure and erythropoesis is discussed in Chapters 4 and 22.

Robert D. Lindeman, M.D.
William O. Smith, M.D.

SELECTED REFERENCES

Bricker, N. S., Morrin, P. A. F., and Kime, S. W., Jr.: The pathologic physiology of chronic Bright's disease: an exposition of the "intact nephron hypothesis." Am. J. Med., 28:77, 1960.

Biber, T. U. L., Mylle, M., Baines, A. D., Gottschalk, C. W., Oliver, J. R., and MacDowell, M. C.: A study of micropuncture and miscrodissection of acute renal damage in rats. Am. J. Med., 44:664, 1968.

Bricker, N. S.: Editorial: on the meaning of the intact nephron hypothesis. Am. J. Med., 46:1, 1969.

10

Maintenance of Body Protein Homeostasis

Victor E. Pollak, M.D., *and Amadeo J. Pesce,* PH.D.

Direct observations relative to the movement of proteins across the glomerular filter, or their reabsorption or secretion by the tubules, require the withdrawing of fluid for analysis from individual nephrons. As yet there are few such observations because nephron puncture studies are difficult to do, and methods for the accurate measurement of minute amounts of plasma proteins are only now being developed. In the two following sections—on the glomerulus as a molecular sieve, and on tubular reabsorption and secretion of proteins—the views presented result largely from investigations in intact animals; thus, our concept of the role of the individual parts of the nephron in the handling of proteins is based mainly on inference rather than direct measurement.

THE GLOMERULUS AS A MOLECULAR SIEVE

The kidney appears to act as a molecular sieve with respect to the plasma proteins. The sieving effect occurs in the glomerulus. Ions and small molecules such as water, glucose, and urea pass through the glomerulus into Bowman's space and proximal tubular lumens, virtually without hindrance, as a function chiefly of the difference between filtration pressure and osmotic pressure. By contrast, plasma protein molecules are retained by the glomerulus most efficiently and to such an extent that the normal urine contains only minimal quantities of protein.

Sieving Coefficient. Inulin, a relatively small molecule (m. w. 5,000) is filtered completely by the glomerulus and is not reabsorbed by the tubules. It is therefore used to measure the glomerular filtration rate, and can serve as a reference substance in considering the permeability of the glomerulus to macromolecules. It has a sieving coefficient of 100 percent (Figs. 10-1 and 10-2).

Albumin is the most abundant plasma protein (m. w. 69,000; molecular radius 35 Å) and has a concentration of 40 grams per liter in the plasma flowing through the glomerulus. The concentration of albumin, measured *by direct puncture* of the first part of the rat proximal convoluted tubules, is about 15 mg./L. or less. In this part of the proximal tubule the inulin concentration is identical to that in the plasma. The true glomerular sieving coefficient for albumin may be expressed as:

$$\left(\frac{[TF_{ALB}] V}{[P_{ALB}]} \div \frac{[TF_{IN}] V}{[P_{IN}]} \right) 100 \quad (1)$$

Where

TF_{ALB} and P_{ALB} = albumin concentration in proximal tubular fluid and in plasma;

TF_{IN} and P_{IN} = inulin concentration in proximal tubular fluid and in plasma;

V = Rate of flow through the proximal tubule. Note that V is common to numerator and denominator.

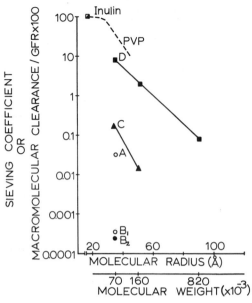

FIG. 10-1. The relationship between molecular radius (and molecular weight) and the sieving coefficient. By definition, the sieving coefficient of inulin is 100.

Three proteins are used in this figure: albumin, M. W. 69,000, molecular radius 35 Å; IgG, M. W. 160,000, molecular radius 50 Å; and, α_2 macroglobulin, M. W. 820,000, molecular radius 90 Å.

Point A is the sieving coefficient in the normal rat, measured in the first part of the proximal convoluted tubule by direct puncture. *Point B_1* is the sieving coefficient in the normal rat, measured indirectly from the urine. *Point B_2* is the sieving coefficient for albumin in healthy man, measured indirectly from the urine. *Line C* is the sieving coefficient for albumin and IgG in a patient excreting a large amount of protein, but in whose glomeruli only minimal changes were found. The sieving coefficient for albumin is greatly increased as compared with the normal. *Line D* is the sieving coefficient derived from studies on a patient with severe proteinuria and a glomerular filtration rate about 5 percent of normal. The glomerulus was highly permeable to albumin and the slope of the line indicates, as compared with Line C, the considerable permeability to proteins larger than albumin.

The interrupted line at the top of the figure is the sieving coefficient for polyvinylpyrrolidone in normal man, and has been replotted from the data of Hardwicke and his colleagues (see Reference), and Hulme, B. and Hardwicke, J.: Proc. Roy. Soc. Med., 59:509, 1966. The measurements from which Points A and B are derived are from unpublished data on the albumin excretion of the normal rat from the authors' laboratory.

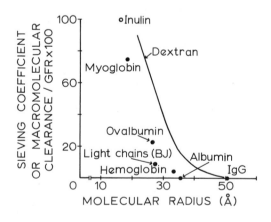

FIG. 10-2. The relationship between molecular radius and the sieving coefficient. By definition the sieving coefficient for inulin is 100. The sieving coefficient for dextran is replotted from the data of Wallenius (see Reference). The sieving coefficients are shown for a number of proteins which have been characterized and studied.

For the rat proximal convoluted tubule, equation (1) is:

$$\left(\frac{[15]\,V}{[40 \times 10^3]} \div (1 \times V) \right) 100 = 0.0375$$

This is indicated by point A in Figure 10-1.

In practice, the sieving coefficient must be measured *indirectly* by a study of the urine rather than the proximal convoluted tubular fluid. In the normal 200-gram rat the albumin excretion is about 0.2 mg. per 24 hours. The glomerular filtration rate is about 1.4 liters per 24 hours. The sieving coefficient for albumin may be calculated as follows:

$$\left(\frac{[U_{ALB}]\,V}{[P_{ALB}]} \div GFR\right) 100 \qquad (2)$$

Where U_{ALB} = concentration of albumin in urine, and

Where $U_{ALB}\,V$ = excretion rate of albumin; for example:

$$\left(\frac{0.2}{[40 \times 10^3]} \div 1.4\right) 100 = 0.00035 \qquad (2)$$

In Figure 10-1 this is represented by point B_1.

If these assumptions are correct (more data are needed), it is reasonable to assume that about 99 percent of the albumin filtered by the glomerulus is reabsorbed by the proximal convoluted tubules. The sieving coefficient for albumin, measured indirectly in this manner, is about 1 percent of that measured directly.

Sieving Phenomenon. In normal man the urine albumin excretion approximates 18 mg. per 24 hours and the inulin clearance 180 L. per 24 hours. Thus the sieving coefficient (equation 2) is:

$$\left\lfloor \frac{18}{40,000} \div (180) \right\rfloor 100 = 0.00025$$

This is close to the figure calculated for the rat and is represented in Figure 10-2 by point B_2.

In disease, large amounts of albumin and of other plasma proteins may appear in the urine and the permeability of the glomerulus to proteins may be greatly increased (Fig. 10-1, points C and D).

The movement of macromolecules through the glomerular capillaries has been studied indirectly by examining the relationship between the size of the macromolecules and their clearance by the kidney into the urine. Polymers (such as dextrans and polyvinylpyrrolidone) of graded molecular size have proved particularly useful. At the top of Figure 10-1 is plotted the re-

sults of a study in healthy subjects, showing the relationship between the sieving coefficient and the molecular radius of polyvinylpyrrolidone. In Figure 10-2, on a different scale, is illustrated the typical curve obtained when dextrans of various molecular radii are infused into dogs. For both macromolecules the sieving coefficient was higher than that for proteins of identical molecular radius—perhaps because proteins are reabsorbed by the normal tubules, whereas the polysaccharide macromolecules are not or because of easier passage through the glomerulus of polysaccharide than of protein molecules of identical radius.

When dextrans were infused (Fig. 10-2), the clearance of molecules with a radius less than 23 Å approached the glomerular filtration rate (i.e., a sieving coefficient close to 100). The clearance fell off sharply as the molecular radius increased from 23 Å to 34 Å; for molecules above 40 Å the clearance was only a minute fraction of the glomerular filtration rate (i.e., a sieving coefficient approaching zero). When heterologous proteins have been injected into experimental animals, curves of a similar general form have been obtained. Clearances of small, relatively low molecular weight proteins such as lysozyme (m. w. 17,000) or myoglobin (m. w. 17,500; molecular radius 19 Å) are relatively high, whereas all but a trace of the high molecular weight proteins such as albumin (molecular radius 35 Å) or IgG (molecular radius 50 Å) are retained by the kidney.

Pore Theory. To explain the molecular sieving phenomenon, two general theories have been proposed. In the *pore theory*, developed fully and in mathematical detail by Pappenheimer, the glomerular capillary walls are envisaged as being perforated by water-permeable pores of molecular dimensions which restrict the passage of proteins and macromolecules. For a glomerular capillary wall containing a homogeneous population of such pores, the limiting size, rate, and amount passing the glomerular capillary wall could be calculated for a series of proteins. According to the pore theory, filtration is the most important factor in transfer of proteins across the glomerular capillaries, but the role of diffusion is not ignored entirely. Indeed, it is suggested that the relative concentration of a protein or macromolecule in Bowman's space is a function of both pore size and diffusion coefficient.

Diffusion Theory. In the *diffusion theory*, the glomerular capillary wall is considered to be a gel that contains fibrils; the entire surface is pictured as being permeable to proteins and macromolecules as well as to water. In contrast to the pore theory, diffusion is considered to be the major mechanism of transfer; hydraulic pressure factors, although important, are assigned a lesser role. All substances are assumed to diffuse across the glomerular capillary wall

at a finite rate that is a function of their intramural diffusion coefficients.

The precise role of the factors favoring restriction or passage of proteins across the glomerulus is not yet clear. The diffusion theory was proposed originally on thermodynamic grounds, and evidence now increasingly favors it. The diffusion theory predicts that protein molecules of any size can be transferred across the glomerular capillary wall; their clearance will approach asymptotically to zero with increasing molecular size or decreasing diffusion coefficient. In fact, virtually all plasma proteins, including minute amounts of large molecular weight proteins, have been found in normal urine.

Glomerular Structure. In assessing the two theories it is useful to recall the unique structure of the glomerular capillary wall. Electron microscopic studies have revealed three distinct layers (Fig. 10-3). The *lamina fenestrata* of the endothelial cells is perforated by many pores of a size too great to restrict passage of any proteins. The *basement membrane*, approximately 3,000 Å

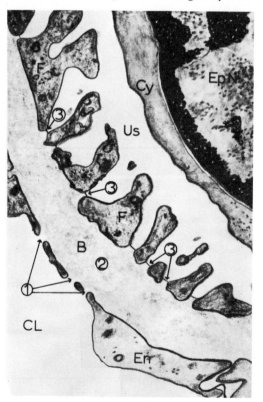

Fig. 10-3. Schematic representation of the human glomerular capillary wall. In the process of transfer from capillary lumen (CL) to urinary space (US), water, solutes and macromolecules must traverse three layers:

(1) The endothelial wall cytoplasm (En), containing numerous fenestrae (1) with a mean diameter of 700 Å. (2) The basement membrane (B), with a mean thickness of about 3,000 Å. (3) The layer of foot process (F) of the epithelial cells (Ep). The foot processes are separated about 250 to 600 Å from each other by slit pores (3), which are lined by a distinct membrane. (Adapted from Picture 8 in Jørgensen, F.: The Ultrastructure of the Normal Human Glomerulus. Copenhagen, Munksgaard, 1966. With the permission of the author and publishers.)

thick in adult man, appears to be a gel-like structure containing fibrils, 30 to 40 Å thick; no pores have been detected in this layer. The third layer, composed of *foot processes,* arises from the trabeculae of the epithelial cells; the foot processes are separated by slit pores, and a distinct membrane has been shown to line the slit.

Many workers now consider that the molecular sieving function of the glomerulus takes place in the basement membrane. If electron-dense macromolecular tracers are injected intravenously, they appear to be restricted in their passage across the basement membrane. Very few molecules or particles—and only the smaller ones—reach the epithelium or the urinary space.

It is still uncertain whether proteins and tracer macromolecules that cross the basement membrane enter thereafter into the foot processes and trabeculae of the epithelial cells or directly into Bowman's space. Studies of serial electron microscopic sections from glomeruli of animals injected with relatively small proteins as tracers (e.g., horseradish peroxidase, m. w. 40,000) indicate that the basement membrane is freely permeable to the tracer, which appears to cross directly into Bowman's space. When a larger protein tracer is injected (e.g., myeloperoxidase, m. w. 160,000), there appears to be an additional permeability barrier or restriction to the passage of the tracer protein, localized at the level of the epithelial slit pore.

Thus, the evidence currently available does not favor one theory to the exclusion of the other. It is perhaps more likely that elements of both theories are correct— namely, that diffusion occurs through the basement membrane as a whole, and that there is a porelike restriction at the junction of the basement membrane and the slit pore.

In proteinuria associated with naturally occurring or experimentally induced disease of the glomeruli, many types of morpho-logical alterations may be seen in the glomerular capillary walls. Gross rupture of glomerular capillaries may occur rarely, resulting in passage of blood into Bowman's space and, presumably, loss of the molecular sieve effect in a few capillary loops of some glomeruli. This may occur in some types of acute hemorrhagic glomerulonephritis and in necrosis of glomerular loops. Recent serial electron microscopic studies suggest that, in certain diseases, actual dissolution of basement membrane may occur in scattered glomerular capillary loops, presumably resulting also in a loss of the sieving effect in those loops.

The glomerular basement membrane is thickened in some disease states. It may appear normal morphologically in other conditions, such as the common idiopathic nephrotic syndrome of childhood or the nephrosis induced in the rat by the amino-nucleoside of puromycin. However, the intravenous injection of tracers indicates that the basement membrane is more permeable in these experimental animals.

TUBULAR ABSORPTION AND SECRETION OF PROTEINS

It is clear that considerable quantities of protein must be reabsorbed by the tubules even when, as in the normal, the concentration of protein in the glomerular filtrate is extremely low. If the albumin concentration in the glomerular filtrate of normal man is 20 mg. per liter (as it may well be), 3.6 grams of albumin must be reabsorbed by the tubules daily from the 180 liters of glomerular filtrate.

Hyaline Droplets. Hyaline, or colloid, droplets have long been recognized in the proximal convoluted tubules. Originally thought to be a manifestation of tubular degeneration and protein secretion, they are now recognized as protein reabsorption droplets. They occur in normal kidneys but are much more obvious in many diseases in which increased glomerular permeability

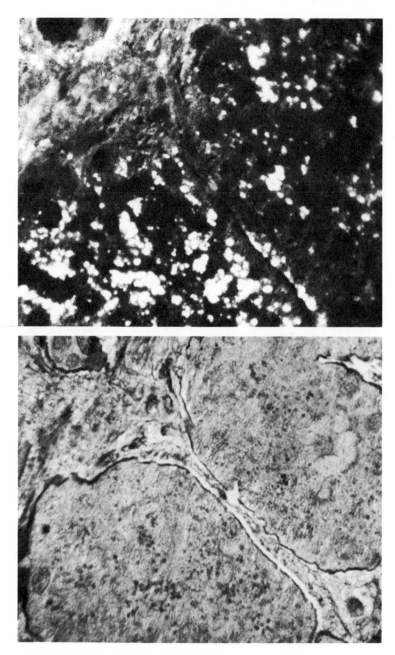

Fig. 10-4. Two proximal tubules from a renal biopsy specimen from a patient with glomerular disease and proteinuria ($\times$ 1,100): (*top*) incubated with a conjugated antiserum directed specifically against human albumin; (*bottom*) the identical section stained with periodic acid-Schiff. There are many droplets containing albumin, particularly in the tubular cytoplasm.

occurs. With the use of specific enzymatic and immunochemical techniques it has been clearly shown that the colloid droplets contain proteins. For example, homologous proteins such as albumin have been demonstrated in colloid droplets of proximal tubules of normal human and rat kidney, and increased numbers of such albumin-containing droplets have been found in proximal tubules from kidneys with abnormal glomerular permeability (Fig. 10-4). Similar droplets may be seen in the visceral epithelial cells of the glomerulus from animals with proteinuria, an observation consistent with reabsorption by pinocytosis of protein from Bowman's space.

Histochemically identifiable proteins have been injected into animals with the objective of studying the course of the protein through the nephron. Extensive studies have been made using horseradish peroxidase, a protein that readily passes into Bowman's space, as a tracer. Within a few minutes of injection, droplets of peroxidase can be detected at the luminal border of proximal tubular cells; the droplets then appear in vacuoles deeper in the cell; still later, chemical and histochemical studies indicate that vacuoles containing droplets of peroxidase become fused with lysosomes in the proximal tubular cells. It is probable that at least some of the absorbed protein is broken down by the cathepsins and other proteolytic enzymes in the lysosomes.

Tubular Protein Reabsorption

The amount of protein that can be reabsorbed by the normal tubules is not known. Clearance experiments suggest that the renal threshold for protein is very low and the quantity of protein that the tubules can reabsorb is small—probably only slightly in excess of that filtered by the normal glomerulus. In conditions with considerable proteinuria the tubular reabsorptive mechanism seems to be overloaded. This can readily be shown experimentally: when horseradish peroxidase is injected, it is re-

absorbed by the proximal convoluted tubules. If the animals are given ovalbumin, the tubular reabsorption droplets become filled with ovalbumin. When horseradish peroxidase is then given to the ovalbumin-treated animals, little or no peroxidase is reabsorbed by the tubules.

It is not clear whether protein reabsorption can occur beyond the proximal convoluted tubules. In proteinuric states the development of smaller numbers of protein reabsorption droplets in the ascending limbs of the loops of Henle and distal tubules suggests that some reabsorption can occur at more distal sites.

Current belief is that reabsorption of proteins by the tubules is nonselective; this implies that all proteins are handled by the tubules in like manner. In truth, there is little direct evidence for this view, save for the observation that reabsorption seems to occur by pinocytosis at the cell membrane.

In the future, it should be possible to examine more effectively the quantitative and the selective aspects of tubular reabsorption, with the use of a combination of clearance, nephron puncture and stop flow techniques, and of immunochemical methods which quantify individual plasma proteins. Further, it may be possible to determine whether plasma proteins are secreted by the tubules (a view for which there is little evidence at present).

The Proteins of Normal Urine

The concentration of protein in normal urine is very low. This has made it difficult to quantify the total amount of protein excreted and to analyze the nature of the proteins in normal urine. In recent years these difficulties have been overcome, particularly as a result of the application of qualitative and quantitative immunochemical techniques to the study of normal urine proteins.

Currently best estimates suggest that the healthy adult excretes between 40 and 150 mg. of protein in the urine each 24 hours. Some of the these proteins are of plasma origin; others apparently derive from the urinary tract.

Immunochemical studies have revealed that normal urine regularly contains a large number of plasma proteins. The distribution of the various plasma proteins in urine differs radically from that in plasma. Consistent with the hypothesis that the normal glomerulus acts as a molecular sieve are the observations that a large proportion of the urine proteins is of low molecular weight, and the high molecular weight plasma proteins are present in normal urine only in trace amounts. Also consistent with this view are observations on the urinary excretion of the hemoglobin-binding protein haptoglobin. This protein can be detected regularly in urine from healthy subjects with haptoglobin type 1-1, in whom the plasma haptoglobin is a monomer of 90,000 molecular weight; it cannot be found in urine from subjects with haptoglobin type 2-2, in whom the haptoglobin exists in plasma as a series of larger, and higher molecular weight, polymers. These findings are in accord also with the hypothesis that protein reabsorption by the tubules is not a selective phenomenon.

Using the immunochemical techniques, the renal clearance of individual plasma proteins has been measured in healthy subjects. If the above hypotheses are correct, a direct relationship between the molecular size (or molecular radius) of the plasma proteins and their renal clearance should be evident (see Fig. 10-2). Such a direct relationship was not found. The glomerular sieving mechanism for polyvinylpyrrolidone is intact in healthy subjects (See 10-1); if the sieving mechanism is intact for proteins as well, it is likely that selective or preferential reabsorption of some proteins occurs, or that some proteins penetrate into the nephron from postglomerular sites.

Tamm-Horsfall Protein. Many proteins that are excreted in normal urine are, no doubt, derived from the kidney and lower urinary tract, from turnover of cells and secretion. Little is known about most of them. An exception is *uromucoid*, a high molecular weight threadlike mucoprotein originally isolated from normal urine by Tamm and Horsfall, by whose names it is often known. It is predominantly the protein found in normal urine, about 40 mg. being excreted per day. It has not been identified in serum but can be found in the urine produced by the isolated perfused kidney. It is secreted by the cells of the ascending limbs of the loops of Henle, the distal tubules and the collecting ducts. Tamm-Horsfall mucoprotein is the major component of the matrix of the casts that form in the lumens of the distal nephron and are excreted in normal and abnormal urine as hyaline casts. Uromucoid is soluble at a pH above 7.0 and precipitates at acid pH. Not surprisingly, these are the solubility characteristics known for the hyaline casts found in normal urine.

Proteinuria and Exercise

Glomerular filtration rate, renal plasma flow and urine flow decrease during heavy physical exercise, but protein excretion in the urine increases manyfold. In one study, for example, the excretion of protein in healthy subjects at rest was 0.035 mg. per minute, whereas after exercise it was 0.21 mg. per minute. Increased urinary excretion of red blood cells and casts also occurs with heavy physical exercise. Urine protein excretion continues to be high after cessation of exercise—and, in fact, may increase in the first 30 minutes of recuperation following exercise.

Recent observations on the nature of the urinary proteins indicate that the increased proteinuria is due almost exclusively to an increase in the amount of plasma proteins excreted. When the findings are compared with those of normal urine, a striking augmentation is noted in the excretion of albumin (m. w. 69,000), of proteins of similar molecular weight, and of proteins of higher molecular weight (ca. 160,000). These findings suggest that a change in glomerular permeability occurs with strenuous exercise. Normal urine does not contain haptoglobin in subjects with plasma haptoglobin type 2-1. After exercise, this large molecular weight ($>$ 100,000) haptoglobin polymer is found in the urine. This observation is consistent with a change in the characteristics in the molecular sieve of the glomerulus.

The mechanism underlying the proteinuria of exercise is not well understood. The most plausible hypothesis was suggested by Poortmans. He proposed that the rise in the release of epinephrine and norepinephrine into the blood during exercise results in constriction of the afferent arterioles of the glomerulus; slowing of the glomerular filtration rate and renal plasma flow would result, thereby permitting a greater diffusion of plasma proteins through the glomerular capillary wall into Bowman's space. This cannot be the entire explanation, for, when studies of the renal clearance of plasma proteins were made, no direct relationship could be demonstrated between the molecular weights of the individual proteins and their renal clearance. An alteration in the capacity of the tubules to reasborb proteins might also be present, such that the tubular maximum reabsorptive capacity is reached for some proteins before it is reached for others. The decreased plasma flow in the peritubular capillaries might well alter the capacity of the tubules to reabsorb protein.

The large number of hyaline and other casts seen in the urine is consistent with an increased secretion of uromucoid by the cells of the distal nephron, as well as a considerable increase in the normal shedding of renal tubular cells into the tubular lumens. The increased number of red blood cells in the urine during heavy exercise seems likely to originate from glomerular or peritubular capillaries. If functional gaps or pores appear in capillaries of a size large enough to permit the passage of red blood cells, it is likely that they would also facilitate the transfer of plasma proteins directly from plasma into Bowman's space or into tubular lumens. Such a phenomenon would indicate that there is at least some heterogeneity of nephron function—that in some nephrons, at least, the molecular sieve may be temporarily bypassed and that postglomerular transfer of plasma proteins into the urine could occur in a few nephrons.

Proteinuria and Posture

In healthy subjects a greater excretion of protein has been demonstrated in the upright than in the recumbent position. Using a sensitive quantitative immunochemical method to estimate the albumin excretion of apparently healthy volunteers, Robinson and Glenn found an excretion rate of 12 μg per minute while their subjects were in the upright position, as compared with 1.1 μg per minute while they were lying down. In these subjects the concentration of albumin (and protein) in the urine in both the recumbent and the upright position was too small to be detected by the qualitative tests in clinical use for the detection of protein.

When the amount of protein in the urine in the recumbent position is insufficient to be detected by the qualitative tests and protein is readily detected in the upright position while standing or during quiet ambulation, orthostatic or postural proteinuria is said to be present. This condition, which is most frequent during adolescence, has long been recognized. However, it remains a source of clinical controversy, for it is still unclear whether or not it is caused by or associated with significant structural abnormality of the kidney. Very limited studies, using quantitative immunochemical methods, indicate that the excretion of albumin in the recumbent position, although very small, exceeds that in healthy subjects, and that the relationship between albumin excretion in the upright position and that in the recumbent position is similar in healthy subjects and those with postural proteinuria. This suggests that in postural proteinuria there is a slight but definite exaggeration of the normal protein excretion in recumbency as well as in the upright position.

Systematic quantitative measurements of proteins other than albumin have not been made in urine and serum of subjects with postural proteinuria, so that data in regard to the interrelationship of the clearance of individual plasma proteins and their molecular size or molecular radius are scanty. Thus, the question of whether or not there is an alteration in the molecular sieve of the glomerulus must await the results of further study.

The mechanism of postural proteinuria is unknown; however, the most plausible explanation is that there is an increased glomerular permeability to protein. This was at one time thought to be a consequence solely of the decrease in glomerular filtration rate and renal plasma flow observed in these patients when they assume the upright position, but there is little evidence that the hemodynamic alterations differ from those in healthy subjects. The demonstration of very minor structural changes in the glomeruli of half the patients with postural proteinuria suggests that there may be a minor alteration in the function of the glomerular capillary wall, facilitating protein transfer into Bowman's space in the recumbent position and reinforced by the decrease in glomerular filtration rate in the upright position.

PROTEINURIA IN DISEASE

In disease, proteinuria may occur when the kidney is normal if there is an unusually high level of proteins in the plasma, particularly proteins of relatively low molecular weight. It may be found when structural and functional changes in the

TABLE 10-1. Types of Proteinuria in Disease

Prerenal, with normal glomerular permeability
Increased glomerular permeability
Renal tubular disorders
Lower urinary tract
Chyluria

glomeruli are associated with an increased glomerular permeability to proteins, or in diseases in which the glomeruli appear to be normal, where there are structural and functional alterations in the tubules. Also, proteinuria may develop when abnormal losses of nonplasma protein arise from cells of the kidney or ureter, lower urinary tract and accessory glands and (at least in theory) from drainage of lymphatics into the urinary tract (Table 10-1).

Proteinuria Associated With Unusually High Levels of Plasma Proteins.

If the level of a plasma protein rises to such a degree that the increased amount crossing the molecular sieve of the glomerulus saturates the reabsorptive capacity of the tubules, proteinuria will be evident. Increased levels of albumin rarely occur in disease states; however, albuminuria can be demonstrated experimentally in animals and man when the plasma albumin level is raised by the infusion of albumin.

In disease, proteinuria occurs particularly when there is an increase in the plasma level of relatively low molecular weight proteins that are normally absent or present in very low concentration (Table 10-2). For example, proteinuria is frequent when hemoglobin circulates in the plasma, as occurs in association with intravascular hemolysis. Initially the hemoglobin binds to the plasma haptoglobin, producing the relatively large hemoglobin-haptoglobin complex, which is retained by the glomeruli. Only when the plasma haptoglobin has been saturated does free hemoglobin circulate. This smaller molecule (m. w. 68,000; molecular radius 33 Å) passes through the

TABLE 10-2. Proteinurias of "Prerenal" Origin

TYPE OF PROTEIN	CLINICAL CONDITION
Infusion of homologous plasma proteins	Experimental or therapeutic
Proteinuria of newborn ruminants: β-lactoglobulin	Physiological
Paraproteinemias—Bence-Jones proteinurias	Multiple myeloma
Hemoglobinuria	Intravascular hemolysis
Myoglobinuria	Muscle destruction
Inflammatory disease with increased low molecular weight proteins in plasma	Inflammation

glomerulus and saturates the reabsorptive capacity of the tubules (see Fig. 10-2). Myoglobin, the related muscle protein, is of low molecular weight (17,500) and small radius (19 Å) and appears in the circulation as a result of injury to or inflammation of muscle tissue. It binds to only a minor degree to other proteins and is readily filtered by the glomerulus; thus, myoglobinuria is the consequence of even a low level of myoglobin in plasma (see Fig. 10-2).

Another example is the light chains (Bence-Jones proteins) of the immunoglobulin molecules (see Fig. 10-2). Light chains are found in normal plasma together with the heavy chains as part of the intact immunoglobulin molecules. There are only minute traces of free light chains present in plasma, usually as monomers or dimers (m. w. 22,000 or 44,000). Light chains are readily filtered by the glomerulus, and minute amounts are found in normal urine. When proliferative disorders of the plasma cells such as multiple myeloma occur, the immune globulins, or their heavy or light chains, or fragments of the chains, are produced in increased amounts and circulate in the plasma. The intact molecules of IgG or IgA, for example, being of relatively high molecular weight (160,000) and large radius (50 Å), do not readily pass the glomerular molecular sieve and rarely appear in the urine, even when the plasma level is greatly increased. In contrast, when the low molecular weight free light chains

are produced in large amounts, they readily pass the glomerular molecular sieve, presumably saturate the tubular reabsorptive mechanism and are excreted in the urine.

Proteinuria Associated With Increased Glomerular Permeability

The urinary excretion of a large quantity of protein is common in patients with diseases affecting the glomeruli, and the protein loss may exceed 20 grams per day or more. Many changes may be seen in the structure of the glomeruli. Capillary loops or whole glomeruli may be obliterated, inflammation with cellular proliferation may occur, the thickness and character of the capillary basement membrane may be altered, and substances may be deposited in the mesangium and on the subendothelial and subepithelial aspects of the basement membrane, and in the basement membrane itself.

Many other structural changes appear to result in a glomerular molecular sieve that is more leaky than the normal. Since it is likely that the tubular capacity to reabsorb filtered proteins does not increase greatly, increased leakiness of the glomerulus or increased filtration of protein will be reflected in increased protein excretion in the urine. If the glomerular ultrafilter becomes blocked or obliterated, the physiological consequence is a decrease in the glomerular filtration rate; proteinuria may become quantitatively reduced or even disappear.

The leakiness of the glomerular ultrafilter

can be expressed physiologically in two ways. Increased permeability results in the filtration and clearance of relatively more protein. Thus, the albumin clearance per unit glomerular filtration rate (i.e., the sieving coefficient) is greatly increased when there is functional damage to the glomerular ultrafilter (see Fig. 10-1, C and D) and may return to within the normal range if that damage can be repaired completely.

Plasma albumin is almost completely retained by the normal glomerular molecular sieve (see Fig. 10-1, A and B) and only trace amounts of proteins larger than albumin appear in the normal urine. Damage to the molecular sieve in disease might be expected to result in a diminished capacity to retain not only albumin but plasma proteins larger than albumin. In diseases associated with structural abnormalities of the glomerulus, the clearance of proteins larger than albumin is often greatly increased, and the urine may contain many serum proteins, including those of large molecular size. The relationship between the molecular size of the plasma proteins and their comparative clearances by the kidney may be used as another index of the leakiness of the diseased glomerular ultrafilter. The clearance by the kidney of significant quantities of proteins of large molecular weight is consistent with a considerable alteration in the molecular sieving action of the glomerulus.

In Figure 10-1 the line originating with point C represents the sieving coefficients for albumin (molecular radius 35 Å) and IgG (molecular radius 50 Å) in a patient with massive proteinuria, normal glomerular filtration rate, and no evident morphological alterations in the glomerular basement membrane. The sieving coefficient for albumin is about 500 times the normal (point B), and that for IgG is one tenth that for albumin.

In Figure 10-1 the line originating with point D is derived from data from a patient with profound damage to the glomerular basement membrane, with resultant severe impairment of glomerular filtration rate and considerable proteinuria. The sieving coefficient for albumin was 8 (i.e., 20,000 times the normal). (See Fig. 10-1.) Larger proteins were also in the urine in abundance. The sieving coefficient for IgG (50 Å) was 2 and that for α_2 macroglobulin (m. w. 820,000; molecular radius 90 Å) was 0.08.

Changes in the glomerulus in disease are often heterogeneous, and it may well be that the glomerular filter is altered in varying degrees from glomerulus to glomerulus. The results of studies on intact kidneys may reflect only the average of a wide diversity of changes in the permeability of the glomerular filter of individual nephrons. Pores or gaps have not been found in the capillary basement membrane of normal glomeruli. Recent evidence suggests that actual dissolution of basement membrane may occur in segmental areas of a few capillary loops of diseased glomeruli at least occasionally. Such gaps in the continuity of the basement membrane could be responsible for the passage into the urine of large molecular weight proteins; their infrequent and segmental distribution emphasizes the probable heterogeneity of physiological alterations that may occur in the glomeruli in disease states.

Proteinuria Associated With Abnormal Renal Tubular Function

Only recently has a particular type of proteinuria been recognized in association with disorders of function of the renal tubules. In addition to the proteinuria, the renal tubular functional abnormalities may include, singly or in various combinations, impairment of reabsorption of water, ions, glucose, uric acid, amino acids and organic acids and impairment of acid excretion.

The proteinuria is of mild degree and, unlike that associated with impairment of glomerular function, rarely exceeds 1 to 2 grams per 24 hours. As indicated previously, the proteins excreted in states of altered glomerular permeability include albumin (m. w. 69,000) as the predominant protein and many plasma proteins larger than albumin. In tubular proteinuria, by contrast, proteins of relatively small size (m. w. 10,000 to 50,000) predominate in the urine. The origin of these proteins was, for a long time, difficult to determine with certainty. Obviously, they might be proteins originating from the kidney itself, particularly from damaged tubules. An alternative explanation is that they are proteins of low molecular weight normally present in plasma and in low concentration. The availability of monospecific antisera directed against many individual low molecular weight plasma proteins has clarified this issue. The proteins excreted in so-called tubular proteinuria have now been shown to originate predominantly from plasma.

Proteins smaller than albumin are filtered more readily by the normal glomerular molecular sieve. Indeed, it is likely that the filtration rate of a small protein such as β-microglobulin (m. w. 13,000) approaches the glomerular filtration rate, at the same time that albumin and large plasma proteins are effectively retained by the glomerulus. Being more readily filtered by the glomerulus, the low molecular weight proteins present to the proximal tubules for reabsorption in amounts that are large compared with those of albumin. In the presence of normal glomerular function, impairment of the ability of the proximal tubules to reabsorb protein will therefore be reflected quantitatively as a relative inability to reabsorb proteins of low molecular weight.

Chyluria

Chyluria is a rare condition arising as a result of fistulous connections between the lymph vessels and the urinary tract; there is a consequent loss of lymph, which drains from the lower limbs and abdominal organs into the urine. Some proteins from the interstitial fluid are probably returned to the bloodstream via the lymphatics; so that the proteins in the urine in chyluria are probably a reflection of the permeability of the nonglomerular capillaries. As might be expected, the proteinuria is very different from that seen in association with glomerular damage. The composition of the urine proteins is very similar to that of plasma, and the clearance of all plasma proteins studied, including α_2-macroglobulin (m. w. 820,000) and β-lipoprotein (m. w. 3,600,-000), is high.

THE METABOLISM OF ALBUMIN AND OTHER PLASMA PROTEINS

It is now clearly recognized that individual plasma protein molecules are continuously being lost, and are continuously replaced by newly synthesized protein molecules. The balance of these two processes determines the actual body pool of a protein at any given time. In normal man it is probable that about one tenth of the mass of circulating plasma protein is broken down each day. Protein molecules, when destroyed, are completely degraded.

Studies of plasma protein metabolism have been made, in the main, by coupling the isolated protein with a radioactive label, most commonly ^{131}I; analysis of the kinetics aims at defining the amounts of protein synthesized and broken down each day, at estimating the size of the intravascular and extravascular (or interstitial) pools of each protein, and at describing the interchanges that take place between the various pools. Precise details of these aspects of protein kinetics are the subject of continued investigation and debate.

The results obtained in any study must be interpreted in the light of the many theoretical and technical factors that may influence the data (Table 10-3). The protein to be studied is altered chemically during the process of isolation from the plasma and may differ in varying degrees

TABLE 10-3. Some of the Factors Affecting the Results of Studies with Isotopic Labeled Proteins

---

Alteration of the protein during isolation

Alteration of the protein during the labeling procedure

Stability of isotopic label

Lack of homogeneity of labeled protein

Reutilization of the isotope label

Mathematical model used

Prolonged or inefficient mixing

Local variations in metabolic rates

Steady state of the subject at the time of study

---

from the native protein. The body recognizes this and metabolizes the altered protein much more rapidly than it does the native protein. Thus, only proteins isolated by the mildest fractionation procedures should be used. The protein molecule may be damaged by the labeling procedure, as occurs when ^{51}Cr is used. Iodination is the most satisfactory method of labeling, provided that the degree of iodination is minimal (not more than one atom of iodine per molecule). The use of damaged protein molecules can be minimized by injecting the labeled protein into another animal that metabolizes them rapidly and, therefore, screens out the most altered of the protein molecules. Reutilization of the isotopic label after catabolism poses a technical problem which may be partially overcome by giving excess amounts of the metabolite (iodine).

Plasma proteins are distributed in two major compartments, the intravascular and the extravascular. The extravascular compartment in fact has many parts, each with its own concentration of protein and rate of transfer from the plasma. Thus, the nature of the mathematical model used to describe plasma protein kinetics in terms of the compartments affects the results, as do the efficiency of mixing, and local variations in metabolic rates. In addition, most observations can be made only under steady state conditions, in which it is assumed that the rate of protein synthesis is equal to the rate of protein catabolism. An analysis of a typical plasma radioactivity decay curve is presented in Figure 10-5.

A recent and representative investigation using iodinated albumin to study albumin kinetics in normal subjects is summarized in Table 10-4. The total albumin mass is 3.9 grams per kg. in women, and 4.7 grams per kg. in men. Of the total mass, 45 percent is intravascular in women, 42 percent in men. Destruction and renewal of albumin appears to occur exclusively in the plasma or intravascular compartment, at a rate (the fractional catabolic rate) of about 8.5 percent of the intravascular albumin mass per day. The rate of albumin synthesis is 0.16 grams per kg. per day.

Albumin biosynthesis (and that of most plasma proteins except the immune globulins) has been shown to occur mainly or exclusively in the liver. Albumin synthesis can perhaps best be studied by making use of a labeled precursor amino acid such as 6-^{14}C-arginine and observing its incorporation into albumin. This method does not necessarily require steady state conditions; it can be used in changing metabolic conditions and can also be applied to sequential studies. Using such a method, in which synthesis and catabolism could be measured independently, Tavill, Craigie, and Rosenoer demonstrated that, over a wide range of rates of synthesis (from 0.05 to 0.30 grams per kg. per day), albumin catabolism and synthesis were equal. They also showed that in subjects with a low albumin pool the infusion of albumin to increase the plasma level resulted in no change in the rate of synthesis.

Fasting has a profound effect in that it results in a decrease in albumin synthesis of about 30 to 50 percent. Malnutrition also leads to a decrease in albumin synthesis; this occurs rapidly as dietary intake falls. A net transfer of albumin into the intravascular pool occurs, and there is a

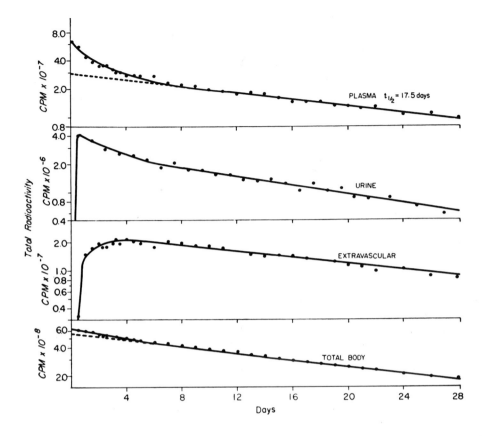

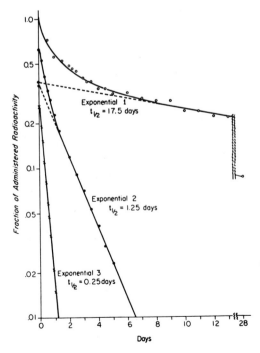

Fig. 10-5. The data obtained in a single human subject after the intravenous injection of a bolus of albumin labeled with [131]I. (*Top*) The curves of plasma and of urine radioactivity were obtained by measurement of timed plasma samples, and of 24-hour urine collections. The total body radioactivity was derived from the amount injected minus the amount excreted in the urine. The extravascular radioactivity was derived by subtracting the total plasma radioactivity from the total body radioactivity. (*Left*) A graphic analysis of the plasma radioactivity decay curve of the same subject. The open circles (top curve) are actual measurements plotted on semilogarithmic paper. The straight terminal portion of this curve was extrapolated back to its intersection with the vertical axis to yield exponential 1. A second curve was derived by subtracting the extrapolated curve from the original curve; exponential 2 was derived from this curve. The values in exponential 3 were obtained from residual values after the calculated data values of exponentials 1 and 2 were subtracted. This is an example of a 3 compartment model for the distribution of albumin. (Beeken, W. L., *et al.*: J. Clin. Invest., *41*:1312, 1962. With the permission of authors and publishers.)

TABLE 10-4. Metabolism of Albumin in Normal Subjects and in
Patients with the Nephrotic Syndrome

	NORMALS*		PATIENTS WITH NEPHROTIC SYNDROME (15)†
	Women (19)	Men (15)	
Plasma albumin concentration (g./100 ml.)	4.3 ± 0.4	4.4 ± 0.2	1.5 (0.9-2.8)
Plasma volume (liters)	2.4 ± 0.3	2.8 ± 0.2	2.9 (1.9-3.8)
Intravascular albumin mass (g./kg.)	1.8 ± 0.3	2.0 ± 0.3	0.71 (0.36-1.28)
Total exchangeable albumin mass (g./kg.)	3.9 ± 0.3	4.7 ± 0.1	1.38 (0.72-2.42)
Ratio (intravascular/total mass) (%)	45 ± 5	42 ± 4	52 (47-60)
Fractional catabolic rate (%/day)	8.5 ± 1.2	8.4 ± 0.5	15 (5-24)
Rate of synthesis (g./kg./day)	0.15 ± 0.02	0.17 ± 0.03	0.20 (0.15-0.29)

* Mean, ± 1 S.D. from Rossing, N.: Clin. Sci., 33:593, 1967
† Mean and range from Jensen, H., Rossing, N., Andersen, S. F., and Jarnum, S., Clin. Sci., 33:445, 1967

fall in the catabolic rate. On low protein diets, low anabolic and catabolic rates have been demonstrated, and the extravascular albumin mass is also decreased. On refeeding, anabolic and catabolic rates can return rapidly to normal.

Patients with the nephrotic syndrome lose large amounts of albumin in the urine. The metabolism of albumin has been studied in 15 such patients by Jensen, Rossing, Andersen and Jarnum (Table 10-4). Their total albumin mass was greatly decreased to 1.38 grams per kg., and a significantly higher proportion (52%) than normal was found in the intravascular compartment. In these patients both the synthetic rate and the catabolic rate were increased.

In normal man about 10 to 15 grams of albumin are catabolized daily. Destruction of plasma proteins has been shown to occur by way of the digestive tract and of the kidney, as well as by catabolism in tissue cells, particularly the liver. In the main because of difficulties in preparation

and labeling discussed above, there is as yet no agreement as to the relative importance of these sites.

A constant leakage of plasma proteins occurs into the secretions of the gastrointestinal tract. There are considerable difficulties in quantifying the magnitude of this loss, but there is little doubt that it is increased in certain types of inflammatory gastrointestinal disease. Perhaps 30 to 70 percent of the total mass of albumin is normally degraded in the gastrointestinal tract, mainly in the small bowel but also in the stomach. The bulk of the amino acids liberated by intestinal catabolism of albumin and other plasma proteins is presumably reabsorbed and reaches the liver via the portal circulation, there to be reutilized for protein synthesis.

As discussed previously, there is much evidence that the glomerular filtrate contains small amounts of albumin and plasma proteins of molecular size smaller than albumin. If the albumin content of the glomerular filtrate is 10 to 20 mg. per liter,

1.8 to 3.6 grams of albumin would be filtered in adult man. Only one hundredth of that amount is excreted in the urine. The remainder is presumably reabsorbed in the proximal tubules which contain large numbers of lysosomes. The protein is absorbed into pinocytotic vacuoles, which then fuse with the lysosomes to form cytophagolysosomes in which the protein is presumably broken down by cathepsins and other proteolytic enzymes. The quantitative importance of this mechanism appears to be relatively small in the normal, but there is some evidence to suggest that in proteinuric states there is a considerable increase in albumin catabolism by the kidney. The amino acids and peptides liberated are conserved and are probably returned to the circulation via the renal lymphatics.

The other major site of albumin catabolism appears to be the liver. When carefully prepared and labeled albumin is used, after screening through another animal, the results suggest that the liver is responsible for about 10 percent of the normal albumin catabolism.

Kinetic studies cannot be applied routinely in clinical medicine; however, the plasma albumin level is observed to be low in many disease states (Table 10-5). It may be low because a large amount of albumin is being lost—in the urine or the gastrointestinal tract or from other sites such as the skin when a large area has been burned. It may be low because the liver is so severely damaged that synthesis clearly is

TABLE 10-5. Some of the Causes of a Low Plasma Albumin Level

Loss
  In the urine
  From the gastrointestinal tract
  From other sites (e.g., skin)
  From intravascular to extravascular pool
Increased fractional rate of catabolism
Decreased synthesis
Genetic analbuminemia

---

decreased. Many other factors such as poor nutrition, fever, tissue breakdown, etc. influence the synthetic and catabolic rates in subtle ways which the physician cannot study at the bedside.

THE CONSEQUENCES TO THE BODY OF THE LOSS OF LARGE AMOUNTS OF ALBUMIN AND OTHER PROTEINS IN THE URINE

When large amounts of protein are continuously lost in the urine, the plasma albumin level is decreased, the plasma cholesterol level becomes elevated, and significant edema occurs. These findings are usually observed when the urine protein loss exceeds 50 to 70 mg. per kg. per day, and, collectively, are known as the nephrotic syndrome. This syndrome occurs in association with a wide variety of disease processes in which the glomeruli are damaged.

In patients with the nephrotic syndrome the plasma albumin level is correlated inversely with the degree of urine albumin (or protein) loss. The low plasma albumin level is a consequence of several factors. It is possible that in some instances excessive loss of albumin occurs into the gastrointestinal tract. An increased fractional rate of catabolism of albumin has been clearly demonstrated in the nephrotic syndrome in man (see Table 10-4), and studies indicate that the renal tubules catabolize considerably more albumin in the animal with experimental nephrotic syndrome than in the normal animal. The rate of synthesis of albumin in the nephrotic syndrome is increased in many subjects; however, malnutrition occurs in many patients with severe nephrotic syndrome, and its effects may well be to decrease the rate of synthesis of plasma albumin.

Another characteristic feature of the nephrotic syndrome is the almost invariable elevation of the plasma cholesterol and phospholipid levels, which are correlated inversely with the plasma albumin level. Plasma triglycerides are elevated particularly when the plasma albumin is very low. The plasma lipids exist almost exclusively in combination with the lipoproteins. In patients with the nephrotic syndrome, significant abnormalities of the plasma lipoproteins occur. In particular, the concentration of the "low" density (D 1.019 to 1.063) and of the "very low" density (D < 1.019) lipoproteins is increased in the plasma. (The density of most proteins such as albumin is about 1.3.)

The cause of the alterations in plasma lipids is still unclear. One explanation suggests that the hyperlipemia may depend on the renal tubular reabsorption of plasma protein and particularly of plasma albumin. Changes in the degree of albuminuria in the experimental rat are accompanied by changes in lipemia of magnitudes such that the relationship between the two variables remains unaltered, and the renal tubular uptake of albumin appears to be related closely to the level of serum lipids.

It is possible that the observed changes may be a reflection of alterations in the rate of hepatic production of the lipids and lipoproteins or in the rate of their disposal, elimination or interconversion. Elevation of the plasma albumin level by the infusion of albumin leads to a decrease in the elevated plasma levels of cholesterol and phospholipids, and similar effects have been described when gamma globulin or polysaccharide molecules such as dextran and polyvinylpyrrolidone have been infused. These observations suggest the possibility that osmotic effects may control hepatic synthesis of plasma lipids. There is some evidence to suggest that the hepatic production of albumin also may be regulated by osmotic effects rather than by albumin concentration per se. Thus, the low oncotic pressure of the plasma, consequent to the decreased plasma albumin level, may be a stimulus to hepatic synthesis of both albumin and lipids.

There are changes in the levels of many plasma proteins in patients with the nephrotic syndrome. It is possible that the rates of hepatic synthesis of many plasma proteins besides albumin are increased. In general, proteins of large molecular size are retained in the plasma by the kidney; their plasma level therefore increases. An example is the large molecular weight (840,000) α_2 macroglobulin, which has the property of binding trypsin and is an inhibitor of α_2 trypsin and plasmin. Smaller proteins are usually excreted in large amounts in the urine, and their plasma level may be decreased. For example, the plasma transferrin (m. w. 90,000) level is usually decreased, leading to a reduction of the plasma iron binding capacity and, possibly, contributing to the increased susceptibility to infection which occurs in these patients. The levels of the thyroxine binding prealbumin (m. w. 61,000) and of the copper binding ceruloplasmin (m. w. 160,000) are also decreased. IgG (m. w. 160,000), which is synthesized in extrahepatic sites, is lost in the urine; the consequent reduction in plasma level may contribute to the susceptibility to infection.

The level of plasma fibrinogen is increased, as are the levels of many coagulation factors; some authors consider that there is an increase in frequency of thrombosis in patients with the nephrotic syndrome. Clotting factor IX, on the other hand, may be lost in the urine in large amounts, and the consequent low plasma level may be associated with the bleeding tendency observed in some patients.

Edema is the characteristic clinical sign observed in the nephrotic syndrome. It occurs almost invariably when the plasma albumin level is decreased to about 16 to 18 grams per liter. As a consequence of the hypoalbuminemia, the plasma oncotic pressure is decreased, leading to an increased leakage of fluid from the intravascular to the extravascular compartment. A low plasma albumin level cannot be the sole cause of the edema, as edema does not occur in patients with analbuminemia in whom the plasma albumin level is usually less than 0.1 grams per liter. In some patients a lowered plasma volume, increased secretion of aldosterone, and increased tubular reabsorption of sodium are other important contributing factors.

ANNOTATED REFERENCES

Baxter, J. H.: Hyperlipoproteinemia in nephrosis. Arch. Intern. Med., *109*:742, 1962.

Berggard, I.: Plasma proteins in normal human urine. *In:* Manuel, Y., Revillard, J. P., and Betuel, H. (eds.): Proteins in Normal and Pathological Urine. p. 7. Baltimore, University Park Press, 1970.

Laterre, E. C., and Heremans, J. F.: Proteins peculiar to normal urine and other secretions. *In:* Manuel, Y., Revillard, J. P., and Betuel, H. (eds.): Proteins in Normal and Pathological Urine. p. 45. Baltimore, University Park Press, 1970. (The two preceding papers give excellent brief accounts, which detail the wide variety of proteins found in urine.)

Chinard, F. P., Lauson, H. D., Eder, H. A., Greif, R. L., and Hiller, A.: A study of the mechanism of proteinuria in patients with the nephrotic syndrome. J. Clin. Invest., *33:* 621, 1954. (The diffusion theory is considered in this paper.)

Dirks, J. H., Clapp, J. R., and Berliner, R. W.: The protein concentration in the proximal tubule of the dog. J. Clin. Invest *43*:916, 1964.

Farquhar, M. G., Wissig, S. L., and Palade, G. E.: Glomerular permeability. I. Ferritin transfer across the normal glomerular capillary wall. J. Exp. Med., *113*:47, 1961.

Farquhar, M. G., and Palade, G. E.: Glomerular permeability. II. Ferritin transfer across the glomerular capillary wall in nephrotic rats. J. Exp. Med., *114*:699, 1961.

Graham, R. C., and Karnovsky, M. J.: Glomerular permeability. Ultrastructural cytochemical studies using peroxidases as protein tracers. J. Exp. Med., *124*:1123, 1966. (The three preceding papers describe some morphological aspects of glomerular permeability.)

Grant, G. H.: The proteins in normal urine. II. From the urinary tract. J. Clin. Path., *12:* 510, 1959.

Hardwicke, J. E., and Soothill, J. F.: Proteinuria. *In:* Black, D. A. K. (ed.): Renal Disease. ed. 2, Chap. 10, p. 252. Philadelphia, F. A. Davis, 1967.

Hardwicke, J. E., Cameron, J. S., Harrison, J. F., Hulme, B., and Soothill, J. F.: Proteinuria, studied by clearance of individual macromolecules. *In:* Manuel, Y., Revillard, J. P., and Betuel, H. (eds.): Proteins in Normal and Pathological Urine. p. 111. Baltimore, University Park Press, 1970. (A very comprehensive modern review which is an excellent starting point for reading)

Katz, J., Sellers, A. L., and Bonorris, G.: Effect of nephrectomy on plasma albumin catabolism in experimental nephrosis. J. Lab. Clin. Med., *63*:680, 1964. (A paper which provides some direct experimental evidence for protein catabolism by the kidney)

Lambert, P. P., Gassee, J. P., and Askenasi, R.: Physiological basis of protein excretion. *In:* Manuel, Y., Revillard, J. P., and Betuel, H. (eds.): Proteins in Normal and Pathological Urine. p. 67. Baltimore, University Park Press, 1970. (An excellent, brief, modern review of the physiological basis of glomerular permeability. It is an excellent starting point for the student interested in this subject.)

McQueen, E. G.: The nature of urinary casts. J. Clin. Path., *15*:367, 1962.

Manuel, Y., and Laterre, E. C.: Tubular proteinuria; biochemical assay. *In:* Manuel, Y., Revillard, J. P., and Betuel, H. (eds.): Proteins in Normal and Pathological Urine. p. 172. Baltimore, University Park Press, 1970.

Pappenheimer, J. R.: Passage of molecules through capillary walls. Physiol. Rev., *33:* 387, 1953. (A full review and theoretical treatment of the subject as it applies to capillaries in general and to the glomerular capillaries)

Poortmans, J., and Jeanloz, R. W.: Quantitative immunochemical determination of 12 plasma proteins excreted in human urine collected before and after exercise. J. Clin. Invest., *47*:386, 1968.

Robinson, R. R., and Glenn, W. G.: Fixed and reproducible orthostatic proteinuria. IV. Urinary albumin excretion by healthy human subjects in the recumbent and upright postures. J. Lab. Clin. Med., *64:*717, 1964.

Robinson, R. R., Lecocq, F. R., Phillippi, P. J., and Glenn, W. G.: Fixed and reproducible orthostatic proteinuria. III Effect of induced renal hemodynamic alterations upon urinary protein excretion. J. Clin. Invest., *42:*100, 1963.

Rossing, N.: The normal metabolism of ^{131}I-labeled albumin in man. Clin. Sci., *33:*593, 1967.

Jensen, H., Rossing, N., Andersen, S. B., and Jarnum, S.: Albumin metabolism in the nephrotic syndrome in adults. Clin. Sci., *33:*445, 1967. (These two papers from the same laboratory provide good comparative data on the normal and the nephrotic.)

Rowe, D. S.: The molecular weights of the proteins of normal and nephrotic sera and nephrotic urine, and a comparison of selective ultrafiltrates of serum proteins with urine proteins. Biochem. J., *67:*435, 1957.

Schultze, H. E., and Heremans, J. F.: Molecular Biology of Human Proteins with Special Reference to Plasma Proteins. Vol. 1, Section IV, Proteins of Extravascular Fluids, Chap. 2, The urinary proteins, p. 670. Section III, The Life of Plasma Proteins, Chap. 1, Synthesis of the plasma proteins, and Chap. 2, Turnover of the plasma proteins, p. 321. Amsterdam, Elsevier, 1966. (This is the most comprehensive brief review available.)

Straus, W.: Occurrence of phagosomes and phago-lysosomes in different segments of the nephron in relation to the reabsorption, transport, digestion and extrusion of intravenously injected horseradish peroxidase. J. Cell. Biol., *21:*295, 1964. (An excellent demonstration of the morphological findings in tubules presented with a protein load)

Tavill, A. S., Craigie, A., and Rosenoer, V. M.: The measurement of the synthetic rate of albumin in man. Clin. Sci., *34:*1, 1968. (An important paper that describes the best approach to the study of protein synthesis)

Wallenius, G.: Renal clearance of dextran as a measure of glomerular permeability. Acta soc. med. upsal., [Supp.] *4:*1954. (An early, detailed paper on the clearance of macromolecules by the glomerulus. It is suitable only for the student who wishes to explore the subject in depth.)

11

Maintenance of
Fluid and Electrolyte Homeostasis

John F. Maher, M.D., *and Karl D. Nolph*, M.D.

INTRODUCTION

Abnormalities in the volume of body fluids and the quantity and concentration of electrolytes therein are frequently encountered in clinical medicine, yet they often are poorly understood by physicians. The healthy kidney compensates for wide variations in intake and extrarenal loss of fluid and electrolytes; when the kidney becomes functionally unable to adjust for these variations, a thorough understanding of fluid and electrolyte physiology is needed to correct resulting imbalances properly.

Serum sodium and potassium concentrations often fail to reflect total body stores. Low serum sodium concentration may occur in spite of increased extracellular fluid volume and total body sodium; conversely, hypernatremia occurs most often with dehydration and often is associated with some decrease in total body sodium. Changes in pH alter serum potassium concentrations without changing total body potassium. Therefore, therapy should be directed primarily not at correcting serum electrolyte concentrations but rather at restoring disordered mechanisms.

NORMAL SODIUM AND WATER METABOLISM

Normal Body Fluid Volumes and Their Measurement

The body is composed mostly of water located predominantly in cells (Fig. 11-1).

Seventy-two percent of lean body mass (body mass excluding fat) is water; body fat is virtually water free. The fat content of the body varies among individuals but averages about 15 percent of body weight. Total body water, then, is about 60 percent of body weight. One third of this is extracellular and two thirds intracellular fluid. Plasma, a portion of the extracellular fluid, represents about 4 percent of body weight. The volume of red blood cells is normally slightly less than that of plasma. In young children, body composition differs; blood volume is greater relative to body weight and extracellular fluid represents 30 percent or more of body weight. Total body water, as a percentage of body weight, decreases with age and with increased weight.

Indicator dilution techniques are used to measure body fluid compartments. A known amount of a substance is injected intravenously and, after equilibration, blood is sampled and the concentration of the indicator substance is determined. From determination of the amount by which the indicator has been diluted, the volume of distribution of the indicator can be calculated using the following formula:

$$\text{Volume} = \frac{\text{Quantity of indicator injected}}{\text{Concentration of indicator per ml. of sample}}$$

Albumin, labeled by radioiodine, or Evans blue dye (T-1824) are the indicators usually used for measuring plasma volume.

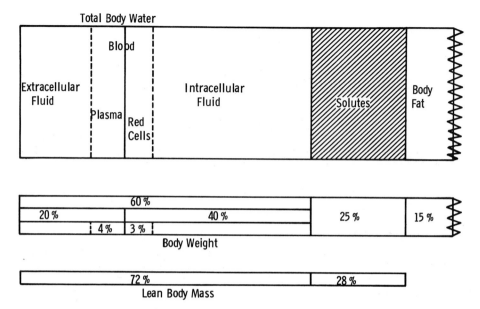

FIG. 11-1. Body composition, shown graphically. The body is made up mostly of water. Solutes in these water compartments and fat make up the remainder of the body mass. The solutes that are in solution approximate 300 mOsm/L. Body fat is obviously a variable quantity from patient to patient.

Blood volume can be calculated if plasma volume and hematocrit are known. Use of the albumin space results in overestimation of plasma volume, because some albumin moves into the extravascular spaces. A more nearly accurate measure of total blood volume is obtained using [51]Chromium-labeled erythrocytes as an indicator of red cell volume, in addition to labeled albumin to measure the plasma volume. Extracellular fluid volume may be estimated using radiosulfate, radiosodium, radiochloride, bromide, thiocyanate or inulin, each having its own theoretical space and distribution volume. Hence, the same "label" should be used to estimate serial changes in the volume of that body fluid compartment. Total body water is measured using tritium- or deuterium-labeled water or radioantipyrine. Intracellular water is calculated as the difference between total body water and extracellular fluid volume.

The fluid compartments are not static but dynamic, with water shifting within compartments and from one compartment to another, under the influence of variables such as posture, exercise and dietary intake. Furthermore, about 8,000 ml. enter and leave the gastrointestinal tract daily (1,500 ml. of saliva, 2,500 ml. of gastric juice, 500 ml. of bile, 700 ml. of pancreatic juice, and 3,000 ml. of intestinal fluid). The glomeruli filter 180 liters of water daily, almost all of which is reabsorbed by the tubules. Equilibration of infused solutes with cell water is very rapid, indicating that large volumes of fluid enter and leave the cells daily. The volume of body fluids is maintained within remarkably close limits by a number of regulatory processes, particularly the thirst and renal concentrating and diluting mechanisms. However, many disorders, particularly renal and gastrointestinal diseases, may cause changes in the volume and composition of body fluids.

Composition of Body Fluids

Each of the various electrolytes and non-electrolytes in normal plasma (Fig. 11-2) contributes to its osmolality (the total solute

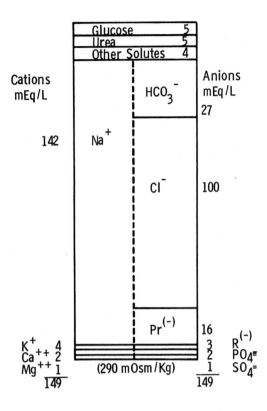

Fig. 11-2. Normal plasma composition. The sum of cations and anions corrected for ionization constants and valences plus nonionized compounds when adjusted for plasma water content equals osmolality (shown in parentheses). Interstitial fluid differs slightly from plasma.

concentration per kilogram of plasma water). Plasma osmolality is normally 285 to 290 milliosmols per kg. or (approximately) per liter of water. A milliosmol equals the molecular weight of a substance in milligrams and is caculated by dividing milligrams of any nondissociable (nonelectrolyte) substance by the molecular weight of that substance. In electrolytic solutions, the ionized cations and anions each contribute to osmolality. Each nonionized cation-anion pair contributes to osmotic pressure or osmolality to the same degree as each ionized cation or anion alone. For this reason, osmolality, measured by the freezing point depression, is less than would be calculated by measuring the concentration of each plasma component, since all electrotype pairs are not completely dissociated. Provided that glucose and urea concentrations are normal and no extraneous solutes such as mannitol are present, plasma osmolality can be approximated by doubling the sodium concentration.

Interstitial fluid (the extravascular, extracellular fluid) is similar in composition to an ultrafiltrate of plasma except that, being protein free, it is modified by the resultant Gibbs-Donnan phenomenon. The Gibbs-Donnan phenomenon refers to the effects of the negatively charged protein molecules upon electrolyte distribution across a semipermeable membrane. The rapid movement of water along osmotic gradients maintains osmolality almost identical in plasma and interstitial fluid. Plasma proteins, because of their large molecular size, contribute little to osmolality but carry many negative charges which must be paired to cations such as sodium. On the protein-free side of the membrane, cation concentration is lower and nonprotein anion concentration higher than in the plasma. Thus, interstitial fluid sodium concentration is approximately 6 percent lower than plasma water sodium concentration and there is a reciprocal increase in the chloride concentration. Plasma osmolality is one to two milliosmoles per kg. of water higher than interstitial fluid because of the combined effects of the plasma protein concentration and the Gibbs-Donnan effect.

The composition of interstitial fluid in specialized areas such as the cerebrospinal fluid, the gastrointestinal tract and the interstitial fluid of the kidney show marked differences, owing to the influence of active transport processes. The blood-cerebrospinal fluid and blood-brain barriers maintain gradients for some solutes such as phosphate and creatinine. Gastric juice has much higher potassium and hydrogen ion concentrations than has plasma. The interstitial fluid of the kidney is composed largely of reabsorbed glomerular filtrate. However, renal medullary interstitial fluid may have sodium and urea concentrations, as well as total osmolality, much higher than those of plasma.

The osmolality of intracellular fluid is assumed to be identical to that of extracellular fluid. The predominant intracellular cations are potassium and, to a lesser extent, magnesium; there is little sodium in cell water. Phosphate and sulfate are the major anions; chloride concentration is negligible. Urea is evenly distributed throughout total body water. Rapid intracellular metabolism of glucose accounts for much lower glucose concentrations within cells than in extracellular fluid.

Maintenance of Normal Fluid Volumes

To maintain health, there must be a balance between fluid intake and loss. The water entering the body is derived from the oral intake of food and liquids or from fluids administered parenterally. Another source of fluid, amounting to about 300 ml. daily, is the water derived from metabolism of carbohydrate, protein and fat. With negative caloric balance, catabolized cells release preformed water that is added to the residual cellular and extracellular fluid.

Water is lost normally from the lungs, skin, bowel and kidneys. Insensible water loss through diffusion and evaporation from the lungs and skin is about 1,000 ml. per day for the average sized adult. Water loss from the lungs can be increased by hyperventilation, intubation, tracheostomy and assisted ventilation; humidification of inspired air reduces this fluid loss. Both an increase in environmental temperature and fever increase cutaneous loss of water. Fluid loss from visible sweating may reach several liters per day. Fecal water normally is only about 100 ml. per day.

The kidney is the organ primarily responsible for maintaining the balance between fluid intake and losses. Urine volume, under normal conditions, may be as low as 500 ml. per day, or even less if the solute load is low. Water diuresis can lead to volumes as high as 20 liters per day, and even higher volumes can be achieved by solute diuresis.

Oral fluid intake ordinarily exceeds urinary output by about 500 ml. per day. Since the difference between gains and losses of water from other sources amounts to a net loss of 500 ml. per day, a balance is achieved. When water balance is maintained, body weight remains constant. Loss or gain of solid tissue is usually accompanied by water loss or gain representing a large fraction of the change in weight.

Maintenance of Sodium Balance

The usual daily dietary sodium chloride intake in this country is highly variable and normally ranges from six to 12 grams (100 to 200 mEq. or 2,300 to 4,600 mg. of sodium). Extrarenal sodium losses are normally less than 10 mEq. per day; the remaining intake must be excreted by the kidney to maintain a constant balance. Since wide variations in intake and extrarenal losses of sodium may occur, effective renal mechanisms exist to vary urinary excretion of sodium from several hundred to as little as one to two milliequivalents per day. Some of these mechanisms (filtered load, aldosterone) are well understood; others ("third factor") are under intensive investigation.

Filtered Load of Sodium. The filtered load of any substance can be determined by multiplying the glomerular filtration rate by its concentration in an ultrafiltrate of plasma. An approximation is made using plasma concentration. The precise ultrafiltrate concentration is estimated by multiplying the concentration of that solute in plasma water by a Gibbs-Donnan factor. As described previously, the sodium concentration in the protein-free ultrafiltrate is lower than that in plasma water. Since the plasma water sodium concentration is higher than measured plasma sodium concentration, corrections for plasma water and Gibbs-Donnan distribution nearly cancel one another. The filtered load of sodium normally is about 26,100 mEq. per day (145 mEq./L. $\times$ 180 per day). Since sodium excretion varies from two to over 200 mEq. per day, it represents from 0.01 to 1.0 percent of the filtered load. The filtered load of sodium is the first factor to be considered in the evaluation of sodium excretion. The renal tubules must reabsorb large quantities of sodium, normally 99 to 99.99 percent of the filtered load of about 18 mEq. per minute. As glomerular filtration rate varies, the tubular reabsorption of sodium does not remain a constant fraction of filtered sodium. A much greater natriuresis may occur with a small increase in filtered load than would be anticipated if the two parameters increased proportionately; conversely, a small decrease in filtered sodium may cause greater sodium retention than would be expected if the two variables showed parallel decreases. When glomerular filtration rate decreases because of loss of nephrons (as occurs in most forms of chronic renal disease), other factors affect the fractional sodium reabsorption so that sodium excretion is usually relatively high.

Renal Tubular Reabsorption of Sodium. The tubule reabsorbs sodium against electrical and concentration gradients by active transport, a process that accounts for most of the oxygen utilization by the kidney (Fig. 11-3). Between 60 and 80 percent of glomerular filtrate is reabsorbed isosmotically in the proximal tubule. Under all conditions, the osmolality of proximal tubular fluid remains nearly identical to that of plasma water. Normally some solutes, such as glucose, are completely reabsorbed in this segment; others, such as urea, are incompletely reabsorbed and some—for ex-

FIG. 11-3. Schematic diagram of the nephron. Normally, the proximal tubule reabsorbs all glucose, most of the sodium, water and phosphate, and some urea. Hydrogen ion and ammonia enter the tubule at many sites. Reabsorption of sodium in the ascending limb of the loop of Henle and distal tubule causes medullary interstitial hypertonicity of as much as 1,200 mOsm/kg. Water diffuses into the hypertonic interstitium from the descending limb of the loop of Henle and, if antidiuretic hormone is present, from the distal and collecting tubules. Distal sodium reabsorption occurs in exchange for potassium and/or hydrogen ion.

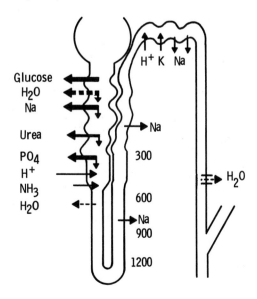

ample, creatinine—are not reabsorbed at all. Under normal conditions, the rate of reabsorption of all solutes approximates the rate of sodium reabsorption. Therefore, sodium concentrations in proximal tubular fluid remain nearly isotonic to plasma. If the proximal tubular fluid contains a large quantity of a nonreabsorbable solute, such as mannitol, the water it obligates will dilute the sodium and lower its concentration.

Sodium is reabsorbed in the ascending limb of the loop of Henle, a water-impermeable segment of the nephron, generating the hypertonicity of the medullary interstitium. In the diluting segment of the distal nephron, sodium reabsorption continues, rendering the tubular fluid hypotonic. In the absence of antidiuretic hormone, the tubular fluid undergoes little modification distal to the diluting segment, and a dilute urine is excreted. If antidiuretic hormone is present in the circulation, the distal nephron becomes water permeable and the final urine is concentrated.

Aldosterone. The adrenal mineralocorticoid hormones increase sodium reabsorption distally in exchange for potassium and hydrogen ions. No water transport occurs by this mechanism. The amount of sodium reabsorbed and potassium secreted depends on the delivery of sodium to this site, the intensity of the aldosterone stimulation, the integrity of the tubular cells and the intracellular potassium and hydrogen ion concentrations. Plasma or extracellular fluid volume depletion, other causes of decreased renal perfusion with resultant renin release, and elevated serum potassium levels are all stimuli for aldosterone release. Thus, aldosterone is the second factor to consider as a major determinant of urinary sodium excretion (For further discussion, the reader is referred to Chapter 16.)

Third Factor. When experimental conditions maintain both a constant glomerular filtration rate and a constant level of aldosterone activity, expansion of plasma or extracellular fluid volume still results in a natriuresis. Therefore, there must be a third factor (or factors) that determines, in part, sodium excretion. Possible explanations for this natriuresis attributed to a third factor are: a natriuretic hormone; an effect on blood viscosity or on tubular diameter; change in peritubular capillary hydrostatic and oncotic pressure or intratubular or interstitial hydrostatic or osmotic pressure; alteration of membrane characteristics; and a redistribution of blood flow from long corticomedullary to short outer cortical nephrons. Attempts have been made to demonstrate an inhibitor of sodium transport in the plasma of saline-expanded animals, using a variety of experimental techniques. A natriuretic hormonal factor that acts on the proximal tubule has not been identified definitely; however, delayed disappearance of saline from micropunctured tubules and increased free water clearance following saline loading have been interpreted as supportive evidence of proximal sodium transport inhibition. Results of cross circulation experiments are consistent with a natriuretic hormone, perhaps of renal origin, which acts on the distal tubule. Such a factor may oppose the action of mineralocorticoids and account for the phenomenon of escape from the salt-retaining effects of aldosterone.

ALTERED SODIUM AND WATER METABOLISM

Sodium ions and their associated anions are the solutes that account for most of the osmolality of extracellular fluid. Loss of sodium from the body isosmotically causes a proportionate decrease in extracellular fluid volume; retention of sodium may result in a proportionate increase in extracellular volume. Serum sodium concentration under these conditions remains normal. Later, compensatory mechanisms operating to restore extracellular fluid volume may superimpose changes in fluid balance and result in hyponatremia or hypernatremia.

Sodium Depletion

Sodium may be lost by any of a variety of mechanisms (Table 11-1). Sodium depletion may result from urinary losses in disorders that impair tubular transport (e.g., obstructive uropathy, pyelonephritis, polycystic kidneys, multiple myeloma, polyarteritis nodosa, diuretic phase of acute renal failure, and, occasionally, glomerulonephritis). Impaired bicarbonate reabsorption due to congenital or acquired disease also may lead to excessive loss of sodium with bicarbonate (renal tubular acidosis). In patients with renal failure, the osmotic diuresis resulting from retained urea and the increased solute load and glomerular filtration rate per residual nephron are associated with a reduced fractional reabsorption of sodium. When salt intake is restricted in such patients, salt depletion can occur. The lack of aldosterone and other mineralocorticoid in adrenal insufficiency also can cause renal salt-losing. Because of failure of the distal exchange (sodium for potassium) mechanism, potassium retention and hyperkalemia is a prominent feature of salt depletion in adrenal insufficiency. Most frequently the cause of renal salt depletion is the administration of diuretics. The mercurial diuretics tend to become ineffective once hypochloremia develops and the thiazide congeners are much less effective when extracellular fluid volume is depleted. The newer, more potent diuretics, ethacrynic acid and furosemide, may continue to induce saluresis in spite of extracellular fluid volume depletion, hyponatremia and reduced glomerular filtration rate.

Renal Response to Sodium Depletion. Salt loss is accompanied by loss of extracellular fluid. The clinical manifestations of such volume depletion include postural hypotension, tachycardia and organ ischemia; severe extracellular fluid volume depletion can result in shock, coma, lactic acidosis, uremia and hepatic failure.

Renal responses to salt depletion occur earlier than other manifestations. Through multiple mechanisms, the decreased extracellular fluid volume diminishes urinary sodium excretion. The decreased volume increases tubular sodium reabsorption by incompletely defined mechanisms currently lumped under the heading of "third factor(s)." Decreased extracellular fluid volume also stimulates aldosterone secretion through the mechanism of increased release of renin and increased angiotensin formation. The aldosterone increases distal tubular reabsorption of sodium in exchange for potassium and hydrogen ions. As volume decreases further, renal blood flow, glomerular filtration rate and filtered load of sodium decrease. Volume depletion also stimulates antidiuretic hormone secretion, which enhances water reabsorption from the distal nephron. The decreased filtered load of sodium may limit the capacity to generate free water in the diluting segment of the nephron and, if insufficient sodium reaches the loop of Henle, the counter-

TABLE 11-1. Causes of Sodium Loss

RENAL	GASTROINTESTINAL	SKIN	DRAINAGE
Renal disease	Vomiting	Sweating	Peritoneal
Diuretic therapy	Diarrhea	Mucoviscidosis	Pleural
Osmotic diuresis	Drainage	Burns	Subcutaneous
Adrenal insufficiency			

current mechanism may be impaired so that neither maximal dilution nor concentration can be achieved. Moreover, the thirst center is stimulated in an attempt to increase fluid intake and restore volume.

If salt loss continues and the patient has access to water, this water is retained and hyponatremia develops. The retained water is not confined to the extracellular space but is distributed in total body water. This results in dilutional expansion of cells.

When sodium is administered without water, an increase in extracellular osmolality occurs. Intracellular water shifts extracellularly to maintain osmotic equilibrium, thereby diluting extracellular and concentrating intracellular fluid. Thus, although sodium is confined extracellularly, it is osmotically effective throughout total body water. The amount of sodium required to restore sodium concentration to normal in the presence of hyponatremia, sodium depletion, and normal total body water is equal to the product of the total body water and the difference between the normal and observed sodium concentrations. For example, at a sodium concentration of 115 mEq. per liter (a decrease in serum sodium concentration of 25 mEq. per liter), a 70-kg. man (estimated total body water, 42 liters) would require 1,050 mEq. of sodium (25 × 42) to raise the serum sodium concentration to normal levels. Of course, losses would have to be stopped or simultaneously replaced in order to achieve the desired net sodium retention.

If total body water is initially reduced, volume depletion will persist after the sodium concentration has been corrected. Repletion with isotonic saline is then necessary to replace the deficit. It is often difficult to determine the relative roles of sodium depletion and water excess in the genesis of hyponatremia. If hyponatremia is due primarily to water retention, attempts to correct hyponatremia by sodium administration may cause increased extracellular fluid volume and potentially life-threatening acute pulmonary edema. Sodium repletion, therefore, can be accomplished best by estimating the sodium deficit from the product of the total body water and the difference between the normal and observed serum sodium concentrations, and administering a fraction of this while monitoring arterial and central venous pressures. Fluid balance may need to be adjusted independently.

Sodium Retention

An excess of total body sodium results in an expanded extracellular fluid volume; this may be associated with a normal, low, or high serum sodium concentration. A clinically detectable increase in extracellular fluid volume is *edema*, which may be localized or generalized. Localized edema may result from diverse causes such as thermal burns, trauma, localized toxins (insect bites), and venous or lymphatic obstruction. Two pressure gradients govern the rate at which edema is formed. The first is the gradient in hydrostatic pressure between the intravascular and interstitial spaces; the second is the osmotic or oncotic pressure gradient resulting from the difference in protein concentration between plasma and interstitial fluid. At the arterial end of the capillary bed, hydrostatic pressure is sufficient to favor movement of fluid into the interstitial spaces against the osmotic pressures exerted by the plasma proteins; at the venous end of the capillary bed, the hydrostatic pressure falls below plasma oncotic pressure so that fluid returns to the intravascular circulation. Any localized change in capillary permeability or in the hydrostatic-osmotic pressure relationships can cause movement of plasma fluid into the interstitium. When localized edema occurs, there is necessarily a transient decrease in extracellular fluid and plasma volumes throughout the remainder of the body. This initiates a renal response which causes sodium retention, restoring normal volumes outside the edematous area; thus, even localized causes of edema should be considered as having generalized effects.

Congestive Heart Failure is characterized by a decreased cardiac output, increased venous volume and pressure, and decreased effective and, often, absolute arterial volume. The increased venous pressure produces an elevated capillary hydrostatic pressure and, hence, increases in interstitial fluid volume. Renal blood flow varies with cardiac output. As it decreases, renin is released, angiotensin is produced and this, in turn, stimulates aldosterone release and resultant sodium retention. Antidiuretic hormone release is stimulated by diminished distention of volume receptors apparently located in the head and atria. As cardiac failure becomes more severe, there is increasing venous pressure and volume, and decreasing cardiac output and renal blood flow; sodium and water retention continue. Administration of sodium, by increasing venous volume, may further aggravate the cardiac failure, and any further decrease in cardiac output only perpetuates a deteriorating clinical condition.

Hepatic Disease may produce increased portal venous pressure, thereby causing increased formation of intraperitoneal fluid (ascites) and contraction of plasma volume. As liver failure progresses, other mechanisms for generalized edema come into play. Inactivation of antidiuretic hormone and of aldosterone by the liver are less effective. Impaired protein synthesis and loss of protein into ascitic fluid lead to hypoproteinemia. The low protein concentration causes a shift of plasma water into interstitial spaces, leading to generalized edema. (For further discussion, the reader is referred to Chapter 22.)

Hypoproteinemia may also result from nutritional deficiency or from protein loss, as in the nephrotic syndrome or in protein-losing enteropathy. Increased degradation of protein and impaired hepatic synthesis may play secondary roles in the *nephrotic syndrome*. Plasma protein depletion re-

duces oncotic pressure and increases fluid loss through the capillaries into the interstitium. This decreases plasma volume, and the kidney responds by salt and water retention. Since the retained sodium and water is not confined to the plasma volume, edema occurs. Under these circumstances, expansion of the plasma volume by colloid administration may promote a diuresis.

Renal failure is characterized by an inability both to conserve sodium and to excrete maximal quantities of salt. Unless the renal circulation is impaired, renal failure must be severe before sodium retention is important clinically. Many of the causes of renal failure, however, involve the renal vasculature and the glomeruli and impair sodium excretion early in their course. Sodium retention may lead to hypertension, congestive heart failure, and decreased renal perfusion and a vicious cycle that further aggravates sodium retention and edema.

Impaired Tubular Reabsorption of Sodium

Diseases that lead to reduced glomerular filtration and enhanced tubular reabsorption of sodium and water cause inappropriate increases in body fluid volumes. Accumulations of sodium and water in concentrations proportionate to that in plasma expand primarily in the interstitial fluid space, causing edema. One of the major therapeutic achievements over the last several decades has been the development of pharmacological agents that block tubular sodium reabsorption and thereby reduce edema formation.

Saluresis and diuresis may be achieved by improving cardiovascular and renal function. Thus, augmentation of cardiac output by digitalis, bed rest, salt deprivation or corrective cardiac surgery, will reverse salt and water retention secondary to congestive heart failure. A low plasma volume due to causes such as nephrotic syndrome or cirrhosis will cause sodium retention. Expansion of plasma volume with blood, albumin, dextran or other plasma expanders or by mobilization of retained fluid by the wrapping or elevating of edematous extremities leads to saluresis. The use of vasodilators such as aminophylline may also achieve a diuresis, owing to increased filtration rate.

Therapeutic diuresis is frequently achieved with drugs that inhibit tubular sodium transport. Osmotic diuresis may result from administration of a nonreabsorbable solute such as mannitol, by elevation of the filtered load of a substance such as glucose above the tubular capacity for reabsorption, or by administration of a nonreabsorbable anion such as sulfate. Because the filtrate remains isosmotic in the lumen of the proximal tubule, the addition of a nonreabsorbable solute obligates water to remain within the lumen to maintain isosmolality. The increased volume augments flow rate of tubular contents, dilates tubules and lowers the tubular fluid concentration of sodium and other solutes, thereby increasing the gradient against which these solutes must be absorbed. These factors all contribute to the saluresis.

Acetazolamide inhibits sodium reabsorption by blocking carbonic anhydrase activity. This is one of the few oral saluretics that has a proximal tubular inhibitory effect, since most of them inhibit distal tubular sodium transport. Chlorothiazide and its numerous analogues inhibit active sodium reabsorption in the diluting segment of the distal tubule; potassium loss is variable. In addition, a mild sodium bicarbonate diuresis results, because thiazides have weak carbonic anhydrase inhibitory effects. Ethacrynic acid and furosemide are two potent saluretics that block sodium transport primarily in the ascending limb of the loop of Henle and the early part of the distal tubule, the diluting segment. Mercurial diuretics also cause a saluresis by inhibiting renal tubular sodium reabsorption. Spironolactone and triamterene prevent the exchange of sodium for potassium and hydrogen in the distal tubule. The former inhibits sodium reabsorption by specifically antagonizing the action of aldosterone, whereas the latter acts directly on the distal tubule exchange site independently of aldosterone activity.

Hyponatremia

Hyponatremia, by definition, represents dilution of extracellular fluid sodium and, except when resulting from an excess of another extracellular solute such as glucose, is associated with a low serum osmolality. Decreased serum concentration may result from sodium loss in excess of osmotically obligated water or sodium depletion with retention of water in an attempt to maintain extracellular volume (as described above); it may also result from primary retention of water in excess of sodium (dilutional hyponatremia).

Serum sodium concentration is also depressed in the presence of hyperlipemia or severe hyperproteinemia. Although the sodium concentration of plasma water is normal, the lipid and protein molecules occupy space. As a result, the measured volume of plasma used in the analytical procedure contains less plasma water and, thus, a decrease in the measured amount of sodium.

Dilutional Hyponatremia

The clinical picture of hyponatremia secondary to water retention differs considerably from that of salt depletion. Coma and convulsions are common and, at least in part, are the result of cerebral edema; no evidence of ischemia of organs such as the skin and kidneys is detectable.

Dilutional hyponatremia may be caused by an extremely high fluid intake, by severe renal failure, by decreased renal perfusion or finally by inappropriate antidiuretic hormone activity. In a normal individual, dilutional hypernatremia does not occur unless water intake exceeds 20 liters per day. With a moderate decrease in renal function, the capacity to excrete water is adequate to handle the usual fluid intake. However, severe renal failure and, obviously, anuria or oliguria limit water tolerance, and dilutional hyponatremia is frequent under these circumstances. Decreased renal perfusion diminishes sodium and water excretion and impairs the capacity to excrete free water.

A marked decrease in renal blood flow may occur from causes such as congestive heart failure, myxedema, vasoconstrictor drugs, and aortic or renal vascular disease. These disorders can be complicated by water retention and dilutional hyponatremia, which often is superimposed upon sodium retention (Fig. 11-4).

Hyponatremia may also result from dilu-

tion of the extracellular fluid by cellular water. This may develop when the osmotic gradient favors the movement of water out of cells. For example, when cells become depleted of potassium, cellular osmolality must decrease. Osmotic equilibrium then demands a shift of sodium into the cell or water out of the cell. Both mechanisms appear to be operative in the hyponatremia of potassium depletion, but the osmotic shift of water appears to be dominant. An increase in extracellular solutes such as mannitol or glucose causes a similar osmotic gradient favoring the movement of water from the cells into the extracellular fluid compartment.

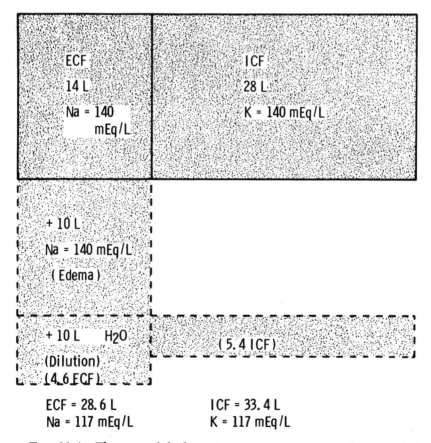

FIG. 11-4. The normal body water compartments may be expanded isotonically (extracellular fluid), causing edema, or hypotonically, causing dilutional hyponatremia. In this example the added water distributes itself proportionately to the volumes of each compartment and dilutes extracellular sodium and intracellular potassium to 117 mEq./L.

Often hyponatremia results from multiple causes. For example, patients with diabetes mellitus may develop hyperglycemia, causing a shift of water from the cells, with expansion of extracellular fluid volume. The hyperglycemia leads to glycosuria, which promotes an osmotic diuresis and renal salt loss. Complications, such as shock, and the therapeutic use of vasoconstrictor drugs may lead to renal ischemia and an inability to excrete free water. Potassium depletion, accompanying diabetic acidosis, may also enhance the development of hyponatremia. It is, therefore, important to identify all possible mechanisms causing the change in sodium concentration in order to correct the underlying abnormality.

Antidiuretic Hormone (ADH). Administration of the antidiuretic hormone, vasopressin, increases the permeability of the distal and collecting tubules to water, causing concentration of the urine. The effect of intravenous aqueous vasopressin lasts about 30 minutes. If fluid is administered while ADH is exerting its effect, dilutional hyponatremia results. Since ADH is metabolized by the liver, impaired metabolism may potentiate water retention in hepatic failure.

Dehydration, with loss of water in excess of solute and a resultant increase in serum osmolality, is the normal stimulus for ADH release. Intracerebral osmoreceptor cells also can be dehydrated and stimulated by infusion of a hypertonic solution such as sodium. There also appear to be volume or "stretch" receptors in the left atrium and, very likely, in the arterial system that are stimulated to cause ADH release when plasma volume is decreased or cardiac output is reduced. ADH is also released by stimuli such as pain and administration of barbiturates, nicotine and morphine. Any one of these stimuli is sufficient to ensure continued release of antidiuretic hormone, in spite of a low serum osmolality. In pituitary or adrenal insufficiency, impaired diluting ability may reflect ADH release associated with low plasma cortisol levels and decreased extracellular fluid volume. Cortisol also may have a direct effect on collecting duct permeability.

A chronic form of inappropriate ADH secretion occurs in some patients with intrathoracic (mediastinal tumors, tuberculosis and bronchogenic carcinoma) or intracerebral diseases. In such instances, where persistent ADH secretion occurs with low plasma osmolality, absence of edema, normal renal, hepatic, cardiac and adrenal function, and no contraction of plasma volume, the term *syndrome of inappropriate secretion of antidiuretic hormone* is applied. Inappropriate ADH activity leads to water retention, increased total body water, cellular swelling with associated cerebral edema and hyponatremia (water intoxication). Expansion of the extracellular fluid volume produces hemodilution and hyponatremia, decreased aldosterone secretion, stimulation of "third factor," and a significant natriuresis. High urine sodium concentration indicates mainly excretion of the sodium intake in a small volume but also reflects some degree of negative sodium balance. The hyponatremia cannot be explained fully by either renal sodium loss or hemodilution owing to retained water. Under these circumstances, hyponatremia is not corrected by sodium administration, since sodium excretion merely increases. To correct the hyponatremia, water intake must be reduced sufficiently to achieve negative fluid balance. These patients present with a low serum sodium concentration and serum osmolality and a urine above minimum osmolality, i.e. greater than 100 milliosmoles per kg. of water. Urine osmolalities as low as 350 to 400 mOsm. per liter are consistent with either inappropriate antidiuretic hormone activity or salt depletion with appropriate stimulation of antidiuretic hormone. However, normal or low values for blood urea nitrogen and normal creatinine clearances help to exclude significant salt depletion.

Hypernatremia

An increase in total body sodium usually is associated with edema, since water must be retained in order to maintain a normal sodium concentration in extracellular fluid. If excessive sodium is administered when there is limited access to water, or if water is lost, hypernatremia will occur. Sodium excesses have occurred when infant's formulas were incorrectly made. Hypernatremia resulting from loss of hyponatric fluids is more frequent. Urinary water loss occurs in congenital and acquired forms of pituitary and nephrogenic diabetes insipidus. Excessive sweating or osmotic diuresis also will produce hypernatremia if insufficient water is replaced.

The most frequent cause of hypernatremia is the use of high protein, high sodium tube feedings for elderly or comatose patients. A high protein diet causes an increase in urea production and, thus, when renal function is not changed, an increase in urea excretion. This produces an osmotic diuresis, with water loss in excess of sodium loss, dehydration and hypernatremia. Another frequent cause of hypernatremia has been peritoneal dialysis using hypertonic glucose. This causes a proportionately greater removal of water than of sodium, resulting in hypernatremia. However, most causes of hypernatremia are associated with salt loss that is associated with proportionately greater water loss. Hypernatremia, therefore, usually indicates dehydration and does not imply an excess of total body sodium. Primary aldosteronism and Cushing's syndrome also may cause mild hypernatremia.

POTASSIUM IN HEALTH AND DISEASE

Sources of Potassium

Exogenous sources of potassium include foods with high protein content, such as meats, dairy products and fruits, and certain drugs, notably salt substitutes and potassium-containing penicillin. The endogenous catabolism of 75 g. of protein releases 33 mEq. of potassium from cell water. Potassium moves from intracellular to extracellular fluid compartments during glycolysis, hemolysis and acidosis. In addition, the osmotic attraction of cellular water to the extracellular fluid carries potassium with it by the process of solvent drag.

Potassium Excretion

Normally, urinary potassium excretion represents only 10 to 20 percent of filtered potassium, indicating net tubular reabsorption. Indeed, all filtered potassium may be reabsorbed in the proximal tubule and all excreted potassium may enter the tubular fluid by active secretion at a distal tubular site. Here, each sodium ion reabsorbed is exchanged for either a potassium or a hydrogen ion, with, under normal conditions, about two potassium ions being secreted into tubular fluid for each hydrogen ion. There appear to be three mechanisms by which potassium secretion at this distal exchange site can be modified.

Delivery of Sodium to Distal Exchange Site. Potassium secretion requires adequate sodium to be delivered to the distal exchange site for reabsorption. If sodium delivery is markedly reduced, as in severe volume depletion, little sodium can be reabsorbed, and little potassium is secreted; by inhibiting sodium reabsorption, using such agents as sulfate, thiocyanate, mannitol or urea, more sodium reaches the distal nephron for reabsorption and more potassium can be secreted in exchange into the tubular fluid.

Aldosterone. The rate of sodium reabsorption in exchange for potassium and hydrogen ion is controlled hormonally by aldosterone. Volume depletion, other causes of decreased renal perfusion with resultant renin release, and elevated serum potassium concentration are all stimuli for aldosterone secretion. The last-named stimulus provides an important mechanism for the elimination of potentially lethal excesses of potassium from the circulation.

Ratio of Available Potassium to Hydrogen Ion. At the distal tubular exchange site, the ratio of potassium to hydrogen ion secretion in exchange for sodium ion is dependent upon the relative availability of each of these ions in the tubular cells. During acidosis, the amount of hydrogen ion available for secretion is increased and, therefore, potassium secretion is suppressed; alkalosis reduces the amount of hydrogen ion available for secretion and, therefore, potassium secretion must be increased.

Similarly, when potassium concentration is decreased, hydrogen ion is preferentially exchanged. This explains why, when sodium load to the distal site remains adequate, a person with hypokalemia and extracellular metabolic alkalosis may excrete a paradoxically acid urine. When serum potassium concentration increases, the ratio of potassium to hydrogen ion secretion also rises; furthermore, since aldosterone secretion is stimulated by hyperkalemia, a resultant increase in the exchange of sodium for potassium occurs.

Hyperkalemia

Potassium is potentially the most toxic substance derived from protein catabolism. If anuria is produced experimentally and the animal is left untreated, death occurs from accumulation of potassium.

Cardiotoxicity. Cardiac arrest or ventricular fibrillation with sudden cardiac death, resulting from hyperkalemia, is well recognized. Below a serum potassium concentration of 6.5 mEq. per liter, the electrocardiogram is often normal; at higher levels, cardiotoxicity is usual but varies in severity. This is so because potassium intoxication is determined by many factors other than the serum concentration. The higher the ratio of extracellular to intracellular potassium, the more rapid the potassium accumulation

and the lower the serum calcium, sodium and pH, the more toxic is a given potassium level. Potassium intoxication may cause anxiety, bradycardia, anesthesia, weakness or paralysis hours or minutes before cardiac arrest; often there are no warning symptoms. However, the electrocardiographic changes usually occur earlier. Depression of the sinus node causes bradycardia. Altered ventricular activation causes high peaked T waves, decreased R and deep S waves, depressed R-ST segments, and occasionally Q waves. Intra-atrial block manifests flat, prolonged and, eventually, absent P waves. A prolonged P-R interval reflects atrioventricular block. Intraventricular block causes a prolonged QRS complex, either diffuse or bundle branch, and may eventually lead to ventricular standstill unless ectopic rhythms occur which culminate in ventricular fibrillation (Fig. 11-5).

Chronic Renal Failure. Because normal potassium homeostasis is maintained by a secretory mechanism that functions far below its capacity (or tubular maximum), hyperkalemia is usually not a serious therapeutic problem in chronic renal failure until the disease is terminal or unless the potassium load to the kidney is increased. Potassium excretion per residual nephron can be increased many fold to compensate for loss of functioning nephrons. This functional adaptation requires aldosterone and adequate sodium reaching the distal nephron. Chronic renal failure provides these circumstances by mechanisms such as osmotic diuresis. Removal of the osmotic load—for example, by dialysis or owing to decreased renal perfusion from salt depletion or heart failure—can impair ability to excrete potassium. Blockade of the sodium-potassium exchange, such as occurs with spironolactone or triamterene therapy, also leads to potassium retention. Usually, however, unless oliguria occurs, potassium is not retained by the kidney sufficiently to cause clinical problems unless supplemental po-

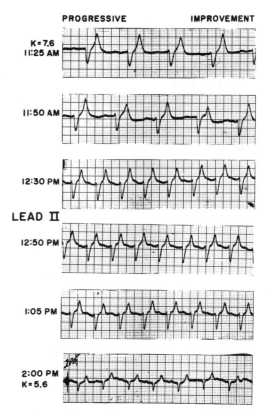

PROGRESSIVE IMPROVEMENT

K = 7.6
11:25 AM

11:50 AM

12:30 PM

LEAD II

12:50 PM

1:05 PM

2:00 PM
K = 5.6

Fig. 11-5. Electrocardiogram of hyper-kalemia. Note bradycardia, flat P waves, PR prolongation, low R wave, widened QRS and high peaked T wave.

tassium is obtained from exogenous sources or endogenous potassium from catabolism or acidosis is markedly increased.

Hypokalemia

Renal conservation of potassium is less complete than is conservation of sodium. Usually, with potassium deprivation, urinary potassium excretion continues at a rate of 10 to 20 mEq. per day or slightly less. Thus, potassium depletion, unlike sodium depletion, can occur with prolonged dietary restriction without increased losses. Causes of potassium depletion include gastrointestinal losses due to vomiting, diarrhea, or drainage and renal losses due to renal tubular acidosis, diuretic therapy, alkalosis, hyperaldosteronism, hypercorticism (Cushing's disease or iatrogenic), and potassium-wasting nephropathy. The last may represent inapparent secondary hyperaldosteronism or the rare Bartter's syndrome

(normotensive hyperaldosteronism). With gastrointestinal losses of potassium, there is often an associated loss of extracellular fluid, leading to secondary hyperaldosteronism and increased renal loss of potassium. Potassium depletion, as indicated by a low cellular potassium, also may occur with rapid tissue anabolism or growth if potassium restrictions are imposed simultaneously. Under these circumstances, total body potassium remains constant as cell mass increases.

Alkalosis contributes to the development of hypokalemia by both cellular and renal mechanisms. A rise in serum pH (or fall in hydrogen ion concentration) causes an intracellular shift of potassium in exchange for hydrogen ion so that serum potassium concentration may be low without total body potassium depletion. In addition, metabolic alkalosis is characterized by development of high serum bicarbonate concentrations. The increased filtered load of bicarbonate is poorly reabsorbed, and this increases delivery of sodium to the distal exchange site where more can be reabsorbed in exchange for potassium and hydrogen. More sodium is also lost in the urine, leading to volume depletion, increased aldosterone secretion and stimulation of the distal exchange mechanisms.

In **diabetic acidosis,** potassium depletion may occur as a result of increased potassium excretion associated with the glucose-induced osmotic diuresis and as a result of a secondary hyperaldosteronism resulting from dehydration and contracted fluid volume. During the acidosis, potassium shifts extracellularly so that the potassium depletion may not be apparent. Once the acidosis is corrected, much of the remaining serum potassium returns intracellularly. Furthermore, potassium moves into cells in association with glucose administration and glycogen formation. Dangerously low levels of serum potassium, with clinical manifestations of hypokalemia, frequently occur in patients with treated diabetic acidosis unless substantial amounts of potassium are given with other therapies.

Effects on the Kidney. Loss of ability to concentrate urine may occur with potassium depletion and is usually attributed to damage to the countercurrent mechanism. Morphologically, a marked vacuolization of proximal tubular epithelium and diffuse interstitial edema, fibrosis and mononuclear cell infiltration may occur. Patients with hypokalemia may also show a decreased glomerular filtration rate, salt-losing, and an increased susceptibility to pyelonephritis.

Role of Chloride in Hypokalemic Alkalosis. Hypokalemic alkalosis tends to be self-perpetuating. Because of metabolic alkalosis with hypochloremia and increased serum bicarbonate concentration, greater amounts of sodium must be reabsorbed with bicarbonate or exchanged distally for potassium or hydrogen ion to prevent sodium loss. This is so because chloride, a readily reabsorbable anion, is reduced in concen-

tration. Increased delivery of sodium to the distal nephron may increase aldosterone-mediated exchange for hydrogen or potassium ions. If potassium supplementation is provided with a poorly reabsorbable anion such as sulfate (or some organic anion such as lactate which is metabolized to bicarbonate), the less diffusible anion impairs proximal sodium reabsorption and sodium homeostasis is maintained only by continued loss of potassium or hydrogen ion through enhanced distal exchange. Administration of chloride, however, allows increased proximal sodium reabsorption, decreasing delivery of sodium to the distal portion of the nephron. If potassium chloride is given, potassium can be retained. If sodium depletion accompanies the hypokalemic alkalosis, as is frequent, sodium chloride administration eventually corrects sodium depletion so that the distal tubule is no longer exposed to the increased aldosterone levels that cause urinary losses of potassium and hydrogen ions. Distal potassium loss also can be interrupted by spironolactone, but this enhances loss of sodium bicarbonate and aggravates any pre-existing sodium depletion.

ANNOTATED REFERENCES

Bartter, F. C., and Schwartz, W. B.: The syndrome of inappropriate secretion of antidiuretic hormone. Am. J. Med., 42:790, 1967. (The clinical features and pathophysiology of this interesting syndrome are outlined by these two pioneer investigators.)

Berliner, R. W.: Renal mechanism for potassium excretion. Harvey Lect., 55:141, 1960. (A lucid summary of the classical experiments demonstrating potassium excretion, relating these to prior studies)

———: Intrarenal mechanisms in the control of sodium excretion. Federation Proceedings, 27:1127, 1968. (This succinct summary of current knowledge is part of a symposium on neural control of body salt and water.)

Bricker, N. S.: The control of sodium excretion with normal and reduced nephron populations. The preeminence of Third Factor. Editorial. Am. J. Med., 43:313, 1967. (A clear presentation of the third factor concept in the handling of sodium by the diseased kidney.)

Bricker, N. S., Klahr, S., Lubowitz, H., and Rieselbach, R. E.: Renal function in chronic renal disease. Medicine, 44:263, 1965. (Of numerous studies of nephron function in renal failure by Dr. Bricker and his associates, this is the most detailed review.)

Davis, B. B., and Knox F. G.: Current concepts of the regulation of urinary sodium excretion. Am. J. Med. Sci., 259:373, 1970. (An up-to-date discussion of intrarenal factors and the extrarenal regulatory mechanisms that govern sodium excretion)

Earley, L. B., and Daugharty, T. M.: Sodium metabolism. New Eng. J. Med., 281:72, 1969. (Current review of sodium, emphasizing renal regulation)

Elkinton, J. R.: Clinical disorders of acid-base regulation: A survey of 17 years diagnostic experience. Med. Clin. N. Am., 50:1325, 1966. (Many years of interest and investigation of acid-base disorders by this lucid writer gives the reader much insight.)

Fuisz, R. E.: Hyponatremia. Medicine, 42:149, 1963. (A detailed and thoroughly referenced review of basic and clinical aspects of hyponatremia)

Kassirer, J. P., and Schwartz, W. B.: The response of normal man to selective depletion of hydrochloric acid. Am. J. Med., 40:10, 1966. (This is one of several carefully controlled studies by these authors, identifying the renal response to chloride depletion.)

Katz, A. I., and Epstein, F. H.: Physiologic role of sodium-potassium-activated adenosine triphosphatase in the transport of cations across biologic membranes. New Eng. J. Med., 278:253, 1968. (A more basic look at the active transport system for sodium and its energy substrate)

Kleeman, C. R., and Fichman, M. P.: The clinical physiology of water metabolism. New Eng. J. Med., 277:1300, 1967. (This review emphasizes factors that control water balance and identifies abnormalities of solute concentration with disordered water regulation.)

Leaf, A.: The clinical and physiologic significance of the serum sodium concentration. New Eng. J. Med., 267:24, 77, 1962. (A medical progress review that considers clinical disturbances affecting the serum sodium concentration from the standpoint of the disordered physiology)

Merrill, J. P., and Hampers, C. L.: Uremia. New Eng. J. Med., 282:953, 1970. (An up-to-date review of uremia, including electrolyte abnormalities)

Morris, R. C., Jr.: Renal tubular acidosis mechanisms: Classification and implications. New Eng. J. Med., 281:1405, 1969. (This discussion clarifies the understanding of renal tubular handling of solutes.)

Nolph, K. D., and Schrier, R.: Sodium, potassium and water metabolism in the syndrome of inappropriate antidiuretic hormone secretion. Am. J. Med., 49:534, 1970. (An extensive metabolic balance study of a patient with dilutional hyponatremia)

Orloff, J. Walser, M., Kennedy, T. J., Jr., and Bartter, F. C.: Hyponatremia. Circulation, 19:284, 1959. (An analysis of the significance and pathogenesis of hyponatremia, with particular reference to the role of antidiuretic hormone)

Pitts, R. F.: Physiology of the Kidney and Body Fluids. Chicago, Year Book Medical Publishers, 1968. (An excellent monograph on renal physiology that provides basic knowledge necessary for the understanding of abnormalities of the regulatory systems)

Ross, E. J., and Christie, B. M.: Hypernatremia. Medicine, 48:441, 1969. (A thorough review, with particular details about the causes of hypernatremia, and including an extensive bibliography)

Schreiner, G. E., and Maher, J. F.: Uremia: Biochemistry, Pathogenesis and Treatment. Springfield, Ill., Charles C Thomas, 1961. (A comprehensive review of uremia and the role of disorders of the electrolytes. The effects of impaired excretory function on organ systems are detailed. The bibliography is extensive.)

Schwartz, W. B., and Relman, A. S.: Effects of electrolyte disorders on renal structure and function. New Eng. J. Med., 276:383, 1967. (The effects of potassium and sodium depletion and hypercalcemia on the kidney are discussed from the functional and morphologic standpoint.)

de Wardener, H. D.: Control of sodium reabsorption. Brit. Med. J., 3:676, 1969. (A lucid presentation of factors that regulate sodium excretion, including the effects of volume expansion)

Welt, L. G., Hollander, W., Jr., and Blythe, W. B.: Consequences of potassium depletion. J. Chronic Dis., 11:213, 1960. (An outline of the abnormalities resulting from potassium depletion)

12

Maintenance of Body Tonicity

Solomon Papper, M.D., and Carlos A. Vaamonde, M.D.

INTRODUCTION

The preservation of the volume and composition of the body fluids requires precise renal regulation of water and solute excretion. Not only must the kidney respond appropriately to the daily ingested load of solutes and water, but it also has to adapt its excretory function to the extrarenal losses of water and solutes occurring in normal life (insensible water loss, sweat, fecal contents) or in disease (excessive sweating, hyperventilation, vomiting, diarrhea, fistulae, etc.). Therefore, when the kidney is diseased fluid homeostasis is in jeopardy.

TERMINOLOGY

We are basically interested in defining two interrelated measures of the concentrating and diluting capacity of the kidney. The first is a measure of *gradient* or *concentration*—that is, the urine concentration, or total solute concentration (urine osmolality, Uosm). The others are *rate-dependent measures:* the minute volume of urine excreted and its solute and water composition relative to plasma—that is, the osmolar clearance (Cosm), the solute-free water excretion or free-water clearance (CH_2O), and the solute-free water reabsorption (T^cH_2O).

In determining urine concentration, one is interested in the number of particles of solute dissolved in a unit volume of urine water (osmolality). This is expressed in milliosmoles per kg. of water (mOsm/kg. H_2O), and the measurement is usually done by freezing point depression osmometry.[*] A "concentrated" urine is characterized by a concentration of total solutes higher in the urine than in plasma. Conversely, the urine is "dilute" when its total solute concentration is less than that of plasma. In the young, healthy adult, plasma osmolality is approximately 280 mOsm/kg. H_2O. Maximum urinary osmolality approximates 1200 mOsm/kg. H_2O, whereas urine may be as dilute as 30 to 35 mOsm/kg. H_2O.

When the urine osmolality (Uosm) is factored by plasma osmolality (Posm), the urine-to-plasma osmotic ratio ($\frac{U}{P}$ osm) is obtained. This is another way of expressing the capacity of the kidney to achieve concentrating or diluting gradients. Thus, the $\frac{U}{P}$ osm ratios vary from 3.5 to 4.5 in maximally concentrated urine to as low as 0.12 in maximally diluted urine.

In the overall formation of urine, no net separation of water from solutes occurs when the urine has the same osmolality as the plasma. Thus, the physiologist also expresses concentrating ability in terms of the volume of water reabsorbed from an isosmotic urine to render it hyperosmotic. This determination is made under conditions of

---

[*] The term osmolarity is commonly used in reference to the osmotic concentration of a solution expressed as osmols of solute per liter of solution. Although in clinical application there is no real difference between osmolality and osmolarity, it is preferable to use the word osmolality when referring to total solute concentration.

hydropenia with or without vasopressin (pitressin) administration. Under these conditions, the concentrated urine can be regarded as consisting of two portions: one that is "isosmotic" to plasma, the other, the amount of pure or "free" water reabsorbed from the isosmotic volume to make it concentrated. This may be expressed as:

(1) V (Urine volume) = isosmotic urine volume — reabsorbed water

In order to know the amount of reabsorbed water, i.e., the concentrating ability, the volume of isosmotic urine must be determined. This is accomplished by answering the question, what volume of urine would have been required to excrete the same solute load isosmotically, i.e., with a urine concentration equal to that of plasma? Or:

(a) Solute excretion (isosmotic)
    = Solute excretion (hyperosmotic)

(b) Isosmotic volume $\times$ Posm
    = V $\times$ Uosm

(c) Isosmotic volume
    $= \dfrac{\text{Uosm}}{\text{Posm}} \times V$

This expression is the same as the "clearance" formula, and the volume of isosmotic urine is referred to as the *osmolar clearance* (Cosm).

For example, let us assume that, during hydropenia, urine volume per minute (V) is 0.5 ml. and Uosm is 1,160 mOsm/kg. H_2O and Posm is 290 mOsm/kg. H_2O (Fig. 12-1, B). According to (c) above, the Cosm is:

$$\frac{1160 \times 0.005}{290} = \frac{580 \; \mu\text{Osm per min.}}{290 \; \mu\text{Osm per ml.}} = 2 \; \text{ml./minute}$$

By substitution in (1) above:

    V = Cosm — reabsorbed water
and,
        Reabsorbed water = Cosm — V
            (Fig. 12-1, B)

Reabsorbed water is usually referred to as *solute-free water reabsorption,* or T^cH_2O,

and urinary concentrating ability may be expressed in these terms.

During hydropenia and vasopressin administration, T^cH_2O will achieve its maximal value only at large rates of solute excretion, i.e., during a solute diuresis (solute or osmotic diuresis is due to restrained water reabsorption occurring proximal to the ascending limb of Henle's loop and to high concentration of relatively nonreabsorbable solute, such as mannitol, urea, sodium sulfate, hypertonic saline, etc.). Under these conditions, V and Cosm increase while Uosm decreases towards isotonicity (Fig. 12-1, D).

In the diluting operation, the urine volume may also be considered as consisting of two portions—one that is isosmotic to plasma (Cosm), the other being free water which is *not* reabsorbed—thus rendering the urine dilute.

(2) V = Cosm + free water (Fig. 12-1, C)

Usually the free water is symbolized as CH_2O *(solute-free water clearance).* T^cH_2O is the negative value of CH_2O and is sometimes referred to as *negative free water clearance* rather than reabsorbed free water. Figure 12-1 depicts graphically the concept of CH_2O and T^cH_2O, with examples of water diuresis, isosmotic urine, and antidiuresis.

The maximal CH_2O and T^cHO reflect, within certain restrictions, the magnitude of sodium reabsorption in the ascending limb of Henle's loop. Therefore, the measurement of these calculated indices of renal function has been used to examine indirectly the characteristics of sodium reabsorption in this segment of the nephron under a variety of normal and abnormal conditions. These indices (Cosm, CH_2O, T^cH_2O), although of physiological and pathophysiological significance, have no practical application in clinical medicine outside of clinical research. The evaluation of maximal or minimal V and Uosm and of Posm is sufficient in most clinical situations.

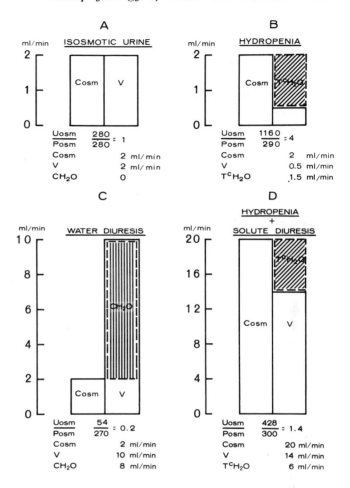

FIG. 12-1. Schematic outline of the relationship between V and Cosm during: (A) isosmotic urine excretion; (B) hydropenia; (C) water diuresis; and (D) hydropenia during solute diuresis. Note that the units in scale C are double those in A and B and the units in scale D are 4 times greater than those in A and B. CH_2O is the amount of water "added" to the urine to make it hypotonic (C). T^CH_2O is the amount of water "removed" from the urine to make it hypertonic (B and D). The values for Uosm, Posm, $\frac{U}{P}$osm, Cosm and CH_2O or T^CH_2O are listed below each example. Note that there is no net separation of water from solutes during excretion of an isosmotic urine (A).

PATHOPHYSIOLOGY OF URINE CONCENTRATION

The Normal Mechanism

It is generally accepted that urine becomes concentrated by the active tubular reabsorption of solutes (sodium) and the consequent passive osmotic equilibration of water. Although water reabsorption occurs along most of the length of the nephron, the final extraction of water in the elaboration of a concentrated urine takes place in the collecting duct. For this to occur, the following conditions are requisite:

(1) The interstitial tissue surrounding the collecting duct must be more concentrated than the urine within the lumen of the collecting duct. The hypertonicity of the medullary interstitium is established pri-

marily by the particulars of the handling of sodium in the loop of Henle. This system is referred to as the countercurrent mechanism and is described below. The particular anatomical spatial distribution of the medullary structures (tubular loops, vasa recta) is known to be related to the capacity of the species (birds and mammals only) to produce a concentrated urine. It is also well recognized that the greater the number of the Henle's loops and the greater the length of the loops, the greater the renal concentrating ability of the species. The medullary circulation (vasa recta) and urea also participate in the maintenance of an hyperosmotic interstitium.

(2) The collecting duct must be permeable to water, so that it can be reabsorbed in accordance with the directional

demands of the concentration gradient. Collecting duct permeability to water requires the action of antidiuretic hormone (ADH, vasopressin).

The Hypertonic Medullary Interstitium. Of the 120 ml. of plasma filtered at the glomerulus each minute, 65 percent is reabsorbed isosmotically in the proximal tubule. Water reabsorption passively follows solute (in particular, sodium) in this segment. In this portion of the nephron, the tubular cells are permeable to water in the absence of ADH.

The reduced volume of isosmotic tubular fluid then enters the loop of Henle in the medulla. Here the concentrating operation is instituted and maintained, serving the needs of conserving water. In the loop, 25 percent of filtered water may be reabsorbed.

The *countercurrent multiplier* (Fig. 12-2) requires the close proximity of the descending limb to the ascending limb of the loop of Henle, as well as their permeability to

sodium, the restricted permeability to water of the ascending limb, and the flow of fluid in opposite directions. The mechanism actually becomes operative by the active transport of sodium *without* water from within the lumen of the ascending limb to the interstitial tissue surrounding the loop of Henle. This process has three important consequences: (a) The fluid leaving the ascending limb and arriving at the distal tubule is hypotonic; (b) the interstitium then becomes hyperosmotic to the fluid in the descending limb, and (c) sodium diffuses passively from the hyperosmotic interstitium into the descending limb.

The mechanism whereby this process results in progressive hypertonicity as the tip of the medulla is approached is known as the "multiplier" effect and it results in the formation of a cortical-medullary osmotic gradient (that is, the cortical tissues are isosmotic while the medullary structures become progressively hyperosmotic toward the papillary tip). This mechanism re-

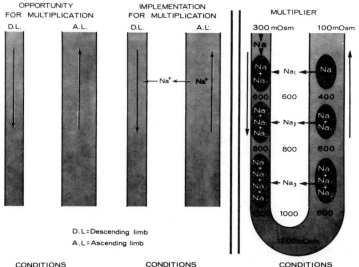

D.L.=Descending limb
A.L.=Ascending limb

CONDITIONS
1) Two channels in close proximity
2) Flow in opposite direction
3) Membranes permeable to Na+, not to H₂O (A.L.)

CONDITIONS
1) Two channels in close proximity
2) Flow in opposite direction
3) Membranes permeable to Na+, not to H₂O (A.L.)
4) Active transport of Na+ out of A.L.

CONDITIONS
1) Two channels in close proximity
2) Flow in opposite direction
3) Membranes permeable to Na+, not to H₂O (A.L.)
4) Active transport of Na+ out of A.L.
5) Continuous flow

FIG. 12-2. Outline of the countercurrent multiplier concentrating mechanism. (Reproduced from Papper, S.: Clinical Nephrology. Boston, Little, Brown and Co., (in press), with permission of the author and the publisher.)

quires that the ascending and the descending limbs be in close proximity and that flow be continuous. Figure 12-2 depicts the essential operation, i.e., sodium being *added* continuously to the interstitium and to the fluid in the descending limb as the fluid proceeds toward the tip, producing increased concentration. Since the descending limb is permeable to water, the fluid within it equilibrates osmotically with the surrounding interstitial fluid (water flows passively out of the descending limb).

For the hyperosmolality to be maintained in the medulla, excess water must be removed from the interstitium; this is accomplished by the vasa recta. This is the *countercurrent exchanger*, which is a *passive* process, depending upon diffusion of solute and water through the walls of the vasa recta capillaries. This system can dissipate the cortical-medullary osmotic gradient if that gradient is not actively maintained by the countercurrent "multiplier" or if there is an accelerated removal of osmotically active solute from the medulla ("washout" effect).

The countercurrent mechanism for the establishment of an hypertonic medullary interstitium is augmented by *urea entrapment*. Through mechanisms not completely defined, urea diffuses from the collecting duct fluid into the interstitium and fluid in Henle's loop. Vasopressin appears to increase the permeability of the collecting duct to urea. This addition of urea quantitatively augments the concentration of solutes in the medullary interstitium. There appears to be considerable "recirculation" of urea through the loops and the vasa recta, which also contributes to conservation of urea as a major medullary solute.

In view of the accumulation of urea in the medullary interstitium at concentrations (during hydropenia) similar to that of the collecting duct fluid, more water is not osmotically "obligated" in the duct lumen for the excretion of urea. It is well known that the urine volume required to excrete a given solute load is less when urea is the principal urinary solute. Thus, urea has a unique role in the urine concentrating process.

Permeability of the Collecting Duct to Water. Antidiuretic hormone (ADH) is required for permeability to water in the distal tubule and collecting duct. ADH is produced in the supraoptic and paraventricular nuclei of the hypothalamus, then stored in the posterior pituitary gland for liberation into the bloodstream in response to stimuli for water conservation. Whereas the proximal tubule is permeable to water in the absence of ADH, ADH is required for permeability to water of the cells of the distal tubule and collecting duct.

In the cortex, during hydropenia the volume of fluid in the distal tubule is reduced by continuous extraction of water (ADH present) from the hypotonic tubular fluid into the isosmotic cortex. The solute concentration of the fluid then becomes isosmotic at the end of the distal tubule.

The achievement of the final urine concentration occurs in the collecting ducts where water leaves the lumen, following the osmotic gradient generated by the countercurrent mechanism. At this level, the final osmotic equilibration takes place; the "final" urine has practically the same osmolality as the structures located deep in the papilla (loops of Henle, vasa recta, interstitium).

Normal Variations in Concentrating Ability

There are some physiological variables that influence the concentrating mechanism.

Age. Infants generally excrete a hypotonic urine. This is related to the relatively large volume of fluid ingested and to the low urea excretion rather than to an immature concentrating mechanism; Uosm reaches adult maximal values in dehydrated infants fed large amounts of protein. Children have values for maximal Uosm similar to those accepted for adults. T^cH_2O values in young adults are between 5 to 7 ml. per minute per 100 ml. of GFR (mannitol diuresis). The aged kidney, in the absence of known renal disease, has decreased concentrating ability, along with generally reduced renal function. The mechanism for this is not known. However, clinically it is an important point, because one cannot assume in the aged the presence of a normal ability to conserve needed water.

Diurnal variations also affect the concentrating mechanisms. Urine flow is lowest and osmolality maximal at night. The cycle appears to be related in part to the solute (sodium) excretion cycle (lowest at night) and to the nocturnal increase in ADH activity.

Enhanced excretion of solutes in normal man results in an increase in urine flow and change in urine osmolality toward that of plasma (Fig. 12-3). If the kidney is concentrating the urine when "solute diuresis" occurs, the urinary concentration declines; a rise in urine osmolality occurs when a solute load is imposed on a kidney that is generating a dilute urine. It is apparent from Figure 12-3 that evaluation of maximal or minimal urine osmolality requires consideration of this important effect of solute excretion on Uosm and V. *Postural natriuresis* through a similar mechanism may be accompanied by a lower maximum Uosm in the hydropenic subject.

Diet may also influence the operation of the concentrating mechanism. Salt deprivation in man will reduce T^cH_2O, apparently by decreasing the amount of sodium available for transport at the concentrating site. The influence of protein and nitrogen metabolism on the concentrating mechanism is apparent from the previous discussion on the role of urea. Low protein diets reduce Uosm and T^cH_2O in man and animals. Finally, osmolality is less in an alkaline than in an acid urine. The mechanism of this effect of pH on the renal concentrating capacity remains unknown.

Consequences of Impaired Concentrating Ability

If the kidney cannot concentrate the urine at all, the urine volume will be determined largely by the filtered load of solute and its reabsorption. In such a setting, water cannot be conserved selectively when there is a need to do so. The water loss may then be regarded as "obligatory" in the sense that it is determined by factors unrelated to any consideration of water homeostasis. Thus, depending upon solute excretion, there may be *polyuria*. If the water loss is not reconstituted by adequate intake, water deficit or *dehydration* may result. Dehydration may stimulate *thirst*. However, thirst in man is determined by many factors (e.g., habit, emotion) in addition to physiological need. The water deficit involves total body water, including the extracellular fluid; therefore, dehydration may result in some contraction of extracellular fluid volume.

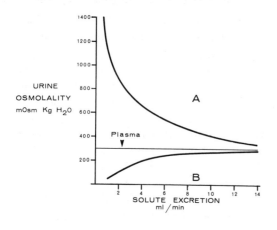

FIG. 12-3. Effect of solute diuresis on urine osmolality during renal concentration (A) or dilution (B) in normal subjects.

In patients with marked impairment in renal function, this results in further reduction in filtration rate. If the pure water loss exceeds the loss of solute, total body water hyperosmolality, including increased serum sodium concentration, results.

Pathophysiology of Impaired Concentration

Reduced concentrating ability occurs when there is an impairment in the establishment of medullary hyperosmolality, and decreased permeability to water in the distal tubule or collecting duct.

Impaired medullary hyperosmolality may occur when any of the circumstances necessary for the establishment of medullary hypertonicity are altered.

Factors that interfere with the delivery of sodium for reabsorption in the loop of Henle may result in impaired concentrating ability. Conceivably, this may obtain when the filtered load of sodium is so reduced (low GFR, hyponatremia) or when the proximal reabsorption of sodium is so increased that less sodium reaches the loop of Henle. It has also been suggested that a low lumenal sodium concentration reaching the concentrating site (during mannitol diuresis) may limit sodium reabsorption by the ascending limb. Theoretically at least, a defect in the capacity of the loop for the transport of sodium could result in failure to create a hypertonic medullary interstitium. For example, some diuretics (furosemide, ethacrynic acid) abolish T^cH_2O in man by blocking sodium reabsorption in the ascending limb of Henle's loop. It is also possible that volume expansion (through any mechanism whatever), may decrease sodium transport at the same site and also decrease medullary hypertonicity. Finally, if the flow of blood in the medulla is increased, the removal of water and solutes from the medullary interstitium may be accelerated ("washout" effect).

Decreased Permeability of the Distal Tubule or Collecting Duct to Water. Decreased permeability to water occurs when there is inadequate production and liberation of ADH. Permeability may be reduced despite the presence of ADH if the cells of the distal tubule and collecting duct are so altered that they fail to respond to the hormone or cannot actually accomplish water reabsorption. In addition, rapid flow of tubular fluid reduces the possibility for osmotic equilibration in the distal tubule and collecting duct (i.e., during osmotic diuresis).

Some Diseases Associated With Impaired Concentrating Ability

Much is known about the normal concentrating mechanism and, from that, one can reason in regard to pathophysiology. Nonetheless, in many diseases associated with impaired concentrating ability, the mechanism(s) of the alteration is not completely understood.

Chronic renal disease regardless of cause, if sufficiently advanced, results in impaired concentrating ability. The precise mechanism(s) whereby this occurs is not established, although at least one important factor is a solute diuresis. There is evidence that there is a decreased number of nephron units in chronic renal disease and and a compensatory increase in GFR per nephron. The latter results in increased filtered load per remaining nephron, insufficient proximal reabsorption and, consequently, delivery of too much tubular fluid to the loop. It is possible that additional causes of inadequate concentrating ability relate to defective responsiveness to ADH as well as to structural inability to reabsorb water. The latter mechanism may be particularly relevant in those renal diseases with predominant medullary involvement, i.e., chronic pyelonephritis.

A group of uncommon clinical diseases (medullary cystic disease, cystinosis, Sjogren's syndrome) is characterized by predominant disease of the medullary structures (selective medullary destruction or fibrosis) that results in polyuria and hyposthenuria in apparent disproportion to the reduction in GFR. Also, amyloid deposits along the collecting tubule were found in a case of nephrogenic diabetes insipidus. The concentration defect appears to be related to a faulty water permeability of the tubular medullary structures.

Hypercalcemic Nephropathy. The earliest renal functional abnormality in hypercalcemic nephropathy is impaired concentrating ability. The early structural changes include damage to the loop of Henle and the collecting duct. It is, therefore, possible that the decreased concentrating ability is due to altered ability of the ascending limb of Henle's loop to transport sodium (T^cH_2O is abolished and CH_2O is reduced) and/or decreased ability of the cells of the collecting duct to reabsorb water or to respond to ADH. Since correction of the hypercalcemia may be followed by reversal of the concentrating defect, one may postulate functional changes in the loop and the collecting duct prior to anatomical damage or healing of the structural abnormality. Hypercalcemic nephropathy may be associated with extensive renal damage, considerable calcium deposition in the parenchyma (nephrocalcinosis) and consequent renal failure.

Hypokalemic nephropathy is a renal lesion with a tendency toward impaired concentrating ability and only little reduction in GFR. The diluting mechanism remains normal. It is difficult to account for the defect in concentrating ability, since the major anatomical lesions consist largely of vacuolization of the cells of the proximal tubule. Perhaps there is altered function in the loop or the collecting duct or both, but this has not been established.

Edematous states, such as cirrhosis of the liver, nephrotic syndrome, and congestive heart failure, are often associated with reversible impairment in concentrating ability. (The situation may be more complicated in the nephrotic syndrome in which, in some instances, there is renal functional deterioration due to renal disease.) Although the mechanism of impaired concentrating ability is not known, two possible explanations have attracted attention. First, these diseases are characterized by greatly increased proximal tubular reabsorption, resulting in decreased delivery of sodium to the loop of Henle. Second, it has been shown that edematous states are associated with a dislocation of renal blood flow from cortex to medulla. It is conceivable that this alteration in intrarenal blood flow, in some way not entirely apparent, limits the ability to establish and maintain a hypertonic medullary interstitium. Although one might expect a shift in blood flow from outer cortex to corticomedullary nephrons (i.e., from relative salt- and water-losing nephrons to salt- and water-retaining nephrons) to enhance concentrating ability, the opposite is observed. In addition, in cirrhosis of the liver, the reduced availability of urea (through malnutrition or decreased synthesis) further reduces maximal Uosm.

Postobstructive nephropathy may be accompanied by decreased concentrating ability. This condition refers to the large diuresis following the relief of urinary tract obstruction. Associated with a solute diuresis due to the high blood levels of urea, there is a defect in sodium reabsorption, and some evidence exists for renal unresponsiveness to ADH.

Acute tubular necrosis in the diuretic phase is associated with abnormal concentrating ability. The large urea load and its osmotic effect may be part of the explanation, but apparently it is not the entire story. A defect in sodium reabsorption, with or without impaired permeability, may also be a factor. This abnormality may persist for several weeks or months, with eventual recovery, or a modest reduction in Uosm may be permanent.

Sickle-cell Disorders. Patients with sickle-cell anemia are unable to concentrate urine normally. Many explanations have been offered, mostly dealing with altered circulation in the medulla. Characteristically, this defect is more pronounced at low urine flows (decreased Uosm maximum) than during osmotic diuresis (T^cH_2O formation may be, indeed, normal or close to normal). The defect is reversible with the transfusion of blood containing normal hemoglobin, suggesting that the concentrating problem probably is related to the abnormal red cells and is not a genetic defect of the kidney. In addition, it has been demonstrated that red blood cells from patients with sickle-cell disease become sickled in hypertonic salt solutions. Thus, it has been postulated that increased sickling in the hypertonic medulla (during hydropenia) causes impaired circulation in the vasa recta secondary to increased viscosity of the blood containing sickled red blood cells. This will decrease the supply of oxygen to the cells of the loop of Henle, resulting in decreased sodium transport and diminished medullary hypertonicity and low Uosm during hydropenia. During osmotic diuresis, when medullary blood flow increases and the tonicity of the medulla is diminished below the hypertonicity of the hydropenic state, the ability of the loop of Henle to reabsorb sodium and of the kidneys to form T^cH_2O improves or approaches normal as a consequence of the diminished sickling and improved medullary circulation.

Diabetes insipidus, no matter what its cause, is characterized by marked reduction or complete absence of ADH. This results in impaired permeability of the distal tubule and the collecting duct to water and, therefore, limited (if any) concentrating ability. Since, even during water diuresis, the medullary structures remain modestly hypertonic to plasma, it is possible that, during severe dehydration (decreased GFR, increased proximal reabsorption), urine osmolality in these patients may increase toward plasma osmolality. A decrease in tubular fluid flow rate,

with increased back-diffusion of water into the interstitium, is the probable explanation.

Nephrogenic diabetes insipidus of the congenital type is a disorder of the kidney making it unresponsive to ADH. Therefore, in this condition, the disturbance in concentrating ability is due to abnormal permeability to water.

Chronic, persistent drinking of water over a protracted period may result in impaired renal concentrating ability which, in some instances, is not reversible for many months. The precise mechanism in man is not known, although there is evidence in rats of a persistent decrease in medullary hyperosmolality.

Malnutrition. Patients with severe malnutrition, in addition to a decreased GFR and RPF, have a decreased concentrating ability (Uosm max and T^cH_2O) with preservation of the diluting capacity. The concentrating defect improves with protein repletion or with exogenous administration of urea. The decreased concentration of urea in the renal medulla appears to be responsible for this defect.

PATHOPHYSIOLOGY OF URINE DILUTION

The Normal Mechanism

Urine becomes diluted by the reabsorption of sodium in the more distal segments of the nephron, where it is reabsorbed without accompanying water. The absence of ADH in the circulating peritubular blood results in reduction in the permeability of the distal tubule and the collecting duct to water. Thus, tubular fluid is being "freed" of its solutes (sodium), resulting in the excretion of a large volume of urine (water diuresis) of low total solute concentration (dilute urine). The concentrating and the diluting mechanism share some common steps and are closely interrelated. Although fewer details of the mechanism of dilution of the urine than of the concentrating mechanism have been described exactly, the essentials of the diluting process have been established.

Sodium Transport in the Diluting Segments of the Nephron. As described previously in the discussion of the concentrating mechanism, the tubular fluid leaving the ascending limb of Henle's loop and arriving at the distal tubule in the cortex is *always* hypotonic, as a result of the active transport of sodium out of the lumen of the ascending limb into the interstitium. This segment of the nephron is not permeable to water, irrespective of the presence or absence of ADH, (i.e., whether the kidney is conserving or excreting water), and has been designated the *medullary diluting segment*. Anatomically it corresponds to the thick outer medullary portion of the ascending limb of Henle's loop. Deep in the medullary structures (at the tip of the papilla), the tubular fluid has become hypertonic to plasma (even during water diuresis, although much less so than during hydropenia), and, as it flows through the ascending limb, its osmolality decreases from hypertonic to isotonic (see Fig. 12-2).

In the cortex, the hypotonic fluid entering the distal tubule at the corticomedullary junction becomes more and more hypotonic, owing to the continuing active extrusion of sodium by the cells of the distal tubule. Water permeability of this segment (designated the *cortical diluting segment*) is dependent on ADH. In its absence, probably only minimal back diffusion of water occurs (from the hypotonic tubular fluid to the isotonic cortical interstitium). Thus, during water diuresis, the volume of the tubular fluid remains practically unchanged during its transit through the cortex.

Finally, the hypotonic tubular fluid enters the collecting duct, also rendered water-impermeable by the absence of ADH, and starts its journey into the medulla. Further reduction in osmolality occurs here (to the maximum diluting capacity—minimal urine osmolality), resulting from reabsorption of sodium in this segment.

Antidiuretic Hormone. The role of ADH is central to the diluting mechanism. In the absence of circulating blood levels of ADH, the permeability of the distal tubule and the collecting duct to water is reduced maximally, resulting in the excretion of a large volume of hypotonic urine (water diuresis). Some degree of back diffusion of water does occur in the absence of ADH, but this in itself does not appear to reduce tubular fluid volume to any significant extent under normal conditions. Experimentally, however, it is possible to demonstrate the production of hypertonic urine in the absence of ADH.

Under normal physiological conditions, inhibition of the release of ADH by the posterohypophysis results from a decrease in the osmolality of the blood perfusing the supraoptic nuclei or an increase in intravascular volume, or both. This decrease in blood osmolality is the result of a positive water balance, usually resulting from water drinking. Conversely, an increase in blood osmolality and/or a contraction of intravascular volume results in the release of ADH by the posterohypophysis (physiological osmotic inhibition or stimulation of ADH release). Other nonosmotic stimuli also affect the release of ADH.

Additional Factors. Any factor that decreases the osmolality of the normally hypertonic medullary interstitium will favor the diluting mechanism by minimizing or eliminating the back diffusion of water. For example, during water and osmotic diuresis, the inner medullary and the papillary hypertonicity are reduced, in spite of continued transport of sodium into the interstitium. An increase in the linear velocity of the tubular fluid that flows through the loop of Henle, a possible increase in the diameter of the lumen of these loops, and an increase in medullary blood flow may be some of the factors responsible for this "wash-out" of the medullary hypertonicity during the diuretic state.

An increase in CH_2O is found after administration of adrenal glucocorticoid. Although an increase in the glomerular filtration rate and consequent increase in the distal delivery of sodium could account for this, there also is evidence that these hormones may have a direct effect on the permeability of certain biological membranes to water. Glucocorticoids physiologically may allow the diluting segments of the nephron to become maximally impermeable to water in the absence of ADH ("permissive" role of the glucocorticoids in water excretion). The precise nature of the mechanism remains unknown.

Normal Variations in Diluting Ability

There are some physiological variables which influence the diluting process.

Age. Infants dilute urine (Uosm) to the same extent as do older children or adults. However, the quantitative values for maximal V and CH_2O during water diuresis are limited by their low glomerular filtration rate.

Circadian variations also occur; the response to water loading is delayed and decreased at night.

Increased solute excretion during water diuresis results in increased V and CH_2O with Uosm rising toward isotonicity (Fig. 12-3, B). Conversely, decreased solute excretion (low sodium diet) may decrease maximal V and CH_2O during water diuresis while further reducing Uosm.

Postural changes are also important. In the upright posture, the response to water loading is reduced, compared to the response in recumbency. Decreased GFR, solute excretion and release of ADH are probably determinant factors.

Consequence of Impaired Diluting Ability

Impaired water diuresis is one of the most common causes of hypoosmolality and hyponatremia observed in clinical practice. Abnormal dilution means that the kidney cannot eliminate solute-free water selectively when there is need to do so. In such a case, the defense of the body water homeostasis has to rely on extrarenal mechanisms (insensible loss, sweat) that are not physiologically equipped to meet these demands.

Impaired water diuresis results in water retention, with absolute or relative hypoosmolality, hyponatremia, hemodilution, and weight gain. If the positive balance of water develops rapidly or is great enough, the syndrome of *water intoxication* may develop. Thus, loss of appetite; nausea or vomiting; irritability, confusion, depression or loss of reflexes; weakness, pyramidal tract signs, stupor or convulsions may appear.

If the kidney cannot dilute the urine at all in the absence of ADH, the urine volume will be determined largely by the filtered load of solute and its reabsorption.

Pathophysiology of Impaired Dilution

Reduced diluting ability occurs when there is: increased permeability to water of the diluting segments of the nephron in the presence or absence of antidiuretic hormone; decreased delivery of water and sodium to the diluting segments; and decreased transport of sodium by the diluting segments.

Increased permeability to water may occur when there is a sustained blood level of circulating antidiuretic hormone increasing the permeability of the distal convoluted tubule and the collecting duct to water. The release of ADH may be appropriate or inappropriate for the stimuli present. *Appropriate release of ADH* for nonosmotic stimuli occurs as a result of hypovolemia secondary to blood loss or extracellular volume depletion, low cardiac output and edematous states, emotional stress and pain (usually transient) and after the use of certain drugs (morphine, nicotine, cholinergic agents, etc.). (Release of ADH in the absence of appropriate stimuli is discussed under the heading Inappropriate ADH Syndrome.)

The increased permeability of the diluting segments to water may occur in the absence of ADH (enhancement of back diffusion of water). This may be secondary to decreased tubular fluid flow, decreased medullary blood flow, or lack of the "permissive" effect of adrenal glucocorticoids.

Reduction in the delivery of filtrate to the distal diluting site may result in impaired diluting ability. This may obtain when the filtered load of sodium (hyponatremia, low glomerular filtration rate) is so reduced or the proximal reabsorption of sodium is so increased (extracellular volume depletion) that less sodium for reabsorption reaches the loop of Henle and the distal tubule.

Decreased sodium transport by the diluting segments of the nephron will decrease their capacity to generate solute-free water. Thus, certain diuretics (e.g., thiazides) block sodium reabsorption by the cortical diluting segment, thereby decreasing CH_2O. Other diuretics (e.g., furosemide, ethacrynic acid, mercurials) block sodium reabsorption by the medullary and cortical diluting segments, resulting in decrease of CH_2O and T^cH_2O. In addition, a diuretic may decrease urinary dilution whenever its marked diuretic and natriuretic effects result in secondary extracellular volume contraction and decreased delivery of filtrate to the distal diluting sites. Massive extracellular volume expansion will also impair dilution by decreasing sodium reabsorption in the diluting segment.

Diseases Associated With Impaired Diluting Ability

Many diseases associated with poor diluting ability may involve the simultaneous operation of several abnormal mechanisms. A major obstacle to a better understanding of these mechanisms has been the difficulty of determining whether or not ADH secretion has been normally supressed by water loading. Development of a sensitive and specific assay for ADH may be of great value in resolving this problem, but, until then, only relative ADH activity can be determined, using the classical physiological criteria. (Fig. 12-4).

Chronic renal disease, regardless of cause, if sufficiently advanced, results in impaired diluting ability. Characteristically, in many of these patients, the diluting capacity persists unaltered or is moderately decreased for longer periods when concentrating ability is grossly impaired. The precise mechanism(s) of the impairment in dilution is not clear but may be related to the decrease in the number of functioning nephrons and the consequent increase in filtered load of solute per remaining nephron. The relative preservation of dilution over concentration may simply result because the absolute change necessary in forming, from isosmotic urine, a maximally concentrated urine is greater than the changes needed to form a maximally diluted urine (see Fig. 12-3). Many patients with chronic renal disease have a normal capacity to generate CH_2O when expressed per 100 ml. GFR, i.e., in relation to the number of remaining functioning neph-

	ANTIDIURESIS	WATER DIURESIS
	ADH ↑	ADH ↓
URINE FLOW	↓	↑
URINE OSMOLALITY	↑	↓
CH_2O	↓	↑
C_{OSM}	NOT ↓	NOT ↑
GFR	NOT ↓	NOT ↑

FIG. 12-4. Analysis of urine flow and urine total solute concentration, together with glomerular filtration rate as an indirect evaluation of ADH activity.

rons. This is in accord with the "intact nephron" hypothesis, rather than the anatomical destruction of the diluting sites. In contrast, there is evidence that the experimentally diseased kidney always generates a greater amount of free water per unit of filtrate than does the normal organ. Perhaps those factors that decrease the back diffusion of water in the absence of ADH are relevant to the relative preservation of dilution. In terminal chronic renal disease with oliguria, the factor limiting water excretion is predominantly the extreme reduction in GFR.

Acute Renal Failure. In oliguric acute tubular insufficiency, the limiting factor for water excretion is the marked reduction in GFR. During the oliguric phase, the urine is isosmotic.

Edematous States. Reversible abnormalities in water excretion are frequently encountered in patients with congestive heart failure, cirrhosis of the liver and nephrosis. The mechanism(s) involved is complex. Impaired water diuresis has been attributed to impaired suppression of ADH during active sodium retention. Decreased inactivation of ADH by the diseased liver also has been implicated, but patients with cirrhosis apparently inactivate vasopressin as well as normals.

In the development of abnormal dilution in edematous states the most important factor appears to be a decrease in the delivery of filtrate to the distal diluting sites. Decreased GFR, with or without hyponatremia, reduces the filtered load of sodium; the decrease in distal delivery of filtrate is further enhanced by the augmented reabsorption of sodium in the proximal tubule, characteristic of these diseases. In addition, increased back diffusion of water resulting from slowly flowing tubular fluid may reduce water excretion. In cirrhosis, water diuresis correlates in general with the severity of the disease. The sickest patients have the poorest responses to water, whereas the majority of patients without ascites or edema have normal, or near normal, diluting capacity.

Adrenal Glucocorticoid Deficiency. Both primary and secondary adrenal insufficiency are characterized by an impaired ability to excrete water normally. The urine is generally hypertonic to plasma and contains a significant amount of sodium. Mineralocorticoid or alcohol administration, high sodium intake and correction of volume depletion do not entirely correct this abnormality. Dilution is rapidly normalized, however, by the administration of glucocorticoids. The exact mechanisms of the glucocorticoid action remain unknown. They may improve water diuresis by inhibiting the back diffusion of water in the diluting segments of the nephron. An abnormality in the release of ADH by the posterohypophysis (glucocorticoids normally inhibiting the release of ADH) has also been postulated. Other factors that may impair dilution in the untreated patient are reduction in GFR and solute excretion and persistence of ADH release by the stimulus of volume contraction. Irrespective of mechanisms, these patients are very sensitive to water intoxication. Thus, dilutional hyponatremia adds to the effect of salt depletion on serum sodium concentration in the patient with untreated adrenal insufficiency.

Hypothyroidism. Primary thyroid deficiency results in a reversible moderate impairment of the diluting capacity in man. Hyponatremia may result in some patients with severe myxedema and in about one half of the patients with myxedema coma. Possible explanations include a decreased GFR, an altered permeability to water and sodium and a reduced sodium transport in the diluting segments of the nephron. This abnormality of water diuresis in hypothyroidism is not corrected by glucocorticoids or alcohol administration; it improves with thyroid replacement.

Inappropriate Secretion of ADH. This syndrome has been reported to occur in a variety of diseases, including malignant tumors (carcinomas of the lung, duodenum and pancreas; thymomas), disorders involving the central nervous system (meningitis, head trauma, brain abscess and tumors, encephalitis, Guillain-Barré syndrome, acute intermittent porphyria), pneumonia, tuberculosis, myxedema and others. The typical features depend upon retention of water resulting from the inappropriate secretion of ADH, secondary hypo-osmolality and hyponatremia, and continued renal excretion of sodium. The urine osmolality is greater than that appropriate for the concomitant tonicity of the plasma. The urine may not be necessarily hypertonic to plasma (although usually it is), but it is certainly not maximally dilute. In addition, there is no clinical evidence of volume depletion or hypotension, and renal and adrenal function are normal. The patient may be asymptomatic, particularly if the serum sodium concentration is above 120 mEq./L. However, when hyponatremia is severe, symptoms of water intoxication usually appear. The syndrome of inappropriate secretion of ADH can be predictably reproduced in normal man by the daily injection of Pitressin tannate in oil while maintaining free access to water. The hyponatremia is usually not corrected by large doses of mineralocorticoid steroids or hypertonic sodium chloride if fluid intake is kept high, although positive sodium balance may occur. It follows from the pathophysiological events that simple fluid restriction is the treatment of choice of this syndrome.

ANNOTATED REFERENCES

Ahmed, A. B., Jr., George, B. C., Gonzales-Auvert, C., and Dingman, J. F.: Increased plasma arginine vasopressin in clinical adrenocortical insufficiency and its inhibiton by glucosteroids. J. Clin. Invest., 46:111, 1967.

Bartter, F. C., and Schwartz, W. B.: The syndrome of inappropriate secretion of anti-diuretic hormone. Am. J. Med., 42:790, 1967. (Review of the subject with emphasis on pathophysiology)

Berliner, R. W., and Davidson, D. G.: Production of hypertonic urine in the absence of pituitary antidiuretic hormone. J. Clin. Invest., 36:1416, 1957. (Elegant demonstration of the production of urine hypertonic to plasma in the absence of ADH)

Berliner, R. W., Levinsky, N. G., Davidson, D. G., and Eden, M.: Dilution and concentration of the urine and action of the anti-diuretic hormone. Am. J. Med., 24:730, 1958. (A detailed description of the concentrating process, including the countercurrent theory)

Bricker, N. S., Dewey, R. R., Lubowitz, H., Stokes, J., and Kirkensgaard, T.: Observations on the concentrating and diluting mechanisms of the diseased kidney. J. Clin. Invest., 38:516, 1959.

Bricker, N. S., Klahr, S., Lubowitz, H., and Rieselbach, R. E.: Renal function in chronic renal disease. Medicine, 44:263, 1965. (A complete analysis of the "intact nephron" hypothesis as applied to renal function in chronic renal disease)

Burg, M. B., Papper, S., and Rosenbaum, J. D.: Factors influencing the diuretic response to ingested water. J. Lab. Clin. Med., 57:533, 1961. (Free water formation depends principally on the sodium load delivery to the diluting segment and on the transport capacity of this part of the nephron.)

Edelman, C. M., Jr.: Maturation of the neonatal kidney. *In*: Becker, E. L.: Proceedings of the Third International Congress of Nephrology. Vol. 3. New York, Karger, 1967. (Review of the functional maturation of the kidney of young infants and comparison to adult indices of renal function)

Eknoyan, G., Suki, W. N., Rector, F. C., Jr., and Seldin, D. W.: Functional characteristics of the diluting segment of the dog nephron and the effect of extracellular volume expansion on its reabsorptive capacity. J. Clin. Invest., 46:1178, 1967. (Characterization of the "medullary and cortical diluting segments" of the dog nephron, with a review of T^cH_2O and CH_2O reliability as indices of sodium reabsorption in the loop)

Epstein, F. H.: Disorders of concentrating ability. Yale J. Biol. Med., 39:186, 1966. (Discussion of renal concentrating ability in diseases characterized by disproportionate, sometimes specific, involvement of the concentrating function: medullary diseases, hypercalcemia, hypokalemia, sickle cell anemia.)

Goldberg, M., McCurdy, D. K., and Ramirez, M. A.: Differences betwen saline and mannitol diuresis in hydropenic man. J. Clin. Invest., *44:*182, 1965. (Evidence is presented in man, suggesting no maximum limit (no Tm) for T^{C}H$_2$O when hypertonic saline is used in place of hypertonic mannitol as the loading solute.)

Gottschalk, C. W., and Mylle, M.: Micropuncture study of the mammalian urinary concentrating mechanism: evidence for the countercurrent hypothesis. Am. J. Physiol., *196:*927, 1959. (Demonstration of the fundamentals of the countercurrent system)

Jamison, R. L., Bennett, C. M., and Berliner, R. W.: Countercurrent multiplication by the thin loops of Henle. Am. J. Physiol., *212:*357, 1967. (Evidence is presented that the thick and the thin segments of the ascending limb of Henle's loop are capable of active sodium transport.)

Klahr, S., Tripathy, K., Garcia, F. T., Mayoral, L. G., Ghitis, J., and Bolaños, O.: On the nature of the renal concentrating defect in malnutrition. Am. J. Med., *43:*84, 1967.

Kleeman, C. R., Czaczkes, J. W., and Cutler, R.: Mechanisms of impaired water excretion in adrenal and pituitary insufficiency. IV. Antidiuretic hormone in primary and secondary adrenal insufficiency. J. Clin. Invest., *43:*1641, 1964.

Leaf, A., Bartter, F. C., Santos, R. F., and Wrong, O.: Evidence in man that urinary electrolyte loss induced by pitressin is a function of water retention. J. Clin. Invest., *32:*868, 1953. (Demonstration of effect of volume expansion (water retention) on sodium excretion—perhaps the most important mechanisms in the pathophysiology of the inappropriate ADH secretion syndrome)

Levinsky, N. G., and Berliner, R. W.: The role of urea in the urine concentrating mechanism. J. Clin. Invest., *38:*741, 1959. (Passive accumulation of urea in the medullary interstitium increases the maximum Uosm that the dog kidney can attain. Uosm and T^{C}H$_2$O are reduced in dog and man on low protein diets.)

Lindeman, R. D., Van Buren, H. C., and Raisz, L. G.: Osmolar renal concentrating ability

in healthy young men and hospitalized patients without renal disease. New Eng. J. Med., *262:*1306, 1960.

Papper, S., and Lancestremere, R. G.: Certain aspects of renal function in myxedema. J. Chron. Dis., *14:*495, 1961.

Papper, S., and Vaamonde, C. A.: The kidney in liver disease. *In:* Strauss, M. B., and Welt, L. G.: Diseases of the Kidney. ed. 2. Boston, Little, Brown & Co., 1971 (In press). (Review of the abnormal renal dilution and concentration capacities in patients with liver disease. Patients with cirrhosis may excrete a "normally" diluted urine (Uosm) less than 100 mOsm./kg. H$_2$O) but have subnormal values for V and CH$_2$O in response to water administration. Increased tubular sodium reabsorption and decreased urea excretion account for the low Uosm.)

Schedl, H. P., and Bartter, F. C.: An explanation for and experimental correction of the abnormal water diuresis in cirrhosis. J. Clin. Invest., *39:*248, 1960. (Impaired response to water administration can be improved by mannitol diuresis. The same authors have demonstrated this effect in congestive heart failure. Thus, increased delivery of solute (sodium) to the distal diluting sites improves diluting ability.)

Stein, R. M., Levitt, B. H., Golstein, M. H., Porush, J. G., Eisner, G. M., and Levitt, M. F.: The effects of salt restriction on the renal concentrating operation in normal hydropenic man. J. Clin. Invest., *41:*2101, 1962.

Strauss, M. B.: Body water in man. The acquisition and maintenance of the body fluids. Boston, Little, Brown & Co., 1957. (A scholarly review of body fluid regulation, with emphasis on volume regulation)

Ullrich, K. J., Kramer, K., and Boylan, J. W.: Present knowledge of the countercurrent system in the mammalian kidney. Prog. Cardiovasc. Dis., *3:*395, 1961. (A detailed description of the countercurrent sytem with historical review of the pioneer work of Kuhn, Hargitay and Wirz)

Vaamonde, C. A., Vaamonde, L. S., Morosi, H. J., Klingler, E. L., Jr., and Papper, S.: Renal concentrating ability in cirrhosis. I. Changes associated with the clinical status and course of the disease. J. Lab. Clin. Med., *70:*179, 1967. (Description of reversible alterations of concentrating ability in cirrhosis)

13

Body Buffering Mechanisms

Edward J. Lennon, M.D.

INTRODUCTION

The hydrogen ion is a small, positively charged, highly reactive particle that exerts profound influences on the rates of biochemical reactions. Virtually all enzymes have maximum activities only within specific, narrow ranges of hydrogen ion concentration ($[H^+]$). The ability to maintain differing $[H^+]$ within cells and on cell surfaces at specific sites in the body permits appropriate enzymes to act in the orderly, stepwise sequences characteristic of the living, metabolizing organism and prevents potentially destructive actions of enzymes at inappropriate sites. For example, gastric peptidases act at the high $[H^+]$ generated within (and tolerated by) the stomach, whereas pancreatic amylase acts at the much lower $[H^+]$ present in the duodenum.

Such gradients in $[H^+]$ at various locations within the body fluids are not static; they are maintained in a dynamic setting in which hydrogen ions are constantly added to and removed from the body by the chemical reactions of metabolism and, since the body is an open chemical system, by exchanges with the environment (intake and excretion).

Because hydrogen ions are so extremely reactive, they are never truly isolated particles in aqueous solution. Instead, they exist as hydronium ions (H_3O^+) or with additional water of hydration. Although it is the hydrated forms of the hydrogen ion that participate in the chemical reactions

to be described, we shall follow the convention of representing these as H^+.

With present knowledge, it is not possible to provide a complete and integrated description of the H^+ content of the body, which would take into account all of the many varying regions of $[H^+]$ within the body water. However, it is possible to view the body as a "black box," by accepting ignorance of the total content, distribution and functional availability of H^+, and to consider the net gains or losses of H^+ to the body over given time intervals. This approach, although incomplete, provides useful insights into the processes that regulate the overall H^+ content of the body and the manner in which these processes are altered by disease states.

DEFINITIONS

Since our concern is the $[H^+]$ within the body fluids, all chemical processes germane to it must result in (a) the addition of hydrogen ions to body fluids, (b) the binding of hydrogen ions to deny their participation in other chemical reactions, or (c) the elimination of hydrogen ions from the body.

An *acid* is a substance containing one or more hydrogen ions that can be liberated in solution (i.e., it competes less effectively for its H^+ than do water molecules). A *base* is a substance that can capture hydrogen ions from solution (i.e., it has a greater affinity for H^+ than do water molecules). Since by definition an acid contains H^+, it is clear that when an acid dissociates it yields not only H^+ but also a substance that has some finite affinity for H^+ which by definition is a base. The base that arises when an acid dissociates is called the conjugate (paired) base of the acid. Strong acids (such as HCl) have weak conjugate bases (such as Cl^-) and are almost completely dissociated in dilute aqueous solutions: $HCl \rightarrow H^+ + Cl^-$. Weak acids (such as H_2CO_3) are only sparingly dissociated in solution and have conjugate bases (such as HCO_3^-) with much greater affinities for H^+: $H_2CO_3 \leftrightarrows H^+ + HCO_3^-$.

Acids may be uncharged compounds (such as HCl and H_2CO_3), cations:

$$NH_4^+ \rightleftarrows H^+ + NH_3$$

or anions:

$$H_2PO_4^- \leftrightarrows H^+ + HPO_4^=$$

Some substances (such as $HPO_4^=$) can behave as either an acid or a base, depending upon the $[H^+]$ of the solution in which they are placed:

$$H_2PO_4^- \overset{+H^+}{\underset{}{\rightleftarrows}} HPO_4^= \overset{-H^+}{\underset{}{\rightleftarrows}} PO_4^=$$

Although the term *alkali* is restricted rigorously to the strong mineral bases (such as NaOH), we shall follow the custom of the current medical literature and use the terms *base* and *alkali* interchangeably.

BUFFERS

When a solution contains both a weak acid and the salt of the weak acid, it exhibits less change in $[H^+]$ on the addition of strong acids or bases than does pure water. Such a solution is said to be buffered. This property is really quite easy to understand. Weak acids are only slightly dissociated; thus a relatively large reservoir of undissociated acid is available to donate H^+ if the $[H^+]$ of the solution is lowered by addition of a strong base. The salts of weak acids, on the other hand, are almost completely dissociated in solution, providing a

large quantity of the conjugate base to react with added strong acids. Consider the buffer pair $H_2CO_3/NaHCO_3$. On addition of a strong acid (such as HCl), the following occurs:

$$HCl + NaHCO_3 \rightarrow NaCl + H_2CO_3$$

The strong acid (HCl), which would have ionized almost completely in water, is converted to a weak and sparingly dissociated acid (H_2CO_3), minimizing the rise in $[H^+]$ of the solution. If a strong base (such as NaOH) were added to the same buffered solution, the reaction would be:

$$NaOH + H_2CO_3 \rightarrow NaHCO_3 + H_2O$$

The strong base (NaOH) is converted to a weaker one (HCO_3^-). In addition, further quantities of H^+ are given up from the undissociated reservoir of H_2CO_3, minimizing the fall in $[H^+]$ of the solution.

In a closed chemical system, the addition of HCl to a solution buffered by $H_2CO_3/NaHCO_3$ would reduce the capacity of the solution to defend $[H^+]$ against additional quantities of HCl (since the HCO_3^- concentration of the solution would be reduced). Similarly, the addition of NaOH would cause a fall in the available quantity of undissociated H_2CO_3, reducing the capacity to buffer subsequent additions of NaOH. Thus, in a closed system, a buffered solution cannot defend $[H^+]$ against continuing additions of acids or bases. Moreover, in a closed system, a solution buffered by a single pair has no defense against the addition of the acid or the salt that together constitute the members of the buffer pair. Fortunately, the living organism is an open system, exchanging with the environment, and elegant mechanisms exist to maintain the concentrations of each of the members of the organism's principal buffer ($H_2CO_3/NaHCO_3$). In addition, other buffer pairs are available to aid in the defense against acute changes in the concentrations of either H_2CO_3 or HCO_3^-.

NORMAL MECHANISMS

The Total Buffer Capacity of the Body

The presence of buffer pairs within the body fluids modulates the changes in [H^+] that occur with sudden increases or decreases in total H^+ content, providing a temporary defense against additions of acids or bases. It is important to re-emphasize that this is a *temporary* defense which leaves the organism vulnerable to subsequent gains of acids or bases and that the ultimate physiological response to gains of acids or bases requires their elimination from the body and restoration of the body's buffer pairs to their normal concentrations.

The buffers of the body fluids are not completely defined. The total volume of water contained in the nonobese adult is equivalent to about 60 percent of body weight. Thus, a 70-kg. adult contains approximately 42 liters of water. Two thirds of this (28 liters) is intracellular. The remaining one third (14 liters) is extracellular and confined, in part, within the vasculature by the oncotic pressure of the plasma proteins. The buffers of the largest fluid compartment—cell water—are least well characterized and probably vary from tissue to tissue. The extracellular, extravascular fluids are buffered almost exclusively by H_2CO_3/$NaHCO_3$. Albumin and phosphate compounds play a significant role in plasma. In red blood cells, the imidazole groups of hemoglobin are also important. Detailed characterizations of the important buffers of whole blood and plasma are available in the references at the end of the chapter.

For the purposes of this chapter, we shall neglect the array of defined and uncertain buffer pairs other than H_2CO_3/HCO_3^-. This seems justifiable since: (a) the H_2CO_3/HCO_3^- buffer pair provides the largest single buffer reservoir in body fluids; (b) the ratio of its members reflects the state of titration of all other buffer pairs; (c) of all body buffers, it is the sole pair for which convincing evidence of independent regulation of the two members has been presented; and (d) it is the only buffer pair in which the two species can be quantitated conveniently.

The concentration of H_2CO_3 in body fluids is a direct function of the partial pressure of gaseous CO_2 (Pa_{CO_2}). The balance between metabolic CO_2 production and pulmonary CO_2 excretion determines Pa_{CO_2} (normally, about 40 mm. Hg). Carbon dioxide is soluble in water, and a small fraction of that dissolved (about 1 part in 800) undergoes hydration, reacting:

$$CO_2 + H_2O \rightarrow H_2CO_3$$

Since all dissolved CO_2 is available for hydration, we shall represent the sum of dissolved CO_2 and actual H_2CO_3 as [H_2CO_3]. The combined solubility-activity coefficient for the conversion of Pa_{CO_2} (mm. Hg) to [H_2CO_3] (mM/L.) is designated S and has a value of 0.03 in blood or plasma at normal body temperature. Thus, [H_2CO_3] (mM/L.) = S·Pa_{CO_2} (mm. Hg), or 0.03 × 40 = 1.2. The normal serum [HCO_3^-] is 24 mM/L. and is maintained by renal HCO_3^- generation, as will be discussed below.

In the experimental animal, the H_2CO_3/HCO_3^- buffer pair can be converted from an open to a closed system if (a) the animal is ventilated mechanically to prevent changes in the rate of CO_2 excretion and (b) the kidneys are removed to prevent renal acid excretion and HCO_3^- generation. Under these conditions, the changes in serum [HCO_3^-] that occur when known quantities of strong acids are infused can be used to calculate the total capacity of the buffers participating in the defense against the acid load. Take, for example, a 20-kg. dog with an extracellular fluid (ECF) volume of 4 liters (20% of body weight) and an ECF [HCO_3^-] of 24 mEq./L. The total HCO_3^- content of the ECF is then 96 mEq. If 100 mEq. of dilute HCl were infused (5 mEq./kg.), and only ECF HCO_3^- acted as a buffer, ECF (HCO_3^-) would fall to zero. In fact, however, in such an experiment, serum [HCO_3^-] falls from 24 mM/L. to about 11.5 mM/L. Thus, ECF HCO_3^- accounts for the buffering of (24 − 11.5) × 4 liters, or 50 mEq. of acid. The remaining 50 mEq. must have been buffered by systems other than ECF HCO_3^- Put another way, the animal behaved "as if" its buffer capacity were that which would exist if HCO_3^-, at the concentration present in ECF, were distributed in a fluid compartment twice the size of ECF volume, or equal to about 40 percent of body weight. As the acid load is increased and the rise in [H^+] of the body fluids becomes more severe, the "apparent" space of distribution of HCO_3^- becomes larger. Presumably, this means that additional buffers come into play as acidemia (increasing blood [H^+]) becomes more severe. The mobilization of alkaline mineral salts from bone probably accounts for a part of this buffer reserve. Experimental studies in the animal and clinical observations in man suggest that death results when acute acid loads exceed about 15 mEq. of acid per kilogram of body weight.

Problems of Measurement

We have implied that [H^+] is measurable directly. Unfortunately, this is not the case. It is possible to measure hydrogen ion *activity* (α [H^+]) in solutions (ignoring, for the moment, a variety of instrumental and theoretical uncertainties that becloud even that measurement). Hydrogen ion activity is a function of the chemical potential of H^+ (i.e., its tendency to enter into chemical reactions). Glass membranes that are selectively permeable to H^+ have been used to construct electrodes that develop an electromotive force (EMF) proportional to α [H^+] in the solution in which they are immersed, such that:

$$EMF = K \cdot \log \frac{1}{\alpha [H^+]}$$

where K = a composite constant. Uncertainties as to the value of the activity coefficient (α) in biological fluids prevent rigorous solution of the equation for [H^+]. Throughout the chapter, we shall sacrifice precision for conceptual clarity and assign α a value of 1.0, thus setting α [H^+] = [H^+].

As Henderson demonstrated, the constituents of the H_2CO_3/HCO_3^- pair can be described according to the mass action law as follows:

$$K \ [H_2CO_3] = [H^+] \ [HCO_3^-]$$

where K is a theoretical constant that varies with the composition of the solution under study and must be corrected for specific solutions (K′). In blood or plasma at body temperature, when [H_2CO_3] and [HCO_3^-] are expressed in mM/L. and a convenient unit for [H^+] is employed i.e., nanoequivalents/liter [10^{-9} Equivalents/liter], K′ has a value of 800. Thus:

$$800 \ [H_2CO_3] \ mM/L. = $$
$$[H^+] \ nanoEq./L. \times [HCO_3^-] \ mM/L.$$

or

$$[H^+] \ nanoEq./L. = 800 \ \frac{[H_2CO_3] \ mM/L.}{[HCO_3] \ mM/L.}$$

Substituting normal values:

$$[H^+] \ nanoEq./L. = 800 \ \frac{1.2}{24} = 40$$

It is clear from Henderson's equation that if any two of the three components are measured, the third can be calculated. In addition, $S \cdot Pa_{CO_2}$ can be substituted for [H_2CO_3]. While Pa_{CO_2} can be measured directly by a variety of techniques, most clinical laboratories measure the serum "total CO_2 content" (T_{CO_2}). In measuring T_{CO_2}, all of the bicarbonate present in a sample of serum is converted to H_2CO_3 by addition of a stronger acid. The H_2CO_3 so generated from HCO_3^- plus the pre-existing H_2CO_3 is then converted to CO_2 gas by shaking in a vacuum. The partial pressure of the liberated CO_2 gas is meas-

TABLE 13-1

Conversion of pH to [H$^+$] NANOEQ./L.				
pH	0.00	0.02	0.04	0.06	0.08
6.9	126	120	115	110	105
7.0	100	96	91	87	83
7.1	79	76	72	69	66
7.2	63	60	58	55	52
7.3	50	48	46	44	42
7.4	40	38	36	35	33
7.5	32	30	29	28	26
7.6	25	24	23	22	21

ured and converted to Tco$_2$ mM/L. The evolved CO$_2$, then, is the sum of serum HCO$_3^-$ + H$_2$CO$_3$. It is important to recall that under virtually all physiological conditions, over 95 percent of what is expressed as Tco$_2$ existed in serum as [HCO$_3^-$]. Thus, Tco$_2$ is a reasonable estimate of serum [HCO$_3^-$]. Adapting Henderson's equation for use with Tco$_2$, one can write:

$$[H^+] = K' \frac{S \cdot Pa_{CO_2}}{Tco_2 - S \cdot Pa_{CO_2}}$$

A second and somewhat more complex unit is employed to express [H$^+$] (or, α [H$^+$]). This system was evolved, in part, to obviate expression of [H$^+$] in awkward units (i.e. [H$^+$] in blood = 0.000000040 Equivalents/liter) and, in part, on the ground that the effects of H$^+$ in chemical systems might correlate better with log $\frac{1}{\alpha [H^+]}$ than with [H$^+$], as suggested by the responses of the glass electrode and of indicator dyes. Hasselbalch adapted Henderson's equation to fit the responses of measuring devices as follows:

Henderson's equation $[H^+] = K' \dfrac{[H_2CO_3]}{[HCO_3^-]}$

First, reciprocals were taken:

$$\frac{1}{[H^+]} = \frac{1}{K'} \times \frac{[HCO_3^-]}{[H_2CO_3]}$$

Then, the equation was put in log form:

$$\log \frac{1}{[H^+]} = \log \frac{1}{K'} + \log \frac{[HCO_3^-]}{[H_2CO_3]}$$

Finally, the convention was adopted:

$$\log \frac{1}{X} \equiv pX$$

Thus:

$$pH = pK' + \log \frac{[HCO_3^-]}{[H_2CO_3]}$$

In blood or serum at body temperature pK' = 6.1. The serious student of acid-base physiology should be familiar with both systems of notation and conversant in employing either. Table 13-1 provides data for interconversion between [H$^+$] and pH over the commonly observed ranges in blood and plasma. The author prefers to utilize [H$^+$] in considering overall regulation of acid-base metabolism.

ENDOGENOUS ACID PRODUCTION

Some 20 moles (20,000 mM) of carbon dioxide are generated each day in the adult as an end product of metabolism. As blood perfuses the tissue sites of CO$_2$ production, CO$_2$ diffuses into the plasma and the red cells. The red cells constitute a "sink" for dissolved CO$_2$, since they contain the enzyme carbonic anhydrase (CA), which greatly accelerates the reaction:

$$CO_2 + H_2O \rightleftarrows H_2CO_3$$

In the peripheral tissues, oxyhemoglobin gives up its oxygen. The imidazole groups of hemoglobin are stronger bases when the oxygen-bearing ferrous groups are in the reduced form. Thus, when H$_2$CO$_3$ forms and dissociates, the H$^+$ is bound to imidazole groups and [HCO$_3^-$] increases within the red cell. Part of the HCO$_3^-$ so generated in the red cell diffuses into the plasma. This is why venous plasma has a somewhat higher [HCO$_3^-$] than does arterial plasma, which tends to limit the increase in venous [H$^+$] resulting from the rise in Pco$_2$. When HCO$_3^-$ diffuses out of the red cell, maintenance of electroneutrality requires that a cation also diffuse out of the cell, or that another anion diffuse into the cell. Since the red cell membrane is relatively impermeable to cations, the latter occurs, and Cl^- enters the cell in exchange for HCO_3^- (the so-called *chloride shift*).

When venous blood perfuses the lungs, this entire process is reversed. Reduced hemoglobin is oxygenated; the imidazole groups then give up their H^+ which react with red cell HCO_3^-, and with HCO_3^- from the plasma diffusing into the cell in exchange for Cl^-. Catalyzed by red cell carbonic anhydrase, H_2CO_3 is converted back to CO_2 and H_2O, the CO_2 diffusing into the alveolar spaces. The overall process results in the potential acid, CO_2, being transported largely as HCO_3^-. Clearly, the quantity of CO_2 produced each day would result in severe and lethal acidemia if pulmonary CO_2 elimination were seriously impaired. The reactions involved in CO_2 transport and excretion are summarized in Fig. 13-1.

A second class of acids (*fixed*, or *nonvolatile* acids) are produced within the body by metabolic processes. Two reactions are involved: When compounds containing neutral sulfur (such as the amino acids methionine and cysteine) are oxidized, the neutral sulfur is converted, step-

wise, to inorganic sulfate ($SO_4^=$). Incident to this oxidation, two hydrogen ions are generated. The overall reaction is:

$$R–S–R \xrightarrow{O_2} urea + CO_2 + H_2O + SO_4^= + 2H^+$$

Since inorganic sulfate is excreted exclusively in the urine, measurement of the 24-hour urine content of inorganic sulfate permits quantitative estimation of the number of hydrogen ions added to the body fluids by this reaction. Adults eating normal diets produce approximately 0.5 mEq. of H^+ per kilogram body weight per day as a result of oxidation of neutral dietary sulfur. Clearly, a diet high in protein will increase acid production from this source.

The second process which results in the endogenous production of fixed acids depends upon urinary losses of intermediary products of metabolism, preventing complete degradation to CO_2 and water. During the combustion of carbohydrates, fats or proteins, organic acids are produced as intermediary steps, for example:

$$glucose \rightarrow 2 \ lactate^- + 2H^+ \rightarrow CO_2 + H_2O$$

If the lactate ions are excreted in the urine (e.g., sodium or potassium salts) hy-

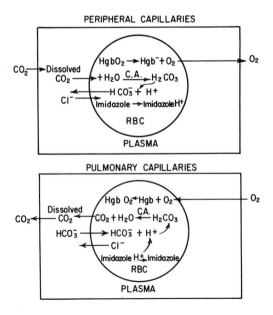

FIG. 13-1. Reactions involved in CO_2 transport and excretion.

drogen ions are left behind in the body fluids and constitute another source of the daily metabolic acid load. Adults eating normal diets excrete approximately 0.5 mEq. of organic acid salts per kg. body weight per day in the urine.

Thus, the oxidation of neutral dietary sulfur and the urinary losses of intermediary organic acid salts result in gains of approximately 1 mEq. of H^+ per kg. per day. The *net* gains of H^+ as a result of metabolism, however, are less because other constituents of the diet, on metabolic degradation, *remove* H^+ from the body fluids. For example, fruits and vegetables contain salts of organic acids (such as K citrate, Na lactate, etc.). Such salts, if absorbed by the intestine and metabolized result in:

$$R\text{-}COO^- K^+ \xrightarrow{O_2} CO_2 + K^+ + HCO_3^-$$

The HCO_3^- so generated reacts with H^+ of the body fluids and is eliminated as $CO_2 + H_2O$.

When the inorganic cations (Na^+, K^+, Ca^{++} and Mg^{++}) and anions ($Cl^- +$ phosphate) contained in normal diets are analyzed, cations exceed anions consistently, but to a variable degree. Electroneutrality requires that the excess inorganic cationic charges in the diet be balanced by anionic charges which, by exclusion, must be organic. These organic anions, to the extent that they can be degraded, represent potential base, because (as shown above) they yield bicarbonate on combustion. The feces also contain an excess of inorganic cations over inorganic anions and therefore contain organic anions. Although they are more likely to represent organic anions secreted into or formed within the lumen of the gut (principally acetate and bicarbonate) rather than unabsorbed organic anions from the diet, they constitute a loss of potential base which must have arisen from neutral precursors, just as do organic acid salts in the urine. The net gain of organic acid salts to the body can be taken as:

diet organic anions minus stool organic anions, mEq. per day. With usual diets, the organic anion content of the diet is about 1 mEq. per kg. of body weight per day and that of the feces about 0.5 mEq. per kg. body weight per day. The net gain of inorganic anion (potential alkali) from the intestine for a 70-kg. male would be 70 minus 35, or 35 mEq. per day. Net endogenous fixed acid production in a 70-kg. male eating a normal diet would then be estimated as:

H^+ from oxidation of organic sulfur	35 mEq.
plus losses of organic acid anions in urine	35 mEq.
	70 mEq.
minus net absorption of diet organic anions	35 mEq.
total net H^+ production	35 mEq.

The components of endogenous fixed acid production are summarized in Fig. 13-2.

If a 70-kg. male had a readily available buffer capacity equal to serum $[HCO_3^-]$ (24 mEq./L.) distributed in a fluid space equivalent to 40 percent of body weight (28 liters), he would be capable of buffering 672 mEq. of acid. If the daily net endogenous fixed acid load (35 mEq. per day) were retained within the body, at the end of a week his buffer capacity would have fallen to 672 minus (35×7), or 427 mEq. Assuming the same apparent space of distribution, serum $[HCO_3^-]$ would then be $\frac{427}{28}$ or about 15 mEq./L. In reality, the serum $[HCO_3^-]$ remains constant from day to day. Thus, it is clear that the quantity of fixed acid generated each day must be precisely balanced by the excretion of an equivalent quantity of acid. This is accomplished by the kidney.

Renal Acid Excretion

Renal tubular cells, like red blood cells, contain the enzyme carbonic anhydrase.

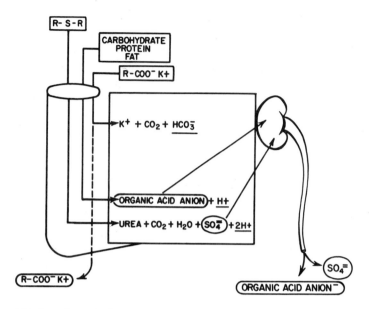

FIG. 13-2. Components of endogenous fixed acid production.

Within the tubule cells, carbonic anhydrase catalyzes the reaction:

$$CO_2 + H_2O \overset{\text{C.A.}}{\rightleftharpoons} H_2CO_3 \leftrightarrows H^+ + HCO_3^-$$

H^+ is secreted into the tubular urine and HCO_3^- is transferred to the blood perfusing the tubules. The kidney has the capacity to establish hydrogen ion gradients, such that, at maximum, $[H^+]$ in tubular urine is 800 times that in peritubular blood. This means that the urine can have a $[H^+]$ as high as 40×800, or 32,000 nanoEq./L. This of itself would do little to achieve excretion of the 35 mEq. of acid generated endogenously each day. The volume of urine required to excrete 35 mEq. of acid at a concentration of 32,000 nanoEq. per L. (0.032 mEq./L.) would be $\dfrac{35}{.032}$, or approximately 1,100 *liters*, nearly ten times the available rate of glomerular filtrate formation! Man would expend his life drinking and urinating (enterprises perhaps no more ignoble than some of his current activities). This is obviated by the buffer capacity of buffer pairs filtered at the glomeruli and an additional buffer formed within the tubule cells —i.e., ammonia (NH_3).

Bicarbonate Reabsorption

The greater part of H^+ excretion by the kidney occurs in the proximal convoluted tubule and is expended in the "reabsorption" of filtered HCO_3^-. At a glomerular filtration rate of 150 liters per day and a serum $[HCO_3^-]$ of 24 mEq./L., some 3,600 mEq. of H^+ must be excreted by the kidneys to prevent losses of HCO_3^- in the urine. According to present concepts, filtered HCO_3^- is not reabsorbed directly across the luminal tubular membrane. Instead, when $NaHCO_3$ is filtered, Na^+ is actively reabsorbed against electrochemical gradients, in exchange for H^+ formed within the tubule cells by the CA-catalyzed hydration of CO_2. The secreted H^+ reacts with HCO_3^- in the tubular urine, and, again catalyzed by CA present at the luminal surface of the proximal tubule cells, the reaction

$$HCO_3^- + H^+ \rightarrow H_2CO_3 \overset{CA}{\rightarrow} CO_2 + H_2O$$

occurs. The CO_2 produced diffuses back into the tubule cell, where it can be reutilized to form additional quantities of H_2CO_3. The $[H^+]$ of the fluid leaving the proximal tubule is altered minimally from that of the original glomerular filtrate.

This step, then, conserves filtered HCO_3^- but does not accomplish *net* acid excretion.

Titratable Acid

As urine proceeds along the tubule to more distal segments, H^+ secretion continues by the same mechanism:

$$(CO_2 + H_2O \overset{CA}{\to} H_2CO_3 \to H^+ + HCO_3^-).$$

With the principal buffer (HCO_3^-) now removed from the urine, $[H^+]$ begins to increase significantly, and the second important filtered buffer pair, $H_2PO_4^-/HPO_4^=$, is titrated. In serum at $[H^+] = 40$ nanoEq./L., $H_2PO_4^-/HPO_4^=$ exists in a ratio of 1/4. This is because K' for the second dissociation (K_2') of $H_3PO_4^= = 160$; thus:

$$[H^+] \text{ nanoEq./L.} = 160 \frac{[H_2PO_4^-] \text{ mM/L.}}{[HPO_4^= \text{ mM/L.}]}$$

At $[H^+] = 40$ nanoEq./L., $[H_2PO_4^-]/[HPO_4^-] = 40/160 = 1/4$. If $[H^+]$ of the tubular urine is raised to 1,600 nanoEq./L., the ratio $[H_2PO_4^-]/[HPO_4^-]$ becomes 1,600/160 or 10/1, trapping H^+ by the reaction:

$$H^+ + HPO_4^= \to H_2PO_4^=.$$

The quantity of acid excreted in the urine by this mechanism is referred to as titratable acid, since it can be quantitated directly by determining the amount of dilute base which must be added to the final urine to return its $[H^+]$ to 40 nanoEq./L. (pH 7.4).

Ammonia Secretion

Finally, the renal tubule cells produce an additional buffer, NH_3, by the reaction:

$$\text{glutamine} \xrightarrow{\text{glutaminase}} \text{glutamic acid} + NH_3$$

Gaseous NH_3 diffuses freely across the cell membrane into the tubular urine, reacting with H^+ to form NH_4^+. The ammonium ion cannot penetrate the lipid cell membrane and is thus trapped and excreted in the final urine.

The net quantity of acid eliminated in the urine, then, is equal to titratable acid plus NH_4^+ minus any HCO_3^- that escapes into the final urine. Normally, NH_4^+ accounts for about two thirds of the acid excreted, and titratable acid for the remaining one third. If body H^+ content is to remain constant, renal acid excretion must equal endogenous acid production. In the 70-kg. man considered previously, it would equal 35 mEq. per day. The mechanisms by which the kidney adjusts net acid excretion to meet the varying demands of acid or base excesses will be discussed under the sections dealing with states of acid ex-

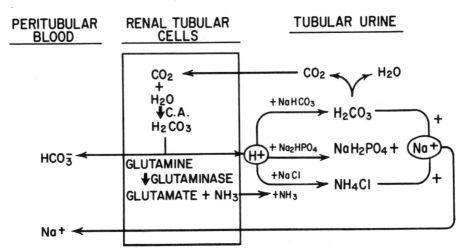

Fig. 13-3. Reactions involved in renal acid excretion.

cess or deficits. The important reactions involved in renal acid excretion are summarized in Figure 13-3.

HYPERCAPNIA AND HYPOCAPNIA

At any steady-state Pa_{CO_2}, CO_2 excretion must equal CO_2 production. Since CO_2 production usually can be assumed to remain constant over a given period of observation, the level of Pa_{CO_2} reflects primarily the efficiency of pulmonary CO_2 elimination.

Hypercapnia (Respiratory Acidosis)

Alveolar ventilation is impaired when the muscles of respiration are weakened or paralyzed (e.g., bulbar poliomyelitis), when airway obstruction is present (e.g., asthma), or when anatomic disruptions cause increases in nonventilated dead space as well as disturbances in the distribution of pulmonary blood flow (e.g., pulmonary emphysema). In each of these instances, Pa_{CO_2} will increase until the gradient for CO_2 diffusion rises sufficiently to compensate for the underlying defect and CO_2 excretion again equals CO_2 production (or, as in asphyxia, death ensues). Since CO_2 is hydrated to the weak acid, carbonic acid, retention of CO_2 produces an acidemia.

Hypercapnia is characterized by an increased Pa_{CO_2}, acidemia, and a compensatory increase in $[HCO_3^-]$ which varies with the duration of hypercapnia. In *acute hypercapnia*, as demonstrated by exposing normal human volunteers to various mixtures of CO_2-in-air, in an environmental chamber, renal acid excretion does not increase significantly. Blood $[H^+]$ increases as a direct, linear function of Pa_{CO_2} such that:

$$[H^+] \text{ nanoEq./L.} =$$
$$0.76 \ Pa_{CO_2} \text{ mm. Hg} + 9.3$$

Serum $[HCO_3^-]$ increases only slightly (by < 3 mEq./L. when Pa_{CO_2} is increased from 40 to 80 mm. Hg). This slight rise in serum $[HCO_3^-]$ can be attributed almost entirely to blood buffering, principally Hgb. Other cell buffers do not appear to participate significantly. The major buffer of the ECF ($H_2CO_3/NaHCO_3$) cannot defend against a rising Pa_{CO_2}. The body's defenses against acute hypercapnia are clearly very limited. Serum $[HCO_3^-]$ values of less than 24 or more than 29 mEq./L. suggest the presence of some additional disturbance in acid-base metabolism.

With *chronic hypercapnia*, as studied in patients with pulmonary diseases and in the dog, blood $[H^+]$ is also a direct, linear function of Pa_{CO_2}, but the slope is less steep and the intercept higher, such that: $[H^+]$ nanoEq./L. $= 0.24 \ Pa_{CO_2}$ mm. Hg $+ 27.2$. At any level of Pa_{CO_2}, serum $[HCO_3^-]$ is higher and blood $[H^+]$ lower after chronic exposure than those values observed during acute exposure. This occurs because of *renal compensation*, which will be discussed below.

Hypocapnia (Respiratory Alkalosis)

Hyperventilation, whether voluntary or involuntary (mechanical ventilation, etc.), causes a fall in Pa_{CO_2} which is ultimately stabilized as the gradient for CO_2 diffusion is reduced. Hypocapnia is characterized by a decrease in Pa_{CO_2}, a fall in blood $[H^+]$ and a compensatory reduction in $[HCO_3^-]$. During *acute hypocapnia*, blood $[H^+]$ is again a direct, linear function of Pa_{CO_2}:

$$[H^+] \text{ nanoEq./L.} = 0.74 \ Pa_{CO_2} + 10.4.$$

This relationship closely parallels that found during acute hypercapnia. When Pa_{CO_2} falls from 40 to 20 mm. Hg ($- 20$ mm. Hg), serum $[HCO_3^-]$ falls, on the average, by 7.4 mEq./L. Renal acid excretion does not change acutely, and only about one third of the fall in $[HCO_3^-]$ can be attributed to buffering by blood. By contrast, when Pa_{CO_2} increases acutely from 40 to 80 mm. Hg ($+ 40$ mm. Hg), serum $[HCO_3^-]$ increases by less than 3 mEq./L. Why is the body better able to buffer against acute increases in Pa_{CO_2}? Studies of the responses of isolated tissues to variations in P_{CO_2} in the bathing media suggest that this represents dynamic buffering as a result of changes in the rate of acid production within cells. Apparently, cells can easily augment acid production but cannot easily shut down their catabolic, energy-yielding (and acid-producing) reactions.

It is not yet clear whether *chronic hypocapnia* results in any further fall in serum $[HCO_3^-]$ in man or whether changes in renal acid excretion play a role in chronic compensatory responses.

Renal Responses to Changes in Pa_{CO_2}

Normally, about 90 percent of filtered $[HCO_3^-]$ is "reabsorbed" in the proximal tubule and the remaining 10 percent in more distal segments. At a glomerular filtration rate of 100 ml. per minute, with a $[HCO_3^-]$ of 20 mEq./L. in the filtrate, 2.4 mEq. of HCO_3^- are filtered per minute and quantitatively reabsorbed. If serum $[HCO_3^-]$ is gradually increased (by infusion of $NaHCO_3$), small quantities of HCO_3^- appear in the urine until the rate of HCO_3^- filtration exceeds 2.7 or 2.8 mEq. per minute. Thereafter, any further increments in filtered HCO_3^- are excreted, demonstrating that the tubules have a maximum limit on HCO_3^- reabsorption ($TM_{HCO_3^-}$). The $TM_{HCO_3^-}$ is not fixed, but varies directly with Pa_{CO_2}. During hypercapnia, the kidney responds by increasing renal acid excretion (ammonium + titratable acid), thereby generating new HCO_3^- and raising the serum $[HCO_3^-]$. Because Pa_{CO_2} is elevated, $TM_{HCO_3^-}$ is also increased, permitting the new, raised levels of HCO_3^- to be maintained. Conversely, a fall in Pa_{CO_2} reduces $TM_{HCO_3^-}$ and results in HCO_3^- excretion.

These effects of Pa_{CO_2} on $TM_{HCO_3^-}$ appear to take place entirely in the proximal convoluted tubules. Secretion of H^+ in all tubular segments depends upon the hydra-

tion of CO_2, catalyzed by carbonic anhydrase. Since CO_2 diffuses freely across cell membranes, it is presumed that P_{CO_2} within cells $= Pa_{CO_2}$. The proximal tubule has a great capacity to secrete H^+, but can do so only against small $[H^+]$ gradients. The distal tubule and the collecting duct have a limited capacity to secrete H^+ but can do so against very large $[H^+]$ gradients. Thus, the level of Pa_{CO_2}, by altering P_{CO_2} and the concentration of H^+ within cells, becomes rate limiting in the high capacity, low gradient, proximal H^+ secretory mechanism but does not affect the low capacity, high gradient, distal system.

METABOLIC ACIDOSIS AND ALKALOSIS

Metabolic Acidosis

Whenever gains of fixed acids or losses of base exceed the rate of acid elimination from the body, metabolic acidosis occurs. This may be the result of metabolic disturbances which increase endogenous acid production (e.g., diabetic ketoacidosis), the ingestion of acid precursors (e.g. methionine, NH_4Cl), abnormal losses of alkaline body fluids (e.g., diarrhea), or a reduced capacity to excrete acid (e.g., chronic glomerulonephritis or renal tubular acidosis). As discussed previously, the abrupt addition of H^+ to the body fluids results first in titration of the body's buffer stores and is reflected by a fall in serum $[HCO_3^-]$. The increase in blood $[H^+]$ is sensed by chemoreceptor cells (located in the aortic arch, the carotid bodies and the respiratory center) which stimulate respiration, thus lowering Pa_{CO_2}. If renal disease is not the cause of the acidosis, the kidneys respond by increasing urinary $[H^+]$, with more complete titration of filtered phosphate. As the duration of metabolic acidosis increases, pulmonary compensation is diminished, probably because of the exhausting respiratory work required. Pa_{CO_2} tends to rise toward normal. The kidneys, on the other hand, gradually increase net renal acid excretion, almost entirely as the result of augmented urinary NH_4^+ content. Unless acid loads are excessive (> 500 mEq. per day), renal acid excretion ultimately reaches —and, for a time, exceeds—acid production, so that the serum $[HCO_3^-]$ rises toward normal.

Thus, *acute metabolic acidosis* is characterized by a sharp reduction in serum $[HCO_3^-]$, accompanied by marked lowering or Pa_{CO_2}, so that blood $[H^+]$ is not greatly increased. *Chronic metabolic acidosis,* in contrast, is identified by a smaller reduction in serum $[HCO_3^-]$ (because of augmented renal acid excretion as NH_4^+), less marked reductions of Pa_{CO_2}, and comparable protection of blood $[H^+]$. In chronic metabolic acidosis, Pa_{CO_2} is a direct, linear function of serum $[HCO_3^-]$, such that:

$$Pa_{CO_2} \text{ mm. Hg} = 1.1 \ [HCO_3^-] \text{ mEq./L.} + 18.3$$

Until recently, the mechanism by which the kidney sensed the presence of systemic acidemia and responded by increasing NH_3 production was puzzling. Although adaptive increases in the enzyme glutaminase, which catalyzes the reaction: glutamine → glutamate $+ NH_3$, have been identified in the rat during induced acidosis, this does not occur in dog or, presumably, man. Studies in vivo and in vitro have shown that renal gluconeogenesis is increased during metabolic acidosis, apparently because of stimulation of a rate-limiting step between α-ketoglutarate and glucose. As a consequence, the conversion of α-ketoglutarate and glutamate to glucose is enhanced. Removal of one of the products of glutamine deamination (glutamate) drives the reaction to the right, increasing the rate of NH_3 production.

Metabolic Acidosis Due to Defective Renal Acid Excretion

Any parenchymatous renal disease impairs the capacity of the kidney to excrete acid. Since acid excretion is a tubular function, diseases that affect the renal interstitium (e.g., chronic pyelonephritis) may exhibit acidemia of a greater degree in relationship to reduction in filtration function than do primarily glomerular diseases (e.g., chronic glomerulonephritis). The fundamental defect in acid excretion in most patients with chronic parenchymatous renal disease is the inability to produce NH_3 in normal quantities. This does not appear to result from any specific cellular defect in NH_3 production, since ammonium excretion may actually be supranormal in terms of the number of surviving nephrons. It appears, rather, that NH_3 excretion falls because the total *number* of functioning nephrons is reduced.

Most patients with chronic azotemic renal disease are able to achieve normal $[H^+]$ gradients between tubular fluid and peritubular blood. When they are acidemic, their urines are characteristically intensely acid. In some patients, however, if serum $[HCO_3^-]$ is raised toward normal, the urine becomes alkaline and a significant HCO_3^- "leak" appears, suggesting a second defect within the kidney. This may not be a defect in tubule cell function, but rather the result of one of the compensatory adaptations to chronic renal failure. Unless sodium intake is (unwisely) overrestricted, patients with chronic renal failure have expanded extracellular fluid volumes. This tends to raise their glomerular filtration rates, improving filtration function.

A second consequence of expansion of the ECF is a reduction in the percent of filtered sodium and water reabsorbed by the tubules. This effect of volume expansion, called *third factor* because it can be dissociated from the effects of glomerular filtration rate and aldosterone, has not been completely explained but appears to involve physical factors (i.e., changes in plasma oncotic pressure, hydrostatic pressure in the peritubular capillaries, and the volume of the interstitial space separating tubules and capillaries) and perhaps also a natriuretic hormone. Since Na^+ must be reabsorbed either with an anion or in exchange for another cation to maintain electroneutrality an increase in the concentration of HCO_3^- in the filtrate of a patient with expanded ECF volume who is rejecting a larger than normal percentage of the filtrate in the proximal tubular segments would cause the distal tubular segments to become flooded with HCO_3^-. Since the capacity of distal tubular segments to secrete H^+ is limited, this may explain the excretion of alkaline, HCO_3^- rich urines.

Renal Tubular Acidosis

A second category of renal disease in which glomerular function is normal but in which specific tubular defects interfere with H^+ secretion is called renal tubular acidosis. This generic term includes many inherited and acquired disorders which may be associated with other tubular transport defects (e.g., glucosuria, phosphaturia, or aminoaciduria).

Renal tubular acidosis can be classified broadly into a proximal and a distal form. In *proximal renal tubular acidosis,* there is an apparent defect in the capacity to generate H^+ in proximal tubular segments. As a consequence, if serum $[HCO_3^-]$ is raised to normal levels, the distal segments receive large loads of filtered HCO_3^-, their low H^+ secretory capacity is exceeded, and the urine is alkaline. If such patients are allowed to develop acidemia, however, the reduced rate of delivery of HCO_3^- to the distal segments unmasks their capacity to develop normal $[H^+]$ gradients, and the urine becomes acid.

The defect in *distal renal tubular acidosis* is the inability to develop normal $[H^+]$ gradients in the distal tubular segments. Since the $[H^+]$ of the filtrate is little altered in proximal segments, the urine is always relatively "alkaline" ($[H^+] < 1525$ nanoEq./ L., pH > 5.8), even when the patient is acidemic. Patients with distal renal

tubular acidosis may be able to cope with modest endogenous acid loads without becoming acidemic. Such patients, however, if challenged with an acid load (NH_4Cl) cannot sufficiently augment renal acid excretion and become acidemic.

Despite markedly reduced rates of renal acid excretion, patients with chronic renal acidosis ultimately achieve stable serum HCO_3^- concentrations at some reduced level. Why don't they develop progressive, lethal acidemia? Two possible answers have been considered. First, it was conceivable that such patients somehow reduced endogenous acid production. This possibility has been excluded. With normal diets, patients with chronic renal disease have normal rates of net endogenous acid production (approximately 0.5 mEq. of acid per kg. of body weight per day).

The second possibility was that patients with chronic renal acidosis were in continuous, positive acid balance. If so, the retained acid must have been neutralized ultimately by slowly available buffers outside of the "HCO_3^- space." This has proved to be the case. The basic mineral salts of bone, chiefly $CaCO_3$, provide the additional source of buffer. The hydroxyapatite crystals of bone mineral in patients with chronic azotemic acidosis are deficient in $CO_3^=$, as compared to normal bone. Calcium balance studies in patients with chronic azotemia indicate daily losses of only 4 or 5 mEq. per day, due entirely to losses in the feces, urinary calcium excretion being less than 1 or 2 mEq. per day. This quantity, even if arising entirely from $CaCO_3$, is inadequate to account for the observed positive acid balances. However, during the continuous remodeling of bone, substitution of $CO_3^=$ deficient hydroxyapatite for normal hydroxyapatite could result in the slow release of additional base without losses of calcium from the skeleton.

A somewhat different mechanism may be operative in experimental (NH_4Cl) acidosis and, presumably, in renal tubular acidosis. When normal subjects are made acidotic by feeding NH_4Cl, there is specific inhibition of tubular calcium reabsorption. The resulting hypercalciuria causes strongly negative calcium balances. The quantities of calcium lost in the urine are sufficient to account for H^+ retention beyond that buffered in the HCO_3^- space. Here, bone mineral breakdown mediated by parathyroid hormone may be contributory, but it is directed primarily at maintaining normal serum $[Ca^{++}]$.

Metabolic Alkalosis

Acid gastric juice is the only body fluid containing high concentrations of H^+ (up to 100 mEq./L.). Protracted vomiting can induce severe alkalosis, serum $[HCO_3^-]$ rising to as high as 60 mEq./L. Concomitant losses of Na^+, K^+, and Cl^-, as a result of their presence in vomitus and because of impaired intake and continued urinary and fecal losses, result in renal responses that tend to sustain the alkalosis (see below). Pulmonary compensatory responses, i.e., increases in Pa_{CO_2}, are limited in man because of the respiratory drive to maintain oxygen delivery.

The Effects of Na^+, K^+ and Cl^- Balances on Renal Acid Excretion

Although the author believes that acid-base balance is regulated specifically and purposefully under normal physiological circumstances, it is clear that other needs of the body may take precedence over $[H^+]$ regulation in disease states. The volume of the extracellular fluid, the *milieu intérieur,* which must carry to and remove from cells their nutrients and waste products, is one such overriding need. The osmotic "skeleton" of the ECF consists of Na^+ (140 mEq./L.), accompanied by Cl^- (104 mEq./L.) and HCO_3^- (24 mEq./L.). The contribution of all other anions (< 12 mEq./L.) is negligible under normal circumstances. In the process of selective re-

absorption of the components of the glomerular filtrate, the renal tubules, in reclaiming Na^+, must either take along an anion (essentially limited to Cl^- or HCO_3^-) or excrete a cation (H^+ or K^+)· If the volume of the ECF is reduced, the kidney is stimulated to increase Na^+ reabsorption, both through suppression of "third factor(s)," increasing tubular Na^+ reabsorption, and by stimulation of aldosterone secretion, increasing the rate of Na^+ reabsorption in exchange for H^+ or K^+ at the distal cation pump.

Consider the forces operative in a patient who has been vomiting large quantities of acid gastric juice. His problems are as follows: He has lost large quantities of H^+ and Cl^-, resulting in a fall in blood $[H^+]$, a rise in serum $[HCO_3^-]$, and a fall in serum $[Cl^-]$; his ECF is depleted in part because of losses of Na^+ in gastric juice—but, more important, because he cannot retain food and for a time continues to excrete sodium in the urine; he has lost significant quantities of K^+, both in vomitus and because of failure of intake in the face of continued renal losses. The renal tubules are presented with a filtrate having a low $[Cl^-]$ and a high $[HCO_3^-]$. They are driven primarily by the "salt hunger" of the body. The primacy of the defense of ECF volume requires that they reabsorb Na^+, either with HCO_3^- or in exchange for an H^+ for K^+. The attempt to restore ECF volume of necessity sustains both the alkalosis and the K^+ depletion. If Na^+ is provided with equivalent quantities of a reabsorbable anion, Cl^- being the only one of physiological significance, a bicarbonate diuresis can occur. The normal quantities of acid produced endogenously ultimately will correct the alkalosis. Clearly, K^+ must also be provided to restore losses. In fact, failure to provide K^+ may delay correction of the alkalosis, since K^+ depletion within tubule cells favors secretion of H^+ in exchange for Na^+ at the distal cation pump.

Adrenal Cortical Hormones and Renal Acid Excretion

The characteristics of sodium reabsorption in the proximal and the distal tubules resemble those of H^+ secretion. Sodium is reabsorbed in the proximal tubule by a high capacity, low gradient system, Na^+ reabsorption ceasing when the peritubular plasma to tubular fluid ratio exceeds 3:2. By contrast, sodium reabsorption by the distal convoluted tubules has a low capacity, but can lower $[Na^+]$ in the urine to < 1.0 mEq./L. Part of Na^+ reabsorption in the distal segments is mediated by a cation pump, in which sodium reabsorption is linked to the excretion of either H^+ or K^+. This pump is stimulated by all adrenal steroid hormones having mineralocorticoid activity, aldosterone being the most potent member of this class.

When aldosterone secretion is excessive, as occurs with autonomous aldosterone-producing adenomas, augmented activity of the distal cation pump results in excessive excretion of both K^+ and H^+, the development of hypokalemia, reduction of blood $[H^+]$, and a rise in serum $[HCO_3^-]$. The degree of metabolic alkalosis induced is not usually severe. As aldosterone-driven Na^+ reabsorption continues, the ECF volume is progressively expanded, and natriuretic forces are activated. Ultimately, the rate of delivery of Na^+ to distal segments exceeds the capacity of these segments to reabsorb Na^+, and sodium intake again is equivalent to sodium excretion (but with an expanded ECF volume). This "escape" from the Na retaining effects of aldosterone also results in increased delivery of HCO_3^- to distal segments, limiting their capacity to accomplish net renal acid excretion.

CLINICAL EVALUATION OF ACID-BASE DISTURBANCES

Many methods have been developed to interpret blood acid-base measurements and to determine whether renal or pulmonary compensatory responses are normal and appropriate. These include the use of the H_2CO_3/HCO_3^- buffer pair, "buffer base," "standard bicarbonate," "base excess," the establishment of confidence limits for expected levels of $[HCO_3^-]$ during acute and chronic changes in Pa_{CO_2}, and expected changes in Pa_{CO_2} during acute and chronic metabolic acid-base disturbances. In addition, there are a host of nomograms. References at the end of the chapter provide access to each of these approaches. Although I have contributed to at least one such method in the attempt to simplify interpretation of acid-base data, reflection leads me to conclude that, while all are useful, all are also susceptible to error. Nothing can substitute for the critical, informed analysis of a carefully elicited history and a carefully conducted physical examination. When supplemented by simple measurements which the physician or nurse can carry out at the bedside or in the ward laboratory (accurate recording of body weight, urinalysis, measurement of tidal volume, etc.), the results from the chemistry laboratory should provide few surprises, and should serve, instead, to quantify clinical conclusions. With this background, any scheme of blood measurements is adequate. I prefer to use that system which has the fewest derived expressions and applies most broadly to total body buffering (rather than the special case of whole blood), namely, the interrelationships of the three parameters of Henderson's equation, $[H^+]$, Pa_{CO_2} and $[HCO_3^-]$.

Since most laboratories report serum total CO_2 content and blood pH, some simple way of deriving the approximate values of $[H^+]$, Pa_{CO_2} and $[HCO_3^-]$ is helpful. Kassirer and Bleich have suggested a clinically useful approach. At pH 7.40, $[H^+]$ in nanoEq./L. is numerically equal to the digits following the decimal point, i.e., 40 nanoEq./L. Over a wide range of pH values, each deviation of 0.01 in pH is equal to a deviation of 1 in [H⁺], but in the opposite direction. That is, when pH rises from 7.40 to 7.48 (+ 0.08), [H⁺] falls by 8, to 32 nanoEq./L. Agreement is precise between pH 7.28 and 7.45. Beyond these ranges, agreement is less precise, but not sufficiently so to destroy clinical utility. Precision can be improved by using the conversion Table 13-1 to obtain a more precise estimate of [H⁺]. Rearranging Henderson's equation, one can write:

$$\mathrm{Pa_{CO_2}} = \frac{[\mathrm{H^+}]\,[\mathrm{HCO_3^-}]}{\mathrm{S \times K}}$$

Using the units employed earlier in the chapter, $\mathrm{S \times K} \cong 25$. As a final simplification, total CO_2 content is substituted for [HCO_3^-] (since over 95 percent of T_{CO_2} arises from HCO_3^-). Then: $\mathrm{Pa_{CO_2}} = \dfrac{[\mathrm{H^+}] \times \mathrm{TCO_2}}{25}$. The small errors of approximation in both directions summate such that calculated $\mathrm{Pa_{CO_2}}$ at any pH between 7.10 and 7.50 does not deviate from true $\mathrm{Pa_{CO_2}}$ by more than 7 percent.

Three postulates are important in interpreting the interrelationships of [H⁺], $\mathrm{Pa_{CO_2}}$ and [HCO_3^-]: First, compensatory responses are *never* complete. Unless blood [H⁺] deviates from the normal, there is no stimulus for renal or pulmonary compensation. Second, lack of *any* compensatory response indicates a complicating acid-base disturbance. There are two exceptions: little, if any, increase in serum [HCO_3^-] is expected in acute hypercapnia and only a moderate increase in $\mathrm{Pa_{CO_2}}$ is expected in acute or chronic metabolic alkalosis. Third, the extent of expected compensation varies with time. Thus, pulmonary compensation in metabolic acidosis becomes less efficient with time, and renal compensation in hypercapnia improves with time. The last of these postulates casts greatest doubt on the

utility of 95 percent confidence limits for expected responses in acute and chronic acid-base disturbances (all obtained in steady states). Clearly, transitional states exist and the duration of any given disturbance is usually not known.

The measurement of serum total CO_2 content alone does not provide sufficient information on which to draw any conclusions. For example, in a given patient, a serum total CO_2 content of 25 mEq./L. would suggest normal acid-base status. However, blood pH might be found to be 7.1 ([H⁺] = 40 + 30, or 70 nanoEq./L.). Using the approximation equation described above:

$$\mathrm{Pa_{CO_2}} = \frac{70 \times 25}{25} = 70 \text{ mm. Hg}$$

As indicated by the concepts presented earlier, this would suggest the presence either of acute hypercapnia or of metabolic acidosis superimposed on chronic hypercapnia. One could could discriminate between these two possibilities only on the basis of the history and physical examination.

The presence of an organic acidosis, such as occurs in diabetic ketoacidosis, lactic acidosis, salicylism, or ethylene glycol or paralydehyde intoxication can be inferred from the serum electrolytes. When the serum [Na⁺] (approximately 140 mEq./L.) is compared to the sum of serum [Cl⁻] + [HCO_3^-] (about 104 + 24, or 128 mEq./L.), [Na⁺] normally exceeds [Cl⁻] + [HCO_3^-] by 6 to 12 mEq./L. The 6 to 12 mEq./L. of "unmeasured anions" is accounted for by the anionic charges on albumin and by the small quantities of $SO_4^=$, $H_2PO_4^-/HPO_4^=$ and organic anions normally present. In chronic azotemic renal disease, retention of phosphate salts and $SO_4^=$ may increase the unmeasured anions to levels seldom exceeding 20 mEq./L. Unmeasured anion concentrations exceeding 12 mEq./L. in patients with normal renal function, or 20 mEq./L. in azotemic patients, indicates the accumulation of organic anions.

Principles of Therapy

The cardinal principle in the treatment of acid-base disturbances is to *correct the underlying defect*. Since CO_2 production cannot be manipulated, the treatment of hyper- or hypocapnia requires that CO_2 excretion be augmented or reduced respectively. Excretion of CO_2 can be increased by relieving airway obstruction (through treatment of infection, use of bronchodilators, intubation, etc.) or by assisting ventilation (using intermittent, positive-pressure breathing, intubation, complete mechanical ventilation, etc.). The excretion of CO_2 can be reduced either by increasing the CO_2 content of inspired air by use of a CO_2-in-air mixture or by rebreathing into a paper bag or by reducing respiratory drive by providing oxygen at increased tensions. It may be necessary on occasion in acute hypercapnia to administer intravenous HCO_3^- to prevent lethal acidemia while other measures are being instituted.

In *acute metabolic acidosis due to overproduction of acids or losses of base*, steps must be taken to restore net acid production to normal by (a) withdrawing the agents that cause increases in endogenous acid production (NH_4Cl, salicylates, etc.), (b) treating of metabolic disturbances that increase endogenous acid production (insulin for diabetic ketoacidosis, treatment of shock or hypoxia causing lactic acidosis, etc.), and (c) preventing of further losses of base (treatment of diarrhea, vomiting in the achlorhydric patient, etc.). While therapy is being instituted for these specific disturbances, intravenous $NaHCO_3$ may be required to raise the serum $[HCO_3^-]$ and to lower the blood $[H^+]$, chiefly to alleviate the exhausting hyperventilation. For example, a 20-kg. child with severe diarrhea might present with a serum total CO_2 content of 4 mM/L. and a blood H^+ of 50 nanoEq./L. (pH 7.30). Pa_{CO_2} would be estimated to be $\dfrac{40 \times 5}{25}$, or 10 mm. Hg. This requires a four-fold increase above normal in ventilation. To raise serum T_{CO_2} ($\cong [HCO_3^-]$) to 15 mM/L. would require a 10 mM/L. increase in $[HCO_3^-]$ throughout the "HCO_3^- space" (equal to 40 percent of body weight). Thus, $20 \times 0.4 \times 10$, or 80 mEq. of HCO_3^- would be required. This is approximately equal to the contents of two commercial ampoules of $NaHCO_3$ (44 mEq./ampoule). This should not be administered in concentrations greater than 284 mEq./L. (i.e., 6 ampoules diluted to 1 liter with water or 5 percent glucose in water) to avoid the irritant effect of very alkaline solutions on veins. In the case of the child considered above, two ampoules of $NaHCO_3$ diluted to 500 ml. with 5 percent glucose and water would be appropriate. Obviously, saline, K^+, etc. would be given in appropriate quantities to correct other deficits. If T_{CO_2} were raised to 15, the same blood $[H^+]$ could be maintained with much less respiratory work, i.e.,

$$Pa_{CO_2} = \frac{50 \times 15}{25} = 30 \text{ mm. Hg}$$

The calculation of the quantity of HCO_3^- to be administered (i.e., [desired T_{CO_2} minus observed T_{CO_2}] $\times$ 0.40 body weight, kg.) is a rough approximation that does not take into account changes in body H^+ content as a result of the balance between H^+ production and H^+ excretion during the time of the infusion. Serial measurements of T_{CO_2} and pH and subsequent adjustments in the rate of the $NaHCO_3$ infusion are required. For example, in idiopathic lactic acidosis (lactic acidosis occurring for unknown reasons, in the absence of shock or arterial hypoxia), rates of acid production are so great that very large quantities of $NaHCO_3$ may be required to prevent lethal acidemia.

In *acute metabolic acidosis due to de-*

creased acid excretion (e.g., oliguria due to acute tubular necrosis), basic mineral salts such as $NaHCO_3$ or $KHCO_3$ cannot be administered, since there is no way for the body to eliminate Na^+ or K^+. Therapy depends upon attempts to reduce endogenous acid production from tissue protein breakdown by providing calories in the form of fat and carbohydrate and by removing H^+ either by hemodialysis or by peritoneal dialysis. During dialysis, the patient's blood is allowed to come to equilibrium with large volumes of fluid, across either a natural (peritoneum) or artificial (cellulose) membrane. Substances that have achieved high concentration in the blood of the uremic patient (such as urea, creatinine, phosphate, H^+, etc.) diffuse into the idealized synthetic fluid and are removed from the body.

Chronic Metabolic Acidosis. In the absence of surreptitious or physician-directed ingestion of substances that increase acid production (NH_4Cl, salicylates, methionine, etc.) or chronically increased rates of endogenous acid production (rare) or base loss (as in chronic diarrhea), chronic metabolic acidosis is most often due to impaired renal acid excretion. Therapy consists of: (1) ensuring that maximal attainable renal function is achieved by looking for and treating infection or obstruction and by providing an adequate sodium intake to attain the salutory effects of mild volume expansion, (2) reducing endogenous acid production by judicious restriction of dietary protein content, and (3) providing exogenous base as $NaHCO_3$ or Na-K salts of organic acids, such as citrate in amounts sufficient to maintain serum $[HCO_3^-]$ at about 20 mEq./L. The amount of base required cannot be predicted a priori because of variable HCO_3^- "leaks" in the urine as serum $[HCO_3^-]$ is raised and differing rates of endogenous acid production among patients. In the average adult, initial provision of 60 mEq. of $NaHCO_3$ per day, with subsequent upward adjustment based on blood measurements, is usually appropriate. If requirements are very large, the daily intake of Na^+ or K^+ incident to therapy may induce edema or hyperkalemia and limit the level of $[HCO_3^-]$ that can be achieved and maintained.

Both *acute* and *chronic metabolic alkalosis* occur chiefly as the result of losses of acid gastric juice. Lesser degrees of alkalosis occur when the kidneys are stimulated to excrete amounts of acid in excess of endogenous acid production (for example, in hypersecretion of aldosterone or as a result of diuretic agents such as thiazide compounds). Treatment consists of taking appropriate measures to stop vomiting or to reduce aldosterone secretion, discontinuing diuretic administration, restoring any degree of volume depletion with saline infusion to permit a bicarbonate diuresis to occur, and restoring associated losses such as K^+.

ANNOTATED REFERENCES

Arbus, G. S., Herbert, L. A., Levesque, P. R., Etsten, B. E., and Schwartz, W. B.: Characterization and clinical application of the "significance band" for acute respiratory alkalosis. New Eng. J. Med., *280*:117, 1969.

Brackett, N. C., Jr., Cohen, J. J., and Schwartz, W. B.: Carbon dioxide titration curve of normal man: effect of increasing degrees of acute hypercapnia on acid-base equilibrium. New Eng. J. Med., *272*:6, 1965.

Brackett, N. C., Jr., Wingo, C. F., Muren, O., and Solano, J. T.: Acid-base response to chronic hypercapnia in man. New Eng. J. Med., *280*:124, 1969.

Van Ypersele de Strihou, C., Brasseur, L., and De Coninck, J.: The carbon dioxide response curve for chronic hypercapnia in man. New Eng. J. Med., *275*:117, 1966. (The four preceding references present descriptions of the experiments which clarified the expected response in man to hyper- and hypocapnia.)

Bates, P. G.: Determination of pH: Theory and Practice. New York, John Wiley & Sons, 1964. (An exhaustive consideration of problems of measurements of (H^+))

Christensen, H. N.: Diagnostic Biochemistry, Quantitative Distributions of Body Constituents and Their Physiological Interpretation. Chaps. 8 and 9. New York, Oxford University Press, 1959.

———: Body Fluids and the Acid-Base Balance. Philadelphia, W. B. Saunders, 1964. (A learning program for students of the biological and medical sciences, with lucid discussions of acid-base physiology)

Edsall, J. T., and Wyman, J.: Biophysical Chemistry. Vol. I, Thermodynamics, Electrostatics and the Biological Significance of the Properties of Matter. Chap. 8. New York, Academic Press, 1958. (The text is a rigorous and complete description of the properties of acids and bases. Chapter 8 is a discussion of problems of measurement of (H^+).)

Goorno, W. E., Rector, F. C., Jr., and Seldin, D. W.: Relation of renal gluconeogenesis to ammonia production in the dog and rat. Am. J. Physiol., *213*:969, 1967.

Kamm, D. E., and Asher, R. R.: Relation between glucose and ammonia production in renal cortical slices. Am. J. Physiol., *218*: 1161, 1970.

Steiner, A. L., Goodman, A. D., and Treble, D. H.: Effect of metabolic acidosis on renal gluconeogenesis in vivo. Am. J. Physiol., *215*:211, 1968. (The three preceding references provide reviews of the relationships among acidosis, renal gluconeogenesis and renal ammonia synthesis.)

Henderson, L. J.: The theory of neutrality regulation in the animal organism. Am. J. Physiol., *21*:427, 1908. (No better description of the $H_2CO_3/NaHCO_3$ buffer system has supplanted this original discussion.)

Kassirer, J. P., and Bleich, H. L.: Rapid estimation of plasma carbon dioxide tension from pH and total carbon dioxide content. New Eng. J. Med., *272*:1067, 1965. (A description of the approximation method for deriving $Paco_2$)

Kassirer, J. P., Berkman, P. M., Lawrenz, D. R., and Schwartz, W. B.: The critical role of chloride in the correction of hypokalemic alkalosis in man. Am. J. Med., *38*:172, 1965.

Schwartz, W. B., Van Ypersele de Strihou, C., and Kassirer, J. P.: Role of anions in metabolic alkalosis and potassium deficiency. New Eng. J. Med., *279*:630, 1968. (The two preceding articles review the role of the chloride ion in acid-base regulation.)

Lemann, J., Jr., Lennon, E. J., Goodman, A. D., Litzow, J. R., and Relman, A. S.: The net balance of acid in subjects given large loads of acid or alkali. J. Clin. Invest., *44*: 507, 1965.

Lemann, J., Jr., Litzow, J. R., and Lennon, E. J.: The effects of chronic acid loads in normal man: further evidence for the participation of bone mineral in the defense against chronic metabolic acidosis. J. Clin. Invest., *45*:1608, 1966. (The two preceding references present evidence that the apparent "bicarbonate space" expands during more chronic acid load.)

Lennon, E. J., and Lemann, J., Jr.: Defense of hydrogen ion concentration in chronic metabolic acidosis: A new evaluation of an old approach. Ann. Intern. Med., *65*:265, 1966. (Includes the equation describing the expected responses of $Paco_2$ to chronic metabolic acidosis)

Lennon, E. J., Lemann, J., Jr., and Litzow, J. R.: The effects of diet and stool composition on the net external acid balance of normal subjects. J. Clin. Invest., *45*:1601, 1966. (A rationale for calculation of net external acid balances with normal diets. The references in this paper give access to the basic theoretical studies of this problem carried out by A. S. Relman and his associates.)

Morris, R. C., Jr.: Renal tubular acidosis. Mechanisms, classification and implications. New Eng. J. Med., *281*:1405, 1969.

Pellegrino, E. D., and Biltz, R. M.: The composition of human bone in uremia. Observations on the reservoir functions of bone and demonstration of a labile fraction of bone carbonate. Medicine, *44*:397, 1965.

Kaye, M., Frueh, A. J., and Silverman, M.: Study of vertebral bone powder from patients with chronic renal failure. J. Clin. Invest., *49*:442, 1970. (The two preceding references present direct evidence that the carbonate content of bone is reduced in azotemic, acidotic patients.)

Pitts, R. F.: Physiology of the Kidney and Body Fluids. ed. 2, Chap. 2. Chicago, Year Book Medical Publishers, 1968. (A review of the mechanisms of renal acid excretion and bicarbonate reabsorption in an excellent and succinct introductory text. Chapter provides 51 references to the original manuscripts which have established the nature of these mechanisms.)

Schwartz, W. B., Jenson, R. L., and Relman, A. S.: The disposition of acid administered to sodium-depleted subjects; the renal response and the role of the whole body buffers. J. Clin. Invest., *33*:587, 1954.

Schwartz, W. B., Orning, K. J., and Porter, R.: The internal distribution of hydrogen ions with varying degrees of metabolic acidosis. J. Clin. Invest., *36*:373, 1957.

Swan, R. C., and Pitts, R. F.: Neutralization of infused acid by nephrectomized dogs. J. Clin. Invest., *34*:205, 1955. (The three preceding references describe acute experiments designed to test the total buffer capacity of the intact organism.)

Welt, L. G.: Clinical Disorders of Hydration and Acid-Base Equilibrium. ed. 3. Boston, Little, Brown & Co., 1970. (An excellent text which presents clinical examples of acid-base disturbances and the approach to their therapy)

Weyer, E. M., and Nahas, G. G. (eds.): Current concepts of acid-base measurement. Ann. New York Acad. Sci., *133*:1, 1966. (This monograph discusses the controversy about systems of nomenclature and methods of evaluating blood measurements of acid-base equilibrium.)

Winters, R. W., Engel, K., and Dell, R. B.: Acid-Base Physiology in Medicine. Cleveland, London Co., 1967. (A self-instructive program and a highly recommended review of the important buffers of the body, for both the neophyte and seasoned clinician)

Appendix A

RENAL FUNCTIONAL TESTS

Six separate renal functions have been selected for discussion as those most frequently quantified: glomerular filtration rate; renal plasma or blood flow; maximum tubular transport capacity; urinary protein excretion; concentrating and diluting ability; and ability to acidify urine. The specific test to be utilized in each category is dependent upon the specific needs. Thus, there are accurate and precise techniques for research and more simple and rapid measures for clinical purposes.

Although somewhat elementary, in all of these tests of renal function, the importance of careful collection of timed urinary specimens, whether they be obtained over a period of 20 minutes or 24 hours, needs to be emphasized early. For example, for collection of a 24-hour urine sample, subjects must be given clear instructions. They must understand that what is sought is all urine formed in a 24-hour period. Thus, at a specified time (say, 8:00 A.M. on arising), they void in the toilet, emptying their bladder. Thereafter, all urine voided up to and including a final voiding at 8:00 A.M. the following morning constitutes the collection. A number of preservatives are available to maintain sterility; 2 cc. of toluene or 20 cc. of 10 percent formalin are satisfactory for most purposes.

Glomerular Filtration Rate

The most useful measure of renal function is an estimate of glomerular filtration rate (GFR). An ideal substance for measuring GFR should meet all of the following criteria: it should be freely filterable, i.e. molecular weight less than 10,000 and not bound to protein; it should not be reabsorbed or secreted by the tubules; it should not be metabolized peripherally, so that venous samples are indicative of arterial concentrations; it should be present in measurable quantities endogenously (if not, it should be nontoxic and inexpensive); and it should be easily, accurately and inexpensively quantified.

Inulin best fulfills the first three criteria, but it remains primarily a research tool, since it requires infusion of an expensive agent and it is a relatively difficult analytical technique. Some of the radioisotope compounds (^{125}I iothalamate, ^{51}Cr edetic acid and ^{60}Co vitamin B-12) are acceptable substitutes for inulin and are more easily and accurately quantified. All of these substances require constant intravenous infusion with collection of simultaneous plasma and urine samples at short intervals.

The following technique is required to obtain sufficiently accurate measures for investigative purposes (Figure III-1). First, a plasma level must be identified which will provide optimal analytical accuracy and precision. If inulin is used, 40 mg. percent is such a level. To achieve this concentration with a loading dose of inulin, the volume of distribution of this substance

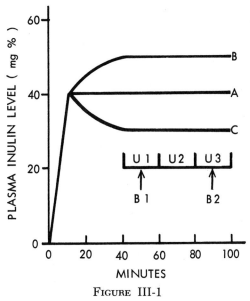

FIGURE III-1

269

must be known. Since inulin also is used as a measure of extracellular fluid volume, an estimate of the volume of distribution would be 20 percent of body weight (i.e., in a 70-kg. man, the volume of distribution is 14 kg. or liters). Since inulin is prepared as a 10 percent solution (10 g. per 100 ml.), a loading dose would require 56 ml. (40 mg. per 100 ml. or 400 mg. per liter × 14 liters = 5,600 mg.) to obtain this level. This is injected slowly over a 10-minute period.

Thereafter, the inulin lost in the urine must be constantly replaced. This requires a reasonable estimate based upon blood urea nitrogen or serum creatinine concentrations or creatinine clearance. If GFR is estimated to be 100 ml. per minute, then 40 mg. per minute would be lost in the urine and 0.4 ml. of the 10 percent solution must be replaced per minute.

A 30- to 45-minute period in the recumbent position is required for equilibration and stabilization of plasma levels. Figure III-1 shows the effect of over- or underestimating the GFR. If the estimate is accurate, the plasma level will follow curve A; if the GFR is over- or underestimated, the plasma levels will follow curves B or C respectively. The subject is hydrated with 20 ml. of water per kg. of body weight about 1 hour before study. Three urine samples are obtained at 20-minute intervals and plasma samples at the midpoint of the first and third periods, extrapolating for the second period. Subjects void spontaneously; catheterization with air or distilled water washes are utilized when problems in complete emptying are encountered.

Chemical methods for the analysis of inulin depend upon acid-hydrolysis of inulin to fructose and measurement of fructose by one of a variety of methods. The diphenylamine method of Harrison, H. E. (Proc. Soc. Exp. Biol., 49:111, 1942) and the resorcinol method of Schreiner, G. E. (Proc. Soc. Exp. Biol. 74:117, 1950) are probably the techniques most frequently used.

For clinical use this technique remains too time-consuming and expensive.

Endogenous clearances of creatinine provide less accurate, but acceptable, techniques for most clinical purposes. Actually, the serum creatinine and serum or blood urea nitrogen concentrations, once they become clearly elevated, provide sufficient clinical information about the status of renal function so that collection of urine for determination of clearance is of limited additional value. Considerable difficulty, however, is encountered in interpreting values which are normal or borderline elevated. Serum creatinine levels are dependent not only upon renal clearance of creatinine but upon body muscle mass and sex. Serum or blood urea nitrogen levels are dependent upon dietary protein and tissue and protein catabolic rates, and intra-individual variations are much greater than those observed with creatinine. The normal BUN:creatinine ratio is 14. Low protein diets lower this ratio; increased protein absorption from the gastrointestinal tract, as seen with GI bleeding, increased tissue catabolism and urinary tract obstruction, increase this ratio. Since daily excretion of creatinine remains fairly constant (1 to 2 g. in normal males; 0.8 to 1.5 g. in normal females), the serum creatinine concentrations are related reciprocally to the creatinine clearances. A substantial amount of urea is reabsorbed from tubular fluid (particularly at flow rates below 2 ml. per minute), so that measurement of urea clearance has largely been abandoned in favor of creatinine clearance methods.

It is important to distinguish between true creatinine and total chromagen techniques when evaluating reported creatinine clearances. About 70 percent of serum chromagen is creatinine; almost all chromagen in urine is creatinine. Most techniques (including those utilizing the Autoanalyzer) measure total chromagen; only those which selectively reabsorb creatinine onto alkaline earth, while other chromagens are removed, and then recover the creatinine from the alkaline earth measure true creatinine clearances (Hare, R. S.: Proc. Soc. Exp. Biol., 74:148, 1950). Since about 30 percent of the true creatinine in the final urine is secreted into the tubular fluid, this measure of GFR averages 30 to 40 percent above inulin clearance. As serum creatinine concentration rises from loss of renal function or from infusion of exogenous creatinine, this percentage increases as does the creatinine to inulin clearance ratio. The total chromagen technique (Bonsnes, R. W., and Taussky, H. H.: J. Biol. Chem., 158:581, 1945) provides an estimate of GFR which is about 30 percent below that provided by the true ·creatinine method. By combining two large opposing sources of error, the total chromagen technique yields a creatinine clearance which usually approximates the inulin clearance; this technique, however, is less precise than the true creatinine method.

Renal Plasma or Blood Flow

Measurement of paraaminhippurate (PAH) clearance is the standard technique to accurately and precisely measure effective renal plasma flow. The total renal plasma flow in man is about 8 percent greater than the renal plasma flow measured by PAH clearances. Renal blood flow can be estimated by dividing the renal plasma flow by one minus the hematocrit. The same considerations described in the conduct of inulin clearances apply for PAH clearances; only the numbers are different. The desired plasma level is 2 mg. percent; the volume of distribution is about 40 percent of body weight; the estimated plasma blood flow is about five times the estimated GFR; and the standard PAH solution comes as a 20 percent solution. This test is described in more detail by H. W. Smith, et al.: J. Clin. Invest., 24:388, 1945.

Phenolsulfonphthalein (PSP) is an aromatic acid which also is removed from the plasma into the tubular fluid by secretion. Although the clearance of PSP is less than that of PAH, its clearance varies directly with PAH and renal plasma flow. The plasma concentrations reached following injection of the usual 1 ml. dose (6 mg.) are far below the concentrations required to saturate tubular transport capacity. Although the test is often referred to as a test of tubular function, it really is a crude index of renal plasma flow which, because of its simplicity, can be used as a clinical estimator of renal plasma flow.

The test is usually performed 40 to 60 minutes after a water load of 10 to 20 ml. of water per kg. of body weight. One ml. of PSP (6 mg.) is injected intravenously and urine samples are collected after 15, 30 and 60 minutes. One to two ml. of 10 percent sodium hydroxide is added to the urine and the total sample is diluted to 1 liter. The red color is compared against a series of standards prepared by diluting various fractions of the quantity of injected dye to 1 liter after alkalinization (e.g., a 10 percent standard represents 0.6 mg. of PSP dye diluted to 1 liter).

Normally, at least 25 percent of the PSP is excreted during the first 15-minute period. This is the period most sensitive to detection of depressed renal function. If all of the dye is administered intravenously, 65 to 80 percent of the dye should be excreted after 1 hour. This technique is useful when analytical techniques are not available; when routine creatinine clearances can be obtained, the PSP test seldom contributes much additional information.

Maximum Tubular Transport Capacity

Measurement of the maximum capacity of the tubules to transport certain substances, either by secretion or absorption, provides a useful index of tubular function or integrity. Because of the complexity of such studies, they are used only for investigative purposes. The maximum tubular transport capacity for a substance (Tm) can be measured by infusing progressively increasing quantities of that substance until the transported load reaches a constant

TABLE III-1. Normal Range of Values

Serum or blood		
Urea		10-40 mg.%
Urea nitrogen (BUN)		5-20 mg.%
Non protein nitrogen (NPN)		20-35 mg.%
Creatinine, true	M	0.7-1.2 mg.%
	F	0.5-1.0 mg.%
Creatinine, total chromagen	M	1.0-1.5 mg.%
	F	0.7-1.2 mg.%
Urine excretions		
Urea and urea nitrogen		Variable
Creatinine, true or total chromagen	M	21-26 mg./kg./day
	F	16-22 mg./kg./day
Clearances		
Inulin		100-140 cc./min./1.73 M^2*
Creatinine, true		130-180 cc./min./1.73 M^2
Creatinine, total chromagen		Same as inulin
Urea		60-90 cc./min./1.73 $M.^2$
PAH		500-700 cc./min./1.73 $M.^2$
PSP		> 25% (15 min.)
		> 40% (30 min.)
		> 55% (2 hr.)
Tubular maximums		
Tm_{PAH} (secretion)		65-78 mg./min.
TmGlucose (reabsorption)		320-400 mg./min.

* Up to age 40 years; thereafter decreases at rate of 1 cc./min./1.73 $M.^2$/year. Similar decreases with age are noted in other clearances. M = men; F = women.

maximum. In studies of the Tm of secreted substances, most notably PAH, where transport characteristics are well documented, the load necessary to exceed Tm can be rapidly established by a priming dose and sustaining infusion. The PAH prime should be calculated to produce a plasma concentration between 20 and 40 mg. percent. This should be injected slowly to avoid vasomotor and autonomic nervous disturbances. With plasma levels in this range, maximum tubular secretion of PAH (T_{mPAH}) can be calculated by quantifying the amount of PAH excreted in the final urine ($U_{PAH}V$) and subtracting the filtered load of PAH (U_{PAH} times GFR measured by inulin clearance). This technique is described in detail in two textbooks on renal physiology (Smith, H. W.: Principles of Renal Physiology. New York, Oxford University Press, 1956; Wesson, L. G.: Physiology of the Human Kidney. New York, Grune and Stratton, 1969).

Urinary Protein Excretion

The measurement of urinary protein excretion provides an evaluation of an entirely independent aspect of renal function. Urinary protein excretion is thought to be dependent upon the balance between glomerular permeability to protein and subsequent tubular reabsorption of the filtered protein. Most heavy proteinurias are associated with significant abnormalities of the glomeruli and with increased glomerular permeability. Dr. Pollak is reluctant to define one specific level of daily urine protein excretion which he would consider separates normal from abnormal protein loss; however, an excretion in excess of 200 mg. per

TABLE III-2. Some Methods Used to Measure Total Protein Content of Plasma, Urine and Other Body Fluids

METHOD	PRINCIPLE	TECHNICAL FEASIBILITY	COMMENTS
Kjeldahl	Chemical; measures nitrogen content of protein	Specialized equipment	Standard reference method; used mainly for preparing standards and for research. In native urine measures (protein + nonprotein) nitrogen.*
Biuret	Chemical; copper complexes with peptide bonds in alkaline solution.	Standard laboratory method. used on autoanalyzer.	Insensitive at low protein concentration. In native urine measures (protein + peptide).
Folin-Lowry	Chemical; biuret and redox	Satisfactory only with pure proteins	Very sensitive; not routine. Interference by compounds such as uric acid
Precipitation (sulfosalicylic acid, trichloracetic acid, or phosphotungstic acid)	Proteins precipitates; measure scattered light or amount of precipitate formed.	Simple and rapid, especially with high protein concentration	Inaccurate; satisfactory as a qualitative test. Some false positive reactions occur and some types of proteinuria may be missed.
Heat and acetic acid	Protein precipitates at pH 5 on heating.	ibid	ibid
Dipstick (Albustix)	Proteins, especially albumin, bind dye, thereby changing the color of the dye.	Simple. One drop of urine needed.	Semiquantitative. Bence Jones proteinuria may be missed. Useful for repeated testing by patient
Refractive index	Physical; protein has higher refractive index than water.	Works if no glucose or other low molecular weight substances present.	Can be useful with some limitations for serum, of little value in urine.
Specific gravity	Physical; protein denser than water	ibid	ibid

* Note: normal plasma contains 13 mg. protein N and at most 0.4 mg. nonprotein N per ml.; urine may contain 10 mg. nonprotein N and 0.015 to 1.5 mg. protein N/ml.

TABLE III-3. Some Methods Used to Concentrate Proteins Present in Dilute Solutions in Body Fluids and Urine

METHOD	PRINCIPLE	TECHNICAL FEASIBILITY	COMMENTS
Precipitation with trichloroacetic acid (TCA) or with perchloric acid (PCA), etc.	The protein is less soluble, possibly because the reagent binds to the protein, thereby changing its structure.	Simple chemicals; quick; ineffective at low protein concentration. Commonly used.	Not all proteins are precipitated; more glycoproteins precipitated with PCA than with TCA. Immunochemical and other properties are altered.
Osmotic concentration against polymer e.g. polyvinylpyrollidone (PVP), carbowax, etc.	The high osmotic pressure outside the dialysis membrane causes water to migrate from the sample.	Relatively simple and quick; uses common chemicals. Can concentrate proteins effectively until osmotic equilibrium occurs. Must be done in the cold.	Considerable protein losses may occur; good reproducibility; selective loss does not appear to be a problem.
Ultrafiltration using collodion membranes	Pressure forces water and salts through the membrane filter, which retains the proteins.	Common inexpensive research equipment; slow procedure; should be done in the cold.	Reasonably reproducible, but large protein losses possible
Ultrafiltration using polymer membranes (Amicon Diaflo)	ibid	Specialized equipment; slow procedure; should be done in the cold.	Highly reproducible. Virtually no protein loss. Excellent for immunochemical analysis. Membranes of varying pore sizes available
Lyophilization	Vacuum distillation of water	Common equipment can take protein to dryness and to highest concentration possible.	Commonly used prior to electrophoresis and immunochemical analysis. Salts are concentrated and retained in urine with high salt concentration.

TABLE III-4. Some Methods for the Measurement of Specific Proteins

METHOD	PRINCIPLE	TECHNICAL FEASIBILITY	COMMENTS
Electrophoresis in dense support media (paper, cellulose acetate etc.)	Proteins migrate according to size and charge.	Simple test. Urine must be concentrated for study.	Common laboratory test. Must be combined with biuret for quantitation. Urine often difficult to study and interpret
Electrophoresis in gel media (starch, acrylamide, etc.)	Proteins migrate according to gel porosity as well as size and charge.	Simple test, more difficult to perform than on dense media; urine must be concentrated.	Common test for some proteins like haptoglobins. Patterns different from those on dense media. Must be done with biuret for quantitation.
Quantitative immunochemical (radial diffusion)	Antigen diffuses into antibody containing gel, forming a ring proportional to antigen concentration.	Simple and reproducible. Antisera commercially available to 20 serum proteins	Reagent used is biological; purity standardization and sometimes identity are problems.
Immunoelectrophoresis	Electrophoretic mobility combined with use of specific antibody to form precipitation	Relatively simple; not quantitative	Can be used to examine many proteins at one time. Good screening method for myeloma
Radioimmunoassay	Labeled and unlabeled antigen compete for same antibody.	Limited at present to very few proteins. Special technique	Most sensitive specific method known. Difficult to use as routine for large numbers of proteins
Bence Jones heat test	Variable solubility of Bence Jones protein on heating	Simple test; but more difficult if there is generalized proteinuria	Relatively poor diagnostic test. Should be replaced by immunochemical analysis.
Dye binding	Among plasma proteins albumin binds certain dyes preferentially, changing their color or fluorescence properties.	Simple test; can be used on autoanalyzer; urine must be dialyzed to remove interfering substances.	Common test. Metabolites influence dye binding. Suitable where sensitivity not accuracy is required
"Salting out"	Albumin more soluble in salt solution than globulins; quantitate with biuret.	Simple procedure; of historical interest only, although still widely used.	Subject to very large error, especially at low plasma protein levels. Does not measure any specific protein.

TABLE III-5. Some Methods Used to Measure the Size of Proteins in Plasma, Urine and Other Body Fluids

METHOD	PRINCIPLE	TECHNICAL FEASIBILITY	COMMENTS
Analytical ultracentrifugation	Proteins denser than water; sediment according to size and shape under high gravitational force	Time consuming; sophisticated instrumentation and mathematical analysis	Reference method. Diagnostic for Waldenström's macroglobulinemia
Gel filtration (Sephadex, Biogel, Sepharose, etc.)	Penetration of the gel matrix is proportional to Stokes radius of the protein.	Instrumentation not commonly available. Analysis is time consuming.	Method is good as a first approximation. Is often used to separate proteins for further analysis.
Acrylamide gel electrophoresis after treatment of protein with detergent	Protein forms negatively charged micelle with detergent. Migration is proportional to penetration of the gel matrix.	Instrumentation not commonly available	New method. Can be used for many samples. With development may become an excellent specialized laboratory method.

day should be considered worthy of clinical evaluation. Table III-2 summarizes the methods available for quantifying urinary protein excretions; Table III-3 presents the methods which can be utilized to concentrate proteins in biological fluids so that they can be more accurately measured; Table III-4 lists some of the methods for the measurement of specific urinary proteins; and Table III-5 presents techniques available for determining protein size.

Tests of Urine Concentrating and Diluting Ability

Drs. Papper and Vaamonde outline some simple tests of maximum urine concentration and dilution. They have excluded tests which have little or no practical clinical application (e.g., C_{H_2O} and $T^C_{H_2O}$ measurements).

Maximal Urine Concentration. This is best done by measuring urine total solute concentration by freezing point depression (osmometry). Where available, this gives the best information and can be done on very little urine (2 ml.), and the results are ready in a matter of a few minutes. A refractometer (which is calibrated in specific gravity units) also provides useful information, requires less training, and can be used quickly with one drop of urine.

Least valuable, except for historic interest, ready availability, and low cost, is the urinometer for determination of specific gravity. *Specific gravity* is dependent not only on the number of particles in solution, but on the mass and density of particles. The main reason specific gravity bears any relation to osmolality is that, in general, the particles in urine are small (salts and urea) and limited in variety. Specific gravity is more influenced than osmolality or refractometry by the presence of heavier particles such as sugar, protein, and x-ray contrast material. Of these, protein is least important quantitatively. "Corrections" of

specific gravity readings are necessary for the presence of glycosuria or proteinuria and for temperature. Thus, for every 0.27 g. of glucose per 100 ml., one subtracts 0.001 from the specific gravity; for every 0.4 g. of protein per 100 ml., one subtracts 0.001 from the specific gravity. Urinometers must be calibrated when new and regularly thereafter, preferably after each use, to be certain they read 1.000 with distilled water; if not, appropriate correction should be made. Failure to calibrate is a major and common source of error. Most urinometers are calibrated at 20°C. and should be read at that temperature. For each 3°C. above or below 20°C., 0.001 must be added or subtracted respectively. For example, urine chilled to 5°C. and tested for specific gravity at that temperature will have to have 0.005 subtracted from the observed specific gravity in order to establish the correct reading at 20°C. Although "simple, easy and available," specific gravity does not provide the precise information sought and does require careful handling. We prefer the cryoscope or the refractometer.

No attempt should be made to concentrate the urine by dehydration in patients with known or suspected azotemia. In general, such individuals cannot concentrate their urine normally, and dehydration is potentially hazardous. With this caution in mind, if the fresh morning urine voided after overnight dehydration (12 to 16 hours of hydropenia) has a carefully determined specific gravity of greater than 1.020 or an osmolality of 800 mOsm per kg. H_2O or more, no further effort is needed to test concentrating ability. This observation will suffice for clinical purposes. If, however, this overnight dehydration does not produce a concentrated urine, and determination of concentrating ability is desired, there are many types of tests available. Two are outlined; in each, a specific gravity of greater than 1.020 or osmolality of

800 mOsm per kg. H_2O or greater can be considered as "normal."

After breakfast on "Day 1," the patient takes no fluids until the test specimen is obtained on "Day 2." On the morning of Day 2 (after 24 hours of hydropenia), his first voided urine is discarded. The concentration of the next voided urine is then tested.

One may eliminate the need for fluid restriction by giving the patient 5 units of vasopressin tannate in oil intramuscularly and determine urinary concentration at any time after 4 hours and within the next 24 hours. Alternatively, 0.5 to 1 unit of aqueous vasopressin may be given slowly intravenously in several minutes and urine collected at half-hour intervals for the ensuing several hours and tested and interpreted as above.

Urine concentration is always higher following dehydration than during vasopressin administration. In addition, urine osmolality is usually higher during prolonged periods of hydropenia (36 to 48 hours) than during shorter ones (12 to 24 hours). Nonetheless, in most clinical situations, the short dehydration test (overnight dehydration) will suffice.

In evaluating results, it should be remembered that many factors may influence maximal concentrating ability independent of any real alteration including: age, diet (sodium, urea), posture, solute diuresis (significant glycosuria, diuretic agents), etc.

Maximal Urine Dilution. Urinary dilution tests are, in general, far less valuable as means of assessing renal function than concentration measurements. The capacity to dilute the urine is best examined by giving an oral water load: 20 ml. of water per kg. of body weight in 15 to 30 minutes. The patient empties his bladder before this and, thereafter, urinates at frequent intervals. The urine flow rate and the osmolality are measured. Normally, within one hour of the ingestion of the water load, there is a marked increase in the urine flow and a marked decrease in urine osmolality. We consider a response to water loading "normal" in the adult when, under these conditions, the maximal urine flow is 8 ml. per minute or greater and the urine osmolality is below 100 mOsm per kg. H_2O. Age, diet, posture and solute diuresis may also influence maximal urinary dilution. In addition, emotional stress and pain (by releasing ADH) may decrease the maximal response to water administration.

Normal Values (in young, healthy adults)

Maximal Renal Concentrating Ability

Uosm maximum (hydropenia)	1200-1300 mOsm/ kg. H_2O
$\frac{U}{P}$ osm maximum (hydropenia)	3.5-4.5
T^CH_2O (hydropenia and mannitol diuresis)	5-7 ml./min./ 100 GFR

Maximal Renal Diluting Ability

Uosm minimum (water diuresis)	30-50 mOsm/ kg. H_2O
$\frac{U}{P}$ osm minimum (water diuresis)	0.12-0.2
V maximum (water and hypotonic solute diuresis)	15-25 ml./min./ 100 GFR
CH_2O maximum (water and hypotonic solute diuresis)	12-20 ml./min./ 100 GFR

Tests of Acid-Base Balance and Urine Acidification

Dr. Lennon has outlined those methods useful in the evaluation of acid-base disturbances and ability to acidify the urine, including steps to be taken in procuring blood and urine samples for these measurements. Normal values also are listed.

I. Acid-base measurements in the blood

A. Methods:

1. *Sample procurement and handling.* For clinical purposes, measurements can be made on arterial or venous blood samples. No important differences in re-

sults pertain if venous blood is properly drawn. Where possible, venous blood should be withdrawn without stasis. If it is necessary to apply a tourniquet, the sample should be drawn *before* releasing compression in order to avoid the "flush" of lactic acid, formed in the tissues during the period of relative ischemia, which occurs on release of tourniquet pressure. Venous blood with an even closer approximation to arterial values can be obtained if the venous sample is "arterialized" by immersing the forearm in warm water for 3 to 5 minutes before sampling. This is not necessary for routine clinical evaluation.

Two blood samples are required for the measurement of pH and total CO_2 content. A sample for pH measurement is drawn into a glass syringe prerinsed with heparin. The syringe is capped, the sample well mixed by rotating the syringe between the palms, and then immersed in ice water for delivery to the laboratory. A second sample, for total CO_2 content, is drawn into a glass syringe prerinsed with mineral oil. This sample is transferred to a centrifuge tube, under oil, and allowed to clot. It is then spun in the centrifuge to separate off the serum.

If the Astrup technique is used, all needed values are obtained from a single, heparinized sample. Aliquots of the sample are placed in tonometers and gassed with two known Pco_2 mixtures. The pH of these two samples, as well as the pH of the original blood, are measured. Using this "CO_2 titration technique" nomograms are available to predict the Pco_2 of the original sample. HCO_3^- can then be calculated.

2. *Measurement of pH.*

Most hospitals now have available temperature controlled, water jacketed micro glass electrodes which permit direct aspiration of the whole-blood sample from the syringe. The syringe should again be rotated between the palms to assure good mixing of the sample before aspiration. A liquid-liquid bridge is then established by immersing the tip of the electrode in contact with a calomel half-cell and pH read directly on a calibrated microvoltmeter. Accurate standards for calibration of the electrode are of course necessary for reliable results.

3. *Measurement of serum total CO_2 content.*

Using the classical Van Slyke manometric technique, a measured sample of serum is drawn into a calibrated pipette and introduced into the chamber of the Van Slyke apparatus. To this is added a small amount of 0.1 N lactic acid to convert all existing HCO_3^- to H_2CO_3. The sample is then shaken under vacuum to release all dissolved CO_2 and H_2CO_3 as CO_2 gas. The pressure of the total gases present is measured at a fixed volume. A small amount of NaOH is then added to absorb all CO_2 and the pressure again measured at the same volume. The difference in the two pressure measurements represents the partial pressure of the liberated CO_2 gas. Knowing temperature, a conversion factor is utilized to convert Pco_2 mm. Hg to total CO_2 content, mM/L. Automated techniques (such as the Autoanalyzer method) which differ in principle but yield similar results are commonly employed in larger laboratories.

4. *Calculation of Pco_2.*

If total CO_2 content and pH are known, Pco_2 can be calculated from the modified Henderson-Hasselbalch equation:

$$pH = pK' + \log \text{ total } CO_2 \text{ content} - SPco_2$$

$$pH = pK' + \log \frac{\text{total } CO_2 \text{ content} - SPco_2}{SPco_2}$$

or, using the conversion table to express pH in terms of $[H^+]$,

$$[H^+] = K'_2 \frac{SPco_2}{\text{total } CO_2 - SPco_2}$$

B. Normal Values:

Comparison of Arterial and Venous Blood Levels of [HCO₃⁻], Pco₂, and [H⁺]
in Normal Subjects.

Study	Number of Observations	HCO_3^- (mEq./liter)	P_{CO_2} (mm. Hg)	H^+ (nanoEq./liter)
Moller (a) (arterial)	100*	23.8	44.2	42.6
Gambino (b) (venous)	55*	27.1	45.2	40.0
Lennon (c) (venous)	120†	27.2 ± 1.7	47.8 ± 4.1	42.1 ± 3.0

* Standard deviations not given for these values.

† Mean ± 1 SD.

(a) Moller, B.: The hydrogen ion concentration of arterial blood. Acta Med. (Suppl. 348), *165*:1, 1959.

(b) Gambino, S. R.; Normal values for adult venous plasma pH and CO₂ content. J. Clin. Path. *32*: 294, 1959.

(c) Lennon, E. J., and Lemann, J., Jr.: Defense of hydrogen ion concentration in chronic metabolic acidosis. Ann. Int. Med., *65*:265, 1966.

II. Acid-Base Measurements in Urine

A. Methods

1. *Sample procurement.*

For simple pH determination, a fresh-voided specimen is adequate. For measurement of acid excretion, a 24-hour urine specimen is collected under mineral oil with phenylmercuric nitrate and thymol added to maintain sterility.

2. *Urine pH.*

This can be approximated by the use of indicator dyes impregnated in paper strips (nitrazine paper, "Dipsticks," etc.). More accurate results are obtained with the glass electrode.

3. *Urine bicarbonate content.*

Specimens having pH values below 5.8 contain no significant quantity of bicarbonate. For specimens having higher pH values, total CO₂ contents are determined, as for blood and Pco₂ and HCO₃⁻ calculated (see above).

4. *Urinary titratable acidity.*

Urinary titratable acid is determined by titrating a measured sample of urine in a water jacketed chamber at 27°C. from its existing pH to the pH of the patient's blood, using 0.1 N NaOH as the titrant. Knowing the quantity of NaOH required, the milliequivalents of titratable acid excreted per 24 hours can be calculated.

5. *Urinary ammonium content.*

The ammonium content of the urine can be measured, using the Conway microdiffusion technique, in which a urine sample is mixed in the outer well of a two-chambered porcelain dish with a strong base to liberate NH₄⁺ as NH₃ gas. Mixing takes place under a sealed cover, and the liberated NH₃ is trapped in an acid in the center well. The NH₃ trapped in the center well reacts with the acid: NH₃ + HA → NH₄A, with a resultant rise in pH. The quantity of NH₃ trapped is determined by back titration with 0.01 N acid, using a colorimetric endpoint. Simpler techniques have been adapted to automated methods.

6. *Net renal acid excretion.*

The net quantity of acid eliminated in a 24-hour urine sample is taken as the sum of urinary titratable acid and urinary ammonium content *minus* any bicarbonate excreted.

B. Normal values

1. *pH.*

Urinary pH varies widely during the day and as a function of endogenous or exogenous loads of acid to be eliminated. The urine is normally acid on arising with pH below 5.8.

2. *Net renal acid excretion.*

Since, in the steady state, endogenous fixed acid must be matched by net renal acid excretion, and endogenous acid production averages 0.5 mEq. per kg. of body wt. per day, it is obvious that renal acid excretion will average 0.5 mEq. per kg. of body wt. per 24 hours. Normally, about two thirds of this is as ammonium and one third as titratable acid.

Appendix B

SUPPLEMENTAL TEXTBOOKS AND MONOGRAPHS

Renal Physiology

Lotspeich, W. D.: Metabolic Aspects of Renal Function. Springfield, Ill., Charles C Thomas, 1959.

Pitts, R. F.: Physiology of the Kidney and Body Fluids. Chicago, Year Book Medical Publishers, 1968.

Smith, H. W.: Principles of Renal Physiology. New York, Oxford University Press, 1951.

Wesson, L. G.: Physiology of the Human Kidney. New York, Grune & Stratton, 1969.

Body Fluids and Electrolytes

Bland, J. H.: Clinical Metabolism of Body Water and Electrolytes. Philadelphia, W. B. Saunders, 1963.

Elkinton, J. R., and Danowski, T. S.: The Body Fluids. Baltimore, The Williams & Wilkins Company, 1955.

Gamble, J. L.: Clinical Anatomy, Physiology and Pathology of Extracellular Fluid. Cambridge, Mass., Harvard Univ. Press, 1947.

Maxwell, M. H., and Kleeman, C. R.: Clinical Disorders of Fluid and Electrolyte Metabolism. Ed. 2. New York, Blakiston Division, McGraw-Hill, 1972.

Strauss, M. B.: Body Water in Man. Boston, Little, Brown and Company, 1957.

Welt, L. G.: Clinical Disorders of Hydration and Acid-Base Balance. Ed. 3. Boston, Little, Brown and Company, 1970.

Acid-Base Physiology

Bates, P. G.: Determination of pH Theory and Practice. New York, John Wiley and Sons, 1964.

Christensen, H. N.: Body Fluids and the Acid-Base Balance: A Learning Program for Students of the Biological and Medical Sciences. Philadelphia, W. B. Saunders, 1964.

Davenport, H. W.: An ABC of Acid-base Chemistry. Ed. 4. Chicago, University of Chicago Press, 1958.

Edsall, J. T., and Wyman, J.: Biophysical Chemistry. Vol. 1, Thermodynamics, Electrostatics and the Biological Significance of the Properties of Matter. New York, Academic Press, 1958.

Peters, J. P., and Van Slyke, D. D.: Quantitative Clinical Chemistry. Baltimore: The Williams & Wilkins Company, 1931.

Weyer, E. M., and Nahas, G. G.: Current Concepts of Acid-Base Measurement. Ann. New York Acad. Sci. *133*:1-274, 1966.

Winters, R. W., Engel, K., and Dell, R. B.: Acid-Base Physiology in Medicine, A Self-instructive Program. Cleveland, London Co., 1967.

Renal Pathophysiology

Becker, E. L.: Structural Basis of Renal Disease. New York, Hoeber Medical Division, Harper and Row, 1968.

Manuel, Y., Revillard, J. P., and Betnel, H.: Proteins in Normal and Pathological Urine. Baltimore, University Park Press, 1970.

Schreiner, G. E., and Maher, J. F.: Uremia Biochemistry, Pathogenesis and Treatment. Springfield, Ill., Charles C Thomas, 1961.

Smith, H. W.: The Kidney. Structure and Function in Health and Disease. New York, Oxford University Press, 1951.

de Wardener, H. E.: The Kidney. An Outline of Normal and Abnormal Structure and Function, Ed. 2. Boston, Little, Brown and Company, 1961.

Renal Pathology

Allen, A. C.: The Kidney: Medical and Surgical Diseases. New York, Grune and Stratton, 1951.

Bell, E. T.: Renal Diseases. Ed. 2. London, Kimpton, 1950.

Heptinstall, R. H.: Pathology of the Kidney. Boston, Little, Brown and Company, 1966.

Clinical Nephrology

Addis, T.: Glomerular Nephritis. New York, Macmillan, 1948.

Black, D. A. K.: Renal Disease. Ed. 2. Philadelphia, F. A. Davis, 1967.

Hamburger, J.: *et al.:* Nephrology. Vols. 1 and 2. Philadelphia, W. B. Saunders, 1969.

Papper, S.: Clinical Nephrology. Boston, Little, Brown and Company, 1971.

Strauss, M. B., and Welt, L. G.: Diseases of the Kidney. Boston, Little, Brown and Company, 1963.

Section Four

Endocrine-Metabolic Mechanisms

Introduction

The endocrine system profoundly affects almost every aspect of human physiology. Hormones control or regulate body water and electrolytes, the growth of both somatic and reproductive tissues, and the mechanisms that convert protein, fat, and carbohydrate into energy. It is these principal functions of hormones and their perturbations which are described in the chapters that follow.

The integrative role which the endocrine system plays in most body functions is assuming new dimensions, as evidence mounts that hormones serve as connecting links between brain function and end-organ responses. On the other hand, the dramatic influence which hormones may have on the mental processes and on nervous system function is only now beginning to receive the attention it deserves.

The field of neuroendocrinology is expanding rapidly. The descriptions of hypothalamic control of pituitary tropic hormone release in the chapters on the thyroid, adrenal cortex, and reproduction reflect this. Probably the synthesis or release of all known anterior pituitary hormones is either stimulated or inhibited by hypothalamic substances (hypophysiotropic hormones) which reach the pituitary through a vascular link. Feedback control by end-organ products is apparently exerted on the hypothalamus by gonadal and adrenal cortical steroids, whereas thyroid hormones act more directly on the anterior pituitary. Evidence also exists that some of the anterior pituitary tropic hormones, acting through a shorter feedback loop, may depress their own hypothalamic releasing factors. The explanation for some of the endocrine syndromes in which too little or too much hormone is secreted undoubtedly lies in hypothalamic function. It is also clear that events recorded in the brain and in the central and autonomic nervous systems will, through afferent pathways, influence hypothalamic endocrine activity. The mechanisms by which the daily life experiences of man may influence endocrine activity and produce disease thus become more apparent.

One of the least-recognized aspects of endocrine disease is the frequency of mild-to-severe psychiatric and emotional disorders which can accompany the major endocrinopathies and complicate management of the patient. These disturbances are particularly common in adrenal cortical hyperfunction (Cushing's disease) or hypofunction (Addison's disease), hyperparathyroidism, hypoparathyroidism, myxedema, and hypoglycemia due to insulinoma.

Marked clinical abnormalities due to excessive hormone emerge when the secreting tissue becomes neoplastic and is no longer susceptible to feedback control. Unbridled hormone secretion may be associated with malignant or adenomatous lesions of the pituitary, adrenal cortex, thyroid, parathyroids, pancreas, or gonads. Many of the resulting clinical entities are discussed in the chapters in this section.

In recent years, some neoplasms of nonendocrine tissues have been found to produce a wide variety of hormones associated with clinical syndromes. These have been referred to as nonendocrine-secreting or paraendocrine tumors and their manifestations as ectopic hormone syndromes. Curiously, neoplasms of the lung and kidney produce ectopic hormones more often than do neoplasms in other tissues. Paraendocrine lung tumors may secrete adrenocorticotropic, antidiuretic, or parathyroid hormones; kidney tumors may secrete parathyroid hormone or erythropoietin. When the tumor is removed entirely, the endocrine syndrome regresses.

Secretion of all of the anterior pituitary tropic hormones (except growth hormone) has been either demonstrated or strongly suspected in various other nonendocrine tumors. Similarly, there is evidence that any of the nonpituitary polypeptide hormones (and some nonpolypeptide hormones) may likewise be secreted by nonendocrine tumors.

Although most nonendocrine tumors are known to secrete only one hormone, multiple hormone production has been reported in a variety of neoplasms. Multiple hormone production may also occur when secreting adenomas are present in several different endocrine or nonendocrine tissues at the same time (multiple endocrine adenomatosis). Thus, the endocrine syndromes produced may be not only striking in their intensity, but also mixed.

The placenta acts, in a sense, as a paraendocrine neoplasm during its period of residence in the uterus. Its emergence as an endocrine tissue, and site of hormone metabolism, which plays a major role in both fetal and maternal physiology, will be evident from the chapter on reproduction.

Perhaps the most prevalent endocrine "disease" today is anovulation induced by oral contraceptives ("the pill"). The extraordinary effectiveness of these hormone analogs in suppressing gonadotropins and ovulation is pointed out in the chapter on reproduction, but their hazardous side-effects, both recognized and potential, are properly emphasized.

The mechanism by which hormone secretion is translated at the cellular level to a tissue response is one of the more fascinating frontiers of biological science. It is now evident that the interaction of a hormone with cell membranes results in the synthesis or accumulation of cyclic 3',5'-adenosine monophosphate (cyclic AMP). This, the "second messenger," then causes changes in enzyme activity or membrane properties which elicit the ultimate physiological effect (e.g., synthesis of another hormone, ion transport by membranes, tissue growth, storage or utilization of energy sources).

Another noteworthy development concerns the prostaglandins. These substances, composed of 20-carbon fatty acids, are present in most tissues. Evidence now suggests that the prostaglandins may regulate endocrine activity in numerous ways, only a few of which have become apparent to investigators. Prostaglandins have tropic hormone-like effects on the adrenal cortex, thyroid, and corpus luteum—effects which seem to correlate with an increase in cyclic AMP. Similarly, insulin secretion is stimulated by both prostaglandins and cyclic AMP. Prostaglandins may therefore act as modifiers of cyclic AMP formation. A recently discovered application of these ubiquitous fatty acids (on an experimental basis) has been the induction of labor or abortion. The uterus is stimulated by the capacity of prostaglandins to produce potent smooth-muscle contraction.

To date, disturbances in cyclic AMP and prostaglandin function have not been tied to mechanisms of endocrine disease, although it is likely that relationships will be demonstrated.

The chapters on endocrinology cover systems and disease mechanisms of major importance. Had space permitted in this text, discussions of less frequent problems would also have been included: for example, carcinoid tumors, hypoglycemia, disorders of sex differentiation, defects of growth and development, and genetic aberrations.

Students wishing to read further on subjects mentioned in this introductory chapter should consult the texts edited by Williams and Bondy (see bibliography). A comprehensive review of the relationships of endocrine and metabolic disturbances to mental function has been presented by Williams in the *Journal of Clinical Endocrinology and Metabolism*, 31:461, 1970. A review of prostaglandins appears in *Recent Progress in Hormone Research*, 26: 139, 1970.

Leonard P. Eliel, M.D.

14

Regulatory Mechanisms of the Pituitary and Pituitary-Adrenal Axis

Don H. Nelson, M.D.

PITUITARY-ADRENAL AXIS

A triad of characteristics makes the hormonal secretion of the adrenal cortex unique in the endocrine system: the hormones secreted are necessary for life; secretion can accelerate rapidly in response to, and as protection against, a variety of stresses; and excessive secretion for a prolonged period produces physiological damage. Thus, a highly sophisticated mechanism of adrenal secretion has evolved to provide the body with adequate hormone at rest, to increase secretions rapidly during stress, and to revert to maintenance levels when the need has passed. This remarkable system is equipped to adjust secretory levels within minutes when needed and yet maintain daily secretion within rather closely defined limits under normal circumstances.

Corticosteroid secretion is regulated chiefly by the hypothalamus, but it may be influenced also by a number of normally occurring substances, as well as pharmacological blocking agents. The level of circulating corticosteroids, in turn, influences the anterior pituitary release of ACTH, the primary regulator of corticosteroid secretion. Pituitary secretion is governed primarily by a hypothalamic substance, corticotropin-releasing factor (CRF), produced in the anterior hypothalamus. Hypothalamic secretion of CRF, in turn, is controlled by nervous impulses from higher centers in the central nervous system and, possibly, by other factors. Thus, constant regulation of adrenal secretion depends upon adequate function of the central nervous system, hypothalamus, pituitary, and the adrenal cortex itself. Defects in production of the specific enzymes involved, tumor formation, or destruction of the necessary hormone-producing tissue at any level can produce a variety of clinical syndromes, including hypopituitarism, Cushing's syndrome, Addison's disease, congenital adrenal hyperplasia, virilization, and hypertension secondary to hyperaldosteronism.

HYPOTHALAMIC CONTROL OF ACTH SECRETION

The demonstration of a portal venous system connecting the hypothalamus to the pituitary gland provided one of the first evidences of a possible hormonal link between the hypothalamus and pituitary. Neurosecretory substances found in the nerve endings of the median eminence are carried by a capillary plexus down the pituitary stalk and, when they reach the anterior pituitary gland, effect the release of anterior pituitary tropic hormones which influence the endocrine glands. Substances that have been isolated from the hypothalamus include corticotropin releasing factor (CRF), luteinizing-hormone releasing factor (LRF), growth hormone releasing factor (GRF), follicle-stimulating-hormone releasing factor (FRF), and thyrotropin releasing factor (TRF). Only the latter has been identified chemically. When specific areas of the hypothalamus that relate to secretion of these releasing factors are destroyed, the pituitary fails to secrete the appropriate tropic substance. As a result, one or more endocrine glands may develop deficiencies, dependent upon the extent of the lesion. In experimental animals, lesions that interfere with the release of ACTH during stress do not necessarily produce adrenal atrophy. Thus, enough ACTH appears to be secreted to maintain adrenal size and a state of relative eucorticism (normal function) in the absence of stress. These animals, however, are unable to increase corticoid secretion in response to stress. Animals with such hypothalamic lesions may actually have enlarged adrenal glands. It is unclear whether this results from hypersecretion of ACTH during the non-stressed periods or whether it is secondary to a factor designated "adrenal weight factor" (which may differ structurally from ACTH). Although specific hypothalamic lesions capable of producing clinical disorders are extremely rare, tumors such as craniopharyngiomas in the suprasellar region may produce hypopituitarism secondary to destruction of the hypothalamus and interference with the blood supply, or compression of the anterior pituitary gland. Similarly, basal skull fractures may interfere with blood supply to the pituitary and, hence, with the circulation carrying releasing factors to the anterior pituitary.

Diurnal Release of ACTH

Diurnal variation in the release of ACTH from the anterior pituitary gland has been well documented. First, corticosteroid plasma levels were observed to be much higher in the early morning than in the late evening. This finding prompted fur-

ther investigations, which have shown that this diurnal variation depends upon a similar morning elevation of ACTH, with the levels decreasing as the day progresses. This diurnal variation in adrenal secretion may be suppressed easily by the exogenous administration of small quantities of corticosteroids or altered by a readjustment to a new dark-light pattern.

Response to Stress

Within a few minutes of exposure to any one of a variety of stresses, including hypoglycemia, tissue trauma, and anoxia, the plasma ACTH level rises sharply. This response presumably is mediated by release of CRF, since it is inhibited by lesions of the hypothalamus. The final pathways by which the signal is transmitted to release CRF have not been described completely. However, the experiments of Egdahl have indicated that an intact nerve supply to an area of trauma is important in stimulating cortical secretion. Severance of the nerve supply to a traumatized limb, for instance, prevents a normal increase in ACTH and corticosteroid secretion. Egdahl also demonstrated that stress results in increased secretion from the pituitary gland when it is isolated from higher neural centers. Probably, then, central nervous system factors inhibit, as well as stimulate, ACTH secretion, and the control of CRF release is mediated chiefly by central nervous system pathways.

Suppression of ACTH Secretion by Corticosteroids

The ability of corticosteroids to suppress ACTH secretion has been clearly established, although the precise site of suppression is unknown. Thus, elevation in corticosteroids apparently prevents secretion of CRF into the portal system and therefore prevents ACTH production. In the normal resting individual, administration of 20 mg. of cortisol a day (or its equivalent, 1 mg. of dexamethasone) will suppress ACTH production and prevent adrenal secretion. Atrophy of the adrenal cortex may result. This so-called feedback mechanism functions to decrease ACTH secretion when there is an increase in plasma corticosteroids—and, conversely, to increase ACTH when there is a decrease in plasma corticosteroid. Probably, the diurnal variation is, to some extent, controlled by the feedback mechanism. Thus, release of ACTH during the early morning hours produces an increase in corticosteroid secretion which inhibits ACTH production for some period of time, with concentrations of both ACTH and corticosteroids reaching low levels in late evening. After midnight the cycle begins again, with increased ACTH release and augmented corticosteroid secretion. The relation of CRF to this diurnal variation is not totally clear. Acute trauma, such as major surgery, produces marked elevations in ACTH, corticosteroids and CRF. Under these circumstances corticosteroids do not inhibit either CRF or ACTH secretion. Thus, the stimulus to CRF production generated by acute stress overcomes the suppressive effect of the corticosteroids.

Administration of corticosteroids in excess of normal secretion, even for a few days, produces marked atrophy of the adrenal cortex. This atrophic gland fails to respond acutely to administration of ACTH; hence, patients being given corticosteroids fail to increase corticosteroid output in response to stress. Because corticosteroids are necessary for protection against severe stress, patients with atrophic glands require increased doses of corticosteroids at times of stress. Although adrenal function usually returns to normal in a few weeks after corticosteroids are withdrawn, an occasional instance of prolonged atrophy has been described. It is most common in patients who have received large doses of corticosteroids for many months. Consequently, they may not secrete sufficient corticosteroids when sub-jected to accidental trauma, surgery, or acute infections, and, unless adequately treated, they may die from adrenal insufficiency. This defect appears to result from long-term inhibition of ACTH and CRF secretion so that, when the exogenous corticosteroid is discontinued, the normal diurnal increase in ACTH fails to return and the adrenal gland remains atrophic. When a period of stress supervenes, the stress response still functions to the extent that plasma ACTH levels are increased, but the adrenal gland is unable to increase corticosteroid secretion because of its atrophic state.

Increased ACTH Production in Adrenal Insufficiency

Adrenal insufficiency is characterized by high ACTH and low corticosteroid concentrations in the blood. The increased plasma ACTH levels are easily suppressed within a few hours by administration of a corticosteroid. Studies in animals have demonstrated that the high levels of ACTH characteristic of adrenal insufficiency require days to develop. This contrasts with the immediate and marked elevation of plasma ACTH which occurs in response to stress.

Metyrapone

This pharmacological agent (2-methyl-1,2 bis(3-pyridyl)-1-propane) is extremely useful in the investigation of abnormalities of the pituitary-adrenal axis. It inhibits 11 beta-hydroxylation, and the adrenal cortex thus fails to produce normal quantities of cortisol under ACTH stimulation and, instead, secretes the 11-desoxy equivalent of cortisol, or substance S (11-desoxy-17-hydroxycorticosterone). As a result of the deficiency in cortisol production, ACTH secretion increases and the production of 11-desoxycortisol rises (Fig. 14-1). The latter compound is biologically inactive and fails to suppress ACTH secretion; since it is measured by most of the standard procedures used for estimation of corticoids in

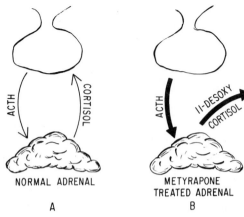

FIG. 14-1. Diagram of effects of metyrapone on the adrenal cortex. (A) The normal relationship between secretion of ACTH by the pituitary to cause secretion of cortisol from the adrenal gland which in turn acts back on the pituitary and/or hypothalamus to suppress ACTH. (B) The effects seen when the adrenal is exposed to metyrapone. Metyrapone acts chiefly as a blocker of 11-hydroxylation in the doses usually administered to a patient, thus resulting in increased secretion of 11-desoxycortisol and decreased secretion of cortisol. As the hypothalamic pituitary system is not suppressed by normal cortisol production, increased ACTH secretion occurs which further stimulates 11-desoxycortisol secretion.

plasma and urine, an increase in corticosteroids after administration of metyrapone may be interpreted as evidence for a normal adrenal-pituitary system. Similarly, an increase in plasma ACTH indicates that normal adrenal secretion has been suppressed. The metyrapone test may be useful in evaluating patients with Cushing's syndrome or hypopituitarism (see below).

CHEMICAL NATURE OF ACTH

ACTH is a polypeptide with a molecular weight of approximately 4,500; but only that part of the chain starting from the NH_2-terminal end has been found to be necessary for full biological potency. A nearly identical amino acid sequence is present in positions 1 to 24 in most animal species, but differences may occur between species in the remaining part of the chain. In addition to the effect of ACTH upon the adrenal cortex, extra-adrenal effects have been demonstrated, including fatty acid release and hyperpigmentation similar to that produced by melanocyte stimulating hormone (MSH). In fact, alpha-MSH, which contains 13 amino acids, closely resembles the first 13 amino acids of the ACTH chain in structure; this similarity probably accounts for the MSH activity demonstrated by ACTH. The effects on fatty acid release from lipid stores and other extra-adrenal effects are produced by very large doses and are considered to be pharmacological rather than physiological effects of the hormone.

ABNORMALITIES OF ACTH SECRETION

Cushing's Syndrome. Cushing's syndrome occurs when the adrenal gland produces excessive amounts of corticosteroids. The condition may be due to tumor or hyperplasia of either the pituitary gland or the adrenal cortex. Some authors prefer to reserve the term *Cushing's disease* for adrenal hyperfunction secondary to excess ACTH secretion. In these patients, excessive pituitary production of ACTH causes adrenal cortical hyperplasia and excessive corticosteroid secretion. A pituitary tumor may be demonstrable; but usually, the pituitary remains normal in size while secreting increased quantities of ACTH, probably secondary to hypothalamic stimuli. It has been suggested that increased secretion may result from microscopic hyperplasia of ACTH-producing cells of the pituitary or from hypothalamic abnormalities, but definitive pathological proof of the etiology is lacking. Although pituitary tumors are relatively rare in Cushing's syndrome, there have been reports of a number of patients who underwent total or partial adrenalectomy and later developed pituitary tumors. Characteristically, such tumors produce greater quantities of ACTH than are associated with untreated Cushing's syndrome, they are large enough to erode the sella turcica, and they appear to produce MSH as well as ACTH in excessive quantities, thus producing deep pigmentation (Nelson's syndrome). In Cushing's syndrome, the increased levels of ACTH show no diurnal variation and are less susceptible to suppression by exogenous administration of normal quantities of corticosteroids. The mechanisms controlling ACTH secretion in these patients have apparently been "reset" so that a higher level of corticosteroid production is necessary to produce ACTH suppression. This leads to supranormal ACTH production and an abnormally elevated level of corticosteroids. The question has arisen, does bilateral adrenalectomy contribute to the development of pituitary tumors? The occurrence of such tumors in patients who have undergone only partial adrenalectomy (or no adrenal surgery at all) suggests that they are a natural consequence of the disease and not secondary to removal of the adrenal glands.

Congenital Adrenal Hyperplasia. This condition is characterized by deficient production of cortisol, thus resulting in overproduction of androgens and, in some instances, other hormones, secondary to increased ACTH production. Although the level of plasma ACTH is elevated in these patients, no primary defect exists in ACTH production; rather, it represents the normal consequence of cortical deficiency leading to an increase in ACTH production.

ACTH Deficiency. A deficiency of ACTH production is usually due to destruction of the pituitary gland or, less often, to hypothalamic injury. A variety of pituitary tumors (chromophobe adenomas mainly) may destroy anterior pituitary tissue; craniopharyngiomas may affect the hypothalamus or the pituitary gland (or both); and various granulomas may interfere with ACTH secretion. In each instance, the result is panhypopituitarism, since other pituitary hormones are also affected. Rarely, there is a specific isolated deficiency in ACTH secretion. Although the exact cause of such a deficiency is unclear, it probably results from an abnormality of CRF production, due perhaps to a lesion of the anterior median eminence similar to that produced experimentally in animals.

ACTH Production by Tissues Other than the Pituitary Gland. In a number of patients with Cushing's syndrome secondary to excessive ACTH production, the source of hormone overproduction has been shown to be not the pituitary gland, but tumors of the lung, pancreas, thyroid, thymus, parotid and prostate. These tumors (which are most commonly oat cell carcinomas of the lung) produce a substance very similar, if not identical, to the ACTH produced by the anterior pituitary gland. The secretion of ACTH by these tumors, however, is independent of normal control; adrenal hyperplasia and Cushing's syndrome result. A few of the patients studied have had benign bronchial adenomas rather than carcinomas, but the latter predominate. The mechanism by which nonpituitary tissue produces a polypeptide such as ACTH is a subject of speculation. Tumors arising in a variety of tissues that do not normally produce hormones may also secrete other polypeptide hormones, including MSH, parathyroid hormone, gonadotropins, antidiuretic hormone, gastrin, erythropoietin and, possibly, thyrotropin and insulin. Hormone secretion by tumors produces clinical manifestations which Liddle has termed "ectopic humoral syndromes."

ADRENOCORTICAL SECRETION

Normal Secretion and Corticosteroidogenesis

Cortisol is the major steroid secreted by the adrenal cortex when it is stimulated by ACTH, but a variety of other steroids are

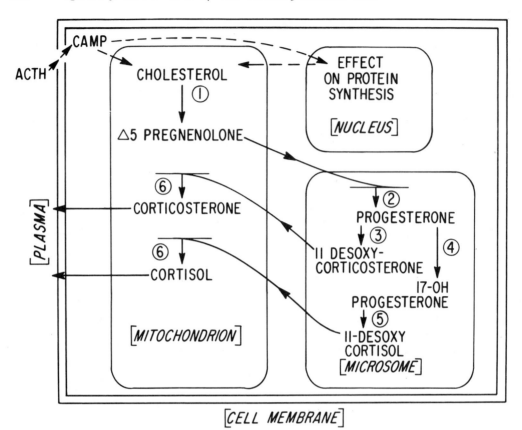

FIG. 14-2. Diagrammatic representation of the action of ACTH on the cell membrane to cause release of cyclic AMP which then, through a mechanism to be established, causes increased conversion of cholesterol to pregnenolone within the mitochondria. Pregnenolone then passes into the microsomes where the 3β-ol-dehydrogenase acts to produce progesterone, and 21- or 17-hydroxylation produces 11-desoxycorticosterone or 17-hydroxyprogesterone. The 11-desoxycorticosterone returns to the mitochondrion where 11-hydroxylation produces corticosterone, while the 17-hydroxyprogesterone undergoes 21-hydroxylation to produce 11-desoxycortisol which is similarly transferred to the mitochondrion for 11-hydroxylation. The chief secretory product, cortisol, or the lesser product, corticosterone, are then secreted out of the mitochondrion and through the cell membrane into the blood perfusing the adrenal cortex.

also secreted in smaller quantities (Fig. 14-2). ACTH acts upon the adenyl cyclase system found in the cellular membrane to convert ATP to 3'-5'cyclic AMP, which is then released into the cell. How cyclic AMP acts remains to be determined, but presumably it stimulates protein synthesis in the cell nucleus and elicits production of a substance or substances that affect the conversion of cholesterol to Δ⁵-pregnenolone by 20-alpha-hydroxylase and or des-

molase located within the mitochondria. Extramitochondrial 3 beta-ol-dehydrogenase is then responsible for conversion of the pregnenolone to progesterone. Progesterone may be hydroxylated either at the 21 position by 21-hydroxylase, or at the 17 position by the microsomal 17-hydroxylase. Corticosterone results from the 11 beta-hydroxylation of 11-desoxycorticosterone or 11-desoxycortisol (substance S) from 21-hydroxylation of 17-OH progesterone. The latter becomes cortisol after 11 beta-hydroxylation. Under normal conditions, the chief secretory product is the substance that has been hydroxylated at the 11, 17, and 21 positions, i.e., cortisol. A normal adult secretes about 20 mg. of this steroid daily. About 2 mg. of corticosterone is secreted daily. Several androgenic steroids that have undergone sidechain cleavage, chief among them dehydroepiandrosterone, are also produced by the adrenal cortex in response to stimulation by ACTH. ACTH is the primary controller of corticoid secretion, with the exception of aldosterone. Control of this steroid is dependent upon angiotensin as well as ACTH production. (This is discussed in greater detail in Chapter 15.)

In addition to effecting increased secretion of corticosteroids, ACTH also influences the growth of the adrenal cortex. Thus, when ACTH is absent, the gland quickly atrophies to less than a quarter of its normal size, whereas excessive stimulation with ACTH increases growth, producing an abnormally large gland. Increased vascularity of the gland has been observed with continued stimulation, and acute administration of ACTH has been associated with increased adrenal blood flow.

Metabolism of the Adrenal Steroids

The adrenal hormones are secreted into the bloodstream where they circulate with a half-life of approximately 1¼ hours. Metabolism occurs chiefly in the liver, where the Δ^4,3-ketone in the A ring is reduced and the hormone is generally conjugated with glucuronic acid to form a glucosiduronate; this highly water-soluble product is rapidly excreted in the urine. Dehydroepiandrosterone may be secreted directly into the blood as the sulfate, and excreted as such in the urine. Whether any significant metabolism occurs at the site of cortisol action has not been determined for certain, but probably no direct

chemical interaction, resulting in an altered molecule, is involved when the hormone produces its effects. The major secretory products may be measured in the urine as 17-hydroxycorticosteroids or 17-ketogenic steroids, as an estimate chiefly of cortisol secretion, or as 17-ketosteroids as an estimate chiefly of androgen secretion.

Effect of Substances Other Than ACTH Upon the Adrenal Gland

Women are known to have larger adrenal glands than men; moreover, the gland becomes smaller after oophorectomy, and its size is restored with estrogen administration. Estrogens have been shown to stimulate ACTH production and 11 beta-hydroxylation, to alter metabolism by increasing 6 beta-hydroxylation by the liver, and to interfere with the action of metyrapone in its inhibition of 11 beta-hydroxylation. The last mentioned effect appears to result from effects of the estrogen that increases 11 beta-hydroxylation (thus, antagonizing the inhibitory effect of metyrapone on this reaction) and induces hepatic mixed function oxidase reactions which result in increased metabolism of the metyrapone. Other effects of estrogen upon the adrenal cortex are stimulation of corticosteroidogenesis (by small amounts of estrogen) and inhibition (by large amounts)—actions that may be antagonized by testosterone.

In addition to alterations in corticosteroidogenesis produced by estrogens and metyrapone, other chemical substances are known to inhibit various aspects of adrenal function. These pharmacological agents, which act chiefly to inhibit secretion, include amphenone (3,3-bis(p-aminophenyl) butanone-2, DDD (2,3 bis(2-chlorophenyl-4-chlorophenyl)-1,1 dichloroethane), aminoglutethimide, and a number of other compounds; they have been tried clinically with limited success as inhibitors of adrenocortical function in patients with adrenal hyperfunction. The sites of inhibition have not been identified clearly for all of these compounds. Aminoglutethimide is of interest because of its ability to interfere with the conversion of cholesterol to pregnenolone.

Cushing's Syndrome

This disease may result not only from adrenal hyperfunction secondary to increased ACTH, but also from development of a primary adrenal tumor. These tumors, benign or malignant, may secrete one or more of the corticosteroids normally produced by the adrenal cortex. Benign tumors usually secrete only one; carcinomas characteristically secrete many. Thus, one may encounter adrenal tumors that specifically produce cortisol in large quantities, androgens in excessive amounts—or, less commonly, one of the intermediates in the synthesis of cortisol such as 11-desoxycortisol (substance S). (The harmful effects of the increased corticosteroids produced are discussed in greater detail below.) When production of corticosteroids becomes excessive, one must determine whether the problem is caused by an adrenal tumor, by overproduction of ACTH by the pituitary, by a tumor of nonendocrine tissue, or (much less often) by accessory adrenal tissue.

Congenital Adrenal Hyperplasia

Congenital adrenal hyperplasia, as the name suggests, is an inherited condition of a recessive nature, resulting in enlarged adrenal glands. The hyperplasia occurs secondary to one of a number of possible defects in corticosteroidogenesis, increasing ACTH secretion.

21-Hydroxylase Deficiency. The most common of the defects in steroid synthesis is in 21-hydroxylation (Fig. 14-2, 5). Instead of cortisol being produced in normal quantities, the production of 17-hydroxyprogesterone is increased. Since this compound has little biological activity, the adrenal gland is overstimulated by ACTH to produce large quantities of this sub-

stance and some corticosterone, but greatly reduced amounts of cortisol. As the result of this increased cortical stimulation by ACTH, excessive androgens (which have no inhibitory effect upon ACTH production) are also produced. These, then, are responsible for the clinical picture typically associated with congenital adrenal hyperplasia—namely, virilization and, in the female infant, pseudohermaphroditism. Clitoral enlargement at birth may result in mistaken sex identification, if a careful examination is not performed. Boys with the condition undergo precocious puberty, with early development of genitalia, facial and body hair growth, and increased muscle mass. Girls fail to undergo normal feminization, and the growth of both sexes is stunted owing to early epiphyseal closure secondary to high androgen levels. This kind of congenital adrenal hyperplasia is characterized by increased excretion of pregnanetriol, the major metabolic product of 17-hydroxyprogesterone. When the condition is recognized, it can be treated adequately by administration of normal quantities of cortisol which suppress ACTH production and, hence, reduce adrenal size and androgen secretion. These patients may live normal lives as long as replacement therapy is carefully maintained. Infants with the disease may tend to lose salt, a result of decreased aldosterone production; but most overcome this condition within a few years and are able to maintain normal sodium balance.

11-Hydroxylase Deficiency. A much less common cause of congenital adrenal hyperplasia is a specific deficiency of 11 beta-hydroxylation (Fig. 14-2, 6). This causes inadequate cortisol and corticosterone synthesis, and excessive 11-desoxycorticosterone and 11-desoxycortisol production. Since 11-desoxycorticosterone is a potent sodium-retaining steroid, these patients develop hypertension, presumably because of salt and water retention. They, too, may be treated by replacement therapy using cortisol, which suppresses the abnormally high ACTH production. Overproduction of androgens also accompanies this syndrome, resulting in virilization.

17-Hydroxylase Deficiency. This condition affects 17-hydroxylation not only in the adrenal glands but also in the ovaries and testes. These patients have a female habitus, fail to undergo normal feminization, have primary amenorrhea, and are hypertensive. They possess the female habitus because, without 17-hydroxylation, they are unable to synthesize testosterone. They cannot undergo normal feminization because this defect also interferes with estrogen production. Without estrogen, of course, menstruation cannot occur. With the inability to produce normal quantities of cortisol, there is an increase in production of 11-desoxycorticosterone and corticosterone (defect at Fig. 14-2, 4). The increased production of 11-desoxycorticosterone is considered largely responsible for the hypertension. Some of these patients may have a concomitant defect in 18-hydroxylation, with very low quantities of aldosterone being produced.

3 Beta-ol-Dehydrogenase Deficiency. An additional defect in corticosteroidogenesis may occur at the point of conversion of pregnenolone to progesterone (Fig. 14-2, 2). These patients produce no biologically active adrenal steroid. If the defect is complete, death is inevitable unless the condition is recognized at birth and corticosteroids are administered.

Androgen-Producing Adrenal Tumors

Although less common than tumors producing Cushing's syndrome, benign or (rarely) malignant tumors of the adrenal cortex may produce only androgens. These tumors are unable to synthesize cortisol and other hydroxylated compounds and thus secrete large quantities of C19 compounds which have androgenic effects. Since these compounds (i.e., dehydroepi-androsterone, 11-hydroxyandrostenedione or etiocholanolone) are relatively weak androgens, large amounts may be produced prior to the appearance of hirsutism or other signs of virilization. This means that high levels of 17-ketosteroids (the easiest measure of androgen metabolites) are found in the urine of almost all patients with androgen-producing tumors of the adrenal cortex. In contrast, gonadal tumors that produce chiefly testosterone (a highly potent androgen) may be associated with normal or only slightly elevated urinary 17-ketosteroids.

Aldosterone

Increased aldosterone production is discussed in Chapter 15. However, we should note briefly here that this steroid is the chief mineralocorticoid produced by the adrenal cortex. Its secretion is not controlled primarily by ACTH, but is influenced by plasma levels of angiotensin as well as ACTH. Excess aldosterone causes sodium and water retention, hypokalemia and hypertension. Hypersecretion of aldosterone is generally present in Cushing's syndrome; in addition, it may be produced by a small benign adrenal cortical adenoma or, less commonly, by hyperplastic glomerular cells. Disappearance of excessive aldosterone production occurs in a few of those patients who have hyperplastic glands when pituitary secretion of ACTH is suppressed by the administration of corticosteroids. In most patients, however, autonomous tumors are responsible for the condition.

Plasma Corticosteroid Concentration

Of the corticosteroids circulating in the blood of man the primary one is cortisol. Although aldosterone is also important, it is present in much smaller quantities, as are the adrenal androgens. Cortisol is present in quantities ranging from approximately 3 to 25 micrograms per 100 ml. of plasma. The wide variation in levels is related to the diurnal variation in ACTH secretion (see above). Cortisol in the circulation is largely bound to a specific globulin called corticosteroid-binding globulin (CBG), or transcortin, and exists only to a very small extent in the free state. Although the exact reason for diurnal variation is unclear, one may assume that this variation allows fairly intensive stimulation of the adrenal gland by ACTH for a period of time each day, while avoiding prolonged, harmful corticosteroid concentrations. The gland is thus kept ready to respond maximally during periods of stress. Diurnal variation is lost during periods of increased secretion or in patients with adrenal atrophy after prolonged treatment with relatively high doses of corticosteroids. A number of other substances can also influence the apparent (if not the biologically significant) level of plasma corticosteroids. Most important among these are the estrogens that stimulate the formation of CBG (as they do a number of other proteins) and may produce extremely high levels of corticosteroids. Although plasma cortisol level, on the average, may be only 10 or 12 micrograms per 100 ml. in the morning, this may increase to 30 micrograms per 100 ml. or more in a patient receiving estrogen therapy. Despite the marked increase in corticosteroid level resulting from the increase in corticosteroid-binding globulin, one sees no evidence of excessive biological effects in patients thus treated. Apparently, then, the corticosteroid bound to CBG is relatively inactive biologically, whereas the free compound continues to be available in normal quantities to exert its effect.

Addison's Disease

Destruction or removal of the adrenal glands results in adrenal insufficiency, or Addison's disease. The important symptoms of adrenal insufficiency result from inadequate production of cortisol and aldosterone. Cortisol deficiency has widespread

effects upon the metabolism of carbohydrate, protein, and fat; aldosterone is mainly responsible for normal sodium retention by the kidney. In the absence of aldosterone, one sees chiefly loss of sodium in the urine, accompanied by loss of fluid volume, hypotension, and an increase in serum potassium. The increased serum potassium probably results from intra- to extracellular shifts of this cation and is influenced by cortisol as well as aldosterone. Similarly, the severe hypotension sometimes resulting from total adrenal insufficiency can be corrected in most cases by administration of large doses of cortisol, although volume repletion may also be necessary. These patients become hypoglycemic when deprived of food, a result of decreased gluconeogenesis, which can be improved by corticosteroid administration. The Addisonian patient, completely deprived of adrenal steroids, becomes hypotensive within 24 to 48 hours and dies in shock if relatively large quantities of corticosteroids are not administered promptly. The Addisonian patient receiving maintenance therapy with normal quantities of corticosteroid will, when subjected to stress (i.e., infection, trauma, or operation), rapidly develop hypotension and shock if increased quantities of corticosteroid are not administered. Thus, small quantities of corticosteroid are necessary for normal function, increased quantities must be given during stress, and prolonged administration of high doses will produce harmful effects because of the ability of cortisol to block anabolic metabolism.

The three levels of corticosteroid action may be referred to as *permissive, protective,* and *pernicious.* The permissive action is illustrated by the effect of normal amounts of cortisol on the maintenance of normal arterial pressure. In contrast, if the Addisonian patient is subjected to stress, 20 mg. of cortisol per day is insufficient to maintain a normal arterial pressure, and the dosage must be raised to 100 mg. or more to prevent hypotension and shock. Thus, during stress, protection is afforded by an increase in the amount of corticosteroid administered. However, very high levels of corticosteroid maintained for a prolonged period (as in Cushing's syndrome) will result in hypertension, muscle atrophy, susceptibility to infection, and other harmful effects which are pernicious and may have serious, if not fatal, results.

If one considers aldosterone in terms of these three levels of action, an analogous situation can be described for the normal individual. Complete removal of aldosterone or other sodium-retaining compounds results in abnormal renal sodium excretion. Normal quantities of aldosterone (approximately 0.2 mg. per day) allow sodium balance to be maintained whether the patient has a daily intake of 50 or 150 mEq. Thus, aldosterone does not determine directly and quantitatively the amount of sodium that is retained, rather, it has an effect on the renal tubules so that they can, through other mechanisms, maintain sodium balance (permissive effect). If, however, sodium intake is decreased to 10 mEq. per day, the Addisonian patient may have difficulty maintaining sodium balance, and the salt-retaining hormone may have to be increased (protective effect). In contrast, if large doses of aldosterone are given, the picture of primary hyperaldosteronism (Conn's syndrome) develops; the patient retains too much sodium, extracellular volume expands, the kidneys lose excessive potassium because of increased sodium and potassium exchange, and the patient becomes hypertensive (pernicious effect).

In addition to the metabolic changes already described, hyperpigmentation is a characteristic of adrenal insufficiency. Because of the general deficiency of corticosteroids, more ACTH is secreted. As mentioned, ACTH has some melanocyte–stimulating property of its own, but MSH probably also is produced in excess in most of these patients, causing an increased pigmentation involving buccal mucosa, palmar creases, extensor surfaces, scars, and areolae.

REGULATION OF GROWTH HORMONE SECRETION

Growth hormone (GH), a protein with a molecular weight of approximately 21,000, is secreted by the anterior pituitary gland in response to stimulation by growth hormone releasing factor (GRF) from the hypothalamus. Growth hormone is relatively species-specific; therefore, only GH obtained from pituitaries of man or other primates is effective in man. Some indication exists, however, that GRF is not species-specific and that the monkey, for instance, will respond to porcine GRF. Although originally thought to affect growth mainly, growth hormone clearly exerts other metabolic effects throughout life. Contrary to what one might expect, the levels of growth hormone are not higher in childhood than in maturity.

Metabolic Effects of Growth Hormones. Growth hormone has numerous effects on intermediary metabolism and affects carbohydrate, fat, and protein synthesis. It accelerates the net transport of amino acids across cell membranes, which may contribute to its anabolic action. It inhibits glucose utilization and increases levels of plasma free fatty acids. Administration of large doses of growth hormone produces glucose intolerance with diabetes mellitus.

Control of Growth Hormone Secretion. A number of conditions that effect ACTH release also elevate plasma growth hormone. Thus, hypoglycemia, exercise, stress, pain, hemorrhage and administration of histamine or pitressin have been shown to increase plasma growth hormone levels within a few minutes. Additional stimulators of growth hormone secretion are arginine (given by infusion), glucagon and estrogens; glucose, in contrast, suppresses growth hormone secretion and, perhaps, secretion of ACTH as well.

Increased Secretion of Growth Hormone in Acromegaly and Gigantism. An abnormal increase of growth hormone secretion during the prepubertal period results in a rapid increase in body growth; patients may achieve heights of seven or eight feet, or more (gigantism). With onset of puberty, a further spurt in growth occurs; then, with closure of the epiphyses of the long bones, height stabilizes. Consequently, oversecretion of growth hormone after epiphyseal closure in adult life cannot produce increased height. Rather, it results in an overgrowth of the membranous bones (acromegaly). This is characterized by enlargement of hands, feet and mandible. Soft tissue growth is also stimulated, producing a heavy, wrinkled facial appearance and further accentuates growth of hands and feet. An increase in growth hormones in either childhood or adult life is due most commonly to a pituitary tumor, often demonstrable by skull x-ray. Associated with acromegaly are the metabolic effects of growth hormone excess, most notably decreased glucose tolerance, some decreased anabolism and, in some patients, elevated serum phosphorus concentration. Pituitary tumors may exert pressure on the optic chiasm, resulting in bitemporal hemianopsia; rarely they may be large enough to exert pressure on the hypothalamus and surrounding neurological tissue. Treatment of these patients consists of ablation of the pituitary gland by radiation and/or hypophysectomy.

Decreased Growth Hormone Secretion. Decreased growth hormone secretion is seen in two forms, either specifically absence of growth hormone alone, or a deficiency associated with hypopituitarism in which all anterior pituitary hormones are decreased or absent. The distribution of specific deficiencies may be either sporadic or familial. Children with this condition have a history of normal gestation and birth weight, but they show diminished rate of growth. Their plasma levels of growth hormone are low, failing to respond normally to hypoglycemia or arginine infusion. The familial type has been shown to be due to an autosomal recessive trait; however, many instances of sporadic, apparently nonfamilial, form have been described.

PANHYPOPITUITARISM

Panhypopituitarism may occur in childhood or in adult life. In childhood, it is characterized by retarded growth and maturation. Tumors of the pituitary region and vascular abnormalities are often responsible, but in some patients no clear-cut cause can be found. In adult life the causes may be similar, although in women panhypopituitarism is commonly associated with postpartum pituitary necrosis (Sheehan's syndrome). This condition usually results from hypotension due to excessive bleeding during delivery. Because the pituitary is enlarged and vascular during pregnancy, it is particularly susceptible to a fall in arterial pressure and, thus, infarcts. In men, chromophobe tumors are a common cause, although other tumors, basilar skull fracture, and granulomatous disorders (i.e., sarcoidosis and Hand-Schüller-Christian disease) may also be responsible.

Patients with hypopituitarism have atrophy of the thyroid, the adrenal glands, and the gonads. Deficiencies in the respective hormonal secretions are apparent, but typical signs of hypothyroidism do not, as a rule, appear until corticosteroids are administered. Characteristically, patients with this condition do well until subjected to some stress, at which time they develop severe adrenal insufficiency and (like the Addisonian patient) require administration of large doses of corticosteroids. Administration of thyroid hormone without corticosteroid increases metabolic rate and provokes an adrenal crisis, requiring the immediate use of corticosteroids.

Hyperpigmentation is not a feature of panhypopituitarism as it is in adrenal insufficiency, because of the lack of ACTH and, probably, MSH. Nor do these patients develop the electrolyte abnormalities observed in Addison's disease. The adrenal gland continues to produce aldosterone because of stimulation by the renin-angiotensin system. Hyponatremia is common; this is not the result of sodium loss but is dilutional. If present before puberty, hypogonadism results in failure to mature sexually; hypogonadism in the adult causes loss of body hair, atrophy of genitalia, decreased libido and, in women, amenorrhea.

Panhypopituitarism can be successfully treated with thyroxin and cortisol and, where indicated, estrogen or testosterone. Diagnosis is reached by demonstration of multiple hormone deficiencies, either of the target organs or the pituitary hormones and lack of normal responses to stimulation by ACTH, metyrapone, or the hypothalamic releasing factors such as TRF.

ANNOTATED REFERENCES

Burgus, R., Dunn, T. F., Desiderio, D. M., Ward, D. N., Vale, W., Guillemin, R., Felix, A. M., Gillessen, D., and Studer, R. O.: Biological activity of synthetic polypeptide derivatives related to the structure of hypothalamic TRF. Endocrinology, 86: 573, 1970. (Description of the probable structure of TRF and the relative activity of various analogs of this structure)

Daniel, P. M.: The anatomy of the hypothalamus and pituitary gland. *In:* Martini, L., and Ganong, W. F. (eds.): Neuroendocrinology. Vol. 1. New York, Academic Press, 1966. (A good description of the hypothalamus and pituitary gland, which is important to an understanding of the relationships between these two organs)

Daughaday, W. H.: The adenohypophysis. *In:* Williams, R. H. (ed.): Textbook of Endocrinology. Philadelphia, W. B. Saunders, 1968. (One chapter in a textbook of endocrinology which describes the chief abnormalities of the pituitary gland as well as its biochemistry and physiology as it might interest those concerned with clinical medicine)

Evans, H. M., Sparks, L. L., and Dixon, J. S. P.: The physiology and chemistry of adrenocorticotrophin. *In:* Harris, G. W., and Donovan, B. T. (eds.): The Pituitary Gland. Vol. 1. Anterior Pituitary. Berkeley, University of California Press, 1966. (Description of the physiology and chemistry of adrenocorticotrophin)

Forsham, P. H., and Melmon, K. L.: The adrenals. *In:* Williams, R. H. (ed.): Textbook of Endocrinology. Philadelphia, W. B. Saunders, 1968. (Chapter from a textbook of endocrinology describing many details about the adrenals not included in this present short chapter)

Goodman, H. G., Grumbach, M. M., and Kaplan, S. L.: Growth and growth hormone. New Eng. J. Med., 278:57, 1968. (An interesting review of growth hormone and its relationship to growth)

Nelson, D. H.: Present status of the problem of iatrogenic adenocortical insufficiency. Anesthesiology, 24:457, 1963. (Description of the adrenal atrophy which occurs following corticosteroids with a probable explanation of the mechanism)

Nelson, D. H., Meakin, J. W., and Thorn, G. W.: ACTH-producing pituitary tumors following adrenalectomy for Cushing's syndrome. Ann. Intern. Med., 52:560, 1960. (Description of the tumors of the pituitary gland which may be found some years following adrenalectomy for Cushing's syndrome and are associated with marked pigmentation of the patient)

Nelson, D. H., and Thorn, G. W.: Diseases of the anterior lobe of the pituitary gland. *In:* Wintrobe, M., Thorn, G. W., *et al.* (eds.): Harrison's Principles of Internal Medicine. ed. 6. New York, McGraw-Hill, 1970. (Short, general description of diseases of the anterior lobe of the pituitary gland, including Cushing's syndrome and other disturbances of ACTH secretion)

Reichlin, S.: Function of the hypothalamus. (Editorial) Am. J. Med., 43:477, 1967.

McCann, S. M., and Dhariwal, A. P. S.: Hypothalamic releasing factors and the neurovascular link between the brain and the anterior pituitary. *In:* Martini, L., and Ganong, W. F. (eds.): Neuroendocrinology. Vol. 1. New York, Academic Press, 1966. (The two preceding articles describe the function of the hypothalamus and the releasing factors involved in stimulation of the pituitary gland.)

15

Regulatory Mechanisms of the Pituitary-Thyroid Axis

Farahe Maloof, M.D.

INTRODUCTION

The major function of the thyroid gland is to synthesize the hormones thyroxine (T_4), triiodothyronine (T_3) and thyrocalcitonin. Thyroxine and triiodothyronine, with their multiple physiological properties, play a major role in controlling many critical bodily functions. (Thyrocalcitonin is discussed in Chapter 16.) Defects in thyroxine synthesis and, rarely, in iodide uptake exist in patients with a wide variety of thyroid disorders such as endemic and sporadic goiter, cretinism, hypothyroidism, hyperthyroidism, biosynthetic abnormalities, nodular goiter and thyroid carcinoma.

The importance of thyroid function becomes evident when one reviews statistics on the incidence of thyroid disease and its frequent association with poor general health and morbidity. For example, endemic or sporadic goiter continues to be a major health problem, even where the disease has been known for generations. The World Health Organization estimates that about 200,000,000 people in the world suffer from goiter. Nodular goiters occur in an impressive number of patients, especially women, even in areas where iodine is not deficient. In women, goiter can develop at the time of puberty or during pregnancy. The mechanism for enlargement of the thyroid in most patients living where dietary iodine is adequate is unknown.

The anterior pituitary gland must secrete optimal and controlled amounts of thyroid-stimulating hormone (TSH) to maintain the normal growth and metabolic functions of the thyroid gland. Any imbalance may lead to significant abnormalities in thyroid growth and function and, conceivably, could cause some thyroid diseases in human beings. The central nervous system plays an important role in maintaining normal thyroid function by producing a neurohumor (TSH-releasing factor, called TRF) in the hypothalamus, which stimulates the synthesis and release of TSH from the anterior pituitary.

HYPOTHALAMIC CONTROL OF TSH

TRF—Isolation and Chemical Structure

The hypothesis that the secretion of TSH from the anterior pituitary is controlled at least partly by neurohumoral agents reaching the pituitary by way of the hypophyseal portal vessels was based upon a series of experiments involving electrical stimulation of the hypothalamus, pituitary stalk transection, and transplantation of the pituitary to sites distant from the hypothalamus. This well-established concept of central nervous system control of the pituitary by neurohumoral substances, transported in the hypophyseal portal blood, provided a firm foundation for subsequent studies in which TRF from the hypothalamus was extracted, assayed and purified. TRF has been identified as a tripeptide, L-pyroglutamyl-L-histidyl-L-proline amide (molecular weight,

362,000). Studies on the effects of components, or of alterations in the tripeptide structure, affirm the necessity of the entire structure for full biological activity. Substitution of asparagine for proline amide yields a compound without TRF activity, and addition of a single methyl group on the amide of pyro (Glu-His-Pro(NH₂)) reduces activity by 95 percent. TRF has been identified not only in hypothalamic extracts but also in the hypophyseal portal blood. Significant TRF-like activity is not detectable in the peripheral circulation of normal animals, but is detectable in that of thyroidectomized-hypophysectomized rats only after exposure to cold. Normal rat serum seems to inactivate TRF.

Control of TRF

TRF activity is greater in the hypothalamus of thyroidectomized rats than in control animals, and this activity is not reduced to subnormal levels by subsequent administration of thyroxine. These data suggest that thyroid deficiency may induce TSH release from the anterior pituitary, partly by an increase in TRF. Whether other factors or alterations may release or inhibit TRF remains in doubt. L-Dopa has been reported to release pituitary LH and growth hormone, possibly by serving as a transmitter for the release of the respective releasing factors. Analogous studies have not been reported for TRF.

Action of TRF

Nanogram quantities of TRF isolated from hypothalamic tissue of oxen, sheep, pigs and human beings stimulate the synthesis and release of TSH from the pituitary in vitro, in the presence of oxygen, calcium and sodium ions. The effect of TRF on TSH release appears not to involve protein or nucleic acid synthesis, since puromycin, cycloheximide and actinomycin D have no inhibitory effect, either in vitro or in vivo. However, growing evidence suggests that cyclic 3′,5′-adenosine monophosphate (cy-

clic AMP) serves as an intracellular messenger to initiate the action of many protein and nonprotein hormones and apparently is involved in the release mechanism of growth hormone, insulin and, more recently, TSH. Isolated TRF from sheep and pigs is effective in releasing TSH in mice, suggesting a lack of species specificity. Synthetic TRF stimulates the release of TSH in mice when given either orally or parenterally; after intravenous administration, it elevates plasma TSH in rats and man.

PITUITARY SECRETION OF TSH

Thyroid stimulating hormone (thyrotropin, TSH) contributes to many aspects of thyroid structure and function: size and vascularity of the gland; height and activity of the follicular epithelium; amount of colloid; biosynthesis and release of thyroxine; and numerous critical biomedical processes in thyroid tissue (e.g., glucose utilization, oxygen consumption, and synthesis of phospholipid, protein, and nucleic acid). These actions vary in time of onset and recently have been thought to result from activation of adenyl cyclase, after the binding of TSH to a receptor site on the cell membrane. The resulting formation of cyclic AMP leads, through an effect on messenger RNA, to the activation of a number of metabolic steps.

Chemistry and Dynamics of TSH

TSH is a glycoprotein (molecular weight, 26,000) secreted by specific anterior pituitary cells (B_1-β_1 basophils). These cells contain granules which have been correlated with TSH secretion. After thyroidectomy, these cells undergo marked hypertrophy with increased endoplasmic reticulum and reduced granularity, demonstrating the high rate of TSH secretion. The pituitary of a normal man weighs about 0.5 g., that of a normal menstruating female about 1.0 g. The TSH content of the normal human pituitary is approximately 300 μg; its secretion rate is 110 μg per day; its metabolic clearance rate is 61 liters per day, and its plasma concentration is about 1.0 to 3.0 ng. per ml. It has been estimated that about 50 percent of the pituitary TSH turns over each day. In contrast, the total luteinizing hormone (LH) content of the human pituitary is 80 μg, which is reported to turn over once per day.

Regulation of TSH Secretion

Release of TSH from the pituitary is stimulated by TRF and by thyroidectomy and is inhibited by locally effective concentrations of thyroxine or triiodothyronine. Whether thyroxine has to be deiodinated to triiodothyronine to be effective remains unclear, but the anterior pituitary is known to have an effective mechanism for deiodinating thyroxine. The inhibitory effect of the thyroid hormones apparently involves both protein and nucleic acid synthesis. Although the thyroid hormones have been localized to parts of the hypothalamus, the evidence that this localization plays a direct role in the feedback control of TSH secretion is circumstantial. Nevertheless, lesions of the thyrotropic area of the hypothalamus increase pituitary sensitivity to feedback inhibition by thyroid hormones.

As early as 16 hours after thyroidectomy in rats, pituitary concentration of TSH falls and plasma TSH rises. In man, total thyroidectomy may lead to an elevated level of plasma TSH within 5 to 7 days. The thyroidectomized animal or human being degrades TSH more slowly than normal, a phenomenon apparently related to the hypothyroid state rather than to removal of the thyroid as a TSH-degrading system; however, the latter remains a possibility.

In the pituitary-adrenal axis, suppressive therapy with adrenal steroids produces prolonged and unpredictable suppression of ACTH release. In contrast, long-term thyroid hormone treatment does not prolong the inhibition of TSH release. Release of TSH occurs in 1 to 5 weeks after various types of thyroid medication are withdrawn.

Reserve Test for TSH

Studies have been devised to determine what factors other than moderate to marked reduction in the concentration of circulating thyroid hormones might increase serum TSH levels. No evidence exists for a diurnal variation in serum TSH, although a possible elevation at 2:00 A.M. has been reported. Acute alterations in blood sugar, intravenous vasopressin, presence of bacterial endotoxins and parenteral administration of glucagon or glucocorticoids have failed to release TSH. The administration of estrogens or progesterone to human beings has been reported to be ineffective. However, data from experiments on rats suggest that estrogens do release TSH. This has been confirmed for man in the laboratory of the author. The serum TSH increases four- to tenfold without contingent detectable changes in the circulating level of total thyroxine, free thyroxine or the thyroxine-binding globulin. The rise in TSH specifically has been confirmed by dilution studies, bioassay and immunological studies. The response appears to be caused by a release of TSH from the pituitary, since it is absent in patients with hypopituitarism and the effect is inhibited by the administration of thyroid hormones.

Thyroid-Releasing Factor. The use of TRF would appear to solve the dilemma of the thyroidologists searching for a reserve test for pituitary TSH. Intravenous administration of synthetic TRF (150 to 1,000 μg) stimulates a fivefold rise in plasma TSH in normal men. Reaching its peak in about 15 to 30 minutes, TSH returns to baseline levels after about 3 hours. Apparently, women respond to TRF more markedly than men do when equivalent doses of TRF are used. The effect of TRF upon the release of TSH in man is blocked by the daily administration of 300 μg of L-thyroxine, and the effect is undetectable in patients with hypopituitarism. Its preliminary effects in normal and hypopituitary man are convincing; however, its validity as a reserve test awaits application in a number of patients with various thyroid, pituitary and hypothalamic disorders.

BIOSYNTHESIS OF THYROID HORMONES

Embryology

The thyroid gland is a bilobed organ which develops as an invagination of the floor of the embryonic pharynx and descends as a cellular stalk to the anterior part of the neck. Then, by a proliferation of its cells, epithelial plates and bands are formed which become the adult organ. Connection with the pharynx (the thyroglossal tract) normally disappears, but its point of origin persists in the adult as the foramen cecum, a dimple on the back of the tongue. These facts of embryonic development explain certain developmental anomalies. Soon after the tenth week of gestation, the thyroid accumulates and binds iodide and forms iodotyrosines and iodothyronines within thyroglobulin. At about this time, one can detect bioassayable TSH in the pituitary and radioimmunoassayable TSH, as well as thyroxine and thyroxine-binding proteins, in the serum.

The thyroid has an abundant blood supply. The normal flow rate of blood through the gland is about 5 ml. blood per g. of thyroid per minute. The blood volume of normal man, 5 liters, passes through his thyroid about once per hour, through his lungs about once per minute, and through his kidneys once in 5 minutes.

Concentration of Iodide

Iodide is actively transported into the thyroid. The follicle cell actively takes up iodide against a substantial electrical gradient and initiates a series of events leading to thyroxine synthesis. The iodide-concentrating mechanism attains saturation when extracellular iodide reaches 5×10^{-4} M. Phosphate bond energy is required, since it is inhibited by anaerobiosis, low temperature and agents such as 2,4-dinitrophenol. It is also inhibited by various cardiac glycosides. Monovalent ions such as thiocyanate and perchlorate can compete with iodide for this concentrating mechanism.

Like the stomach, mammary and salivary glands, the thyroid gland is able to concentrate iodide from the blood. However, the thyroid uptake mechanism is the most efficient, accumulating up to 40 times the serum level of iodide. Under appropriate conditions, the concentration gradient can reach levels of 500:1. Only the thyroid uptake is responsive to TSH, and only the thyroid incorporates iodine into thyroid hormone.

Iodide Transport Defect. Patients with an iodide transport defect are hypothyroid and have a goiter; laboratory studies show a low concentration of serum thyroxine, and ^{131}I uptake is low. The defect involves not only the thyroid, but the salivary gland and the gastric mucosa also. Analysis of the thyroid in vitro reveals it to be hyperplastic histologically, its iodine content is low, and the concentration of ^{131}I in the thyroid is the same as in the medium (normally 10 to 20 times greater). In one patient, large doses of inorganic iodide led to an increase in the synthesis of thyroxine, a decrease in goiter size, and a rise in the basal metabolic rate to normal. Although the locus of the defect is unknown, an abnormal carrier protein of the plasma membrane of the follicular cell has been postulated as the cause.

Biosynthesis of Thyroid Hormones

Once iodide is accumulated by thyroid tissue, it is rapidly oxidized to some higher valence which serves to iodinate thyroglobulin-bound tyrosine. This, then, appears to be the initial step in thyroxine biosynthesis. Although the sequence of events involved in this biosynthetic reaction is not known precisely, the factors involved are: the thyroid peroxidase; a source of hydrogen peroxide; an iodinating intermediate postulated to be sulfenyl iodide species (SI⁺); and a suitable acceptor molecule, namely, peptide-linked tyrosine, in the thyro-globulin molecule.

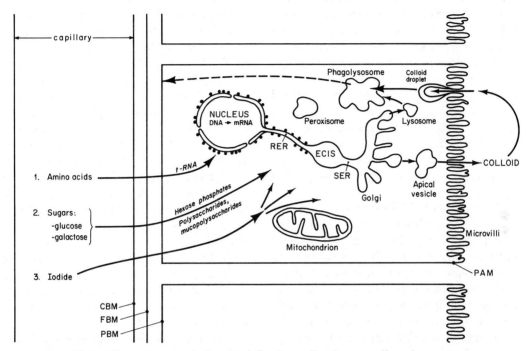

FIG. 15-1. Ultrastructure of thyroid follicular cell: cbm, capillary basement membrane; fbm, follicular basement membrane; pbm, plasma basement membrane; pam, plasma apical membrane; mv, microvilli; nuc, nucleus; nl, nucleolus; RER, rough endoplasmic reticulum; SER, smooth endoplasmic reticulum; ECIS, endoplasmic cisterna; GA, Golgi apparatus; AV, apical vesicle; CD, colloid droplets; Lys, lysosomes; Plys, phagolysosomes; mt, mitochondria.

Requirements for the biosynthesis, storage and secretion of thyroxine:

I. Biosynthesis of Thyroglobulin   *Site*

     (a) amino acids   RER and SER

     (b) sugar—glucosamine ⎫

                mannose ⎬   RER and SER

                galactose ⎫

                fucose   ⎬   Golgi apparatus

                sialic   ⎭

     (c) –S–S–bonds   Endoplasmic

         (disulfide exchange enzyme)   reticulum

     (d) iodination

            peroxidase   Endoplasmic reticulum

            source of H_2O_2 (?) Peroxisomes

            iodinating intermediate

II. Storage of thyroglobulin   Follicular lumen

III. Proteolysis of thyroglobulin   Phagolysosomes

IV. Deiodination of iodotyrosines   (?)

The peroxidase, found in the endoplasmic reticulum of the cytoplasm, appears to be a cytochrome b type hemoprotein. The final characterization awaits purification. The source of peroxide is unknown. The postulated iodinating species (a sulfenyl iodide) is gaining scientific support; but direct confirmation is yet unavailable. A series of studies on thyroglobulin reveal that the conformational structure of this large glycoprotein is essential for optimal iodination. About 90 percent of body iodine resides in the thyroid gland, chiefly as organic iodine, and is stored in thyroglobulin. This large iodine pool (5.0 to 7.0 mg.) turns over slowly, about one percent per day.

The epithelial follicular cells probably take up thyroglobulin from the colloid by pinocytosis, and within these cells thyroid hormones are released by degradation of thyroglobulin. Possibly lysozymes coalesce with the droplets in the cytoplasm, releasing hydrolytic enzymes which degrade thyroglobulin to liberate thyroxine, triiodothyronine, monoiodotyrosine and diiodotyrosine. Deiodination of iodotyrosines takes place in the follicular cells, and the iodothyronines are secreted into the parafollicular capillaries along with minute quantities of thyroglobulin and, possibly, iodotyrosines.

Defect of Thyroid Hormone Biosynthesis

Several types of goiter associated with defects in thyroid hormone biosynthesis are now recognized: those in which the thyroid is unable to convert inorganic iodide into an organic form (iodide organification defect); partial organification defect associated with deaf-mutism (Pendred's syndrome); those in which the thyroid is unable to couple mono- and diiodotyrosine into triiodothyronine or thyroxine (coupling defect); those in which there is a general tissue lack of iodotyrosine deiodinase (dehalogenase defect); and a group characterized by the secretion of an abnormal iodinated peptide into the serum (serum iodopeptide defect) (Fig. 15-1 and Table 15-1).

Iodide Organification Defect. Complete and total defects are rare. This syndrome is characterized by a hypothyroid patient who has a goiter (which may be quite large) and a low circulating thyroxine level. The thyroid iodide uptake is high (50 to 80%) and reaches a peak rapidly, within 2 to 4 hours. Thereupon, it is rapidly and almost completely dischargeable by the administration of thiocyanate or perchlorate. Analysis of the thyroid tissue reveals it to be hyperplastic, with 99 percent of the iodine content present as inorganic iodide. The tissue is unable to iodinate tyrosyl residues in vitro. Postulated defects might occur in the thyroid peroxidase, the source of peroxide, the iodinating intermediate, or the acceptor molecule, thyroglobulin. A patient studied recently in the laboratory of the author has been found with this defect due to a lack of peroxidase activity.

Pendred's Syndrome (Congenital Goiter and Nerve Deafness). In these patients, either the goiter is apparent at birth or it becomes so within the first decade or two of life. Eighth-nerve deafness is present from birth or develops during childhood but vestibular function is normal (although somewhat impaired) and tympanic membranes are intact. Intelligence and growth are usually normal. Whether this disease is inherited remains unclear from the available family studies. The patients usually have normal radioactive iodine uptakes and normal circulating levels of thyroxine. However, they have a positive perchlorate or thiocyanate discharge test. They do not have exceptionally large goiters. Inhibition of thyroid iodide uptake after the perchlorate discharge test persists much longer than in normal patients, suggesting another potential abnormality.

Defect in Coupling of Mono- and Diiodotyrosine. These patients may or may not

TABLE 15-1. Congenital Defects in the Biosynthesis of Thyroxine

	CLINICAL	GOITER	SERUM T_4	THYROID ^{131}I UPTAKE	SALIVARY/ PLASMA ^{131}I RATIO	SCN or ClO_4 DISCHARGE	SERUM MIT	SERUM DIT	SERUM BII
Concentration defect	Hypothyroid	++	↓	↓	N ~ 1.0	−	−	−	−
Organification defect	Hypothyroid	++++	↓	High early peak rapid release	N	−	−	−	−
Pendred's syndrome	Euthyroid Deaf (VIII)	++	N	N	N	+	−	−	−
Coupling defect	Euthyroid	++	N or ↓	N or ↑	N	neg.	−	−	−
Dehalogenase defect	Hypothyroid	++	↓	N or↑ rapid release	N	neg.	+	+	−
Iodopeptide defect	Hypothyroid	++	N or↑	N or↑	N	neg.	−	−	+

T_4 = thyroxine

MIT = monoiodotyrosine

DIT = diiodotyrosine

BII = butanol nonextractable iodine

N = normal

have an abnormally low level of circulating thyroid hormone. The thyroid uptake of ^{131}I may also be high, but the ^{131}I is not dischargeable by perchlorate or thiocyanate. The thyroid tissue contains labeled mono- and diiodotyrosine but little detectable triiodothyronine or thyroxine. The defect seems to be in the conformational structure of thyroglobulin and not in a postulated coupling enzyme.

Iodotyrosine Dehalogenase Defect. In these hypothyroid patients, serum thyroxine concentration is low; and, thyroid uptake of ^{131}I may be high, with a rapid turnover. After ^{131}I administration, there is a high serum concentration of ^{131}I-labeled mono- and diiodotyrosine, whereas normally little or none is detectable. After intravenous administration of ^{131}I-labeled mono- or diiodotyrosine to such patients, a large quantity of these iodotyrosines is excreted unchanged in the urine, providing evidence for a generalized tissue defect in dehalogenation. Tissue obtained at surgery is hyperplastic histologically and demonstrates no dehalogenating activity. This is one of the two enzymatic defects which have been demonstrated in thyroid tissue. Administration of large doses of iodide has stimulated synthesis of a normal amount of thyroxine.

Abnormal Serum Iodopeptides. This defect occurs in a heterogeneous group of patients with goiter and, characteristically, with elevated serum concentration of acid-butanol-insoluble iodine, which leads to a high serum protein-bound iodine. Such patients usually have a normal or elevated thyroid retention curve of labeled ^{131}I. In some patients, the acid-butanol-insoluble material, the iodinated protein, resembles albumin immunologically.

Tissue obtained at surgery reveals little or no thyroglobulin, but variable amounts of low molecular weight iodoproteins are present. The abnormality seems to be defective synthesis or structure of thyroglob-ulin or an abnormality in the proteolysis of thyroglobulin. Iodination of serum albumin in the thyroid apparently accounts for the iodinated albuminlike material in serum.

CIRCULATING THYROID HORMONES

The major circulating thyroid hormones are thyroxine (T_4) and triiodothyronine (T_3), which exist in serum at a concentration of 4 to 11 μg percent (about 10^{-7} M) and 150 to 250 ng. percent (3×10^{-9} M), respectively. Measurements of the serum-free thyroxine reveal it to be 0.8 to 2.5 ng. per 100 ml. Free triiodothyronine is about 1.5 ng. per 100 ml. Approximately two thirds of the circulating T_3 is secreted by the thyroid and one third is derived from the deiodination of T_4.

KINETICS OF THYROXINE AND TRIIODOTHYRONINE METABOLISM

In the blood, T_4 and T_3 are largely bound to protein. Thyroxine is firmly associated with plasma-binding proteins, such that the free plasma hormone constitutes only one three thousandth of the total circulating level. T_3 is less firmly bound (one third as strongly as T_4) and the free hormone is one hundredth of the total. The major binding proteins are thyroxine-binding globulin (TBG), thyroxine-binding prealbumin (TBPA), and albumin; these account for 60, 30, and 10 percent, respectively, of the thyroxine-binding capacity of plasma. (The kinetics of thyroid hormone metabolism is summarized in Table 15-2.)

EXCESS SYNTHESIS AND SECRETION OF T_4 AND T_3

Hyperthyroidism

Accumulated data have been interpreted to incriminate either the hypothalamic-pituitary area or an autonomously hyper-

TABLE 15-2. Kinetics of Thyroxine and Triiodothyronine Metabolism

	T_4			T_3	
	NORMAL	HYPER-THYROID	HYPO-THYROID	NORMAL	HYPER-THYROID
Space (liters)	9.4	9.1	7.5	43	45
Turnover rate (% per day)	10.6	24.0	9.8	52	84
Clearance rate (liters per day)	1.0	2.1	0.7	22	37
Organic iodide pool (μg I)	508	1544	107	81	–
Daily hormonal disposal (μg I per day)	53.6	359	10.7	60	336
T/2 (days)	6.4	2.7	9.4	1.5	–

functioning thyroid as the prime abnormality in hyperthyroidism; however, these data have been neither direct nor substantial. Proponents of the hypothesis implicating autonomous thyroid hyperfunction note that hyperthyroidism can occur in patients with decreased function of the anterior pituitary and that serum TSH concentrations (measured by bioassay or radioimmunoassay) are not elevated in most patients with hyperthyroidism, presumably because TSH is suppressed by excessive levels of circulating thyroid hormone. However, a few patients with pituitary tumors and hyperthyroidism have had elevated levels of serum TSH (confirmed by bioassay and recently by radioimmunoassay). Irradiation of the pituitary of such patients has ameliorated the hyperthyroidism.

Hyperthyroidism is generally considered to be caused by a diffusely hyperplastic thyroid, independent of pituitary function, an autonomy reflected by the failure of T_3 to suppress thyroid activity. Since the serum TSH concentration is usually not elevated, investigators have searched for an additional source of thyroid stimulation. A bioassay for plasma TSH, based upon the release of thyroid radioactivity in animals pretreated with ^{131}I, has revealed a long-acting thyroid stimulator (LATS) in the plasma of patients with thyrotoxicosis. LATS has a slower and more sustained effect than TSH, giving a maximum bioassay response at 8 to 16 hours, instead of 2 to 3 hours as for TSH. The activity of LATS is associated with the 7S gamma globulin of serum, is neutralized by antisera to human gamma globulin but not by antiserum to TSH, and is not suppressible by thyroid hormone. Perhaps the major argument for LATS as a causative factor in hyperthyroidism is its presence in neonates with hyperthyroidism, born of mothers with LATS activity in the serum. The disease in the neonate seems to subside concomitantly with a decrease in LATS activity and, at times, without specific antithyroid activity. These data demonstrate that LATS may cross the placenta and stimulate thyroid activity.

It has been stated as late as 1967 that, despite the wealth of literature on the role of stress in hyperthyroidism, and the full documentation that neural factors are concerned with the regulation of the pituitary axis, the accumulating data in regard to LATS leaves disappointingly little room for a neuroendocrine-thyrotropic hormone theory of hyperthyroidism. This statement may need to be altered in view of the clear demonstration that suppressibility of ^{131}I uptake bears no relationship to the presence or absence of LATS-like activity in the serum of hyperthyroid patients.

Suppression Test for Hyperthyroidism. The susceptibility or resistance of an endocrine target organ to suppression forms the basis of a commonly used test for assessing its function. Hence, abnormal suppressibility is the cardinal feature of a hyperactive endocrine organ. The test consists of the administration of varying doses of the biosynthetic hormone of the target organ and the measurement of alterations in that organ's functional state. For the thyroid, changes in the ^{131}I uptake by the thyroid and the circulating level of thyroxine are measured. Decrease in the uptake of ^{131}I and in the circulating level of thyroxine of over 50 percent is considered normal after the administration of physiological doses of triiodothyronine (75 μg per day) for 10 days. The test period for thyroid is longer than that for most other endocrine target organs, because circulating thyroxine has a longer half-life than most other hormones.

An explanation of the abnormal T_3 suppression test in hyperthyroid patients with thyroid hyperplasia is not clear. Whether this is due to an inability to suppress an autonomously functioning hyperplastic thyroid or an inability to suppress an abnormally functioning hypothalamic-pituitary mechanism is difficult to ascertain at present. The final answer to this important problem awaits a refinement in the radioimmunoassay of serum TSH, epecially at the supposedly normal levels, an analysis of the role of the hyperthyroid gland in inactivating TSH and a precise determination of the effects of excess thyroxine or triiodothyronine on the biosynthesis and secretion of TSH, possibly leading to an abnormal form of TSH which is biologically active but not assessed by current radioimmunoassay techniques.

Hyperfunctioning Hyperplastic Thyroid (Graves' Disease). Hyperthyroidism can be defined as a metabolic state resulting from the increased production and secretion of either thyroxine or triiodothyronine, or both. The disease can occur in patients with a diffusely hyperplastic thyroid or with hyperfunctioning single or multiple thyroid adenomas. Excessive secretion of thyroid hormone produces numerous clinical features related to increased metabolic activity, such as nervousness, tremor, tachycardia, sweating, heat intolerance, fatigue, and weight loss in spite of increased appetite. Although an enlarged thyroid is usually present, this is not invariable. Exophthalmos is observed in most patients. The secretion rate and the metabolic disposal rate of thyroxine and/or triiodothyronine are increased. A usual feature in hyperthyroidism is a high radioiodine uptake by the thyroid. A small number of patients with hyperthyroidism have normal uptake values; these values, however, are not reduced by the administration of T_3 (75 to 100 μg per day for 10 days). The lack of suppressibility of either an elevated (or normal) [131]I uptake or the elevated serum level of T_4 or T_3 is a prime feature of the disease. In some patients the uptake and turnover of ^{131}I may be so rapid that a normal value at 24 hours reflects a markedly hyperfunctioning gland.

Hyperfunctioning Thyroid Nodule or Nodules (Plummer's Disease). Hyperthyroidism may result from increased synthesis and secretion of thyroxine or triiodothyronine from single or multiple thyroid nodules. This disorder can be most precisely demonstrated in the patient with a single hyperfunctioning nodule; ^{131}I uptake is high over the nodule and low or absent in the opposite lobe. This disorder (Plummer's disease) differs from Graves' disease in the following ways: the disease occurs in older persons; the symptoms are more subtle and less explosive; malignant exophthalmos and pretibial myxedema are rarely present, and, usually, no LATS activity is measurable in the serum. However, the elevated uptake of ^{131}I in this disease is not suppressed by the exogenous administration of T₃—findings similar to those in Graves' disease. In Plummer's disease, pituitary TSH may be suppressed by the hyperfunctioning nodule, since there is atrophy of the contralateral lobe, and its low ^{131}I uptake may be corrected by the exogenous administration of bovine TSH. Pituitary TSH may also be suppressed in a patient with a hyperfunctioning thyroid nodule who is not hyperthyroid and has normal circulating levels of T₄ and T₃.

Triiodothyronine Hyperthyroidism. This disorder has been observed in a small number of patients, but the number and type can be determined precisely only when the assay for triiodothyronine is specific, accurate and reproducible. The syndrome consists of hyperthyroidism (evidenced by an elevated BMR), a normal level of circulating thyroxine, normal thyroxine-binding proteins, and an elevated circulating triiodothyronine level. Radioiodine uptake may be elevated or normal, but it is nonsuppressible by the exogenous administration of thyroid hormone.

Effect of Excess Thyroid Hormones on Other Endocrine Function

Thyroid hormones may have profound effects on the function of other endocrine systems, thus reflecting the interdependency of endocrine function upon systemic functional homeostasis. Excessive secretion of thyroid hormones may affect the secretion, clearance, and metabolism of androgens, estrogens, progesterone, and adrenal steroids (Table 15-3).

Hypermetabolism Associated With Neoplasia

Abnormalities in thyroid function that are the result of malignancy are rare. How-

TABLE 15-3. Effect of Thyroid Hormones upon Other Endocrine Function

Androgen Metabolism

Decreased metabolic clearance rate of testosterone

Normal metabolic clearance rate for androstenedione

Decreased conversion of testosterone to androstenedione; increased conversion of androstenedione to testosterone

Normal plasma production rate of testosterone

Increased plasma testosterone; suppressible by the administration of estrogen

Increased androsterone and decreased etiocholanolone excretion in urine

Estrogen Metabolism

Increased estrogen production rate in men (from a normal of approximately 60 μg per day to 150 to 250 μg per day)

Estradiol is metabolized to 2-hydroxyesterone rather than estriol

Progesterone Metabolism

Increased in favor of the 5α metabolites such as 5α-pregnenediol and 3α-hydroxy, 5α-pregnone.

Adrenal Steroid Metabolism

Increased cortisol production rate, but also increased excretion, hence plasma levels are normal

Increased 11-ketonic deviation of cortisol, the products of which are less active in inhibiting ACTH release, thereby producing increased ACTH in plasma

ever, certain patients with tumors of trophoblastic origin present laboratory and, occasionally, clinical evidence of hyperthyroidism. Laboratory data indicative of thyroid hyperfunction have been recorded for at least 13 patients with hydatidiform mole, 14 women with choriocarcinoma, and one man with embryonal carcinoma of the testes. All of these patients had a high level of urinary chorionic gonadotropin excretion. A few had clinical evidence of hyperthyroidism, including an enlarged thyroid with a bruit; however, most patients had an increase in several tests of thyroid function without clinical signs of hyperthyroidism. These tests included the serum protein-bound iodine (PBI), butanol extractable iodine (BEI), thyroxine, BMR and ^{131}I uptake. The thyroxine-binding globulin capacity for thyroxine was markedly increased (80 to 100 μg %) in patients with hydatidiform mole. The elevated thyroid uptake of ^{131}I in one patient was normally suppressible with 50 μg of T_3 for only 4 days.

Elevations of TSH in the serum and tumor tissue have been found in some patients by bioassay but not by human TSH radioimmunoassay. This is reminiscent of an extract of human placenta with TSH-like activity, which has low biological activity and reacts poorly in the human TSH radioimmunoassay system. Chemotherapy of the metastatic trophoblastic disease or removal of the mole usually leads to clinical and chemical amelioration of the hyperthyroidism.

DECREASED THYROXINE AND TRIIODOTHYRONINE SYNTHESIS AND SECRETION

Primary Thyroid Myxedema

A relative or absolute deficiency of circulating thyroid hormone (hypothyroidism) may result primarily from the spontaneous or iatrogenic destruction of thyroid tissue, disorders of thyroid hormone biosynthesis or the inhibition of thyroxine biosynthesis by iodide in certain patients with thyroid disorders and secondarily from TSH deficiency. A primary decrease in the synthesis and secretion of thyroxine leads to a compensatory oversecretion of TSH, goiter formation (if viable thyroid tissue is available) and the peripheral effects of thyroxine deficiency. Peripheral effects consist of decreased metabolism, bradycardia, cold sensitivity, thickening of the skin by mucinous edema, intellectual deterioration, mental and physical lethargy and impaired hearing. Physical examination usually reveals a striking delay in relaxation of tendon reflexes. Primary hypothyroidism is associated with a decreased serum T_4 and ^{131}I uptake, with elevated serum TSH and, in most patients, with circulating thyroid autoantibodies. The uptake usually fails to rise in response to the daily administration of bovine TSH for 3 days.

Elevated levels of serum TSH have been found in a large number of patients with primary thyroid myxedema, making the radioimmunoassay of TSH a useful diagnostic test. The elevated serum TSH concentration is due to accelerated secretion of TSH from the pituitary (average 688 μg per day) and a prolonged half-life of serum TSH (about 85 min.). The elevation in serum TSH levels may range widely, from 6 to 250 ng. per ml. in response to a variable secretory rate of TSH from the pituitary (ranging from 260 to 15,380 μg per day). The variability, no doubt, stems from several factors: some patients do not have complete lack of thyroxine; some patients with low thyroxine levels may synthesize adequate amounts of triiodothyronine under specific conditions, which prevents serum TSH from being as high as one would expect from the serum thyroxine level; or severe and prolonged myxedema may lead to myxedema of the pituitary, causing decreased synthesis and secretion of TSH. The concept of a hypothyroid pituitary has been based upon the results of bioassay experiments for TSH and has been confirmed by the radioimmunoassay for TSH. This concept can be extended to involve other pituitary hormones, since some patients with myxedema have impaired release of growth hormone, gonadotropins and ACTH. These abnormalities are usually reversible with thyroid treatment.

Hypothyroidism (Secondary to Pituitary Failure)

In hypothyroidism secondary to pituitary or hypothalamic failure, the clinical picture usually differs somewhat from that of primary thyroid myxedema. Skin and hair changes may be less severe than in primary hypothyroidism. Decreased pigmentation, especially of the areola of the breasts, may be an impressive feature, supposedly the result of decreased ACTH. Arterial pressure is usually normal or low, and blood cholesterol is less elevated. Circulating antibodies to thyroid antigens are more often positive in patients with primary myxedema. Plasma TSH measured by radioimmunoassay should be low or nondetectable, and the low ^{131}I uptake by the thyroid should respond to exogenous administration of TSH. Patients with pituitary hypothyroidism usually have a deficiency of other pituitary hormones and their respective target-organ hormones, although pituitary hypothyroidism occasionally occurs secondary to an isolated TSH deficiency. In spite of what should be a clear distinction between the two disorders, clinical differentiation may sometimes be difficult, since patients with a primary thyroid deficiency may have myxedematous hypofunction of the pituitary or its target organs, or both. The use of thyroid-releasing factor as a reserve test for TSH should prove helpful.

Effect of Thyroid Upon Elevated Serum TSH

When thyroid hormone is administered to patients with hypothyroidism, the elevated serum TSH concentration returns toward normal. The time required to accomplish the reduction may vary considerably, depending upon the type and dosage of thyroid hormone administered.

Prehypothyroid State

An elevated serum TSH in the presence of apparent euthyroidism (both clinically and metabolically) as assessed by a normal serum T_4 and BMR suggests a prehypothyroid state. The syndrome may be seen in patients with Hashimoto's thyroiditis or in those who have been previously treated for hyperthyroidism by ^{131}I or possibly by surgery. After subtotal thyroidectomy for hyperthyroidism, serum TSH may rise transiently and then decline to normal as the thyroid remnant undergoes compensatory hypertrophy and thyroid hormone secretion becomes adequate. The ^{131}I-treated gland seems unable to make the same growth response, and high levels of serum TSH are required to maintain the same thyroid hormone output from the damaged thyroid. The subsequent course of events is variable: some patients eventually develop clinical hypothyroidism with a low serum thyroxine; others may remain euthyroid with elevated serum TSH and normal levels of thyroxine; still others may remain euthyroid with low levels of serum thyroxine. In the last situation, the euthyroid state may be due to a normal level of circulating triiodothyronine. Such patients have a decreased thyroid reserve, as evidenced by an impaired response to exogenous TSH stimulation and an abnormal perchlorate discharge study.

Elevated serum TSH in hyperthyroid patients previously treated by ^{131}I or surgery is a sensitive indicator of the potential subsequent development of hypothyroidism. Routine assays for TSH in such patients may obviate the disabling consequences of hypothyroidism—especially in children, in the reproductive woman, and in the aged.

Syndromes Associated With Primary Thyroid Myxedema

Three syndromes associated with primary hypothyroidism and possible hypothalamic pituitary derangement other than for TSH merit discussion.

Amenorrhea-Galactorrhea. This syndrome has been noted in postpartum women who previously menstruated normally. Lactation in the absence of breast-feeding and amenorrhea were their main complaints, averaging 28 months in duration.

Pituitary tumor has been excluded on the basis of normal visual field examination and negative roentgenograms of the sella turcica. Usually a well-documented history and laboratory evidence of primary thyroid failure have been available, including an elevated level of TSH by radioimmunoassay. Low urinary gonadotropin excretory patterns as well as hypoestrogenic vaginal smears have been reported. All patients have responded rapidly and completely to thyroid therapy. The mechanism for the galactorrhea associated with this syndrome is unknown; its elucidation must await sensitive assays for serum prolactin. However, TRF secretion, which is presumably increased in myxedema, has been reported to release prolactin from pituitary cells in culture.

Isosexual Sexual Precocity. Although untreated hypothyroidism usually delays sexual maturation, precocious puberty has been reported in a small number of hypothyroid children (18 cases up until 1967, 15 girls and 3 boys). This disorder has been associated with Down's syndrome in 4 patients. Age at onset of the hypothyroidism varies from 1 to 10 years, and the precocious puberty appears to follow this by one or more years. Clinical features in both sexes consist of advanced genital development. In girls there is precocious breast development, enlarged labia minora and increase in vaginal estrogenization and menstruation; galactorrhea may be present. Pubic and axillary hair is usually scant or absent in spite of other evidence of advanced sex organ maturation. An enlarged penis is uniformly present in boys. The sella turcica is frequently enlarged. Urinary gonadotropins (by bioassay) are not elevated as one might expect. All the symptoms and signs of sexual precocity regress with thyroid therapy.

Van Wyck and Grumbach have called this fascinating entity a "hormonal overlap" syndrome. They have suggested that thyroid deficiency leads to a hypersecretion of multiple heterologous tropic hormones, such as follicle-stimulating hormone (FSH) and LH, in addition to TSH; however, gonadotropin levels are not increased (at least by current methods of assay). No evidence exists for increased ACTH release, although hyperpigmentation may be a part of the disorder. Another possible explanation for the syndrome arises from observations in hypothyroid animals, namely, that the half-life of estrogen is prolonged and that the gonads show an increased sensitivity to gonadotropins. Hypothyroidism in normal menstruating women may lead to amenorrhea, oligomenorrhea or metropathia. Those with amenorrhea have decreased urinary pregnanediol excretion and a nonsecretory endometrium (on biopsy) during the luteal phase, suggesting defective luteal function.

Inappropriate Secretion of the Antidiuretic Hormone. Occasionally, in a patient with myxedema, the findings of hyponatremia, natriuresis, and the production of urine hypertonic to serum have led to the diagnosis of inappropriate secretion of the antidiuretic hormone (ADH), vasopressin. These findings have occurred in the absence of dehydration, hypotension, azotemia and renal or adrenal disease. Levels of antidiuretic hormone remain to be determined in such patients.

End-Organ Refractoriness to Thyroid Hormones

One family with three siblings has been reported with deaf-mutism, stippled epiphyses, goiter, elevated levels of circulating total and free thyroxine, and elevated total triiodothyronine; yet, clinically they were euthyroid. Absence of hyperthyroidism, delayed maturation of bone, and failure to respond to the administration of up to 1,000 μg of T_4 or 375 μg of T_3 daily suggest selective tissue resistance to thyroid hormone.

ALTERATION IN PLASMA-BINDING PROTEINS

Thyroxine-Binding Globulin (TBG)

TBG is rich in carbohydrate and has a serum concentration of 2 mg. per 100 ml. and a binding capacity for T_4 of 20 μg percent. It binds tetra- and triiodinated thyronines, but does not bind deaminated thyronines such as tetraiodothyroacetic acid. The binding affinity of TBG for T_4 is extremely powerful. Binding of T_3 by TBG is only one third as strong as binding of T_4. Its binding capacity is increased by estrogen administration, during pregnancy, in myxedema and in acute intermittent porphyria. Binding capacity is decreased by androgens, by high doses of corticosteroids, in debilitated patients, in thyrotoxicosis, and by drugs such as diphenylhydantoin which compete with thyroxine for binding.

Increased Thyroxine-Binding Globulin. An elevated serum thyroxine level secondary to an increased TBG capacity for T_4 can lead to an erroneous diagnosis of hyperthyroidism. The abnormality appears to be hereditary, being transmitted as a simple mendelian autosomal dominant trait. Clinical findings include an elevated serum PBI or thyroxine without symptoms of hyperthyroidism, a normal BMR and ^{131}I uptake and a low T_3 resin uptake. The thyroxine-binding capacity for thyroxine is increased to levels of 50 μg per 100 ml, whereas the T_4-binding capacity of prealbumin is decreased. This results in an elevated level of serum T_4 and a decreased proportion of free T_4, but a normal concentration of free T_4.

Decreased Thyroxine-Binding Globulin. A decreased serum thyroxine level secondary to decreased TBG capacity for T_4 may lead to an erroneous diagnosis of hypothyroidism. This abnormality also appears to be hereditary and is transmitted as a mendelian X-chromosome-linked dominant trait. Clinically, one sees a euthyroid patient with a normal ^{131}I uptake, a normal BMR, and a low serum T_4 owing to a low thyroxine-binding capacity for T_4 and a high T_3 resin uptake. Such patients have pronounced abnormalities in the peripheral metabolism of thyroxine. The volume of distribution and the fractional rate of ^{131}I-labeled thyroxine are increased; hormonal clearance rates are markedly augmented. However, these data have been interpreted to mean that the daily disposal of hormone by both excretory and degradative routes is normal. Prolonged administration of synthetic estrogens or adrenal corticosteroids has failed to alter significantly the binding capacity of TBG.

Thyroxine-Binding Prealbumin (TBPA)

Thyroxine-binding prealbumin is poor in carbohydrate, rich in tryptophan and has a serum concentration of 30 mg. percent and a binding capacity of 300 μg percent. It has one binding site per molecule and binds tetraiodinated thyronines (such as T_4) and deaminated thyronines but not triiodinated thyronines (such as T_3). The binding capacity of TBPA is increased in patients given androgens and corticosteroids; it is decreased immediately postoperatively and in acute and chronic illness, thyrotoxicosis, and patients treated with estrogens or drugs that compete for binding, such as salicylates, penicillin or dinitrophenol.

Analbuminemia

This rare disorder, inherited as an autosomal recessive trait, is characterized by a small concentration of albumin in the serum (100 μg per 100 ml.). Such patients have marked elevations in the concentrations of other serum proteins. The binding capacity of albumin for T_4 is about 2×10^5. Studies reveal a normal PBI and an increased capacity of TBG and TBPA to bind. Unexplained is an increase in the dialyzable thyroid hormone fraction, in view of the increased TBG capacity. An increase in the concentration of serum free fatty acids in such patients might explain this discrepancy.

METABOLIC DERANGEMENTS ASSOCIATED WITH THYROID CARCINOMA

Follicular Adenocarcinoma (Metastatic)

Hyperthyroidism has been noted in a small number of patients with metastatic follicular adenocarcinoma of the thyroid. A review of 13 such patients (in 1964) revealed a sex ratio of 10 women to 3 men, consistent with that of most thyroid diseases. The ages ranged from 38 to 66. Nine patients had undergone thyroidectomy 18 years (on the average) before the hyperthyroidism appeared. The clinical features resembled those of ordinary hyperthyroidism, with two exceptions: no patient had a goiter at the time that hyperthyroidism was detected (but one had a palpable thyroid with an elevated uptake of ^{131}I), and none had malignant exophthalmos.

Medullary Carcinoma

Medullary carcinoma of the thyroid arises from the parafollicular cells, which are said to be derived from the ultimobranchial bodies. These cells do not concentrate iodide or synthesize thyroxine, but they do synthesize thyrocalcitonin, which lowers serum calcium. Of 33 patients, 26 were women and 7 men; their ages at the time of surgery ranged from 17 to 69 years (75% were beyond the fourth decade). The duration of the disease before operation ranged from 1 to 12 years, except in one case with a 32-year preoperative history. Although thyroid carcinoma is not known to be hereditable, the medullary variety does occur in some families with a dominant autosomal inheritance. This tumor has attracted great interest recently because of the diverse endocrinological syndromes with which it may be associated.

Calcitonin-Secreting Medullary Thyroid Tumor. The main clinical features of this condition include: goiter, acne, Marfan features, neuromas of tongue and eyelid, prominent lips, pigmentation on hands, feet, and around the mouth, proximal myopathy, loose joints, and episodic flushing. In such patients, the plasma concentration of calcitonin is elevated from 100 to 100,000 pg. per ml. (normal, 0 to 100 pg. per ml.). In tumor tissue, sometimes 5,000 times as much calcitonin may be found as in normal thyroid tissue. Despite the increased tissue and serum thyrocalcitonin, most patients are normocalcemic and normophosphatemic, although patients with hypocalcemia and hypophosphatemia have been reported. Generally, the tumor seems to be under the same control as normal thyroid tissue: calcium infusion leads to a dose-dependent increase in the serum level of calcitonin, and EDTA infusion leads to increased serum calcitonin. Response to glucagon infusion varies; but in certain patients glucagon infusion produces increased levels of serum calcitonin, whereas in normal patients glucagon infusion results in decreased serum calcitonin levels. Although most patients with this syndrome are normocalcemic, about 20 percent may have hypercalcemia associated with increased levels of parathormone, owing to concomitant parathyroid hyperplasia or adenoma.

Pheochromocytoma. Pheochromocytoma has been found in association with thyroid carcinoma; in 11 of 17 such patients the thyroid cancer was of the medullary type. The significant features of this relationship include the frequent incidence of bilateral adrenal tumors and the frequent family history of pheochromocytoma. A few patients have parathyroid adenoma, suggesting the multiple endocrine adenoma syndrome (somewhat different from that involving the pituitary, parathyroid and pancreas).

Cushing's Syndrome. Nine of 11 patients with thyroid carcinoma and Cushing's syndrome had medullary carcinoma and two had a papillary thyroid carcinoma. In the two with papillary carcinoma, the Cushing's syndrome preceded the diagnosis of cancer. In the other patients, the tumor preceded the onset of Cushing's syndrome. These thyroid tumors appear to have ACTH-like activity.

Prostaglandin Secretion. Appreciable amounts of prostaglandinlike activity were present in tumor tissue from 4 of 7 patients with medullary carcinoma; raised blood levels of prostaglandin were detected in two. It has been postulated that the diarrhea seen in patients with medullary carcinoma might be related to an excess of prostaglandin.

By way of summation, the reader is referred to Table 15-1, which is designed to indicate the general level of function in a variety of diseases involving the thyroid gland.

ANNOTATED REFERENCES

Benhamon-Glynn, N., El Kabir, D. J., Roitt, I. M., and Doniach, D.: Studies on the antigen reacting with the thyroid stimulating immunoglobulin (LATS) in thyrotoxicosis. Immunology, *16*:187, 1969. (Immunological studies of LATS)

Burgus, R., and Guillemin, R.: Hypothalamic releasing factors. Ann. Rev. Biochem., *39*: 499, 1970. (Current biochemical review of the hypothalamic releasing factor)

Edelhoch, H.: The structure of thyroglobulin and its role in iodination. Recent Prog. Hormone Research, *21*:1, 1965. (Review of the structure and function of thyroglobulin)

Hoch, F. L.: Thyrotoxicosis as a disease of mitochondria. New Eng. J. Med., *266*:446, 498, 1962. (Good review of action of thyroxine)

Ingbar, S. H., and Freinkel, N.: Regulation of the peripheral metabolism of the thyroid hormones. Recent Prog. Hormone Research, *16*:353, 1960. (Review of the regulation of the peripheral metabolism of thyroxine)

McKenzie, J. M.: The long-acting thyroid stimulator: its role in Graves' disease. Recent. Prog. Hormone Research, *23*:1, 1967. (Review of the development and role of LATS)

————: Humoral factors in the pathogenesis of Graves' disease. Physiol. Rev., *48*:252, 1968. (Review of the etiological factors in Graves' disease)

Maloof, F., and Soodak, M.: Intermediary metabolism of thyroid tissue and the action of drugs. Pharmacol. Rev., *15*:43, 1963. (Good review of the intermediary metabolism of the thyroid as of 1963)

Martini, L., and Ganong, W. P.: Neuroendocrinology. Vols. 1 and 2. New York, Academic Press, 1967. (Excellent text of neuroendocrinology)

Means, J. M., DeGroot, L. J., and Stanbury, J. B.: The Thyroid and its Diseases. New York, McGraw-Hill, 1963. (Textbook of clinical thyroidology)

Odell, W. D., Wilber, J. F., and Utiger, R. D.: Studies of thyrotropin physiology by means of radioimmunoassay. Recent Progr. Hormone Research, *23*:47, 1967. (Excellent review of the radioimmunoassay of TSH)

Oppenheimer, J. H.: Role of plasma protein in the binding and metabolism of the thyroid hormones. New Eng. J. Med., *278*:1153, 1962. (Review of the plasma-binding proteins of the thyroid hormone)

Pitt-Rivers, R., and Trotter, W. R.: The Thyroid Gland. London, Butterworth, 1964. (Textbook encompassing clinical, physiological, and biochemical thyroidology)

Rawson, R. W.: Modern concepts of thyroid physiology. N. Y. Acad. Sci., *86*:311, 1960. (Review of current concepts of thyroid physiology as of 1960)

Robbins, J., and Rall, J. E.: Thyroid hormone transport in circulation. Recent Progr. Hormone Research, *13*:161, 1957. (Review of the concepts in thyroid hormone transport)

Schally, A. V., et al.: Hypothalamic neurohormones regulating anterior pituitary function. Recent Progr. Hormone Research, *24*:497, 1968. (Review of the hypothalamic releasing factor)

Silver, S.: Radioactive Nuclides in Medicine and Biology. Ed. 3. p. 1. Philadelphia, Lea and Febiger, 1968. (Review of the tests of thyroid function)

Stanbury, J. B., Wyngaarden, J. B., and Fredrickson, D. S.: The Metabolic Basis of Inherited Disease. Ed. 2. p. 215. New York, McGraw-Hill, 1966. (Review of the defects in synthesis of thyroxine)

Taurog, A.: Thyroid peroxidase and thyroxine biosynthesis. Recent Progr. Hormone Research, 26:189, 1970. (Current review of the thyroid peroxidase)

Woeber, K. A., Sobel, R. J., Ingbar, S. H., and Sterling, K.: The peripheral metabolism of triiodothyronine in normal subjects and in patients with hyperthyroidism. J. Clin. Invest., 49:643, 1970. (Study of the peripheral metabolism of triiodothyronine)

Wolff, J.: Transport of iodide and other anions in the thyroid gland. Physiol. Rev., 44:45, 1964. (Good review of the concentrating mechanism of iodide)

16

Endocrine Control Mechanisms of Electrolyte and Water Metabolism

Frederic C. Bartter, M.D.

BODY WATER

Total body water depends ultimately upon the sum of the intake water and the water reabsorbed from the renal tubules minus the sum of the filtered water and the insensible water loss (sweat and lungs). Control of total body water may be attributed to interplay of these four variables.

Water Intake

Thirst control depends upon osmolality of blood in the hypothalamic region. Under conditions in which the subject has access to water and is conscious, water intake is controlled normally by thirst. Evidence from animal experiments suggests that thirst, in turn, depends upon the osmolality of the blood bathing the hypothalamico-hypophyseal system. For example, the experimental infusion of hypertonic saline into the arterial supply of this region produces excessive drinking in goats. Because it is the tonicity of this fluid which determines its thirst-promoting property, it is clear that this stimulus reflects in fact the need for "free water" (see below). Normally, the need for water and the need for free water occur at the same time, but, in view of this mechanism, a subject might still be thirsty if, for any reason, he was loaded with solute, so that fluid volume was normal but plasma osmolality was elevated because of *relative* insufficiency of water.

Renal Water Reabsorption

Renal tubular reabsorption of water, independent of solute (sodium), is almost entirely dependent upon the concentration of circulating antidiuretic hormone (ADH, vasopressin). This is apparent from the comparison of the quality of urine excreted with normal secretion of ADH and that excreted without it. In the former situation, osmolality may reach 1,000 mOsm per kg., for example. This means that more than two thirds of the water filtered at the glomerulus has been reabsorbed during the secretion of urine. In the latter instance, without ADH, urinary osmolality may be, for example, 50 mOsm per kg., a finding that suggests that almost all the urine freed of solute in its passage through the tubules is being excreted. ADH produces its effects by increasing the permeability of distal tubules and collecting ducts so as to allow more back-diffusion of water freed of its solute in the ascending limb of the loop of Henle. The osmotic force outside the distal convoluted tubules is that of normal plasma; the osmotic force outside the collecting ducts is that of nondiffusible solutes (sodium and chloride) in the renal medulla, normally present in concentrations three to four times those of the interstitial fluid surrounding the distal convoluted tubule. At both sites water is not reabsorbed in the absence of ADH but is reabsorbed in its presence.

Antidiuretic Hormone (ADH). Control of ADH secretion depends most importantly on the osmolality of the plasma bathing the supraoptico-hypophyseal system. Increased plasma osmolality (total amount of solute per molecule of water) in this area leads to an increase in the secretion of ADH, just as it does to an increase of thirst. It is not known whether the thirst stimulus is the same as the stimulus for the secretion of ADH; probably, it arises with separate neurogenic impulses.

Strong afferent stimuli can override the stimulus of osmolality to produce ADH. Pain, for example, can stimulate the secretion of ADH in spite of hypotonicity of plasma. This relationship is important in the patient who has undergone a major operation and may continue to secrete ADH after his serum has become hypotonic. If such a patient receives dextrose and water, for example, the water is retained, and plasma hypotonicity may result.

An important mechanism for the control of ADH secretion depends upon some function of intravascular volume, which may, of course, reflect total extracellular fluid volume. Loss of intravascular volume as the result of hemorrhage, for example, leads to increased production of ADH, even when tonicity of the plasma is normal.

Pain and hypovolemia are, therefore, two stimuli that are independent of the tonicity stimulus and of the greatest physiological importance. Accordingly, they have been included among the causes of "inappropriate" secretion of ADH. Thus, hypersecretion or sustained secretion of ADH resulting from these stimuli, and independent of plasma osmolality, is called inappropriate secretion of ADH because it can continue in the presence of the hypotonicity which, under normal circumstances, would prevent it.

The secretion of ADH is inhibited by alcohol; although this relationship has not found important clinical application, it may be useful in differential diagnosis—for example, between the syndromes in which ADH is produced from the supraoptico-hypophyseal system, on the one hand, and those in which it is produced, under no known control, from a tumor such as a bronchogenic carcinoma.

Diabetes Insipidus. In pituitary diabetes insipidus, the supraoptico-hypophyseal system is unable to secrete or release ADH (vasopressin) into the general circulation. As a result, free water which is generated in the ascending limb of the loop of Henle is not reabsorbed in the distal convoluted tubule unless the patient receives vasopressin exogenously. The collecting ducts also remain less permeable to water than they are under the influence of vasopressin. The consequence of the former defect is the continued delivery through the collecting ducts into the urine of large volumes of urine of low osmolality. The disordered physiology of the collecting ducts does not contribute to the syndrome because reabsorption of water at this site normally represents only that small volume which raises distal tubular fluid from isotonicity with plasma to hypertonicity. Even such transport of water as does occur in the collecting ducts may be sufficient to lower the osmolality of medullary fluid by "washout." Such washout, which is thus an indirect effect of ADH deficiency, limits the tonicity of urine which can be produced immediately upon administration of vasopressin. Thus, the full hypertonicity of urine ultimately attainable by the use of vasopressin cannot be established until enough time has elapsed to allow for reaccumulation of medullary solute sufficient to permit maximal concentration of the urine resulting from the influence of vasopressin on the collecting ducts.

The untreated patient with diabetes insipidus, unable to reabsorb water normally, excretes a large amount of dilute urine. The immediate result can be seen in the rising serum tonicity and sodium concentration. This, in turn, may have its normal effect in inducing thirst; and, if sufficient water is available, the patient may drink enough to compensate fully for the water loss, restoring normal plasma tonicity and establishing a new steady state requiring the constant ingestion of large volumes of water for its maintenance. The treatment of diabetes insipidus in this manner with water alone does not leave the kidney in a physiologically normal condition, as medullary hypertonicity is frequently diminished or abolished. In the patient with diabetes insipidus who is treated with water alone, response to other stimuli (besides plasma osmolality) that can induce secretion of ADH is not predictable. A powerful pharmacological stimulus to ADH release, nicotine, may release some ADH and allow transitory urinary concentration in a patient with this syndrome. If a patient with diabetes insipidus is subjected to quiet standing (a procedure which must be carried out with care to avoid syncope), there is generally a rise of urinary osmolality, which may even exceed the osmolality of plasma. The physiological implications of such change cannot be ascertained without direct measurement of plasma ADH concentration: in some patients, plasma ADH concentration does indeed rise under the stimulus of hypovolemia in spite of the nonresponsiveness to hypertonicity. In other patients, however, the marked fall in glomerular filtration rate, which results from the hypovolemia of quiet standing, may allow moderate concentration of the urine in the absence of ADH. In this instance the sequence of events presumably is: (1) fall in glomerular filtration rate; (2) decrease in percentage of proximal fluid delivered to distal sites; (3) limitation in formation of free water; and (4) increase of tonicity of urine by withdrawal of water into medullary hypertonic fluid in spite of absence of ADH.

Syndrome of Inappropriate ADH Secretion. This is a disorder of sustained ADH secretion not responding to osmolality of plasma perfusing the supraoptico-hypophyseal system. When long-acting vasopressin is administered to a normal subject, the sequelae depend entirely upon the water intake. Thus, if the water intake is continued low, no sequela is apparent. If, on the other hand, there is a relatively high intake of solute-free water, a striking chain of sequelae results.* As the water is retained body weight at first rises. Because water is being retained in relative excess of solute or sodium, serum hypotonicity and hyponatremia develop. Hypertonicity of the urine, however, persists in spite of the hyponatremia because water reabsorption is maintained by the exogenous vasopressin. With expansion of extracellular fluid volume, renin secretion and aldosterone secretion are suppressed; and proximal tubular reabsorption of salt and water is also suppressed. Since expansion of extracellular fluid volume also increases the glomerular filtration rate and, thus, the filtered sodium load, these changes result in a loss of sodium in the urine, an event that further aggravates the hyponatremia.

Certain tumors, such as "oat-cell" bronchogenic carcinomas, can reproduce all the features of this complex that can be produced experimentally with vasopressin and water. Furthermore, the tumor contains a substance indistinguishable by physiological and radioimmunological tests from arginine vasopressin. Because the serum hypotonicity and hyponatremia are unable to diminish vasopressin release into plasma, the syndrome has been termed that of "inappropriate" secretion of ADH. It is

---

* A strict analysis of the requirements can be made if the total dietary solute load destined for urinary excretion is divided by the total water intake minus the insensible water loss. If this figure (mOsm per kg.) is lower than the osmolality of serum, the syndrome will be produced.

easy to recognize this syndrome when all of its features are present but it is important to note that all these features may be modified in a patient in whom the basic physiological derangement is present.

As noted above, if the water intake is limited, either voluntarily or therapeutically, hyponatremia will gradually disappear as insensible water loss lowers body water; a liter a day may be lost in this way. If, in contrast, sodium intake is kept very low, urinary sodium may fall to equal intake sodium even when hyperexpansion of body fluid volume is maintained by an excessive water intake. Indeed, for any given sodium intake, a steady state may be reached at which urinary sodium equals the dietary intake, in spite of continued overexpansion of body fluids, hyponatremia, and plasma hypotonicity.

In view of the original definition of the syndrome as one in which secretion of ADH (or an analogous substance from a tumor) continues in spite of hypotonicity of plasma, it is reasonable to include in it other syndromes in which the secretion of ADH may be responding to another known stimulus. Thus, the hyponatremia often found in the postoperative state may result from a combination of sustained secretion of ADH because of operative trauma and excessive intake (or infusion) of water. Similarly, the hyponatremia often found after acute hemorrhage when water is drunk or infused, may result from a "volume" stimulus to secretion of ADH. For these two syndromes we have retained the term "inappropriate," defined strictly as inappropriate vis-a-vis an abnormally low tonicity of plasma.

The hypotonicity of plasma found in some patients with Addison's disease (see below) may depend in part upon sustained secretion of ADH because of "volume" stimuli.

Volume Control

An abnormality of water metabolism exists in Addison's disease which provides an example of predominance of volume over osmolality control. Abnormalities in patients with primary adrenal insufficiency result from an insufficiency of cortisol, on the one hand, and aldosterone insufficiency on the other. Whereas the deficiency in cortisol leads to decreased glomerular filtration rate, which limits the amount of salt and water delivered to the proximal tubule and thus ultimately the formation of free water, the deficiency of aldosterone production results in a limitation of sodium reabsorption at all renal tubular sites. The important clinical consequences depend upon the limitation of sodium reabsorption at two sites. In the ascending limb, the formation of free water is limited, and the return of sodium to the medulla is also limited by the same process. This does not allow medullary solute concentration to reach its normal high values, thus ultimately limiting the concentration of the urine. In addition, at the distal convoluted tubule a deficiency of aldosterone limits the exchange of sodium-for-potassium or hydrogen. As a consequence of these abnormalities, the patient with Addison's disease is not able to excrete normally a load of free water; consequently he develops hyponatremia and hypotonicity of the plasma. Such hypotonicity is not able to decrease normally the production of vasopressin; consequently, the secretion of vasopressin is sustained. Thus, in Addison's disease, the secretion of antidiuretic hormone, inappropriate as regards serum tonicity, responds to hypovolemia and hypotension. It has not been established whether another type of control of vasopressin secretion depends directly upon the plasma cortisol. Some results, indeed, suggest that the sustained secretion of antidiuretic hormone may be lowered rapidly by the administration of cortisol. As cortisol produces a prompt water diuresis by increasing the glomerular filtration rate and perhaps by decreasing the reabsorption of water in the distal tubules, the

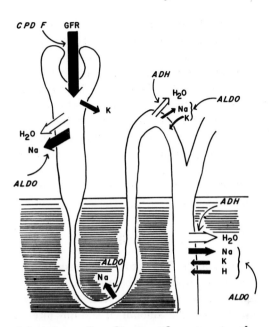

FIG. 16-1. A representation of the sites of action of the hormones on water and electrolyte control in the nephron. Cortisol (CPD F) increases the glomerular filtration rate (GFR). Aldosterone (ALDO) causes sodium (Na) reabsorption in the proximal tubule, the loop of Henle (against a concentration gradient), and in exchange for the secretion of potassium (K) and hydrogen (H) ions in the distal tubule and in the collecting duct. In the proximal tubule water is reabsorbed isosmotically with sodium; in the distal tubule and collecting duct permeability to water is controlled by antidiuretic hormone (ADH).

decrease of secretion of antidiuretic hormone may be a secondary effect. (These interrelationships between adrenal cortical hormones and ADH and renal function are presented diagrammatically in Fig. 16-1.)

"Free Water" Excretion

A water load is excreted in the urine by decreasing ADH secretion in response to a reduction in plasma osmolality. This response is often used to assess the normal responsiveness of the supraoptico-hypophyseal system, and it is assumed that patients who show a normal response of free water excretion to a water load do not suffer from the syndrome of inappropriate ADH secretion. Conversely, many investigators have assumed that the failure to excrete normally a water load constitutes *prima facie* evidence for the failure of ADH secretion to be diminished normally. Whereas a normal response does indeed make it most unlikely that the syndrome is present, it is important to note that a "false positive" result may occur in a number of conditions in which there is not a sustained secretion of ADH. These are conditions in which formation of free water in the ascending limb is limited by extensive re-

absorption of sodium and water in the proximal tubule. Under these circumstances, a failure of water diuresis after a water load may indicate only that the water load was not able to decrease the excessive proximal tubular reabsorption. This situation is seen in patients with cirrhosis and ascites, in whom proximal reabsorption may account for a very large proportion of the filtered sodium, and urinary sodium may approach zero. The "false positive" results in this condition can often be distinguished from the true positive results in a patient with the syndrome of inappropriate secretion of ADH by measurement of the osmolality of the urine during the water clearance. There is often a small amount of free water excreted during the test, as shown by the decrease of osmolality of urine to values below that of serum. In a truly positive test, in which there is sustained inappropriate secretion of ADH, urinary osmolality frequently stays above that of plasma throughout the usual 4 to 5-hour collection period.

The same condition of "false positive" water test because of extensive proximal reabsorption of sodium and water may occur in patients with cardiac failure and edema whose urinary sodium is very low. The excessive proximal reabsorption may limit the volume of water diuresis in such a patient and, thus, a "positive" test need not establish the diagnosis of inappropriate secretion of ADH. In some patients with severe cardiac failure, however, there is, in fact, also a sustained secretion of ADH in spite of hyponatremia; the secretion of ADH is probably responding to a decrease in *effective* blood volume secondary to the cardiac failure.

BODY SODIUM

Control of body sodium must depend ultimately upon the balance between sodium intake, and sodium reabsorbed by the tubules less the amount of sodium filtered; fecal sodium normally remains low. The control of sodium intake may reflect sodium appetite; hormonal control, if any, is not established.

Sodium Filtration

The filtered load of sodium is dependent upon the glomerular filtration rate and is partly dependent upon carbohydrate-active steroids (Fig. 16-1). Because some 14 mEq. of sodium are filtered each minute in the glomeruli, tubular reabsorption of sodium is quantitatively the most important variable. The filtered load of sodium is dependent upon glomerular filtration rate (GFR) which, in turn, is partly dependent upon the action of carbohydrate-active steroids. When the normal concentrations of carbohydrate-active steroids are diminished, as in adrenocortical or anterior pituitary insufficiency, a decrease in glomerular filtration rate regularly ensues. This change is dependent upon the carbohydrate activity of the steroids, as it may be rapidly restored when this steroid action is reinstituted by replacement therapy, even when this is done with steroids having no sodium-retaining activity. The net

effect of such deficiency upon sodium balance is to allow a state of balance in spite of the diminished ability of the tubules to absorb sodium.

Tubular Reabsorption

Reabsorption of sodium and water by the proximal tubules is probably dependent, in part, upon the action of sodium-retaining steroids such as aldosterone (Fig. 16-1). This has been shown directly by the effect of aldosterone in increasing the rate of decrease in the width of a segment of fluid between oil drops in the proximal tubule, and indirectly by the demonstration that sodium-retaining steroids can decrease the maximal free water clearance (whereas their action on the distal tubule clearly is such as to produce an increase in free water clearance).

In part, the reabsorption of sodium and water in the renal tubule is independent of sodium-retaining steroids. This part has been studied extensively by measurements of sodium reabsorption under the conditions where maximum concentrations of sodium-retaining steroids are sustained exogenously. The control of such reabsorption is commonly attributed to "third factor," which refers to reabsorption which represents a change in that fraction of the filtered load which is reabsorbed and is not dependent upon steroids. Proximal reabsorption is rapidly decreased by expansion of ECF volume. For example, that a portion of the salt and water lost in the urine was liberated from the proximal tubules can be shown both by direct micropuncture evidence and by indirect evidence based on the increased formation of free water as isotonic fluid is delivered to distal sites.

Reabsorption of sodium in the ascending limb of Henle's loop and in the distal convoluted tubule is partly dependent upon the action of sodium-retaining steroids. This action removes sodium from a section of the tubule which is relatively impermeable to water and consequently is responsible for the production of free water. The anion reabsorbed with this sodium is largely chloride.

Reabsorption of sodium in the distal convoluted tubule is also under the influence of sodium-retaining steroids where much of the sodium is reabsorbed not with chloride ion but rather in exchange for potassium or hydrogen ion. In this action, which further conserves filtered sodium in the body, there is also an important contribution to the ability of the tubules to secrete potassium and hydrogen into the urine.

BODY POTASSIUM

Control of body potassium depends upon the potassium intake minus the sum of the urinary and fecal potassium. Extracellular fluid potassium control depends upon the sum of the potassium intake and potassium released from cells minus the sum of the urinary potassium, the fecal potassium, and the potassium entering cells. Potassium intake is under no known hormonal control; however, its secretion into the gastrointestinal tract is partly under control of sodium-retaining steroids which promote the exchange of potassium for sodium.

Renal Function

The potassium that is filtered at the glomerulus is largely reabsorbed in the proximal tubules; accordingly, most of the potassium appearing in the urine is secreted across the distal tubular lumen. This process requires luminal sodium, so that potassium is exchanged, directly or indirectly, for sodium across the tubular wall. The exchange would be considered indirect if the sequence of events is, in fact, reabsorption of sodium, establishment of a potential gradient with the lumen negative to the peritubular fluid and entry of potassium into the lumen in response to the electro-

chemical gradient thus established. This distal tubular exchange of sodium for potassium is strongly under the control of sodium-retaining steroids, in the absence of which a defective exchange can lead to potassium retention and hyperkalemia (Fig. 16-1).

Cellular Potassium Loss

Loss of intracellular potassium depends upon the electrical gradient between intracellular and extracellular fluid and an active exchange with sodium. (An electrical potential 20 mv more negative than the actual one on the inside of the cell would be necessary to prevent loss of potassium from cells by electrical forces alone.) The return of potassium to the cell is dependent upon a "sodium pump" which excludes sodium ions from the intracellular space in exchange for extracellular potassium. Whereas a similar pump in the renal tubule and the unicellular membrane (such as toad bladder) is clearly dependent in part upon the action of sodium-retaining steroids, it has not been established whether body cells in general require such steroids for normal activation of the pump. An excess of such steroids in the whole organism produces hypokalemia, as discussed above. The consequent loss of potassium from cells, with partial replacement with sodium and hydrogen ions, may be secondary to the electrochemical gradient for potassium thus established.

Carbohydrate-active steroids themselves can induce a rapid loss of potassium from cells, resulting in a transient hyperkalemia before the potassium is lost in the urine. This very early event following the administration of steroids is independent of the sodium-retaining action of steroids.

The entry of potassium into cells as discussed above is dependent mostly on the action of an energy-requiring cation-exchanging pump, whose steroid dependence for body cells is not known.

BODY HYDROGEN ION

Hydrogen Ion Balance

Body hydrogen ion balance depends upon the sum of the "acidity" of intake, the metabolic production of hydrogen ion from endogenous sources, and endogenous production from the loss of hydroxyl ions, less the sum of the urinary and pulmonary losses. Extracellular hydrogen ion concentration depends upon these factors in addition to the loss of hydrogen ion from cells less the intracellular entry of hydrogen ion.

The "acidity" of the intake ash is a term derived from the sum of the fixed dietary anions less the fixed cations of the diet (all expressed in milliequivalents). The assumption that this measures the hydrogen ion intake depends upon the assumption that all dietary cation is excreted as the hydroxide or as neutral salt with a dietary anion, and all dietary anions as the acid or neutral salt with a dietary cation. The acidity of the ash of an *ad libitum* diet is not known to depend upon hormonal control.

Metabolic Release

Metabolic release of hydrogen ion from endogenous sources includes a similar process when body constituents are involved; and, in addition, hydrogen ions produced from partial catabolism of endogenous constituents, such as the hydrogen ions released when methionine is converted to sulfuric acid or phospholipids to phosphoric acid or neutral carbohydrates to ketoacids. The conversion of substances to carbon dioxide represents a special case. Since the carbon dioxide is hydrated to form carbonic acid, the dissociation into hydrogen and bicarbonate furnishes a supply of hydrogen ions from body water, since the corresponding hydroxyl ion is bound to carbon dioxide.

Any process that promotes catabolism of body tissue may provide a source of hydrogen ions; gluconeogenesis from protein is partially controlled by carbohydrate-active steroids; lipolysis is partially controlled by catecholamines, insulin, and prostaglandins. Insulin may also serve to remove hydrogen ions indirectly by promoting the complete oxidation of ketoacids to carbon dioxide.

Loss of Hydroxyl Ions

Body hydrogen ion concentrations are increased by the loss of hydroxyl ions dissociated from body water to the external environment.

Thus, loss of hydroxyl ions from the gut occurs whenever alkaline secretions are excreted in the feces, as with any form of diarrhea. Feces are normally alkaline and, thus, even normally represent a small continued loss of hydroxyl ions representing a net gain of hydrogen ions to the body. Whereas fecal loss of hydroxyl ion is not known to be under direct hormonal control, it has been suggested (but not proved) that sodium-retaining steroids may promote the reabsorption of sodium in exchange for hydrogen. However, any hormonal disorder associated with diarrhea has the net effect of increasing total body hydrogen ions.

Loss of hydroxyl ions from the renal tubule appears as an increase in urinary bicarbonate and, thus, will appear whenever there is a defect in the rate of tubular hydrogen ion secretion (which allows reabsorption of filtered bicarbonate) or in the hydrogen ion gradient ("distal" renal tubular acidosis) which can be achieved between lumen and plasma. This function of tubular epithelium, which is entirely dependent upon the exchange secretion of sodium ion from lumen to plasma, is strongly dependent upon the action of sodium-retaining steroids, probably in both the distal and proximal convoluted tubules, as well as in the collecting ducts. Parathyroid hormone also induces a prompt alkalinization of urine with loss of bicarbonate. Accordingly, the ultimate action of this hormone may be to increase total body hydrogen ion concentration.

Tubular secretion of hydrogen ion is partly controlled by sodium-retaining steroids. On a controlled regimen, sodium-retaining steroids can be shown to promote hydrogen ion loss into the tubule; this effect is marked under conditions where sodium-for-potassium exchange is limited (e.g., potassium depletion). The urine becomes acid (or urinary titratable acidity and ammonium increase). As noted below, this explains the paradoxical aciduria (with alkalosis) found with overdosage of desoxycorticosterone or early in primary aldosteronism.

Pulmonary loss of hydrogen depends on control of respiration. The hydration of carbon dioxide and the dissociation of carbonic acid promote effective dissociation of water into hydrogen ions, which increase the body hydrogen ion concentration, and hydroxyl ions, which are "sequestered with" the carbon dioxide as bicarbonate. The process of elimination of carbon dioxide by the lungs, requiring as it does a dehydration of carbonic acid, eliminates free hydrogen ions. Whereas the control of respiration is altered in a number of endocrine diseases, there is probably no direct hormonal control of respiration.

$$CO_2 + H_2O \underset{lung}{\overset{tissue}{\rightleftarrows}} H_2CO_3 \rightleftarrows H^+ + CO_2 \cdot OH^-$$

Addison's Disease: Lack of Sodium-Retaining Hormone

Adrenocortical insufficiency may result from destruction of the adrenal cortex by tuberculosis, a virus, antibodies, or other agents. The disorder which results represents a deficiency not only of cortisol production but also of aldosterone. The renal effect of cortisol deficiency has been discussed: to the extent that it lowers the glomerular filtration rate, it lowers the filtered sodium load and thus may serve to limit the effects on the body economy of the defects in the tubular reabsorption of sodium. It is also possible that the defect in cortisol production results in an inability of the tubules to excrete water above and beyond the defect that results from sustained inappropriate secretion of ADH (see above, and Fig. 16-1).

Effect on Proximal Tubule. To the extent that proximal tubular sodium and chloride reabsorption depend upon the action of sodium-retaining steroids, such reabsorption will be defective in Addison's disease, not only because of the lack of aldosterone but also because of the lack of the moderate sodium-retaining activity of cortisol. There results an increase of delivery of sodium from the proximal to the distal convoluted tubule; without replacement, much of this is lost in the urine. The resulting hypovolemia with hypotension may become the gravest physiological disorder in the untreated patient.

Effect on Ascending Limb. To the extent that sodium reabsorption in the ascending limb is dependent upon the action of sodium-retaining steroids, sodium retention is affected at this site as well in the patient with Addison's disease. Such a defect has three results: in the first place, it limits the amount of sodium transported to the medullary areas responsible for the ultimate urinary hypertonicity in dehydration. Thus, the upper limit of urinary osmolality is considerably reduced in the untreated patient with Addison's disease. Secondly, it limits the production of free water which is normally accomplished by the extraction of sodium in the ascending limb which is relatively impermeable to the "freed" water which remains in the tubule. This limits the ability of the kidneys to excrete a free water load; consequently, if more free water (water relatively in excess of solute) is ingested by the patient with Addison's disease, this water is retained and contributes to further serum hypotonicity and hyponatremia. In the third place, the failure of reabsorption of sodium contributes to the ultimate loss of sodium in the urine.

Effect on Distal Tubule. Whereas some sodium is reabsorbed in the distal convoluted tubule together with chloride, other important fractions of the distal tubular sodium are reabsorbed in exchange with potassium or hydrogen. A limitation in these functions in the patient with Addison's disease who lacks aldosterone characteristically results in hyperkalemia. This may be attributable, ultimately, not only to ingested potassium but also to some loss of potassium from cells as sodium enters cells in reponse to a less than optimal control of the sodium-for-potassium exchange mechanism. The distal defect may result also in acidosis if the patient is taking an acid ash diet, giving rise to more hydrogen ions than the distal exchange mechanism can accommodate.

Secondary Increase of Renin Production. The lowering of arterial pressure or of pulse pressure or of intravascular volume in Addison's disease stimulates the juxtaglomerular apparatus of the afferent arterioles to produce renin in excess. The angiotensin thus generated by the excessive renin serves to support the arterial pressure to some extent and to prevent even more serious hypotension. When Addison's disease results from destruction of the adrenal cortex, the angiotensin II resulting from this interaction is, of course, unable to stimulate further production of aldosterone from the adrenal.

In anterior pituitary insufficiency, with limited or abolished production of ACTH, the failure of cortisol production may effectively lower the glomerular filtration rate and may, together with the sustained secretion of ADH, limit the ability to excrete free water. Whereas it might be supposed that a normal release of renin would lead to a normal control of aldosterone secretion from the zona glomerulosa in this form of

Addison's disease, this is not the case, and aldosterone production is limited. Response of aldosterone production to salt deprivation is also limited, so that hypovolemia and hypotension not infrequently appear. In some patients with pituitary Addison's disease, there is excessive loss of sodium in the urine; in spite of this, the increase in aldosterone production which normally follows loss of sodium may be slight or absent. Such sodium loss provides clinical evidence that sodium retention by the renal tubules depends not alone upon aldosterone but also upon the presence of cortisol, which has much weaker molar activity on sodium transport but which, of course, is normally present in much larger quantities than aldosterone. The failure of aldosterone secretion to rise to excessive values in the hypotensive patient with pituitary Addison's disease provides clinical evidence that the control of aldosterone secretion by the zona glomerulosa depends, in part, upon ACTH activity.

Primary Aldosteronism: Autonomous Excess of Sodium-Retaining Steroids

In primary aldosteronism, resulting from an adenoma of the adrenal cortex, overproduction of aldosterone is autonomous vis-a-vis known physiological controls, with the important exception of that control exerted by the serum or cellular potassium. Even with tumor, potassium depletion limits the production of aldosterone. The syndrome of primary aldosteronism also may be produced by hyperplasia of all adrenal cortical tissue. This syndrome is indistinguishable from that resulting from adenoma; whereas its existence suggests that the overproduction may not be autonomous but may depend rather upon an unknown tropic influence, this syndrome as regards electrolyte and water metabolism is identical with that resulting from adenoma.

Effect on Proximal Tubule. In the presence of excessive quantities of aldosterone, it is likely that proximal tubular reabsorption is stimulated to exceed the normal rate.

With the expansion of extracellular fluid volume, which is a normal and immediate effect of a sodium-retaining steroid, however, two processes are set in action which produce a fractional decrease in proximal reabsorption of sodium and a relative increase in amount of sodium delivered to distal sites. First, with expansion of extracellular fluid volume, glomerular filtration rate rises. And secondly, such expansion activates the mechanism generally referred to as "third factor" whereby the proximal tubular reabsorbate is decreased in spite of maximal activity of sodium-retaining steroids. Since proximal reabsorption constitutes some four fifths of the total reabsorption of sodium, this decrease in proximal reabsorption supplies sodium to distal sites, exceeding their capacity to reabsorb it, and prevents any sustained retention of sodium beyond that required initially to produce the expansion of extracellular volume.

Effect on Ascending Limb. Excess of sodium-retaining activity at the ascending limb produces the reverse of those changes noted for the patient with Addison's disease: increased sodium transport, increased generation of free water in the tubule, and a decreased amount of sodium delivered to more distal sites. In fact, the proximal rejection of sodium and water (described above) is such as to exceed the capacity of the distal tubule, so that this action cannot produce excessive retention of sodium over any prolonged period. Accordingly, unless maximal free water formation is measured, there may be no clinical evidence of the excessive action of sodium-retaining steroids upon the ascending limb.

Effect on Distal Tubule. In the patient with primary aldosteronism and sustained excessive secretion of sodium-retaining steroids, the distal tubule is stimulated toward excessive exchange of sodium-for-potassium or -hydrogen. Actual exchange of sodium-for-potassium requires adequate supplies of sodium in the tubular lumen and this is present because of the mechanism outlined

above. Accordingly, hypokalemia regularly ensues because of the augmented sodium-for-potassium exchange. In a similar fashion, the sodium-for-hydrogen exchange is augmented by the action of the steroid and finds adequate luminal sodium to take part in the actual exchange. Accordingly, the immediate effect of an excess of sodium-retaining steroids is to produce an acid urine as the plasma is left relatively more alkaline (paradoxical aciduria). In the patient with primary aldosteronism which has continued for more than a short time, these processes (that of potassium exchange and that of hydrogen exchange) are complicated by the development of hypokalemia as a net loss of hydrogen from the body is induced. In hypokalemia, when exchange of sodium-for-hydrogen is further enhanced by relative deficiency of potassium to compete for the same transport mechanism, the secretion of ammonia (NH_3) may become greatly enhanced. Under these circumstances, whereas the net secretion of hydrogen into the urine is continuously augmented and alkalosis of plasma is maintained, the urine ceases to be acid and becomes alkaline as ammonia enters from the tubules and is hydrated. A second tubular defect of potassium deficiency may be found in the pitressin-resistant polyuria with urinary osmolality consistently below that of plasma. If water loss is thus excessive, relative to the sodium loss, hypernatremia may result.

Secondary Decrease of Renin Production. As the production of renin may be lowered by sodium loading, especially when accompanied by administration of sodium-retaining steroids or by other factors which expand extracellular fluid volume, renin production is effectively reduced in the patient with primary aldosteronism. This lowering is best appreciated by measurements while the patient is standing, a maneuver which normally further increases renin production. This secondary reduction in renin production has provided a valuable aid in the diagnosis of primary aldosteronism; but it has no obvious physiological consequences. The hypertension occurs in spite of the suppression of renin production, and aldosterone production continues to be excessive. The lowering of renin does provide further evidence that a tropic factor which is *not* angiotensin stimulates the adrenal cortex to produce aldosterone in the syndrome of primary aldosteronism with hyperplasia of all adrenal cortical tissue.

Syndrome of Juxtaglomerular Hyperplasia (as Autonomous Excess of Renin)

In the syndrome of juxtaglomerular hyperplasia with normal blood pressure, hyperaldosteronism, and hypokalemic alkalosis, the effect of angiotensin in elevating arterial pressure is reduced to one tenth to one hundredth of that in the normal subject. The origin of this defect, which may be part of an inheritable disease, is not known. The result, however, is a continuous, autonomous overproduction of renin resulting in a continuous production of angiotensin. The angiotensin, in turn, can stimulate the adrenal cortex to produce aldosterone in excess. Whereas neither the high circulating angiotensin concentrations nor the high circulating concentrations of aldosterone are able to produce hypertension, the aldosterone has its usual effect on the renal tubules in stimulating excessive sodium-for-hydrogen and sodium-for-potassium exchange, thus inducing hypokalemic alkalosis. Physiologically, the syndrome represents a sustained autonomous overproduction of renin in association with extremely low responsiveness of the blood vessels to angiotensin.

CALCIUM

Calcium Balance

Control of total body calcium depends upon the sum of net gastrointestinal absorption and tubular reabsorption less the filtered calcium. Control of serum calcium represents the sum of gastrointestinal absorption and tubular reabsorption minus the filtered calcium plus that which is lost from bone less the calcium entering bone.

Gastrointestinal Tract. Gastrointestinal absorption of calcium depends partly upon parathyroid hormone (PTH). This has been shown both indirectly from clinical studies of diseases and directly by measurement of calcium absorption with and without PTH in the whole organism as well as the isolated gut sac. Vitamin D apparently is more important that PTH in control of intestinal absorption of calcium; and considerable absorption occurs independently of both these agents.

Kidney. Tubular reabsorption of calcium depends partly upon PTH, which increases reabsorption. It is also likely that thyrocalcitonin (TCT), in some species, has an effect in decreasing tubular reabsorption of calcium. The action of PTH can be shown experimentally by careful measurement of the reabsorption at constant filtered calcium and sodium loads before and after the hormone is administered. It is probably responsible for the relative hypocalciuria in hyperparathyroidism and for the relative hypercalciuria (low threshhold) in hypoparathyroidism.

Bone. Loss of calcium from bone depends upon the interaction of PTH and TCT. PTH increases bone resorption by osteoclasts and osteoblasts; in this action, it probably promotes the formation of cyclic 3'5'-AMP as "second messenger." TCT prevents the resorption of bone, not only by interfering with the action of PTH by mechanisms not yet clarified but also by a more direct action since it can be shown to act in an in-vitro system not containing PTH.

Control of Parathyroid Hormone (PTH)

Control of PTH is largely an inverse function of serum ionized calcium. Whereas all conditions which increase serum phosphate (except hypoparathyroidism) produce increased PTH secretion, it has not been shown that phosphate can stimulate the parathyroids independently of its action in depressing serum ionized calcium. In addition, magnesium ion concentration may exert a moderate degree of control upon PTH secretion. Such an effect can be overcome experimentally by changes in serum ionized calcium.

Phosphate Reabsorption. An important additional action of PTH is its effect in decreasing renal tubular reabsorption of phosphate. This can be shown readily in the animal or patient lacking PTH by measuring filtered and urinary phosphate under steady conditions before and during PTH administration. The immediate increase in urinary phosphate produced by the hormone is accompanied by an equally prompt increase in the urinary excretion of cyclic AMP and bicarbonate. It has not been shown how these three actions are interrelated. In pseudohypoparathyroidism (failure of kidney and bone to respond to PTH), however, where phosphaturia does not occur with PTH the increase in cyclic AMP is also lacking.

Thyrocalcitonin (TCT)

Control of TCT secretion is largely a direct function of the serum ionized calcium. Convincing experimental evidence exists indicating that glucagon may be capable of stimulating the release of TCT without elevating serum ionized calcium. If this constitutes an independent control mechanism, its physiological significance has not been determined.

Magnesium-Calcium Competition at Gut and Tubular Membrane. Calcium metabolism is influenced by magnesium metabolism in at least two sites. Thus, factors that increase gastrointestinal absorption of magnesium reduce calcium absorption, perhaps by competition for binding sites. Somewhat in favor of this explanation is the reverse phenomenon, which can also be demon-

strated, to wit: decreased reabsorption of magnesium may be induced by factors that tend to increase gastrointestinal calcium absorption. In an analogous fashion, factors which increase renal tubular magnesium absorption appear to limit calcium absorption. To the extent that it can be tested experimentally, it appears that the reverse also applies to renal tubules.

Sodium-Calcium Competition at Tubular Membrane. There is extensive evidence that the renal tubular reabsorption of sodium and calcium are closely linked so that almost all measures which increase the renal sodium clearance also increase renal calcium clearance. An important implication of this relationship is that calcium clearance can be measured only under conditions in which the corresponding dynamics for sodium are known and held constant. It is clear that not all tubular reabsorption of sodium shares a common mechanism with that of calcium: this applies to the sodium-for-potassium exchange in the distal tubules. It is also likely that certain renal tubular cells can change reabsorption of calcium without a corresponding change in sodium reabsorption. The portion of calcium reabsorption which is promoted by the action of PTH may thus be independent of any influences upon the reabsorption of sodium.

Double Feedback Loop

Control of plasma calcium ion concentration is subject to the action of two interrelated "feedback loops" involving bone, TCT, and PTH (Fig. 16-2). Thus, a decrease in plasma calcium ion concentration results in prompt increase in PTH secretion; this, in turn, prompts bone resorption by osteoclasts and osteoblasts, an action which tends to restore the serum ionized calcium. As plasma calcium ion concentration rises, secretion of TCT is stimulated which, in turn, inhibits further resorption of bone, thus limiting a continued rise of ionized calcium. The increase in secretion

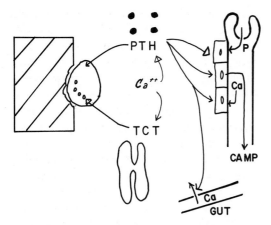

FIG. 16-2. The control and actions of parathyroid hormone (PTH) and of thyrocalcitonin (TCT). An increase in calcium ion (Ca⁺⁺) inhibits ↟ PTH secretion and stimulates ↑ TCT secretion. TCT inhibits ↟ bone resorption (*left*) while PTH stimulates ↑ bone resorption. In addition, PTH has three known effects on the kidney tubule (*right*): inhibition of phosphate (P) reabsorption, stimulation of secretion of cyclic AMP (CAMP), and stimulation of calcium (Ca) reabsorption. PTH also promotes calcium absorption from the gut (*lower right*).

of PTH may also increase the serum ionized calcium concentration by increasing renal tubular reabsorption of filtered calcium.

Hypoparathyroidism

Idiopathic hypoparathyroidism is a condition in which there is unexplained paucity or lack of parathyroid tissue. It must be carefully distinguished from the condition known as pseudohypoparathyroidism, in which the parathyroid tissue is present in excess; but the renal and, generally, the osseous tissues are unable to respond normally to PTH. The lack of parathyroid tissue produces a syndrome which is explained by the known physiological actions of the hormone.

Effect on Gut. Absence of PTH limits absorption of dietary calcium such that for any given calcium intake the contribution of the gut is less than it would be in the presence of PTH. Clinically, this effect can be partly overcome by administration of large doses of vitamin D, whose action is, in part, independent of parathyroid hormone.

Effect on Tubule. Renal clearance of calcium is increased in hypoparathyroidism relative to the normal state. This effect which produces a relatively high clearance of calcium ion is reflected in the low threshold for calcium characteristic of this condition. The patient with hypoparathyroidism may thus show hypercalciuria even in the presence of hypocalcemia.

Effect on Bone. Absence of PTH removes the important element of bone resorption from the feedback loop by which calcium ion concentration is regulated. Thus, as calcium ion is decreased because of the failure of absorption or of tubular reabsorption, the parathyroid-induced release of calcium from bone cannot occur. Whereas it is likely that TCT secretion is also decreased in the presence of hypocalcemia, any contribution of such decrease to the maintenance of serum calcium ion concentration of hypoparathyroidism is clearly inadequate as the clinical condition invariably results in hypocalcemia.

Hyperparathyroidism

In hyperparathyroidism almost the precise opposite of the abnormalities found in hypoparathyroidism may be observed, with a single exception: an excess of TCT may, in some patients, "protect" the bones against the excessive PTH.

Effect on Gut. In hyperparathyroidism there is relative hyperabsorption of calcium, such that a greater fraction of the intake is absorbed than in the normal. The clinical result of this disorder ranges widely because the normal absorption of calcium in the absence of PTH ranges widely among individuals and among families.

Effect on Tubule. The renal tubules reabsorb relatively more of the filtered calcium ion in the patient with hyperparathyroidism than in the normal. Thus, the renal clearance of calcium ion is relatively low in this condition, and the patient may show normal or indeed low urinary calcium in the presence of hypercalcemia. The second effect on the tubule is that produced by the action of PTH in decreasing phosphate reabsorption. Clinically, this results in the pathognomonic signs of hypophosphatemia and greatly increased renal clearance of phosphorus.

Effect on Bone. In hyperparathyroidism, with a great excess of PTH, bones may show rapid resorption resulting in thin bones and osteitis fibrosa cystica; the bones are thinner than normal in spite of obvious compensatory bone formation, as shown by the wide osteoid borders and marked osteoblastic activity.

Secondary TCT Production. All the effects that we have considered for PTH—increased gastrointestinal absorption, tubular reabsorption, and bone resorption—have the ultimate effect of increasing the serum ionized calcium concentration. The increased serum ionized calcium is the most effective stimulus for the secretion of TCT. Therefore, in some patients with hyperparathyroidism, secretion of TCT may be greatly augmented; and it is likely that, in some patients, this increase is sufficient to limit the resorption of bone to such an extent that the bone disease is not clinically apparent. Thus, secondary overproduction of TCT may play a part in the very common syndrome of hyperparathyroidism without bone disease.

With the increased sensitivity of tests for PTH secretion it has become apparent that many patients with hyperparathyroidism who show indeed a sustained hypersecretion of PTH may develop only hypercalciuria but no hypercalcemia, hypophosphatemia, or osteitis fibrosa cystica.

Such patients, who often present with a history of kidney stone formation, may be detected by demonstrating that PTH secretion, while only slightly in excess, is sustained (and lacks circadian variability), and is autonomous as regards such stimuli as hypercalcemia, which normally suppress PTH secretion. The presence of bone disease may be demonstrable by such technics as microradiography by which it may be shown that there is excessive resorption of bone.

Medullary Carcinoma of Thyroid (Primary TCT Excess)

With the discovery that the cells of the medullary thyroid carcinoma produce TCT, a syndrome characteristic of excess of TCT was sought in patients with this disorder. Whereas one might anticipate that such patients would develop hypocalcemia with hypophosphatemia, changes which can be induced with exogenous TCT, it is rarely that a patient is found with this combination.

Secondary PTH Excess. As already noted, the most important stimulus to the production of PTH is decreased serum ionized calcium concentration. Probably all patients with medullary carcinoma of the thyroid in whom TCT is liberated into the circulation in excess respond to a transient hypocalcemia by overproduction of PTH. In a number of such patients, parathyroid adenomas have been found, and accompanying hyperparathyroidism may thus obliterate most of the effects of the excessive TCT.

In some patients with medullary thyroid carcinoma, osteomalacia is produced. The sequence of events which would explain this complication may be as follows: excess of TCT induces hypocalcemia; hypocalcemia induces increased PTH secretion; PTH increases phosphate clearance and lowers serum phosphate concentration; the ion product of a low-normal serum calcium and low serum phosphate concentration is inadequate for normal calcification of osteoid tissue; and thus, the bones develop osteomalacia.

REFERENCES

Bartter, F. C.: Regulation of the volume and composition of extracellular and intracellular fluid. Ann. N. Y. Acad. Sci., *110* (Part II): 682, 1963.

Bartter, F. C., and Fourman, P.: The different effects of aldosterone-like steroids and hydrocortisone-like steroids on urinary excretion of potassium and acid. Metabolism, *11*:6, 1962.

Bartter, F. C., and Schwartz, W. B.: The syndrome of inappropriate secretion of antidiuretic hormone. Am. J. Med., *42*:790, 1967.

Talmage, R., and Munson, P. (eds.): Parathyroid Hormone and Thyrocalcitonin (Calcitonin). Proceedings of the Fourth Parathyroid Conference. New York, Excerpta Medica Foundation, 1971.

Verney, E. B.: Antidiuretic hormone and the factors which determine its release. Proc. Roy. Soc. London B. *135*:25, 1947-48.

17

Metabolism and Energy Mechanisms

Donald B. Martin, M.D.

GENERAL FEATURES OF METABOLISM

The most extraordinary feature of intermediary metabolism is that it works! Not only does it work, but it operates with remarkable efficiency. Consider how variable the calories one expends are: from about 60 calories per hour (sleeping) to 600 calories per hour (doing heavy work); then contrast this with the complete lack of attention given to the amount and type of food consumed. One can only be amazed by the relative stability of body weight.

In order to assure both a ready supply of immediate energy as well as a constant replenishment of depleted energy stores, a complex, but efficient, method of energy exchange between stationary and mobile forms of energy is required. This must be accomplished for several types of foodstuffs and during periods of fasting as well as feeding (Fig. 17-1). Each type of food is available to the body both in storage form or as a mobile carrier of energy. The latter can be converted to a common set of metabolites and fed into a common "generator" which supplies and maintains all tissues with the universally used instant energy source, adenosine triphosphate (ATP). ATP is used in all mammalian cells for physical work, biosynthetic processes, generation of electrical energy, and so forth.

The storage forms of energy seem simple enough on initial inspection, but the regulation of their amount, and the factors which induce their synthesis and breakdown are quite complex. For all three foodstuffs (proteins, carbohydrates, and fats), the arc of synthesis and the arc of breakdown are usually under hormonal control. It is at the storage level of metabolic regulation that instant and fine controls are needed to take up any extra energy substrate (furnished by food intake) and to provide a mobile energy carrier in times of food deprivation or stress. The storage form of fat is by far the most efficient. Not only does it provide more than twice as many kilocalories per gram as protein or carbohydrate (9 vs. 4) but also it is stored in a water-free form. In contrast, glycogen needs water of hydration for its formation, adding undesirable weight per unit energy stored.

For energy to be transported from the respective storage depots to other tissues to be metabolized, mobile energy carriers for each foodstuff are required. In addition, at this mobile energy carrier level, digested foodstuffs are "fed into" the scheme. Water solubility is of prime importance in this phase of energy transfer. This poses no problem for the simpler forms of carbohydrate and protein (glucose and amino acids). However, for transfer of fat energy, solubility in aqueous solutions is an enormous obstacle. Although ketone bodies are readily water-soluble, they play a minor role in normal energy balance, but in protracted starvation and diabetic ketoacidosis they represent an important energy source. For the free fatty

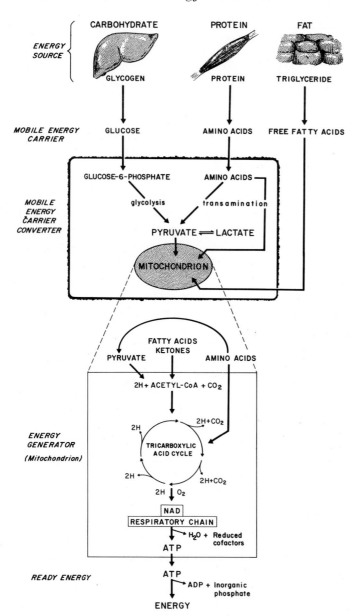

Fig. 17-1. The metabolic pathways for each type of food through the various energy interrelationships.

acids released from adipose tissue, and for the chylomicra and other lipids taken up by the gut and released from the liver postprandially, linkage to protein (as lipoproteins) is essential to insure water solubility.

Once the mobile carrier energy has reached its cell of destination, it must be converted to a substrate that can be utilized in the cell mitochondrion to generate ATP. For some metabolites, glucose, and certain amino acids in particular, this means conversion to pyruvate which, in turn, can be cycled into the tricarboxylic acid (Krebs or citric acid) cycle. Other amino acids are transaminated and utilized by the mitochondrion. Free fatty acids are oxidized in the mitochondrion to acetyl-CoA and metabolized in the tricarboxylic acid cycle.

From the metabolic control point of view, these initial cellular reactions of this energy conversion step (prior to mitochondrial oxidation) are usually not under hormonal control. Rather, they tend to be modified at key steps by metabolite repression or stimulation. Thus, the levels of critical metabolites affect (positively or negatively) the rates of enzymatic reactions which, in turn, influence the degree to which a given pathway is utilized. This type of control is common at enzymatic steps where different enzymes catalyze the forward and reverse directions of the metabolic process. These unidirectional enzymes at key metabolic crossroads are examples of another mechanism by which metabolic processes are regulated.

Once the mobile energy source has been converted (prepared) to a form acceptable for utilization in the mitochondrion, the substrates are oxidized by a complex coupled series of reactions in the respiratory chain of the mitochondrion to carbon dioxide and water, with the generation of ATP from ADP and the concomitant reduction of NAD to NADH. This process requires oxygen. All three foodstuffs ultimately contribute their major share of energy production through these mitochondrial reactions. This emphasizes the central role of mitochondrial oxidative phosphorylation as the final common pathway of energy production in the mammalian cell. Under certain abnormal conditions, the formation of ATP that depends upon oxygen utilization can become "uncoupled"; that is, extra oxygen is utilized for a given amount of ATP formed. This is called uncoupled oxidative phosphorylation and is found with the use of certain drugs, for example, dinitrophenol.

An alternative mechanism for the generation of ATP exists, in which glucose metabolism in effect stops at the level of energy conversion (prior to mitochondrial oxidative phosphorylation) and yields pyruvate and lactate. This more "primitive" energy system, widely found in the lower forms of life, offers the major advantage that oxygen is not needed. This system of glucose metabolism is called anaerobic glycolysis (in contrast to aerobic glycolysis, described above). It is relatively inefficient, yielding only 52 kilocalories of energy and 2 moles of ATP for each mole of glucose utilized (vs. 686 kilocalories and 38 moles of ATP with aerobic glycolysis). Whereas it is the sole source of energy in a number of bacterial systems, its principal function in mammalian systems is to provide energy for muscular work during periods of excessive muscular activity when oxygen supply is outstripped. Although the pyruvate and lactate formed in this circumstance are not used by muscle, they are not "lost." They are released into the circulation, transported to the liver, and therein metabolized. Muscle also possesses a backup form of high-energy phosphate in the form of creatine phosphate which acts as an energy reservoir, keeping the ATP as "fully charged" as possible, and ready to provide energy for muscular work.

ORGAN PHYSIOLOGY

Muscle, liver, and adipose tissue will be discussed in detail in this section and their metabolic relationships contrasted during the fed and fasting states. Other tissues will not be discussed in detail. Omission of a detailed discussion of other tissues does not mean to imply a lack of importance but rather a relative lack of ability to modify their metabolic processes in times of fasting or stress. The gut and hemic cells are good examples of this; the nervous system is, as well (with the notable exception of the situation of protracted fasting). Since the kidney closely resembles, meta-

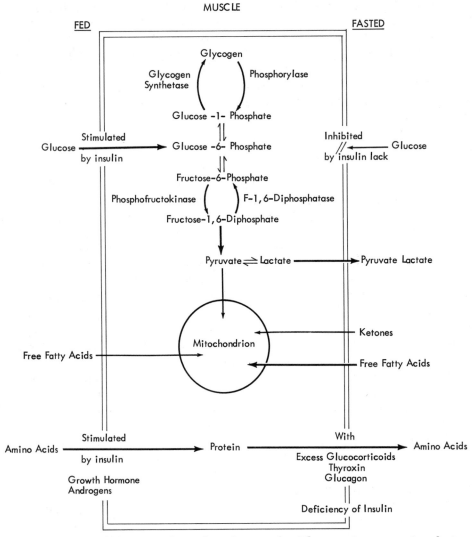

FIG. 17-2. Metabolic interrelationships for muscle. The reactions occurring during the fed state are represented on the left; those for the fasted state appear on the right.

bolically, the liver, it will not be discussed separately.

Muscle

Muscle, comprising about 50 percent of total body mass, is by far the major consumer of energy. During the resting state, for example, metabolism of muscle accounts for 25 percent of oxygen consumption. Before discussing the fed and fasting states and their effects on muscle metabolism, it should be noted that of the storage forms of the three foodstuffs mentioned above, glycogen and triglyceride are present in relatively small amounts and are relatively unaffected by change in metabolic status. Moreover, amino acids play a small role in energy-yielding substrates in muscle, although muscle is known to contain various transaminases.

Muscle Metabolism: Fed State. During the *fed* state (Fig. 17-2), glucose is taken up by muscle, a process governed largely by the level of circulating insulin.

Glucose is taken by muscle and phosphorylated to glucose-6-phosphate. From this metabolic turning point, energy can flow from glucose to glycogen, or to pyruvate via the Embden-Meyerhof pathway. In point of fact, both can occur during periods of excess caloric intake, with the Embden-Meyerhof pathway metabolizing most of the glucose taken up. The flow of energy from glucose-6-phosphate to pyruvate is governed by the activity of the enzyme phosphofructokinase. This enzymatic step exemplifies the two types of control mechanisms described under energy conversion in the preceding section. Phosphofructokinase is a unidirectional enzyme, converting fructose-6-phosphate to fructose-1,6-diphosphate; a different enzyme, fructose-1,6-diphosphatase, catalyzes the reverse reaction. In addition, the activity of phosphofructokinase is under metabolite control. It is inhibited by increased amounts of ATP and citrate and stimulated by ADP. Thus, with continued glucose degradation and mitochondrial oxidative phosphorylation, a gradual build-up of these two inhibitors occurs with diminished activity of this pathway. The oxidation of fatty acids by muscle mitochondria provides more of these inhibitory metabolites. For these reasons, muscle prefers fatty acids as fuel, even in the fed state. Because of its bulk, muscle is the principal reservoir for protein. Unlike carbohydrate and fat, the storage and the utilizable forms of protein are apparently the same. The various proteins in muscle, both structural and enzymatic, are constantly being broken down and renewed, in good part under hormonal control. Insulin, growth hormone, and androgen all favor the protein synthesis; and, when excessive amounts of thyroxine and glucocorticoids are present, protein breakdown is accelerated. Insulin deficiency also accelerates protein breakdown.

Muscle Metabolism: Fasting State. During *fasting*, muscle takes up much less glucose for several reasons. The first is the lower level of circulating insulin. Another is related to mitochondrial metabolism of the increased amounts of free fatty acids induced by caloric deprivation. Phosphofructokinase is inhibited, as explained above. In addition, the increased amounts of acetyl-CoA, formed from fatty acid oxidation, inhibit the enzyme which converts pyruvate to acetyl-CoA (pyruvate dehydrogenase). This further inhibits the Embden-Meyerhof pathway. The unoxidized pyruvate leaves the cell and is transported to the liver where it is metabolized. During fasting, protein is broken down into amino acids and released into the circulation largely as a result of the lower level of circulating insulin. These amino acids are picked up by the liver where they are used for gluconeogenesis. In addition, in pathological states of excess amounts of circulating adrenal corticoids and thyroxine, protein is broken down in a more accelerated fashion with consequent depletion of muscle protein stores and increased levels of circulating amino acids.

Liver

Although representing a very small percent of body mass, the liver is an extraordinarily important regulator of energy metabolism (Fig. 17-3). Before discussing hepatic metabolism in the fed and fasting states, several features unique to liver should be described. The liver is the first organ to receive foodstuffs that have been absorbed from the gut. In addition, insulin and glucagon from the pancreatic islets must traverse the liver before entering the systemic circulation. The "quick control" which modifies hepatic metabolism is usually accomplished by metabolite inhibition or stimulation at points where the enzymes are unidirectional (allowing for unidirectional control). Nevertheless, hormonal effects are important even though they are not rapid in onset and do not occur in small hormone doses, as is the case with muscle and adipose tissue. The

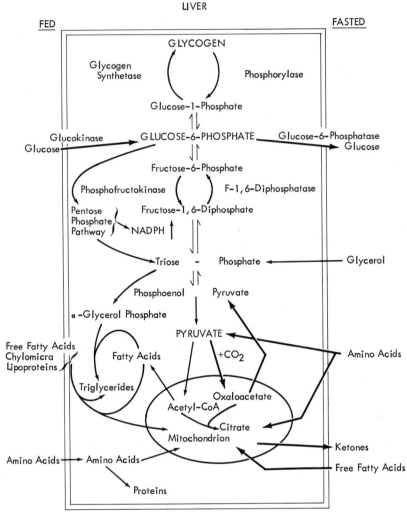

Fig. 17-3. Metabolic interrelationships for liver. The design of the scheme is the same as for muscle (Fig. 17-2).

actions of insulin and glucagon on the liver are still a subject of controversy. Data from experiments with isolated perfused livers indicate that insulin lowers hepatic glucose output (and glucagon the reverse), independent of an action on glycogen. The locus of these effects has not been defined.

The effects of glucocorticoids on inducing hepatic enzymes (of gluconeogenesis) have been well described, but these effects require time and metabolism of the liver cells before they are observed.

Regulation of glycogenolysis in response to epinephrine or glucagon is extremely rapid and can be demonstrated in broken hepatic cell preparations.

The enzymes of glycogen synthesis and degradation, and their hormonal control, serve as good examples of different ways in which metabolic processes can be regulated. The enzyme of glycogen formation, glycogen synthetase, differs from that of glycogen degradation, phosphorylase, allowing differential control (Fig. 17-3). Both enzymes exist in active and inactive forms. A special family of enzymes that adds a phosphate group are termed kinases. The kinase that adds a phosphate to phosphorylase activates (increases the activity of) the enzyme. A different kinase adds a phosphate to glycogen synthetase and has the opposite effect of rendering it "inactive." This is an example of a similar, rather simple, enzymatic mechanism having opposite metabolic effects. The phosphorylase kinase enzyme, in turn, is regulated by a similar enzymatic mechanism, and this enzyme is called phosphorylase kinase kinase! Recent evidence suggests that phosphorylase kinase kinase is the same enzyme as the glycogen synthetase kinase. This permits the turning off of glycogen synthesis and the turning on of glycogenolysis through the action of a single enzyme. This is of special interest in that this enzyme, which is capable of controlling both glycogen synthesis and breakdown, is under the control of cyclic 3',5'-AMP.

The compound cyclic 3',5'-AMP, first characterized about 15 years ago, has become widely accepted as an intracellular "second messenger." It is now implicated in the release of most pituitary polypeptide hormones, in the release of insulin by the beta cell, in glycogen formation and breakdown, and in numerous other phenomena. With reference to glycogen, for example, epinephrine and glucagon increase the amount of cyclic AMP which inhibits glycogen synthetase and activates phosphorylase. Both actions increase the amount of available glucose in the circulation.

Hepatic Metabolism: Fed State. In the *fed* state, the liver takes up glucose to form glucose-6-phosphate (Fig. 17-3). Several hepatic enzymes change glucose to glucose-6-phosphate, including a specific glucokinase and a less specific hexokinase. Interestingly, the specific glucokinase has a low affinity for glucose (needs a higher ambient sugar level before it acts). In addition, its activity diminishes with fasting or insulin lack. In contrast, hexokinase activity does not change during altered nutritional or hormonal states, and is less specific, phosphorylating other hexoses (mannose, galactose, fructose, etc.).

Once glucose-6-phosphate is formed, the metabolic options are more complex in liver than in muscle. Not only can glucose be transformed into glycogen via glucose-1-phosphate and to pyruvate by the Embden-Meyerhof pathway, but it has two additional fates of metabolic significance. First, it is metabolized in the pentose phosphate pathway, re-entering the Embden-Meyerhof pathway at the triose phosphate level. This pathway is an important generator of a reduced cofactor (NADPH) that is essential for biosynthetic reactions. Therefore, in the fed state, these reductive hydrogens are available and are used in the synthesis of long-chain fatty acids, cholesterol, and in other synthetic reactions. The other reaction unique to liver (and kidney) is the dephosphorylation of glucose-6-phosphate (by the enzyme glucose-6-phosphatase) to form free glucose, which can be released into the circulation as a (mobile) energy source. This glucose, released by the liver, constitutes the principal source of glucose during periods of fasting and is absolutely essential for maintaining function of the central nervous system.

In liver, the Embden-Meyerhof pathway functions much as it does in muscle, the major control point being the activity of the unidirectional enzymes operating between fructose-6-phosphate and fructose-1,6-diphosphate.

The fate of pyruvate in liver during the fed state is more complex than in muscle. As in muscle, pyruvate can enter the tricarboxylic acid cycle and form lactate. In addition, pyruvate is decarboxylated to acetyl-CoA, which forms citrate from oxaloacetate in the mitochondrion. Acetyl-CoA and citrate are thought to provide the substrate for the extramitochondrial synthesis of long-chain fatty acids. The mechanism for the transfer of these compounds from the inside to the outside of the mitochondrion is poorly understood. The acetyl group is almost surely not transferred as the coenzyme A derivative, since this compound does not penetrate the mitochondrial cell wall, but rather as the carnitine derivative, which does. Moreover, the appropriate enzymes are present for the exchange of the CoA and the carnitine parts of the molecule. It should be emphasized that the synthetic pathway for long-chain fatty acids differs from the pathway of degradation. Fatty acid synthesis occurs outside the mitochondrion and oxidation within; and the enzymes of the two processes differ. The major difference lies in the enzyme acetyl-CoA carboxylase which forms malonyl-coenzyme A, the key intermediate of the synthetic pathway; this is absent from the pathway of fatty acid oxidation. A separate pathway for synthesis and oxidation offers another mechanism for control. Acetyl-CoA carboxylase is markedly stimulated by citrate, a by-product of glycolysis in times of carbohydrate excess. It is inhibited by palmityl coenzyme A, which is present in excess in times of increased triglyceride breakdown. The other important fate of pyruvate is to form phosphoenolpyruvate (via oxaloacetate), a compound which can be transformed to glucose; this is important in the fasting, not in the fed state.

In the fed state, the liver does metabolize fatty acids in the mitochondrion. The sources of these fatty acids are the breakdown (turnover) of endogenous triglycerides, and free fatty acids. Chlyomicra from the intestinal absorption of fatty foods can serve as another source. The intermediary reactions involving chylomicra and lipoproteins occur in the liver, but their details remain obscure.

The uptake and metabolism of amino acids by liver in the fed state proceed in a manner analogous to that of muscle. The proteins formed are, of course, very different from those in muscle. Serum albumin, for example, is produced in the liver.

Hepatic Metabolism: Fasting State. In the *fasting* state, liver metabolism changes dramatically. The metabolic efforts of the liver change from uptake and storage of glucose (as glycogen), lipids (as triglyceride), and amino acids (as protein) to mobilization (i.e., the liver becomes the prime source of glucose for the body). All metabolic events in the liver during fasting bend to this goal. The two major fuels during fasting are free fatty acids from adipose tissue and hepatic triglyceride breakdown, and amino acids from the liver and muscle. Quantitatively, smaller amounts of fuel come from glycerol (from triglyceride breakdown in adipose tissue and liver), and from lactate which enters via pyruvate and comes from the metabolism of muscle and the erythrocyte. In addition, hepatic glycogen is a source of fuel during short periods of fasting.

The metabolism of the increased amounts of free fatty acids circulating during fasting provides the energy required for gluconeogenesis. Moreover, their increased levels provide a mechanism by which the process is controlled. The coenzyme A derivatives of the long-chain fatty acids inhibit phosphofructokinase, diminishing the activity of the Embden-Meyerhof pathway. They also inhibit pyruvate dehydrogenase and citrate synthetase, blocking the entry of pyruvate into the tricarboxylic acid cycle. This latter inhibition is complemented by the fact that the acetyl-CoA, derived from fatty acid oxidation, acts as a cofactor in the step from pyruvate to oxaloacetate, which, in turn, can be further metabolized to glucose. The acetyl-CoA cannot itself be metabolized to oxaloacetate and is therefore not a precursor of glucose. Thus, fatty acids cannot be transformed into glucose. For this reason, the body must rely on the complicated process of gluconeogenesis.

The increased amounts of amino acids from protein breakdown in muscle and liver during fasting provide the substrate for gluconeogenesis via transamination.

The other metabolic event in liver during fasting which is of great importance is ketogenesis. Part of the explanation of ketogenesis is inherent in the discussion of gluconeogenesis. Excessive amounts of fatty acids, by way of their coenzyme A derivatives, inhibit the formation of citrate from acetyl-CoA and oxaloacetate. Furthermore, they inhibit acetyl coenzyme A carboxylase, the rate-controlling enzyme in

fatty acid synthesis. Thus, two major exits of acetyl-CoA are blocked, leaving ketogenesis as the only open pathway.

Adipose Tissue

Adipose tissue, the principal repository of the energy storage form of fat, comprises a percentage of the total body weight ranging from a few percent in a starved person or thin, well-trained athlete to as high as 50 percent in obese persons. Other types of lipid (i.e., cholesterol, phospholipids, sphingolipids) are usually assigned a structural function and are present in insignificant amounts in adipose tissue.

Adipose Tissue Metabolism: Fed State. In the *fed* state (Fig. 17-4), each major food-

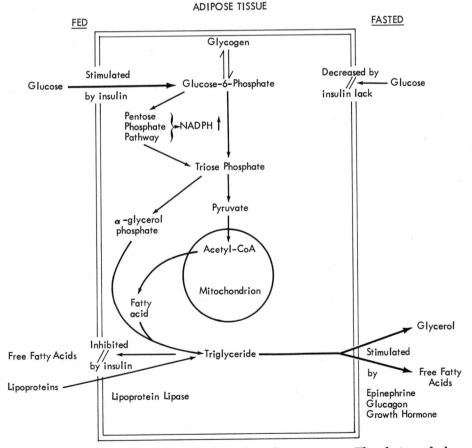

FIG. 17-4. Metabolic interrelationships for adipose tissue. The design of the scheme is the same as for muscle (Fig. 17-2).

stuff contributes substrate to adipose tissue. Amino acids are taken up by adipose tissue but play no significant role in energy-exchange mechanisms. In contrast, glucose plays an enormous role as an energy source. In adipose tissue as in muscle, the prime governor of sugar uptake is the hormone insulin. During periods of carbohydrate caloric excess, there is an increased circulating level of insulin. This provides a mechanism whereby carbohydrate calories can be taken up in times of "plenty" and stored as triglycerides. The higher circulating insulin plays another critical regulatory role—the inhibition of the release of free fatty acids. In fact, the level of insulin at which this inhibition occurs is lower than the level necessary for stimulation of glucose uptake. The fate of glucose within the cell resembles that in muscle. It is phosphorylated to glucose-6-phosphate, can proceed to glycogen, or to pyruvate via the Embden-Meyerhof pathway. In addition, in the fed state glucose can be degraded via the pentose phosphate pathway, providing reduced pyridine cofactors (NADPH) required for fatty acid synthesis. Another extremely important by-product of the Embden-Meyerhof pathway is alpha-glycerol phosphate, the carbohydrate moiety of the triglyceride molecule. All the alpha-glycerol phosphate used in triglyceride synthesis must come from glucose, since adipose tissue cannot rephosphorylate the glycerol released from breakdown of triglyceride. The pyruvate formed is decarboxylated in the mitochondria into acetyl-CoA, and the citrate and acetyl-CoA presumably exit to return to the cytoplasm for fatty acid synthesis (similar to liver).

Adipose tissue can take up lipoproteins, although the exact mechanism is unclear. The enzyme lipoprotein lipase plays a major role in this function, although its exact locus remains a point of controversy (the cell membrane versus the capillary adjacent to the adipose tissue cell). Clearly, the activity of this enzyme changes greatly with altered nutrition and hormonal influence. The enzymatic activity is decreased in starvation and increased with feeding. In turn, the enzymatic activity is increased in the presence of insulin and decreased by catecholamines. The relative contribution of lipids taken up in this manner to the lipids synthesized from glucose is unclear.

Adipose Tissue Metabolism: Fasting State. In the *fasting* state, metabolism in adipose tissue shifts dramatically, largely owing to a decrease in circulating insulin. With relative insulin lack, glucose uptake ceases (and therefore the production of alpha-glycerol phosphate, NADPH, and long-chain fatty acids diminishes) and the inhibition of free fatty acid release decreases. In addition to the free fatty acids, glycerol is also released. The free fatty acids provide the liver with a source of energy for gluconeogenesis and muscle with a substrate for the energy of muscular work. The glycerol is a glucose precursor in liver and enters the gluconeogenic pathway at the triose-phosphate level.

Lipolysis proceeds under the control of a number of lipases (not to be confused with the lipoprotein lipase described above), at least one of which is under hormonal control. A number of hormones and pituitary peptides release free fatty acids from adipose tissue in vitro, presumably by activating the tissue lipase via cyclic AMP. Their physiological role is not clear. The functions of insulin in the inhibition, and of epinephrine and glucagon in the stimulation, of lipolysis are clear; they probably play the major physiological regulatory role.

ALTERED STATES AND DISEASES

Starvation

One can best understand and correlate metabolism with its hormonal control by examining the physiological changes induced by fasting. In the sections above, the forms of energy transformation in various tissues were discussed. The physiological adaptation to starvation, of course, revolves around the availability of a reservoir of calories. Obviously, the circulating metabolites alone (i.e., glucose, amino acids, ketone bodies, free fatty acids) can hardly be expected to contribute significantly to caloric balance. For example, if one were to calculate the calories available as extracellular glucose in an average man with a blood sugar of 70 mg.% and a normal extracellular fluid volume, and multiply this by the factor of 4 kilocalories per gram, one would arrive at an unimpressive figure of somewhat over 80 kilocalories. Similarly scanty amounts are found when the same calculations are applied to extracellular amino acids and free fatty acids. The storage forms of fuel provide the major source of calories during the period of food deprivation (i.e., glycogen, protein and fat). An inspection of these three caloric depots makes it readily apparent that glycogen can supply only a relatively small amount of fuel. The total amount of glycogen in muscle and liver calculated for an average percent of tissue content amounts to only about 1,000 calories (and it was noted earlier that muscle glycogen is not totally exhausted during fasting). This then leaves protein and fat (as triglycerides) as the major fuel sources. With excess caloric intake one can add more and more to the adipose tissue reservoir. However, extra dietary protein does not increase muscle mass but is metabolized and excreted in the urine as urea. Of these two major fuel sources, protein is critical as machinery, and metabolic efforts are made to spare it. Depot fat, on the other hand, has very little function other than to serve as the primary energy source during starvation— minor exceptions being structural function in certain specialized areas (for example, periorbital fat). As noted above, stored fat

has a further advantage because of its high calorie-to-weight ratio and the absence of water hydration, which gives it greater calorie-to-weight "efficiency."

A well-nourished man, fasting over a 24-hour period, expends approximately 1,800 calories. For this he burns about 75 grams of protein, primarily from muscle, about 160 grams of triglycerides from adipose tissue, and about 180 grams of carbohydrates provided by the liver (and kidney) through gluconeogenesis (Fig. 17-5). The principal tissue that consumes glucose, of course, is the central nervous system. It requires approximately 150 grams per day (which is burned into carbon dioxide and water). The remainder of the glucose is used by such tissues as the hemic cells and, to some extent, muscle, which burn some glucose but convert most of it to lactate. This lactate, in turn, is returned to the liver which can metabolize it to form more glucose through gluconeogenesis. In this way, part of the energy of glucose utilized by these cells can be recycled. All the other tissues (heart, skeletal muscle, etc.) use either free fatty acids or ketone bodies as their source of calories.

The principal organ adaptation is that of the liver. This is so because it is the primary source of newly formed glucose from precursors such as amino acids (largely alanine), pyruvate, lactate, and glycerol. In the fasting state, the liver derives its energy from the oxidation of fatty acids.

What are the hormonal control mechanisms which operate during fasting? Because of the number of hormones involved, of course, the answer is complex. The principal difficulty lies in trying to ascribe primacy to the action of the hormone. However, if one were pressed to pick a single hormone as central to the physiological adaptation to starvation, it would be insulin. Insulin plays a major role in both energy breakdown and build-up. It is capable of inhibiting free fatty acid release from

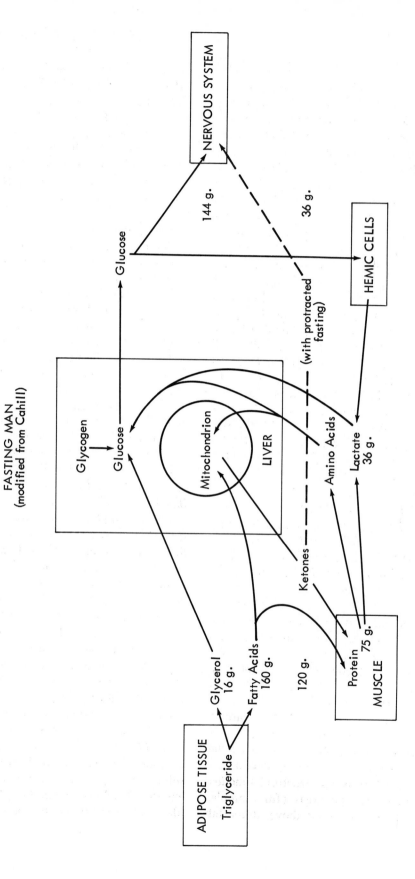

FIG. 17-5. Metabolic interrelationships during fasting in man. The scheme. (Modified from Cahill, G. F., Jr.: New Eng. J. Med., 282:668, 1970)

adipose tissue at very low concentrations. In addition, evidence suggests that insulin may regulate protein breakdown and the release of amino acids into the circulation. On the side of energy accumulation, insulin has an important function in governing sugar uptake by adipose tissue and by muscle. Thus, the insulin level regulates the sugar supply into both muscle and fat, and the release of critical fuels during starvation.

Another hormone that may play a critical role in the adaptation to starvation is glucagon. Glucagon not only provokes brisk hepatic glycogenolysis but also augments gluconeogenesis. Thus, it is receiving increasing recognition as a participant in body function, the state of starvation being no exception. Certainly other hormones, such as glucocorticoids, growth hormone, and catecholamines, are also important in fasting. For example, in the absence of adequate adrenal function, gluconeogenesis is impaired, inducing profound hypoglycemia. Yet, these hormones function more "permissively," lacking the primary role of insulin (and perhaps glucagon).

The amount of protein breakdown day by day during starvation has several physiological implications. Of clinical importance is that protein breakdown can be spared by supplying modest amounts of carbohydrate to serve as metabolic fuel for the nervous system. Referred to as the so-called *protein-sparing ability* of glucose in starvation, this provides the physiological rationale for including glucose in intravenous fluids postoperatively. Another is the obligatory protein breakdown occurring in the fasting state which places a limitation upon the duration of total caloric deprivation to which an obese patient can be subjected. Recent evidence has shown, however, that as the fast is prolonged (above 3 to 4 weeks), body metabolism further adapts to reduce protein breakdown. Specifically, with protracted fasting the central nervous system can adapt to the use of ketone bod-

ies as fuel (betahydroxybutyrate and acetoacetate). This permits the nervous system to metabolize less glucose, necessitating diminished gluconeogenesis, and requiring the utilization of smaller amounts of glucose precursors (notably amino acids), thereby sparing protein breakdown.

Therefore, fasting man has two principal sources of fuel, adipose tissue fat and muscle protein. Initial reserves of glycogen are quickly exhausted. As the fast is prolonged, the principal user of carbohydrate energy, the central nervous system, further adapts by consuming ketone bodies as fuel and, hence, decreases the demands for structural protein as an energy source.

Diabetes Mellitus

The pathophysiological discussion of starvation serves as a proper introduction to that of diabetes mellitus.

The pathophysiological parallels between diabetes mellitus and starvation are striking, but there are some important differences. These parallels are logical when one considers diabetes mellitus as *intra*cellular starvation due to insufficient insulin. Thus, the metabolic events during fasting (discussed above) also occur in this disease. In muscle, glucose utilization diminishes with a subsequent decrease of activity in the Embden-Meyerhof pathway. The metabolic needs of muscle are met by increased utilization of free fatty acids and/or ketone bodies. Smaller amounts of amino acids are taken up, and increased amounts are released. In adipose tissue, glucose uptake and utilization also decrease, coincident with a decline in the activity of the Embden-Meyerhof pathway (with a decreased formation of alpha-glycerol phosphate) and the pentose phosphate pathway (with a reduction in the amount of reductive hydrogens formed). Triglycerides are broken down into free fatty acids and are released in increased amounts into the circulation.

Hepatic adaptation is striking; metabolism reverts from glucose utilization to glucose production. The influx of amino acids and free fatty acids, coupled with a decrease in the breakdown of glucose, reverses the direction of the Embden-Meyerhof pathway toward glucose synthesis (and release). As a result of this increased formation of free fatty acids, the formation and release of ketone bodies increase. Why excessive ketone bodies are formed in this situation is not completely clear. Earlier data seemed to offer a reasonable explanation; i.e., the level of oxaloacetate in diabetes mellitus was reported to be reduced. With less oxaloacetate, enzymatic activity of citrate synthetase would be diminished, whereas the enzyme forming phosphoenolpyruvate would not. This explained the metabolic flow in the direction of phosphoenolpyruvate and glucose. However, more recent information refutes this, demonstrating little or no decrease in the amount of oxaloacetate. Thus, more recent theories suggest that the enzyme citrate synthetase is directly inhibited. In either case, there is a resultant diminished conversion of oxaloacetate to citrate. The possible metabolic fate of acetyl-CoA produced by fatty acid oxidation is thereby limited. Lipogenesis is decreased because of the diminished amounts of citrate, NADPH, and alpha glycerol phosphate. The fate of the remaining acetyl-CoA therefore resides with the ketone bodies—hence, the close interplay between gluconeogenesis and ketogenesis.

This raises an old argument in the field of diabetes, concerning whether overproduction or underutilization of glucose is the main culprit; does the intense ketogenesis stimulate overproduction of glucose, which, in turn, contributes to the hyperglycemia? The answer is that the hyperglycemia results partly from increased hepatic glucose output and partly from underutilization. Substrate supplied to the liver from the periphery certainly plays an important role in regulating glucose production and output, but also important are the regulatory effects of hormones (especially insulin

and glucagon) on hepatic glucose production and release.

An interesting interplay between gluconeogenesis and ketogenesis is seen in veterinary medicine. Cows with bovine ketosis demonstrate ketogenesis and excessive gluconeogenesis in association with *hypo*glycemia, rather than with *hyper*glycemia as is observed in diabetes mellitus. This syndrome is found in lactating cows, usually but not always at the time of parturition. Most significant is that it occurs at a time of very high milk production, requiring large amounts of glucose as a precursor for the lactose of the milk. Unfortunately for the cow, the glucose is derived essentially in toto from gluconeogenesis, since the ruminant does not absorb the breakdown products of cellulose, per se, but rather the short-chain fatty acids resulting from bacterial fermentation in the rumen. The demand for glucose can be so great that hypoglycemia is induced (with cell starvation). The resultant mobilization of fatty acids with subsequent ketosis can overwhelm the cow's acid-base balance, producing ketoacidosis, dehydration, coma, and death. (The proper treatment is intravenous administration of glucose!)

In diabetes mellitus, of course, there is intense hyperglycemia, but the concept of intracellular starvation, with its metabolic consequences, is still pertinent. The chemical and clinical findings in uncontrolled diabetes mellitus are quite predictable if the foregoing discussion is kept in mind (Fig. 17-6). Thus, one finds, in addition to the hyperglycemia, increased levels of circulating free fatty acids and amino acids (secondary to insufficient insulin) and, as a result, increased ketone bodies, lactate, and urea. There are two important physiological consequences of these abnormalities. The first, osmotic diuresis, is produced by the urine glucose excretion induced by the high blood sugar. As a result, there is a profound urinary loss of electrolytes, causing depletion of extra- and intracellular

DIABETES MELLITUS

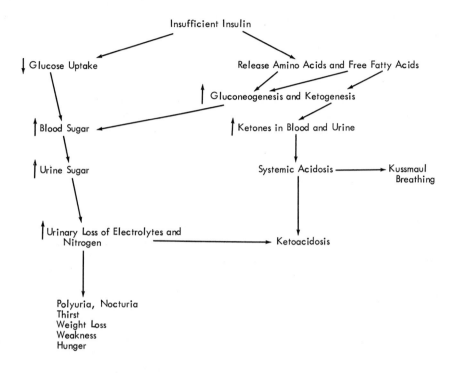

Fig. 17-6. Metabolic relationships in diabetes mellitus.

fluids and, thus, dehydration, weakness, and weight loss. The second consequence is the striking increase in ketone bodies, upsetting the systemic acid-base balance. Systemic acidosis can supervene, and in severe cases collapse, coma, and death in ketoacidosis follow.

There is a clinical variation of this syndrome, where the two processes—hyperglycemia and osmotic diuresis with fluid and electrolyte depletion, versus ketogenesis—can be separated. This is the syndrome of hyperglycemic nonketotic coma. In this situation, there is intense hyperglycemia, with polyuria, dehydration, electrolyte depletion and excessive thirst. As long as the patient can keep up a reasonable fluid intake, coma is averted. Usually, however, a point comes when, because of dehydration and weakness, oral fluid replacement is unable to match urinary losses.

At this juncture profound dehydration ensues, with partial circulatory collapse and sharp increase in blood sugar. Coma supervenes. The patient therefore presents with intense hyperglycemia, with blood sugar levels of 800 to 2,000 mg.%, but without ketoacidosis. A precise explanation of this lack of ketoacidosis is lacking. Several tentative explanations have been offered. One hinges on the fact that the level of insulin required for glucose uptake by adipose tissue is an order of magnitude higher than the level of insulin required for inhibition of fatty acid release. Although there are only a few reports of insulin levels measured in this clinical situation, they are in the range that might separate these effects, allowing inhibition of fatty acid release, but not permitting glucose uptake. Another explanation emphasizes the final dramatic increase in blood sugar, secondary to partial circulatory collapse and poor renal perfusion. This extraordinarily high blood sugar level is thought to allow enough glucose entry in the cells by a mass-action effect to "turn off" ketogenesis. There are animal experiments showing clearing of ketone bodies by the artificial elevation of blood sugar to the very high levels found in this syndrome. A third explanation is related to measurements in this syndrome of blood cortisol and growth hormone, which are found to be elevated to a much smaller degree than in the "usual" patients with ketoacidosis. The lower levels of these lipolytic factors might explain the lack of ketoacidosis. To what degree any or all of these explanations play a role is not clear. The clinical syndrome is, therefore, well defined, but its pathophysiology remains to be precisely elucidated. Happily, the pathological process can plateau at much less severe levels of symptomatology, depending in good part on the amount of available insulin, and also depending on such factors as stress, metabolic needs of the body, and body weight.

What is the hormonal basis of diabetes mellitus? Since the pioneering work of Von Mering and Minkowski, and that of Banting and Best, the pancreas and the beta cell of the pancreatic islets have been thought to be central to the pathogenesis of diabetes mellitus. Yet with the discovery and understanding of the mechanisms of action of a number of other hormones, each hormone has, in its turn, also been thought to play a critical pathophysiological role. Catecholamines, for example, can cause hyperglycemia by inducing glycogenolysis and a decrease in insulin release by the beta cell; indeed, patients with tumors secreting excessive amounts of catecholamines (pheochromocytomas) can demonstrate varying degrees of hyperglycemia. Glucocorticoids in high doses can produce hyperglycemia secondary to the induction of protein breakdown, increased circulating amino acids, increased free fatty acids, and

an increase in gluconeogenesis (and possibly induction of resistance in peripheral tissues to the action of insulin). Patients with adrenal tumors or those who receive high doses of exogenous glucocorticoids can have hyperglycemia. Similarly, growth hormone can produce hyperglycemia; patients with increased circulating growth hormone secondary to pituitary tumors (acromegaly) also can have hyperglycemia. Although during ketoacidosis (and stress) all these hormones have been found to be elevated, at times of proper metabolic balance no consistent patterns have emerged to suggest that any of these hormones are elevated in a meaningful way. Thus, no proof exists that these hormones are causative in diabetes mellitus. Glucagon, the newest hormone to be appreciated as an important metabolic regulator, recently has been thought to contribute significantly, but more data are needed to test the validity of this hypothesis.

A discussion of the pathophysiology of diabetes mellitus inevitably leads to a consideration of insulin and pancreatic beta cell physiology. The development of a reliable, reproducible, accurate method for measuring insulin in blood (a radioimmunoassay) has led to a veritable explosion of physiological data.

The initial impact of the radioimmunoassay was to confirm the earlier data obtained by bioassay of the insulin content of the blood and the pancreas of so-called *juvenile* (also termed ketosis-prone or insulin-requiring) diabetics. Both approaches indicated that this type of diabetes was characterized by insulin deficiency, these patients having little, or no, insulin in their blood or pancreas. This severe deficiency of insulin initiates the pathophysiological process noted in the sections above in a reasonably predictable manner.

Another striking finding noted (using the radioimmunoassay) was that the other type of diabetic—the so-called *maturity onset* (also termed insulin-independent or ketosis-resistant) diabetic—had measurable blood insulin. In fact, on initial inspection of the data, the insulin appeared to circulate in greater amounts in this type of diabetic patient than in normal individuals. To explain this apparent paradox of hyperglycemia in the face of hyperinsulinemia, the concept of tissue resistance to the action of insulin was offered. Several hypotheses were suggested to explain this resistance. Free fatty acids are known to inhibit the action of insulin on muscle both in vivo and in vitro. Since free fatty acids are elevated in the pertinent clinical situations (maturity-onset diabetes mellitus, acromegaly, hyperadrenocorticism, and pheochromocytoma), they were thought to play a pathogenetic role. Other insulin antagonists were suggested to explain this apparent paradox, one being an albumin derivative (termed *synalbumin*). Although antagonism to the action of insulin certainly occurs, it is thought to be relatively unimportant in the pathogenesis of diabetes mellitus. Other explanations have implicated the pancreatic beta cell. Thus, the insulin that is elevated in the diabetic patient has been suggested to be different from normal insulin. This hypothesis has special appeal in the light of developments in the area of beta cell insulin synthesis. It is now acknowledged that the two-chain insulin molecule has a single-chain precursor (termed proinsulin) and that the peptide connecting the two chains is cleaved and released into the circulation with the hormone. The proinsulin, although physiologically much less active than insulin, still reacts with the insulin antibody used in the usual radioimmunoassay to a variable, but significant, degree. Therefore, if the diabetic patient had a supranormal amount of proinsulin in his blood, one would predict a higher level of immunologically measured insulin. Preliminary work suggests that the mechanism converting proinsulin to insulin is not faulty, but more work must be done to confirm this. The insulin or proinsulin in

the diabetic patient has never been characterized chemically. Therefore, the possibility that the molecule is different in diabetes remains open to question.

Mounting evidence suggests that the major defect in the maturity-onset diabetes is the same as that in juvenile diabetes (i.e., too little insulin). Apparently, the difference between the two types lies in the degree of insulin deficiency. The seeming paradox noted above is resolved when one realizes that the levels of insulin which appear to be above normal are, in fact, less than normal for that particular blood sugar level. Obesity aggravates or precipitates the condition because these patients have resistance to circulating insulin at elevated levels. If one compares two obese patients, one with diabetes mellitus and one without, the diabetic releases insufficient insulin and has a lower blood insulin level. Thus, all diabetics have diminished insulin production. This finding of diminished or inadequate insulin output for a given blood sugar stimulus has permitted the concept of the "sluggish" beta cell.

Thus, in the maturity-onset diabetic patient, enough insulin apparently circulates to allow some glucose uptake and some inhibition of free fatty acid release from adipose tissue. Because of this, the metabolic catastrophe of the insulinoprivic ("juvenile") diabetic patient is avoided. The patient with maturity-onset diabetes mellitus, however, lacks enough available insulin to increase glucose uptake in the tissues that are insulin-responsive, or to inhibit free fatty acid release from adipose tissue in normal amounts. The increased circulating free fatty acids not only inhibit the action of insulin in the periphery but stimulate hepatic gluconeogenesis. This augments hepatic glucose output (and ketone bodies as well) with resultant further increase in blood sugar. Whether the lack of insulin and/or excess of glucagon also play an important role directly in the liver remains a moot point. A patient in this pathophysiological state can continue on in this manner for months or years. Should an additional stress supervene (e.g., weight gain, infection, surgery), the balance can be tipped negatively, with hyperglycemia and ketonemia overwhelming the body insulin reserve and with subsequent clinical deterioration.

Patients with diabetes mellitus of either "type" can develop certain microvascular complications involving the eye, the nervous system, and the kidney. Although a detailed clinical and pathological description of these complications is beyond the scope of this chapter, it should be stated that they involve small vessels in tissues that are non-insulin-sensitive. An enormous controversy exists regarding the effect of "normalizing" blood sugar in an attempt to prevent or delay these microangiopathic complications. The argument, unfortunately, has little ultimate practical meaning. All diabetics have abnormal blood sugars intermittently, even within a given day, despite the most zealous attention to the details of their diet and insulin administration. The problem, therefore, realistically hinges on the question of whether lowering the blood sugar "toward normal" or "as normal as possible" has an effect on the rate of development or degree of these pathological processes. The answer to this question is not known. The frequency of these complications even in clinics that stress "tight" control of blood sugar is very high and roughly the same as in other clinics where control is not stressed. Many experts in the field, therefore, do not feel it is in the patient's best interest to emphasize "tight" control to the point of making the diabetic regimen oppressive, especially if repeated episodes of hypoglycemia are a by-product of an overzealous attempt at control. This is not to say that elevation of blood sugar above normal may not be playing a role in the pathogenesis of these complications. This is unknown and hopefully will be the basis

of more basic and clinical research in this area. In addition, it will be of extraordinary interest to see if true normalization of blood sugar, by such techniques as pancreatic transplant or some yet-to-be-devised servomechanism whereby blood sugar can be precisely controlled, will have an effect on the rate of frequency and degree of vascular complications.

ANNOTATED REFERENCES

Cahill, G. F., Jr.: Starvation in man. New Eng. J. Med., 282:668, 1970. (Excellent review which forms the basis of the material presented in the section on starvation in the text above)

Coleman, J. D.: Metabolic Interrelationships Between Carbohydrates, Lipids, and Proteins. *In:* Bondy, P. K.: Diseases of Metabolism. p. 88. Philadelphia, W. B. Saunders, 1969. (An exhaustive discussion of the various enzymatic pathways and their interrelationships)

Jeanrenard, B.: Adipose tissue dynamics, revisited. Ergebnisse Physiol., 60:58, 1968. (An excellent review of adipose tissue physiology and biochemistry)

Krebs, H. A.: Bovine ketosis. Veterinary Rec., 78:187, 1966. (Excellent review by an acknowledged expert on the interrelationship between gluconeogenesis and ketogenesis)

Kronfeld, D. S.: Excessive gluconeogenesis and oxaloacetate depletion in bovine ketosis. Nutrition Rev., 27:131, 1969. (A very good precis of the problem of possible interrelationship of the regulation of gluconeogenesis and the availability of oxaloacetate)

Lehninger, A. L.: Bioenergetics. New York, W. A. Benjamin, 1965. (A very good primer on energy interrelationships)

Levine, R., and Haft, D. E.: Carbohydrate homeostasis. New Eng. J. Med., 283:175, 237, 1970. (Well-written general review of carbohydrate physiology and biochemistry)

Levine, R.: Mechanisms of insulin secretion. New Eng. J. Med., 283:522, 1970. (Summary of the status of our understanding of the physiology of insulin secretion)

Randle, P. J., and Morgan, H. E.: Regulation of glucose uptake in muscle. Vitamins and

Hormones, *60*:58, 1968. (An excellent review of the regulation of glucose metabolism in muscle)

Sawin, C. T.: The Hormones. Endocrine Physiology. Boston, Little, Brown and Company, 1969. (A very readable short text on all aspects of endocrinology)

Sutherland, E. W.: On the biological role of cyclic AMP. JAMA, *214*:1281, 1970. (An excellent short review of the actions of the compound by the principal investigator in the fields)

Weber, G., *et al.*: Regulation of gluconeogenesis and glycogenolysis: Studies on mechanisms controlling enzyme activity. Advances in Enzyme Regulation, *5*:257, 1967. (A good review of hepatic adipose physiology)

Williams, R. H.: Textbook of Endocrinology. Philadelphia, W. B. Saunders, 1968. (The chapters on the pancreas and on hypoglycemia are especially pertinent, but the whole text is an excellent general review of endocrinology and metabolism.)

18

Endrocrine Mechanisms
of Reproduction

Mortimer B. Lipsett, M.D.

INTRODUCTION

Reproductive biology encompasses many diverse disciplines—from population dynamics, pregnancy, contraception, through gonadal physiology, and biochemistry of the steroid and protein hormones, to molecular biology. Since each of these topics has been the subject of extensive monographs, coverage in this chapter must necessarily be selective, brief, and somewhat didactic. In a rapidly expanding area of science such as reproductive biology, new hypotheses quickly supplant old ones, and what is fact today may be error tomorrow. With this disclaimer, I shall present fact and informed opinion concerning several aspects of *human* reproductive endocrinology with reference to other species only when evidence from the human studies is not decisive.

It may seem that to discuss reproductive endocrinology without considering sexual behavior in depth is to omit an integral, important, and certainly interesting aspect of the field. But sexual behavior is a sphere in which man clearly differs from all other species. In animals, sex drive and behavior are instinctual and depend on appropriate hormonal signals. In man, sex behavior is conditioned by other stimuli and is largely independent of these hormones. We shall therefore ignore the behavioral aspects of sex which have only recently been considered a legitimate topic

for study and concentrate instead on those physiological mechanisms wherein the biologist has been able to quantify response and develop sound concepts.

To the physician the word *sex* denotes differentiation at many levels. To describe the sex of an individual, one should be able to specify genetic sex, gonadal sex, hormonal sex, sex of internal and external genitalia, sex of rearing, and sex of orientation. One or several of these may be discordant, raising difficult therapeutic problems. For the individual to assume a mature and stable role in society, it is necessary for the physician to diagnose and evaluate deviations from the normal pattern and to recommend therapy. Concepts and techniques now at hand make this possible in every instance.

THE HORMONES

Definitions

The terms *androgens, estrogens,* and *progestins* are operational definitions for classes of compounds that exert distinctive biological effects. Thus, any compound that causes growth of the prostate and seminal vesicles in the castrate male is an androgen. Similarly, estrogens cause growth of the uterus and hyperplasia of the endometrium, and progestins transform the endometrium to a secretory type and maintain pregnancy after castration. The steroid chemist has synthesized hundreds of steroids with one or more of these actions and several are in therapeutic use.

The corresponding hormones circulating in the blood are testosterone, estradiol, and progesterone. However, during pregnancy estriol is the principal circulating estrogen and is present in amounts several-fold greater than estradiol. Whether these are the active intracellular hormones is not clear yet and will be discussed subsequently. In addition to these hormones, there are a number of steroids secreted by the endocrine glands that have no intrinsic biological activity but are converted by the liver and other tissues into the hormones. An example of this is the prehormone, androstenedione, secreted predominantly by the adrenal cortex. This prehormone has little intrinsic androgenicity in man but is androgenic because of its transformation to testosterone after secretion.

Gonadal Steroids

As a prelude to the study of the pathophysiology of those endocrine glands important in reproductive biology, the hormones themselves should be considered briefly. The biosynthesis of steroids by the adrenal cortex, testis, ovarian follicle and corpus luteum has the same basic pattern (Fig. 18-1). Either plasma cholesterol or cholesterol synthesized from acetate by the gland is the sterol converted by mitochondrial enzymes to the first steroid, pregnenolone. Subsequent steps leading to the synthesis of testosterone and estradiol are directed by microsomal enzymes. Steroid biosynthesis also occurs in the placenta but the pathways may differ from those in the gonads and adrenal cortex.

Although the testicular Leydig cells and ovarian follicle cells have the same enzymes, the steroid secretory patterns differ. The testis aromatizes androgens (inserts double bonds) poorly and little, if any, estrogen is synthesized and secreted. In the ovary, although androstenedione and testosterone are the proximate precursors of estrone and estradiol, respectively, little of each androgen is secreted normally. Aromatization of the A-ring of the steroids to produce estrogens proceeds by a series of steps in the ovary. The process of aromatization is irreversible in vivo. The ovary and placenta have the highest aromatizing activity, but the testis, adrenal cortex, and liver are capable of carrying out this transformation to a limited extent.

The principal precursors of urinary 17-ketosteroids (excreted metabolites of androgens) in men and women are the adrenal cortical steroids, dehydroepiandrosterone and its sulfate. Since the liver and other tissues are able to hydroxylate the steroid molecule at several sites, many metabolites of testosterone have been identified. These metabolites are present in small amounts and are of chemical interest, but have little physiological significance. Following reduction of the 3-ketone and oxidation of the 17-β-hydroxyl group of testosterone, the resulting isomeric ketosteroids are conjugated primarily with glucosiduronic acid and to a lesser extent with sulfuric acid. This conjugation reduces metabolic activity and renders the steroids water-soluble so that they are excreted by the kidney.

The principal metabolites of estradiol are estrone, estriol (16-β-hydroxyestradiol) and 2-hydroxyestrone. These steroids, too, are conjugated at the 3-position with glucosiduronic acid; conjugation with sulfuric acid and other compounds also occurs. Over 30 different metabolites of estradiol (all of minor physiological interest) have been identified in urine. Estrone, estradiol, and estriol can be measured although they are normally excreted at rates of less than 10 μg per day. However, 2-hydroxyestrone is easily destroyed; and since this may be the major pathway of metabolism in some patients, it should be recognized that the measurement of the classical urinary estro-

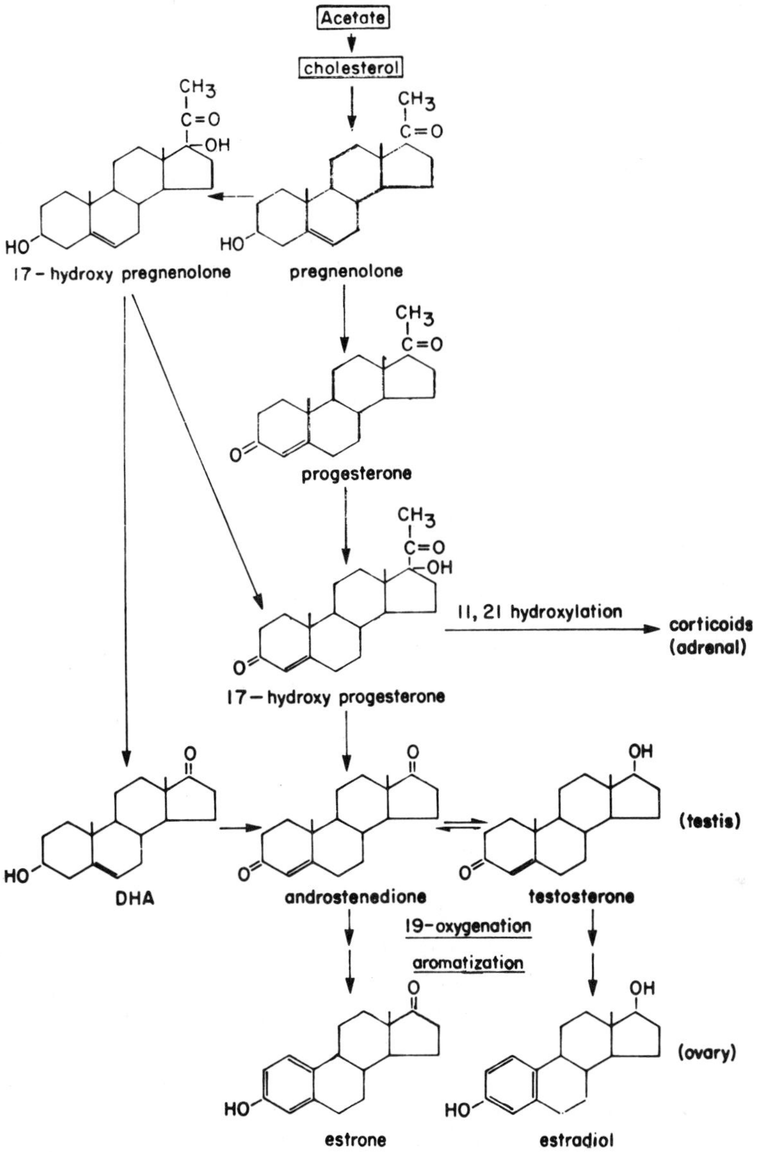

FIG. 18-1. Synthesis of testosterone.

gens may not always accurately reflect estrogen secretion rates.

Protein Hormones

The four protein hormones that are of immediate relevance are: follicle-stimulating hormone (FSH), luteinizing or interstitial cell-stimulating hormone (LH), human chorionic gonadotropin (HCG), and human placental lactogen (HPL). The first two are synthesized and secreted by the anterior pituitary gland in response to hypothalamic-releasing factors. FSH is necessary for ovarian follicle growth; in the testis, it initiates and maintains seminiferous tubule function in conjunction with androgen. LH stimulates testosterone synthesis by the Leydig cells, initiates ovulation and is responsible for maintenance of the corpus luteum during the menstrual cycle. HCG and HPL are synthesized by the placenta, by tumors originating from trophoblastic cells, and rarely by other cancers. HPL has a molecular weight of 20,000 to 30,000 and is quantitatively the most important protein synthesized by the placenta late in pregnancy. This hormone is thought to be the most critical hormone concerned with fetal growth and development, exerting its influence upon the fetus through alterations in maternal metabolism, notably conversion to metabolism of fat by sparing glucose for fetal consumption. It has activities resembling those of growth hormone and prolactin; in fact, it has considerable immunological and structural similarity to growth hormone.

HCG is the placental gonadotropin (m. w. 60,000) whose presence in increased amounts in the urine is responsible for positive pregnancy tests. HCG, FSH, and LH may be split into two polypeptide chains, one of which is common to the three hormones. Specificity is apparently conferred by the other nonidentical polypeptide. HCG is a glycoprotein containing 30 percent of carbohydrate. At the terminal end of each carbohydrate chain is sialic acid which confers specific properties on HCG. A high content of sialic acid is associated with a long half-life of the molecule; thus HCG has a half-life of 10 to 20 hours in the circulation. By contrast, FSH and LH have relatively small amounts of sialic acid and are removed quickly from the blood. HCG has the important function of maintaining corpus luteum function and high rates of progesterone secretion during the first 6 to 8 weeks of pregnancy; its role later in pregnancy is unknown.

Prostaglandins

The prostaglandins are 20-carbon fatty acids containing a cyclopentane ring and are synthesized in vivo from 20-carbon essential fatty acids by cyclization and oxidation. General descriptions of the effects of the prostaglandins are difficult

since each of the 15 to 20 naturally occurring prostaglandins may have a different effect. For example, two of these, PGF (prostaglandin F) and PGE (prostaglandin E), stimulate uterine contractions whereas other prostaglandins have opposite effects. Further, the hormonal status of the uterus influences the response to the prostaglandins. Prostaglandins may be involved in the regulation of corpus luteum function by several mechanisms: alterations of blood flow, stimulation of steroidogenesis, and competition with LH for receptors.

Since prostaglandins have been isolated from many tissues, it is difficult to characterize them as hormones. However, such intriguing findings as the correlation of seminal fluid prostaglandin content with male fertility and the effects of prostaglandins on fallopian tube motility and on ovarian secretory activity make their study an exciting area of research.

Blood Levels and Transport

The development of relatively simple methods for measuring small amounts of steroid or protein hormones has initiated the recent advances in reproductive endocrinology. These radioligand assays for proteins, polypeptides, steroids, and various small molecules have assumed such importance in medicine that an understanding of the principles is necessary.

A radioligand assay requires a binding substance, generally a protein, with high binding affinity and specificity for the ligand which is chemically similar or identical to the substance to be measured. Pure ligand must be labeled at high specific activity, e.g., tritium-labeled estradiol or iodine-131-labeled LH. When labeled ligand, $^{\ast}L$, is added to a solution containing the binding protein, B, the equilibrium, $B + {^{\ast}L} \rightleftarrows B{^{\ast}L}$ is reached, and a characteristic association constant is 10^8 to 10^{10}. Since labeled ligand can be displaced from the complex by unlabeled material, the radioactivity not bound to protein will be proportional to the amount of unlabeled material present. The assayist must then separate *L from B*L and there are adequate and simple ways of doing this. The sorts of binding proteins that have been used in these measurements are: antibodies directed against the protein hormones or against the steroids used as haptenes coupled to a protein, plasma transport binding proteins such as cortisol-binding globulin or testosterone-estradiol-binding globulin (TeBG), specific intracellular cytoplasmic steroid receptor proteins and membrane-bound polypeptide hormone receptors. The necessity for the sensitivity achieved by these methods is emphasized by the low plasma concentrations of the protein and steroid hormones (Table 18-1).

In plasma, estradiol and testosterone are bound to TeBG, a β-globulin with an association constant for these steroids of about 10^9. This binding protein has moderate specificity, complexing structurally related, but not necessarily biologically active, steroids. By contrast albumin binds steroids but the Ka is only 10^5. At physiological concentrations, most testosterone and estradiol is bound to the specific binding protein and only a small fraction can be considered to be either free or albumin-bound in plasma. Because of the relatively tight binding, protein-bound hormone cannot be metabolized by the cell nor can it enter cells of target tissues. Since the level of binding protein may be altered by drugs and hormones, knowledge of binding as well as of total steroid hormone concentration may be necessary to appreciate fully alterations in steroid secretion and metabolism.

MECHANISM OF ACTION

This exciting frontier of endocrinology has seen the greatest progress in defining the way the sex steroids act. The steroids enter the cytoplasm of steroid-responsive tissues and are retained by a specific cytosol steroid-binding protein of high affinity. This process has been studied most carefully for the estrogens, and such cytosol receptors have been found in the following estrogen-responsive tissues: vagina, uterus, hypothalamus, anterior pituitary, carcinogen-induced rat mammary carcinomas, and some human breast cancers. These receptor proteins have biological specificity in contrast to the structural specificity of the plasma binding proteins. Thus, any compound that is an estrogen binds to the cytosol-binding receptor but may not bind to TeBG.

Because of the presence of cytoplasmic binding receptors, estradiol is retained within estrogen-responsive cells for pro-

TABLE 18-1. Plasma Steroid and Protein Hormone Concentrations (ng./100 ml.)

	MEN	WOMEN		
		FOLLICULAR PHASE	LUTEAL PHASE	PREGNANCY THIRD TRIMESTER
Testosterone	700	35	40	100
Estradiol	2	4	15	100
Progesterone	30	40	1,500	15×10^3
LH	15	120 }Peak level	15	—
FSH	15	30 }	12	12
HCG	0	0	0	100×10^3
HPL	0	0	0	600×10^3
Estriol				20×10^3

longed periods without further metabolism. The protein-estradiol complex is then transported to the nucleus where further alterations in the complex take place. This process presumably initiates transcription of new information from the DNA resulting in synthesis of those structural proteins and enzymes necessary for cell growth.

When testosterone exerts its effect on prostate and seminal vesicles, it too is retained by the cells because of a cytoplasmic binding protein. However, after entering the nucleus, it is reduced at the $\Delta^{4,5}$-double bond to 5α-dihydrotestosterone. The nuclear 5α-reductase is present only in androgen-responsive tissues. Presumably the dihydrotestosterone then initiates transcription of DNA with resulting growth and increase in function of the responsive cells.

These are the barest outlines of the intersection of molecular biology and endocrinology. But it is here that mechanisms of disease will be explored and solutions found.

TESTIS

Fetal Activity

The fetal testis plays an important role during embryogenesis. Leydig cells are prominent by the twelfth week of develop-

ment and secrete an androgen that stimulates the wolffian duct to form the vas deferens, epididymis, and seminal vesicles and induces male external genital development—fusion of the genital folds to form the scrotum, lengthening of the genital tubercle and fusion of the folds to form the phallus. It is probable but not proved that this androgen is testosterone. In addition to the androgen, the fetal testis produces a substance of unknown structure that suppresses müllerian duct structures which in the female are the anlagen of the uterus, the fallopian tubes and the upper one third of the vagina. This substance probably acts in concert with testosterone to effect normal wolffian duct differentiation (Fig. 18-2). In the absence or failure to respond to testicular secretions, the male develops a female phenotype.

Patient 1. An 18-year-old girl was referred because of primary amenorrhea. She was tall and well developed, and there was an absence of axillary and pubic hair. There were masses in the labia, the vagina was short and ended in a blind pouch; cervix and uterus could not be palpated. The karyotype was 46 XY. Laparotomy revealed epididymis, testes with abundant Leydig cells and disordered spermatogenesis, vas deferens in the labia, and an absence of müllerian derivatives.

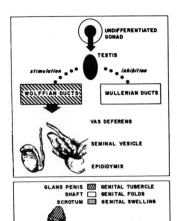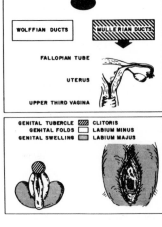

FIG. 18-2. Gonadal and genital development. Note that internal genitalia develop from separate primordia present in both sexes and external genitalia develop in a continuous transformation of anlage common to both sexes. (Federman, D. D.: New Eng. J. Med. 277:351, 1967. With permission)

It is known from many studies that plasma testosterone concentrations in these women are in the normal male range and that these patients have a genetic defect which prevents their responding to testosterone. This congenital insensitivity to androgen, which existed in fetal life, is a classical experiment of nature, revealing the role of fetal androgen in development and providing a striking demonstration of the many different aspects of sex. Although a normal fetal testis was present, the fetal androgen was ineffective and phenotypic development was therefore female. However, the inducer of müllerian duct regression was secreted and was active so that the uterus and fallopian tubes did not differentiate. This syndrome is known as testicular feminization.

The patient then was a genetic and gonadal male, had male internal genitalia, female external genitalia, and was female in sex of rearing and orientation. These patients are well-adjusted women who are unhappy only about their amenorrhea and their inability to have children. A clinical note—the physician should always refer to these patients as women for they are women in the most important ways—sex of rearing and of orientation. The fact that they are genetic, gonadal, and hormonal males is of little concern. Should the physician disclose any of these last facts to the patient he may do irreparable psychological harm.

Pubescence

During childhood, the testis has questionable endocrine activity although in several experimental animals it is possible to demonstrate the existence of a testicular-hypothalamic interplay before sexual maturation. The mechanism of pubescence in many remains unknown. Through childhood FSH and LH are secreted in low amounts with a slight predominance of FSH. During pubescence, the levels of these hormones increase, but the rate of increase of LH is three- to four-fold whereas

FSH increases about 1.5 times, so that by maturity urinary LH is greater than urinary FSH. At about age 8, the seminiferous tubules begin to enlarge and develop a lumen and subsequently Sertoli cells appear. Six months to 2 years before the first clinical evidence of pubescence, the Leydig cells differentiate from mesenchymal cells under the influence of increasing LH levels, and plasma testosterone begins to increase toward the normal adult levels.

Leydig Cells. The adult testis has two functions, production of the microgamete (the sperm) and secretion of testosterone. Each function is separately controlled by the gonadotropins in a relatively independent manner. The Leydig cells comprise less than five percent of the total testicular volume in adults. Thus, testicular volume is directly related to seminiferous tubule development. Since seminiferous tubule maintenance requires FSH, relatively simple clinical observations and a few laboratory tests may suggest important clinical inferences.

Patient 2. A 6-year-old boy was referred because of precocious virilization. Increased height, muscular development, facial and pubic hair, and phallic growth confirmed the impression of increased androgenic effect. The testes, however, were small. Since testicular volume had not increased with other evidence of pubescence, this could not be true precocious puberty due to early initiation of gonadotropin secretion (FSH and LH). Rather, the cause of the virilization was most likely secretion of an androgen from an unusual source, either an adrenal cortical tumor, congenital adrenal hyperplasia, or a testicular tumor.

In adult men, testosterone is secreted by the Leydig cells at a rate of about 7 mg. daily, which maintains a plasma testosterone concentration of about 0.7 μg/100 ml. Testosterone is not a 17-ketosteroid although its principal metabolites are. Since the urinary 17-ketosteroids resulting from testosterone secretion constitute only 25 percent of the total urinary 17-ketosteroids, this measurement cannot be an adequate index of testosterone production. Further, since the analysis of urinary 17-ketosteroids is relatively imprecise, small changes are unreliable.

Patient 2. (cont'd). The boy's urinary 17-ketosteroid excretion was 3.5 mg./24 hr. But his plasma testosterone level was 0.35 μg/100 ml., sufficient to account for his virilization. To achieve this testosterone level in a 6-year-old boy, a testosterone secretion rate of about 1 mg. daily would be required. This would yield only 0.5 mg. of 17-ketosteroids, an amount within the error limits of the assay. Thus, the apparent paradox of high plasma testosterone levels and low urinary 17-ketosteroids is easily explained. Since patients who are virilized by adrenal tumors or congenital adrenal hyperplasia almost invariably have high urinary 17-ketosteroids, the working diagnosis became a tumor of the testis.

Normally, Leydig cell activity is strictly dependent upon pituitary LH although endogenous or exogenous HCG can also stimulate the Leydig cells. The secretion of LH is related to Leydig cell activity by a negative feedback system. When free testosterone levels in plasma fall, hypothalamic receptors are activated and LH-releasing factor is secreted, causing the pituitary to release LH thereby returning plasma testosterone levels toward normal. Conversely, a high dose of androgen will suppress LH secretion and plasma testosterone will decrease.

Patient 2 (cont'd). Since one possible cause of the virilization was the secretion of HCG (placental protein hormone with LH activity) by a trophoblastic tumor, plasma and urinary HCG levels were measured and found to be low. The patient was given a synthetic androgen in doses high enough to inhibit LH secretion in normal men and thereby cause a 90 percent decrease in plasma testosterone. His testosterone level did not change and his low urinary gonadotropins were unaltered. Since functional endocrine tumors are usually autonomous and maintain secretory activity irrespective of normal control mechanisms, it was probable that the virilization was the result of a tumor of the Leydig cells. The testes were explored even though no tumor was palpable. At surgery, a small benign interstitial cell tumor was found and the testis

removed. One week later, plasma testosterone concentration was 0.04 μg/100 ml. and marked regression of virilization was noted 6 months later.

Spermatogenesis. This process is as complex as that of stem cell differentiation and proliferation in the bone marrow. Waves of spermatogenesis begin at different areas within the same tubule so that casual inspection gives an impression of haphazard maturation. The kinetics of spermatogenesis have been studied in detail, and in man development of the mature sperm from the most primitive spermatogonium takes about 75 days. The unique feature about this maturation process is the large segment of 20 to 30 days occupied by the premeiotic prophase of the spermatocyte. The duration of each step in the sequence of spermatogenesis is fixed and neither drug nor alteration of hormonal environment has altered this timing. If FSH is withdrawn, as after hypophysectomy in the rat or development of a pituitary tumor in man, the spermatogonia slowly disappear, the tubules hyalinize and the changes then become irreversible. However, if FSH secretion remains low from childhood, then the tubules remain in their immature state and can be induced to mature at any time by administration of human FSH and androgen.

Control of FSH secretion in men is unknown. In general, FSH and LH increase and decrease in parallel; however, experimental situations have been described in which levels of these hormones vary independently. Recent experiments present good data suggesting that some component of the seminiferous tubule secretes a substance that regulates pituitary FSH secretion.

Secondary Hypogonadism. *Patient 3.* A 21-year-old man was referred because of sexual immaturity. He had not entered pubescence; there was no beard, the voice was high-pitched, the testes and penis were small, the prostate was not palpable. All tests of pituitary function were normal except that plasma and urinary FSH and LH levels were in the range found in the prepubescent child. The diagnosis then was hypogonadotropic hypogonadism.

In general, pubescence begins by age 12, and rarely pubescence does not begin until age 20 or later. If spontaneous pubescence occurs, then the diagnosis must have been only delayed pubescence. At some point, however, a diagnosis of hypogonadotropic hypogonadism must be entertained. These patients cannot be distinguished from those with delayed puberty except by the duration of the problem. There is an interesting variant of hypogonadotropic hypogonadism in which the defect is associated with anosmia (Kallman's syndrome); this sensory modality should be tested in any patient, man or woman, with delayed pubescence. Parenthetically, the relationship between the sense of smell and reproductive function is important in many animals.

Patient 3 (cont'd). Because it was imperative for psychological reasons to induce virilization in this 21-year-old man, he was given injections of testosterone enanthate, a long-acting preparation, 200 mg. every 2 weeks. On this regimen he virilized completely, but his testes remained small. He married at age 24, and 3 years later inquired about the possibility of restoring fertility. The measurements of gonadotropins were unchanged. Testicular biopsy showed a few primary spermatogonia and no Leydig cells. There was no tubular fibrosis. The testosterone was stopped, and the patient was given FSH, as human menopausal gonadotropin, 250 IU three times weekly, and HCG, 1,000 IU three times weekly. Ten weeks later sperm were present in the ejaculate, and after 16 weeks the sperm count had reached 30,000,000 per mm.[3] The patient's wife became pregnant soon after.

In the previously unstimulated testis, FSH will produce full maturation of the seminiferous tubule in the presence of androgen. The process of maturation involves differentiation of Sertoli cells, progression of spermatogenesis through meiotic division to mature sperm. Concurrently, elastic fibers become visible in the tunica propria

of the tubule and their presence is a sign that pubescence was initiated.

In the hypogonadotropic states, it is usual to have FSH and LH depressed simultaneously whether the cause be idiopathic, as in Case 3, genetic, as in Kallman's syndrome, or a result of pituitary destruction. Monotropic deficiency of FSH has not been recognized as yet. The reverse situation, a relative deficiency of LH, has been accorded the title of "fertile eunuch syndrome." In these patients pubescence fails because of inadequate stimulation of Leydig cell secretions. However, the testes enlarge owing to seminiferous tubule development, and spermatogenesis is complete in a few tubules. Since these men have little libido or potentia, they are fertile in name only. When HCG is given, masculinization ensues and the sperm count improves.

One of the most frequent symptoms of pituitary tumors in men is loss of libido owing to low LH levels and consequent decreased testosterone secretion. Since only gonadotropin production by the pituitary may be impaired, other pituitary function tests may be normal. Following removal of the tumor and consequent hypopituitarism, androgen deficiency may be replaced by either HCG or a synthetic androgen, and FSH deficiency may be overcome by the use of human menopausal gonadotropin when fertility is desired. In contrast to patients with hypogonadotropic hypogonadism in whom testicular function may always be restored, if pituitary function in normal adult men is interrupted for a prolonged period, irreversible hyalinization and fibrosis of the seminiferous tubules occur.

Primary Hypogonadism. *Patient 4.* An 18-year-old boy reported for a preinduction physical examination. On casual inspection he was noted to be without facial hair and to have mild gynecomastia. Pubic hair was sparse and the phallus was normal, but the testes were firm and pea-sized. The appearance of gynecomastia and small firm testes alerted the physician to the possibility of Klinefelter's syndrome, the most common cause of primary hypogonadism.

Patients with Klinefelter's syndrome, as described in 1942, are men with gynecomastia, aspermatogenesis, increased numbers of normal Leydig cells in clumps and variable eunuchoid features. In the subsequent years, increasing knowledge of variants of the syndrome and its pathogenesis have changed the definition somewhat. It is accepted now that any man with more than one X chromosome in any tissue has Klinefelter's syndrome. Thus, karyotypes such as 47 (XXY), 48 (XXYY), 48 (XXXY) all define the disease. Rarely, mosaicism may produce an XXY karyotype in the testis only, so that there is disturbance of testicular function only.

The clinical features of over a thousand patients with this relatively common syndrome have been reviewed. Among the most frequent signs were small testes, azoospermia (no sperm in the ejaculate), and impaired spermatogenesis. Gynecomastia and features of eunuchoidism appeared less frequently. When gonadotropins were measured, plasma or urinary LH and FSH were usually high.

The histological picture of the testis is essentially normal before pubescence, but the onset of pubescence is associated with hyalinization of the seminiferous tubules and greatly decreased numbers or absence of germinal cells. Elastic fibers do not appear in the tunica propria. The Leydig cells appear hyperplastic and are often clumped. In spite of this, plasma testosterone levels tend to be reduced and respond poorly to HCG. It has been concluded that the hyperplasia is due to continued LH stimulation, but that there is an intrinsic Leydig cell defect that prevents adequate testosterone synthesis.

Patient 4 (cont'd). Plasma testosterone was increased from 0.09 μg/100 ml. to only 0.21 μg/100 ml. after 4 days stimulation with 4,000 IU of HCG daily. Testicular biopsy showed

hyalinization and fibrosis of most of the tubules and a few intact tubules with progression to the spermatocyte stage. The karyotype was 47 (XXY). Because of inadequate virilization, the patient was placed on a depo-testosterone preparation. This caused an increase in beard growth and some deepening of the voice, but the patient never shaved more than twice a week. Curiously, and without adequate explanation, some eunuchoidal patients with Klinefelter's syndrome seem to be resistant to what are usually adequate virilizing doses of androgen.

Primary testicular insufficiency may range from anorchia to instances of spermatogenic arrest. We should note that the presence of normal male internal and external genitalia is adequate evidence that fetal testicular function was normal. Thus, the patient with anorchia suffered the loss of testes after organogenesis had been completed and before birth. The etiology of this syndrome is unknown.

The common cause of male infertility is spermatogenic arrest. This is descriptive of many probably unrelated disorders resulting in inadequate sperm maturation because of arrest of the process at any of several levels. This may not be absolute, and seminal fluid analysis may reveal either azoospermia or low numbers of normal active sperm. The cause is unknown and treatment with gonadotropins has generally proved unsuccessful. It is important to remember that men with sperm counts as low as 10 to 20 million per ml. may still be fertile so that a diagnosis of absolute sterility should never be made when sperm are present. Further, evaluation of such men should include at least three semen analyses at 5-day intervals, since sperm counts may fluctuate widely.

THE OVARY

The adult ovary is a complex organ with continually shifting relationships between specific organelles and cell types throughout the menstrual cycle. Thus, each compartment within the ovary must be considered separately. They are: the follicle, containing the ovum, the granulosa cells and theca interna cells; the corpus luteum; and the interstitium, comprised of stromal cells, and hilus (interstitial) cells which resemble the testicular Leydig cells.

The steroid-synthetic capacities of each compartment have been studied, and the in vitro information can be summarized briefly as follows: the interstitium (stromal cells) is active in the synthesis of the C_{18}-steroids such as androstenedione, dehydroepiandrosterone, and testosterone; the corpus luteum synthesizes chiefly progesterone but is also active in estrogen synthesis; the theca interna cells during their phase of rapid development, 4 to 6 days before ovulation, are the important source of estrogens; and the granulosa cells may subsequently contribute to estrogen synthesis.

These activities may be brought into focus by consideration of the events of the normal menstrual cycle (Fig. 18-3). The menstrual cycle has been divided into follicular and luteal phases by the LH peak, and the luteal phase is considered to end with the onset of menstrual bleeding. Towards the end of the luteal phase, the plasma concentration of FSH begins to rise and continues to the peak in the early follicular phase. It is probable that this early increased secretion of FSH is necessary for follicular development, since, if it is prevented, the menstrual cycle is postponed. It has been shown in the monkey that follicular growth begins at this time. Throughout the follicular phase the level of LH rises slowly. During this time, the dominant follicle has been increasing in size and, a few days before the LH surge, the theca interna cells begin to proliferate and take on the appearance of secretory cells. Under the influence of LH acting perhaps in concert with FSH, the theca interna cells increase their secretion of estradiol. As this rises to a peak, it triggers

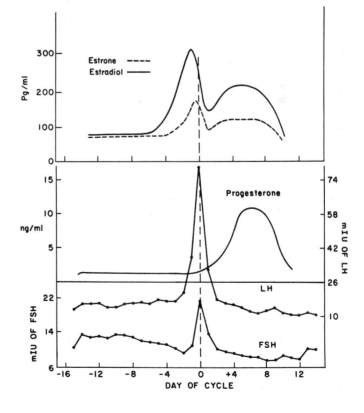

Fig. 18-3. Hormonal patterns in the normal menstrual cycle. The follicular phase lasts from Day — 16 to Day 0; Day 0 is the day of the LH peak. The luteal phase lasts from Day 0 to the onset of menses which usually begins on day + 14.

the release of LH, probably through the LH-releasing factor, which then causes ovulation 18 to 36 hours later. The events associated with ovulation induce luteinization of the granulosa cells. Granulosa cells, removed from the preovulatory follicle and placed in tissue culture, undergo functional and morphological changes compatible with luteinization. This suggests that, once the granulosa cells have matured sufficiently, luteinization is not hormonally directed but is a consequence of follicular rupture. At the time of rupture and follicular disorganization, the secretion rate and plasma level of the estrogens fall, to increase again as the corpus luteum begins to function.

The plasma progesterone concentration remains low until the LH surge and then rises slowly for 2 to 3 days. With full luteinization and development of the corpus luteum, the rate of increase of plasma progesterone is accelerated and it reaches a peak 5 to 8 days after the LH surge. As the progesterone again reaches normal levels the uterine endometrium sloughs and signals the onset of a new cycle.

The control of corpus luteum function varies widely among species so that the following comments apply only to women. It is now probable that maintenance of the normal corpus luteum life span of 14 ($\pm$ 1) days depends upon the continuous presence of LH. In a limited series of studies, it has been found that the life of the corpus luteum cannot be greatly prolonged by LH although HCG will extend it. Since HCG is secreted by trophoblastic cells of the implanted blastocyst 8 to 10 days after ovulation, this would constitute the signal for continuation of corpus luteum function in pregnancy.

In several animals such as the guinea pig and sheep, the presence of a normal uterus causes cessation of corpus luteum function and the start of a new estrus cycle. When the uterus is removed, or when pregnancy ensues, corpus luteum life span is extended.

That this effect is local was shown by removal of one uterine horn in the sheep with consequent prolongation of function of the ipsilateral corpus luteum. It has been suggested that prostaglandins may be the uterine factors that regulate corpus luteum function by their effects on local blood flow. It is doubtful that uterine control of luteal function is important in man.

AMENORRHEA

In either primary or secondary amenorrhea, the central nervous system, the anterior pituitary gland, or the ovary may be primarily involved. Although the causes may be different, investigation of the patient is similar and necessitates methodical evaluation of each gland. Thus, the integrity of the hypothalamic-pituitary system must be assessed, the adequacy of gonadotropin secretion measured, gonadal function evaluated, and the genetic milieu determined. It should be clear that these are dependent variables; e.g., gonadal dysgenesis due to genetic defects causes low estrogen secretion with consequent high gonadotropin secretion.

Primary Amenorrhea. The largest proportion of women with this disease have a 45 (XO) karyotype resulting in the typical Turner's syndrome. In the absence of either a second X or a Y chromosome, gonadal differentiation does not occur. In the absence of fetal gonadal hormones, a female phenotype develops. At the expected time of puberty, plasma LH and FSH levels are elevated in the absence of normal feedback controls. The gonads in such patients demonstrate undifferentiated stromal tissue and an absence of follicles. With sequential estrogen and progestin therapy, normal menstrual cycles and development of secondary sex characteristics can be produced. The wide variety of associated anomalies in Turner's syndrome affecting skin, muscle, skeleton, kidneys and cardiovascular system have no ready explanation but give the clue to the existence of Turner's syndrome in the prepubertal girl.

There are many other variants of intersexes that result in gonads without ova. Whether these gonads were destined to be ovaries or testes can only be surmised from the karyotype—a gonad in a woman is an ovary only when ova are present. However, in the absence of follicles, estrogen production remains low and the high gonadotropin levels are again observed.

Ovariectomized women and postmenopausal women have estrone and estradiol in urine and plasma. The facile explanation is that adrenal cortex secretes estrogens, since suppression of adrenal cortical function diminishes estrogen levels. Although it is true that the adrenal cortex does secrete a very small amount of estrogen, most of the estrogen is derived from conversion in the liver of androstenedione secreted by the adrenal cortex. Thus, estrogen production will be increased when androstenedione secretion is high or the rate of conversion to estrogen is increased. The former situation occurs in feminizing adrenal cortical carcinoma in man; the latter is the cause of some instances of postmenopausal bleeding due to endometrial hyperplasia.

Primary amenorrhea may also be caused by failure of the normal process of pubescence, as in the male. Differentiation of delayed pubescence from hypogonadotropic hypogonadism may also be difficult in women, but it is exceedingly rare for pubescence to occur spontaneously after age 18 in women. Anosmia or hyposmia may be a clue to hypothalamic dysfunction. Treatment with estrogen and progestin to achieve sexual maturation should be initiated by age 16. The question of subsequent fertility can be assessed by proving the presence of ovarian follicles. If they are present, then the chances are good that adequate stimulation and ovulation can be achieved later with human menopausal gonadotropin.

Secondary Amenorrhea. The preponderance of women in this category have "psychogenic" or "hypothalamic" amenorrhea, although it is clear that these two general terms do not necessarily imply the same pathophysiological mechanism. Hence, the length of the adjectives is inversely related to our knowledge of this condition. It seems probable, however, that emotional stress of many kinds can suppress the hypothalamic center responsible for cyclic variations in gonadotropins, leaving only constant low levels of gonadotropin secretion. In over 50 percent of such patients, menses resume spontaneously within a year. However, some have prolonged amenorrhea and induction of ovulation may be necessary to restore menses and fertility.

Amenorrhea is an early sign of pituitary disease due usually to a tumor. Since only gonadotropins may be affected initially and for long periods of time, other tests of pituitary function may not be useful. Further, some small intrasellar tumors may not deform the sella turcica so that secondary amenorrhea is the only clue to their presence. These patients cannot be distinguished from those with psychogenic amenorrhea except by the progress of the disease.

Diseases of the ovaries cause amenorrhea. Follicular atresia may occur earlier than usual so that a postmenopausal ovary can be present in a woman under 40. The high plasma and urinary gonadotropins confirm this diagnosis.

Excess androgen production (see below) will suppress gonadotropin secretion and thereby induce amenorrhea. The other clinical manifestations will direct the diagnostic efforts.

Finally, the most common cause of secondary amenorrhea is pregnancy. Insistence on a pregnancy test has made many a diagnosis early that would have become apparent later.

Induction of Ovulation. Two methods of inducing ovulation have become available in the past several years. As in men, injections of human menopausal gonadotropin (primarily FSH) will promote development of the cells surrounding the germ cell derivatives; thus, follicular growth occurs in response to FSH. The adequacy of follicular development can be assessed by measuring urinary estrogen excretion derived presumably from the secretory activity of the differentiating theca interna cells. At the appropriate time an injection of HCG is given to induce ovulation by simulating the LH surge of the menstrual cycle. When this is successful, fertilization, nidation and pregnancy can result without further medical intervention. Since the amount of FSH given is less than precise, on occasion more than one follicle will mature and superovulation and multiple pregnancies occur.

The second agent that induces ovulation is a very weak synthetic estrogen, clomiphene citrate. This compound acts as a peripheral antagonist of estrogens at any receptor. When a woman who is producing some estrogens receives clomiphene citrate, the clomiphene competes with estrogen at hypothalamic receptors, thereby inducing the release of FSH and LH, and often initiates follicular development with a subsequent ovulatory menstrual cycle. Occasional hyperstimulation with this agent has also resulted in multiple pregnancies.

Ovarian Virilizing Syndromes. Since testosterone and testosterone prehormones are intermediates in the ovarian biosynthesis of estrogens, the occasional role of the ovary in virilizing syndromes is understandable. There are a variety of ovarian tumors, benign and malignant, that cause virilization owing to secretion of testosterone. Hyperplasia of the stromal cells of the ovary is one of the rarer causes of virilization.

Patient 5. A 32-year-old woman was referred because of increased beard growth, minimal balding, deepening of the voice, acne, and amenorrhea of 6 months' duration. Physical examination disclosed a hirsute woman with increased facial hair, temporal hair recession, loss of female fat contours, and clitoral

hypertrophy. There were no signs of Cushing's syndrome. Pelvic and abdominal examination were normal.

The severe virilization suggested that the most likely cause was a tumor of either the adrenal cortex or ovary. Urinary 17-ketosteroids were 12 mg./24 hr. Plasma testosterone was 0.4 μg/100 ml., a value well within the normal male range. With this information, the ovaries were examined at surgery and a small yellowish tumor was found. The pathological diagnosis was benign arrhenoblastoma; the ovary was removed, and the patient resumed menses 2 months later.

In virilized women, normal urinary 17-ketosteroid excretion associated with male plasma testosterone levels places the disease in the ovaries. When adrenal cortical hyperplasia or tumor causes virilization, 17-ketosteroid excretion is high, owing to the secretion of androstenedione and dehydroepiandrosterone and their subsequent conversion to testosterone. Ovarian virilizing tumors have the capacity to synthesize and secrete testosterone so that the urinary ketosteroids may be the same as in normal man. Tumors of the ovary can cause a high urinary ketosteroid excretion, however, so that this evidence cannot be used to localize the disease to the adrenal cortex.

Mild virilization consisting usually of hirsutism alone frequently accompanies secondary amenorrhea, either idiopathic or due to the polycystic ovary syndrome. In each instance it has been shown that testosterone is both secreted by the ovary and produced peripherally from ovarian androstenedione. In only a minority of these patients is adrenal cortical hyperactivity of primary importance in the genesis of the increased androgens.

Polycystic ovary syndrome (Stein-Leventhal syndrome) is another cause of hirsutism and infertility. Women with this disease have the onset of amenorrhea and mild hirsutism at any time after the menarche through the third decade of life. The hirsutism is generally mild, but in a few patients appreciable beard growth and balding may occur. The ovaries are large and have multiple large cysts lying below a thickened capsule. The cysts are surrounded by luteinized theca cells. On the basis of the luteinization and some bioassay data, it was suggested that LH secretion was inappropriately high. This has been confirmed recently by radioimmunoassay, although the events leading to the alteration in LH secretion remain unknown.

Hirsutism in many women is a familial trait which is unassociated with any significant endocrine abnormality except for menstrual irregularities. The 17-ketosteroid excretion is normal or slightly elevated. Most patients with mild hirsutism have increased testosterone production rates and increased plasma testosterone levels (Table 18-2). To understand the origin of the testosterone, one must consider the origin of testosterone in normal women. About 50 percent of testosterone is secreted, 15 percent is derived from peripheral conversion of dehydroepiandrosterone, and the remainder results from transformation of androstenedione secreted by ovary and adrenal cortex. In women with hirsutism or polycystic ovaries, testosterone production rates are higher and more testosterone is secreted. In addition, androstenedione secretion rates are also increased so that there is an additional increment of plasma testosterone

from this prehormone. In these syndromes, the ovary is usually the source of the increased androgen secretion.

THE FETO-PLACENTAL UNIT

Steroid Metabolism

The efforts of Diczfalusy and his collaborators during the past 15 years have clarified many of our concepts of sources and routes of steroid metabolism. This knowledge is of more than theoretical interest because measurements of steroid and protein hormones can yield information about the health of fetus and placenta.

Estriol excretion increases from 10 μg per day in the luteal phase to 30 mg. daily during the third trimester of pregnancy (see Fig. 18-4). Other estrogens increase proportionately. Estriol has its origin mainly from dehydroepiandrosterone sulfate, 70 to 80 percent of which is secreted by the fetal adrenal cortex, the remainder coming from the maternal adrenal cortex. In the anencephalic fetus, whose adrenal cortex is atrophic, urinary estriol is low. Since estriol is easily measured in the urine at these high levels, it is feasible to follow women *sequentially* and thus look for alterations in fetal adrenal cor-

TABLE 18-2.   Testosterone Production Rates

	Plasma T (μg/100 ml.)	MCR$_T$* (L./24 hr.)	BPR$_T$† (μg/24 hr.)	Plasma Δ‡ μg/100 ml.	MCRΔ (L./24 hr.)	BPR‡ (μg/24 hr.)	$_p\Delta T$§	T from Δ‖ (μg/24 hr.)
Normal women	0.04	600	240	0.15	2000	3000	0.04	120
Hirsute women	0.11	1000	1000	0.28	2000	5600	0.04	225

T = testosterone.

* MCR = metabolic clearance rate or the virtual volume of plasma cleared of steroid per day.

† BPR = blood production rate of testosterone or androstenedione, or the amount of testosterone or androstenedione, entering the bloodstream per day.

‡ Δ = androstenedione.

§ $_p\Delta T$ = fraction of blood androstenedione converted to blood testosterone.

‖ T from Δ = the amount of testosterone entering the circulation from peripheral conversion of plasma androstenedione.

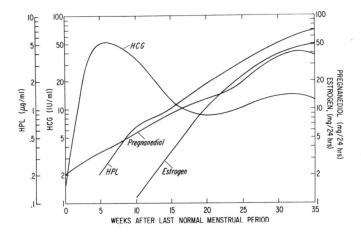

Fig. 18-4. Pattern of hormones throughout pregnancy. Human placental lactogen (HPL) and HCG are given per ml. of serum, estrogen and pregnanediol as urinary excretion per day.

tical activity indicating fetal disease. The large variability among women with respect to estriol excretion during the third trimester of pregnancy makes occasional determinations of little use unless the excretion is less than 3 to 4 mg. per day. Below this level and depending on the laboratory, fetal death is indicated.

Patient 6. A 23-year-old juvenile diabetic, taking 40 U of NPH insulin daily, was referred for her first pregnancy. At the 30th week of gestation, estriol excretion was 11 mg. per day, the mean being 20 mg. This could indicate retardation of fetal growth, but the wide scatter about the normal mean made such deductions hazardous. Weekly urinary estriol excretion continued to increase so that by the 34th week, when fetal viability seemed assured, it was 17 mg. per day. At this point, urinary estriol measurements were made biweekly. In the 36th week, there was a fall from 21 to 16 mg. per day. This was repeated for the next 3 days and successive values of 16, 14, and 10 mg. were obtained. Since this suggested fetal distress, a cesarean section was performed with delivery of a 5-pound infant. The placenta showed evidence of segmental infarction.

In diseases such as diabetes and toxemia of pregnancy where placental insufficiency may result in decreased nutrition of the fetus and subsequent fetal distress, sequential estriol determination can signal these events before conventional signs such as alterations of fetal heart rate will be apparent. Here, then, is an outstanding example of the adaptation of fundamental knowl-

edge to a clinical test and its translation to normal medical practice.

Progesterone is synthesized by the placenta. Curiously, the placenta cannot perform 17-α-hydroxylation, necessary for biosynthesis of androgens and estrogens. Since the corpus luteum and the placenta produce progesterone, urinary pregnanediol, the major metabolite of progesterone, might be expected to reflect contributions from both sources. The corpus luteum is the major source of progesterone during the first 6 weeks of pregnancy but its function remains low thereafter. If luteectomy is done during the first 6 weeks of pregnancy, there is an almost 100 percent incidence of abortion; when performed after 20 weeks, abortion does not occur. Since, after 8 weeks' gestation, progesterone is synthesized almost exclusively by the placenta, a drop in progesterone excretion will signify a change in placental function. Death of the fetus does not immediately alter progesterone excretion, only estrogen excretion decreases.

Human placental lactogen (HPL) is secreted primarily into the maternal circulation, and by the 32nd week of pregnancy its blood level may be as high as 0.7 μg per 100 ml. It can be detected in maternal plasma as early as the 6th week; and at term, the placenta secretes 300 to 1,000 mg. daily, suggesting that HPL is the principal placental protein synthesized during the last trimester. Of particular significance is the close correlation between placental weight and HPL blood levels. Because of this, plasma HPL levels give information concerning the state of the placenta. For example, low HPL concentrations have indicated retarded fetal growth in patients with diabetes or hypertension. Reduced HPL, associated with bleeding during pregnancy, indicates a threatened abortion. In some patients, subsequent stable·or rising levels have given reassurance about the functional integrity of the placenta.

With differentiation of trophoblastic cells, the blastocyst becomes ready for implantation. At this time, HCG is synthesized by the syncytiotrophoblast and is detectable in urine 9 days after ovulation. This timing is critical, since 7 to 8 days after ovulation corpus luteum synthesis of progesterone begins to decline. It is probable that, in man, HCG is the luteotropic hormone responsible for continued function of the corpus luteum.

HCG levels in blood and urine increase rapidly, reaching a peak of 50 IU per ml. 6 to 8 weeks after ovulation (Fig. 18-4). They slowly decline to levels of about 10 IU per ml. and remain relatively constant throughout pregnancy. Serial measurement of HCG has not been rewarding clinically. It does tend to be elevated in twin pregnancies, but the normal variations are so wide that the predictive value is low.

CONTRACEPTIVES

This topic is immediately relevant to any discussion of reproductive biology. The complexity of processes of spermatogenesis, maturation of the ovum and ovulation, sperm and ovum transport, fertilization, and nidation offers countless critical points where interference would prevent pregnancy. A variety of mechanical means have been used to prevent fertilization, and the intrauterine device was foreshadowed 1,000 years ago by the insertion of an intrauterine pebble to prevent pregnancy in camels.

The use of the "pill" is, of course, the most important example of hormonal contraception. The commercially available oral contraceptives contain an estrogen, either ethinyl estradiol or the 3-methyl ether of ethinyl estradiol, mestranol, and a progestational agent. Although increasing doses of estradiol can induce the LH surge and thereby cause ovulation, the combination of a greater than physiological dose of estrogen and a progestational agent will effectively suppress both gonadotropins. Thus, when a normally cycling woman is given oral contraceptives, sequential measurements of LH and FSH show a steady low level of each hormone with elimination of the midcycle peaks.

These agents are remarkably effective, the failure rate being less than 0.5 percent. This means that there is less than one pregnancy for every 200 woman-years of exposure. The effectiveness of the agents must be balanced against their side effects. Based on retrospective studies, users of the pill are at increased risk for venous thrombosis and for death from pulmonary embolism. The absolute risk is small, and certainly no greater than that to be expected from the maternal mortality resulting from unplanned births due to the use of less effective contraceptive measures such as the diaphragm. The thromboembolic effects are due to the estrogenic component of the "pill" and one study suggests that the lower the dose of estrogen, the fewer the episodes of thromboembolism.

The oral contraceptives cause a large number of alterations of the normal physiological state; the significance of these is not clear. For example, even at the lowest effective dose of estrogen, plasma proteins such as ceruloplasmin, thyroxine-binding globulin, and cortisol-binding globulin are increased. This is good biological evidence that 50 μg of ethinyl estradiol is a greater than physiological amount of estrogen.

The level of renin substrate, the β-globulin, which releases angiotensin, is uniformly increased; and recent studies suggest a significant percentage of women (as high as 30 percent or more) may become hypertensive. Plasma triglycerides increase and glucose tolerance is reduced — whether these changes will ultimately accelerate arteriosclerosis or diabetes remains to be seen. From a review of these additional effects, it is apparent that even small excesses of the sex steroids may exert widespread effects.

Another example of hormonal contraception is the "minipill" containing any of several progestational agents at low doses. During therapy, the hypothalamic-pituitary-ovarian system appears normal by measurement; FSH and LH peaks appear and ovulation occurs. The reason for the effectiveness of low-dose progestins is not clear. Progesterone alters the characteristics of the cervical mucus, changing it from a profuse, watery, low-viscosity fluid favorable for sperm penetration to a scanty, thick secretion that either prevents sperm penetration or decreases sperm viability. Recent decisions about the continued clinical trial of these agents indicate the profound philosophical problems encountered by such regulatory agencies as the Food and Drug Administration. Some progestational agents that were in clinical trial were noted to cause mammary tumors in the beagle, a dog that is particularly susceptible to development of mammary carcinoma. Although the relevance to the human experience is unknown, clinical trials were stopped. However, estrogens have been known for 40 years to cause cancer in susceptible strains of mice and to produce mammary carcinoma in rats. Thus, all the current oral contraceptives contain one agent carcinogenic for at least one animal. There is no evidence that these agents are carcinogenic in man nor is there evidence to the contrary.

The prostaglandins (PGE_1, PGE_2 and PGF_{2a}) have been shown to be effective abortifacients in the first and second trimester of pregnancy. The mechanism is apparently the stimulation of uterine contractions. Similarly, the prostaglandins have been used to induce labor in women at term. The other effects of prostaglandin such as inhibition of progesterone secretion have not been studied adequately as yet. It should be clear, however, that any compound that prevented synthesis of progesterone for 3 days during the first 6 weeks of pregnancy would terminate that pregnancy, since maintenance of the endometrium of pregnancy requires progesterone.

ANNOTATED REFERENCES

Baird, D. T., Horton, R., Longcope, C., and Tait, J. F.: Steroid pre-hormones. Perspectives Biol. Med., *11*:384, 1967. (Discussion of endogenous steroids without specific biological activity)

Bardin, C. W., and Lipsett, M. B.: Testosterone and androstenedione blood production rates in normal women and women with idiopathic hirsutism or polycystic ovaries. J. Clin. Invest., *46*:891, 1967.

Bardin, C. W., Ross, G. T., Rifkind, A. B., Cargille, C. M., and Lipsett, M. B.: Studies of the pituitary-Leydig cell axis in young men with hypogonadotropic hypogonadism and hyposmia: Comparison with normal men, prepuberal boys, and hypopituitary patients. J. Clin. Invest., *48*:2046, 1969. (Discussion of secondary hypogonadism with anosmia, Kallman's syndrome)

Channing, C. P.: Influence of the *in vivo* and *in vitro* hormonal environment upon luteinization of granulosa cells in tissue culture. Recent Progr. Hormone Res., *26*:589, 1970. (Fascinating tissue culture study which shows granulosa cells undergoing functional and morphological changes compatible with luteinization)

Federman, D. C.: Abnormal sexual development; a genetic and endocrine approach to differential diagnosis. Philadelphia, W. B. Saunders, 1967. (Excellent review of syndromes resulting from the dysgenetic gonad and anorchia)

Gorski, J., Taft, D., Shyamala, G., Smith, D., and Notides, A.: Hormone receptors: Studies on the interaction of estrogen with the uterus. Recent Progr. Hormone Res., 24:45, 1968.

Jensen, E. V., and Jacobson, H. I.: Basic guides to the mechanism of estrogen action. Recent Progr. Hormone Res., 18:387, 1962.

Heller, C. G., and Clermont, Y.: Kinetics of the germinal epithelium in man. Recent Progr. Hormone Res., 20:545, 1964. (Detailed discussion of spermatogenesis)

Hertz, R., Odell, W., and Ross, G. T.: Diagnostic implication of primary amenorrhea. Ann. Intern. Med., 65:800, 1966. (Discussion of the pathological evaluation of such a patient)

Jost, A.: Problems of fetal endocrinology: The gonadal and hypophyseal hormones. Recent Progr. Hormone Res., 8:379, 1953. (Discussion of the role of the fetal testis during embryogenesis)

Karolinska Symposia on Research Methods in Reproductive Endocrinology. Immunoassay of Gonadotropins. Diczfalusy, E. (ed.), Stockholm, 1969.

Karolinska Symposia on Research Methods in Reproductive Endocrinology. Steroid Assay by Protein Binding. Diczfalusy, E. (ed.); Geneva, 1970. (The above two references detail the techniques involved in immunoassay and bioassay of hormones.)

McCullagh, E. P., Beck, J. C., and Schaffenburg, C. A.: A syndrome of eunuchoidism with spermatogenesis, normal urinary FSH and low or normal ICSH: ("fertile eunuchs"). J. Clin. Endocrinology, 13:489, 1953.

Paulsen, C. A., Gordon, D. L., Carpenter, R. W., Gandy, H. M., and Drucker, W. D.: Klinefelter's syndrome and its variants: A hormonal and chromosomal study. Recent Progr. Hormone Res., 24:321, 1968. (Review of 1,000 patients with this disease)

Penny, R., Guyda, H. J., Baghdassarian, A., Johanson, J., and Blizzard, R. M.: Correlation of serum follicular-stimulating hormone (FSH) and luteinizing hormone (LH) as measured by radioimmunoassay in disorders of sexual development. J. Clin. Invest., 49:1847, 1970. (Particular emphasis is directed to primary amenorrhea.)

Ramwell, P. W., and Shaw, J. E.: Biological significance of the prostaglandins. Recent Progr. Hormone Res., 26:139, 1970. (Discussion of the physiological effects of the prostaglandins)

Ross, G. T., Cargille, C. M., Lipsett, M. B., Rayford, P. L., Marshall, J. R., Strott, C. A., and Rodbard, D.: Pituitary and gonadal hormones in women during spontaneous and induced ovulatory cycles. Recent Progr. Hormone Res., 26:1, 1970. (Excellent discussion of the hormonal changes during the normal menstrual cycle)

Ryan, K. J., and Smith, O. W.: Biogenesis of steroid hormones in the human ovary. Recent Progr. Hormone Res., 21:367, 1965. (Discussion of the steroid-synthetic capacities of each compartment of the ovary)

Salhanick, H. A., Kipnis, D. M., and Vande Wiele, R. L.: Metabolic effects of gonadal hormones and contraceptive steroids. New York, Plenum Press, 1969. (An excellent review of the gamut of side effects from oral contraceptives)

Saxena, B. N., Emerson, K., Jr., and Selenkow, H. A.: Serum placental lactogen (HPL) levels as an index of placental function. New Eng. J. Med., 281:225, 1969. (Role of human placental lactogen in pregnancy and fetal growth)

Southren, A. L.: The syndrome of testicular feminization. Adv. Metabolic Disorders, 2:227, 1965.

Vande Wiele, R. L., Bogumil, J., Dyenfurth, I., Ferin, M., Jewelewicz, R., Warren, M., Rizkallah, T., and Milkhail, G.: Mechanisms regulating the menstrual cycle in women. Recent Progr. Hormone Res., 26:48, 1970. (This report shows that maintenance of the corpus luteum depends upon LH.)

Wilson, J. D., and Gloyna, R. E.: The intranuclear metabolism of testosterone in the accessory organs of reproduction. Recent Progr. Hormone Res., 26:309, 1970. (The above three references show the important interrelationships of intracellular biochemical alterations by endocrine hormones and cell function.)

Yoshimi, T., Strott, C. A., Marshall, J. R., and Lipsett, M. B.: Corpus luteum function during early pregnancy. J. Clin. Endocrinology, 29:225, 1969. (Study detailing the role of the corpus luteum during pregnancy)

Appendix A

Each chapter in this section provides a list of references and monographs (each having excellent bibliographies). In this Appendix are presented important textbooks which provide broad and comprehensive discussions of the clinical problems involved in endocrinology and metabolism. Those textbooks listed under "General Endocrinology and Metabolism" have exceptionally well-selected references on every aspect of endocrinology and metabolism; those textbooks in the "Specialized Areas of Endocrinology and Metabolism" have more detailed and specialized references.

General Textbooks

Bloodworth, I. M. B., Jr.: Endocrine Pathology. Baltimore, The Williams & Wilkins Company, 1968.

Bondy, P. K.: Duncan's Diseases of Metabolism. Ed. 6. Two volumes. Vol. 1. Genetics and Metabolism. Vol. 2. Endocrinology and Nutrition. Philadelphia, W. B. Saunders, 1969.

Hsia, D. Y-Y: Inborn Errors of the Metabolism. Part I. Clinical Aspects. Part 2. (with Inouye, T.) Laboratory Methods. Chicago, Year Book Medical Publishers, 1966.

Lisser, H., and Escamilla, R. F.: Atlas of Clinical Endocrinology. Ed. 2. St. Louis, C. V. Mosby, 1962.

Stanbury, J. B., Wyngaarden, J. B., and Fredrickson, D. S.: The Metabolic Basis of Inherited Disease. Ed. 2. New York, Blakiston Division, McGraw-Hill, 1966.

Tepperman, J.: Metabolic and Endocrine Physiology. An Introductory Text. Chicago, Year Book Medical Publishers, 1962.

Williams, R. G.: Textbook of Endocrinology. Ed. 4. Philadelphia, W. B. Saunders, 1968.

Specialized Textbooks

Albright, F., and Reifenstein, E. C., Jr.; The Parathyroid Glands and Metabolic Bone Disease. Baltimore, The Williams & Wilkins Company, 1948.

Ciba Foundation Study Group No. 27: The Human Adrenal Cortex. Boston, Little, Brown and Company, 1967.

Eisenstein, A. B.: The Adrenal Cortex. Boston, Little, Brown and Company, 1967.

Federman, D. D.: Abnormal Sexual Development. Philadelphia, W. B. Saunders, 1967.

Jackson, W. P. U.: Calcium Metabolism and Bone Disease. Baltimore, The Williams & Wilkins Company, 1967.

Jolliffe, N.: Clinical Nutrition. Ed. 2. New York, Hoeber Medical Book, Harper and Brothers, 1962.

Lissak, K., and Endroczi, E.: The Neuroendocrine Control of Adaptation. New York, Pergamon Press, 1966.

Locke, M.: Control Mechanisms in Developmental Processes. New York, Academic Press, 1968.

Means, J. H., DeGroot, L. J., and Stanbury, J. B.: The Thyroid and Its Diseases. Ed. 3. New York, Blakiston Division, McGraw-Hill, 1963.

Schettler, G.: Lipids and Lipidoses. New York, Springer-Verlag, 1967.

Werner, S. C.: The Thyroid. Ed. 2. New York, Hoeber Medical Book, Harper and Row, 1962.

Wohl, M. G., and Goodhart, R. S.: Modern Nutrition in Health and Disease. Ed. 4. Philadelphia, Lea and Febiger, 1968.

Appendix B

TESTS OF THYROID FUNCTION

Basal Metabolic Rate (Normal: ± 10%).

The action of thyroid hormone upon the tissues of the body is difficult to assess. Although the BMR is probably the best test available, the wide range of normality, the marked likelihood of error in determination, and the limited reproducibility between laboratories and technicians limit

375

its usefulness. It can be helpful in diagnosis when measurements of some of the new thyroid function tests have been altered by various medications. A small percentage of hyperthyroid patients may have BMR values within the normal range.

Radioactive Iodine (^{131}I) Uptake (Normal ^{131}I uptake by the thyroid in 24 hours: 20 to 50%)

This represents the proportion of the dose of ^{131}I administered which is accumulated by the thyroid, and is a function of the relative clearances by the thyroid (33 ml. per min.) and kidney (25 ml. per min.). It is a measure of the avidity of the thyroid gland to concentrate iodide and to iodinate organic moieties. The peak uptake is achieved normally in 24 hours. It may be achieved earlier (2 to 6 hours) in patients with hyperthyroidism and later (48 to 72 hours) in patients with renal disease. The test should not be performed in those patients who have recently ingested other isotopes (e.g., arsenic, technetium, etc.) which are used for scanning studies of other organs, since such isotopes are taken up by the thyroid and distort the results. ^{131}I uptake is high in patients with hyperthyroidism, prolonged iodine deficiency, and dyshormonogenesis. The uptake is low in patients with hypothyroidism, subacute thyroiditis, or an iodide concentration defect. Exogenous iodide or thyroid hormone administration will depress uptake for a variable period, depending upon the functional activity of the thyroid. Alterations in the uptake of ^{131}I due to unknown sources of exogenous iodide may be verified by tests for 24-hour urinary excretion and iodide. The normal urinary iodide excretion is about 150 μg per day in the New England area; however, this may vary elsewhere, depending upon the diet. A random urine collection examined for iodide and creatinine content may suffice for this evaluation by using the following formula:

$$\text{Urine Iodide } (\mu\text{g per 24 hours}) = \frac{\text{Iodide } (\mu\text{g/L.})}{\text{Creatinine } (\text{g./L.})} \times K$$

where K is the average creatinine excretion in g. per 24 hours for the individual.

Perchlorate of Thiocyanate Discharge Test

This test is useful for detecting abnormalities in intrathyroidal organification of iodide such as in Hashimoto's thyroiditis or other intrathyroidal defects in organification. The test consists of giving a tracer solution of ^{131}I (with or without 500 μg carrier iodide), measuring activity over the thyroid or several hours until a maximum is reached, and then administering either 400 mg. of sodium perchlorate or 1.0 g. of thiocyanate by mouth. Sequential measurements are made of the radioactivity over the thyroid for the next hour. A discharge of ^{131}I greater than 10 percent is abnormal and suggests impairment of iodination.

Thyroid Hormone Blood Levels

Tests which measure the concentration of circulating thyroid hormone vary directly with the quantity of hormone secreted by the thyroid and the quantity and binding avidity of TBG and less critically, TBPA and albumin.

Protein Bound Iodine (PBI) (Normal: 4 to 8 μg %). The measurement consists of precipitation of serum proteins by Somogyi's reagent, digestion of these proteins with the consequent liberation of inorganic iodide which is measured colorimetrically in a chemical reaction involving the reduction of a ceriarsenous acid mixture. Since the test determines the quantity of iodine in thyroid hormone, a misleading value may result from the presence of exogenous iodine. The PBI measures T_4, T_3 iodinated dyes, thyroglobulin, other iodinated proteins, and excess iodide. The precipitated proteins consist largely of the thyroxine-binding proteins to which thyroxine or triiodothyronine is loosely and noncovalently bound. In abnormal thyroid conditions, such as chronic thyroiditis and dyshormonogenesis, iodinated proteins are released into the circulation, in which the organic iodide moiety is peptide linked, and this, too, is measured in the PBI test. The PBI test also indiscriminately measures iodides, taken in the form of medication, which become attached to serum proteins. In fact, any PBI over 20 μg percent should suggest iodide contamination. Test values are falsely elevated by medications affecting protein-binding, such as estrogens, and decreased by androgens and drugs which compete with thyroxine for the thyroxine-binding protein (e.g., salicylates or diphenylhydantoin). Heavy metals (e.g., mercury, gold) interfere with the test, producing low values.

Butanol-Extractable Iodine (BEI) (Normal: 2.6 to 6.5 μg %). The BEI test measures iodine which is nonpeptide linked to serum proteins and which is butanol-soluble in an acid medium. Acid butanol extracts T_4, T_3, and iodinated dyes (the latter producing an artifactually elevated BEI. The test is also affected by the same drugs as is the PBI.

Butanol Nonextractable Iodine (BII) (Normal: 0.5 to 1.6 μg %). BII can be measured chemically and represents peptide-linked organic iodine which is butanol-insoluble. It may be elevated in patients with hyperthyroidism, Hashimoto's thyroiditis, and dyshormonogenesis.

Serum Thyroxine (Normal: 4 to 11 μg %). Since this analysis is not contingent primarily upon iodine content, it usually is not affected by exogenous iodine. The test is performed in one of two ways: either by competitive protein-binding (Murphy-Pattee method) or by column chromatography. The former method is preferred, since the chromatographic analysis is altered by the presence of iodoprotein and organic iodine contrast material in the serum. In the Murphy-Pattee method, T_4 is measured by extracting the thyroxine from the patient's serum with ethanol and then quantitating its ability to displace radioactive thyroxine in the patient's serum. Thyroxine values are determined from a standard curve. Medications affecting thyroxine-binding proteins (e.g., estrogens, androgens, steroids) will alter the measurement of serum thyroxine, as in the BEI. Nevertheless, for routine assays this test should replace the PBI and BEI.

Free Thyroxine Concentration (Normal: 0.8 to 2.6 ng. %). This test measures the very small concentration of thyroxine, supposedly metabolically active, present in the circulation (normally, about 0.1% of the total). The test is performed by incubating the patient's serum with a specific amount of purified radioactive thyroxine in a dialysis bag. The displaced radioactive T_4 then passes through a dialyzing membrane over a 16-hour period. The labeled material in the dialysate is then further purified to obtain primarily labeled T_4. The amount of this activity is compared to the total radioactivity initially placed in the dialysis bag. This fraction multiplied by the value of the total serum thyroxine represents the concentration of free thyroxine. The test for free thyroxine is unaffected by exogenous iodine and is the only test which does not produce false values for those patients who are taking estrogens. However, elevations in the free thyroxine concentration occur in euthyroid patients during acute or chronic illness or immediately following surgery. In such instances, the total thyroxine is usually normal, excluding the possibility of hyperthyroidism.

Total Triiodothyronine (Normal: 150 to 250 ng. %). This is a relatively new test and the precision of the measurement remains to be determined. Serum is extracted with ethanol to remove T_4 and T_3, which are then separated by column and paper chromatography. The T_3 is measured by competitive binding displacement. It may be elevated in patients with symptoms of hypermetabolism and normal levels of circulating thyroxine.

Free Triiodothyronine (Normal: 1 to 3 ng. %). The validity of this test, which is also new, remains to be evaluated in various clinical conditions.

Thyroxine-binding Proteins

These capacities are measured by conventional (TBPA) and reverse-flow electrophoresis (TBG) using barbital buffer, pH 8.6. Normal values are as follows:

	AGE (YEARS)		
	0-12	30-40	70-80
TBG* μg/100 ml.	26	22	24
TBPA	70	180	80

T₃ Resin Uptake (Normal: 25 to 35%)

This test consists of adding labeled T_3 and a sample of serum to a sponge resin and mixing. A competition exists between the binding capacity of the resin and the thyroxine-binding proteins for the added T_3. Test values are elevated when the blood concentration of thyroid hormone is increased as in hyperthyroidism or when the concentration of these plasma proteins which bind thyroid hormone is decreased by physical or chemical means. The value is depressed in conditions leading to a decrease in the level of circulating thyroid hormone (e.g., hypothyroidism) or in conditions leading to an increase in the concentration or binding capacity of the binding proteins (e.g., estrogens). It is an indirect assessment of the binding protein and the concentration of free thyroxine. Although inorganic iodides do not alter the test, organic iodides may. Because the results are altered by a variety of illnesses and drugs, the test is impractical for general use except to evaluate the true nature of an elevated serum T_4 and, possibly, to confirm the diagnosis of hyper-

thyroidism (elevated T_3) or substantiate the effect of estrogens (low T_3 uptake).

Tanned Red Cell Agglutination Test (TRC)

The TRC measures the titer of antibodies present in plasma that are directed against human thyroglobulin. The presence of antibodies in a dilution greater than 1:20 is abnormal. Titers above 1:20 are observed in about 70 percent of chronic thyroiditis patients. This test, coupled with a fluorescent antibody technique for measuring circulating antibodies to thyroid cytoplasmic and colloidal antigens, is abnormal in approximately 90 percent of patients with chronic thyroiditis.

Radioimmunoassay of TSH (Normal: 0 to 5 ng./ml)

The radioimmunoassay for human TSH is based upon a double antibody precipitation technique. It is an excellent test for the differential diagnosis of primary thyroid myxedema and pituitary myxedema, being elevated in the former condition and nonexistent or very low in the latter. However, in long-standing primary thyroid myxedema, it may not be as high as one would expect, suggesting pituitary myxedema. Radioimmunoassay of TSH is the best test for detecting an early potential prehypothyroid state in patients previously treated for hyperthyroidism by surgery or [131]I. It may be elevated for 6 to 12 months at a time when the circulating level of thyroxine is normal.

Reserve Test for TSH

The isolation and synthesis of a thyroid-releasing tripeptide from the hypothalamus has led to a valuable assay for the reserve capacity of the pituitary to release TSH. Administration of 200 to 800 μg TRH intravenously to normal men produces a rise in radioimmunoassayable serum TSH from levels of 3 miu per ml. to 15 to 20 miu per ml. The effect appears to be specific and constant in normal men. It is ineffective in patients with hypopituitarism and the effect is inhibited by the prior administration of thyroxine or triiodothyronine.

Long-Acting Thyroid Stimulator (LATS)

This substance is measured by bioassay and involves administration of the patient's serum to a mouse whose thyroid has been prelabeled with ^{131}I. The ^{131}I released by the thyroid is measured at 2 and 9 hours and contrasted with suitable controls. A significant increase in the level at 9 hours over that at 2 hours is considered positive for LATS activity. It is reported to be present in about 80 percent of hyperthyroid patients. (This is an interesting research tool, but at present is of doubtful value as a routine diagnostic measure.)

Tests Assessing Pituitary-Thyroid Homeostasis

Suppression Test. This test consists of administering triiodothyronine (25 μg every 8 hours for 10 days), measuring ^{131}I thyroid uptake and serum T_4, and then comparing the results with control values for ^{131}I and T_4 prior to the test. *Normal suppression* constitutes a decrease of ^{131}I uptake to less than 10 percent or less than half of the control value or a decrease in serum T_4 to below the normal levels. *Abnormal suppression* indicates active thyrotoxicosis, latent hyperthyroidism, a hyperfunctioning nontoxic adenoma, or persistence of hyperthyroidism after long-term antithyroid treatment.

TSH Stimulation Test. This test involves daily intravenous administration of 10 units of bovine TSH for 3 days. If thyroid function is normal, ^{131}I uptake rises, indicating ability to form new hormone, and an increased T_4 indicates enhanced release of preformed hormone. This test is used clinically to differentiate primary from secondary myxedema. It is generally negative in the former and positive in the latter patients. It may also be useful in assessing the thyroid glandular function during thyroid hormone administration. The test shows no response with decreased thyroid reserve; and it is also used to evaluate reason for lack of ^{131}I activity in the opposite thyroid lobe which has been suppressed by a hyperfunctioning single adenoma.

The following tables of endocrine functional tests should be of help in the differential diagnosis of thyroid (Table IV-1) hypercalcemic (Table IV-2), and adrenal (Table IV-3) diseases.

TABLE IV-1. Thyroid Function Studies in Disease*

	¹³¹I Uptake 20-55%	T₄ 4-11 μg%	FT₄ 0.8-2.6 mg.%	T₃ Uptake 25-35%	TBG 16-24 μg%	TSH < 5 ng./ml.	BMR -10 to +10	TRC 7 1/20	T₃ Supp.	TSH Stim.	TRH Response	Total T₃ 140-220 mμ
Normal values												
Iodine Deficiency	↑	N↓	N↓	N↓	N	N↑	N↓	–	N	▪	–	–
Iodine Excess												
Iodide	→↓	N	N	N	N	N	N	–	–	–	–	–
Organic Iodine	→	N↑	N↑	N↑	N	N	N	–	–	–	–	–
Iodide Myxedema	→	→	→	→	N↑	↑	→	N↑	–	–	–	→
Estrogen	N	↑	N	→	↑	N↑	N	–	–	–	N↑	–
Androgen	N	→	←	←	→	N	N	–	–	–	–	–
Cortisone	→	→	←	←	→	N↓	N	–	–	–	N	–
Heparin	N	N↑	←	←	–	N	–	–	–	–	–	–
Diffuse Toxic Goiter												
Active	←N	←N	←N	←N	↓N	N	←N	N←	Abn	–	Abn	–
Inactive	N	N	N	N	N	N↑	N	N←	N	–	–	–
Latent	N↑	N	N	N	N	N	N	N←	Abn	–	–	–
Active Ophthalmopathy (nontoxic)	N	N	N	N	N	N	N	N↑	Abn	–	–	–
T₃ Thyrotoxicosis	N↑	N	N	N↑	N	N	←	–	Abn	–	–	←
Thyrotoxicosis Factitia												
(1) T₄ or Desiccated Thyroid	→	←→	←→	←N	N→	–	←	–	–	←¹³¹I	–	N↑
(2) T₃	→	←→	←→	N↑	N→	–	←	–	–	←¹³¹I	–	←
Functioning Adenoma												
Toxic	N↑	←N	←N	←N	N	–	←N	–	–	Abn	–	←N
Nontoxic	N	N	N	N	N	–	N	–	N, Abn	–	–	N
Multinodular Goiter (toxic)	↑N	↑N	↑N	↑N	N	–	N↑	–	Abn	–	–	N↑

Myxedema												
Primary (Thyroid)	↓	↓	↓	↓	N ↑	↑	↓	↑ N	—	negative	exaggerated	↓
Secondary (Pituitary)	↓ N	↓	↓	↓ N	N	↓ (N)	↓	N ↑	—	↑ ^{131}I, T_4	negative	↓ N
Tertiary (Hypothyroidism)	↓ N	↓	↓	↓	N	↑ (sl)	↓	N	—	↑ ^{131}I, T_4	exaggerated	↓ N
Prehypothyroid State	N ↓	N	N	N	N	↑	N	N ↑	N	variable	?	N ↓
Hashimoto's Thyroiditis												
Euthyroid	N ↓	N	N	N	N	N ↑	N	↑	N	variable	—	N
Hypothyroid	↓ N	↓	↓	↓	N	↑	↓	↑	N	variable	—	↓
Subacute Thyroiditis	↓	↑ N	↑ N	↑ N	N	N	↑ N	↑ N	—	Abn	—	?
Genetic												
Increased TBG	N	↑	N	↓	↑	N	N	—	—	—	—	—
Decreased TBG	N	↓	↓	↑	↓	N	N	—	—	—	—	—

* ^{131}I uptake, 24-hour uptake; T_4, thyroxine; FT_4, free thyroxine, T_3 uptake, triiodothyronine; TBG, thyroid-binding globulin; TSH, thyroid-stimulating hormone; BMR, basal metabolic rate; TRC, tanned red blood cell uptake; T_3 supp., triiodothyronine suppression test; TSH stim., thyroid-stimulating hormone stimulation test; TRH response, thyrotropic-releasing hormone response; total T_3, total triiodothyronine. For details of test see this Appendix and text of Chapter 15. N = normal; ↓ = decreased; ↑ = increased; Abn = abnormal.

TABLE IV-2.   Alterations of Plasma Calcium (Ca), Phosphorus (P), and Alkaline Phosphatase (Alk. Ptase), Urine Excretion of Calcium (Ca) and Tubular Reabsorption of Phosphate (TRP) in the Normal Subject and Diseased Patient.

DISEASE	PLASMA			URINE	
	Ca	P	Alk. Ptase	Ca	TRP
Normal	9-11 mg.%	3.2-4.2 mg.%	3-12K-AU	125-250 mg./24 hr	90-95%
Hyperparathyroidism					
Primary	↑	↓ or N	N or ↑	N or ↑	↓ or N
Secondary					
Renal (azotemic)	↑ or N	↑	↑	↓	↓
Vitamin D deficiency	↓ or N	↓	↑	↓	↓
Renal rickets	↓	N or ↑	↑	↓	
Hypoparathyroidism	↓	↑	N	↓	↑
Vitamin D intoxication	↑	N or ↑ or ↓	N or ↑	↑	N or ↓
Sarcoidosis	↑	N or ↓	N	↑	N or ↓
Milk-alkali syndrome	↑	N	N	N	N or ↓
Paget's disease	N or ↑	N	↑	↑	N
Metastatic bone disease	↑	N	N	↑	N

TABLE IV-3.   Adrenal Functions in the Normal and Various Adrenal Pathophysiological Conditions

Condition	PLASMA CORTISOL (µg %)	PLASMA ACTH (mU %)	24-HOUR URINE EXCRETION			17-OH-CS ACTH STIM.	URINE EXCR. Metyrapone
			17-KS (mg.)	17-OH-CS (mg.)	Aldosterone (µg)		
Normal	(8 a.m.) 9-32 (4 p.m.) 2-18	0.05-0.5	♂ 6-21 ♀ 4-17	♂ 5-23 ♀ 3-15	2-20	↑	↑
Hypercorticism							
Adenoma	↑	↓	↓N	↑	N↓	↑, no ↑	no ↑
Hyperplasia	N↑	↑	N↑	↑	N↑	↑	↑
Carcinoma	↑	↓	↑	↑	N↓	no ↑	no ↑
Hypocorticism	↓	↑	↓	↓	↓	no ↑	no ↑
Hypopituitarism	↓	↓	↓	↓	N-↓	↑	no ↑
Primary Aldosteronism	N	N	N	N	↑	N	
Congenital adrenal hyperplasia (enzymatic defects)							
21-Hydroxylase	↓	↓	↑	↓	↓	↑	
11-Hydroxylase	↓	N		↓	↓		
17-Hydroxylase		N	↓	↓	↓		
18-Hydroxylase	N	N	N		↓	N	N

Section Five

Gastrointestinal Mechanisms

Introduction

At the turn of the century gastro-enterology appeared to have superb prospects for future development. In the preceding century Beaumont had provided excellent physiological descriptions of the viable gastric mucosa, Virchow had given us much of the anatomical picture of gastrointestinal diseases, Billroth had successfully resected stomachs and Pavlov had demonstrated that the vagus nerves regulate gastric and pancreatic secretion. Shortly thereafter, Starling and Edkins founded endocrinology with their discoveries of the two hormones, secretin and gastrin. Inexplicably, our knowledge about gastrointestinal function in health and disease grew very slowly over the next half century.

The past two decades have seen a resurgence of interest, investigation and insights. The modern medical center has its gastrointestinal laboratories, where sophisticated methods are used to aid the clinician. The field is exploding, and one can sense the excitement of current advances and future promise. The chemical structures of gastrin, secretin and cholecystokinin have been identified, and immunoassay of these substances as a clinical test of disease is in the offing. The concept of a broken mucosal barrier in ulcer disease has been established. The interaction of neural and hormonal factors regulating gastric, pancreatic and biliary secretion is unfolding. The role of gastrin and secretin in the Zollinger-Ellison syndrome has been delineated. In regard to the small intestine, recent developments in pathophysiology include information about selective enzyme deficiencies leading to malabsorption of carbohydrates, the relation of steatorrhea to disturbed micelle formation, the application of the triple-lumen tube in studies of fluid and electrolyte transport in man, new

insights and information about cholera and the use of biopsy instruments to diagnose mucosal diseases. Sophisticated electromyography has been employed to characterize motor disorders of the gut. Powerful tools of biochemistry—chromatography, fluorometry, radioisotopes, ultracentrifugation and the like—have been used to delineate disturbances in biliary metabolism and other forms of liver disease. There are descriptions of genetic and immunological diseases of the gastrointestinal tract. Bacteriological and viral factors in gut disorders, pinocytosis and inverse pinocytosis, and the relationship of intestinal transport to hematological disturbances are subjects about which we have new information. These and many other areas of pathophysiological investigation are described in the ensuing four chapters of this book. Gastroenterology is no longer the poor relative of medicine. The dark ages are past.

The four chapters that follow are devoted to considerations of disordered mechanisms of disease in the gastrointestinal tract, including the liver. For emphasis and completeness, two additional areas are considered: pathophysiological aspects of pancreatic and mesenteric vascular diseases.

The pancreas is a complex organ, having both exocrine and endocrine functions. It is difficult to approach diagnostically, therapeutically and even investigationally, owing to its inaccessible location. Exocrine pancreatic function involves production and secretion of enzymes into the intestinal lumen. These enzymes mediate intraluminal digestion (preparation for absorption) of fat, protein and carbohydrate. The pancreas also secretes a large volume of alkaline fluid essential for the buffering of acidic gastric juice, thereby rendering intestinal pH suitable for enzyme and bile salt activity. Pancreatic secretion is stimulated by the intestinal hormones cholecystokinin (CCK) and secretin, which are released when acid or food comes in contact with the mucosa of the proximal small intestine.

Disordered exocrine function is characterized by deficient pancreatic secretion into the intestine and can be caused by mechanical obstruction (gallstone, tumor at the common biliary and pancreatic orifice, spasm of the sphincter of Oddi secondary to drugs such as morphine and other opiates) or loss of pancreatic exocrine glandular substance (chronic pancreatitis, cystic fibrosis, surgical extirpation and, rarely, extensive replacement by malignant tissue). Relative deficiency of pancreatic secretion and defective digestion may follow surgical procedures in which the pancreas and its ducts are left intact (gastrectomy with gastroenterostomy) but food bypasses both the site of CCK and secretin release and the segment of gut into which pancreatic juice is secreted.

Theoretically, pancreatic exocrine deficiency can lead to severe malabsorption secondary to impaired intraluminal digestion with severe malnutrition. However, this rarely occurs, because alternative mechanisms allow some digestion and absorption of fat, protein and carbohydrate even after total pancreatectomy.

In the Zollinger-Ellison syndrome, pancreatic digestive function may be impaired when massive volumes of gastric acid enter the duodenum and overwhelm the buffering capacity of pancreatic and intestinal fluid. In this case the pH of the intestinal contents remains acidic, inhibiting the activity of pancreatic enzymes and bile salts.

Pancreatic exocrine function is clinically assessed by the measurement of fecal fat (and the ratio of triglyceride to fatty acid), and by stimulating pancreatic secretion with secretin. The secretin test is performed on a fasting patient, employing a double lumen nasogastric tube; one lumen drains gastric contents (to reduce acid stimulation of endogenous secretion release)

while the other drains secretions from the upper small intestine. Secretin is injected parenterally, and the intestinal collection after secretin is compared with control values for secretion of fluid, bicarbonate and enzyme. In a normal individual, secretin induces a brisk increase in secretion of all three. In many patients with chronic pancreatitis there is a reduced output of bicarbonate and enzymes, whereas patients with pancreatic carcinoma often respond to secretin with impaired volume secretion but have normal bicarbonate and enzyme concentrations (presumably as a result of intrapancreatic ductal obstruction by tumor). Unfortunately, these changes are not invariable and therefore are not reliable in the diagnosis of the individual patient with pancreatic disease.

Mesenteric vascular diseases constitute a commonly occurring and usually life-threatening group of disorders. In the past decade they have begun to receive the deserved attention of clinical investigators. These entities are difficult to diagnose and often end in death. The patient with a mesenteric vascular disease is typically elderly and has an associated cardio-vascular disorder, such as generalized arteriosclerosis, congestive cardiac failure controlled with digitalis, and shock.

The mesenteric circulation has certain physiological characteristics that are readily disturbed with coexistent cardiovascular disease. There is, for example, a countercurrent exchange for oxygen in the small vessels of the villi that permits shunting of the gas from arteriole to venule. This mechanism results in oxygen bypassing the capillary bed at the tips of the villi. Normally, the countercurrent does not deprive the distal villi of much oxygen, but, when mesenteric blood flow is reduced, the loss of oxygen exaggerates tissue hypoxia and leads to mucosal necrosis, starting at the villous tips.

The small vessels of the mesenteric circulation can develop microthrombi during persistent low flow states, thereby further

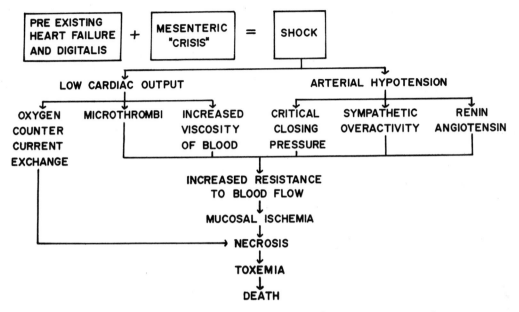

Fɪɢ. V-1. Pathophysiological mechanisms in nonocclusive mesenteric ischemia.

obstructing tissue perfusion. As blood flow is reduced, the viscosity of blood tends to rise. As the driving head of pressure in the arteries declines, as in shock, the critical closing pressure of small vessels in the mesenteric circulation is approached and blood flow is restricted. Norepinephrine, dopamine, vasopressin and angiotensin II are released during severe circulatory stress states; these vasoactive mediators constrict the precapillary segment of the mesenteric circulation. Furthermore, cardiac glycosides, which are often being administered to the elderly patient in shock or heart failure, are direct constrictors of mesenteric arterioles. The sequence of pathophysiological events leading to severe bowel ischemia and demise is shown schematically in Figure V-1. Unfortunately, many patients go undiagnosed before death, and various forms of treatment have been unsuccessful.

In the following section on disordered mechanisms of gastrointestinal disease we have categorized major disorders under the major recognized normal functions of the digestive system: secretion, motility, absorption and hepatic metabolism. In America today, gastrointestinal diseases constitute an annual multibillion dollar economic loss, striking people often in their most productive period. The family physician, the gastroenterologist and the surgeon care for millions who suffer some of the most common symptoms and diseases of the modern American: constipation, diarrhea, ulcers, cirrhosis, and colitis. Before rational diagnosis and therapy of gastrointestinal diseases can be instituted, comprehension of their disordered mechanisms is essential.

T. E. Bʏɴᴜᴍ, ᴍ.ᴅ., and
Eᴜɢᴇɴᴇ D. Jᴀᴄᴏʙsᴏɴ, ᴍ.ᴅ.

19

Motility

James Christensen, M.D.

Study of the structure and function of the operating systems of gastrointestinal motor functions is difficult because nerve and muscle cells are so small, the systems vary among species, and the gut is relatively inaccessible. Small wonder, then, that disorders of gastrointestinal motor function, though common, are poorly understood. However, a recent surge of interest has led to such advances in knowledge that the pathogenesis of some of these disorders may now be surmised; others remain enigmatic.

THE ANATOMY OF GASTROINTESTINAL MUSCLE

The *muscularis propria,* arranged in an outer longitudinal layer and an inner circular layer throughout the gut, consists wholly of smooth muscle except at the beginning and the end of the gut.

Pharynx and Esophagus. Through the pharynx and upper esophagus, striated muscle forms a continuous sheet, with a thickened band at the pharyngoesophageal junction, the cricopharyngeus. The striated muscle receives nerves from the glossopharyngeal nerve and the pharyngeal and esophageal branches of the vagus. In man, the striated muscle occupies both outer and inner muscle layers in the proximal third of the esophagus and interdigitates with smooth muscle in the middle third, but the length and the level of the transitional segment vary. Most other common mammals have striated muscle throughout most of the esophagus. Both outer and inner esophageal muscle layers are separately continuous with the longitudinal and circular layers of the stomach, without thickening at the gastroesophageal junction, in man. The smooth muscle segment receives branches from the esophageal plexus, encircling the distal half or third of the esophagus.

Stomach. In the stomach, the longitudinal muscle layer is incomplete, lying in two broad bands centered along the lesser and the greater curvatures. Internal to the complete circular coat lies a third, the incomplete oblique layer. It invests the fundus and extends over the body in broad anterior and posterior bands, thinning along the lesser and greater curvatures. The circular layer gradually thickens distally in the antrum. A concentration of fibrous tissue at the pylorus separates the circular coats of stomach and duodenum. Some fibers of the longitudinal layer bridge this connective tissue segment. Gastric smooth muscle receives branches of the abdominal vagi and perivascular nerves from the celiac plexus.

Small Intestine. In the small intestine, the two layers of smooth muscle are truly longitudinal and circular, rather than helical, as once was claimed. Though apparently separate, the two layers are joined by oblique muscle bundles. The only change in the muscle along the bowel is a progressive caudal reduction in wall thickness and bowel circumference. The small bowel receives nerves from the abdominal sympathetic plexuses and the vagi.

Colon. In the colon, the arrangement of

the longitudinal layer varies among species. In cats the longitudinal layer covers the entire circumference uniformly. In man it lies mainly in three longitudinal bands, the taeniae coli, except in the appendix and rectum. Between adjacent taeniae, the longitudinal muscle is very thin. The taeniae maintain a constant position relative to the mesentery, one taenia always lying along the mesenteric attachment. In the anal canal, the circular coat thickens, forming the internal anal sphincter, and the longitudinal coat merges into the striated external anal sphincter. The colon is said to receive vagal branches as well as extrinsic nerves from the abdominal sympathetic plexuses and pelvic nerves.

Ultrastructure. In microscopic appearance, gastrointestinal striated muscle resembles somatic striated muscle, but it has received little serious study. The microscopic anatomy of smooth muscle is better known. In these small mononuclear cells, approximately 100 to 200 microns by 5 to 10 microns, the cytoplasm appears homogeneous (hence "smooth"), but faint striations occur; presumably these are contractile protein filaments. These pass through dark regions—areas of densely packed filaments lying randomly in the cytoplasm and all along the cell wall. These filaments are probably actin. Myosin has been found chemically, but myosin filaments cannot be seen by ordinary light microscopy. A sliding filament mechanism probably operates in smooth muscle, but no details are known. Smooth muscle, once viewed as syncytial, actually consists of separate cells intimately joined by special structures called nexuses. A nexus, the localized fusion of plasma membranes of adjacent cells, can be either a simple abutment of extensions from adjacent cells across the intercellular space or a bulbous projection of one cell into another. Partial fusion of the membranes of the two cells occurs at the nexus. The function of the nexus is physiological, not mechanical,

though it has mechanical stability. It may provide a low-resistance electrical shunt between cells. The extracellular space in smooth muscle contains connective tissue elements giving mechanical linkage throughout the tissue.

THE NERVES OF THE GUT

Gastrointestinal nerves are still imperfectly understood, owing to their extreme complexity. The classic view will be summarized here, although it is certainly incomplete.

Extrinsic Nerves. The vagi distribute parasympathetic preganglionic fibers from cells of the brain-stem vagal nuclei and they also distribute sympathetic fibers from branches of the cervical sympathetic ganglia. The vagi also carry special somatic efferent nerves passing without synaptic interruption to striated muscle. The vagal preganglionic parasympathetic fibers pass to the esophagus, stomach, small bowel, and proximal colon, making synapse with neurons of the myenteric and submucosal plexuses. These plexuses innervate smooth muscle with postganglionic fibers. The sympathetic fibers pass directly to the muscle. Postganglionic sympathetic fibers also reach the muscle from cell bodies in the celiac, superior and inferior mesenteric ganglia. Each ganglion distributes its branches with the branches of the associated artery. Beyond the splenic flexure the colon receives its parasympathetic preganglionic nerves from pelvic nerves. The postganglionic parasympathetic cholinergic neurons are excitatory, being opposed by the inhibitory postganglionic sympathetic neurons. Visceral afferent nerves in the vagus, arising from the unipolar cells of the nodose ganglion, serve the esophagus and stomach. Sympathetic visceral afferent fibers also serve esophagus and stomach, small bowel, and proximal colon.

Plexuses. The synaptic relationships of the extrinsic nerves to the enteric plexuses are unclear. These plexuses—the subserous, the myenteric, the submucosal, and the mucous plexuses—interact extensively, independent of the extrinsic nerves. They constitute a separate enteric nervous system. Nerve bundles connect plexuses in adjacent gut layers.

The *subserous plexus* is not prominent except at the mesenteric attachment of the small bowel. This plexus contains ganglia and nerve bundles.

The *myenteric plexus,* between the longitudinal and circular muscle coats, is composed of three networks in the same plane. The primary plexus of large nerve bundles linking ganglia of various sizes contains the secondary plexus, a closer network of smaller nerve bundles with scattered neurons. This secondary, or interfascicular, plexus is continuous with a tertiary plexus, or autonomic ground plexus. The tertiary plexus is a network of small bundles of very fine fibers penetrating the muscle coats. The myenteric plexus varies in density; its neurons are sparse in the esophagus and much denser in the stomach. Neurons are concentrated along the mesenteric border of the small bowel and under the taeniae in the colon. Other small stellate or fusiform cells, the interstitial cells of Cajal, lie in the meshes of the myenteric plexus. These cells may be neurons, but they have not been clearly identified as such. Their function is controversial.

The *submucosal plexus* contains several tiers of networks of nerve bundles linking small ganglia. This plexus is probably absent from the esophagus, and, in all the gut, it is less conspicuous than the myenteric plexus.

The *mucous plexus,* an ill defined nerve network without neurons, is an extension of the submucosal plexus into the mucosa.

Enteric neurons are classified morphologically as Dogiel Types I, II, and III, on the basis of the number and size of axons and dendrites. This classification has little morphological meaning (since many intermediate forms occur) and no established functional significance. Neurons and nerve fibers are incompletely enveloped by Schwann cells, which are probably syncytial rather than separate and form no definite capsule about ganglia.

Recent observations indicate that the classic view of the nerves of the gut must be changed. Cholinesterase and catecholamines can now be identified histochemically in nerve fibers and cells, allowing identification of these structures as cholinergic or adrenergic. Most (but not all) neurons in the myenteric plexus are cholinesterase positive, and none contains norepinephrine. Catecholamine-containing nerve fibers form prominent meshworks about ganglion cells, with relatively few entering the muscle layers except, perhaps, in the colon. This suggests that adrenergic fibers act on secondary parasympathetic neurons as well as on muscle. Although differences between organs and species in this pattern have been explored only superficially, there appears to be great variation. Another important observation is the demonstration of a new type of enteric neuron, the nonadrenergic inhibitory neuron. The evidence for its existence is physiological; its morphology and chemistry are unknown.

Receptors. Drug effects on autonomic nerves and smooth muscle are often used to explain actions of autonomic nerves on smooth muscle. The effects of the major peripheral autonomic neurohormones (acetylcholine and norepinephrine) and their congeners and antagonists can be explained by assuming the existence of drug receptors. Receptors are postulated active sites in nerve and muscle cells interacting selectively with corresponding active parts of the drug molecule. The pertinent autonomic receptors are the muscarinic and nicotinic cholinergic receptors and the alpha- and beta-adrenergic receptors which interact with the two established autonomic neurohormones, acetylcholine and norepinephrine. Other substances, histamine and serotonin especially, with corresponding theoretical receptors, are less clearly established as autonomic neurohormones.

Muscarinic cholinergic receptors elicit responses in which, in the presence of physostigmine, acetyl-beta-methacholine and acetylcholine are more potent than carbachol. These responses are selectively opposed by atropine and similar drugs, but not by nicotine and similar drugs. Such receptors occur in smooth muscle cells and in nerves. *Nicotinic cholinergic receptors* are invoked to explain responses in which carbachol, acetylcholine, and nicotine are more potent than acetyl-beta-methacholine, all in the presence of physostigmine. These responses are selectively opposed by nicotine, hexamethonium and *d*-tubocurarine. Such receptors occur in the striated muscle end-plate, in autonomic ganglion cells and in motor nerve fibers and sensory endings. *Alpha-adrenergic receptors* bring about responses in which the order of potency of catecholamines is first norepinephrine, then epinephrine, and last isopropylnorepinephrine (isoproterenol). These responses are selectively opposed by phentolamine, phenoxybenzamine, and similar agents. Such receptors occur in smooth muscle cells and, probably, in cholinergic postganglionic nerves. *Beta-adrenergic receptors* are invoked to explain responses in which the order of potency of catecholamines is isopropylnorepinephrine (isoproterenol), epinephrine, and norepinephrine. These responses are selectively opposed by dichloroisoproterenol, pronethalol, and propranolol. Such receptors occur in smooth muscle cells. Many other agents (gastrin, secretin, bradykinin, angiotensin, vasopressin, substance P, morphine, and the prostaglandins, among others) affect gastrointestinal motor function, but their actions cannot be so carefully defined in terms of the above receptors. Some may act directly on nerves or muscle, others indirectly through release of neurohormones from their sites of storage.

In the gut, stimulation of muscarinic and nicotinic cholinergic receptors generally excites contractions, whereas beta-adrenergic receptors generally inhibit them. Stimulation of alpha-adrenergic receptors excites contractions in some parts of the gut and inhibits them in others.

INTEGRATION OF CONTRACTION IN SMOOTH MUSCLE

In somatic striated muscle, grading and integration of contraction requires a motor unit in which one motor neuron controls a fixed number of muscle cells. Gastrointestinal striated muscle may have such an arrangement, but not smooth muscle. In smooth muscle, motor nerves lie diffusely scattered among the cells. There are no specialized motor nerve terminals. Neurohormones are released over long segments of terminal nerve fibrils. Precise spatial relationships between nerve and muscle remain obscure because of limitations in anatomical technique. In general, grading and integration of contraction in smooth muscle depend upon: (1) extrinsic nerves, (2) local reflexes, (3) circulating hormones, and (4) the membrane electrical properties of smooth muscle.

Extrinsic Nerves. The influence of extrinsic nerves is difficult to evaluate. The necessary techniques, nerve section and stimulation, suffer from limitations: motor nerves reach the gut through multiple routes; "normal" rates of nerve stimulation are not clearly known; and most extrinsic nerves may be mixed bundles of sensory and motor, sympathetic and parasympathetic fibers. It is virtually impossible to trace single extrinsic nerve fibers to their terminations in the enteric ganglia. Any single fiber must contact myriad ganglion cells: the vagi contain only a few thousand parasympathetic motor fibers as they enter the abdomen, fibers destined to reach many millions of ganglion cells in the stomach and the small bowel. There are no comparable estimates for the sympathetic motor nerves. Except for the esophagus, gut movements continue after section of extrinsic nerves. Extrinsic nerves are viewed as exerting a general modulation of local mechanisms, largely through centrally mediated reflexes.

Local reflexes, both excitatory and inhibitory, modify contractions. They constitute a major control mechanism. Such reflexes are initiated by movement, stretch and chemical stimuli. The sensing structures are unidentified anatomically and incompletely characterized physiologically. The exact pathways are generally undefined. Most reflexes can be described only in terms of stimulus and response.

Hormones. Circulating hormones produce more general or remote effects than local reflexes. Some systemic hormones (e.g., epinephrine) have general effects, whereas gastrointestinal hormones (e.g., gastrin) seem to have a more sharply localized action. Some of the latter hormones may have specific roles in gastrointestinal movement. Gastrin, which contracts the gastroesophageal sphincter and retards gastric emptying, may also be partly responsible for the gastrocolic reflex.

Electrophysiology. The cell membrane electrical properties of smooth muscle constitute a major integrative mechanism. The spontaneous activity of gastrointestinal smooth muscle is related to fluctuations in cell transmembrane potential that reflect ion movement across the membrane. These fluctuations are partly myogenic, independent of the innervation. Besides being linked to contractions, these membrane electrical phenomena have another function by virtue of the nexuses, the low-resistance intercellular shunts. Cells spread over relatively great distances all act together, electrically.

The *resting membrane potential* is the voltage drop across the cell membrane when the membrane is electrically stable. In smooth muscle, the resting potential is 50 to 80 mv. The inside is negative to the outside. This wide range reflects differences among tissues as well as inaccuracies in measurement. (The technical problems of measurement are great.) Transmembrane potentials in smooth muscle are unstable. They vary with the tension and the state of activity of the muscle, making it difficult to determine actual resting potential. Ion distributions underlying the resting membrane potential in smooth muscle are unknown, owing to uncertainty in regard to intracellular ion concentrations and membrane permeabilities for ions. Not only does resting potential fluctuate slowly; it is interrupted frequently by transient depolarizations associated with activity, transients varying in form among tissues. The variability of these transients and our ignorance of their genesis make classification difficult except on the basis of form. This criterion distinguishes two kinds of such transients—action potentials (or action spikes), and slow waves.

During spontaneous or induced contractions, smooth muscle generates rapid, transient, spikelike depolarizations, *action potentials.* Their amplitude approaches the value of the resting membrane potential. They precede and seem to trigger contraction, they can spread, and they seem to be of an "all or none" character. Compared to nerve action potentials, action potentials of smooth muscle are slow. The rate of rise is less than 20 volts per second, duration at half-maximum amplitude is up to 30 milliseconds. They tend to occur in bursts or periods. The patterns vary among tissues, with tension, and with general level of activity. Spikes are myogenic; they persist after denervation by a variety of methods.

Their occurrence depends on resting membrane potential. Spikes occur with low resting potentials more commonly than with high ones. The ionic shifts responsible for these spike potentials are unknown.

Slower transients, called *slow waves*, occur in some but not all smooth muscles. Slow waves are slow fluctuations in membrane potential, apparently conducted over very long distances. Although, within a tissue, they cycle with a regular periodicity and are uniform in configuration, period and configuration differ among organs and among species. In the organs where they do occur, they are constantly present. In the gut, they have been described only in the distal part of the stomach and in the small bowel and colon. The form, velocity and direction of their apparent propagation are different and characteristic for each organ. Although the slow waves in these organs occur with great regularity, their period and the direction of their apparent movement can be modified. They are often accompanied by bursts of action potentials, which occur on the peak of each slow wave, at the period of maximal depolarization. It appears that the two phenomena result from two separate processes: the first, the slow wave generator, cannot elicit contraction, but it can trigger the second, the spike generator, which is contraction-coupled. Ionic shifts underlying slow wave generation are unknown. Since slow waves appear to pace contractions and to move without decrement throughout a whole viscus or segments of it, they seem to integrate activity throughout the whole organ.

THE PHARYNX AND ESOPHAGUS

To begin a swallow, one first closes the entrance to the gut by sealing the lips, the nasopharynx, and the glottis. Orderly contraction of the tongue and pharynx propels a bolus toward the gullet with the structures contracting in a fixed sequence so that the contraction seems to sweep continuously, with acceleration, from the base of the tongue to the pharyngoesophageal or upper esophageal sphincter. The anatomical counterpart of this sphincter is the cricopharyngeus and adjacent circular esophageal and pharyngeal muscles. This 3-cm. segment is closed at rest, exerting a variable pressure in the lumen, about 18 to 60 cm. H_2O above atmospheric pressure. Just after contraction begins in the upper pharynx, the sphincter relaxes so that it is open when the pharyngeal peristaltic sweep arrives. Having remained open for less than 1 second, the sphincter closes, as the pharyngeal contraction reaches it, in a forceful contraction lasting a few seconds. It then relaxes to its resting tension.

The esophageal body, which begins at the sphincter, is flaccid at rest. Luminal pressure is subatmospheric, reflecting intrapleural pressure. Contraction begins at the rostral end just after closure of the upper esophageal sphincter and sweeps toward the stomach. The strength of this contraction is less in the proximal than in the distal half of the esophageal body. Intraluminal pressure produced by the contraction seldom exceeds 30 cm. H_2O proximally, 60 cm. H_2O distally. The contraction accelerates as it moves, velocity being about 2 to 4 cm./sec. The length of the contracted segment is about 5 cm. in the upper esophagus, about 12 cm. in the lower.

The gastroesophageal junction is closed at rest, creating a barrier between the hypobaric esophageal lumen and hyperbaric gastric lumen. This lower esophageal sphincter is a functional unit 2 to 4 cm. long. It has no clear anatomical counterpart. Its closure tension exerts a luminal pressure about 10 cm. H_2O above gastric fundal pressure. In a swallow, this sphincter opens less than 2 seconds after the upper sphincter opens, well before the esophageal peristaltic contraction arrives. Luminal pressure in this zone falls to gastric fundal pressure. After 3 to 5 seconds, when the peristaltic wave arrives, the sphincter closes. In the rostral half of the sphincter, the tension of closure transiently exceeds resting tension for a second or two. Transient distention of the esophageal body can cause peristalsis below the stimulus, called secondary peristalsis, and relaxation of the lower esophageal sphincter.

This complex series of events is controlled by a combination of central and peripheral nervous mechanisms, still inadequately understood. Contractions of buccal muscles, tongue, pharynx and upper esophagus, the buccopharyngeal phase of swallowing, are initiated by excitation of afferent nerves distributed widely throughout the pharynx, the larynx, and the upper esophagus. These afferent nerves bear upon a swallowing center in the reticular formation between the caudal pole of the facial nucleus and the rostral pole of the inferior olive. When activated, the center produces a stereotyped spatiotemporal pattern of contraction in the responding striated musculature through actions on motor neurons of the trigeminal, facial, and hypoglossal nuclei, and the nucleus ambiguus. Sequential firing of neurons in these centers establishes sequential contractions of striated muscle, leading to the progression of contraction through the pharynx and proximal esophagus.

How the upper esophageal sphincter is controlled is not clear. Some believe that luminal closure at this level is passive, the result of elasticity of the muscle, and that luminal opening in swallowing is also passive, the mechanical effect of elevating the larynx. Others claim that closure represents tonic contraction of the muscle and that opening results from inhibition of the motor nerves. There is good evidence to support the latter hypothesis.

Control of the smooth muscle segment of the esophagus is obscure. This muscle could be controlled by the swallowing center. Peristalsis, however, can occur in the isolated smooth muscle segment, so the extrinsic innervation must play only a supportive role in esophageal movement. In the isolated segment of opossum esophagus composed wholly of smooth muscle, a complex series of reflex responses to stretch occurs. These reflexes resemble those occurring in vivo in man. Initiation of localized distention excites a brief circular muscle contraction just rostral to the stretch, often progressing retrograde. This is a direct response of the muscle, not a nervous reflex. As long as the stretch is maintained the esophagus shortens, owing to isolated contraction of the longitudinal muscle, a local reflex involving cholinergic ganglion cells. About 1 second after termination of the stretch, a brief circular muscle contraction occurs just caudal to the stretch, often progressing down the full length of the organ, at other times involving the entire caudal segment at once. This response, resembling secondary peristalsis in man, results from a noncholinergic motor nerve mechanism. In contrast to the striated muscle, the smooth muscle segment has a high degree of autonomy, yet the peristaltic wave passes over the junction of striated and smooth muscle unbroken. No one knows how this occurs.

What keeps the gastroesophageal sphincter closed? Some believe that closure is passive, crediting the diaphragmatic hiatus and phrenoesophageal ligament, which enclose this region in most animals. Yet the opossum, an animal whose gastroesophageal junction lies far below the diaphragm, has a clear sphincter at the junction. Most authorities believe that closure represents sustained active contraction of the smooth muscle, but sustained electrical activity has not been shown here. If closure is active, either local or central neuromechanisms, or both, could be responsible. Sphincter closure is weak under general anesthesia and after vagal section, and very weak or absent in the isolated organ, suggesting influences of central motoneurons, presumably in the dorsal motor nucleus. The sphincter is opened by swallowing and esophageal distention, possibly through centrally mediated reflexes. Whether opening represents central or peripheral inhibition of a neural closure mechanism, or activation of separate muscle-inhibiting nerves, is unclear. It is also unknown whether these systems use cholinergic, adrenergic or nonadrenergic inhibitory nerves.

DISORDERS OF ESOPHAGEAL MOTOR FUNCTION

Motor disorders of the esophagus are better defined than those elsewhere in the gut because the esophageal process is brief, stereotyped, and accessible to study. Thus, the impression that motor disorders are commoner in the esophagus than in other viscera may simply reflect a better understanding of the esophageal process. Moreover, most motor disorders of the esophagus are explained on the basis of nerve disease—which may reflect the fact that the esophagus is mostly controlled by nerves, whereas myogenic mechanisms dominate the operation of the other viscera.

In **achalasia** (or **cardiospasm**), peristalsis in the esophageal body is lost and the gastroesophageal sphincter fails to open with swallowing. In Europe and North America this idiopathic entity usually afflicts young adults and is apparently not associated with other disorders. It resembles a disorder found in Brazil and associated with Chagas' disease, developing long after the initial infection with *Trypanosoma cruzi*. Moreover, a similar and rare disorder occurs in dogs. The atony and dilatation of the esophageal body in idiopathic achalasia were once thought to be passive, the result of the obstruction at the gastroesophageal junction. Authorities now view this as the active result of disease in the esophageal body. The entire esoph-

ageal body is variably dilated and often elongated. The pharynx and upper esophageal sphincter act normally, but peristalsis initiated by swallowing passes only a few centimeters down the esophagus. Below this point contractions may be feeble or nonprogressive. Weak nonpropulsive contractions sometimes occur spontaneously. These often appear on barium swallow as transient narrow rings, shallow and randomly distributed, which are especially prominent in the distal half of the organ. The lower esophageal sphincter is closed, producing symmetrical tapering to a point at the gastroesophageal junction. The sphincter may open a little spontaneously but not with swallowing. Resting luminal pressure in the sphincter is at least normal and is usually increased.

Achalasia is the result of denervation. In both idiopathic achalasia and achalasia of Chagas' disease the number of ganglion cells is reduced, the deficiency being greater in Chagas' disease. There may also be some extrinsic denervation. Postvagotomy achalasia occurs rarely after vagotomy in man. Vagal section and destruction of the nucleus ambiguus in dogs produces esophageal abnormalities resembling achalasia, but the canine esophagus consists entirely of striated muscle. The concept of cholinergic postdenervation hypersensitivity is invoked to explain the effect of acetyl-beta-methacholine (Mecholyl) on the esophagus in achalasia. This agent, given subcutaneously in sufficient dose, produces cholinergic responses because it is slowly destroyed by cholinesterases. In achalasia, smaller doses than those required to cause other cholinergic effects will produce strong, repeated contractions of the esophagus. The dilated esophagus narrows, the walls are apposed. This response is sometimes used diagnostically to distinguish achalasia from cancerous obstruction at the gastroesophageal junction. Muscle strips from the distal esophagus of patients with achalasia respond poorly to ganglion cell stimulants, also suggesting ganglion cell depletion. Beta-adrenergic receptors, which are inhibitory, seem to function normally in such strips. The cause of neuronal degeneration in idiopathic achalasia is a mystery, as is the cause of the achalasia of Chagas' disease.

In **diffuse esophageal spasm,** swallowing causes powerful simultaneous uncoordinated movements of the esophageal body. This is usually most prominent in the distal two thirds of the organ, but it may involve the whole esophagus. The two sphincters operate normally. The force of the contraction varies. Weak contractions may cause only dysphagia, but strong contractions are quite painful. Symptoms are aggravated by ingestion of cold foods or liquids and by nervous tension. Nonperistaltic responses to swallowing occur in about 10 per cent of swallows in healthy young subjects, and the proportion gradually increase with age. In most nonagenarians nearly all responses are nonperistaltic. The abnormality acquires the status of a disease only when the nonperistaltic contractions occur with a frequency disproportionate to the patient's age, or when they produce symptoms of dysphagia or odynophagia. The disorder, whether symptomatic or not, causes distortion of the radiographic silhouette which radiologists refer to as the corkscrew esophagus, pseudodiverticulosis, curling, or tertiary contractions, according to the gross pattern of the contractions.

The pathogenic process may be degeneration of local neurons, an idea supported by the positive response of some patients to methacholine. Also, some patients present a disorder intermediate between diffuse esophageal spasm and achalasia; a patient has been described whose disorder appeared to evolve from the one to the other. The kinds of nerves involved must be somewhat different from those involved in achalasia. Perhaps inhibitory nerves are more affected than excitatory ones. As in achalasia, there is no identifiable cause for neural dysfunction. The disease may be viewed as an exaggeration of the aging process.

The gastroesophageal sphincter shows two kinds of abnormal function, **hypertension,** and **relaxation,** or **chalasia of the sphincter.** In the hypertensive sphincter, resting tension is abnormally increased but sphincter opening is normal. Although this is often associated with diffuse esophageal spasm, it may occur alone. Other patients may have no detectable sphincter; the zone is always relaxed. This may be called chalasia, a term used mostly in pediatrics. In premature and full-term neonatal infants the sphincter is often not demonstrable. Esophageal peristalsis is also disorganized in such infants, but normal esophageal body and sphincter function soon develop. These abnormalities are viewed as the consequence of neuronal immaturity. In adults, the absence of sphincter closure is usually a troublesome disease, allowing free reflux of gastric contents into the esophagus, leading to esophagitis and sometimes stricture. This is so often associated with hiatus hernia that a causal relationship is inferred. Proponents of this view say that the sphincter fails because it is displaced from the support of the diaphragmatic hiatus and because the normal gastroesophageal angle is obliterated. However, many patients with hiatus hernia have, in fact, an effective sphincter and, conversely, the sphincter may fail even though there is no demonstrable hernia. A disturbance of the innervation seems a likely explanation for both of these abnormalities of the gastroesophageal sphincter, but there is no direct evidence for this.

Pulsion diverticula may also result from an esophageal motor disorder. Pulsion diverticula occur mainly adjacent to the sphincters. These diverticula are saccular with narrow mouths, and they tend to grow. The *epiphrenic diverticulum* arises from the region of the lower esophageal sphincter and commonly occurs in achalasia. *Zenker's diverticulum* arises from the pharynx just above the upper esophageal sphincter. Though pulsion diverticula are usually considered the result of a primary weakness of the pharyngeal or esophageal wall, this does not account for their usual parasphincteric location. They could also result from abnormal sphincter closure. Closure tension of the lower esophageal sphincter is increased in some patients with achalasia. Some patients with Zenker's diverticulum show delayed opening of the upper esophageal sphincter.

In most patients with **progressive systemic sclerosis,** or **scleroderma,** the esophagus shows abnormal peristalsis and the lower sphincter fails to close. The pharynx and upper esophageal sphincter act normally but esophageal contractions are weak and often simultaneous. The weakness is explained by focal replacement of the muscle by fibrous tissue. Both the weakness and the histologic lesion are more severe in the smooth muscle, but they may also occur in the striated muscle. If coordination of esophageal movement is a function of nerves, then one cannot account for the nonpropulsive simultaneous contractions of the esophageal body often seen in scleroderma on the basis of fibrous replacement of muscle. However, no esophageal neural lesion has been described in scleroderma.

Esophageal dysfunction in **diabetes mellitus** is common, although not usually symptomatic. Pharyngeal contractions are weak; the esophageal responses are feeble and often simultaneous, and lower esophageal sphincter closure tension is reduced. These abnormalities are attributed to visceral neuropathy, but there is little anatomical or physiological evidence to support this. In fact, the concept of visceral, or autonomic, neuropathy remains very poorly defined.

Esophageal dysfunction also accompanies a variety of neuromuscular disorders. In **myasthenia gravis,** the striated musculature is weak, producing weakness in the upper esophageal sphincter. In **myotonic dystrophy,** closure tension of the upper sphincter is reduced, contractions are feeble and often simultaneous in the pharynx and esophageal body, and the gastroesophageal sphincter is normal. After **bulbar poliomyelitis,** pharyngeal and proximal esophageal contractions are weak or absent, but the upper esophageal sphincter is normal. Characteristically, the last named disease involves the nucleus ambiguus but spares the dorsal motor nucleus of the vagus which, some believe, controls the sphincter. Motor abnormalities have also been described in **amyotrophic lateral sclerosis, familial dysautonomia,** and **multiple sclerosis,** among other entities.

Why is dysphagia so rare when esophageal dysfunction is apparently so common? Man eats in the upright position, hence, the esophagus can empty by gravity even when the propulsive mechanism is abnormal. Generally, dysphagia occurs only when the disturbance of the mechanism is such that the descent of food into the stomach is obstructed. Failure of the pharynx or upper esophageal sphincter produces oropharyngeal dysphagia, a sense of suprasternal obstruction with aspiration or nasal reflux of the bolus. Failure of the esophagus or lower esophageal sphincter produces esophageal dysphagia, a sense of retrosternal obstruction.

THE STOMACH

The stomach stores food, mixes it and regulates its delivery to the bowel. The three anatomical divisions of the stomach are: the fundus, lying cephalad to the gastroesophageal junction, the corpus, extending from the gastroesophageal junction to the incisura angularis, and the antrum, extending from the incisura angularis to the pylorus. These divisions are somewhat arbitrary, however, and do not correspond to the two gastric motor functions, receptive relaxation and emptying.

Gastric Filling. After a fast, the stomach contains about 50 ml. of fluid. After a meal it may contain more than a liter. We know very little about the process of filling, but the popular concept is expressed in the term *receptive relaxation.* Although the process may be viewed as a passive one (i.e., an infolded empty stomach simply unfolding until a maximal volume is reached), it appears, in fact, to be active; the diameter of the stomach accommodates to the volume in such a way as to keep wall tension rather constant over a wide range of volumes. This function resides mostly in the fundus and less in the gastric body. The function has not been assigned to a single muscle layer, but the arrangement of the innermost oblique layer suggests that that layer could be chiefly responsible. Studies in cats indicate that receptive relaxation is caused by the intrinsic nonadrenergic inhibitory nerves. These nerves seem to be controlled by central mechanisms under the influences of vagal afferent nerves from the pharynx and esophagus.

Gastric Emptying. This process is accomplished by tonic and peristaltic contractions. *Tonic contractions* are slow, transient rises in wall tension, lasting about one minute, and are more prominent in the fundus than in the antrum. Their function is not clear, but probably they slowly move gastric contents by reducing the volume of a single segment of the stomach. It is tempting to think that they move contents from the reservoir (fundus) into the pump (antrum) but this has not been proved. *Peristaltic contractions* are localized contractions of the circular muscle layer, moving smoothly along the antrum. The advancing faces of contracting and relaxing fibers define a contracted segment 1 to 2 cm. wide. The contractions appear to arise about the middle of the body of the stomach as very shallow indentations and move toward the pylorus, with acceleration and a progressive increase in depth. Rarely they fade before reaching the antrum,

occasionally they pass as narrow rings to the pylorus, but usually they terminate with antral systole, a simultaneous contraction of the terminal antral segment and the pylorus. When they are regular at their highest rate, their frequency is constant. In man, these contractions occur maximally at three times per minute; other species have different characteristic frequencies. Moreover, in man, when they are slower than three times per minute, they appear at intervals of some multiple of 20 seconds. When they are shallow, as in the body of the stomach, some matter is forced ahead while other matter escapes back through the contraction ring. As contractions deepen toward the antrum, the proportion of material pushed forward progressively increases.

Net gastric emptying has no fixed rate; it is graded and is controlled by several mechanisms. Rate of emptying varies with the volume of residual gastric content. Also, it can be reduced by the action of at least three types of duodenal receptors: acid, fat and osmoreceptors, each sensing the concentration of their respective excitants in the gastric effluent and appropriately reducing the rate of emptying by way of reflexes and, perhaps, hormones. The reduction is probably accomplished mainly through a reduction in stroke volume of the antral pump. Gastric emptying is also affected by systemic stimuli such as somatic and visceral pain, and by hypoglycemia. The role of hormones is under debate. Gastrin, which excites gastric muscle, and a poorly described inhibitor, enterogastrone, may have some role in the physiological regulation of gastric emptying.

Although grading of stroke volume results probably from changes in the depth of the antral stripping wave, emptying could also be graded by changes in the frequency of peristalsis. Antral peristalsis is paced by a carrier, the *gastric slow wave,* which establishes frequency, velocity, and direction of peristalsis. The gastric slow wave consists of an omnipresent initial potential and a second potential appearing only during contractions. The spikelike initial potential arises almost imperceptibly at an unidentified pacemaker about two thirds of the way up the stomach and passes toward the pylorus with constant amplification and acceleration. The pacemaker starts a new initial potential every 20 seconds. During peristalsis, a second electrical deflection appears (the second potential), phased by the initial potential with a constant phase lag. It appears as a burst of spikes or fused spikes, and it is the electrical correlate of contraction. The frequency, velocity and direction of propagation of antral peristalsis depend upon the slow wave initial potential. After gastric transection, the slow waves above the level of transection continue unaffected, but below the transection they are slower and often move retrograde, appearing to arise from a pyloric pacemaker. The gastric slow wave can be affected by autonomic neurohormones. Acetylcholine, given by injection into a small artery supplying the stomach, enhances the second potential and its accompanying contraction, and can induce a premature initial potential if delivered in a critical susceptible period. Catecholamines, given by a similar route, inhibit second potentials and contractions; large doses cause rapidly repeated spikes resembling initial potentials. Excitation of both alpha- and beta-adrenergic receptors can produce these effects. The gastric initial potential continues with great decrement for a few centimeters into the duodenum, apparently inducing duodenal contractions at that period in the cycle when each pump stroke discharges food into the duodenum.

DISORDERS OF GASTRIC MOTOR FUNCTION

Disorders of the motor function of the stomach are less well described than those of the esophagus. Among patients with **diabetes mellitus,** gastric emptying may be delayed to a variable extent. This is usually taken as evidence of autonomic neuropathy, though the concept is poorly defined. Occasionally gastric emptying is so severely retarded that vomiting occurs, leading to difficulty in diabetic control.

In treating peptic ulcer, modern gastric surgeons usually perform **vagotomy** with many varieties of gastric operations. Vagotomy alone delays gastric emptying. The addition of pyloroplasty (commonly called a drainage procedure) partly reverses this effect. In pyloroplasty, the incised pylorus is sewn in such a way as to widen the diameter of the pyloric channel. But pyloric channel diameter is a minor factor in antral stroke volume, except when the channel is scarred by an ulcer. It is reported that vagotomy produces chaotic spread of the antral initial potential, with many retrograde sequences arising from the pyloric pacemaker. Pyloroplasty depresses this pyloric pacemaker. After vagotomy and pyloroplasty, antral slow waves are antegrade but somewhat slower than normal.

After various kinds of gastric resection, emptying is widely held to be accelerated, as suggested by the term **dumping syndrome.** Nothing definite is known of the changes in emptying after subtotal gastric resections.

Vomiting is a complex reflex response to excitation of unidentified receptors in the pharynx, stomach and other viscera, and to emotions and dizziness. The multiple afferent nerves bear on a medullary vomiting center near the respiratory center and the vagal motor nuclei. Responsible efferent nerve tracts to the gut are vagal, in part, but many other pathways must be involved in the whole response. First there is inhibition of gastric peristalsis and small bowel movement, followed serially by an increase in small bowel tonus, contraction of the gastric antrum and relaxation of the gastroesophageal sphincter. Then contraction of the diaphragm and abdominal wall expels gastric contents. These events are difficult to study in man because sensing elements passed by mouth are expelled in the response.

THE GALLBLADDER AND BILE DUCTS

The movements of the biliary system, a neglected area of study, probably involve four functionally distinct parts: the gallbladder, the valve of Heister, the bile ducts and the sphincter of Oddi. Both the vagi and the splanchnic nerves carry both excitatory and inhibitory nerves to these structures. The vagi mostly excite, the splanchnics mostly inhibit. Intramural ganglion cells occur throughout the biliary system, in the adventitia, the muscle, and the submucosa. These are connected by nerve bundles in irregular plexuses. Afferent nerve fibers from these viscera travel with both the vagi and the splanchnic nerves.

Filling of the gallbladder is commonly viewed as a passive yielding of the wall to the low pressure of the entering bile, but, in fact, the physiology of this process is unknown. Emptying of the gallbladder is accomplished by contraction of the wall. Current opinion holds this to be mainly the response to a hormone, cholecystokinin-pancreozymin, liberated from the duodenal mucosa in response to fats and to some amino acids in the lumen. This view leaves unexplained the functions of the local nerves. The older literature emphasized the nerves in the emptying process. All told, our understanding of the interaction of the hormones, nerves, and muscles in both filling and emptying is deficient.

The valve of Heister is actually a spiral arrangement of the cystic duct about whose function little is known, certainly not enough to justify the implications of the term "valve." It may provide a barrier to emptying of the gallbladder, since a pressure gradient can develop between the gallbladder and the common bile duct in unanesthetized dogs but not in anesthetized animals or cadavers.

The common bile duct is mostly fibrous in dogs; it contains more smooth muscle in man. Peristalsis is thought not to occur in this tube but concrete evidence is sparse. The little reliable information available suggests that this tube is, at best, a passive conduit; yet this view leaves unexplained the presence of smooth muscle in the wall.

The sphincter of Oddi is a ring of muscle buried in the duodenal wall and surrounding the exit of the common bile duct. Students have argued whether closure at this level is produced by passive pinching of the duct by the duodenal wall or whether closure is accomplished by an active sphincter. Probably an active sphincter is responsible, for the muscle at this site is pharmacologically distinguishable from duodenal muscle. Alpha-adrenergic receptor stimulation excites contractions, and morphine is a potent excitant. Also, after excision or incision of the sphincter in man, barium sulfate may freely enter the biliary system during roentgenographic examination of the small bowel, a phenomenon not seen in normal persons. A clear description of nervous and hormonal control of the sphincter and its normal operation has not yet come forth.

MOTOR DISORDERS OF THE BILIARY SYSTEM

The pain of **cholecystitis** and **cholelithiasis** is often attributed to abnormal contractions somewhere in the biliary system. The gallbladder empties poorly when stones are present. The relationship between the inflammation and stones, on the one hand, and the motor disturbances, on the other, is conjectural.

Biliary dyskinesia is a vague diagnostic term applied to pain, like that of cholecystitis, which persists after cholecystectomy or occurs in the absence of radiographically demonstrable biliary tract disease. The term indicates that those who use it believe that a motor abnormality is responsible for the pain. No convincing evidence exists to support this idea.

THE SMALL INTESTINE

The varieties of movement in the small bowel have long been a source of argument. No clear picture exists, largely because the events are complex and the methods of observation are limited.

Segmenting contractions, stationary localized circumferential contractions of the circular muscle one to two centimeters wide, predominate. Segmentation may occur in apparently random fashion in time and space, or may be ordered in time to occur regularly as rhythmic segmentation.

Pendular movements are defined poorly and seen too rarely to allow any consensus about them. Some suggest that they represent rhythmic segmentations occurring alternately in adjacent segments.

Peristalsis is also controversial. Defined as a moving ring of contraction, peristalsis certainly occurs in the isolated intestine, but it is less clear in the normal gut in vivo. Some claim to have seen it; others never see it. If not actually rare, it is, at least, much less often recognizable than segmentation. The whole problem can be viewed as one of integration and differentiation of the activity of many units whose basic function is segmentation. These segments, one to two centimeters long, may be aligned side by side, or they may overlap. If such units contract completely at random in time and space, they accomplish optimal internal circulation or mixing of contents. The opposite possibility, complete integration, would constitute apparent peristalsis, the *peristaltic rush* seen at times in various animal preparations. This would produce optimal propulsion. It seems likely that the degree of integration of segmentation is never at one extreme or the other, but somewhere in between, to produce the various patterns seen. The small bowel thus differs from the esophagus and gastric antrum where complete integration in time and space is always present to produce peristalsis.

As in the stomach, many **integrating mechanisms,** not yet clearly understood, operate in the small bowel. The classic local reflex is the peristaltic reflex of Bayliss and Starling, a contraction rostral to, and inhibition caudal to a point of local stimulation. Centrally mediated reflexes also operate: movement in the entire small bowel in situ is inhibited by manipulation of one isolated segment and by somatic and visceral pain. Currently, the most interesting integrating mechanism is a myogenic one, the small bowel slow wave.

Electrical records from small bowel muscle show slow waves and spikes. The slow waves are near-sinusoidal oscillations of resting membrane potential occurring in concert in millions of adjacent muscle cells. Slow waves are constant, spikes are intermittent. When spike bursts occur, they do so at the period of maximum slow wave depolarization and just preceding contraction. The slow wave is generated by the longitudinal muscle layer but passes into the circular layer through muscle bridges. A single slow wave appears to move toward the colon as a ring along the intestine, reaching all points on the circumference of the coronal section of the gut simultaneously. Slow wave frequency and apparent velocity are highest in duodenum, lower in jejunum and lowest in ileum. Frequency declines in widely spaced steps, being fairly constant for variable lengths of gut between steps. Slow waves apparently trigger spikes. Nerves have little effect on the characteristics of the slow waves. Spikes, on the other hand, are nerve-modulated. Cholinergic agents excite spike bursts, anticholinergic agents inhibit them, serotonin excites them, and stimulation of alpha- and beta-adrenergic receptors inhibits them. Extrinsic nerve stimulation variably excites or depresses spike bursts, depending on the frequency of stimulation.

The slow wave is a pacemaker, a carrier, upon which hang other integrating activities. It provides two important governors of small bowel contractions—one spatial, the other temporal. The spatial one is polarity: probably the constantly aboral movement of the slow wave somehow assures that contents flow always aborad. The temporal one is rate-limiting for the frequency of rhythmic segmentation: when any given segment of the intestine is excited to contract rhythmically at its maximum rate, it does so at exactly the frequency of the omnipresent slow wave. Since slow wave frequency declines along the gut, the proximal intestine can contract maximally at a faster rate than the distal intestine. This frequency gradient of contraction also may be related to aboral propulsion.

DISORDERS OF SMALL BOWEL FUNCTION

Ileus is the prolonged absence of small bowel contractions. It represents the disappearance of contractions under the influence of local and centrally mediated reflexes. The slow waves probably persist, however, because, in animal experiments, the slow waves disappear only when the muscle is dead. In **small bowel infarction,** mapping the slow wave at the margin of the infarcted segment may be useful in determining the limits of resection.

If movements of the small bowel accomplish propulsion, why is obstruction not universal in **regional enteritis?** In this disease the muscular wall is infiltrated by a chronic inflammatory process. The wall is grossly rigid and thick, and the lumen is narrowed. This process occurs over sharply demarcated segments of variable length, especially in the terminal ileum. Although

many patients with regional enteritis do develop small bowel obstruction, others do not. The propulsive force of the last few centimeters of normal bowel above the stenotic segment may be sufficient to propel matter down a rigid-walled segment, the success of the effort depending upon the balance between the propulsive force of the muscle contractions and the resistance of the pipe.

Reversed segments of the small bowel produce partial obstruction readily, even when they are short. Mostly as an experimental procedure, but occasionally therapeutically, a short segment of small bowel is resected and replaced in continuity in the reversed position. Even if very short, such segments slow down transit, probably because they always retain their original polarity.

THE COLON

The contractile repertoire of the colon, like that of the small bowel, is imperfectly described. Most often the colon exhibits *segmentation*, being divided into segments by lumen-occluding narrow contractile rings. These segments are the familiar haustra. Some believe that these rings are fixed in position by localized anatomical or physiological specialization, but, in fact, their positions probably vary from time to time. Segmenting contractions retard propulsion by creating resistance to flow. Segmentation occurs in the sigmoid colon about 50 per cent of the time; in the cecum it is much less frequent. Regional variations in occurrence and force of segmentation may check flow of stool through certain regions of the colon. *Propulsion* in the colon is accomplished by withdrawal of all segmentation below the fecal mass and the application of a force above the fecal mass. The nature of this force is unknown. The consequent intermittent movement of the fecal mass into an empty segment of colon is called the *mass movement*. Mass movement occupies only a few seconds, and recurs two to three times daily. Rarely, a moving ring of contraction, *peristalsis*, occurs in the colon. Considering the rarity of peristalsis, it is difficult to assign this process any function in normal colonic movement.

Colonic movement is depressed by sleep, usually increased by fear, and variably affected by other emotional stimuli. Eating excites both mass movements and segmentation; this is the gastrocolic reflex. Whether this is neurogenic or a humoral response is a matter of controversy.

Colonic movement is probably affected by extrinsic and intrinsic nerves, but the details are confusing, and no complete picture emerges. Studies of drug effects have not cleared things up. Muscarinic cholinergic stimulants actually inhibit contractions in vivo, but the cholinesterase inhibitor, prostigimine, increases segmentation. Muscarinic antagonists such as atropine do not clearly depress segmenting contractions in vivo. Ganglionic stimulants applied in vitro inhibit spontaneous contractions. Adrenergic drugs have not been studied well.

Slow waves in the colon have unique characteristics. They are generated in the circular muscle and seem to pass around and around; their frequency varies somewhat, being slower in cecum and ascending colon than elsewhere. Slow waves can be altered by drugs affecting the autonomic nervous system, but slow waves are clearly myogenic. Their function in the colon is obscure, but they probably exert some sort of control over segmentation.

DISORDERS OF COLONIC MOTOR FUNCTION

In simple **diarrhea** and **constipation**, segmentation appears to be at fault. Diarrhea accompanies greatly diminished colonic segmentation; constipation accompanies normal or increased segmentation. These observations fit with the view that segmentation provides resistance to flow. The more basic processes leading to diarrhea and constipation, then, are those governing segmentation, and these are still unknown.

In **diverticulosis** also, segmentation is at fault. The primary event is hypertrophy of localized rings of smooth muscle in the rectosigmoid colon. The cause of the hypertrophy is unknown; some authorities blame it on a low-residue diet. In persons with hypertrophy alone (the *prediverticular state*), eating excites powerful rhythmic contractions of this segment through the gastrocolic reflex. Pain may result either from the extreme force of these contractions or from transient obstruction of the colon and gaseous distention above the sigmoid colon. The extremely high intraluminal pressures produced by the hypertrophic rings causes herniation of the mucosa through the circular muscle layer at the weakest points, the places where arteries penetrate the muscle.

Functional bowel syndrome and **irritable colon syndrome** are vague diagnoses indiscriminately applied to patients who have abdominal pain, diarrhea, or constipation, without objective evidence of disease. The uncertainties and gaps in our knowledge of colonic physiology make it quite possible that these reflect, in part, various motor dysfunctions of the colon. In a related entity loosely called the laxative colon, constipation persists despite long-continued laxative use, and the distal colon lacks haustra. Some evidence suggests that excessive use of laxatives damages the neurons of the intrinsic plexuses.

Hirschsprung's disease, or congenital aganglionic megacolon, is a congenital stenosis of a segment of colon, usually rectum, with great dilatation of the colon proximal to the stricture. In the constricted segment, intrinsic plexus cells are absent. Although adrenergic fibers are present in this segment, they are morphologically abnormal. The dilated segment of the colon has a normal neural histology. The spatial correspondence of sustained contraction and absent ganglion cells suggests that such cells are normally inhibitory to the rectum. Experimental evidence supports this concept.

ANNOTATED REFERENCES

Burn, J. H.: The Autonomic Nervous System for Students of Physiology and Pharmacology. ed. 2. Philadelphia, F. A. Davis, 1965. (A synoptic review of the anatomy and pharmacology of the autonomic nervous system)

Burnstock, G., and Holman, M. E.: Effects of drugs on smooth muscle. Ann. Rev. Pharmacol., 6:129, 1966. (A comprehensive and critical review of the ways in which drugs may affect processes in smooth muscle)

———: Smooth muscle: autonomic nerve transmission. Ann. Rev. Physiol., 25:61, 1963. (A comprehensive and critical treatment of interrelationships between nerves and smooth muscle)

Burnstock, G., Holman, M. E., and Prosser, C. L.: Electrophysiology of smooth muscle. Physiol. Rev., 43:482, 1963. (A critical discussion of the significance of electrical events in smooth muscle)

Christensen, J., Caprilli, R., and Lund, G. F.: Electric slow waves in circular muscle of cat colon. Am. J. Physiol., 217:771, 1969. (The first detailed study of the subject)

Christensen, J., and Lund, G. F.: Esophageal responses to distention and electrical stimulation. J. Clin. Invest., 48:408, 1969. (The first experimental study of these reflexes in an appropriate animal model)

Code, C. F.: Handbook of Physiology. Section VI, Alimentary Canal, Vol. IV, Motility. Washington, D. C., American Physiological Society, 1968. (Currently the most complete and detailed collection of reviews on the structure and function of the nerve and muscle apparatus of the gut. Appropriate chapters deal with clinical applications. The emphasis throughout is on recent observations, with an attempted synthesis of data.)

Code, C. F., Creamer, B., Schlegel, J. F., Olsen, A. M., Donoghue, F. E., and Ander-sen, H. A.: An Atlas of Esophageal Motility in Health and Disease. Springfield, Ill., Charles C Thomas, 1958. (A profusely illustrated manual, presenting, clearly and succinctly, the results of early studies on the movements of the esophagus by manometry. Reproductions of manometric records show patterns from normal patients and those with the common esophageal motor disorders of clinical importance. Abnormalities are presented largely in qualitative, not quantitative, terms and manometric techniques have subsequently been refined.)

Code, C. F., Hightower, N. C., and Morlock, C. G.: Motility studies of the alimentary canal in man: review of recent studies. Am. J. Med., 13:328, 1952. (The only recent major attempt to interpret manometric studies from the small intestine)

Daniel, E. E., and Chapman, K. M.: Electrical activity of the gastrointestinal tract as an indication of mechanical activity. Am. J. Dig. Dis., 8:54, 1963. (A detailed and critical review of the technique and interpretation of the recording of gross electrical events from the gut. It provides a synthesis of information up to that time, aimed at physicians.)

Daniel, E. E., Sehdev, H., and Robinson, H.: Mechanisms for activation of smooth muscle. Physiol. Rev.: (Supp.), 42(5):228, 1962. (A discursive treatment of the cellular processes in smooth muscle leading to activation of the contractile apparatus)

Hunt, J. N.: Gastric emptying and secretion in man. Physiol. Rev., 39:491, 1959. (An authoritative review of studies on net gastric emptying; much of the data comes from man)

Ingelfinger, F. J.: Esophageal Motility. Physiol. Rev., 38:533, 1958.

———: The esophagus. Gastroenterology, 41:264, 1961.

———: The esophagus. March 1961 to February 1963. Gastroenterology, 45:241, 1963. (These, with the Kramer reference, constitute a critical and continuing review of the nature and control of esophageal movement in health and disease.)

Kewenter, J.: The vagal control of the jejunal and ileal motility and blood flow. Acta physiol. scand. (Supp.), 257(65):1, 1965. (A published thesis reviewing a series of experiments on this subject)

Kock, N. G.: An experimental analysis of mechanisms engaged in reflex inhibition of intestinal motility. Acta physiol. scand. (Supp.), *164*(47):1, 1959. (A published thesis reviewing a series of experiments on this subject)

Kosterlitz, H. W., and Lees, G. M.: Pharmacologic analysis of intrinsic intestinal reflexes. Pharmacol. Rev., *16*:301, 1964. (An authoritative and critical review of reflex mechanisms and an interpretation of the effects of drugs on these reflexes)

Kramer, P.: The esophagus. Gastroenterology, *49*:439, 1965.

Lerche, W.: The Esophagus and Pharynx in Action: A Study of Structure in Relation to Function. Springfield, Ill., Charles C Thomas, 1950. (A very detailed, well illustrated review of anatomy dealing extensively with the older literature)

McLennan, H.: Synaptic Transmission. Philadelphia, W. B. Saunders, 1963. (An authoritative monograph which provides an overview of the exceedingly complex, rapidly expanding and highly specialized subject of transmission at neural synapses. A similar and nearly simultaneous monograph was published by Eccles.)

Martinson, J.: Studies on the efferent vagal control of the stomach. Acta physiol. scand., (Supp.), *255*(65):1, 1965. (A published thesis reviewing a series of experiments on this subject)

Norberg, K. A., and Hamberger, B.: The sympathetic adrenergic neurone. Acta physiol. scand., (Supp.) *238*(63):1, 1964. (A published thesis reviewing a series of experiments on this subject)

Pick, J.: The Autonomic Nervous System. Philadelphia, J. B. Lippincott, 1970. (A recent textbook, highly detailed and well illustrated)

Prosser, C. L.: Conduction in nonstriated muscles. Physiol. Rev. (Supp.), *42*(5):193, 1962. (A discursive and technical review of integration of electrical events in smooth muscle)

Truelove, S. C.: Movements of the large intestine. Physiol. Rev., *46*:457, 1966. (The most complete and the only recent review of this complex subject)

20

Mechanisms of Gastric and Pancreatic Secretion

Horace W. Davenport, D.SC.

In this chapter, the general physiological aspects of secretory defects will receive primary consideration. The diseases caused by abnormalities of gastric and pancreatic secretion may arise from either abnormal production or composition of any of the several secretions or abnormal responses of the digestive tract to those secretions. Defective controlling mechanisms may cause abnormal secretion through either excessive or insufficient stimulation. Alternatively, the relationship among the modes of control themselves may be deranged, or the secreting cells may be malfunctioning as a result of hypertrophy or hyperplasia, degeneration, or an inability to respond to the controlling mechanisms. On the other hand, normal mucosa of the digestive tract may succumb to attack by its own secretions if these are excessive or if they arrive at the wrong place. Drugs, other extrinsic agents, or some intrinsic influence may lower mucosal defenses. Finally, disease may arise from the pathophysiological responses of the digestive tract to its own normal secretions. These responses may alleviate or excerbate the effects of "auto-attack" and sometimes obscure the true nature of the underlying process. All of these possibilities will be illustrated below.

THE SECRETIONS

The oxyntic glandular area of the gastric mucosa secretes four components: (1) an isotonic acid of approximately 150 mN H^+, 155 mN Cl^-, and 5 mN K^+, (2) a collection of precursors of several proteolytic enzymes active at low pH (here simply called pepsinogen), (3) several complex mucoproteins dissolved in a small volume of fluid closely resembling an ultrafiltrate of plasma, and (4) the intrinsic factor. Pancreatic juice has two components: (1) an aqueous phase consisting of an isotonic solution of Na^+, HCO_3^- and Cl^-, with the concentration of HCO_3^- approaching 100 mM during rapid secretion, and (2) an enzymatic component containing amylolytic and lipolytic enzymes, the inactive precursors of proteolytic and other hydrolytic enzymes, e.g., elastase, ribonuclease, and phospholipase. Bile has two major components: (1) the bile salts, whose functions are described in another chapter, and (2) an aqueous phase with an electrolyte composition and bicarbonate content resembling that of the aqueous component of pancreatic secretion. Because of this resemblance and the fact that both the biliary and pancreatic components are subject to the same control mechanisms, statements about the aqueous component of pancreatic juice apply to that of bile, as well. Normal compositional variations in these secretions and their causes—a subject beyond the scope of this chapter—are discussed in standard textbooks of gastroenterological physiology.

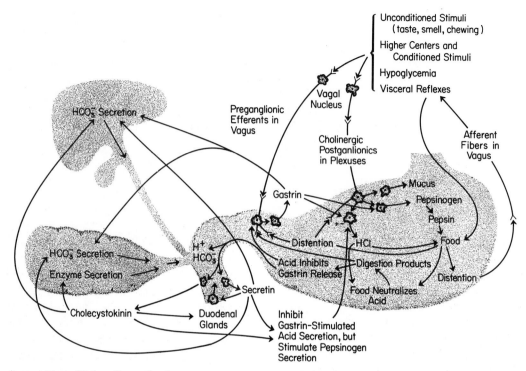

FIG. 20-1. Control of gastric, pancreatic and hepatic secretion by nerves and hormones.

CONTROL OF GASTRIC AND PANCREATIC SECRETIONS (Fig. 20-1).

The efferent pathway for central nervous regulation travels through preganglionic fibers in the vagus nerve. These fibers synapse with postganglionic cell bodies in the gastric intramural plexuses, and postganglionic fibers from the cell bodies innervate cells secreting acid, pepsinogen, and mucus. Vagal postganglionic fibers also innervate cells in the pyloric glandular mucosa which secrete the hormone gastrin. All the postganglionic fibers are cholinergic. Gastrin exerts its major secretory effect by stimulating secretion of acid and pepsinogen. Moreover, it acts synergistically with acetylcholine liberated by vagal impulses. The combined effect of gastrin and cholinergic nervous stimulation far exceeds the effect of either alone.

Distention of the stomach by food stimulates sensory endings in the stomach wall whose afferent fibers travel centrally in the vagus nerve. Distention also stimulates endings of neurons that lie entirely within the intramural plexuses and constitute the afferent limb of a reflex arc confined to the plexuses. Gastric distention, operating through both long and short reflex pathways, stimulates secretion of acid and pepsinogen by the oxyntic glandular mucosa and also the release of gastrin into the blood from the pyloric glandular mucosa.

Acid in contact with the pyloric glandular mucosa inhibits the release of gastrin. Because food neutralizes and dilutes acid, at the beginning of digestion of a meal gastrin-stimulated secretion is copious. Then, as the acidity of gastric contents in contact with the pyloric glandular mucosa rises, gastrin release slows and finally stops altogether; acid secretion dies away.

When the acid of chyme enters the duodenum, the hormone secretin is released into the bloodstream. Secretin stimulates pancreatic and hepatic secretion of bicarbonate-containing fluid, as well as gastric secretion of pepsinogen. Secretin inhibits gastrin-stimulated acid secretion. The products of protein and fat digestion in duodenal contents release the hormone cholecystokinin into the blood (chemically identical with the hormone formerly called "pancreozymin") from the small intestine. Cholecystokinin strongly stimulates pancreatic secretion of enzymes, weakly stimulates secretion of bicarbonate-containing fluid by pancreas and liver, competitively inhibits gastrin-stimulated acid secretion and, like secretin, stimulates secretion of pepsinogen.

There are two ways in which this control network can go awry: either through aberrations in nervous regulation, or by hormonal imbalance. Circumstantial evidence suggests that nervous influences, acting through the vagus as their efferent pathway, produce some disorders attributed to excessive secretion. For example, rats immobilized by being wrapped in wire gauze, monkeys punished for failing to solve insoluble problems, men harrassed and frustrated, and patients with severe injuries have all been observed to develop "stress" ulcers of the gastric mucosa. Ulcers result from failure of mucosal defense to withstand attack, and the assumption is common that stress breaks down normal defenses in the face of an overwhelming gastric attack — the attack consisting of excessive acid secretion mediated by excessive vagal activity. Although secretion may equal or surpass some standard of normality in such circumstances, no direct evidence is available that excessive vagal activity is responsible. True, in the normal feeding cycle vagal activity waxes and wanes, and this is partly responsible for increases and decreases in secretion. But it remains to be proved that sustained vagal activity in the interdigestive period maintains the secretion responsible for peptic ulceration. That vagotomy or administration of anticholinergic drugs reduces secretion during stress proves nothing, for cholinergic nervous stimulation is synergistic with hormonal stimulation, and alteration of the level of one of the synergists profoundly modifies the response of the target cells to the other.

Basal Secretion. To set a standard of normality, gastric secretion in normal subjects was measured.* Fasting human beings secrete acid at some basal rate, maintained either by spontaneous activity of secretory cells or by some residual level of stimulation. In contrast, gastric secretion in dogs is absent under basal conditions, indicating that their cells are not spontaneously active. On the basis of these findings, human gastric cells are assumed also to lack spontaneous activity. Consequently, basal secretion is attributed to ongoing stimulation, nervous or hormonal—a conclusion supported by the ability of atropine and other anticholinergic drugs to reduce basal secretion.

Since the extent of basal secretion depends not only upon the strength of stimulation but also upon the ability of the subject's cells to respond, basal secretion is compared to the maximum secretory ability of the stomach as a clue to the relative intensity of stimulation responsible for basal secretion. The subject's maximum secretory capacity is measured by giving the dose of stimulating drug which is thought to evoke a maximum response. However, the measured capacity of a single subject can vary, depending upon his condition, particularly his state of hydration and his K^+ balance, as well as upon the dose of drug used.

Stimulated Secretion. Most persons secrete at a maximal rate for their prevailing conditions when histamine acid phosphate is subcutaneously injected in a dose of 0.04 mg./kg. body weight; that dose is routinely

---

* Clinically, basal secretion is usually measured by collecting gastric juice from the subject before he has eaten breakfast. Every clinic develops its own technique and accumulates its own data to establish standards.

used for the *augmented histamine test.* Troublesome side effects of histamine must be prevented by prior administration of an antihistaminic drug thought to exert a minimal effect on gastric function. Gastric juice is collected every 15 minutes and peak secretion is usually observed in the second and third periods after histamine injection (between 15 and 45 minutes). Peak output is expressed as rate of acid production, i.e., mEq. H⁺ per hour. Table 20-1 compares representative peak output values with basal secretory data obtained in the same clinic. Many other clinics have reported similar results.

Because it has fewer side effects, the histamine analog betazole (Histalog) is often substituted for histamine. Although its peak action occurs later and is more prolonged, the peak responses to subcutaneous injection of the dihydrochloride (1.7 mg./ kg. body weight) duplicates that obtained in the augmented histamine test. Table 20-2 compares data on the basal secretion and Histalog-stimulated acid secretion of normal subjects and patients with ulcers

either of the duodenum or of the oxyntic glandular area of the stomach (the latter called gastric ulcers). Although the data are internally consistent, two aspects of the tests unfortunately preclude comparison with outside data. First, the gastric juice samples were titrated only to pH 3.5, so that any buffered acid was missed;* and, second, the selected dose of Histalog (0.5 mg./kg. body weight) was not high enough to elicit responses equal to those in the augmented histamine test. Nevertheless, the results accurately define the two populations studied and indicate the range of variability.

Generally, men secrete more acid than women, patients with duodenal ulcers secrete more than normal subjects, and patients with ulcers of the oxyntic gland area secrete less acid than normal subjects. The rate of acid secretion tends to

* For analysis of the technique and meaning of titrations of gastric juice, *see* Moore, E. W., and Scarlata, R. W.: Gastroenterology, 49:178, 1965; Moore, E. W.: Gastroenterology, 54:501, 1968; and Makhlouf, G. M., Blum, A. L., and Moore, E. W.: Gastroenterology, 58:345, 1970.

TABLE 20-1. Basal Gastric Secretion Compared with Peak Secretory Response to the Augmented Histamine Test in 20 Normal Subjects.*

		VOLUME (ml./hr.)	ACIDITY (mN)	ACID SECRETION (mEq./hr.)
Basal Secretion†				
Men	Mean ± SD	38.7 ± 23.0	29.8 ± 21.7	1.3 ± 1.6
	Range	10–93	0–71	0–6.2
Women	Mean ± SD	40.6 ± 38.8	20.3 ± 18.3	1.1 ± 1.8
	Range	2–140	0–54	0–6.0
Augmented Histamine Test‡				
Men	Mean ± SD	214.9 ± 91.5	94.7 ± 35.1	21.6 ± 13.8
	Range	38–406	10–135	0.3–50.2
Women	Mean ± SD	138.2 ± 68.7	87.2 ± 32.8	12.3 ± 9.0
	Range	23–304	18–130	0.4–32.0

* From Baron, J. H.: Gut, 4:136, 1963.

† One-hour collection after overnight fast.

‡ Two times the maximum 30-minute response after subcutaneous injection of histamine acid phosphate, 0.04 mg./kg.

TABLE 20-2. Basal and Peak Gastric Acid Secretion in Normal Subjects,
Patients with Duodenal Ulcer, and Patients with Gastric Ulcer of All Ages°

	NO. SUBJECTS		BASAL (mEq./hr.)†	HISTALOG‡ STIMULATION (mEq./hr.)
Normal Subjects				
Women	615	Mean ± SD	2.44 ± 2.85	11.64 ± 7.62
		Range	0–17.1	0–48.4
Men	634	Mean ± SD	1.33 ± 2.00	7.53 ± 5.20
		Range	0–15.0	0–24.7
Duodenal Ulcer Patients				
Men	787	Mean ± SD	5.29 ± 4.63	19.91 ± 9.70
		Range	0.1–31.7	1.0–60.4
Women	245	Mean ± SD	2.87 ± 3.14	13.42 ± 6.87
		Range	0.1–26.2	1.2–36.5
Gastric Ulcer Patients				
Men	148	Mean ± SD	1.45 ± 1.99	9.68 ± 6.97
		Range	0–14.8	0.1–35.5
Women	81	Mean ± SD	1.00 ± 1.49	7.95 ± 5.96
		Range	0–8.4	0.6–23.1

° From Grossman, M. I., Kirsner, J. B., and Gillespie, I. E., Gastroenterology, 45:14, 1963. (Reproduced by permission)

† Titrated to pH 3.5.

‡ Dichloride dose, 0.5 mg./kg. body weight.

fall in persons over 50 and is usually low in patients with gastric cancer (these two groups are not represented in Table 20-2).

Because normal variation is so great, often the physician cannot easily categorize a specific individual squarely in one or another population. Nevertheless, he must adopt some standard of normality. Sometimes he does this intuitively; sometimes he applies an informal analysis of accumulated experience. And seldom are sophisticated statistical methods more helpful. To illustrate this problem (Figure 20-2), the cumulative frequency of basal secretion in men, ages 20 to 49, was plotted for two groups: 387 control subjects and 434 patients with duodenal ulcer. For the control group, an upper limit of normal was calculated which would yield a 0.975 probability that 5 percent of the subjects had secretory rates above the upper limit. From these data,

the upper limit of normal was calculated to be 6.5 mEq. per hour. Yet, over 70 percent of the patients with duodenal ulcer secreted at a lesser rate. Thus, if a single individual is observed to secrete at the rate of 5 mEq. per hour, in what category does one place him? Results of the Histalog test on such a patient would pose the same dilemma.

In some clinics gastric juice is collected continuously by suction for 12 hours, beginning at 8:00 or 8:30 P.M. while the patient tries to sleep. Table 20-3 lists representative values for normal subjects.

Pentagastrin, the commercially available derivative of gastrin, contains the active N-terminal amino acid sequence of gastrin and possesses its physiological properties. Pentagastrin, subcutaneously or intramuscularly injected (6 µg./kg. body weight) or intravenously infused (6 µg./kg. per hr.)

TABLE 20-3. Overnight Gastric Secretion in Normal Subjects*

	NO. SUBJECTS		VOLUME† (ml./12 hrs.)	ACIDITY (mN)	ACID SECRETION (mEq./12 hrs.)
Men	33	Mean	643	30	20
		Range	173–1,188	1–90	0.4–93
Women	17	Mean	460	27	14
		Range	148–903	3–72	0.6–68

* From Levin, E., Kirsner, J. B., Palmer, W. L., and Butler, C. Arch. Surg., 56:345, 1948. Reproduced by permission.

† Juice collected between 8:30 P.M. and 8:30 A.M.; subjects not fasted before test.

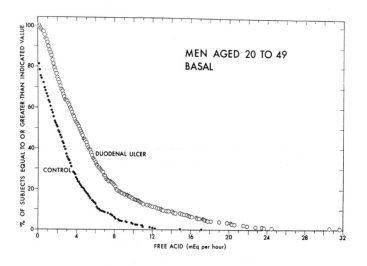

MEN AGED 20 TO 49
BASAL

DUODENAL ULCER

CONTROL

FREE ACID (mEq per hour)

FIG. 20-2. Cumulative frequency of the secretion of acid by 387 control subjects and by 434 patients with duodenal ulcer, plotted against the rate of acid secreted in mEq. per hour. Because titration was carried only to pH 3.5, just the ionized acid was measured; this component is sometimes called "free acid" (as labeled on the abscissa). (Grossman, M. I., Kirsner, J. B., and Gillespie, I. E.: Gastroenterology, 45:14, 1963. Copyright 1963 by Williams and Wilkins Company, and reproduced by permission.)

evokes the same response as 0.04 mg. of histamine acid phosphate or 2 mg. of Histalog per kilogram of body weight.

Peptide-secreting adenomatosis* is a disease of secretion caused by excessive hormonal stimulation. Its cardinal features are acid hypertension, recurrent ulceration, diarrhea, and the presence of a tumor usually, but not invariably, in the pancreas. The tumor contains and secretes large quantities of gastrin, and some have been found to contain and secrete secretin. Both

hormones are peptides, hence the name peptide-secreting adenomatosis. The following "case history"† will provide an illustration of this disease.

For 2 years a male patient, 50 years old, had passed as many as 20 loose, pale stools a day, with daily fecal output ranging from 3 to 5 liters. He had lost weight and had periods of extreme muscular weakness. He complained of retrosternal pain unassociated with meals but relieved by alkali.

* The syndrome was described by Zollinger, R. M., and Ellison, E. H., Ann. Surg., 142:709–728, 1955, and it bears their names. However, we use the term peptide-secreting adenomatosis, in observance of the trend to abandon eponyms and in recognition of the variations of the disease that have been described.

† This "composite" case, written to illustrate the course of a typical case of pancreatic tumor, was drawn from several sources, notably Rawson, A. B., England, M. T., Gillam, G. G., French, J. M., and Stammers, F. A. R.: Lancet 2:131, 1960, and Sircus, W., Brunt, P. W., Walker, R. J., Small, W. P., Falconer, C. W. A., and Thomson, C. G.: Gut 11:197, 1970.

His arterial pressure was 95/65 mm. Hg. Plasma Na$^+$ was 125 and K$^+$ 2.0 mEq. per hour; 24-hour stool collections contained 100 to 400 mEq. of Na$^+$ and 100 to 300 mEq. of K$^+$. On a daily fat intake of 70 g., fecal fat was 15 to 20 g., mostly soaps. Radiological examination following a barium swallow showed hypertrophic gastric folds and duodenal and jejunal ulceration. In repeated tests basal acid output was 8.0 to 14.5 mEq. per hour. Twelve-hour overnight volume was 2,500 ml., containing 246 mEq. of H$^+$. Peak acid output in response to histamine acid phosphate (0.04 mg./kg.) was 350 ml. per hour, containing 37 mEq. of H$^+$. However, peak output after pentagastrin administration (6 μg./kg.) was only 15 mEq. per hour. Analysis of plasma for gastrin by radioimmune assay gave 4 ng./ml. whereas normal values never exceed 0.075 ng./ml. Bioassay of plasma indicated secretinlike activity.

Laporotomy revealed a 10-g. tumor in the head of the pancreas; it was excised. The tumor contained a high concentration of a gastric secretagogue, not histamine, and a substance which was found to stimulate secretion of bicarbonate-containing fluid, but not of enzymes, in the pancreas of an anesthetized cat. After surgery, diarrhea ceased immediately. In two tests, basal secretion was 11 and 12 mEq. of H$^+$ per hour; 12-hour overnight secretion was 900 ml., containing 61 mEq. of H$^+$. Peak responses in two augmented histamine tests were 400 and 600 ml. per hour, containing 53 and 60 mEq. of H$^+$.

Comment: Circulating gastrin from the tumor produced constant stimulation; hence the high basal and overnight secretion. The fact that the patient was severely dehydrated and hypokalemic may explain why secretion was not even greater. Because secretin inhibits gastrin-stimulated acid secretion, circulating secretin probably kept acid secretion below its potential level. Circulating secretin probably accounts for

the patient's small response to pentagastrin injection. Gastrin stimulates protein synthesis in the gastric mucosa, and the patient's high level of gastrin probably caused the gastric hypertrophy and hyperplasia. Ability to secrete acid is roughly proportional to the parietal cell mass; therefore, the augmented histamine test yielded high values both before and after surgery. That the value was not higher preoperatively can be attributed to dehydration induced by loss of Na$^+$ in the stool and to a greatly reduced total exchangeable K$^+$ mass, of which hypokalemia is a reflection.

When a normal man digests a meal, the pH of the duodenal bulb drops from 6 to 2 as chyme spurts from the stomach, toward the end of gastric emptying. Earlier in the digestive and emptying processes, the pH of chyme is not so low, for gastric contents have not yet been acidified. At mid-duodenum, the prevailing pH is above 6, with only occasional and brief falls to 4, as acid chyme is propelled downward from the duodenal bulb. After this, the pH always exceeds 6. When, as in the example above, the stomach secretes acid at a high rate in the interdigestive period, no food is present to dilute and neutralize the acid, upper intestinal pH is low, and mid-jejunal pH may be 2.5. Continuous perfusion of the intestine by acid accounts for ulceration and mucosal edema. Acid in the duodenum strongly stimulates intestinal motility; contraction of indurated bowel causes pain.

Infusion of acid into the duodenum and jejunum causes outpouring of fluid from the mucosa; this, together with increased motility, accounts for the diarrhea. In addition, the low intraluminal pH precipitates glycocholate bile salts and inactivates lipase, so that the micellar phase of fat digestion is inadequately prepared. This defect (more fully described in Chapter 21) promotes steatorrhea which, in turn, aggravates the diarrhea.

Loss of sodium in the stool shrinks the extracellular fluid volume, and reduction of plasma volume impairs cardiac output. Loss of potassium reduces intracellular K^+ mass and results in skeletal and cardiac muscular weakness.

Cells die if their cytoplasm becomes acidic. However, the mucosa of the digestive tract is exposed to acid, and cells are destroyed only in pathological conditions. Clearly, protective mechanisms must be present, which differ in the upper intestine, the body of the stomach, and the pyloric antrum.

Gastric Emptying. Because the upper intestine is exposed to acid coming from the stomach, the amount and concentration of intestinal acid depends upon the rate of gastric emptying and the concentration of acid in chyme leaving the stomach. The rate of gastric emptying is directly proportional to the square root of the volume of gastric contents. Since the volume is greatest at the beginning of digestion, when the meal has just been swallowed, the rate of emptying is greatest then, as well. Some chyme flows from the stomach through the pyloric sphincter as each peristaltic wave passes over the terminal antrum, one stroke every 22 seconds. Early in emptying, each stroke delivers less than 10 ml. to the duodenum. However, early in emptying, the acidity of gastric contents is low. Late in emptying, acidity of gastric contents is high, but, because the volume remaining in the stomach is small, the amount of acid delivered per stroke is also small. The rate of gastric emptying is further controlled by a negative feedback regulation of gastric motility. Presence of acid in the duodenum slows gastric emptying so that the amount of acid delivered is closely adjusted to the duodenum's capacity to cope with the acid. In contrast, duodenal motility is stimulated by acid, thereby dispersing acid quickly over a large area of mucosa where it can be diluted and neutralized. Possibly some derangement of this motor mechanism can cause overly rapid delivery of acid to the duodenum.

Duodenal Buffering. Intestinal acidity is reduced in three ways: (1) acid is diluted and neutralized by bicarbonate-containing secretions of pancreas and liver; (2) it is diluted and neutralized by fluid coming from the mucosa; and (3) it is absorbed through the mucosa into interstitial fluid and blood. Acid in the duodenum stimulates the release of secretin which, in turn, stimulates the pancreas and liver to secrete their alkaline fluids. Secretin is released at a rate proportional to the amount, not the concentration, of acid in the duodenum, thus effectively adjusting the quantity of juice capable of neutralizing acid to the quantity of acid to be neutralized. The duodenal and jejunal mucosae have a large surface area, highly permeable to water and substances of a molecular weight below 100. Consequently, not only can H^+ diffuse into it, but with almost equal ease HCO_3^- ions can diffuse from interstitial fluid into the lumen. Provided that blood flow through the capillaries of the mucosa is sufficiently rapid, acid diffusing into the mucosal interstitial spaces can be neutralized and removed before it harms the mucosal cells. In the absence of pancreatic juice and bile, acid in the duodenum is neutralized almost as rapidly as during normal secretion, with the pH of upper intestinal contents averaging only one pH unit below normal. Only when the protective mechanisms are deranged or overwhelmed by acid is the integrity of the mucosa compromised.

Gastric Mucosal Barrier. In the body of the stomach, acid, instead of being neutralized and diluted, is contained. Here acid comes into contact with only a small gastric mucosal surface, the surface-to-volume ratio being nearly the minimum possible. Moreover, the mucosa has no villi and just a few blunt microvilli. Acid in contact with the oxyntic glandular mucosa diffuses into it with remarkable slowness. The impenetrability of oxyntic glandular mucosa to acid is called the gastric mucosal barrier, the chief defense of the stomach against digestion by its own secretions. In contrast to the intestinal mucosa, the oxyntic glandular mucosa is likewise very slightly permeable to Na$^+$ and other ions of the interstitial fluid. Just as the gastric mucosal barrier keeps H$^+$ ions within the lumen, it also keeps the components of interstitial fluid within the mucosa.

The pyloric glandular mucosa that lines the gastric antrum has properties lying between those of the oxyntic glandular mucosa and the duodenum. It, too, has a small surface-to-volume ratio, and, though somewhat more permeable to acid and Na$^+$ than the oxyntic glandular mucosa, it is far less permeable than the duodenal mucosa.

Peptic Ulceration. In the duodenum and upper jejunum, ulceration occurs when normal defenses must cope with an abnormally large quantity of acid. Gastric mucosa is damaged when insufficient defenses face a normal amount of acid. When the gastric mucosal barrier to H$^+$ is broken, acid can diffuse into the mucosa, with the consequences depicted in Figure 20-3.

Acid diffuses slowly through the normal barrier and rapidly through the broken barrier, releasing histamine and probably other injurious substances. Histamine, in turn, stimulates acid secretion so that, concomitant with the rapid diffusion of acid into the mucosa, more acid appears in the lumen. Consequently, the concentration and the amount of acid in the lumen may not fall. If this addition of acid to the lumen is not recognized, one may fail to understand that acid is simultaneously disappearing. Histamine dilates mucosal capillaries and increases their permeability, promoting filtration of plasma into interstitial spaces. Edema and high interstitial pressure increase lymph flow. Interstitial fluid is filtered across the mucosa, and, as modified by the filtration process, it enters the gastric lumen. Acid also destroys mucosal capillaries, with destruction especially pronounced when mucosal capillary hydrostatic pressure is raised by concurrent cholinergic stimulation. Acid stimulates intramural plexuses, and gastric motility is enhanced through a cholinergic reflex. Action of acid on intramural plexuses may also contribute to stimulation of secretion of pepsinogen.

The gastric mucosal barrier may be abnormally permeable for reasons at present totally unknown. In some patients with ulcers of the oxyntic glandular area of the stomach, the gastric mucosa appears to be more permeable to acid than does that of normal subjects. One apparent character-

FIG. 20-3. Scheme showing the pathophysiological consequences of back diffusion of acid through the gastric mucosal barrier. (Davenport, H. W.: Physiology of the Digestive Tract. Ed. 3. Chicago, Year Book Medical Publishers, 1971. Copyright 1971 by Year Book Medical Publishers, and reproduced by permission.)

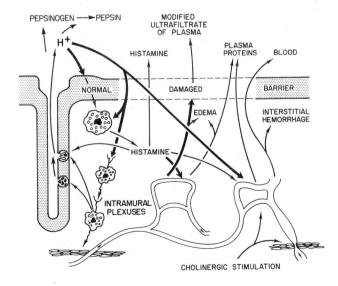

istic of this condition is the secretion of less acid (see Table 20-2) and more Na+ than is measured in normal subjects. Before the concept of barrier function was applied to this disease, these patients were thought to secrete weakly acidic juice. But this raised two questions: First, how can the gastric mucosae, which (according to most gastroenterologists) secrete acid at a variable rate but in a fixed composition, switch to secreting juice of a different composition? And, second, how can weakly acidic juice cause ulceration of the oxyntic glandular mucosa when, under normal circumstances, a strongly acidic juice cannot? If inadequacy of the barrier comes first, the sequence of events diagrammed in Figure 20-3 explains the course of the disease. The mucosa does, in fact, secrete an acid juice of normal composition, but before the acid is aspirated many of its H+ ions have diffused back into the mucosa through the broken barrier, to be replaced by Na+. Subnormal defense, and not abnormal attack, is responsible.

The exact nature and locus of the gastric mucosal barrier are unknown; however, barrier function must be performed at least in part by the apical plasma membrane of the surface epithelial cells. With their rapid cycle of desquamation and renewal, normally about half a million of these cells are shed each minute and replaced by an equal number of new surface epithelial cells. The new cells arise by mitosis in the region of the necks of the glands and migrate to the mucosal surface. Some flaw in the process of differentiation and replacement may be responsible for inadequate barrier function. Excessive adrenal cortical hormones, either exogenously administered or endogenously secreted, delay replacement of desquamated or excised surface epithelial cells. A subtle derangement in cell structure, in mode of attachment of the cells to one another, in the plasma membrane, or in the composition of mucus syn-

thesized by the cells may some day be identified.

Destruction of Gastric Mucosal Barrier. The gastric mucosal barrier is also destroyed by commonly ingested substances, the most important being acetylsalicylic acid (aspirin) and ethanol, alone or together.

Because *acetylsalicylic acid* is an organic acid with a pKa of 3.5, more than half its molecules in solution remain un-ionized below pH 3. Below pH 2, over 97 percent are undissociated. On the other hand, above pH 5 more than 97 percent of its molecules in solution are ionized. Acetylsalicylic acid is fat-soluble in the un-ionized state and relatively fat-insoluble in the ionized state. It is a well established physiological principle that un-ionized, fat-soluble substances can easily penetrate the lipoprotein layer that forms the plasma membrane of cells, whereas ionized, fat-insoluble substances cannot. Therefore, acetylsalicylic acid diffuses rapidly from acid gastric contents into the cells of the gastric mucosa, but the diffusion of acetylsalicylate in neutral gastric contents is much slower. A sufficient amount of acetylsalicylic acid in the gastric mucosa kills the surface epithelial cells and accelerates their rate of desquamation, an event that occurs when the luminal concentration of acetylsalicylic acid in acid solution reaches 5 mM (5 mM is the acid concentration of one 5-grain aspirin tablet dissolved in about 370 ml. of gastric contents). In physiological terms, acetylsalicylic acid in acid solution, but not in neutral solution, destroys the gastric mucosal barrier. Once acetylsalicylic acid has opened the gates of mucosal defenses, H+ ions can pour into the mucosa, with all the consequences depicted in Figure 20-3. The concentration of acid in gastric contents necessary for this insult to the mucosa lies between 25 and 65 mN. Acetylsalicylic acid is especially potent in affecting mucosal capillaries; when acid diffuses into mucosa damaged by acetylsalicylic acid, gastric mucosal bleeding is more frequent and copious than after damage by any other agent.

Ethanol, a small molecule soluble in both water and fat, is rapidly absorbed through the gastric mucosa. Neither an acid nor a base, its rate of absorption is entirely independent of the acidity of gastric contents, depending only upon its concentration in the lumen. At a luminal concentration greater than 10 percent (w/v)—equal to the alcohol concentration in light sherry—it, too, destroys the gastric mucosal barrier. Although the effect of ethanol alone is virtually independent of the presence of acid, the combination of ethanol, acetylsalicylic acid, and hydrochloric acid is particularly devastating. Ethanol at a concentration of only 4 percent (w/v)—equal to that in beer—has no effect upon the mucosal barrier, but when combined with 10 mN HCl it doubles the virulence of aspirin.

Because a lipid layer constitutes a major component of the gastric mucosal barrier, exposure to detergents such as bile salts and lysolecithin can disrupt the barrier. Bile salts, present in intestinal contents in the interdigestive phase, increase in both concentration and quantity in the duodenum after feeding, when the gallbladder contracts and the enterohepatic circulation is accelerated. Lysolecithin is also present in duodenal contents, sometimes in high concentration. Lecithin secreted in bile and ingested in food is converted to lysolecithin by phospholipase A of pancreatic juice. Under normal circumstances the downward gradient of intestinal motility sweeps bile salts and lysolecithin to the lower small intestine, where they are absorbed. However, when duodenal contents are regurgitated into the stomach, both may come into contact with the gastric mucosa (another example of normal contents of the digestive tract being in the wrong place). Patients with gastric ulcers along the lesser curvature have a far higher concentration of bile salts in their gastric contents than do normal subjects, both fasting and after a meal. Similar data on the concentration of lysolecithin in the stomach are unavailable. Bile salts in the stomach compromise the integrity of the gastric mucosal barrier in man as well as in experimental animals, but whether their presence in the stomach is a cause of gastric ulceration or an effect of disordered motility due to ulceration remains in question.

Back Diffusion of Acid. The aphorism "no acid, no hemorrhage" is justified by the fact that acid secretion and acid back diffusion are necessary but not sufficient causes of mucosal bleeding. Other factors, many poorly understood, affect mucosal susceptibility to bleeding. The prerequisites for this condition are: the presence of acid on the surface of the mucosa, acid penetration of the mucosal barrier and the existence of a diffusion gradient and a pathway through which diffusion can occur. Since the concentration of acid within mucosal cells and interstitial fluid is essentially zero, the gradient is determined entirely by the concentration of acid in the lumen. Under normal circumstances, the upper limit of acid concentration is set by the maximum concentration at which it is secreted, approximately 150 mN. At this concentration, H^+ ions diffuse so slowly through the normal mucosal barrier that they do no damage. If the concentration of HCl in the lumen is raised to 300 to 450 mN, it diffuses 2 to 3 times more rapidly through the barrier, destroys the mucosal defenses and causes hemorrhage. However, the rate of diffusion of acid through a normal barrier can also be increased by making the acid fat-soluble. The aliphatic acids, acetic through hexanoic, are sufficiently water-soluble that a high concentration is possible in gastric contents, and they are sufficiently fat-soluble to diffuse through the gastric mucosal barrier much faster than the H^+ ions of HCl. If the barrier is broken by some other agent (e.g., acetylsalicylic acid), H^+ ions can diffuse more rapidly along a lower gradient into the mucosa and cause hemorrhage. Thus, a low concentration of acid (say 25 mN) in the lumen may initiate bleeding in a mucosa already made vulnerable by acetylsalicylic acid. However, the fact that the concentration of acid is low in a sample of gastric contents contaminated with blood does not mean that the low concentration of acid was the original offender.

The concentration of acid recovered from the stomach is determined not only by the concentration and amount of acid secreted but also by the loss of acid through neutralization and back diffusion. In a bleeding stomach, acid not only has disappeared by diffusion into the mucosa but has been diluted and neutralized by bicarbonate-containing fluid (derived from interstitial fluid) and by mucosal blood. Consequently, the low concentration of acid found in the gastric contents of patients with gastric ulcer is no evidence that the stomach has secreted acid at a low concentration.

Once acid has penetrated the mucosa, its ability to cause bleeding is affected by mucosal blood flow. Adequate blood flow through mucosal capillaries is required to ensure neutralization and removal of acid, and, when blood flow is greatly reduced, as by a large dose of vasopressin, an otherwise innocuous concentration of acid may cause hemorrhage. Reduced blood flow has often been thought responsible for a breach in mucosal defenses, but proof is lacking.

When acetylsalicylic acid in acid solution breaks the mucosal barrier, mucosal blood flow increases and remains high after removal of the salicylate-containing fluid from the lumen, yet the mucosa is particularly susceptible to hemorrhage. Electronmicrographs of stomach tissue dam-

aged by acetylsalicylic acid show these vessels to be exposed by destruction of the overlying cells. If resistance to venous outflow is large, capillaries and venules, especially those immediately under the surface epithelial cells, are grossly dilated, a condition readily inviting their damage. Because of the passage of venous drainage of the mucosa through the gastric muscular wall, each contraction impedes venous drainage, and small vessels upstream are engorged. Cholinergic stimulation provokes a more forceful gastric muscular contraction, perhaps accounting for the fact that cholinergic stimulation enhances the bleeding tendency of the mucosa, particularly after damage by acetylsalicylic acid.

Although acid may be required to initiate bleeding, it is not necessarily the cause. In many experimental situations in which the gastric mucosal barrier may be broken, with acid diffusing into the mucosa, bleeding fails to occur. For example, when the oxyntic glandular mucosa is treated with *p*—chloromercuribenzoate (a compound which reacts with thiol groups), appropriate concentrations of the compound break the barrier as effectively as 20 mM acetylsalicylic acid in acid solution. Yet bleeding rarely occurs when the mucosa is subsequently bathed with 100 mN HCl.

Achlorhydria, (the inability of the stomach to secrete acid) is physiologically unimportant in itself, for peptic digestion is nonessential to gastrointestinal function. However, gastric atrophy sufficient to prevent the stomach from secreting the intrinsic factor always includes inability to secrete acid. Absence of intrinsic factor results in pernicious anemia; thus, determination of the presence of achlorhydria is an important step in diagnosis of that disease.

Intrinsic Factor. The intrinsic factor is the only organic constituent of either gastric or pancreatic secretion essential for life. Vitamin B-12, with a molecular weight of 1,357, is the largest molecule absorbed intact by the gut. Intrinsic factor, a mucoprotein secreted by human oxyntic cells, binds vitamin B-12 and its co-vitamin derivatives. The complex moves with chyme to the terminal ileum where, as the first step in absorption, the complex is taken into the epithelial cells by pinocytosis. In gastric mucosal atrophy the stomach loses its ability to secrete, first, acid, then pepsinogen and, finally, intrinsic factor. When the vitamin then fails to be absorbed in sufficient quantity, the major result is failure of red cells to mature. Consequently, after the patient with gastric atrophy has exhausted his body stores of the vitamin, pernicious anemia ensues. The deficit can be overcome by parenteral administration of the vitamin or by feeding massive amounts. Orally taken, a small fraction of the vitamin is absorbed without intrinsic factor, and the daily requirement of 1 μg. is satisfied. If intrinsic factor is secreted, its combination with vitamin B-12 is exposed to many perils on its trip through the alimentary canal, until it reaches the ileal mucosa. Other mucoproteins compete with intrinsic factor for the vitamin, but, having bound the vitamin, they are incapable of assisting its absorption. Organisms living in the small intestine may purloin the vitamin for their own uses. Finally, disease or resection of the terminal ileum may deprive the body of the site of absorption.

Mucus. To assess the role of mucus in gastrointestinal disease, one must know its function. Beyond the fact that mucus is known to lubricate the surface of the gut, nothing is certain. The layer of mucus on the surface of epithelial cells cannot be an effective barrier to diffusion, for ions and even large molecules diffuse through a mucous gel almost as quickly as through an unstirred layer of water. Thus, the acid concentration at the bottom of a mucus layer can equal the concentration at the top of the layer. Furthermore, acid coagulates mucus. Since mucus does not inhibit pepsin, the presence of a mucus layer cannot explain the usual immunity of the stomach to peptic digestion. Another enzyme, phospholipase A, penetrates the mucus layer to destroy the gastric mucosal barrier. If phospholipase A can reach the barrier, there is no reason why pepsin cannot, too. Gastric mucus is an extraordinarily complex mixture, undergoing subtle changes in composition in some conditions, such as prolonged salicylate administration, but whether these changes have any physiological significance is unknown.

Pepsin performs no indispensable digestive function. Hence its deficiency or total absence cannot cause disease. None of the problems of patients with complete gastrectomy or gastric atrophy can be attributed to absence of pepsin; protein is adequately digested by pancreatic and intestinal proteolytic enzymes. At the other extreme, excessive pepsin secretion has not been shown to produce any pathological effects. Pepsin does not digest the normal gastric mucosal barrier, and it may not contribute to attack by other agents. Addition of a high concentration of pepsin does not exacerbate the damage to canine gastric mucosa by aspirin in acid solution. Pepsin may, however, aggravate the damage caused by excessive acid in the upper intestine. When the duodenum of an experimental animal is perfused with a solution resembling acid gastric juice, the presence of pepsin somewhat intensifies the destruction of duodenal and jejunal mucosa.

Pancreatic Enzymes. In the absence of salivary and gastric enzymes, starch, protein and fat must be digested by pancreatic and intestinal enzymes. Intestinal enzymes are found on the surface of intestinal epithelial cells, where they are accessible to substrates in luminal contents, and within the cytoplasm of the cells. Because about 250 g. of intestinal epithelial cells are desquamated and themselves digested each day, large amounts of proteolytic and lipolytic enzymes contained within the cells are available in the intestinal lumen. Therefore, in pancreatic deficiency, the major foodstuffs can still be digested and absorbed, albeit imperfectly. The deficit in protein digestion varies greatly from patient to patient and even in the same patient from day to day, but a person with no pancreatic proteolytic enzymes may digest and absorb about 25 to 75 percent of his protein intake. Owing to the insolubility of fat in water, lipid must undergo an elaborate series of delicately adjusted physical and chemical transformations before it can be digested and absorbed, and the complexity of this process renders it particularly vulnerable. Yet, patients with no functioning pancreatic tissue may absorb between 60 and 90 percent of ingested fat. Because steatorrhea is so obvious when it occurs and so troublesome in its consequences, failure of fat absorption receives much attention. Still, one should not forget that the victim of steatorrhea remains capable of digesting and absorbing some fat. (Defects in fat digestion and absorption leading to steatorrhea and diarrhea are discussed in Chapter 21.)

Soy beans and egg white contain compounds that inhibit the pancreatic enzyme trypsin. Rats fed these inhibitors develop massive hypertrophy of the pancreas and secrete enormous amounts of pancreatic enzymes. Because the enzymes are not themselves completely digested and absorbed, the rats develop a deficiency of essential amino acids that are carried away by the enzymes lost in the stool, as well as negative nitrogen balance. Such excessive pancreatic secretion has not been identified in human subjects.

ANNOTATED REFERENCES

Augur, N. A., Jr.: Gastric mucosal blood flow following damage by ethanol, acetic acid, and aspirin. Gastroenterology, 58:311, 1970. (This paper describes the hyperemia that follows damage to the gastric mucosa and relates the hyperemia to the tendency of the mucosa to bleed.)

Baron, J. H.: Lean body mass, gastric acid, and peptic ulcer. Gut, 10:637, 1969. (This paper contains the standard data on the secretory capacity of the stomach; it and its references should be consulted for the definition of "normal values.")

Castle, W. B.: Intrinsic factor and vitamin B_{12} absorption. In: Code, C. F. (ed.): Handbook of Physiology. Sec. 6, Alimentary Canal, Vol. 3. Washington, D. C., American Physiological Society, 1968. (This is the authoritative review of the role of the stomach in vitamin B-12 absorption.)

Davenport, H. W.: Gastric mucosal injury by fatty and acetylsalicylic acids. Gastroenterology, 46:245, 1964. (This paper establishes the fact that it is the fat-soluble forms of aspirin and fatty acids in acid solution which damage to gastric mucosa.)

————: Fluid produced by the gastric mucosa during damage by acetic and salicylic acids. Gastroenterology, 50:487, 1966. (This paper describes the pathophysiological consequences of increased gastric mucosal permeability; these include back-diffusion of acid, increased sodium output, increased capillary permeability and plasma protein output.)

————: Physiological structure of the gastric mucosa. In: Code, C. F. (ed.): Handbook of Physiology. Sec. 6, Alimentary Canal. Vol. 2. Washington, D. C., American Physiological Society, 1967. (This is an attempt to integrate the knowledge of the permeability characteristics of the stomach; although it was written in 1965, not much must be altered now.)

————: Gastric mucosal hemorrhage in dogs. Effects of acid, aspirin, and alcohol. Gastroenterology, 56:439, 1969. (This paper describes quantitatively the relations between acid, aspirin and alcohol leading to mucosal hemorrhage. The concepts developed apply to the human stomach.)

————: Back diffusion of acid through the gastric mucosa and its physiological consequences. In: Glass, G. B. J. (ed.): Progress in Gastroenterology. Vol. 2. New York, Grune & Stratton, 1970. (This paper summarizes current knowledge of the consequences of a broken gastric mucosal barrier.)

Davenport, H. W., Warner, J. A., and Code, C. F.: Functional significance of the gastric mucosal barrier to sodium. Gastroenterology, 47:142, 1964. (This paper is the first to describe the physiological consequences of a broken gastric mucosal barrier to sodium and hydrogen ions, and its references cover the genesis of the idea of the mucosal barrier.)

Horowitz, M. I.: Mucopolysaccharides and glycoproteins in the alimentary tract. *In:* Code, C. F. (ed.): Handbook of Physiology. Sec. 6, Alimentary Canal. Vol. 2. Washington, D. C., American Physiological Society, 1967. (This review summarizes current knowledge of the chemistry of the mucous substances in the gastrointestinal tract.)

Hunt, J. N., and Wan, B.: Electrolytes of mammalian gastric juice. *In:* Code, C. F. (ed.): Handbook of Physiology. Sec. 6, Alimentary Canal. Vol. 2. Washington, D. C., American Physiological Society, 1967. (This review describes the variations in composition of gastric juice, and it summarizes the several theories attempting to explain these variations. Its references are useful as a guide to the older literature.)

Jacobson, E. D.: Secretion and blood flow in the gastrointestinal tract. *In:* Code, C. F. (ed.): Handbook of Physiology. Sec. 6, Alimentary Canal. Vol. 2. Washington, D. C., American Physiological Society, 1967. (This is an authoritative review, with valuable references on the relation between blood flow and function of the digestive tract.)

Janowitz, H. D.: Pancreatic secretion of fluids and electrolytes. *In:* Code, C. F. (ed.): Handbook of Physiology. Sec. 6, Alimentary Canal. Vol. 2. Washington, D. C., American Physiological Society, 1967. (This review summarizes current knowledge of pancreatic secretion of fluids and electrolytes, and its references are a guide to the older literature.)

Lagerlöf, H. O.: Pancreatic secretion: Pathophysiology. *In:* Code, C. F. (ed.): Handbook of Physiology. Sec. 6, Alimentary Canal. Vol. 2. Washington, D. C., American Physiological Society, 1967. (This review deals with the disturbances of pancreatic function found in diseased states, and it should be read by all who are beginning their study of pancreatic pathophysiology.)

Lipkin, M., and Bell, B.: Cell proliferation. *In:* Code, C. F. (ed.): Handbook of Physiology. Sec. 6, Alimentary Canal. Vol. 5. Washington, D. C., American Physiological Society, 1968. (This review covers the important subject of the turnover of cells of the digestive tract and the factors influencing the turnover.)

Max, M., and Menguy, R.: Influence of adrenocorticotropin, cortisone, aspirin, and phenylbutazone on the rate of renewal of gastric mucous cells. Gastroenterology, 58: 329, 1970. (This paper demonstrates that factors known to decrease the ability of the stomach to protect itself may act through influencing rate of cell renewal.)

Rhodes, J., Barbardo, D. E., Phillips, S. F., Rovelstad, R. A., and Hofmann, A. F.: Increased reflux of bile into the stomach in patients with gastric ulcer. Gastroenterology, 57:241, 1969. (Reflux of intestinal contents may be a factor promoting gastric ulceration; this paper contains the primary data on the occurrence of reflux in patients with gastric ulcers.)

Ruckley, C. V., and Sircus, W.: Tests for gastric secretory function and their clinical applications. *In:* Glass, G. B. J. (ed.): Progress in Gastroenterology. Vol. 2. New York, Grune & Stratton, 1970. (This review by experienced clinicians summarizes current knowledge of gastric secretory tests and their reliability.)

Smith, G. P., and Brooks, F. P.: Brain, behavior, and gastric secretion. *In:* Glass, G. B. J. (ed.): Progress in Gastroenterology. Vol. 2. New York, Grune and Stratton, 1970. (This paper reviews what little is known about the influence of nervous factors on gastric secretion; it is valuable chiefly for pointing out that much that is currently believed has little scientific foundation.)

GENERAL REFERENCES

Davenport, H. W.: Physiology of the Digestive Tract. Ed. 2. Chicago, Year Book Medical Publishers, 1966. (This is the standard introductory textbook of gastroenterological physiology, and it is the point of departure for those studying the digestive tract.)

Grossman, M. I. (ed.): Gastrin. Berkeley, University of California Press, 1966. (This is a survey of the knowledge of gastrin at the dawn of the modern age; so far there is no more modern equivalent.)

Ivy, A. C., Grossman, M. I., and Bachrach, W. H.: Peptic Ulcer. Philadelphia, Blakiston, 1950. (This is a massive survey of what was and what was not known about the subject in 1950; those studying the subject should begin here.)

Skoryna, S. C. (ed.): Pathophysiology of Peptic Ulcer. Philadelphia, J. B. Lippincott, 1963. (This book summarizes the knowledge of the basis of ulcer disease as of 1962, and it is valuable for its references and for its demonstration of how little real thought has been applied to the subject.)

Wolf, S.: The Stomach. New York, Oxford University Press, 1965. (This book contains historical material that should interest every student of gastroenterology.)

21

Absorption

Konrad H. Soergel, M.D., *and Alan F. Hofmann,* M.D.

The absorptive capacity of the human intestinal tract far exceeds the usual daily metabolic requirements of the body. This functional reserve capacity permits rapid correction of metabolic deficits by sudden increases in oral intake; however, the evidence for physiological autoregulation of absorptive mechanisms is scant. Excluding calcium and iron, metabolites undergo absorption with little regard to the body's needs.

Many variables affect net absorption of a particular substance, including its intraluminal concentration; digestion and solubilization; contact time with the absorbing surface; integrity of the absorbing cells; blood and lymph flow, and, when applicable, the rate of endogenous secretion of this same substance into the intestinal tract. All must be considered before a complete picture of intestinal absorption in health and disease emerges.

GENERAL CONSIDERATIONS

Digestion

Carbohydrates and proteins can be absorbed only after complete enzymatic hydrolysis to monosaccharides and amino acids, respectively. Fat (triglyceride) requires, in addition, physical dispersion of the water-insoluble products of enzymatic hydrolysis by bile acids. Although fat-soluble vitamins are absorbed without enzymatic hydrolysis, they must be dispersed by bile acids.

Digestive activity occurs both intraluminally and at the lumen surface. Intra-

luminally, digestion is carried out by salivary, gastric and pancreatic enzymes, as well as bile acids. Neurohumoral regulation of digestive secretions is complex. However, the hormone cholecystokinin-pancreozymin (CCK), which is released from the proximal small intestine and causes pancreatic secretion and gallbladder contraction, is the most important mediator of intraluminal digestion. At the luminal surface, digestion is carried out by enzymes within the luminal surface of the absorbing cells.

Surface Area

Anatomically the small intestine is a pliable 280-cm. cylinder. Were its lining smooth, the surface area would be 2,000 cm.^{2}; however, the spiral or circular valvulae conniventes and villi increase the surface area to about 54,000 cm.^{2}. The microvilli covering the epithelial cells cause a further expansion of total surface area to nearly 2 million cm.^{2}. The surface area per unit length of gut decreases from jejunum to colon because of loss of valvulae conniventes in the ileum and the absence of villi in the colon. The surface area of the colon is one fiftieth that of the small intestine.

Intestinal Epithelium (Fig. 21-1)

The *absorptive cells* of the small intestinal mucosa are derived from undifferentiated cells in the crypts of Lieberkuehn. The latter cells proliferate rapidly (mean generation time, 24 hours) and exhibit secretory activity. Absorptive cells con-

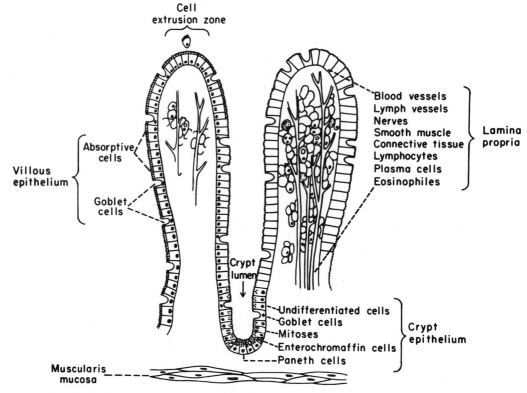

Fig. 21-1. Schematic representation of small intestinal crypt and villi. (Code, C. F. (ed.): Handbook of Physiology. Section 6, Vol. 3, p. 1127. Washington, D. C., Am. Physiol. Soc., 1968. By permission of the publisher)

stantly differentiate from crypt cells and migrate steadily towards the lumen. They appear at the villus base within 24 hours, reach the villus tips in 5 to 7 days, and are then shed into the intestinal lumen. It has been estimated that 20 to 50 million epithelial cells are lost from the human intestine each minute. The absorptive cells continue to mature during their upward migration and do not acquire their full complement of enzymes—and, presumably, their full absorptive capacity—until they reach the upper half of the villus. *Goblet cells,* present in the epithelium of both crypts and villi, secrete mucus continuously by a merocrine process. Two additional cell types are found only in the crypts: *Paneth cells,* with prominent secretory granules which are discharged into the crypt lumen by mixed apocrine and mero-

crine secretion; and *enterochromaffin (Kulchitsky) cells,* containing serotinin and probably a kallikrein-type enzyme, which secrete into the blood rather than the crypt lumen.

The high turnover rate makes intestinal epithelium particularly vulnerable to alterations in cell proliferation. Ionizing radiation, radiomimetic cytotoxic drugs, and starvation cause mitotic inhibition in the intestinal crypts, resulting in inadequate replacement of the senescent cells lost from the villus tips, mucosal atrophy and shortening of both crypts and villi. A different disturbance exists in nontropical (celiac) sprue. The absorptive cells are shed abnormally fast, causing a compensatory increase in undifferentiated cell proliferation and epithelial cell migration rate and resulting in a characteristic mucosal appearance of flattened or absent villi and elongated crypts.

The complex arrangement of the intestinal epithelium has obscured our knowledge of its absorptive function. Probably, however, the crypt epithelium fails to make extensive contact with intestinal contents, due to the narrow diameter of the crypt orifices (100 μ) and the steady secretory flow from the crypts. Although a positive correlation exists between absorptive capacity and the state of functional maturation of the epithelial cells, it remains poorly defined. Moreover, very little is known about composition, volume and regulation of the secretions discharged by goblet and undifferentiated crypt cells. This secretion, the succus entericus (which, by definition, includes the desquamated epithelial cells), accounts for about 60 percent of fasting jejunal contents and nearly all of the fasting ileal contents; the rest represents unreabsorbed gastric, biliary, and pancreatic secretions.

After extensive jejunal resections, the ileum adapts to the sudden drop in absorptive surface by increasing the number of cells per villus, the villus height and the absorptive capacity. Detailed studies in the rat have shown that villus size is modified by a villus growth factor present in pyloric and duodenal secretions and by a villus reducing factor in ileal contents.

Mucosal Contact Time

All other factors being equal, absorption of any substance increases with the time it is in contact with the absorbing surface. If one considers the small intestine as a collapsible tube 280 cm. long, the time spent by a bolus within this organ depends upon the mean velocity of bolus propulsion. During fasting, intestinal contents move at 1.5 to 2.0 cm. per minute. When an average meal is then ingested, little or no change ensues. Thus, a bolus of food must reach the terminal ileum 2½ to 3 hours after leaving the stomach. Within physiological

limits, variations in intestinal flow *rate* elicit no changes in aborad flow *velocity*. The flow rate in the midjejunum is 2.0 to 3.5 ml. per minute after a small meal; however, the intestine can accommodate flow rates up to 7 to 8 ml. per minute by increasing its diameter, without change in flow velocity. This factor, plus the large reserve capacity of the small intestine for absorption, tends to prevent malabsorption from occurring, even when greatly increased amounts of food and drink enter the small intestine or when intestinal transit velocity is increased.

Intestinal Blood Flow

Blood supplies the absorptive cells with oxygen and energy-producing substrates, provides the water entering the intestinal lumen in response to osmotic gradients, and transports absorbed water and water-soluble substances into the general circulation. Sight of a meal increases intestinal blood flow, which rises further during digestion and absorption. The intestinal blood flow, expressed as a fraction of the cardiac output, increases, indicating a redistribution of flow to the splanchnic circulation. A reverse relationship (variations in blood flow affecting absorption) undoubtedly exists, but is probably of no physiological significance. In dogs, intestinal blood flow must be halved before the rate of glucose absorption decreases. To what extent, and how, intestinal mucosal blood flow and absorption are linked in normal human physiology remain unexplained. The major absorptive role of intestinal lymphatics is to transport absorbed lipids in the form of emulsion droplets (chylomicra).

Intestinal Bacterial Flora

More than 60 bacterial species have been isolated from the human intestine. About 15 percent of samples obtained from the stomach, duodenum, jejunum, and proximal ileum during fasting are sterile. The rest contain viable bacteria in counts up to 10^5 per ml. This flora is predominantly gram-positive and microaerophilic *(Streptococcus viridans,* Lactobacillus, Staphylococcus). In the distal ileum, the bacterial flora changes in composition and increases greatly in number (up to 10^9 per ml.). Wet stool has a colony count of about 10^{10} per g.; bacteria account for about one third the dry fecal weight. The flora in distal ileum and colon is predominantly gram-negative and anaerobic (*Bacteroides, Lactobacillus bifidus*); in addition, enterobacteriae (*E. coli, A. aerogenes*), Proteus, Pseudomonas, *Streptococcus faecalis,* aerobic lactobacilli, and Staphylococcus are usually present. Small numbers of fungi, mainly *Candida albicans,* can be found at all levels of the intestinal tract. The mechanism operative in preventing bacterial proliferation in the proximal small intestine is mainly mechanical cleansing by the rapid passage of intestinal contents; however, bile acids, gastric acidity, and the secretory immunoglobulin IgA may also inhibit growth.

The bacterial flora of the proximal small intestine does not detectably alter nutrients or bile acids. Any increase in colony counts is generally due to replacement of the normal flora by gram-negative, mainly anaerobic, organisms of the type found usually only in the distal ileum and colon.

The normal bacterial flora of ileum and colon does not affect absorption directly; rather, it acts upon a variety of substances present in intestinal contents. Bacterial alterations and subsequent metabolic fate of certain substances entering the colon in health and disease are summarized in Table 21-1.

The normal intestinal flora assumes pathological importance when increased substrate enters the colon, causing bacteria to form metabolites in toxic quantities, and when stasis of small intestinal contents permits bacterial proliferation—for example, with diverticula, blind loops, gastrojejuno-colic fistula or disturbances of motil-ity. In this situation the bacteria may cause vitamin B-12 deficiency by competing with the host for this vitamin. In addition, bacterial alteration of bile acids reduces their concentration in solution, with consequent impairment of fat digestion.

Although colonic bacteria may synthesize the B vitamins, vitamin K and folic acid, this is of no nutritional significance to the host.

MECHANISMS OF INTESTINAL MEMBRANE TRANSPORT

Basic Considerations

Absorption represents the transport of water and solutes from intestinal contents which have a variable composition across a membrane—the intestinal epithelium—to the body fluids which have a relatively fixed composition. The rates of absorption and secretion of a substance for a given intraluminal concentration are determined by the properties of the membrane, which vary along the intestine. Conversely, membrane characteristics can be deduced from determining transport rates under defined conditions. Classification of the complicated transport processes across the intestinal epithelium and other biological membranes is based upon the solute and water movement across artificial, simple membranes. Transport across such simple membranes follows the laws of physical chemistry, derived especially from irreversible thermodynamics. Although application of these definitions and principles derived from in vitro experiments to the intact intestine obviously represents an oversimplification, it provides a uniform manner of describing biological observations.

Membrane Structure

Permeability of the epithelial cell wall depends upon its lipid structure, the presence of water-filled channels (pores) and the action of carrier molecules. A double

TABLE 21-1. Metabolism of Selected Compounds by Colonic Bacteria

	COMPOUND	TYPE OF TRANSFORMATION	PRODUCT	FATE	COMMENTS
Fatty acids	Oleic acid	Hydration	10-hydroxy-stearic acid	probably poorly absorbed	May induce colonic secretion of Na^+
Bile acids	Conjugated bile acids	Deconjugation	Free bile acids and glycine or taurine	see below	Many bacterial species have deconjugating enzymes.
	Cholic acid	7α-dehydroxylation	Deoxycholic acid	poorly absorbed	If absorbed, conjugated with glycine or taurine, BE-EHC
	Chenodeoxycholic	7α-dehydroxylation	Lithocholic acid	very poorly absorbed	Small fraction absorbed, conjugated, sulfated (H); BE
Carbohydrates	Glucose Lactose	Oxidation and reduction	Organic acids: acetic, lactic, formic, and propionic.	probably poorly absorbed	Cause osmotic diarrhea; low intraluminal pH may contribute. If absorbed, expired or metabolized to CO_2.
	Raffinose		Gases: CO_2 and H_2	absorbed	Expired
Amino acids	Glycine	Oxidative deamination	CO_2	absorbed	Expired
	Tryptophan	Deamination	Indole	probably poorly absorbed	Oxidized to indoxyl (H); sulfated to form indican. 2 molecules of indican may condense to form an indigo dye.
		Decarboxylation	Tryptamine	absorbed	Oxidized (H) to indole acetic acid (RE)
	Histidine	Deamination	5-imidazole acrylic acid (urocanic acid)	absorbed	Converted to formino-glutamic acid
Bile pigments	Bilirubin diglucuronide	Deconjugation and reduction	Urobilinogens	poorly absorbed	BE-RE
Urea	Urea	Hydrolysis	Ammonia and CO_2	absorbed	Reconverted to urea (H)–RE; also utilized for protein synthesis

Abbreviations: H, hepatic, i.e., conversion carried out by hepatic enzymes; BE, biliary excretion; RE, renal excretion; EHC, undergoes enterohepatic circulation. Absorption is defined as an amount of magnitude sufficient to detect urinary or biliary excretion of a compound or its metabolite.

Comments: This listing is incomplete, but all of the transformations shown have been found to occur in man. Others, not shown, occur in vitro and probably occur in vivo.

layer of polar lipid molecules forms the "backbone" of the membrane (cell wall) which is impermeable to ions and water. Un-ionized solutes can penetrate the membrane in direct relation to their lipid solubility. In addition, measurements of membrane permeability suggest the presence of water-filled "pores" which can be penetrated by water and water-soluble solutes, depending upon the particle size relative to that of the channel, and on the friction between particle and channel wall. Fixed electrical charges may be present in the channel wall which affect the mobility of ions within the channel. The existence of these channels has been postulated solely on the basis of permeability measurements which indicate a pore radius of about 4 Å (0.4 mμ) in the jejunum; this apparent radius decreases progressively in the ileum and colon. Some solutes penetrate the membrane more rapidly than predicted for this model. *Carrier molecules* have, therefore, been postulated to exist within the cell wall. The solute is believed to combine with a specific site on the carrier, causing the carrier to become mobile within the membrane. The substrate is thereby delivered to the opposite side of the membrane, where it dissociates from the carrier. This process is highly specific for certain solutes, exhibits a maximum velocity and is susceptible to inhibition when a second substrate competes for binding sites on the carrier.

Pathway of Absorption

Absorption progresses from the intestinal contents to the blood along the following pathway: (1) a thin unstirred layer of fluid; (2) the "fuzzy coat" (glycocalix) covering the microvilli (the site of surface digestion); (3) the cell membrane (site of aqueous pores and lipid backbone); (4) the cytoplasm (source of metabolic energy and site of certain biotransformations); (5) the serosal and/or lateral cell wall (where sodium ion is actively extruded); (6) the

lateral intercellular space; (7) the basement membrane; and (8) the wall of capillaries and lymphatics. When discussing water and solute transport in conceptual terms, this pathway is often considered to be a single cell wall. It is further assumed that all cells in contact with luminal contents are functionally homogeneous.

Transport Processes

Since the molecular arrangement of cell membranes is not known, since no carrier molecule has been isolated from mammalian intestinal epithelium, and since the steps that provide metabolic energy for active transport have not been fully elucidated, precise molecular description of the transport mechanisms for solutes across the epithelial cell must be presently tentative. Nonetheless, certain unifying concepts are emerging.

A major activity of the epithelial cell is the maintenance of a fixed, low intracellular sodium concentration by pumping sodium ion into the lateral intercellular spaces; thus, a local region that is hyperosmolar is created. This localized area of high sodium ion concentration induces an osmotic water flow from the cell. As water flows, it carries with it those solute molecules that are small enough to pass through the pores of the lateral cell wall. Thus, water movement is passive in response to active sodium pumping at the lateral cellwall, and this osmotic flow of water transports solutes by convection. Sodium is intimately linked to two steps in the absorption of most amino acids and sugars: First, the entry of these solutes into the absorbing cell requires the presence of sodium ions in the mucosal solution. Second, active transport of sugars and amino acids causes their intracellular accumulation, resulting in concentrations higher than those in luminal contents. Since this "uphill" accumulation can occur only while sodium is actively extruded through the lateral cell wall, sodium-coupled transport is emerging as a major mechanism of intestinal transport.

Both solvent drag and active transport are processes that can move solute against a concentration gradient—that is, to a compartment with higher solute concentration. Processes such as simple or facilitated diffusion move solute to the compartment with lower solute concentration.

Active Transport and Facilitated Diffusion. Active transport is defined operationally as net absorption against a concentration or electrical gradient, or both; it requires metabolic energy and is assumed to involve a carrier molecule in the membrane which associates reversibly with the actively transported solute. As noted, there is increasing evidence that uphill movement of solute is coupled with downhill movement of sodium ion into the cell. Active transport creates a serosal-to-mucosal concentration gradient; because of the gradient, the solute diffuses passively back into the mucosal solution. In this case, the net transport rate equals the active transport rate minus the rate of passive back diffusion. Thus, the net absorption of actively transported substances, particularly of inorganic ions, will decrease with increasing diffusion permeability of the membrane. When a solute is absorbed down a mucosal-serosal concentration gradient without energy expenditure, but requiring interaction with a carrier to gain entry into the cell, the process is termed *facilitated diffusion.*

The interaction between solute and carrier conforms to the reaction kinetics of an enzyme (the carrier) with a substrate. This relationship can be expressed in terms of first-order (Michaelis-Menten) kinetics by considering the rate of active transport or facilitated diffusion as the velocity of an enzymatic reaction. Figure 21-2, A, illustrates the facilitated diffusion of D-fructose from the human jejunum. In Figure 21-2, B, the data have been translated into a Lineweaver-Burk plot, which allows the calculation of the maximal absorption rate of this sugar (V_{max}) and of the concentration at which half the maximal transport rate occurs (that is, at which the carrier is half-saturated with the substrate (K_m)). In general, substrates with

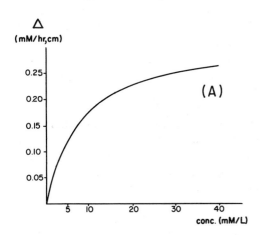
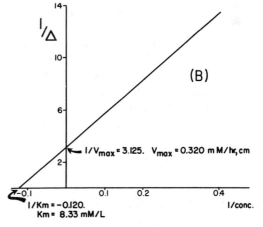

FIG. 21-2. Absorption of D-fructose from the human jejunum during continuous perfusion (9 ml./min). The test solutions contained D-fructose in varying concentrations, a nonabsorbable marker substance, and NaCl to an osmolality of 285 mOsm/kg. Samples for analysis were obtained 15 cm. and 45 cm. distal to the site of infusion, and absorption of D-fructose was calculated from changes in D-fructose concentration relative to that of the marker. (A) Plot of mean concentration vs. absorption rate (Δ). (B) Lineweaver-Burk plot of the same data.

high affinity for the carrier (low K_m) exhibit a low V_{max}.

Active transport of single ions generates electrical membrane potentials, but so many factors influence the potential difference during intestinal absorption that potential difference measurements have shed little light on the mechanism of active ion transport.

Movement of Solutes by Osmotic Flow of Water; Solvent Drag. When unequal concentrations of solute exist on either side of a membrane, osmotic forces are generated that cause bulk flow of water through the "pores" of the membrane. When solute particles are too large to enter the "pores," they exert the full *theoretical* osmotic pressure (as measured in an osmometer) on the membrane. With small solutes, a fraction of the molecules will hit the "pore" entry in dead center and thus traverse the "pore" channel. When, for example, every third solute particle enters the "pore," the so-called reflection coefficient of this solute is 2 out of 3, or 0.67. The *effective* osmotic pressure created by this solute then equals 0.67 times the theoretical osmotic pressure. Bulk flow through an essentially cylindrical channel is analogous to flow through a capillary (Poiseuille's Law); this means that increases in "pore" diameter will have more marked effects on bulk flow (flow varies with the fourth power of pore radius) than on the process of diffusion (which varies with the square of the area). As water passes through the "pores," it sweeps solute along, a process termed *solvent drag.*

Osmotic permeability decreases progressively from duodenum to rectum. Thus, the jejunum is best suited for rapid equilibration of osmotic gradients generated by (for example) ingestion of a hypertonic meal. In patients with nontropical sprue, pore size is diminished, thereby interfering with bulk flow, solvent drag and diffusion of small molecules and ions that penetrate normal pores.

Simple Diffusion. The force that drives this process is the concentration gradient, which causes net transport of the solute until concentration equilibrium is established between the mucosal and the serosal compartment. *Water-soluble substances* (e.g., urea, potassium or glycerol) diffuse through water-filled "pores" in the membrane if the solute molecules or ions are sufficiently small to pass through the "pore," i.e., if they have a molecular weight greater than 150. Besides "pore" size, factors that influence diffusion rates are pore length and friction caused by pore configuration. For electrolytes, electrical forces in the membrane also influence diffusion. Thus, an anion, such as chloride, will be attracted to the positively charged side of a membrane across which a potential difference exists. Fixed negative charges in the "pore" act to slow diffusion of anions and accelerate cation diffusion through the channel.

Lipid-soluble substances such as fatty acids and cholesterol, diffuse through the lipid bilayer of the membrane. For lipids that ionize, only uncharged species of such molecules can diffuse into the membrane. Accordingly, the rate of their movement across the membrane depends upon the dissociation constant of the acid or base in relation to the pH of the solutions on either side of the membrane.

Ion Exchange. Ion exchange, i.e., a stoichiometric exchange of ions of like charge between luminal content and mucosal cell, is an important determinant of ion concentration in gut contents. In the stomach, hydrogen ion of gastric contents may be exchanged for sodium ion. In jejunum and ileum, sodium from the lumen probably is exchanged for hydrogen ion formed in the absorptive cells. In ileum and colon, luminal chloride is exchanged for cellular bicarbonate. At present, the molecular basis for these exchanges, their overall importance in ion movements during digestion, and their regulation are not understood.

Pinocytosis, the engulfing of fluid by the cell membrane which then pinches off to form a vacuole containing the imbibed fluid, has not been demonstrated in adults. In the newborn, however, pinocytosis occurs and is important.

Secretion and Blood to Lumen Transport. Under some conditions, the rate of secretion of electrolytes and water by the intestine increases and becomes an important factor in the pathogenesis of diarrhea. Secretion from the small intestine has been induced experimentally by cholera toxin or intestinal obstruction and observed clinically in patients with cholera, sprue and bacterial diarrhea; it is probably an important mechanism in viral diarrhea. Secretion from the large intestine has been induced by dihydroxy bile acids during perfusion of the human colon.

In diseases associated with alterations in the structure of the fine lymphatics, ulcerative or hyperplastic mucosal disease, and in gastrointestinal hypersensitivity states, considerable transudation of plasma proteins may occur. If proteins pass into the proximal gut in these conditions of "protein-losing enteropathy," they will be digested to amino acids which are absorbed; if proteins are lost in the colon, they escape digestion, and fecal nitrogen loss will be increased. If the rate of transudation exceeds the capacity of liver to synthesize protein, hypoproteinemia and consequent edema may occur.

ABSORPTION OF WATER AND WATER-SOLUBLE SOLUTES

Fluid and Electrolyte Movement During Digestion

Water and salt ingested orally represent only a fraction of the amounts that the intestine must absorb to preserve homeostasis. Most of the absorptive load comes from endogenous secretions (Fig. 21-3). The succus entericus contributed by the jejunum and, in much smaller quantities, by the ileum, contains 0.3 g.% of proteins, 0.5 mEq./L. of calcium, and visible mucus, in addition to the major electrolytes. With food intake, the load increases by the amount ingested—and mainly by a two- to fourfold augmentation of salivary, gastric, biliary and pancreatic secretions. By the

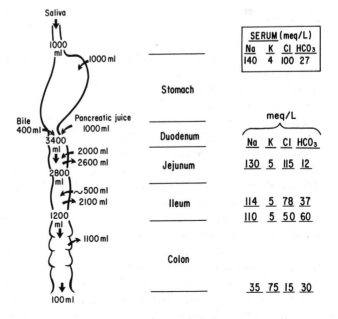

FIG. 21-3. Secretion, absorption, intraluminal flow rate (ml./day), and composition of fasting gastrointestinal tract contents in man. The numbers represent approximate values.

time an ordinary meal reaches the ileum, however, the total increment in luminal flow amounts to only one third of the original meal volume. Most of the meal and the digestive secretions stimulated by it are absorbed in the jejunum.

Flow rates in the proximal jejunum after a large meal range up to 7 ml. per minute. The small intestine can distend to accept flow rates of 8 ml. per minute. At higher flow rates, the luminal diameter remains constant and flow velocity increases, causing a reduction in time available for absorption. In fact, "overflow diarrhea" occurs when jejunal flow rate reaches about 10 ml. per minute. Clinical examples are asiatic cholera, in which the proximal small intestine secretes excessive amounts of an isotonic electrolyte solution, and some patients with gastrin-producing pancreatic islet cell tumors (Zollinger-Ellison syndrome), in which excessive gastric acid production is associated with increased alkaline pancreatic secretions (to neutralize the acid). In both diseases, the absorptive capacity of water and electrolytes is essentially preserved; when diarrhea occurs, it is due to overloading of the intestinal tract with endogenous secretions.

Electrolyte Movements

From jejunum to colon, the electrolyte composition of intestinal contents during fasting deviates progressively from those of serum and interstitial fluid (Fig. 21-3). Some predictions can thus be made about the presence and efficiency of active transport along the intestinal tract. The sodium concentration in jejunal contents is 10 mEq. per liter below that of serum, a difference widened to 30 in the ileum and to 105 to 140 mEq./L. in the colon. Hence, sodium is absorbed against chemical concentration gradients and, in addition, against an electrical gradient, since the polarity of the potential difference tends to inhibit cation absorption. These data suggest that sodium is absorbed actively and with increasing efficiency from jejunum to

colon. The aboral decrease in intestinal permeability, which progressively restricts passive diffusion of sodium from the blood back into the intestinal lumen, probably also contributes.

In contrast to sodium, potassium appears to be passively distributed between blood and intestinal lumen. In the jejunum, concentration of chloride is slightly greater than that in blood, whereas the concentration of bicarbonate is less. In the ileum and colon, the concentration of chloride falls and that of bicarbonate rises reciprocally.

The nature of the active transport processes that maintain the lumen-to-blood concentration gradients of sodium, chloride, and bicarbonate is the subject of continued debate. It has not yet been possible to assess quantitatively the following factors: the continued secretion of water and electrolytes by the crypt epithelium; the existence of pore channels that permit ions to diffuse back along the concentration gradients established by active transport; and the lack of correlation between intestinal potential difference and the rate of net sodium absorption.

The most likely mechanisms for sodium, chloride, and bicarbonate transport are as follows: in the jejunum, luminal contents are slightly acid, and sodium is absorbed in exchange for actively secreted hydrogen ion produced by absorptive cells; hydrogen ion then reacts with bicarbonate in the lumen to form CO_2 and H_2O, accounting for the low bicarbonate concentration of jejunal contents. In the ileum, bicarbonate is secreted in exchange for chloride, in addition to sodium and hydrogen ion exchange. According to this scheme, sodium and chloride transport does not generate a potential difference. In the colon, the exchange of chloride for bicarbonate continues, while sodium appears to be directly and actively absorbed. Colonic sodium absorption and potential difference rise in the presence of increased blood levels of aldosterone.

In patients with congenital chloridorrhea the ileal and colonic exchange of chloride and bicarbonate is defective, resulting in diarrhea, high chloride concentrations in stool water, and metabolic alkalosis owing to decreased bicarbonate loss with persistent H⁺ excretion (mainly as ammonium ion) in stool.

Passive Water Movement. Water movement across the intestinal mucosa is entirely passive, occurring only in response to osmotic gradients; bulk flow of water (together with solutes) occurs until the osmotic gradient disappears. This mechanism efficiently maintains the isotonicity of intestinal contents with respect to plasma (~ 280 mOsm./kg.), with one exception: the osmolality of stool water may reach 400 mOsm./kg. This probably results from continuing production of solute—mainly volatile fatty acids—in the colon, with insufficient time for osmotic equilibration across the relatively impermeable colonic and rectal mucosa. Active solute absorption is accompanied by water absorption, even in the presence of opposing osmotic gradients. In fact, when water absorption is plotted against solute absorption from isotonic solutions, a linear correlation is obtained. This coupling of water to active solute absorption represents a form of "local osmosis." The solute, after crossing the cell wall with a carrier, raises the osmotic pressure on the serosal side of the membrane, causing bulk flow of water from the lumen. These considerations led to the formulation of a model for solute and water transport by the epithelial cell consisting of three compartments separated by two membranes in series: the first membrane has a low permeability to solutes and water, and the second is highly permeable. An attempt to relate this hypothesis to the morphology of the mucosal cell is shown in Figure 21-4. Here, sodium is extruded into the lateral extracellular space through the membrane with low permeability, creating a standing osmotic gradient which causes bulk water flow from the cell and progressive widening of the intercellular space toward the serosal pole. Basement membrane and capillary walls provide the

Fig. 21-4. Pathway of glucose and sodium absorption incorporating the three-compartment model of solute and water transport. A, glucose-sodium-carrier complex in luminal cell wall; B, (Na⁺K)-activated, Mg-dependent adenosine triphosphate hydrolase providing energy for active sodium extrusion from cell; C, sodium-carrier complex in lateral cell wall; 1, lateral (and basal) cell wall: small "pore" radius, hence, large osmotic gradients; 2, basement membrane and capillary wall; large "pore" radius, hence, small osmotic but larger diffusion gradients. (The lateral intercellular space, located between barriers 1 and 2, is the middle compartment of Curran's three-compartment model.)

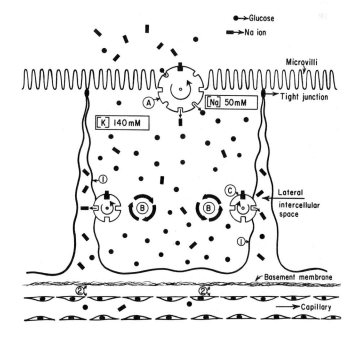

second (high permeability) diffusion barrier.

Carbohydrates

An American adult consumes about 350 g. of carbohydrates daily, consisting about 60 percent of starch, 30 percent of sucrose, and 10 percent of lactose. After digestion, this amount of carbohydrate yields about 2,000 millimoles of monosaccharides, by far the largest contribution to the total exogenous solute load presented to the intestine.

Amylopectin, the main component of starch, is composed of 1,4 α-linked glucose chains, connected by 1,6 α-branching points. Amylose, the minor component, is a straight chain of 1,4 α-linked glucose molecules. Salivary and pancreatic amylase hydrolyze 1,4 α-glucose-glucose links but show little or no activity against either 1,4 α-links located near the end of the glucose chain or those adjacent to the 1,6 α-branching points. Therefore, the products of starch digestion are two- and three-unit pieces of straight glucose chains (maltose and maltotriose.) and α-limit dextrins, i.e., the intact 1,6 α-branch points with short 1,4 α-linked glucose chains attached.

Oligosaccharidases. The digestion of dietary starch is nearly complete in the first 10 to 20 cm. of jejunum. Since α-amylase is the only intraluminal enzyme capable of carbohydrate breakdown, and since carbohydrates are absorbed as monosaccharides, the final step in carbohydrate digestion must occur on the surface of the absorbing membrane. Indeed, a number of oligosaccharidases have been identified in the glycocalix (fuzzy coat) and the microvillus membrane of the small intestinal mucosa. Four or five *maltases* split two- and three-unit 1,4 α-glucose chains, one of which has *sucrase* (invertase) activity and another α-dextrinase (*isomaltase*) activity. The latter hydrolyzes the 1,6 α-glycosidic links of the α-limit dextrans. In addition, one *lactase* and one *trehalase* are present.

Located outside the entry barrier of the absorbing cells, these enzymes hydrolyze dietary oligo- and disaccharides somewhat faster than monosaccharides can be absorbed. About 75 percent of the monosaccharides liberated are immediately transported into the cell without ever appearing in intestinal contents. The proximity of the sites of oligosaccharide digestion and monosaccharide uptake provides a kinetic advantage to the absorption of sugars arriving at the intestinal mucosa as di- or oligosaccharides. Monosaccharides liberated on the cell surface are absorbed more rapidly than monosaccharides present in intestinal contents. It seems likely, therefore, that ingested sucrose will elevate blood glucose levels just as rapidly as ingested glucose.

Oligosaccharidase activity in biopsy samples of intestinal mucosa rises from the level found in the duodenum to a peak in the distal jejunum, falling to low levels in the distal ileum.

In normal subjects, lactase is least active, maltase most active, and sucrase and α-dextrinase are intermediate. The activities of jejunal maltase and sucrase rise 2 to 5 days after a high sucrose or fructose diet is begun, whereas lactase activity remains unaltered by a high lactose or galactose diet for up to 5 weeks. Diets high in glucose do not affect intestinal oligosaccharidase levels.

Oligodisaccharidase Deficiency. Lack of these enzymes is a frequent cause of carbohydrate malabsorption. Transient, acquired deficiencies attend acute and chronic diseases of the small intestine, such as bacterial or viral enteric infections and idiopathic and tropical sprue. Lactase activity may then be nearly undetectable, while the other enzymes may be reduced to 25 percent of their normal activity. The clinical result is lactose intolerance, with increased diarrhea, acidic stools, and flatulence after ingestion of milk and ice cream.

The high activity of maltase, which actually represents four or five different enzymes, explains why isolated maltase deficiency has not been described. The enzymes responsible for sucrase and α-dextrinase (isomaltase) activity are closely related, so that deficiency of one usually, if not always, accompanies deficiency of the other. Patients with congenital sucrase-isomaltase deficiency have diminished maltase activity, as expected, but enough enzyme remains to ensure adequate digestion of maltose and α-limit dextrine. This rare disorder can be managed adequately by a diet low in sucrose and amylopectin.

A marked decrease or loss of intestinal lactase activity is very common. In fact, intestinal lactase activity decreases after weaning in all mammalian species except man. Lactase deficiency is found in 10 percent of Caucasians and in 95 percent of Negroes, Orientals, and American Indians. Although overall, persistent lactase activity correlates well with regular milk ingestion after weaning, this does not explain the absence in some Caucasians. The usually mild symptoms of lactase deficiency can be entirely eliminated by simple dietary adjustments.

Monosaccharide Absorption. The common hexoses, D-glucose, D-galactose, and D-fructose, as well as the pentose, D-xylose, are absorbed more rapidly from jejunum than from ileum (absorption rate per unit length of intestine). They exhibit a V_{max} (maximum absorption velocity), which suggests that hexose and pentose absorption is carrier-mediated (see Fig. 21-2). Active transport of glucose and galactose is readily demonstrated; however, most glucose absorption occurs because jejunal concentrations are usually higher than concentrations in serum. Because glucose absorption is rapid, carrier-mediated transport must be involved, but, since glucose moves from a higher to a lower concentration, it is unclear what fraction of total absorbed glucose is transported by energy-requiring active transport and what part by facilitated diffusion. If small intestinal content equilibrated with blood, one could calculate that only one gram of glucose

would escape absorption daily. Thus, the role of active glucose transport is to accelerate glucose absorption, especially in the jejunum.

In congenital glucose-galactose malabsorption (a rare syndrome) active absorption of glucose and galactose is absent. Affected infants have profuse diarrhea and fail to thrive until all dietary carbohydrates except fructose have been eliminated. Whether the defect arises from the absence of the glucose-galactose receptor on the carrier or from uncoupling of the carrier transport from its energy source remains unclear.

Studies in vitro have clearly shown that active monosaccharide absorption requires the presence of sodium. (Figure 21-4 includes a simplified presentation of the sodium-glucose interrelationship.) The carrier in the luminal cell wall has binding sites for sodium and glucose; the presence of sodium in the mucosal solution increases the carrier affinity for glucose. The sodium concentration gradient across the luminal cell wall causes the sodium-glucose-*carrier complex* to diffuse toward the inside of the cell, with subsequent dissociation of sodium from the carrier. This reduces the binding affinity for glucose so that glucose dissociates from the carrier also and accumulates intracellularly against a concentration gradient. The low intracellular sodium concentration is maintained by active extrusion of sodium ions across the lateral cell walls. Glucose diffuses from the cell toward the interstitial space and capillaries along its concentration gradient. The net sodium transport occurring in a stochiometric relation to glucose transport is believed to represent only a small fraction of total sodium absorption. In this scheme, the energy for active glucose absorption is provided by the mechanism that maintains intracellular sodium concentration at the low level of approximately 50 mM per liter.

The concept of a carrier with multiple specific receptor sites explains several other observations. For instance, active amino acid transport also has an absolute requirement for sodium, and the existence of a single carrier complex with specific receptor sites for sugars, amino acids and sodium has been postulated.

Glucose absorption measurably decreases when the intraluminal concentration of sodium is less than 50 mEq. per liter. Since the sodium concentration of small intestinal contents is usually higher, this ion has little or no influence on sugar absorption in vivo.

Hormones regulate sugar absorption to a limited extent. The active absorption of glucose and, interestingly, amino acids is increased in insulin-deficient diabetes, returning to normal with insulin administration.

D-Xylose, a pentose that is rapidly absorbed from the jejunum, is widely used as a test substance for assessing the functional integrity of the proximal small intestinal mucosa. A test dose is given orally and urinary excretion is measured. Impaired excretion suggests a diffuse abnormality of the duodenal and jejunal muosa, such as that occurring in idiopathic or tropical sprue and in Whipple's disease.

Consequences of Carbohydrate Malabsorption. Malabsorption of sugars is caused by reduced disaccharidase activity or by reduced ability to absorb monosaccharides. Either defect can represent a congenital lesion or develop in the course of acquired small intestinal disease, including acute infectious diarrhea. Increased quantities of carbohydrate delivered to the terminal ileum and colon are fermented by the resident bacterial flora to gas and organic acids; very little mono- and disaccharide appear in the stool.

The concentration of organic acids— mainly acetic, propionic, *n*-butyric and lactic acid—in stool water is about 50 mEq. per liter in normal subjects on a carbohydrate-free diet. This rises to 120 to 170 mEq. per liter with a regular diet. These short-chain organic acids are produced by bacterial fermentation of about 10 grams of glucose per day in healthy subjects eating ordinary food. Excessive production of these acids owing to carbohydrate malabsorption has the following consequences: (1) the conversion of one molecule of glucose into two or three smaller molecules increases the solute concentration of colonic contents. The resulting water bulk flow into the colon, though too slow to achieve complete osmotic equilibration, increases the stool volume. This is a form of *osmotic diarrhea* in which the organic acids constitute a large part of the total solute in stool water. By contrast, in overflow diarrhea, stool water contains largely inorganic ions. Hence, with equal stool volumes, osmotic diarrhea causes less sodium loss than does overflow diarrhea. (2) Part of the organic acids react with the bicarbonate of colonic contents, causing the bicarbonate in stool water to diminish or disappear; stool water then becomes acid (pH less than 6.0). It follows that the discharge of acid stools does not represent a net acid loss to the body but rather a net loss of bicarbonate, which can result in metabolic acidosis. (3) Acetic acid in high concentrations may aggravate the existing diarrhea by interfering with absorption and stimulating secretion of water and electrolytes by the colon.

Colon bacteria also form *hydrogen gas* from carbohydrate. After a 6-hour fast, hydrogen gas is produced at a rate of 0.6 ml. per minute in the colon. After a 24-hour fast the rate falls to 0.05 ml. per minute. One hour after a few grams of lactose have entered the terminal ileum and colon, up to 4 ml. of hydrogen gas are produced per minute. Production of hydrogen gas increases not only during carbohydrate malabsorption but also when large quantities of certain nonabsorbable oligosaccharides (e.g., raffinose in shell beans) are ingested.

Proteins and Amino Acids

Dietary protein is diluted two- to three-fold by the 170 g. of protein which reaches the intestine with desquamated epithelial cells and digestive and intestinal secretions each day. Most of the dietary protein is absorbed in the proximal small intestine. However, in contrast to sugar absorption, the jejunum and the ileum have roughly the same capacity for amino acid transport.

Intraluminal Digestion. Pepsins, trypsin and chymotrypsins are endopeptidases which attack peptide bonds inside protein molecules. Pancreatic carboxypeptidases are exopeptidases which hydrolyze peptide bonds adjacent to the terminal carboxyl group of proteins and peptides. The result of intraluminal protein digestion is a mixture of amino acids and small peptides.

Surface Digestion and Absorption. Protein is absorbed mainly in the form of free L-amino acids. The small amounts of intact di-, tri- and tetrapeptides entering the absorptive cells are either hydrolyzed within the cells or appear unchanged in the portal venous blood.

Numerous specific peptide hydrolases are present in the intestinal epithelial cells, many of which have been localized to the microvillus membranes, presumably exterior to the entry barrier of the cell wall. The free amino acids liberated by these surface digestive enzymes are then transported into the epithelial cells. Intestinal peptide hydrolases are distributed similarly to the oligosaccharidases, their activity rising progressively from duodenum to jejunum and falling to low levels in the terminal ileum. In diseases affecting the intestinal mucosa, their activity is reduced. Clinically important congenital or acquired deficiencies of peptide hydrolases have not yet been reported.

Amino acid absorption shares several features with glucose absorption, including carrier-mediated transport which proceeds along a "downhill" concentration gradient in vivo. Studies in vitro indicate that this transport is active, because the substrate is found to accumulate within the epithelial cells and because the observed absorption rates exceed those expected for facilitated diffusion. In both, amino acid and glucose absorption, sodium is required. Moreover, the sites of the end stage of digestion and the entry mechanism into the cell on the microvillus membrane are closely associated for both, carbohydrates and proteins.

A grouping of the amino acids into different transport systems has been proposed on the basis of competitive inhibition studies of cellular uptake in vitro: (1) monoamino-monocarboxylic acids, including tryptophan, (2) basic amino acids (ornithine, lysine, arginine and cystine), (3) imino acids (proline and hydroxyproline) and perhaps, (4) glycine.

Congenital Defects in Amino Acid Transport. Studies of several rare congenital disorders of renal and intestinal amino acid transport help to validate this scheme of amino acid transport groups. In *Hartnup's disease*, the intestinal uptake and renal tubular reabsorption of group 1 amino acids is low. The patients present with cerebellar ataxia, a rash and, frequently, mental retardation. These features, along with increased urinary excretion of indican, are presumably due to the malabsorption of tryptophan, the precursor of nicotinamide (vitamin B-6). In the *blue diaper syndrome*, an unexplained isolated defect in intestinal absorption of tryptophan occurs. An oxidation product of urinary indican turns the diaper blue. Patients with *cystinuria* have a similar but not identical transport defect of group 2 amino acids in the intestine and the renal tubules.

Pinocytosis. During the first few days of life, the intestine absorbs gamma-globulin molecules by pinocytosis, allowing the newborn infant to acquire passive immunity by absorbing antibodies contained in the colostrum. Later in life, intact protein is not absorbed by the gut.

Iron

Absorption of iron is regulated by the amount of total body iron and is modified by various intraluminal factors. The average adult body contains about four grams of iron, of which half is in hemoglobin; the rest is stored in parenchymatous organs and the reticuloendothelial system as ferritin, and in cytochrome enzymes, myoglobin, and peroxidases. The body loses iron in desquamated cells from skin, intestine, lungs and kidneys and, in women, with menstrual bleeding. Net iron absorption, occurring mainly in the duodenum, is 1 to 3 mg. per day.

When the body is overloaded with iron, the amount of ferritin within the intestinal epithelial cells increases; in iron deficiency, it decreases. Part of the iron taken up by the columnar epithelial cells enters the circulation, where it is bound by transferrin; the rest remains in the epithelial cells, combines with apoferritin, and re-enters the intestinal lumen when the sensecent cells are shed from the villus tip. In iron deficiency, mucosal uptake rises and the proportion of mucosal iron entering the body becomes greater. Iron overloading evokes opposite changes. In *idiopathic hemochromatosis,* the mechanism normally controlling iron absorption and, hence, stabilizing metabolic iron balance goes awry. Iron continues to be absorbed despite greatly increased iron stores in the body.

Inorganic iron is absorbed in the ferrous form. Absorption of inorganic iron is decreased by conditions that favor oxidation to the ferric form—such as absence of gastric hydrochloric acid—or precipitate or sequester iron, such as the presence of phosphate, carbonate and plant phytates. Ascorbic acid, a reducing agent, augments iron absorption.

Most dietary iron is present in organic compounds, mainly hemoglobin, ferritin, and myoglobin, and phytates and siderochromes in plants. The absorption of hemoglobin iron is unaffected by the intraluminal factors influencing the absorption of inorganic iron. Probably hemoglobin iron enters the mucosal cell while still attached to the porphyrin ring. (For a further discussion of iron absorption the reader is referred to Chapter 23.)

Vitamin B-12

Specific acceptor sites exist in the glycocalix (fuzzy coat) of the ileum which bind the intrinsic factor-vitamin B-12 complex before vitamin B-12 enters the mucosal cells. Malabsorption of this vitamin results from resection of the distal ileum or bypass by an ileocolostomy, from disease of the ileum such as regional enteritis (Crohn's disease) and, frequently, from tropical and nontropical sprue. In these clinical conditions, malabsorption of a vitamin B-12 test-dose given orally is not corrected by the addition of intrinsic factor. This measure serves to differentiate between vitamin B-12 malabsorption due to absence of acceptor sites and vitamin B-12 deficiency caused by lack of gastric intrinsic factor production.

Folic Acid

This vitamin exists in food as pteroyl polyglutamate. Its absorption requires a peptidase, commonly known as folate conjugase, to break down the oligopeptide to pteroylglutamic acid. The monoglutamate is absorbed, probably by an active transport mechanism, in the duodenum and upper jejunum. Folic acid deficiency is often the earliest sign of diffuse intestinal mucosal disease at this site, particularly in tropical sprue.

Calcium

This ion is actively absorbed by the duodenum and proximal jejunum. One of the factors influencing its uptake is the amount of a specific calcium-binding protein present in the cytoplasm of the absorbing cells. Synthesis of this protein requires certain metabolically active derivatives of vitamin D. Inadequate absorption of calcium accompanies diffuse diseases of the proximal intestinal mucosa or may be caused by vitamin D deficiency due to insufficient dietary intake, impaired absorption secondary to disturbances of micelle formation (see below) or resistance to the metabolic actions of this vitamin (in familial vitamin-D-resistant rickets).

ABSORPTION OF LIPIDS

Fat

Triglyceride is a major source of calories in man because of its high caloric value per unit weight (9 cal. per g.). Triglyceride is insoluble in water, and its absorption in man entails enzymatic hydrolysis by pancreatic lipase, solubilization of the lipolytic products by bile acids, uptake by the cell membrane, re-esterification to triglyceride, and aggregation and emulsification to form chylomicra. Thus, fat absorption resembles protein and polysaccharide absorption in that intraluminal digestion is necessary. However, it differs in that surface digestion is not involved, and intracellular resynthesis and emulsification are essential steps. Because efficient fat absorption depends upon so many sequential and interrelated steps, fat malabsorption is more common than carbohydrate or protein malabsorption. Furthermore, fat is more resist-

ant to bacterial destruction, and measurement of unabsorbed fat is simple and accurate. Consequently, the term steatorrhea is often used as a synonym for malabsorption—which is understandable but incorrect.

Fat Digestion: Chemical and Physical Events. Fat digestion is the sequence of chemical and physical events that transforms a water-insoluble molecule (triglyceride) into molecules dispersed in water in a diffusible form. Normal fat digestion relies upon three processes: (1) the secretion of pancreatic lipase and bicarbonate evoked by the hormones cholecystokinin (CCK), (also called pancreozymin) and secretin; (2) the hydrolysis of the outer triglyceride bonds by pancreatic lipase to yield two molecules of fatty acid and one molecule of 2-monoglyceride; and (3) the solubilization of these lipolytic products in polymolecular aggregates by bile acid molecules (Fig. 21-5).

Ingested fat is emulsified partly in the stomach, and emulsification is continued in the small intestine, especially by bile (a concentrated detergent solution of bile acids and lecithin). Pancreatic lipase, an interfacial enzyme, adsorbs to the oil-water interface. The enzyme has positional specificity and attacks only the 1 and 3 ester bonds of the triglyceride molecule.

The lipolytic products, with their low

FIG. 21-5. The physical and chemical events in fat digestion. Triglyceride (*left*) is cleaved by pancreatic lipase to yield two molecules of fatty acid and one molecule of 2-monoglyceride (*right, middle*). These lipolytic products are dispersed by bile acids as a mixed micelle, shown in cross section. The bile acid molecules coat the outside of the micelle apposing their hydrophobic backs to the paraffin chains of the lipolytic products. The center of the micelle is liquid hydrocarbon and can dissolve cholesterol and fat-soluble vitamins. The bile acid molecules and the lipolytic products are believed to be rapidly exchanging between other micelles.

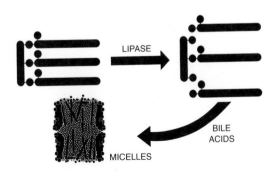

solubility in water, are probably present as bilayers at the oil-water interface. Bile acids desorb the lipolytic products from the oil-water interface and disperse them in the aqueous phase of intestinal content. The bile acid molecule has a hydrophobic and a hydrophilic side and, thus, resembles a typical detergent molecule; such a molecule is termed amphipathic (possessing both properties). Amphipathic molecules have a low molecular solubility, and, when the saturation point is exceeded, they aggregate to form small polymolecular clusters called micelles. Micelles are formed over a narrow concentration range, called the critical micellar concentration. Bile acid solutions, above their critical micellar concentration of about 2 mM, have a striking ability to solubilize fatty acid and monoglyceride. In the bile acid-lipolytic product micelle, a small bilayer of lipolytic products is ringed by the bile acid molecules which have their hydrophobic backs apposed to the paraffin chains of the fatty acid moiety (Fig. 21-5). The paraffin chains are in a liquid state, and the center of the micelle acts as a solvent for other lipids such as cholesterol and fat-soluble vitamins. Dispersion of lipolytic products into micelles by bile acids is the dominant *physical* event in fat digestion.

Because the concentration of bile acids during digestion is about 10 mM, 80 per cent of the bile acids are in micellar form. When intestinal content is isolated during digestion, the water-clear aqueous micellar phase is found to contain bile acid and lipolytic products in about equal proportions by weight. An oil phase, containing insoluble dietary triglyceride as well as diglyceride is also present, and lipolytic products are distributed between the two phases according to their partition coefficients. Intestinal content during fat digestion is thus analogous to a liquid partition system in a separatory funnel (Fig. 21-6).

Enterohepatic Circulation of Bile Acids. Dispersion of lipolytic products in micellar form by bile acids accelerates their transport from the oil-water interface to the surface of the jejunal cell. Bile acids, however, are poorly absorbed by the jejunum, and the greater part passes to the ileum, where they are efficiently absorbed.

The amount of bile acids in the human body, called the bile acid pool, may be measured by isotope dilution. Normally, about 5 to 10 millimoles (2 to 4 g.) are present. If fecal bile acids are measured, man can be shown to excrete about 1 millimole daily. By the use of perfusion techniques, it has been determined that 5 to

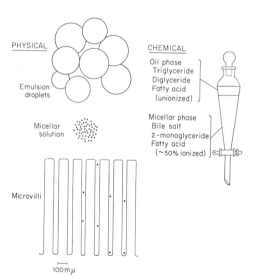

FIG. 21-6. Schematic representation of the physical and chemical states of triglyceride and its lipolytic products during fat digestion. An oil phase composed of higher glycerides and protonated, i.e., nondissociated, fatty acid is in equilibrium with an aqueous, micellar phase of bile acid, monoglyceride, and fatty acid; the fatty acid is partially ionized at jejunal pH. Nonpolar lipids such as fat-soluble vitamins and cholesterol are partitioned between the two phases. The emulsion droplets, micelles (black dots), and microvilli are drawn approximately to scale.

10 millimoles of bile acids are secreted into the duodenum during each meal. Obviously, intestinal absorption via the portal system is extremely efficient. Furthermore, peripheral blood levels are extremely low, so that hepatic clearance of bile acids from portal blood must be equally efficient. Thus, the pool of bile acids circulates once or twice with each meal, is efficiently absorbed by the intestine and is cleared from portal blood by the liver. This recycling process involving the liver and intestine is called the enterohepatic circulation. It renders 15 to 60 millimoles of bile acids available for digestion despite a daily synthesis of only 0.5 to 1.0 millimole. The intactness of enterohepatic circulation determines the size of the bile acid pool, which, in turn, is the major determinant of jejunal bile acid concentrations.

Disturbances in Fat Digestion. Disturbances in fat digestion are the most common cause of fat malabsorption and are due to either lipase deficiency, which impairs hydrolysis of triglycerides, or bile acid deficiency, which impairs micelle formation, or both. Recent studies relating lipase output to steatorrhea indicate that lipase output must be reduced to one tenth of normal before steatorrhea appears. Thus, steatorrhea caused by lipase deficiency occurs only with severe pancreatic disease or as a congenital isolated lipase deficiency. In patients with lipase deficiency, micelle formation is unimpaired, but the unhydrolyzed triglyceride remains in the oil phase. Since the presence of a micellar phase permits fat-soluble vitamin absorption, the major consequence of lipase deficiency is reduced fat absorption, and a caloric deficit. Moreover, since lipase deficiency usually is associated with a reduction in other pancreatic enzymes, there is frequently an associated malabsorption of protein.

Bile acid deficiency is more commonly a cause of disturbed fat digestion than is lipase deficiency. A lack of bile acid is caused by reduced secretion or by precipitation or sequestration in the intestinal lumen. To maintain normal concentrations of bile acids, hepatic secretion of a large bile acid pool must be unimpaired. This process, in turn, is influenced by both hepatic and intestinal factors. Bile acid deficiency may result from enterohepatic or intraluminal causes. Enterohepatic factors include ileal dysfunction (causing bile acid malabsorption and a reduced bile acid pool), hepatic disease (causing reduced secretion) and biliary obstruction or external fistula. Intraluminally, bile acids may be precipitated by cationic molecules (e.g., the antibiotic neomycin or the anion exchange resin cholestyramine) or diluted by excessive secretion. Reduced bile acid concentration in the syndrome of intestinal stasis and bacterial overgrowth is discussed below.

Lipolysis is unimpaired in bile acid deficiency. In the absence of micelles, an oil phase composed chiefly of lipolytic products is in equilibrium with an aqueous phase containing a very low concentration of fatty acid in molecular solution. Fat absorption occurs more slowly, but, if intestinal surface and transit are normal, bile acid deficiency seldom causes severe steatorrhea, i.e., fecal fat loss above 30 g. per day. In the absence of a micellar phase, fat-soluble vitamins are poorly absorbed; hypoprothrombinemia is often an early sign of bile acid deficiency. In patients with partial ileal resection, jejunal bile acid concentrations may be normal, because a compensatory increase in bile acid synthesis occurs, counteracting the effects of reduced absorption. In the patient with a large ileal resection and reduced bile acid concentrations, steatorrhea is caused by maldigestion and reduced surface area.

Uptake and Chylomicron Formation. When a water-clear micellar solution of taurine conjugated bile acids and lipolytic products, such as that present in digestion, is perfused into the jejunum of a healthy subject, its fatty acid and monoglyceride are readily absorbed, but its bile acids are not. Thus, uptake of the intact micelle seems unlikely. The present view is that lipolytic products move into the epithelial cell by simple adsorption and diffusion through the lipid portion of the cell membrane. Studies in healthy human subjects have shown that fat absorption is largely completed by mid-jejunum, whereas most bile acids are absorbed in the ileum. Indeed, the low rate of jejunal absorption of bile acids helps maintain their micellar concentration well above the critical level.

Electron microscopy of intestinal mucosa during fat digestion and absorption shows emulsion droplets in the lumen and fat globules in the cellular endoplasmic reticulum, just below the terminal web. No lipid is seen in or between the microvilli, because the micelles cannot be visualized by electron microscopy. The appearance of oil droplets in the endoplasmic reticulum suggests that this organelle contains the re-esterification enzymes. Experiments with specifically labeled triglycerides have shown that the 2-ester bond is not hydrolyzed during passage through the mucosal cell. It is now well-established that the major biochemical pathway for triglyceride resynthesis is acylation of absorbed monoglyceride by absorbed fatty acids. In this monoglyceride acylation pathway, the fatty acids are activated with coenzyme A by means of a thiokinase, and resynthesis is considered to be mediated by a multi-enzyme complex.

Although the triglyceride droplets are readily observed by electron microscopy to move laterally in the mucosal cell and pass into the intercellular spaces, the biochemical correlates of these morphological events are poorly defined. An emulsifying layer, consisting chiefly of phospholipid that contains some free cholesterol, overlies the surface of the oil droplet.

A lipoprotein, synthesized in the mucosal cell, is also associated with the surface of the chylomicron; however, its significance and relationship to other serum lipoproteins are unknown. At present, the evolutionary advantages of resynthesis of lipolytic products to triglyceride and of transport into lymphatics rather than vascular absorption of chylomicron triglyceride are unclear.

Disturbances in Uptake and Chylomicron Formation. Loss of surface, such as occurs in mucosal disease such as nontropical sprue, should reduce the rate of uptake of micellar lipid, but there is little experimental evidence for this in human disease. When jejunal uptake is reduced because of maldigestion or mucosal disease, the site of fat absorption moves into the ileum, where the uptake of lipolytic products appears to induce formation of esterifying enzymes.

Defects in intracellular processing of fat, as well as chylomicron formation, are rare and poorly understood. A specific defect in esterifying activity has not been demonstrated, although reduced esterifying activity occurs in generalized mucosal disease. Chylomicra accumulate in the cell in children with the rare disease of β-lipoproteinemia or in animals poisoned with inhibitors of protein biosynthesis, but a detailed biochemical explanation for these disturbances, which are readily detectable morphologically, is lacking.

Combined Defects. The disturbances causing fat malabsorption in certain common clinical conditions are complex and poorly understood. In nontropical sprue there is decreased release of CCK, causing impaired gallbladder contraction and reduced bile acid secretion. Concentration of bile acids frequently falls below the critical micellar concentration. Lipolytic products in the aqueous phase in greatly reduced concentrations are presented to a mucosa with decreased surface area and reduced esterifying capacity. In post-gastrectomy steatorrhea, multiple causes of impaired fat digestion are present: poor emulsification of fat in the stomach; impaired CCK release or impaired gallbladder contraction, or both, consequent to vagotomy; dissynchronization of CCK release and gastric emptying; and bacterial overgrowth which causes bile acid deconjugation.

Medium-Chain Triglycerides. Medium-chain triglyceride is fat which is produced synthetically by esterifying glycerol with coconut oil fatty acids of predominantly eight carbon chain length. The metabolism of medium-chain fatty acids (C8) differs in many ways from that of long-chain fatty acids (14 to 18 carbon atoms) during digestion and absorption for several reasons. Thus, for a given amount of pancreatic lipase, medium-chain fatty acids are hydrolyzed more readily. In addition, the lipolytic products of medium-chain fatty acids—octanoic acid, and 2-mono-octanoin—are sufficiently water-soluble to pass into the aqueous phase even when bile acids are absent. The 2-monoglycerides of medium-chain fatty acids are isomerized extremely rapidly to the 1-isomer, which is readily cleaved by lipase; thus, absorption is chiefly in the chemical form of fatty acid. Moreover, for a given concentration, medium-chain fatty acid is absorbed more rapidly than long-chain fatty acid. The medium-chain fatty acids are not converted to a coenzyme-A derivative in the cell nor are they resynthesized to triglyceride. They are not incorporated into chylomicra but leave the mucosal cell by the portal system, where they are bound, in part, to serum albumin. Medium-chain fatty acids are bound less firmly to serum albumin than are long-chain fatty acids; on reaching the liver, they are rapidly oxidized and are not esterified to form complex lipids. Thus, medium-chain fatty acids have the physical and gustatory characteristics of long-chain fatty acids; however, after hydrolysis, their absorption and subsequent metabolism resemble those of a rapidly metabolized nutrient such as glucose.

Principles of Therapy for Fat Maldigestive and Malabsorptive Conditions. In pancreatic lipase deficiency, preparations of porcine pancreas are fed by mouth.

However, success is modest because these preparations vary widely in enzyme activity, much of the ingested enzyme is destroyed by gastric acidity, and the dose required to reduce steatorrhea requires ingestion of many tablets with each meal, entailing considerable expense. Medium-chain triglyceride has benefited children with pancreatic insufficiency due to cystic fibrosis. Results of attempts to prepare fatty acids or monoglyceride in palatable form have generally been unpromising.

In bile acid deficiency, bile acid preparations are of limited value, because, in conditions associated with bile acid malabsorption, diarrhea occurs. In liver disease, bile acid ingestion causes elevated blood levels, with increased pruritis; in patients with biliary fistula, failure of bile acid conservation means that large doses must be given to simulate normal digestive conditions (e.g., 2 to 8 g. per meal), and bile acids are gastric irritants. Bile acid deficiency is of great clinical significance in children with biliary atresia; substitution of medium-chain triglycerides (which do not require micellar dispersion by bile acids for absorption) for long-chain triglycerides markedly increases growth rate.

In disturbances in chylomicron release, substitution of medium-chain triglyceride (absorbed in unesterified form in blood) for long-chain triglyceride in blood will decrease steatorrhea, and fat droplets will disappear from the mucosal cell. Unfortunately, medium chain triglyceride is an expensive caloric source.

Fate of Unabsorbed Fat. Unabsorbed fatty acids are hydrogenated by colonic bacteria to form saturated fatty acids or hydrated to form hydroxy fatty acids. A number of bacterial species are now known to convert oleic acid to 10-hydroxy stearic acid. In vitro, the enzymes responsible for this conversion are induced by the presence of fatty acids in the culture medium. Since the structure of 10-hydroxy stearic acid resembles that of ricinoleic acid, the dominant fatty acid of castor oil, it has been proposed that 10-hydroxy stearic acid has cathartic properties, i.e., that it will induce colonic secretion of water. Colonic bacteria are not thought to metabolize or synthesize long-chain fatty acid de novo to any significant extent.

Absorption of Lipids Other Than Fat

Lecithin (phosphatidyl choline) is an important constituent of bile and a major lipid in certain foods, such as egg yolks. The mucosal cells synthesize this essential constituent of the chylomicron surface readily by several routes. There is no evidence that absorption of biliary or dietary lecithin influences the rate of triglyceride absorption. A human pancreatic phospholipase hydrolyzes the 2-ester linkage of lecithin to give a 1-lysolecithin. This hydrolysis is assumed to occur at the oil-water interface; lysolecithin, being water-soluble, moves to the aqueous phase. Lysolecithin is absorbed, presumably by a mechanism similar to that described for lipolytic products, and is reacylated in the cell. The amount of chylomicron phospholipid is uninfluenced by dietary phospholipid; by contrast, the major source of the fatty acids of chylomicron triglyceride is dietary triglyceride.

Cholesterol absorption has drawn considerable attention because of the relationship between hypercholesterolemia and atherosclerosis. The paraffin chains of the lipolytic products in the micelle form a liquid hydrocarbon core in which cholesterol can dissolve. Dietary cholesterol, together with cholesterol from bile and sloughed cells, is partitioned between the micellar phase and the oil phase. In contrast to other lipolytic products, cholesterol is virtually insoluble in water; and micelles must be present for it to pass from the oil to the aqueous phase and be absorbed. Endogenous cholesterol is actively synthesized in the cell and, with absorbed cholesterol,

is partially esterified before being incorporated into chylomicra. Despite our understanding of the molecular events in cholesterol absorption, we are extremely ignorant about the factors influencing the efficiency of absorption. It seems likely that cholesterol absorption increases with cholesterol intake. Since cholesterol and phospholipids are synthesized ubiquitously within the body, lack of absorption has no metabolic significance.

Fat-soluble vitamins (A, D, E, K) are insoluble in water and require micellar dispersion for absorption. Malabsorption of fat-soluble vitamins occurs when intestinal bile acid concentration is reduced below the critical micellar concentration and, probably, also when generalized mucosal disease is present.

Absorption of Bile Acids

Bile acids are synthesized in the liver from cholesterol and may be considered as its water-soluble excretory product. In human beings, the two primary bile acids are cholic acid (a trihydroxy acid) and chenodeoxycholic acid (a dihydroxy bile acid). The carboxylic acid group of these bile acids is conjugated before secretion into bile, i.e., is linked to the amino group of glycine or taurine in an amide bond (Figure 21-7). Thus, the ionizable group of the conjugated bile acids in bile is either the carboxyl group of glycine ($pK \cong 4$) or the sulfonic acid group of taurine ($pK \cong 2$). Both are much stronger acids than the carboxylic acid group of the unconjugated bile acid ($pK \cong 6$). The lower pKa of conjugated bile acids has important physiological consequences. Conjugated bile acids are fully ionized at jejunal pH and, thus, cannot be absorbed by dissolution in the lipid membrane of the mucosal cell (Fig. 21-8). As a result, jejunal absorption is limited. Conjugation also influences solubility in water at acidic pH. The unconjugated bile acids are insoluble below pH

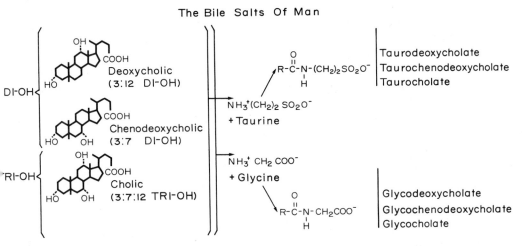

FIG. 21-7. The predominant bile acids of human bile. Deoxycholic acid (*top*) is produced in the colon by bacterial 7α-dehydroxylation of cholic acid, absorbed, conjugated in the liver with glycine or taurine, and recycled with the other primary bile acids. Bacteria also act upon 7α-dehydroxylate chenodeoxycholic acid to form lithocholic acid, but lithocholic acid precipitates from solution and is present in only trace concentrations in bile. Glycine-conjugated bile acids predominate, and, in health, the bile acids in bile are completely conjugated.

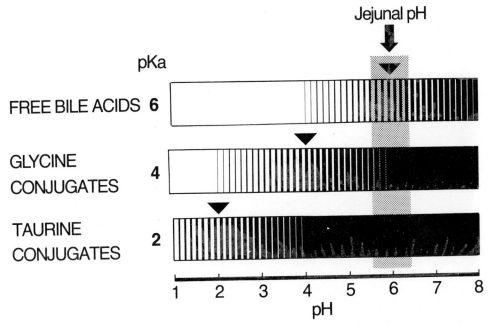

FIG. 21-8. Relationship between pK_a and the state of ionization at jejunal pH for the three classes of bile acids. At a pH value equivalent to the pK_a (black inverted triangle), the bile acids are half ionized. At higher pH values, the fraction in nonionized form (white) decreases, and the absorption rate decreases, since only the nonionized form is adsorbed. At jejunal pH, free bile acids are absorbed most rapidly, and only limited absorption of glycine-conjugated bile acids occur; no significant absorption of taurine-conjugated bile acids occurs.

6; glycine conjugates are soluble down to pH 4; and taurine conjugates are completely resistant to acid precipitation.

Formation of Secondary Bile Acids. Bile acids entering the intestine are exposed to intestinal and bacterial enzymes. Although bile acids are completely resistant to intestinal enzymes, bacterial enzymes in the large intestine hydrolyze the amide bond, converting the conjugated bile acid to a free or unconjugated bile acid. They also remove the 7α-hydroxy group from the steroid nucleus, converting cholic acid ($3\alpha,7\alpha,12\alpha$-trihydroxy) to deoxycholic acid ($3\alpha,12\alpha$-dihydroxy) and chenodeoxycholic acid ($3\alpha,7\alpha$-dihydroxy) to lithocholic acid (3α-hydroxy). These bile acids are called secondary bile acids. Lithocholic acid is insoluble and is not absorbed. However, some deoxycholic acid is reabsorbed and returns to the liver, where it is conjugated with glycine or taurine to recycle with the primary bile acids. Thus, normal bile contains the glycine and taurine conjugates of the two primary bile acids, cholic and chenodeoxycholic acid, and of the secondary bile acid deoxycholic acid (Fig. 21-9).

Enterohepatic Cycling and Regulation of Synthesis. The intestine conserves bile acids efficiently by the cumulative effect of active ileal transport and passive small and large intestinal absorption. Bile acid transport mechanism is located in the terminal ileum. It transports trihydroxy acids more

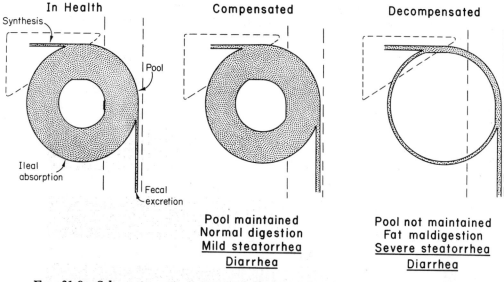

FIG. 21-9. Schematic representation of the enterohepatic circulation of bile acids in health (*left*), in clinical conditions causing mild bile acid malabsorption (*center*), or severe bile acid malabsorption (*right*). In health, efficient intestinal conservation maintains a large bile acid pool, and the small fecal loss is balanced by hepatic synthesis. (In this figure, jejunal and colonic absorption are not shown, but these also occur.) With mild bile acid malabsorption (*center*), increased hepatic synthesis maintains the bile acid pool and fat digestion is unimpaired. Increased passage of bile acids into the colon causes diarrhea which responds to cholestyramine. With severe bile acid malabsorption (*right*), the bile acid secretion falls despite increased hepatic synthesis. Fat maldigestion occurs; this plus the reduced surface area cause steatorrhea. In such conditions, diarrhea has been postulated to be caused by the secretory effect of the hydroxy fatty acids formed by bacterial hydration of unabsorbed oleic acid, and the diarrhea may respond to removal of long-chain fatty acid from the diet.

readily than dihydroxy acids, and it may transport many other amphipathic anions as well. Active ileal transport of bile acids is the intestinal factor most important in conserving the bile acid pool.

Jejunal absorption of bile acids is passive, involving predominantly the least polar of the bile acid conjugates, the glycine dihydroxy bile acids. These are absorbed in uncharged form by dissolution in the membrane, as described for acidic compounds. Free bile acids—i.e., unconjugated bile acids, especially the dihydroxy acids—deoxycholic acid and chenodeoxycholic acid—are still more lipophilic. Perfusion experiment in man have demonstrated that they are absorbed still more rapidly than any of the conjugated bile acids. However, free bile acids are absent in health, appreciable concentrations being measured in the small intestinal lumen only in disease associated with stasis and bacterial overgrowth. In these rare conditions, in which bacteria proliferate in the small intestine, bile acid deconjugation occurs. The resulting free bile acids are rapidly absorbed, as well as precipitated from solution. The result is intraluminal bile acid deficiency, causing impaired micellar dispersion.

Bile acids are returned to the liver in portal blood, bound, in part, to serum albumin. The liver efficiently clears bile acids from portal blood, but the factors responsible for this uptake remain undefined. The uptake capacity of the liver appears to be several times the absorptive capacity of the small intestine, and ingestion of bile acids by mouth raises peripheral blood levels only slightly.

A negative feedback mechanism regulates hepatic synthesis of bile acids from cholesterol. Whenever the enterohepatic circulation of bile acids is interrupted, as in biliary fistula or ileal dysfunction, hepatic synthesis increases. In man, the rate of synthesis may increase from the normal millimole per day to as high as 15 millimoles per day. Recent evidence suggests that the converse is also true, namely, that if bile acids are fed, increasing the return to the liver, synthesis decreases.

The major changes in the bile acid molecule mediated by bacteria include cleavage of the amide bond (deconjugation), dehydroxylation, and oxidation of hydroxy groups. Over 25 bile acids have been identified in human feces. Deconjugation and oxidation of hydroxyl groups to keto groups are carried out by a great variety of aerobic and anaerobic bacteria, but only a few carry out dehydroxylation. Since destruction of the steroid nucleus of bile acids has never been shown to occur in man, the measurement of fecal bile acids is considered to reflect hepatic synthesis accurately.

Colonic Secretion. Deoxycholic acid, at concentrations above 1 millimole, induces secretion of water and electrolytes by the human colon. Chenodeoxycholic acid has a similar effect, but only at somewhat higher concentrations; cholic acid has no secretory effect at 10 millimoles. Thus, bacterial 7α-dehydroxylation converts the pharmacologically inert cholic acid into the potent secretory bile acid deoxycholic acid. However, chenodeoxycholic acid, with its moderate secretory activity, is converted to lithocholic acid, which is insoluble and precipitates from solution. It seems likely that the net effect of bacterial alteration in the colon is to reduce the concentration of bile acids in solution. This in turn, permits maximal colonic absorption of water.

Bile Acid Malabsorption. Bile acid malabsorption occurs when the capacity of the ileal transport mechanism is reduced because of ileal disease, resection or bypass or when bile acids are bound in the lumen so strongly that ileal uptake is impaired. At present, the only agents demonstrated to cause significant malabsorption of bile acids are cholestyramine and neomycin. Congenital absence of the ileal transport mechanism has not been observed.

Bile acid absorption is assessed directly by radioisotopic labeling of the bile acid pool and then determining the rate of isotope excretion in feces. Patients with bile acid malabsorption excrete 50 to 90 percent of the label in 24 hours, whereas healthy individuals excrete less than 20 percent. Since bile acid malabsorption almost invariably is associated with increased hepatic synthesis, bile acid malabsorption may be inferred when fecal bile acids are increased.

The consequences of bile acid malabsorption follow from principles already discussed: increased hepatic synthesis, reduced jejunal bile acid concentrations when increased hepatic synthesis cannot compensate for decreased return, and increased passage of bile acids into the colon.

In patients with limited ileal resection, a greatly increased hepatic synthesis fully compensates for decreased return, and fat digestion is unimpaired. Although these patients show little steatorrhea, a watery diarrhea results from the high concentration of bile acids in colonic contents. Administration of cholestyramine, which binds bile acids in the intestinal lumen, is an effective treatment for the diarrhea of these patients. In some patients, the increased passage of bile acids into the colon appears to inhibit bacteria carrying out 7α dehydroxylation, since deoxycholic acid may be absent from feces.

In patients with larger ileal resection, hepatic synthesis is unable to compensate for decreased hepatic return, and bile acid secretion is decreased. The concentration of bile acids in jejunal content is decreased, causing reduced concentrations of micellar lipid. Fat maldigestion, together with the decreased surface area, causes fat malabsorption. Such patients have severe steatorrhea as well as diarrhea. The unabsorbed fatty acids are converted by colonic bacteria to hydroxy fatty acids, which may induce colonic secretion of electrolytes and water. Bile acid malabsorption seems un-

important as a cause of diarrhea, since bile acid concentrations in colonic contents are low. Substitution of medium-chain triglyceride for long-chain triglyceride reduces steatorrhea, since medium chain fatty acids do not require bile acids for absorption. Diarrhea is ameliorated in addition, probably in relation to decreased formation of hydroxy fatty acids.

In patients with bile acid malabsorption, a marked increase in the synthesis of glycine-conjugated bile acids occurs, because in man, dietary taurine is rate-limiting for bile acid conjugation, whereas glycine for bile acid conjugation is readily synthesized via multiple pathways. This increase in the synthesis of glycine-conjugated bile acids is signaled by an increase in the glycine/taurine ratio of biliary bile acids from 3 (in health) to 10 to 20 in patients with ileal resection. Recent evidence suggests that glyoxylate (OHC-COOH) may be an important precursor of bile acid glycine under such circumstances, and that glyoxylate may then accumulate and be converted to oxalate, causing hyperoxaluria and, in some instances, the formation of oxalate kidney stones. Taurine administration reduces the synthesis of bile acid glycine and induces a striking fall in urinary oxalate excretion in this syndrome of bile acid malabsorption and hyperoxaluria.

MALABSORPTION AND DIARRHEA

Definitions

Malabsorption is a historic (but still useful) term that refers to less efficient absorption of a major nutritional component, owing to some pathological state. It is signaled by an increased quantity of a given substance or its breakdown products in the stool. However, increased fecal excretion of a given substance may not result solely from malabsorption of ingested materials, for fecal excretion of any substance represents the balance between total input (food, drink and endogenous secretion) and

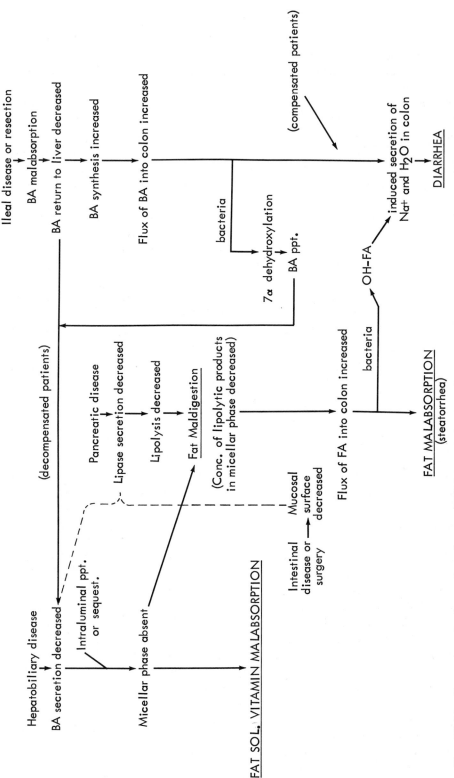

FIG. 21-10. Summary of relationships between abdominal disease, digestion, and absorption. Multiple causes of steatorrhea, diarrhea, or both are seen; these may or may not involve bacterial transformations. The sequence of events in patients with small ileal resection is described by the column on the right. The events in patients with large ileal resection (decompensated patients) are complex, since they may include decreased bile acid secretion, fat maldigestion, and bacterial production of hydroxy fatty acids. The dotted line indicates the decrease in enzyme and bile acid secretion observed in patients with decreased mucosal surface and consequent cholecystokinin deficiency, such as occurs in sprue.

absorption. Indeed, endogenous secretions of water and electrolytes constitute the major source of increased fecal output. In the case of bile acids, endogenous secretion is the only source of increased fecal output. The term *malabsorption syndrome*, although widely used clinically, is not a diagnosis and should be replaced whenever possible by the name of the specific disease (e.g., tropical sprue), accompanied by identification of the particular substance malabsorbed (e.g., carbohydrate, fat). *Diarrhea*, defined as increased fecal loss of water and electrolytes, is characterized by a stool weight greater than 200 g. per day. Since stools consist chiefly of water (80 to 90 percent), increased stool weight represents chiefly increased fecal water. To the patient, diarrhea means either bulky or watery stools and, usually, increased fecal frequency. Diarrhea may occur alone or in association with increased fecal loss of other nutrients such as fat (steatorrhea) or nitrogen (azotorrhea).

Mechanisms

Increased fecal losses occur when the absorptive load exceeds the absorptive capacity.

Increased Absorptive Load. Absorption overload, leading to osmotic diarrhea, occurs in patients given excessive amounts of tube feeding; the situation is analagous to the ingestion of an osmotic cathartic such as magnesium sulfate. More commonly, increased absorptive load is caused by an increased secretion of water and electrolytes from the stomach, small intestine or colon, as has been discussed.

Decreased Absorption. Decreased absorptive capacity may result from intraluminal digestive disturbances due to reduced enzymatic activity which may result from decreased exocrine secretion, lack of conversion of proenzyme to active enzyme, enzyme inhibition or denaturation or reduced concentrations of bile acid. The diminished absorption may be caused by

surface digestive disturbances that involve digestive enzymes on the luminal cell surface and may be produced by decreased enzyme activity per cell, decreased number of cells per unit area, or decreased intestinal length. The enzymes that carry out surface digestion are predominantly oligosaccharidases and oligopeptidases. Mucosal absorptive disturbances that affect uptake, transport through the cell, exit from the cell, and transport from the mucosa in lymph or blood constitute a third possible mechanism of decreased absorption. Examples are: a generalized reduction in permeability, as occurs in nontropical sprue; reduced transport of specific substances, as in glucose-galactose malabsorption or bile acid and B-12 malabsorption in ileal disease; loss of absorptive area because of intestinal resection; disturbances in biosynthetic processes such as those required for chylomicron formation; and disturbed exit from the cell or intracellular spaces because of obstruction to lymph or blood flow.

Systemic Effects

Malabsorption, as defined above, may have any of the following consequences: water and electrolyte depletion; caloric deficiency, causing inadequate weight gain in children and weight loss in adults; signs of deficiencies of specific vitamins and minerals. In addition, unabsorbed substances may interact with intestinal bacteria, and the products formed may influence intestinal function or they may be absorbed, causing metabolic disturbances. In some instances, unabsorbed substances may alter the bacterial flora or induce bacterial enzymes.

ANNOTATED REFERENCES

Books

Code, C. F. (ed.): Handbook of Physiology. Section 6, Alimentary Canal, Vol. III (Intestinal Absorption) and Vol. V (Bile Acid Absorption). Washington, American Physiological Society, 1968. (Authoritative sources of information on intestinal absorption up to the time of their publication)

Smyth, D. H. (ed.): Intestinal absorption. Brit. Med. Bull., 23:205, 1967. (A collection of lucidly written review papers representing the British schools of thought)

Ugolev, A. M.: Physiology and Pathology of Membrane Digestion. New York, Plenum Press, 1968. (A highly original comprehensive statement of a Russian author's concept of the role of membrane digestion; more recent investigations have confirmed the importance of surface digestion in carbohydrate and protein absorption but have cast doubt on the view that pancreatic enzymes are adsorbed to the mucosa and contribute significantly to digestion.)

Review Articles

Crane, R. K.: Intestinal absorption of sugars. Physiol. Rev., 50:789, 1960. (The beginning of the modern era of research in sugar absorption)

Danielsson, H.: Present status of research on catabolism and excretion of cholesterol. Advances, Lipid Res., 1:335, 1963. (Summary of factors influencing cholesterol and bile acid metabolism.)

Diamond, J. M., and Wright, E. M.: Biological membranes: the physical basis of ion and non-electrolytes selectivity. Ann. Rev. Physiol, 31:581, 1969. (The latter half of this brilliant interpretive review deals with factors influencing the rate of passive absorption of hydrophobic compounds.)

Dietschy, J. M.: Mechanisms for the intestinal absorption of bile acids. J. Lipid Res., 9: 297, 1968. (A lucid discussion of the mechanisms of bile acid absorption from the intestine. Whether absorption of intact micelles occurs, as proposed here, remains uncertain.)

Fordtran, J. S.: Speculations on the pathogenesis of diarrhea. Fed. Proc., 26:1405, 1967. (A superb analysis of the pathophysiology of various forms of diarrhea)

Fordtran, J. S., and Dietschy, J. M.: Water and electrolyte movement in the intestine. Gastroenterology, 50:263, 1966. (An imaginative effort to interpret water and salt transport on the basis of membrane physiological concepts)

Gray, G. M.: Carbohydrate digestion and absorption. Gastroenterology, 58:96, 1970.

(A first-rate review, mainly of carbohydrate digestion, which corrects some persistent errors in the current literature.)

Hofmann, A. F.: Physico-chemical approach to the intraluminal phase of fat absorption. Gastroenterology, 50:56, 1966. (A classification and discussion of disturbances in fat maldigestion based on disturbances in micelle formation.)

————: The syndrome of ileal disease and the broken enterohepatic circulation: choleretic enteropathy. Gastroenterology, 52:752, 1967. (Algebraic treatment of enterohepatic circulation: prediction of compensated and decompensated states in patients with bile acid malabsorption)

Hofmann, A. F., and Small, D. M.: Detergent properties of bile salts: correlation with physiological function. Ann. Rev. Med., 18: 433, 1967. (Condensed review of bile acid physical chemistry and physiology)

Mao, C. C., and Jacobson, E. D.: Intestinal absorption and blood flow. Am. J. Clin. Nutr., 23:820, 1970. (A lucid, brief review of the interrelationship between blood flow and absorption)

Matthews, D. M., and Laster, L.: Absorption of protein digestion products: a review. Gut, 6:411, 1965. (A review of amino acid and peptide absorption based mainly on observations made in vitro)

Schanker, L. S.: On the mechanism of absorption of drugs from the gastrointestinal tract. J. Med. Pharm. Chem., 2:343, 1960. (Relationship between state of ionization and absorption rates)

Scheline, R. R.: Drug metabolism by intestinal micro-organisms. J. Pharm. Sci., 57:2021, 1968. (A review of bacterial transformations of drugs in vitro and in vivo)

Schultz, S. G., and Curran, P. F.: Coupled transport of sodium and organic solutes. Physiol. Rev., 50:637, 1970. (A classical review of the evidence for the sodium-gradient hypothesis of carrier-mediated sugar and amino acid transport)

Senior, J. R.: Intestinal absorption of fats. J. Lipid Res., 5:495, 1964. (Imaginative review of a decade of progress)

Small, D. M.: A classification of biological lipids based upon their interaction in aqueous systems. J. Am. Oil Chem. Soc., 45:108, 1968. (A remarkable attempt to relate the behavior of lipids in water to their molecular structure and biological function)

Tabaqchali, S.: The pathophysical role of small intestinal bacterial flora. Quadrennial Review of the 4th World Congress of Gastroenterology (Supp. 6), p. 137, 1970. (Review of pathophysiology of consequences of bacterial proliferation in the small intestine)

Thier, S. O., and Alpers, D. H.: Disorders of intestinal transport of amino acids. Am. J. Dis. Child., 117:13, 1969. (Clinically oriented description of disorders of intestinal and renal tubular amino acid absorption)

Experimental Papers

Altmann, G. B., and Leblond, C. P.: Factors influencing villus size in the small intestine of adult rats as revealed by transposition of intestinal segments. Am. J. Anat., 127: 15, 1970. (A description of the intraluminal factors that regulate villus size and, thereby, the absorptive surface area. Observations in man, though less complete, are consistent with these findings.)

Arnesjo, B., Nilsson, A., Barrowman, F., and Borgstrom, B.: Intestinal digestion and absorption of cholesterol and lecithin in the human. Scand. J. Gastroent., 4:653, 1969. (Studies demonstrating that phospholipid is hydrolyzed to l-lysolecithin and absorbed in this form)

Borgstrom, B., Dahlqvist, A., Lundh, G., and Sjovall, J.: Studies of intestinal digestion and absorption in the human. J. Clin. Invest., 36:1521, 1957. (The first demonstration that digestion and absorption of a meal is almost completed in the jejunum)

Borgstrom, B., Lundh, G., and Hofmann, A. F.: The site of absorption of conjugated bile salts in man. Gastroenterology, 45:229, 1963. (Experimental evidence for ileal absorption of bile acids in man)

Curran, P. F., and MacIntosh, J. R.: A model system for biological water transport. Nature (London), 193:347, 1962. (A brief description of a hypothetical model, to account for the coupling of water to solute transport. This model has since received broad support from physiological and morphological studies, as well as from thermodynamic considerations.)

Danielsson, H., Eneroth, P., Hellstrom, K., Lindstedt, S., and Sjovall, J.: On the turnover and excretory products of cholic and chenodeoxycholic acid in man. J. Biol. Chem., 238:2299, 1963. (Studies in man, using ¹⁴C cholic and ³H chenodeoxycholic acid which define the origins of deoxycholic and lithocholic acid in feces)

Dillard, R. L.: Eastman, H., and Fordtran, J. S.: Volume-flow relationships during the transport of fluid through the human small intestine. Gastroenterology, 49:58, 1965. (Application of the dye-dilution method to the perfused small intestine which allows calculations of flow velocity and luminal radius)

Eneroth, P., Gordon, B., Ryhage, R., and Sjovall, J.: Identification of mono- and dihydroxy bile acids in human feces by gasliquid chromatography and mass spectrometry. J. Lipid Res., 7:511, 1966. (Identification of individual secondary bile acids)

Fordtran, J. S., Soergel, K. H., and Ingelfinger, F. J.: Intestinal absorption of D-xylose in man. New Eng. J. Med., 267:274, 1962. (The absorption of D-xylose from the human small intestine as related to mucosal contact time and distance from the ligament of Treitz)

Garbutt, J. T., Heaton, K. W., Lack, L., and Tyor, M. P.: Increased ratio of glycine to taurine conjugated bile salts in patients with ileal disorders. Gastroenterology, 56:711, 1965. (Increased glycine/taurine conjugated bile acid ratio in patients with bile acid malabsorption because of ileal disease)

Hofmann, A. F., and Borgstrom, B.: Physicochemical state of lipids in intestinal content during their digestion and absorption. Fed. Proc., 21:43, 1962. (First paper providing evidence for the formation of micellar phase during fat digestion)

Kopp, W. L., Trier, J. S., MacKenzie, I. L., and Donaldson, R. M., Jr.: Antibodies to intestinal microvillous membranes. I. Characterization and morphologic localization. J. Exp. Med., 128:357, 1968. (Conclusive demonstration of the acceptor for the intrinsic factor-vitamin B-12 complex in the glycocalix of the ileal mucosa)

Levitt, M. D., and Ingelfinger, F. J.: Hydrogen and methane production in man. Ann. N. Y. Acad. Sci., 150:75, 1968. (Methane and hydrogen gas are produced almost entirely in the colon.)

Lindstedt, S.: The turnover of cholic acid in man: bile acids and steroids 51. Acta physiol. scand., 40:1, 1957. (First description of bile acid kinetics in man by isotope dilution)

Love, A. H. C., Mitchell, T. G., and Phillips, R. A.: Water and sodium absorption in the human intestine. J. Physiol. (London), *195:* 133, 1968. (Maximum absorptive capacity in man for sodium and water during total intestinal perfusion)

Mattson, F. H., and Beck, L. W.: The specificity of pancreatic lipase for the primary hydroxyl groups of glycerides. J. Biol. Chem., *219:*735, 1956. (Demonstration of the positive specificity of pancreatic lipase)

Mattson, F. H., and Volpenhein, R. A.: The digestion and absorption of triglycerides. J. Biol. Chem., *239:*2772, 1964. (Studies in rats showing that triglyceride is absorbed mainly as fatty acid and 2-monoglyceride)

Shefer, S., Hauser, S., and Bekersky, I.: Feedback regulations of bile acid biosynthesis in the rat. J. Lipid Res., *10:*646, 1969. (Infused bile acids suppress hepatic synthesis of bile acids.)

Simmonds, W. J., Hofmann, A. F., and Theodor, E.: Absorption of cholesterol from a micellar solution: intestinal perfusion studies in man. J. Clin. Invest., *46:*874, 1967. (Perfusion studies showing that micellar monoglyceride is rapidly absorbed from the human jejunum, whereas taurocholate is not absorbed.)

Soergel, K. H.: Flow measurements of fasting contents in the human small intestine. *In:* Demling, L., and Ottenjann, R. (eds.): Motility of the Gastrointestinal Tract. p. 81. Stuttgart, Georg Thieme Verlag, 1971. (An analysis of velocity and rate of flow in the human small intestine during fasting and after a meal)

Soergel, K. H., Whalen, G. E., and Harris, J. A.: Passive movement of water and sodium across the human small intestinal mucosa. J. Appl. Physiol., *24:*40, 1968. (Observed differences in mucosal permeability to diffusion and osmotic (bulk) flow fit Curran's 3-compartment model, which contains 2 barriers of different "pore" size in series.)

Turnberg, L. A., Bieberdorf, F. A., Morawski, S. G., and Fordtran, J. S.: Interrelationships of chloride, bicarbonate, sodium, and hydrogen transport in the human ileum. J. Clin. Invest., *49:*557, 1970.

Turnberg, L. A., Fordtran, J. S., Carter, N. W., and Rector, F. C., Jr.: Mechanism of bicarbonate absorption and its relationship to sodium transport in the human jejunum. J. Clin. Invest., *49:*548, 1970. (Support of the hypothesis that Na^+ and Cl^- absorption in the small intestine occurs by exchange with actively secreted H^+ and HCO_3^- is presented in these 2 papers.)

22

Hepatic Mechanisms

William W. Faloon, M.D.

The normal liver is uniquely equipped to perform a variety of biochemical and physiological functions, many of which are still inadequately understood. Certain inborn errors affect a single function; in generalized liver disease several functions are lost simultaneously. One can best understand the physiological alterations that occur by considering the anatomical position of the liver, as well as the general types of function it usually serves.

Because the liver is situated astride the blood flow from the intestinal tract, hepatic cells are the first to modify material entering the body through the intestinal tract, with the exception of certain materials carried by lymphatics. Hepatic cells act not only on food, drugs and foreign compounds that are ingested but also on a number of endogenous materials formed in the intestinal tract itself (enterohepatic circulation). Since the liver receives blood both from the portal venous system draining the splanchnic area and from the hepatic artery, changes in blood flow of either type may alter hepatic function.

The diversified functions of the liver may be grouped into four general categories: uptake, or transfer of compounds from blood into liver cells; synthesis, metabolism and conjugation and, related to this, conversion of certain compounds and subsequent storage of derived or newly formed compounds (amino acids, carbohydrate, lipids, and vitamin); transfer across hepatic cell to the bile canaliculus, with excretion into the bile or transfer from the cell to the blood in the hepatic veins; and phagocytosis carried out by the Kupffer cells.

If one remembers that there is very little biochemistry or physiology in the body unrelated to or unaffected by hepatic function, the multiplicity of the signs and symptoms of liver disease is understandable. The patient with liver disease may not only have jaundice but may also have low prothrombin activity, deficiency of bile acids, decreased serum albumin with abnormal globulin fractions, disorders of sodium and potassium metabolism, and hypoglycemia.

In this chapter, emphasis is placed on the disorders and physical changes most commonly seen in patients. As a general principle, hepatic function is disturbed when blood flow is altered, when specific activities of the liver cell are reduced or destroyed, when chemical agents inhibit various metabolic functions or compete with the usual materials for metabolism, and when excretion is blocked either chemically or mechanically.

BILIRUBIN METABOLISM AND JAUNDICE

In the normal individual, bilirubin formation and removal proceed smoothly through the steps outlined in Figure 22-1. Approximately 85 percent of the bilirubin formed is derived from conversion of the heme of hemoglobin to bilirubin within the reticuloendothelial cells of the various organs, the remainder being derived from other

BILIRUBIN METABOLISM

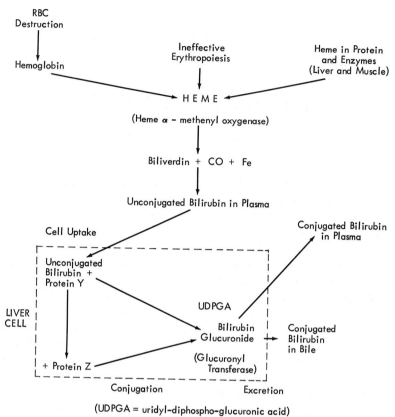

FIG. 22-1. Bilirubin metabolism

sources. This 15 percent, called early-labeled bilirubin, may account for some of the elevation in serum bilirubin observed in liver disease. Some of the early-labeled pigment appears to be associated with erythropoiesis; the remainder originates primarily in the liver itself. Bilirubin formation from heme, either of red cell or other origin, results in the release of carbon monoxide, measurement of which may reflect bilirubin formation. The occurrence of jaundice represents failure to remove or to excrete the bilirubin that is formed. Obviously, excessive bilirubin formation also can produce this clinical sign by overwhelming the clearing capacity. Furthermore, an inborn or induced error can occur at nearly all of the known steps in bilirubin metabolism, and many examples of such errors have been identified. These are usually classified according to the mechanisms involved, into five categories: (1) overproduction of bilirubin, (2) impaired hepatic uptake, (3) intracellular conjugation defects, (4) defects in bilirubin glucuronide excretion, and (5) hepatocellular disease.

Overproduction of Bilirubin. Increased destruction of red cells in certain diseases may yield more bilirubin than can be cleared from the circulation and metabolized by the liver. The occurrence of this phenomenon is not confined to hemolytic diseases; it is also occasionally observed when blood escapes from the circulation into the tissues, as in hematomas or infarcts. Resultant hemolysis and removal of heme provides more bilirubin than can be cleared, particularly if liver function is depressed.

In patients with liver disease or with ineffective red cell manufacture, bilirubin may be derived from materials that have not been contained in circulating red cells. This is sometimes called "shunt" hyperbilirubinemia and occurs in conditions such as thalassemia. The other chief source of bilirubin is its direct production from heme, as opposed to hemoglobin. Heme is contained in tissue proteins and in enzymes found in the liver itself.

When the binding of bilirubin by albumin in the serum is impaired, bilirubin may be deposited in certain tissues. This impairment assumes particular clinical importance in newborn infants, in whom a type of brain damage known as kernicterus may follow. The condition, which attends hemolytic disease in infants, is aggravated by certain drugs (e.g., vitamin K, sulfonamides) administered to the pregnant mother near term or to the newborn child. These drugs compete with unconjugated bilirubin for binding by albumin, and, when albumin binding is inadequate, bilirubin penetration of the brain is enhanced.

Impaired Hepatic Uptake. Ordinarily bilirubin is transferred from plasma to the liver cell quite rapidly. Although the mechanism has not yet been delineated, apparently it depends chiefly upon permeability of the plasma membrane and the presence of two non-albumin proteins which have been identified in liver cell cytoplasm and are designated Y and Z proteins. Both are basic proteins, but Y protein binds more bilirubin and bromosulfophthalein and (BSP) dye than does Z. The Z protein appears to bind these materials when their concentrations on Y attain critical levels. How these proteins contribute to the unconjugated hyperbilirubinemias in the human remains unclear. Among animals, the mutant Southdown sheep appears to have a defect in the transfer of a number of organic ions from plasma into the liver cell. This may be the animal analogue of a human disease not yet characterized. The failure of Y and Z proteins to develop may play a role in Gilbert's syndrome or in neonatal jaundice in humans. In other conditions, damage to these proteins by certain chemical agents, drugs, infection, or malnutrition may also account for elevation of indirect bilirubin.

Intracellular Conjugation Defects. Within the liver cell, bilirubin is prepared for excretion by conjugation with glucuronic acid. The responsible enzyme, glucuronyl transferase, catalyzes the transfer of glucuronic acid from uridyl-diphospho-glucuronic acid (UDPGA) to a number of compounds, including bilirubin. Glucose serves as the source of glucuronic acid, and glucuronyl transferase activity is greatest in the endoplasmic reticulum of the liver cell. Bilirubin glucuronide is water-soluble and can be excreted into the bile as well as excreted in the urine if it is returned to the blood in certain types of jaundice ("regurgitation"). Known as direct-reacting bilirubin, bilirubin glucuronide represents bilirubin that has been metabolized by the liver.

Inhibition of glucuronyl transferase is best typified by the syndrome seen in the breast-fed infants of mothers who excrete pregnane-3 alpha, 20-beta-diol in their milk. When breast-feeding is stopped, the hyperbilirubinemia disappears—usually within 3 to 12 days. A related syndrome appears to occur in overtly healthy mothers who have infants with severe nonhemolytic, unconjugated hyperbilirubinemia. The maternal serum contains an unidentified factor that inhibits glucuronyl transferase activity in vitro. The factor appears to reach its highest concentration in the mother before delivery and disappears after birth of the infant. This transient familial nenonatal hyperbilirubinemia has been called the Lucey-Driscoll syndrome and may be due to a steroid with properties resembling those of pregnane-3-alpha, 29-beta-diol. Administration of drugs such as novobiocin may inhibit transferase. A defect in the enzyme may occur in hypothyroidism.

Glucuronyl Transferase Deficiency. An animal model for deficiency of glucuronyl transferase has been found in the Gunn rat. Unconjugated bilirubin accumulates in its serum, bilirubin is absent from the urine, and the bile is virtually colorless. In the human, a defect in glucuronyl transferase resembling that in the Gunn rat is known as the Crigler-Najjar syndrome. In these infants, kernicterus frequently occurs and survival is rare in severe cases. Recent studies indicate that this syndrome may take two forms. In type I, the bilirubin elevation is greater and brain damage is frequent. The bile is almost colorless, containing only traces of unconjugated bilirubin. Phenobarbital, which is able to induce increased glucuronyl transferase activity, has no effect. In type II, affected infants have less severe hyperbilirubinemia (8 to 22 mg./100 ml.), show no kernicterus, and have pigmented bile containing bilirubin glucuronide. When phenobarbital is administered to these patients, jaundice often disappears completely in 5 to 12 days.

The development of glucuronyl transferase may be delayed in the newborn infant. As a result, jaundice due to unconjugated hyperbilirubinemia is common in the first few days of life, after which it subsides. Possibly, this syndrome is the result, not of transferase deficiency alone but also of enhanced bilirubin production resulting from the reduced life span of red cells in the newborn. In addition, steroids of fetal or placental origin or a reduced transfer of bilirubin into the hepatic cell may be responsible for this "physiological" jaundice.

Unconjugated hyperbilirubinemia is also seen in Gilbert's syndrome, which may have many causes. Both glucuronide formation in vivo and liver transferase activity in vitro are usually normal, and impaired transfer of bilirubin from plasma into hepatic cell has been suggested. Other mechanisms that have not yet been proved or eliminated include compensated hemolysis, increased production of non-red-cell heme and, perhaps, continuing damage from viral hepatitis. Jaundice in Gilbert's syndrome is usually accentuated by fasting, exercise, fever, pregnancy, use of oral contraceptives and other drugs and alcohol consumption.

Defects in Bilirubin Glucuronide Excretion. After conjugation with glucuronic acid, bilirubin is normally excreted into the bile quite rapidly. An active transport system is believed to be responsible, and a large functional reserve is probably present. Thus, severe hemolysis in the presence of normal liver function is known to be associated with unconjugated hyperbilirubinemia, whereas hemolysis in patients with coexisting liver damage is associated with significant conjugated hyperbilirubinemia.

The human disorder that appears to typify hepatic excretory dysfunction is the Dubin-Johnson syndrome. This inherited abnormality is characterized by a reduced capacity to excrete certain organic anions, including BSP, iodopanoic acid and bilirubin; the result is a hyperbilirubinemia of conjugated bilirubin. In the liver cells of these patients, a black-brown pigment accumulates in the hepatic lysosome. The pigment is a compound derived from epinephrine metabolites whose biliary excretion is also impaired. Evidence of cholestasis or obstruction to the flow of bile is absent in such individuals and plasma bile salt concentrations are normal, suggesting a specific defect in bilirubin excretion dissociated from bile salt excretion.

Certain drugs, such as the C-17-alkylated anabolic steroids and oral contraceptive agents, reduce the capacity of the liver to excrete certain anions and may aggravate an existing reduction in excretory capacity for bilirubin. This may explain the jaundice of the third trimester of pregnancy in some women who also develop jaundice when given oral contraceptive agents.

Drug-induced cholestasis appears to differ from the pure excretory defect for bilirubin; i.e., in patients with cholestasis, bile salt formation or secretion is altered and serum bile acid elevations are common.

Hepatocellular Disease. Viral hepatitis and cirrhosis are the most commonly observed causes of jaundice. In a given individual, a defect in uptake or conjugation may be shown to predominate, as opposed to a defect in hepatic cell bilirubin excretion. Both conjugated and unconjugated bilirubin is found in the serum, and the most reasonable view of the jaundice in parenchymal cell disease is that all of the steps involved may contribute. Accordingly, there may be (a) increased hemolysis or increased bilirubin synthesis from non-red-cell sources, (b) impaired hepatic uptake of bilirubin from plasma as a result of membrane or protein damage, and (c) reduced glucuronyl transferase activity within the diseased liver cell or impaired excretion of conjugated bilirubin into the bile. Added to these is the possibility that the inflammatory process may mechanically obstruct the flow of bile in the canaliculi.

Reasonably, one should expect that obstruction from common duct stones or tumors of the biliary tract would be the conditions in which elevation of conjugated (direct) bilirubin alone would occur. In actual practice, although direct bilirubin frequently predominates, both types of bilirubin are usually elevated. One cannot rely upon the relative elevations of "indirect" (unconjugated) bilirubin and "direct" (conjugated) bilirubin in the patient with common duct stone, biliary tract tumor, hepatitis or cirrhosis. Regurgitation of conjugated bilirubin from the obstructed biliary radicle or from the liver cell would explain the increase in conjugated serum bilirubin. An effect of obstruction upon the intracellular metabolism of bilirubin or even upon its uptake by the liver cell must be postulated, however, to explain the elevation of unconjugated bilirubin.

Thus far, therapeutic management of jaundice has revealed little about the physiology of hyperbilirubinemia. Relief of the obstruction restores the level of both types of bilirubin to normal and at approximately the same rate. Similarly, abstinence from alcohol yields no predominant change in one fraction of bilirubin over the other. The inborn errors and their treatment appear to offer the best clues to the understanding of jaundice. For example, phenobarbital, known to elicit an increase in enzyme activity, reduces serum bilirubin levels in certain infants with Crigler-Najjar syndrome and in other newborn infants who develop physiological neonatal hyperbilirubinemia. Our expanded knowledge of hemolysis and bilirubin transport has allowed rational therapy of hyperbilirubinemia in newborn infants. However, the lessons learned from drug-induced hyperbilirubinemia have shed little light on the physiological disturbances in bilirubin metabolism, particularly those seen in the common types of liver disease—cirrhosis and hepatitis.

PORTAL HYPERTENSION

A variety of mechanisms can produce increased pressure in the portal veins. The most common is intrahepatic fibrosis associated with cirrhosis, both alcoholic or "postnecrotic" in type. The pressure often rises to four or five times normal, with complications that may be life-threatening. Portal hypertension may be transient in patients who have "alcoholic hepatitis" associated with extreme fatty infiltration and swelling of the liver. Increased pressure may result from multiple thrombi of the intrahepatic venules, with minimal alteration of liver function, or from thrombosis of the extrahepatic portion of the portal vein as a result of trauma, infection or cavernous transformation. Either primary or metastatic neoplastic tumor in the liver may compress the blood vessels and raise the portal vein pressures. A rare condition associated with portal hypertension, the Budd-Chiari syndrome, consists of thrombosis of both the inferior vena cava and the hepatic vein above the liver. This phenomenon is now also reported occasionally in patients using contraceptive medication. In tropical countries, schistosomiasis is the leading cause of portal hypertension due to fibrosis of the intrahepatic vascular system. In some tropical areas veno-occlusive disease of the liver is common and is suspected to be caused by certain alkaloids present in native "bush tea." Possibly other toxins produce similar hepatic vascular changes. Occasionally, portal hypertension occurs in patients with arteriovenous fistulas, causing increased flow of blood into the portal system and consequent hypertension.

Portal hypertension leads to vascular changes, because it is necessary for blood in the splanchnic area to find a new route to the systemic circulation. Various collateral channels are opened, the most common being the abdominal and periumbilical veins (easily visible on the abdominal wall), esophageal varices and the hemorrhoidal vessels in the rectum. In addition, the spleen is often enlarged, owing to portal obstruction and increased splenic pressure. Portal pressure may be accurately reflected by wedged hepatic vein pressures measured through a venous catheter.

When the hepatic blood flow in portal hypertensive states decreases, the hepatic extraction or uptake of certain constituents of portal blood likewise decreases. These compounds then reach the systemic circulation in abnormally high concentration and may be deleterious. The best example is ammonia, which is formed in the colon and

is ordinarily removed by the liver to be used in urea synthesis. However, when diverted into the systemic circulation by way of the collateral vessels, ammonia may produce neurological changes (to be discussed more fully under hepatic encephalopathy).

Similarly, metabolism of carbohydrates and certain fatty acids may be disturbed in patients with cirrhosis and portal hypertension. In cirrhosis, above-normal concentrations of carbohydrates and fatty acid are found in peripheral veins after oral loads. The same may be true of compounds that undergo enterohepatic circulation, such as bile salts and urobilinogen. When collateral circulation circumvents a cirrhotic liver and the organ cannot remove endogenous or exogenous compounds, even when they are delivered via the hepatic artery, a number of chemical and physiological changes may take place. For example, high levels of circulating estrogen may lead to gynecomastia; circulating vasodilator compounds may produce vascular "spiders" and palmar erythema and elevated bile salt levels in the serum may produce itching.

The presence of portal hypertension may also contribute to the localization of the excessive extravascular fluid in patients with cirrhosis. Excess fluid is deposited at the site of the highest venous pressure (and lowest tissue pressure)—usually the portal area in such patients. This will be discussed in greater detail under the pathological physiology of ascites formation. In cirrhosis, not only is the flow of blood through the liver obstructed, but the flow of lymph may also be decreased, thus necessitating an increase in thoracic duct lymph flow.

The propensity of large veins in the esophagus to bleed under increased pressure is a major hazard for the patient with cirrhosis. The mechanism that initiates bleeding from these varices is not well understood. In some patients, the reflux of acid from the stomach into the esophagus, which erodes the mucosa over the vessels, may be the cause. However, bleeding also occurs in some achlorhydric patients.

PATHOLOGICAL PHYSIOLOGY OF ASCITES FORMATION

In the patient with severe liver disease, fluid frequently accumulates in the abdominal cavity; ankle edema may or may not be present. Because low serum albumin concentrations and increased portal venous pressure are associated with such ascites formation, the traditional view has been that fluid accumulation is initiated by one or both of these factors. This concept of ascites formation, outlined in Figure 22-2, is often referred to as "outflow block," since venous drainage is reduced from the liver. The obstruction to flow is followed by engorgement of the lymphatic spaces within the liver. The splanchnic blood volume is also thought to be increased, owing to increased portal venous pressure. Fluid may then leak from the liver capsule and from the lymphatics and capillaries of the splanchnic area into the peritoneal cavity, forming ascites. When the lymphatics leading out of the peritoneal cavity are able to drain this fluid, ascites may not become clinically apparent. If increased inflow or decreased outflow of fluid occurs, abdominal swelling may appear.

Renal excretion of sodium and water is consistently reduced in patients with ascites. In the traditional view, trapping of blood in the splanchnic space reduces the so-called "effective" plasma volume, thus stimulating the renin-angiotensin-aldosterone mechanism. Enhanced aldosterone levels may then produce marked sodium retention by the renal tubule.

The sequence of events described above leaves certain clinical observations unexplained. Numerous patients with elevated

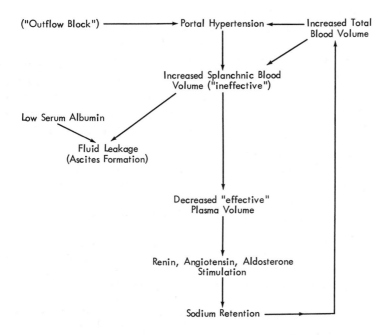

"CLASSICAL" VIEW OF ASCITES FORMATION

FIG. 22-2. "Classical" view of ascites formation.

portal venous pressures have no ascites. In one study, venous pressure in a group of patients without ascites did not differ significantly from that in a similar group who had ascites when portal pressures were measured at the time of portal shunt operation. Similarly, when portacaval shunt is created surgically, with marked reduction in portal venous pressure, fluid retention apparently is relieved only rarely. In some patients, the ascites accumulates for the first time after surgery. In others, the ascitic fluid disappears but ankle edema becomes marked postoperatively. Ability to excrete sodium in the urine is not regularly enhanced after portacaval shunt even when salt-poor human albumin is given to raise the serum albumin concentration. The effect that a portacaval shunt, with relief of the portal hypertension may have upon renin values, has not yet been studied definitively. In a few patients, creation of a shunt reduces urinary aldosterone excretion; in others, aldosterone levels fail to return to normal. In regard to those patients in whom aldosterone levels decrease and urinary sodium excretion is enhanced, it is unclear whether these phenomena are coincidental and related primarily to improvement in liver function or whether they are related directly to the surgical shunt.

It is noteworthy that, after portacaval shunt, the plasma volume in cirrhotic patients does not return to normal, thus suggesting that factors other than the portal hypertension increase plasma volume.

The role of hypoalbuminemia as the "trigger" mechanism in fluid retention is also difficult to demonstrate convincingly. The concentration of serum albumin does not appear to correlate with the presence or absence of ascites. Total body albumin may be normal or increased in some individuals with ascites, and the lowered serum concentration may be due partly to dilution of albumin by the increased extracellular fluid and plasma volume. Human serum albumin given intravenously to maintain serum albumin levels at or above normal produces little or no change in ascites. Diuresis of ascites may occur spontaneously with long-term treatment, frequently without a significant increase in serum albumin concentration. When diuretic drugs induce loss of ascites and edema, the concentration of serum albumin frequently rises. This suggests that loss of the excessive extracellular fluid permits concentration of albumin in the remaining fluid volume, with a return to normal values in the plasma.

Loss of ascites and edema is regularly related to increased renal excretion of sodium and water. Although this may occur spontaneously or consequent to therapeutic measures, it is worth noting that adrenalectomy and the administration of compounds such as amphenone, which inhibit the synthesis of aldosterone, also result in sodium excretion and the disappearance of ascites. These findings emphasize the importance of sodium and water retention in ascites formation.

An Alternate Mechanism of Ascites Formation

The above observations lead to consideration of another mechanism of fluid retention in cirrhosis. Certain fluid-retention characteristics of cirrhotic patients with ascites are given in Table 22-1. The

TABLE 22-1.   Characteristics of the Cirrhotic Patient with Ascites

Marked renal sodium retention
Increased plasma volume
Increased extracellular fluid volume
Decreased salivary and sweat sodium
Increased aldosterone secretion and urinary excretion
Increased plasma renin
Normal or decreased renal plasma flow and glomerular filtration
Decreased diuresis after water-loading

fact that reversal of the sodium retention is associated with loss of ascites, even without change in portal pressure may be a crucial factor in ascites formation. The association of sodium retention with increased plasma renin and increased aldosterone secretion suggests that these events occur without feedback suppression of renin. Ordinarily, the excessive aldosterone and sodium retention would shut down renin production. The question of what stimulates renin production in the cirrhotic patient remains unanswered.

The levels of plasma renin ordinarily rise after: (a) a decrease in plasma volume; (b) hypotension; (c) sodium depletion; and (d) changes in renal hemodynamics related to intrarenal blood flow. In the cirrhotic patient, blood volume remains high even after a portacaval shunt, persistent hypotension is absent (although hypertension is rare in cirrhotic patients), and the total body sodium is increased. Thus, a change in renal hemodynamics is the most likely possible stimulus for renin production in cirrhosis. The extreme hemodynamic instability of the cirrhotic patient has been documented and is characterized by a decrease in mean renal blood flow and a reduction in perfusion of the renal cortex. The abnormalities were most marked in patients with the most severe renal failure associated with cirrhosis. Other studies confirmed the elevation of plasma renin in patients with sodium retention, which was found to be particularly high in cirrhotic patients with evidence of renal failure. A surprisingly low plasma renin substrate (angiotensinogen) was found in some patients, but there was no evidence of consistent hepatic failure to remove renin from blood.

POSTULATED MECHANISM FOR ASCITES FORMATION

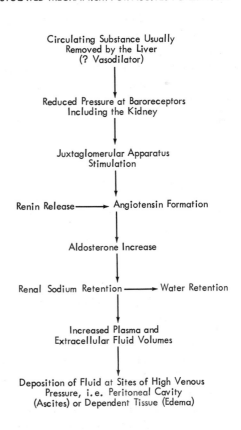

FIG. 22-3. Postulated mechanism for ascites formation.

Circulating Substance Usually
Removed by the Liver
(? Vasodilator)

Reduced Pressure at Baroreceptors
Including the Kidney

Juxtaglomerular Apparatus
Stimulation

Renin Release ⟶ Angiotensin Formation

Aldosterone Increase

Renal Sodium Retention ⟶ Water Retention

Increased Plasma and
Extracellular Fluid Volumes

Deposition of Fluid at Sites of High Venous
Pressure, i.e. Peritoneal Cavity
(Ascites) or Dependent Tissue (Edema)

On the basis of these many observations, which suggest an abnormality in the control of the renin-angiotensin-aldosterone system, it seems reasonable to suspect the existence of a humoral substance that may stimulate renin production either by a systemic effect or by specific stimulation of the juxtaglomerular apparatus.

A mechanism for ascites formation, incorporating these observations, has been proposed (Fig. 22-3). According to this concept, a vasodilator substance may exist that would stimulate baroreceptors related to the juxtaglomerular apparatus, resulting in renin release. The existence of such a substance should not be surprising, if one reviews the evidence of vasodilatation in patients with severe cirrhosis, i.e., increased cardiac output with decreased peripheral resistance, peripheral vascular changes such as erythema of the palms, spider angiomata, warm skin even when the blood pressure is low, portal-to-pulmonary-vein shunting, and pulmonary arteriovenous shunts. In addition, the cirrhotic patient has a reduced response to vasoconstrictor agents such as angiotensin and norepinephrine. In early studies, a vasodepressor material (VDM) had been demonstrated in patients with severe liver disease. This material was later believed to be ferritin. More recently, a peripheral and splanchnic vasodilator peptide derived from normal intestinal tissue has been described. It is reasonable to believe that this type of compound might participate in stimulating the release of renin and, thus, in initiating the events that lead to fluid retention.

PATHOLOGICAL PHYSIOLOGY OF RENAL FAILURE IN SEVERE LIVER DISEASE

In severe liver disease, a type of renal failure frequently develops which is not associated with specific pathological changes in the kidney. This condition, which some call "renal circulatory failure,"

is almost always accompanied by refractory ascites. Coma, bleeding and jaundice may or may not be present. Hyponatremia, hyperkalemia, hypochloremia, elevated blood urea and elevated serum creatinine are found consistently, in addition to severely impaired liver function. A progressive decrease in urinary output and a falling blood pressure that responds only transiently to vasoconstrictor agents such as norepinephrine or metaraminol characterize the course of the syndrome. Generally, the specific gravity of the urine consistently exceeds 1.010, and abnormal cellular elements are present in the urine. Maintenance of kidney function is also evident in the marked retention of sodium, since a nonfunctioning tubule would fail to reabsorb sodium. The osmolarity of the urine may be higher than that of the plasma, water is poorly excreted, and both the glomerular filtration rate and the estimated renal plasma flow are low. The extraction ratio for para-aminohippurate (PAH) usually remains normal, and the kidney shows no specific lesion histologically. Thus, although such patients may die of renal failure, their kidneys can be transplanted successfully into compatible recipients, with subsequent normal function.

Although occasionally patients recover from renal circulatory failure in severe liver disease, prognosis, on the whole, is very poor. Various measures, including infusions of albumin, plasma or ascitic fluid and intravenous administration of hypertonic glucose or manitol, have been employed in an attempt to improve urinary flow but without success. The available data indicate that renal vascular resistance is markedly increased, with poor perfusion of the cortical portions of the kidney. After portacaval shunt in a patient with apparent terminal renal failure, it was reported that renal function improved, pressure returned to normal, and both aldosterone excretion and plasma renin declined. Evidence in regard to a possible relationship between the portacaval shunt and the improvement in renal function must be considered inconclusive until more patients have been studied.

Although this syndrome usually occurs spontaneously or without known association with therapeutic measures, in some instances renal failure is precipitated by hemorrhage or by the vigorous use of diuretic agents. Renal failure in advanced liver disease appears to be one end of a spectrum, perhaps representing an extreme instance of the same abnormality that initiates the accumulation of ascitic fluid and edema. In a study of 30 patients with ascites, 19 of whom had impaired renal function with creatinine clearance below 80 ml. per minute, peripheral venous renin was highest in patients with marked ascites; an inverse correlation was found between plasma renin and creatinine clearance. Patients with little or no ascites had low plasma renin activities and normal creatinine clearance. Infusion of dopamine increased the para-aminohippurate clearance and, in 10 of 13 patients, renin activity decreased. These studies also suggest that a humoral substance may alter blood flow in the kidney and stimulate the production of angiotensin.

PATHOLOGICAL PHYSIOLOGY OF HEPATIC ENCEPHALOPATHY

Patients with severe liver disease often develop a syndrome of confusion and neurological changes, progressing to stupor and coma. Although several factors appear to play a role, almost all such episodes result from excessive amounts of ammonia reaching the brain. The chemistry of these neurological changes is poorly understood but is probably related to interference with energy metabolism in the brain. However, the mechanisms by which ammonia is formed and reaches the brain are now well delineated.

In human beings the major site of ammonia formation is the colon, where bacterial urease produces ammonia and carbon dioxide from urea. (Ammonia as used herein refers to ammonium-nitrogen measured in blood.) Other nitrogenous materials, such as amino acids, are also potential sources of ammonia. In addition, some ammonia is derived from the kidney, but, except in conditions of potassium depletion, the amount is relatively small compared to that produced in the intestine. When ammonia is absorbed, it is conveyed by the portal vein blood to the liver, where it is converted to urea in successive steps involving ornithine, citrulline and arginine. In general, urea is a nontoxic compound, chiefly excreted by the kidney, but it also serves as the main source of ammonia in the intestinal tract. This continuing cycle is harmless except when liver disease is severe, when marked shunting of the blood around the liver occurs, or when urea concentrations are increased in the blood and extracellular fluid. In addition, an inborn error (e.g., ornithine transcarbamylase deficiency) may lead to excessive ammonia in the blood after birth. Such infants develop ammonia intoxication but have histologically normal livers.

In the patient with cirrhosis or severe hepatitis, the mechanism for ammonia removal is inadequate, and, thus, ammonia reaches the brain in increased concentration. The effect upon cerebral metabolism is reflected in the syndrome known as hepatic encephalopathy (Fig. 22-4).

Although other metabolic alterations leading to hepatic encephalopathy have been suggested, only ammonia intoxication has the necessary biochemical and clinical correlation for wide acceptance. Other factors that have been suggested as coma-producing agents include certain short-chain fatty acids, indole or skatole compounds, certain amino acids and the occurrence of hypoglycemia. Precipitation of encephalopathy has also been associated with the injudicious use of certain analgesic or sedative drugs.

Ammonia intoxication is usually associ-

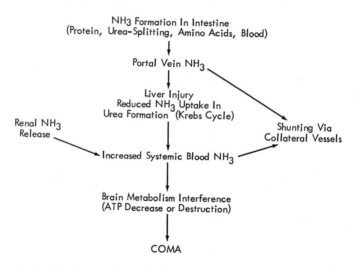

MECHANISM OF AMMONIA

RELATED HEPATIC COMA

NH3 Formation In Intestine
(Protein, Urea–Splitting, Amino Acids, Blood)

Portal Vein NH₃

Liver Injury
Reduced NH₃ Uptake In
Urea Formation (Krebs Cycle)

Renal NH₃
Release

Shunting Via
Collateral Vessels

Increased Systemic Blood NH₃

Brain Metabolism Interference
(ATP Decrease or Destruction)

COMA

FIG. 22-4. Mechanism of ammonia-related hepatic coma.

ated with elevated levels of arterial ammonia, the correlation being less definite with venous ammonia. Ammonia levels rise with oral administration of ammonia salts, urea, amino acids, protein, whole blood or red cells. Certain clinical changes may also increase the ammonia load. These include potassium deficiency, administration of diuretic agents, the presence of bacteria in the small or the large intestine, ureterocolic anastomosis and even vigorous exercise. Obviously, patients who have increased shunting of blood around the liver, such as those with surgically produced portavacal anastomoses, are exposed to increased risk of hepatic coma. In some individuals, spontaneously occurring shunts produce the same effect.

Unfortunately, no specific treatment is yet available to counteract the effect of ammonia upon the central nervous system. Therefore, reversibility of hepatic encephalopathy depends upon maintaining life until the injured liver can remove the ammonia present in the blood and tissues. The methods of therapy now available are

designed to reduce or eliminate the ammonia presented to the liver. For example, reducing the intake of protein or ammonia-containing compounds reduces the ammonia load that the liver must remove. Even more marked is the salutary effect of poorly absorbed antibiotics, such as neomycin, which reduce the bacterial flora, particularly in the colon. These agents depress ammonia formation, and this is consistently followed by a reduction in blood ammonia levels and reversal of the encephalopathy when the liver disease is otherwise stable. Potassium replacement in the potassium depleted patient reduces ammonia production by the kidney; blood ammonia levels may be reduced in some patients when potassium replacement alone has been carried out.

The recurrence of chronic encephalopathy in certain individuals has stimulated the search for agents that could be used instead of or in addition to neomycin. Although production of antibodies to urease has been reported, this has not been found to be clinically useful. Acetohydroxamate compounds have been employed to block urease activity but clinical experience has been limited. Recently, lactulose, a disaccharide that is not normally split by human intestinal mucosal enzymes but is split by intestinal bacteria, has been employed. Lactulose lowers the pH in the colon, so that ammonia formed therein tends to remain and the ammonia from blood and extracellular fluid may shift back into the colon. Surgical measures intended to shunt intestinal contents around the area of high bacterial concentration in the colon (colon bypass) have been employed in certain individuals with chronic hepatic encephalopathy, but this entails a significant risk from the surgical procedure.

More drastic measures have been used to remove ammonia and other toxic substances in patients with fulminating hepatic necrosis, such as that occasionally occurring in infectious or drug-induced hepatitis. These measures include exchange transfusion, circulation of blood through invitro pig liver preparations, and cross-circulation between the patient and baboons or human volunteers. When effective, all of these measures remove ammonia in large quantities, affording at least temporary clinical improvement.

PATHOLOGICAL PHYSIOLOGY OF CHOLESTEROL AND BILE ACID ABNORMALITIES IN LIVER DISEASE

In the normal individual, cholesterol is derived from food and from synthesis in the liver and intestine. The greater part of cholesterol is converted into bile acids, which are excreted into the biliary tract to be employed in digestive and absorptive processes and subsequently reabsorbed in the terminal ileum. (The role of bile acids in intestinal absorption is discussed elsewhere in this section.) Although errors in the metabolism of hepatic cholesterol and bile acid are observed in several diseases,

they are most prominent in two conditions: biliary cirrhosis and gallstone disease. In the syndrome known as primary biliary cirrhosis, no abnormality of the bile duct is recognizable, whereas secondary biliary cirrhosis is attended by long-standing obstruction due to gallstones, tumor, or stricture of bile ducts. Similar chemical changes are observed in both situations.

In primary biliary cirrhosis a distinctive pattern of changes in lipid and cholesterol metabolism is usually found. The serum lipids rise markedly, with totals sometimes exceeding 2,000 mg. percent, and the serum cholesterol concentration generally exceeds 500 mg. percent. When the cholesterol levels are elevated for several months, xanthomata and xanthelasma of the skin frequently occur. In some individuals, similar cholesterol deposits in the coronary arteries may lead to clinically evident heart disease. Despite the marked elevations in lipids, the serum is usually clear rather than milky. As the patients develop more marked liver damage and various nutritional changes occur, the serum lipids and cholesterol fall toward normal. Although this change may represent liver failure, it is sometimes erroneously interpreted as a response to treatment.

A striking feature in some patients with biliary cirrhosis is severe itching of the skin, associated with elevated concentrations of the bile acids in serum and skin. Surgical drainage of the bile to the outside, thus preventing reabsorption and recycling of the bile acids, lowers the concentrations of serum lipid, cholesterol and bile acid. The use of cholestyramine, a resin that binds bile acids, produces a similar effect by preventing their reabsorption. Unfortunately, neither of these procedures alleviates the basic disturbance in primary biliary cirrhosis, which appears to be caused by failure of normal bile acid excreation. The site of this metabolic error remains unclear, and there are presently no reliable chemical studies of biliary cholesterol, lecithin, and bile salts in primary biliary cirrhosis.

Another error in cholesterol and bile acid metabolism may result in gallstone formation. In order to maintain solubility of cholesterol in bile, bile salts and lecithin must be present in adequate concentration. In patients with cholesterol stones, this relationship appears to be disturbed. This may allow precipitation of cholesterol, the chief constituent of gallstones in most humans. Appropriate studies have been performed on normal subjects, on Caucasian patients with gallstones, and on Indians in the Southwest United States, in whom the incidence of gallstone formation is high. In Indian women who had not yet formed stones, as well as in Caucasian patients with stones, the ratio of bile acids to cholesterol in gallbladder and hepatic bile was consistently reduced. The data showed concentrations that, plotted by the triangular coordinate technique, fell outside the range necessary for micelle formation. Thus, hepatic bile in these patients was supersaturated with cholesterol. Gallbladder bile in patients who already have formed gallstones appears to have two phases: a solid phase, composed of cholesterol crystals, and a liquid phase, of bile saturated with cholesterol. Additional studies of the total bile acid pool size (the amount of bile acid circulating in the liver, the biliary tree and the intestine, plus a small amount in the blood) reveal a reduced pool in patients with gallstones. Patients with gallstones produce less cholic acid than do normal individuals. The cholic acid half-life, although somewhat lower in the patients with gallstones than in those without, is not significantly different. Studies of the individual bile acids in the bile of normal subjects and patients with gallstones revealed no significant difference in the ratio of trihydroxy to dihydroxy bile acids (cholic to deoxycholic and chenodeoxycholic). Although the incident that causes precipitation of cholesterol from the supersaturated solution is unknown, these studies suggest that, in certain patients, gallstones result from an error in liver function, with failure to maintain a normal bile salt pool. The frequent appearance of gallstones in several members of the same family suggests that an inheritable trait may be responsible in such instances.

As one might expect, bile acid metabolism is disturbed in patients with cirrhosis or hepatitis, owing to interference with production, uptake from the blood, conjugation with taurine or glycine, and re-excretion into the bile. In jaundiced patients serum bile acid levels commonly are elevated. In contrast to patients with obstructive jaundice, those with parenchymal disease may have more dihydroxy bile acid (chenodeoxycholic) than trihydroxy acid (cholic). A decrease in deoxycholic and lithocholic acid in bile has been observed in cirrhotic patients.

DISTURBANCES IN METABOLISM OF CARBOHYDRATE, PROTEIN, FAT AND OTHER NUTRIENTS

Carbohydrates. The liver plays a central role in carbohydrate metabolism. Disturbances of glucose regulation, although relatively common, appear in such mild forms that no clinical effects are apparent in most patients. In general, the patient with cirrhosis has difficulty storing administered glucose as glycogen, and does so relatively slowly. As a result, the glucose tolerance curve is frequently higher than normal and is accompanied by elevated serum insulin levels after administration of a standard glucose load. Hypoglycemia may result from poor mobilization of glucose; tolbutamide injected intravenously causes a greater fall in blood sugar in cirrhotic patients than in normal subjects while the level of plasma insulin rises higher in the former. In this respect, patients with cirrhosis resemble patients with diabetes mellitus of the adult-onset type. For these individuals, existence of hepatogenous diabetes has been suggested. In acute fulminating liver destruction, hypoglycemia may occur, but it is usually masked by the routine administration of intravenous glucose. Hypoglycemia may be related to failure of the liver to destroy insulin as well as to deficient glycogen storage. In patients who have had portacaval anastomosis, the shunting of blood around the liver may also reduce inactivation of insulin.

Certain primary disorders of carbohydrate metabolism may be associated with the development of cirrhosis. Type IV glycogenesis (Anderson's disease) is a rarely found deficiency of the branching enzyme amylo-1,4,1,6-transglucosidase. It produces an abnormal glycogen, and cirrhosis may occur. Galactosemia due to deficiency of galactose 1-phosphate uridyl transferase produces a type of cirrhosis usually found in young individuals. Hereditary fructose intolerance is due primarily to a deficiency of hepatic fructose 1-phosphate aldolase; hypoglycemia follows administration of fructose. Cirrhosis and liver failure become evident subsequently. Other, even more rare, alterations in carbohydrate metabolism are associated with disease of the liver. With the exceptions noted above, most of these errors result in hypoglycemia.

Protein. Although protein deficiency is generally believed to be important in the development of liver disease, errors in protein metabolism resulting from established hepatic disorders are relatively few. As noted above, nitrogen metabolism plays a significant role in the development of hepatic coma. In addition, patients with liver disease may lose excessive amino acid in the urine, and their amino acid patterns, in both serum and urine, may be abnormal.

As the liver disease improves, these changes are usually reversed. In the patient with liver disease, protein digestion and absorption are normal, and the protein requirement for maintenance of nitrogen equilibrium does not differ significantly from normal.

Fat. As in the case of carbohydrate and protein, the liver is responsible for several steps in the metabolism of fat. The liver receives lipids chiefly in the form of triglycerides and chylomicra and normally takes up fat as fatty acids, which may be: (a) used for energy; (b) reused in forming triglycerides which may be stored or released as lipoprotein into the circulation; (c) converted to phospholipids; or (d) used to form cholesterol esters. An additional function is the synthesis of liver lipid from smaller carbon compounds. As one should expect, this function of the liver can be disturbed by a wide variety of pathological situations, including inborn errors, chemical injury with alcohol or carbon tetrachloride, nutritional aberrations (starvation or obesity), diabetes, thyrotoxicosis and viral infections. In most of these conditions, the liver responds by developing or retaining more than the normal amount of fat; in some cases (but not all), this is accompanied by increased release of lipids. Thus, the alcoholic patients may have hyperlipemia and a fatty liver. In advanced portal cirrhosis, serum cholesterol is low and esterification of cholesterol is diminished, whereas, in biliary cirrhosis, hypercholesterolemia and hyperlipidemia are the rule. Changes in lipoprotein fractions, as studied by electrophoretic separation, may sometimes characterize portal cirrhosis (decreased pre-beta-lipoprotein) or obstructive jaundice (increased beta and alpha fractions). Our inability to use lipid changes as diagnostic clues reflects the numerous lipid metabolic functions of the liver as well as the technical difficulty in measuring their parameters.

Vitamins. In patients with liver disease, metabolism of vitamins may be disturbed by the inability of the body to store vitamins or by excessive release when liver cells are damaged. Although abnormally high vitamin levels occasionally result, more often the levels are low because of prior poor nutritional intake. In patients with progressive or severe liver injury, the signs of deficiency may persist despite large therapeutic doses of vitamins. This reflects either substrate deficiency or the inability of the liver to convert the vitamin into the necessary active compounds. In certain individuals with acute liver disease such as hepatitis, elevated serum vitamin B-12 concentrations reflect release from intrahepatic stores. This finding may be used as an indication of liver injury, and the higher levels usually persist until the disease is resolved.

Trace Metals. Except perhaps for iron and copper, the alterations in trace metals in liver disease appear to be related to low intake or poor absorption. The patient with hemochromatosis shows excessive iron stores in many tissues of the body, particularly the liver and pancreas. A controversy exists whether this is secondary to the liver disease or whether the primary error is excessive absorption and abnormal metabolism of iron, leading to its deposition and subsequent deleterious effects on tissue. Copper deposition is an etiological factor in Wilson's disease (hepatolenticular degeneration). In patients with this abnormality, a deficiency of serum ceruloplasmin is present, and copper is deposited in the tissues of the central nervous system, the liver and the eyes. Removal of copper by the use of penicillamine produces beneficial effects in most such patients. Here too, as in the case of iron, there is some controversy: does the copper cause liver damage, or does pre-existing (perhaps congenital or neonatal) liver disease cause the abnormalities in ceruloplasmin and copper metabolism?

ANNOTATED REFERENCES

GENERAL REFERENCES

Bondy, P. K.: Duncan's Diseases of Metabolism. ed. 6. Philadelphia, W. B. Saunders, 1969. (A well written chapter covering the fundamental physiology and biochemistry of bilirubin metabolism is presented by N. I. Berlin, P. D. Berk and R. B. Howe (Chap. 12).)

Schiff, L. (ed.): Diseases of the Liver. ed. 3. Philadelphia, J. B. Lippincott, 1969. (This book, which is a collection of treatises on various aspects of the disorders of the liver, provides excellent comprehensive discussions of ascites, hepatic coma, biliary cirrhosis, jaundice and bile acid metabolism. The authors of these sections are: Chap. 4, Bile salts and hepatobiliary diseases (Bile Acid Metabolism), J. B. Carey; Chap. 8, Jaundice, L. Schiff and B. H. Billing; Chap. 11, Ascites, W. H. J. Summerskill; Chap. 12, Hepatic coma, C. S. Davidson and G. J. Gabuzda; Chap. 20, Biliary cirrhosis, W. T. Foulk and A. H. Baggenstoss.)

Spiro, H. M.: Clinical Gastroenterology. New York, Macmillan, 1970. (This text presents a primarily clinical approach to the problems of liver disease. Although it does not delve deeply into physiology, it is comprehensive in its treatment of the various signs, symptoms and complications of hepatic disorders.)

TEXT REFERENCES

Bilirubin Metabolism and Jaundice

Arias, I. M., and Gartner, L. M.: Production of unconjugated hyperbilirubinemia in full-term newborn infants following administration of pregnane-3 (alpha) 20 (beta)-Diol. Nature, 203:1292, 1964.

Coburn, R. F., Williams, W. J., White, P., et al.: The production of carbon monoxide from hemoglobin in vivo. J. Clin. Invest., 46:346, 1967.

Crigler, J. F., and Gold, N. I.: Effect of sodium phenobarbital on bilirubin metabolism in an infant with congenital, nonhemolytic, unconjugated hyperbilirubinemia and kernicterus. J. Clin. Invest., 48:42, 1969.

Gartner, L. M., and Arias, I. M.: Formation, transport, metabolism and excretion of bilirubin. New Eng. J. Med., 280:1339, 1969. (This is an excellent review of the clinical syndromes and pathological physiology of disturbed bilirubin metabolism. It is clearly written, with due regard for individuals who are not completely familiar with the field.)

Levi, A. J., Gatmaitan, Z., and Arias, I. M.: Two hepatic cytoplasmic protein fractions, Y and Z, and their possible role in the hepatic uptake of bilirubin, sulfobromophthalein, and other anions. J. Clin. Invest., *48:* 2156, 1969.

Robinson, S. H.: The origins of bilirubin. New Eng. J. Med. *279:*143, 1968. (This is a good discussion of the fundamental chemistry and the clinical applications of bilirubin formation and is particularly helpful in regard to bilirubin from nonhemoglobin sources.)

Portal Hypertension

Boyer, J. L., Hales, M., and Klatskin, G.: Intrahepatic portal vein thrombosis—a primary cause of portal hypertension. 71st Annual Meeting of the American Gastroenterological Association, May, 1970.

Leevy, C. M., and Britton, R. C.: The hepatic circulation and portal hypertension. (Proceedings of a Conference) Ann. N. Y. Acad. Sci., *170:*1, 1970. (A collection of papers dealing with nearly all aspects of the portal circulation and portal hypertension.)

Viallet, A. Joly, J. G., Warleau, D., and Lavoie, P.: Comparison of free portal venous pressure and wedged hepatic venous pressure in patients with cirrhosis of the liver. Gastroenterology, *59:*372, 1970.

Ascites Formation

Barnardo, D. E., Strong, C. G., and Baldus, W. P.: Failure of the cirrhotic liver to inactivate renin; evidence for a splanchnic source of reninlike activity. J. Lab. Clin. Med., *74:*495, 1969.

Barnardo, D. E., Summerskill, W. H. J., Strong, C. G., and Baldus, W. P.: Renal function, renin activity and endogenous vasoactive substances in cirrhosis. Am. J. Dig. Dis., *15:* 419, 1970.

Chey, W. Y., and Shay, H.: Observations on the mechanism and treatment of ascites in hepatic cirrhosis. Am. J. Med. Sci., *244:* 1, 1962.

Eisenmenger, W. J., and Nickel, W. F.: Relationship of portal hypertension to ascites in Laennec's cirrhosis. Am. J. Med., *20:*879, 1956.

Epstein, M., Berk, D. P., Hollenberg, N. K., Adams, D. P., Chalmers, T. C., Abrams, H. L., and Merrill, J. P.: Renal failure in the patient with cirrhosis. Am. J. Med., *49:*175, 1970.

Faloon, W. W., Eckhardt, R. D., Cooper, A. M., and Davidson, C. S.: The effect of human serum albumin, mercurial diuretics and a low sodium diet on sodium excretion in patients with cirrhosis of the liver. J. Clin. Invest., *28:*595, 1949.

Fisher, E. R., and Hellstrom, H. R.: Evaluation of indices of juxtaglomerular cells and zona glomerulosa in human and experimental cirrhosis. J. Lab. Clin. Med., *67:* 199, 1966.

Gabuzda, G. J., Cirrhosis, ascites and edema. Gastroenterology, *58:*546, 1970.

Habif, D. V., Randall, H. T., and Soroff, H. S.: The management of cirrhosis of the liver and ascites with particular reference to the portacaval shunt operation. Surgery, *34:* 580, 1953.

Laragh, J. H., and Ames, R. P.: Physiology of body water and electrolytes in hepatic disease. Med. Clin. N. Am. *47:*587, 1963.

Laragh, J. H., Cannon, P. J., Bentzel, C. J., et al.: Angiotensin II, norepinephrine, and renal transport of electrolytes and water in normal man and in cirrhosis with ascites. J. Clin. Invest., *42:*1179, 1963.

Lieberman, F. L., Denison, E. K., and Reynolds, T. B.: The relationship of plasma volume, portal hypertension, ascites and renal sodium retention in cirrhosis: the overflow theory of ascites formation. Ann. N. Y. Acad. Sci., *170:*202, 1970.

Lieberman, F. L., Ito, S., and Reynolds, T. B.: Effective plasma volume in cirrhosis with ascites. Evidence that a decreased value does not account for renal sodium retention, a spontaneous reduction in glomerular filtration rate (GFR), and a fall in GFR during drug-induced diuresis. J. Clin. Invest., *48:* 975, 1969.

Orloff, M. J.: Pathogenesis and surgical treatment of intractable ascites associated with alcoholic cirrhosis. Ann. N. Y. Acad. Sci., *170:*213, 1970.

Pecikyan, R., Kanzaki, G., and Berger, E. Y.: Electrolyte excretion during spontaneous recovery from the ascitic phase of cirrhosis of the liver. Am. J. Med., *42:*359, 1967.

Said, S. I., and Mutt, V.: Polypeptide with broad biological activity: isolation from small intestine. Science, *169:*1217, 1970.

Schorr, E., Zweifach, B. W., and Furchgott, R. F.: Hepatorenal factors in circulatory homeostasis; influence of humoral factors of hepatorenal origin on vascular reactions to hemorrhage. Ann. N. Y. Acad. Sci., 49:571, 1948.

Schroeder, E. T., Eich, R. H., Smulyan, H., Gould, A. B., and Gabuzda, G. J.: Plasma renin level in hepatic cirrhosis, relation to functional renal failure. Am. J. Med., 49:186, 1970.

Summerskill, W. H. J.: Disorders of water and electrolyte metabolism in liver disease. Am. J. Clin. Nutr., 23:499, 1970.

Renal Failure

Baldus, W. P., Feichter, R. N., and Summerskill, W. H. J.: Kidney and cirrhosis, I. Clinical and biochemical features of azotemia in hepatic failure. Ann. Intern. Med., 60:353, 1964.

Barnardo, D. E., Summerskill, W. H. J., Strong, C. G., and Baldus, W. P.: Renal function, renin activity and endogenous vasoactive substances in cirrhosis. Am. J. Dig. Dis. 15:419, 1970.

Koppel, M. H., Coburn, J. W., Mims, M. D., Bolstein, H., Boyle, J. D., and Rubini, M. E.: Transplantation of cadaveric kidneys from patients with hepatorenal syndrome. Evidence for the functional nature of renal failure in advanced liver disease. New Eng. J. Med., 280:1367, 1969.

Lieberman, F. L.: Functional renal failure in cirrhosis. Gastroenterology, 58:108, 1970.

Schroeder, E. T., Numann, P. J., and Chamberlain, B. E.: Functional renal failure in cirrhosis, recovery after portacaval shunt. Ann. Intern. Med., 72:923, 1970.

Hepatic Coma

Conn, H. O.: A rational program for the management of hepatic coma. Gastroenterology, 67:715, 1969.

Faloon, W. W., and Evans, G.: Precipitating factors in the genesis of hepatic coma. N. Y. J. Med., 70(23):2891, 1970.

Levin, B., Oberholzer, V. G., and Sinclair, L.: Biochemical investigations of hyperammonemia. Lancet., 1:170, 1969. (A specific defect in the hepatic urea biosynthetic pathway is described in three patients. This is an example of an inborn error which produces the same picture seen in adults with cirrhosis and ammonia intoxication.)

Shear, L., and Gabuzda, G. J.: Potassium deficiency and endogenous ammonium overload from kidney. Am. J. Clin. Nutr., 23:614, 1970.

Summerskill, W. H. J., and Wolpert, E.: Ammonia metabolism in the gut. Am. J. Clin. Nutr., 23:633, 1970.

Walker, C. O., and Schenker, S.: Pathogenesis of hepatic encephalopathy with special reference to the role of ammonia. Am. J. Clin. Nutr., 23:619, 1970.

Cholesterol and Bile Acid

Admirand, W. H., and Small, D. M.: The physiochemical basis of cholesterol gallstone formation in man. J. Clin. Invest., 47:1043, 1968.

Small, D. M., and Rapo, S.: Source of abnormal bile in patients with cholesterol gallstones. New Eng. J. Med., 283:53, 1970.

Thistle, J. L., and Schoenfeld, L. J.: Lithogenic bile among young Indian women. Lithogenic potential decreased with chenodeoxycholic acid. New Eng. J. Med., 284:177, 1971.

Turnberg, L. Z., and Grahame, G.: Bile salt secretion in cirrhosis of the liver. Gut., 11:126, 1970.

Vlahcevic, Z. R., Bell, C. C., Buhac, I., Farrar, A. T., and Swell, L.: Diminished bile acid pool size in patients with gallstones. Gastroenterology, 59:165, 1970.

Vlahcevic, Z. R., Bell, C. C., and Swell, L.: Significance of the liver in the production of lithogenic bile in man. Gastroenterology, 59:62, 1970.

Vlahcevic, Z. R., Buhac, I., Bell, C. C., and Swell, L.: Abnormal metabolism of secondary bile acids in patients with cirrhosis. Gut., 11:420, 1970.

Carbohydrate, Protein, Fat and Nutritional Pathologic Physiology

Berkowitz, D.: Glucose tolerance, free fatty acid and serum insulin responses in patients with cirrhosis. Am. J. Dig. Dis., 14:691, 1969.

Gabuzda, G. J., and Shear, L.: Metabolism of dietary protein in hepatic cirrhosis; nutritional and clinical considerations. Am. J. Clin. Nutr., 23:479, 1970.

Leevy, C. M., Thompson, A., and Baker, H.: Vitamins and liver injury. Am. J. Clin. Nutr., *23*:493, 1970.

Lieber, C.: Metabolic effects produced by alcohol in the liver and other tissues. Advances Intern. Med., *14*:151, 1968.

Prasad, A. S., Oberleas, D., and Rajasekaran, G.: Essential micronutrient elements, biochemistry and changes in liver disorders. Am. J. Clin. Nutr., *23*:581, 1970.

Samaan, N. A., Stone, D. B., and Eckhardt, R. D.: Serum glucose, insulin and growth hormone in chronic hepatic cirrhosis. Arch. Intern. Med., *124*:149, 1969.

Sherlock, S.: Carbohydrate changes in liver diseases. Am. J. Clin. Nutr., *23*:462, 1970.

Stormont, J. M., and Waterhouse, C.: Effect of variations in previous diet on fasting plasma lipids. J. Lab. Clin. Med., *61*:826, 1963.

Appendix A

Classical References

Secretion

Babkin, B. P.: Secretory Mechanisms of the Digestive Glands. Ed. 2. New York, Hoeber, 1950.

Bayliss, W. M., and Starling, E. H.: The mechanism of pancreatic secretion. J. Physiol. (London), 28:325, 1902.

Beaumont, W.: Experiments and Observations on the Gastric Juice and the Physiology of Digestion. New York, Dover, 1959.

de Reuck, A. V. S., and Cameron M.P.: The Exocrine Pancreas, Normal and Abnormal Functions. Boston, Little, Brown and Company, 1962.

Dreiling, D. A., Janowitz, H. D., and Perrier, C. V.: Pancreatic Inflammatory Disease. New York, Hoeber, 1964.

Gregory, R. A.: Secretory Mechanisms of the Gastro-Intestinal Tract. London, Edward Arnold, 1962.

Ivy, A. C., Grossman, M. I., and Bachrach, W. H.: Peptic Ulcer. Philadelphia, Blakiston, 1950.

Pavlov, I. P.: The Work of the Digestive Glands. Ed. 2. London, Griffin, 1910.

Motility

Alvarez, W. C., and Zimmerman, A.: Movements of the stomach. Am. J. Physiol., 84:261, 1928.

Bayliss, W. M., and Starling, E. H.: The movements and innervation of the large intestine. J. Physiol. (London), 26:107, 1900.

Bayliss, W. M., and Starling, E. H.: The movements and innervation of the small intestine. J. Physiol. (London), 24:99, 1899; 26:125, 1901.

Cannon, W. B.: The Mechanical Factors of Digestion. London, Edward Arnold, 1911.

Chaudhary, N. A., and Truelove, S. C.: Human colonic motility. Gastroenterology, 40:1, 1961.

Code, C. F., Creamer, B., and Schlegel, J. F.: An Atlas of Esophageal Motility in Health and Disease. Springfield, Ill., Charles C Thomas, 1958.

Code, C. F., Hightower, N. C., and Morlock, C. D.: Motility of the alimentary canal in man. Am. J. Med., 13:328, 1952.

Grace, W. J., Wolf, S., and Wolff, H. G.: The Human Colon. New York, Hoeber, 1951.

Ingelfinger, F. J.: Esophageal motility. Physiol. Rev., 38:533, 1958.

Youmans, W. B.: Nervous and Nuerohumoral Regulation of Intestinal Motility. New York, Interscience, 1949.

Absorption

Verzar, F., and McDougall, E. J.: Absorption from the Intestine. London, Longmans, Green & Co., 1936.

Wilson, T. H.: Intestinal Absorption. Philadelphia, W. B. Saunders, 1962.

Liver

Davidson, C. S.: Liver Pathophysiology. Boston, Little, Brown & Co., 1970.

McMichael, J.: The portal circulation. J. Physiol., 75:241, 1932.

Patek, A. J., and Ratnoff, O. D.: The natural history of Laennec's cirrhosis. Medicine, 21:207, 1942.

Schiff, L.: Diseases of the Liver. Ed. 3. Philadelphia, J. B. Lippincott, 1969.

Sherlock, S.: Diseases of the Liver and Biliary System. Ed. 4. Philadelphia, F. A. Davis, 1968.

Appendix B

Laboratory Tests for Disturbed Gastrointestinal Functions

TESTS OF GASTRIC SECRETION (GASTRIC ANALYSIS)

Basal Acid Output (BAO)

Technique. Patient on no drugs, fasted for 12 hours prior to test. Pass nasogastric tube, position tip in most dependent portion of stomach under fluoroscopic monitoring. Place patient in supine position, somewhat on left side. Patient must expectorate all saliva. Continuous gentle aspiration of all gastric contents. Initial recovery ("residual volume") is set aside, after recording volume measurement. Collect all subsequent gastric contents for 90 minutes, divided into six 15-minute specimens. Discard first two 15-minute specimens. Measure volume of last four 15-minute specimens and determine acidity as hydrogen ion concentration (mEq./L.) by glass pH electrode. BAO is calculated by multiplying the volume (liters) of each specimen by its acidity (hydrogen ion concentration in mEq./L.) and adding up the products. The answer is given as mEq. per hour.

Normal range. 1.3 to 4.2 mEq. per hour.

Interpretation. Many normal individuals have zero BAO. A patient cannot be considered achlorhydric unless he fails to produce acid after stimulation (see below). An elevated BAO may be found in gastric outlet obstruction, "retained antrum" syndrome, Zollinger-Ellison syndrome and some (but not all) duodenal ulcer patients.

Comment. An important test in the diagnosis of a relatively uncommon disease (Zollinger-Ellison syndrome). Not reliable in the diagnosis of duodenal ulcer, gastric ulcer or gastric cancer. (See below.)

Stimulated Gastric Secretion

Maximal Histamine Test

Technique. Prepare patient as outlined above for basal acid output. Administer an intramuscular dose of antihistamine followed by subcutaneous administration of 0.04 mg. of histamine per kg. of body weight. Collect four 15-minute specimens of gastric contents and measure volume and hydrogen ion concentration as outlined above for BAO. Take the two highest 15-minute output values (peak half hour output), multiply by 2, express in mEq. per hour. This is maximal acid output (MAO), a value similar to peak acid output (PAO).

Normal range. 6 to 40 mEq. per hour (representative mean—20 mEq. per hour).

Interpretation. Patients with no acid output after maximal histamine stimulation have histamine-fast achlorhydria (i.e., true achlorhydria), often have pernicious anemia, may have gastric cancer, cannot have duodenal ulcer.

Low acid output is common with gastric ulcer. High outputs occur in duodenal ulcer, and Zollinger-Ellison syndrome. A BAO that is high and 60 percent or greater of MAO indicates Zollinger-Ellison syndrome.

Comment. There is great overlap in range of values for normals, duodenal ulcer, gastric ulcer and gastric cancer; the test cannot be relied upon in the diagnosis of these conditions.

Maximal Histalog Test

Comment. Betazole (Histalog) 1.5 mg. per kg. of body weight is administered

I.M. instead of histamine; it avoids the necessity of prior administration of antihistamine. Otherwise, technique, normal range and interpretation are the same as the maximal histamine test.

Insulin Test for Gastric Vagal Innervation (Hollander test)

Technique. Prepare patient as outlined above for basal acid output technique. Draw blood for fasting blood sugar. Perform BAO as outlined above. Administer 0.2 units of regular U-80 insulin/kg. body weight I.V. Collect gastric content specimens as outlined above for MAO, but obtain eight 15-minute specimens. Draw blood for blood sugar determinations every 30 minutes after insulin administration. Determine acid concentration (mEq./L.) for each gastric specimen.

Interpretation. A positive response (indicating vagal innervation to the stomach) is an increase in acid concentration for two or more successive 15-minute specimens of 20 mEq./L. over the basal level, or 10 mEq./L. if the basal level is zero. A negative response indicates that vagal nerve supply to the stomach has been completely interrupted. A test cannot be said to be negative unless the blood sugar determination falls to below 50 mg.% after insulin administration.

Comment. This test is done to determine if a prior surgical procedure accomplished a complete vagotomy, so that the stomach is left with no vagal innervation. These patients often have partial gastric resections or, at least, pyloroplasty; both (if functioning properly) cause gastric contents to empty into the intestine more rapidly and impair the ability to collect gastric contents adequately by naso-gastric tube suction. This gives a greater chance for false negative results.

Significant hypoglycemia (even to levels required for success with this test) may endanger elderly patients.

TEST OF PANCREATIC SECRETION

Secretin Stimulation

Technique. Patient on no drugs, fasted for 12 hours prior to test. Pass double lumen tube, position tip (distal lumen) in third portion of duodenum and proximal lumen in most dependent portion of stomach, under fluoroscopic monitoring. Continuous gentle aspiration of all gastric contents (discard) and all duodenal contents. Collect basal duodenal contents for two 20-minute specimens. Administer secretin 1.0 unit/kg. body weight and collect all duodenal contents for four 20-minute specimens. Determine volume, bicarbonate concentration and enzyme (usually amylase) concentration on all duodenal specimens.

Normal range.

Volume—99 to 450 ml.
Bicarbonate—60 to 140 mEq./L.
Amylase—1 to 20 units/ml.

Interpretation. "Quantitative" deficiency (low volume and total quantity, normal concentrations) is characteristic of pancreatic duct obstruction (e.g., pancreatic cancer). "Qualitative" deficiency (normal volume, low concentrations of bicarbonate and enzymes) is characteristic of chronic pancreatic inflammation (e.g., chronic pancreatitis).

Comment. Recovery of duodenal contents is rarely complete, often 20 percent or more being lost in the best circumstances.

The test is not reliable in the diagnosis of acute pancreatic inflammation or cancer of the tail; extensive malignant infiltration gives results similar to chronic inflammation. Cancer is usually far advanced when quantitative secretory abnormalities are detectable.

TESTS OF INTESTINAL ABSORPTION

Most tests of absorption can be classified as follows: balance tests, in which the fecal output of an unabsorbed substance or its metabolites is expressed as fraction of input; or tolerance tests, in which absorption of a test substance is inferred from the appearance of the substance or its metabolic product in the blood or urine. The most reliable test of nutrient absorption is the fat balance test, which is valuable for establishing the presence of malabsorption, even if it gives no clue to its mechanism. With the development of new techniques as well as advances in the understanding of absorptive mechanisms, it now seems possible to classify other tests of absorption according to the scheme proposed for malabsorption: (1) tests of intraluminal digestion; (2) tests of surface digestion; (3) tests of mucosal uptake; and (4) tests of bacterial overgrowth. This section does not cover the diagnostic approach to a patient with malabsorption, which, in addition to the tests discussed, will generally include appropriate endoscopy, x-ray examination, and small intestinal biopsy.

Tests for Presence of Malabsorption: Classical Balance Test

Here fecal excretion of fat (and occasionally nitrogen) is measured while the patient adheres to a defined diet. Although unabsorbed food constitutes the main source of fecal fat and nitrogen, endogenous secretion, as well as bacterial production and destruction, may contribute. In addition to these factors, practical problems attend the collection of a representative fecal sample: (1) it may be difficult to obtain a complete collection during the test interval; and (2) defecation may be irregular, with stool excretion in a particular 24-hour period not representing the mean daily output.

Measurement of fecal water, indicated by fecal weight, is essential for defining the severity of diarrhea. The significance of increased fecal water differs from that of increased fecal fat and nitrogen, in that endogenous secretions are its predominant source.

To reduce the error of fecal sampling, 48- or 72-hour stool collections are the present practice. However, in diarrheal conditions, a 24-hour sample is usually adequate. In the future, measurement of fat absorption will probably feature the administration of radioactive fat together with a nonabsorbable reference marker. By comparing the ratio of marker to radioactivity in the test meal with that in the stool, the fraction of absorbed radioactive fat can be calculated. In addition, a procedure such as this indicates the true absorption of the administered fat, since endogenous secretions are not measured. In principle, the use of labeled test substances administered together with a marker should permit the development of a variety of useful absorption tests.

TESTS OF DIGESTION

Tests of Pancreatic Exocrine Function

Since intraluminal digestion requires pancreatic enzymes and bile acids, tests of intraluminal digestion are concerned first with pancreatic exocrine function. To test this, one determines bicarbonate or enzymes in duodenal aspirates after intravenous administration of secretin or CCK, respectively. Alternatively, one measures enzyme concentration and extent of lipolysis in jejunal content after a test meal. Pancreatic disease severe enough to cause fat malabsorption is almost invariably associated with reduced bicarbonate concentration or reduced volume output after intravenous administration of secretin. To detect isolated lipase deficiency (an extremely rare congenital condition), measurement of individual enzymes in duodenal aspirates after intravenously given CCK is necessary. Total pancreatic enzyme output in response endogenous stimulation may be determined by duodenal perfusion with amino acids and a nonabsorbable marker. Measurement of marker and enzyme concentration in samples aspirated from the distal duodenum permits calculation of the rate of enzyme secretion. Comparison of enzyme output after endogenous stimuli (acting via CCK-release) with that evoked by exogenous hormone permits the diagnosis of impaired CCK release, as has been observed in nontropical sprue. Fecal chymotrypsin measurements are a simple but rather unreliable test of pancreatic exocrine insufficiency.

Tests of Fat Digestion

To assess fat digestion, i.e., micellar dispersion as well as lipolysis, a standard liquid meal containing carbohydrate, fat and protein is fed. Jejunal content is then collected for several hours and pooled by 20- or 30-minute intervals.

Lipase concentration is measured, as well as the ratio of free fatty acids to total fatty acids, to give an indication of lipolysis. The sample of intestinal content is ultracentrifuged to determine the lipid content of the aqueous micellar phase. This actually measures the partition coefficient of the lipolytic products between the subnatant aqueous micellar phase and the oil phase. The concentration of bile acids in the aspirate is a measure of bile acid secretion into the duodenum relative to dilution. The resulting value is compared with the concentration of bile acids in the micellar phase to provide information on the fraction of bile acids in solution. Bile acids are examined chromatographically to detect free bile acids, indicating bacterial deconjugation.

Tests of Surface Digestion

The most commonly employed test of surface digestion is the lactose tolerance test. Absorption rate may be monitored qualitatively by changes in blood glucose levels, or by the appearance of diarrhea; ^{14}C-lactose may be used and absorption assessed qualitatively by appearance of $^{14}CO_2$ in breath. In principle, similar tests for oligopeptide digestion could be developed. Disaccharidases can be measured quantitatively in intestinal biopsy samples obtained by peroral intubation.

TESTS OF MUCOSAL UPTAKE

Tests of mucosal uptake employ material that requires neither enzymatic digestion nor micellar dispersion. Oral tolerance tests used to assess absorption are of limited value, because a flat blood tolerance curve does not distinguish delayed gastric emptying from decreased absorption or increased metabolic disposition of the test substance or its metabolic product. Tests based on urinary recovery are further subject to the vagaries of renal excretion. Investigators, recognizing the value of tests of mucosal absorptive function, have attempted to circumvent these problems in several ways: (1) before the test, body stores are saturated with a loading dose of the test substance in order to block tissue uptake. For vitamin B-12, for example, nonradioactive vitamin B-12 is given parenterally before oral administration of the labeled vitamin. (2) Substances are chosen for which a constant fraction of the amount absorbed is excreted into the urine, e.g., xylose. For any tolerance test based on renal excretion, urine is collected in a period sufficiently long, so that the amount recovered allows a reasonable estimate of the fraction of the test substance absorbed and is little influenced by the absorption rate.

TESTS OF BACTERIAL OVERGROWTH

Samples of small intestinal content are obtained by intubation and cultured areobically and anaerobically. Intestinal aspirates can be examined by thin layer chromatography for the presence of free bile acids. In the future, diagnostic procedures will likely include assay of specific bacterial enzymes in intestinal aspirates, as well as breath analysis to detect volatile products produced by rapid bacterial degradation of ingested substances.

USEFUL DIAGNOSTIC TESTS

The tests summarized here are informative and reliable, and based on substantial clinical experience. In the text several tests which are discussed are based upon limited experience at present. Other tests, e.g., the vitamin A tolerance tests, are popular for screening but do not distinguish digestive from absorptive disturbances. Tests of pancreatic function are discussed in the introduction to the gastrointestinal section, the Schilling test for vitamin B-12 absorption is described in Chapter 23.

Stool Weight

Normally, less than 200 g. of feces are excreted per day. Measurement of stool weight is essential for establishing the presence of diarrhea. Paint cans have proven to be convenient for stool collections and are being used increasingly.

Stool Fat

Normal stool fat is less than 7 g. per day in subjects on a 100-g. fat diet, as determined by methods based on saponification and titration of liberated fatty acids. Reportedly the fraction of fecal fat in the form of triglyceride increases in patients with steatorrhea due to defective lipolysis (pancreatic insufficiency). It may be as-

sessed by thin-layer chromatography, provided that stools are frozen immediately after collection and lipid extraction is carried out promptly. Tests based on microscopic appearance of oil droplets stained with fat-soluble dyes are insensitive and subject to sampling errors.

Oral D-Xylose Test

When 5-hour urinary excretion of D-xylose is measured after administration of a 25-g. oral dose, normal excretion is > 5.0 g.; questionably subnormal excretion is 3.0 to 5.0 g.; and diffuse abnormality or resection of jejunum is < 3.0 g. "Falsely" low values are obtained with bacterial overgrowth in the proximal small intestine (bacterial utilization of D-xylose), ascites (distribution of absorbed D-xylose into an expanded extracellular space), renal disease (D-xylose clearance from the blood equals 60 to 80% of the glomerular filtration rate), and urinary retention (excreted D-xylose retained with residual urine in the urinary bladder).

Lactase Assay

By tolerance test (50 g. of lactose by mouth; maximum rise in blood sugar 20 mg.%), and by direct mucosal enzyme assay (normal = 1 Unit [1 mole lactase hydrolyzed per minute] per gram of mucosa, at 37° C). The tolerance test is unreliable in the presence of diabetes mellitus and may show falsely low results when gastric emptying is delayed.

REFERENCES FOR TESTS

Christiansen, P. A., Kirsner, J. B., and Ablaza, J.: D-Xylose and its use in the diagnosis of malabsorptive states. Am. J. Med., 73: 521, 1969.

Jover, A., and Gordon, R. S., Jr.: Procedure for quantitative analysis of feces with special reference to fecal fatty acids. J. Lab. Clin. Med., 59:878, 1962.

Newcomer, A. D., and McGill, D. B.: Disaccharidase activity in the small intestine: prevalence of lactase deficiency in 100 healthy subjects. Gastroenterology, 53:881, 1967.

Thompson, J. B., Su, C. K., and Welsh, J. D.: Fecal triglycerides. II. Digestive vs. absorptive steatorrhea. J. Lab. Clin. Med., 73:521, 1969.

Wollaeger, E. E., Comfort, M. W., and Osterberg, A. E.: Total solids, fat, and nitrogen in the feces. Gastroenterology, 9:272, 1947.

CLINICALLY USED TESTS OF LIVER FUNCTION

TEST	NORMAL VALUE	COMMENT
To determine the presence of liver disease and its course		
Serum Bilirubin		Elevation of unconjugated bilirubin, rarely seen by itself, may occur in Dubin-Johnson syndrome. Conjugated bilirubin may predominate in hemolytic disease and is usually found when liver function is also impaired.
Unconjugated (Direct, or 1 minute)	0.1 to 0.3 mg./100 ml.	
Conjugated (Indirect, or value of 30 min. figure less 1 min. bilirubin)	0.3 to 0.5 mg./100 ml.	
Bromsulfalein Removal (BSP)	0 to 5% retention in 30 min.	Removal is decreased (retention in blood therefore increased) in cardiac failure or febrile patients. Will be elevated in nearly all jaundiced patients. Determination not necessary in presence of bilirubin elevation.
Enzymes		SGOT and SGPT can be elevated in myocardial infarcts and congestive heart failure. LDH will be elevated in myocardial infarct and hemolytic disease much more than in liver disease. In hepatitis, SGOT and SGPT usually over 400 u., in obstruction usually under 300; lesser elevations occur in cirrhosis. Alkaline phosphatase elevated more than 3 times normal in obstruction; in parenchymal disease elevation is usually less.
Serum glutamic oxalacetic transaminase (SGOT)	10 to 40 units	
Serum glutamine pyruvic transaminase (SGPT)	10 to 40 units	
Lactic dehydrogenase (LDH)	100 to 350 Berger-Brody units	
Alkaline phosphatase	2.0 to 4.5 Bodansky units	
Urine Urobilinogen	0.3 to 1.0 Ehrlich units per 2 hours (1 to 3 P.M.)	Urine urobilinogen reflects bilirubin reaching the intestine where bacteria change it to urobilinogen. Failure of liver to remove absorbed urobilinogen allows increased urinary excretion.
Prothrombin Time	Less than 2 second deviation from control value	Prothrombin decreased due to failure of liver synthesis (parenchymal disease) or poor absorption (obstructive jaundice or malabsorptive disease).
Other biochemical studies of less value or less frequently used		
Total serum protein	6.0 to 8.0 g./100 ml.	For protein changes, see references.
Paper electrophoresis or protein		
Albumin	50 to 60%	
Alpha-1 globulin	4.2 to 7.2%	
Alpha-2 globulin	6.8 to 12%	
Beta globulin	9.3 to 15%	
Gamma globulin	13 to 23%	
Serum copper	90 to 120 μg./100 ml.	Serum copper decreased (10 to 70 μg./100 ml.) and low serum ceruloplasmin (under 10 mg./100 ml.) is found in Wilson's disease (hepatolenticular degeneration).
Serum ceruloplasmin	20 to 40 mg./100 ml.	
Serum iron	60 to 190	Serum iron elevated in hemochromatosis.

CLINICALLY USED TESTS OF LIVER FUNCTION (Continued)

Test	Normal Value	Comment
Cephalin Flocculation	0 to 2+	Cephalin flocculation and thymol turbidity elevation reflect serum protein changes seen with parencyhmal disease and only rarely with obstruction.
Thymol Turbidity	0 to 4 units	
Serum Cholesterol Total Esters	 150 to 250 mg./100 ml. 60 to 75% of total	Total cholesterol usually decreased in severe parenchymal disease, increased in biliary cirrhosis and in some patients with obstructive jaundice or recent alcohol intake (Zieve's syndrome).
Serum vitamin B-12	200 to 800 mg./100 ml.	Serum B-12 markedly increased in hepatitis or other inflammatory liver disease.
Blood ammonia	50 to 150 μg./ml. (arterial) (venous value lower than arterial)	Ammonia values useful in distinguishing stupor or coma due to liver disease from that due to drugs or trauma, or in recognizing Wernicke's syndrome (normal ammonia).

To differentiate the cause of jaundice

SGOT

SGPT

Serum bilirubin fractionation

Alkaline phosphatase

Urine urobilinogen

(If prolonged) prothrombin response to vitamin K given parenterally

Cephalin Flocculation or Thymol Turbidity

Cholesterol and esters

Protein and electrophoresis

Smooth muscle antibodies

Antimitochondrial antibodies

For discussion of tests for differentiation of jaundice, see above or see references.

Other procedures and tests of liver function

Cholecystography
oral
intravenous
transhepatic

In the presence of significant liver disease or obstructive jaundice, oral or intravenous cholangiograms rarely are helpful owing to inability of the liver to remove and concentrate the contrast compound. Transhepatic cholecystography is usually helpful only to those with much experience in doing it.

CLINICALLY USED TESTS OF LIVER FUNCTION (Continued)

TEST	COMMENT
Hepatic scan with radio-isotopes Gold (Au 198) or Technetium sulfate (Tc 99m)	Scans are helpful in metastatic neoplasm, cysts or hepatic abscesses, but confusion is common because vascular abnormalities in parenchymal disease may simulate defects due to metastatic tumor.
Needle biopsy of liver	Needle biopsy should be done in co-operative patients without hemorrhagic tendency. Useful in hepatitis, cirrhosis, fatty liver, metastatic tumor, drug-induced liver damage or cholestasis.
Splenoportography	Splenoportography may be helpful in distinguishing intrahepatic from extra-hepatic portal vein obstruction.
Esophagoscopy	Use of the flexible fiberoptic esophago-scope makes this a very convenient procedure for finding varices.
Others, peritoneoscopy, paracentesis, umbilical vein angiography	See references for discussion of these.

References for Liver Function Tests:

Bouchier, I. A. D.: Clinical Investigation of Liver Function. London, Blackwell Scientific Publications, 1969.

Conn, H. O.: Is hepatic scanning overrated? Gastroenterology, 54:135, 1968.

Evaluation of Liver Function in Clinical Practice. Indianapolis, Indiana, Eli Lilly and Company, 1965.

Schiff, L.: Diseases of the Liver. Ed. 3. Philadelphia, J. B. Lippincott, 1969.

Hematological Mechanisms

Introduction

Study of the blood of man is a fascinating aspect of physiology, since its ready availability for sampling permits research in depth not only for the field of molecular biology but also for the transient changes which occur in the active, dynamic state of life. Blood is the river of life flowing through the vascular channels and carrying cellular nutrients. Its contents are complex, and the study of blood has become very specialized. Some investigators study only the red cells, whereas others discard the cells in favor of the plasma proteins.

The explosive accumulation of information in two aspects of hematology and the research potential for investigating a third have not been summarized completely in the chapters to follow. Blood cells, being relatively easy to obtain from human sources, have served as models for the study of other cells. Between 1965 and the present a number of reviews have been written on the physiology and disorders of the cell membrane. The demonstration that virtually all of the lipid in the mature erythrocyte can be found in the membrane indicated that alterations in lipid metabolism should be investigated for red blood cell defects. Hence, a wide variety of morphological abnormalities of red cells, which had long been recognized, have provided us with new clues about red blood cell pathophysiological alterations and can now be directly linked with alterations in the physical state of the red cell membrane. In-vitro phenomena such as osmotic and mechanical fragility, although they do not identify precisely the red cell defect, nevertheless permit ways of identifying alterations in the red cell membrane which lead to premature erythrocyte destruction by a final common pathway. The cell shape, the contents and physical properties of the cell membrane contribute to the normal biconcave disc shape. All conditions which lead to spherocytosis result in a geometrically rigid cell which adapts poorly to the deformation required for movement through the microcirculation. Fragmentation in vivo may occur when portions of the red cells stick to either epithelial cells or areas of fibrin mesh, resulting in tearing of fragments from the cell. These cell fragments are then abnormally trapped in the reticuloendothelial system, especially the spleen. This mechanical deformation results in random hemolysis and the condition *microangiopathic hemolytic anemia*. The spleen serves as a filter for erythrocytes with inclusion bodies or damaged membrane structures. The erythrocytes pass adjacent to the splenic adventitial cells when moving through the cord into the sinus lumen. They enter the basement membrane finestra and then squeeze between the endothelial cells of the sinuses. Should they contain an inclusion body, this is extruded with part of the cell membrane during this passage. This is the so-called "pitting" function of the spleen. A mature cell which undergoes this membrane loss is unable to synthesize a new membrane, loses its biconcave shape and becomes spherocytic. Another important consequence of cell membrane research was information concerning the involvement of the leukocyte membrane in phagocytosis. The answer to the difference in the nature of phagocytic ability appears in the nature of the cellular plasma membrane. Phagocytosis of neutrophils can be blocked by the removal of surface sialic acid. Due to the prevention of glycolysis (at the phosphohexose isomerase level) lysophosphatides play a role in leukocyte metabolism. During phagocytosis there is a striking stimulation of the incorporation of lysolecithin. The biochemical properties of proteins, derived from isolated membranes of washed human platelets, have afforded an opportunity to further characterize the role of platelets in expending energy during the initiation of blood clotting. A dynamic, active membrane in platelets may be an important feature, not only in hemostasis, but also in the initiation of intravascular clotting (specifically, intracardiac and arterial thrombi). The interruptions in the platelet membrane act as sites for the dendritic bridges which function in holding the clot stroma together.

A second area of rapidly accumulating information is concerned with the relationship of the complement-kallikrein system to plasminogen activation. Human α-2 macroglobulin can complex and inhibit in vitro proteolytic activity of three enzymes important in maintaining normal blood flow (*i.e.*, plasmin, thrombin, and plasma kallikrein). In contrast, C1 inactivator has been shown to inhibit C1 esterase, plasmin, kallikrein, chymotrypsin and a plasma permeability factor. The α-2 macroglobulin-plasmin complex may also serve as a plasminogen activator. The kinin system is activated by factor XII-a (active Hageman factor), thereby establishing an interrelationship between the intravascular coagulation, fibrinolytic, complement and kinin systems. In hereditary angioneurotic edema, there are permanently reduced levels of kallikrein inhibitor. Increased levels of spontaneous esterolytic activity (presumably bradykinin) have been observed in patients with the carcinoid syndrome during active flushing.

On the research horizon evidence is appearing which is linking the autonomic nervous system with the complex interaction of blood proteins comprising the clot-forming to clot-lysing systems which permits unimpeded circulation of blood. Vascular endothelium, a hitherto unknown factor in the initiation of platelet plugging and thrombosis, may contain many of the same proteins identifiable in the platelet. A platelet protein, thrombasthenin, bears a similar relation chemically to the muscle protein, actomyosin. Furthermore, the vasoactive response to neurogenic stimulation may also influence the endothelial-blood cell interface reaction as well as the intravascular cellular contents.

The origin of the cellular elements of the blood is a question which has long troubled the hematologist. In the human embryo, blood cells are formed in the numerous blood islands of the yolk sac. Later, hematopoiesis occurs in blood islands located in the embryonic mesoderm. Those cells which differentiate centrally become the primitive blood cells whereas those located peripherally become the endothelial walls. These primitive blood cells bear some resemblance to a mature histiocyte in that they possess nucleoli, basophilic cytoplasm, and a spongy chromatin network. Therefore, a common ancestor cell for both the endothelial and circulating cells of the blood suggests that some residual function may occur in the walls of the blood vessels themselves. Advances in our understanding of protein synthesis and the methodology for isolating and purifying protein substances should afford a rapid accumulation of knowledge in this heretofore unrecognized area of investigation. Our future understanding of the pathophysiology of the blood tends to focus on the molecular ecology of the blood cells, the circulation of these molecules within the cells, and the mode of transport and release of these substances into the surrounding media. The substances which regulate the growth, differentiation, release and destruction of the blood cells are now being explained.

James W. Hampton, M.D.

23

Erythropoiesis

Walter H. Whitcomb, M.D.

THE ERYTHRON

A Concept

The erythrocyte originates from a stem cell located within bone marrow and differentiates into a cell having the capacity to synthesize hemoglobin. Within 3 to 5 days, the red cell precursor (normoblast) undergoes further differentiation and division. It is capable of synthesizing deoxyribonucleic acid (DNA), ribonucleic acid (RNA), lipid, heme and proteins and has an electron transfer system, a functional Krebs tricarboxylic acid cycle and the Embden-Meyerhof and hexose monophosphate pathways. The cell then loses its nucleus (and is no longer capable of DNA and RNA synthesis) and enters the vascular compartment. The young erythrocyte (reticulocyte) matures and remains intravascular for approximately 120 days; then, with its energy supply exhausted, it is removed by the reticuloendothelial system of the spleen, the liver and the bone marrow.

This cell system, termed the erythron by Castle and Minot, is a dynamic one. The erythroid marrow maintains a population of 25×10^{12} circulating erythrocytes which contain 750 g. of hemoglobin. With maximal stimulation the erythroid marrow is capable of increasing the production rate six- to eightfold and can produce a volume of red cells equivalent to that contained in one-half pint of whole blood per day. Normally, aged red cells are removed from the blood at a rate of about 1.25 percent per day. If, in disease, the red cells are removed from the circulation at an increased rate, production must increase in order to maintain the optimum number of circulating cells. For example, if the red cell survival time is reduced to one fourth of normal (from 120 days to 30 days), production must increase to four times normal. If, however, the survival time is reduced to one tenth of normal (from 120 days to 12 days), the maximal production rate (6 to 8 times normal) is exceeded and a deficit in circulating red cells ensues (anemia). Pathophysiological states of the erythron can therefore be broadly categorized as due to insufficient production, excessive destruction, or a combination of these. Conversely, a pathological state may be associated with excessive production of red cells, resulting in an increased circulating red cell mass. An outline of pathophysiology of the erythron based upon the concepts of insufficient production, increased destruction, and increased production is presented in Table 23-1.

The Erythrocyte—A Transport for Oxygen

The principal function of the erythrocyte is to provide a transport vehicle for hemoglobin, a respiratory pigment that can accept, transport and deliver oxygen to the tissues. In order to accomplish this, the mature erythrocyte must also provide an environment in which the hemoglobin can be maintained in its functional reduced state. Oxygen is accepted in the pulmonary capillaries and is transported, bound to hemoglobin, to the tissue capillaries, where it is released. The pressure of oxygen in the capillary must be sufficiently great to

TABLE 23-1. Outline of Pathophysiology of Erythron

I. Insufficient Production of Erythrocytes

    A. Defects in general controlling mechanisms

        1. Erythropoietin deficiency—renal disease, erythropoietin antibody, possible inactivators

        2. Endocrinopathies of pituitary, thyroid, adrenals, and gonads

    B. Defects in cell proliferation and maturation

        1. Impaired hemoglobin synthesis, due to

            a. Defective synthesis—i.e., congenital hemoglobinopathies and thalassemia; toxins, i.e., lead

            b. Nutritional deficit—iron; Vitamin B-12; folic acid; protein

        2. Defective erythropoiesis—bone marrow failure

            a. Aplastic anemia—congenital (Fanconi's anemia); acquired, due to drugs and toxins

            b. Red cell aplasia—congenital (Blackfan-Diamond); acquired, due to drugs

            c. Refractory anemia—secondary to systemic disease, i.e., infection, neoplasms, etc.

II. Increased Destruction of Erythrocytes

    A. Intracorpuscular abnormality

        1. Inherited abnormalities—hemoglobinopathies; hereditary spherocytosis; hereditary non-spherocytic hemolytic anemias with enzyme deficiency; thalassemia

        2. Acquired abnormalities—paroxysmal nocturnal hemoglobinuria, lead poisoning, nutritional deficit (iron, B-12, folic acid)

    B. Extracorpuscular abnormality—erythrocyte environment

        1. Acquired, associated with antibodies—isoantibodies (transfusion reaction); autoantibodies; drugs (alpha-methyldopa)

        2. Acquired, without antibodies—chemical and physical agents; infectious diseases (i.e., malaria, bacterial sepsis); neoplasms; splenomegaly; microangiopathic states

III. Increased Production of Erythrocytes

    A. Appropriate compensatory mechanism in response to

        1. Increased destruction of erythrocytes (if pathophysiological mechanism affects only the erythrocyte segment of the erythron)

        2. Tissue oxygen deficit—relative insufficiency in oxygen transport

            a. Environmental oxygen deficit (altitude)

            b. Cardiorespiratory insufficiency

            c. Abnormal hemoglobin with increased oxygen affinity

    B. Inappropriate

        1. Excessive erythropoietin activity—associated with tumors of kidney, liver, cerebellum, adrenals, uterus, and lung

        2. Excessive proliferation of erythroid cells or all cells of marrow (myeloproliferative disease) without increased erythropoietin activity, i.e., polycythemia vera.

permit oxygen to leave the capillary and diffuse through tissue. The pressure of oxygen in arterial blood is about the same throughout the body, whereas the pressure of oxygen in venous blood varies, depending on the amount of oxygen utilized by a given tissue. Thus, the quantity of oxygen delivered to a given tissue depends on the amount of oxygen in inspired air, the functional state of the lung, the cardiac output, the hemoglobin concentration, the affinity of the hemoglobin for oxygen and the vascular distribution of blood to a given tissue. The mean capillary oxygen tension —and, hence, tissue oxygen tension—can be influenced by any one of these factors. In the event of oxygen deprivation, compensatory mechanisms occur, i.e., increased number of active tissue capillaries; a shift in the oxygen-hemoglobin dissociation curve, providing for the release of oxygen at lower pressure; increased cardiac output, and an increased number of circulating red cells and, hence, hemoglobin. A functional concept of anemia must therefore consider the adequacy of tissue oxygen supply. For example, a normal individual who moves to a high altitude has an inadequate number of red cells for adequate tissue oxygenation and develops the symptoms of hypoxia.

Anemia. The symptoms of anemia depend upon its severity, its rate of development and its duration and, moreover, upon the ability of other physiological mechanisms to compensate for the reduction in circulating red cells, principal among which is the status of the cardiovascular system. Signs and symptoms common to all anemias are pallor and bruising, shortness of breath, increased heart rate, palpitation, angina and heart failure, irritability, dizziness, fatigue, headache and tinnitus. In general, these effects are the result of the decreased oxygen-carrying capacity of the blood. However, it is important to realize that, if the anemia develops at a slow rate, remarkable compensation and adaptation may occur, with the appearance of few, if any, symptoms in the absence of physical exertion.

Normally, blood contains about 15 g. of hemoglobin per 100 ml., and each gram of hemoglobin carries 1.36 ml. of oxygen when completely saturated. Arterial blood carries 21 ml. of oxygen; venous blood contains about 15 ml. of oxygen per 100 ml. (6 ml. have been delivered to the tissues). If the hemoglobin concentration falls below 5 g. per 100 ml., less than 6 ml. of oxygen per 100 ml. of circulating blood is available for the tissues, and major adjustments in other systems must occur. Cardiac output remains relatively constant until the hemoglobin level falls to about 7 or 8 g. per 100 ml., but then it increases in proportion to the hemoglobin deficit. Blood flow becomes redistributed to the most vital centers. Renal blood flow may be reduced by 50 percent, and cutaneous flow is also markedly decreased. Liver function is usually fairly well maintained in severe anemia, provided that the disease process causing the anemia has no direct effect on the liver per se.

Certain clinical signs and symptoms may suggest specific types of anemia. For example, dysphagia, brittle nails and spooning of nails may occur in severe iron deficiency anemia. A lemon-yellow tint to the skin, atrophy of the papillae of the tongue, delirium, and subacute combined degeneration of the spinal cord are associated with vitamin B-12 deficiency. Petechiae occurring at the base of hair are characteristic of vitamin C deficiency (scurvy), and associated dermatitis suggests nutritional deficiency. Jaundice may be associated with hemolytic anemia. However, although one may suspect the cause of the anemia from the clinical presentation, further study of the erythron is necessary to establish the etiology precisely. Conversely, anemia "discovered" in relatively asymptomatic patients is a sign that in itself needs to be explained, e.g., iron deficiency anemia may be traced to chronic blood loss and depletion of iron stores as a result of a carcinoma of the cecum which, for the moment, is asymptomatic.

Erythremia. In contrast to anemia, inappropriate increase in red cell production resulting in erythremia is commonly associated with a ruddy complexion and pruritus. Symptoms may be similar to those of anemia, i.e., the cardiovascular stress may be manifest as chest pain or fullness, increased heart rate, shortness of breath, claudication; the effect on the central nervous system may result in irritability, light-headedness, fatigue, tinnitus. Gastrointestinal complaints may be related to the increase in blood volume and viscosity, with sludging in small vessels and a propensity to thrombotic phenomena. Oxygen transport is in excess of tissue demand. Like anemia, the complaints of erythremia (too many erythrocytes) may be referable to several organ systems and to one of several remediable etiologies and, therefore, need to be explained.

Because the signs and symptoms of anemia and erythremia are manifestations of multiple system involvement, the determination of the exact pathophysiological derangement (the diagnosis) depends, for the most part, upon laboratory techniques. A brief outline of the "tools" available to the physician follows:

An estimate of the concentration of erythrocytes can be obtained by an enumeration of red cells per cubic millimeter of blood (red cell count) and the determination of the percent of blood that is cellular by centrifugation (hematocrit). An estimate of the hemoglobin concentration can be determined chemically and expressed as grams per 100 ml. of blood. The volume of all circulating red cells and of plasma can be measured, using radioisotopic labeling of an aliquot of the patient's blood and applying a dilution formula (red cell and plasma volumes). The rate of red cell production can be estimated by counting the proportion of young red cells on a blood smear (the reticulocyte count) and expressing the result as a percent or relating this proportion to the red cell count for an expression of "absolute" reticulocyte count. The rate can also be estimated by injecting a radioactive material that is taken up by the developing red cell (e.g., radioactive iron (^{59}Fe), which is incorporated into cellular hemoglobin) so that newly formed circulating cells are labeled. The survival time of the red cell can be measured, either by injecting a radioactive material that becomes incorporated into the developing cell or by tagging a sample of the circulating cells and, in either case, then measuring by radiation detectors the proportion of labeled cells remaining in the circulation over a period of time.

Examination of a blood smear microscopically provides information in regard to size, shape and hemoglobin content of the individual cells and is perhaps the single most important simple procedure to perform. For example, pale cells (hypochromic) suggest defective hemoglobin formation which could be due to iron deficiency or other defects in hemoglobin synthesis; large cells (macrocytes) suggest defects in nucleic acid and DNA synthesis, such as occurs with vitamin B-12 and folic acid deficiency; spherical cells (rather than biconcave discs) suggest an abnormal red cell which can be expected to be destroyed (hemolyzed) early in its life span. Small cells suggest increased proliferation of cells, with inadequate hemoglobin synthesis; i.e., iron deficiency, thalassemia, pyridoxine responsiveness.

Additional information concerning the specific defect in the cell can be obtained by more sophisticated chemical methods, both qualitative and quantitative, employed in regard to the hemoglobin and its components, the cellular enzyme systems and the reaction of the cells to certain electrolyte concentrations, chemicals and antibodies. In addition to the study of the circulating red cell, estimates of the integrity of the erythroid marrow—the organ of production—can be made by direct microscopic examination and by the use of radioactive materials. A sequence of inquiry becomes apparent, i.e., the patient's symptoms, the patient's physical manifestations of the disease, and the use of laboratory tests selected to gain evidence for or against known biochemical and physiological defects.

REGULATION OF ERYTHROPOIESIS

Erythropoietin

Inasmuch as the erythron occupies a central position in the organism's mechanisms of respiration, one might anticipate that a major factor influencing the control of the erythron would be the relation between oxygen supply and the metabolic requirements (Fig. 23-1). Indeed, the major

FACTORS INFLUENCING THE CONTROL OF ERYTHROPOIESIS

FIG. 23-1. The interrelationship of oxygen transport, blood volume, the erythropoietin mechanism and the endocrine influence on erythroid cell proliferation. USC, unipotential stem cell; PrNB, pronormoblast; BNB, basophilic normoblast; PNB, polychromatophilic normoblast; ONB, orthochromatic normoblast; R, reticulocyte.

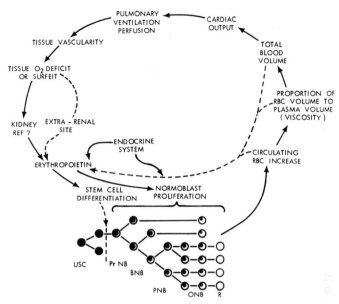

stimulus to erythroid cell production is tissue oxygen deficit; moreover, in the presence of oxygen excess, the rate of erythropoiesis is diminished or ceases entirely. Red cell production is now thought to be regulated by the erythropoietic stimulatory factor (ESF) erythropoietin, which is produced or activated by the kidney in response to tissue oxygen requirements. A proportional relationship has been demonstrated between the degree of induced hypoxia or anemia and the level of erythropoietic activity in the plasma and urine. In man, plasma and urine erythropoietic activity increases to maximum between 24 and 60 hours of induced hypoxia and then decreases to near normal levels within 4 to 5 days. The reason for this wave of plasma ESF activity rather than a sustained level is not clear, but recent evidence suggests that the shift in the oxyhemoglobin dissociation curve results in the delivery of 20 to 30 percent more oxygen prior to any actual increase in the number of circulating red cells. This compensatory mechanism may permit the erythropoietin levels to return to a level not measurable by the bioassay methods currently employed. How-

ever, sustained levels of ESF are seen in chronic severe anemia, presumably because the capacity of other compensatory mechanisms has been exceeded.

The principal source of erythropoietin in man is the kidney. That other sites of origin exist is apparent from the observation that erythropoietin activity can be detected in the plasma of certain anephric individuals. Evidence has been presented which suggests that ESF is formed by the enzymatic action of subcellular fractions of renal tissue (renal erythropoietic factor) on a globulin substrate (plasma precursor) of plasma or serum; however, complete agreement on this point is lacking. Although erythropoietin has not been completely characterized chemically, it resides in the mucoprotein fraction of plasma.

The principal mode of action of ESF is to induce differentiation of an erythroid precursor cell. The chain of biochemical events initiated by erythropoietin in the target cell and leading to erythroid cell differentiation remains to be defined. Erythropoietin increases iron incorporation into hemoglobin, and the enhanced incorporation of iron in heme requires DNA-dependent RNA synthesis. Recent reports indicate that ESF stimulates DNA and stroma synthesis and glycolysis. In summation, the production rate of erythroid cells is governed for the most part by erythropoietin which, in turn, is elaborated in response to differences of tissue oxygen requirements and the delivery of oxygen to the tissue. In a larger context, the erythropoietin system may be viewed as one facet of the total mechanism having to do with blood volume regulation. The role of the kidney in plasma volume regulation is discussed elsewhere in this volume (see Chapters 3 and 11).

Anemias resulting from abnormalities of the erythropoietin mechanism thus far have been related to inadequate production and inactivation. In the former, renal disease provides the best example. The cause of the anemia of renal disease is complex and involves defective production and, to a much smaller extent, increased destruction of red cells, both of which may be caused, in part, by toxins associated with renal failure. Current evidence indicates that erythropoietin activity is decreased in these patients, but whether this decrease is due only to insufficient production of ESF or to inactivation by inhibitors as well remains to be clarified. That erythropoietin antibodies can occur in certain diseases (e.g., thymoma) seems to be established. In this instance, the combination of ESF and the antibody prevents the biological activity of ESF in an assay animal.

Erythremia resulting from abnormalities of the ESF system may be categorized as inappropriate ESF activity, in contrast to the erythremia resulting from hypoxia and an appropriate increase in ESF (Table 23-1—III, A and B). The former is most commonly associated with tumors of the kidney (hypernephroma, cysts), the liver (hamartoma, hepatoma,) the cerebellum (cerebellar hemangioblastoma), the adrenals (adrenal carcinoma, pheochromocytoma), the uterus (fibromata) and the lung.

Whether the tumor secretes erythropoietin per se or stimulates the formation of ESF remains to be clarified. It has been shown that increased pressure on the rabbit kidney, such as occurs in hydronephrosis, results in increased ESF formation.

In these pathological states involving the ESF mechanisms, size, shape and color of the red cell are usually normal (normocytic, normochromic).

The Endocrine System

Other hormones may also be viewed as regulators of erythropoiesis (see Fig. 23-1 and Table 23-1—I, A.2). Hypofunction of the pituitary thyroid, testes or adrenals results in diminished red cell production, and hyperfunction of the thyroid or the adrenal glands is associated with increased rates of erythropoiesis. Administration of testosterone and certain analogues induces erythremia in animals and man, and various estrogenic preparations induce anemia. The mechanisms of action of these hormones have not been completely elucidated but must involve effects on metabolism and oxygen requirements of the organism, "growth-stimulating" effects, and specific effects on erythroid cellular function which at present are not defined. Awareness of the effect on metabolism and tissue oxygen requirements has provided a stimulus for the investigation of the interrelationship of testosterone, estrogen and thyroxin (and analogues) and erythropoietin production.

It is now clear that one of the mechanisms by which testosterone stimulates erythropoiesis is increased elaboration of erythropoietin, presumably by a direct renal effect. Administration of fluoxymesterone to eunuchoid males or females as treatment for neoplasms results in increased levels of urinary erythropoietin and a concomitant increase in circulating red cells. Evidence is accumulating that suggests a direct effect on the erythroid cell also. There is a clear correlation between the hypermetabolism and increased oxygen requirements of hyperthyroidism and an increased rate of erythropoiesis. (Erythremia does not occur, because of a mild hemolytic process that offsets the increased production.) One would expect erythropoietin levels to be increased; however, reports at present are conflicting. Circumstantial evidence in favor of this concept is provided by the fact that patients with hypothyroidism and decreased erythropoiesis respond to hypoxia and bleeding with an increase in erythropoiesis.

In general, the anemia of endocrine deficiency (ovarian deficiency excluded) is characterized by normal appearing erythrocytes and hemoglobin levels of about 10 g. per 100 ml. Usually the symptoms and signs of the primary disease predominate and the anemia is defined by laboratory methods. The anemia is corrected by hormone replacement therapy over a period of weeks. It is important to realize that clinically nonapparent endocrine deficiency may be the cause of anemia of obscure etiology. Further, the endocrine dysfunction may have other effects that result in more than one operating mechanism. For example, the diminished vitamin B-12 absorption associated with myxedema (hypothyroidism) may result in a syndrome that suggests pernicious anemia (vitamin B-12 deficiency).

Negative Feedback

It is clear that the absence of erythropoietin (induced by anti-ESF antisera) in normal animals is accompanied by the cessation of erythropoiesis; less clear, however, is the fact that, when the red cell life span is shortened, erythropoiesis may increase markedly in the absence of anemia and tissue oxygen deficit. This suggests that control mechanisms other than erythropoietin are operative. It has been postulated that a negative feedback mechanism (an inhibitor to erythropoiesis) related to aged RBC's may exist; thus, when aged RBC's are removed from the circulation,

erythropoiesis may proceed. Recent reports indicate the presence of erythropoietic inhibitory activity in the plasma of animals transfused to excess with RBC's, in erythremic man returning from high altitude and in animals that have been exposed to high oxygen. Also, various protein preparations, from urine, kidney and other tissues, with the capacity to inactivate erythropoietin have been described. Exploration of the control of erythropoiesis (negative feedback) is developing but more work is required before its clinical significance can be assessed. This concept is illustrated in Figure 23-1 by the broken lines, indicating that the ESF activity is diminished in the presence of excess blood volume, RBC or oxygen transport.

CELLULAR PROLIFERATION

Cellular Differentiation

Erythropoiesis ceases when circulating erythrocytes are present in excess. Those cells to the right of the unipotential stem cell (see Fig. 23-1) mature and enter the circulating blood and are not replaced. Examination of bone marrow five days after transfusion of excess RBC's reveals no erythroid cells. If the excess red cells are removed, or if the animal is made hypoxic or is given erythropoietin, erythropoiesis is resumed.

Current concepts hold that the erythroid cell line differentiates from a unipotential stem cell, the latter to be distinguished from those stem cells that are destined to differentiate into granulocytes (white blood cells) and megakaryocytes (platelet forming cells). The unipotential stem cell differentiates into the most immature form of the red cell that can be identified morphologically—the pronormoblast. The young erythrocytes then develop by nearly imperceptible gradations through an orderly sequence, arbitrarily divided (on the basis of morphological and staining charac-

teristics) into basophilic, polychromatophilic, and orthochromatic normoblasts,
reticulocytes and, finally, mature erythrocytes.

Although the differentiating erythroid
cells undergo well-recognized morphological changes, the process is continuous.
The pronormoblast is a cell characterized
by a large nucleus containing a nucleolus
and having dark-blue cytoplasm. During
differentiation and maturation, the nucleolus disappears (basophilic normoblast) and
the nucleus contracts, becomes gradually
pyknotic and is located eccentrically in the
cell; the cytoplasm changes from dark blue
to red (polychromatophilic and ortho-
chromic normoblasts) as the concentration
of hemoglobin increases. The pyknotic
nucleus is extruded and the young red cell
(reticulocyte) is released into the circulating blood. It has been estimated that 10
to 15 percent of the cells die at various
stages of development. About 3 mitotic
divisions occur during the evolution from
pronormoblast to normoblast. The generation of the cell (from completion of one
division to the completion of the next)
comprises 4 stages: (1) mitosis, G_0; (2)
postmitotic rest period, G_1; (3) period of
DNA synthesis, S; and (4) premitotic rest
period, G_2. The duration of this cell cycle
has been estimated to be 20 to 30 hours for
basophilic and polychromatic cells. The
time required for evolution to occur has
been estimated to be 30 hours for pro-
normoblasts, 12 to 64 hours for basophilic
normoblasts, 8 to 25 hours for polychromatophilic normoblasts, and 19 hours for orthochromatic normoblasts, or a total of about
3 to 5 days.

It is apparent that the erythron is not a
self-maintaining population of cells; rather,
the existence of the erythrocyte precursors
is dependent upon the differentiation of a
stem cell in response to physiological requirement of the organism which are
mediated by erythropoietin.

The stem cells, on the other hand, are
considered to be a population of cells that
are self-maintaining, i.e., when depleted by
the action of erythropoietin, they are
capable of replenishing their number. The
mechanism that limits the growth of the
stem cell compartment, in the absence of
the depleting effect of ESF, remains unknown. Self-regulation in other cell systems
has been attributed to inhibitors of cell
growth that are contained within the cells
themselves and reach a critical level in
dense cell populations. Knowledge of
regulation of this compartment is important
to our future understanding of defective
erythropoiesis and, hence, of anemia of obscure etiology.

Examples of diseases in which the stem
cell and/or its immediate progeny (the
erythroid, myeloloid and megakaryocyte
precursors) may be defective are polycythemia vera and aplastic anemia. Current evidence indicates that the marrow
of patients with polycythemia vera does not
respond to erythropoietin as normal marrow does when studied in an in vitro culture system. Patients exhibit erythremia
without increased levels of ESF and also
have increased numbers of platelets and
white cells for no apparent reason (Table
23-1—III, B.2). In this instance it seems
that the stem cell fails to recognize the
normal inhibition to growth and proliferates
autonomously. Patients with aplastic anemia may present with anemia, decreased
platelets (thrombocytopenia) and decreased white cells (leukocytopenia)
(Table 23-1—I, B.2). Erythropoietin levels
are frequently very much increased. In
this group of diseases one might postulate
either that the stem cells are unable to
respond to erythropoietin and those factors
stimulating platelet and white cell production or that the stimulating factors are in
some way defective.

Hemoglobin Synthesis

The production of normal erythrocytes
requires multiplication and maturation of

a highly differentiated series of cells having a unique intracellular physiology that requires optimal concentrations of proteins, vitamins and minerals and also, precise interactions of these components. Derangements of this balance result in structural alterations which may be recognizable microscopically. Two major biosynthetic pathways in the developing erythroid cells are heme and globin synthesis (Fig. 23-2).

Heme Synthesis. The initial step in heme biosynthesis is the condensation of glycine and succinate. Succinyl-coenzyme A derived from the tricarboxylic acid cycle combines with glycine to form α-amino, β-keto adipic acid which is then decarboxylated to form Δ-aminolevulinic acid. The formation of Δ-aminolevulinic acid from succinyl-CoA and glycine requires pyridoxal phosphate, coenzyme A, ferrous iron and the rate-limiting enzyme Δ-aminolevulinic acid synthetase (ALA synthetase). Two molecules of Δ-aminolevulinic acid condense to form porphobilinogen, a reaction that is catalyzed by the enzyme Δ-aminolevulinic acid dehydrase. Porphobilinogen is then converted to uroporphyrinogen III, the reaction being catalyzed by porphobilinogen deaminase and cosynthetase. Uroporphyrinogen III is transformed to coproporphyrinogen III by the enzyme uroporphyrinogen decarboxylase. Protoporphyrinogen IX is formed by the highly specific enzyme coproporphyrinogen oxidative decarboxylase and is then converted to protoporphyrin IX by protoporphyrinogen oxidase. (Uroporphyrinogen III and coproporphyrinogen III are autooxidized to uroporphyrin III and coproporphyrin III, and the latter compounds can be demonstrated in both nucleated and non-nucleated erythrocytes.) Iron is then inserted into protoporphyrin IX by the enzyme heme synthetase to form heme.

Regulation of heme biosynthesis is influenced by the availability of those enzymes and cofactors necessary for the activation of glycine and succinate (vitamin B-12, vitamin B-6, folic acid), by the rate-limiting enzyme ALA synthetase and by the availability of iron. The critical step is the formation of Δ-aminolevulinic acid, which is controlled by ALA synthetase. Recent evidence suggests that the activity of this enzyme is enhanced by erythropoietin. Conversely, feedback inhibition has been demonstrated, i.e., heme inhibits ALA synthetase when present in high concentration. Since pyridoxal phosphate is a coenzyme for ALA synthetase, the availability of pyridoxine (vitamin B-6) also may be limiting. It is well known that hemoglobin synthesis is reduced by iron depletion and that the addition of iron to red cells will increase hemoglobin synthesis (by an increase in polysome production). Heme influences cellular protein synthesis. Heme stimulates globin synthesis by promoting polyribosome formation and stabilization, possibly by more rapid translation of messenger RNA and by promoting the assembly of hemoglobin.

Protein synthesis within the developing red cell is thought to be similar to protein synthesis occurring in other cells (see Fig. 23-2). The genetic information required for the synthesis of the specific protein is coded in the deoxyribonucleic acid (DNA) of the cell nucleus. The DNA molecule is a double-stranded helix composed of alternating units of deoxyribose and a phosphate compound bound to each of the sugar units by the base pairs, adenine and thymine or guanine and cytosine. The genetic information coded in DNA is transferred by complementary base pairs of messenger RNA, which differs from DNA in that there is a single stranded helix, the sugar is ribose, and the base thymine is replaced by uracil; this RNA leaves the nucleus, carries the genetic information derived from the DNA of the gene, and directs the synthesis of the specific protein by the ribosomes in the cytoplasm. Transfer RNA, with an affinity solely for a specific amino acid, delivers the amino acids

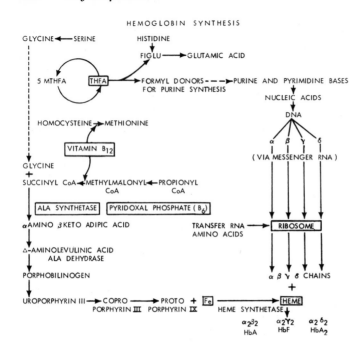

HEMOGLOBIN SYNTHESIS

FIG. 23-2. The metabolic pathways of heme and globin synthesis in which vitamin B-12, folic acid, vitamin B-6 and iron participate. THFA, tetrahydrofolic acid; 5M-THFA, N^5 methyl THFA; ALA synthetase, Δ-aminolevulinic acid synthetase.

to the ribosomes. The binding of the amino acid in transfer RNA requires a specific activating enzyme and energy in the form of ATP. The specific site of location of the amino acids on the ribosome is determined by the relationship of the order of the base pairs of the messenger RNA to the order of those of the transfer RNA. The ribosome moves along the messenger RNA template, and a polypeptide composed of precisely ordered amino acids is formed. The completed polypeptide chains are designated α, β, γ, and δ. Paired polypeptide chains form the various hemoglobins: 2 alpha and 2 beta chains combine to form hemoglobin A; 2 alpha and 2 gamma chains combine to form hemoglobin F; 2 alpha and 2 delta chains combine to form hemoglobin A_2.

This series of steps in the synthesis of hemoglobin requires control and coordination. The globin portion of the hemoglobin molecule may be regulated by the rate of amino-acid-binding to RNA during the assembly of chains on the ribosome or by the rate of chain release from the ribosome. It may be that structural alterations of the chains also alter synthesis, because the

rate of synthesis of abnormal hemoglobin is usually less than that of normal hemoglobin.

Of greater importance clinically is the genetic control of hemoglobin chain synthesis. In hemoglobin A ($\alpha_2^A\beta_2^A$), the α and β chain synthesis is controlled by pairs of genes at different genetic loci. The genotype of hemoglobin A may be expressed as

$$Hb\alpha^A/Hb\alpha^A\text{-}Hb\beta^A/Hb\beta^A,$$

where the superscript denotes the gene controlling the synthesis of the chain. The gene that affects the rate of a chain's synthesis is located near the gene that affects the structure of that chain. This close proximity of genes results in the inheritance of rate and structural mutations as alleles. (An allele is one of two or more forms of a gene that occurs at a given locus on a chromosome.)

Normally, during fetal development, α chains combine with ϵ chains to form hemoglobin Gower, with γ chains to form hemoglobin F, with β chains to form hemoglobin A and with δ chains to form hemoglobin A_2. In adult life, hemoglobin A ($\alpha_2\beta_2$) is the major component of hemoglobin, with hemoglobin A_2 ($\alpha_2\delta_2$) comprising 2.5 percent of the total. Thus, during fetal development β and δ chain synthesis remains at a low level until birth, when γ chain synthesis decreases and β and δ chain synthesis becomes fully activated.

The significance of the genetic control lies in the potential for various combinations of normal and abnormal chains, which may result in hemoglobin with a variety of structural and functional abnormalities (Table 23-2). Individuals who inherit one abnormal gene will make two hemoglobins —one normal and one abnormal. An example could be normal adult hemoglobin, hemoglobin A ($\alpha_2^A\beta_2^A$) and the abnormal hemoglobin, hemoglobin S ($\alpha_2^A\beta_2^S$). (The genotype may be written as $Hb\alpha^A/Hg\alpha^A$- $Hb\beta^A/Hb\beta^S$.) Two abnormal genes of the same type may be inherited, e.g., $Hb\alpha^A/Hb\alpha^A$- $Hb\beta^S/Hb\beta^S$, in which case only abnormal hemoglobin S ($\alpha_2^A\beta_2^S$) is formed (Table 22-2). Two abnormal genes of different type but at the same locus may occur, e.g., $Hb\alpha^A/Hb\alpha^A$-$Hb\beta^S/Hb\beta^C$, and only hemoglobin

S ($\alpha_2^A\beta_2^S$) and hemoglobin C ($\alpha_2^A\beta_2^C$) are formed. Two abnormal genes of different type and at different loci may also occur, e.g., $Hb\alpha^A/Hb\alpha^G$-$Hb\beta^A/Hb\beta^C$, with the formation of four hemoglobins.

In addition to the genes that regulate chain structure, certain genes regulate the rate of synthesis of chains without altering their structure (Thalassemia, Table 23-2). If by a mutation in the regulator gene controlling the rate of β chain synthesis, that rate is diminished, the activity of its allelic gene increases. Varying amounts of γ and δ chain production may occur, with cells having variable proportions of Hb A ($\alpha_2\beta_2$), Hb F ($\alpha_2\gamma_2$) and Hb A_2 ($\alpha_2\delta_2$). If the allelic gene directs synthesis of an abnormal hemoglobin such as S, the only β chains formed are those determined by the β^S gene; little Hb A ($\alpha_2^A\beta_2^A$) is formed and Hb S ($\alpha_2^A\beta_2^S$) is unimpeded.

Gene mutations resulting in structural alterations of the chains (amino acid substitutions) result in clinical disease if they alter the spatial or electronic configuration of the hemoglobin molecule. Normally, the molecule is spheroidal in shape, measuring

TABLE 23-2. Examples of Abnormal Hemoglobins, with Physiological Effects*

HEMOGLOBIN	AMINO ACID SUBSTITUTION	FUNCTIONAL DERANGEMENT	MORPHOLOGY OF RBC	CLINICAL PHYSIOLOGICAL EFFECT
S	$\beta6$ glu $\to$ val	Structural change with tactoid formation	Sickled cells	Increased viscosity $\to$ thrombosis Hemolytic anemia
M Boston	$\alpha58$ his $\to$ tyr	Increased stability methemoglobin	Normal	Cyanosis, erythremia in some instances
Chesapeake	$\alpha92$ arg $\to$ leu	Increased O_2 affinity	Normal	Erythremia
Kansas		Decreased O_2 affinity	Normal	Cyanosis, mild anemia
Zurich	$\beta63$ his $\to$ arg	Unstable	Normal, but inclusions after drugs	Hemolytic anemia and methemoglobinemia after drug ingestion
Thalassemia	None	Decreased chain synthesis	Hypochromia Target cells	Hemolytic anemia

* Adapted from Harvey, A. M., *et al.*: The Principles and Practice of Medicine. Ed. 17 Table 4, Chapter 55, p. 522, New York, Appleton-Century-Crofts, 1968.

approximately $64 \times 55 \times 50$ Å, and has a molecular weight of 64,458. It is composed of 4 heme groups and 4 sub-units of globin, the two α and the two β chains which are coiled and arranged in a tetrahedral array. The 4 heme groups are positioned in separate crevices on the surface of the molecule. The configuration is maintained by various noncovalent forces, important among which are the bonding forces (van der Waals) between the internal hydrophobic side chains of each amino acid. The stability of hemoglobin is largely dependent on the interaction of the globin molecule and the heme group that forms hydrophobic bonds with the internal side chains of the globin. A further stabilizing force is the interaction of the α and the β chains. The contours of the folded or coiled chains are complementary, though showing few actual sites of contact. The interior of the chains is composed of nonpolar residues which are in contact with their neighbors by van der Waals forces; the side chains that are ionizable lie at the surface of the subunits. The heme groups, lying in widely separated crevices, are suspended between two histidine residues. The iron of the heme is attached directly to the proximal histidine residue. The distal histidine residue lies opposite (but is not attached to) iron, to provide the site of combination with molecular oxygen.

Structural changes may result in hemoglobins that are unstable or have an altered oxygen affinity or have increased stability or have aberrant physical properties that change the shape of the cell (Table 23-2). For example, hemoglobin Zurich (Table 23-2) is an unstable hemoglobin resulting from the substitution of arginine for the heme-linked histidine in the β chain. Individuals having this hemoglobin have mild hemolytic anemia which is exacerbated if they are exposed to certain drugs that induce precipitation of hemoglobin within the cell, leading to increased rates of cell destruction.

Altered oxygen affinity may be related to the steric configuration of the α and the β chains which is critical to the role of hemoglobin in oxygen transport. During oxygenation, the chains move together. Amino acid substitution in this local region of the molecule interferes with heme-heme interaction, with the result that oxygen affinity is increased. Hemoglobin Chesapeake serves as an example (Table 23-2). The release of oxygen to the tissue is impaired, and the patient develops erythremia as a compensatory mechanism to provide adequate oxygen delivery. In contrast, Hb Kansas possesses a decreased oxygen affinity, with an associated decrease in red cell production and consequent cyanosis. Hemoglobin M is a hemoglobin that is so stable that the normal erythrocyte-reducing enzymes are ineffective. In hemoglobin M, the heme-linked histidines are replaced by tyrosine, allowing trivalent iron to form a stabilizing intramolecular bond which prevents reduction by methemoglobin reductase. Oxygenation is impaired, and mild erythremia may result as a compensatory mechanism analogous to that occurring in patients with Hb Chesapeake.

Sickle hemoglobin presents the classic example of a marked change in physical properties. Substitution of valine for glutamic acid in position 6 of the β chain results in an intramolecular ring, with valine in position 1 of the same chain. During deoxygenation the change in the spatial relationship of the α and the β chains results in deformation of the cells, which appear microscopically to be sickle-shaped. The cell is rigid and fragile and tends to adhere to other cells. This change in the red cell results in a disease characterized by hemolytic anemia and thrombus formation in small vessels. Patients present with a history of chronic anemia and jaundice (associated with hemolysis), beginning in childhood. They often have severe abdominal pain and limb pain and, also, pneumonia-like episodes, owing to repeated pulmonary thromboses. Increased red cell destruction occurs, with deoxygenation and infection.

Thalassemias. Alterations in the rate of chain synthesis without structural changes in the chain are now recognized as responsible for a group of diseases, called the thalassemias, in which partial or complete suppression of one of the chains occurs. Failure to synthesize β chains results in diminished hemoglobin formation in each cell, and the cells are pale (hypochromic). The relative excess of α chains may precipitate in the cell and form inclusion bodies. Thus, although the chains per se may be normal, their disproportion results in a structurally altered red cell that is hypochromic and has a shortened survival time.

At present, more than 100 different abnormal hemoglobins have been described, and, as stated earlier, an individual may inherit more than one abnormal hemoglobin. Examples of abnormal hemoglobins, together with the amino acid substitution, functional derangement and physiological expression are presented in Table 23-2. The biochemical defect appears to determine the altered physiology, and it is evident that patients having a hemoglobinopathy may present with anemia or erythremia. The erythrocyte may be normal in appearance or it may be markedly changed in size, hemoglobin content and configuration. Rarely the clinical syndrome is sufficiently characteristic to suggest the hemoglobinopathy, e.g., homozygous sickle-cell disease. Red cell morphology may assist in providing clues, but the diagnosis depends on further study of the hemoglobin by chemical methods.

Nutritional Factors

Vitamin B-12. The importance of vitamin B-12 and folic acid in hemoglobin and protein synthesis has been mentioned previously. Vitamin B-12 participates in the biosynthesis of methionine from homocysteine (Fig. 23-2). This reaction involves the release of N^5-methyl-tetrahydrofolate to available tetrahydrofolate (THFA). Vitamin B-12 is required for normal propionate metabolism as a coenzyme to methylmalonyl Coenzyme A isomerase in the transformation of propionyl CoA to succinyl CoA. Increased levels of methylmalonate in urine are attributed to the defect in this conversion and are associated with B-12 deficiency. In regard to protein metabolism, B-12 may affect the conversion of ribonucleotides to deoxyribonucleotides, or it may have the direct effect of making labile groups available from precursors such as formate, the α carbon of glycine or the β carbon of serine. Although disagreement exists in regard to the mechanism, the fact remains that DNA synthesis in marrow erythroid cells is deranged in patients with B-12 deficiency.

Vitamin B-12 is synthesized by microorganisms and not by mammalian cells. The dietary intake (3 to 30 μg. per day) is in the form of animal protein, hence deficiency states (with rare exceptions) are the result of impaired absorption. Normally, the absorption of B-12 in the ileum is dependent upon the presence of Castle's intrinsic factor, a complex mucoprotein (m.w. 50,000) secreted by the parietal cells in the fundus of the stomach. Intrinsic factor preferentially binds B-12 in the acid milieu of the stomach, and the complex is absorbed by specific receptor sites in the microvilli of the brush border. The absorbed B-12 is transported by a beta globulin (transcobalamin II) in the plasma. The total body stores of B-12 are estimated to be about 4 mg., of which 1 mg. resides in the liver. Because the daily requirement is only about 1 μg., a period of 3 to 10 years is required for B-12 deficiency to become manifest if absorption fails to occur. Vitamin B-12 deficiency and anemia may occur then, from inadequate dietary intake by food faddists who avoid meat for prolonged periods of time; from malabsorption resulting from lack of intrinsic factor or due to disease of the ileum, where the B-12–intrinsic-factor complex is absorbed; or from inadequacy of the amounts available for absorption, because of utilization of B-12 by intestinal bacteria in blind loops of bowel or diverticula. Pernicious anemia results from B-12 deficiency caused by lack of intrinsic factor. Individuals who have had surgical removal of the gastric fundus can be expected to develop anemia when the stores of B-12 become depleted. Impaired absorption may occur with any disease of the small intestine that affects the ileum, including tropical sprue, regional enteritis, tuberculous enteritis, ileal resection, lymphoma and Whipple's disease.

Folate. The biological function of folic acid is dependent upon its conversion to N^5-methyl-tetrahydrofolate (Fig. 23-2). Its biochemical role is that of a coenzyme in a number of reactions in which single carbon atoms are transferred to or built into intermediates. The metabolic systems dependent upon folic acid coenzymes are the conversion of serine to glycine the synthesis of methionine, thymine and purines, and the degradation of histidine and purine. A deficiency of tetrahydrofolate (THFA) may result from folic acid deficiency or B-12 deficiency which impairs the transformation of N^5-methyl-tetrahydrofolate (5M-THFA) to THFA. This relationship provides an explanation for the correction by large amounts of folic acid, of anemia due to B-12 deficiency. THFA is necessary for the transformation of histidine to glutamic acid. In this reaction histidine is metabolized to formiminoglutamic acid (FIGLU) which, in turn, yields its formimino group to tetrahydropteroylglutamic acid, resulting in the formation of 5-methyl-tetrahydrofolic acid (5M-THFA). This reaction provides an explanation for the increased excretion of formiminoglutamic acid (FIGLU) in the urine of folate-deficient subjects and in about one third of B-12-deficient subjects, after histidine administration.

Folic acid occurs naturally as polyglutamic folate in yeasts, vegetables, liver, fruits, grains, nuts and eggs. Most diets contain more than 1,000 µg. per day; the minimal requirement is about 50 µg. per day. Folic acid is absorbed in the jejunum and is present in serum as 5M-THFA (5.9 to 21 µg./ml.). Total stores of N^5-methyltetrahydrafolic acid are 5.0 to 10 mg., of which 3.5 to 7.5 mg. are stored in the liver. Because the daily requirement of folic acid is about 100 times that of B-12, the stores of folic acid are depleted more readily (in weeks) than are those of B-12 (in years) when dietary intake is inadequate or absorption is defective. Deficiency states result from inadequate dietary intake, inadequate absorption associated with intestinal disorders or the use of certain drugs (diphenylhydantoin, oral contraceptives, methotrexate) that inhibit absorption and, finally, increased requirements associated with increased RBC production rates (e.g., in hemolytic anemias, pregnancy, childhood and hyperthyroidism).

In contrast to Vitamin B-12, the supply of folic acid and its utilization by the erythroid precursors is affected more readily by relatively acute disease processes. In fact, relative folic acid deficiency is now recognized as being second only to iron deficiency as a cause of anemia.

In the absence of B-12 or folic acid, effective red cell production and red cell survival are decreased. In normal erythropoiesis, cytoplasmic RNA gradually decreases and has almost disappeared by the time hemoglobin appears. In B-12 and folate deficiencies, RNA disappears more slowly and persists into the stages during which hemoglobinization is well advanced. The derangement of DNA synthesis is believed to result in prolonged resting phases between mitoses. The resulting cell is designated the megaloblast, since it is larger than normal size and has an excessive amount of cytoplasm, with disproportionate aging of the nucleus.

Because appearance of the erythroid cells (megaloblasts) and the erythrocytes (macrocytes) are the same in B-12 and in folic acid deficiency, additional chemical and functional tests are required to distinguish between the two. Vitamin B-12 absorption can be evaluated by the oral administration of radioactive B-12 and subsequent measurement of the radioactive B-12 appearing in the plasma, liver and stool. If, in addition to the oral radioactive B-12, a parenteral injection of vitamin B-12 is given, the plasma B-12 levels are increased sufficiently to exceed the renal threshold, and the excretion of the radioactive B-12 in the urine can then be measured. If absorption is diminished, the amount of B-12 in the stool is increased, while B-12 in the plasma, liver, and urine is decreased. Vitamin B-12 deficiency can also be determined by direct measurement of the serum B-12 level and by the measurement of methylmalonate in the urine. Folic acid deficiency may be evaluated by the direct measurement of serum folate and by the administration of physiological doses of folic acid. The importance of distinguishing between deficiency of vitamin B-12 and folic acid deficiency lies in the awareness of the neurological disase which may develop if B-12 deficiency is not recognized.

Vitamin B-6. The availability of vitamin B-6 may influence hemoglobin synthesis in certain diseases. Vitamin B-6 is the class name for several naturally occurring forms including pyridoxine, pyridoxal phosphate and pyridoxamine. The 5-phosphate of pyridoxal is the coenzyme of ALA synthetase, hence deficiency would be expected to result in decreased heme synthesis. Although deficiency of B-6 is not associated with anemia in man, certain anemias (sideroblastic) do improve with the administration of pyridoxine. Certain drugs, such as isonicotinic acid hydrazine, which is used in the treatment of tuberculosis, inhibit pyridoxal phosphate and may induce anemia.

Protein. The significant contribution of the nutritional factors, vitamin B-12, folic acid, vitamin B-6 to normal protein synthesis is clear (Fig. 23-2). Less clear, however, is the effect of dietary protein deprivation on erythropoiesis. The reason for this is that nutritional deficiencies in man are almost invariably multiple and hence the singular effect of protein deprivation is difficult to separate from the effects of deprivation of B-12, folic acid, iron and other nutrients. Anemia does occur in protein-deprived animals, and it is generally agreed that protein intake affects the erythron in man. The mechanism may affect the synthesis of the cell directly, or it may affect the rate of differentiation and proliferation through the erythropoietin mechanism. It is known that the metabolic rate decreases during protein deprivation and, hence, oxygen requirements are less. In this context, the lower levels of ESF may be viewed as a physiological adjustment to the reduction in oxygen requirement.

Iron. The crucial role of iron in normal hemoglobin synthesis is more readily appreciated if one recalls that this element is a major structural component in the hemoglobin molecule and, in a larger sense, is one of the most important biocatalytic elements in enzymology. Inasmuch as 6.3 g. of hemoglobin, containing 21 mg. of iron, are synthesized and degraded in each 24 hours (3 million erythrocytes per second), inadequacy of the supply of iron to the developing erythroid cells becomes apparent, in a short period of time, in diminished erythropoiesis and the formation of small erythrocytes that have decreased hemoglobin concentration. (Evidence has been presented that suggests that cell size is determined by the duration of time between cell divisions and that this interval is determined in part by a critical level of hemoglobin concentration.) The supply of iron to the erythroid cell is dependent upon adequate iron transport and iron stores; and adequate stores are dependent upon the availability of iron and its absorption, as well as on the loss of iron from the body. In the absence of iron loss the iron content of the body remains relatively constant.

Normally, in the U. S. the typical man, who consumes 2,500 calories per day, ingests about 15 mg. of iron, of which about 1.0 mg. (10 to 15%) is absorbed. The range of absorption has been estimated to be from 0.5 mg. in the iron-replete individual to 3.0 mg. in the iron-deficient person. Loss of iron occurs through occult blood in the intestine (0.2 to 0.5 mg.), urine excretion (0.1 mg.), desquamation of skin (0.2 to 0.3 mg.) and desquamation of gut epithelium. The net obligatory loss of iron is therefore estimated at about 1.0 mg., and normal absorption can keep pace. To compensate for menstrual blood loss (0.5 mg. of iron daily), the mature woman normally must absorb 0.7 to 2.0 mg. per day; during pregnancy (with its requirement of 440 to 1,050 mg. of iron), she must absorb about 3.5 mg. of iron per day in order to compensate for the loss. In the man, iron loss beyond the obligatory loss of 1.0 mg. results from abnormal bleeding. If 1.0 ml. of erythrocytes contain 1.0 mg. of iron, a small amount of blood (6 to 10 ml.) lost each day can result in negative iron balance and depletion of iron stores, for which the maximal absorption capacity of dietary iron of 3.0 mg. cannot compensate. Defective hemoglobin formation and anemia ensue. The sequence of events is: iron deficiency, depletion of iron stores, decreased saturation of the transport protein, decreased delivery to the erythroid cell, impaired hemoglobin synthesis, and anemia. It follows that the first evidence of iron deficiency is diminished iron stores (which can be evaluated by appropriate

staining of marrow) and that the change in red cell size and anemia are late manifestations of iron deficiency.

Absorption. Control of iron absorption is relative and not absolute, since excessive soluble inorganic iron presented to the gut can be absorbed. Normally, the mechanism is responsive to body needs, with greater amounts being absorbed in the iron-deficient state and lesser amounts when the stores are replete. Absorption is greatest in the proximal portion of the small intestine, in contrast to B-12 and folic acid. Luminal factors that influence but do not "control" iron absorption are: hydrochloric acid of gastric juice, the presence of ascorbic acid, and the formation of soluble or insoluble iron complexes (phosphates, phytates). Evidence in regard to the influence of gastric juice on iron absorption is at present conflicting, although a protein factor that binds iron has been described. It is clear, however, that iron absorption is impaired in individuals who have undergone removal of 80 percent of the stomach. The intestinal mucosa does play an important role in regulating iron absorption when the dietary intake of iron is within a physiological range. The control is apparent in relation to iron stores and the rate of erythropoiesis. Absorption is increased in iron deficiency, with or without accompanying anemia, and when the rate of erythropoiesis is accelerated, as in hemolytic anemia (increased destruction of erythrocytes). The messenger to the mucosal cell remains unknown, although it is postulated that the concentration of iron within the mucosa to some extent regulates the absorption of iron by the cell. This "messenger" iron is thought to be incorporated into the cell at an early stage of development, while the cell is in the crypts of Lieberkuhn, and, as the cells migrate along the villi, they may accept additional iron. Iron transfer across the cell is believed to be the result of a metabolic process. A portion of the absorbed iron is rapidly delivered to the plasma and a portion is found as ferritin. Increased rates of absorption are associated with the presence of relatively small amounts of ferritin iron; decreased rates of absorption are associated with larger amounts of ferritin, suggesting that a deviation of iron to ferritin prevents excessive absorption. The level of plasma iron may influence the concentration of iron in the young crypt cell and, hence, may influence iron absorption. In addition to these possible mechanisms, the luminal iron concentration also affects the absorption rate.

Iron-binding. Iron entering the plasma is transported by an iron-binding protein (each molecule binds two atoms of iron) that has the electrophoretic mobility of a β_1 globulin. At least 14 molecular species have been identified, the most common having a molecular weight of 92,000 to 93,000. This transport protein is normally one-third saturated with iron, the quantity of iron ranging from 60 to 200 μg. per 100 ml. of plasma. Plasma iron concentration represents the balance between iron delivered to plasma from the gut and hemoglobin breakdown and iron removed from the plasma by heme biosynthesis, cell metabolism and deposition in storage sites. This dynamic interchange is demonstrated by the fact that from 27 to 32 mg. of iron enter and leave the plasma pool during a 24-hour period. Of the incoming iron, about 80 percent is transported to the erythron and 20 percent to iron storage sites.

Iron (500 to 1,000 mg.) is stored intracellularly as ferritin (a water-soluble protein, apoferritin, combined with iron and having ferric hydroxide phosphate micelles at the periphery and iron aggregates at the center of the molecule) and hemosiderin (large aggregates of ferritin molecules, together with porphyrin and other pigments) in liver, spleen, bone marrow and muscle. Radioactive iron tracers reveal that iron is first incorporated into ferritin and later into hemosiderin and, during iron deple-

tion, is first mobilized from ferritin. An estimate of iron stores can be obtained from observing the hemosiderin granules in smears of bone marrow stained with Prussian blue. The incorporation of iron into ferritin is reported to be dependent upon oxidative reaction requiring ATP and ascorbic acid, and the release of iron from ferritin is postulated to be dependent upon the reduction of ferritin by the enzyme xanthine oxidase.

Kinetics. The kinetics and pattern of distribution of iron entering the body can be summarized as follows: Ingested iron salts are absorbed quickly, and an increase in plasma iron (transferrin-bound) can be detected within 30 minutes and reaches peak values within 2-1/2 to 5 hours. If the iron is injected intravenously, it combines with transferrin immediately. The time required for 50 percent of the iron to leave the plasma is about 60 to 90 minutes. This plasma iron clearance, depicted graphically, describes a curve comprised of about three exponential functions—the reason being that iron is leaving the plasma (transferrin) transport system and later may recirculate from labile iron pools. As iron leaves the plasma, it can be detected within the marrow in a matter of minutes. (The incorporation of iron in the developing cell and into hemoglobin is known to require only 6 to 8 minutes.) The iron then appears in newly formed erythrocytes in the circulating blood over a period of one to 10 days, depending upon the rate of erythrocyte production.

Iron deficiency, the most common cause of anemia, results in defective hemoglobin synthesis, decreased cell proliferation, and erythrocytes that are small, pale and misshapen, with a shortened survival time. Similar morphological characteristics are seen in other conditions that affect heme biosynthesis, e.g., thalassemia, the sideroachrestic anemias and pyridoxine-responsive anemias. Iron deficiency anemia may be identified by direct vizualization of iron in hemosiderin within the marrow and direct measurement of the iron transport proteins in serum and the proportion of iron attached to the protein (serum iron binding capacity and serum iron). Having determined that the cause of the anemia is iron deficiency, the next step is to determine why iron stores are depleted. This may result from dietary inadequacy, malabsorption (disease of stomach or small intestine), iron loss (usually chronic blood loss), iron requirements greater than available iron and, rarely, inadequate transport protein.

THE MATURE ERYTHROCYTE

Metabolism and Survival

The mature cell is a biconcave disc, seven microns in diameter, with a soluble core and a membrane, 75 Å thick, which is comprised of lipids, protein, ions and water. The classical concept postulates a bimolecular layer of lipids covered internally and externally by a layer of protein. The membrane lipids (phospholipids, free cholesterol and glycolipids) are not synthesized in the RBC; however, some of the lipids do exchange with corresponding plasma compounds. The nonhemoglobin membrane proteins are less well defined, owing, in part, to the complexities of protein fractionation (and the different methods used) and to their interractions with each other. Important functional ligands of the protein are the membrane SH-groups which, when reacted with specific chemicals, result in alteration of membrane function. An important characteristic of the protein is its high concentration of glycoproteins, which accounts for the acidic nature of the surface. It has been suggested that cells agglutinate more readily when the surface charge and the heme electrostatic repulsive forces are disrupted by agents such as antibodies. Many enzyme activities have been reported to be associated with the membrane, including some of the glycolytic enzymes such as glyceraldehyde phosphate dehydrogenase and phosphoglycerate kinase. Despite incomplete knowledge in regard to the structure and chemical components of the membrane, the concept and certain functions attributable to the membrane are helpful in understanding hemolytic mechanisms. The boundary between the "membrane" and the interior of the cell is viewed as being dynamic—functionally related to and dependent on cellular metabolism.

The red cell is unique in that by the time it has matured and left the bone marrow, it is the only cell that survives without nucleus, ribosomes, mitochondria, an intact Krebs cycle, cytochrome system, or the capacity for oxidative phosphorylation. Despite these losses in metabolic machinery, the cell survives for approximately 120 days, during which time it travels a distance of 100 miles between the heart and various tissues. Its success in withstanding the buffeting received by a free cell in constant motion is determined in large measure by its unique shape; a biconcave disc is deformable. A sphere is relatively rigid, and any factor that causes spherocytosis will decrease deformability and, hence, survival. The normal cell can withstand great bending forces without irreversible change; however, tangential stress (stretching) leads to an increase in membrane tension. Stretch resulting in an increase of surface area greater than 10 to 15 percent causes hemolysis. Maintenance of the biconcave disc shape and deformability is dependent upon the energy metabolism of the cell.

Metabolism. An understanding of the concepts concerning the glycolytic pathways of the mature erythrocyte provides a basis for considerations of normal red cell aging (senescence) and of certain genetic and acquired metabolic defects that result in abnormal shortening of the life span of the erythrocyte, which is manifested as hemolytic anemia (Table 23-1, II A).

Energy requirements are met by the

metabolism of glucose to lactate through two principal pathways, the Embden-Meyerhof anaerobic glycolytic pathway (EMP) and aerobic oxidative hexose monophosphate shunt (HMP) (Fig. 23-3).

In the hexose monophosphate shunt, 10 percent of the glucose is metabolized, providing 25 percent of the potential energy by generation of NADPH. Regeneration of NADPH by glycolysis is important to the methemoglobin reductase system which reduces methemoglobin to hemoglobin. The maintenance of functional hemoglobin is thus dependent on this glycolytic pathway (HMP). The HMP pathway is limited by glucose-6-phosphate dehydrogenase (G-6-PD) activity and by the formation of reduced glutathione (GSSG).

In the Embden-Meyerhof pathway 1 mole of glucose is catabolized to 2 moles of lactate, with the generation of 2 moles of adenosine triphosphate (ATP). ATP serves mainly as substrate for the Mg^{++}, K^+, and Na^+-activated ATP-ase reaction, which is essential for the active transport of potassium into the cell and sodium out of the cell by the cell membrane, thus preserving chemical intracellular gradients and, also, the molecular configuration of the cell membrane.

The rate-limiting steps in the EM pathway have been identified as the reactions catalyzed by hexokinase (HK), phosphofructokinase (PFK) and pyruvate kinase (PK). Hexokinase is inhibited by its product, glucose-6-phosphate (G-6-P), PFK is inhibited by ATP, and PK activity is limited by the concentration of its substrate, phosphoenolpyruvate (PEP). Inorganic phosphate also serves to regulate glycolysis, inasmuch as the inhibitory effect of glucose-6-phosphate and ATP can be overcome by phosphate.

The glycolytic generation of ATP may be diminished by a metabolic bypass in the conversion of glyceraldehyde-3-phosphate (G-3-P) to lactate. G-3-P is phosphorylated to 1,3-diphosphoglycerate (1,3-DPG), which may then be converted to 3-phosphoglycerate (3-PG), with the generation of 1 mole of ATP, or mutated to 2,3-diphosphoglycerate (2,3-DPG), which may then be returned to the main pathway as 3-phosphoglycerate without the generation of the energy-rich ATP. This shunt, called the Rapaport-Luebering shunt, has been considered to be wasteful of a high percentage of glucose in the EM pathway.

2,3-DPG. The role of 2,3-diphosphoglycerate (2,3-DPG) in the economy of the erythrocyte has been poorly understood. Recent evidence suggests that this material may be a major determinant of hemoglobin affinity for oxygen. This important relationship and its significance in total oxygen transport in the body is discussed in detail in Chapter 8.

Normal Senescence. Normal aging of red cells is associated with denaturation of protein enzymes and a 50-percent decrease in enzyme concentration which cannot be replenished. These enzymes can also be modified by drugs, nutrients and the toxic effects of certain diseases. More specifically, when the concentrations of the rate-limiting enzymes hexokinase (HK) and pyruvate kinase (PK) are compared in old and in young erythrocytes, a decrement of 75 to 80 percent is found. With the decrease in HK a concomitant decrease in ATP also occurs; the membrane function is affected; intracellular sodium and water increase; the biconcave shape becomes spheroidal; and the cells are more susceptible to osmotic lysis. Initial changes in membrane deformability occur when ATP is 70 percent of the initial value.

Intracorpuscular Abnormalities — Inherited. In the HMP pathway, the enzymes that may be deficient owing to hereditary factors include glucose-6-phosphate dehydrogenase and 6-phosphogluconate dehydrogenase. In the EM pathway they include pyruvate kinase, triose phosphate isomerase, hexokinase, glucose phosphate isomerase, phosphoglycerate kinase and diphosphogly-

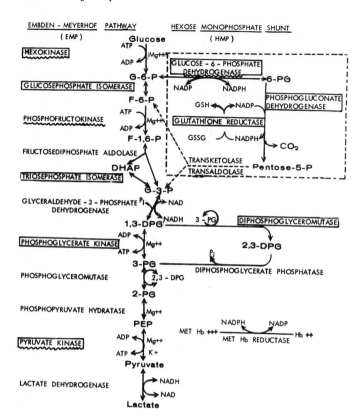

FIG. 23-3. Metabolism of the erythrocyte, glycolytic pathways. Enzymes underlined ᔕᔕᔕ are those which are rate-limiting. Enzymes enclosed ⬚ are those which have been found deficient due to hereditary factors. G-6-P, glucose-6-phosphate; 6-PG, 6-phosphogluconate; F-6-P, fructose-6-phosphate; F-1,6-P, fructose-1,6-phosphate; DHAP, dihydroxyacetone phosphate; G-3-P, glyceraldehyde-3-phosphate; 1,3-DPG, 1,3-diphosphoglycerate; 2,3-DPG, 2,3-diphosphoglycerate; 3-PG, 3-phosphoglycerate; 2-PG, 2-phosphoglycerate; PEP, phosphoenolpyruvate.

ceromutase. Nonglycolytic enzymes that may be deficient are glutathione reductase, adenosine triphosphatase, glutathione peroxidase and glutathione synthetase.

The diseases that are the result of hereditary deficiencies of glycolytic and nonglycolytic enzymes are referred to as the hereditary nonspherocytic hemolytic anemias (Fig. 23-3; enzymes enclosed in boxes). They share the characteristics of a hemolytic anemia resulting from an intrinsic defect of the erythrocyte: the erythrocytes are not spherocytic; there is no hemoglobin variant (i.e., structural defects in hemoglobin chains or defective synthesis), and there is evidence for genetic control of the disease. Glucose-6-phosphate dehydrogenase (G-6-PD) deficiency was the first to be described and serves as an example. It is an inherited sex-linked disorder, occurring in all races. Two

forms of the disease occur. The first is associated with chronic hemolytic anemia and occurs primarily in individuals with Northern European ancestry. When G-6-PD is low, glutathione (GSSG) cannot be maintained in reduced form (GSH) because of impaired ability to reduce NADP in the HMP pathway, and the reduction of methemoglobin (Met Hb^{+++}) is impaired. In the second form, which has a relatively high incidence in Negroes, and peoples of Mediterranean ancestry, the hemolytic anemia is not noticeable until the individual is exposed to an oxidative agent (e.g., fava beans, primaquine, sulfonamides). Hemolysis then occurs and affects the more senescent cells (because with aging, G-6-PD level decreases) which are unable to withstand the oxidant stress.

Abnormal ion transport in the red cell is also recognized as a feature of several forms of hemolytic anemia. Accumulation of sodium in excess of potassium loss is associated with a gain in cell water and swelling of the cell. Univalent cations move slowly across the cell membrane in response to the concentration gradients of cell and plasma. In order to maintain an appropriate level, the energy-dependent membrane-localized cation pump returns sodium from the cell to the plasma and potassium from plasma to cell. The membrane ATPase is activated by Mg^{++}, Na^+, and K^+. The pump accelerates glycolysis and ATP production, presumably by providing ADP for the phosphoglycerate kinase reaction and inorganic phosphate for the hexokinase and phosphofructokinase reactions (Fig. 23-3). In hereditary spherocytosis, an intrinsic RBC membrane defect is inherited as an autosomal dominant in which there is a leakage of sodium into the cell, which activates the sodium pump, thus requiring increased glycolysis. In addition to the decrease in surface area and an increase in permeability to sodium, the membrane is also deficient in both cholesterol and phospholipid and the lipids are unstable, with inordinate lipid loss occurring with energy depletion. The result is a spherocytic, rather than biconcave, cell which, owing to loss of deformability, is destroyed in the normal spleen. Surgical removal of the spleen allows RBC survival of sufficient duration to prevent anemia.

Intracorpuscular Abnormalities—Acquired. Paroxysmal nocturnal hemoglobinuria is a disease that provides an example of an acquired, rather than a familial, intrinsic defect of the RBC. The major manifestation of the abnormality is increased susceptibility to destruction in the presence of complement and in the absence of demonstrable antigen-antibody reaction. There is also an associated deficiency of the enzyme acetylcholinesterase.

Extracorpuscular Abnormalities — Acquired; Associated with Antibodies. Survival of normal red cells with no intrinsic defect may be threatened by alterations in the cellular environment. The cause of cell destruction is then considered to be extrinsic to the cell and may be the presence of antibodies or other factors such as infection, chemical agents, or diseases that affect small blood vessels. Antibodies may have a detrimental effect on the cell by activation of the complement system. Serum complement is the name for a system of 9 proteins which react sequentially after activation of antibody at the cell surface. Components C'8 and C'9 are involved in the final cytolytic step of the sequence which causes a membrane defect 80 to 100 Å in diameter. The effect is a disruption of ionic equilibrium due to bypass of the Na^+-K^+ pump resulting in the exodus of K^+ from the cell. The oncotic pressure of hemoglobin draws excess water into the cell, resulting in osmotic lysis and the release of hemoglobin. Certain antigen-antibody reactions may cause increased destruction in the absence of complement and of direct membrane damage. In this instance, destruction is believed to be caused by interaction of the antibody on the cell surface with the macrophages in the reticuloendothelial system.

The RBC may also be destroyed by complement lysis that may occur as a result of reactions to drugs such as quinine and quinidine, in which an antigen-antibody reaction occurs between the drug and an antibody to a drug-protein complex (previous sensitization). In this instance, the RBC is uninvolved in the primary reaction but is damaged by the complement.

Immunological destruction of RBC may be caused by autoantibodies found in association with diseases that affect lymphatic tissue, such as lymphatic leukemia, malignant lymphoma, or systemic lupus erythematosus, and with the administration of particular drugs, such as α-methyldopa. In 50 percent of instances no cause can be determined. The etiology of the autoimmune hemolytic anemias is, for the moment, obscure; however, several varieties of immunoglobulins are involved. They have been divided into those that react at 37°C. (warm antibodies, IgG) and those that react below 37°C. (cold antibodies, IgM).

Detection of antibodies coating RBC's is accomplished by the Coombs' test, which employs rabbit antihuman globulin serum. This serum causes agglutination of RBC's coated with gamma globulin antibody. Specific antiglobulin serums can be produced which react characteristically with specific globulins on the cell surface. Further study of the antibody can be carried out by elution of the antibody from the cell surface.

Extracorpuscular   Abnormalities — Acquired, Without Antibodies. Infectious agents induce damage by various mechanisms, depending upon the etiologic agent —e.g., the malaria parasites produce increased osmotic and mechanical fragility by their presence in the cell; bacterial infections may induce hemolysis by direct enzymatic action on the red cell membrane, (e.g., *Cl. welchii* produces an α-lecithinase that attacks the lipoprotein structure).

Mechanical injury may exceed the tolerance of normal RBC's and result in a shortened life span. Hemolytic anemia of significant degree may follow the replacement of the aortic valve with a ball-valve prosthesis, and mild hemolysis has been reported in association with aortic valvular stenosis. Diseases that affect the small vessels also may be accompanied by hemolytic anemia. An example is the hemolytic-uremic syndrome, in which fragmentation of erythrocytes occurs as a result of the deposition of fibrin strands and platelets in the capillaries of the lungs and the kidneys. Irregularities in the endothelium of vessels also have been observed to cause arrest and deformation of erythrocytes. This form of hemolytic anemia has been termed microangiopathic and often is recognized by the appearance of fragments of RBC's in the blood smear.

A number of chemical agents may induce hemolytic anemia, and these have various modes of action. Some, such as arsenic, lead and benzene, induce a hemolytic anemia that is not enzyme-dependent, in contrast to the oxidant drugs such as primaquine. Lead poisoning is accompanied by a hemolytic process that results from defective heme synthesis (Fig. 23-2) at various levels of the pathway (ALA synthetase, ALA dehydrase, coproporphyrin decarboxylase and heme synthetase), a block prior to the formation of coproporphyrinogen III and impairment of the conversion of protoporphyrin IX into heme. The cell formed is hypochromic and microcytic, similar to the cell formed in the presence of iron deficiency.

In summary, the survival of the red cell is dependent largely on its unique shape, which allows deformation to occur in the smallest vessels and is, in turn, dependent on the energy derived from the metabolism of glucose in the EM and HMP pathways, together with the membrane enzymes that regulate the ionic concentration within the cell and the integrity of the structural components of the cell. Congenital deficiencies of enzymes, defects in hemoglobin synthesis and defects of the membrane (i.e., intrinsic defects) may result in shortening of the life span. Extrinsic factors present in the environment of the cell also may interfere with normal synthesis or damage the membrane sufficiently to cause changes in the shape, a decrease in deformability and shortening of life span.

The onset of hemolytic anemia may be rapid or gradual, depending upon the cause. When it is rapid, the symptoms are weakness, malaise, headache, irritability, and pain of the back, abdomen and extremities; gastrointestinal symptoms—anorexia, nausea and vomiting—may occur. Physical findings are the signs of anemia and, in addition, varying degrees of jaundice, depending upon the quantity of heme breakdown products formed and the capacity of the liver to remove the pigment from the blood and excrete it into bile. Because the clinical findings are similar in many of the hemolytic anemias, the history of past events is of particular importance. An approach is (1) to determine the possibility of an underlying or associated disorder, i.e., infection, chemical exposure (lead), sensitivity to drugs, disease of the liver or kidneys, disease of the hematopoietic system such as leukemia or malignant lymphoma; (2) to evaluate the morphological appearance of the RBC on blood smear, i.e., sickled cells suggest sickle cell disease; target cells suggest thalassemia, lead intoxication, severe iron deficiency, hemoglobin C disease; spherocytes suggest hereditary spherocytosis and may also be present in the immune hemolytic anemias; burr cells suggest disease of the liver, kidney, microangiopathic disease states, autoimmune disease and metastatic malignancy (if the erythroid marrow is able to respond appropriately to the increased destruction, reticulocytes are markedly increased); (3) to evaluate the hemoglobin by electrophoretic or other chemical means; (4) to determine the presence of antibody on the RBC by serological tests; and (5) to determine the possibility of an enzyme deficiency by specific enzyme assay.

Erythrocyte Destruction

The removal of the senescent or injured red cells is a function of the reticuloendothelial system (RES) and its mononuclear macrophages, except for the rare occasion of acute intravascular hemolysis. The sequence of this operation is: injury (i.e., lesion of RBC due to aging, biochemical defect or extrinsic agent); sequestration (i.e., the selective filtration of the injured cells and concentration within the vascular spaces of the RES); and hemolysis by enzymatic digestion within macrophages, after phagocytosis. Hemolytic mechanisms may be summarized as: pathological increases in membrane permeability, leading

to swelling and sphering, fragmentation; and phagocytosis. Fragmentation of red cells occurs randomly in normal as well as in abnormal cells, and can occur without loss of hemoglobin. The net effect is a decrease in the ratio of surface area to volume of the fragment and of the remaining portion of the RBC.

Reticuloendothelial System. The principal organs of the reticuloendothelial system are the spleen, the liver and bone marrow. The hepatic RES is a sinusoidal organ in which blood passes directly into the macrophage-lined sinusoids and thence to the venous outflow. In contrast, the arterial vessels of the spleen enter the pulp, where the artery is surrounded by a sleeve of loose reticular tissue packed with small lymphocytes which are immunologically competent. This sleeve also accommodates ovoid collections of lymphocytes (follicles which grossly are the malpighian corpuscles) that have germinal centers and are the sites of antibody formation. The sleeve, or sheath, is surrounded by a marginal zone consisting of fine meshed reticular connective tissue which receives the terminal arterioles and also contains venous sinuses and free cells such as macrophages, lymphocytes and plasma cells. The function of this marginal zone is the filtration of blood. The spleen thus presents a greater hazard to the RBC's than does the liver, although the blood circulating through the liver (35% of blood volume per minute) is about 7 times the volume circulating through the spleen. The irregular and narrow passages of the spleen constitute a mechanical hazard and detain the RBC's, exposing them to macrophages and to immunologically competent cells— an environment that permits the entire series of steps, from antigen encounter to antibody production. The hematocrit of spleen blood is high, viscosity is increased, Po_2 is decreased, and glucose concentration is decreased. These conditions favor spherocyte formation. The passages through this filter measure about 3 microns (RBC diameter, 7 microns), and the normal healthy cell is believed to be capable of passing through a tube having a diameter of 2.8 to 3.6 microns without permanent injury. The smallest diameter of the extrasplenic circulation is said to be 4 microns. The RBC membrane retains deformability until Po_2 values decrease to 30 mm. Hg (normal venous Po_2, 38 to 42 mm. Hg). Values of Po_2 are presumed to reach this level during stasis of the RBC in the spleen.

Whether sequestration occurs in the liver or spleen is determined by the degree of damage sustained by the erythrocyte rather than a selective affinity for specific types of injury. Mildly injured cells (i.e., those exposed to non-complement-fixing antibodies, low doses of oxidant drugs, etc.) are sequestered in the spleen. Similarly, it is the degree of antibody coating rather than the type that determines the site of sequestration. A high ratio of antibody to RBC mass favors hepatic sequestration. The hepatic RES is capable of clearance of agglutinated erythrocytes and severely injured nonagglutinated erythrocytes. The terminal event for effete cells and fragments is erythrophagocytosis by the macrophages and intracellular digestion of all red cell components.

Bilirubin. Hemoglobin is not reutilized as such, the component parts are metabolized separately. The protein fraction is returned to the general pool, iron is returned to the iron pool and storage sites, and the porphyrin fraction is not reused. In the RE cell the methene bridge of the heme is ruptured, and bilirubin is formed, which is then transported in plasma to the liver, where it is conjugated with glucuronic acid. The bilirubin glucuronide is excreted by the biliary system into the intestine, where it is degraded by intestinal bacteria to several compounds known collectively as fecal urobilinogen. Fifty percent of the urobilinogen is reabsorbed and recirculated, to appear in bile and urine as well as feces.

The measurements of bilirubin and bilirubin glucuronide in serum and urobilinogen in urine and feces provide indices of hemoglobin degradation and, thus, of red cell destruction.

ANNOTATED REFERENCES

Dacie, J. V., and Worlledge, S. M.: Auto-immune hemolytic anemias. *In:* Brown, E. B., and Moore, C. V.: Progress in Hematology. Vol. VI, p. 82. New York, Grune & Stratton, 1969. (A review article providing a perspective of the auto-immune hemolytic anemias)

Greenwalt, T. J., and Jamieson, G. A.: Formation and Destruction of Blood Cells. Philadelphia, J. B. Lippincott, 1970. (A monograph reflecting current knowledge of the formation and destruction of blood cells)

Harris, J. W., and Kellermeyer, R. W.: The Red Cell. rev. ed., published for the Commonwealth Fund. Cambridge, Mass., Harvard University Press, 1970. (A complete treatise concerning the chemistry and physiology of the red cell)

Harvey, A. M., *et al.:* The Principles and Practice of Medicine. ed. 17. New York, Appleton-Century-Crofts, 1968. (A textbook of internal medicine emphasizing the approach to clinical problems rather than disease entities)

Jandl, J. H. (guest ed.): Symposium on disorders of the red cell. Am. J. Med., *41:* 657, 1966. (A symposium by several noted authors which provides a brief perspective of red cell disorders)

Krantz, S. B., and Jacobson, L. O.: Erythropoietin and the Regulation of Erythropoiesis. Chicago, University of Chicago Press, 1970. (A reference for current knowledge of the regulation of erythropoiesis)

Leavell, B. S., and Thorup, O. A., Jr.: Fundamentals of Clinical Hematology. ed. 2. Philadelphia, W. B. Saunders, 1966. (A concise textbook of hematology providing an overview of diseases of the blood)

Miescher, P. A., and Jaffe, E. R. (eds.): Nutritional anemias. Seminars Hemat., 7:1, January, 1970. (A series of articles concerning each of the major nutritional factors important to erythropoiesis)

———: The red cell membrane. I. Seminars Hemat., 7:249, July, 1970.

——: The red cell membrane. II. Seminars Hemat., 7:355, October, 1970. (A series of articles defining the state of knowledge concerning the red cell membrane; composition and structure)

——: Haemolytic anemias. Seminars Hemat., 7:10, 1969. (An issue of this quarterly journal containing several review articles on hemolytic anemias)

Valentine, W. N.: The normal metabolism of human erythrocytes and hemolytic anemia due to inborn errors. *In:* Plenary Session Papers. p. 160. XII Congress, International Society of Hematology, 1968. (An overview of erythrocyte metabolism and inborn errors)

Weed, R. I.: The cell membrane in hemolytic disorders. *In:* Plenary Session Papers. p. 81. XII Congress, International Society of Hematology, 1968. (A review of the red cell membrane in disease states)

Wintrobe, M. M. (ed.): Clinical Hematology. ed. 6. Philadelphia, Lea & Febiger, 1967. (A reference textbook of hematology)

Weatherall, D. J., and Clegg, J. B.: The control of human hemoglobin synthesis and function in health and disease. *In:* Brown, E. B., Moore, C. V.: Progress in Hematology. Vol. VI, p. 261. New York, Grune & Stratton, 1969. (A review article devoted to the mechanisms controlling hemoglobin synthesis)

24

Leukopoiesis

James W. Hampton, M.D.

INTRODUCTION

Granulocytes, produced from primordial stem cells in the bone marrow, are released into the circulation in the mature form. However, these circulating leukocytes do not remain long within the vascular space, but rather migrate in and out of the tissues.

The role of infection in stimulating leukopoiesis is the subject of a separate chapter of this textbook (Chapter 32). Therein, inflammation is considered and the mechanisms promoting phagocytosis, the primary function of granulocytes, is discussed. The extravascular migration of leukocytes makes it difficult to determine their lifespan, and, as yet, the physiological mechanisms which regulate their production and release from marrow stores is unknown. Nevertheless, the use of radioactive labels has made it possible to measure neutrophil survival time in the blood and effective neutrophil production and turnover.

"LIFESPAN" OF LEUKOCYTES

Neutrophilic granulocytes are formed in the bone marrow as a result of mitotic cell division by three morphologically identifiable and concatenated cell types: the myeloblast, the promyelocyte, and the myelocyte. As cells divide they progress through these early stages of maturation (i.e., changes in cytoplasm and granules with,

finally, cessation of cellular division); their cytoplasmic maturation, however, continues and nuclear maturation begins. The latter process, continuing through two additional morphological stages, the metamyelocyte (called juvenile in the peripheral blood) and band neutrophil, finally culminates in the mature, segmented neutrophilic granulocyte, which is released from the marrow into the circulation. In normal man only bands and segmented neutrophils are found in the blood in significant numbers. However, in response to appropriate stimuli (i.e., inflammation) large numbers of them, including bands and a few metamyelocytes as well, are released into the blood. Because large numbers of the three most mature neutrophil forms are present in the marrow and can be released into the blood as needed, they are referred to collectively as the marrow granulocyte reserve. After release from marrow, neutrophils spend a brief sojourn intravascularly, during which time they may circulate freely or adhere to the walls of small vessels. From the latter site they may either reenter the circulation or migrate into the tissues, where they perform their primary phagocytic functions, die and are eliminated from the body. This sequence of events (Fig. 24-1) provides a convenient schema for a detailed discussion of granulocyte kinetics and granulocytopoiesis. Because of uncertainty as to the duration of cell survival in the tissue phase the precise duration of neutrophil lifespan remains to be determined. Moreover, although the above outline appears to occur in all animals thus far studied, the quantitative aspects may

* The author deeply appreciates the editorial comments and suggestions of Dr. John W. Athens, who has contributed so much to the development of this important field of Hematology, and also is indebted to him for contributing the illustrations for this chapter.

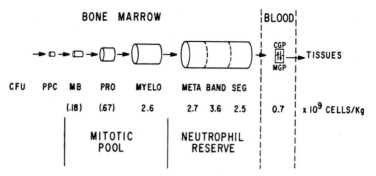

FIG. 24-1. The sequence of maturation of a neutrophil, its peripheral circulation and migration to the tissues is depicted. The relative size of marrow compartments with actual value taken from Donahue *et al.* (1958). The value for the blood neutrophil pool is from Cartwright *et al.* (1964). CFU, colony forming unit; PPC, primitive precursor cell; MB, myeloblast; PRO, promyelocyte; MYELO, myelocyte; META, metamyelocyte; BAND, band neutrophil; SEG, segmental neutrophil. (Gordon, A. S. (ed.): Regulation of Hematopoiesis. p. 114. New York, Meredith Corp., Appleton-Century-Crofts, 1970. With permission)

vary considerably from one species to another. Wherever possible human physiology will be emphasized in this discussion; but in some instances clinical studies have been impossible to perform, and inferences have been drawn from animal experiments.

STEM CELL COMPARTMENT

Evidence has been offered to support the concept that a pluripotential stem cell is the precursor for the myeloid, erythroid, and megakaryocytic series of cells. Thus, the Philadelphia (Ph₁) chromosome, characteristic of adult chronic myelocytic leukemia, is present in nucleated erythroid cells, granulocytic cells, and megakaryocytes of such patients. This finding has been interpreted as being compatible with the hypothesis that chronic myelocytic leukemia is the result of an abnormality in a pluripotential stem cell compartment. In addition, splenic colonies that form after lethal doses of irradiation and marrow infusion in mice are probably derived from a single precursor cell, since a single colony sometimes contains erythroid, granulocytic, and megakaryocytic forms.

The precise identity of the pluripotential stem cell is unknown, but the reticulum cell and the small lymphocyte are prime suspects. That infused leukocytes protect irradiated animals from death suggests that normal blood may contain such stem cells. Since only the normal circulating lymphocyte can be stimulated to grow and divide, it or some component of the lymphocyte population seems the most likely candidate.

CONTROL MECHANISMS

Several lines of reasoning suggest that the neutrophil system is maintained under some form of regulatory control. First, circulating blood neutrophil concentration remains remarkably constant in the normal individual. Secondly, the system responds predictably to certain stimuli including endotoxin, etiocholanolone or cortisol injection, leukocytopheresis, infections, and obliteration of the marrow mitotic pool by administration of vinblastine or nitrogen mustard. And thirdly, blood neutrophil concentration returns to normal levels following withdrawal of these provoking stimuli. Although we can observe the phenomenon of mobilization of neutrophils from the circulation to tissues, the distribution of blood neutrophils in marginal and circulating sites, release of neutrophils from the marrow reserve to the blood, the rate of cell production in the mitotic pool, and the inflow of stem cells into the mitotic pool, very little is known about leukocyte control at any of these possible sites.

Some control of neutrophil kinetics may occur at the sites of blood cell egress from the circulation. Thus, it seems likely that endothelial injury (either by physical, chemical, or other means) occurs continuously. This most likely promotes sticking of leukocytes to vascular endothelium and subsequent migration into tissues. The channels through which the neutrophils are lost somehow "sense" the number of cells leaving the circulation and, in some manner, signal the marrow to release more cells. Although this concept seems plausible, there is little information available concerning the sites through which the neutrophils disappear. Indirect evidence has been offered from the fact that neutrophils appear in saliva, urine, pulmonary secretions and those areas constantly exposed to bacterial, chemical and, perhaps, even physical injury. Thus, it is common clinical knowledge that neutrophils rapidly accumulate at sites of tissue damage.

Perhaps the best studied control point to date is release of cells from the marrow. Endotoxin, or a similar substance released from damaged tissue, has been emphasized as a regulator of marrow cell release. However, in a variety of experimental studies a neutrophil-releasing factor has been identified: from an extract of a specific mouse mammary tumor; or the plasma of dogs during the leukopenic and recovery phase that occurs after vinblastine, nitrogen mustard, or endotoxin injection; or in human plasma after endotoxin injection. In other studies, rat femurs perfused with plasma having a low leukocyte content resulted in increased marrow cell release. As increased numbers of mature leukocytes accumulated in the perfusate,

further leukocyte release was suppressed. On this basis a "negative" feedback control mechanism, mediated by mature neutrophils, which may regulate marrow cell release, was postulated.

Control of cell production rate by some independent means seems likely from studies of vinblastine injection into dogs, since marrow recovery began before neutrophil-releasing activity appeared in the blood and before correction of the blood leukopenia began. It seems that cell production may increase by: shortening of cell generation time; stimulation of resting cells to enter active cell cycle; reduction of postulated normal cell death; or increased inflow from the precursor stem cell compartment.

In chronic myelocytic leukemia (CML) an increase in neutrophil production and turnover occurs; this appears to result from increased mitotic compartment size rather than from increase in cell division rate. A factor present in plasma from patients having CML may induce stem cells to enter the myeloid compartment. Urine from such individuals has been the best source of this factor which is necessary for the growth of myeloid marrow colonies in vitro. Marrow lymphocytes may be multipotential hemopoietic stem cells; this is supported by reduction in marrow lymphocytes 6 hours after injection of leukocytosis-inducing factor into rats. Many other substances have been reported to alter neutrophil production, but, as yet, probably only neutrophilia-inducing factor can be accorded some physiological significance.

CIRCULATING LEUKOCYTES

Normal Leukocyte Pool Size and Distribution

From the earliest studies of inflammation the concept was established that neutrophils could be freely circulating or marginated along the walls of small blood vessels at sites of tissue injury. This concept was then extrapolated to the normal in an attempt to explain the brief increase in circulating neutrophils, without a "shift to the left," that occurs with exercise or excitement. Quantitative measurement of the size of the blood neutrophil mass and its distribution was first accomplished following the development of the in-vitro method for labeling circulating neutrophils with diisopropylfluorophosphate (DFP[32]). During a 1-hour incubation period with autologous blood, the DFP[32] binds to neutrophils in high concentration, and to erythrocytes and thrombocytes at a much lower concentration. An aliquot of the labeled blood is saved for neutrophil count and neutrophil radioactivity measurement; the remainder is returned to the subject. After allowing for equilibration within the vascular system, another blood sample is withdrawn and the blood neutrophils are isolated and their radioactivity measured. As the DFP[32] does not disappear from or damage the circulating neutrophils, the size of the total blood granulocyte pool (TBGP) can be calculated by means of the isotope dilution principle.

In normal men a mean value of 70×10^7 cells per kg. (Table 24-1) was obtained, and an average of 44 percent of the injected, labeled neutrophils could be accounted for in the circulating granulocyte pool (CGP). The latter can be calculated

TABLE 24-1. The size of the total blood granulocyte pool (TBGP), circulating granulocyte pool (CGP), marginal granulocyte pool (MGP), half disappearance time ($T_{1/2}$) and granulocyte turnover rate (GTR) in normal man[*]

	MEAN	95% LIMITS
TBGP $\times 10^7$/kg.	70	14-160
CGP $\times 10^7$/kg.	31	11-46
MGP $\times 10^7$/kg.	39	0-85
$T_{1/2}$ in hrs	6.7	4-10
GTR $\times 10^7$/kg./day	163	50-340

[*] TBGP, CGP, and MGP values were obtained in 109 normal males. $T_{1/2}$ and GTR values were obtained in 56 of these 109 subjects.

by multiplying the total blood volume by the blood neutrophil concentration.

Lack of evidence that DFP[32] damages or elutes from neutrophils led to the suggestion that the difference between the measured TBGP and the calculated CGP represents cells "marginated" along the walls of small blood vessels (MGP). This assumption was supported by studies in which the proportion of injected neutrophils recovered in the CGP was increased from 44 to a mean of 78 percent (54 to 100%) immediately following exercise or epinephrine injection. Granulocytes, then, can be mobilized from intravascular sites rather than from the unlabeled marrow granulocyte reserve and epinephrine and exercise can shift or redistribute leukocytes. The size of the MGP measured by the above technique varies considerably from one individual to another as well as within the same individual. Presumably lability in distribution of cells between the CGP and MGP accounts for the neutrophilia seen in association with paroxysmal tachycardia or convulsive seizures.

A diurnal variation in blood neutrophil concentration has been reported with a 50 percent rise in neutrophil concentration occurring in the night. The time of peak granulocyte count varied widely and could not be related clearly to meals or other activity. Another explanation suggested for this highly variable diurnal variation is intravascular shift of neutrophils.

Normal Leukocyte Turnover and Survival Time

After measurement of TBGP with DFP[32] labeled cells, serial measurements of blood neutrophil radioactivity and thus blood neutrophil disappearance time can be made.

When blood granulocyte disappearance time was measured by this means in normal subjects in a steady state, radioactivity decreased exponentially with an average half-time of disappearance ($T_{1/2}$) of 6.7 hours (Table 24-1). In contrast to erythrocytes, neutrophils leave the blood in a random fashion and do not have a finite blood survival time. Measurement of blood granulocyte disappearance rate with radiochromate-labeled cells gives $T_{1/2}$ values very similar to those obtained with DFP[32]. If the estimates of TBGP size are approximately correct and if neutrophils leave the blood predominantly in a random fashion, the blood granulocyte turnover rate (GTR) can be easily calculated from the first-order decay equation and the two measured parameters, TBGP and $T_{1/2}$:

$$GTR = \frac{0.693 \times TBGP}{T_{1/2}}$$

In subjects, either normal or abnormal, who are in a steady state the blood GTR describes the number of neutrophils delivered to and removed from the blood during a given time interval and thus is a measurement of effective granulocytopoiesis (Table 24-1).

Leukocyte Turnover and Survival Time in Disease

Abnormalities of TBGP size are found in patients with neutrophilia and neutropenia; and, in general, the correlation between TBGP size and blood neutrophil concentration has been good. However, in a few situations blood neutrophil concentration does not reflect TBGP size adequately because of intravascular redistribution of cells (Fig. 24-2).

Normal TBGP Size. TBGP size and the approximately equal distribution of cells in circulating and marginal sites in normal subjects are illustrated in Fig. 24-2, A. Neutrophilia without concomitant increase in TBGP size has been observed transiently following exercise of short duration or epinephrine injection (Fig. 24-2, B). Whether a more sustained redistribution or "shift neutrophilia" occurs is uncertain. A sustained, mild neutrophilia and normal TBGP size has been observed in four patients, two of whom subsequently developed granulomatous disease. The other two patients, one with essential hypertension and the other with chronic anxiety, may have had chronic "stress" or "shift leukocytosis." If real, this entity may be not unlike "stress erythrocytosis" which has been explained on the basis of contracted plasma volume (see Chapter 3). Several patients with neutropenia of borderline degree and with normal TBGP size have been observed (Fig. 24-2, C). Such a cell redistribution appears to explain the reduced neutrophil concentration seen several hours after endotoxin injection.

Expanded TBGP Size. As expected, most

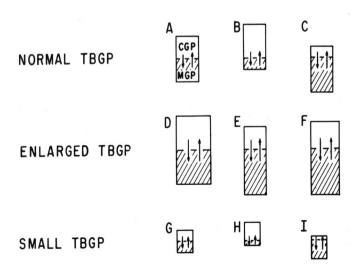

Fig. 24-2. Peripheral blood neutrophil concentration may not always reflect the total blood granulocyte pool (TBGP). Each rectangle represents a possibility in the ratio of the circulating granulocyte pool (CGP) and the marginal granulocyte pool (MGP). For a discussion of these various possibilities, A through I, refer to text. (Gordon, A. S. (ed.): Regulation of Hematopoiesis. p. 1152. New York, Meredith Corp., Appleton-Century-Crofts, 1970. With permission)

NORMAL TBGP

ENLARGED TBGP

SMALL TBGP

patients with neutrophilia (e.g., chronic infection, chronic myelocytic leukemia, myelofibrosis, polycythemia vera, Hodgkin's disease) have expanded TBGP with approximately even distribution of cells between marginal and circulating sites (Fig. 24-2, D). However, in some patients with myelofibrosis or polycythemia vera, with increased TBGP size, the preponderant increase was in marginated cells. In such patients, the blood granulocyte count did not give a good indication of the TBGP size (Fig. 24-2, F). Occasionally, some patients with normal blood granulocyte counts have modestly elevated TBGP size or a "masked granulocytosis" (Fig. 24-2, E); this entity is consistently demonstrable in patients with mild, induced skin inflammations.

Decreased TBGP Size. Diseases associated with neutropenia are difficult to study because large blood samples are required to obtain sufficient neutrophils to measure accurately the radioactivity. However, some patients with neutropenia have a reduced TBGP size in which the cells are distributed equally between circulating

and marginal sites (Fig. 24-2, G). Other patients with reduced TBGP size have most of their cells in the CGP (Fig. 24-2, H), and some patients have an apparent neutropenia but the TBGP may be normal (Fig. 24-2, I).

BONE MARROW LEUKOCYTES

Marrow Mitotic Compartment

The bone marrow contains myeloblasts, promyelocytes, and myelocytes, cells which are capable of mitotic division as judged by direct observation in cell cultures and by virtue of their uptake of tritiated thymidine. Cells in this compartment mature from myeloblast to promyelocyte to myelocyte. The number of cell divisions during each morphological stage is unknown. However, from studies of the complex blood radioactivity curve obtained after the intravenous injection of DFP[32], it has been suggested that there must be at least 3 divisions at the myelocyte stage. If a single division were postulated, then, for the myeloblast and promyelocyte stages, a minimum of 5 cellular divisions must occur

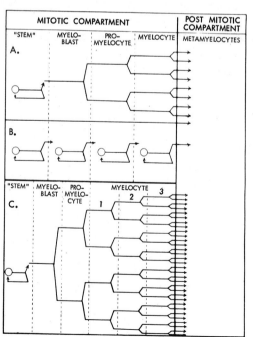

Fig. 24-3. An important consideration in leukopoiesis is the mitotic compartment and mode of cell division. Although schemes A and B have been proposed, the best scheme would appear to be C, since the myelocyte compartment is at least four times the size of the preceding promyelocyte pool. (Boggs, D. R., Athens, J. W., Cartwright, G. E., and Wintrobe, M. M.: J. Clin. Invest., 44: 643, 1965)

during myelopoiesis (Fig. 24-3, C). The scheme of myeloid mitosis and proliferation currently proposed (Fig. 24-3, C) suggests that the major locus of neutrophil production in man is at the myelocyte stage. This is supported by other evidence demonstrating that the myelocyte pool is at least 4 times the size of the preceding promyelocyte pool (see Fig. 24-1).

The cellular output by the mitotic pool must at least equal the blood granulocyte turnover rate. Whether cell production is greater than this, in order to compensate for a sizable component of premature cell death in the marrow (i.e., ineffective granulocytopoiesis), is unknown. Myelocyte production in the dog has been calculated to be about double the measured inflow of cells into the metamyelocyte compartment, suggesting considerable ineffective granulopoiesis in this species. No adequate studies of marrow granulocyte production rate (MPR) in disease are available that would indicate ineffective neutrophil production pathologically.

In the steady state, the efflux of cells from the mitotic pool should be equal to the cellular inflow plus the cell birth rate, assuming no cellular death, within the pool. Since inflow from the stem cell compartment is quantitatively insignificant compared with mitotic pool efflux, measurement of cell birth rate in the mitotic pool approximates marrow production rate (MPR). Theoretically, mitotic pool cellular birth rate can be estimated either from mitotic indices or from other data obtained after labeling with tritiated thymidine (^{3}HTdR).

Leukapheresis:

A Test of Leukocyte Reserve

Leukapheresis studies in dogs represented the first demonstration of a marrow "storage" pool of neutrophils that could be mobilized to replace cells removed from the blood. Removal of circulating leukocytes in a number greater than 4.5 blood volumes in several hours was required to produce a leukopenia. Slower removal rates do not always yield neutropenia. The storage pool is not dependent upon intact marrow production, as shown in dogs treated with vinblastine to suppress mitotic activity. They were able to mobilize large numbers of neutrophils even though the mitotic pool was almost completely destroyed. Similar studies, in man, have confirmed these findings and led to the concept of a granulocyte reserve.

Autoradiographic studies, following the injection of tritiated thymidine (^{3}HTdR) into dogs and man, have demonstrated labeling of cells in the mitotic compartment but no labeling in the more mature forms initially. The subsequent appearance of the label, first in metamyelocytes, then in bands and, finally, in segmented neutrophils, demonstrates that neutrophils move through the neutrophil reserve on a first-in, first-out basis (Fig. 24-1).

Athough estimates of bone marrow cell content have been made in small animals by direct volume measurement and cell content, this method obviously is not feasible in man. Nevertheless, estimations of marrow cell content are possible. The time lag between the injection of radiophosphate or tritiated thymidine and the appearance of the labeled cells in circulating blood allows an estimate of the time required for cells to mature and to be discharged from the neutrophil reserve. Minimum values range from 4 to 6 days, while the peak of radiophosphate and tritiated thymidine radioactivity curves occurs at 6 to 8 days. Thus, it has been concluded that the marrow neutrophil reserve contains approximately a 4- to 8-day supply of cells at normal utilization rates. From this, and the daily blood granulocyte turnover rate, the cellular content of the MGR can be calculated (Table 24-2). Total marrow nucleated red cell mass can be measured by ^{59}Fe distribution in bone marrow and the entire myeloid mass can be quantified from the ratio of myeloid to erythroid cells. By this means,

TABLE 24-2. Estimates of Marrow
Granulocyte Reserve

METHOD OF CALCULATION	MGR IN CELLS $\times 10^9$/kg.
From ^{3}HTdR curve (4 to 8 days) and in-vitro DFP^{32} GTR	6.5-13
From in-vivo DFP curve (8 to 14 days) and in-vitro DFP^{32} GTR	13-23
^{59}Fe technique	8.8

the size of the several myeloid compartments can be calculated from the marrow differential counts.

A discrepancy exists between estimates of MGR obtained in human beings by means of DNA labels (4 to 8 days) as compared with the in-vivo DFP^{32} method (8 to 14 days). As similar measurements in normal dogs give about the same values for MGR with both methods (about 5 days), it seems likely that the differences in clinical data reflect differences among the "normal" human subjects studied. Thus, ambulatory, healthy adult men were used in the DFP^{32} studies while "hematologically normal" subjects with tuberculosis or carcinoma were used in the DNA-labeling studies and the ^{59}Fe study. The time lag between the injection of radioactive labels and the appearance of tagged cells in the blood reflects the time required for the completion of cell division and the period of traversing the nondividing reserve. The marrow reserve is therefore thought to be a 4- to 8-day supply of cells at normal utilization rates. From these studies in patients and normal subjects it seems likely that the marrow granulocyte reserve may vary in size (and/or that the maturation time may vary) in disease, since the transit time from myelocyte to blood may be only 48 hours in a patient with severe infection as compared with a range of "normal," 95 to 144 hours.

Leukocytes in Tissues

Following injury, leukocytes are mobilized from the blood into tissues. Such a cellular inflow into exudates is clearly evident within 2 hours after injury. There is no evidence for return of neutrophils from the tissues to blood, and their ultimate fate once they have left the blood is presumed to be death, or migration from tissues into body secretions (saliva, urine, pulmonary secretions, gut), where they may yet perform a protective function.

LEUKOPOIESIS IN DISEASE

Chronic Neutrophilia

Measurement of blood neutrophil turnover rate (effective granulocytopoiesis) in patients with neutrophilia (as a result of chronic infection, polycythemia vera, chronic myelocytic leukemia, or a variety of other diseases) is normal or increased, with values ranging up to 12 times that of the normal. In these chronic neutrophilic diseases the GTR was increased but proportionately less than the TBGP size. The $T_{1/2}$ values were normal or prolonged rather than short. In those patients with immature circulating neutrophils (chronic myelocytic leukemia, myelofibrosis, and myeloid metaplasia) very prolonged $T_{1/2}$ values (25 to 90 hours) were found as compared with values of less than 18 hours in patients with neutrophilias in which mainly mature cells are present. The prolonged $T_{1/2}$ values in the former are thought to reflect the inability of immature cells to leave the blood readily and/or also represent recirculation of immature cells to the bone marrow and spleen. In chronic, steady-state neutrophilias, in which mature cells are present (infection and inflammation), it appears that increased numbers of cells are made available to the tissues by *increasing* TBGP size rather than by shortened transit through the blood. However, in many patients the number of cells supplied appears to exceed the demand, and neutrophilia with an increased TBGP size and a prolonged $T_{1/2}$ results. Clearly, large numbers of cells migrate into inflammatory sites.

A different kinetic picture is present with neutrophilia resulting from adrenal corticosteroid administration. Steroids induce a neutrophilia which results from a real increase in TBGP size (without significant redistribution of cells), and the $T_{1/2}$ is moderately prolonged. As a result, the granulocyte turnover rate (GTR) is normal. It has been noted that in patients on steroids there is decreased migration of neutrophils into induced inflammations, and resistance to bacterial infections may be compromised by virtue of this. Presumably, some abnormal mechanism of neutrophil disposal is activated in these patients. In any case, it should be remembered that although steroids will often increase the blood neutrophil concentration this may not be beneficial to the patient.

Acute Neutrophilia:

The Mechanism of Acute Infection

As one might expect from clinical observations, neutrophils are mobilized from the blood into inflammatory sites in several hours. In dogs with induced pneumonia, there is both an increase in TBGP size and a rapid turnover of blood neutrophils beginning within 4 hours after injection of pneumococci into the bronchi. If the infection can be controlled, the release of cells from the marrow into the blood decreases and the neutrophilia and disturbed kinetics gradually return to normal. When the infection is overwhelming, the demand for neutrophils exceeds the supply (TBGP plus marrow granulocyte reserve) and neutropenia ensues.

Neutropenic States

Neutropenia may result either from decreased cell production or from excessive cell destruction. In patients with neutropenia associated with hypoplastic marrows (i.e., patients treated with irradiation or chemotherapy) the TBGP was small and

the $T_{1/2}$ normal. As a result the granulocyte turnover rate was decreased. Similar results have been observed in patients with aplastic anemia, multiple myeloma, acute leukemia, and drug-induced neutropenia. Dogs made neutropenic with busulfan or melphalan also had decreased production (GTR) characterized by normal $T_{1/2}$ values. In contrast, patients with cirrhosis may have neutropenia as a result of increased cell destruction in an enlarged spleen. In such patients short $T_{1/2}$ values have been observed. An apparent decrease in susceptibility to infection and restoration toward normal has followed splenectomy in several patients. It also seems likely that neutropenia may result from ineffective granulocytopoiesis; i.e., intramedullary cell destruction analogous to the anemia of thalassemia major or pernicious anemia.

Two familial agranulocytosis syndromes have been described. One, a benign form, is inherited as a dominant and was accidentally discovered in a healthy adult. In contrast, a neutropenia transmitted as an autosomal recessive and found in offspring of consanguineous parents is associated with a disease syndrome and has been termed infantile genetic agranulocytosis. Repeated infections occur in spite of peripheral monocytosis and a diffuse increase in gamma globulins. Death usually occurs at an early age. Dicentric chromosomal and chromatid breaks have been reported to occur in cell cultures in this second type.

The Leukemias

The chief centers for blood cell formation and destruction belong to the reticuloendothelial system, otherwise known as hematopoietic tissues. The multipotential stem cell (or primary reticulum cell) may give rise to any of the cells ordinarily found in a hematopoietic organ. Leukemias are thought to represent a disorder of multifocal origin, since patients with these disorders usually show early clinical evidence of systemic involvement. Since the leukemias represent an unrestrained proliferation of cells, they are considered neoplastic conditions.

Leukemias are classified according to the predominant cell component. The similarity between the myeloproliferative diseases includes progressive proliferation of one or more of the cellular constituents of the bone marrow that are derived from a common multipotential undifferentiated precursor cell. Tumors of the lymphopoietic tissue may also involve the marrow both by spread to the marrow and by the activation of lymphoid foci within the marrow itself. The classification of leukemias ranges from undifferentiated, stem-cell leukemias to highly differentiated chronic myelocytic and chronic lymphocytic leukemias. The recognition of the regulatory mechanisms in leukemia is poorly understood. In some instances there appears to be an actual "inhibitory" factor present in the leukemic serum which has been termed a chalone (after the Greek word for bridle). These inhibitory serum factors may be present in high concentration early in the disease, and gradually fall as the leukemic picture is controlled.

Leukemoid Reactions

Leukemoid reactions are marked elevations in the leukocyte count (and/or an increase in immaturity of circulating leukocytes) in the circulating blood as a consequence of a severe and overwhelming sepsis, a chronic granulomatous condition, or a recovery response to marrow suppression by a drug. In leukemoid reactions serum or urine lysozyme levels are not increased and alkaline phosphatase content of peripheral blood leukocytes is usually normal or elevated. The neutrophil responses are unaffected by the usual hormonal factors, and the leukocytosis tends to disappear when the underlying process has been resolved.

ANNOTATED REFERENCES

Athens, J. W.: Neutrophilic granulocyte kinetics and granulocytopoiesis. In: Gordon, A. S. (ed.): Regulation of Hematopoiesis. p. 1143. Appleton-Century-Crofts, New York, 1970. (A detailed analysis of the experimental evidence for leukocyte pools and their alterations in disease)

————: Granulocyte kinetics in health and disease. In: Human Tumor Cell Kinetics. Bethesda National Cancer Institute Monograph, 1969. (A review of the leukocyte kinetics)

Bierman, H. M., Kelly, K. H., Bryon, R. L., Jr., and Marshall, G. J.: Leucapheresis in man. I. Hematological observations following leucocyte withdrawal in patients with nonhematological disorders. Brit. J. Hematol., 7:51, 1961. (A study of the dynamics of leukapheresis and leukocyte kinetics)

Boggs, D. R., Athens, J. W., Cartwright, G. E., and Wintrobe, M. M.: Leukokinetic studies. IX. Experimental evaluation of a model of granulopoiesis. J. Clin. Invest., 44:643, 1965.

Boggs, D. R.: The kinetics of neutrophilic leukocytes in health and disease. Seminars Hemat., 4:359, 1967. (A review of leukocyte kinetics)

Bond, V. P., Fliedner, T. M., Cronkite, E. P., Rubini, J. R., and Robertson, J. S.: Cell turnover in blood and blood forming tissues studied with tritiated thymidine. In: Stohlman, F., Jr. (ed.): The Kinetics of Cellular Proliferation. p. 188, New York, Grune & Stratton, 1959. (Mitotic pools in leukocyte proliferation are described.)

Cartwright, G. E., Athens, J. W., and Wintrobe, M. M.: The kinetics of granulopoiesis in normal man. Blood, 24:780, 1964.

Craddock, C. G., Jr., Perry, S., and Lawrence, J. S.: The dynamics of leukopoiesis and leukocytosis as studied by leukopheresis and isotopic techniques. J. Clin. Invest., 35:285, 1956. (Experimental data on physiological mechanisms of leukopoiesis)

Cronkite, E. P., and Vincent, P. C.: Granulocytopoiesis: Symposium on myeloproliferative disorders of animals and man. 1969. (Review of granulocytopoiesis)

Cutting, H. O., and Lang, J. E.: Familial benign chronic neutropenia. Ann. Intern. Med., 61:876, 1964. (Experimental quantitation of marrow cells)

Donohue, D. M., Reiff, R. H., Hansen, M. L., Belson, Y., and Finch, C. A.: Quantitative measurement of the erythrocytic and granulocytic cells of the marrow and blood. J. Clin. Invest., 37:1571, 1958. (Experimental evidence for control of leukopoiesis)

Hersch, E. M., and Bodey, G. P.: Leukocytic mechanisms in inflammation. *In:* de Groff, A. C., and Coeger, W. P. (eds.): Annual **Rev. Med.**, *21*:105, 1970. (Review of role of leukocytes in inflammation)

Kostmann, R.: Infantile genetic agranulocytosis. Acta pediat. scand. (Suppl. 105), *45*:1, 1956. (Clinical studies of leukocyte disorders)

Mauer, A. M.: Diurnal variation of proliferative activity in human bone marrow. Blood, *26*:1, 1965. (Evidence for normal variation in leukopoiesis)

Novikoff, A.: Lysosomes in physiology and pathology of cells: Contribution of staining methods. *In:* de Reuk, A. V. S., and Cameron, N. P. (eds.): Ciba Foundation Symposium on Lysosomes, p. 36, 1963. (Review of function of leukocytes)

Patt, H. M., and Maloney, M. A.: Kinetics of neutrophil balance in the kinetics of cellular proliferation. *In:* Stohlman, F., Jr. (ed.): The Kinetics of Cellular Proliferation. p. 201. New York, Grune and Stratton, 1959. (Role of leukocytic regulation by kinetics)

Perry, S., Craddock, C. G., Jr., and Lawrence, J. S.: Rates of appearance and disappearance of white blood cells in normal and in various disease states. J. Lab. Clin. Med., *51*:501, 1958. (Review of leukemia as a condition of altered leukocyte kinetics)

Rabinovitch, M.: Phagocytosis: The engulfment stage. *In:* Miescher, P. A., and Jaffe, E. R. (eds.): Seminars in Hematology, *V*: 134, 1968. (Review of the principal role of leukocytes, phagocytosis)

Stryckmans, P. A., Cronkite, E. P., Fache, J., Fliedner, T. M., and Ramas, J.: DNA synthesis time of erythropoietic and granulopoietic cells in human beings. Nature, *211*: 717, 1966. (Experimental evidence for mitotic maturation)

Thomas, E. D., Plain, G. L., and Thomas, D.: Leukocyte kinetics in the dog studied by cross-circulation. J. Lab. Clin. Med., *66*:64, 1965. (Regulation of leukocytes in experimental model)

Whang, J., Frei, E., III., Tiio, J. H., Carbone, P. P., and Brecher, G.: The distribution of the Philadelphia chromosome in patients with chronic myelogenous leukemia. Blood, *22*:664, 1963. (Genetic markers in leukemia)

Zeulzer, W. W.: Myelokathexis—a new form of chronic granulocytopenia. New Eng. J. Med., *270*:699, 1964. (A new concept of reduced circulating neutrophils)

25

Hemostasis

Oscar D. Ratnoff, M.D.

In vertebrates, blood circulates in a closed circuit of conduits, the blood vessels. Elaborate devices have evolved to minimize blood loss after disruption of these channels. Vasoconstriction, an immediate response to vascular injury, is usually transient and ineffective, but blood platelets quickly adhere to the injured endothelium and, if the vascular break is minor, these cells will stanch the flow of blood. At the same time the blood clots, making the platelet mass more compact and forming a plug that seals larger wounds. If bleeding occurs in a closed space such as the forearm, the back pressure of blood accumulating extravascularly helps to stem further loss; the rapid development of a black eye after trivial injury illustrates how bleeding progresses where back pressure is minimal. In the special case of the uterus, contraction of extravascular muscles helps to minimize hemorrhage after childbirth by narrowing vascular lumens. Failure of any of these mechanism—a relatively common occurrence—leads to undue blood loss or even death from exsanguination.

PHYSIOLOGY AND PATHOLOGY OF THE BLOOD CLOTTING MECHANISM

Blood clotting reflects the conversion of plasma *fibrinogen* (Factor I) to an insoluble network of fibers, the *fibrin* clot. Fibrinogen is a protein with a molecular weight of 340,000 and is synthesized in the parenchymal cells of the liver; it is composed of a long spindle with a nodule at each end and a third in the middle. Each molecule contains three pairs of polypeptides, the α(A), β (B) and γ chains, held together by disulfide bonds.

The conversion of fibrinogen to fibrin is brought about by *thrombin,* a protease which is evolved during the clotting process and hydrolyzes arginyl-glycine bonds in each of the α (A) and β (B) chains (Fig. 25-1). Two pairs of polypeptide fragments, fibrinopeptides A and B respectively, are

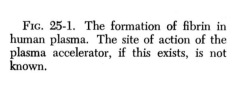

FIG. 25-1. The formation of fibrin in human plasma. The site of action of the plasma accelerator, if this exists, is not known.

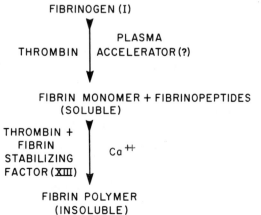

FIBRINOGEN (I)

THROMBIN | PLASMA ACCELERATOR (?)

FIBRIN MONOMER + FIBRINOPEPTIDES (SOLUBLE)

THROMBIN + FIBRIN STABILIZING FACTOR (XIII) | Ca^{++}

FIBRIN POLYMER (INSOLUBLE)

released, and what remains—monomeric units of fibrin—polymerizes, both side-to-side and end-to-end, to form the insoluble fibrin clot. Visible clotting requires the release of fibrinopeptide A; if fibrinopeptide B is cleaved selectively, clotting does not take place. Calcium ions greatly accelerate polymerization. Additionally, plasma speeds fibrin formation through an unknown mechanism.

Fibrin derived from purified fibrinogen has poor tensile strength and dissolves in such agents as 5M urea or 1 percent monochloroacetic acid, as if its molecules were loosely bound. Clots formed from normal plasma, in contrast, are stronger and do not dissolve in these agents. Such fibrin has been bonded chemically by *fibrin-stabilizing factor* (Factor XIII), a plasma enzyme that forms amide links between the γ-carboxyl group of glutamine and probably the β-carboxyl of asparagine in one fibrin monomer and the epsilon-amino group of lysine in another. Fibrin-stabilizing factor, a protein with a molecular weight of 300,000, functions only when "activated" by thrombin and when calcium ions are present. It provides fibrin with the strength needed to hold together the edges of a wound.

The active site of thrombin, in common with some other proteases, contains the sequence glycyl-aspartyl-seryl-glycyl-glutamyl-alanine. Thrombin generates during clotting, from its precursor *prothrombin* (Factor II). Two mechanisms have been distinguished through which thrombin forms, the extrinsic and intrinsic pathways of blood clotting. When shed blood comes into contact with particles of tissue, thrombin evolves rapidly. The agent or agents responsible, known generically as *tissue thromboplastin,* are lipoproteins localized to microsomes. Since tissue thromboplastin is not a constituent of blood, it is said to initiate clotting via the extrinsic pathway (Fig. 25-2). Thromboplastin contains a heat-labile protein portion and a more stable phospholipid, both needed for thrombin formation. The protein fraction interacts, in the presence of calcium ions, with a plasma protein, *Factor VII,* yielding a product that "activates" a second plasma protein, Stuart factor (Factor X). Activated Stuart factor, which may be a proteolytic enzyme, forms a potent *prothrombin-converting principle,* in the presence of another plasma component, *proaccelerin* (Factor V). This, in turn, separates thrombin (which has a molecular weight of 32,000)

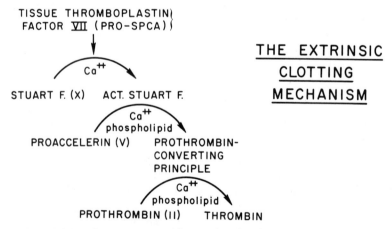

TISSUE THROMBOPLASTIN⎱
FACTOR Ⅶ (PRO-SPCA)⎰

↓

Ca⁺⁺

STUART F. (X)   ACT. STUART F.

Ca⁺⁺
phospholipid

PROACCELERIN (V)   PROTHROMBIN-
CONVERTING
PRINCIPLE

Ca⁺⁺
phospholipid

PROTHROMBIN (II)   THROMBIN

THE EXTRINSIC
CLOTTING
MECHANISM

Fɪɢ. 25-2. The extrinsic pathway for the formation of thrombin. Omitted from the diagram are inhibitors of various steps and the alteration induced by thrombin in proaccelerin. The phospholipid needed for the formation and action of the prothrombin converting principle is derived for the most part from tissue thromboplastin.

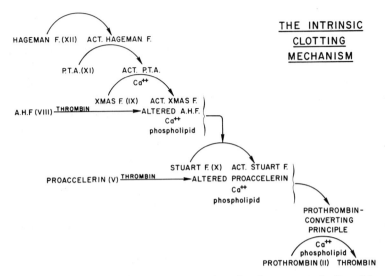

Fig. 25-3. The intrinsic pathway for the formation of thrombin. Omitted from the diagram are inhibitors of various steps. The phospholipid needed is furnished by platelets and by the plasma itself. (P.T.A., plasma thromboplastin antecedent; XMAS F., Christmas factor; and AHF, antihemophilic factor)

from its parent molecule, prothrombin, whose molecular weight is 68,000. Thrombin formation occurs only in the presence of calcium ions and phospholipids, the latter furnished by tissue thromboplastin, as if activated Stuart factor and proaccelerin interacted upon the surface of phospholipid micelles. The effect of proaccelerin is enhanced if it is first altered by thrombin.

Cell-free plasma, prepared so that it is not contaminated with tissue, nonetheless clots in glass tubes, through a process described as the intrinsic pathway of clotting (Fig. 25-3). Thrombin formation takes place as the result of a sequence of enzymatic steps, beginning with the "activation" of *Hageman factor* (Factor XII), a protein with a molecular weight of about 80,000. This alteration can be brought about by contact with glass or other negatively charged surfaces. Glass is foreign to the body, but substances as diverse as collagen, sebum, crystalline sodium urate or calcium pyrophosphate may activate Hageman factor under normal or pathological conditions. The nature of the activation process is not understood; possibly, Hageman factor

changes from a water-soluble to a hydrophobic form. Once activated, Hageman factor not only promotes clotting but also may initiate reactions leading to the inflammatory response, to the activation of plasminogen to form plasmin, a plasma proteolytic enzyme (see Chapter 26), and to a change in the first component of complement (C1) to its active form, C1 esterase ($C\bar{1}$). Its role in clotting is to change a second substance, *plasma thromboplastin antecedent* (PTA, Factor XI), from an inert to a clot-promoting form. This factor, whose molecular weight is 165,000, is probably a protease; its function is the activation of a third factor, *Christmas factor* (PTC, Factor IX), a step requiring calcium ions. Activated Christmas factor then interacts with *antihemophilic factor* (Factor VIII), a glycoprotein with a molecular weight greater than 2,000,000, to form an agent that can activate Stuart factor. This step requires calcium ions and phospholipids and is accelerated if antihemophilic factor has first been altered by thrombin. Once Stuart factor has been activated, the subsequent steps in the intrinsic pathway are the same as those of the extrinsic pathway. In the intrinsic pathway, platelets and the plasma itself furnish the needed phospholipids.

The mechanisms described are relevant to clotting as studied in the test tube. In life, coagulation is probably initiated, in different situations, by exposure of blood to injured tissue or to an activator of Hageman factor such as collagen or sebum. Other stimuli may bring about clotting under special circumstances. Thus, some snake venoms are clot-promoting. The venom of the Russell's viper has components that activate Stuart factor and enhance the action of proaccelerin; venoms of the Malayan pit viper and Southern copperhead snake convert fibrinogen to fibrin. Certain strains of *Staphylococcus aureus* produce a "coagulase" that induces thrombin formation by acting upon a prothrombinlike component of plasma, which perhaps explains the localization of staphylococcal infections. Pancreatic trypsin also brings about clotting, through actions upon Stuart factor and prothrombin; the physiological significance of this property is uncertain, since plasma inhibits trypsin readily.

As might be expected, plasma possesses potent inhibitors that limit the clotting process, a safeguard against thrombosis. The best defined of these agents are those that block thrombin and the activated forms of Hageman factor, PTA and Stuart factor. Additionally, activated clotting factors may be inactivated by passage through the liver or the reticuloendothelial system.

Diagnosis of Disorders of Blood Coagulation

The blood clotting mechanism may be inadequate because of a functional deficiency of one or more clotting factors, as the result of either defective synthesis or excessive utilization. Alternatively, coagulation may be impaired by circulating inhibitors or by the proteolytic action of plasmin (Chapter 26). The degree of the defect in hemostasis is not related to the intensity of the abnormality measured in the test tube. A severe deficiency of Hageman factor is usually not accompanied by any discernible defect in hemostasis, whereas a deficiency of antihemophilic factor, resulting in a comparable prolongation of the time elapsing until blood clots in vitro, is associated with a serious bleeding tendency.

In studying the patient with hemostatic disturbances, inquiry must be made as to whether bleeding has been life-long or of recent origin, and whether it has appeared de novo or during the course of another disorder, such as hepatic or renal disease. It is important to learn whether the patient has ingested (or has been exposed to) possibly injurious agents, has recently (or concurrently) undergone infection or has subsisted on a deficient diet. The type of bleeding sustained by the patient is diagnostically helpful. *Ecchymoses*—that is, black and blue marks—are commonplace in many forms of hemorrhagic disease, whereas *petechiae*—pin-point dull red spots that do not blanch on pressure—are unusual in purely coagulative abnormalities but are frequent in disorders of platelets or blood vessels. Bleeding from the umbilicus at birth suggests a defect in the extrinsic pathway of clotting or in fibrin formation. One must determine whether there is a family history of bleeding, or whether the patient is the offspring of a consanguineous union, a hallmark of a rare autosomal recessive trait.

Differential diagnosis of disorders of blood clotting is almost entirely a laboratory exercise. A functional deficiency of any factor involved in the intrinsic pathway of clotting, or the presence of inhibitors of this pathway, may almost always be detected by measurement of the *partial thromboplastin time*. In this test, citrated plasma is mixed with calcium ions and phospholipid, and the clotting time—the interval until clotting takes place—is measured. Sometimes kaolin or diatomaceous earth is added to the mixture to ensure activation of Hageman factor. The test will detect almost all cases of functional deficiency of factors participating in the intrinsic pathway but gives normal values in patients with deficiencies of Factor VII or fibrin-stabilizing factor. Much less sensitive is measurement of the *clotting time of whole blood;* in at least one third of patients with classic hemophilia or Christmas disease (deficiencies of antihemophilic factor and Christmas factor, respectively) the clotting time of whole blood is normal. Another screening test for defects in the intrinsic pathway is measurement of *"prothrombin consumption"* or *"serum prothrombic activity."* During clotting, the conversion of prothrombin to thrombin proceeds slowly; clotting occurs when enough thrombin has formed to change fibrinogen to fibrin. Even in normal blood, appreciable prothrombin remains an hour after whole blood has been placed in glass tubes. In patients with defects of the intrinsic pathway, the formation of thrombin may be impeded, so that serum may contain excessively high amounts of prothrombin. This test is relatively insensitive, giving normal results in patients with partial deficiencies of the factors participating in the intrinsic pathway. Abnormal prothrombin consumption is also found in patients with qualitatively abnormal or decreased numbers of platelets.

Defects in the extrinsic pathway of clotting are readily detected by measuring the *one-stage prothrombin time,* that is, the clotting time of a mixture of citrated plasma, tissue thromboplastin and calcium ions. The one-stage prothrombin time is so short that the simultaneous contribution of the intrinsic pathway to the formation of thrombin is normally negligible. The one-stage prothrombin time is abnormally long whenever a functional deficiency of any factor participating in the extrinsic pathway exists or when fibrin formation is impeded. It is also long when an inhibitor of the pathway is present, but it is normal in deficiencies of Hageman factor, PTA, Christmas factor, antihemophilic factor or fibrin-stabilizing factor.

Specific localization of deficiencies of clotting factors other than fibrinogen or fibrin-stabilizing factor is carried out by the technique of "cross matching." In brief, a small amount of the plasma to be tested is mixed with a plasma known to be deficient in a single factor. The mixture is then tested by whatever method is appropriate for the factor under study—modifications of the partial thromboplastin time for assay of Hageman factor, PTA, Christmas factor or antihemophilic factor, and modifications of the one-stage prothrombin time for measurement of Factor VII, Stuart factor, proaccelerin or prothrombin.

Quantitative measurement of plasma fibrinogen is readily performed by converting this protein to fibrin by the addition of thrombin and assaying the fibrin chemically; other methods, although simpler, lack accuracy. Defects in the rate of formation of fibrin are detected by measuring the *thrombin time,* the time elapsed until clotting, after thrombin has been added to plasma. Hypofibrinogenemia may be suspected when the partial thromboplastin time, the one-stage prothrombin time or the thrombin time is long; the blood is incoagulable in afibrinogenemia, the absence of detectable fibrinogen in plasma. A useful test for *deficiency of fibrin-stabilizing factor* is determination of the solubility of a plasma clot in 5M urea or 1 percent monochloroacetic acid.

This list of available procedures is, of course, incomplete. There are, among others, tests for the detection of circulating anticoagulants—agents in plasma which abnormally delay clotting—and for substances in *serum* antigenically related to fibrinogen (Chapter 26).

Hereditary Disorders of Blood Clotting

Classic hemophilia, a functional deficiency of antihemophilic factor (Factor VIII), is a life-long bleeding disorder limited almost exclusively to males and inherited as an X chromosome-linked trait. All daughters of hemophiliacs are carriers of the trait; one half of a carrier's sons are hemophiliacs, and one half of her daughters are carriers. The severity of the defect varies among different families, but the disorder is usually of about the same intensity among affected individuals within a single family. Two groups of hemophiliacs have been differentiated immunologically. As detected by heterologous antibodies against antihemophilic factor, all hemophiliacs appear to synthesize normal amounts of a functionally incompetent form of this factor. Among these patients, a small group of mildly affected hemophiliacs can be distinguished in whom the abnormal antihemophilic factor can also be detected by tests employing homologous circulating anticoagulants against this substance; this is not the case in most hemophiliacs. As would be anticipated from these considerations, the carriers of hemophilia have normal amounts of antihemophilic factor in their plasma, as determined immunologically. Their titer of antihemophilic factor, measured functionally, varies widely, but, on the average, it is about half that of normal individuals.

The deficiency of functional antihemophilic factor in the plasma of patients with classic hemophilia impairs the intrinsic pathway of clotting, so that the formation of thrombin is abnormally slow. In severe cases, the clotting time of whole blood is prolonged and prothrombin consumption is poor. In milder cases, these tests may give normal results, but the partial thromboplastin time is almost always abnormally long. Diagnosis rests upon specific assays for antihemophilic factor. Affected individuals have titers of less than 0.40 units of activity attributable to antihemophilic factor per ml. of plasma; on the average, normal plasma contains one unit per ml. In contrast to von Willebrand's disease, the bleeding time is normal.

The symptoms of classic hemophilia are not well explained. In severe cases, the patients may have apparently spontaneous hemorrhages into the skin, muscles, soft tissues, central nervous system and gastrointestinal and urinary tracts and may bleed profusely after injuries, surgery or dental extractions. Bleeding into joint spaces, particularly of the knee, hip and elbow, are especially common, and permanent, crippling damage may ensue. In milder cases, bleeding may occur only after injury or surgery, making diagnosis perplexing. Ordinarily, trivial scrapes do not bleed profusely, perhaps because the injured tissues initiate clotting by the extrinsic mechanism. Why hemophiliacs bleed into joints is unclear, but possibly tissue thromboplastin is not readily available.

Therapy, besides local measures to improve hemostasis and minimize tissue injury, consists of temporary correction of the patient's coagulative defect by the transfusion of concentrates of normal plasma rich in antihemophilic factor. Since half of transfused antihemophilic factor disappears from the circulation within 12 hours, repeated infusions are needed.

Von Willebrand's Disease. A second inherited deficiency of antihemophilic factor occurs in von Willebrand's disease (vascular hemophilia), inherited as an autosomal dominant trait. The basic nature of the abnormality in von Willebrand's disease is not understood. The deficiency in antihemophilic factor, usually not as great as in the severer forms of classic hemophilia, apparently reflects a true incapacity to synthesize this substance. Transfusion of hemophilic plasma into patients with von Willebrand's disease is followed by a significant increase in the recipient's titer of antihemophilic factor, maximal after 4 to 8 hours, as if hemophilic plasma contained either a precursor of antihemophilic factor or an agent stimulating its synthesis.

Besides the deficiency in antihemophilic factor, patients with von Willebrand's disease have an unexplained defect in hemostasis manifest in the laboratory by a long bleeding time and often by diminished adhesiveness of platelets to glass. The bleeding tendency and the defect measured in the laboratory vary from patient to patient, even within a family, and not every patient has all of the abnormalities noted. In most cases, bleeding is mild, occurring only after injury or surgical procedures. Menorrhagia is common, but childbirth is usually unaccompanied by hemorrhage, since the titer of antihemophilic factor rises during pregnancy, often to normal levels. This observation has led to the prophylactic use of oral contraceptive agents in female patients.

Christmas disease, which is clinically similar to classic hemophilia and is inherited in the same way, results from a functional deficiency of Christmas factor (Factor IX). A few patients with Christmas disease synthesize a nonfunctional variant of Christmas factor; however, most of them do not seem to synthesize this protein. Although the titer of Christmas factor in the plasma of carriers varies greatly, on the average it is about one half that of normal individuals.

Hereditary Disorders of Fibrinogen. Two distinct forms of hereditary disorders of fibrinogen have been described. In *congenital afibrinogenemia*, the plasma contains no detectable fibrinogen so that the blood is incoagulable. Inherited as an autosomal recessive trait, this disorder is remarkable in that affected individuals bleed only when subjected to injury, beginning with bleeding from the umbilicus at birth. In these patients, thrombin formation is unimpaired, and platelet aggregation and adhesion to wounds are usually normal. The freedom from spontaneous bleeding is in sharp contrast to severe hemophilia or Christmas disease, in which the formation of thrombin is excessively slow.

A rare but instructive abnormality, transmitted in an autosomal dominant way, is *congenital dysfibrinogenemia,* a state in which the patients synthesizes a functionally defective fibrinogen, usually in normal amounts. The disorder ordinarily comes to notice because the one-stage prothrombin time is prolonged. Analysis of the cause for this abnormality demonstrates that the patient's fibrinogen clots excessively slowly upon the addition of thrombin, nearly always because the polymerization of fibrin is defective. Several different abnormalities have been described, distinguished by immunological and functional tests, each named after the city of origin. In one case, fibrinogen Detroit, an arginine residue in the N-terminal part of the α (A) chain of fibrinogen was replaced by a neutral amino acid, probably serine, and the protein contained less carbohydrate than normally. Unexpectedly, patients with dysfibrinogenemia do not ordinarily have a bleeding tendency, although exceptions have been reported. In a few patients, operative wounds have dehisced, and in others, a thrombotic tendency is present. In one family, affected individuals had a fibrinogen that clotted more rapidly than normally and exhibited a high incidence of thromboembolic phenomena.

Deficiency of Functional Fibrin-Stabilizing Factor (Factor XIII). Fibrin formation may also be impeded by this hereditary anomaly. Individuals lacking fibrin-stabilizing factor may have a severe bleeding tendency, beginning with umbilical bleeding after birth and often ending with central nervous system hemorrhage. Why the patients bleed is not clearly evident, since the rate of fibrin formation, tested in vitro, is normal. In some instances, the patients' wounds break down repeatedly and bleed afresh, suggesting that the formed fibrin easily disrupts, since it lacks its normal chemical bonding. The plasma of patients with fibrin-stabilizing factor deficiency contains material antigenically related to this substance, as if a nonfunctional form is synthesized. Two groups of patients have been discerned. In some families, the defect is inherited as an autosomal recessive trait, and consanguinity is frequent. In others, the disorder is limited to men and consanguinity is rare, but proof of X chromosomal inheritance is lacking.

Other Familial Disorders of Blood Clotting. A number of such disorders have been described, all inherited in an autosomal recessive manner. *Hageman trait,* the hereditary absence of Hageman factor (Factor XII) is usually asymptomatic, the defect being detected by chance observation of a much prolonged clotting time. Suitable tests demonstrate a defect in the intrinsic pathway of coagulation. The defect in homozygous *PTA deficiency* (Factor XI) is usually incomplete, so that clotting time and partial thromboplastin time may be only marginally prolonged. Although some patients with PTA deficiency are asymptomatic, others have a mild bleeding tendency, and a few deaths have been reported. In both Hageman trait and PTA deficiency, the defect is apparently a failure to synthesize the missing protein, since no material antigenically related to Hageman factor or PTA can be detected.

In four autosomal recessive traits, the one-stage prothrombin time is prolonged, and the disorders must then be differentiated by specific testing. In all, the degree of the bleeding tendency varies widely from family to family, a difference not explained entirely at this time. In *Stuart factor deficiency* (Factor X) and *prothrombin deficiency* (Factor II) some individuals appear to synthesize a nonfunctional form of the factor, whereas others have no detectable material in their blood related to Stuart factor or prothrombin. In *Factor VII deficiency* and *parahemophilia* (the hereditary absence of proaccelerin), the plasma probably does not contain antigens related to

the missing factors. Since autosomal recessive traits occur in both sexes, it is not surprising that menorrhagia and bleeding at childbirth are common problems.

Acquired Disorders of Blood Clotting

The number of clinical situations in which a clotting abnormality is present is legion, but certain syndromes are of peculiar interest because they illustrate pathophysiological mechanisms.

Four plasma clotting factors—prothrombin, Stuart factor, Factor VII and Christmas factor—are synthesized only if vitamin K is available. Although small amounts of vitamin K are furnished by diet, the bulk of this material is synthesized by intestinal bacteria. Natural vitamin K is lipid-soluble and is absorbed only if bile salts are present in the gut. Once absorbed, it is utilized in the parenchymal cells of the liver for production of the four factors, but it does not become an integral part of their structure. Synthesis of the four factors appears to involve two stages. First, a functionally inert polypeptide is produced, after which, in some unknown way, vitamin K converts it to a functional form before its release from the ribosomes. In the absence of vitamin K, the incomplete protein may enter the bloodstream, where it acts as a competitive inhibitor of the vitamin K-dependent factors. Any interruption of this chain of events results in a deficiency of the vitamin K-dependent clotting factors.

Deficiencies of the Vitamin K-Dependent Clotting Factors. The newborn infant's only source of vitamin K is milk; if this supply is inadequate, a severe depletion of the vitamin K-dependent factors may ensue, resulting in hemorrhagic disease of the newborn. Administration of small amounts of vitamin K to the mother just before childbirth, or to the newborn infant will prevent this disorder. If the infant survives, the disorder is self-limited because the intestinal flora soon supply adequate amounts of vitamin K. Sterilization of the gut with antibiotics, as may take place in preparation for surgery, will cut off this source. In this situation deficiency of the vitamin K-dependent factors is fostered by restriction of dietary intake of the vitamin. Impaired absorption of vitamin K because of *obstructive jaundice*, in which bile salts do not reach the duodenum, is responsible for the bleeding tendency that accompanies this disorder. The deficiency of vitamin K-dependent clotting factors is readily corrected by parenteral administration of the vitamin, greatly reducing the risk of surgery.

Impaired absorption may also reflect intrinsic disease of the bowel, such as sprue, in which lipid-soluble agents are poorly absorbed. Parenchymal hepatic disease is commonly associated with deficient synthesis of the vitamin K-dependent factors; in this case, the parenteral administration of the vitamin is usually without benefit. Finally, the administration of certain anticoagulant drugs, notably those related to coumarin or phenindione, interferes with synthesis of the vitamin K-dependent factors, presumably by competitive inhibition. The clinical picture in all of these states resembles that of classic hemophilia except that the onset is usually sudden. The diagnosis should be suspected whenever the one-stage prothrombin time is prolonged, since this test measures, in part, the concentrations of Factor VII, Stuart factor and prothrombin. Suitable analysis will also reveal a depression in the concentration of Christmas factor. In otherwise normal individuals in whom the one-stage prothrombin time is prolonged and multiple deficiencies of the vitamin K-dependent factors are detected, the question of surreptitious ingestion of coumarinlike compounds must come to mind; here the mechanism of disease goes beyond the alterations in hemostasis, for these patients exhibit evidence of severe emotional disturbances.

Intravascular Clotting. When a massive amount of tissue thromboplastin is injected intravenously, the circulating blood clots and the animal dies because the flow of blood is obstructed. When thromboplastin is injected slowly, the animal survives and its blood is incoagulable. The mechanisms underlying these events are easily understood. Injected thromboplastin generates thrombin via the extrinsic pathway. A large bolus of thromboplastin induces the explosive formation of thrombin and intravascular clotting. Lesser amounts of thromboplastin bring about much slower and, perhaps, incomplete fibrin formation, so that the blood may contain soluble intermediates of fibrin formation—either monomers or soluble polymers of fibrin, or complexes of fibrin monomer and fibrinogen. These products are rapidly removed from the circulation by the reticuloendothelial system. As a result, the animal's circulating fibrinogen is depleted and its blood will not clot. At the same time, other clotting factors may be decreased in concentration, particularly antihemophilic factor and proaccelerin, which are inactivated by thrombin. The platelet count falls, perhaps because the platelets are clumped by thrombin. Sporadically, evidence of intravascular activation of plasminogen, the precursor of the plasma proteolytic enzyme plasmin, may be detected. At the same time, the serum (that is, the liquid phase of blood after defibrination) contains material antigenically related to fibrinogen and fibrin, either soluble fibrin or fibrinogen-fibrin complexes, or the products of the digestion of fibrin by plasmin (see Chapter 26). These fibrinogen-fibrin related antigens may be responsible for the anticoagulant properties of blood that appear after the infusion of thromboplastin.

These experimental observations find their counterpart in many clinical situations.

In *amniotic fluid embolism*, amniotic fluid and its contaminants enter the maternal bloodstream during parturition; lanugo hairs and fetal epithelial cells are demonstrable in maternal blood. The patient may die suddenly of respiratory failure. More usually, an insidious hemorrhagic tendency appears, with bleeding from the placental site, venepuncture wounds, the gastrointestinal tract and elsewhere; unless vigorous treatment is instituted, death from exsanguination may result. Studies of maternal blood demonstrate depletion of fibrinogen and other clotting factors, thrombocytopenia, the presence of an anticoagulant interfering with fibrin formation and, inconstantly, excessive fibrinolysis. These changes, similar to those seen in animals injected with thromboplastin, presumably result from initiation of the extrinsic clotting pathway by procoagulant agents in amniotic fluid.

In the same way, intravascular clotting may follow *envenomation* by poisonous snakes, the mechanism varying from species to species. For example, Russell's viper venom activates Stuart factor and alters proaccelerin so as to make it more effective; the venom of the tiger snake behaves like activated Stuart factor, and the venoms of the Malayan pit viper and the Southern copperhead snake clot fibrinogen directly.

Many other hypofibrinogenemic syndromes have been attributed to intravascular coagulation. In *massive transfusion of incompatible blood,* the hemolyzed erythrocytes presumably initiate the clotting process. The hypofibrinogenemia complicating *premature separation of the placenta* has been ascribed to the thromboplastic effect of placental tissue, which may gain entrance to the maternal circulation; in another view, hypofibrinogenemia results from depletion of plasma fibrinogen in the formation of the retroplacental clot.

In other cases, damaged vascular endothelium is thought to be the source of procoagulant material. This is most clearly seen in *purpura fulminans,* in which widespread vasculitis and thrombosis leads to patchy, usually superficial gangrene. The

hypofibrinogenemia and other evidences of intravascular clotting observed in *heat stroke,* in severe *sepsis* (such as occurs after self-induced abortion) or in the *Waterhouse-Friderichsen syndrome* have similarly been attributed to vascular injury. Tumor cells or secretions may be the source of the procoagulant that induces hypofibrinogenemia in *leukemia* or *carcinoma,* particularly of the prostate. Perhaps similar mechanisms are responsible for the hypofibrinogenemia observed in association with *giant cavernomatous hemangioma.*

The view that hypofibrinogenemia is secondary to intravascular clotting may be fortified by inducing remission by the administration of heparin, a powerful anticoagulant. For example, the hypofibrinogenemia found in some patients with *intrauterine retention of a dead fetus* is temporarily corrected by administration of heparin. The detection of fibrinogen-fibrin related antigens in serum also supports the diagnosis of intravascular clotting. In experimental animals, the induction of intravascular clotting may lead to fragmentation of erythrocytes, which take on a bizarre appearance, as if sheared by strands of fibrin. Demonstration of such erythrocytes in the blood of patients with hypofibrinogenemia has been taken as evidence of intravascular coagulation.

Microangiopathic Hemolytic Anemia. The observations discussed above have suggested to some that other syndromes, not necessarily associated with hypofibrinogenemia, may be associated with intravascular clotting. The abnormal erythrocytes characteristic of the syndromes known collectively as *microangiopathic hemolytic* anemia, including *thrombotic thrombocytopenic purpura,* the *hemolytic uremic syndrome* and *eclampsia,* have been ascribed to intravascular clotting secondary to vascular damage. Similar erythrocytic changes may be seen in *acute glomerulonephritis, malignant hypertension* and the hemolytic anemia complicating *cardio-*

vascular prosthesis. Needless to say, caution is necessary before accepting the view that these syndromes are, in fact, complicated by intravascular clotting.

Hepatic Disease

Patients with chronic hepatic disease, such as cirrhosis of the liver or carcinoma, commonly have a bleeding tendency, a complication which may also accompany the severer forms of acute hepatitis. The pathogenesis of bleeding may be complex. Commonly, a deficiency of the vitamin K-dependent clotting factors is present. In most cases this is the result of impaired synthesis by hepatic parenchymal cells and does not respond to parenterally administered vitamin K. Less often, intrahepatic obstruction to the flow of bile may impair absorption of the vitamin, and in such cases, parenterally administered vitamin K may correct the defect. Deficiencies of Stuart factor, Factor VII and prothrombin can be detected by an abnormally prolonged one-stage prothrombin time. A deficiency of proaccelerin, also synthesized in the liver, is frequent; this abnormality also lengthens the prothrombin time. Unusually, patients with hepatic disease may have hypofibrinogenemia. In some cases this has been attributed to deficient synthesis of fibrinogen by the liver; in others, it may result from intravascular coagulation. In patients with portal hypertension, the platelet count is often reduced, probably because these cells are sequestered in the enlarged spleen. Thrombocytopenia may also result from alcoholism, either as a direct toxic effect of alcohol or because of concomitant folic acid deficiency. Further complicating matters, patients with chronic hepatic disease often have evidence of increased plasma fibrinolytic activity (see Chapter 26). This may be one of several reasons why clotting of plasma upon the addition of thrombin is delayed in patients with hepatic disease. Thus, the bleeding syndrome of patients with disorders of the liver has no simple explanation, and each case must be studied if a knowledge of the mechanisms involved is important for the patient's care.

Circulating Anticoagulants

An occasional but important complication of hemophilia is the development of circulating anticoagulants directed against antihemophilic factor. These agents, detected in as many as 10 percent of patients or more, are apparently IgG autoantibodies which inactivate antihemophilic factor. In such patients, transfusions of antihemophilic factor are without benefit, and bleeding is therefore more difficult to control.

Rarely, similar anticoagulants may appear de novo in apparently normal adults; these have also been observed in some patients with systemic lupus erythematosus, after penicillin reactions or after childbirth. Since the patients lack functional antihemophilic factor, they have symptoms similar to those of severe hemophilia, and, unless the anticoagulant disappears, they may ultimately succumb to hemorrhage.

Circulating anticoagulants have also been detected in some patients with Christmas disease or hereditary deficiencies of PTA, fibrinogen, Factor VII or proaccelerin. The nature of these anticoagulants has not been well studied, but presumably they are autoantibodies.

Another type of anticoagulant, seen most frequently in patients with systemic lupus erythematosus, interferes with the formation of the prothrombin-converting principle. In such cases, the one-stage prothrombin time is prolonged, particularly if measured with diluted tissue thromboplastin. Hemorrhagic symptoms are often minimal in this syndrome.

DISORDERS OF BLOOD PLATELETS

The Role of the Platelets in Hemostasis

Platelets are small cytoplasmic fragments of megakaryocytes, large multinucleated cells found, in the adult, primarily in the red marrow. Maturation of megakaryocytes to the point at which they shed platelets into the circulation takes as long as 5 to 7 days. The maturation process is stimulated by one or more agents in plasma.

Normal peripheral blood contains 150,000 to 350,000 platelets per mm.[3] About one third of the total platelet mass is not in the peripheral circulation but is sequestered in the spleen. The average platelet survives about 10 days; the bulk are lost through senescence, although normally a small number may be destroyed at random, presumably because they are utilized in hemostasis. The aged or injured platelets are apparently removed from the circulation by the reticuloendothelial system, primarily in liver and spleen.

When vascular endothelium is damaged, platelets rapidly accumulate at the point of injury, adhering to exposed subendothelial collagen. Once this takes place, platelets discharge their cytoplasmic granules, releasing adenosine diphosphate (ADP) and other intracellular constituents. ADP, derived from platelets and probably also from injured endothelium, induces loose aggregation of fresh platelets to those adherent to the wound. Platelet phospholipoprotein (principally that in the platelet membrane) becomes available for local thrombin formation, which is brought about by contact of blood with thromboplastin (furnished by injured tissue) and by activation of Hageman factor by exposed collagen. Platelets also contain proaccelerin and antihemophilic factor, derived from circulating plasma, and these may intensify the evolution of thrombin. Thrombin stimulates local deposition of fibrin, binding the platelets into a hemostatic plug sufficient to stop bleeding from small wounds. Thrombin also induces further platelet aggregation and degranulation, with the release of intracellular components including ADP—the phenomenon of viscous metamorphosis. The action of

thrombin on platelets is complex, involving participation of intracellular proteins functionally similar to plasma fibrinogen and fibrin-stabilizing factor. Further, it brings about contraction of an intracellular protein, thrombasthenin, a substance with adenosine triphosphatase activity similar to that of muscle actomyosin. This contraction of platelets makes the platelet plug more compact and, in the test tube, is responsible for clot retraction, in which the clot shrinks, expressing the serum within its meshes. Platelet fibrinogen and calcium ions are needed both for platelet aggregation and for clot retraction.

When blood or platelet-rich plasma comes into contact with glass, some of the platelets adhere to the glass surface. This effect, perhaps related to that induced by collagen, depends upon the presence in plasma of fibrinogen and perhaps other substances. The significance of platelet adhesion to glass, a substance foreign to the body, is not apparent, but decreased adhesiveness is a useful test for the presence of abnormal platelets or of von Willebrand's disease.

Tests of Platelet Function

Platelets should be counted in the blood of every patient suspected of a hemostatic defect, and a smear should be examined to see if they are morphologically abnormal. Since platelets are essential for the phenomenon of clot retraction, qualitative or semiquantitative tests of this process must be performed. Qualitative abnormalities in the clot-promoting properties of platelets may be detected because the clotting time measured in plastic tubes is prolonged, or because platelets behave defectively in the prothrombin consumption test or the thromboplastin generation test of Biggs and Douglas. In these procedures, the normal conversion of prothrombin to thrombin is impeded if platelets fail to release adequate amounts of clot-promoting phospholipids. Simple tests have been devised to measure the capacity of platelets to adhere to glass (a property said to be defective in von Willebrand's disease) and to aggregate upon the addition of collagen, ADP, epinephrine or thrombin.

The bleeding time—that is, the time elapsing until bleeding stops from an incised wound—is prolonged when the platelet count is depressed below about 80,000 per mm.[3] A long bleeding time is also found in patients with thrombocythemia (with platelet counts above 800,000 per mm.[3]) or with qualitative abnormalities of platelets. Other causes of a long bleeding time are von Willebrand's disease and the dysproteinemias.

Thrombocytopenia

Individuals in whom the platelet count is less than 100,000 per mm.[3] may exhibit a bleeding tendency in rough proportion to the degree of thrombocytopenia. They bruise readily and may bleed from the nose, the gingiva or the gastrointestinal or genitourinary tracts. Petechiae are commonplace, both on the skin and on mucous membranes. Bleeding into the central nervous system, often with lethal outcome, is particularly frequent in patients with acute leukemia or aplastic anemia but is unusual in idiopathic thrombocytopenic purpura.

Examination of aspirates of bone marrow is essential in determining the cause of thrombocytopenia in almost every case. The aspirate may demonstrate the presence of abnormal cells, such as are found in leukemia, metastatic carcinoma or Gaucher's disease. In other cases, no abnormal cells may be seen but megakaryocytes may be absent, as in aplastic anemia. In contrast, in idiopathic thrombocytopenic purpura or the thrombocytopenia accompanying systemic lupus erythematosus, sensitivity to drugs or miliary tuberculosis, normal numbers of megakaryocytes may be present, but often the cytoplasm of these cells is scanty. In still other cases, such as agnogenic myeloid metaplasia, myelofibrosis, aplastic anemia or the early stages of leukemia, no marrow may be obtainable by aspiration, and in these instances surgical biopsy of the marrow should be performed in an effort to reach a diagnosis.

Thrombocytopenia may be the result of several different processes (Table 25-1). In some cases, the rate of formation of platelets is abnormally low. Sometimes, as in hypoplastic or aplastic anemia, which may occur idiopathically or after exposure to ionizing radiation, drugs or toxins, megakaryocytes are few in number or absent, often in association with depression of other marrow elements. In other cases, the marrow may be infiltrated with abnormal cells which may crowd out the normal elements and, in this way, lead to anemia and thrombocytopenia. In still others, the normal stimuli to platelet maturation may be lacking or suppressed, as in the megaloblastic anemias of vitamin B-12 or folate deficiencies. Suppression of normal marrow activity may account for some instances of thrombocytopenia accompanying infection.

Thrombocytopenia may also be due to excessively rapid destruction of platelets, most clearly seen in individuals sensitive to certain drugs. The list of offending agents is long. Of these, Sedormid, quinine, quinidine and digitoxin have been studied most intensively. The most likely explanation for their effects is that the patient acquires circulating antibodies against the drugs. Upon subsequent administration of the medication, antigen-antibody complexes form within the circulation and injure the platelets through the process of immune adherence.

A similar explanation has been offered for the pathogenesis of idiopathic thrombocytopenic purpura. Two types of this disorder can be delineated—an acute self-limited purpura, usually lasting no more than several months, and a more chronic

TABLE 25-1. Mechanisms Responsible for Thrombocytopenia*

Disorders in which production of platelets is probably reduced

Hypoplasia or aplasia of megakaryocytes
  Depression of marrow by ionizing radiation
  Depression of marrow by drugs or toxins (chloramphenicol, benzene, gold salts)
  Idiosyncrasy to drugs
  Congenital hypoplastic anemia
  Fanconi's familial anemia
  Congenital thrombocytopenia with absent radii
  Aplastic anemia with thymoma
  Agnogenic myeloid metaplasia or myeloid fibrosis
  Idiopathic aplastic anemia
  Congenital thrombocytopenia

Infiltration of marrow by abnormal cells
  Leukemia and other lymphomas
  Metastatic carcinoma
  Multiple myeloma
  The histiocytoses

Ineffective production of platelets
  Deficiency of vitamin B-12
  Deficiency of folate
  Iron-deficiency anemia (inconstant)
  Azotemia
  Hyperthyroidism (?)
  Infections
  Alcoholism
  Excessive Prednisone dosage
  Congenital absence of plasma factor needed for platelet production
  Familial thrombocytopenia
  Congenital thrombocytopenia with eczema and repeated infections (Wiskott-Aldrich
    syndrome) (?)

Disorders in which the life span of the platelets is probably decreased

Sensitivity to drugs
  quinine
  quinidine
  digitoxin

Iso-immunization to platelets
  transfusion
  pregnancy

Systemic lupus erythematosus

Idiopathic thrombocytopenic purpura

Lymphomas
  lymphosarcoma
  Hodgkin's disease
  chronic lymphatic leukemia

---

* The reader will note that the same basic disease process may, in different situations, result in thrombocytopenia through different mechanisms.

TABLE 25-1.   Continued

Hemolytic anemias

Microangiopathic anemias
    thrombotic thrombocytopenic purpura
    hemolytic-uremic syndrome
    eclampsia

Intravascular coagulation
    amniotic fluid embolism
    heat stroke
    carcinoma

Extracorporeal circulation

Infections
    miliary tuberculosis

Sarcoidosis

Chronic cor pulmonale

Disorders in which platelets are sequestered

Splenomegaly
    congestive splenomegaly
    Gaucher's disease
    miliary tuberculosis
    agnogenic myeloid metaplasia
    Hodgkin's disease
    sarcoidosis

Congenital hemangiomatosis

Kaposi's sarcoma

Dilution of platelets

By transfusion of platelet-poor blood

Disorders in which the pathogenesis of thrombocytopenia is unclear

Infections

Onyalai

Thermal burns

Kwashiorkor

Macroglobulinemia and other dysproteinemias

Paroxysmal nocturnal hemoglobinuria

Congenital erythropoietic porphyria

Envenomation by the brown recluse spider

form which may persist for many years. In either case, the marrow contains normal numbers of megakaryocytes which, however, lack the budding platelets normally found at the periphery of their cytoplasm. The plasma of patients with idiopathic thrombocytopenic purpura contains an agent that will induce thrombocytopenia when introduced into normal individuals either by transfusion or across the placenta. The agent has characteristics of an autoantibody of the IgG type. A similar explanation has been offered for the thrombocytopenia accompanying systemic lupus erythematosus, miliary tuberculosis, lymphosarcoma and other disorders.

The life span of the platelets may also be decreased because of excessive peripheral utilization of platelets, the most dramatic examples being the syndromes of intravascular clotting and the thrombocytopenia that follows the use of extracorporeal circulatory apparatus. A similar pathogenesis may explain the thrombocytopenia that is a feature of widespread vascular damage such as occurs in thrombotic thrombocytopenic purpura, the hemolytic-uremic syndrome or eclampsia.

In some cases, the number of circulating platelets is reduced because these cells are sequestered within an enlarged spleen. Under these conditions, platelet production and life span are usually normal, and splenectomy often brings about remission of the thrombocytopenia. Other lesions in which a wide meshwork of small vessels is present, such as giant cavernomatous hemangioma or Kaposi's sarcoma, may be complicated by thrombocytopenia.

In still other cases, the platelet count is lowered by dilution. When patients with severe hemorrhage are transfused with many units of stored blood, the platelet count falls because the infused blood is deficient in viable platelets. Other clotting factors deficient in stored blood may also decrease in concentration.

Thrombocytopenia during the course of infection may come about in different ways. For example, in rubella the platelets are destroyed prematurely, perhaps by circulating autoantibodies. In infections in which vascular damage is prominent, such as bacterial sepsis, rickettsial disease or certain other hemorrhagic fevers, thrombocytopenia may result from consumption of platelets in intravascular clotting. In still other infections—for example, rare cases of acute hepatitis—thrombocytopenia may result from marrow aplasia. Almost certainly,

in many infectious processes several different mechanisms may be operative.

This lengthy classification does not encompass all the disorders in which thrombocytopenia has been detected. Prominent among those in which the pathogenesis is unclear are various forms of hereditary thrombocytopenias. Although some of these have been ascribed to deficient formation from megakaryocytes, in others, no such explanation is forthcoming. Most interesting is the rare congenital thrombocytopenia that responds to the transfusion of normal plasma, as if patients with this disorder lacked a stimulus needed for platelet formation.

Thrombocytosis and Thrombocythemia

A transient rise in the platelet count above normal is common during the periods after surgical procedure, severe hemorrhage or childbirth, or as an accompaniment of inflammatory states. A more sustained rise is more likely to be associated with iron deficiency anemia, malignancy or one of the myeloproliferative syndromes — polycythemia vera, chronic myeloid leukemia or agnogenic myeloid metaplasia. The thrombocytosis is apparently in response to some unknown stimulus to platelet production; the platelet life span is usually normal. Thrombocytosis may also follow splenectomy or atrophy of the spleen, as if platelet survival were enhanced by the absence of an organ important for the removal of these cells. Although, usually, the rise in the platelet count after splenectomy is transient, it may persist for many years. When the platelet count is elevated to 800,000 per mm.^{3} or more—a condition called thrombocythemia—the patient may exhibit a bleeding tendency. Although many explanations for this paradox have been offered, none is satisfactory.

Qualitative Abnormalities of Platelets

Inherited qualitative defects in platelet function are uncommon; they serve, however, to fortify our views concerning the role of these cells in hemostasis. In *thromb-asthenia,* or Glanzmann's disease, a bleeding tendency is correlated with impaired clot retraction, a long bleeding time and impaired aggregation of platelets by ADP and thrombin, but the platelet count is usually normal. In *thrombopathic purpura,* or von Willebrand-Jurgen's syndrome, the platelets lack clot-promoting activity, behaving as if they cannot release their phospholipid. The bleeding time is prolonged and aggregation of platelets by ADP and thrombin may be impaired but clot retraction is normal. A third group of patients, with a condition called *thrombo-pathia,* usually have only a mild bleeding tendency. Again, the bleeding time is long and aggregation of platelets by collagen or by small amounts of ADP may be impaired, but clot retraction and clot-promoting activity are normal.

An acquired qualitative platelet abnormality may contribute to the bleeding tendency of patients with *uremia.* Platelet aggregation by collagen is also depressed by the *ingestion of aspirin,* which may aggravate the bleeding tendency in patients with other forms of hemostatic abnormality.

DISORDERED HEMOSTASIS DUE TO VASCULAR PATHOLOGY

When the vascular wall is damaged, bleeding may occur after minor trauma or apparently spontaneously. Hemorrhage may follow damage to the endothelial surface or to the supporting structures within and around the blood vessels, and it is especially likely to occur where intravascular pressure is relatively high, as in the small blood vessels of the feet and ankles.

Hereditary Disorders of Connective Tissue. Inadequate vascular support is a prominent feature of several of the hereditary disorders of connective tissue. In the *Meekrin-Ehlers-Danlos syndrome,* the col-

lagen fibrils appear to be bound together abnormally, so that they slide over each other more easily than normally. In this disease, inherited as an autosomal dominant trait, affected individuals may have hypermobility of joints, hyperelasticity of skin, a blue cast to the sclerae, dislocation of the lens of the eye, and diaphragmatic hernia. The patients bruise readily and may have subcutaneous hematomas, presumably because their blood vessels lack the support of normal collagen. In some families, spontaneous rupture of large arteries is a frequent and often lethal event. In *osteogenesis imperfecta,* usually inherited as an autosomal dominant trait, the skin is deficient in collagen. A patient with this disorder may bruise easily and have other hemorrhagic manifestations; the platelets are said to function abnormally, contributing to the bleeding diathesis. In *pseudoxanthoma elasticum,* an autosomal recessive trait, visceral, retinal, joint and cutaneous bleeding is commonplace. In this disorder, which is characterized by the presence of waxy papules in the folds of the skin, one basic defect appears to be a degeneration of the elastic fibers of connective tissue surrounding the blood vessels.

Scurvy. A diffuse lack of vascular support is also a prime feature of scurvy—a disease, once extremely common, caused by protracted dietary deficiency of vitamin C. Hemorrhages into the skin, the gingiva and the joints are important signs. These phenomena have been attributed to a loss of vascular support as the result of defective synthesis of connective tissue. Recently, study of the platelets in scurvy has suggested that these cells are qualitatively defective, contributing to the bleeding tendency.

Senile Purpura. A loss of connective tissue support of blood vessels, due to atrophy of collagen associated with aging, is thought to be the basis for senile purpura, the dark purple spots seen on the skin of elderly people. In this harmless but sometimes frightening problem, cutaneous hemorrhages are most likely to be present on the extensor surface and radial border of the forearms, on the backs of the hands and in the region where eyeglasses press upon the face. Characteristically, the lesions are sharply demarcated, unlike ordinary bruises, and last for days or weeks, leaving a residue of brownish pigmentation. Bleeding is thought to occur because the vessels are readily torn by shearing strains upon the skin. Similar lesions are seen in some individuals treated with corticosteroids, perhaps because the structure of collagen is altered. In other patients receiving steroids, typical ecchymoses may appear spontaneously or after minimal trauma. These lesions are indistinguishable from those of *simple purpura,* an unexplained tendency to bruise readily which is present almost entirely in women. Simple purpura—or "the devil's pinches"—is a cosmetic problem, but has no other significance.

Anaphylactoid purpura (allergic or Henoch-Schönlein's purpura)—a relatively common disorder in which bleeding may be an important element—is a systemic disease involving the skin, the mucous membranes, the joints, the gastrointestinal tract, the kidneys, the central nervous system and the heart. The basic lesion is a diffuse angiitis, in which the arterioles, capillaries and vessels are surrounded by neutrophils, mononuclear cells and eosinophils; the vascular walls may be completely necrotic. The vessels may be plugged by aggregations of leukocytes and platelets. Particularly in children, anaphylactoid purpura may follow closely upon infection, often by the hemolytic streptococcus. In other cases, the syndrome appears to be initiated by the administration of one or another drug. In many ways, the lesions of anaphylactoid purpura suggest an immunologic process; IgG immunoglobulins and $\beta_1 C$ (the third component of complement) have been demonstrated in renal lesions in some cases but not in skin. In common with many vascular lesions, evidences of local intravascular clotting have been described.

Autoerythrocyte Sensitization. The complexity of the mechanisms responsible for alterations in hemostasis is best exemplified by an unusual disorder of women — *autoerythrocyte sensitization*—which is characterized by recurrent crops of painful ecchymoses that have an inflammatory component. Because the lesions can be reproduced by the intracutaneous injection of erythrocytic stroma, the disorder was at first attributed to autosensitization to red blood cells. More recently it has become apparent that women with this disease uniformly have severe emotional disturbances, suggesting that neural influences may be important. Autoerythrocyte sensitization serves as a reminder that our focus upon biochemical alterations may teach us about the nature of disease but leaves us short of the mark when we study illness in the patient.

ANNOTATED REFERENCES

Biggs, R., and Macfarlane, R. G.: Human Blood Coagulation and its Disorders. Ed. 3. Philadelphia, F. A. Davis, 1962. (A comprehensive review of clinical and laboratory diagnosis and therapeutic measures for bleeding disorders. Although considerably dated, it provides a guideline for understanding the interpretation of certain laboratory tests of hemostasis.)

———: Treatment of Haemophilia and Other Coagulation Disorders. Philadelphia, F. A. Davis, 1966. (Emphasis on the collaborative therapy of patients with hemophilia and related disorders)

Cartwright, G. E.: Diagnostic Laboratory Hematology. Ed. 4. New York, Grune and Stratton, 1968. (Review of methods used in student laboratories and doctor's offices for the diagnosis of hematological disorders)

Davie, E. W., and Ratnoff, O. D.: The Proteins of Blood Coagulation, *In:* Neurath, H. (ed.): The Proteins. Ed. 2. Vol. 3, p. 359.

New York, Academic Press, 1965. (A review of the chemical analyses of proteins known to interact in the processes of blood coagulation and fibrinolysis)

Hardaway, R. M., III: Syndromes of Disseminated Intravascular Coagulation. Springfield, Ill., Charles C Thomas, 1966. (Review of the clinical evidence for disseminated intravascular clotting by a surgeon)

Hardisty, R. M., and Ingram, G. I. C.: Bleeding Disorders. Investigation and Management. Philadelphia, F. A. Davis, 1965. (A short monograph on the laboratory investigation of a bleeding diathesis)

Hougie, C.: Fundamentals of Blood Coagulation in Clinical Medicine. New York, McGraw-Hill, 1963. (The well-established facts that form a rational basis for diagnosis and therapy of disorders of hemostasis for practitioners)

Johnson, S. A., and Greenwalt, T. J.: Coagulation and Transfusion in Clinical Medicine. Boston, Little, Brown and Co., 1965. (A monograph on the clinical management of congenital and acquired bleeding disorders)

Laki, K. (ed.): Fibrinogen. New York, Marcel Dekker, 1968. (A recent comprehensive compilation of essays in the field of fibrinogen investigation, covering developments up to 1968—when several advances in our understanding of the molecular variants were reported)

McKay, D. D.: Disseminated Intravascular Coagulation: An Intermediary Mechanism of Disease. New York, Harper and Row, 1965. (Review of the theoretical and experimental evidence for the unusual but ubiquitous syndromes of disseminated intravascular clotting)

McKusic, V. A.: Heritable Disorders of Connective Tissue. Ed. 3. St. Louis, C. V. Mosby, 1966. (A review of the genetic diseases, with discussion of the hemorrhagic disorders attributed to connective tissue defects)

Marcus, A. J., and Zucker, M. B.: The Physiology of Blood Platelets. New York, Grune and Stratton, 1964. (A comprehensive review of the biochemistry, microscopy and methods of study of blood platelets, by contributors working in the field)

Margolius, A., Jr., Jackson, D. P., and Ratnoff, O. D.: Circulation anticoagulants. A study of 40 cases and a review of the literature. Medicine, 40:145, 1961. (Comprehensive summary of circulating anticoagulants

studied at the Johns Hopkins and Western Reserve Universities)

Miescher, P. A., and Koffler, D.: Schönlein-Henoch's and related syndromes. In: Miescher, P. A., and Muller-Eberhard, H. J. (ed.): Textbook of Immunopathology. Vol. 1, p. 359. New York, Grune and Stratton, 1968. (A short statement on the clinical, pathogenetic and therapeutic aspects of Schönlein-Henoch's purpura)

Owen, C. A., Jr., Bowie, E. J. W., Didisheim, P., and Thompson, J. H.: Jr.: The Diagonsis of Bleeding Disorders. Boston, Little, Brown and Co., 1969. (A narrative review of laboratory methods in the diagnosis of bleeding syndromes)

Poller, L. (ed.): Recent Advances in Blood Coagulation. London, J. and A. Churchill, 1969. (A collection of essays on the current developments in the investigation of hemophilia and the use of anticoagulants)

Quick, A. J.: Hemorrhagic Diseases and Thrombosis. Ed. 2., Philadelphia, Lea and Febiger, 1966. (A review of the laboratory diagnosis of hemorrhagic diatheses and anticoagulant therapy for the treatment of thrombosis)

Ratnoff, O. D.: Bleeding Syndromes. Springfield, Ill., Charles C Thomas, 1960. (Practical information on bleeding syndromes, written for the practicing physician)

Ratnoff, O. D. (ed.): Treatment of Hemorrhagic Disorders. New York, Harper and Row, 1968. (Assembly of views on the treatment of hemorrhagic disorders)

Ratnoff, O. D.: Disordered hemostasis in hepatic disease. In: Schiff, L. (ed.): Diseases of the Liver. Ed. 3. Philadelphia, J. B. Lippincott, 1969. (A review of the pathophysiology of disorders of hemostasis caused by hepatic disease)

Ratnoff, O. D.: Hereditary disorders of hemostasis. In: Stanbury, J. B. Wyngaarden, J. B., and Frederickson, D. S. (eds.): The Metabolic Basis of Inherited Disease. Ed. 3. New York, McGraw-Hill (In press). (A review of the hereditary diseases of clotting, platelet function and vascular abnormalities)

Salzman, E. W., and Britten, A.: Hemorrhage and Thrombosis. A Practical Clinical Guide. Boston, Little, Brown and Co., 1965. (A monograph on the clinical management of congenital and acquired bleeding disorders)

Seegers, W. H. (ed.): Blood Clotting Enzymology. New York, Academic Press, 1967.

(A limited review of the prothrombin activating system in the terminology that perpetuates the hypothesis that derivatives of the prothrombin structure will accelerate the autoconversion of prothrombin to the enzyme thrombin. Other contributors are somewhat limited in number.)

Tocantins, L. M., and Kazan, L. A. (eds.): Blood Coagulation, Hemorrhage and Thrombosis. Methods of Study. New York, Grune and Stratton, 1964. (Comprehensive review of laboratory methods for clinical and basic investigation of the hemostatic mechanisms)

26

Fibrinolysis

Victor J. Marder, M.D., *and Sol Sherry,* M.D.

INTRODUCTION

Maintenance of a circulating plasma which is continuously primed for appropriate fibrin formation and lysis, yet simultaneously protected against inappropriate stimulation of either the coagulation or the fibrinolytic process is understandably complex. The fibrinolytic system controls and regulates the enzymatic degradation of both circulating fibrinogen and intra- and extravascular deposits of fibrin. This control is mediated by the plasminogen-plasmin proteolytic enzyme system, composed of precursors, activators and inhibitors which are constantly available and utilized under physiological conditions. Differences in their concentrations and activities in the thrombus, as opposed to the circulating blood, are basic to an understanding of the physiological control of fibrinolysis; alterations of any of the reacting or modifying components may lead to a variety of pathological states.

COMPONENTS OF THE FIBRINOLYTIC SYSTEM

Enzymes and Inhibitors

Plasminogen is found most abundantly in plasma, where its concentration has been estimated at 0.1 to 0.2 mg. per ml. and its activity at 2.5 ± 1.0 caseinolytic units per ml.; however, it is also present in smaller amounts in all body fluids and secretions. Plasminogen concentration in areas of inflammation correlates with the fibrinogen concentration, and significant amounts are incorporated in the meshwork of fibrin clots. The site or sites of plasminogen production remains obscure. On the basis of immunofluorescent studies, it has been suggested that plasminogen is synthesized in the bone marrow eosinophil and then transported to the circulation and tissues when needed; however, many investigators consider the liver to be the major site of production and catabolism of this circulating component. Highly purified preparations of plasminogen that closely resemble the native material (for example, high solubility at neutral pH) have been obtained by a variety of chemical precipitation and chromatographic techniques, and physicochemical studies indicate that it is a monomeric protein with a molecular weight of about 81,000.

Plasmin is an endopeptidase which hydrolyzes susceptible arginine and lysine bonds in proteins at neutral pH and acts upon most synthetic substrates and proteins susceptible to digestion by trypsin. Plasmin digests fibrinogen and fibrin at about the same rate and also hydrolyzes other naturally occurring proteins, including the coagulation Factors V and VIII, serum complement components, ACTH, growth hormone, and glucagon. Alpha casein is the substrate of choice for laboratory assay because of its high solubility and sensitivity to plasmin digestion. Plasminogen is converted to plasmin enzymatically, probably by the splitting of a single arginine-valine bond, resulting in two peptide chains held together by a single disulfide bond. Techniques are available for the preparation of highly purified plasmin comparable in purity to that achieved for plasminogen. However, the instability of plasmin requires the presence of stabilizing agents such as glycerol or lysine in such preparations. Recent studies suggest that plasmin has a molecular weight of 75,410, or slightly lower than plasminogen's.

Plasminogen Activators. The biological activators of plasminogen are also proteolytic enzymes, in that they are capable of hydrolyzing arginine and/or lysine bonds. However, they are highly specific for the plasminogen substrate and have little demonstrable effect on proteins that are digested by plasmin. Plasminogen activators can be found in trace amounts in all body fluids, including the plasma and urine (urokinase) and in most body tissues. It is likely that tissue activators are concentrated in the cellular lysozomal granules and in the vascular endothelial cells of most organs. The type derived from vascular cells is soluble and presumably diffuses into the circulation in response to vasoactive stimuli. The lysozomal variety is difficult to extract and does not lend itself well to purification, but considerable progress has been made in the purification and characterization of the plasminogen activator from pig heart and rabbit kidney. Urokinase has been purified and crystallized, and measurements indicate that it is a single polypeptide chain of molecular weight 53,000, with an active enzyme center capable of hydrolyzing arginine and lysine bonds. Urokinase is probably a single molecular species, but limited proteolytic digestion such as may occur during purification results in active fragments of smaller size which are still capable of plasminogen activation. Studies in vitro show that human urokinase and milk activator or adrenal tissue activator are immunologically dissimilar, whereas urokinase and activator harvested from human kidney tissue culture are identical. In vivo studies, in man, fail to show de-

tectable urokinase in peripheral venous plasma in situations in which there is evidence of increased levels of circulating plasminogen activator. These observations are consistent with the view that urokinase is produced by the kidneys independent of circulating plasma activator levels. Its function is presumably to keep the urinary tract free of fibrinous deposits.

Proactivator. There is little evidence as yet for the existence of a proactivator in plasma as a separate and distinct entity. Originally invoked to explain the two-step activation of plasminogen to plasmin by the bacterial substance streptokinase, the "proactivator" in this instance probably is plasminogen or even plasmin itself. The latter reacts with streptokinase in a 1:1 molar ratio to form an "activator" complex, which is a potent activator for human plasminogen as well as for those animal plasminogens ordinarily resistant to the reaction of streptokinase alone.

The presence of a completely intrinsic plasma activation system is supported by observations of increased fibrinolytic activity, presumably generated by Hageman factor activation. This concept of an endogenous source of plasma activator would provide evidence for the existence of a plasma proactivator (or activator-inhibitor complex) and would help explain the mechanism for induction of fibrinolysis at sites of vascular and tissue injury. However, the data relating Hageman factor activation to enhanced fibrinolytic activity are based primarily on observations of euglobulin precipitates under unphysiological conditions (low pH and ionic strength); consequently, this aspect needs further clarification.

Inhibitors. Plasma contains antiplasmin activity sufficient to inactivate ten times the available plasmin. The two types of antiplasmin that have been described most likely represent at least two separate alpha globulin components, an α-1-globulin (slow-acting inhibitor, probably identical with α-1-antitrypsin) and an α-2-macroglobulin (immediate-acting inhibitor). Platelets also contain anti-plasmin activity of the immediate-acting type, although it has not been sufficiently purified for characterization. In addition, plasma probably contains components that inhibit the activation of plasminogen, but these are not well characterized or even clearly delineated from the antiplasmin activity of plasma.

Fibrinogen and Fibrin Degradation Products

Physicochemical Properties. Degradation products of fibrinogen and fibrin can appear in the systemic circulation as the result of enzymatic digestion of circulating fibrinogen or fibrin clot dissolution. These derivatives have striking anticoagulant effect and, when present in sufficient concentrations, contribute to the severe hemorrhagic disturbances seen in these clinical states. Interest in the structure and function of these degradation products developed independently in several laboratories, with simultaneous studies of physicochemical, physiological and clinical aspects of the subject proceeding in parallel. After prolonged digestion of plasmin, fibrinogen is degraded into at least five definable degradation products which can be individually purified and assayed. Two of these, Fragments D and E, react with anti-fibrinogen antiserum and contain entirely distinct antigenic determinants of the parent fibrinogen molecule. Crude plasmin digests of fibrinogen or fibrin containing Fragments D and E interfere with normal clotting of fibrinogen by their ability to delay and interfere with the polymerization of fibrin. Of greater interest, however, is the fact that partially digested fibrinogen has a greater anticoagulant effect than such "terminal" digests, suggesting that intermediate degradation products are the most potent anticoagulant derivatives. These derivatives are called Fragments X and Y and have molecular weights of approximately 240,000 and 155,000, respectively, as compared to a molecular weight of 320,000 for fibrinogen and 85,000 and 50,000 for Fragments D and E, respectively. The derivatives of fibrinogen and fibrin degradation are physicochemically quite similar, except for the absence in the fibrin derivatives of the N-terminal fibrinopeptides; the latter have been split off previously by thrombin during the formation of the fibrin monomer molecules from fibrinogen.

It has been postulated that the enzymatic hydrolysis of fibrinogen or fibrin is punctuated by sudden large alterations of the molecule as it splits into the smaller derivatives. The first noticeable change is the conversion of fibrinogen to Fragment X, which is clottable by thrombin and about 20 percent smaller than fibrinogen. This degradation is associated with the release of several small "minor" products from the C-terminal end of the molecule. Fragment X presumably splits unevenly into a Fragment Y molecule and a single Fragment D molecule, following which the Fragment Y is itself split unevenly to a second Fragment D and a single Fragment E molecule. Thus, two Fragment D and one Fragment E molecules are derived from each parent fibrinogen molecule. Fibrinogen is considered to be a dimer consisting of three pairs of polypeptide chains of about 50,000 to 70,000 molecular weight, and it is not yet clear whether Fragments D and E represent individual chains or contain parts of each chain. Fragment Y (molecular weight 155,000) is approximately half the size of fibrinogen, but it is probably not an exact half of fibrinogen, since its parent molecule is Fragment X (molecular weight 240,000) and not fibrinogen.

Anticoagulant Properties. Degradation products are regularly detected in the serum of patients with fibrinolytic syndromes, and it is generally accepted that the intermediate degradation products are the main factors causing the bleeding associated with these disorders. In purified systems, Fragment X is slowly clotted by thrombin, yet it is a potent anticoagulant molecule; Fragment Y has the greatest anticoagulant effect in a thrombin time system. This effect is mediated by the formation of large molecular weight complexes of degradation products with fibrin monomers, which interfere with the polymerization of the monomer units into the fibrin clot. Furthermore, the clots that do form in the presence of degradation products are of a defective and weakened structure. Other clotting functions also are disrupted by these large degradation products, such as thromboplastin generation and thrombin activity, and platelet function also can be impaired by the small peptides released during degradation. Expectedly, patients with bleeding due to circulating degradation products usually manifest abnormalities in a number of clotting tests (partial thromboplastin time, prothrombin time) as well as in tests of hemostasis (bleeding time).

Detection in Serum. The detection of elevated levels of degradation products in the blood, with or without abnormal bleeding, depends primarily upon the persistence of fragments or non-clottable complexes in the serum, after clotting of the blood sample. Serum rather than plasma is the preferred test material, since plasma fibrinogen reacts in most test systems and yields false positive results. Fragment X and some complexes may be partially removed during clot formation, but sufficient quantities usually remain in the serum to be detected by the assay systems. Tests of clotting functions or hemostasis directly assess the anticoagulant effect of the degradation products, but they are nonspecific, since deficiencies of clotting factors or platelets can produce similar abnormalities. In addition, the complexes of degradation products and fibrin or Fragment X monomer are precipitated by cooling at

4° ("cryofibrinogens"), providing further indirect evidence of fibrinolysis. However, proof that abnormal bleeding or cryoprecipitation is owing to degradation products usually requires the demonstration of their presence by specific assays which, with the exception of staphylococcal clumping, utilize antifibrinogen antibody.

The most commonly used immunological tests are the tanned red cell hemagglutination inhibition immunoassay (TRCHII), immune precipitation, and latex particle flocculation tests. Each assay has been modified by individual investigators, as illustrated by the variations of latex tests now in use or under development. One type (Fi test) depends upon the clumping of antibody-coated latex particles by degradation products that react with the antifibrinogen antibody. Although moderately sensitive, this test is the least reliable for assays on clinical material, and a modified version utilizes fibrinogen-coated latex particles that are aggregated by dilute antifibrinogen antiserum. The latter is preincubated with test material containing degradation products, neutralizing the antifibrinogen antibody and rendering it incapable of agglutinating the fibrinogen-coated latex particles. This modified assay system is more sensitive and will probably find wider use in clinical circumstances. Immunoelectrophoresis or double diffusion (Ouchterlony) studies in agar gel using antifibrinogen antiserum are relatively insensitive techniques, but they are of value in distinguishing specific degradation products in the test material. The flocculation test is basically the same as the Ouchterlony, except that immunoprecipitation can be achieved rapidly by direct mixture of the test material with the antiserum in a cavity plate. It is usually positive in patients with titers of degradation products sufficient to cause abnormal bleeding. The TRCHII is the most sensitive of the immunological tests and is basically the same as the modified latex agglutination assay described above, with fibrinogen-coated tanned erythrocytes substituted for fibrinogen-coated latex particles. The staphylococcal clumping test utilizes strains of *Staphylococcus aureus* that are clumped by fibrinogen and certain of its derivatives or complexes. Both the staphylococcal clumping test and the TRCHII are of high sensitivity in clinical situations and correlate well in individual determinations.

There is some evidence to suggest that fibrin degradation products have a strong tendency to polymerize with circulating fibrinogen, and this observation has prompted an intensive search for new detection systems utilizing a plasma rather than a serum substrate. These assays (protamine sulfate and ethanol precipitation) presumably rely upon the dissociation of the components of the complex, with subsequent spontaneous clotting of the monomers into macroscopic gels.

PHYSIOLOGICAL MECHANISM FOR FIBRINOLYSIS

Basic Considerations

Since the action of plasmin on fibrin produces fibrinolysis, it was initially assumed that the regulation of plasmin levels was the mechanism utilized by the organism for controlling and regulating fibrinolysis in circulating blood. This concept implied that fibrinolytic activity of circulating fluids, as measured by dilute whole blood or euglobulin clot lysis, was equivalent to plasmin activity. Biochemical considerations have made such an explanation unlikely, since plasma contains large amounts of inhibitor that inactivate considerable increments in circulating plasmin. Furthermore, plasmin has no particular specificity for fibrin, hydrolyzing fibrinogen and numerous other plasma proteins of biological significance. It became apparent that the organism utilized a more complex control mechanism when observations in vivo during certain physiological

(e.g., intense exercise) and pharmacological states (e.g., after intravenous nicotinic acid) showed striking increases in plasma fibrinolytic activity without measurable plasmin formation. The factor responsible for such heightened circulating fibrinolytic activity was identified as plasminogen activator (Fig. 26-1), as opposed to increased circulating free plasmin, which is associated with digestion of fibrinogen as well as other plasma proteins, resulting in severe coagulation defects often with a hemorrhagic diathesis. Further insight into the mechanism of fibrinolysis in vivo is afforded by the two-phase concept for plasminogen (Fig. 26-1). The soluble form in plasma serves as a mobile reserve, to be deposited in fibrin whenever and wherever the latter forms, and the gel-phase plasminogen is present in the thrombus. In the absence of the naturally occurring inhibitors, activation of soluble phase plasminogen would lead rapidly to hyperplasminemia and the digestion of large amounts of circulating fibrinogen. Activation of gel-phase plasminogen, however, leads only to the formation of plasmin within the thrombus, resulting in the local dissolution of fibrin without affecting circulating proteins such as fibrinogen.

Thus, regulation of the level of plasminogen activator provides the body with a specific mechanism for fibrinolysis. Under physiological circumstances and following an appropriate stimulus, activator is transiently released into the circulation and directly raises the clot dissolving activity of plasma by its ability to activate gel phase plasminogen, without invoking the consequences of increased plasma proteolysis. The dissolution of a preformed thrombus or fibrin deposit is relatively slow, since, under the usual clinical circumstances, the activator must diffuse into the clot and mediate lysis in a progressive manner. However, much more rapid fibrinolysis occurs when fibrin formation proceeds in the presence of activator and plasminogen.

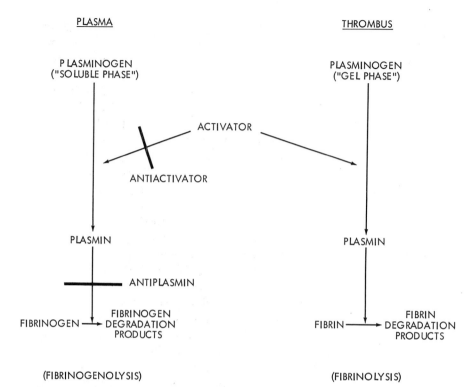

PLASMA   THROMBUS

PLASMINOGEN   PLASMINOGEN
("SOLUBLE PHASE")   ("GEL PHASE")

ACTIVATOR

ANTIACTIVATOR

PLASMIN   PLASMIN

ANTIPLASMIN

FIBRINOGEN   FIBRIN
FIBRINOGEN → DEGRADATION   FIBRIN → DEGRADATION
PRODUCTS   PRODUCTS

(FIBRINOGENOLYSIS)   (FIBRINOLYSIS)

FIG. 26-1. Simplified comparison of the fibrinolytic systems in circulating plasma and in a thrombus. In this context, "fibrinolysis" is synonymous with clot dissolution. The heavy lines indicate sites of inhibitory activity.

Under the latter circumstance, both constituents are incorporated throughout the fibrin meshwork and, with the widespread formation of plasmin, prompt dissolution may take place. Most measurements of plasma fibrinolytic activity utilize fibrin substrates formed in the presence of plasminogen and activator and reflect primarily the activator content of the biological fluid.

Undoubtedly a number of factors influence the normal fibrinolytic process in vivo, one of which is the age of the fibrin. Early observations demonstrated that, after 5 to 7 days in situ, experimental clots in animals usually became quite resistant to lysis by thrombolytic agents, and in general, this has been the experience of most clinical investigators using such agents in man. The mechanism for this increased resistance has not been clarified entirely. A major consideration is the fibrin stabilization reaction, in which thrombin-activated Factor XIII (fibrin stabilizing factor) catalyzes a transamidation between fibrin polymers, bonding them together chemically so as to provide a stronger structural integrity for the clot, which renders it more resistant to lysis. Additional factors which may influence fibrinolytic phenomena remain to be clarified, but recent studies suggest that sex hormones may suppress plasminogen activator activity both in the endometrium and in the circulating blood.

Agents that promote fibrinolysis clinically, such as urokinase and streptokinase, are plasminogen activators and have been under intensive trial recently for the treatment of pulmonary and peripheral arterial embolism and deep venous thrombosis, as compared with standard heparin-coumadin anticoagulant therapy. Theoretically these lytic agents should produce more rapid dissolution of the thrombus, since anti-coagulants act only to prevent further clot accumulation. Early results of several studies appear to support this hypothesis. Other agents presently under clinical trial for treatment of thrombosis are the snake venom derivatives, which enzymatically clot circulating fibrinogen, thereby causing a profound hypofibrinogenemia. These clots quickly lyse and a relatively simple and safe anticoagulant situation is produced. Although such agents may prove better anticoagulants than heparin, they would not be expected to enhance the lysis of preformed clots, and therefore would be less advantageous than the fibrinolytic agents.

Circulating Fibrinolytic Activity

Microcirculation. To fulfill its role in biological repair most efficiently, the organism provides a means by which high concentrations of activator are made available locally at the immediate site of fibrin deposition. This activator is concentrated in the endothelial cells of blood vessels and is more readily soluble and diffusible than that present in lysosomal granules of all tissue cells. Present evidence indicates that some activator is released continuously from endothelial cells, but this is strikingly enhanced by a variety of vasoactive stimuli, particularly vasodilatation, and through such a mechanism the local concentration of activator at sites of injury, trauma or inflammation is raised significantly. Observations made on blood taken from the systemic circulation poorly reflect the changes occurring at local sites of fibrin deposition, since activator is primarily released in the microcirculation, and that which enters the systemic circulation is diluted in a large circulating pool and then rapidly cleared. Clearance is achieved by plasma inhibitors and an active hepatic clearing mechanism, which reduces the half-life of activator in the circulation to approximately 13 minutes. Thus, it is essential to consider the dynamics of fibrinolytic phenomena in the microcirculation independent of those occurring in the systemic circulation, where activity is considerably lower. This is demonstrated by the significant venous-arterial differences in plasminogen activator concentration across all organ circulations, including the pulmonary, renal, mesenteric and gastric circulations. In addition, fibrinolytic activity can be augmented readily through the release of further increments of plasminogen activator in response to various vasoactive stimuli, particularly ischemia, anoxia, exercise, and pharmacological agents such as epinephrine, acetylcholine, histamine, nicotinic acid, and pyrogens. This high activity in the microcirculation probably plays an important physiological role in maintaining the patency of the capillary bed, since fibrin formed in the presence of high concentrations of plasminogen activator dissolves quickly. The dynamic state of fibrinolysis in vivo is substantiated by two types of observations. First, there is a rapid release of large quantities of fibrin degradation products into the systemic circulation following an acute episode of disseminated intravascular coagulation, and, second, rats pretreated with epsilon aminocaproic acid suffer extensive fibrin deposits in the microcirculation after injection of thrombogenic serum, whereas control animals show no fibrin deposits in the capillary bed.

Systemic Circulation. The systemic circulation has considerably less fibrinolytic activity than does the microcirculation, since the relatively small endothelial surface area of the large blood vessels contributes little plasminogen activator to the circulation. In addition, the activity demonstrable in the systemic circulation is continuously inactivated and cleared. As a consequence, the systemic circulation is ill prepared to deal with the problem of fibrin deposition and thrombus formation. Fibrin deposition can be initiated by a variety of pathological changes, most important of which are alterations in the vascular endothelium that remove its thrombus-inhibiting properties, circulatory stasis such as the impaired venous return observed in congestive heart failure, and the release of clot-promoting substances into the local or systemic circulation. Once initiated, a thrombus undergoes growth or resolution, depending upon the interplay between two opposing reactions—continued fibrin formation and fibrinolysis. In the microcirculation, heightened fibrinolytic activity shifts the balance in favor of resolution, but the low activity in the general circulation allows continued thrombus growth. This low systemic activity contributes significantly to the magnitude and consequences of thromboembolic diseases as well as to its initial appearance in the patient.

Although the titer of fibrinolytic activity in the systemic circulation is low, the measurable amounts probably play a role in the natural, slow dissolution of thrombi. As they form, thrombi incorporate plasminogen activator and plasminogen from the circulating blood, and the subsequent activation of plasminogen in close proximity to the clot fibrils favors spontaneous resolution. Thus, sterile blood clotted at 37° usually undergoes complete liquefaction in 7 to 10 days, and a high proportion of pulmonary emboli slowly resolve with only anticoagulant therapy.

ABNORMAL STATES OF FIBRINOLYSIS

Pathological Fibrinogenolysis (Hyperplasminemia, "Primary Fibrinolytic Disorders")

Three mechanisms exist for the production of pathological fibrinogenolysis. First, inordinate amounts of plasminogen activator may be administered for therapeutic purposes or released from endogenous sources. The latter include activator-rich neoplastic tissue, such as metastatic prostatic carcinoma, or the activator-rich vascular endothelium of many organs in response to the profound stimuli of severe trauma, anoxia, shock or extensive surgery. Plasminogen activation temporarily overwhelms normal plasma inhibitory mechanisms, and free plasmin circulates for significant periods of time. Second, impaired clearance of activator from the circulation may appear in patients with hepatic cirrhosis, especially during portocaval shunting procedures. Such individuals cannot adequately handle plasminogen activator released in normal amounts, which ordinarily do not produce significant hyperplasminemia. Third, proteolytic enzymes other than plasmin that are capable of degrading fibrinogen may be liberated by leukemic cells and produce a similar hyperplasminemic state.

Patients with sustained increase in circulating proteolytic activity have a marked increase in fibrinolytic activity as measured by blood or euglobulin clot lysis. In addition, circulating fibrinogen and other clotting factors are enzymatically destroyed, often causing a severe hemorrhagic diathesis. Blood clotting studies show delayed clotting of whole blood or plasma by all tests and an abnormal, shaggy appearance of the blood clot. These clot abnormalities are mainly the result of circulating fibrinogen degradation products, and they normalize as the plasma inhibitory properties limit and then eliminate the hyperplasminemia. The spontaneous appearance of pathological fibrinogenolysis, severe enough to cause sustained hypofibrinogenemia, is rare, and cases of "hypofibrinogenemia" and "fibrinolysis" usually are instances of disseminated intravascular coagulation (DIC) with accelerated local fibrinolysis (Fig. 26-2). In the true case of hyperplasminemia, the obvious and effective treatment of choice is the administration of a fibrinolytic inhibitor, but when given to patients with disseminated intravascular coagulation, such inhibitors can result in the rapid development of multiple large thrombi. In patients with excessive fibrinogenolysis and certain instances of local

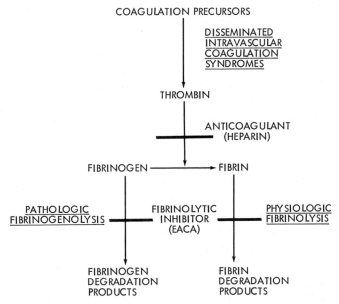

Fig. 26-2. Rationale of therapy in disseminated intravascular coagulation and pathologic fibrinogenolysis. Solid bars indicate inhibitory effect of the anticoagulant (usually heparin) or antifibrinolytic agent (e.g., epsilon aminocaproic acid). Activation of the coagulation system in the circulating blood could be initiated by a variety of stimuli, including obstetric accidents, carcinoma, hemolysis and endotoxinemia associated with hypotension.

fibrinolysis, inhibition can be accomplished by the administration of agents such as epsilon aminocaproic acid (EACA), Trasylol, p-aminocyclohexane carboxylic acid (AMCHA) and p-aminomethylbenzoic acid (PAMBA). These compounds have had great usefulness in the management of post-prostatectomy bleeding, where inhibition of the local action of urokinase in the genitourinary tract significantly improves hemostasis.

Excessive Fibrinolysis (Secondary Fibrinolysis)

It is now generally accepted that most instances of acute spontaneous hypofibrinogenemia are caused by disseminated intravascular coagulation (DIC), also referred to as consumption coagulopathy and acute defibrination syndrome. This syndrome can be initiated by many diseases, most of which are characterized by the release of thromboplastinlike material into the circulation, leading to an excessive deposition of fibrin in the microcirculation of many organ systems. As a result of thrombi present in the brain, kidneys, skin and gastrointestinal tract, the patient may manifest coma, anuria, azotemia, skin

necrosis and intestinal ulceration. Rapid local fibrinolysis evokes a physiological response which is teleologically desirable for maintaining the patency of the microcirculation (Fig. 26-2). The reopening of these vascular channels is associated with the release of excessive amounts of soluble fibrin degradation products into the systemic circulation. The degradation products may accumulate, achieving concentrations exceeding that of the circulating fibrinogen, and serious derangements of coagulation and hemostatic function may ensue, with the clinical manifestations of a generalized hemorrhagic diathesis. Since the fibrin degradation products appear as a secondary response to fibrin deposition, the bleeding problem has sometimes been referred to as a secondary fibrinolytic disorder, in which fibrinolysis is localized to the capillary bed, with little or no evidence of heightened fibrinolytic activity in the systemic circulation. Hyperplasminemia and fibrinogenolysis are usually absent, except in those mixed disorders in which tissue thromboplastin and tissue plasminogen activator may be liberated simultaneously into the bloodstream.

In its classical appearance, the laboratory

picture of the DIC syndrome with secondary local hyperfibrinolysis is associated with consumption and deficiency of coagulation factors such as platelets, fibrinogen, and Factors V and VIII, consumption of fibrinolytic components such as plasminogen, and the disruption of normal clotting by the circulating fibrin degradation products. The latter is manifested by incoagulable blood or severely impaired clotting, usually with the formation of a small clot (low fibrinogen) that does not lyse (low plasminogen) and with circulating complexes of fibrin monomers and degradation products that may exhibit a tendency to precipitate at 4° (cryofibrinogen).

The primary form of treatment is heparin administration, to curtail the underlying intravascular coagulation, followed by transfusions with plasma and platelets, to replace consumed clotting factors. On occasion, the serious hemorrhagic complications of the circulating degradation fragments may necessitate additional therapy with fibrinolytic inhibitors to curtail the rapid release of these products. However, the administration of such inhibitors should not be undertaken without prior administration of heparin; otherwise serious or even fatal thrombosis may be precipitated.

A similar type of defibrination state has been described in which a less dramatic and frequently more protracted process (subacute form) is associated with a chronic disease such as carcinoma. Although the syndrome may become quite serious, with multiple thrombi and even extensive hemorrhage, the usual case may be much less dramatic and the laboratory findings less severely deranged. The most consistent finding has been the presence of increased amounts of circulating degradation products, and heparin has proved useful as a diagnostic, as well as a therapeutic tool.

Decreased Fibrinolytic Activity

The continuously active and readily stimulated normal fibrinolytic mechanism in vivo presumably plays an important role in limiting the extent of fibrin deposition. It would therefore be expected that impaired fibrinolysis would favor the growth of newly formed thrombi. Unfortunately, current methods for the assay of circulating fibrinolytic activity discriminates only between enhanced and normal levels and is not sufficiently sensitive to distinguish between small amounts of normal activity and abnormally low levels. Such a distinction should be an important goal, since it may provide a better understanding of the predisposition to and the course of thromboembolic disease in a variety of clinical conditions. Of particular importance are the occurrence of venous thrombosis during immobilization, in which the perfusion of inactive muscles may not provide sufficient activator to prevent thrombosis in the deep veins; its occurrence in congestive heart failure, in which the resolution of pulmonary emboli appears to be delayed, and its occurrence in malignant disease and obesity, in which some evidence suggests an impaired fibrinolytic mechanism.

There are several potential mechanisms for the production of reduced fibrinolytic activity, such as inadequate production and release of plasminogen activator, accelerated mechanisms for the clearance of activator from the circulation, severe deficiency of plasminogen, and excessive amounts of activator inhibitor. However, there is relatively little documentation of patients with such hereditary or acquired abnormalities. Such observations may help to distinguish between those factors initiating thrombosis and those that influence its subsequent course. In this regard, the administration of inhibitors of plasminogen activation, such as epsilon aminocaproic acid, has not been associated with new thrombus formation, although it may retard lysis and promote continuation of an underlying thrombotic states. Severe deficiency of plasminogen has been described only during intense thrombolytic states induced by the administration of plasminogen activators such as streptokinase and urokinase. When rethrombosis occurs under these circumstances, fibrin resolution is impaired significantly and the thrombosis is relatively resistant to subsequent fibrinolytic therapy.

ANNOTATED REFERENCES

Alkjaersig, N., Fletcher, A. P., and Sherry, S.: The mechanism of clot dissolution by plasmin. J. Clin. Invest., 38:1086, 1959. (Experimental work lending much fundamental support to current views of fibrinolysis)

Ashford, T. P., Weinstein, M. C., and Freiman, D. G.: The role of the intrinsic fibrinolytic system in the prevention of stasis thrombosis in small veins. An electron microscopic study. Am. J. Path., 52:1117, 1968. (Anatomic study demonstrating the role of the fibrinolytic system of small vessels in preventing local fibrin deposition)

Bang, N. U., Fletcher, A. P., Alkjaersig, N., and Sherry, S.: Pathogenesis of the coagulation defect developing during pathological proteolytic ("fibrinolytic") states. III. Demonstration of abnormal clot structure by electron microscopy. J. Clin. Invest., 41: 935, 1962. (Electron microscopic demonstration of abnormal clot structure formed in the presence of fibrinogen degradation products)

Iatrides, S. G., and Ferguson, J. H.: Active Hageman Factor: A plasma lysokinase of the human fibrinolytic system. J. Clin. Invest., 41:1277, 1962. (Experimental evidence supporting the concept that activation of Hageman factor can activate fibrinolysis)

Marder, V. J., Matchett, M. O., and Sherry, S.: Detection of serum fibrinogen and fibrin degradation products. Comparison of six techniques using purified products and application in clinical studies. Am. J. Med., 51:71, 1971. (Comparative study of tests for detecting serum degradation products, with clinical application)

Marder, V. J., Shulman, N. R., and Carroll, W. R.: High molecular weight derivatives of human fibrinogen. I. Physiocochemical and immunological characterization. Marder, V. J., and Shulman, N. R.: II. Mechanism of their anticoagulant activity. J. Biol. Chem., 244:2111, 2120, 1969. (Physicochemical studies of the intermediate and final fibrinogen degradation products, demonstration of their anticoagulant action by complex formation and description of a concept of fibrinogen fragmentation by plasmin)

Robbins, K. C., Summaria, L., Hsieh, B., and Shah, R. J.: The peptide chains of human plasmin. Mechanism of activation of human plasminogen to plasmin. J. Biol. Chem., 242: 2333, 1967. (Biochemical analysis of the structure and activation of the pivotal enzyme, plasmin)

Sherry, S., Alkjaersig, N., and Fletcher, A. P.: Fibrinolysis and fibrinolytic activity in man. Physiol. Rev., 39:343, 1959. (Comprehensive review, which first formulated many of the currently accepted concepts of fibrinolysis)

Verstraete, M., Vermylen, C., Vermylen, J., and Vanderbroucke, J.: Excessive consumption of blood coagulation components as cause of hemorrhagic diathesis. Am. J. Med., 38:899, 1965. (Clinical description of the defibrination syndromes, with special emphasis on the cause of the associated hemorrhagic diathesis)

Warren, B. A.: Fibrinolytic activity of vascular endothelium. Brit. Med. Bull., 20:213, 1964. (Basic support for the effective fibrinolytic process in the microcirculation)

Appendix A

Classical references have been cited in this section for those scholars who wish to read the initial and definitive proof of some widely accepted facts in our knowledge of human blood.

Erythropoiesis

Barcroft, J.: The significance of hemoglobin. Physiol. Rev., 4:329, 1924.

Belcher, E. H., and Courtenay, V.: Studies of Fe[59] uptake by rat reticulocytes in vitro. Brit. J. Haemat., 5:268, 1959.

Bloom, W., and Bartelmez, G. W.: Hematopoiesis in young human embryos. Am. J. Anat., 67:21, 1940.

Brewer, G. J., and Dern, R. J.: A new inherited enzymatic deficiency of human erythrocytes: 6-phosphogluconate dehydrogenase deficiency. Am. J. Human Genet., 16:472, 1964.

Custer, R. P.: Studies on the structure and function of bone marrow. J. Lab. Clin. Med., 17:951, 960, 1932.

Dacie, J. V., and White, J. C.: Erythropoiesis with particular reference to its study by biopsy of human marrow. J. Clin. Path., 2:1, 1949.

Daland, G. A., Heath, C. W., and Minot, T. R.: Differentiation of pernicious anemia and certain other macrocytic anemias by the distribution of red cell diameters. Blood, 1:6, 1946.

Daughaday, W. H., Williams, R. H., and Daland, G. A.: The effect of endocrinopathies on the blood. Blood, 3:1342, 1948.

Davidson, C. S., Murphy, J. C., Watson, R. J., and Castle, W. B.: Comparison of the effects of massive blood transfusions and of liver extract in pernicious anemia. J. Clin. Invest., 25:858, 1946.

Downey, H.: The megaloblast-normoblast problem. J. Lab. Clin. Med., 39:837, 1952.

Ebert, R. V., and Stead, E. A., Jr.: Demonstration that in normal man no reserves of blood are mobilized by exercise, epinephrine and hemorrhage. Am. J. Med. Sci., 201:655, 1941.

Ebaugh, F. G., Jr., Emerson, C. P., and Ross, J. F.: The use of radioactive chromium 51 as an erythrocyte tagging agent for the determination of red cell survival in vivo. J. Clin. Invest., 32:1260, 1953.

Erslev, A.: Humoral regulation of red cell production. Blood, 8:349, 1953.

Finch, C. A., Hegsted, M., Kinney, T. D., Thomas, E. D., Roth, C. E., Hoskins, D., Finch, S., and Fluhorty, R. G.: Iron metabolism, the pathophysiology of iron storage. Blood, 5:983, 1950.

Gardiner, F. H., and Pringle, J. C., Jr.: Androgens and erythropoiesis. Arch. Int. Med., 107:846, 1961.

Gilmour, J. R.: Normal haematopoiesis in intra-uterine and neonatal life. J. Path. Bact., 52:25, 1941.

Gordon, A. S.: Hemopoietine. Physiol. Rev., 39:1, 1959.

Gurney, C. W., Goldwasser, E., and Parr, C.: Studies on erythropoiesis. VI. Erythropoietin in human plasma. J. Lab. Clin. Med., 50:534, 1957.

Heath, C. W., and Daland, G. A.: The life of reticulocytes. Arch. Int. Med., 46:533, 1930.

Herbert, V.: Minimal daily adult folate requirement. Arch. Int. Med., 110:649, 1962.

Jacob, H. S., and Jandl, J. H.: Glutathione in the regulation of the hexose monophosphate pathway. J. Biol. Chem., 241:4243, 1966.

Jacobson, L. O., Gurney, C. W., and Goldwasser, E.: The control of erythropoiesis. Adv. Int. Med., 10:297, 1960.

Jaffe, E. R.: The reduction of methemoglobin in human erythrocytes incubated with purine nucleosides. J. Clin. Invest., 38:1555, 1959.

Jandl, J. H., et al.: Clinical determination of the sites of red cell sequestration in hemolytic anemias. J. Clin. Invest., 35:842, 1956.

Maximow, A. A.: Relation of blood cells to connective tissues and endothelium. Physiol. Rev., 4:533, 1924.

Michels, N. A.: Erythropoiesis. Haematologica, 45:75, 1931.

Osler, W.: A clinical lecture on erythraemia. Lancet, *1*:143, 1908.

Pauling, L.: Abnormality of hemoglobin molecules in hereditary hemolytic anemias. The Harvey Lectures, Academic Press, New York, 1955.

Stohlman, F., Jr., Rath, C. E., and Rose J. C.: Evidence for a humoral regulation of erythropoiesis. Blood, 9:721, 1954.

Whitcomb, W. H., and Moore, M. Z.: The inhibitory effect of plasma from hypertransfused animals on erythrocyte iron incorporation in mice. J. Lab. Clin. Med., *66*:641, 1965.

Wintrobe, M. M.: Blood of normal men and women. Bull. Johns Hopkins Hosp., 53:118, 1933.

Leukopoiesis

Athens, J. W., Raab, S. O., Haab, O. P., Boggs, D. R., Ashenbrucker, H., Cartwright, G. E., and Wintrobe, M. M.: Leukokinetic studies. X. Blood granulocyte kinetics in chronic myelocytic leukemia. J. Clin. Invest., *44*:765, 1965.

Cabot, R. C.: The lymphocytosis of infection. Am. J. Med. Sci., *145*:335, 1913.

Craddock, C. G., Jr.: Studies of leukopoiesis. The technique of leukophoresis and the response of myeloid tissue in normal and irradiated dogs. J. Lab. Clin. Med., *45*:881, 1955.

Damashek, W., and Colmes, A.: The effect of drugs in the production of agranulocytosis with particular reference to amidopyrine sensitivity. J. Clin. Invest., *15*:85, 1936.

Dougherty, T. F., and White, A.: Influence of hormones on lymphoid tissue structure and function. The role of the pituitary adrenotrophic hormone in the regulation of the lymphocytes and other cellular elements of the blood. Endocrinology, *35*:1, 1944.

Downey, H., and McKinlay, C. A.: Acute lymphadenosis compared with acute lymphatic leukemia. Arch. Int. Med., *32*:82, 1923.

Gordon, A. S.: Some aspects of hormonal influences upon the leukocytes. Ann. N. Y. Acad. Sci., 59:907, 1955.

Kline, D.: DNA P[32] studies of white cell formation. Distribution and life span. *In:* F. Stohlman, Jr. (ed.): The Kinetics of Cellular Proliferation. p. 142. New York, Grune & Stratton, 1959.

Menkin, V., and Kadish, M. A.: Presence of the leukocytosis-promoting factor in the circulating blood. Arch. Path., *33*:193, 1942.

Paul, J. R., and Bunnell, W. W.: The presence of heterophile antibodies in infectious mononucleosis. Am. J. Med. Sci., *183*:90, 1932.

Reznikoff, P.: The etiologic importance of fatigue and the prognostic significance of monocytosis in neutropenia (agranulocytosis). Am. J. Med. Sci., *195*:627, 1938.

Tullis, J. L.: Prevalence, nature and identification of leukocyte antibodies. New Eng. J. Med., *258*:569, 1958.

Wintrobe, M. M.: Diagnostic significance of changes in leukocytes. Bull. N. Y. Acad. Med., *15*:223, 1939.

Hemostasis

Aas, K. A., and Gardiner, F. H.: Survival of blood platelets labelled with Chromium[51]. J. Clin. Invest., 37:1251, 1958.

Ackroyd, J. F.: The pathogenesis of thrombocytopenic purpura due to hypersensitivity to Sedormid (allyl-isopropyl acetylcorbormid). Clin. Sci., 7:249, 1949.

Aggeler, P. M., *et al.:* Plasma thromboplastin component (PTC) deficiency: A new disease resembling hemophilia. Proc. Soc. Exp. Biol. Med., 79:692, 1952.

Brinkhous, K. M.: A study of the clotting defect in hemophilia. The delayed formation of thrombin. Am. J. Med. Sci., *198:* 509, 1939.

Brinkhous, K. M., Langdell, R. D., Penick, G. D., Graham, J. B., and Wagner, R. H.: Newer approaches to the study of hemophilia and hemophilioid states. JAMA, *154:* 481, 1954.

Conley, C. L., Hartmann, R. C., and Morse, W. I., II.: The clotting behavior of human "platelet-free" plasma: evidence for the existence of a plasma thromboplastin. J. Clin. Invest., 28:340, 1949.

Davie, E. W., and Ratnoff, O. D.: Waterfall sequence for intrinsic blood clotting. Science, *145*:1310, 1964.

Duckert, F., Jung, E., and Shmerling, D. H.: A hitherto undescribed congenital haemorrhagic diathesis probably due to fibrin stabilizing factor deficiency. Thromb. Diath. Haemorrh., 5:179, 1960.

Gaardner, A., Janesen, J., Laland, S., Hellem, A., and Owren, P. A.: Adenosine diphosphate in red cells as a factor in the adhesiveness of human blood platelets. Nature, *192:* 531, 1961.

Harrington, W. J., Sprague, C. C., Minnich, V., Moore, C. V., and Duboch, R.: Immunologic mechanisms in idiopathic and neonatal thrombocytopenic purpura. Ann. Int. Med., *38:*433, 1953.

Hougie, C., Barrow, E. M., and Graham, J. B.: Stuart clotting defect. I. Segregation of an hereditory hemorrhagic state from the heterogenous group heretofore called "stable factor" (SPCA, proconvertin, Factor VII) deficiency. J. Clin. Invest., *36:*485, 1957.

Laki, K., and Gladner, J. A.: Chemistry and physiology of the fibrinogen-fibrin transition. Physiol. Rev., *44:*127, 1964.

Luscher, E. F.: Retraction activity of the platelets: Biochemical background and physiological significance in blood platelets. Henry Ford Hospital Symposium. p. 445. Boston, Little, Brown and Company, 1961.

Macfarlane, R. G.: An enzyme cascade in blood clotting mechanisms and its function as a biochemical amplifier. Nature, *202:* 498, 1964.

McKee, P. A., Coussons, R. T., Buckner, R. C., Williams, G. R., and Hampton, J. W.: Effects of the spleen on canine Factor VIII levels. J. Lab. Clin. Med., *75:*391, 1970.

Nilsson, I. M., Blomback, M., and von Francken, T.: On an inherited autosomal hemorrhagic diathesis with antihemophilic globulin (AHG) deficiency and prolonged bleeding time. Acta med. scand., *159:*35, 1957.

Osler, W.: On a family form of recurring epistaxis associated with multiple telangiectasis of the skin and mucous membranes. Johns Hopkins Hospital Bull., *12:*333, 1901.

Otto, J.: An account of a hemorrhagic disposition existing in certain families. *In:* Major, R. H. (ed.): Classic Descriptions of Disease. p. 522. Springfield, Ill., Charles C Thomas, 1945.

Patek, A. J., Jr., and Taylor, F. H. L.: Hemophilia. II. Some properties of a substance obtained from normal human plasma effective in accelerating the coagulation of hemophilic blood. J. Clin. Invest., *16:*113, 1937.

Prichard, R. W., and Vann, R. L.: Congenital afbrinogenemia. Am. J. Dis. Child, *88:*703, 1954.

Quick, A. J., and Collentine, G. E.: The role of vitamin K in the synthesis of prothrombin. Am. J. Physiol., *164:*716, 1951.

Ratnoff, O. D., Davie, E. W., and Mallett, D. L.: Evidence that activated Hageman factor in turn activates plasma thromboplastin antecedent. J. Clin. Invest., *40:*803, 1961.

Rosenthal, R. L., Dreskin, D. H., and Rosenthal, N.: New hemophilia-like disease caused by deficiency of a third plasma thromboplastin factor. Proc. Soc. Exp. Biol. Med., *82:*171, 1953.

Seegers, W. H., and Alkjaersig, N.: Comparative properties of purified human and bovine prothrombin. Am. J. Physiol., *172:*731, 1953.

Spaet, T. H., *et al.:* Reticuloendothelial clearance of blood thromboplastin by rats. Blood, *17:*196, 1961.

Tocantins, L. M.: Platelets and the structure and physical property of blood clots. Am. J. Physiol., *114:*709, 1936.

Van Creveld, S., Ho, L. K., and Veder, H. A.: Thrombopathia. Acta haemat., *19:*199, 1958.

Fibrinolysis

Albrechtsen, O. K.: Fibrinolytic activity of human endometrium. Acta endocrinol., *23:* 207, 1956.

Alkjaersig, N., Fletcher, A. P., and Sherry, S.: The mechanism of clot dissolution by plasmin. J. Clin. Invest., *38:*1086, 1959.

————: Pathogenesis of the coagulation defect developing during pathological plasma proteolytic (fibrinolytic) states. J. Clin. Invest., *41:*917, 1962.

Astrup, T.: Activation of a proteolytic enzyme in blood by animal tissue. Biochem. J., *50:* 5, 1951.

Barnhart, M. I., and Riddle, J. M.: Cellular localization of profibrinolysin (plasminogen). Blood, *21:*306, 1963.

Bidwell, E.: Fibrinolysins of human plasma. Biochem. J., *55:*497, 1953.

Blomback, B., and Blomback, M.: Purification of human and bovine fibrinogen. Arkiv. Kemi., *10:*415, 1956.

Duguid, J. B.: Thrombosis as a factor in the pathogenesis of coronary atherosclerosis. J. Path. Bact., *58:*207, 1946.

Ham, T. H., and Curtis, F. C.: Plasma fibrinogen response in man: Influence of the nutritional state, induced hyperpyrexia, infectious disease and liver damage. Medicine 17:413, 1938.

Hampton, J. W., Mantooth, J., Brandt, E. M., and Wolf, S.: Plasma fibrinogen patterns in patients with coronary atherosclerosis. Circulation, 34:1098, 1966.

Johnson, A. J., Fletcher, A. P., McCarty, W. R., and Tillett, W. S.: The intravascular use of streptokinase. Ann. N. Y. Acad. Sci., 68:201, 1957.

Johnson, S. A., and Schneider, C. L.: Existence of antifibrinolysin activity in platelets. Science, 117:229, 1953.

Kaplan, M. H.: Nature and role of the lytic factor in hemolytic streptococcal fibrinolysis. Proc. Soc. Exp. Biol. Med., 57:40, 1944.

Kline, D. L.: The purification and crystallization of plasminogen (profibrinolysin). J. Biol. Chem., 204:949, 1953.

Kwaan, H. C., and McFadzean, A. J. S.: Plasma fibrinolytic activity induced by ischemia. Clin. Sci., 15:245, 1956.

Milstone, H.: A factor in normal human blood which participates in streptococcal fibrinolysis. J. Immunol., 42:109, 1941.

Norman, P. S.: Antiplasmins. Fed. Proc., 25: 63, 1966.

Reid, D. E., Weiner, A. E., and Roby, C. C.: Presumptive amniotic fluid infusion with resultant postpartum hemorrhage due to afibrinogenemia. JAMA, 152:227, 1953.

Robbins, K. C., et al.: Further studies on the purification and characterization of human plasminogen and plasmin. J. Biol. Chem., 240:541, 1965.

Sherry, S.: Participation of fibrin in activation of fibrinolytic enzyme of plasma. J. Clin. Invest., 33:966, 1954.

Tillett, W. S., Edwards, L. B., and Garner, R. L.: Fibrinolytic activity of hemolytic streptococci. Development of resistance to fibrinolysis following acute hemolytic streptococcus infections. J. Clin. Invest., 13:47, 1934.

Wessler, S.: Experimentally produced phlebothrombosis in study of thromboembolism. J. Clin. Invest., 32:610, 1953.

Appendix B

Normal values for a variety of hematological studies are presented in the following three tables. Table VI-1 presents the normal differential proportion of cells in the bone marrow; and Table VI-2 presents normal hematological indices of the circulating blood. (Table VI-2A, normal hematological values, and Table VI-2B, values for chemical tests used in the differential diagnosis of anemias). Table VI-3 presents the normal values for a variety of tests of coagulation (general, vascular, platelet, and clotting factors).

TABLE VI-1. Normal Bone Marrow Differential

	PERCENT
Normoblasts	7-32
Pronormoblasts	1-8
Neutrophils (PMN)	7-30
Metamyelocytes	13-32
Myelocytes	5-19
Promyelocytes	1-8
Myeloblasts	0-5
Lymphocytes	3-17
Eosinophils	0-1
Basophils	0-1
Plasma cells	0-1

TABLE VI-2.

A. Normal Hematological Values

Hematocrit	45.0
Males	$45.0 \pm 7\%$
Females	$38.0 \pm 5\%$
Newborn	49-54%
Children	35-49%
Hemoglobin	
Males	15 ± 2 g. %
Females	13 ± 2 g. %
Newborn	16.5-19.5 g. %
Children	11.2-16.5 g. %

Leukocytes	5,000-10,000/mm.[3]
PMN	3,000-5,000/mm.[3] (54-62%)
Bands	150-400/mm.[3] (3-5%)
Lymphocytes	1,500-3,000/mm.[3]
Monocytes	258-500/mm.[3]
Eosinophils	50-250/mm.[3]
Basophils	15-50/mm.[3]
M.C.H. (mean corpuscular hemoglobin)	29 ± 2 mg.
M.C.V. (mean corpuscular volume)	87 ± 5 cu. micra.
M.C.H.C. (mean corpuscular hemoglobin concentration)	$34 \pm 2\%$

B. Chemical Tests for Anemias

Serum folic acid assay	3-16 nanograms/ml.
Serum B-12 assay	200-900 pg./ml.
Urine methylmalonate	Negative
Serum iron	50-150 μg %
Total iron-binding capacity	250-400 μg %
Percent saturation	35-40%
Hemoglobin electrophoresis	A—adult type
Fetal hemoglobin	$< 2\%$ of total
Haptoglobin	< 125 mg. %
Methemoglobin	0-8% total Hgb.
Autohemolysis test	
with glucose	0.3-0.7% hemolysis
without glucose	0.4-4.5% hemolysis
with ATP	0.0-0.8% hemolysis
Osmotic fragility	
without incubation	
initial	0.50% saline
complete	0.30% saline
with incubation	
initial	0.65% saline
complete	0.40% saline
ALA synthetase (urine)	2-4 mg./24 hrs.

TABLE VI-3. Tests of Hemostasis.

Vascular Tests	Normal values
Petechiometer	0-5
Rumple-Leed	none
Bleeding time (Duke)	< 3 min.

Platelets	
Count	200-400,000/mm.^{3}
Adhesiveness	> 20%
Factor 3	10-15 secs.

Screening tests	
Silicone clotting time	25-35 mins.
Clot retraction	1-2 hrs.
Whole clot lysis	> 12 hrs.
Prothrombin time (Quick)	15 secs.
Partial thromboplastin time	< 50 secs.
Euglobulin lysis time	< 60 mins.
Fibrin split products	0-5 μg %
Thromboplastin generation test	< 10 secs. in 6 mins.

Specific clotting factor assays	
Fibrinogen (I)	200-400 mg. %
Prothrombin (II)	150-350 u.
Accelerin (V)	50-150%
Convertin (VII)	50-150%
Antihemophilic (VIII)	50-150%
Christmas (IX)	50-150%
Stuart (X)	50-150%
P.T.A. (XI)	50-150%
Hageman (XII)	50-150%
Plasma transglutaminase (XIII)	8-20 u./ml.

Section Seven

Neuromuscular Mechanisms

Introduction

Health is manifested by a behavior of bodily systems that achieves and maintains a comfortable interaction or relationship with the environment. J. B. S. Haldane stated this principle when he said that progress in medicine depends on understanding how the human organism adapts to changes in the environment. Thus, the healthy person increases his red blood count when living at altitude but not at sea level.

Polycythemia developed at sea level spells disease. The bodily mechanisms required to increase the number of circulating red blood cells, however, are identical in health and disease. So it is with other bodily systems. For example, the difference between infection and mere exposure to microbes depends on a neat balance of activation and restraint of immunological and other defense mechanisms. Inadequate modulation of immunological function appears to be responsible for the exaggerated immune behavior of certain connective tissue diseases. Disease, then, may reflect too much or too little of certain adaptive functions, resulting in essentially inappropriate physiological behavior.

A distorted balance of regulatory functions can at times be attributable to a genetic error, particularly the failure of appearance of certain enzymes, or an incorrect sequence of amino acids in the molecule. Likewise, nutritional disturbances may impair the development of proper visceral controls. So may a wide variety of experiences, including overload of the system, as from climatic extremes, injury, infection or perhaps social and psychological pressures, especially during periods of growth and development.

Proper balance of bodily functions is maintained in part by accommodative behavior of the endocrine glands, largely under the governance of the central nervous system. Other functions are regulated more directly by neural mechanisms. In the chapters that follow, the authors have attempted to relate what is known of neural regulation and control of visceral processes to the manifestations of disease. Any regulatory system requires the interaction of activating and restraining forces with a feedback of some sort. The graceful movements of a ballerina, a pianist or a champion athlete depend more on the modulating restraint of inhibitory neurons from cerebellum, red nucleus and basal ganglia than they do on the activation of the Betz cells, the prime movers of the corticospinal tract. As the painful, useless skeletal muscle contractions of tetanus infection or strychnine poisoning are attributable to blocking or inactivation of the normal modulating influence of an inhibitory network, so the almost ceaseless gastric secretion of HCl and proteolytic enzymes characteristic of duodenal ulcer or the sustained elevation of blood pressure by initially normal arterioles may reflect the failure of the normal balance of autonomic excitatory and inhibitory influence.

Inhibitory pathways in the central nervous system have only recently come under serious scrutiny. Much has been learned through study of the inhibitory neurotransmitters, glycine and GABA. The latter has been identified, not only in association with Purkinje cells of the cerebellum, the sole output of that organ, but elsewhere in the central nervous system and even peripherally in the walls of arteries and arterioles. One may infer from this and other work an elaborate inhibitory network responsible for the modulation of visceral behavior as it is for the function of skeletal muscles.

When Charles Richet, in 1886, first combined chloral hydrate and glucose to make what he called chloralose he noted that, while it dulled consciousness, in enhanced visceral responsiveness. Neurophysiologists have taken advantage of this property of the drug to identify autonomic pathways in anesthetized animals.

As more and more has been learned about excitatory and inhibitory influences, about facilitatory and inhibitory regulation of synaptic transmission, the concept of the reflex nature of bodily regulation has given way to a concept of neural interaction in which virtually all parts of the nervous system are interconnected so that local perturbations may have widespread effects. Rich interconnections between somatic sensory, visceral sensory and the effector neurons of all sorts have been discovered that link many zones of the central nervous system, including thalamus, hypothalamus and limbic cortex with the frontal lobes. The extent of interrelatedness of all of these structures in the formulation of organismal behavior not only has led to the discarding of the too simplistic reflex concept of regulation but has made it clear that the somatic and visceral pathways are not two systems after all but a single system with different kinds of neuronal hookup in a state of continuous dynamic interaction.

The rapidly growing understanding of the way the nervous system relates to all bodily structures has made it increasingly attractive to consider such disorders as pathological aggressiveness, alcoholism, epilepsy and hemolytic anemia as resulting from a defective balance of excitatory and inhibitory mechanisms. At present it appears that the fault lies most often with inadequacy of smoothly regulatory inhibition.

With the newer knowledge of the vast central ramifications of the autonomic system and the fact that visceral effector neurons can be found as "high" in the cerebral hemispheres as those that activate skeletal muscles, the principal difference between somatic and autonomic effector nerves is evident only after they leave the cord, where autonomic nerves synapse before acting on the effector organ while somatic motor nerves do not. Furthermore, peripheral neural plexuses, such as are found in the gut and other viscera, are uniquely characteristic of autonomic innervation. The peripheral synapse of autonomic nerves seemed to have very little functional significance until the recent discovery of interneurons in mammalian autonomic ganglia. The implication of this discovery is that further visceral and vascular regulatory activity is possibly peripheral to the central nervous system.

The neural plexuses that invest many visceral and vascular structures endow them with greater versatility and range of function than the skeletal muscles and enable them to perform and to adapt within limits even when isolated from the body and suspended in an artificial medium. Doubtless a capacity for automaticity and autoregulation in various organs was in part responsible for the fact that the widely ramified representation of autonomic nerves in the central nervous system was overlooked for so long a time.

Recent studies with the techniques of operant conditioning have shown that both somatic and visceral function can be modulated through "learning." Perhaps doors have been opened to new therapeutic measures through training maneuvers.

Stewart G. Wolf, M.D.

27

Neural Integration

Robert B. Livingston, M.D.

INTRODUCTION

When John Farquar Fulton arrived at Yale University to establish the Laboratory of Physiology, he was invited to address the local medical society on a subject of his choice. He suggested the hypothalamus. The Program Chairman asked, "The hypowhat?" Professor Fulton replied, "You know, that small patch of nervous tissue at the base of our brains which regulates our eating, drinking, sexual functions, and sleep." "By Jove!" came the startled response, "That *is* marvelous! And, in the correct sequence, too!"

The scope of central nervous system regulation, revelation of which was startling in Fulton's day, is considerably more familiar now, and better understood. There remains, however, an immense territory that beckons for discovery, and what is known still continues to be awe inspiring.

Old ideas are being eroded. Emerging concepts are strictly formative. The aim of this chapter is to provide a comprehensive view of the nervous system as a dynamic mechanism, responding with normal and sometimes deranged somatic and visceral regulations, and responsible for a full range of corresponding subjective reactions. The main aim is to assemble a theoretical framework to explain the organization and operation of the nervous system as a whole in health and disease. Mind-body unity, structural-functional unity, and a transactional conception of holistic performance in an evolutionary and developmental context are pervading assumptions.

We exist as integrated concatenations; we need to comprehend what induces and achieves this integration. Mental processes are not to be left aside. Subjective experiences, and conscious processes in general, can become contributory forces in many instances of regulation and should not be neglected or considered as passive epiphenomena.

SOME GENERAL FEATURES OF BRAIN ORGANIZATION RELATING TO REGULATION

Complexity, Interdependence and Transactions

Progress in understanding how the nervous system regulates somatic and visceral activities continues to reveal greater complexity than was hitherto imagined. There appear to be more and more feedback loops capable of finer tuning of regulatory systems. The following example of nervous system control of endocrine functions illustrates this general conceptual trend.

An Example of Interdependence of Parts. It was initially presumed that endocrine glands were independent of one another and of the nervous system. Later it was discovered that the hypophysis generated a dominating influence on endocrine glands. The proposition was popularized that the pituitary was the "conductor of an endocrine orchestra." Still later it was recognized that the hypophysis is itself influenced by the levels of hormonal products secreted by certain endocrine glands, and, secondly, is itself affected by actions of the hypothalamus. The hypothalamus was found to discharge "releasing hormones" into the portal circulation. These were amplified by the adenohypophysis in the discharge of hormones into the general circulation whereby they activated target endocrine glands, gonads, thyroid, etc. Hormones released by the target glands returned to affect the hypophysis and the hypothalamus.

More recently, hormonal controls generated by the hypothalamus have been found to include inhibitory influences on adenohypophyseal secretion, holding in check background secretory rates and counteracting the effects of specific releasing hormones. Thus, hypothalamic control is more regulatory than simply triggering the release of hypophyseal hormones. Direct innervation of endocrine glands is being explored for its possible facilitatory and inhibitory roles in the control of endocrine secretion, and also the influences of secretory products and the actions of near and remote parts of the nervous system on the hypothalamus.

At each step along the way the nervous system and endocrine mechanisms have been found to be more interdependent than was originally conceived. The limits of this increasing complexity are nowhere in sight. Comparative studies indicate that endocrine and nervous systems have been evolving in intimate juxtaposition, and probably interacting just as intimately, from the time of their simultaneous evolutionary start. Neurosecretory and neurohormonal regulation appear to belong to the same line of biological developments that led to the signaling between nerve cells and at neuroeffector junctions by release of neurotransmitter substances.

Transactional Mechanisms. This neuroendocrine example can be extended to the interpretation of practically any other aspect of nervous system regulation. The emerging picture is one of multiple, mutually interdependent parts in simultaneous action. These are called transactional mechanisms and they are far more complex and subtle than interactions. The whole is a far more refined regulatory system than the sum of its parts.

The nervous system is the most complicated mechanism known to man. It works, generally speaking, and in health and disease it functions in ways that are often predictable in spite of the fact that many of the intervening mechanisms have not yet been revealed. Regularities of systems transactions recur at so many different levels of complexity that there must exist some superordinate principles, shaped by evolution. How did these principles emerge and what is their nature? Chance and selection, operating throughout evolution, have produced a staggering variety of self-preserving, self-reproducing systems, all cast by chance and hammered out and tempered on the hard anvil of selection. Those evolutionary lines that did by chance survive had that capacity built in by selection; from outward appearance this overall phenomenon appears to be goal-seeking.

Of the abundance of organisms surviving and reproducing, many tend to reproduce enough variations among their descendants so that there are organisms capable of surviving and reproducing in spite of changes in the environment, and enough organisms capable of surviving in different environments. Thus, the environment gets in on the act of selection, and organisms, because of success by chance, appear to be "seeking" survival. Emergent awareness of self and of dangers and opportunities in the environment became an early evolutionary advantage in organisms with nervous systems. Selection pressures have eventually yielded fuller self-consciousness and efforts at individual and group survival. Evolution is still going on. Perhaps still greater self-awareness will come to our rescue yet.

Successful recipes for self- and species-preservation have become widely shared by many different forms of life. From embryogenesis until death these mechanisms regulate detailed goal-seeking processes and overall behaviors oriented toward survival of the individual and survival of the species. They constitute evolutionarily created, not externally introduced, *teleological mechanisms*. An important consequence of evolutionary pressure is that surviving species tend to perfect their capabilities for survival in a given context —one that may be altered to their detriment or demise.

Environmental pressures pay off for variety and abundance of living forms. Biological selection pressures pay off for yield of systems that will serve successfully for many different forms of life, and provide recurrent regularities throughout many different levels of biological complexity, even including psychological, social and cultural complexity. We agree with Peter Medawar that these are not products of biological complexity—they are integral expressions of biological complexity itself. The severity and universality of evolutionary forces shaping life make it practical for us to infer that there are teleological commitments in each of the biological systems and subsystems that we are trying to understand. If we cannot perceive the goals to which these mechanisms are committed, it is because we haven't looked at them correctly.

Evolution of Nervous System Mechanisms

Control of appetite-satiety, approach-avoidance, activity-inactivity, and offense-defense is observed in single-celled organisms such as paramecia. This forces us to recognize that such discriminative and integrative properties are basic to protoplasm, and to bear in mind that such non-nervous controls might contribute to pathophysio-

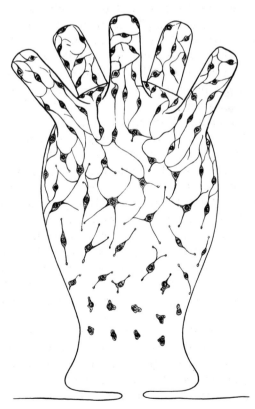

FIG. 27-1. Hydra bud developing its syncytial nerve net which serves to regulate appetite, feeding behavior, satiety and such behavior as letting go the perch to drift to a better location. The nerve net is without ganglia, so that the intricately integrated activity must be by some "democratic process" depending upon relative excitability among presumably equivalent neurons. (Bullock, T. H., and Horridge, G. A.: Structure and Function in the Nervous Systems of Invertebrates. San Francisco, W. H. Freeman and Company. Copyright © 1965)

logical processes in higher organisms in ways that compete with or complement nervous system controls.

Nerve Nets. Simple multicelled organisms like the hydra have nerve nets that are entirely decentralized. Regulation in the hydra has, literally, no headquarters. There is not even any directionality to the synapses; hence, impulses can flow from any neuron to any other with which it has connections. Yet the hydra shifts from torpor to activity, groping with its tentacles to capture free-swimming daphnia and to stuff them into its ostium which the nerve net obligingly opens like a beggar's purse to receive this provender.

The nerve net performs by apparently "democratic processes" according to shifting levels of excitability distributed among neurons that are reacting to visceral needs. Visceral needs create some kind of "popular demand" that culminates in apparently purposive behavior. The hydra's behavior must involve complicated sequential activities among multiple, mutually interdependent parts. Waves of coordinated activity must sweep through the hydra's nerve net in fulfillment of this succession of transactional performances, or else hydra would not survive.

Reticular Formations. We may draw a functional parallel in regulatory activity between the syncytial nerve net of the hydra and the synaptically directional reticular formation in higher organisms. This parallel can be readily appreciated in the regulation of respiration, thirst, appetite and temperature controls in mammals. In each case, the regulatory control is built into the chassis of the nervous system in the form of a reticulum of nerve cells that are responding to their own individual and sometimes specialized visceral needs. By some kind of "democratic process" among these neurons, and by their "popular demand," coordinated behaviors are organized and executed for the restoration, maintenance and well being of the whole organism.

Regardless of whatever modulations of these actions are introduced by other inputs and by feedback loops, the nexus of the regulatory performance is achieved by a diffusely organized system of neurons, a reticular formation.

Jellyfish emerged early in evolution, bearing two important new cytological contributions: directional synapses and the means for transforming local, decremental action potentials into decrementless spikes that conduct impulses over long distances. More sophisticated organisms have stretched and plaited neurons "invented" by the jellyfish into functional organizations that include a great variety of reticular and laminar systems of great intricacy.

Origin of Somatic Mechanisms. The larvae of certain filter-feeding tunicates developed transient motor and sensory mechanisms capable of exploration and selection of a suitable perch for the adult's sessile life. Following the larval stage this specialized motor and sensory apparatus degenerated. Some descendants of tunicates took advantage of the propulsive and discriminative powers of the tunicate larvae by retaining the larval apparatus. This marks the beginning of somatic as contrasted with visceral nervous mechanisms. Chordates and vertebrates in general utilized this advantage as an additional means to obtain visceral satisfactions. From what we know at present, somatic nervous systems are still subservient and devoted to securing fulfillment of visceral needs!

Embryological Emergence of Regulatory Mechanisms

All neurons and glia, primary elements of nervous system regulation, are derived from a single layer of ependymal cells which lines the primitive neural tube. Priority in both phylogeny and ontogeny is given to the development of diffusely projecting neurons which in the human nervous system constitute the extensive gray and white reticular zones which are visible all along the neuraxis but are most fully developed throughout the brain stem and especially toward its cephalic end. The reticular neurons form a matrix, literally a mother tissue, a framework within which later developing neurons migrate to their final destinations. Within this matrix the more differentiated nuclei and more easily recognized pathways make their appearance. Reticular organization provides a scaffolding in the subventricular and intermediate zones of the expanded neural tube upon which the rest of the brain takes form.

At the cephalic end of the brain stem, thalamic nuclei make their appearance as though crystallizing out of a reticular substrate. Reticular neurons pushed aside in this process remain within the thalamus as the intralaminar nuclei, periventricular nuclei, the centrum medianum and, on the outside margins of the thalamus, the nucleus reticularis which covers the thalamus like a cowl, anteriorly, dorsally and laterally. Thus, reticular remnants of the original matrix maintain functional continuity with the mesencephalic reticular formation. By means of these residual reticular pathways through the thalamus behavioral arousal takes place. This is the pathway by which the mesencephalic reticular formation activates the cerebral hemispheres.

Ventrally, the hypothalamus is practically completely composed of gray reticular formation, also in direct continuity with reticular formation of the midbrain and thalamic regions.

Reticular Scaffolding. Reticular scaffolding makes its appearance prior to the more differentiated features of nervous systems and it continues to develop *pari passu* with them throughout phylogeny and ontogeny. This priority of development of the reticular formation provides reticular regulatory control and participation all along the neuraxis.

The large, well differentiated nuclei and tracts which are emphasized in textbooks and teaching as the main stations and circuits of the central nervous system are more akin to immigrants, evolutionary and embryological latecomers who have settled down to participate in an already integratively successful population of reticular neurons. This may account for the finding by Hess that specific functional representations in the hypothalamus and along the brain stem do not follow anatomically discrete nuclei and well marked pathways. Instead, they hover among these structures like a cloud smoothly adumbrating a more detailed landscape. Electrophysiological, pharmacological and thermal influences and the effects of lesions all indicate that functional systems extend in configurations different from those implied by traditional anatomical maps and probably conform more specifically to as yet undisclosed characteristics of the underlying reticular matrix.

Input and Output Channels

Signals entering the central nervous system are conveyed by way of an estimated few tens of millions of input fibers. These come from distant receptors and from visceral and somatic sense organs and receptors throughout the body. Impulses leaving the central nervous system are confined to an estimated few millions of outgoing fibers, somatic and autonomic. These provide central control over glands and smooth, cardiac and skeletal muscles. Thus there is already a convergence of perhaps ten-to-one between the number of input and output channels.

The number of neurons that are wholly contained within the central nervous system is difficult to appreciate. This number is traditionally estimated to be 10 billion nerve cells (10^{10}), but recent estimates of the number of cells in the cerebellar cortex suggest that there may be 10^{11} cells in the granular layer alone! This warrants our revising upward the estimated total

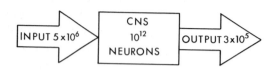

Fig. 27-2. Input neurons constitute an estimated few tens of millions of neurons. Output neurons are 10-fold or so less in number. Central neurons, into whose expanse incoming signals enter and from which emerge output signals to muscles and glands, are some tens to hundreds of thousandfold more numerous compared to the total of input and output neurons, totaling an estimated 100 billion neurons.

of central neurons to 10^{12}. Hence, for every input or output channel, there are tens of thousands to perhaps hundreds of thousands of central channels. This leaves entirely out of account whatever influence may be exerted by the even more numerous glia. There are estimated to be 10 times as many glia as neurons. It is known that glia can modulate neurons, but there is as yet no clear idea about how glia may contribute to the control of central circuitry.

Incoming signals entering this almost unlimited pool of neurons are segregated among ascending and descending pathways according to different sensory modalities and various stimulus features. They are many times represented and re-represented in various ways for comparison with built-in references such as thermoregulatory set-points, with genetically endowed and idiosyncratically acquired memory stores, and with information entering through other input channels. The central neurons accessible to input signals provide both the territories and the mechanisms for differentiation and abstraction of incoming messages, for the general function of *analysis*.

Output of signals is achieved by a reverse flow, an unimaginably vast convergent condensation of impulses from widespread incoming and central pathways, from genetically organized reflexes and other built-in motor programs and acquired motor patterns as well as from currently incoming sensory signals. All of these activities ultimately focus on the "final common path," the visceral and somatic motor apparatus that influences glands and muscles. This pluripotential convergence provides for coordination and cooperation among signals coming from many sources and serves the general function of *synthesis*.

Analysis and Synthesis. It should not be supposed that sensory mechanisms are limited to analysis. Sensory differentiation and abstraction is simply not at all possible without analysis. Yet it is also clear that acts of perception indubitably include important aspects of synthesis. "Mom," for the infant, is made up of olfactory, taste, visual, auditory, and somesthetic cues synthesized into a whole that is further enriched according to an historical context of past experiences and expectations and the characteristics of past and present infant needs and yields of satisfactions.

Neither should it be supposed that motor mechanisms are limited to synthesis. Motor acts simply cannot occur without some degree of synthesis. Even the crudest mass action, supposing every output channel to be firing, would require a fantastically convergent synthesis from the billions of central neurons and the tens of millions of input neurons to the few million output possibilities. But motor activities also clearly involve analysis—analysis and extraction from previous action programs and their outcomes in order, unconsciously and consciously, to compose an alternative and perhaps an improved contemporary performance.

Both of these opposite kinds of processes are taking place simultaneously and often in parallel channels as commonplace phenomena throughout the nervous system. Synaptic switching to synthetic convergence or analytic divergence is undoubtedly controlled, as are all nervous system regulations, by genetically endowed and acquired (according to past payoffs) visceral satisfactions and dissatisfactions. Later, we shall explore ways by which such controls may be acquired.

Horizontal and Vertical Organization

Segmental Reflex Coordinations. Because the nervous system is stretched out lengthwise, it is easier to cut across than to slice vertically. This has given rise to an exaggerated emphasis on horizontal organization of central nervous mechanisms.

It is well known that there are quite direct horizontal segmental reflex connections involving one or only a few synapses and relatively less direct contributions crossing to the opposite side of the spinal cord. It is also well established that there are numerous parallel paths involving many interneurons, the most central of which are diffusely projecting, the reticular formation of the spinal cord. This zone is anatomically and physiologically continuous with brain stem reticular formation. These pauci- and pluri-synaptic paths ultimately converge on both somatic and autonomic motor nuclei.

What is less often emphasized is that incoming sensory fibers branch immediately on entry into the spinal cord, before ven-

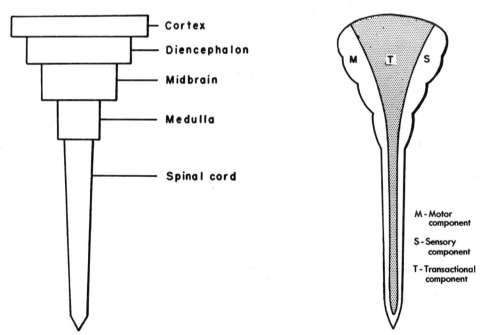

FIG. 27-3. It has been traditional to emphasize the horizontal organization of integrative processes. Partly this is due to the investigation of reflexes and partly because it is easier to cut across the nervous system rather than longitudinally in interfering with regulatory processes. In reality the nervous system is as thoroughly stitched together vertically as horizontally. The two figures are both correct and should be considered as representative of two aspects of one system. Transactional mechanisms are considered those such as spinal and brain stem reticular formation, hypothalamus and limbic system. (Livingston, W. K., Haugen, F. P., and Brookhart, J. M.: The vertical organization of function in the central nervous system. Neurology, 4:485, 1954)

turing into synaptic relations, and send tributaries upstream and downstream into several other segments as well as into the segment of their entry, thus influencing several segments simultaneously. Thus the nervous system is stitched together vertically as well as horizontally.

A conspicuous fraction of incoming sensory fibers send branches that ascend and descend in the dorsal columns. The ascending branches may travel all the way to the lower brain stem before synapsing. The dorsal column fibers give off frequent collaterals to the spinal reticular formation and intermediate nuclei along the way. By this means, first order sensory neurons contribute to near and far regions of the spinal cord and brain stem.

Interlimb Reflex Coordinations. Sensory signals, in addition to contributing to segmental reflexes, contribute to long interlimb reflexes of the spinal cord and to reflexes which link spinal and cranial performances. These long intersegmental systems follow two distinctly different paths. The first, called *propriospinal,* is a diffusely projecting, multisynaptic system that ascends and descends in widely distributed fashion and projects to both sides of the spinal cord. It crosses and recrosses the cord as it relays upward and can bypass successive contralateral hemisections of the spinal cord. A faster and more discrete path is followed by a second system, the *spino-bulbar relay* system, which transmits directly to the medulla oblongata and from

SPINO-SPINAL & SPINO-CRANIAL COORDINATION

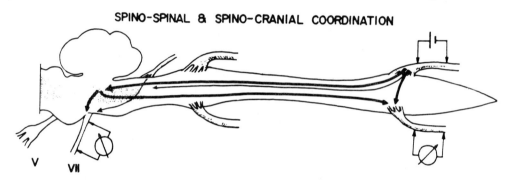

CRANIO-CRANIAL & CRANIO-SPINAL COORDINATION

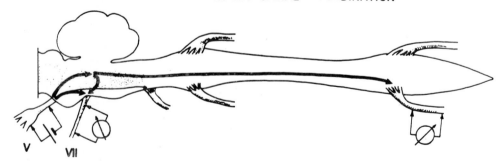

Fig. 27-4. Among longitudinal regulatory systems are spino-bulbar relay and brain stem-bulbar relay pathways which originate from sensory input and contribute to motor outflow at all spinal and brain stem levels. This bulbar relay projection system is to be contrasted with more diffusely projecting, locally integrating spinal and brain stem projections, also longitudinal, also serving the length of the neuraxis, but which do not depend on bulbar relay. Bulbar relay correlates the final outcome of influence by both systems with remote parts of the neuraxis and with the vital bulbar centers. (Shimamura, M.: Longitudinal coordination between spinal and cranial reflex systems. Exp. Neurol., 8:505, 1963)

there relays impulses to all spinal and cranial motor nuclei. The importance of this bulbar relay is that it allows longitudinal control to come under the influence of the bulb which is a quintessentially convergent area that receives signals from many widespread regions of the nervous system. The bulb also contains the vital centers for cardiovascular and respiratory regulation and mechanisms for general inhibitory control over all central nervous system output. Its communications must have a high priority.

Cranial sensory input contributes to a similar diffusely projecting, multisynaptic pathway that sweeps throughout the neuraxis to reach all cranial and spinal motor nuclei and to a bulbar relayed system that reflects back upstream to affect cranial motor outflow and downstream to affect spinal motor outflow at all levels. Both the diffusely projecting and bulbar relayed systems provide motor activation at all levels of the neuraxis. These longitudinal control systems participate in posture and locomotion and contribute to the background activation of motor systems underlying pyramidal and extrapyramidal activities.

Embryological Organization of Sensory and Motor Systems. The nervous system originates as a flat plate which by genetically controlled processes of cell division, cell migration, and preordained cell death, forms a shallow groove. This closes over by exuberant growth from the sides, growth that spills back laterally from the top to form the neural crest. From this crest are derived dorsal root ganglia, sympathetic ganglia and the adrenal medulla. Closure of the tube proceeds headward and tailward from near the middle region.

Cells in the dorsal (posterior) half of the neural tube, dorsal to an indentation in the wall of the tube, the sulcus limitans, are dedicated predominantly to sensory functions. Cells lying ventral (anterior) to that sulcus are involved mainly in motor activities. This dorsal/ventral : sensory/motor

organization of the spinal cord and brain stem can be followed into the forebrain. Cortex lying anterior to the central sulcus is mainly motor, governing body parts in relation to each other and to the gravitational field. Increasingly complicated representations extend forward of the precentral gyrus. Prognostications relating to a wide latitude of behavioral options for the future seem to be the preoccupation of association areas of the frontal lobe. Cortex lying posterior to the central sulcus is mainly sensory, representing somesthetic, visual and auditory modalities and their individual and conjoined associations.

The basal ganglia, which contribute mostly to motor performance and probably especially to stereotypical acts and imitative behavior, are located generally anterior and lateral to the thalamus, which is devoted mainly to sensory relay and sensory association functions. The hypothalamus, which lies ventrally perforce of its junction with the hypophysis, the glandular part of which is derived from an outpouching in the roof of the mouth, is the focus of visceral and neuroendocrine synthesis. The hypothalamus lies in the centralmost position for transactions among the frontal lobes, limbic system, neuroendocrine pathways and the midbrain reticular formation.

Visceral and Branchial Representations. At spinal levels, visceral mechanisms are sandwiched between dorsal somatic *sensory* and ventral somatic *motor* controls. Visceral regulatory systems form a nearly continuous longitudinal column which lies adjacent to the sulcus limitans, from lower sacral to lower cervical segments. At lowermost brain stem levels, first order neurons in the dorsal columns relay and arch ventrally, decussating, to ascend to the thalamus via the medial lemniscus on the opposite side. Separation and ventral displacement of the dorsal column-lemniscal system exposes the spinal canal to the surface and elevates visceral regulatory mechanisms to the floor of the fourth ventricle. Here visceral nuclei are joined by cell

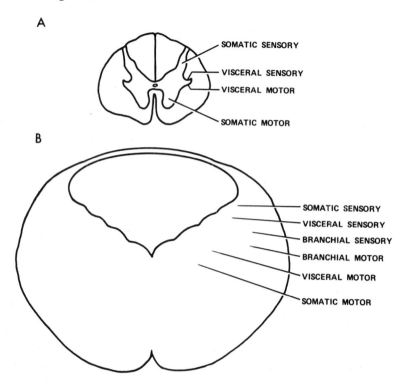

Fig. 27-5. Comparability between spinal somatic and visceral sensory and motor centers and those in the brain stem, including the addition of branchial regulatory components, can be seen in this highly schematized representation of spinal (A) and bulbar (B) organizations.

clusters that serve face, throat and neck structures derived from branchial arches.

Sensory systems thus become dorsolaterally located, with somatic, branchial and visceral elements arranged in that order toward the midline. Motor systems, visceral, branchial, and somatic are arranged in reciprocal order continuing ventromedially. Juxtaposition among somatic, visceral and branchial input and output systems provides functional integrity. Additionally, they are all embedded in reticular formation. This ensures smooth and powerful lower motor neuron performance of vital functions by synthesis of respiratory and cardiovascular controls, mechanisms for vocalization, facial, neck and shoulder gestures and mimicry, laughing, crying, coughing, sneezing and vomiting.

At mesencephalic levels, visceral regula-tory systems continue to remain central-most, coursing alongside the aqueduct, including both central gray and white reticular formation. Included in this region is a very important projection representing the convergent synthetic center for frontal lobe, limbic and hypothalamic outflows, the *frontal-limbic midbrain area.* From this location, visceral regulatory systems can be traced cephalically along the walls and floor of the third ventricle. Above diencephalic levels, visceral representations are expanded enormously into the limbic system, which is the head ganglion of visceral experience and expression, of emotion, feeling, mood.

The Limbic System. Limbic structures form a double ring of cortical and subcortical tissues that are perched on either side of the cephalic brain stem, like a pair of jaunty crowns. There they encircle the

Fig. 27-6. The limbic system includes a ring of cortical and subcortical structures that encircles the stem of each hemisphere. It therefore lies between brain stem and the neocortex and serves in important regulatory capacities linking these two levels of the neuraxis. The limbic system is phylogenetically old and is functionally related to experiences and expressions of mood, feeling and emotion. (MacLean, P. D.: The triune brain, emotion, and scientific bias. *In:* The Neurosciences, Second Study Program. New York, Rockefeller University Press, 1970)

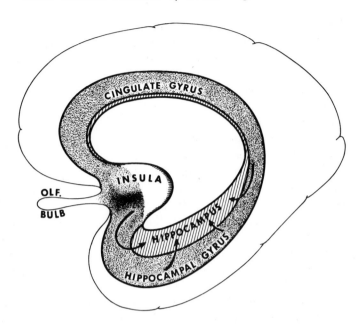

stalks of the two cerebral hemispheres. In embyronic development, the limbic system begins to develop before other parts of the hemispheres.

Like Troy, the limbic system is built upon and includes ancient structures, the evolutionarily original cortex, archipallium or entopallium, a slim circlet of two layered cortex which traces the same course around the hilus of the hemisphere. An expanded region of this ancient cortex is the hippocampus. The bulk of cortical substance of the limbic system is paleocortex or mesopallium, more primitive structurally and functionally and older than neocortex or neopallium. The latter makes its evolutionary debut with mammals. It appears on the outer surface of the lateral ventricles as these form outpouchings from the forward end of the third ventricle, passing through the double ring of limbic structures via the foramina of Monro. Thus neocortex becomes applicated on the outside of the hemisphere.

Because neopallium is increasingly expanded in higher mammals, it increasingly hides the limbic system from view. But the hidden limbic system structures can be palpated by reaching one's fingers into the

cleft between the hemispheres and then tracing all the way around the hemisphere with the fingertips pressed against the outpouching stalk of the hemisphere. In this location, they press against the ring of limbic structures that encircles the base of each hemisphere.

The limbic system thus forms a continuous medial boundary for the entire hemisphere. It was originally referred to by Broca as the *ourlet,* or hem, like the hem of a woman's gown, and later, more formally, as the grand limbic lobe, limbic meaning marginal. This strategic location of the limbic system astraddle the brain stem and portal guardian to the hemispheres, like the Colossus at Rhodes, enables the limbic system to monitor traffic and participate in all brain stem-hemispheric transactions. As a student declared, "The cerebral hemispheres are harnessed to the brain stem by a limbic yoke."

Integration by Analysis and Synthesis

The two principal functional processes, analysis and synthesis, referred to above, have their anatomical counterparts in laminar and reticular organizations. Although the nervous system usually exhibits both of these kinds of organization and function together, a predominantly laminar organization favors analysis and a predominantly reticular organization favors synthesis. The most familiar and pure form examples of these categories are cortex and reticular formation.

Analytic Functions—Mainly Sensory. Sensory systems strongly favor laminar organization. This is seen in the olfactory bulb, retina, substantia gelatinosa and its continuation in the trigeminal root, the cochlear nuclei, both colliculi, the geniculates and the cortical receiving areas, olfaction in paleocortex, and somesthesis, vision and audition in limbic paleocortex and neocortex. As signals ascend sensory pathways, they are subjected to a succession of analytic operations. Hubel and Wiesel have demonstrated in the visual system that more and more complex and abstract features of visual stimuli are detectable as neuronal units are examined proceeding centralward from retina to visual cortex and visual association areas. Cortex is organized for widespread display of representations—representations for sensory analysis of the outside world as relayed inward from distance and body surface receptors, and representations for motor synthesis of the body parts in relation to each other, to the spatial world and to the direction and force of gravity. Sensory cortex is particularly exquisitely organized for analytic functions, for selective differentiation and abstraction.

Synthetic Functions — Mainly Motor. Motor systems predominantly display reticular organization. Two exceptions are motor and cerebellar cortex, both of which contribute special analytic functions involved in motor programming and coordination. Both contribute powerfully to downstream synthetic steps. Elsewhere, throughout the basal ganglia, brain stem and the extrapyramidal system in general, cellular architecture is typically reticular. Reticular organization provides neuronal accessibility to a wide variety of different kinds of inputs. Convergent synthesis is characteristic of integrative activities of both upper and lower motor neurons.

The reticular formation is a prime example of reticular organization. It is both diffusely receiving and diffusely projecting. Dendrites of the reticular formation in both spinal cord and brain stem reach out radially toward a variety of sources of converging signals. When thin cross sections of gray and white reticular formation at brain stem levels are held up to a light source, a radiating "sunburst effect" is seen, caused by reflections from the innumerable outreaching dendritic processes of reticular neurons. Microscopists, impressed by the details of this phenomenon, refer to the midbrain reticular formation as an "isodendritic core," meaning that the reticular dendrites are characteristically radially distributed, universally outward reaching.

Dendrites of the largest reticular cells are stocky, tortuous and many-branched. They radiate widely like the snake-heads of an aroused medusa. These dendrites make contact with a wide variety of collaterals branching off from each of the several sensory, motor and central pathways. This arrangement provides reticular neurons with multimodality information from sensory, motor and central sources. These same gigantic cells possess extremely long axons that branch repeatedly and usually course up and down both sides of the neuraxis, sometimes reaching all the way from above the diencephalon to the spinal cord, giving off collaterals and axonal terminations to sensory, motor, relay and association nuclei belonging to many different functional systems.

There are also many intermediate and small nerve cells in the reticular formation. Although we tend to be attracted by larger neurons, perhaps the vast majority of reticular units are small cells with short dendrites and short, unmyelinated axons. They obviously do business locally. Some are so short that they extend less distance than the

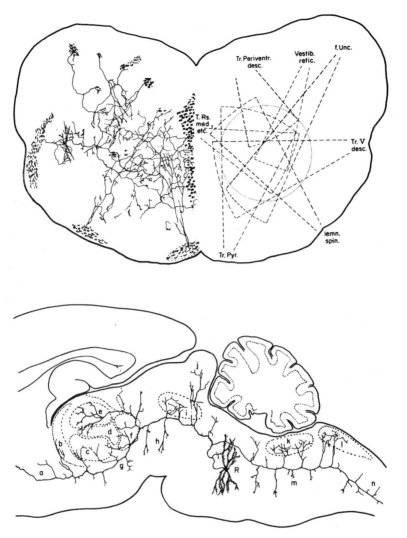

Fig. 27-7. The reticular formation lies medially all along the neuraxis. (Top) Invading collaterals from both sensory and motor pathways entering the regions of dendritic fields of brain stem reticular neurons. (Bottom) A single large reticular neuron. Its localized dendritic display makes contact with numerous input sources and its very extended axon passes upstream and downstream along the neuraxis with distribution of terminals to many different sensory and motor and central ganglionic constellations. The convergence of inputs and widely distributed output characterize synthetic functions involved in central regulatory processes. (Scheibel, M. E., and Scheibel, A. B.: Anatomical basis of attention mechanisms in vertebrate brains. *In:* The Neurosciences. New York, Rockefeller University Press, 1967)

length of an action potential! They probably function as local "on" and "off" signals and as very local switching devices. Perhaps their role is one of sustaining background levels of excitability among neurons within their limited scope.

Altogether, reticular arrangements can be seen to provide remarkable advantages for synthesis through the gathering in of various signals by reticular dendrites, and the further synthesis at each of the various points of reticular axonal distribution whereby recipient cells can be given a synthetic view of an extended range of nervous system activities, near and remote. It is worthwhile to look upon each region under discussion in terms of its inputs and outputs and more particularly in regard to whether it is organized in a laminar way and may be functioning predominantly for analysis or whether it is organized in a reticular way and may be functioning predominantly for synthesis.

PATHOPHYSIOLOGICAL MECHANISMS AFFECTING NERVOUS SYSTEM REGULATION

In the space available, only a few shorthand examples of pathophysiological processes can be introduced. Examples are selected from among the most frequently presenting and most important symptomatic and regulatory disorders encountered in medicine. Acquaintance with these examples may provide a cognitive framework for the reader whereby other pathophysiological processes affecting regulation may be schematized.

Pathophysiological Processes Involved in Pain Perception

Pain may be considered the subjective accompaniment of a set of bodily responses aimed at protection and withdrawal from noxious and destructive stimuli. Pain contributes a perceptual impetus for immediate avoidance behavior, and for anticipation and escape from disagreeable situations in the future. But pain is much more than this definition implies. Pain makes many misleading contributions such as referred pain and phantom limb pain and it can become an overwhelming disorder in its own right. It can become a kind of physiological juggernaut that disrupts normal somatic and visceral regulatory processes and dominates all aspects of perception, judgment and action.

Pain is one of the most common of all symptoms presented to the physician. Pain can regulate subjective existence in a way perhaps more compelling than any other experience. It can do so in one incandescent instant, it can be prolonged indefinitely, and it can increase in severity with time. Severe pain may be a different experience from what we normally encounter and what we can remember realistically. Prolonged severe pain can radically distort all of a person's reactions and social relations, destroy his motivations and erode any of the values he has attached to continuing to live. An understanding of pain mechanisms is sought therefore to provide a more humane, empathic and understanding communication with patients who are enduring pain, and a more progressive, physiological therapy for relieving their suffering.

There is one essential principle to remember: *pain is a subjective experience* that is not to be directly inferred from the presence or absence of apparently suitable stimuli or even from the presence or absence of reflex changes or detectable activities in peripheral or central pathways. No one knows how and where pain is perceived. Whatever is sufficient at that level may be all that is necessary for the full scale reality of pain to be experienced, regardless of what may or may not be happening along incoming "pain pathways." Usually there is a presumptively sufficient stimulus in evidence and certain intermediate pathways may be demonstrated as being activated. But pain is still no more and no less than a subjective experience and, like many other vivid subjective experiences, may not be publicly shareable. The only evidence the doctor may have that a patient is experiencing pain is that patient's own testimony. Such unconfirmed testimony is not binding on the physician, unfortunately, the physician may be sorely mistaken ignoring that testimony and by assuming that the patient is malingering. This turns out in time to be an occurrence more frequent than the converse.

Acute injury may be accompanied by widespread alterations in visceral and somatic regulations: one has the "wind knocked out," sphincter control may be lost, or a limb may become numb or paralyzed. Less frequently, a transient paraplegia or quadriplegia may occur. Consciousness may remain intact throughout. Pain may appear at the outset, or it may be hours or days in coming on. It has been established that this widespread pathophysiological commotion of the spinal cord can occur with no more than peripheral nerve injury. Evidently, a strong barrage of incoming impulses is capable of disrupting normal communications among widespread regions.

Long-lasting abnormal sensory input is usually accompanied by long-lasting reflex disorders, manifested by heightened muscle tension, spasms and spontaneous jerking, excessive dryness of the skin or more usually excessive sweating, and changes in blood flow and temperature of the affected part. Thus, both somatic and visceral regulations are upset. Long-continued severe pain may be associated with pronounced "trophic changes" such as abnormally heavy pigmentation, continuous dripping of perspiration, excessive rate of hair growth and altered hair texture, clawlike appearance of nails, being greatly arched longitudinally and laterally with a marked reduction in the finger pad and loss of the sulcus between nail and pad. There may be disappearance of nearly all subcutaneous fat and thin, parchmentlike changes in glabrous skin. There may be pallor, duskiness, frank cyanosis, or redness and mottling of both hairy and glabrous skin. Return of color following elevation may be slowed. It is highly likely that some of these reflex changes, particularly ones involving increased skeletal muscle tension and vascular spasm, may themselves generate additional sources of pain. The origins of the patient's symptoms thus become multiplied. A feed-forward vicious-circle disorder has become manifest.

Causalgia. Severe intractable pain of long duration may have astonishing symptoms. These were identified as "causalgia" by S. Wier Mitchell in reference to the high incidence of burning sensations. The patient with causalgia may suffer various kinds of boring, gnawing, rasping, bursting, squeezing, vicelike, lancinating, and lightning pains as well as excruciating burning sensations. The painful part may be less painful if wetted with a cold cloth or sponge or submerged in cold water. Even a faint breath of air across a sensitive region may give rise to paroxysms of pain. Some patients are so extraordinarily sensitive that the least vibration or noise worsens their symptoms. This suggests that disturbances in central regulation may have reached at least to bulbar and pontine levels. Sometimes the opposite extremity, uninjured, gives rise to "mirror image" symptoms of a similar but less intense kind. There may also be interference with regulatory reflexes on the side contralateral to the injury. This indicates a central nervous "spill over," affecting circuits crossing to the opposite side.

Patients with causalgia are rare, but their symptoms and the spread of their reflex abnormalities may provide valuable insight into the pathophysiological processes involved in "minor causalgias" ranging all the way from major causalgias to instances of "normal pain." Successful lines of treatment for causalgia may have wide application. Obviously, severe cases of causalgia are difficult to treat. Therapy should be addressed to attempting to restore normal input signals from the affected part, for it is observed that "normal input conditions normal output." Return to normal input will tend to restore the exaggerated reflexes to their usual regulatory levels and thereby decrease the number of sites of origin of painful incoming signals. Return to normal input will also

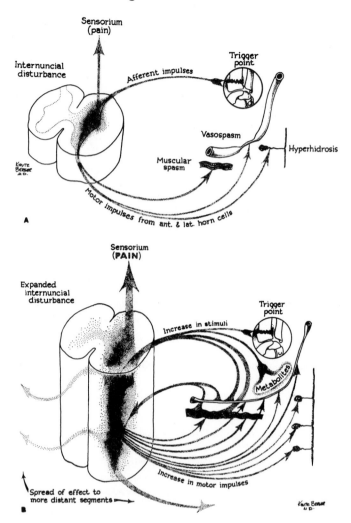

FIG. 27-8. (A) Reflex repercussions experienced promptly following injury (in this case a bone fracture). Normally the pain and reflex disturbances subside spontaneously following reduction of initiating stimuli. (B) The reflex repercussions experienced in relation to chronic pain syndromes, such as causalgia, and to a lesser degree in many clinical disorders including peptic ulcer, angina pectoris, etc. Reflex changes may include increased sudomotor activity, skeletal muscle spasm, vasomotor constriction (to the point of tissue hypoxia) and smooth muscle spasm of other kinds that may initiate secondary sources of pain perception. In some cases these secondary sources, peripheral and central, may be recalcitrant to treatment even after the initiating cause has been corrected. (Bonica, J. J.: The Management of Pain. Philadelphia, Lea & Febiger, 1953)

tend to allow subsidence of whatever higher centers have begun to become self-sustaining sources of abnormal pain perception.

Treatment may need to begin with the patient controlling submersion of his affected extremity into quiet, cold water. After this has become tolerable, the water may be stirred slowly and ultimately used as a whirlpool bath. Then the patient may be able to begin self-palpation and gentle massage underwater. Eventually, warm water massage and dry massage by a therapist may be tolerated. Procaine injections into especially sensitive "trigger points" and perhaps also into the appropriate sympathetic ganglia may accelerate rehabilitation. Incre-

mental progress is indicative of a favorable outcome from even very refractory cases. Trophic and reflex changes ameliorate along with the disappearance of abnormal symptoms.

Pathophysiology of Pain Syndromes. The central pathophysiology in severe prolonged pain is like that in severe transient pain with its accompanying central and reflex disturbances, except that the pain continues and may worsen. The mystery is, why does every case not subside in the way that pain normally does? The difference may lie in the failure of subsidence of firing from peripheral trigger points. It may lie in persistence of an initial central (perhaps segmental) disturbance, which continues to keep reflexes and sensory relays exalted beyond normal levels of activity. There is evidence that the pattern of firing of fast and slow conducting peripheral nerve fibers contributes to a "gating" of late-arriving impulses into channels that will affect higher center interpretations relating to pain. Less is known about how much centrifugal controls affecting sensory transmission (see Regulation of Perceptual Processes, p. 591) may alter receptors and central relays in relation to this pathophysiological process.

Obviously, central pathophysiology must include local disturbances of reflexes involving one or more segments, presumably implicating exaggerated reticular and interneuronal activation. If the pain persists in severe form, the increased central excitatory state seems to spread to achieve increasingly powerful local excitation and upward to involve reticular zones at higher levels. This progression may mount slowly, reaching at least as high as the lateral part of the mesencephalic reticular formation. As symptoms persist, the hyperactive sites seem to become more deeply ingrained, habituated or committed. A kind of slow powder burn establishes itself probably in the reticular formation and ascends progressively to higher levels. After a while the contributions from the original site of injury may be only a part of what the patient is experiencing and interference at that level alone may not provide full relief. But if the process can be interrupted or slowed, it may wind down progressively even though it had been highly agitated for a long time.

Alumina cream, aluminum hydroxide, which is a soothing restorative on the surface of a gastric ulcer, is a powerful chronic irritant when applied to the nervous system. Small amounts of this substance applied to cortex in experimental animals yield epileptic foci after some days or weeks. Application of the same material to the dorsal columns (peripheral nerve fibers in transit in the central nervous system, traditionally not a pain pathway) yields a chronic hyperesthesia together with hyperactive and hyperirritable visceral and somatic reflexes. Perhaps something similar occurs with peripheral nerve injury.

The key to therapy is always to try to reduce or interrupt this feed-forward progression as soon as it is recognized. Early treatment is obviously advantageous. Treatment of any and all sources of pain, primary and secondary, is indicated. It is eminently worthwhile conscientiously to try to detect all of the minute individual sources of disturbance, including sites in skin, deep fasciae, spastic muscles and other places which on even gentle stimulation may worsen the pain. If these are too tender to approach, they can usually be "babied up to" by cooling the approach or infiltrating it with a local anesthetic.

When individual trigger points are adequately infiltrated with local procaine and gently massaged, some of the near and even remote sources of pain may be immediately relieved. Oddly enough, this relief may long outlast the duration of the local anesthetic. It is as if temporary intervention to interrupt the "vicious circle" would permit the entire disturbance to partially or completely subside. It may subside only to return again, but repeated treatments are often increasingly successful in stepwise fashion. Successful treatment achieves a pattern of recovery that is not unlike the normal symptomatic and reflex subsidence in spontaneous recovery from intense transient pain.

The interpretation of pain syndromes may have wide implications for exaggerated regulatory processes in general. So W. K. Livingston wrote in 1943:

Doubtless, in many clinical syndromes there may be a combination of somatic, visceral and psychic irritants, each contributing to the central process, whether they represent primary or secondary factors. Once a vicious circle is established, the process tends to become self-sustaining and it is increasingly difficult to stop. The activities of the regulatory centers within the spinal cord and brain, which normally are so beautifully synchronized, are thrown out of step with one another. If the irritant foci are removed early enough the central process subsides and improvement becomes possible. If the process is permitted to continue for a long time, the habit pathways are cut deeper and it becomes more difficult to re-establish normal function, even after the original irritants are gone.*

The Symptom Complex Known as Epilepsy

Epilepsy is a symptom, not a disease. Epileptic seizures may be caused by a wide variety of disorders: genetic, traumatic, chemical, infectious, neoplastic or degenerative. The diagnosis may have to be based largely on history because seizures may be infrequent enough not to be seen or recorded. Appropriate EEG abnormality, if present, is pathognomonic. Seizures interfere with cerebral regulatory processes and provide an important means for mapping the central representation of somatic and visceral functions; knowledge of the cerebral location of these functions can be of help in determining the site of origin of the seizure activity. For purposes of this discussion, epilepsy can be divided into four categories: focal epilepsy, grand mal seizures, petit mal seizures, and psychomotor epilepsy.

Focal Epilepsy. In focal epilepsy the abnormal synchronous, rhythmically discharging population of neurons remains localized. If the focus is in motor areas of neocortex, there may be involuntary jerkings of limited parts of skeletal musculature on one side of the body such as myoclonic jerking of the right thumb, face, or foot, or deviation of the eyes to the contralateral side, interruption of speech or emission of jargon speech, as examples. If the focus is in sensory areas there may be numbness or tingling of parts of the body, flashes of light or visual hallucinations, buzzing, ringing, music, or sounds resembling speech.

* Livingston, W. K.. Pain Mechanisms. New York, The Macmillan Company, 1943.

The remarkable feature of neocortical local seizures is that they are singularly detached from the person and from his intimate feelings, mood and emotion. They remain "neutral" events, as if imposed from outside the "self." This is in contrast with limbic system seizures which may be focal or more widespread and which are ordinarily accompanied by intimate involvement of the person. Even so, limbic seizures are ordinarily recognized by the patient as abnormal happenings. Focal limbic seizures may be associated with olfactory hallucinations, usually a disagreeable or repulsive odor.

Focal seizures may not remain localized on every occasion; the electrical perturbation may progress or march to involve mirror image structures on the opposite side of the body and engulf the whole forebrain in what is called a grand mal seizure.

Grand Mal Seizures. Seizures that result in generalized EEG and bodily disturbances may originate practically anywhere in the forebrain. There may be some vague prodromal sensations or motor effects, such as twitchings, and these may be noticed some hours in advance of a seizure. Generally, however, there is only a very brief aura of some few seconds' duration, followed by an abrupt loss of consciousness, an explosive outcry, severe tonic breathholding and hypertension, giving way gradually to rhythmic tonic-clonic discharges that rack the whole body and may be associated with profuse salivation and perhaps lip or tongue biting, profuse perspiration and, often, loss of bladder or bowel sphincter control. The tonic-clonic phase becomes slower, perhaps irregular and then ends, abruptly followed by a quick deep sigh and somewhat labored respiration. Breathholding may have been so prolonged that a considerable cyanosis may develop. There is absence of consciousness for the entire event and some retrograde amnesia for a few moments before the outcry. There is usually marked confusion, interference with thinking, bewilderment, fatigue, muscle soreness, and drowsiness and lassitude following the seizure. Seizures coming close upon one another may call for heroic measures to rescue the individual from *status epilepticus.* After grand mal seizures there may be reflex signs characteristic of upper motor neuron dysfunction.

Petit Mal Seizures. This epileptic discharge shows a characteristic spike and wave pattern occurring at intervals of 3 per second on the electroencephalogram. These signs are quite generalized and are thought to originate from abnormal foci in the diffusely projecting (intralaminar) nuclei of the thalamus. There is usually an abrupt loss of consciousness, without warning. In contrast to grand mal seizures, there is an absence of motor manifestations except for flicking of the eyelids or twitching of some facial muscles and staring into space. The individual, although unable to respond to commands or remember events that transpire during the attack, may continue walking, riding a bicycle, or doing some other familiar act. After 2 to 15 or 30 seconds, consciousness is abruptly and fully restored and the patient promptly resumes what he was doing with no more than a short gap or "absence" in his stream of recollected consciousness.

Petit mal is more frequent in childhood and adolescence. Although benign, there may be so many such episodes (from a few up to hundreds per day) that they can interfere with the individual's daily activities. A young patient may have enough cumulative losses of consciousness and disturbances of mental processes that he can readily fall behind in schoolwork. As the patient matures, the petit mal may disappear or give way to grand mal seizures.

The main pathophysiological involvement in petit mal is in the regulation of consciousness; there is an abrupt switching off and on of consciousness that corresponds closely to the onset and disappearance of the spike and wave pattern on the electroencephalogram, as if a curtain had been abruptly dropped down on the stage of consciousness. This is presumably because of an interference in the stream of impulses from cephalic brain stem to forebrain which are necessary to the normal waking state. There is typically no loss of postural tone or control and little or no visceral effects, showing the independence of this neocortical arousal system from basal gangliar, limbic, hypothalamic and lower brain stem mechanisms.

Psychomotor Epilepsy. Seizures originating in the limbic system and associated temporal lobe structures and the medial surface of the hemisphere are conspicuously different from the preceding epilepsies in that they seem to engage the patient's proper self, his intimate personal being. There is more interference or disturbance of consciousness than outright loss, and J. Hughlings Jackson, who first described such a case in a young physician, referred to the disorder as *dreamy states.* The patient may be unresponsive and may fail to remember the episode but he still may be able to perform complicated actions while the seizure is progressing. For example, Jackson's first patient performed a physical examination, correctly diagnosed lobar pneumonia and ordered the patient to bed all the while that he was having a psychomotor seizure. He had no memory for what he had done and, on checking back on his actions, he was relieved to find that he had done everything correctly even though without laying down memory stores for the events.

On other occasions (in that patient and in other patients) there are experienced distortions of consciousness such as: perceiving objects to be farther away than they are, or nearer; strange objects and persons seeming familiar (déjà vu), and familiar objects and persons seeming strange (jamais vu); colors and surfaces seeming especially entrancing and life seeming more than usually tenuous. The resemblance of these psychic shifts to experiences with tranquilizers, hallucinogens, and psychotomimetic drugs is evident. The pathophysiological processes are not yet understood in close relation except for the fact that both the psychomotor epilepsy and the chemical agents seem to have their main effects localized in limbic and temporal lobe structures.

The patient may appear to be practically conscious, for he is usually able to perform acts of dexterity and habitual skill and may engage in a simple, perhaps vague and distracted, conversation. Obvious motor manifestations may be absent or there may be staring, lip licking and smacking, chewing and swallowing, all characteristic responses elicited by stimulation of the amygdala and associated temporal lobe structures. As might be expected, psychomotor seizures may begin with an olfactory aura. They may also begin with shifts in visual size, shape or distance relations, visual memories, auditory reminiscences, and feelings of love, fear, apprehension or anger.

Another main category of seizure activity generated in this region is centered around interferences with visceral regulations. A seizure may begin with epigastric queasiness or frank abdominal pain and be associated with detectable signs of gastrointestinal hypermotility. Cortex implicated is on the island of Reil, continuous with limbic cortex and functionally a proper part of it. There may be episodes of cardiac arrhythmia, transient cardiac arrest, hypertension and marked flushing or pallor of the face or extremities, marked perspiration, respiratory changes, etc.

The limbic circuit, from orbito-medial surface of the frontal lobe looping up over the corpus callosum, around the splenium of the corpus callosum posteriorly and forward again along the medial and inferior surfaces of the temporal lobe, rejoining the orbital surface by way of insular cortex, forms a ring of somatic and visceral representations within which territory psychomotor seizures may remain localized where

they began. Or they may march around, involving extensive regions of the limbic system or the whole of it, but generally staying within the boundaries of the limbic system itself. Limbic structures have the lowest threshold for seizure discharge among all brain structures and abundant connections throughout the system. Psychomotor epileptic events are intimately personal, not detached as in neocortical seizures. Limbic system seizures not only involve parts of the patient's body in abnormal discharge but implicate his innermost self although the patient can usually recognize that he is experiencing an abnormal attack.

Because the limbic system is the seat of emotional experience and expression, seizures in this region are apt to reflect an emotional component, a group of visceral responses composed into a meaningful sector of behavior. The accompanying subjective experience is correspondingly realistic. The patient at the onset of a seizure may become unaccountably enraged with little or no apparent provocation and proceed to mount a vicious personal assault that is well directed and violent. Another patient may become sexually aggressive during limbic seizures, or disrobe in public. Another may exhibit great elation and joy, or grief and despair without cause.

Probable Pathophysiology of Epilepsy. The pathophysiology of epilepsy derives from fundamental processes of excitability of neurons and populations of neurons. Disorder may arise from increased local excitability or from reduced inhibition, probably from both. Nobody understands the means by which normal circuits avoid seizure activity. The potentiality is always there. Anyone can have a seizure elicited by modest electrical or chemical stimulation.

It is known that neurons and other excitable tissues, muscles and glands, deprived of their innervation become hyperexcitable. Normal excitability is restored when these tissues are reinnervated. Since such *sensitization of denervation* may occur in places where there is no known inhibitory process, it may be supposed that hyperexcitability may be the usual outcome of any process that interrupts innervation to some excitable tissue. This may explain epilepsy following cortical injury and the subsequent scarring which may interfere further. Note that the process of sensitization of denervation may have important implications for many different kinds of regulatory disorder.

Following local injury, then, there is likely to be sensitization of denervation. Moreover, the loss of normal input to that region will relieve cells there from the desynchronizing influence of normally invading impulses. There would then be a reduced likelihood of desynchronization in a region of already heightened excitability, in the presence of natural tendencies for nerves to go into synchronous discharge, and in the likely absence of local inhibitory influences.

Once an epileptic focus is generated, its manifestations depend on what functions are represented in that particular locale, whether the discharge will march beyond that immediate region and what circuits ultimately become involved. Focal epilepsy, by definition, tends to remain localized. Petit mal is thought to originate in the thalamic intralaminar nuclei which sustain consciousness in the forebrain. Its abnormal discharge tends to decouple brain stem and forebrain and thereby transiently to interrupt the stream of consciousness. Psychomotor epilepsy begins and may be confined to the low-threshold limbic structures that are so involved in the processes of mood, feeling and emotion. Perceptual distortions, particular memory recall, emotional crises and alterations in the sense of relatedness to other persons and objects are characteristic features which can be accounted for by seizures confined to the limbic system. Focal epilepsy, petit

mal and psychomotor seizures if sufficiently violent can pass into grand mal seizures by presumably invading enough neocortex or by causing generalization of seizure activity by involving the cephalic brain stem reticular formation.

Disturbances of Respiratory and Cardiovascular Regulations

It is well known that the vital centers for cardiovascular and respiratory regulation, without which self-supported life cannot persist, are located in the bulbar and pontine reticular formation. It is there that reflexes essential for regulation of respiration, blood pressure and cardiac rhythm have their "segmental" reflex transactions. Although regulation at the bulbar and pontine levels is essential for survival, higher levels of integration are necessary for existence in a more active than vegetative state, for upright stance and more elaborate exertions.

It appears that the reticular formation exerts a role in these visceral regulations from spinal levels to and including the cephalic brain stem. Spinal levels, for example, exert influences in the correct reciprocal direction for the maintenance of rhythmic respiration and for the support of arterial pressure, but without commanding enough power to sustain respiration. If the cord is transected below the level of phrenic outflow but above the highest sympathetic outflow, rhythmic respiration is maintained by the diaphragm working alone; after a period of spinal shock, blood pressure returns to normal levels and becomes capable of pressor and cardioacceleratory responses following stimulation of the splanchnic nerve. The chronic spinal dog can also compensate by vasoconstriction following blood loss by virtue of responses of spinal centers to hypoxia. If the isolated cord is now destroyed, the pressure again falls to shock levels although there is some recovery due to sensitization of denervation of sympathetic ganglia and recovery of intrinsic tone in arterioles. Regulation, however, is almost totally absent.

Bulbar neurons in the reticular formation contribute importantly to vasoconstrictor activity, sufficiently to maintain adequate arterial pressure. In the bulb, pressor regions are generally cephalad and lateral to depressor influences as elicited by direct stimulation. Pressor tone is not dependent on afferent influx. Depressor influences are dependent on afferent influences and are affected by reduction of vasoconstrictor tone and slowing of the heart. Sympathetic cholinergic vasodilator fibers to skeletal muscles are controlled by pathways that bypass the bulbar cardiovascular centers.

Chronic high mesencephalic transections in cats, in contrast to transections at lower mesencephalon and pontine or bulbar levels, show highly integrated rage responses, not as complete as those obtained in hypothalamic and decorticate animals, but with sufficient cardiac acceleration and vasoconstriction to support the general response. At mesencephalic levels there is a relay of sympathetic vasodilator influence in the basal parts of the superior colliculi which is normally activated from hypothalamic or limbic levels. A parallel system invokes vasoconstriction in vessels of the skin and abdominal viscera, a projection that also bypasses the bulbar reticular formation.

Direct hypothalamic stimulation in unanesthetized, unrestrained dogs yields responses of the left ventricle similar to responses exhibited in the same animals when exercising. Lesions in the mesencephalic central gray eliminated normal ventricular responses to exercise and to eating. Thermoregulatory mechanisms closely related to cardiovascular regulation include hypothalamic regions for heat loss, in the anterior hypothalamus and preoptic area, and for heat production and conservation in more caudal parts of the hypothalamus. The eliciting of heat loss by direct stimula-

tion results in cutaneous vasodilation, visceral vasoconstriction, sweating, postural exposure of ventral areas, and panting. Heat conservation activation is expressed by increased skeletal muscle tension, shivering, postural guarding of ventral surfaces, piloerection, and release of calorigenic hormones by hypothalamic influence on the hypophysis.

It should be emphasized that at this diencephalic level there is already an integration of somatic and visceral mechanisms for a more elaborate goal-directed response (exercise, heat regulation) instead of simply an influence on cardiovascular regulation. Similar hypothalamic controls are exerted on respiratory functions. Control of respiration for gesture vocalization is introduced at mesencephalic levels, but for speech and sustained vocalization as in singing, there is need for control from neocortex.

Cortical regions influencing respiratory and cardiovascular responses include limbic cortex (quite widely effective and powerful in its influence) and neocortex (presumably limited to sensorimotor areas in the middle of the hemisphere). Limbic influences include the orbital surface of the frontal lobes which embody an afferent field of representation of the vagus and the outflow of which transits the median forebrain bundle and may include relays in the hypothalamus. More lateral temporal lobe elements of the limbic system which influence respiratory and cardiovascular effects bypass the hypothalamus altogether, passing directly by way of the mesencephalic reticular formation.

Although carotid sinus reflexes are mediated at bulbar levels, they can be modulated, inhibited or enhanced by stimulation of the amygdala or hippocampus. Mesencephalic influences of a like kind are obliterated by thalamic or high mesencephalic transections, indicating that the mesencephalic effects are probably exerted on fibers of passage from limbic or neocortical centers or were dependent on an upward-bound circuit going to those regions. Neocortical cardiovascular responses include vasoconstriction and skeletal muscle sympathetic cholinergic vasodilation.

Knowledge of higher levels of control of visceral activities and visceral reflexes is still in its infancy even though important pioneering work was initiated by Schiff and others before the analysis of cortical representation of skeletal motor regulation had begun. It is still only a surmise that these higher level controls are organized to accommodate "specialized psychomotor engagements" of the autonomic and somatic nervous systems. One can speculate that neocortical motor cortex may enjoin autonomic and somatic activities to enable the pitcher to hurl the ball and the batter to swing at it with confident synchrony of respiratory, cardiovascular, sudomotor, postural and voluntary musculature. One of the most urgent aspects still to be investigated further is evidence that cardiac irregularities so often associated with anesthetics and manipulation of the heart and great vessels may be fostered by anesthetic or physiological sensitization of high level autonomic controls.

Regulation of Perceptual Processes

It is evident that our individual past experiences shape our present perceptions. There are centrifugal nerve fibers which travel in a reverse course parallel to sensory input pathways and modify incoming sensory signals all the way from sense receptors to cortex. This system is subject to conditioning and learning. It can drastically modify the sensory signals upon which we depend for perception of the world around us. It can vastly reduce or augment, narrow or broaden, the central display of sensory signals in our brains in ways that are reflected in correspondingly vastly altered behavior. As an example, yogi during meditation are reported by Indian workers to exhibit little or no auditory evoked potentials recorded at the scalp. Zen Buddhists during meditation are reported by Japanese workers to have steady high level auditory evoked potentials recorded at the scalp. When not in meditation, each group displays normal auditory evoked potentials, that is, they begin with relatively high amplitude responses to novel stimuli and with repetition of the stimulus they show habituation to a lower, fluctuating level. The differences are notable only during meditation. Since the objective of yogi meditation is to eliminate intrusion of the outside sensate world and the objective of Zen meditation is to openly accept sensations but remain passive to them, it would seem from electroencephalographic evidence that in considerable measure the two styles of meditators are succeeding physiologically in their subjective aims.

This, and much other evidence from experimental studies in animals and clinical studies in man, suggests that there is considerable conditionable control over the distribution and amplitude of sensory signals. Subjective testimony and outward behavior confirm that these changes are associated with real differences in perceptual experience.

In animals, it has been shown that all sense modalities possess centrifugal controls, that these modulate distance, and cutaneous, visceral and body wall receptors (roughly 10% of cutaneous receptors appear to have efferent controls, and other end-organs change excitability in response to circulating catecholamines). In the auditory pathway, for example, there are centrifugal projections extending downward from auditory cortex to medial geniculate, thence to inferior colliculus and continuing on to serve auditory relay nuclei downward to and including the cochlear nuclei and finally the organ of Corti itself. The last centrifugal projections end on individual hair cells. There is a parallel diffusely projecting descending circuit that travels via reticular formation and also exerts influences on ascending auditory signals. It has long been known that the central nervous system controls gates to afferent input such as the middle ear muscles, the pupil, efferent control of muscle spindle afferents, etc. The least thoroughly recognized centrifugal sensory controls remaining are those in the visual system in mammals. In man some 10 percent of the optic nerve fibers are efferent to the retina, but their function in man is unknown. All other systems, olfactory, vestibular, somesthetic, are confirmed and known at least in a preliminary way to be effective in the shaping of incoming sensory data.

In view of the powerful conditionability of this centrifugal system and its shift in influence in accordance with expectations and purposes, it is likely that perceptions may be strongly modulated according to an individual's past experiences and his direction of attention and intentions. There exists the likelihood that cultural differences in language, custom and human values may be deeply embedded in these mechanisms of central control of perceptual processes. If so, we each wear an idiosyncratic physiological lens system that is invisible to us because it affects the very circuits on which we depend for knowledge of ourselves and the world around us; we have no other avenue for knowledge. Our idiosyncratic lens system may be exposed to verification only when we trade observations with others whose lens systems, invisible to them, differ from our own perforce, by reason of their having experienced different backgrounds, different expectations and different purposes.

Until now, science has provided only two explanations for discordant testimony between persons witnessing the same event: either one or both of the parties is lying, or one or both of the parties is psychologically incapable of maintaining testimony in correspondence with percep-

tions. The black robes of law, the green table of diplomacy and the bloody fields of battle are populated with individuals who believe righteously that the opposition is either lying, or psychologically undisciplined. A third possibility, revealed by studies carried out on central regulation of sensory receptors and sensory transmission systems, is that one and the same event may be experienced by two or more people entirely differently. Since our past experiences, expectations and purposes cannot be equivalent, present perceptions are not likely to be equivalent except in territories of generous overlap of experience. Thus, the same event, as detected by sense data reaching consciousness in two different people may be experienced as two distinctly different events. This needs to be taken into consideration by physicians.

Naturally, differences in perceptual processing will be greatest when the gulf of differences in past experiences is greatest. Hence, cross-cultural communications may be fundamentally more difficult than we have hitherto supposed. It is easy to recognize that groups of individuals similarly reared may take uniform (mutually reinforcing) exception to the testimony of other groups differently reared. Because of central control of perception, they will tend to confirm the rectitude of their own group's way of perceiving events and to join in condemning any contradictory testimony. At root, many of man's most urgent and difficult problems may be foundering on this biological system of perceptual regulation.

It is likely that when men come to grips with this evidence, they will have a more reverberant ado about the changes implied in their relations with one another than men had in the 16th century when they came to recognize, contrary to common sense, that the earth traveled around the sun, or in the 19th century, when they learned that they were not a divine, separate creation, but related and interdependent with other members of the plant and animal kingdom.

At present, one can only conjecture about the possible role of such mechanisms in pain, apprehension, and conflict, and the importance for the physician to know that his patients may need special consideration for what they suffer and for how they comprehend and respond to his ministrations.

Storage of Memories

Perhaps the most important regulatory mechanism of higher nervous systems relates to the capacity for acquiring new skills in perception, judgment and behavior. How this is accomplished is being studied very intensively. From what has been discussed above, it is evident that something is already known about mechanisms for wakefulness, consciousness and direction of attention. It is also apparent that some experiences, engaging certain regions of the brain in activity, generate in us satisfaction, other experiences dissatisfaction. The lateral aspect of upper mesencephalic levels have intimately to do with pain, fear and avoidance behavior. The entire limbic system has to do with various aspects and various intensities of satisfaction. Neocortical fields, much of the thalamus, most of the basal ganglia and the cerebellum are all different, neutral in respect to affect.

It is conjectured that events with strong

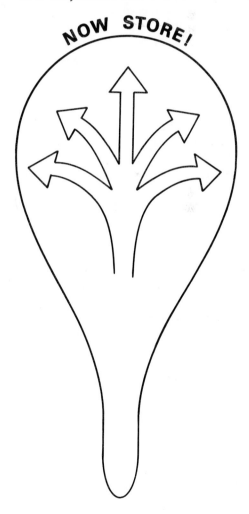

Fig. 27-9. This is an entirely hypothetical representation of how the nervous system may initiate storage of information. Recently activated circuits would increase the likelihood of repeating similar fixing patterns as a result of a generalized *now store* order generated by limbic system excitation of the cephalic brain stem reticular activating system. The latter has diffuse projections to the entire hemisphere. The *now store* order would store all experiences following satisfaction or dissatisfaction. Those regularly recurring experiences, or especially vivid ones, would be especially stably imprinted.

affective tone arouse a discharge of a general sort to effect memory storage among all recently active circuits. In this way recurrent events leading regularly to satisfaction or dissatisfaction will be stored in the relevant circuits, and events of an indifferent sort will not be stored. This would provide for selective economization in storage and retrieval of useful prognostications for the guidance of forthcoming behavior. The nature of this "now store" command is being investigated. Perhaps it is exerted through the outflow of the limbic system into the frontal-limbic midbrain area from whence it could be delivered via the reticular activating system throughout the entire forebrain.

The virtue of a generalized "now store" order, as against previous theories of learning which invoke specific pathway selection for strengthening of connections during the learning process, is that the generalized order will provide that *all recently active circuits* will be strengthened. Those that were only adventitiously active during a rewarding or punishing event will tend to lose strength of connections because they are not regularly recurrent in relation to biologically meaningful events. Any event that is regularly recurrent in relation to biologically meaningful events will be stored.

Much remains to be done. There has been so far only a little progress toward understanding regulatory processes in health and disease. There has been only a modest gain toward understanding the nature of human nature. The physician is in a key position to contribute to man's well being on an individual basis, and collectively, insofar as knowledge goes, and by analyzing and sharing his clinical experience he can contribute to man's further understanding of himself and of the many opportunities he has for constructive adaptation among all men throughout this magnificent spaceship earth.

Of all the wonders, none is more wonderful than man who has learned the art of speech and wind-swift thought—and of living in neighborliness.

Sophocles

ANNOTATED REFERENCES

Adey, W. R., Segundo, J. P., and Livingston, R. B.: Corticofugal influences on intrinsic brainstem conduction in cat and monkey. J. Neurophysiol., 20:1, 1957. (So much is going on in the brain stem reticular formation, in terms of intrinsic activity and traffic up and down, that it seemed worthwhile to investigate whether the cortex had any influence. It does, but only selected cortical areas, and these contribute apparently unique corticofugal contributions.)

Ames, A., Jr.: The Nature of Our Perceptions, Prehensions and Behavior. Princeton, N. J., Princeton University Press, 1955. (Ames was a lawyer who became interested in painting and, later, perception. He made such fundamental demonstrations in relation to perception that John Dewey declared them to be the most important contributions to philosophy and human understanding in the 20th century. Ames showed that perception always involves contributions by the perceiver based on his past experiences, expectations and purposes. The problem now is to discover how neurophysiological processes are involved in bringing past experiences, expectations and purposes to bear on the perceptual act.)

Bishop, G. H.: Natural history of the nerve impulse. Physiol. Rev., 36:376, 1956. (A masterful description of neuron physiology in an evolutionary context)

Bonica, J. J.: The Management of Pain. Philadelphia, Lea & Febiger, 1953. (A thoughtful compendium on pain and its treatment by local anesthesia, nerve block, spinal block, etc. The first few chapters are especially pertinent to the problem of the regulation of pain and its pathophysiology.)

Bullock, T. H., and Horridge, G. A.: Structure and Function in the Nervous Systems of Invertebrates. Vols. 1 and 2. San Francisco, W. H. Freeman, 1965. (The most valuable treatise on invertebrate nervous systems, beautifully written, illustrated and documented. Since invertebrates conspicuously outnumber and out-vary vertebrates, there are a great many wonderful examples for further exploration of how nervous regulation is accomplished.)

Dell, P., and Olson, R.: Projections thalamiques, corticales et cérébelleuses des afférences viscérales vagales. Compt. Rend. Soc. Biol., *145*:1084, 1951.

Dell, P. and Olson, R.: Projections "secondaires" mésencéphaliques, diencéphaliques, et amygdaliennes des afférences viscérales vagales. Compt. Rend. Soc. Biol., *145*:1089, 1951. (Dell and Olson are among the best authorities on central projections of visceral afferents. They distinguish between the classical [lateral] ascending pathways which they call "primary" and the extraclassical [medial] ascending pathways which they call "secondary." The latter have longer latency, slower conduction and are more vulnerable to anesthetics, although of course they are phylogenetically and embryologically first to provide higher nervous system regulation and may be just as primary in terms of functional contributions.)

Folkow, B.: Nervous control of blood vessels. Physiol. Rev., *35*:629, 1955. (Hardly anyone knows as much as Folkow about circulatory integration, peripheral and central. This is a substantial review by a great authority.)

Galambos, R.: A glia-neural theory of brain function. Proc. Nat. Acad. Sci., *47*:129, 1961. (Galambos caught everyone by surprise with this article which pointed out that the glia, for all anyone knew, might have just as significant a role as any other cells in the central nervous system in controlling excitability and switching connections. Assumptions about what neurons do in contrast to glia were called into question. [See Kuffler and Nichols, below, for a thorough physiological review of glia functions.])

Galambos, R.: Neural mechanisms of audition. Physiol. Rev., *34*:497, 1954. (This is the first review to emphasize the descending centrifugal control systems as affecting auditory processes.)

Galambos, R., Myers, R. E., and Sheatz, G. C.: Extralemniscal activation of auditory cortex in cats. Am. J. Physiol., *200*:23, 1961. (The authors transected the classical auditory path but still got "normal appearing" evoked cortical potentials of no or only slightly shorter latency than in intact animals. These evoked potentials must depend on medial, reticular conduction pathways, and the paper demonstrates that this system is vulnerable to barbiturates but not to chloral hydrate.)

Herrick, C. J.: George Ellett Coghill, Naturalist and Philosopher. Chicago, University of Chicago Press, 1949. (Don't fail to read this one. Coghill [and Herrick] made great contributions to American Neurology. The through-thinking of processes and implications of evolution, embryology and development, perception and consciousness are exemplary. Modern electrophysiological techniques need to be combined with these anatomical and behavioral studies.)

Hockman, C. H., Livingston, K. E., and Talesnik, J.: Cerebellar modulation of reflex vagal bradycardia. Brain Res., *23*:101, 1970. (Electrical stimulation of the fastigial lobe of the cerebellum will inhibit carotid sinus nerve reflexes, but this modulation apparently depends on long circuiting up through the limbic system.)

Hockman, C. H., Talesnik, J., and Livingston, K. E.: Central nervous system modulation of baroceptor reflexes. Am. J. Physiol., *217*:1681, 1969. (Direct stimulation of mesencephalic central gray and limbic structures [amygdala and hippocampus] can inhibit or potentiate baroceptor reflexes. The mesencephalic effects, however, are not obtainable after high decerebration, hence they must be dependent on limbic integrity.)

Hubel, D. H., and Wiesel, T. N.: Receptive fields and functional architecture of monkey striate cortex. J. Physiol., *195*:215, 1968. (The beautiful sequence of experiments on visual physiology in cats by these two authors has now been extended to the monkey. There are some physiologically significant differences which should contribute to improved perceptual capabilities in the monkey, ones that are presumably closer to our own.)

Jackson, J. H.: On a particular variety of epilepsy ("intellectual aura"), one case with symptoms of organic brain disease. Brain, *11*:179, 1888. (This is the earliest description, and probably the best, of temporal lobe seizure, now called psychomotor epilepsy. The patient was a young physician who was a good observer. His experiences interpreted by the great master, Jackson, are historical gems. The patient turned out to have a cyst in the temporal lobe just beneath the olfactory cortex [uncus].)

Jasper, H. H., Ward, A. A. Jr., and Pope, A. (eds.): Symposium on basic mechanisms of the epilepsies. Boston, Little, Brown, and Co., 1969. (Some of the world's best authorities share insights into the mechanisms of epileptic seizures.)

Klüver, H., and Bucy, P. C.: An analysis of certain effects of bilateral temporal lobectomy in the rhesus monkey, with special reference to "psychic blindness." J. Physiol., 5:33, 1938. (This is another classical paper. Amputation of the anterior regions of the temporal lobes bilaterally, in monkeys, revealed the "temporal lobe syndrome" of increased tameness [lack of usual fear responses], orality, sexuality and changes in cognitive capacities later shown to be devastating defects in new memory storage.)

Kuffler, S. W., and Nicholls, J. G.: The physiology of neuroglial cells. Ergebn. Physiol., 57:1, 1966. (Kuffler and Nicholls present the most comprehensive and authoritative analysis of the physiology of glia available to date. This has stimulated great interest in the possible roles of these cells in central processes of nervous regulation.)

Livingston, R. B.: Central control of receptors and sensory transmission systems. In: Handbook of Physiology. Section I, Neurophysiology. Vol. 1. Chapter 31. Washington, D.C., American Physiological Society, 1959. (A review of central control of perceptual processes, a field that was just then emerging. A more up-to-date review is needed but that will have to be confined to a modality at a time or it will require a big book because the information has greatly expanded and the findings are both complicated and fascinating.)

————: How man looks at his own brain. In: Psychology: a Study of a Science. vol. 4. Biologically Oriented Fields. pp. 51-99. New York, McGraw-Hill, 1962. (This attempts to bridge the gap between neurophysiology and psychology. It's a pity that teaching and professional careers in these two fields are still so disparate. The momentum of 2,400 years of thinking that mind and brain are separate entities is still carrying us along.)

Livingston, W. K.: Pain Mechanisms. New York, The Macmillan Company, 1943. (This is a small and delightful book to read. Its clinical interpretations are challenging. Neurophysiologists are still finding mechanisms which help account for the clinical varieties of pain syndromes, and there is obviously a great deal of headway yet to be made.)

Livingston, W. K., Davis, E. W., and Livingston, K. E.: "Delayed recovery" in peripheral nerve lesions caused by high velocity projectile wounding. J. Neurosurg., 2:170, 1945. (This study of high velocity missile wounds demonstrates that peripheral nerve injury alone is sufficient to disorganize sensory and motor projection systems at least transiently. Physiological discombobulation, if you will. Can this be sufficient to cause longer lasting difficulties?)

Livingston, W. K., Haugen, F. P., and Brookhart, J. M.: The vertical organization of function in the central nervous system. Neurology, 4:485, 1954. (This is a think piece designed to call attention to the vertical systems that organize and regulate nervous functions. When this idea of vertical integration is coupled with the more traditional idea of a hierarchy or horizontal levels of integration, the nervous system begins to hang together properly.)

MacLean, P. D.: Contrasting functions of limbic and neocortical systems of the brain and their relevance to psychophysiological aspects of medicine. Am. J. Med., 25:611, 1958. (MacLean is another "great" in the tradition of Herrick, Cogdill and his immediate mentors, Papez, Cobb and Fulton. Here he compares neocortical and limbic systems and draws inferences of value in the practice of medicine. MacLean finds separate regions of the limbic system for self-preservation [centered in the amygdala] and preservation of the species [centered in the septum], thus giving a nervous system account for two very fundamental biological processes.)

————: The limbic system ("visceral brain") and emotional behavior. Arch. Neurol. Psychiat., 73:130, 1955. (The case for limbic representation of emotional experiences and expressions is here well established.)

————: Psychosomatic disease and the "visceral brain." Psychosomat. Med., 11:338, 1949. (This was MacLean's first essay on improving the Papez hypothesis for an anatomical substrate for emotions [see Papez below] and adding physiological and psychoanalytic interpretations to that. It is remarkable how prescient MacLean was at this time. Although he has been fulfilling the objectives outlined since 1949, others, too, have been finding that paper a germinal source and forecast as well.)

———: The triune braine, emotion, and scientific bias. *In:* The Neurosciences Second Study Program. pp. 336-349. New York, Rockefeller University Press, 1970. (He keeps getting better and better every time, and the information is improving, too. This includes anatomical evidence for quite direct somatic as well as visceral representation in the limbic system, and a panoramic view of visceral integration.)

Magoun, H. W.: Caudal and cephalic functions of the brain stem reticular formation. Physiol. Rev., *30:*459, 1950. (A master who opened up revolutionary opportunities between neurophysiology and the behavioral sciences. Here Magoun suggests that there might be a central regulation of sensory input as a parallel to mechanisms for motor control, about which he and his associates have made so many contributions. Soon, in his laboratories and elsewhere, papers began to appear which demonstrated in a variety of modalities central control of sensory transmission.)

———: The Waking Brain. Springfield, Ill., Charles C Thomas, 1958. (On sabbatical leave at the National Institute of Neurological Diseases and [then] Blindness, Magoun wrote this slim but pithy summary of brain mechanisms underlying behavior. Compare it with Sherrington's classical *Integrative Actions of the Nervous System* to see how the nervous system is coming together into a whole that includes mechanisms for higher mental processes.)

Mitchell, S. W., Morehouse, G. E., and Keen, W. W.: Gunshot Wounds and Other Injuries of Nerves. Philadelphia, J. B. Lippincott, 1864. (During and following the Civil War, Mitchell, Morehouse and Keen observed many soldiers with peripheral nerve injuries. Some of their patients suffered bizarre pain symptoms, now classical examples of causalgia, a name contributed by Mitchell.)

Mountcastle, V. B. (ed.): Medical Physiology. vols. 1 and 2. St. Louis, C. V. Mosby Company, 1968. (This is a most worthy text with superb treatment of the nervous system. Mountcastle, the senior editor, has done extremely fine work on sensory systems, but the whole two volumes shine.)

Nauta, W. J. H.: Hippocampal projections and related neural pathways to the midbrain in the cat. Brain, *81:*319, 1958. (Nauta, a wonderfully articulate and gracious authority, rescued us from being unable to visualize and trace finely myelinated fibers which make up about 95 percent of the connections in the nervous system. Nauta and others working with his staining techniques and later improved versions of them have been busily tracing the brain's hook-up systems in a number of mammals, and more broadly, in reptiles, amphibians and birds. The cumulative work is monumental. Here Nauta himself traces limbic projections to the midbrain, what he now calls the frontal-limbic midbrain area.)

———: Some brain structures and functions related to memory. Neurosciences Res. Prog. Bull., *2(5):*1, 1964. (Same Nauta editing a Neurosciences Research Program Work Session on brain mechanisms involved in memory. In this monograph appears the first theorizing that memory may be stored in recently active circuits as a result of a generalized "Now Print!" order.)

Nauta, W. J. H., and Kuypers, H. G. J. M.: Some ascending pathways in the brain stem reticular formation. *In:* Reticular Formation of the Brain. pp. 3-30. Boston, Little, Brown, & Company, 1958. (Nauta and Kuypers illustrate and interpret systems of collateral engagement of reticular neurons and the associated ascending [mainly medial] pathways.)

Olds, J.: Hypothalamic substrates of reward. Physiol. Rev., *42:*554, 1962. (Before Olds and Milner [1954] showed that there are centers for positive reinforcement in the brain, hypothalamic and limbic, it was argued that motivation could be accounted for solely on the basis of drive reduction, relief from disagreeable sensations and experiences. That can still be argued, but much less persuasively, because animals will approach and cross an electrified grid to reach a means for central stimulation; they will learn a maze with no other "reward" than central stimulation, etc.)

Papez, J. W.: A proposed mechanism of emotion. Arch. Neurol. Psychiat., *38:*725, 1937. (This is the paper that launched a thousand experiments and anticipated a working junction between neurophysiologists and psychologists that was ultimately brought about by MacLean and Magoun and others. The argument is primarily anatomical and very shrewd.)

Penfield, W. G., and Jasper, H. H.: Epilepsy and the functional anatomy of the human brain. Boston, Little, Brown, & Company, 1954. (Penfield and Jasper, neurosurgeon and neurophysiologist, provide here an authoritative resource for those interested in the epilepsies and what can be learned from them concerning brain organization and regulation.)

Scharrer, B.: General principles of neuroendocrine communication. pp. 519-538. *In:* The Neurosciences, Second Study Program. New York, The Rockefeller University Press, 1970. (Berta Scharrer, remaining member of a great husband and. wife team of Scharrers, here summarizes the field of neuroendocrinology with special emphasis on emergent riddles and opportunities for further incisive investigation. Although the mechanisms already revealed are complicated beyond belief, it makes good integrative sense through her interpretations.)

Scheibel, M. E., and Scheibel, A. B.: Anatomical basis of attention mechanisms in vertebrate brains. *In:* The Neurosciences. pp. 577-601. New York, Rockefeller University Press, 1967. (A wife and husband team [in the order of authorship presented here] who grace our generation in the tradition of the great Santiago Ramón y Cajal. They are systematically tracing out single fiber connections using silver impregnation techniques which complement those of Nauta to depict populations and bundles of fibers.)

Shimamura, M., and Akert, K.: Peripheral nervous relations of propriospinal and spinobulbo-spinal reflex sytems. Jap. J. Physiol., *15*:638, 1965. (Shimamura pioneered in defining two types of long interlimb reflex systems, propriospinal and bulbar relay, which operate along the axis of the brain stem as well, linking head to body and body to head as they should be linked by regulatory systems. Here he and the Swiss anatomist, Akert, contribute insight into the functional connections between peripheral nerves and these longitudinal reflex systems.)

Whorf, Benjamin Lee: Language, Thought and Reality. Selected writings edited by John B. Carroll. New York, John Wiley and Sons, 1956. (Whorf was an amateur anthropologist interested in language functions and particularly in the conceptual binding properties of different languages. Whorf has been widely criticized by contemporary anthropologists, who do not spare each other either, but his main thesis that there are different logics of thinking which depend on upbringing and in which language forms an important part as fulcrum or as a manifestation of the distinctions, seems to withstand at least my [amateur] reading of the criticisms.)

28

Autonomic Control of Cardiovascular Function

H. Page Mauck, Jr., M.D.

INTRODUCTION

The complex interrelationships involved in the integrated neural regulation of the circulatory system offer a fascinating and important as well as difficult area for study. The central nervous system is in a continuous state of activity as it exerts control over multiple organ systems, thus attesting to its important regulatory influence, but also presenting obstacles to its quantitative evaluation. Difficulties encountered in recording the intensity of neural discharge have precluded a precise description of neural afferent and efferent impulse traffic and synaptic routing. The complicating effects of general anesthesia and administration of artificial stimuli, often of unphysiological intensity, have further complicated the problem of interpreting experimental work. Thus, only a qualitative assessment of cardiovascular effects is feasible at present. With these limitations in mind, integrated neural activity pertaining to cardiovascular function will be discussed with respect to disease.

INTEGRATED NEURAL CONTROL

The cardiovascular system utilizes many control mechanisms in adjusting to the highly variable demands made upon it. Control may be at the cellular level, as with the intrinsic contractile activity of smooth muscle, at the tissue level, with its local neural network and its locally produced humoral agents, or at various levels of the nervous system.

The entire complex of controls appears to be arranged in a hierarchy, each level being capable of definitive activity in response to information received locally or from below and responsive also to information from higher integrative levels, including those in the forebrain that interpret human relationships and life experiences. The central interactions, therefore, involve centripetal as well as centrifugal effects. Thus, afferent neural discharges to both the vasomotor and the cardiac acceleratory regions of the medulla oblongata may strikingly alter neural activity in the more rostral regions of the brain, whereas changes in neural activity within the cerebral cortex, the hypothalamus and the basal ganglia produce significant variation in the sensitivity of reflex mechanisms.

According to traditional concepts, peripheral reflex control is held to be mediated through diffuse and uniform efferent neural discharge to all vascular beds; however, newer evidence clearly demonstrates the presence of selectivity of control. Thus, although vascular responses to afferent information may, under some circumstances, be generalized, the patterns of reaction may also be discrete and adaptive to specific demands. The neural pathways that regulate vasomotor behavior vary from one vascular bed to another. Presently available evidence indicates, for example, that central neural influences have little to do with the behavior of cerebral vessels, whereas skin blood flow, in contrast, is extremely sensitive to neural changes. Muscle blood flow, also highly responsive to neural regulation, is controlled by another central mechanism, different from the one that regulates heat loss and, hence, cutaneous circulation.

Locally produced vasoactive substances such as histamine, bradykinin and serotonin may affect vascular smooth muscle directly. Depending upon circumstances, their activity may reinforce, override, or be overridden by neural effects. This is an extremely complex set of interrelationships that appears to be elegantly adaptive.

Centrally mediated vasoconstrictor effects are important in determining regional and organ blood flow. Thus, changes in the resistance ratio between arterioles and venules constitute a major determinant of capillary pressure and filtration exchange and changes in arteriolar resistance are responsible for major alterations in arterial inflow. The thick arteriolar smooth muscle wall possesses a high degree of autonomous myogenic activity, which operates independently of neural regulation and renders it capable of exhibiting a significant degree of resting tone. The larger veins (except, possibly, those located deep in the skeletal muscle of the extremities) possess little local myogenic activity but are well-endowed with sympathetic nerve endings. They function primarily as capacitance vessels. The thin venous wall and large lumen permit only small changes in resistance to flow but exert significant effect on total blood capacity and, therefore, venous return to the heart and cardiac output. Shunt vessels, located in the skin and concerned with temperature regulation, likewise demonstrate essentially no local myogenic tone but constrict in response to sympathetic stimuli appropriate to their function. Under usual circumstances, the degree of arteriolar vasoconstrictor activity is determined by the tissue needs of that

particular organ and through the vasodilator effects produced by local metabolites, the elaboration of which may or may not result from neuronal discharges. It becomes apparent, however, that despite the importance of locally mediated vascular tone, arterial pressure would decline precipitously and tissue perfusion to vital organs would be severely compromised if these local needs were permitted to override general control of total vascular resistance. The centrally mediated vasoconstrictor nerves, therefore, permit a level of general vasoconstriction which assures that arterial pressure is maintained at adequate levels. The overall circulatory restrictions imposed by the central nervous system, though differing in various tissues, permit redistribution of immediately required nutritional blood supply to any one organ from other organs.

Effects of Exercise

In the resting state the central nervous system is called upon to maintain arterial pressure and regulate the distribution of tissue blood flow; moreover, it plays an equally important role in circulatory adjustments to sudden changes in tissue needs by virtue of its rapidity of response. For example, during strenuous exercise, muscle blood flow in an extremity may increase from resting levels of 3 ml. per 100 grams of muscle per minute to 40 to 50 ml. per 100 grams of muscle per minute; total muscle blood flow may reach 15 liters per minute, and cardiac output 20 liters per minute. Under these circumstances, the neural cardiovascular mechanisms must play a major role in both maintaining cardiac output and properly adjusting the distribution of flow. Older concepts held that neural circulatory responses evoked with the onset of muscular effort are initiated through local mechanisms resulting from vasodilatation induced by tissue metabolites. Such a mechanism produces an initial small decrease in arterial pressure while exercising muscles squeeze blood out of the capacitance vessels to increase venous return, making it possible to increase the force of myocardial contraction through the Starling mechanism. Actually, in a healthy person who is exercising, the need for increased cardiac output is anticipated by cortical mechanisms, so that rate and stroke volume may be enhanced through sympathetic discharges without any preliminary recourse to the Starling effect of stretching the heart muscle fibers. Supramedullary regions, within both cortical and subcortical areas of the brain, are activated to increase muscle blood flow, arterial pressure and heart rate, not only concomitant with the onset of motor activity at the beginning of exercise but also in anticipation of muscle activity. Such cortically integrated responses give way to simple reflex adjustments as exercise proceeds, so that, in sustained exercise, tissue perfusion is dictated by local metabolic needs. Supramedullary control is particularly important, however, to anticipate requirements in emergency situations and competitive athletic events that require intense effort for brief periods.

Emotional Effects

Many clinical and experimental observations attest to the profound effects of human emotions upon the cardiovascular system. Changes in heart rate and arterial pressure, cutaneous blanching and flushing, and increased sweating of the palms and soles are cardinal manifestations of the human response to anger, rage, anxiety and fear.

Studies in anesthetized decorticate cats have clearly demonstrated a combination of somatic and autonomic effects, including pupillary dilatation, baring of the teeth, spitting, snarling, arching of the back, and changes in heart rate and arterial pressure, following mild sensory stimulation. The response, originally termed *sham rage* by Bard, is clearly a well-integrated pattern of behavior, characterized by activation of both the somatic and autonomic nervous system. More recently, a neural basis for a well organized defense reaction has been documented through the demonstration of a sympathetic vasodilator pathway in skeletal muscle of mammals, which has its origin in the cortex, with relays in the hypothalamus and without synaptic connection to the bulbar vasomotor regions.

Stimulation of certain hypothalamic areas produces active vasodilatation of skeletal muscle, with renal, splanchnic and cutaneous vasoconstriction, minimal change in arterial pressure, and increase in heart rate. Recent studies in unanesthetized cats observed during actual fighting have shown similar cardiovascular responses. The neural mechanism responsible for skeletal muscle vasodilatation is unclear, although cholinergic fibers, beta-adrenergic receptors, and release of histamine by sympathetic dilating fibers have been considered. In studies on human subjects, endeavoring to solve difficult mathematical problems, a cardiovascular reaction has been observed, consisting of increases in arterial pressure, heart rate and cardiac output, and decreases in renal and cutaneous blood flow. In addition, there was increased sweating of the hands as well as enhanced blood flow in the hand and forearm. Such studies demonstrate that significant circulatory changes may occur as part of a well organized response of the organism to threats from the environment.

In contrast to such hyperdynamic responses, certain circulatory alterations of an opposing type, namely bradycardia and lowered arterial pressure, are familiar during sudden fright and in emotional fainting. Similar cardiovascular behavior has been induced in animals by stimulation of the limbic system or visceral brain in specific areas of the cingulate gyrus.

Thus, there is a well-established neurophysiological basis for many of the circulatory changes associated with environmental stress. Whether these responses are activated frequently to levels of intensity that produce actual threats to life, as in extreme bradycardia or profound hypotension, can be satisfactorily resolved only by more careful clinical documentation.

HIGHER CENTRAL NEURAL CONTROL

The intermediolateral cells of the spinal cord are the initial neurons of a preganglionic double neuronal autonomic "final common pathway" to visceral structures. These cells are responsive not only to peripheral sensory input, but also to modulation from higher neural stations located in cortical and subcortical regions. Rather than suddenly being called into play, they exercise their regulatory function in a dynamic way by being in a state of virtually continuous activity.

Although the influence of higher neural centers upon spinal neurons as it pertains to cardiovascular function has long been recognized, identification of the specific neural centers and descending pathways involved has evolved slowly, and only recently has experimental evidence established the regulatory role of discrete neural regions and fiber tracts.

Medullary Centers

The medulla oblongata is a major integrative area in the hierarchical arrangement of neurovascular controls. Pressor responses are obtained from the stimulation of interconnected neuronal groups which occupy an extensive region of the lateral reticular formation in the upper dorsal two thirds of the medulla oblongata, whereas the depressor effects result from stimulation of a portion of the medial reticular formation in the lower half. Both groups of cells are tonically active and integrate all basic vasoregulatory function. Recent evidence indicates that afferent impulse

traffic from the aortic and carotid mechanoreceptors passes through the midline depressor region and proceeds to the pressor regions located more laterally. Afferent fibers from mechanoreceptor sites have been traced to the nucleus solitarius and into regions of the medullary reticular formation, suggesting a site for the neural pathway for ascending impulses to higher cortical levels. Cardiac accelerator neurons are located in the same general areas in which the vasomotor cells are found but are not as well localized and cannot be identified as a cluster.

Supramedullary Centers

Supramedullary regions are capable of producing significant autonomic effects on cardiovascular function. Electrical stimulation of specific regions of the cerebral cortex (for example, gyrus proteus, orbital and sigmoid gyri, and portions of the rostral limbic system, including the cingulate and subcallosal gyri and insula) may evoke pressor and depressor effects as well as tachycardia and bradycardia.

The close functional interrelationships of the autonomic and somatic divisions of the central nervous system at the cortical level have been clearly demonstrated, in that electrical stimulation of the cortex results in simultaneous skeletal movement of the lower limbs and marked increase in muscle blood flow.

In general, the anterior half of the cortex exerts considerable autonomic control on cardiovascular function, whereas the posterior half is devoid of significant influence. Electrical stimulation of rostral regions of the limbic system (frequently referred to as the visceral brain) may elicit either marked bradycardia, together with arterial hypotension and respiratory arrest or, in a closely adjacent area, opposite responses.

Hypothalamus

The role of the hypothalamus as a major center for integration of autonomic responses is clearly established and, together with the septal nuclei and lateral preoptic region, it constitutes a nodal area in integration of autonomic activity in the limbic system. In addition, axons passing from the amygdala and septum converge and impinge upon discrete cells located predominantly in the ventromedial hypothalamus, with a small number of fibers passing into dorsal and lateral areas.

Little is known about the descending fiber tracts from these higher neural areas, whether they synapse in the medulla oblongata, or pass directly to the intermediolateral cells of the spinal cord. Reciprocal modulation of activity probably occurs at all higher neural stations, with the resulting integrated discharge projecting to spinal neurons.

REFLEX CONTROL

It is necessary that the various neural and chemical factors influencing cardiovascular function be subject to a high degree of integrated control. For example, chemoreceptors located in the carotid and aortic bodies are sensitive to alterations in blood gases and pH. Also, alterations in pressure in the arterial and venous circuits and the heart affect receptors, eliciting changes in cardiac output, systemic resistance and venous capacity. These responses are dependent upon neural mechanisms that respond to signals from these receptors by increasing or decreasing efferent neural impulse traffic to the heart and blood vessels. The reflexes that originate from receptors within the cardiovascular system are termed *intrinsic* whereas those from receptors located outside of the cardiovascular system are termed *extrinsic*. The intrinsic, or cardiovascular, reflexes function to prevent sudden large changes in vascular capacity that would lead to rapid and significant alterations in blood volume. The importance to the organism of the extrinsic reflexes has not been so clearly

defined, but they appear to participate in responses observed clinically, such as shock resulting from severe extremity pain or the introduction of a needle into a peripheral artery.

Traditionally, the major cardiovascular reflexes have been considered as isolated stimulus–response sequences, mediated by simple neural pathways responsible for specific regulatory functions at a subconscious level. This rather circumscribed view of reflex function implies the presence of sensory receptors, afferent neurons, one or more central neurons comprising a center, and an efferent motor neuron for the effector organ. However, this traditional concept of a simple neural arc is oversimplified and inconsistent with the findings of more recent experimental work. It is now known that the entire central nervous system maintains a state of dynamic equilibrium which is readily disturbed by afferent neural discharge from any peripheral site. The neural input projects not only to the spinal cord but to all levels of the central nervous system, and the resulting neural efferent discharge likewise represents the integration of neural activity at all levels.

High Pressure Mechanoreceptors

The arterial mechanoreceptor (frequently termed stretch receptor, baroreceptor or pressor receptor) reflexes play a dominant role in the maintenance and control of arterial pressure and are in a state of continuous activity. The receptors, activated by tissue stretch, are located at the bifurcation of the carotid arteries and in the aortic arch. Afferent neural pathways lead from the carotid receptor sites through the glossopharyngeal nerve and from aortic receptors through the vagus nerve, and relay in medullary vasomotor centers in the reticular formation for integration of the response. The efferent limb of the reflex involves the vasomotor nerves and both sympathetic and parasympathetic cardiac efferent fibers. Again, it must be emphasized that afferent mechanoreceptor neural activity ascends to neural levels above the vasomotor center and significant modification in efferent neural discharge to the periphery may occur as a result of modulation from descending neural activity from higher integrative levels. Such evidence clearly supports the concept of the central nervous system as a diffuse interacting and interconnected system rather than a simple reflex arc.

Early studies have clearly demonstrated that the mechanoreceptor reflexes fail to operate at arterial pressures below 60 mm. Hg and maximally activate at pressures above 200 mm. Hg. The curve of response is sigmoid, with maximal efficiency of the response in the normal range of arterial pressure. Recently, the reflex has been shown to be more sensitive to pulsatile pressure than to static pressure, and also to the rate of change (dp/dt) of arterial pressure. Thus, both total pressure and pulse pressure are important determinants of the degree of mechanoreceptor stimulation. It is important to emphasize that decline in arterial pressure does not stimulate the mechanoreceptors and that hypotension produces a positive reflex by decreasing afferent discharge, which leads to decreased vasodilatation and cardiac inhibition.

Elevation of arterial pressure by stretch of the arterial wall at mechanoreceptor sites, therefore, increases mechanoreceptor afferent neural discharge to the medulla. The response is inhibition of sympathetic peripheral vasoconstrictor tone and cardiac sympathetic efferents, with resultant bradycardia and a small decrease in atrial and ventricular contractility. Enhanced parasympathetic discharge to the heart also augments bradycardia and decreases myocardial contractility.

The mechanoreceptor reflex produces highly significant changes in total systemic resistance, with effects on both the arterial and venous circuits. The most striking changes are observed in the arterioles of the splanchnic bed, kidney, muscle and skin, as well as in the veins of the splanchnic vascular bed. The effects upon skeletal muscle and cutaneous veins are relatively insignificant.

Abnormal function of the arterial mechanoreceptors may produce serious clinical manifestations. For example, some individuals exhibit unusual sensitivity of the carotid sinus, so that minimal pressure in the neck—even the wearing of a tight collar —may elicit severe reflex bradycardia and hypotension. Less frequently, these signs may result from a cervical rib, dilatation of the carotid artery or a tumor encroaching upon the carotid sinus region. Patients with orthostatic hypotension experience dizziness, loss of vision and syncope upon assuming the upright position. This condition results from inadequate peripheral vasoconstrictor tone and frequently results from diseases involving autonomic pathways—for example, tabes dorsalis and diabetes mellitus. It is also observed with central lesions in the medulla and hypothalamus, and, when no overt pathological lesion is encountered, the process is called idiopathic.

Low Pressure Mechanoreceptors

Cardiovascular reflexes may arise from mechanoreceptors located in the walls of the superior and inferior venae cavae at the junction with the right atrium, in the region of the tricuspid valve, in pulmonary arteries and veins, along the posterior left atrial wall, in the ventricles, and in the coronary arteries. The intensity of afferent neural discharges varies with the phase of the cardiac cycle. These impulses travel in the vagi to higher neural centers separately from aortic depressor fibers. The effects are the same as those of the aortic and carotid sinus reflexes, namely, bradycardia and peripheral vasodilatation. Mechanoreceptors increase afferent vagal discharge in response to changes in atrial muscle tone such as may occur with atrial systole or atrial diastolic filling. Those located in ventricular musculature increase their neural firing rate during ventricular systole.

Although the exact role of the cardiac mechanoreceptors has not been clearly defined, they have been implicated in certain clinically observed physiological adjustments. Increased atrial systolic force and tachycardia associated with exercise may activate atrial receptors, thereby increasing vagal activity and modulating the degree of tachycardia. Release of antidiuretic hormone is elicited through low-pressure mechanoreceptors in the left atrium, their afferent limb traveling in the vagus nerve. Other low pressure mechanoreceptors control renin release at the juxtaglomerular level of the kidney. Their location has not yet been determined precisely, but the efferent limb is contained in sympathetic nerves.

Low pressure mechanoreceptors may be stimulated by the injection of a variety of pharmacological agents into various regions of the heart. For example, when veratrum is injected into the coronary arteries or the left ventricle, severe bradycardia and hypotension are induced without alterations in respiration; when it is injected into the pulmonary artery, it elicits bradycardia and hypotension, accompanied by respiratory arrest and perhaps a variety of cardiac arrhythmias. Left-sided injections activate only low-pressure receptors in the left side of the heart, whereas the pulmonary arterial injection activates receptors in the pulmonary artery as well as those in the coronary artery and left ventricle. These studies have certain important clinical implications. For example, in cardiac catheterization, passage of the catheter into a cardiac chamber may impinge upon the endocardial wall and provoke bradycardia, hypotension or arrhythmias. These effects may be abolished by parasympathetic blockade with atropine. Also, the cardiac arrest or arrhythmias associated with hypotension and respiratory alteration following pulmonary embolus may result from activation of pulmonary mechanoreceptors through acute elevations of pulmonary arterial pressure.

Chemoreceptors

Arterial chemoreceptor reflexes represent a major cardiovascular adaptation to oxygen deprivation. Unlike the arterial mechanoreceptors, which are continuously active, the chemoreceptors probably have minimal influence in the resting state or even with mild exercise. Chemoreceptors reside in the carotid body, but the origin of the occipital artery and the aortic body are composed of highly vascular groups of epithelioid cells that are activated by decrease in Po_2, increase in Pco_2, or decrease in pH. Diminution in blood flow to arterial chemoreceptors may also evoke responses, as do a variety of drugs. The chemoreceptor reflex, although usually described as an isolated neural reflex arc, represents in the intact organism only one of many sensory inputs that are continuously responding to signals from the periphery and transmitting information to higher neural centers, so that the cardiovascular response evoked represents the integrated effect of these inputs. Therefore, the reflex in the intact animal must be considered not as an isolated phenomenon but rather as a single component of a broader neural adaptation affecting the heart and peripheral vasculature as well as respiratory activity.

The primary circulatory effects of the chemoreceptors are best observed in animals in whom ventilation is maintained constant by artificial means. Under these circumstances, the cardiovascular responses to carotid chemoreceptor stimulation induced by perfusion with hypoxic blood consist of bradycardia, decreased cardiac output, decreased ventricular contractility, and elevation of arterial pressure associated with increased resistance in the extremities and splanchnic and renal areas. The cardiac response is mediated by the vagus nerve; the vasomotor is the result of enhanced sympathetic discharge. Also, the cardiac response may be evoked by modest increases in arterial P_{CO_2} or decreases in pH, and these changes increase the sensitivity of the chemoreceptors to oxygen lack. The primary response clearly represents an adaptation by the organism to conserve oxygen utilization by diverting blood flow from less critical areas to the heart and brain. In contrast, if the animal breathes spontaneously, perfusion of the carotid sinus with hypoxic blood results in an immediate increase in respiratory rate and total ventilation through direct effects upon the respiratory center, increased heart rate and insignificant change in arterial pressure. The differences observed between artificial and spontaneous ventilation may be attributed to secondary effects resulting from the increase in respiration, and possible mechanisms include activation of pulmonary stretch receptors through increased inflation and arterial hypocapnia. It is known that lung inflation produces increase in cardiac rate through a reflex with its afferent limb located in the vagus, while sympathetic efferents concomitantly produce peripheral vasodilatation. The site of the pulmonary receptor responsible for this reflex is unknown. This reflex has been applied clinically in determining effective adrenergic blockade by using digital plethysmography to assess inhibition of vasoconstriction during a deep sigh.

In the normal human being, the primary effect of vasoconstriction and bradycardia is overridden by tachycardia and vasodilatation produced by increased respiratory activity, and the overall response observed is the result of components of both effects. Clinically, the influence of chemoreceptor activity may be observed in patients with cyanotic congenital heart disease, in whom episodes of respiratory arrest are frequently associated with severe bradycardia and hypertension. Institution of artificial ventilation rapidly elicits tachycardia and normotension. Furthermore, chemoreceptor reflexes contribute to the maintenance of the circulation when blood volume is depleted. Following hemorrhage, afferent neural discharge from the carotid and aortic bodies is increased, exciting the bulbar vasoconstrictor centers to increase peripheral vasoconstriction. Therefore, the reflex acts synergistically with the carotid and aortic mechanoreceptors, which, by decreasing discharge to the vasopressor center, decrease its tonic inhibition.

Diving Reflex

Upon submersion, diving animals elicit striking reflex adjustments of the cardiovascular system, consisting of cessation of respiration, bradycardia, decreased cardiac output and pronounced vasoconstriction in all vascular beds except the heart and the brain. The reflex is designed to distribute oxygen reserves to these vital organs. The circulatory responses appear to be the result of both reflex mechanisms and alterations in blood gas tensions. The sensory stimulus initiating the reflex originates in the nares and is dependent upon the integrity of the trigeminal nerve, and the efferent vagal component producing bradycardia may be abolished by either vagotomy or atropine. Within a few seconds after emergence from water, marked tachycardia ensues, followed by a less rapid decrease in peripheral vasoconstriction, permitting arterial pressure to gradually return to normal levels. It has been suggested that the tachycardia and decreased vasoconstriction result from activation of pulmonary stretch receptors stimulated by the return of respiration following emergence from water. The diving reflex, although frequently considered a single neural response, actually represents only one component of a widespread cardiovascular adjustment to submersion. The response is initiated by cortical inhibition of the respiratory center as a result of sensory stimulation of receptors by the dousing of the face with water. The ensuing apnea decreases arterial Po_2 and pH and elevates Pco_2 thereby activating the arterial chemoreceptors which in turn lead to bradycardia and systemic vasoconstriction. The functional significance of such a reflex adjustment is to conserve limited oxygen reserves during the period of submersion while maintaining relatively normal perfusion through the cerebral and coronary vascular beds. It is apparent that the diving reflex is quite distinct from the "defense reaction," which is characterized by tachycardia, increased cardiac output and decreased peripheral resistance in skeletal muscle.

The diving reflex has been considered to be active in a number of diverse clinical conditions. For example, some have attributed bradycardia and vasoconstriction of the newborn to this reflex, and it has been postulated as a cause of sudden death in myocardial infarction. Divers also exhibit bradycardia and peripheral vasoconstriction.

Temperature Regulation

The human being maintains a relatively constant body temperature by balancing heat loss against metabolic heat production. Heat loss occurs through radiation, convection and vaporization and is influenced primarily by vasodilatation of cutaneous blood vessels and increased sweating. These responses are activated by increased afferent neural discharge to anterior hypothalamic areas from cutaneous thermal receptors as well as by direct excitation of specific neurons by the warm blood supplying other hypothalamic areas. Thus, when the temperature of blood perfusing the hypothalamus rises above 37° C., the vasoconstrictor response to stimulation of cutaneous cold receptors is markedly suppressed. Both central and peripheral thermal drives are important in adaptation to heat and cold.

Vascular responses to increased tempera-

ture have been studied best in the hand and forearm. The responses are mediated through sympathetic discharge and, therefore, cannot be elicited in the sympathectomized limb. The exact mechanism responsible for vasodilatation of the forearm is not very clear, but direct effects of sympathetic cholinergic dilators or sympathetic cholinergic discharge to sweat glands causing release of a local humoral substance are possible mediators. Exposure of an extremity to cold may evoke strong reflex vasoconstriction in contralateral extremities, which is abolished by sympathetic blockade. This response may result from cold stimulation of skin receptors as well as from stimulation of hypothalamic centers by cooled blood.

Following severe vasoconstriction of the fingers resulting from intense cold, vasodilatation supervenes. This response results, at least in part, from a local axon reflex, considered to result from stimulation of local sensory nerve fibers, which, in turn, act antidromically upon nearby blood vessels. This reflex is not dependent upon the integrity of the spinal reflex arc or higher neural centers.

Therefore, it is apparent that the diffuse projections of afferent neural discharge from cutaneous thermal receptors to the posterior hypothalamus, with probable integration at all intervening neural levels, make the older concept of a massive and diffuse reflex no longer tenable. In the control of heat loss through alterations of vasoconstrictor tone, the body maintains a hierarchy of neural responses, with central effects dominating peripheral effects.

NEUROEFFECTOR FUNCTION

The efferent neuroeffector system of both sympathetic and parasympathetic preganglionic fibers, as well as parasympathetic postganglionic fibers, is mediated by release of acetylcholine. It is also possible that sympathetic postganglionic fibers to sweat glands may be cholinergic. Acetlycholine, the chemical transmitter released by neural impulses at the preganglionic neuronal synapse and also at the myoneural junction of somatic muscle, resides in terminal neuronal vesicles. This neurohumoral compound acts upon the cell membrane of muscle cells to produce depolarization associated with sodium ion influx.

The parasympathetic nervous system influences circulatory function primarily through vagal effects upon cardiac pacemaker activity, conduction and myocardial contractility and, possibly, through alterations in resistance in coronary vessels.

The sympathetic efferent neuroeffector system in relation to cardiovascular control is much more complex. Recent studies have clarified the nature of sympathetic innervation of vascular smooth muscle, the organization of the sympathetic postganglionic axon, and the mechanism responsible for impulse transmission from the nerve terminal to the smooth muscle receptor site. Anatomically, the peripheral adrenergic system consists of a reticulum of slender processes, which enlarge at various points into presynaptic bags or varicosities. The latter structures pass in close proximity to vascular smooth muscle cells. The peripheral sympathetic neurotransmitter norepinephrine and adenosine triphosphate and protein as well, are present in granular vesicles located within the varicosities. The major portion of norepinephrine is granule-bound within the vesicle, but a smaller portion is mobile and may be transported. Norepinephrine and storage granules are manufactured within the cell bodies of the neuron and pass by axoplasmic flow to the presynaptic terminals. Electron microscopic studies have demonstrated spaces between the terminal nerve processes and the nearest vascular smooth muscle cell. This is in contrast to skeletal muscle, in which transmission occurs through the motor end plates. The transmission between nerve and smooth muscle occurs by release of

transmitter from the vesicles, which, in turn, passes to the membrane of muscle cells. This discontinuous process at the neuroeffector site has been termed *synapse en passant*. Norepinephrine is released continuously from granules, and the major portion is inactivated through deamination by the enzyme monoamine oxidase. Arrival of a neural impulse at the varicosity depolarizes it and releases norepinephrine, which passes into the synaptic cleft. Whether the transmitter is released directly from the granule or has an intermediate location from which it can be delivered is unclear. Norepinephrine, once released, may combine with either the alpha- or the beta-adrenergic receptor located on or in the surface membrane of the smooth muscle cell. The norepinephrine-receptor combination is spontaneously reversible, permitting the substance to pass back, upon release, into the synaptic cleft, where it may be taken up once more by an active process into the terminal varicosity and storage granules. A major portion of norepinephrine is metabolized to biologically inactive compounds by the enzyme catechol-o-methyl transferase, while a small component is discharged into the bloodstream and transported throughout the body. The beta-adrenergic receptor is now considered to be the enzyme adenyl cyclase or a closely related substance that possibly acts upon it and, in turn, catalyzes the conversion of adenosine triphosphate to cyclic AMP. It is now considered that beta-adrenergic activity is related to the ability of this enzyme to alter the intracellular level of cyclic AMP, either by increasing or, perhaps, by decreasing membrane activity.

CENTRAL NERVOUS SYSTEM CARDIAC CONTROL

The control of cardiac performance is mediated by highly integrated responses involving neural, humoral, metabolic and hemodynamic factors. Under different circumstances, these factors may act either antagonistically or synergistically.

The influence of respiration on the circulation emphasizes the continuous and significant effects of other organ functions upon cardiovascular function. Respiratory action can affect significantly the circulation through neural, humoral, or mechanical means. Blood gas and pH alterations represent important humoral influences on cardiac performance; and changes in venous return to the heart, as a result of intrathoracic pressure shifts during various phases of the respiratory cycle, are mechanical effects inducing changes in cardiac output. The neural effects either are reflex, involving the vasomotor center directly, or arise from stimuli having origin in the respiratory center, with secondary effects on the medullary regions.

Significant changes in cardiac sympathetic and vagal cardioinhibitory traffic, as well as in peripheral sympathetic efferent nerves, occur with respiration. It is well known, for example, that the heart rate accelerates during inspiration and decelerates during expiration, a response easily documented clinically by a sinus arrhythmia. This neural response disappears in dogs following cardiac denervation produced either by autotransplantation or by bilateral vagotomy. Although parasympathetic effects are important, modest electrical stimulation of certain mesencephalic areas strongly increases cardiac sympathetic activity and may also abolish the response. Thus, it is most likely that sinus arrhythmia in man results from interplay between both branches of the autonomic nervous system. A major neural mechanism for the increase of heart rate with inspiration appears to be through activation of pulmonary receptors with lung inflation and concomitant decrease in vagal impulse traffic to the heart. This reflex also effects small decreases in sympathetic efferent discharge to the periphery and, thus, may play a role in the decline in blood pressure frequently observed during inspiration.

The autonomic nervous system exerts significant influence in man, on both cardiac rhythm and performance. However, its role in regulating coronary blood flow is less clear. Most evidence suggests that coronary blood flow, like cerebral blood flow, is controlled predominantly by local vasodilator metabolites. Other factors of importance include aortic pressure, heart rate and myocardial compression of coronary vessels; moreover, the coronary vasculature does have significant autonomic innervation. Alpha- and beta-receptors are present in the coronary circulation and are capable of producing vasoconstriction and vasodilatation. The vagus nerve also exerts control on coronary blood flow, and vagal stimulation produces vasodilatation by direct effects. Increased sympathetic preganglionic neural impulses, transmitted in the white rami of thoracic outflow, have been produced by experimental occlusion of the coronary artery in cats and clearly demonstrate significant neural control of the coronary circulation. Difficulties in assessing the contribution of neural factors in the regulation of coronary blood flow in man arise primarily from attempts at extrapolation of information obtained from isolated experimental preparations. From these studies we are unable to identify the effects of heart rate, myocardial contractility and myocardial metabolic activity in the intact animals. Therefore, although the autonomic nervous system clearly exerts significant influence on the coronary circulation, a quantitative assessment of its clinical role is not yet available.

The sympathetic and parasympathetic divisions of the autonomic nervous system contribute significantly to atrial and ventricular myocardial contractility and are critical in adjustments of cardiac performance to environmental needs. The vagus distributes broadly to the myocardium and conducting system and sends dense neural projections to the atria and nodal tissue and lesser numbers to the ventricles. Although traditional concepts have considered ventricular muscle to be devoid of vagal control, parasympathetic efferent discharge produces not only significant depression in atrial contractility but also modest depression of ventricular contractility.

The normal heart is richly supplied with sympathetic nerve endings and norepinephrine, which is stored in discrete granules within the myocardium. Sympathetic impulses arriving at the myocardial nerve terminals effect a release of neurotransmitter from the granule, which results in a positive inotropic effect. The norepinephrine, present at the cardiac sympathetic nerve endings, may be synthesized from tyrosine or dopamine, or it may be derived from the uptake of circulating norepinephrine.

The morphology of the nerve endings in the myocardium has been clarified by a number of recent studies. Motor end-plates analogous to those observed in skeletal muscle are not present in smooth or cardiac muscle, and direct penetration of the sarcoplasm by free nerve endings has not been observed. Newer histochemical techniques have demonstrated that the postganglionic sympathetic fibers enter an autonomic ground plexus, and that axons travel together in a single Schwann cell. The space between the Schwann cell and cardiac muscle is occupied by ground substance which may act as a diffusing medium for the transmitter from the axon to the muscle cell. Present evidence suggests that terminal components of autonomic nerves, which may comprise both autonomic divisions, are invested in the same ground plexus and enclosed together with cardiac cells in a common envelope. The envelopes are arranged as small subunits in such fashion that increased neural discharge in one or more subunits would not necessarily affect subunits in other areas. Positive inotropic

effects could occur not only by liberation of increasing quantities of catecholamines within each subunit but also by the recruitment of additional subunits. Synchrony and asynchrony of contraction may be explained on this basis. It must be recognized, however, that it is not possible to distinguish between cholinergic and adrenergic nerve endings in the myocardium, and a model such as the one described (though appealing from a functional viewpoint) must be subject to further experimental documentation.

Cardiac Rhythm and Conduction

Although the heart may maintain rhythm and conduction without neural control, both are continuously under autonomic influence. The tonic activity of both autonomic divisions is demonstrated clearly by observation that bilateral section of the vagus nerves increases heart rate, whereas sympathectomy decreases heart rate. Furthermore, control of heart rate, myocardial contractility, and, probably coronary blood flow act in concert in a broad neural adaptation to the needs of the entire organism.

The sino-atrial node is located beneath the epicardium at the junction of the superior vena cava and right atrium and lies closely adjacent to nerves and ganglia which abound in surrounding areas. The atrioventricular node is located beneath the right atrial endocardium, anterior to the coronary sinus ostium and slightly superior to the insertion of the septal leaflet of the tricuspid valve. It contains no ganglia, although many neural structures surround it, particularly in regions adjacent to its posterior margin. Nerve endings do not terminate directly on conducting or nonconducting myocardial cells but end at a distance. The firing rate of the sino-atrial node is influenced by sympathetic discharge from both right and left cardiac nerves, with the right predominating at physiological heart rates. Both alpha- and beta-receptors are present and affect function in the sino-atrial node. Parasympathetic control of the sino-atrial node is mediated by bilateral vagal discharge, with the right-sided influence predominant, and increased vagal discharge depresses the pacemaker by slowing the rate of depolarization. Again, by reciprocal effects on the rate of depolarization in the transmitter tissue, conduction through the atrioventricular node is increased by sympathetic and decreased by parasympathetic discharge, with the left dominating the right vagus.

Intracellular study of single cardiac cells has demonstrated clearly that pacemaker cells undergo spontaneous depolarization, whereas nonpacemaker cells do not. Spontaneous cellular depolarization is not dependent upon neural discharge, since it also occurs in denervated hearts. It is well established, however, that the rate of depolarization is greatly influenced by neural mechanisms. An increase in sympathetic efferent discharge increases the slope of depolarization of all pacemaker tissue, whereas increased parasympathetic discharge decreases it. Furthermore, pacemaker cells' varying responses to varying intensities of neural discharge or the cell membranes' differences in sensitivity to a given intensity would be capable of producing individual variations in the slope of diastolic depolarization between cells and, thus, lead to a shift in the pacemaker site. This change could produce a number of cardiac arrhythmias, including atrial or ventricular tachycardia.

Supramedullary Effects. Supramedullary regions of the central nervous system exert considerable influence on cardiac rate and conduction through alterations in the intensity of sympathetic and parasympathetic discharge. Experimental evidence supporting this relationship has been derived largely from studies involving electrical stimulation of cortical and subcortical re-gions. These studies, however, are deficient in that they neither correlate the rhythm disturbances with concomitant changes in the cardiac sympathetic or vagus nerve traffic nor identify the autonomic division mediating the response. Furthermore, the stimuli are of artificially high intensity. Despite these objections, a certain degree of specificity of higher autonomic control upon heart rate and conduction is apparent.

Stimulation of various regions of the cortex, basal ganglia, midbrain and pons evokes predominantly sympathetic or parasympathetic responses, whereas certain other areas demonstrate interplay of both autonomic components. Thus, stimulation of the motor and the premotor cortices elicits increases and decreases of heart rate and arterial pressure from closely adjacent neural regions. In contrast, stimulation of deeper cortical structures produces predominantly parasympathetic effects, consisting of marked sinus bradycardia and sinus arrest, complete heart block, atrial flutter, and fibrillation. Stimulation of portions of the hypothalamus and mesencephalon produces marked increases in heart rate and ventricular arrhythmias. The latter responses are usually associated with increases in arterial pressure; however, information concerning changes in other cardiovascular variables, such as cardiac output, systemic resistance and distribution of blood flow, are usually not available from these experimental studies.

A number of clinical observations indicate that pathological changes in the myocardium may occur with a variety of diseases of the brain. The pathological lesion termed focal myocytolysis consists of an area within the muscle fiber, devoid of tissue, leaving an empty sarcolemmic sheath, muscle nuclei, lipofuscin pigment and sparse histiocytes. Also, patients with intracranial disease, particularly subarachnoid hemorrhage, show striking electrocardiographic changes. Thus, central nervous system dysfunction produces pronounced changes in cardiac function, but the exact location of the higher neural centers involved and the pathways of mediation are not yet clearly defined.

Serious arrhythmias may occur as a result of heart disease. In acute myocardial infarction, for example, occlusion of the right posterior coronary artery frequently elicits sino-atrial arrest, A-V dissociation and varying heart block. Since cholinergic ganglia and adrenergic and cholinergic nerve fibers lie closely adjacent to the sino-atrial and atrioventricular nodes, it has been suggested that an alteration of autonomic discharge in these neural structures (as a result of anoxia or acidosis) may produce these arrhythmias. The frequent abolition or attenuation of the rhythm disturbances with atropine supports overriding parasympathetic activity as a causative mechanism. However, alternative mechanisms exist, capable of enhancing parasympathetic discharge in this clinical situation, and include activation of mechanoreceptors as a result of distention of intracardiac chambers or extreme fear sufficient to evoke overwhelming cortical vagal discharges.

Cardiac Failure

As the myocardium progressively deteriorates in disease, with diminished capability for increasing speed and force of contraction, cardiac output is less able to increase in response to tissue demand. Therefore, it becomes necessary for the heart to utilize compensatory mechanisms for the maintenance of adequate function. Three mechanisms are available for this purpose. The heart may increase its myocardial mass through hypertrophy, in order to increase the number of active contractile units. Obviously, this mechanism is slow in development and is limited by the ability of coronary blood supply to maintain normal oxygen requirements relative to the increased muscle mass. Secondly, the Frank-Starling mechanism may be utilized, leading

to an increase in ventricular end-diastolic volume and an increase in systolic tension and stroke volume. This mechanism prevents a decline in output which would otherwise occur as the myocardial contractile state is depressed. As this response is initiated, an increase in blood volume occurs through the retention of salt and water, with a resultant elevation of pulmonary venous pressure. The latter response may produce dyspnea and congestion. The third compensatory mechanism consists of an increase in speed and force of muscle contraction from any ventricular end-diastolic muscle length which may be mediated by enhanced sympathetic efferent discharge. This mechanism, acting together with neural vasoconstrictor adjustments in the periphery, permits a rapid overall circulatory compensation.

Norepinephrine located in storage granules of the myocardium becomes seriously depleted in the failing heart. The deficiency results from both inadequate synthesis of the hormone and a failure of binding. The deficiency in synthesis is most likely due to a defect in the rate limiting enzyme system tyrosine hydroxylase, which is necessary for the biosynthesis of norepinephrine. Thus, the quantity of neurotransmitter released by nerve stimulation is markedly reduced at a time when the failing heart has a serious need for a mechanism to increase its myocardial force and velocity of contraction. It must be recognized, however, that the myocardium is not entirely dependent upon norepinephrine stores for its intrinsic state of contractility. Studies in the denervated heart have clearly shown that myocardial contractile function can be maintained at adequate levels despite a norepinephrine reduction equivalent to that observed in the failing heart. The increased levels of circulating catecholamines observed in heart failure may be a mechanism that helps to sustain myocardial contractility when the transmitter is depleted at the neuroeffector site.

In addition to myocardial alterations, profound changes occur in the peripheral circulation in congestive heart failure. Arteriolar and venous constriction is enhanced by sympathetic efferent discharge, an effect most clearly demonstrated with muscular exercise. In addition, the degree of stiffness of the arteriole is increased, probably through increased tissue fluid, and this limits the arteriole's ability to undergo local metabolic dilatation.

Peripheral sympathetic arteriolar constriction produces a more effective distribution of the limited cardiac output to vital organs, such as the heart and brain, with relatively high metabolic requirements at rest and in exercise. The arteriolar constriction is reflected in decreased blood flow to the cutaneous, renal, splanchnic and skeletal muscle circulations. With the onset of exercise, the subject in heart failure shows adequate perfusion of the cerebral and coronary circulation, minimally adequate perfusion of exercising skeletal muscle, and a marked decrease of blood flow to all other regions.

Venous pressure is elevated in congestive heart failure, as a result of sympathetic venoconstriction, and this mechanism improves right ventricular diastolic filling, thereby permitting the depressed myocardium to increase cardiac output by increasing diastolic volume and stroke volume.

In states of diminished myocardial contractility, therefore, the sympathetic nervous system demonstrates highly efficient and rapidly acting adjustments that maintain arterial blood pressure at adequate levels while distributing limited but adequate blood flow to regions where the needs are the greatest.

ANNOTATED REFERENCES

Adams, D. B., Baccelli, G., Mancia, G., and Zanchetti, A.: Cardiovascular changes during naturally elicited fighting behavior in the cat. Am. J. Physiol., 216:1226, 1969. (A concise discussion of cardiovascular changes elicited during fighting, with additional data from studies in unanesthetized cats)

Bard, P.: Regulation of the systemic circulation. In: Mountcastle, V. B. (ed.): Medical Physiology. Ed. 12, Vol. I, Chap. 2. St. Louis, C. V. Mosby, 1968. (A comprehensive review of the integrated neural control of the circulation)

Benzinger, T. H.: Heat regulation: homeostasis of central temperature in man. Physiol. Rev., 49:671, 1969. (A critical review of current knowledge of temperature regulation in man)

Berne, R. M.: Regulation of coronary blood flow. Physiol. Rev., 44:1, 1964. (An excellent review of the difficult subject of coronary vascular control)

Braunwald, E., Ross, J., and Sonnenblick, E. H.: Mechanisms of contraction of the normal and failing heart. Boston, Little, Brown & Co., 1967. (A comprehensive review of the physiology of cardiac muscle and the factors influencing cardiac performance)

Connor, R. C. R.: Myocardial damage secondary to brain lesions. Am. Heart J., 78:145, 1969. (A discussion of neurological disease relative to its effects upon the heart)

Cooper, T.: Terminal innervation of the heart. In: Randall, W. C. (ed.): Nervous Control of the Heart. p. 130. Baltimore, Williams & Wilkins, 1965. (A clear review of current knowledge of the nervous innervation of the heart from the morphological point of view)

Folkow, B., Heymans, C., and Neil, E.: Integrated aspects of cardiovascular regulation. In: Hamilton, W. F., and Dow, P. (eds.): Handbook of Physiology. Section 2, Circulation, Vol. III, p. 178. Washington, D. C., American Physiological Society, 1965. (A comprehensive review of cardiovascular regulation)

Hoffman, B. F., and Cranefield, P. F.: The physiological basis of cardiac arrhythmias. Am. J. Med., 37:670, 1964. (A clear, concise discussion of the neural control of cardiac rhythm)

Kontos, H. A., and Patterson, J. L., Jr.: Reflex control of the circulation. In: Gordon, B. L., Carleton, R. A., and Faber, L. P. (eds.): Clinical Cardiopulmonary Physiology. p. 63. New York, Grune & Stratton, 1969. (A clear discussion of the vascular reflexes from the traditional viewpoint)

Lofving, B.: Cardiovascular adjustments induced from rostral cingulate gyrus. Acta physiol. scand., 53(Supp. 184):1, 1961. (A thorough review plus excellent experimental studies emphasizing the importance of sympathetic inhibitory pathways from cortical and subcortical structures)

Mauck, H. P., Jr., and Hockman, C. H.: Central nervous system mechanisms mediating cardiac rate and rhythm. Am. Heart J., 74:96, 1967. (A review of experimental work on cortical and subcortical control of heart rhythm)

von Euler, U. S.: Some factors affecting catecholamine uptake, storage, and release in adrenergic granules. Circ. Res., 21 (Supp. 3):5, 1967. (A concise review of an important subject)

Wolf, S.: The bradycardia of the dive reflex. A possible mechanism of sudden death. Trans. Am. Clin. Climat. Assoc., 76:192, 1964. (An interesting discussion suggesting the dive reflex as a mechanism of sudden death in a variety of clinical states)

Zelis, R., Mason, D. T., and Braunwald, E.: A comparison of the effects of vasodilator stimuli on peripheral resistance vessels in normal subjects and in patients with congestive heart failure. J. Clin. Invest., 47:960, 1968. (An excellent study elucidating the mechanisms responsible for peripheral vasoconstriction in congestive heart failure)

29

Autonomic Control of Gastrointestinal Function

Ove Lundgren, M.D.

INTRODUCTION

Much information is available concerning the effects of direct stimulation of regional parasympathetic and sympathetic nervous pathways in the gastrointestinal tract of experimental animals, and the physiological and pathophysiological mechanisms underlying neuromuscular control have also been examined thoroughly. Far less information, however, is available concerning higher central nervous system controls in man, which are responsible for emotionally significant experiences affecting motility and blood flow. Therefore, any detailed neurophysiological analysis of such disorders is impossible at present. This chapter is concerned with neuromuscular control of motility and blood flow by the autonomic nervous system. (For further discussion on motility and secretion, the reader is referred to Chapters 19 and 20, respectively.)

ANATOMICAL CONSIDERATIONS

The Intramural Nerve Plexuses

The most conspicuous nervous structures of the gastrointestinal tract are the two intramural nerve plexuses: the myenteric (Auerbach) and the submucosal (Meissner). The myenteric plexus is located between the circular and the longitudinal layers of the muscularis propria. Both plexuses consist of ganglion cells and intermingled unmyelinated nervous fibers that form a dense meshwork. The nerve cells and axons making up the plexuses may be classified as efferent, afferent, or associative.

The *efferent* (motor) nerves emanate from the sympathetic and the parasympathetic nervous systems. The sympathetic nervous pathways, reaching the gastrointestinal wall along the vasculature, are, by classical definition, mainly postganglionic, whereas the parasympathetic nervous outflow (vagus and pelvic nerves) is preganglionic. The fibers of the latter synapse with postganglionic neurons in the intramural plexuses. The number of intramural nerve cells is far larger than the number of preganglionic nerve fibers, indicating a considerable divergence of the preganglionic axons. The postganglionic parasympathetic fibers innervate the gastrointestinal musculature. There is reason to believe that some efferent neurons in the intramural plexuses are part of local reflex arcs that innervate vascular smooth muscle (producing, for example, reflex vasodilatation) and the muscularis propria (producing, for example, a peristaltic reflex).

The *afferent* (sensory) neurons of the intramural plexuses consist, in part, of nerve cells that terminate at mucosal receptors. They constitute the first neurons in the local reflex arcs. There also appear to be neurons within mesenteric neurovascular bundles of the intramural plexuses that project centripetally. In fact, most fibers in the vagal and pelvic nerves are afferent. Association neurons connect two or more nerve cells in one or the other plexus.

Innervation of the Muscularis Propria and Vasculature

According to classical concepts, the muscularis propria of the gastrointestinal tract is innervated by both sympathetic and parasympathetic fibers. Recent histochemical investigations suggest, however, that the sympathetic innervation of the muscularis is sparse as compared to the myenteric plexus, where the dense adrenergic fibers apparently synapse with nonadrenergic ganglion cells. In contrast, the parasympathetic innervation of the muscularis propria is abundant. These cholinergic nerves may innervate directly only certain key smooth muscle cells, which electrically influence adjacent cells indirectly through low-resistance junctions (so-called nexa) between cells. The autonomic effector in the intestine consists of a hexagonal bundle of closely-packed smooth muscle cells. Neuromuscular junctions of the type known from striated skeletal muscles have not been found in the intestinal muscularis, and the true nature of the intestinal autonomic neuromuscular junction remains unknown.

Sympathetic innervation of vascular smooth muscle exists in all the circulatory beds of the gastrointestinal tract. The vessels piercing the muscularis propria and the vascular network of the submucosa are innervated by adrenergic nerve fibers which form a characteristic network of varicose nerve fibers (sympathetic ground plexus) on the outer surface of the tunica media. This nervous network controls directly the adjacent vascular smooth muscle and indirectly the rest of the smooth musculature of the media by low-resistance nexa. The vessels of the basal parts of the mucosa also are innervated by sympathetic fibers, whereas the superficial mucosal parts almost completely lack any adrenergic innervation.

Parasympathetic innervation of the gas-trointestinal vasculature is not so clearly understood. It was once believed that all vascular beds were innervated by parasympathetic nerve fibers which induced effects opposite to those of the sympathetic nervous system. However, the small intestinal vessels (and probably also those of the stomach) lack any direct extrinsic parasympathetic control. The colonic vasculature is probably innervated by parasympathetic nervous fibers from the pelvic nerves, but these nerve fibers are not cholinergic and their intramural distribution is unknown.

MOTILITY

Stomach

Three types of contractions occur in the stomach and are most often observed after a meal. *Peristalsis,* a complex movement involving both the circular and the longitudinal gastric smooth muscles, consists of a wave of contraction preceded by a relaxation that starts in the fundus region and moves toward the antrum. Only two out of three peristaltic waves reaching the antrum create an intraluminal pressure high enough to empty a portion of the gastric contents into the duodenum. In doing so, it appears that the gastric antrum, the pylorus, and the proximal duodenum behave as a well coordinated functional unit. Thus, peristalsis serves two functions: mixing and propulsion. *Terminal antral contraction* involves contraction of the gastric antrum against a closed pyloric sphincter. It mixes the gastric contents by moving the antral contents in a retrograde direction. As the stomach empties, there is a slow, continuous contraction of the whole stomach, reducing the size of the stomach without changing the intragastric pressure.

In the following discussions, inhibition of motility refers to a decrease of rate and/or amplitude of contraction; with respect to the stomach, this is accompanied by a concurrent decrease of rate of gastric emptying. Relaxation refers to an increase of volume resulting from a decrease in smooth muscle tonus without change in intraluminal pressure.

Parasympathetic Control. In the experimental animal, efferent electrical stimulation of the vagus may induce either contraction or relaxation of the gastric muscularis propria. There exist two sets of nerve fibers in the vagi, which differ functionally and electrophysiologically. Stimulation of the low-threshold (comparatively thick) fibers induces contractions and increased gastric wall tonus, a response that can be blocked by atropine. This indicates that the transmitter substance at the effector cell is acetylcholine. On visual inspection, the stimulation of the low-threshold vagal fibers is seen to induce strong peristaltic waves moving from the fundus toward the pylorus. Stimulation of the high-threshold (comparatively thin) fibers, in contrast, induces gastric wall relaxation which can be

blocked neither by atropine nor by alpha or beta-adrenergic blocking agents. The chemical nature of the transmitter involved in this response is unknown. The vagal relaxation of the stomach occurs mainly in the corpus–fundus part and may increase gastric volume two- or threefold. This relaxation is accompanied by increased gastric secretion of hydrochloric acid and pepsinogen as well as augmentation of gastric blood flow. These secretory and vascular effects can be reduced by atropine. Teleologically, this response pattern (relaxation, increased secretion, augmented blood flow), elicited by stimulation of the high-threshold vagal fibers, seems to be very appropriate during gastric digestive work.

The effects of the vagal stimulation described above can be abolished by ganglionic blocking agents, since the vagal fibers are preganglionic. However, vagal control of gastric function may be arranged in

PARASYMPATHETIC CONTROL

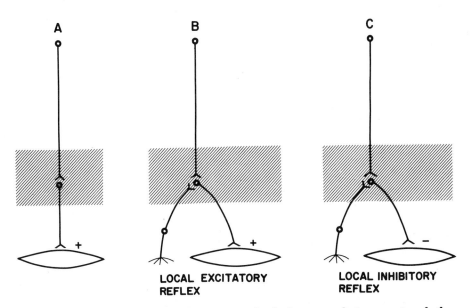

FIG. 29-1. The parasympathetic control of the muscularis propria of the gastrointestinal tract schematically illustrated. Shaded areas denote intramural nerve plexuses. The net effects indicated on the figure (+, contraction; —, relaxation) are the result of an activation of the preganglionic parasympathetic fibers facilitating the synapses in (B) and (C).

two fundamentally different ways (Fig. 29-1): The preganglionic vagal nerve fibers may synapse with one postganglionic fiber, which, in turn, innervates the gastric smooth muscularis propria (Fig. 29-1, A). The preganglionic gastric fibers may facilitate reflex arcs that may induce contractions, relaxations, or more complex muscular movements such as peristalsis (Fig. 29-1, B and C).

Sympathetic Control. Electrical stimulation of gastric sympathetic fibers produces inhibition of gastric contractions and a relaxation of the smooth musculature. These effects can be blocked by adrenergic blocking agents. Adrenergically mediated gastric relaxation differs from parasympathetically induced relaxation in that return to prestimulatory control volume occurs within 1 to 2 minutes in the former, as compared to 10 to 30 minutes after vagal stimulation. Furthermore, the magnitude of the increase in gastric volume induced by sympathetic stimulation seems to be dependent upon nervous activity within vagal excitatory fibers and/or within

local excitatory reflex arcs. Thus, sympathetic stimulation causes a marked gastric relaxation during concomitant pronounced vagal gastric tonus, whereas a very slight volume increase is seen during an identical sympathetic stimulation in the absence of such vagal tonus. It has also been shown that adrenergic fibers may inhibit local excitatory reflexes (Fig. 29-2, A). These observations are consistent with the view that the sympathetic influence on gastric motility exerts mainly, if not exclusively, an inhibitory effect on intramural ganglia (Fig. 29-2, A and C).

Reflex Control. The neural effects described above are elicited normally by means of reflexes. Thus, vagally mediated reflex gastric *contractions* can be provoked by sight, smell or taste of food and, possibly, also from gastric mechanoreceptors (Fig. 29-1, A and B). Painful stimuli of skin or internal organs and intestinal distention are known to inhibit gastric motility and tonus through sympathetic connections. This reflex inhibition, seen after intestinal distention, has been termed the intestino-gastric inhibi-

SYMPATHETIC CONTROL

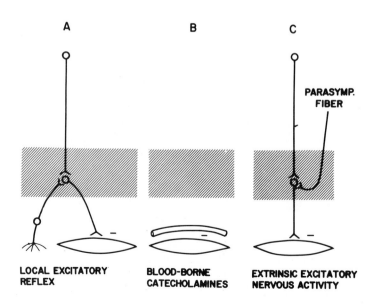

A   B   C

PARASYMP. FIBER

LOCAL EXCITATORY REFLEX

BLOOD-BORNE CATECHOLAMINES

EXTRINSIC EXCITATORY NERVOUS ACTIVITY

FIG. 29-2. The sympathetic control of the muscularis propria of the gastrointestinal tract schematically illustrated. Shaded areas denote intramural nerve plexuses. The net effects indicated on the figure (—, relaxation) are the result of an activation of the sympathetic postganglionic nervous fibers inhibiting the synapses in (A) and (C).

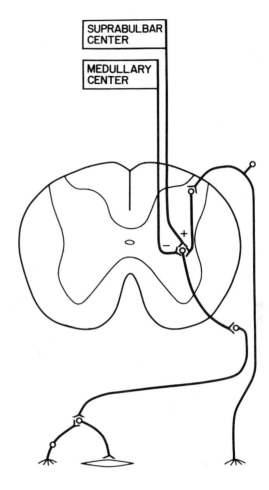

FIG. 29-3. The central nervous control of the intestino-intestinal inhibitory reflex schematically illustrated. The figure is based mainly upon the ·experimental investigations of Johansson and co-workers. The afferent as well as the efferent fibers run within sympathetic nerves. Neither the number of spinal neurons of the reflex nor the exact anatomical localization of the supraspinal control of the reflex are known.

tory reflex (Fig. 29-3). Similar reflexes are known to produce inhibition of small and large intestinal contractility. Such reflexes can be elicited from pain receptors and mechanoreceptors that are located within the mucosa. They contribute to the painful distention of the gut following abdominal surgery. Since these sympathetic reflexes act at the ganglion, the net increase of gastric volume and net decrease of rate of emptying depend on the current excitatory activity within the vagal fibers and/or local excitatory reflexes (Fig. 29-2, A and C). In so-called *receptive relaxation,* vagal fibers are concerned with actual gastric relaxation. In the experimental animal, this reflex relaxation is elicited by esophageal or antral distention (by a balloon), by

touching the pharyngeal mucosal surface, by spontaneous swallows, or during sham feeding. This mechanism prepares the stomach to receive food, involving as it does an increased rate of gastric secretion as well as increased mucosal blood flow.

The most thoroughly studied of the reflexes controlling gastric motility are those induced from receptor sites located in the duodenum and upper jejunum. From these receptors gastric motility and emptying are reflexly decreased. The most potent inhibitory effects are elicited by certain fatty acids, low pH and hyperosmolar intestinal content. The true nature of these mechanisms is not known. Hormones such as enterogastrone have been said to be liberated by the fatty acids. However, the fast inhibitory response seen in man and in the experimental animal upon duodenal feeding of fatty acids and HCl suggests that the reflex is nervously mediated—at least in part.

There are conflicting opinions regarding the possible nervous pathways of these reflexes. Some report that the reflex inhibitions of contractions and reflex relaxations are abolished by vagotomy; others claim that the reflexes are sympathetically mediated. These opposing views possibly may be reconciled by the above-mentioned observation that a sympathetically mediated effect is dependent upon the current nervous activity within vagal excitatory fibers and/or local excitatory reflex arcs.

Pathophysiological Aspects. The physiological controlling mechanisms described above may regulate the rate of gastric emptying in such a way as to keep the pH and/or osmolarity of the duodenal contents within certain limits. The importance of these regulatory mechanisms for the organism as a whole is most clearly demonstrated by the pathological conditions that develop when gastric emptying is too fast or too slow. Thus, various surgical procedures may disturb the delicate mechanisms controlling gastric motility and emptying. For example, a gastric resection with removal of the pylorus and duodenal bulb is frequently performed for peptic ulceration. Such an operation increases gastric emptying approximately twofold and produces in certain patients a so-called *dumping syndrome.*

The dumping syndrome usually appears within half an hour following a meal. The initial symptoms usually are intestinal, i.e., epigastric discomfort, a sensation of fullness and distention, and nausea. Somewhat later, certain circulatory and general symptoms appear, including tachycardia, pallor, tiredness, weakness and dizziness. The basic physiological disturbance is not entirely clear but certainly it includes increased rate of gastric emptying. Since the gastric contents enter the small intestine only moderately diluted by gastric secretion, the abnormally high osmolarity of the intestinal contents causes a rapid transfer of large volumes of fluid from blood to the intestinal lumen. This, in turn, distends the small intestine and disturbs body electrolyte and water balance. The two latter factors are believed to induce most of the symptoms of dumping.

The increased rate of gastric emptying after gastric surgery may be attributable, in part, to the removal of portions of the stomach and duodenum normally involved, through local reflexes, in the control of gastric emptying contractions. Other reflexes may also be disturbed by such surgery. Thus, a vagovagal receptive relaxation is elicited from distention-sensitive receptors in the antral mucosa, probably decreasing the rate of gastric emptying. It has, however, been convincingly demonstrated that the inhibitory reflex of gastric motility that can be elicited from receptive sites in the duodenum and the jejunum is intact even after partial gastrectomy.

Vagotomy is another surgical procedure for peptic ulceration which is directed at decreasing the secretion of gastric juice. Ligation of vagal fibers also affects gastric motility, and often the patient initially develops symptoms of gastric retention (e.g., distention after meal; nausea, vomiting). This is probably the result of inhibition of gastric motility and slowing of rate of gastric emptying in the absence of vagal excitatory nervous activity. In fact, gastric retention is so often seen after vagotomy that some drainage operation (e.g., pyloroplasty or gastroenterostomy) must be performed to provide a normal rate of emptying.

Acute gastric dilatation is a condition characterized by enlargement of the stomach in the absence of any apparent organic obstruction. The stomach may be enormously distended by swallowed gas and secretions which fill the entire abdominal cavity. It is most often seen 1 to 3 days after intra-abdominal surgery, but it may also be observed during certain infections, emotional stress and other conditions.

Sympathetic inhibition of gastric as well as intestinal motility occurs postoperatively by means of the reflex illustrated in Figure 29-3. In some instances, the sympathetic inhibition of gastric motility and emptying is prolonged, and swallowed air and gastric secretions as well as regurgitated bile and pancreatic juice may distend the stomach considerably. One cannot, however, exclude the possibility that the vagally induced receptive relaxation also is of importance in the pathogenesis of acute gastric dilatation, since it has been demonstrated that this reflex can be elicited by distention of the gastric antrum. Acute gastric dilatation is potentially fatal because of the great loss of fluid and electrolytes into the stomach. Prompt decompression of the stomach contents by nasogastric tube drainage and restoration and maintenance of water and electrolyte balance are the essentials of therapy.

Small Intestine

The small intestine exhibits three different types of movement, which can also be observed in the denervated gut. *Segmentation* may be defined as a localized contraction, predominantly of the circular muscle layer of the small bowel wall. It divides the intestinal contents into small segments by constricting the lumen. *Pendulum movement* is a motility pattern akin to segmentation. It can be described as a local contraction in which the active part of the small intestine moves to and fro, thus mixing the intestinal contents without propelling it. *Peristalsis,* a local reflex, is elicited by mucosal receptors sensitive to distention. It involves both longitudinal and circular smooth muscle layers. As in the stomach, it consists of a wave of contraction preceded by relaxation. This complex muscular movement is programmed by the intramural

plexuses to propel the intestinal contents aborally. Thus, in contrast to the two other types of intestinal movements, peristalsis is propulsive.

Parasympathetic Control. Stimulation of vagal fibers in the experimental animal augments small intestinal motility through a direct effect upon the muscularis propria (Fig. 29-1, A) and/or by facilitating local excitatory reflexes (Fig. 29-1, B). The strength of the contractions as well as the tone of the muscularis are increased during vagal stimulation. This vagal effect is blocked by atropine and, hence, is cholinergic.

Sympathetic Control. Electrical stimulation of the regional sympathetic fibers promptly inhibits spontaneous intestinal motility and the intestinal contractions evoked by stimulation of the vagal fibers. A concurrent intestinal relaxation is also observed. These adrenergic effects, which in the cat seem to be more pronounced in the ileum, are probably mediated through intramural synapses (Fig. 29-2, A and C). Catecholamines, released from the adrenal glands, induce the same effects on intestinal motility as described above (Fig. 29-2, B) and seem to be of particular importance for the sympathetic control of jejunal motility.

Reflex and Central Nervous Control. The best documented extrinsic intestinal reflex is the *intestino-intestinal inhibitory reflex,* the intestinal equivalent to the intestino-gastric inhibitory reflex described on page 618. This reflex is elicited by the distention of one intestinal segment which promptly decreases rhythmicity and intestinal tonus of another segment (Fig. 29-3). Concomitantly, a reflex vasoconstriction occurs in the intestinal and renal vasculatures. The minimum pressures needed for the activation of the mucosal mechanoreceptors of this reflex are well below those normally seen in the contracting small intestine.

The intestino-intestinal inhibitory reflex is controlled by central nervous structures in a way similar to that described for the somatomotor flexor reflex. Electrical stimulation of medullary centers inhibits the reflex, and apparently there are certain suprabulbar nervous structures that facilitate the reflex (Fig. 29-3). Such a control system of an autonomic reflex arc may be a principal and important mechanism of central nervous control of gastrointestinal motility.

Inhibition of intestinal motility by sympathetic fibers can also be elicited reflexly from receptor sites elsewhere than in the intestinal mucosa. Thus, reflex inhibition of motility and relaxation of the small intestine can be evoked by cutaneous pain, mechanical or chemical peritoneal irritation, or distention of the ureter or the urinary bladder.

Pathophysiological Aspects. Derangement of the propulsive motility of the small intestine causes some of the most commonly observed disorders of the small intestine. Obviously, the passage of food through the small intestine may proceed too slowly or too fast. A pronounced nervous inhibition of small intestine motility and, hence, decreased propulsion are seen in *paralytic ileus,* in which the small intestine is dilated by gas and fluid, without mechanical obstruction. This distention gives a very characteristic radiological picture of dilated loops of intestine with layerings of fluid. Paralytic ileus is most often seen in patients with diffuse peritonitis and can be interpreted as an exaggerated form of the inhibition of intestinal motility observed after all intra-abdominal surgery. It may also be caused by various painful conditions such as renal attacks, torsion of the testis or blunt abdominal trauma.

Intestinal motility in, for example, a severe case of peritonitis is reflexly inhibited from receptors in the peritoneum (see above). The absence of stirring and propulsion of the intestinal contents, possibly associated with some as yet unknown circulatory derangement of the intestinal

mucosa, may decrease the rate of absorption from lumen to blood to such an extent that the intestine becomes distended with nonabsorbed gas and fluid. This dilatation, in turn, further exaggerates the inhibition of intestinal motility by means of the intestino-intestinal inhibitory reflex. Since there is also an intestino-colonic inhibitory reflex, colonic motility also is reflexly decreased.

Paralytic ileus is, like acute gastric dilatation, a potentially fatal condition because of the induced disturbance of water and electrolyte balance. It has been treated physiologically—by spinal anesthesia, abolishing the sympathetic inhibition of intestinal motility. This treatment is, however, only symptomatic and proper therapy should be directed to the underlying illness also (peritonitis, renal calculus, etc.).

A more subtle disturbance of small intestinal motility is manifested in the radiological picture of malabsorption with dilatation of the gut lumen and segmentation of swallowed barium. Such patterns, potentially rapidly reversible, have been described in association with emotional stress.

Hypermotility of the small intestine and at times, hypomotility may lead to a decreased transit time of the intestinal contents, sometimes resulting in diarrhea. The causes of this common condition are numerous and include infections, poisonings and allergic conditions. It is not known, however, to what extent nervous mechanisms are responsible for this altered intestinal motility.

Colon and Rectum

The colon has two major jobs: removal of salt and water from the intestinal content, and propulsion of the residue for elimination by defecation. The former is accomplished mainly by annular contractions of the colonic wall, resembling segmentation in the small intestine. This type of nonpropulsive motility is usually called *haustration,* and it is easily recognized in a radiological examination of the large bowel by its accordionlike appearance. Propulsive activity involves coordinated movements of circular and longitudinal muscles which differ on the two sides (ascending and descending) of the colon. A progressive squeezing motion occurs on the right side, periodically assisted by a caudally progressive ringlike contraction, while the left side assumes a shortened, narrowed sustained contractile posture resembling a more or less rigid tube.

Parasympathetic Control. The proximal colon is controlled by vagal fibres in a manner similar to that of the small intestine (see above). Electrical stimulation of the pelvic nerves in the experimental animal causes a powerful and sustained colonic contraction, a response that cannot be blocked by atropine. However, the chemical nature of the transmitter involved is not known. This nervously induced colonic motility involves both the circular and the longitudinal smooth muscle layers. It is accompanied by an increase of colonic blood flow and secretion. The pelvic nerve control of colonic motility may be exerted through one or both of the mechanisms illustrated in parts A and B of Figure 29-1.

Sympathetic Control. Electrical stimulation of the regional sympathetic fibers promptly inhibits spontaneous colonic motility and the colonic contractions evoked by stimulation of the vagal or pelvic nerve fibers (Fig. 29-2, A and C).

Reflex and Central Nervous Control. The ascending part of the colon is being filled continuously by the fluid chyme entering through the ileocecal valve. Water and electrolytes are absorbed mainly in the proximal colon, giving the consistency of feces to the contents in distal colonic parts. Three to four times a day, part of the contents of the proximal colon is swiftly transported through the transverse into the descending colon and the rectum. This gross peristaltic movement, usually referred

to as *mass movement* or *gastrocolic reflex*, may be elicited by smoking, by food entering the stomach, or even by the thought of eating. Recent studies suggest that the hormone gastrin may be involved in this mechanism.

The muscular movements required to transport colonic contents from the cecum to the rectum constitute the familiar gastrocolic reflex. In this motility pattern, each portion of the colon has its own contractile characteristic. A systolic contraction occurs in the cecum, followed by shortening and narrowing of the ascending colon; a sequential wave of circular muscle contraction passes distally producing an effect midway between peristalsis and stripping in the transverse colon. On the left side and extending down to the rectum, such propulsive movements are not evident; instead, a sustained contraction of longitudinal and circular muscles occurs, so that this segment becomes a stationary semirigid open tube, through which the contents of the proximal parts are squeezed to the colon by the propulsive movements on the right side. This complex and effective progression occurs with a periodicity that varies widely, both from person to person and with time in a given individual. The nature of the defecated stool (watery, soft, formed, hard or scybalous) is dependent upon the relative predominance, in duration, of propulsive versus desiccating contractile activity.

Rectal distention by feces evokes a sense of fullness and an urge to defecate. This *defecation reflex* can be inhibited consciously by higher brain centers. Defecation is a very complex propulsive movement, involving a contraction of the colon and rectum and a relaxation of the anal sphincters, as well as so-called straining movements of the diaphragm and abdominal wall muscles.

Afferent as well as efferent nervous fibers of the defecation reflex run in the pelvic nerves, severance of which abolishes the voluntary control of the reflex. Reflex contraction of the rectum, together with a reciprocal relaxation of the anal sphincters, can, however, occur in the absence of extrinsic nerves. Such autonomous rectal contraction is comparatively weak, and defecation is difficult without laxatives. A medullary center for the control of defecation as well as inhibitory areas in the hypothalamus have been postulated from animal experiments. Central nervous control of this reflex may be organized in a manner similar principally to that of the intestino-intestinal inhibitory reflex (Fig. 29-3).

Colonic motility is profoundly altered as part of an individual's reaction to emotionally-charged situations. Either constipation or diarrhea may result, with or without associated symptoms or signs such as pain, mucous discharge or bleeding.

Pathophysiological Aspects. The importance of the intramural nerve plexuses for the proper coordination of colonic motility is demonstrated by the clinical condition called *congenital megacolon* (Hirschsprung's disease). This disorder is characterized by an enormous enlargement of the colon proximal to a contracted rectum or sigmoid colon. Colonic dilatation causes a progressive abdominal enlargement. The basic lesion is a congenital absence of myenteric ganglion cells in the contracted bowel. In the dilated colonic portion, however, the intramural nerve plexuses are intact. The presence of the myenteric plexus is apparently a prerequisite for a coordinated relaxation of the large bowel. Thus, the aganglionic bowel segment acts as a functional obstruction, causing the findings described above.

Dilatation and redundancy of the colon may occur without structural lesions and as a result of altered higher neural control, especially accompanying emotional depression. Such constipation constitutes a major problem in psychiatric institutions.

Rectal constipation, or inertia, is characterized by retention of feces in the rectum (and often in the distal colon). Prolonged retention in the rectum results in excessive absorption of water from the feces, which then becomes hardened (impaction).

In rectal constipation, the passage of chyme through the colon and into the rectum is normal, but the defecation reflex fails. This, in all probability, results from a continuous nervous inhibition of the defecation reflex over a period of months or years. Such inhibition of defecation may be caused by a disruption, during emotional tension, of the complex pattern of conditioned reflex behavior of defecation. The defecation reflex may also be inhibited by pain or fear of pain, particularly if the pain is evoked by defecation itself, as in anal fissures and hemorrhoids. The eliciting stimuli act through supraspinal centers controlling the excitability of reflex synapses in the spinal cord or in the intramural plexuses (Fig. 29-3) and/or by a nervous inhibition at the effector cells. Correction of bowel habits through proper training is often successful in the treatment of the psychosomatic type of rectal constipation.

The haustral markings apparent in x-rays of the barium-filled colon indicate the presence of nonpropulsive, drying contractions. The colon arranges itself into a series of small compartments, created by the simultaneous contraction of circular muscles and the taeniae coli which are attached to them at right angles. As adjacent segments contract alternately, a kneading motion squeezes the colonic contents. Water is removed and reabsorbed by the rich vascular network from all levels of the colon. Excessive to-and-fro desiccating movements reduce bulk and make hard stool masses which may be difficult to evacuate. Moreover, the contractions themselves, when intense, may be painful and produce the clinical picture of *spastic constipation.*

Very high pressures may be generated in these haustral segments, produced by the simultaneous contraction of circular muscles and of the taeniae coli. It has been suggested that the pathogenesis of *diverticulosis* involves excessive contractile activity and hypertrophy of the taeniae coli. As a result, diverticular outpouchings occur at the site of entrance of blood vessels into the weakened colonic wall; mucosal herniation might be expected as a consequence of unduly prolonged periods of increased intraluminal pressure.

Haustral markings indicative of desiccating contractions are not evident during propulsive contractile activity. Persistent propulsive activity without adequate desiccating contractions may lead to a watery *diarrhea*, a situation frequently resulting from inflammation and infectious agents or other causes of injury to the colon, especially the impairment of water absorption by cathartics or bacterial toxins. In these situations, the normal propulsive colonic activity is exaggerated and prolonged at the expense of desiccating contractions that reabsorb precious water and salts.

Sometimes, periods of excessive segmental contraction may alternate with exaggerated propulsive activity, resulting in pain and discomfort of alternating constipation and diarrhea, variously called *spastic colon, functional colitis,* or *mucous colitis.*

BLOOD FLOW

The complex vasculature of the gastrointestinal tract may, from an anatomical and physiological point of view, be described in terms of parallel- and series-coupled vascular sections. The parallel-coupled sections are well defined and consist of the vascular circuits in the different layers of the gastrointestinal wall. The series-coupled (consecutive) vascular sections are composed of the parts described below.

The key section of any circulation is its exchange vessels, the true capillaries. It is across their thin endothelial walls that the all-important exchange takes place between intra- and extravascular compartments. The vascular bed proximal to the capillaries consists of the precapillary sphincters and the arterioles. The sphincters usually do not

contribute much to total regional peripheral resistance; however, they are of paramount importance, since they control the size of capillary area and, thus, the mean diffusion distance and time for transcapillary exchange. The average sphincter tone can be estimated by the determination of the capillary filtration coefficient (CFC) (Fig. 29-4). Distal to the capillaries are the venules and small veins, which may profoundly affect the pre- to postcapillary resistance ratio. This ratio is one of the main determinants of mean hydrostatic capillary pressure and, hence, of the rate of the hydrodynamic fluid exchange across the capillary wall. Changes in the capacitance vessels alter regional blood volume without significantly influencing regional resistance to blood flow. Reactions within the above-mentioned consecutive vascular sections can be followed by continuously registering tissue volume (Fig. 29-4) or weight.

Vascular smooth muscle is controlled by myogenic, nervous and local chemical factors. The vascular smooth muscle tone present in a vascular bed during resting conditions (and in the absence of any autonomic nervous influence) is referred to as basal vascular tone. This inherent tone, which varies among the different vascular beds of the body, is predominantly a consequence of precapillary myogenic activity, moderately reinforced by the distending force of the transmural pressure. This factor exerts an excitatory influence on myogenic activity and implies, for example, that an increased arterial pressure augments basal vascular tone, i.e., increases the flow resistance. This is one of the explanations for the autoregulation of blood flow (i.e., relative constancy of flow during wide variations of arterial perfusion pressure) seen in the small intestinal vascular bed. However, in many physiological situations, the myogenic autoregulatory mechanism is more or less overruled by local chemical and nervous factors. Local chemical factors usually have a dilatatory action and consist of metabolites accumulated during aug-

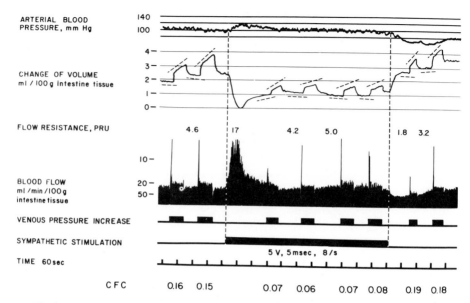

FIG. 29-4. Effect of electrical stimulation of the regional sympathetic vaso-constrictor fibers on arterial pressure, blood flow, tissue volume and capillary filtration coefficient (CFC) of the small intestine of a cat. Capillary filtration coefficient was determined from the slow, continuous increase of tissue volume (filtration slope, indicated by dotted line), registered by a plethysmograph after a sudden increase of venous outflow pressure. (Folkow, B., *et al.*: The effect of the sympathetic vasoconstrictor fibres on the distribution of capillary blood flow in the intestine. Acta physiol. scand., *61*:458, 1964. By permission)

mented tissue metabolism. The nervous influence on the vasculature is mediated by sympathetic vasoconstrictor fibers and, in some compartments, parasympathetic dilator mechanisms.

Stomach

Besides its obvious function (i.e., storage), the stomach is predominantly a secretory organ, a fact that is clearly reflected in the vascular architecture of the mucosa. Thus, there is an extensive capillary network surrounding the secretory cells of the crypts, and a major portion (70 to 80%) of the gastric blood flow is diverted to the mucosa. Mucosal blood flow increases during secretion, either by neural or hormonal mechanisms.

Parasympathetic Control. Stimulation of the low-threshold vagal fibers, which causes a contraction of the gastric muscularis pro-

pria, has no noticeable effect on total gastric blood flow. On the other hand, activation of the high-threshold vagal fibers, which induces corpus–fundus relaxation (receptive relaxation), increases secretion and total gastric flow. High-threshold vagal stimulation relaxes not only the resistance vessels but also the precapillary sphincters, increasing the capillary surface area available for exchange. Such a nervous stimulation might also increase gastric capillary permeability.

Sympathetic Control. Stimulation of the sympathetic fibers produces a characteristic response pattern, which has been studied most thoroughly in the small intestine (Fig. 29-4). The gastric response is qualitatively identical to the response of the intestine; quantitatively, however, differences may exist between the two organs. There is a drastic initial reduction of blood flow and a constriction of the capacitance vessels (decrease of tissue volume, Fig. 29-4) upon stimulation. However, when the stimulation is prolonged, blood flow again increases and, within 3 to 4 minutes, reaches a new steady-state flow level (Fig. 29-4). This phenomenon has been named *autoregulatory escape* from the vasoconstrictor fiber influence. The tissue volume, recorded plethysmographically, remains constant throughout, except for the phasic decrease obtained initially when the blood is expelled as a result of constriction of the capacitance vessels. This indicates that there is no net transcapillary exchange of fluid (i.e., mean hydrostatic capillary pressure is not altered significantly as compared to control). The CFC is markedly reduced, however, reflecting a diminished surface area available for transcapillary exchange. The comparatively small increase of regional vascular resistance, in the face of a drastic reduction of CFC, suggests a redistribution of the intramural blood during neurally mediated vasoconstriction.

Reflex and Central Nervous Control. The gastric vasodilatation observed upon vagal stimulation is usually associated with increased acid secretion. Augmentation of gastric secretion and blood flow may be evoked by emotion as well as by the sight, smell or taste of food. There also exist local intramural reflex arcs that increase the rate of gastric secretion and, probably, mucosal blood flow upon distention and local chemical stimulation of the stomach during the so-called gastric phase of secretion.

Vagal effects on gastric secretion and blood flow have been evoked from various central nervous structures, but it is impossible at present to translate this neurophysiological information into psychological terms. It is well established, however, that emotions are reflected in changes of gastric function. For example, it was demonstrated by Wolf and Wolff on their fistulous subject Tom, that persistent feelings of anxiety were associated with increased gastric secretion and mucosal blood flow, lasting for weeks.

Reflex and central nervous control of the gastrointestinal circulation, as mediated through regional sympathetic vasoconstrictor fibers, has been studied mainly in the small intestine, the most accessible organ for studies of gastrointestinal blood flow. It seems reasonable, however, to assume that the vasculatures of the stomach and small and large intestine behave in a qualitatively similar fashion, although quantitative differences probably exist. Consequently, the following description of sympathetic reflex and central nervous control applies to all vascular beds of the gastrointestinal tract.

Decreased mechanoreceptor activity (elicited by lowered pressure in the carotid sinus) and increased chemoreceptor activity (induced by increased Pa_{CO_2}, or decreased pH or Pa_{O_2}, singly or in any combination) produce reflex sympathetic vasoconstriction through bulbar vasomotor centers. A similar response is seen in the experimental animal upon stimulation of the somatic pressor afferents in peripheral nerves. These fibers probably conduct pain stimuli from the body surface. Visceral pain, in contrast, evokes a fall in arterial pressure and an increase in blood flow as a result of inhibition of sympathetic activity.

Changes in the regional vasoconstrictor neural outflow to the gastrointestinal tract have been observed on stimulation of central nervous structures. One particular area is in the anterior hypothalamus, from which the so-called defense (alarm) reaction can be elicited. This reaction, coupled to a typical behavioral response characterized by attack and flight, includes a well defined autonomic response pattern, including, among other things, an increased neurogenic drive to the heart, skeletal muscle vasodilation, and vasoconstriction in most other circulations, including that of the gastrointestinal tract. It has been proposed that this integrative hypothalamic area is triggered in human beings during stressful conditions. Thus, Tom, the subject studied by Wolf and Wolff, reacted in fear with an alarm reaction, including a blanching of the gastric mucosa.

Pathophysiological Aspects. Peptic ulcer, either duodenal or gastric, is one of the most important of gastrointestinal diseases. Generally speaking, it results from the inability of the gastroduodenal mucosa to withstand the digestive action of the gastric juice. Two main groups of causative agents are usually discussed: factors that induce an excessive secretion of gastric juice, particularly hydrochloric acid, or factors that cause a decline of mucosal tissue resistance. The duodenal ulcer, in contrast to the gastric ulcer, is usually associated with an increased rate of gastric secretion.

Among several factors that may lessen tissue resistance to peptic ulceration is one that has been called vascular insufficiency. This is usually discussed in terms of a mucosal ischemia, possibly caused by a nervous vasoconstriction such as might be induced from the defense area mentioned above. However, it seems improbable to this author that a nervous vasoconstriction *alone* could cause tissue damage. It seems possible that such a mechanism, when occurring together with other local disturbances, may precipitate tissue damage.

Small Intestine

The small intestine is predominantly an absorptive organ. The vascular architecture of the intestinal villi includes an extensive capillary network, situated just below a monolayered intestinal epithelium. As in the stomach, the major fraction (75 to 90%) of the intestinal blood flow reaches the mucosa-submucosa. No extrinsic parasympathetic vasodilator fibers have as yet been demonstrated in the small intestine. However, a local nervous vasodilatation, evoked by mechanical stimulation of the intestinal mucosa, has been demonstrated and may be responsible, in part, for the functional

hyperemia of the small intestine, observed during digestive work. Sympathetic control of the circulation of the small intestine is very similar to that of the stomach. Figure 29-4 illustrates the effects of electrical stimulation of regional sympathetic vasoconstriction fibers on the consecutive vascular sections of the small intestine.

Pathophysiological Aspects. Shock is a clinical condition characterized by a low arterial pressure, pallor, sweating, nausea, restlessness and confusion. This clinical picture reflects a circulatory insufficiency, the causes of which are numerous (see Chap. 4).

The relationship between shock and the intestinal circulation remains unresolved at present. Some authors consider the intestinal vascular bed the key to irreversible shock; others believe the intestinal circulation has little relevance for the genesis of shock. Definitive conclusions cannot be drawn concerning the importance of sympathetic effects upon vessels in explaining necrosis and ulcerations of the intestinal mucosa in experimental animals dying in shock. To judge from experiments in which the regional sympathetic vasoconstrictor fibers have been stimulated (Fig. 29-4), it seems reasonable to assume that the overall intestinal flow resistance is not markedly enhanced during the acute compensatory phase after hemorrhage, whereas superficial mucosal blood flow may be considerably decreased, owing to a redistribution of flow toward deeper intestinal layers. Furthermore, the reduction in venous return to the heart may be partly compensated for by the neurally mediated vasoconstriction of the intestinal capacitance vessels.

Colon

The colon not only has storage and desiccating functions but, like the stomach, is also a secretory organ, producing, mainly in the distal half, a mucoid secretion that presumably serves to lubricate the inspissated fecal contents in association with defecation. A major portion (80 to 90%) of total colonic blood flow is diverted to the mucosa-submucosa in accordance .with functional needs. The constrictor fiber control of colonic vessels is similar to that of the small intestine and the stomach. As to the parasympathetic supply, direct stimulation of the vagal fibers distributed to the proximal part of the colon does not affect colonic blood flow. Activation of the pelvic nerves, on the other hand, induces a transient, marked vasodilatation in the distal parts of the colon, particularly in its mucosal-submucosal parts. This initial increase in blood flow, which often reaches maximal values, is followed by a more prolonged but somewhat less intense vasodilatation, probably evoked by a concurrent colonic secretory activity. The last-mentioned nervous effects are abolished by atropine and, hence, contain a cholinergic link. The initial, marked vasodilatation is not diminished by the adminstration of atropine, and the chemical nature of the transmitter involved is not known. This parasympathetic nervous outflow in the pelvic nerves is reflexly activated during defecation. No experimental work has been done in regard to the central nervous control of colonic blood flow. However (as discussed above under *Motility*), colonic functions are markedly influenced by emotionally stressful events, responses to which are integrated in central connections of the autonomic nervous outflow.

No clinical disorder seems to be positively attributed to neurogenic changes in colonic blood flow. However, it has been repeatedly demonstrated that emotions influence colonic function and mucosal blood supply, evidently by means of nervous pathways. Thus, Grace, Wolf and Wolff found that "life situations provocative of abject fear and dejection were associated with colonic hypofunction of most of the large intestine with pallor, relaxation, lack of contractile activity" whereas "feelings of anger, resentment and hostility or of anxiety and apprehension were found to be associated with hyperfunction of the colon." The onset and exacerbations of colonic disorders such as irritable colon and ulcerative colitis are often associated with emotional conflicts, sometimes referred to as stress. A similar relationship between illness and emotions is known to exist in the stomach. It may be more than a coincidence that gross mucosal disturbances and even ulcerative lesions are almost exclusively confined to the only two compartments of the gastrointestinal tract whose vascular beds, besides their control by sympathetic vasoconstrictor fibers, are exposed to the powerful vasodilator influence of their parasympathetic supply. The possibility exists that a chain of events similar to those producing peptic ulceration may occur in the colonic mucosa also.

ANNOTATED REFERENCES

Almy, T. P.: Experimental studies on the irritable colon. Am. J. Med., *10*:60, 1950. (An interesting review of the author's work on the irritable colon syndrome)

Berger, T.: Studies on the gastric emptying mechanism. Acta chir. scand. (Supp.), *404*: 1, 1969. (A recent study with new techniques on gastric emptying in normal humans and after partial gastrectomy)

Christoffersson, E.: Studies on intestinal and circulatory reactions in provoked dumping attack. Acta chir. scand. (Supp.), *349*:1, 1965. (A pathophysiological study of dumping)

Code, C. F. (ed.): Handbook of Physiology. Section 6, Alimentary Canal, Vol. IV. Washington. The American Physiological Society, 1968. (*The* book as regards the physiology of gastrointestinal motility)

Davenport, H. W.: Physiology of the digestive tract. Chicago, Year Book Medical Publishers, 1971. (An outstanding introduction to the physiology of the digestive tract)

Dragstedt, L. R., Montgomery, M. L., Ellis, J. C., and Matthews, W. B.: The pathogenesis of acute dilatation of the stomach. Surg. Gynec. Obstet., 52:1075, 1931. (A classical paper on acute gastric dilatation)

Folkow, B.: Regional adjustments of intestinal blood flow. Gastroenterology, 52:423, 1967. (A review of the investigations on intestinal blood flow performed at the author's laboratory, including a discussion of the functional organization of the intestinal vascular bed)

Grace, W. J., Wolf, S., and Wolff, H. G.: The Human Colon. New York, Paul B. Hoeber, 1951. (A fascinating monograph on colonic function in humans)

Hultén, L.: Extrinsic nervous control of colonic blood flow and motility. Acta physiol. scand. (Supp.), 335:1, 1969. (A comprehensive review and study of colonic motility and blood flow)

Jacobson, E. D.: The circulation of the stomach. Gastroenterology, 48:85, 1965. (A thorough review of the literature on the subject together with an interesting discussion of the relationship between gastric blood flow and peptic ulcer)

Jansson, G.: Extrinsic nervous control of gastric motility. Acta physiol. scand. (Supp.), 326:1, 1969. (A careful and thorough investigation of the parasympathetic and sympathetic control of gastric motility)

Jansson, G., Lundgren, O., and Martinson, J.: Neurohormonal control of gastric blood flow. Gastroenterology, 58:425, 1970. (A brief review of the subject together with some thoughts on the pathogenesis of peptic ulcer)

Johansson, B., Jonsson, O., and Ljung, B.: Supraspinal control of the intestino-intestinal inhibitory reflex. Acta physiol. scand., 63:442, 1965. (An interesting study of the central nervous control of the intestino-intestinal inhibitory reflex. Figure 29-3 of this chapter is based on this and the following paper.)

———: Tonic supraspinal mechanism influencing the intestino-intestinal inhibitory reflex. Acta physiol. scand., 72:200, 1968. (See note to the preceding reference.)

Machella, T. E.: Mechanism of the post-gastrectomy dumping syndrome. Gastroenterology, 14:237, 1950. (A classical paper on the dumping syndrome)

Norberg, K. A.: Adrenergic innervation of the intestinal wall studied by fluorescence microscopy. Int. J. Neuropharmacol., 3:379, 1964. (The adrenergic innervation of the gut mapped by use of the ingenious fluorescence technique of Hillarp)

Texter, E. C., Jr.: Small intestinal blood flow. Am. J. Dig. Dis., 8:587, 1963. (A review article including a fairly extensive discussion of the small bowel in shock)

Truelove, S. C.: Movements of the large intestine. Physiol. Rev., 46:457, 1966. (A good physiological review with pathophysiological aspects)

Wangensteen, O. H.: Intestinal Obstructions. Springfield, Ill., Charles C Thomas, 1955. (The well known monograph on surgical disorders of intestinal motility)

Wolf, S.: The Stomach. New York, Oxford University Press, 1965. (A recent monograph on gastric function by one of the investigators who made the classical studies on the fistulous subject Tom)

Youmans, W. B.: Nervous and Neurohormonal Regulation of Intestinal Motility. New York, Interscience Publishers, 1949. (An outstanding monograph on the subject)

30

Neural Control of Skeletal Muscle

William J. Crowley, Jr., M.D.

MECHANISMS OF DEVELOPMENT AND MAINTENANCE OF NORMAL MUSCLE BULK

Development and maintenance of muscle bulk depend upon many different factors, including inheritance, stimulation of development by the motor nerve and by somatotrophic, androgenic and perhaps other hormones, adequate nutrition and optimum exercise.

Disease Mechanisms Producing Atrophy

Meager muscle bulk, or atrophy, may result from absence of one or more of the factors mentioned above or may be a consequence of disease. Examples of conditions producing atrophy include denervation atrophy, myositis and disuse.

Denervation Atrophy. The motor nerve exerts an influence, called the trophic factor, on muscle fibers that is indispensable for development and maintenance of structure, metabolism, excitability and contractile properties of muscle. Acetylcholine, the neuromuscular junction transmitter, released both spontaneously (to produce miniature end-plate potentials) and as a result of nerve impulse, has been postulated to be the trophic agent but probably is not. A radioactive isotope of phosphorus injected into the anterior horn cell region of the spinal cord is transported along the motor axons by axonal flow. Eventually, it appears within the muscle fiber, suggesting that acetylcholine is not the only substance transmitted from nerve to muscle. Muscle, when deprived of its nerve supply,

atrophies markedly—owing mainly to loss of sarcoplasmic protein, including energy-producing enzyme systems. There is early sarcolemmal nuclear proliferation and increased muscle fiber deoxyribonucleic acid (DNA) content. Myofibrillar protein is retained for a relatively longer period, but it, too, eventually diminishes, with concomitant loss of contractility. In the histological examination of incompletely denervated muscle, atrophic fibers may be grouped according to their distribution in the motor unit. The appearance of islands of normal muscle fibers surrounded by seas of atrophied fibers and chains of sarcolemmal nuclei is particularly characteristic of chronic denervating processes in which some amount of reinnervation has taken place. Motor neurons destroyed by diseases such as paralytic poliomyelitis cannot regenerate, thus, recovery is limited and occurs only to the extent possible as a result of terminal axon sprouting. In denervation atrophies due mainly to axonal damage—for example, peripheral neuritis—astonishing regrowth may occur.

Atrophy in Myositis. In myositis, muscle fibers undergo necrosis and phagocytosis and are replaced, to a degree, by fibrous tissue and fat. Fiber regeneration may be active but, if this regeneration cannot keep pace with the destructive process, muscles will atrophy. The extent of fiber regeneration can be gauged by the amount of restoration of muscle bulk that takes place in treatment of some myositis patients with corticosteroids.

Disuse Atrophy. Muscles can be prevented from contracting by immobilization (for example, casting) or tenotomy and in isolated spinal cord preparations. The disuse atrophy which results is characterized by loss of myofibrillar protein and increase in collagenous protein (fibrosis). Although the late picture of disuse atrophy may be indistinguishable from the late stages of denervation, there are differences in time sequence and in degree and types of changes.

Other Mechanisms. Lesions in the cerebral cortex, especially those occurring early in life, may be associated with underdevelopment or atrophy of the contralateral half of the body, including the muscles. This seems to be more marked with lesions occurring in the parietal cortex than with lesions occurring elsewhere in the cortex and is probably the result of trophic influences as yet undisclosed. Finally, sickness can produce nonspecific wasting, resulting, in part, from inactivity and, probably, in part from liberation of toxins or competition for available metabolites.

Disease Mechanisms
Producing Hypertrophy

Pathological enlargement of muscles, as contrasted with physiological enlargement due to work (exercise), may occur in gigantism, hemihypertrophy, localized limb hypertrophy and pseudohypertrophic muscular dystrophy. In the first condition, excessive growth hormone is present. In localized or lateralized hypertrophy, there is an association with neurofibromatosis, arteriovenous malformations, Wilms' tumors and familial occurrence, which suggests that trophic nerve factor, blood supply or other growth-stimulating influences may be excessive. In pseudohypertrophic muscular dystrophy, the calf and deltoid (and other) muscles may be enlarged, in part because of increased fat and fibrous tissue content but also because of unusually enlarged individual muscle fibers. Such muscles are weak

compared to healthy muscles of comparable bulk but may be strong compared to other muscles in the patient's body. Compensatory hypertrophy by work may be one cause of the enlargement of these muscle fibers, and there may be other factors as yet unknown. Similarly, the stimulus for production of fibrosis in dystrophic muscle is not known.

MECHANISMS OF DEVELOPMENT AND MAINTENANCE OF NORMAL MUSCLE TONE

The extremely slight, smooth elasticity observed by an examiner when flexing or extending a limb of a relaxed normal subject constitutes normal tone. The anatomical structures and physiological processes underlying normal muscle tone are as yet imperfectly understood. Reduced to simplest terms, this phenomenon involves skeletal muscle innervated by anterior horn cells whose firing is controlled by local and remote influences.

Local Influences

When the tendon of a normal muscle is struck forcibly, a brisk reflex contraction of that muscle is elicited. This reflex can be abolished by section of the sensory as well as of the motor nerve fibers supplying the muscle. Afferent impulses arising in muscle spindles, which are found in all skeletal muscles, are essential to the generation of deep tendon reflexes. Muscle spindles are complex sensors of muscle length and rate of lengthening and run parallel to the skeletal muscle fibers (Fig. 30-1). Each spindle is a fusiform structure, consisting of a lymph-filled fibrous capsule, attached at each end to the perimysium of adjacent muscle fibers and containing from 2 to 8 specialized intrafusal muscle fibers. There are two types of intrafusal muscle fibers: nuclear bag fibers with distended equatorial regions filled with sarcolemmal nuclei, and thin nuclear chain fibers within which the

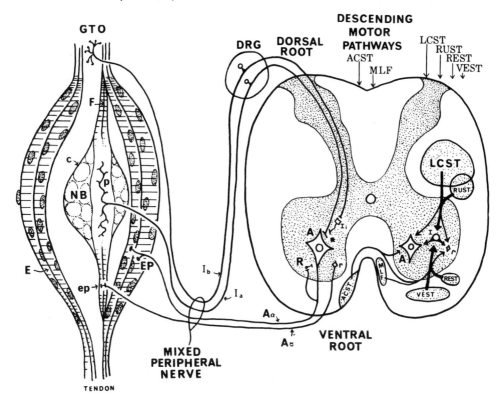

FIG. 30-1. Semi-diagrammatic representation of the muscle spindle and its related neural pathways.

A	Alpha motor neuron (anterior horn cell)	γ	Gamma motor neuron
Aα	Axon of alpha motor neuron	GTO	Golgi tendon organ
Aγ	Axon of gamma motor neuron gamma efferent, fusimotor fiber)	I	Interneuron
		Ii	Inhibitory interneuron
ACST	Anterior corticospinal tract	LCST	Lateral corticospinal tract
C	Capsule of muscle spindle	MLF	Medial longitudinal fasciculus
DRG	Dorsal root ganglion	NB	Nuclear bag region of intrafusal muscle fiber
E	Extrafusal muscle fiber (skeletal muscle)	P	Primary (annulospiral) nerve ending
		R	Renshaw cell
EP	Neuromuscular junction (end plate) of alpha motor neuron axon on skeletal muscle	REST	Reticulospinal tract
		RUST	Rubrospinal tract
		VEST	Vestibulospinal tract
ep	Neuromuscular junction (end plate) of gamma motor neuron axon on intrafusal muscle fiber	I_a	Group I_a primary spindle afferent fiber
		I_b	Group I_b Golgi tendon organ afferent fiber
F	Intrafusal muscle fiber		

sarcolemmal nuclei are arranged in single file. Intrafusal muscle fibers, both types of which are smaller and shorter than ordinary extrafusal muscle fibers, are found in most spindles. The extremities of each intrafusal muscle fiber are striated muscle in-

nervated by thin motor axons that conduct impulses at the speed of group A gamma fibers and are called fusimotor or gamma efferents. Stimulation of gamma efferents evokes contraction of the intrafusal muscle fibers, an action that stretches the equatorial region of the fiber but does not directly shorten or increase the tension of the muscle as a whole. Two types of specialized sensory nerve endings are found at the equatorial region of intrafusal muscle fibers, each with a characteristic location and disposition of its branches. These are the primary (annulospiral) and the secondary (flower-spray) endings with their large myelinated axons, I_a and II afferents, respectively. Elongation of the muscle—and, consequently, of the spindles—distorts the sensory nerve endings, thus generating impulses that produce excitatory postsynaptic potentials (EPSP's) on the somas of anterior horn cells (alpha motor neurons) innervating the same muscle. If these afferent facilitatory impulses bombarding the alpha motor neuron are sufficiently intense, they may initiate a conducted action potential in it. This, in turn, provokes contraction, which shortens the muscle, reduces the distortion of the spindles and diminishes the afferent facilitatory impulses. The muscle spindles can shorten and resume firing, even in a shortened muscle, through the action of the gamma efferents, the stimulation of which causes intrafusal muscle fibers to contract, thus stretching the central sensory region again. The deep tendon reflex thus has an afferent component, perhaps mediated largely by nuclear bag fibers, and two efferent components which enable elicitation of the reflex over wide ranges of muscle length and tension. However, the segmental reflex arc serves a function far more important than the clinically useful deep tendon reflex phenomenon. It is the mechanism by which muscle tone is established, perhaps mediated largely by nuclear chain fibers. In addition, voluntary movement is facilitated through coordinated functioning of both types of spindle fibers.

An additional sensory organ that influences this segmental reflex arc is the Golgi tendon organ—a high-threshold tension sensor, located at the musculotendinous junction in series with muscle fibers. Group I_b afferent fibers from the Golgi tendon organ enter the spinal cord and synapse with inhibitory neurons, which, in turn, make contact with the presynaptic endings of I_a afferent terminals. These act to depolarize the fine presynaptic nerve terminals and, thus, reduce their ability to liberate the chemical transmitter substance which excites alpha motor neurons. This is an example of presynaptic inhibition. Golgi afferents also stimulate motor neurons supplying antagonists of the muscle in question, producing facilitation of their discharge (the inverse myotatic reflex).

Alpha motor nerve fibers give rise to a recurrent collateral, which returns to the spinal cord anterior horn of its origin to synapse with an inhibitory interneuron, the Renshaw cell. This, in turn, synapses with the original anterior horn cell giving rise to the recurrent collateral.

A motor neuron is activated when the spatial and temporal summation of converging synaptic depolarizations on its dendrites and soma creates sufficient current flow to depolarize the critical initial segment of the soma to threshold for development of a self-regenerating action potential which is propagated down the axon. The synaptically induced, localized decremental depolarizations of dendrites and soma are termed excitatory postsynaptic potentials (EPSP's). The membrane of the soma and dendrites is stabilized and prevented from depolarizing through the effects of synaptic contact from other cells—for example, Renshaw cells, which produce inhibitory postsynaptic potentials (IPSP's). Another method of inhibition involves reduction of the ability of presynaptic nerve terminals to liberate transmitter substance, as with Golgi afferents—an effect termed presynaptic inhibition. Theoretically, presynaptic inhibition could be exerted at both excitatory and inhibitory synapses. A motor neuron may thus receive excitatory and inhibitory postsynaptic potentials or these influences may be reduced by presynaptic inhibition. The functioning of this reflex system may be illustrated by the simple act of flexing the elbow against gravity. The messages to contract originate in the contralateral precentral gyrus and are conducted by the corticospinal and other descending motor pathways to the appropriate segments of the cervical spinal cord. Some of these descending impulses cause depolarization of alpha motor neurons supplying the biceps muscle, causing them to discharge. This initiates contraction in the biceps muscle but also initiates two processes that tend to inhibit further biceps alpha motor neuron discharges. These processes are recurrent collateral (Renshaw) inhibition and biceps muscle shortening, reducing the stretch on the muscle spindles. The latter reduces the discharge rate in I_a afferents and thus diminishes the net depolarization of biceps alpha motor neuron membranes. At this point, further biceps muscle contraction would stop were it not for the depolarization of gamma motor neurons created by other descending impulses. Gamma motor neuron discharges produce intrafusal muscle fiber contraction, stretching the spindle and permitting I_a afferent messages to continue. In this manner, biceps alpha motor neuron discharges continue, and biceps muscle continues to contract until the desired degree of flexion is achieved and descending upper motor neuron influences diminish.

Remote Influences

It is apparent that increased muscle tone and movement can occur as a result of facilitation as well as decreased inhibition of alpha and gamma motor neuron discharge. By the same token, decreased muscle tone and movement can occur as a result of less facilitation as well as increased inhibition of alpha and gamma motor neuron discharges. The degree to which a given mechanism may result in the observed phenomenon is not always readily determinable. Facilitatory influences appear to orig-

inate in the propriospinal system of the spinal cord, vestibular nuclei, tegmentum (reticular formation) of the mesencephalon, pons and rostrolateral medulla, cerebellar deep nuclei and cerebral cortex. Inhibitory influences originate in the ventromedial bulbar reticular formation, cerebellar cortex, subthalamic nucleus, the rostral cingulate gyrus, corpus striatum and cerebral cortex. Normal muscle tone—the semivoluntary muscle contractions that support the limbs and body against gravity and the anticipatory postures that serve as the ground from which voluntary movement emerges—depends upon normal integration of the functions of the excitatory and the inhibitory centers described above and the connections between these structures and the local spinal segments. Influences originating in these centers descend to the spinal level by way of the corticospinal, vestibulospinal, rubrospinal, olivospinal, reticulospinal and tectotegmentospinal tracts in the lateral and ventral white matter of the spinal cord and terminate on spinal interneurons, gamma motor neurons and alpha motor neurons. It is incorrect to think in terms of distinct pyramidal and extrapyramidal systems, each responsible for independent functions. For example, parkinsonian rigidity, a common example of a manifestation of "extrapyramidal" disease, is greatly reduced by section of the pyramidal tract. It is certainly true that lesions of the basal ganglia result in patterns of disturbance of tone, posture and movement different from those produced by lesions in the cortex or corticospinal pathway. For simplicity of conceptualization, it is convenient to think in terms of the contrasting pyramidal and extrapyramidal disorders.

Disease Mechanisms Producing Hypotonia

Excessive floppiness of the limbs, the tendency for joints to be hyperextensible and diminished damping of deep tendon reflexes (pendularity) constitute hypotonia. It is noted when segmental reflex arcs are interrupted, as in peripheral neuropathies, radiculopathies, myelopathies and flaccid paralysis accompanying other neuromuscular disorders. Hypotonia is also present when facilitating influences upon motor neurons are decreased or inhibitory influences are increased, as in sleep and coma and in cerebellar disturbances. In acute spinal cord transections or following any acute brain or brain stem lesion that affects motor systems, muscle tone and deep tendon reflexes in the affected portion of the body may be decreased or absent, at least transiently. This effect is attributed to loss of tonic excitatory influences upon spinal motor neurons. The recovery and subsequent increase in muscle tone and deep tendon reflexes have been attributed to postdenervation hypersensitivity to excitatory transmitter substance of central facilitatory neurons.

Disease Mechanisms Producing Hypertonia

Muscle tone is evaluated best when the patient is relaxed, warm and comfortable. However, children, elderly people and demented or anxious individuals may be unable to relax. Often, movement of the patient's limb by the examiner is not truly passive but is anticipated and assisted or resisted by semivoluntary contractions of the patient's muscle (paratonia). The resistance encountered under these circumstances should not be interpreted as resting muscle tone. Furthermore, movement may be inhibited by pain and joint changes.

Spasticity. Initial resistance to passive elongation of a muscle followed by sudden collapse of the resistance is known as spasticity. It results from lesions of the pyramidal tract. The pyramidal tracts appear to inhibit brain stem and spinal influences responsible for facilitating muscle tone. Lesions of the pyramidal tract "release"— i.e., permit to become active—facilitatory influences of these lower centers. These stimuli, carried chiefly by the vestibulospinal, reticulospinal and propriospinal pathways, produce facilitation of the gamma motor neurons. Pathological facilitation of gamma efferents produces pathological increase in spindle sensitivity to muscle lengthening (passive or active) and increased spindle afferent drive to alpha motor neurons. Attempted muscle elongation results in brisk reflex muscle contraction. The increase in muscle tension, in an effort to overcome the reflex contraction and continue to elongate the muscle, activates the Golgi tendon organs. This results in reflex inhibition of the motor neurons supplying the elongating muscle and reflex stimulation of the motor neurons supplying antagonists of the elongating muscle. These mechanisms produce the classical "clasp-knife reaction," in which passive muscle stretch is met initially with resistance (muscle stretch reflex) and, later, with collapse ("lengthening reaction") of this resistance, owing to activation of Golgi tendon organ inhibition. Spasticity may be mild (barely detectable) to marked (difficult to overcome). It is responsible for exaggerated deep tendon reflexes, clonus, scissoring gait and snapping into place or jiggling of the knee while walking (spastic gait). In human beings, spasticity from cerebral lesions is most marked in the gastrocnemius-soleus, quadriceps, adductors, finger flexors and biceps brachii (antigravity muscles) and is most readily detected in these muscles in mild cases.

Decerebrate Rigidity. This state was originally described by Sherrington in midcollicular transection of the mesencephalon of experimental animals. In human beings, it consists of rigidity of the extensor and adductor muscles of the legs, plantar flexion of the feet and toes, and extension, internal rotation and adduction of the upper extremities. Often, opisthotonos is also present. Like spasticity, opisthotonos results from a release of brain stem facilitatory drive of gamma efferents, chiefly vestibulospinal; however, Golgi inhibition is not as apparent.

Alpha Rigidity. Carnivores submitted to experimental anoxic decerebration (by bilateral occlusion of carotid and vertebral arteries) develop severe rigidity which is not abolished by dorsal root section and does not depend upon fusimotor hyperactivity—and is therefore unlike spasticity and decerebrate rigidity. It is the result of direct facilitation of alpha motor neuron discharge, presumably because of loss of anterior lobe vermis influence (perhaps loss of fastigiovestibular inhibition).

Extrapyramidal Rigidity. This is defined as resistance to passive elongation of the muscles which may vary from slight to marked and is uniform throughout the extent of elongation of the muscle. There is no clasp-knife phenomenon such as occurs in spasticity. Extrapyramidal rigidity is attributed to lesions in the substantia nigra (Parkinson's disease), caudate nucleus (Huntington's chorea), putamen and globus pallidus (Wilson's disease, kernicterus, dystonia musculorum deformans, etc.). Rigidity may often be abolished by lesions in the medial globus pallidus, the nucleus ventralis lateralis of the thalamus, or the pyramidal tracts, by dorsal root section or by infiltration of the muscle with dilute procaine. This indicates that, at least in some instances, rigidity is mediated by the corticospinal tracts and depends upon activation of the muscle spindles. It seems apparent that central dopaminergic pathways (particularly the nigrostriatal pathway) must be important in the production of parkinsonian rigidity, since the administration of L-dopa to parkinsonian patients tends to overcome it. Another indication that central monoamine pathways are important in the production of extrapyramidal rigidity is the fact that the administration of drugs that interfere with central monoamine synaptic transmission—for example, reserpine, phenothiazines, monoamine oxidase inhibitors and tricyclic antidepressants—may also produce extrapyramidal rigidity.

Rigidity Resulting from Failure of Local Inhibition of Motor Neuron Firing. Tetanus, strychnine intoxication and the rare, spontaneously occurring stiff-man syndrome are examples of muscular stiffness and spasm resulting from failure of local inhibition of motor neuron firing. This failure is probably due to specific defects in the functioning of inhibitory interneurons such as the Renshaw cell.

Myotonia is defined as the inability to relax promptly after a vigorous muscle contraction. It is typically more marked in distal muscles, worse after a period of rest and less bothersome after repeated elicitation. It is observed in myotonia atrophica, myotonia congenita, paramyotonia (cold-induced myotonia) and in association with hyperkalemic periodic paralysis. It is also seen in laboratory animals (goats) and in poisoning with 2,4-dichlorophenoxyacetic acid (2,4-D) and other chemical substances. Whether or not myotonia is produced by the same basic mechanism in each of these conditions is not known. Forceful muscle contraction elicited by voluntary effort, electrical stimulation or percussion evokes, after the stimulus has subsided, repeated depolarization of muscle membrane and consequent contractions. Properties of the sarcolemma, sarcoplasm, sarcoplasmic reticulum and myofibrils are being investigated in these disorders. It is clear that sarcolemmal excitability is increased and that nonphysiological stimuli are adequate to provoke depolarization and contraction. Perhaps this is caused by changes in the resting membrane potential, critical membrane depolarization required to elicit an action potential (threshold), ionic conductances or electrical impedance. Junctional transmission, excitation-contraction coupling and myofibrillar contraction do not appear to be the principal sites of defect.

MECHANISM FOR DEVELOPMENT AND MAINTENANCE OF MUSCULAR STRENGTH

In the normal individual, muscular strength is a function of muscle mass, training and effort. Exercise produces changes in muscle which increase the strength of contraction and the endurance for repetitive contractions. The quantitative changes that appear depend, to a degree, upon the type of exercise. Very brief, intensive training appears to increase the amount of glycolytic activity of muscle fibers, particularly the white (fast, Type II) fibers. Prolonged intensive training increases myofibrillar and sarcoplasmic protein. Weight lifting emphasizes the former, running and swimming the latter. Increasing activity of enzymes and cofactors of oxidative metabolism, increased myoglobin and increased numbers of mitochondria are produced by prolonged training. Increased cross-sectional areas of muscle fibers (red more than white) and increased numbers of capillaries (more so in white than in red) are also produced by prolonged training. These changes result in increased contractile machinery and increased capability to supply energy to this machinery.

Disease Mechanisms Producing Muscular Weakness

Muscular weakness is a common symptom which may result from lesions in the anatomical locations or physiological processes outlined in Table 30-1.

Neuropathies. The motor neuron, as the final common pathway from the central nervous system to muscle, is of paramount importance in the integration of segmental and extrasegmental facilitatory and inhibitory influences. Motor nerve impulses are the only useful means of evoking muscular contraction. Loss or dysfunction of the motor neuron, from whatever cause, results in weakness or paralysis. The distribution of the muscle weakness depends upon the site and extent of nerve disease. In motor neuron disorders such as amyotrophic lateral sclerosis, weakness becomes generalized. In spinal cord disorders such as paralytic poliomyelitis or spinal cord tumor, weakness is determined by the extent of spinal cord involvement. In nerve root disorders, weakness is seen in the muscles innervated by a particular nerve root. In mononeuritis, the denervated muscles are those supplied by a single peripheral nerve. In polyneuritis, the distal musculature is usually involved first and most severely, but the condition may intensify to produce generalized flaccid paralysis. In most peripheral neuropathies, these changes are accompanied by sensory and trophic disturbances.

Normal Physiology of the Neuromuscular Junction

The neuromuscular junction consists of the terminal arborization of the motor axon, the specialization of the sarcolemma underlying the motor nerve terminal (the sole

TABLE 30-1. Anatomical and Physiological Processes Resulting in Muscle Weakness

---

Destruction of anterior horn cells or failure of anterior horn cells to discharge

Inadequate effort
Upper motor neuron lesions
Lower motor neuron lesions
  Amyotrophic lateral sclerosis
  Poliomyelitis
  Other brain stem and spinal cord disorders

Nerve root disorders

Degenerative disorders of the spine and intervertebral disks
Neoplasms

Peripheral nerve disorders

Neuromuscular junction disorders

Myopathies
  Hereditary
  Acquired
    Inflammatory myopathies
    Dyshormonal myopathies
    Metabolic myopathies
    Toxic myopathies

Disuse

Diseases of bone, joints or tendons

Systemic illness

plate) and the intervening synaptic cleft. Acetylcholine is formed in the motor nerve terminal from choline and acetyl CoA and stored in uniform quantities in small vesicles near the junctional axolemma. Periodically, single vesicles spontaneously discharge their content of acetylcholine into the synaptic cleft. The liberated acetylcholine associates with receptors located in the sarcolemma, causing local subthreshold depolarizations, called miniature end plate potentials (MEPP's). Motor nerve action potentials produce synchronized release of the acetylcholine from many vesicles. This saturates the sarcolemmal receptors and produces depolarization of the entire sole plate, called end plate potential (EPP), which is suprathreshold for generation of conducted muscle fiber action potentials. Acetylcholinesterase enzymes located in the sarcolemma cause rapid hydrolysis of acetylcholine and repolarization of the sole plate to make it ready for the next motor nerve impulse.

Neuromuscular Junction Disorders

Myasthenia gravis is characterized by fluctuating weakness, increased by repetitive muscle contraction and relieved by rest and the administration of cholinesterase inhibitors. The underlying defect in myasthenia gravis appears to be progressive decrease in the quantity of acetylcholine released from motor nerve terminals in response to repetitive voluntary motor nerve impulses. Declining transmitter release results in progressively increasing neuromuscular blockade during sustained or repetitive muscle contractions (myasthenic fatigue). An analogous phenomenon is observed in the EMG during repetitive motor nerve stimulation; the amplitude of indirectly evoked muscle responses progressively declines (myasthenic decrement). Muscle membrane excitability, muscle contractility, sensitivity of junctional sarcolemma to acetylcholine and acetylcholinesterase activity appear to be normal. The

reason for low acetylcholine output by nerve terminals is not known. Some defect in storage or packaging of acetylcholine in the nerve terminal appears to be involved. Synaptic vesicles in the motor nerve terminal appear to contain less than the normal amount of acetylcholine. This is reflected in the lowered amplitude of miniature end plate potentials and by the lowered amplitudes of end plate potentials. The total amount of acetylcholine within the nerve terminal and the release mechanism itself appear normal. An association of myasthenia gravis with thymic disorders—for example, thymic neoplasm (thymoma) and thymic hyperplasia and inflammation—has long been known. Some defect in thymus activity is almost certainly involved in the genesis of myasthenia gravis, perhaps liberation of a toxic substance from a reactive or inflamed gland or from a thymic neoplasm. Antibodies which bind to thymocytes, skeletal muscle striations and sarcolemmal nuclei are also frequently present, but their role is not understood. The presence of these antibodies, the association of myasthenia gravis with other autoimmune disorders (e.g., thyroiditis) and the relief of symptoms by corticosteroid or corticotrophic therapy have suggested that myasthenia gravis is an autoimmune disorder. Overtreatment of myasthenics with cholinesterase inhibitors such as neostigmine may produce depolarizing and desensitizing neuromuscular blockade, which may significantly increase the patient's weakness and produce unpleasant autonomic side effects. This is called cholinergic crisis. The same manifestations accompany administration of some anesthetic adjuvants and intoxication by organic phosphorus insecticides (e.g., parathion) and some nerve gases. The toxic effects of some of these agents can be reversed by the administration of pralidoxime; parasympathetic hyperactivity accompanying most of these intoxications is overcome with atropine. Nondepolarizing neuromuscular blockade is produced by curarelike drugs which combine with receptor sites on the muscle membrane. The drugs themselves do not depolarize the receptor sites but prevent acetylcholine from reaching its receptor site to produce depolarization. Patients with myasthenia gravis are extraordinarily sensitive to extremely low doses of curare, an effect which may be used in a provocative diagnostic test for myasthenia in questionable cases.

Myasthenic Syndrome. The Eaton-Lambert syndrome is an uncommon disorder, characterized by weakness of the proximal limb muscles, which is overcome, to a degree, by sustained or repeated muscle contractions. In contrast to myasthenia gravis, acetylcholine release is insufficient at the beginning of voluntary or indirectly evoked muscle contractions. The initial weakness is overcome by progressive augmentation of acetylcholine release during sustained voluntary contraction or repetitive evoked contractions. The mechanism of release of acetylcholine from presynaptic vesicles in response to motor nerve impulse appears to be defective. The number of vesicles discharging their acetylcholine rather than the amount of acetylcholine contained in each vesicle appears to be decreased. This results in lowered amplitude of initial end plate potentials produced by motor nerve stimulation, but amplitude and frequency of miniature end plate potentials appear to be normal. Repeated stimulation results in augmentation of the number of vesicles discharging their contents with each succeeding impulse, creating augmentation of the amplitude of evoked muscle responses as junctional transmission is improved. This facilitation of junctional transmission with repeated stimulation is detected with an evoked potential electromyogram and constitutes the basis for diagnosis of this condition. Guanidine is often effective in relieving weakness in this syndrome whereas standard cholinesterase inhibitors may be less so.

Botulism. Botulinus toxin appears to coat the motor axon terminals to prevent acetylcholine liberation. Botulinus antitoxin binds the circulating toxin and guanidine helps augment acetylcholine release, but the mortality from intoxication with this substance is still virtually 100 percent.

Miscellaneous Neuromuscular Junction Poisons. Hemicholinium and triethylcholine inhibit acetylcholine synthesis in the neuromuscular junction and produce failure of junctional transmission after prolonged stimulation in experimental preparations. High concentrations of these drugs produce postjunctional blockade.

Myopathies

Weakness in myopathies may be the result of decreased numbers of muscle fibers, as occurs in all muscular dystrophies and inflammatory myopathies, or to impairment of the ability of the muscle fibers to contract. The latter disorder may result from loss of sarcomeres through necrosis and phagocytosis or shrinkage of fibers with loss of myofibrils and sarcoplasm. Replacement of myofibrils by noncontractile substances such as central cores, nemaline bodies, vacuoles, fat and hyaline material (loss of cross striations, indicative of loss of actin and myosin structures and interrelationships) may also be responsible. Muscle fiber contractility may be impaired, owing to an inadequate milieu for muscle excitation and/or contraction, such as occurs in thyrotoxicosis, myxedema, adrenal dysfunction, hypokalemia, hyperkalemia, hypercalcemia and failure of energy-yielding metabolism.

Thyrotoxic Myopathy. Muscular weakness often occurs in thyrotoxicosis and is associated histologically with simple fiber atrophy and fatty infiltration especially in proximal muscles. In electron micrographs of muscle obtained from patients with thyrotoxic myopathy, abnormal, swollen mitochondria and focal dilatation of the transverse tubular system have been observed.

Similar changes have been induced experimentally by administration of thyroxine. Also, administration of excessive thyroxine is thought to uncouple oxidative phosphorylation in mitochondria, resulting in a decrease in adenosine triphosphate (ATP) formation, an increase in oxygen utilization and an increase in heat production. Excessive thyroxine also appears to decrease creatine phosphokinase (CPK) activity resulting in a decrease in formation of creatine phosphate. Since creatine phosphate is an energy-storage molecule which donates a phosphoryl group to adenosine diphosphate (ADP) to form ATP when needed, ATP concentration is thus further reduced. Muscle contraction is known to involve actin, myosin and ATP; reduction of the supply of ATP decreases the ability of muscle to contract. Diminished CPK activity in thyrotoxic muscle results in increased creatinemia with increased creatinuria (200 to 500 mg. per 24 hours) and decreased uptake by muscle of exogenously administered creatine (decreased creatine tolerance). Creatinine is the inactive metabolite of creatine which is formed in the muscle and excreted through the kidney at a relatively constant rate. In thyrotoxicosis, the production of creatinine and its excretion in the urine are both decreased to less than one gram per 24 hours.

The pathogenesis of the muscular dystrophies is not known. As in most inherited disorders, some enzymatic defect may be postulated but this defect is yet to be identified. Certain characteristics of dystrophic muscle have been identified, such as the ease with which creatine phosphokinase escapes through the sarcolemma (which accounts for the extremely high concentrations of this enzyme in the sera of afflicted individuals and carriers of the genetic trait). The resting membrane potential in dystrophic muscle is reduced, but muscle excitability remains relatively normal. Similarly, the etiology of polymyositis is not known.

In the past, it was classified among the collagen-vascular disorders, but recent electron micrographs reveal virus particles in muscle cells, raising the question of viral etiology for some cases of this condition.

The Role of Potassium in Normal Muscle Functioning

The resting membrane potential of nerve and muscle is maintained by the active exclusion of sodium ions from within the fiber by the sodium pump, a process involving oxidative metabolism. An integral part of the functioning of the sodium pump is to maintain a high concentration of potassium within the fiber against its concentration gradient but in approximate accordance with the electrical potential gradient across the membrane. In fact, an approximation of the resting membrane potential can be determined experimentally if the concentrations of potassium ions inside and outside of the membrane are known. Depolarization and hyperpolarization of the membrane may be brought about by altering the external concentrations of potassium significantly. The resting membrane is quite impermeable to sodium ions and relatively indifferent to significant alterations in serum sodium concentrations.

Weakness Associated With Hypokalemia. Lowered serum potassium ion concentration is found in renal diseases, excessive use of diuretic agents, certain gastrointestinal disorders, in association with thyrotoxicosis, and in hyperaldosteronism. In these disorders, lowered serum potassium levels are associated with weakness or paralysis. Abnormal potassium metabolism is associated with muscular weakness in three inherited conditions known as the familial periodic paralyses. The most common variety is hypokalemic periodic paralysis, in which attacks of muscle weakness are ushered in by a fall in serum potassium ion concentration, most of which is absorbed by the muscle fibers. Despite the demonstrated rise in intracellular and fall in extracellular potassium ion concentration, the membrane potential of paralyzed muscle does not rise. In fact, it remains constant or depolarizes slightly. The propagated muscle action potential decreases in amplitude and increases in duration at the beginning of an attack of paralysis, but decrease in strength of muscle contraction occurs earlier and is more marked than changes in the action potential. This suggests that some alteration of depolarization-contraction coupling or of the contraction process itself is involved in the muscle weakness. This is reinforced by the observation that the weakness persists despite the return of serum potassium ion concentration to normal or above normal, with treatment. Central vacuoles are commonly found in muscle fibers of patients with hypokalemic periodic paralysis. Some observers consider these to represent dilated sarcoplasmic reticulum, but there is no universal agreement on this point. Glycogen may be deposited in these vacuoles, suggesting that carbohydrate metabolism is disordered and, in fact, may be the principal pathogenic defect in hypokalemic periodic paralysis. Muscle fiber degeneration associated with this vacuolar change is undoubtedly the reason for the irreversible muscular weakness and wasting which appear in some patients with hypokalemic periodic paralysis. The administration of diuretics, potassium, or spironolactone greatly improves the condition or totally prevents attacks.

Weakness Associated With Hyperkalemia. Hyperkalemic periodic paralysis is characterized by attacks of muscle weakness associated with elevated serum potassium ion concentrations. Membrane potential is decreased during attacks of hyperkalemic periodic paralysis, owing, in part, to mechanisms described above, but it is doubtful that the increase in extracellular potassium ion concentration is sufficient to account for the degree of membrane depolarization observed in this condition. As the membrane potential falls during an attack, muscle fibers become irritable, fire spontaneously and demonstrate myotonia; soon they become inexcitable. The mechanism of paralysis in normokalemic periodic paralysis is not known. Both of these conditions, which are considered similar, if not identical, to Gamstorp's disease (adynamia episodica hereditaria), are aggravated by the administration of potassium. Hyperkalemic periodic paralysis may be treated with intravenous calcium gluconate during the attack, whereas normokalemic paralysis is best treated with 9α-fluorohydrocortisone. Although many of these conditions may subside spontaneously later in life, occasionally they are associated with the development of proximal muscle weakness and wasting of a progressive myopathy.

MECHANISMS OF MOVEMENT

In human beings, voluntary movement appears to originate in the cerebral cortex, especially the frontal lobe. Impulses originating in cortical pyramidal neurons (including Betz cells from the precentral gyrus) descend with the corticospinal and corticobulbar tracts through the internal capsule and cerebral peduncle; the tracts decussate in the brain stem and terminate on alpha motor neurons and interneurons in the brain stem and spinal cord. The frontal motor cortex initiates appropriate voluntary movements in response to sensory information arising from the periphery, integrated with memories of previous experience stored in association cortex and motivated by impulses from limbic structures and elsewhere. This is not to imply that other structures and pathways do not play a role in voluntary movement. In fact, the frontal motor cortex is really the highest level of control of movement, exerting its influence over lower centers, each of which performs its own function. These functions seem to assume a progressively more complex reflex pattern as the level of mediation of the reflex ascends from the caudal to the rostral end of the nervous system. For example, the segmental monosynaptic reflex responsible for the muscle stretch reflex is, in one sense, uncomplicated. The triple flexion reflex in response to nocioceptive stimuli is mediated over a number of spinal segments and is correspondingly more complex. The tonic neck reflex, righting reflex and reflex walking are examples of progressively more complex reflex mechanisms mediated at progressively higher brain stem (or brain) levels and employing the lower reflexes in their expression. Some of these reflexes are observed in normal subjects only under special circumstances, but the reflex mechanisms function below the level of consciousness to facilitate voluntary movements initiated by the frontal motor cortex.

Disease Mechanisms Producing Hypomotility

Since muscular weakness is one of the principal causes of hypomotility, the differential diagnosis and the mechanisms responsible for the two manifestations may be similar. However, hypomotility may occur in the absence of muscular weakness. For example, hypokinesia, bradykinesia, lack of associated movements, infrequent blinking and masked facial expression occur in Parkinson's disease. These are negative signs, resulting from the absence of normal activity in neurons destroyed by disease. Lesions in the substantia nigra, the nigrostriatal pathway and other central dopaminergic pathways no doubt play an important role in the development of these manifestations of Parkinson's disease, since treatment with L-dopa frequently leads to gratifying improvement in them. However, one of the complications of L-dopa therapy is the production of akinetic attacks in which the patient stands for long periods without being able to move at all. This suggests that nigrostriatal pathway dysfunction is not exclusively responsible for hypokinesia in Parkinson's disease and directs attention

to lesions elsewhere—for example, in the globus pallidus. Experimentally, profound hypokinesia may result from extensive bilateral lesions in the globus pallidus. Hypokinesia may also be seen in emotional disorders such as hysterical paralysis, depression and catatonic schizophrenia. Transmission at central monoamine synapses is implicated in the pathogenesis of at least one of these disorders (depression). It is known that agents that exhaust the central nervous system content of serotonin (e.g., reserpine) may induce depression, whereas agents that increase the central nervous system content of catecholamines (e.g., monoamine oxidase inhibitors) are beneficial in the treatment of depression.

Disease Mechanisms Producing Involuntary Movements

Some involuntary movements (tremor, chorea, athetosis, ballismus, torsion dystonia and other dyskinesias, nystagmus and clonus) occur because activity in the surviving neurons is unopposed or unmodified by activity in other neurons that have been destroyed or disconnected by disease process. In a sense, the function of certain nervous system elements is released from control or inhibition by damaged cells, producing abnormal involuntary movements which, together with increased muscular tone, represent positive neural phenomena. Another way in which neurons may react to injury is to become irritable, with discharges occurring largely independent of synaptic influences. In tetany, fasciculations and, perhaps, epilepsy and myoclonus, these irritative phenomena appear to be present.

Ballismus is a movement disorder characterized by abrupt, violent flinging movements of the extremities, the upper extremities especially. Pathologically, it was associated with lesions in the subthalamic nucleus. The most common cause of such lesions is vascular occlusion—for example, in patients with diabetes mellitus. Experimentally, destruction of 20 percent of the subthalamic nucleus in primates results in contralateral hemiballismus and chorea. On the basis of the known connections of the subthalamic nucleus, it is postulated that the net effect of this group of neurons is to inhibit the globus pallidus. In the terms mentioned above, hemiballismus is a release phenomenon of the globus pallidus.

Chorea. Characteristically, choreic movements are extremely quick, often very slight graceful movements recurring repeatedly, especially in distal or facial muscles. They occur in their classical form in Huntington's disease, in which degeneration of the corpus striatum and cerebral cortex is found. They are also seen in senile progressive chorea, in which degeneration of the corpus striatum is unaccompanied by degeneration of the cerebral cortex and, hence, dementia does not occur. Choreic movements occur in Sydenham's (rheumatic) chorea, in which no consistent pathology has been demonstrated. Choreiform movements may also be seen with infarcts in the region of the red nucleus or the subthalamic nucleus, and occasionally in hepatic encephalopathy and in Wilson's disease.

Athetosis is characterized by sinuous movements of the extremities, especially distally, in which slow flexion alternates with hyperextension, most particularly in the digits, with hyperextension appearing to be the predominating tendency. Although this condition is described as mobile spasm, there is a tendency toward assumption of athetoid posturing of the hands extended against gravity and toward postural spasms of the limbs when the body is suspended off the ground. Athetosis may be regarded as the midportion of a spectrum of movements which merges, at one extreme, with chorea and, at the other, with torsion dystonia. Pathologically, lesions are found most consistently in the corpus striatum, especially the putamen, but by no means are the lesions limited to this region. Disorders in which degeneration of the puta-

men is particularly noted include congenital double athetosis, calcification of the cortex and basal ganglia and Wilson's disease. Occasionally, athetosis may be seen in disorders in which the globus pallidus appears to be primarily involved—for example, Hallervorden-Spatz disease and kernicterus. Athetosis may occasionally be seen in demyelinating disorders such as the Pelizaeus-Merzbacher syndrome.

Torsion Dystonia (Dystonia Musculorum Deformans). The identifying feature of this condition is the tendency to a spasmodic twisting and turning of the spine and limbs; the neck and back are hyperextended and the face and distal portions of the extremities are relatively spared. Movements come in attacks, either rapid or slow in their onset and subsidence, are intensified by willed movements with which they interfere and subside in sleep. Choreic and athetoid movements may be present to a greater or lesser degree, suggesting once again the close relationships between these three manifestations. Occasionally this rare disorder occurs in families, in which case pathological changes occur in the putamen and thalamus, globus pallidus, substantia nigra, subthalamic nucleus and cerebral cortex. It has been produced in humans by surgical lesions in the parvicellular area of the centromedian nucleus of the thalamus. Occasionally it may occur as a result of encephalitis or cerebral anoxia, or in the course of tumors or Wilson's disease.

Mechanisms of Disturbances of Posture. Mechanisms responsible for muscular tone are also responsible for posture. The decerebrate state has been discussed in a previous section (Hypertonia). Another characteristic disturbance of posture occurs in Parkinson's disease, in which there is a tendency toward progressive flexion, as if the patient intended to roll up in a ball. This is the prototype of Denny-Brown's pallidal syndrome, in which tonic antigravity influences appear to be pathologically diminished. Presumably, this is caused by lesions in the globus pallidus, although lesions in the substantia nigra, corpus striatum, ansa lenticularis and cerebral cortex are also found in Parkinson's disease. The opposite disturbance of posture, Denny Brown's striatal syndrome, is typified by dystonia musculorum deformans, in which the principal lesions are found in the corpus striatum. Experimentally, torsion dystonia can be produced by destructive lesions in the centromedian nucleus of the thalamus. Spasmodic torticollis, considered by many to be an incomplete form of dystonia musculorum deformans, has been produced by destructive lesions in the mesencephalic tegmentum. Both of these postural syndromes have been abolished by lesions in the nucleus ventralis lateralis of the thalamus.

Tremors are traditionally considered to be of five varieties: essential, metabolic, anxiety, Parkinsonian, and cerebellar. Oscillations or vibrations of the head, trunk and extremities are natural phenomena occurring in all individuals. In normal persons, they may be of such low amplitude as to be imperceptible except on very close observation—for example, with the aid of biomedical instrumentation. If the oscillations are noticeable to visual inspection, they are called tremor. Normal persons tremble—*i.e.*, the oscillations increase in amplitude—under certain conditions, including anxiety, fear, anger, fatigue and cold (shivering). With the possible exception of shivering, these tremors are usually simple vertical oscillations of low amplitude, occurring at a frequency of between 8 to 12 per second. They are best seen in the extended finger tips with the arms held straight out from the body, in which case they are called *static* or *postural* tremor. These tremors are also seen during movements of the limbs, in which case they are called *action tremor*. Action tremor is usually of constant frequency and amplitude throughout the full range of movement but it is intensified when an attempt is made to perform

some movement requiring precise control. Some persons tremble without the predisposing conditions mentioned above and to this extent, they are not normal. If postural or action tremor is the only abnormality present, the disorder is called *essential tremor*. If essential tremor occurs in several members of a family, it is called *familial tremor*. Essential or familial tremors are usually mild, but occasionally they are so intense as to be disabling. Tremor occurs in many patients with thyrotoxicosis, pheochromocytoma and severe anxiety states. These tremors resemble essential tremor in that they are postural or action tremors and occur in the same general frequency range (8 to 12 per second). There is a pharmacological synergism between thyroxine, epinephrine and norepinephrine. Epinephrine is thought to increase the amplitude of postural tremor, in part by stimulating peripheral beta-adrenergic receptors; thyroxine increases the physiological response to epinephrine. These three hormones also influence the rate of development and decay of muscle tension in response to a single nerve impulse ("twitch tension") and alter (often decrease) the smoothness of maintained muscular tension during repetitive nerve impulses short of completely fused tetany. Propranolol decreases the amplitude of tremors induced by intravenously administered epinephrine as well as in patients with thyrotoxicosis and those with severe anxiety.

Two experimental substances, tremorine and its active metabolite, oxotremorine, have been developed for the purpose of investigating mechanisms of tremor. Like epinephrine, they have significant effects upon peripheral beta-adrenergic effectors which can be blocked by the administration of propranolol. In addition, they increase the liberation of acetylcholine in the central nervous system. Thus, they may exert a portion of their tremorogenic effect at the level of the basal ganglia, the reticular formation or even in the spinal cord, perhaps at the level of the Renshaw cell. Parkinsonian tremor is described as a pill-rolling tremor at rest, abolished by voluntary movement. Parkinson's disease has been subdivided into three distinctive types on the basis of antecedent or associated illness and clinical and histopathological manifestations. Postencephalitic parkinsonism follows an attack of encephalitis and is characterized by dyskinesias such as oculogyric crises; lesions are found in the substantia nigra, locus ceruleus, globus pallidus and ansa lenticularis. Arteriosclerotic parkinsonism is associated with onset at a later age; the predominating manifestation is rigidity, and the putamen especially is involved pathologically. Paralysis agitans, idiopathic parkinsonism, is unassociated with a history of encephalitis or manifestations of cerebrovascular disease. The onset is at an earlier age and tremor may be the predominating manifestation. Pathological lesions are seen in the globus pallidus and ansa lenticularis, and hyaline neuronal inclusions, Lewy bodies, are prominent in the substantia nigra and locus ceruleus. Treatment with L-dopa has sometimes been disappointing in regard to suppression of parkinsonian tremor and also because it tends to evoke dyskinetic movements in a significant number of patients. This suggests that parkinsonian tremor may be only indirectly (if at all) produced by lesions in the nigrostriatal dopaminergic pathway, which is presumed to be responsible for other manifestations such as rigidity and bradykinesia. Although the locus of origin of parkinsonian tremor is not known, surgical lesions in the nucleus ventralis lateralis and the nucleus ventralis anterior of the thalamus relieve the tremor. Tremors resembling parkinsonian tremor have been produced by experimental lesions in the substantia nigra, but it has been argued that such lesions also involve portions of the mesencephalon and the brachium conjunctivum, which are certainly important in the genesis of tremor.

Cerebellar tremor, also called ataxic or intention tremor, is said not to be present at rest or with posture holding, but it becomes manifest with voluntary movement and is exaggerated near the end point of intended movement. Lesions of the cerebellar cortex, the deep cerebellar nuclei and the superior cerebellar peduncle produce ipsilateral intention tremor as well as ataxia, dysmetria, dyssynergia, dysdiadochokinesia, dysrhythmia, hypotonia and dysarthria. Lesions of the red nucleus, which receives projections from the contralateral deep nuclei of the cerebellum, produce effects similar to those mentioned above but on the side of the body contralateral to the side of the injury. In addition, lesions of the red nucleus produce choreiform movements and, occasionally, myoclonus. Physiological data indicate that the net effect of cerebellar Purkinje cells upon the cerebellar nuclei to which they project is inhibitory. The net effect of activation of the cerebellar nuclear cells is excitatory upon the centers to which they project—for example, Deiter's nucleus, brain stem reticular formation, red nucleus and thalamus. Thus, the effect of activation of the cerebellar cortex must be to inhibit these projection centers. The cerebellum receives important projections from spinal cord pathways mediating unconscious proprioception. Coordinated movements which depend upon integration of these proprioceptive impulses with cortical and subcortical influences are disturbed by cerebellar lesions. Cerebellar tremor has been abolished by lesions in the nucleus ventralis lateralis and ventralis anterior, which indicates that impulses necessary for production of cerebellar tremor must pass through these nuclei on their way to the cortex.

The distinctions between cerebellar, parkinsonian and essential tremors are questioned on experimental as well as clinical grounds. Experimentally, tremors may be produced by lesions in many regions of the nervous system. For example, cerebellar as well as parkinsonian tremors can be produced by lesions in the brachium conjunctivum. From the clinical point of view, parkinsonian tremor is often not abolished during voluntary movement, certain types of cerebellar tremors are present at rest, and essential tremors share in characteristics of the other two varieties. All of these tremors have been successfully abolished by surgical lesions in the ventral lateral and ventral anterior nuclei of the thalamus, which funnel cerebellifugal (dentato-rubro-thalamic) and pallidofugal (nigro-striato-pallido-thalamic) influences to frontal lobe cortex. These thalamic nuclei must be important way stations in the production of tremor if not the site of its origin. Unlike clonus, these tremors are unaffected by dorsal rhizotomy and are abolished by spinal cord transection. It appears that tremor results from failure of damaged neural circuits to achieve precisely the proper muscle tone or to smoothly synergize movements, with reciprocal overcompensation by counterbalancing influences. Tremor rhythm may be determined by the duration of inhibitory postsynaptic potentials, particularly in spinal cord reticular neurons. Whether these tremors originate in the cerebral cortex, thalamus, basal ganglia, cerebellum, or the spinal cord (for example, the propriospinal system) remains to be seen.

Convulsive Phenomena. Epilepsy and convulsive phenomena involve a profound disturbance of the controlled sequential discharge of neurons that underlies normal central nervous system function. Failure of inhibition and excessive and avalanching synchrony of discharge of variable populations of neurons are characteristic of the disturbance of neuronal function in epileptic seizures. In experimental as well as in naturally occurring epileptic foci, membranes of individual neurons are depolarized, the threshold for firing is reduced, and the local population of epileptic neurons fires synchronously. In some instances the dendrites of epileptic neurons are stunted or atrophic, and there is a failure of inhibitory influences to reach them. Surrounding the epileptic focus may be a zone of neurons with increased membrane polarization and resistance to repetitive firing, the inhibitory surround. An epileptic focus may remain localized without clinical accompaniment, detectable only by electroencephalogram (EEG) or electrocorticography (ECoG) as an area from which spikes can be recorded. Then, for reasons poorly understood, the epileptic focus may spread by recruitment of neighboring neurons into the epileptic pool—perhaps as a result of failure of surround inhibition—and by projection of impulses to central structures and other synaptic terminals of the axons of epileptic neurons. The local spread may be accompanied by a subjective experience, the aura, or other partial manifestations that indicate the beginning of the seizure. If the epileptic focus is located in the precentral cortex, this centrifugal concentric spread is manifested by a jacksonian march. If the focus is in the temporal lobe, the patient may experience, among other things, hallucinations, perceptual distortions, recurrent memories, stereotyped movements or affective changes, all fragments or caricatures of normal temporal lobe functioning. Projection to central structures, i.e., the diencephalon or rostral midline mesencephalon, may be followed by secondary projections of synchronized discharges to the cortex at large through the nonspecific thalamocortical projection system, accompanied by loss or alteration of consciousness. Another consequence of central projection may be the caudal transmission of epileptic discharges to brain stem and spinal cord motor neurons. This usually occurs in two phases, accompanied by characteristic EEG (or ECoG) changes. The first, or tonic, phase consists of vigorous extension of the limbs and trunk, accompanied, in the EEG, by 8 to 10 spikes per second recorded from all cortical and subcortical areas. These represent soma spikes, neuronal action potentials. The second, or clonic phase, begins as a fine quivering of the muscles, which gradually evolves into distinct contractions separated by progressively lengthening relaxation periods, until the seizure ends, after about five minutes, when another contraction fails to follow a period of relaxation. The EEG during this phase shows slow waves, probably inhibitory polarization in dendritic fields, which intersperse spikes or brief bursts of spikes with a frequency of approximately 1 to 6 per second. Following his seizure, the patient is frequently flaccid and comatose, the EEG being, at the same time, flat and featureless (asynchronized neuronal discharge). Gradually the patient regains consciousness and the EEG regains its usual appearance. It has long been believed, although increasingly questioned, that epileptic seizures may originate centrally as opposed to focally in the cortex. These idiopathic, familial, centrencephalic or primarily generalized seizures are postulated to begin in the intralaminar or reticular nuclei of the thalamus or mesencephalic reticular formation and project rostrally and caudally through the pathways described above. Usually, rostral and caudal projections occur simultaneously, but occasionally primarily rostral projection occurs to produce altered consciousness with little motor concomitant. This may occur in the absence variety of petit mal attacks. The EEG correlate of this type of seizure is a rhythmical wave and spike pattern of 3 per second, accentuated by hyperventilation. Primarily caudal projection may occur in the akinetic or myoclonic forms of petit mal (in which a 3-per-second wave and spike pattern may also be seen) and in myoclonus and tonic spasms which may not be associated with any specific EEG abnormalities or alteration of consciousness. Tonic spasms may be seen in brain stem anoxia (anoxic, ischemic and histo-

toxic), compression of the brain stem, transtentorial or foramen magnum herniation and in tetany, tetanus and poisoning by medications such as strychnine, picrotoxin and camphor. Epileptic seizures are, in fact, manifestations of underlying dysfunctions, which may be either primary in the central nervous system or a reflection of a systemic process. Effective treatment involves identification, as far as possible, of the underlying cause and therapy specifically directed toward its correction. Treatment of epileptic seizures may be directed toward suppressing the epileptic focus (phenobarbital), preventing the transsynaptic spread (diphenylhydantoin), or increasing inhibitory tone (diones and succinimides).

Tetany.   Neuromuscular irritability (cramps, fasciculations, carpopedal spasms, Chvostek's sign and convulsions) accompanies lowered serum concentrations of ionized calcium. Normal total serum calcium concentration is 10 mg. per 100 ml. (2.45 mM per liter). Approximately 50 percent (1.18 mM per liter) of total serum calcium is ionized and is diffusible across excitable membranes. The remainder is bound to proteins, especially to albumin, or complexed with phosphates, citrates, etc. Calcium ions are in equilibrium with undissociated, bound and stored (in the skeleton) calcium, the relative concentrations determined in part by parathyroid hormone levels and serum pH. When the serum level of ionized calcium falls below 1 mM per L. at pH 7.4, the above-described manifestations begin to appear. Excitable membranes of both nerve and muscle cells develop increased excitability in the presence of lowered serum calcium concentration. This excitability results from a decrease in membrane resistance, an increase in sodium permeability and decreased accommodation to depolarizing currents. The way in which calcium produces these changes is not known. Certainly calcium and other divalent cations improve non-synaptic membrane stability. The net result is a reduction of the resting membrane potential and elevation of the critical membrane potential required for development and propagation of an all-or-nothing action potential. Spontaneous impulses occur when the excitability is thus increased, apparently in central and peripheral neurons and in muscle fibers. Low serum calcium levels inhibit synaptic transmission (including neuromuscular junction) by inhibiting the release of synaptic transmitters. Spontaneous activity of muscle cells produces fibrillations that can be recorded electromyographically. However, fasciculations, cramps, spasms and convulsions are abolished by curare, indicating that neural activity is responsible for their production. Fasciculations persist in tetany following nerve section or block, indicating a peripheral nerve origin for these phenomena. Tonic–clonic convulsions in tetany are abolished by dorsal root section and by spinal cord transection. This indicates that brain and spinal reflex arcs are required to produce the tonic and clonic spasms. Metabolic or respiratory alkalosis produces tetany, in part by reducing the ionized calcium content of the blood. Total serum calcium may be normal in this condition. However, carbon dioxide is also a potent stabilizer of excitable membranes and hypocapnia results in vasoconstriction, so that tissue anoxia may be responsible for some aspects of respiratory alkalosis.

Nystagmus. In the conscious subject, the position and voluntary movement of the eyes is normally under the control of the eye movement centers in the frontal and the occipital lobe—responsible, respectively, for voluntary and pursuit eye movements. This control is mediated through centers in the pretectal region, midbrain and pons, the eye muscle nuclei being connected by the medial longitudinal fasciculus. Some involuntary control of eye position and movement is exerted by the labyrinth and perhaps other structures—for example, the cervical muscle spindles. When the labyrinth, eighth cranial nerve, vestibular nuclei or medial longitudinal fasciculus is injured, its influence upon eye position and movement is decreased or lost. The net effect of tonic vestibular influence upon eye muscles is to turn the eyes conjugately in the contralateral direction. When the impulses from one labyrinth are interrupted, the eyes tend to deviate toward the involved side, driven there by the opposite normal labyrinth. In the unconscious subject, in the absence of frontal and occipital lobe influences, the eyes tend to become tonically deviated to the side of the lesion. In the conscious subject, the cortical gaze centers exert sufficient influence to restore the eyes to midline, but an imbalance persists which requires periodic readjustment. This imbalance results in phasic eye movements called nystagmus. Labyrinthine nystagmus is characterized by two phases—a slow phase, induced by insufficiency of labyrinthine influences, and a fast phase induced by cortical restorative influences. The fast phase, by which the nystagmus is named, is in reality the "normal" phase of the nystagmus. Unfortunately, phasic nystagmus does not always accurately localize or even lateralize a lesion in the labyrinth–eye muscle pathway. In an effort to improve the accuracy of diagnostic localization of lesions in patients with brain stem, eighth nerve or vestibular lesions, tests such as caloric stimulation have been developed. Lavage of the external auditory canal of the right ear with cold water normally produces nystagmus, fast component to the left (tonic deviation of the eyes to the right in the unconscious subject). Nystagmus induced by vestibular lesions may be horizontal, diagonal or rotatory, but vertical nystagmus is virtually always indicative of a lesion of the brain stem.

Pendular eye movements seen in darkroom workers and mine workers (miner's nystagmus) are not of vestibular origin but are an exaggeration of the normal scanning eye movements that underlie the physiology of vision. Miner's nystagmus is induced by inadequate illumination for the development of macular fixation. Scanning eye movements are also present in macular degeneration, particularly that occurring early in life. Opticokinetic nystagmus (elevator or railroad car nystagmus) is nystagmus evoked by trains of moving objects passed in a vertical or horizontal direction before the eyes of a subject. It is constituted by momentary fixation and pursuit of one object in the train, followed by movement of the eyes to fixate upon and follow the next object, and so on. Lesions in the posterior half of the cerebral hemisphere are most likely to cause decrease or absence of opticokinetic nystagmus in moving the objects from the normal side toward the side of the lesion.

Palatal nystagmus (palatal myoclonus) is an involuntary movement involving the soft palate and, frequently, the external nares, tongue, pharyngeal constrictors, the tiny muscles in the middle ear and the diaphragms. The involuntary movement is a rhythmical contraction of these muscles with a frequency of approximately 2 to 4 per second. Pathologically, it is associated with lesions in the central tegmental tract and inferior olivary nucleus. It may be seen with limited brain stem infarctions, multiple sclerosis, encephalitis and other disorders. Unlike so many of the other involuntary movements, it persists in sleep.

Myoclonus. Extremely abrupt, brief, occasionally very powerful disorganized muscle contractions occurring involuntarily are known as myoclonus. Myoclonic jerks may occur normally, for example, when one is just dropping off to sleep, at which time they are caused presumably by a brief alteration of the influence of the brain stem reticular formation upon spinal cord motor neurons associated with falling asleep. Myoclonus may also be seen in many diseases, especially those with neuronal degeneration. In the latter instance,

it is often associated with convulsive seizures. Examples of these disorders include progressive myoclonus epilepsy, dyssynergia cerebellaris progressiva, Tay Sachs disease, and so on.

Fasciculations. These brief involuntary contractions of small portions of a muscle actually represent the approximately synchronous depolarization and contraction of all muscle fibers contained in a single motor unit. These twitchings may be sufficient to contract a small (for example, interphalangeal) joint. In the EMG, they can be differentiated only with great difficulty from voluntary motor unit potentials. They may occur in otherwise normal individuals, especially under conditions of stress or fatigue or following unaccustomed exercise or excessive coffee drinking. Under these conditions, they may involve more than a single motor unit, occur rapidly and repetitively in the same muscle and are sometimes called myokymias. Fasciculations may indicate motor neuron disease, for example, amyotrophic lateral sclerosis, in which case they rapidly become associated with other signs of the disorder such as weakness, wasting and pathological reflexes. It appears that the nerve impulse responsible for fasciculations arises near the neuromuscular junction, since sectioning the motor nerve or blocking it with local anesthetic does not abolish them. They also may be seen in radiculopathies, occasionally in peripheral neuropathies and, not uncommonly, in the presence of neuromuscular junction depolarizing agents such as succinylcholine.

ANNOTATED REFERENCES

Cooper, I. S.: Involuntary Movement Disorders. New York, Hoeber Div., Harper & Row, 1969. (A remarkable monograph presenting the accomplishments and theories of an innovative neurosurgeon)

Denny-Brown, D.: The Basal Ganglia and Their Relation to Disorders of Movement. New York, Oxford University Press, 1962.

(A monograph containing the observations and theories of a neurologist who has had considerable influence upon the thinking of a generation of American neurologists and neurophysiologists)

Eccles, J. C.: The Physiology of Synapses. New York, Springer-Verlag, 1964. (One of several excellent monographs dealing with synaptic transmission by a famous Australian neurophysiologist who won the Nobel Prize for physiology in 1963)

Eccles, J. C., Ito, M., and Szentagothai, J.: The Cerebellum as a Neuronal Machine. New York, Springer-Verlag, 1967. (Further evidence of Eccles' genius, this work was accomplished after he had already won the Nobel Prize.)

Elmqvist, D., Hoffman, W. W., Kugelberg, J., and Quastel, D. M. J.: An electrophysiological investigation of neuromuscular transmission in myasthenia gravis. J. Physiol. (London), *174*:417, 1964. (A work which has shaped our current concept of the pathophysiology of myasthenia gravis)

Elmqvist, D., and Lambert, E. H.: Detailed analysis of neuromuscular transmission in a patient with the myasthenic syndrome sometimes associated with bronchogenic carcinoma. Mayo Clin. Proc., *43*:689, 1968. (An analysis of the pathophysiology of the Eaton-Lambert syndrome)

Erlanger, J., and Gasser, H. S.: Electrical Signs of Nervous Activity. ed. 2. Philadelphia, University of Pennsylvania Press, 1968. (Reprinting of the work for which the authors received the Nobel Prize in 1944)

Granit, R.: Receptors and Sensory Perception. New Haven, Yale University Press, 1955 (2nd printing, paperback, 1962). (Lectures dealing with some of the work for which the author received the Nobel Prize in 1967)

Guth, L.: "Trophic" influences of nerve on muscle. Physiol. Rev., *48*:645, 1968. (A very concise statement of the influences of nerve upon skeletal muscle)

Gutmann, E.: The Denervated Muscle. Prague, Czechoslovak Academy of Sciences, 1962. (A somewhat more detailed analysis of the trophic effects of nerve upon muscle by a pioneer in the field)

Hodgkin, A. L.: The Conduction of the Nervous Impulse. Springfield, Ill., Charles C Thomas, 1964. (A monograph dealing with the physiology of nerve membranes, including work for which the author shared the Nobel Prize in 1963)

Hornykiewicz, O.: Dopamine (3-hydroxytryptamine) and brain function. Pharmacol. Rev., *18*:925, 1966. (An exhaustive review of the anatomy and neurochemistry of central nervous system dopaminergic synapses)

Katz, B.: Nerve, Muscle and Synapse. New York, McGraw-Hill, 1966. (A very readable monograph dealing with membrane phenomena and synaptic transmission by the recipient of the Nobel Prize for physiology in 1970)

McArdle, B.: Metabolic and endocrine myopathies. *In:* Walton, J. N. (ed.): Disorders of Voluntary Muscle. ed. 2, Chap. 18, p. 607. Boston, Little, Brown & Co., 1969. (A tidy summary of this topic with an excellent bibliography)

Martin, J. P.: The Basal Ganglia and Posture. London, Pitman Medical Publishing Co., 1967. (Speculations regarding the functions of the basal ganglia from the clinical point of view)

Mettler, F. A.: Cortical subcortical relations in abnormal motor functions. *In:* Yahr, M. D., and Purpura, D. P. (eds.): Neurophysiological Basis of Normal and Abnormal Motor Activities. p. 445. New York, Raven Press, 1967. (Functions of the basal ganglia from the physiological point of view)

Pearson, C. M.: Polymyositis and related disorders. *In:* Walton, J. N. (ed.): Disorders of Voluntary Muscle. ed. 2, Chap. 15, p. 501. Boston, Little, Brown & Co., 1969. (Excellent summary of this important subject by the leading contributor in the field today)

Sherrington, C. S.: The Integrative Action of the Nervous System. ed. 2. New Haven, Yale University Press, 1947 (paperback). (This is included mainly for its historical value, a classic in neurophysiology by the founding father of a school of neurophysiologists and recipient of the Nobel Prize for physiology in 1932.)

Shy, G. M., Wanko, T., Rowley, P. T., and Engel, A. G.: Studies in familial periodic paralysis. Exp. Neurol., *3*:53, 1961. (Probably the most complete and technically the best article dealing with this subject)

Thesleff, S., and Quastel, D. M. J.: Neuromuscular pharmacology. Ann. Rev. Pharmacol., *5*:263, 1965. (An excellent review of neuromuscular pharmacology)

Truex, R. C., and Carpenter, M. B.: Human Neuroanatomy. ed. 6. Baltimore, Williams & Wilkins, 1969. (A very readable and up-to-date neuroanatomy text containing much clinical correlation)

Waldstein, S. S.: Thyroid-catecholamine inter-relations. Ann. Rev. Med., 17:123, 1966. (A good review of the pharmacology of thyroid and adrenal hormones and their interaction)

31

Learning Mechanisms

Leo V. DiCara, PH.D.

CONDITIONING AND LEARNED RESPONSES IN VISCERAL CONTROL

Classical conditioning is important to the study of psychophysiologic disorders because it is assumed to provide the basic model for emotional learning. In the Russian laboratories, however, classical conditioning has had much broader significance, partly because of Pavlov's particular theories of the neurophysiological events behind conditioning, but also because of the diverse experimental observations that the Russian investigators have made. In general, the Russians have been more deeply aware of the importance of higher cortical integrative activity in visceral and somatic activity than have their American counterparts. The Russians have concentrated on three aspects of conditioning: interoceptive conditioning, semantic conditioning, and theories of neurosis and personality that arise from Pavlov's work. In experiments on interoceptive conditioning, the Russian investigators have shown that all internal organs are under neural control and can have their functions altered by conditioning. More recently, American investigators have demonstrated that visceral responses also can be directly modified by trial-and-error learning, a form of learning much more flexible than classical conditioning (see below).

AUTONOMIC NERVOUS SYSTEM

Peripheral Aspects

The peripheral organization of the autonomic nervous system ensures the autonomy and automatism of the various visceral functions. Central connections ensure that visceral control is adaptive to external conditions and responsive to the highest levels of neural integration. Peripheral organization is found to some extent in the effector organ itself, as in smooth muscle or glands, as well as autonomic ganglia and plexuses.

The peripheral autonomic nervous system consists of a sympathetic, or thoracolumbar, division and a parasympathetic, or craniosacral, division. The division is based on anatomical, pharmacological and physiological grounds, the most important of which include the facts that the anatomical distributions of the nerve fibers in the two divisions are distinct from each other, and that synaptic transmission in the two divisions is accomplished, for the most part, by the action of different neurotransmitters. The effects of the two divisions on the organs they innervate are often antagonistic to each other.

Ganglia of the sympathetic division lie in close proximity to the spinal cord (Fig. 30-1). The sympathetic fibers leave the cord through the anterior roots of the spinal nerve and pass into the sympathetic chain, where they synapse. From here, fibers travel in two directions; some pass into visceral sympathetic nerves that innervate the internal organs, and the others return through the gray ramus back into the spinal nerve. The latter fibers then travel throughout the body along the spinal nerves. Since the synaptic connections in the sympathetic division are made near the spinal cord, the

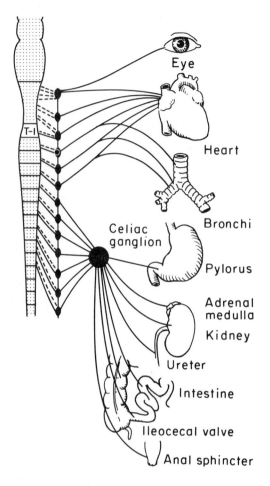

Eye

Heart

Bronchi

Celiac
ganglion

Pylorus

Adrenal
medulla

Kidney

Ureter

Intestine

Ileocecal valve

Anal sphincter

T-I

FIG. 31-1. Oversimplified anatomy of peripheral sympathetic nervous system.

preganglionic fibers are short and the post-ganglionic fibers are long.

Although parasympathetic fibers may be found in several cranial nerves, the greatest number travel in the tenth cranial (vagus) nerve (Fig. 30-2). The cell bodies of the preganglionic neurons are in the brain stem or sacral cord, and their fibers usually run all the way to the organ that they innervate, where they synapse with the postganglionic neurons. This is quite different from the sympathetic system, whose postganglionic cells are located in the sympathetic ganglia at relatively great distances from their respective target organs.

The parasympathetic division contains two anatomically distinct outflow sections, the cranial and the sacral. The cranial out-

flow comprises visceral motor fibers from several brain stem nuclei (e.g., vagus and glossopharyngeal), whose projections reach postganglionic neurons in the walls of the viscera. The sacral outflow contains visceral efferents from lower cord nuclei which project directly to these postganglionic neurons in pelvic organs.

Several neurotransmitters are involved in visceral regulation. Acetlycholine is apparently the transmitter between preganglionic and postganglionic sympathetic neurons. Several excitatory and inhibitory autonomic neurotransmitters have been identified, including acetylcholine, norepinephrine, serotonin and histamine.

The neurotransmitter is synthesized within the neuron and is stored in small "packages" or quanta at the nerve terminal,

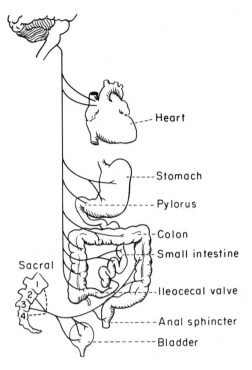

FIG. 31-2. Oversimplified anatomy of peripheral parasympathetic nervous system.

to be released upon excitation of the neuron. The sympathetic and parasympathetic divisions work together to achieve a great variety of reflex and generally adaptive patterns of response. The neurotransmitter substances act on the target tissue in a localized and precise fashion by way of "receptors" discretely sensitive to one or another transmitter. A neuron that elaborates norepinephrine at its terminal may, for example, either constrict or dilate an arteriole, depending on whether its connection at the vessel wall is an alpha or a beta receptor.

Sympathetic and parasympathetic divisions affect various organ functions quite differently. Thus, sympathetic activation prepares the body for action, raises arterial pressure, increases heart rate, dilates bronchi, and releases glucose from the liver. Parasympathetic activation, for the most part, results in conservation of energy although, in the gastrointestinal tract, stimulation is associated with increased acid secretion and motility. However, the sym-

pathetic and parasympathetic systems do not operate independently of one another. Normally, impulses are transmitted continuously through the fibers of both divisions, thereby allowing for maintenance of tonic and dynamic activity. This allows each system to exert both facilitating and inhibiting effects, thus modulating the activity of the organ concerned. Psychosomatic disorders have been viewed as reflecting a disturbed balance of autonomic function or as representing inappropriate patterns of autonomic response.

Central Aspects of the Autonomic Nervous System

Central regulation of autonomic function is discussed in Chapter 27. It is important to emphasize that interactions among frontal, limbic, hypothalamic, thalamic and cerebellar structures regulate responses integrated at brain stem, spinal and peripheral levels. The nature of the effector function depends, therefore, on myriad potential influences from circuitry concerned with individual past experience, learning, attitudes, values, goals and aspirations. An appreciation of the complexity of the central integrative process helps to explain the question, "Why does one person react to "stress" in a fashion widely different from another person?"

Certain areas of the neocortex have a direct cardiovascular function, as demonstrated by techniques of electrical stimulation and ablation. Thus, stimulation of the motor cortex can effect renal vasoconstriction severe enough to produce nephron necrosis. These changes depend on intact renal nerves, suggesting that cortical autonomic control may be capable of activating the renin-angiotensin system involved in hypertension.

Connections between higher integrative levels and visceral effector pathways in the brain stem are concentrated in the hypothalamus. In addition, the hypothalamus provides access to hormonal regulators by way of "releasing" hormones transported through portal veins from the hypothalamus to the anterior hypophysis, which is attached to the hypothalamus by the infundibulum. In this way, the hypothalamus controls the secretion of most anterior pituitary hormones and, thereby, many bodily metabolic functions.

Interest in the connections between the hypothalamus and superior limbic structures, often called the visceral brain, has been intense since Papez' classical studies and his hypothesis .that activity in this system provides the autonomic basis for emotional experience. Thus, electrical stimulation in certain parts of the limbic system can produce fully integrated patterns of behavior in the awake animal that seem identical with the behavior normally occurring in situations in which the animal is required to fight or to defend itself. These results of brain stimulation in animals have been confirmed clinically in man by observing patients who are conscious during certain brain operations. In such patients, for example, electrical stimulation in or near the amygdala is reported to result in feelings of fear. Other lines of research converge and support the notion that the limbic system is involved in emotional and visceral behavior.

LEARNING AND CONDITIONING AND VISCERAL PATHOPHYSIOLOGY

Learning is the outcome of experience, and it is the adaptive use of experience in the life of the organism. Conditioning is the description of very elementary types of learning. The extent to which any creature learns and the kinds of learning it executes are the result both of the capacity for learning in that animal and the opportunities for learning presented by the environment. In recent years, there have been a number of attempts to derive rational methods of symptomatic treatment of certain disorders such as tics, and phobias, based on the assumption that these symptoms are learned patterns of behavior.

Pathological Model Systems

A major feature of the pathophysiology of most, if not all, psychosomatic conditions is profound disturbance in vegetative function. These dysfunctions include excessive lability and range as their primary characteristics, and exaggerations or disturbances of integrative patterns of autonomic mobilization responses and homeostatic mechanisms are also importantly involved.

It has been postulated that chronic diseases such as peptic ulcer, hypertension, hyperthyroidism, ulcerative colitis, and rheumatoid arthritis result when disturbances in the activity of the autonomic nervous system are prolonged.

In order to understand better the natural history of such diseases, recent experiments have attempted to produce pathological model systems in animals by special conditioning procedures. Except for work by

Gantt and Masserman in the United States, the major part of previous work on this problem has been reported by Soviet scientists. In some of these experiments the investigators have been able to produce symptoms of cardiovascular pathology by conditioning. For example, sustained hypertension accompanied by histologically verified subendocardial necrotic foci, frequently resulting in fatal congestive failure, has been observed in dogs who had developed an experimental neurosis as a result of conditioning procedures in which mutually competitive responses were trained to the same stimulus. Furthermore, it has been reported that not only do conditioned disturbances in higher nervous function give rise to disorders in cardiovascular regulation, but also that the reverse holds true; namely, that experimentally induced and naturally occurring cardiovascular disorders have been found to give rise to concomitant parallel changes in those central nervous system structures participating in cardiovascular regulation. On the basis of such results and of experience with natural and other experimental vasoneuroses as well as the evidence of psychogenic influence on arterial pressure, a corticovisceral theory of hypertension has been formulated.

Experimental Neurosis and Individual Differences

The origins of experiments to produce experimental neurosis and pathophysiological effects date back to Pavlov, who noticed in his studies on conditioned reflexes that certain dogs behaved unpredictably and were difficult to handle. Systematic studies disclosed that experimental neuroses and altered physiology could be produced by conflict between emotions during conditioning or by requiring the animal to establish too fine a sensory discrimination as well as by extremes in any of the aspects of the training procedure.

Recent experiments have confirmed and extended Pavlov's original demonstration. In summary, these experiments indicate that several procedures can lead to physiological disturbances. Exposure to severe and unexpected pain, emotional excitement in a life-threatening situation, and subjection to a conflict between strong innate drives and learned reactions can lead to behavioral and physiological disturbances.

Pavlov observed that the basic temperament and autonomic reactivity of his dogs were important in determining the degree of susceptibility to, as well as the type and severity of, "neurotic" disturbance. When animal subjects with different temperaments are exposed to identical stresses, they tend to develop different types of disorders. In a somewhat analogous situation, Lacey and Lacey have clearly demonstrated that human beings who were experimentally stressed showed consistent response tendencies which differed among individuals. This consistency of autonomic reactivity or autonomic constitution could be the product of a number of different mechanisms which might predispose the individual, as clinical observation suggests, to specific types of behavior. Thus, the same objective conditions may impose different amounts of stress on different individuals, and the responses of different organ systems may vary with its intensity. In addition, genetic differences may predispose to different responses in different individuals. Moreover, experiences in infancy may affect the physical growth and the physiological activity of specific systems, providing a basis for subsequent differences in susceptibility. Finally, differences in the response of visceral systems may be attributable to the effects of classical conditioning by early learning experiences.

Instrumental Learning of Visceral and Glandular Responses

Previous work in the Soviet Union has shown the surprising degree to which a number of glandular and visceral responses under the control of the autonomic nervous system can be modified by classical conditioning. The traditional view has been that these responses can be modified only by an inferior form of learning called Pavlovian, or classical, conditioning and not by the superior form called trial-and-error, or instrumental, learning, employed in the training of skeletal responses. The difference between the two forms of learning is illustrated in Figure 31-3.

In classical conditioning, the reinforcement must be by an innate or unconditioned stimulus that already elicits the specific response to be learned, for example, the sight and smell of food, which elicits a salivatory response. In trial-and-error or instrumental learning, the reinforcement, called a reward, has the property of strengthening any immediately preceding response. Therefore, the possibility for reinforcement is much greater than is the

case in classical conditioning, since a given reward may reinforce any one of a number of different responses and a given response may be reinforced by any one of a number of rewards. The old belief that trial-and-error learning is possible only for voluntary responses mediated by skeletal muscles and, conversely, that the involuntary responses mediated by the autonomic nervous system can be modified only by classical conditioning has been used as the main argument for the notion that instrumental learning and classical conditioning are two basically different phenomena with different neurophysiological substrates. Recent experiments, however, have disproved this belief by the demonstration that visceral responses can be modified by trial-and-error learning. These experiments have deep implications for theories of learning, for psychosomatic medicine and, possibly, for an increased understanding of homeostasis.

A major problem encountered in research

PAVLOVIAN OR CLASSICAL CONDITIONING

After several pairings the initially neutral CS substitutes for the UCS, producing the UCR

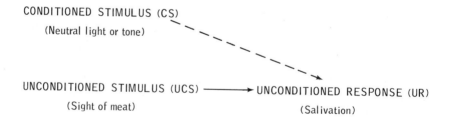

CONDITIONED STIMULUS (CS)

(Neutral light or tone)

UNCONDITIONED STIMULUS (UCS) ——→ UNCONDITIONED RESPONSE (UR)

(Sight of meat)　　　　　　　　　　　(Salivation)

INSTRUMENTAL OR TRIAL AND ERROR LEARNING

After several reinforcements of the CR, the CS serves as a signal to perform
the learned response (CR) in order to obtain reinforcement

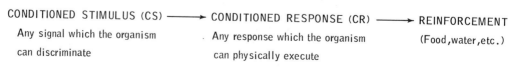

CONDITIONED STIMULUS (CS) ——→ CONDITIONED RESPONSE (CR) ——→ REINFORCEMENT

Any signal which the organism　　　Any response which the organism　　(Food, water, etc.)

can discriminate　　　　　　　　　can physically execute

FIG. 31-3. Difference between classical conditioning and instrumental or trial-and-error learning.

on the instrumental modification of visceral responses is that the majority of such responses are altered by voluntary responses such as tensing of specific muscles or changing the rate or pattern of breathing. One way to circumvent this problem, at least in animals, is to abolish skeletal muscle activity by curarelike drugs, such as *d*-tubocurarine; these interfere pharmacologically with the transmission of the nerve impulse to the skeletal muscle but do not affect autonomically mediated responses.

Curarized subjects cannot breathe and must be maintained on artificial respiration, and, because they cannot eat or drink, the possibilities for rewarding them in a training situation are somewhat limited. It is well known, however, that training of instrumental skeletal muscle responses can be accomplished either by using direct electrical stimulation of rewarding areas of the hypothalamus or by allowing escape from and/or avoidance of mildly noxious electric shock. Recent experiments in which these techniques have been used with curarized animals have shown that, by instrumental procedures, either increases or decreases can be produced in visceral responses such as heart rate, intestinal motility, arterial pressure, vasomotor responses, urine formation and contractions of the uterus. Several investigators have reported similar instrumental learning of heart rate and arterial pressure in man and have started to employ these techniques in therapy of certain clinical cardiovascular disorders.

Other experiments have shown that visceral learning can be quite specific, so that learned changes in heart rate do not involve systematic changes in intestinal contractions and learned changes in intestinal contractions do not involve systematic change in heart rate. Similarly, learned changes in renal blood flow and rate of urine formation have been found to be independent of changes in arterial pressure, heart rate or peripheral vasomotor responses. In another experiment, a marked dissociation of learned arterial pressure responses and heart rate was obtained in experimental animals but not in yoked controls. Taken as a group, these results indicate that visceral learning in rats is not mediated by a generalized reaction such as a shift in overall level of parasympathetic or sympathetic arousal. The striking degree of specificity to which a visceral response can be trained is shown by the results of an experiment in which opposite vasomotor responses were learned in the two ears of the curarized rat.

The demonstration that visceral and glandular responses are subject to instrumental learning and that this learning can have significant behavioral and physiological consequences means that it is theoretically possible for learning to produce the tendency, experimentally demonstrated by Lacey and Lacey, for individuals to respond to stressful situations with consistently ordered hierarchies of visceral responses that vary among individuals. It is also theoretically possible that such learning can be carried far enough to create an actual psychosomatic symptom. Presumably, genetic and constitutional differences among individuals would also explain the differences in response of the different visceral systems. Moreover, it seems probable that, under circumstances that can reward different psychosomatic symptoms, the subject will be most likely to learn the one to which he is innately most predisposed.

The effectiveness of the technique used with animals—objectively recording glandular and visceral responses and then modifying them by immediately rewarding, first, slight and, then, progressively greater changes—suggests that this kind of procedure might be used to produce therapeutic learning in patients with psychophysiological disorders.

A New View of Homeostasis

The reactions through which the constancy of the internal environment is maintained are more complex and indirect than are the ordinary chemical and physical adjustments. They are essentially physiological reactions regulated through the visceral nerves. The traditional view has been that the processes that tend to maintain homeostasis are carried out automatically. They involve both relatively simple reflex mechanisms and integrating mechanisms of higher orders, the most important of which are located in the hypothalamus. According to classical views, the automatic regulation of the routine necessities for maintenance of homeostasis permits the functions of the brain that subserve manual skills, intelligence, imagination, and insight to find their highest expression. On the other hand, we now know that visceral responses are not always involuntary but can be learned in the same way as skeletal muscle responses. This raises the fundamental question as to whether or not the capacity for instrumental learning of visceral responses is in itself important to the maintenance of the homeostasis of the internal environment.

Since skeletal muscles operate mainly upon the external environment, the ability to learn responses that bring rewards such as food, water, or escape from pain has survival value. To what extent does regulation of the internal visceral responses by the autonomic nervous system have a similar adaptive value? Will an animal with experimentally altered internal homeostasis perform the instrumental tasks necessary to correct the imbalance? In a two-part experiment designed to test this question, Miller, DiCara and Wolf injected antidiuretic hormone (ADH) into rats if they selected one arm of a T maze and isotonic saline if they selected the other. One group of normal animals, loaded in advance with water by stomach tube, learned to select the saline arm; choice of the ADH arm would have prevented the secretion of excess water required for restoration of water balance. By contrast, a group of rats

with diabetes insipidus, and loaded in advance with hypertonic NaCl by stomach tube, regularly chose the ADH arm. For this group the homeostatic consequences of the injections were reversed; ADH caused concentration of urine, thus promoting excretion of the excess NaCl, whereas isotonic saline would have merely produced further imbalance. The control rats received neither water nor NaCl and exhibited no preference in the maze.

These instrumental responses aimed at correcting internal imbalance appear to be no less adaptive than those learned skeletal responses that are directed toward altering the animal's relationship to its external environment.

Taken together with the other evidence that glandular and visceral responses can be instrumentally learned, these data raise the question of the degree to which animals normally learn the various autonomic responses required to maintain proper homeostasis. Whether or not such learning could contribute to homeostasis in ordinary day-to-day life experience would depend on the range of sensitivity or inertia of the control mechanism. Thus, a very small drift in the quantity of a physiological variable may be quickly compensated for under ordinary circumstances. But where such a control system may display greater inertia and require a larger drift before corrective change is triggered, visceral learning, rewarded by a return to homeostasis, might be available as a back-up mechanism.

ANNOTATED REFERENCES

Bykov, K. M.: The Cerebral Cortex and the Internal Organs (W. H. Gantt, translator and editor). New York, Chemical Publishing Co., 1957. (Presents evidence that glandular and visceral responses, controlled by the autonomic nervous system, can be modified by classical conditioning)

DiCara, L. V.: Learning in the autonomic nervous system. Sci. Am., 222:30, 1970. (Review of evidence that visceral responses can be modified by trial-and-error learning)

Figar, S.: Conditioned circulatory responses in men and animals. *In:* Hamilton, W. F. (ed.): Handbook of Physiology. Section 2, Circulation, Vol. 3, p. 1991. Washington, D. C., American Physiological Society, 1965.

Gantt, W. H.: Cardiovascular component of the conditioned reflex to pain, food and other stimuli. Physiol. Rev. (Supp.), *40*(4): 266, 1960.

Lacey, J. I., and Lacey, B. C.: Verification and extension of the principle of autonomic response stereotypy. Am. J. Physiol., *71:* 50, 1958.

Miminoshvili, D. I., Magakian, G. O., and Kokaia, G. I.: Attempts to obtain a model of hypertension and coronary insufficiency in monkeys. *In:* Utkin, I. A. (ed.): Problems of Medicine and Biology in Experiments on Monkeys. p. 103. New York, Pergamon Press, 1960. (The above four references present evidence in the experimental animal, as well as in man, that symptoms of cardiovascular pathology can be produced by conditioning.)

Gellhorn, E.: Autonomic Imbalance and the Hypothalamus. St. Paul, University of Minneosta Press, 1957.

and

Gellhorn, E., and Loofbourrow, G. N.: Emotions and Emotional Disorders. New York, Harper & Row, 1963. (Two of the most comprehensive and critical textbooks available on the subject of autonomic imbalance and psychosomatic disorders)

Haymaker, W., Anderson, E., and Nauta, W. H.: The Hypothalamus. Springfield, Ill., Charles C Thomas, 1969. (A comprehensive treatise on the hypothalamus, with emphasis on anatomy and *function* and the basis of *clinical syndromes*)

Hoff, E. C., Kell, J. F., and Carrol, M. N.: Effects of cortical stimulation and lesions on cardiovascular function. Physiol. Rev., *43*:68, 1963.

Ingram, W. R.: Central autonomic mechanisms. *In:* Field, J. (ed.): Handbook of Physiology. Section 1, Neurophysiology, Vol. 2, p. 951. Washington, D. C., American Physiological Society, 1960. (Review of literature indicating that autonomic function is influenced by structures at all levels in the central nervous system)

MacLean, P. D.: Psychosomatics. *In:* Field, J. (ed.): Handbook of Physiology. Section 1, Neurophysiology, Vol. 3, p. 1723. Washington, D. C., American Physiological Society, 1960. (Comprehensive review concerning mechanisms through which emotion and stress alter function leading to psychosomatic illness)

Miller, N. E., DiCara, L. V., and Wolf, G.: Body fluid homeostasis and learning in rats: T-maze learning reinforced by manipulating anti-diuretic hormone. Am. J. Physiol., *215:* 684, 1968.

Papez, J. W.: A proposed mechanism of emotion. Arch. Neurol. Psychiat., *38*:725, 1937.

Suggested Review Articles on Learning Mechanisms

Astrup, C.: Pavlovian Psychiatry. Springfield, Ill., Charles C Thomas, 1965.

Holland, B. C., and Ward, R. S.: Homeostasis and psychosomatic medicine. *In:* Arieti, S. (ed.): American Handbook of Psychiatry. Vol. 3. New York, Basic Books, 1966. (The above two references are suggested as excellent review articles, in addition to others already cited, concerned with learning and autonomic mechanisms in relation to pathophysiology.)

Appendix

The following references are added for those scholars wishing to read further into the interrelationships of neural and somatic function.

Gellhorn, E.: Principles of Autonomic-Somatic Integrations—Physiological Basis of Psychological and Clinical Implications. Minneapolis, University of Minnesota Press, 1967.

Alexander, F.: Psychosomatic Medicine, Its Principles and Applications. New York, The Chemical Publishing Co., 1957.

Raab, W. (ed.): Ischemic Heart Disease. Springfield, Ill., Charles C Thomas, 1966.

Wolf, S., and Goodell, H.: Harold G. Wolff's Stress and Disease. Ed. 2 (edited and revised). Springfield, Ill., Charles C Thomas, 1968.

Greenfield, N. S., and Sternbach, R.: Handbook of Psychophysiology. New York, Holt, Rinehart and Winston (in press).

Field, J., Magoun, H. W., and Hall, V. E.: Handbook of Physiology. Section 1: Neurophysiology. Washington, D. C., American Physiological Society, 1960.

Section Eight

Immunological Mechanisms

Introduction

Much awareness of the science of immunology (in contexts other than allergy and resistance to infection) has developed concurrent with successful application of the fundamental principles of this once aloof science toward solution of tissue transplantation problems. However, most of the immunological principles useful in organ transplantation had evolved from previous studies concerned with resistance to infection. Continuing progress toward application of these immunological principles to the solutions of problems in transplantation, allergy, infectious diseases, and malignancy (to name but a few) are practical goals of contemporary immunology. In this section we shall apply concepts of immunology to other clinical disciplines by emphasizing the relationships between mechanisms of immunity and mechanisms of disease. We hope that this approach will provide sufficient insight into a discipline still in a process of rapid evolution to permit the student to "bridge" the gap between fundamental immunobiology and modern clinical medicine. The scope of modern immunobiology is indeed broad (Table VIII-1); and it should be clear, even from a cursory glance at this table, that the entire field of immunology cannot be covered in the available space. The topics chosen were selected to provide an overview of the content of immunobiology.

In the classical sense, immunity is synonymous with resistance to infection. However, immunology has become much more than the study of resistance to infection. It is, rather, the study of the mechanisms which tend to maintain uniqueness of "self" through recognition of and rejection of "nonself" at the cellular and molecular levels.

Because of the recent rapid expansion of immunology, the science has literally outgrown its vocabulary. Definitions given in this text were adopted for their working value, but students are warned that in immunology synonyms do not always have synonymous meaning, and many authors use the technical terms of immunology differently. For this reason alone, comprehension of recently published findings in immunology may be difficult. There is no doubt that the group of principles and mechanisms which comprise immunology is complex; but they are not really complicated. With this orientation, one should expect a limited degree of difficulty in the initial study of immunology, and he should be alert for the emergence of genuine biological principles.

Traditionally, the components of immunology have been categorized into two major overlapping groups, both of which afford protection against infection. One of these, called *innate immunity*, is made up of all of the anatomical and physiological barriers against infection. Included are the integrity of the integument, bacteriostatic and bacteriocidal activity of secretions, ciliary action of respiratory epithelium, gastric acidity, motility of the intestine, effects of inflammatory responses, body temperature, phagocytic elements of the reticuloendothelial system and even the circulation of the blood. The inflammatory responses are also components of innate immunity. All of these factors tend, in rather obvious ways, to prevent or control infection nonspecifically (i.e., regardless of the nature of the potentially infectious agent). Similarly, the components of innate immunity tend to reject, or otherwise react with, foreign material other than infectious agents and complicate delivery of blood and other therapy to patients.

A second category of barriers against infection concerns the elements of *adaptive*

TABLE VIII-1. Scope of Contemporary Immunobiology

SUBJECT OF STUDY	CONTENT
Resistance to infection	1. Innate barriers 2. Adaptive barrier to recurrent or persistent infection 3. Resistance to specific pathogens 4. Immunosuppressive action of pathogens
Fundamental relationships in cellular and humoral immunity	1. Cells and molecules basic immunological function 2. Relation of structure to function of the immunological apparatus 3. Development of cells and molecules, tissues and organs of immunological apparatus 4. Identification and measurements of cells and molecules of immunological apparatus 5. Immunological phenomena in relationship to cellular biology 6. Relation at the cellular and molecular levels of structure to function in the immunology system 7. Defects in development of immunity, their classification, recognition and analysis as "Experiments" of Nature 8. Immunological tolerance, specific negative adaptation
Tissue transplantation	1. Antigenic relationships of host and graft 2. Immune responses in graft rejection 3. Inhibition of graft rejection
Malignant adaptation and immunity	1. Antigenic features of malignant cells 2. Tumor immunity 3. Chemical carcinogens and immunity 4. Malignancy in genetic and therapeutic immunodeficiency 5. Immunodeviation 6. Attempts to modify malignant adaptation by immunological means
Mechanisms of immunology injury	1. Self-recognition and autotolerance 2. Models of autoimmunity 3. Autoimmune phenomena in human disease 4. Relation of immunodeficiency, autoimmunity, virus infection and aging
Allergy	1. Drug and chemical 2. Humoral and cellular immunology 3. The IgE system 4. Atopy 5. Prophylaxis and therapy

immunity. Some of these mechanisms will be introduced as basic principles and then applied clinically. A discussion of the evolution of adaptive immunity should provide a further degree of insight into the biological value of adaptive immunity. Principles of adaptive hypersensitivity and application of these to clinical problems in tissue transplantation and "autoallergy" will be discussed; and the complement system, an amplifier of innate as well as adaptive immunity and hypersensitivity, is presented.

Adaptive immunity differs from innate

immunity in that adaptive immunity is inducible, it manifests a high degree of specificity, and it is the system of primary concern with regard to immunological hypersensitivity. In response to initial exposure of deeper tissues to foreign substances (called antigens or allergens), immunologically "competent" tissues form antibody and/or become hypersensitive. These "adaptive" reactions of an individual to antigen or allergen comprise the primary immune response. Similar reactions which occur subsequent to a second exposure to the same (or chemically similar) antigen or allergen are called the secondary immunological responses. These secondary responses differ from primary immune responses in that the secondary ones develop more rapidly and, in quantitative terms, are more intense. Since primary immune responses often appear (superficially) to be transient, secondary responses are also called anamnestic (memory) responses.

The development of immunity and immunological hypersensitivity subsequent to exposure to antigen or allergen occurs through the same or extremely similar mechanisms. The essential difference between these two phenomena is that immunity is associated with beneficial effects (increased resistance to infection or suppression of malignancy) whereas hypersensitivity is associated with deleterious effects (allergic conditions and graft rejection).

There are at least two identifiable kinds of adaptive immunity and, in parallel, two kinds of immunological hypersensitivity: cell-mediated (or thymus-dependent), and humoral (or thymus-independent) immunity or hypersensitivity. In general, cell-mediated immunity involves increased resistance to many infectious agents, which have antigens to which a patient has been exposed previously. Cell-mediated hypersensitivity (also called delayed hypersensitivity) is related to pathological reactivity

to otherwise bland substances (in this context called allergens rather than antigens). The same is true of humoral immunity and the immediate hypersensitivities.

While these simplified distinctions are useful in defining the terms, immunity and hypersensitivity may not occur separately. The relative increase in resistance to infection with a given pathogen which results from immunity to the antigens of such an agent is usually accompanied by hypersensitivity to these same antigens (allergens). Clinically, this information is often put to practical use by assuming that hypersensitivity to antigens of a given agent (which can be detected by skin tests) is evidence of resistance to infection by agents bearing the same antigens. This is an assumption which is generally useful, but not infallible. It is also possible, for example, that a patient or experimental animal could be hypersensitive to a pathogen-associated allergen which is not crucial to that pathogen's ability to cause disease. In such cases, hypersensitivity may in fact render the patient less resistant to infection. Similarly, patients may have adaptive increases in resistance to infection without demonstrable hypersensitivity.

Several forms of adaptive immunity are classified on the basis of the source of immunizing antigen or sensitizing allergen. *Active immunity* results subsequent to exposure of immunologically competent tissues of normal individuals to antigen. If an antigen (or allergen) is encountered as a result of normal exposure to the environment (e.g., the result of frank or inapparent infection, exposure to dietary antigens, inhalation of organic debris, etc.), the resulting immunity (or hypersensitivity) is called natural active immunity (or hypersensitivity). In contrast, when the antigen (or allergen) is provided in a vaccine, the resulting immunity is called artificially induced immunity or simply artificial active immunity. Immunity may also be acquired *passively*. Patients are passively immunized with elements of the thymus-independent system by providing them with serum or serum gamma globulin from an actively immunized patient or animal. In human beings, this also occurs naturally through transplacental passage of maternal gamma globulin to fetus. Passive immunization is always transient and may be contraindicated when the patient demonstrates hypersensitivity to allergens in the serum or the gamma globulin to be transferred. However, passive immunization has the advantage of delivering some of the elements of immunity immediately. This is of particular importance when the disease to be controlled is produced by toxins which must be neutralized immediately (as in diphtheria or tetanus). In such cases active immunity (which requires a finite induction period) may not develop fast enough to protect the patient.

Passive immunity can also be achieved by transplantation of lymphoid tissues or cells from actively immunized donors. This allows transfer of elements of thymus-dependent as well as thymus-independent immunity. When immunity is thusly transferred by "adopting" a previous immunity already acquired by a donor, it is called *adoptive* immunity. Adoptive immunity may be similar to active immunity (with respect to permanency), but the immunological barriers against transplantation of cells are more difficult to overcome than are the barriers which complicate serotherapy. In addition to the problem of graft rejection (more precisely host versus graft rejection) there is the possibility of reaction of the engrafted lymphoid tissue against antigens of the recipient (graft versus host reaction). Adoptive immunity is appropriately a research tool which is currently being explored as a special therapeutic procedure.

Several areas of applied clinical immunology are not discussed under these specific immune mechanisms. Thus, such immunological problems as "allergy," "asthma," and atopy—still incompletely understood from a fundamental viewpoint —are not discussed in depth in this Section. However, by understanding the basic concepts discussed under inflammation, adaptive immunity, "autoallergy," immunological amplification, and even the evolution of immune mechanisms the student will develop a clearer insight and overview for handling these "thorny" clinical problems.

William A. Cain, Ph.D.

32

Inflammatory Mechanisms and Fever

Arthur R. Page, M.D.

INTRODUCTION

Inflammation, a basic pathophysiological response to injury, is present in some form in all multicellular animals. Injury triggers a complex series of interdependent reactions that serve to bring phagocytes into the area to defend against invasion by microorganisms and also set the stage for repair of the injury through proliferation of connective tissue cells and collagen production.

The majority of injuries result only in a local inflammatory response; a major injury due to trauma or virulent infection results in systemic signs of inflammation characterized by fever and leukocytosis. The discussion that follows outlines in greater detail what is presently known about the mechanisms behind these responses.

THE LOCAL INFLAMMATORY RESPONSE

Local inflammatory response is characterized by a series of interrelated events that, once triggered, proceed in a regular sequence independent of the initiating stimulus or the organ involved. The response can be divided into four stages: increased vascular permeability; neutrophil exudation; mononuclear cell exudation; and cellular proliferation and repair. Although all stages are present sequentially in all inflammatory reactions, the underlying initiating event may influence the development of the various stages.

INCREASED VASCULAR PERMEABILITY

Increased vascular permeability, the first phase of the inflammatory response, begins immediately after the initiating event and is readily demonstrable by the leakage of colloidal dyes through the walls of small venules near the capillary-venule junction. Electron microscopic studies show that leakage of colloidal material occurs between endothelial cells. Under normal conditions, the junction between endothelial cells is filled with a substance that is altered during inflammation, allowing passage of high molecular weight material.

When white blood cells are specifically attracted to the area by chemotactic stimuli, they escape from the vessels by the same route between endothelial cells, widening the spaces and allowing the passive escape of a small number of red blood cells. The local increased vascular permeability may last only a few minutes, indicating that the vascular lining may be quickly restored once the chemical mediators are withdrawn. However, under most circumstances, these substances are produced continuously for at least 24 hours following the initiating event, allowing for continued exudation of plasma as well as of cells.

Histamine, serotonin (5-hydroxytryptamine) and the kinins have been shown to be direct chemical mediators of increased vascular permeability. All three carry a net positive charge and may act by combining with acidic mucopolysaccharides or glycoproteins in the blood vessel wall. Serotonin, active in rodents, does not play this role in human beings or other mammals; however, free histamine and the kinins are probably involved in all inflammatory reactions.

Histamine is present in most tissues of the body, concentrated in the basophilic granules of tissue mast cells and circulating basophils. It is stored in these specialized cells in a complex with heparin. Release of histamine occurs through several different mechanisms. Mechanical disruption during trauma to tissues rich in mast cells, such as the dermis, will release histamine. Antigen-antibody reactions can initiate histamine release in two ways: Interaction of an antigen with cytotropic antibody results in direct histamine release without activation of the complement system, and combination of antigen with complement-fixing antibody can activate the complement system, producing anaphylatoxin. Another mechanism resulting in histamine release in the absence of antigen-antibody interaction occurs subsequent to anaphylatoxin activation by proteolytic enzymes. Finally, neutrophil granules contain an arginine-rich polypeptide that has histamine-releasing activity. This peptide is important in maintaining vascular permeability during the inflammatory response.

Platelets. Blood platelets, which are rich in histamine, participate in the mechanisms just described. An additional antigen-mediated mechanism has been defined for platelets whereby sensitized lymphocytes reacting with specific antigen produce a platelet-histamine-releasing factor. Neither circulating antibody nor complement is required for this reaction.

Kinins. The kinins are small arginine-containing polypeptides that are released from a plasma alpha-globulin, kininogen, by certain proteolytic enzymes. Two enzymes in human plasma, kallikrein and permeability factor/dilute (PF/dil), both of which exist in inactive forms and are activated by Hageman factor, have been shown to produce kinins from plasma kininogen. In vitro, Hageman factor, or Factor XII in the blood clotting system, can be converted from an inactive to an active enzyme by adsorption on a wide variety of different materials. It is probable that, in the body, Hageman factor initiates inflammatory reactions by activating the kinin-producing enzymes.

The complement system also has been implicated in activating plasma kallikrein, so that activation of the complement system produces both kinin- and histamine-releasing substances.

Neutrophils also contain enzyme capable of producing kinins from kininogen and, again, probably are important in maintenance of increased vascular permeability during an inflammatory response.

These interactions are summarized in Figure 32-1.

Neutrophil Exudation

Neutrophil exudation, the second phase of the local inflammatory reaction, starts 30 to 45 minutes after the initiating event, reaching a maximum after 6 to 8 hours, then declining in intensity in the later stages. Neutrophil exudation is dependent upon the production of specific chemotactic factors which attract neutrophils to the inflammatory site. The neutrophils respond first by sticking to the blood vessel wall and then move through the vessel wall by squeezing between endothelial cells (diapedesis); finally, they penetrate the basement membrane and enter the tissue interstitial spaces. These cells actively phagocytize and kill microorganisms and furnish an important barrier against bacterial invasion. In addition, they release products that maintain the inflammatory response by increasing vascular permeability and attracting more neutrophils and eventually set the stage for mononuclear cell exudation.

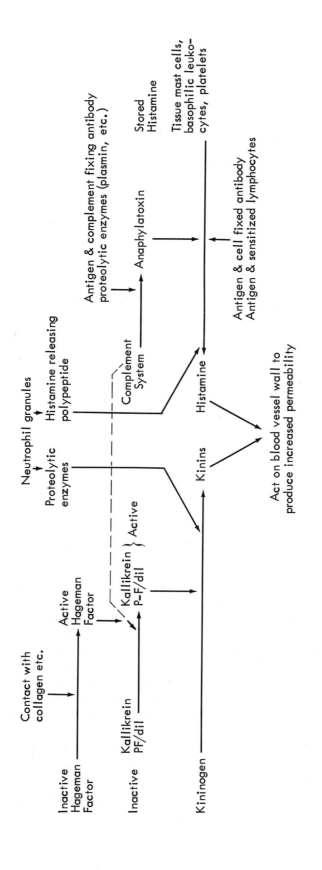

Fig. 32-1. Some mechanisms for the production of increased vascular permeability during inflammation.

Chemotaxis. Factors chemotactic for neutrophils (i.e., chemical factors which attract neutrophils) are generated in fresh plasma by clotting and by activation of the complement system. At least three such factors generated by the complement system have been identified and studied extensively. The factor of major importance is activated by antigen-antibody interaction and is a small peptide split-product of $C'5$. Another such factor is the macromolecular complex of $C'5$, $C'6$ and $C'7$. The complement system may be important for neutrophil chemotaxis even in the absence of antibody-antigen reactions. Thus, neutrophil exudation is absent in patients with glomerulonephritis, whose plasma contains an inhibitor of the complement-generated chemotactic factors, and patients with $C'2$-deficiency also have reduced neutrophil exudation following simple injury. The plasma fibrinolytic enzyme plasmin has been shown to produce chemotactic split-products from purified $C'3$, and it seems probable that many small peptides released from plasma or tissue proteins by proteolytic enzymes also will be shown to be chemotactically active. Serum from freshly clotted blood contains heat-labile neutrophil chemotactic factors that are different from the complement-derived factors. These heat-labile factors may be by-products of the clotting sequence, but they have not been studied in detail.

Neutrophil granules are a source of chemotactically active polypeptides that may play a role in continued leukocyte exudation, and some bacteria also produce chemotactic factors, not requiring serum, plasma or antibody-antigen interaction for their activity.

The intensity and duration of the neutrophil exudation are directly proportional to the amount of liberated chemotactic factors. In allergic inflammation produced by complement-activating antibody-antigen interaction, the neutrophil exudation may continue for 24 to 48 hours, and in bacteria-induced inflammation, neutrophil exudation continues, until all invading organisms are killed and digested, even though this may require several days. The concentration of large numbers of neutrophils may lead to abscess formation. An abscess is a cavity produced by tissue necrosis and filled with an exudate rich in neutrophils. Large abscesses may require drainage before healing can occur.

Mononuclear Cell Exudation

The third stage of the inflammation response is mononuclear cell exudation. The first mononuclear cells enter the simple lesion 4 hours after the initial stimulus, and the response reaches a peak after 16 to 24 hours. Mononuclear cells reaching the lesion during the early hours are of the same size and shape as lymphocytes, but studies of their rapid turnover and electron-microscopic morphology indicate that most belong to a special *monocytic* population produced in bone marrow. As the inflammatory lesion progresses, cells take on the appearance and characteristics of large mononuclear macrophages.

Unlike the circulating neutrophils, which respond instantly to an inflammatory stimulus, the circulating mononuclear cells must first be stimulated to produce new messenger-RNA and protein before they can participate in the inflammatory reaction. This "triggering" of new messenger-RNA and protein production is initiated by presently unidentified substances, which are released into the circulation from the inflammatory lesion during the first hour of injury. Thus "armed," the mononuclear cells are attracted by the chemotactic substances released by neutrophils from the inflammatory site.

Another source of chemotactic stimuli for these "armed" mononuclear cells is produced by immune small lymphocytes that have interacted with antigen in a delayed hypersensitivity-type reaction. In reactions of this type, the mononuclear cell response is prolonged and intensified.

Repair

The final stage of the inflammatory response is resolution and repair. Within 18 hours, fibroblasts in the area are already beginning to synthesize new DNA in preparation for division, and after 48 to 72 hours, fibroblast proliferation reaches its peak. Depending upon the issue involved, regeneration of organ-specific cells may also take place. Some organs—e.g., the skin and liver—have tremendous capacity for regeneration, whereas others—e.g., the central nervous system—have no such capacity.

Fibroblasts actively produce acidic mucopolysaccharides and collagen fibers during the repair stage. The formed collagen produces a strong scar, and it has been postulated that the increased acid-mucopolysaccharide production is important in resolving the inflammatory reaction by combining with and neutralizing chemical mediators at the site.

The final stage may end in three ways: complete repair or restoration with or without scar tissue; abscess formation; or granuloma formation.

When tissue necrosis is extensive and neutrophil exudation is prolonged by the presence of virulent microorganisms, the end stage may be an abscess, or bag of pus cells, walled off by fibroblasts and collagen fibers. These lesions may never completely resolve unless they are surgically drained.

Granuloma formation results when the inflammatory lesion is produced by microorganisms (which survive inside of phagocytic cells) or by vegetable or mineral fibers not digested by phagocytic cells. Under these circumstances, mononuclear cells surround the area and produce a fibrous capsule around the foreign material. The center of the mature granuloma will contain the offending material, often within multinucleated giant cells formed by the cytoplasmic fusion of many macrophages. In this manner, undigestible microorganisms such as the tubercle bacillus can be walled off and contained for many years even though living organisms are inside the granuloma.

Rebuck Skin Window

The cellular events associated with the local inflammatory response can be studied in patients by the use of the Rebuck skin window technique. A small superficial abrasion is made on the skin and a sterile coverslip is taped over the area. The coverslip is changed at 2-hour intervals for 24 hours and then stained and studied microscopically. Normally, during the earlier hours the predominant cells are neutrophils. Later, mononuclear macrophages predominate. This technique is widely used as a clinical research tool and for the diagnosis of diseases such as those manifested by intracellular collection of mucopolysaccharides, lipid, etc. In many of these diseases, the macrophages at the inflammatory site contain characteristic storage granules.

Phagocytosis and Killing

The primary function of an inflammatory reaction is to bring neutrophils and macrophages into a local area where they can phagocytize and kill the invading microorganisms. Most virulent microorganisms are able to invade because they have some defense against phagocytosis. For instance, pneumococci have a capsular polysaccharide and the streptococci have a cell wall protein (M-protein), that protects them from phagocytosis. In turn, the body reacts by producing specific antibody that reacts with the external protective coating and allows the bacteria to be phagocytized within 2 days. In the presence of large amounts of antibody alone or small amounts of antibody and complement, the cell will be able to phagocytize the virulent bacteria. During activation of the complement system, many molecules of $C'3$ are deposited on the bacterial surface. Phagocytes, in turn, have receptors for C'3 molecules that enable them to phagocytize bacteria coated with these molecules. C'5 is required for optimal phagocytosis of some microorganisms. This process of coating particles to obtain optimal conditions for phagocytosis is termed opsonization.

In the process of phagocytosis, bacteria are engulfed by the cellular pseudopod, and external cellular membrane surrounding the organism is pinched off and internalized. The phagocytized particle or organism is entrapped within a membrane-lined sac called a phagocytic vacuole (Fig. 32-2). The phagocytic cells contain many small granules or lysosomes within their cytoplasm. These are essentially small sacs filled with digestive enzymes. As soon as phagocytosis begins, a process called de-

granulation occurs. The lysosomes approach phagocytic vacuoles and the lysosomal membranes fuse with the membrane of the phagocytic vacuole so that lysosomal enzymes are spilled into the phagocytic vacuole, forming phagolysosomes (Fig. 32-3). Thus, potentially toxic bactericidal enzymes are concentrated in the phagolysosome and are kept from harming the host cell, since the phagocytic vacuole is functionally exterior to the cell.

During phagocytosis, the cell becomes very active metabolically. Although anaerobic glycolysis furnishes the energy needed for phagocytosis, the killing of microorganisms is dependent on oxidative metabolism. During phagocytosis, the marked increase in oxygen consumption is associated with peroxide production. The

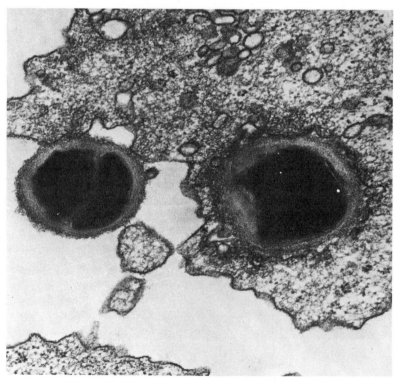

FIG. 32-2. Electron micrograph demonstrating the process of phagocytosis. One organism has just been ingested and lies within a phagocytic vacuole. The other organism is adherent to the external membrane of the neutrophil and is being surrounded by pseudopods. (The electron micrograph was provided by Dr. James White.)

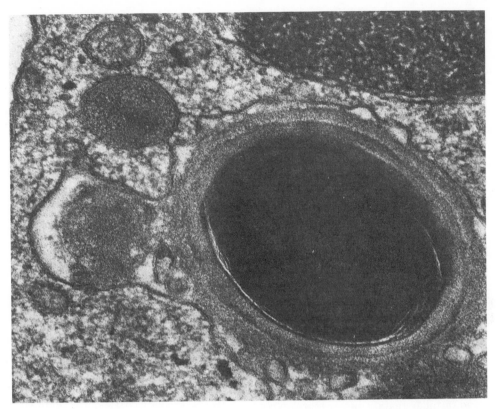

FIG. 32-3. Electron micrograph demonstrating degranulation. The bacterium lies within a phagocytic vacuole and two neutrophil granules (lysosomes) are ready to dump their contents into the phagocytic vacuole. (The electron micrograph was provided by Dr. James White.)

peroxide, in combination with peroxidase and iodide ions, iodinates bacterial proteins and kills bacteria.

SYSTEMIC MANIFESTATIONS OF INFLAMMATION

Fever and leukocytosis are the usual systemic accompaniments of inflammatory reactions that are not successful in completely localizing and neutralizing bacterial invaders.

Fever

It has long been argued that fever is an adaptation that has value to the body by raising the body's temperature above that optimal for growth of pathogenic bacteria. It is questionable, however, whether the body's cells are at an advantage at a higher temperature in defending against microorganisms. Although fever therapy was once used (e.g., for syphilis), its effectiveness was never established.

The mechanism of fever production following injection of gram-negative bacterial endotoxin has been extensively studied and has been a useful model that now seems to explain most fevers associated with infection or inflammatory or allergic reactions. Endotoxin produces fever by stimulating certain cells to release a protein (endogenous pyrogen) into the circulation. This protein acts on the central nervous system to produce fever. Neutrophils and mononuclear macrophages can be stimulated to produce this protein in vitro under a wide variety of stimuli, such as exposure to bacterial endotoxin, phagocytosis of bacteria or even of inert particles, and a change in the ionic strength and oxygen tension of the medium. Neutrophils and macrophages within inflammatory exudates contain preformed endogenous pyrogen that can be released. However, circulating normal neutrophils do not contain this protein prior to stimulation of its synthesis.

Cells other than neutrophils and macrophages may also be capable of producing pyrogen in the intact animal, since animals, including man, respond with fever to endotoxin even when these cell types are absent. Pyrogen production may be a generalized function of cells of the reticuloendothelial system.

Fever following exposure to bacteria, viruses, antigens and sterile inflammatory reactions is mediated through the production of endogenous pyrogen.

Leukocytosis

Leukocytosis is another systemic manifestation of inflammation but is more selective than fever. As a general rule, bacterial infections produce leukocytosis with an increase in the number of circulating neutrophils. Viral infections usually are not associated with leukocytosis and may even be accompanied by mild leukopenia.

Leukocytosis following intravenous injection of bacterial endotoxin results from the release of mature cells and band forms from the marrow. The peak response occurs from 4 to 6 hours after the injection. Endotoxin stimulates cortisone release from the adrenal gland, and cortisone injection produces the same pattern of leukocytosis. That cortisone is the mediator of endotoxin-induced leukocytosis has not been proved, but the leukocytosis accompanying infection with bacteria is more than simple release of cells from the marrow. Marrow is also stimulated to produce more cells, and the mechanism for this stimulation is believed to be through the release of substances that specifically stimulate neutrophil production.

Acute Phase Proteins

Another systemic manifestation of inflammation is the production of certain proteins by the liver. Blood concentrations of fibrinogen, alpha-2-glycoproteins, C-reactive protein, some clotting factors and some complement components are increased, owing to release of some factor or factors from inflammatory lesions that stimulate their synthesis.

Erythrocyte Sedimentation Rate. The erythrocyte sedimentation rate (ESR) has evolved as a useful clinical tool for the detection of inflammation. The rate at which red cells settle out of plasma is a function of the size of the cell. Under normal conditions, the cells remain as individual units because the negative charges on their membranes keep them from sticking together. However, when red blood cells are heavily coated with fibrinogen, they clump together and settle faster (rouleaux formation). Under most circumstances, an increased ESR is a function of increased blood fibrinogen, although increased gamma globulin or anemia also can cause increased ESR, and polycythemia inhibits the ESR.

C-Reactive Protein. This protein is a liver product, not normally present in plasma, which is produced during acute inflammation. It was discovered originally because it formed a precipitin reaction with pneumococcal capsular polysaccharide. It is now detected in the clinical laboratory by its reaction with an antibody. In general, this protein appears in plasma within a few hours after the onset of infection and disappears shortly after the infection is under control; in mild inflammation it does not appear at all.

Albumin. The hepatic synthesis of serum albumin is depressed during inflammatory reactions so that serum electrophoretic patterns usually show slightly reduced albumin and elevated alpha-2-globulin and gamma globulins during chronic inflammation.

Anemia

Microcytic hyperchromic anemia is another systemic manifestation of chronic inflammation. This is not a true iron-deficiency anemia; rather it is a defect in ability to utilize iron for hemoglobin synthesis. Bone marrow stores of iron are increased, in contrast to their decrease in true iron deficiencies secondary to chronic blood loss or inadequate diet.

INTERFERONS

Interferons are proteins that are produced by virus-infected cells and protect other cells from infection by virus. Unlike antibodies, interferons are not specific for the virus. Many cellular types can be induced to produce interferons in tissue culture. In vivo their site of production depends upon the nature of the infection.

These proteins are undoubtedly important in limiting viral infections, and in many ways their production is similar to the production of endogenous pyrogen. Thus, live virus, inactivated virus, synthetic double-stranded nucleic acids or bacterial endotoxins induce cells to produce and release interferon. Interferon then acts upon non-infected cells by inducing the production of a second protein which inhibits virus multiplication. The finding that a synthetic double-stranded RNA, poly-inosine-cytidine (Poly-I-C), is effective as an inducer of interferon and will protect animals against some virus infections when given at the proper time has raised the hope of an effective therapy for human viral infections. Recent work in animals indicates that interferons may be effective antitumor agents, at least in the case of virus-induced tumors.

ANTI-INFLAMMATORY DRUGS

In many clinical situations the symptoms and disability produced by an inflammatory reaction are the main concern of the patient and physician. These situations arise when the cause of the inflammatory reaction is an antibody-antigen interaction in which either the antigen itself is not harmful (e.g., production of asthma and hayfever by plant pollens) or the etiologic agent is not known but the chronic inflammatory reaction is causing disability (e.g., rheumatoid arthritis). Therapy of these conditions is, at least in part, designed toward limiting the inflammatory reaction.

Several drugs having some anti-inflammatory activity are available. The more effective drugs have more than one mechanism for reducing inflammation. The most potent agents in use are adrenal corticosteroids. These drugs block both inflammatory edema and cell immigration; however, because they are so effective, they also increase susceptibility to infection. Other actions rendering them undesirable for long-term use are growth retardation, osteoporosis, peptic ulcers, potassium depletion, hypertension and vasculitis. Hence, these drugs should be used only as a last resort.

Antihistaminic drugs are the only effective drugs whose site of action is known with confidence. These drugs competitively inhibit histamine and effectively block inflammatory edema due to histamine alone. They are effective only in acute allergic reactions where histamine is released, as in hayfever. Antihistaminics are of limited usefulness since they do not inhibit other mediators of edema.

Acetylsalicylic acid (aspirin) is by far the most commonly used drug. It is effective in reducing inflammatory edema, but its mechanism of action remains unknown. 6-Mercaptopurine (6-MP), a nucleic acid analogue, depletes monocyte precursors in the bone marrow; when used in relatively low dosages, it has a specific effect on preventing mononuclear cell exudation. It has been used to treat chronic active hepatitis, but its effectiveness in that disease may be related to some mechanism other than its anti-inflammatory activity.

DISEASES ASSOCIATED WITH ABNORMALITIES OF INFLAMMATION

Inflammatory Edema

Hereditary angioneurotic edema is discussed in detail in Chapter 34. This disease is characterized by repeated life-threatening episodes of inflammatory edema, especially when the trachea is involved. The patients lack a serum alpha-2-globulin which is a natural inhibitor of C'1 esterase and also of the kinin-producing enzymes, kallikrein and PF/dil. Attacks reportedly have been treated by infusion of plasma containing the normal inhibitor.

Abnormalities of Leukocyte Migration

Patients with severe neutropenia are unable to produce a normal inflammatory response to infection. Not only do they fail to concentrate neutrophils in the local area, but mononuclear cell mobilization is impaired, owing to a lack of chemotactic factors normally produced by neutrophils in the inflamed area. Ability to respond normally to injury is not strictly correlated with the white blood cell count, since some patients with severe chronic neutropenia can release cells from their marrow when stimulated and show a normal local inflammatory response even though circulating neutrophils are below 500 cells per ml. Moreover, some patients (those with classical cyclic neutropenia) mobilize large numbers of activated circulating monocytes at the time they are neutropenic, and these cells are able to participate in the inflammatory response without neutrophils being present. In general, however, total lack of neutrophils for more than 2 weeks is incompatible with survival.

Patients in diabetic acidosis, although they may have increased levels of circulating neutrophils, are unable to mobilize these cells at the inflammatory site. Correction of the acidosis also corrects the inflammatory defect. These patients have increased susceptibility to bacterial infection.

Patients with acute or chronic glomerulonephritis associated with low serum complement also have defective neutrophil migration. This is due to a circulating inhibitor that interferes with the response to complement-generated chemotactic factors. Patients with a congenital deficiency of C'2 also show defective neutrophil migration, owing to a lack of generation of the complement-dependent chemotactic factors. These patients do not have increased susceptibility to infection.

The most common cause of defects in neutrophil and mononuclear cell migration is the use of natural and synthetic corticosteroids, 6-mercaptopurine or azathioprine as immunosuppressve agents in treatment following transplantation and for systemic autoimmune or connective tissue diseases such as systemic lupus erythematosus. These drugs may reduce inflammatory edema, limit neutrophil exudation and prevent mononuclear cell exudation, but they also predispose the patient to generalized bacterial, viral or fungal infections.

Defects of Phagocytosis

Patients with defects in their ability to phagocytize microorganisms have been reported, but in every case this has been the result of deficiency in serum factors, either antibody or some complement component.

Chronic Granulomatous Disease

A group of diseases have been defined that are characterized by a defect in intracellular killing of microorganisms in which the neutrophil can phagocytize bacteria normally but fails to kill certain bacteria such as staphylococcus and *E.coli*. Clinically, these patients have repeated acute and chronic infections with these organisms although they have had no problems with organisms such as the streptococcus and pneumococcus. The defect in killing results from the inability of these neutrophils to generate hydrogen peroxide. A similar defect can be produced by incubating normal cells under anaerobic conditions. The ability of these cells to kill only certain bacteria is explained by the fact that all bacteria produce some hydrogen peroxide, but bacteria that produce catalase can detoxify peroxide before it reacts with myeloperoxidase and halide ions to produce the bactericidal effect. Catalase-negative organisms essentially "commit suicide" in the abnormal neutrophil by producing the peroxide that the cell can use to kill them. Normal neutrophils produce excess peroxide that cannot be detoxified by the catalase in the bacteria and can therefore kill catalase producers as well.

Two distinct genetic forms of this disease have been identified. The first form, involving multiple male members of families, is characterized clinically by repeated attacks of lymphadenitis, pneumonia and osteomyelitis. These male patients have all died of infectious complications before reaching puberty. The terminal illness is often caused by infection with antibiotic-resistant fungi. Sex-linked inheritance has been firmly established by showing that neutrophils from the patients' mothers, maternal grandmothers and 50 percent of the female siblings have an intracellular bactericidal defect intermediate between that of patients' cells and control cells. Further, it has been shown that in the female carrier, 50 percent of the cells are normal and 50 percent are defective when studied for their ability to reduce nitroblue tetrazolium dye. All the cells of the patients are defective in nitroblue tetrazolium reduction because of the inability of the cell to produce peroxide. This histochemical reaction can be used as a quick screening test for the laboratory diagnosis of the disease.

The second form of this disease involves women and is clinically less severe, with many patients surviving into adult life. This form is apparently inherited as an autosomal recessive gene. The neutrophils from female patients are identical with those of the male patients in bacterial killing and also in their metabolic defects. The metabolic defects include a failure of stimulation of oxygen consumption, peroxide production and glucose metabolism through the hexose monophosphate pathway during phagocytosis. The female patients are deficient in glutathione peroxidase, an enzyme that catalyzes conversion of peroxide to water and oxidation of reduced glutathione. The enzyme defect in the males has not yet been clearly identified, but it is now clear that more than one enzyme defect can result in the same metabolic defect, indicating that a series of closely linked enzyme reactions are necessary for the production of hydrogen peroxide in response to phagocytosis. Patients with no myeloperoxidase have a similar but less severe intracellular killing defect and a more benign clinical disease.

ANNOTATED REFERENCES

Becker, E. L.: The relation of complement to other systems. Proc. Roy. Soc. Biol., 173:383, 1969. (A short discussion of implied interaction between the clotting and complement systems)

Becker, E. L., and Austen, K. F.: Mechanisms of immunologic injury of rat peritoneal mast cells. J. Exp. Med., 124:379, 397, 1966. (Studies of the mechanism(s) of immunological induction of histamine release. Similarities between complement-dependent and independent pathways are demonstrated.)

Dias, D. A., Silva, W., and Lepow, I. H.: Complement as a mediator of inflammation. J. Exp. Med., 125:921, 1967. (Characterization of a pathway leading to production of anaphylatoxin. Biological activities of this anaphylatoxin, derived from interactions of the first four components of complement, are carefully demonstrated.)

Douglas, S.: Disorders of phagocytic function. Blood, 35:851, 1970.

Downey, H.: Reactions of blood and tissue cells to acid colloidal dyes. Anat. Rec., 12:429, 1917. (Original observations demonstrating the formation of phagosomes)

Henson, P. M.: Release of vasoactive amines from rabbit platelets induced by sensitized mononuclear leukocytes and antigen. J. Exp. Med., 131:287, 1970 (Studies on the mechanism of leukocyte-dependent release of vasoactive amines from platelets)

Hirsch, J. G.: Cinemicrophotographic observations on granule lysis in polymorphonuclear leukocytes during phagocytosis. J. Exp. Med., *116:*827, 1962. (A classic paper demonstrating lysosome degranulation during phagocytosis. The process is shown in 57 time-lapse photographs.)

Holmes, B., Page, A. R., and Good, R. A.: Studies of the metabolic activity of leukocytes from patients with a genetic abnormality of phagocytic function. J. Clin. Invest., *46:*1422, 1967. (The essential relationships among phagocytosis, degranulation, glucose oxidation, hydrogen peroxide formation and intracellular bactericidal activity)

Holmes, B., Park, B. H., Malawista, S. E., Quie, P. G., Nelson, D. L., and Good, R. A.: Chronic granulomatous disease in females. A deficiency of leukocyte glutathione peroxidase. New Eng. J. Med., *283:*217, 1970.

Keller, H. U., and Sorkin, E.: Chemotaxis of leukocytes. Experientia, *24:*641, 1968. (A concise survey of leukocyte-chemotaxis and leukocyte-chemotactic factors, including definition of terms and methods for studying chemotaxis (103 references))

Lay, W. H., and Nussenzweig, V.: Receptors for complement on leukocytes. J. Exp. Med., *128:*991, 1968. (Demonstration of differences of affinity for antigen-antibody-complement complexes by neutrophils, lymphocytes and macrophages)

Menkin, V.: Modern views on inflammation. Int. Arch. Allerg., *4:*131, 1953. (Early studies on the chemical mediators of inflammation)

Mechinikoff, E.: Lectures on the Comparative Pathology of Inflammation (F. Starling and E. Starling, translators). Lender, Kegan, Paul, Treach, Trubnef and Co., 1873. (Early observations including the discovery of phagocytosis)

Nordlund, J. J., Root, R. K., and Wolf, S. M.: Studies on the origin of human leukocyte pyrogen. J. Exp. Med., *131:*727, 1970. (Systematic study on mechanisms of endogenous leukocyte pyrogen activation)

Page, A. R.: Inhibition of the lymphocyte response to inflammation with antimetabolites. Am. J. Path., *45:*1029, 1964. (Studies on the dynamics of lymphocytes in inflammation)

Page, A. R., and Good, R. A.: A clinical and experimental study of the function of neutrophils in the inflammatory response. Am. J. Path., *34:*623, 1957. (Original observations demonstrating the role of neutrophils in inflammation and the stepwise nature of an inflammatory response)

Quie, P. G., Messner, R. P., and Williams, R. C.: Phagocytosis in subacute bacterial endocarditis. Localization of the primary opsonic site to the Fc fragment. J. Exp. Med., *128:*553, 1968. (A thorough study demonstrating critical activities of the Fc fragment of IgG antibody in opsonization)

Sbarra, A. J., and Karnovsky, M. L.: The biochemical basis of phagocytosis. J. Biol. Chem., *234:*1355, 1959. (Studies on metabolic changes associated with phagocytosis)

Seegers, W., and Janoff, A.: Mediators of inflammation in leukocyte lysosomes. J. Exp. Med., *124:*833, 1966. (Description of two lysosome-derived mediators of inflammation)

Symposium on the chemical biology of inflammation. Biochem. Pharmacol. (Special Supp.), March, 1968. (A collection of research reports and lectures on the components of inflammation)

Symposium on vasoactive peptides. Fed. Proc., *27:*49–99, 1968. (A series of short papers on the nature of the kinins and other vasoactive peptides)

Tyrrell, D. A. J.: The role of interferon in the response to infection. Proc. Roy. Soc. Med., *62:*297, 1969. (A very short review of the subject)

Volkman, A., and Gowans, J. L.: The production of macrophages in the rat. Brit. J. Exp. Path., *46:*50, 1965. (A study of the inflammatory response demonstrating the origin of certain macrophages)

Ward, P. A.: Chemotaxis of mononuclear cells. J. Exp. Med., *128:*1201, 1968. (Studies on components of chemotaxis showing differences between the response of neutrophils and mononuclear cells)

33

Adaptive Immunity

Robert A. Good, M.D., and Joanne Finstad, M.S.

INTRODUCTION

The fundamental role of immunity can be viewed as the preservation of the integrity of the individual. In its ultimate expression, capacity to mount an immune response must facilitate perpetuation and elaboration of the germ plasm; and in this instance, such facilitation involves as a basic premise prevention of invasion by injurious organisms or cells from both external and internal sources. Modern concepts of immunity have developed from earlier studies that were focused on pragmatic efforts to understand and foster resistance to viral and bacterial infection or to minimize injurious consequences of such infection. This goal continues to be a cornerstone of modern immunobiology, and as such pervades much of immunobiological inquiry. The modern perspective, however, considers resistance to exogenous infection to be but one function of an adaptive process of far broader scope. It is the purpose of this chapter to reflect briefly upon immunobiology in its broader aspects and to consider the impingement of this upon the present-day practice of medicine.

We will reflect on the perturbations of adaptive immunity, focusing especially on inborn errors in which deficiencies of development of the systems have been demonstrated. We will consider as well the fundamental aspects of ontogenetic and phylogenetic development of the lymphoid systems and their immunological functions, define the divisions of labor within the lymphoid systems, and attempt to analyze certain basic bulwarks of the bodily defense especially in the perspective permitted by primary and secondary immunodeficiency diseases of man. Finally, we will make an effort to provide an insight into fundamental relationships of immunity to malignancy.

HOST-PARASITE RELATIONSHIPS

Specific Adaptive Barrier to Recurrent or Persistent Infection. In spite of continuing function of the set of innate nonadaptive and nonspecific adaptive barriers (described in the Introduction to this Section and in the chapter concerned with inflammatory mechanisms), lack of capacity to achieve adaptive immunity leaves man vulnerable to recurrent and persisting infection with a wide variety of microorganisms. Indeed, absence or even severe crippling of the adaptive immune responses in all mammals is incompatible with long survival in the sea of bacteria, fungi, viruses and protozoa that share ecological niches with the phylogenetically more recent forms. The adaptive immune responses (Table 33-1) which represent barriers to exogenous microbial invasion include ability both to produce and to secrete specific antibodies and ability to develop populations of specifically reactive lymphoid cells in response to antigenic stimulation. The fundamental process of immunity, namely, ability to achieve specific responses to antigens and organisms never before encountered by the individual, is not yet fully understood.

TABLE 33-1. Adaptive Immune Responses

1. Produce circulating antibody on initial antigenic stimulation

2. Develop delayed allergy and cellular immunity on exposure to antigen

3. Recognize as foreign and reject foreign cells, tissues and organs

4. Exhibit memory of prior immunological system

5. Possess systems of cells which respond specifically with proliferation and differentiation to antigenic stimulation.

Further, the existence of this capacity in the general absence of adaptive specific reactivity against host constituents – the self-nonself discriminations so fundamental to immune reactions—remains enigmatic in fundamental terms. Considerations of the basis of these processes and discriminations in relation to known adaptive genetic and differentiative mechanisms continue to provoke the immunobiologist and to be the focus of much of this theoretical effort even in most recent writings. That adaptive mechanisms of the host are both discriminatory and anticipatory has been established; and it is clear that without such adaptations survival of individual members of more recent phylogenetic forms is impossible.

Resistance to Specific Pathogens. Subdivisions within the adaptive immunological systems have been clearly indicated by clinical studies for many years. For example, untreated immunologically deficient patients, such as those having multiple myeloma, are very susceptible to infections with the pneumococcus and certain other pyogenic pathogens of high-grade virulence. By contrast, they seem to have no unusual deficiency in ability to resist many viruses, fungi, or intracellular bacterial pathogens. On the other hand, untreated patients suffering from Hodgkin's disease, a malignancy of the reticulum cells, have long been known to be inordinately susceptible to tuberculosis, fungus infections or infec-

tions with any of a variety of viruses. Similarly, children with different kinds of primary immunodeficiencies differ from one another in the kinds of infection to which they express inordinate susceptibility. Thus, it is apparent in considerations of clinical disease that a high degree of specialization must exist within the bodily defenses. Some aspects of this specialization can now be understood; much further work needs be done, but it is clear that extensive effort in this direction will provide a most useful approach to classification of both microorganisms and host defenses. Such a classification will reflect fundamental interactions at a genetic level which have been taking place for hundreds of millions of years.

The clearest evidence of these fundamental relationships between forms evolving in intimate relationships to one another can be seen with certain viral infections in which long-standing associations of host and virus have resulted in mutual adaptations that permit regular invasion of host and persistent parasitism without damage to either organism or host. Thus, for example, in the case of the herpes virus, parasitism by variants of this virus class can be demonstrated in most vertebrate hosts. There can be no question that interaction between these viruses and their vertebral hosts have been in process for hundreds of millions of years, and that all major branches of vertebrates in which studies have been carried out are infected with one or another of these agents. Hosts carrying such viruses (such as man, infected with the herpes simplex virus) may be healthy for many years, showing only the minor annoyance of "cold sores" from time to time, as an expression of the continuing infection. The same organism, however, infecting a closely related species (e.g., mouse, rabbit, monkey, or chimpanzee) may produce devastating and rapidly fatal infection. If, however, a primary host of the virus experiences perturbation of his immunological defense system (as may occur during lymphoreticular malignancy or as a consequence of immunosuppressive therapy), or if infection with the herpesvirus occurs during the immediate neonatal period or in an aged person, the same agent may produce overwhelming and fatal infection.

With many pathogens the capacity of host phagocytic cells to kill ingested organisms by a particular enzymatic mechanism is vital to survival, whereas with still other pathogens specific components of the complement system have been found to be vital to the bodily defense. One of the major concerns of clinical immunobiology, then, must be a persistent effort to sort out, understand, and hopefully to manipulate capacities for resistance to specific pathogens.

Immunosuppressive Actions of Pathogens. Shortly after the turn of this century, von Pirquet recognized that host defense mechanisms can be profoundly affected by infection. He discovered that patients showing hypersensitivity to the tubercle bacilli temporarily lose this hypersensitivity during and immediately following an attack of measles. For many years little was made of this relationship, but more recent studies have established that rubeola, rubella, almost all oncogenic viruses, and many other viral and bacterial agents are able to interfere with one or another of the specific host defense mechanisms. This ability often is revealed by depression of a specific component of host's immunological defense. Such influences of infectious agents on the immune reactions have been demonstrated both in vivo and in vitro. It seems reasonable to postulate that in such relations one also may be seeing evidence of an ongoing interaction of basic adaptive genetic potential of the microorganisms on the one hand and specific adaptive defenses of the vertebrate host on the other. Consequently, in viewing the adaptive immunological mechanisms of the host in relation to invasive and destructive potential of the innumerable kinds of microorganisms one must always consider the relationships in a framework of ongoing interactions between multiple adaptive systems.

Practical Prophylactic and Therapeutic Usefulness of Adaptive Resistance. Among the most powerful of all tools of modern medicine is the ability to prevent serious disease by active immunization. Almost unbelievable prevention of morbidity, mortality, crippling and suffering has been accomplished by application of active immunization to large populations. The discipline of immunology, indeed, first appeared as a reflection of Jenner's insight which led him to attempt to vaccinate against smallpox by inducing cowpox in susceptible persons. It is now known, of course, that cowpox represents an attenuated virus infection. The legacy from Jenner's insight included four major contributions: the first practical use of the positive adaptive component in immunity; the first association of hypersensitivity in resistance to infection; the first use of an attenuated virus in active immunizations; and the first mention and association of viral interference with resistance to infection in man. The achievement of immunity by this means, and the general acceptance of the methodology combined with modern methods of distribution and communication, now promise that the earth can be freed of smallpox as a disease during the present decade. Similar promise exists for poliomyelitis. Both killed and live polio virus vaccines are effective and in large population have achieved most impressive herd immunity. Beginning use of attenuated live measles and rubella vaccine also may lead to similar salutary achievements in preventative medicine. Listed in Table 33-2 are the viral infections against which effective live attenuated virus immunization is available.

Active immunization against the exotoxins of diphtheria and tetanus organisms

TABLE 33-2. Attenuated Viruses
Used in Immunization

1. Vaccinia
2. Polio 1, 2, 3
3. Rubeola
4. Rubella
5. Yellow fever
6. Mumps

has also been established as widely applicable achievements in the prevention of disease. Striking figures demonstrating the progressive reduction in severe disease from infections with bacteria producing these exotoxins can be attributed to this form of active immunization. Similarly active immunization with an attenuated form of the tubercle bacilli has been effective in preventing spread of tuberculosis among susceptibles in many parts of the world. Immunization with other bacterial and viral vaccines has been somewhat less effective.

Specific passive immunizations have also been found clinically applicable. With this approach outstanding success in prevention of measles, tetanus, diphtheria, rabies, pertussis and now even chickenpox has been achieved. Although not nearly so generally applicable nor so effective as are active immunizations, passive immunizations reflect practical consequences of immunological inquiry which will surely continue to be important in human and veterinary medicine.

Diagnostic measures reflecting specific cellular or humoral responses to antigens contained in or produced by microorganisms have also proved of greatest practical value. These analyses have facilitated diagnosis and aided in directing therapy and management of human disease. Routine clinical and public health laboratories and special government- and WHO-sponsored laboratories can carry out literally hundreds of different kinds of immunological analyses. In the same way they can provide specific antigens for skin tests which can facilitate diagnoses and foster clinical management of many diseases of man and animals.

HUMORAL IMMUNITY

Fundamental Relationships in Cellular and Humoral Immunity. The immunological responses and specific immunological defenses can be divided into two primary groups, those of cellular and those

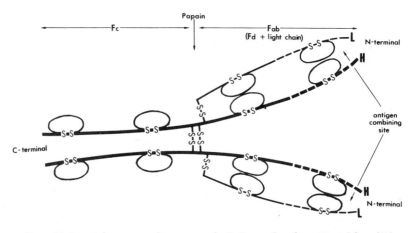

FIG. 33-1. Schematic diagram of IgG molecule. Variable (V)-regions of heavy (H) and light (L) chains are shown as dashed lines. Solid lines represent (c) constant regions.

of humoral nature. All immune responses of more recent vertebrates (phylogenetically) appear to be functions of the lymphoid cell systems. However, in some of these the specificity resides at the surface of circulating lymphocytes and in others at the combining site of immunoglobulin molecules.

In all placoderm-derived vertebrates the molecules of the five major immunoglobulin classes have the same basic structure and characteristics, the analyses of which were made possible by the availability of myeloma proteins in man and mice. The basic units of all known immunoglobulins are comprised of 4 polypeptide chains, two light chains and two heavy chains (Fig. 33-1). Both light and heavy chains comprise a variable N-terminal region and constant C-terminal portion. The specific combining site for antigen is found in one of the repetitive loops of the variable or V-region. Light chains and heavy chains are covalently bonded, being held together by interchain disulfide linkages. Heavy chains and light chains are, as their name implies, of very different molecular sizes. Light chains of all immunoglobulins are of two fundamental classes designated kappa

and lambda chains. Heavy chains are distinguishable from light chains also by their antigenic specificity, and in man heavy chains of each of 5 major classes can be distinguished one from the other on an immunological basis. Heavy chains of different classes have antigenically specific characteristics.

Within several of the 5 major immunoglobulin classes subgroups have been defined which again can be distinguished one from the other by virtue of distinct antigenic characteristics. The molecular distinctions of the several immunoglobulins are simply presented in Figure 33-2. The macromolecular 19S IgM immunoglobulins in man are pentamers in which the basic subunit has sedimentation characteristics of 7S. In more phylogenetically primitive forms, and occasionally in human diseases where antibody formation is maximally stimulated (as, for example, in trypanosomiasis), large amounts of 7S IgM may be found in the circulation. The IgM molecules possess large amounts, approximately 10 percent, of carbohydrate which is attached to the carboxy terminal constant region of the molecules. The most abundant of the immunoglobulin molecules in

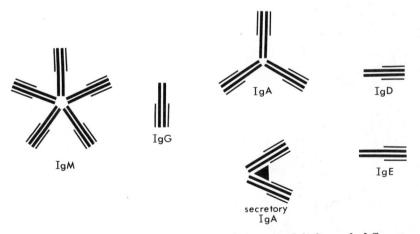

FIG. 33-2. Schematic comparison of immunoglobulins of different classes. Heavy chains are shown as thick lines, light chains as narrow lines. Note the basic similarity of immunoglobulin structures (the 2 H-chain plus 2 L-chain unit).

the human body are those of the IgG fraction. For man, IgG molecules comprise at least 4 subclasses, and distinct functional specialization has already been described from some of the subclasses. For example, IgG_2 is the subclass which contains most of the antibodies against polysaccharide antigens. IgG_4, unlike the other immunoglobulins of IgG class, does not readily fix complement—even in the aggregated form. Probably of greatest importance in the ultimate function of immunoglobulin molecules is their distribution. IgM immunoglobulins in man are largely confined to the intravascular compartment. IgG molecules are widely distributed in blood, lymph and interstitial fluid—the total space for IgG being approximately 2.2 times that of the plasma volume.

IgA immunoglobulins are of two major subclasses and are found in circulation primarily as monomers comprising 2 light and 2 heavy chains. Small amounts of dimeric IgA may also be present in the circulation. IgA immunoglobulins are selectively concentrated in glandular, urinary tract and gastrointestinal secretions, but in the secretions they are present in a special form in which dimers of the basic IgA molecule are combined with a specific additional component, termed the "Transport Piece."

Although any of the immunoglobulin molecules can be found at cell surfaces, it is IgE class of molecules in man and many mammals which can specifically attach to certain cells, especially mast cells. Normally only small amounts are present in the circulation. The rates of degradation and distribution of the several classes of immunoglobulin molecules are primarily a function of the so-called Fc of C-terminal part of the molecule (Fig. 33-1). Thus, transport of IgG across the placenta from maternal plasma to fetal plasma requires a transport mechanism dependent upon the integrity of the Fc component of the molecule. Under usual conditions of aggregation or distortion, after combination with antigen both IgM and IgG efficiently engage the complement cascade, a biological amplification system capable of

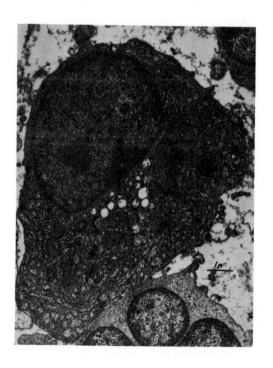

Fig. 33-3. Electron micrograph of human plasma cell. Note the highly developed and prominent rough endoplasmic reticulum, Golgi apparatus (appearing as clear vesicles), and eccentric nucleus, all characteristic of these cells which synthesize and secrete antibody.

facilitating primitive defense responses, including reactivity of blood vessels, coagulation of blood, inflammation and phagocytosis. (The reader is referred to Chapter 34 for a detailed discussion of complement.)

By contrast neither IgA nor IgE seems capable of activating the complement system under usual conditions. In its special "transport" form IgA comprises the major component of a local antibody system. IgE represents the immunoglobulin system involved in atopy (immediate hypersensitivity) but its raison d'être in bodily defense is still enigmatic. Little is known of the highly variable IgD but antibodies of IgD class have been described.

All immunoglobulins are produced and secreted by lymphoid cells. Cells with morphological features of lymphocytes and plasma cells are capable of both synthesis and secretion of immunoglobulins (Fig. 33-3). Many lymphoid cells, however, seem readily capable of synthesizing but

are not differentiated to a point where they are capable of efficient secretion of immunoglobulin molecules.

The Bursa of Fabricius. Only in birds is the organ specialized for differentiation of the antibody and immunoglobulin secreting cells known with certainty. The specialized site for differentiation of the cells of this B-cell system is in the bursa of Fabricius (Fig. 33-4) where cells capable of producing IgM and IgG differentiate sequentially. Lymphoid cells in the bursa first become capable of IgM synthesis, and then later switch to capacity for IgG synthesis. Later renewal of the two, IgM-producing and IgG-producing cell lines, is transferred to the bone marrow and the bursa itself involutes. Involution of the bursa coincides with sexual maturation.

The immunoglobulin producing and secreting cells develop from stem cells located first in the blood islets of the yolk sac and later, successively, in fetal liver and then, bone marrow. Proliferating within

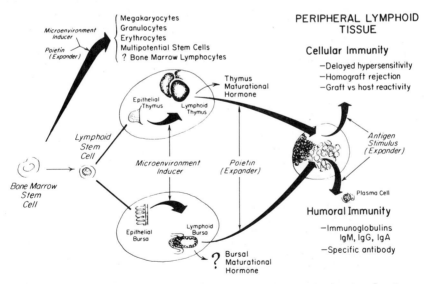

FIG. 33-4. Developmental basis of the two component adaptive lymphoid systems: bursa-dependent (B-cells) and thymus-dependent (T-cells). Both cellular systems differentiate from stem cells which might also differentiate into any of the other bone marrow derived cells. (Good, R. A., *et al.: In:* Birth Defects: Original Article Series, The Immunological Deficiency Disease in Man. vol. 4, p. 17, 1968. With permission of the publisher, The National Foundation, N. Y., 1968)

the bursa these cells become capable of immunoglobulin production, but by the time they leave the bursa most lymphocytes are not specialized enough for immunoglobulin secretion. Such cells may be called B_1; and their development into immunoglobulin-secreting cells occurs later. Lymphocytes and plasma cells capable of immunoglobulin secretion may be referred to as B_2 cells (Table 33-3). Specialization of B cells not only into cellular subpopulations capable of both synthesis and secretion of different specific molecular immunoglobulin classes but also with respect to ability to synthesize (but not to secrete) immunoglobulin molecules seems certain. The subclasses of cells are specialized even with respect to distribution in the peripheral lymphoid organs and tissues (e.g., lymph node and spleen). In many mammals, including man, the B_1 cells are located in the far cortical areas of the lymph nodes, sites where germinal centers preferentially appear. The B_2, or secretory B cells, by contrast are located preferentially in the medullary cords of lymph nodes and in the red pulp of the spleen. Both B_1 and B_2 cells almost certainly have both class-specific and subclass-specific surface markers which will permit identification of these

lymphocytes from other kinds of lymphocytes, as well as from one another. For example, in the mouse, the antibody prepared against myeloma cells PC_1 will attach to secretory B cells but probably not to the B_1 cells; and several investigators have described markers (i.e., C'3 receptor) which may help identify the B_1 cells.

In mammals, reptiles, amphibians and fishes a system of B cells specialized for immunoglobulin and antibody production and secretion can be identified. In none of these forms, however, has an organ morphologically equivalent to the bursa of Fabricius of birds been identified. Although areas of gut-associated lymphoid tissue (GALT) having certain similarities to the bursa of birds have been located in each of the major phylogenetic deviations, for none of these has convincing evidence been presented that establishes its role as the differentiation site equivalent to the bursa of Fabricius. However, the studies of the bursa and its role have proceeded apace and have improved definition of the bursal development and functions sufficiently to permit early identification of bursal equivalents (Table 33-4). Appropriate study in fishes, amphibians, reptiles and mammals using these characteristics to define the

TABLE 33-3. B-Cell System

CELL	LOCATION	CHARACTERISTICS
Prebursal	Blood islets of yolk sac, fetal liver, bone marrow	Not immunologically competent, capable of traffic to differentiative site in bursa or equivalent
Postbursal B_1 cell	Medulla of bursa, far cortical areas of lymph node—germinal centers periarteriolar accumulations in malpighian corpuscles of spleen. Efferent lymph following antigenic stimulation. Thoracic duct lymph, circulating blood	Capable of synthesis of ab, but not specialized for secretion of large amounts of ab. C3 receptor of Nussenzweig. Probably specific antigen in all animals
Postbursal B_2 or S cell	Bone marrow, medullary cords of lymph nodes, red pulp of spleen, lamina propria of GI tract and secretory glands, interstitial tissue of bone marrow, occasionally peripheral blood or lymph	Secretory lymphocyte, plasma cells. Neg. for Nussenzweig marker. Specialized for both production and secretion of the several immunoglobulin classes. PC_1 antigenic marker in mouse. Probably specific antigens in other animals

TABLE 33-4. Bursa and Bursal Equivalent Sites

1. First site in B lymphoid system to possess immunoglobulin forming cells.

2. Most likely a lymphoepithelial site.

3. First site to have lymphoid cells forming readily demonstrable μ chains.

4. Later, also the first site to have cells forming γ chains.

5. Only site to have appreciable number of cells forming both μ and γ chains in large amounts at the same time. Thus cytoplasm stains specifically for both IgG and IgM.

6. A site where antigenic stimulation does not alter either the number or time of appearance of μ or γ containing cells and the follicles producing these cells.

bursal site should permit definitive identification of bursal equivalents.

THE ROLE OF THE THYMUS

T-Cell System. Counterpoint to the B cell system is the system of T cells, responsible for the cell-mediated immunities. This population of lymphoid cells is derived from primitive stem cells under the differentiative influence of the thymus. Stem cells from the blood of islets of the yolk sac (or later from fetal liver and then bone marrow) travel to the thymus where they come under the inducing and differentiating influence of the epithelial component of the thymus. Under conditions of extraordinary, apparently wasteful proliferation, the stem cells develop first into thymocytes which have surface markers and then into post-thymic cells (T_1). Both thymocytes and post-thymic cells in mice possess a characteristic antigenic marker, theta (θ), by which they can be distinguished from lymphoid cells of the B population and other mononuclear elements. Already several classes of post-thymic lymphocytes have been defined (largely from experimental studies in mice); but it seems likely that the same distinctions will hold for other mammals and for man as well (Table 33-5).

In chickens, in which the two primary lines of T and B lymphocytes can be manipulated separately by early removal of the thymus or the bursa, there is no question that these lines of cells are distinct from one another and that each represents

TABLE 33-5. T-Cell System

CELL	LOCATION	CHARACTERISTICS
Prethymic	Yolk sac, fetal liver, bone marrow	Theta-negative, not immunologically competent. Radiosensitive. Migrate to thymus and bone marrow
Post-thymic T_1	Fetal liver, spleen of neonatal mouse. Peripheral lymphoid tissue	Not immunocompetent. Sensitive to humoral inductive influence of thymus. Theta-positive
Post-thymic T_2	Blood, thoracic duct lymph, peripheral lymphoid tissues, thymus dependent areas. Recirculating in blood and lymph	Theta-positive, immunocompetent long-lived recirculating cell
Post-thymic T_3	Peripheral lymphoid system, blood inflammatory exudates	Not recirculating, theta-positive, produces lymphokines, selectively migrate to inflammatory site
Post-thymic T_4	Spleen, lymph node, marrow	Radioresistant, theta-positive, memory cell capable of engaging radiosensitive, nonspecific marrow cells to achieve delayed allergic reaction and cellular immunity

a specialized line directed in its development by a central lymphoid organ. Similarly in mice, rats and man sufficient evidence is at hand to permit the conclusion that at least two separate main lines of lymphocytes exist. Among the cells called lymphocytes in the peripheral blood of man, however, a minimum of 5 distinct, functionally separate cells can be identified. As more is learned of specialization among the lymphoid cell populations, this list will certainly be considered a minimum.

The B-cell system exercises its specific influences by virtue of the immunoglobulin products of its cells at least some of which act at long range from the cells as antibodies capable of engaging several biological amplification systems (Table 33-6). The basis of B-cell specificity is, however, clearly the immunoglobulin molecules of the several classes; and it seems virtually certain that the antigen receptors through which the cells of this class can be specifically stimulated by antigen comprise the same immunoglobulin molecules. Thus, antigenic stimulation of B-cells is thought to be accomplished by specifically stimulating antigen combining with antigenic receptors at the cellular surface thereby launching a process of proliferation and differentiation. Thus far, the basis for the specific reactivity of T-cells has not

TABLE 33-6. Biological Amplification Systems Used by Immune Reactions

1. Inflammation: via chemotactic factors of complement system, chemotactic lymphokines, MIF, kinin and kallikrein systems

2. Vascular reactivity: through anaphylatoxins from complement system, lymphokines, kallikrein and kinin systems and histamine and SRS-A from mast cells activated by IgE or IgG and antigen

3. Coagulation via complement system, kallikrein and kinin system possibly lymphokines

4. Phagocytosis: through immune adherence via complement system, MIF, or macrophage activating factors

5. C3 opsonization via complement cascade, etc.

TABLE 33-7. Views Concerning Nature of T-Cell Receptor

1. Production but not secretion of all different forms of immunoglobulin

2. IgX

3. Light chains, especially kappa chains attached to or embedded in the membranes

4. V. regions of light chains in the membrane

5. A primitive immunoglobulin such as Y. S. Choi's primordial immunoglobulin of chicken

6. Finstad's agglutinators of primitive invertebrates

been established. Most investigators believe it to be an immunoglobulin molecule possessing light chain—heavy chain structure bound firmly to the surface of T-cells. The several views regarding the nature of the surface receptor which accounts for specificity of the T-cell system are summarized in Table 33-7.

T-cells are not secretory cells and their specificity is engaged in immune responses only when these lymphocytes encounter antigen at their surface, to which they react specifically. The T-lymphocytes by virtue of this specificity are responsible for a major component of the bodily defense. The small T-lymphocytes are capable of producing several biologically active substances which can effect certain physiological processes, seem to be able to kill nonspecifically and can engage at least some of the primitive reactive systems of the body as biological amplification systems. Table 33-8 lists

TABLE 33-8. Currently Recognized Lymphokines

1. Migration Inhibitory Factor

2. Lymphotoxic Factor

3. Skin-reactive Factor

4. Chemotactic Factor

5. Mitogenic Factor

6. Interferon

7. Antibody

those products of lymphocytes which have been demonstrated to be produced following specific antigenic stimulation of these cells.

Distribution of T- and B-Lymphocytes in the Peripheral Lymphoid System. In a typical mammalian lymph node, T-cells are found preferentially in the deep cortical areas while B-lymphocytes are preferentially located in the far cortical regions, germinal centers and in the medullary cords. These regions are not exclusive one of the other, but the bulk of T-cells in the node are in the T-cell regions and vice versa. In the spleen the B-cell population is located both in certain para-arteriolar accommodations in the malpighian corpuscles and in the red pulp. By contrast, T-cells predominate in the periarteriolar regions of the spleen and diffusely in the malpighian bodies. B-cells specialized for secretion of IgA dominate the lamina propria regions of the small and large bowel, lamina propria of intestinal glands of external secretion and a number of other subepithelial locations. Both B- and T-cells are found in the loose connective tissue, efferent lymph from regional nodes, thoracic duct lymph and in the peripheral circulation.

Functions of T-Cells. As has been established from crucial studies in a wide variety of experimental animals and from studies of patients with selective immunodeficiencies, T-cells in mammals are essential for life. They represent the specific component of a major bulwark of the bodily defense against fungi, certain viruses and the facultative intracellular bacterial pathogens such as the tubercle bacillus, atypical acid fast organism, Salmonella, Listeria, Brucella and many others. The known functions of T-cells include development of delayed allergic reactions, a major component in recognition and rejection of solid tissue allografts, initiation of graft-versus-host reactions and probably a major contribution to surveillance against the development of malignancy (Table 33-9).

TABLE 33-9. The Thymus-Dependent Cell Responses

1. Ability of host to develop delayed allergy
2. Ability of host to reject skin allografts
3. Ability of cells to initiate graft-versus-host disease
4. Ability of cells to respond in vitro by blast transformation and/or replication when stimulated with:
   a. phytohemagglutinin (PHA)
   b. allogeneic cells
   c. antigen to which the host has received prior stimulation

The T-cells respond in vitro with blast cell transformation and replication to stimulation with phytohemagglutinin or with allogeneic cells in mixed leukocyte cultures. They also respond in vitro with blast transformation and replication to stimulation with antigens to which the host from which they are derived had been sensitized previously. Summarized in Table 33-10 are those tests by which T-cell function can be assessed clinically. The T-cells comprise a major component of the very small lymphocyte population so that

TABLE 33-10. Tests for Adequacy of T-Cell Functions

1. Quantitative response of cells in whole blood to PHA—in vitro transformation
2. Dose response analysis of PHA responsiveness in vitro
3. In vitro response of lymphocytes to allogeneic, irradiated or mitomycin-treated lymphocytes
4. Development of delayed allergy to ubiquitous antigens, SK-SD, mumps, Candida, trichophyton, PPD
5. Development of contact allergy to 2-4 dinitrochlorobenzene
6. Small lymphocyte count
7. Capacity to reject allograft of skin
8. Presence of an abundant cell population in deep cortical areas of lymph node following antigenic stimulation
9. Vigorous defense against fungi, virus and facultative intracellular bacterial pathogens

a crude morphological indication of their adequacy or deficiency can be obtained by a carefully executed count of the numbers of circulating small lymphocytes.

Viewed in another way, the T-lymphocyte population makes up a major component of the long-lived recirculating small lymphocyte population and a major component, if not the great bulk, of the readily mobilizable lymphoid pool that can be exhausted by extracorporeal irradiation, thoracic duct drainage or treatment with antilymphocyte serum.

Several tests are available by which the adequacy or inadequacy of B-cell function can be assessed clinically (Table 33-11). This antibody-producing population of cells represents a major bulwark of the bodily defense against the high grade encapsulated pyogenic pathogens. It also contributes in a major way to elimination of foreign red blood cells and hematopoietic elements, prevention and elimination of certain virus infection, and possibly through

TABLE 33-11. Tests to Evaluate B-Cell Function

1. Quantitation of levels of all major immunoglobulin classes by radial immunodiffusion or radioimmunoassay.

2. Evaluation of fractional catabolic rate and/ or synthesis rates for individual immunoglobulins

3. Antibody concentration to antigens widely distributed in nature, e.g., isohemagglutinins, ASO, Schick test, antiviral antibodies etc.

4. Quantitation of antibody response to killed polio virus vaccine—Pasteur Institute

5. Quantitation of antibody response to diphtheria and tetanus toxoids

6. Antibody responses to polysaccharide antigens from pneumococcal, meningococcal and Hemophilus polysaccharides

7. Concentration of IgA and analysis of form of IgA in saliva

8. Quantitation of immunoglobulin subclasses

9. Specific identification of bacteria causing frequent pneumonia, sepsis, conjunctivitis and meningitis. The organisms that particularly plague patients with B-cell defects include pneumococci, streptococci, *Hemophilus influenzae,* meningococci and *Pseudomonas aeruginosa.*

the local antibody system prevention of stimulation with many antigens that cross react with those of the host.

T- and B-Cell Interactions. Synergistic actions of the T- and B-cells in reconstituting immunological capacity of fatally irradiated mice have been studied by many investigators. T-cells and B-cells may interact to facilitate responses to certain antigens, and these findings have suggested that T-cell–B-cell interactions are obligatory to the achievement of antibody production. Indeed, one of the most popular views of the distinction between self and nonself proposes that tolerance is accomplished when T-cells or B-cells are stimulated by antigen directly, and that an immune response is achieved when a different signal derives from T-cell–B-cell interaction. (Although hypotheses based on these findings present interesting possibilities, abundant evidence has been forthcoming indicating that T-cell and B-cell interactions are not obligatory to antibody synthesis or immunoglobulin production and secretion.)

IMMUNOLOGICAL DEFICIENCY DISEASES OF MAN

The extraordinary interplay between clinic and the basic laboratory in development of the modern immunobiological perspective is seen particularly well when one considers the primary immunodeficiency diseases of man (Table 33-12). Human diseases have been defined and studied in terms of which system of lymphoid cells (either T-cell development or B-cell development or both) is primarily defective.

Patients with X-linked recessive form of primary immunodeficiency, a rare disease, lack plasma cells and B-lymphocytes and often fail to produce even the smallest amounts of antibody even after repeated antigenic stimulation. Such patients, by contrast, develop cell-mediated immunities very well. Small lymphocyte counts are

TABLE 33-12. Classification of Primary Immunodeficiency Disorders

TYPE	SUGGESTED CELLULAR DEFECT		
	B-CELLS	T-CELLS	STEM CELLS
Infantile X-linked agammaglobulinemia	+		
Selective immunoglobulin deficiency (IgA)	some +		
Transient hypogammaglobulinemia of infancy	+		
X-linked immunodeficiency with hyper-IgM	+	?	
Thymic hypoplasia (pharyngeal pouch syndrome, DiGeorge's syndrome)		+	
Episodic lymphopenia with lymphocytoxic antibody		+	
Immunodeficiency with or without hyperimmuno-globulinemia	+	+ (sometimes)	
Immunodeficiency with ataxia-telangiectasia	+	+	
Immunodeficiency with thrombocytopenia and eczema (Wiskott-Aldrich syndrome)	+	+	
Immunodeficiency with thymoma	+	+	
Immunodeficiency with short-limbed dwarfism	+	+	
Immunodeficiency with generalized hematopoietic hypoplasia	+	+	+
Severe combined immunodeficiency			
(a) autosomal recessive	+	+	+
(b) X-linked	+	+	+
(c) sporadic	+	+	+
Variable immunodeficiency (common, largely un-classified)		(sometimes)	

normal or nearly normal. Delayed allergic reactions seem to develop normally after appropriate challenge with such antigens as 2-4 dinitrochlorobenzene. Responses to phytohemagglutinin stimulation of lymphocyte in vitro are vigorous, and lymphocyte populations in deep cortical areas of lymph nodes and periarteriolar components of spleen appear normal. However, the far cortical areas of lymph nodes are poorly populated (Fig. 33-5). Further, germinal centers are lacking in both lymph nodes and spleen. All immunoglobulins are lacking or are almost lacking. The thymus, which has been repeatedly studied, cannot be distinguished from the thymus of a normal child or from the thymus which shows accidental involution in immunologically normal person suffering from those chronic or acute infections which frequently plague these children.

Counterpoint to these patients with isolated primary B-cell deficiency are those rare patients born without a thymus (DiGeorge syndrome), who lack T-cells but possess a well-developed B-cell system. (Recently a similar genetic abnormality has been defined in the athymic nude mice.) Such patients and mice appear usually to develop plasma cells, far cortical areas of lymph nodes, germinal centers and immunoglobulins normally. They produce antibodies to stimulation with many antigens quite well, but are almost completely lacking in very small lymphocytes and in lymphocytes of the deep cortical areas of

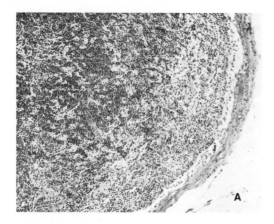

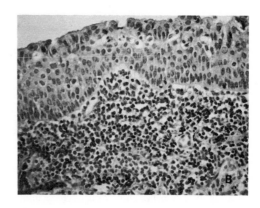

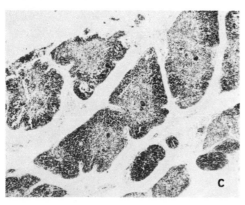

FIG. 33-5.   (A) Cortical region of a lymph node from a patient with X-linked recessive immune deficiency disease (Bruton-type agammaglobulinemia). Note the sparsity of cells and absence of lymphoid germinal centers in the far cortical space. (Compare with Fig. 33-6.) (B) Tonsil of same patient. Again note absence of germinal center and plasma cells. (C) Thymus of same patient, which is essentially normal. (Good, R. A., *et al.: In:* Birth Defects: Original Article Series. vol. 4, p. 17, 1968. With permission of the publisher, The National Foundation, N. Y., 1968)

the lymph node, and are deficient in the periarteriolar populations of lymphocytes in the spleen. Such patients cannot reject allografts of skin or other tissues and fail to develop delayed allergic responses. They generally lack cells in peripheral blood that can respond in vitro with blast transformation and proliferation to phytohemagglutinin, allogeneic cells or antigens to which host has previously been stimulated.

Patients with the isolated B-cell and T-cell deficits differ strikingly in the pattern of infections that cause them the greatest difficulty. Those with primary B-cell immunodeficiency, unless treated with gamma globulin, experience a succession of life-threatening infections with extracellular encapsulated pyogenic pathogens, such as pneumococci, *Hemophilus influenzae*, streptococci, meningococci and *Pseudomonas aeruginosa*. Recurrent otitis, sinusitis, and skin infections also occur with great frequency. By contrast, BCG immunization, primary tuberculosis, virus vaccination, many viral infections, fungus infections, and systemic infections with enterobacteria are either well handled or do not occur with inordinate frequency. Recurrent sinus infection, respiratory disease and persistent spruelike gastrointestinal disease persist in such patients, probably as a result of the inadequacy of the local antibody system even after administration of large amounts of serum gamma globulin.

Patients who lack T cells, but possess B cells in normal state of development and in normal numbers, present a strikingly different lymphoid morphology (Fig. 33-6). Such patients often die of infection with atypical acid-fast organisms, vaccinia virus, fungus infections, or infections with enterobacteriaciae.

Immunological Deficiency Diseases of Man

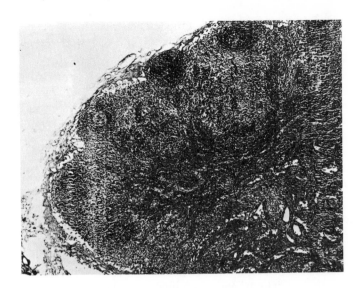

FIG. 33-6. Lymph node from patient born without a thymus (DiGeorge syndrome). Note the presence of germinal centers in the far cortical region. Plasma cells are present near the germinal centers and in the medulla. Lymphocytes in the T-region (deep cortical area or corticomedullary junction) are sparsely distributed. (Cooper, M. D., *et al.: In:* Birth Defects: Original Article Series. vol. 4, p. 378, 1968. With permission of the publisher, The National Foundation, N. Y., 1968)

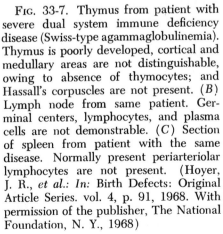

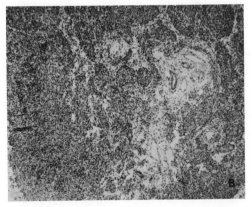

FIG. 33-7. Thymus from patient with severe dual system immune deficiency disease (Swiss-type agammaglobulinemia). Thymus is poorly developed, cortical and medullary areas are not distinguishable, owing to absence of thymocytes; and Hassall's corpuscles are not present. (*B*) Lymph node from same patient. Germinal centers, lymphocytes, and plasma cells are not demonstrable. (*C*) Section of spleen from patient with the same disease. Normally present periarteriolar lymphocytes are not present. (Hoyer, J. R., *et al.: In:* Birth Defects: Original Article Series. vol. 4, p. 91, 1968. With permission of the publisher, The National Foundation, N. Y., 1968)

Severe Dual System Immunodeficiencies. In patients with the most extreme form of severe dual system immunodeficiency T-cell functions as well as B-cell functions are virtually absent. The thymus is regularly present but is underdeveloped, lacks Hassall's corpuscles, and is made up entirely (or almost entirely) of stromal-epithelial cells. All peripheral lymphoid areas (and selective sites for both T- and B-cell populations) are sparsely populated, or not populated at all, with lymphoid cells. Forms of this disease have been transmitted either as an autosomal recessive trait or as an X-linked recessive disorder. Such patients often have persistent and invasive infections with Candida and regularly die during the first or early part of the second year with infections caused by disseminated BCG infection, generalized vaccinia, overwhelming Hecht's pneumonia due to measles virus, fungus infection, or infection with either high grade encapsulated pathogens or lower grade enterobacteriaciae. Often they can develop neither cellular nor humoral immunity although considerable variability in phenotypic expression of the basically inherited disorder has been recorded. This variability permits, in some instances, a degree of T-cell and/or B-cell development in individual patients whereas in other family members virtually no development of either system may be seen. (Figure 33-7 illustrates the characteristics of the lymphoid tissues of these patients.)

Ataxia-Telangiectasia. Characteristically, immunological deficiencies are observed in patients having what was once considered primarily a neurological disease, ataxia-telangiectasia. Such patients regularly show deficiencies of cell-mediated immunities, in approximately 60 to 70 percent deficiencies of IgA production, and even a higher incidence in serum IgE deficiency. Such patients are prone to experience frequent sinopulmonary disease, the frequency of which correlates well with combined IgA and IgE deficiency. This familial disease is generally considered autosomal recessive; but recently questions have been raised, since the incidence of consanguinity is not as high as might be anticipated for the usual autosomal inheritance, and among involved sibships too many children have been affected. Consequently, a basis in isoimmunization has been proposed, and we have considered the possibility that the syndrome reflects a slow virus infection permitted by the host's genetic makeup. Finstad has proposed that isoimmunization with a homologue of the T-cell specific theta antigen of Reif could account for both the central nervous system and immunological maldevelopment. (Figure 33-8 shows the scleral telangiectases of a patient with ataxia-telangiectasia.)

Wiskott-Aldrich Syndrome. In this disease of X-linked inheritance, the affected boys express an abnormality of platelet structure, thrombocytopenia, and consequent purpura with a propensity to bleed.

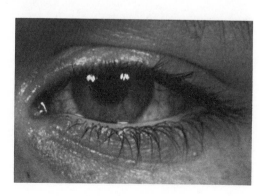

Fig. 33-8. Scleral telangiectases, characteristic of patients with ataxia-telangiectasia (similar vascular defects are widely distributed in these patients).

In addition, a persistent eczema, which is most difficult to treat, is regularly present and associated with elevated circulating IgE levels. Increased susceptibility to infection with viral, fungal and bacterial pathogens is associated with a unique broadly based immunodeficiency (Fig. 33-9). The immunodeficiency associated with Wiskott-Aldrich syndrome was first shown to be a deficiency of IgM immunoglobulins the result of inability to produce antibody to certain antigens—especially those of polysaccharide nature. In addition, patients with this syndrome show a progressive deficit of cell-mediated immune responses. Frequently a reduction in numbers of lymphocytes in peripheral blood occurs, and associated is a progressive decline of lymphocyte responses in vitro to phytohemagglutinin, allogeneic cells and antigens to which the host has been stimulated. Changing peripheral lymphoid tissue morphology reflects the progressive immune deficiency.

Thus far, the basic inherited polypeptide or enzyme deficiency which underlies this disease remains undefined and the condition at present is lethal. Fudenberg and his associates have claimed to be able to treat this immune deficit effectively by giving Lawrence's transfer factor from normal persons. Bach has also described a patient whom he treated with a combination of large doses of cyclophosphamide and marrow transplantation from an HL-A matched sibling donor (see Chap. 35). The result was long-lasting chimerism and partial correction of the immunodeficiency with most salutary clinical improvement.

Common Variable Form of Immunodeficiency. One of the most frequent clinical forms of immunodeficiency, presently termed the common variable form of immunodeficiency, is the disease which used to be referred to as acquired agammaglobulinemia (also termed hypogammaglobulinemia, sporadic late occurring immunodeficiency, and a variety of forms of dysgammaglobulinemia). Extensive experience has revealed a number of families in which an autosomal recessive inheritance appears to be operating. In several families, however, in which the disease appears sporadically, an apparently dominant hereditary characteristic has been transmitted. Doubtless, several ultimately distinguishable entities remain grouped in this wastebasket classification of immunodeficiencies. However, there can be no doubt that within this group great variability of disease is seen from time to time, not only in the same patient but also from patient to patient within the same family. It is for this reason that the concept of dys-

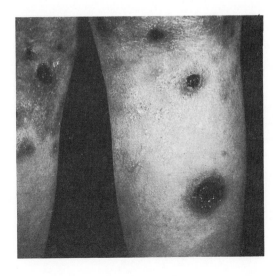

FIG. 33-9. Herpes simplex lesions on a patient with Wiskott-Aldrich syndrome. In immunologically competent individuals, herpes lesions are not associated with such overwhelming disease. (St. Gene, J. W., Jr., Prince, J. T., and Burke, B. A.: New Eng. J. Med., 273: 229, 1965. Reproduced with permission of the publisher.)

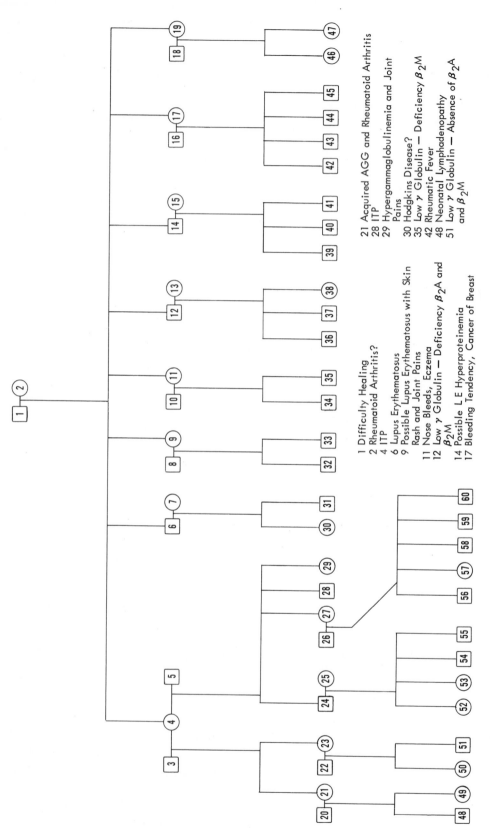

FIG. 33-10. Genetic pedigree of a family with multiple expressions of immunological disease. Note that diseases categorized as immune deficiency, autoallergy, and malignancy are represented. (Modified slightly from Wolf, J. K., Gokcen, M., and Good, R. A.: J. Lab. Clin. Med., 61:230, 1963)

1 Difficulty Healing
2 Rheumatoid Arthritis?
4 ITP
6 Lupus Erythematosus
9 Possible Lupus Erythematosus with Skin Rash and Joint Pains
11 Nose Bleeds, Eczema
12 Low γ Globulin — Deficiency β₂A and β₂M
14 Possible L E Hyperproteinemia
17 Bleeding Tendency, Cancer of Breast

21 Acquired AGG and Rheumatoid Arthritis
28 ITP
29 Hypergammaglobulinemia and Joint Pains
30 Hodgkins Disease?
35 Low γ Globulin — Deficiency β₂M
42 Rheumatic Fever
48 Neonatal Lymphadenopathy
51 Low γ Globulin — Absence of β₂A and β₂M

gammaglobulinemia has been discarded. Patients having at one time a disease designated as dysgammaglobulinemia (abnormality of one or more of the five gamma globulin classes) may have panhypogammaglobulinemia later. Similarly, one patient in a family may have severe panhypogammaglobulinemia, but another family member may demonstrate a dysgammaglobulinemia, and a third family member may have had transient hypogammaglobulinemia of infancy. In some large sibships immunological deficiency with hypogammaglobulinemia may be but one manifestation of an immunological perturbation expressed in other members of the family as autoallergic hematological diseases (e.g., Coombs' positive hemolytic anemia, rheumatoid arthritis, regional enteritis, etc.). The family history of one such family is presented in Figure 33-10. Mesenchymal and autoallergic diseases and phenomena have occurred far too frequently among patients with this form of immune deficiency (and their relatives) to be explained solely by chance.

Persisting Virus Infection and Immunodeficiency. As mentioned above, one form of immunodeficiency may be produced by acute or persisting infection with a variety of viruses. A striking example of a viral etiology of immunological deficiency is represented by in-utero infection with the rubella virus. During the active phase of the persisting infection which developed from exposure to the rubella virus in utero this agent usually produces a rather striking dysgammaglobulinemia in which IgM levels may be very high while IgG levels and production are very low. Such patients occasionally have circulating lymphocytes unresponsive to phytohemagglutinin, to allogeneic cells or to antigen to which the host has been stimulated. Five such patients have been described whose persistent hypogammaglobulinemia, involving all the immunoglobulins, seemed attrib-

utable as did other congenital abnormalities to early in utero infection with the rubella virus.

Transient Hypogammaglobulinemia of Infancy. Full term human infants are regularly born with apparently full responsiveness of all populations of lymphoid cells and with IgG levels derived almost entirely from the mother, usually slightly higher than those of the mother. As the child assumes immunological responsibility, the maternal IgG declines in the circulation and gradually IgM, IgG, IgA and IgE produced by the child increase to adult levels, usually during the first two years of life. Occasionally a child shows persistent deficiency of all immunoglobulins or particularly of IgG and assumption of immunological responsibility is delayed. This may lead to the transient hypogammaglobulinemia of infancy which differs from the other primary immunodeficiencies in that gradually immunological vigor and immunoglobulin levels within the normal range may be achieved. In our experience, however, the so-called transient hypogammaglobulinemia of infancy is often only the expression in one member of a family of the common variable form of immunodeficiency. Such children, although certainly returning toward normal immunological vigor, may demonstrate, later in life, immunoglobulin concentrations and immunological capacity somewhat lower than normal.

Isolated Absence of IgA. Many persons have been described whose serum and tissues lack IgA. In some the deficiency of IgA is most profound and no IgA can be demonstrated even when most sensitive techniques are used. Patient with deficiencies of IgA frequently, but not always, lack a vigorous local antibody system; consequently, they suffer from recurrent viral and bacterial infections of respiratory tract and bowel. Such patients have been most extensively studied and have been found to be inordinately susceptible to respiratory and gastrointestinal diseases, to develop autoallergic diseases and to express autoallergic phenomena in high frequency. It seems clear to us, however, that perturbations of the local antibody system regularly are associated with disease. One major reason for the local antibody system may be to inhibit effective antigenic stimulation of the other immunological systems by antigens closely related to and cross reactive with the host's own antigenic constituents.

Secondary Immunodeficiencies. Immunodeficiencies not infrequently occur secondarily to malignancy of the lymphoid system (Table 33-13). In patients with myeloma as in those with Bruton's agammaglobulinemia, major trouble with infections is attributable to the encapsulated high grade pathogens (e.g., pneumococcus). By contrast, such organisms cause much less trouble in patients with Hodgkin's disease. These patients, however, are more vulnerable than normal to infections with tubercle bacilli, fungi and certain viruses. These patterns of susceptibility to different pathogens as in primary immunodeficiency diseases reflect the basic perturbations of the different immunological systems.

PHYLOGENETIC DEVELOPMENT OF IMMUNITY

The immunological process so elaborately developed in mammals and birds has been evolved over more than 300 million years with the development of the vertebrates. Many invertebrates seem to lack true adaptive immunity; and in none of the invertebrates have all of the fundamental characteristics of adaptive immunity been fulfilled. Indeed, although extensive studies have been performed to link both adaptive responses and specific serological reactions of invertebrates with immunological processes of vertebrates, no immunoglobulin molecule of any of the classes found among vertebrates has yet been described in any invertebrate. Further, no invertebrate has been found to have a thymus, true lymphocytes or plasma cells of the sort characteristic of all vertebrates. By contrast, all major lines of vertebrates from the most primitive ostrachoderm-derived cyclostomes to man can be shown to possess adaptive immunity, lymphocytes and immunoglobulin molecules resembling those of mammals. Thus, although it is certain that the basic origins of adaptive immunity will ultimately be traced in the invertebrates, it

TABLE 33-13. Secondary Immunodeficiencies

MALIGNANCY	IMMUNE DEFICIENCIES
1. Multiple myeloma	Abnormality of B_2 cells. Failure of normal immunoglobulin synthesis and antibody production
2. Hodgkin's disease	Malignancy of the reticulum cells. Progressive defect in function of T-cell population. Little or no deficiency of B-cell function
3. Chronic lymphatic leukemia	Both T- and B-cell functions are defective early in the course
4. Acute "lymphatic" leukemia	Neither B- nor T-cell functions deficient early in the disease. Deficits in platelets, granulocytes and red cells
5. Acute myelocytic leukemia	(Same as for acute "lymphatic" leukemia)
6. Cytotoxic treatment of patients of groups 1-5	Marked deficiencies of both T- and B-cell function secondary to immunosuppressive cytotoxic therapy in patients of all groups

is equally clear that the immunological systems, as we know them in mammals, appeared in the most primitive vertebrate lines. The major developments have been elaborated through the eons of history with the elaboration of the vertebral forms.

Immunity in Invertebrates. Several comparative immunobiologists have, however, described recognition of foreignness of skin xenografts and possibly allografts, a form of transplant rejection, and even "immunological memory" in the annelid earthworms. The bases of this process, however, have not yet been defined in convincing cellular or humoral terms. Such studies, carried out to full immunobiological definition, are of greatest importance, since they may elucidate basic origins of either or both cellular and/or humoral immunity of more recent forms.

The hemolymph of many invertebrates contains agglutinators which show specificity to foreign red blood cells. When the purified agglutinators have been analyzed, however, they have not been found to share primary, secondary, tertiary or quaternary structural features with any known vertebrate immunogobulin. The agglutinators were found to be comprised of complex polymers of varying multiplicities of basic units with M.W. of 25,000 to 30,000 linked by noncovalent bonds. They have now been sufficiently purified and some have been found to have a single N-terminal amino acid, promising that extensive amino acid sequences can be carried out which should establish whether or not the specificity of the agglutinators for different red blood cells relates in any way to the basis of specificity of immunoglobulin molecules.

Vertebrate Immunoglobulins. The mod-

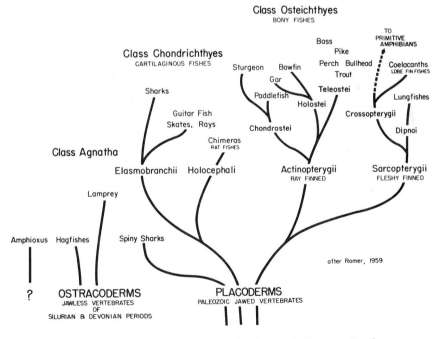

Fig. 33-11. Simplified schematic of the evolutionary development of the fish. (Finstad, J., and Good, R. A.: Phylogenetic studies of adaptive immune response in lower vertebrates. *In:* Smith, Miescher, and Good (eds.): Phylogeny of Immunity. p. 173. Gainesville, University of Florida Press, 1966)

ern representatives reflecting the most primitive vertebrate deviations in phylogeny are the cyclostomes which derive from the ancient ostrachoderms apparently along two separate pathways (Fig. 33-11). Both hagfishes and lampreys are readily available for modern immunobiological analysis and both possess the elements of adaptive immunity. Antibody production, delayed allergic responses and allograft rejection are readily demonstrable in the lamprey. The lamprey and hagfishes have cells which look like lymphocytes and those of the lamprey possess numerous markers by which mammalian lymphocytes are distinguished. Neither species possesses a thymus during adult life, but the lamprey, during embryonic life, has theliolymphocytes arranged in a relationship to the pharyngeal pouch epithelium that permit the diffuse lymphoepithelium to be considered a protothymus. As with the invertebrates, at no stage of development is a true thymus distinguishable as an organ. There is no spleen as such, and these primitive fishes do not have lymph nodes or bone marrow. Thus, they achieve all the elements of adaptive immunity with none of the organs that have been so intimately associated with the immunological process in mammals. Although it has been claimed that both lamprey and hagfish possess an IgM type immunoglobulin molecule, this has not been our experience. The only immunoglobulin we have been able to find in the lamprey is a 9-11S molecule comprising four subunits of 70,000 molecular weight held together by noncovalent bonding. No light-chain–heavy-chain differentiation was present in the purified lamprey immunoglobulin molecules. The antibody activity was greatest when the molecule was in tetrameric form.

The subunits in lamprey immunoglobulin has been considered to be similar to a μ chain, but recent studies reveal that its circular dichronic patterns distinguish it sharply from the μ chains of phylogenet-

ically more recent forms. This protein is therefore more similar to an unusual avian protein which is present in the serum of agammaglobulinemic chickens. These proteins also have similar molecular weights and sedimentation constants. Lamprey immunoglobulin, a primordial immunoglobulin molecule, is comprised of polypeptide chains of approximately 70,000 M.W., and is tetrameric. It possesses no light chains and the subunits are held together by noncovalent linkages. Subunits of the primordial immunoglobulin of chicken are immunologically related to μ heavy chains, whereas the dissociation characteristics of these molecules are strikingly like those of the lamprey immunoglobulin.

The Fishes. Beginning with the most primitive elasmobranchii and extending to include all vertebral forms a true thymus of lymphoepithelial nature is present. The thymus in all forms develops from epithelial primordial just as do mammalian thymuses. Indeed, in all the fishes, amphibians, reptiles and birds the thymus is so similar to the mammalian thymus that it is often difficult to distinguish examples of one from the other. The thymus often contains large epitheloid cells and myoidzellen. Further, in all forms studied Hassall's corpuscles have been distinguishable. Figure 33-12 compares the thymus of several primitive vertebrates, a cyclostome, the lamprey; an elasmobranch, the guitarfish; a chondrostean, the paddlefish; and aves, the chicken; with the normal human thymus. A further similarity has been found in the regular involution of the thymus of all vertebrates with advancing age. In the elasmobranchs the spleen has appeared as an organ with lymphoid structure. In addition, malpighian corpuscles and plasma cells are present and all of these structures are found in all phylogenetically distal forms. Lymph nodes were first identifiable in the amphibians, but true complex lymph node structure is present in the egg-laying mammals and is not seen

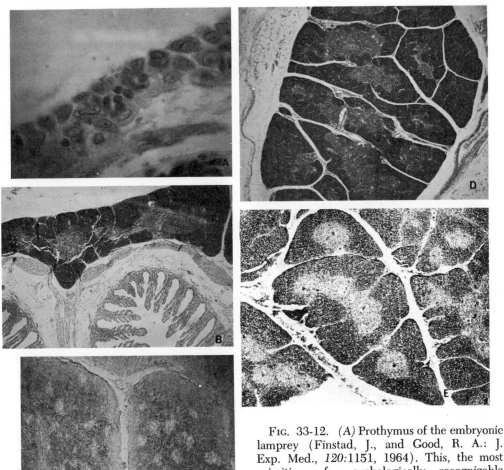

FIG. 33-12. (A) Prothymus of the embryonic lamprey (Finstad, J., and Good, R. A.: J. Exp. Med., *120*:1151, 1964). This, the most primitive of morphologically recognizable thymuses, is composed of phanyngeal pouch epithelial cells and the darker theliolymphocytes. (B) Guitarfish thymus. (Papermaster, B., *et al.: In:* Good, R. A., and Gabrielsen, A.: The Thymus in Immunobiology. New York, Harper & Row, 1964) (C) Paddlefish thymus (Finstad, J., and Good, R. A.: Lab Invest., *13:* 490, © 1964, The International Academy of Pathology) (D) Chicken thymus. (E) Human thymus. Note the basic similarity of histological structure in thymuses shown in (B) through (E). All are pharyngeal pouch derived and lymphoepithelial, all are organized into cortex and medulla and all have Hassall's corpuscles. This figure therefore demonstrates the constancy of thymic structure through hundreds of millions of years of evolution, suggesting that, once this fundamental component of adaptive immunity evolved, little alteration was permitted by selection.

in any of the reptilian forms or in birds. Plasma cells appeared in clearly identifiable form in chondrosteans and can be identified in all more distal phylogenetic deviations. The separation of humoral and cellular immunity antedated in phylogeny the appearance of true plasma cells, since lymph-

ocytes with ultrastructural characteristics suggesting proplasmacytes have been found in them most primitive forms of sharks and rays.

Evolution of Modern Immunoglobulin Diversity. Studies of the molecular evolution of immunoglobulin molecules have been pressed in several laboratories, including our own, and a reasonably clear picture of the evolutionary sequence can be presented.

True mammalian type immunoglobulins are present in the circulation of all forms ultimately derived from the ancient placoderms. In all elasmobranchs, chondrosteans, holosteans, marine and fresh water teleosts, the immunoglobulins possess light and heavy chain subunits. The heavy chains are strikingly similar to μ chains of mammals. Heavy and light chains are linked by disulfide bonds. Carbohydrate content represents approximately 10 percent of the molecule and is attached to one of the constant regions of the heavy chain. A 7S subunit is frequently, if not always, demonstrable and after repeated immunization in many species can be found to possess antibody activity. The heavy chains can be considered to be of μ type. From the most primitive elasmobranchs to man, a constant molecular size (conforming to the basic formula VCCCCC, where C represents a constant region of some 12,000 M.W. and V the variable region of approximately 10,000 M.W.) has been preserved. The light chains of elasmobranch fit the formula VC as is the case in mammals and the amino acid sequences in the N-terminal portion of light chains and heavy chains reveal striking homologues one with another and with the kappa light chains of mouse and man. Except for variability of the molecular weight of the polymerized macromolecule in the different primitive fishes (some being tetramers and some pentamers) the molecular form of the IgM type of immunoglobulin molecule has been remarkably constant through the eons of history. The 7S immunoglobulin of fishes possesses heavy chains which are identical immunologically and physicochemically with the heavy chains of the macromolecular immunoglobulin form.

A second major immunoglobulin class is first distinguishable in the lung fishes (dipnoid derivatives of the actinopterygii line leading to amphibians) where, for the first time, immunoglobulin molecules possessing heavy chains distinct from μ type appeared. The heavy chains of the immunoglobulin newly present in lungfishes have a low molecular weight (approximately 38,000) similar to IgG molecules of certain reptilian and avian forms. By contrast, amphibian and mammalian IgG type immunoglobulins have a molecular weight of approximately 50,000.

Whereas phylogenetic modifications of the IgM-type molecule involved only polymer size, with the evolution of an IgG molecular type considerable variability occurred in the number of C regions and the carbohydrate content. This variability probably has influenced greatly the distribution of antibody molecules in the body and allowed capacity to engage the more primitive biological amplification systems by virtue of recognition mechanisms located in the C-terminal, Fc portion of the antibody molecule.

N-terminal amino acid sequence studies of immunoglobulin molecules from primitive fishes to man have revealed that even in the so-called variable portion of the immunoglobulin molecules considerable constancy exists.

Careful analyses and comparison of N-terminal amino acid sequences have revealed evidence of a genetically based mechanism for insertion and deletion of single amino acids in phylogeny which may have most important consequences in development of a system of molecules having the extraordinary variability demanded of immunoglobulin molecules basic to the bodily defense. Much work must be done to define the phylogenetic origin of additional class and subclass heterogeneity of the immunoglobulin molecules, but progress already made in phylogenetic analysis of the origins of the two basic immunoglobulin classes promises that the phylogenetic approach can be most helpful in understanding this complex cellular-molecular system.

Lack of understanding of the molecular basis of cell-mediated immunity in man and mammals makes difficult comparable analysis of the evolutionary basis for heterogeneity in this other major bulwark of the bodily defense. It seems certain, however, that as soon as the molecular basis of cellular immunity has been defined phylogenetic analysis will be as revealing with this system as it has been for immunoglobulins and antibodies.

RELATIONSHIP OF MALIGNANT ADAPTATION AND IMMUNITY

The application of long-lasting immunosuppressive measures to prevent allograft rejection in man has brought to the front reflections on the relationship between immunity and malignancy. When immunosuppressive therapy was first found to be effective in preventing renal allograft rejection in man, patients inadvertently received renal transplants with epithelial tumors of donor origin. In several instances the epithelial tumors became widely disseminated. In several other instances, the tumors were completely rejected when the immunosuppressive therapy which permitted their establishment was discontinued. Thus, with these allografts of tumor the most powerful antitumor therapy imaginable was achieved simply by permitting the host's immune responses to come to bear on these foreign malignant cells.

Patients being treated with intensive immunosuppressive therapy in an effort to prevent allograft rejection have developed de novo malignancies far more frequently than other patients of the same age. These malignancies have involved the lymphoid system, reticulum cells, connective tissue and several epithelial systems including stomach, colon, ovaries, uterus and skin. In contrast to the transplanted malignancies those developing de novo have not

regularly regressed when immunosuppressive treatment has been discontinued.

Similarly, patients with various kinds of immunodeficiency have developed cancer far too frequently to be explained by chance. Studies on the occurrence of malignancies in patients with immune deficiency diseases suggest that defects primarily of humoral immunity predispose to leukemias whereas defects depressing immunological vigor of both immunity systems predispose to solid tissue malignancies of both the lymphoreticular system and other rapidly replicating systems.

The findings of such a very high incidence of malignancy in patients with primary immunodeficiencies, whatever its ultimate basis, fulfills in a dramatic way a prediction made by Lewis Thomas in his original statement of the concept that the immunity systems, especially cell-mediated and allograft immunity, represent a major defense against development of cancer. Since Thomas' original statement this concept of immunosurveillance has been extensively promulgated by Burnett and studied experimentally by many investigators.

Whether immunosurveillance is a major defense against malignancy or not, there can be no question that depression of host immune response opens the door to transplantation of malignant disease and de novo development of malignancy. Thus, in very real sense the immunity mechanisms, particularly cell-mediated mechanisms, stand as a major defense against cancer. If the basis of this defense is not to be realized in the concept of immunosurveillance, it may be found in the defenses against infection or against stimulation of many tissues by foreign antigens not permitted when the immune system is intact.

It is now clear as well that most if not all tumors possess at their surface tumor-specific antigens which can stimulate host immune responses. In a number of clinical and experimental systems a delicate balance between effective host immunity to eliminate the tumor cells and capacity of the tumor cells to achieve an immunodeviation which protects them from such elimination has been detected. These same delicate relationships have now been found to exist in long-lasting chimeric states involving normal cells in experimental animals and man. It seems certain from this background that imaginative manipulation of tumor antigenicity and host cellular and humoral immune responses can permit the immunological adaptations to be brought to bear with sufficient vigor to facilitate elimination of the minimal residual tumor cells remaining after effective chemotherapeutic or surgical elimination of the bulk of tumor cells.

Already clinical application of those principles in immunotherapy of tumor is being attempted in several quarters. Perhaps more important, these studies of the many interfaces between immunity and malignancy may lead to more nearly complete understanding of the very raison d'etre of both processes.

IMMUNITY AND AGING

Numerous studies in both animals and man reveal clear evidence that with aging a progressive loss of immunological vigor occurs. This decline is seen with cell-mediated and with humoral immune responses. In birds, the bursa of Fabricius involutes with sexual maturation. In all animals from primitive fishes to man, thymic involution is noted the initiation of which seems to coincide with sexual maturation. The progressive involution of immunological vigor with age is also associated with increasing frequency of auto-allergic phenomena, increased susceptibility to many microbial pathogens and increasing incidence of cancer. Indeed, one theory of the aging process relates to the perturbations of immunological function with age. It seems likely that extensive studies of this apparently programmed involution of the lymphoid system to aging and of the diseases that occur so frequently in aging will lead to better understanding of aging and its concomitants and possibly to their ultimate manipulation as well.

ANNOTATED REFERENCES

Alexander, J. W., and Good, R. A. (eds.): Immunobiology for Surgeons. Philadelphia, W. B. Saunders, 1970. (A concise text containing the important principles and concepts of immunobiology)

August, C. S., Rosen, F. S., Filler, R. M., Janeway, C. A., Markowski, B., and Kay, H. E. M.: Implantation of a foetal thymus restoring immunological competence in a patient with thymic aplasia (DiGeorge's syndrome). Lancet, 2:1210, 1968. (A case report: restoration of immunological competence in a patient with congenital thymic aplasia)

Barandun, S.: Das Antikorpermangelsyndrome. Basel-Stuttgart, Benno Schwabe & Co., Verlag, 1959. (A monograph describing antibody deficiency syndromes)

Bretcher, P., and Cohn, M.: A theory of self-nonself discrimination. Science, 169:1042, 1970. (A concisely presented theory attempting to account for self recognition and immunological tolerance)

Bruton, O. C.: Agammaglobulinemia. Pediatrics, 9:722, 1952. (A case report: initial observation of humoral immunological deficiency)

Burnet, F. M.: Immunological aspects of malignant disease. Lancet, 1:1171, 1967. (Discussion of multiple myeloma, immunological surveillance and allogeneic inhibition)

Cleveland, W. W., Foge, B. J., Brown, W. T., and Kay, H. E. M.: Foetal thymic transplant in a case of DiGeorge's syndrome. Lancet, 2:1211, 1968. (Similar to August *et al.*, above)

Cochrane, C. G.: Immunologic tissue injury mediated by neutrophilic leukocytes. *In:* Dixon, F. J., and Kundel, H. G. (eds.): Adv. Immunology. Vol. 9. p. 97, New York, Academic Press, 1968. (A review describing the roles of antigen, antibody complement, and leukocytes in allergic tissue destruction)

Cone, L., Uhr, J. W.: Immunological deficiency disorders associated with chronic lymphocytic leukemia and multiple myeloma. J. Clin. Invest., 43:2241, 1964.

Fahey, J. L., Scoggins, R., Utz, J. P., and Szwed, C. F.: Infection, antibody response and gamma globulin components in multiple myeloma and macroglobulinemia. Am. J. Med., 35:698, 1963. (Original article documenting immune deficiency and susceptibility to infection in patients with B-cell malignancy)

Fudenberg, H., Good, R. A., Goodman, H. C., Hitzig, W., Kunkel, H. G., Roitt, I. M., Rosen, F S., Rowe, D. S., Seligmann, M., and Soothill, J. R.: Primary immunodeficiencies. Report of WHO committee. Pediatrics, 47:927, 1971.

Good, R. A.: Immunologic reconstitution: The achievement and its meaning. Hospital Practice, 4:41, 1969. (Lucid description of the significance and interpretation of implications of successful therapeutic lymphoid tissue transfer)

Good, R. A., and Bergsma, D.: Immunologic Deficiency Diseases in Man. Birth Defects. Original Article Series. Vol. 4. 1968, National Foundation Press. (Descriptions of primary and secondary immune deficiency diseases)

Good, R. A., Finstad, J., and Gatti, R. A.: Bulwarks of the bodily defense. In: Mudd, S. (ed.): Infectious Agents and Host Reactions. Chap. 4. p. 76. Philadelphia, W. B. Saunders, 1970. (Identification of the major mechanisms of bodily defense, and characterization of the normal functions of these mechanisms)

Good, R. A., and Fisher, D.: Immunobiology. Connecticut, Sinauer Assoc. Inc., 1971.

Good, R. A., and Gabrielsen, A. E. (eds.): The Thymus in Immunoboliology. New York, Hoeber-Harper, 1964. (Proceedings of 1962 symposium on immunobiology; an extensive basic text on the subject of immunobiology much of which is still current)

Kagan, B. M., and Stiehm, E. R. (eds.): Immunological Incompetence. Chicago, Yearbook Medical Publishers, 1971.

Kincade, P. W., Lawton, A. R., Bochman, D. E., and Cooper, M. D.: Differentiation of immunoglobulin class heterogeneity: Effects of antibody mediated suppression of IgG synthesis in chickens. Proc. Nat. Acad. Sci., 67:1918, 1970. (Direct demonstration of

central lymphoid organ role in IgM to IgG switch)

Krivit, W., and Good, R. A.: Aldrich's syndrome. J. Dis. Child., 97:137, 1959. (Extensive evaluation of morphological, immunological, and endocrinological parameters in Wiskott-Aldrich Syndrome)

Landy, M., and Braun, W. (eds.): Immunological Tolerance. New York, Academic Press, 1969.

Lawrence, H. S., and Landy, M. (eds.): Mediators of Cellular Immunity. New York, Academic Press, 1969. (Enumeration and characterization of cellular and subcellular factors functional in cell mediated immunity)

Mitchison, N. A.: Immunocompetent cell populations. In: Landy, M. and Braun, W. (eds.): Immunological Tolerance. p. 125. New York, Academic Press, 1969.

Müller-Eberhard, H. J.: Development of the complement system. In: Kagan, B. M., and Stiehm, E. R. (eds.): Immunologic Incompetence. p. 73. Chicago, Yearbook Medical Publishers, 1971.

Parrott, D. M., de Soussa, M. A. B., and East, J.: Thymus dependent areas in the lymphoid organs of neonatally thymectomized mice. J. Exp. Med., 123:191, 1966. (Original description of thymus-dependent region of lymph nodes and spleen)

Peterson, R. D. A., Cooper, M. D., and Good, R. A.: The pathogenesis of immunologic deficiency diseases. Am. J. Med., 38:579, 1965. (Early attempt at integration of clinical and experimental observations toward definition of the immunological deficiency diseases)

Smith, R. T., Good, R. A., and Miescher, P. A. (eds.): Ontogeny of Immunity. Gainesville, University of Florida Press, 1967.

Smith, R. T. and Good, R. A. (eds.): Cellular Recognition. New York, Appleton-Century-Crofts, 1969. (The two preceding references are proceedings of workshops. Both contain valuable data and discussions on the indicated subjects.)

Smith, R. T., and Landy, M. (eds.): Immunological Surveillance. New York, Academic Press, 1970.

Smith, R. T., Miescher, P. A., and Good, R. A. (eds.): Phylogeny of Immunity. Gainesville, University of Florida Press, 1966. (Proceedings of a workshop; contains description of major steps in the evolution of immunological competence.)

Teague, P. O., Yunis, E. J., Rodey, G., Fish, A. J., Stutman, O., and Good, R. A.: Autoimmune phenomena and renal disease in mice. Role of thymectomy, aging and involution of immunologic capacity. Lab. Invest., 22:121, 1970. (Demonstration of role of thymus in aging)

Stutman, O.: Detection of post-thymic cells in mouse hemopoietic tissues. Fed. Proc., 30: 529, 1971. (abstr.) (Identification of a subpopulation of T-cells, T_2)

Turk, J. L.: Delayed hypersensitivity. Amsterdam, North Holland Pub. Co., 1967. (Extensive monograph covering all aspects of delayed hypersensitivity)

Walford, R. L. (ed.): The Immunologic Theory of Aging. Baltimore, The Williams and Wilkins Company, 1969. (Extensive monograph on changes in immunological reactivity with aging)

Yunis, E. J., Stitman, O., Fernandes, G., Teague, P. O., and Good, R. A.: The Thymus autoimmunity and the involution of the lymphoid system. *In:* Sigel, M., and Good, R. A. (eds.): Proc. Symposium on Tolerance, Autoimmunity and Molecular Aging. (In press)

34

Mechanisms Involving the Complement System

Richard J. Pickering, M.D., *and Ann E. Gabrielsen,* PH.D.

INTRODUCTION

In the late 1800's a labile factor was observed in normal blood which, in the presence of specific antibody, would kill bacteria or lyse erythrocytes. Although the potential importance to host defense of a serum substance capable of killing bacteria was recognized, the complexity of the system and deficiency of appropriate tools severely limited investigations. Rapid advances in recent years in protein separation and identification, use of radioisotopes and other developments have provided the means for functional and chemical identification of the component parts of the system.

The complement system is one of the important effectors of the inflammatory reaction through its release of factors which increase capillary permeability, cause smooth muscle contraction, and influence polymorphonuclear leukocyte migration. In addition, it plays specific roles in some histamine-release reactions, viral neutralization, and bactericidal and cytolytic systems. Its participation in these areas of host reaction may be beneficial as an aid to elimination of invading organisms or in the destruction of the host's own cells which have been altered by mutation or viral transformation. In contrast, its participation in inflammatory reactions in nonregenerative tissue, such as the kidney, may actually lead to a destructive process which will ultimately be detrimental to the host.

The complement system is multicomponent, consisting of 11 distinct, sequentially interacting serum proteins and at least 3 functionally interacting inhibitors. Originally the term "complement" applied to the serum factors bringing about lysis of erythrocytes sensitized with antibody. The identification of components depended upon two criteria: absence of the component impaired the capacity of the otherwise intact system to produce hemolysis of sensitized erythrocytes; and substitution of the component in question restored hemolytic function.

For many years 4 such components, C1, C4, C2, and C3, were recognized. However, since 1960 C1 has been found to be a complex molecule consisting of 3 subunits, and C3 has been dissected into at least 6 factors. Most of the work involved in the delineation of this system has been carried out in two species—the guinea pig and man; and, although some known differences between these two species exist, the basic concepts apply to both.

NOMENCLATURE

The functional components of complement are numbered C1, C2, C3, C4, C5, C6, C7, C8 and C9. With one exception this indicates the sequence of their interaction with each other: C4, however, reacts in the classical hemolytic sequence between C1 and C2. C1 consists of 3 subcomponents, C1q, C1r, and C1s, which are believed to exist in serum combined with each other and with calcium ion.

C2 and C4 of the classical functional definition are now known to be single factors functionally and chemically. Classical hemolytic C3 (to be referred to as C3* in order to distinguish it from the single component C3), however, is the cumulative function of 6 chemically and functionally distinct components, C3, C5, C6, C7, C8 and C9.

A detection system for the individual components consists of immune lysis of cells, usually sheep erythrocytes (E) sensitized with antibody to antigens of the cell surface (A) to form EA. Complement components which have been activated specifically and combined to the cell-antibody indicator particle are represented as EAC$\overline{1}$, EAC$\overline{14}$, EAC$\overline{142}$ . . . to EAC$\overline{1\text{-}9}$. These cell intermediates are used for detection of one or more of the components, since cytolysis will occur only when these intermediates have been provided with the remaining complement factors. The release of hemoglobin from the cells provides a convenient material for spectrophotometric assay of complement activity.

CHARACTERISTICS OF COMPLEMENT PROTEINS

The 11 complement proteins account for approximately 10 percent of the globulin fraction of human serum; more than half of this is beta 1C globulin (C3). Though much has been done to simplify and clarify the terminology, terms such as 11S protein (C1q), beta 1C globulin (C3), beta 1E globulin (C4), and beta 1F globulin (C5) still appear in the literature and provide points of confusion. These immunochemical terms apply to the isolated proteins; but, since they are also functional units in the complement sequence, they should also be given the appropriate numerical designation.

All of the complement proteins isolated to date are macromolecules with molecular weights ranging from 70,000 to 400,000. Information concerning the chemistry of all of the complement components is not yet available; however, a summary of some of their properties is presented in Appendix Two of this Section.

FUNCTIONS OF THE COMPLEMENT SYSTEM

Although immune cytolysis, associated with activation of C1 through C9, has been the hallmark of complement function and is undoubtedly important, it is not the only biological function of the system. Nor do some of the other functions require all of the components through C9. However, a full grasp of the events which occur during the activation of the complement system, with attendant interaction of all the proteins in the system, is likely best attained by considering all of the events from the initial activation of C1 to the terminal event of cytolysis. First, however, it may be useful to consider cytolysis.

The functional lesion in the cell membrane has been characterized though its chemical nature is not known. It was shown in the late 1950's that tumor cells, acted upon by antibody and complement, lost ions and small molecules (such as free amino acids and ribonucleotides) and some proteins through a cell membrane which was altered but apparently unruptured, as judged by phase and electron microscopy. The membrane, however, retained a degree of discrimination against movement of large molecules. Another criterion of activation of the terminal portion of the sequence through C9 is the presence of round 80- to 100-Å electron-dense areas on the membrane of erythrocytes lysed by antibody and complement. Since it has been concluded that these "lesions" depend on the activation of the final components, they have been used to indicate C8 and C9 activation on isolated membrane surfaces such as endotoxin.

C1

When immunoglobulin (IgM or IgG) becomes altered, either specifically by combination with an appropriate antigen or nonspecifically by aggregation with heat, for example, it becomes capable of interacting with C1q. The site of interaction with C1q is on the heavy chain of the immunoglobulin molecule and presumably becomes available upon a specific structural alteration of the latter. (For the sake of later discussion, it is important to mention here that at least one subclass of human IgG, IgG-4, and three other human immunoglobulin classes, IgA, IgD, and IgE, do not activate complement via C1.)

Although this interaction between C1q and immunoglobulin may serve other as yet unidentified biological functions, it is now considered to be a major mechanism whereby the complement system is activated. A simplified scheme representing the present concept of complement component interactions and the resultant biological activities is presented in Figure 34-1.

In this scheme one might consider each component as having one activation site and, once activated, two combining sites: one, which combines with the previously active component or the membrane, and a second which, either alone or in specific steric relationship to previously activated components, is able to react with the next complement component.

C1q. C1q has been considered individually because of the functional characteristics described. However, it does not exist as such in human serum but rather is part of a large macromolecule with two other C1 subcomponents, C1r and C1s. The structural and functional integrity of C1, therefore, depends upon the presence of optimum amounts of calcium ion and the presence of all three subcomponents. Thus, chelation of calcium with ethylenediaminetetraacetic acid (EDTA) or other chelating agents produces instability and separation of the subcomponents from each other, effectively preventing participation of C1 in the hemolytic sequence. In addition, the absence of one of the subcomponents (e.g., C1r) also allows for separate existence of the remaining two subcomponents and interferes with or prevents the usual sequence of events leading to immune cytolysis.

C1r. The mechanisms of interaction within the C1 molecule are even less well understood than those which involve combination of C1q with immunoglobulin. However, there is evidence that C1r has proteolytic and esterolytic activity which, following binding of C1q by immunoglobulin, facilitates conversion of C1s to its active state.

C1s. C1s is the smallest of the subcomponents of C1 and is believed to exist in the native state as a proesterase. Once activated by C1r (or by any of several other proteolytic enzymes such as trypsin or pepsin) C1s becomes capable of hydrolyzing synthetic esters and also assumes its specific capacity to interact with its natural substrates C4 and C2 on a particle surface or in the fluid phase. If C1 becomes bound to an antigen (on an erythrocyte or an insoluble particle or membrane), subsequent components will, on activation, bind directly to the surface provided or indirectly to a previously bound complement component. However, if C1s becomes activated in the fluid phase in the absence of the normal C1 esterase inhibitor (see below) this enzyme may then activate its natural substrates, C4 and C2. Since the activated state of these components is of short duration in the absence of an appropriate binding site, this fluid phase interaction leads to destruction of C4 and C2 with some activation of C3, but likely no other later components. (An example of

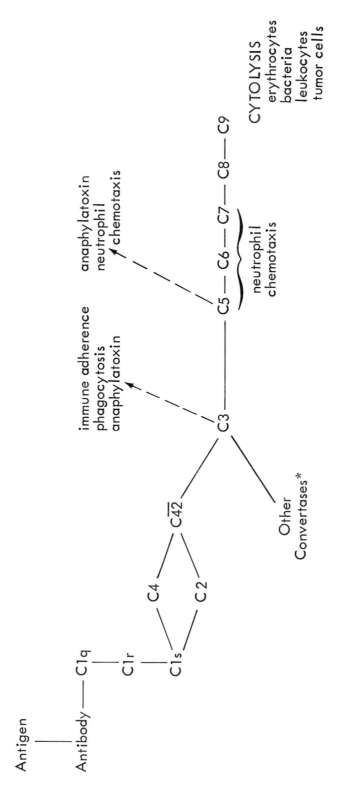

FIG. 34-1. Simplified scheme of complement component interactions and products.

*trypsin, plasmin, thrombin, endotoxin, cobra venom factor

this interaction will be discussed when we consider hereditary angioedema, a disease in which the normal C1 esterase inhibition is absent.)

It is important to note that the binding of C1 to immunoglobulin is not irreversible. For example, the dissociation of C1 from the immune complex can be enhanced by increasing the ionic strength of the environment. This effect appears to be an equilibration phenomenon, since the combination of C1 with the immune complex is favored by low ionic strength. The capacity of activated C1 to move from one complex to another while retaining its ability to interact with the later-acting components implies that one C1 molecule can activate several natural substrate molecules. Examples of other amplification reactions will be considered at other steps of the complement reaction.

Inhibitor of C1 Esterase

The serum of several species, including man and guinea pig, contains a protein which inhibits the fluid phase conversion of the proesterase C1s to its active state and inhibits the interaction of C1 esterase with its substrates including C4, C2 and synthetic esters. This protein, an acid-labile alpha-2-globulin, is separable from the inhibitors of trypsin and plasmin. Such an inhibitor may be functional in localizing the effects of antigen-antibody activation of complement to specific surfaces or membranes. This may serve to increase the efficiency of the surface reaction, protect the organism from harmful effects of massive fluid-phase enzymatic reactions and, perhaps, conserve these proteins.

The mechanism of C1 esterase inhibitor function is not completely clear, but there is evidence for a combination of the inhibitor with the esterase, since there is stoichiometric loss of inhibitor activity during in-vitro inhibition of C1 esterase.

C1 esterase inhibitor also inhibits other chain reaction systems including the coagulation and the kinin systems. Although this may represent a useful functional bond among these three effector systems, the significance of such a link is not understood.

The C$\overline{42}$ Complex

Activation of C4 and C2 by EAC$\overline{1}$ leads to formation and binding of the C$\overline{42}$ complex to the membrane surface. The resulting EAC$\overline{142}$ is very unstable, owing to decay and release of C2. The remaining EAC$\overline{14}$ complex is not only stable but capable of interacting again with fresh C2 to form a new functional EAC$\overline{142}$ complex.

On the basis of experiments showing the increased efficiency of binding and stability of C2 following treatment with an oxidizing agent, iodine, and the reversal of these hemolytic properties of C2 by parachloromercuribenzoate, it has been proposed that alteration of sulfhydryl groups plays some role in the activation and function of C2.

The production of this EAC$\overline{142}$ complex appears to be the last step requiring the presence of activated C1. Thus, a C$\overline{42}$ complex (C3 convertase) is capable, even in the absence of C1, of interacting with C3. Specific in-vitro conditions favor the formation of a stable but reversible combination of native C4 and C2 which provides for efficient conversion to active C$\overline{42}$ by C1 esterase. The biological advantage of such an affinity, if it occurs in vivo, is immediately apparent.

The final and probably most important point is that an activated C$\overline{42}$ complex is capable of activating many C3 molecules and thus represents the second example of an amplification step within the complement sequence.

During the activation of C4 and C2 by C1 esterase, small molecular fragments are cleaved from the parent molecules, the larger fragments providing the components of the enzyme C3 convertase.

C3

The beta 1C globulin (C3) has been a major focus of complement investigation because of its relative stability and high concentration in serum. For these reasons it has been used widely as a measure of serum complement concentration and as a marker for complement binding in tissue through the use of immunofluorescent techniques. In addition, C3 seems to be a pivotal component biologically. It is apparently susceptible to activation by enzyme systems similar to, but distinct from, the C3 convertase, a role which may be of importance pathogenetically. As noted in Figure 34-1, C3 not only generates a number of biological activities itself but is essential to the functions of the components from C5 on. Thus, in several ways, understanding of C3 seems to be crucial to our understanding of the entire system and its physiological and pathological roles. The activation of C3 and its binding to appropriate receptor sites results in a number of phenomena:

Interaction With C5. Activation of C3 by $EAC\overline{142}$ or $EAC\overline{42}$ leads to binding of C3 to the cell membrane and the formation of $EAC\overline{1423}$ or $EAC\overline{423}$. This complex has the capacity to cleave and activate C5 enzymatically. It loses its capacity to interact with C5, owing to the spontaneous decay of C2. However, the $EAC\overline{143}$ or $EAC\overline{43}$ intermediate remains intact and can be activated once again through the addition of fresh C2. Although the $EAC\overline{42}$ complex is capable of cleaving and activating large numbers of C3 molecules, only a fraction of those molecules actually bind to the membrane and, in turn, only a small percentage of those actually bound to the membrane contribute to the functions of the intermediate cell formed.

C3 Peptidase Activity. The formation of the $EAC\overline{1423}$ or $EAC\overline{423}$ intermediate produces a peptidase, the activity of which may be demonstrated by hydrolysis of peptides containing aromatic amino acid residues. From quantitative measurements, it appears that a special relationship between the $C\overline{42}$ and C3 molecules is necessary for the generation of this peptidase activity; that is, a C3 molecule must be bound either directly to or directly adjacent to a $C\overline{42}$ complex.

Immune Adherence Agglutination Activity. This fascinating phenomenon is one in which indicator particles such as bacteria or erythrocytes (and perhaps other antigens), carrying on their surface at least C4 and C3, adhere to specific receptor sites on primate erythrocytes, polymorphonuclear leukocytes, or dog platelets. It appears that this phenomenon serves to enhance the sticking of such sensitized particles to cell membranes and thus facilitates engulfment by phagocytic cells.

Fragments of C3 With Other Biological Activities. During the enzymatic cleavage of C3 by C3 convertase or other enzymes, at least two major fragments are produced. The large fragment is the portion which attaches to the membrane (or an appropriate surface receptor); the low molecular weight peptide produces smooth muscle contraction and alters capillary permeability in skin of experimental animals. This anaphylatoxin activity is inhibited by antihistamine. In addition, a small fragment derived by plasmin cleavage of C3 is one of the split products from complement proteins which attract polymorphonuclear leukocytes across membranes in vitro.

Virus Neutralization. Under certain conditions viruses are not neutralized by antibody in the absence of complement components. Depending upon the virus used and the assay techniques, neutralization can be detected following binding of antibody, C1 and C4. The degree of neutralization is enhanced by adding C2 and maximum neutralization occurs in the presence of C1, C4, C2, and C3. Maximum neutralization by serum deficient in C5 or C6 suggests that C5 through C9 serve little or no function in complement-induced neutralization of selected viruses in spite of the fact that terminal components may be activated and leave their 80- to 100-A mark on the virus envelope.

Thus, the activation and binding of C3 to specific receptor sites leads to generation of a number of functions, including enhanced polymorphonuclear leukocyte phagocytosis, increased vascular permeability, smooth muscle contraction, and directed polymorphonuclear leukocyte migration, all of which are integral parts of the inflammatory reaction. In addition, complement-induced virus neutralization occurs in the presence of antibody C1, C4, C2, and C3.

C3 Inactivator. Another naturally occurring inhibitor affects this second point of amplification within the complement system. This is the C3 inactivator, a heat-stable beta globulin, which appears to interfere with cell-bound C3 functions but does not affect native C3. The biological role and significance of this inactivator is not known.

Alternative Modes of Activation of C3

In addition to C3 convertase a number of enzymes including trypsin, plasmin and thrombin, as well as specific bacterial endotoxins and a factor derived from cobra venom, lead to alterations in C3 which are similar to, if not identical with those produced by the complement component-derived C3 convertase. For example: bacterial endotoxin interaction with serum leads to formation of a reagent capable of binding C3, producing biologically active fragments and probably carrying the sequence through to the activation of C8 and C9 as indicated by the presence of 80-100-Å electron-dense areas on the endotoxin surface. This reaction is of interest because it appears to occur without interaction with C1, C4, or C2, or by a mechanism which involves these components in a fashion strikingly different from the classical system described earlier. It has been proposed

that this interaction of bacterial endotoxins with the complement system may be responsible for some of the in-vivo changes occurring in endotoxin shock.

Another biological product which has been used extensively to study the functions of the complement system is a factor isolated from cobra venom. It has long been known that cobra venom, when added to normal serum, impairs the hemolytic activity of complement. In recent years it has been possible to separate the "complement-inactivating" protein from the many other toxins in this lethal parotid secretion. This has enabled investigators to study the interaction of this single protein with functionally pure complement components in vitro and has also provided a tool for the investigation of some aspects of complement function in the intact animal. First, the action of cobra venom factor (CVF) appears to be with C3, and its effect is ultimately an inactivation of C3 plus C5, C6, C7, C8 and C9. There is no apparent effect on C1, C4 and C2. Secondly, the CVF reacts with the purified complement components only in the presence of a heat-labile serum protein which combines with CVF to form a stable cobra venom factor convertase (CVFC) capable of interacting with C3. This CVFC reacts directly with C3 (and perhaps with C5 and later components) to produce fragments from C3 and C5 which bring about biological activities similar to, if not identical with those discussed above.

In addition, it has been observed that if unsensitized erythrocytes are present during the reaction of CVF with serum, these erythrocytes may be lysed through complement activation in spite of the fact that there is no antibody, C1, C4 or C2 bound to the cell surface. This "passive membrane injury" appears to result from an extensive activation of terminal complement components near enough to the membrane surface to result in cell lysis. Thus, the capacity of cobra venom to "inactivate" complement results from a reaction which leads to the production of an enzyme similar to C3 convertase. The loss of C3 and later components results from the specific cleavage of these molecules, leading to activation and release of biologically active fragments, followed by degeneration of the large fragments in the absence of specific binding sites.

The cobra venom factor can be given to animals intravenously or by the intraperitoneal route to reduce the animals' C3, C5, C6, C7, C8 and C9 in order to study the responses of such a depleted animal in a number of experimental models. Conclusions concerning the role of complement in experimental glomerulonephritis based on use of CVF will be referred to in the section on experimental models.

C5

Activation of C5 depends upon enzymatic cleavage of this molecule by preformed $EAC\overline{423}$. A small fragment released during this interaction has both anaphylatoxic and chemotactic properties. Once again, there is a possible amplification step at this stage, since C5 molecules activated by $EAC\overline{423}$ complexes demonstrate a limited capacity to transfer from one receptor site to another. C5 will also yield biologically effective small fragments when activated by the endotoxin complex, CVFC, or trypsin. An interesting and perhaps the most important aspect of C5 is its unique ability to participate (apparently in the absence of bound antibody C1, C4, C2 or C3) in nonimmune cytolysis resulting from activation of C5-C9. Several in vitro examples of this phenomenon have been described, and the hemolysis in paroxysmal nocturnal hemoglobinuria may be the result of this type of "nonimmune" reaction.

C6 and C7

Variable behavior of these components under different experimental conditions has made analysis of their interaction with each other and adjacent reacting components extremely difficult. Under physiological conditions, human C5, C6 and C7 seem to have an unusual affinity for each other, showing a tendency to form a reversible protein complex. This characteristic, plus the observation that under the same conditions C5 combines poorly with $EAC\overline{1423}$ in the absence of C6 and C7, gave rise to the concept that these components act as a functional unit. Under specific circumstances the activation of C5, C6 and C7 may lead to the formation of a large molecular complex which attracts polymorphonuclear leukocytes across a millipore filter membrane in vitro. Since there is also abundant evidence that C5, C6, and C7 do function sequentially in many experimental circumstances, the questions surrounding the significance of the "functional unit" versus sequential interaction in vivo remain unanswered.

C8 and C9

Activation and binding of C8 by a complex carrying all of the required components up to C7 leads to an alteration of the membrane which allows a very gradual release of intracellular contents. As discussed previously, activation of C9 increases the rate of lysis and is responsible for the membrane "lesions" previously discussed. In addition, it has been reported that an iron chelating agent, 1,10-phenanthroline, mimics the lytic activity of C9, perhaps indicating some relationship of iron with complement cytolysis. Thus, the activation of the terminal complement components, either as a result of a $C\overline{42}$ derived C3 convertase, or through an alternative convertase, results in the production of fragments or complexes which seem to influence several of the basic components of the inflammatory reaction, including phagocytosis, altered capillary permeability, smooth muscle contraction, and polymorphonuclear leukocyte movement. With the activation of C8 and especially C9, a cytolytic step occurs which may be functional in direct lysis of foreign cells such as bacteria or the host's own cells which may have been altered in some way by aging, viral infection or other cellular derangements.

COMPLEMENT ABNORMALITIES IN MAN

Congenital

These include deficiencies of individual complement components and defects in other systems which lead to deficiencies in serum complement activity.

Hereditary Angioedema (hereditary angioneurotic edema). This disorder is characterized by a sudden onset of localized edema which may involve any part of the body. The edema usually increases rapidly over a period of hours, is nonpruritic and nonpitting and lasts for 12 to 72 hours, after which it subsides spontaneously. In some instances it is preceded by a faint pink serpiginous rash. The episodes of edema are often apparently unprovoked, but are occasionally preceded by trauma or emotional stress. When the edema is in the gastrointestinal tract it causes abdominal pain with vomiting and occasionally leads to dehydration. Laryngeal edema may lead to severe respiratory embarrassment and death if not relieved by tracheotomy. This disease, as one of its names implies, was once considered to be related solely to emotional instability. In 1963, however, Donaldson and Evans described the deficiency of the serum inhibitor of C1 esterase in patients with this disorder and showed that the defect was inherited as an autosomal dominant. Later reports described two genetic forms of the disorder: one, in which the C1 esterase inhibitor is absent, and the other in which the inhibitor protein is present but nonfunctional. Free esterase activity, which appears in the serum (a phenomenon normally prevented by C1 esterase inhibitor) during attacks of edema

in affected individuals, is thought to be responsible for the fluid phase enzymatic cleavage of C4 and C2 resulting in the reduction of serum concentration of these components during these episodes. Recently, a small peptide with some kininlike activities has been found in the serum of these patients and it is proposed that this may be a fragment derived from C4 or C2 as a result of the enzymatic cleavage. How all of these changes relate to the edema, its localized nature, and the periodicity of the attacks is not yet understood.

Although there is no generally accepted form of therapy for the progressive edema, there is some evidence that administration of fresh frozen plasma may limit the extension of the edema. It has also been reported that treatment of one or two patients with epsilon-aminocaproic acid decreased the frequency of attacks of edema. However, because of the very unpredictable nature of this disorder in terms of frequency and duration of attacks, forms of therapy on a small scale are difficult to evaluate.

C2 Deficiency. The first report of a deficiency of C2 in man appeared in 1960, and shortly thereafter 14 similar cases were found among Swiss army recruits. This deficiency is inherited as an autosomal recessive trait, with heterozygotes demonstrating approximately one half the normal C2 activity in serum. The near-absence of serum C2 appears to be a result of defective synthesis. Although these patients show a deficiency in the capacity to bring about cytolysis of sensitized erythrocytes, their serum is able to produce near-normal immune adherence activity. The reason for this apparent discrepancy is not totally understood, but appears to be related to the ability of a single $C\overline{42}$ complex to activate large numbers of C3 molecules and thereby make efficient use of the small amount of available C2. In addition, these sera, deficient in hemolytic C2 activity, have near-normal bactericidal complement activity in vitro if no exogenous antibody is added to the system. These observations suggest that under conditions that approach those in vivo, namely, in undiluted blood using endogenous antibody, the effects of the C2 deficiency are minimized. These have been proposed as bases for the observation that none of these individuals deficient in C2 has recurrent infections as might be expected in the face of such an apparently severe deficiency.

C5 Deficiency. Investigation of a patient with recurrent infections led to the discovery of a familial deficiency of serum phagocytosis-enhancing factor(s). Although the serums of these patients have normal hemolytic complement activity and C5 concentration, their inability to enhance phagocytosis was corrected by providing purified human C5. As yet there is no explanation for this discrepancy between hemolytic activity and phagocytosis, another biological function apparently dependent upon a complement protein; but such dissociation suggests a heterogeneity of function within what otherwise seems to be a homogeneous molecular population.

While the remaining complement system abnormalities are likely related to congenital defects, they have not been shown to be either congenital or genetically transmissible.

C3 Deficiencies. Two forms of C3 deficiency have been reported. A number of individuals within a healthy blood donor population were found to have extremely reduced serum C3 function. Paradoxically, these individuals had substantial amounts of serum $\beta 1C$ globulin. The basis for this discrepancy between hemolytic activity and protein concentration is not clear. The complement system defect in these people is not limited to C3 but is associated with significant reductions in the hemolytic activity of several components.

The second disorder associated with reduced serum C3 activity was found in a patient who had a history of recurrent in-

fections. This reduction of C3 is the result of the absence of a heat-labile beta pseudoglobulin which apparently protects C3 from conversion to an inactive form, C3b. When this patient was given fresh-frozen plasma, relatively large amounts of active native C3 appeared in his serum; the level subsided gradually while the proportion of inactive C3b increased. His extremely rapid catabolic rate for native C3 was substantially reduced following infusion of fresh-frozen plasma, thus supporting the concept of a protective phenomenon. The exact characteristics of the C3 protective protein have not yet been ascertained.

C4 Deficiency. Employing mass screening, the group of Japanese workers who described the first of the C3 deficiencies (see above) have since found three individuals who were severely deficient in serum hemolytic C4 activity. Once more they observed an incongruity between function and protein concentration by showing that these sera have near-normal amounts of what is believed to be C4 protein. In addition, these sera contain a potent C4 "inactivator" which may be responsible for the reduced amount of hemolytically active C4 available. Why the hemolytically inactive C4 molecules are not removed from the circulation as they are in hereditary angioedema has not been explained. The effect of this defect on complement-related biological functions other than hemolysis has not yet been defined.

Acquired Abnormalities

These abnormalities include a variety of reasonably common disorders in which serum complement is depleted. Generally, acquired complement aberrations are assumed to be secondary to in-vivo immunological activation and depletion. In a number of instances, this assumption is supported by the return of complement levels to normal as the disease process is resolved. There are, however, instances in which a prior defect cannot be excluded; indeed, only the most meager data on both the range and the incidence of complement deficiencies in the general population are available. This problem is of particular relevance in patients with chronic renal disease and persistent hypocomplementemia.

Three major techniques have been used to evaluate the status of the complement system in man, including an assay of hemolytic activity (total hemolytic activity or hemolytic activity of individual components) or immunochemical assessment of one or more of the complement proteins such as C3 (β1C globulin) in serum. A second method involves the detection of complement proteins in tissue, using immunofluorescence as an indicator of abnormal activation and binding of these components. C3 has been studied most extensively using this latter technique, and its presence has often been taken as evidence for involvement of all complement components. There is, however, evidence that C3 may be present in tissue in the absence of other components, indicating a need for caution in extrapolating from data on a single component. Radioisotopes have also been used to study the metabolism of purified complement proteins.

With reference to the assay of complement or complement component concentrations in serum, relatively little attention has been given to diseases in which there is increased serum complement activity or protein concentration. In some *acute* inflammatory processes (e.g., myocardial infarction, bacterial infections, renal allograft rejection, and juvenile rheumatoid arthritis) the complement system appears to react as an acute phase system. This characteristic has not been explored sufficiently to shed any light on the mechanisms involved.

Of the major disorders which have been studied using the above techniques, only renal diseases, some of the hemolytic anemias, and transplantation will be discussed.

Acute Post-streptococcal Glomerulonephritis (APGN). As early as 1914, transient decreases in serum complement activity were noted in association with streptococcal infections. That the association was with post-streptococcal *nephritis* rather than streptococcal infection per se did not become evident until more than 25 years later when the link of streptococcal infection to acute nephritis was clarified. During the acute phase, the total hemolytic activity of the serum is usually substantially reduced (to 30% or less of normal) as a result of extremely small amounts of C3* (C3 through C9). Emphasis has been on C3 (β1C) as the limiting factor; indeed, provision of normal amounts of β1C completely restored hemolytic function of some of these sera. More recently, C5 deficiencies have also been noted in APGN and there may well be other defects in the C3* complex as other components are studied. The early components may also be low in some of these patients; C2 may occasionally be as low as 25 percent of normal, and C1 and C4 concentrations are not infrequently found to be 50 to 60 percent of normal. These low levels of complement function and protein concentration are transient and generally return to normal somewhere between 1 and 6 weeks following the onset of the symptoms of acute nephritis. C1, C4 and C2 precede C3* and β1C globulin in this return to normal. Patients thought to have APGN, who have abnormalities for more than 2 months, should be considered for renal biopsy to exclude chronic glomerular disease. There are occasionally patients who may have normal serum complement levels, even during the acute phase of a relatively severe episode of APGN; but, generally, the severity of disease correlates well with the degree of reduction of serum complement activity.

Depletion of serum complement in patients having APGN was one of the observations which supported the hypothesis that this glomerular disease might have an immunological basis. The development of immunofluorescent techniques made it possible to demonstrate deposits of IgG and β1C in the glomeruli of these patients, suggesting that an immunological complement-fixing reaction was occurring in the kidney at a rate in excess of the body's capacity to synthesize the complement proteins involved. There is, however, some evidence to suggest that the increased activation of C3 is not sufficient to bring about the degree of depletion observed in the serum, and it has been proposed that there is a reduction in the rate of synthesis of C3 as well. Evidence accumulating from studies of patients with APGN (as well as in experimental animals) suggests that the kidney, by virtue of its properties as a filtering mechanism, becomes an object of indirect insult. In this view, immune complexes composed of IgG antibody plus an antigen are formed elsewhere in the body, bind complement components either in the circulation or following deposition in tissue and then, in some way, become sequestered in the glomeruli. It is presumably these complexes which appear on immunofluorescent microscopy as subepithelial nodules or as linear interrupted deposits along the basement membrane. The activation and binding of complement components in these areas may be responsible for alterations in capillary permeability and the accumulation, perhaps as a result of the immune adherence phenomena, of polymorphonuclear leukocytes in the region of the immune deposits. It is proposed that the polymorphonuclear leukocytes and/or platelets which accumulate as part of this inflammatory reaction may, through release of intracellular enzymes, be responsible for the glomerular basement membrane injury which results in proteinuria. The associated proliferative response on the part of the glomerular elements, plus the local edema, lead to decreased glomerular perfusion, oliguria, and, sometimes, anuria.

Systemic Lupus Erythematosus Glomerulonephritis. Many patients with systemic lupus erythematosus (SLE), especially in the childhood age groups, develop a glomerulonephritis which can, as a result of persistent glomerular inflammatory activity, lead to progressive glomerular destruction and chronic renal failure. Although the kidney is not the only site of such destructive inflammatory activity, the glomerular destruction represents the major threat to life in these patients. There is usually a marked reduction of total hemolytic complement in the acute phase of the disease when nephritis is present. Although the degree of reduction is similar to that observed in patients with acute post-streptococcal glomerulonephritis, there is a definite difference in the components which appear to be involved. Both groups show reduced C3* and C3 protein, but SLE patients have, in addition, depletion of C1q protein, and hemolytic C1, C4 and C2. One report indicated that C5 protein concentration may be normal, even in the presence of abnormalities of earlier components. In general, patients found to have SLE nephritis have been treated with steroids (with or without azathioprine), and it is not clear whether spontaneous changes in the complement profile would occur. However, following initiation of therapy levels of all affected complement components usually do return to normal within 2 to 8 weeks.

Immunofluorescent studies of glomeruli in this disease reveal a spectrum of changes. Some patients have lesions similar to those described for APGN: nodular or granular deposition of IgG and β1C globulin (C3) in subepithelial regions in the glomeruli. The nodules tend to be slightly smaller than those observed in acute post-streptococcal glomerulonephritis. These patients have a predominantly proliferative disease. Some SLE patients have appreciable thickening of the glomerular basement membrane, and in these instances there is often deposition of IgG and β1C globulin in subendothelial areas and/or within the membrane. The pattern of immune deposition here is more reminiscent of chronic glomerulonephritis with a pronounced membranous component. In spite of the similarity of the lesions of some of these patients to those of APGN, the etiology is clearly different: in the case of post-streptococcal disease (particularly in children), the process ordinarily resolves with no sequelae; in SLE, the process is progressive and, if unchecked, rapidly destroys the kidney.

There is another major difference between the two diseases: in SLE the nature of the antibody deposited in the kidneys has been established in some instances as anti-DNA. In this disease material presumably derived from cell nuclei, including free DNA, may enter and persist for a time in the circulation. Consequently, IgG and IgM antibodies develop against this free nuclear material and presumably form immune complexes which are, in all likelihood, cleared by the reticuloendothelial system. These antibodies are, at least in part, responsible for the positive fluorescent antinuclear antibody and the LE cell phenomenon which are the hallmarks of this disease. The capacity of the immune complex involving DNA and anti-DNA to activate complement may be an important factor in determining the extent of glomerular damage. This is suggested by the observation that the anti-DNA antibody from patients who have SLE nephritis is much more efficient in its capacity to activate the complement system than the anti-DNA obtained from SLE patients without nephritis. The reasons for this variation in efficiency of complement activation are unknown.

It is important to recognize that this emphasis on immunological activation of the complement system as a prominent factor in the development of glomerular disease, to the apparent exclusion of other effector systems, is, of course, a compartmentalized view for purposes of simplification. It is well known, for example, that immune complexes produce extensive changes in a number of other systems, including platelets and the clotting system, as well as the kallikrein-kinin system. Any or all of these systems may be involved in the inflammatory reaction in the glomeruli.

Chronic Glomerulonephritis. This term is a broad one, referring to long-standing progressive glomerular destruction of varied etiology. Within this group is a subgroup characterized by low serum complement activity. In the pediatric age group, as high as 30 percent of patients with chronic glomerulonephritis have this abnormality; in adults it is much less common. With respect to complement components, this defect is more isolated than that observed in either APGN or SLE: it is largely restricted to C3*. Its most singular characteristic, however, is its persistence. Low complement levels have been documented for several years in some patients, though in others the levels may fluctuate substantially. Multiple forms of immunosuppressive therapy, including prednisone in combination with azathioprine or cyclophosphamide, have been tried, without apparent effect on the glomerular lesion, and with minimal change in the serum C3* levels. In one study, the extent of fibrin deposit in the glomeruli prompted the therapeutic use of anticoagulants—once again, without consistent improvement in the glomerular disease, although serum complement levels frequently returned to the low normal range.

Immunofluorescent studies of the glomeruli rarely show the subepithelial "humps" of IgG and β1C globulin found in both APGN and SLE. Membrane involvement is more frequent, and the immune deposits tend to be subendothelial in location. Again there tends to be a spectrum, from proliferative to membranous disease, paralleled roughly by a shift in the locus of the immune deposits from the epithelial to the endothelial side. There is another characteristic which distinguishes the glomerular lesions of these patients: in many patients there is no detectable IgG in the presence of substantial amounts of β1C globulin, suggesting either that IgG is obscured by extensive deposits of C3, or that C3 activation and binding occur in the kidney even in the absence of IgG. This observation, plus the very selective deficiency of C3*, suggests complement activation by a pathophysiological mechanism similar to that proposed for endotoxin (see above). Studies of the metabolism of C3 in these patients suggest that at least two mechanisms may be involved: increased rate of catabolism of C3, presumably resulting from immunological activation, and a decreased synthesis rate of this protein.

Thus, to summarize this group of glomerular diseases, it may be said that all appear to be characterized by reduced amounts of one or more complement component activities and the presence of IgG and/or β1C globulin in the glomeruli. Serum complement component profiles differ, chiefly in the clear depression of C1, C4 and C2 in SLE and the sparing of these same components in chronic glomerulonephritis. In APGN, the evidence is clearest for involvement of the terminal components (C3*), but there is often depletion of C2 and sometimes depletion of C4 and C1. There is also an immunopathological spectrum. Subepithelial deposition of IgG and β1C is typical of APGN (but also noted in SLE and chronic hypocomplementemic nephritis when the disease is proliferative). In SLE and chronic glomerulonephritis, there may also be intramembranous and subendothelial deposition of IgG and/or β1C; most of these patients are considered to have membranous disease. In SLE and APGN, some patients have linear deposition of IgG and β1C. Many of these would not be considered to have renal disease on clinical grounds. Linear deposition in another clinical context has been associated with anti-glomerular-basement-membrane antibodies, but these have not been demonstrated in either systemic lupus erythematosus or acute post-streptococcal glomerulonephritis.

Glomerulonephritis in Goodpasture's Syndrome. This disorder is characterized by intermittent episodes of pulmonary hemorrhage, which may be massive, accompanied by a usually relentless, destructive glomerulonephritis. Unlike the diseases discussed above, this disorder is not associated with reduced serum complement activity. However, there is a significant accumulation of IgG and β1C in, or at least along, the glomerular basement membrane. The pattern is ribbonlike and linear, without any apparent evidence of granularity, and appears to coincide with the distribution of glomerular basement membrane. Immunohistochemically, this disease is virtually identical to certain experimental models, and it is in a large measure these models which have led to the hypothesis that the glomerular injury in these patients is brought about by antibody directed against antigens present in the glomerular basement membrane.

It is surprising that a process which leads to local deposition of large amounts of C3 does not lead to any significant change in the serum concentration of this or any other complement component measured so far. This is especially puzzling, since the experimental counterpart of the nephritis of Goodpasture's disease is associated with altered serum complement in at least one phase of the disease. These observations suggest that we have very naive concepts of the events which occur in patients and experimental animals during the course of these illnesses.

Acquired Hemolytic Anemias. That hemolysis in some acquired hemolytic anemias might be mediated by serum factors was demonstrated by Donath and Landsteiner in 1904. They found that red cells of patients with paroxysmal cold hemoglobinuria could be lysed by a hemolysin or antibody in their own serum. It was not until the 1940's, when the Coombs technique became available, that gamma globulins were actually shown to be present on the red cell surface in patients with acquired hemolytic anemias. Since the hallmark of complement activity through the years has been in-vitro hemolysis, it was only a short step to the implication of complement in some of these hemolytic disorders (see also Chapter 35).

For this discussion it is useful to think of two categories of hemolytic processes in which complement participates: Firstly, autoimmune disorders in which antibody against erythrocyte membrane antigens is present and presumably activates the complement system from C1 through C9 to produce hemolysis. This group is made up predominantly of the "warm antibody" hemolytic anemias in which gamma globulin (usually IgG) is bound to the membrane of unlysed cells obtained from the patients during hemolytic episodes. The presence of IgG on the cell membrane is demonstrated by agglutination of the erythrocytes by anti-human-IgG. The availability of antibody against human C4 and C3 made it possible to show that in many instances these erythrocytes also carry these complement proteins in addition to immunoglobulin. During hemolytic episodes the serum of some of these patients may contain reduced total hemolytic activity and C3 protein.

Second, paroxysmal nocturnal hemoglobinuria (PNH), a disorder in which there is an abnormality of the erythrocyte membrane and a process of complement activation leading to episodes of hemolysis apparently without the participation of antibody. In vitro, erythrocytes from patients with PNH show a tendency to undergo extensive hemolysis in the presence of diluted normal human serum (pH 6.5), and this hemolysis is substantially enhanced by fluid phase activators of the first component of complement. Although the first three components of complement (C1, C4 and C2) appear

to be essential for hemolysis, they do not become bound to the cell membrane as they do in typical antibody-mediated hemolysis. Instead, small amounts of activated C1 esterase apparently lead to sufficient $C\overline{42}$ formation to allow activation and binding of C3 to the surface of these abnormal erythrocytes. This process could lead, in vivo, to activation of the later components and to the production of hemolysis. Certainly in the in-vitro activation, C3 can be shown to bind to the erythrocyte membrane, and these erythrocytes can subsequently be lysed by the addition of partially purified terminal complement components. In addition, using agglutination methods, acid-treated PNH erythrocytes have C3 bound to the cell surface, whereas gamma globulin and C4 are absent. This susceptibility of PNH cells to hemolysis by fluid phase activation of complement is not unique to these cells, but PNH cells are significantly more sensitive to complement lysis than normal cells.

Serum levels of total complement and C3 in these patients during both quiescent and hemolytic periods are normal. This suggested that whatever increased complement activation is involved in the hemolytic crisis, it is not enough to deplete the plasma C3 pool. A recent study has shown slight increases in the fractional catabolic rate of C3 in PNH patients during periods of active hemolysis. The same investigators found that patients who had "warm antibody" hemolytic anemias, and whose erythrocytes carried gamma globulin, C4 and C3, also had increased fractional catabolic rates which were generally somewhat greater than those of the PNH patients. In contrast to PNH patients, several of those with warm antibody hemolytic anemia also have reduced serum levels of C3 and total complement activity. Patients with "cold antibody" hemolytic anemia or hemolytic anemia associated with α-methyldopa administration had normal serum complement and C3 levels and normal or near normal fractional catabolic rates and synthesis rates. These findings support the concept that complement participates in a number of hemolytic processes in vivo. The available evidence suggests that the activation of C8 and C9 occurs as a result of at least two different mechanisms, exemplified clinically and experimentally in "warm antibody" hemolytic anemias and classic immune hemolysis on one hand and paroxysmal nocturnal hemoglobinuria and CVF-induced hemolysis on the other.

Renal Allograft Rejection Reactions. The major reaction underlying the rejection of allografts (tissue or organs from genetically disparate members of the same species) depends upon the cellular limb of the immune system, the function of which is independent of complement. However, there are some instances in which antibody and complement likely play a significant role in the rejection of organ allografts such as the kidney. The first of these situations arises in individuals who, because of multiple transfusions, hemodialysis, or previous organ transplants, have developed cytotoxic antibodies directed against antigenic determinants of cell surfaces. These host antibodies will, in the event of implantation of an organ carrying the appropriate antigens, combine with the antigen. Activation of complement and an acute inflammatory reaction characterized by an accumulation of polymorphonuclear leukocytes, platelets, and erythrocytes within the lumens of small arterioles and capillaries will lead to endothelial injury and impairment of blood flow. Similar conditions may arise if the organ transplantation is carried out across the ABO blood group or other tissue antigen barriers in which preformed iso-antibodies are present in the recipient's plasma. In these instances, reductions in total serum complement and C2 have been observed. A kidney exposed to these conditions will often fail to function and may show the above changes within minutes following implantation.

The second instance in which antibody may play a significant role is in the very slow, long-term rejection reaction. The recipient very gradually becomes immunized to donor antigens as a result of shedding of these antigens from the implanted organ. In these instances the process is insidious and usually is not associated with any reduction in the total concentration of complement components in the serum. Further evidence suggesting the participation of antibody and complement in the slow chronic rejection reactions in a transplanted organ that has been essentially "accepted" is the finding of IgG, IgM and C3 deposited in the glomeruli.

Finally, there is a group of patients in whom an acute rejection episode occurs after the transplanted kidney has been "accepted" for an extended period. Decreases in total complement and C2 activity during this type of rejection crisis suggest participation of complement in the reaction. In these three instances, the rate of complement participation in the reaction appears to determine whether or not detectable changes occur in the serum. In the first and last instances, the change is so rapid that the rate of synthesis perhaps does not adjust with sufficient rapidity to replenish the supply of complement proteins; in the second instance, failure to observe changes in the serum complement may reflect relatively undisturbed synthesis/catabolism rates.

Several groups have studied total serum complement and/or complement component levels in patients during allograft rejection episodes. Emphasis in the earliest investigations was on depression of C2, but in later work lability of C3 and C4 levels and increases in catabolism of these proteins were also noted. Although these changes are clearly not seen in all rejection episodes, it is difficult on the basis of present data to conclude that these occur only in the antibody-mediated type of episode noted above. The patient groups studied have been heterogeneous in terms of duration of the transplant and the acuteness or chronicity of the rejection process. In addition, histology of the rejecting kidney usually has not been correlated with the complement information. One of the most interesting questions, which cannot be answered yet, is whether component fluctuations will be seen even if the cellular component of the rejection is clearly dominant over the humoral.

There are a number of other disorders in man in which there appear to be abnormalities in metabolism of complement indicated by alterations in serum complement activity or complement protein concentrations and/or the deposition of complement proteins in injured tissue. These include rheumatoid arthritis, plasma cell hepatitis, amyloidosis, and rheumatic heart disease, to mention only a few. Discussion of these has been omitted, in part because of the limitation of space, but also because they have received less investigative attention than those diseases described above.

Experimental Models

There is one additional type of evidence pertaining to the role of complement in disease, alluded to in the sections on clinical illnesses, namely, the experimental models. These are the bridge, in a sense, between the in-vitro systems (where the investigator has the greatest control) and patients where the physician observes the results of pathogenetic mechanisms he can manipulate minimally, if at all.

Immune Complex Disease. Glomerulonephritis has been induced in rabbits by repeated parenteral administration of a foreign protein, such as bovine serum albumin (BSA), in a dose calculated to maintain serum antigen/antibody ratio optimum for complex formation and complement activation. Under these conditions the animals develop progressive destructive glomerular inflammation which resembles the reaction seen in chronic

membranous glomerulonephritis. By light microscopy the glomeruli show basement membrane thickening with a variable degree of endothelial proliferation and neutrophil accumulation. Gamma globulin and C3 are localized as granular deposits in the glomerular capillary walls by fluorescent microscopy. Serum complement changes do not seem to correlate with the deposition of C3 in the glomeruli. The metabolism of complement components in these animals has not been studied, nor has it been possible to deplete complement long enough to ascertain whether the full pathological picture would develop in its absence. Thus, in this chronic glomerulonephritis model, the participation of complement is implied by its presence in the immune deposits in the kidney and by the analogy of this glomerular reaction with the Arthus type of inflammation.

A single injection of heterologous protein in rabbits produces a short-lived glomerulonephritis which heals spontaneously and thus resembles APGN in man. Following injection of the heterologous protein three phases of loss from the serum occur: equilibration with the total extracellular fluid; catabolic elimination; and immune elimination facilitated by the production of antibody. The glomerulonephritis appears just prior to or very early in the phase of immune elimination when antibody and antigen are present in an optimum ratio for pathogenic immune complexes to form in the circulation. During this phase, serum complement levels fall substantially below normal and this depletion involves early as well as late components. This complement reduction is transient, lasting 2 or 3 days. The glomerular lesions are characterized by deposits of antigen, gamma globulin and C3 in a fine granular discontinuous pattern along the glomerular capillary walls.

In many ways the alterations observed in this disorder resemble those of APGN in man, with the exception of the complement components lost from the serum; i.e., early and late components here but predominantly C3* in APGN. This may be a reflection of a difference in the type of antigen used rather than a difference in the basic mechanism of glomerular injury.

Anti - Glomerular - Basement - Membrane Disease (Nephrotoxic Nephritis). This is the experimental model mentioned earlier as the immunohistochemical counterpart of Goodpasture's nephritis. This experimental model is one in which antibody against glomerular basement membrane is produced in a heterologous species and then injected back into the donor species. This results in a two-stage lesion. At first, the nephrotoxic antibody combines with the basement membrane antigens, activating complement rapidly enough to deplete serum hemolytic activity temporarily. This complement alteration occurs in nephrectomized animals, indicating that the antibody combines with an equivalent antigen in other parts of the body. The second phase of this disorder occurs as a result of the host's antibody response to the heterologous immunoglobulin. This host antibody combines with the heterologous antibody (nephrotoxic immunoglobulin) persisting in the glomeruli, producing a second phase of complement activation and inflammation. Depletion of serum complement during this phase depends on the presence of renal tissue. Complement components have not been studied in detail in this experimental model.

That complement participates in the production of lesions in stage one of this model seems likely, but its main role seems to be an influence on polymorphonuclear leukocyte accumulation within the glomeruli. Experimental manipulation to deplete an animal of complement with CVF or polymorphonuclear leukocytes with nitrogen mustard prior to injection of anti-GBM antibody reduces glomerular injury but does not eliminate it. Immunofluorescent studies of the glomeruli in this model show uniform ribbonlike staining along

capillary walls throughout the entire glomerular tuft, of antigen (nephrotoxic immunoglobulin), antibody (host-immunoglobulin) and host C3. Since this disease can be induced in animals deficient in C5 and C6, it appears that complement's role is limited to functions served by the first four components.

ANNOTATED REFERENCES

Carpenter, C. B.: Immunologic aspects of Renal Disease. Ann. Rev. Med., *21*:1, 1970. (Complement levels in APGN and SLE and observations of complement in graft rejection and C3 synthesis rate)

Daniels, C. A., Borsos, T., Rapp, H. J., Snyderman, R., and Notkins, A. L.: Neutralization of sensitized virus by purified components of complement. Proc. Nat. Acad. Sci., *65:* 528, 1970.

Linscott, W. D., and Levinson, W. E.: Complement components required for virus neutralization by early immunoglobulin antibody. Proc. Nat. Acad. Sci., *64*:520, 1969. (The last two references are original articles on the role of complement components in viral neutralization)

Donaldson, V. H., and Evans, R. R.: A biochemical abnormality in hereditary angioneurotic edema. Am. J. Med., *35*:37, 1963. (Original observation of primary defect in hereditary angioneurotic edema)

Gewurz, H., Clark, D. S., Finstad, J., Kelly, W. D., Varco, R. L., Good, R. A., and Gabrielsen, A. E.: Role of the complement system in graft rejections in experimental animals and man. Ann. N. Y. Acad. Sci., *129*:673, 1966. (Complement changes during graft rejection)

Gewurz, H., Pickering, R. J., Clark, D. S., Page, A. R., Finstad, J., and Good, R. A.: The Complement System in the Prevention, Mediation and Diagnosis of Disease and its Usefulness in Determination of Immunopathogenetic Mechanisms. *In:* Immunologic Deficiency Diseases in Man. Birth Defects Original Article Series IV, 396, 1968. (Review and original observations on complement profiles and alternative mechanisms of complement activation)

Herdman, R. C., Pickering, R. J., Michael, A. F., Vernier, R. L., Fish, A. J., Gewurz, H., and Good, R. A.: Chronic glomerulonephritis associated with low serum complement activity (Chronic Hypocomplementemic Glomerulonephritis). Medicine, 49:207, 1970. (A review article)

Humphrey, J. H., and Dourmashkin, R. R.: The lesions in cell membranes caused by complement. Adv. Immunol., 11:75, 1969. (Review of the nature of complement-mediated membrane damage)

Koffler, D., Agnello, V., Carr, R. I., and Kunkel, H. G.: Variable patterns of immunoglobulin and complement deposition in the kidneys of patients with systemic lupus erythematosus. Am. J. Path., 56:305, 1969. (Immunofluorescent studies of kidney from patients with SLE)

Kohler, P. F., and ten Bensel, R.: Serial complement component alterations in acute glomerulonephritis and systemic lupus erythematosus. Clin. Exp. Immunol., 4:191, 1969. (Observation and discussion of complement levels in APGN and SLE)

Lachmann, P. J., and Thompson, R. A.: Reactive lysis: the complement mediated lysis of unsensitized cells. J. Exp. Med., 131:643, 1970. (Original observations of complement activation by nonimmunological mechanism)

Müller-Eberhard, H. J.: Chemistry and reaction mechanisms of complement. Adv. Immunol., 8:2, 1968.

————: Complement. Ann. Rev. Biochem., 38: 389, 1969.
(The above two references present thorough reviews of mechanisms of complement component action and interaction, and refer to original papers concerning acquired hemolytic anemias and primary complement deficiencies)

Nelson, R. A., Jr.: The role of complement in immune phenomena. In: Zweifach, B. W., Grant, L., and McCluskey, R. T. (eds.): The Inflammatory Process. p. 819. New York, Academic Press, 1965.

Nomenclature on Complement. Immunochemistry, 7:137, 1970.

Osborn, T. W. B.: Complement or Alexin. London, Oxford University Press, 1937. (A scholarly review of the knowledge of complement as of 1937)

Petz, L. D., Fink, D. J., Letsky, E. A., Fudenberg, H. H., and Müller-Eberhard, H. J.: In vivo metabolism of complement. J. Clin. Invest., 47:2469, 1968. (Study of complement levels and catabolism of C3 in several diseases)

Pickering, R. J., Wolfson, M. R., Good, R. A., and Gewurz, H.: Passive hemolysis by serum and cobra venom factor: A new mechanism inducing membrane injury by complement. Proc. Nat. Acad. Sci., 62:521, 1969. (Original observation of nonimmune terminal component activation)

Porter, K. A., Andres, G. A., Calder, M. W., Dossetor, J. B., Hsu, K. C., Rendall, J. M., Seegal, B. C., and Starzl, T. E.: Human renal transplants. II. Immunofluorescent and immunoferritin studies. Lab. Invest., 18: 159, 1968. (Observation and discussion of complement and antibody participation in glomerulonephritis and graft rejection)

Ratnoff, O. D.: Some relationships among hemostasis, fibrinolytic phenomena, immunity, and the inflammatory response. Adv. Immunol., 10:145, 1969. (Broadly based review of subjects named in the title)

Ruddy, S., and Austen, K. F.: Inherited abnormalities of the complement system in man. Prog. Med. Genet., 7:69, 1970. (Review of genetically determined complement related diseases including C-2 and C-5 deficiency in man)

Tojo, T., and Friou, G. J.: Lupus nephritis: varying complement fixing properties of immunoglobulin G antibodies to antigens of cell nuclei. Science, 161:904, 1968. (Comparison of antibodies from SLE patients with and without kidney disease)

Torisu, M., Arata, M., Sonozaki, H., and Majima, H. C′3c deficient human sera with a βIc precipitin line. J. Immunol., 99:629, 1967.

Alper, C. A., Abramson, N., Johnston, R. B., Jr., Jandl, J. H., and Rosen, F. S.: Studies in vivo and in vitro on an abnormality in the metabolism of C3 in a patient with increased susceptibility to infection. J. Clin. Invest., 49:1975, 1970. (This and previous reference include observation and discussion of C3 deficiency)

Torisu, M., Sonozaki, H., Inai, S., and Arata, M.: Deficiency of the fourth component of complement in Man. J. Immunol., 104:728, 1970.

Unanue, E. R., and Dixon, F. J., Experimental glomerulonephritis: immunological events and pathogenetic mechanisms. Adv. Immunol., 6:1, 1967. (Review of components of immune renal disease)

35

Autoallergic Mechanisms and Transplantation

J. Wesley Alexander, M.D., SC.D.

At first glance, transplantation and auto-immunity appear to be widely divergent fields of interest. However, the principles of pathophysiology and therapy in transplantation and in the autoimmune diseases are similar enough to warrant simultaneous consideration. Both involve expressions of immunological injury and their control by immunosuppressive therapy.

MECHANISMS OF IMMUNOLOGICAL INJURY

Role of Antibody

Five classes of immunoglobulin are synthesized by plasma cells and large and medium-sized lymphocytes. Of these, only IgG and IgM appear to play important roles in immunological injury associated with transplantation and the autoimmune diseases. IgM and three of the four subclasses of IgG can activate the complement sequence necessary to initiate immune adherence, release of vasoactive substances, including anaphylatoxins, opsonization for phagocytosis of foreign molecular complexes and cells, and enzymatic damage of cell membranes, which can result in cellular death. The degree of complement-mediated damage is related to the density of the involved antigens on the cell membrane and to the type of antibody reacting with these antigens. Specific antibody can also inactivate enzymes and may increase the removal of particulate matter by the reticu-

loendothelial system, even in the absence of complement.

Immunological damage caused by sensitized thymus-dependent lymphocytes (sometimes called killer cells) is initiated by the interaction of the antigen on target cells, such as those of a transplanted organ, and specific receptor sites on the cell membranes of the lymphocytes. The resulting release of enzymes and other cytotoxic substances from the lymphocytes can cause death of target cells. The exact biochemical sequence in this process is not known.

Synergistic action of damage caused by circulating antibody and sensitized thymus-dependent lymphocytes has been observed, but the mechanism of this synergism has not been well defined. In contrast, specific IgG antibody has been shown to have a very important protective role in both tumor grafts and normal tissue grafts. This process, known as efferent immunological enhancement, occurs when: the cell membrane is resistant to complement lysis; the specific antibody binding with the antigenic determinant sites on the cell membrane does not activate the complement sequence (as in the case with IgA, IgD, IgE and IgG-4 antibodies); the density of antigens on the surface of the cells capable of stimulating an immune response is low; or the antigens are located improperly to cause activation of a sufficient amount of complement to cause lysis. Complete binding of antibody to antigenic receptor sites on the

x Antigenic Site

⬭ Antibody Molecule (IgG)

c' Complement

A. Cytolysis

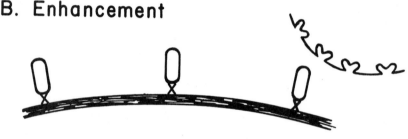

Cell Membrane

c'-Mediated Damage
to Membrane

B. Enhancement

C. Cell-Mediated Damage

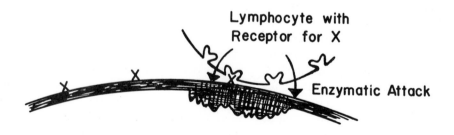

Lymphocyte with
Receptor for X

Enzymatic Attack

FIG. 35-1. Diagrammatic sketch showing the relationship between cell-mediated cytolysis, antibody-mediated cytolysis and enhancement. Note how the presence of antibody may help the cell to survive. (Alexander, J. W., and Good, R. A.: Immunobiology for Surgeons. Philadelphia, W. B. Saunders, 1970)

cell surface prevents their interaction with thymus-dependent lymphocytes, and the cell becomes protected from immunological cell-mediated damage. The differences and similarities between cell-mediated damage, cytolysis and enhancement are shown diagrammatically in Figure 35-1.

TRANSPLANTATION

It was not until the beginning of the 20th century that Alexis Carrel developed the surgical techniques necessary for transplantation of organs by vascular anastomoses. In the most elegant of his studies he demonstrated that grafts between unrelated animals of the same species could be performed with satisfactory early function. However, good early functional results were always followed by unrelenting deterioration and eventual death of the grafted tissues. The processes responsible for this phenomenon remained somewhat of an enigma until the classical work of Sir Peter Medawar and his colleagues in the early 1940's showed that immunological mechanisms were responsible for the rejection process. With the knowledge of the importance of genetic identity in transplantation as evidenced by long-term survival of skin isografts (grafts between two individuals with identical genetic constitutions), Murray and his co-workers performed the first successful human renal transplant between identical twins in 1954. Xenogenic (cross species) renal grafts had previously been attempted in man as early as the beginning of the 20th century, and renal allografts (grafts between individuals of the same species) had also been tried in man, but all failed. After Murray's success, several further attempts to transplant kidneys of allogenic origin in man were made during the next few years; these also failed. In 1959, Schwartz and Dameshek demonstrated the immunosuppressive potency of 6-mercaptopurine (6-MP), which led to successful long-term maintenance of transplanted allogenic organs. Increasing scientific interest soon centered around transplantation biology, which has already benefited every medical discipline.

The greatest experience with transplanted organs has been with the kidney. Since the problems in transplantation biology are similar for all types of vascularized tissue grafts, the following discussion will be directed to the more extensive clinical and laboratory experience with renal transplants.

Mechanisms of Rejection

The active process of destruction of a surgically successful graft is called immunological rejection. Three clinically distinct types of rejection will be described.

Immediate or Hyperacute Rejection. It is generally accepted that the greater the genetic disparity between the donor and the recipient, the more vigorous will be the immunological reaction against a newly grafted organ. With extreme differences, such as may be observed in xenografts, naturally occurring preformed antibodies may already be present in the recipient. Even when the disparity of histocompatibility antigens is not so great, preformed antibodies may be present in the recipient against donor antigens because of past sensitization resulting from pregnancy, a previous graft or antigens received with blood transfusions. Regardless of the cause for existence, hyperacute rejection occurs within minutes or hours of reconstitution of the blood supply when a kidney graft is transplanted to an individual with preformed antibodies.

Hyperacute rejection is caused by the interaction of circulating antibody and histocompatibility antigens of the endothelial membranes of the grafted organ, which initiates activation of the complement sequence with platelet aggregation in the small vessels. The clotting mechanism is activated by platelet lysis and the effect of complement on plasmin. Endothelial membranes are damaged, resulting in diapedesis of neutrophils and mononuclear phagocytes into the interstitial tissues. Cortical blood flow becomes markedly reduced as a consequence of vasospasm and microthrombi, and, in extreme instances, acute cortical necrosis may ensue. The extensive ischemic damage to the organ usually precludes recovery of function. Hyperacute rejection occurs, therefore, because of the fixation of a large amount of antibody by the grafted organ, with associated complement-mediated damage. The damage caused by this reaction is extremely difficult to control, both clinically and in the laboratory animal, but, more important, it can be avoided by the exclusion of donor-recipient combinations in which the recipient has preformed antibodies against the donor's tissues. (Some of the basic features of this type of rejection are shown in Fig. 35-2, A.)

Acute Rejection. In a healthy unmodified unrelated recipient, a transplanted allogenic renal graft functions well initially when a technically satisfactory surgical procedure has been accomplished. However, the renal graft rather abruptly ceases to function approximately 6 days after implantation as the result of acute immunological rejection (Fig. 35-2, B). This process is heralded by a brief but rapidly increasing degree of renal functional abnormality that is associated with swelling and edema, interstitial hemorrhage and a profound accumulation of pyroninophilic cells in and around small vessels, particularly the venules and capillaries. With the progressive vascular damage, reduction and, finally, ablation of tubular blood flow occur, with subsequent ischemic necrosis. Fibrinoid necrosis of vascular walls is also observed, often associated with accumulation of immunoglobulin complexes. In addition to pronounced mononuclear cell infiltration, nonspecific accumulations of neutrophils may be a prominent feature. In untreated patients, acute rejection eventually results in the death of the graft because of unrelenting progression of vascular obliteration and thrombosis, interstitial hemorrhage and infarction. The lesions in acute rejection are the direct result of immunological assault by the host, which is directed against the histocompatibility antigens located on the cell membranes of the new graft. Since the endothelial surfaces of the graft are continually exposed to the bloodstream, they are the first and most severely affected.

The small recirculating thymus-dependent (TD) lymphocytes appear to play the major role in acute rejection processes. In the presence of appropriate antigen, sensitized or immune TD cells are transformed into blastlike forms that are capable of enzymatic destruction of allogenic target cells when exposed to them in vitro. In addition to the well recognized role of the cell-mediated immune response in acute rejection, there is increasing evidence that circulating antibody, synthesized in response to the stimulus of a new graft, may also result in acute rejection. Cytotoxic antibodies often appear in the circulation coincidentally with, or shortly after, acute rejection episodes, and cytotoxic antibody can be eluted from rejected grafts. In transplanted hearts and lungs in both man and animals, acute vascular injury may be seen with little or no evidence of lymphoid cell infiltration, especially when the recipient is receiving immunosuppressive drugs. The endothelial damage is caused by activation of large amounts of complement rather than by the antibody itself.

Acute rejection often may be reversed by intensive immunosuppressive therapy; however, far-advanced lesions invariably result in permanent damage to the organ. The intensity and timing of the pathological changes during acute rejection episodes depend not only upon the degree of disparity of histocompatibility between recipient and host, but also upon the vigor with which the recipient can mount an

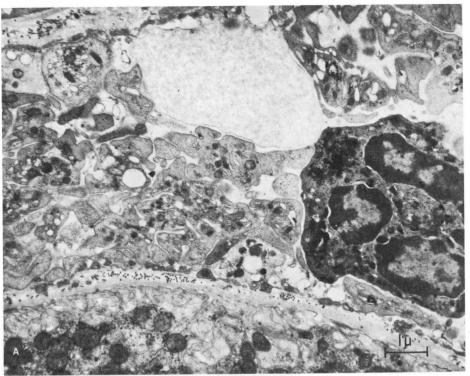

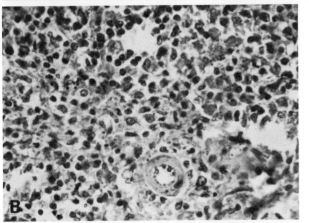

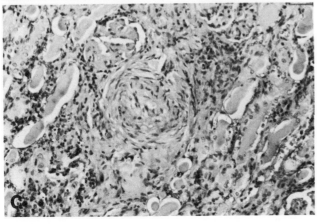

Fig. 35-2. Structural changes associated with different types of rejection. (A) Electron micrograph of a biopsy specimen, removed from a dog kidney 5 hours after allogenic transplantation, showing the features of the immediate type of rejection. The host had been presensitized with specific donor antigens. Complete obliteration of the vessels in the figure is caused by a completely obstructing platelet thrombus. Degranulation and extensive fusion of the platelets are evident. (B) Microscopic appearance of a dog kidney late in acute rejection. There is extensive infiltration of the interstitial tissues with mononuclear cells and extensive hemorrhage ($\times$ 160). (C) Microscopic appearance of a dog kidney showing vascular lesions often found in chronic rejection. The artery in the center is almost completely obliterated ($\times$ 63). (Adapted from Alexander, J. W., and Good, R. A.: Immunobiology for Surgeons. Philadelphia, W. B. Saunders, 1970)

immunological response and the relative degree to which cell-mediated and antibody-mediated immunity is stimulated. The presence of specific antibody against antigens in the graft is also an important component of immunological adaptation.

Chronic Rejection. Still another type of rejection is sometimes recognized in patients receiving immunosuppressive therapy. This condition is associated with chronic progressive renal damage; it is less common than acute rejection episodes and is much less responsive to immunosuppressive agents. The lesions of chronic rejection are characterized by accumulation of subendothelial deposits of amorphous material in the glomeruli and arterioles, destruction of the internal elastic lamina, a chronic obliterative process in the vessels, and patchy fibrous obliteration of nephron units (Fig. 35-2, C). The patchy destruction occurs because of an unequal involvement of the affected vessels supplying the units.

Histocompatibility Testing

Histocompatibility antigens can be defined as complex chemical configurations on cell membranes common to all of the individual's nucleated cells. Both major and minor histocompatibility antigens exist. In man, these antigens are inherited through alleles on a single chromosome at what appears to be two subloci. At the present time, more than 12 twelve distinct antigens have been recognized by international nomenclature, but many more exist. The histocompatibility locus has been studied more completely in the mouse and is known as the H-2 locus. In the rat, it is the AgB locus, and in the chicken, it is the B locus. In man, the major blood group antigens (A, B and AB) also act as major histocompatibility antigens. Accumulating data have already shown that as the antigenic relationship (genetic relationship) between donor and recipient becomes

closer, the vigor of the rejection episode will be diminished and it will be easier to obtain prolonged graft survival and good function with immunosuppressive therapy. This has been particularly well demonstrated in sibling-to-sibling human grafts, but the principle is also clearly important in other types of interfamilial grafts and even in grafts between unrelated individuals (cadaveric grafts). Table VIII-7 (see Appendix, this section) shows why histocompatibility testing is essential in the selection of donor-recipient combinations from sibling groups. As expected from mendelian concepts, one fourth of randomly selected siblings will show identity of histocompatibility antigens, whereas another one fourth potentially have four antigenic differences. Even though the increased effectiveness of immunosuppressive regimens and the extenuating circumstances involving availability of a donor organ for a given recipient may at times be overriding considerations, provision of a recipient with an organ from a well matched donor is of great importance in obtaining long-term graft function (Fig. 35-3). Methods for histocompatibility testing are still far from being exact because of many laboratory variables, and the antigenic expression of inherited histocompatibility antigens may differ in density on cell surfaces of the various organs. Because of this, antigens weakly expressed on lymphocytes (the cell type most often used in histocompatibility typing) may sometimes be strongly expressed on the kidney or other organ.

Immunological Adaptation

Prolonged graft survival in an allogenic host requires manipulation of the immunological apparatus in order to minimize or negate an effective immune response. Early graft survival during the first few months is highly dependent upon effective immunosuppression, but truly long-term survival of a graft more probably depends upon a

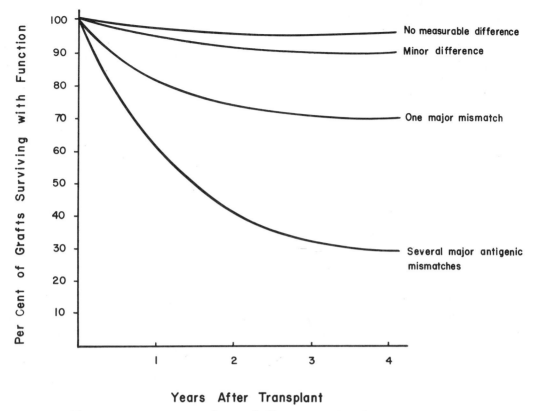

FIG. 35-3. Approximate survival rate of allogenic renal grafts in treated, good risk, young patients according to the degree of histocompatibility similarity. (Alexander, J. W., and Good, R. A.: Immunobiology for Surgeons. Philadelphia, W. B. Saunders, 1970)

complex interaction of three basic processes: enhancement, true immunological tolerance and adaptation of the graft.

Enhancement. In contrast to the damaging effect of antibody which can cause cytolysis by activating large amounts of complement at antigenic receptor sites on cell membranes, antibody paradoxically may protect the graft from the damaging effects of cell-mediated immune recognition by binding to antigenic receptor sites without activating complement (see Fig. 35-1). This process was first observed in tumor experiments when it was found that administration of tumor-specific antibody to an animal before implantation of the tumor resulted in an accelerated or enhanced rate of tumor growth (hence, the name enhancement). Enhancing antibody (IgG) activates minimal amounts of complement either because it is incapable of initiating the cascade reaction (IgG-4) or because the antigenic receptor sites are sufficiently sparse to prevent two molecules of IgG from simultaneously complexing to a single C'1 complex (see Chap. 34 on complement). In general, the presence of antibody without activated complement does not damage the cell. Enhancement of organ grafts can be produced either by active immunization or by the administration of antibody. The mechanism seems to be peripheral rather than central, and the effect can be increased by nonspecific immunosuppressive therapy.

Enhancement with specific antibody

against transplantation antigens as the sole means of graft protection has been successful in certain laboratory animals, and the principle has been applied at least once in man. This technique appears to be a promising approach to the prevention of acute rejection, but there is insufficient accumulated experience to recommend clinical usage. Much more important than its early effect, enhancement seems to play a predominant role in the long-term survival of whole organ grafts. Recent studies have shown that most, if not all, organ recipients develop both circulating antibodies and cell-mediated immunity against antigens of the graft. At present, it appears that the relative responses of humoral immunity (causing enhancement) and cellular immunity (causing damage) are the most significant factors influencing long-term graft acceptance.

Tolerance. Immunological nonresponsiveness can also be achieved in a normal animal by the administration of specific antigen. In 1942, Felton demonstrated that a large dose of pneumococcal polysaccharide given to an adult mouse could result in specific loss of reactivity to that antigen. This type of tolerance occurs in parabiotic unions between two animals and is often referred to as high-dose tolerance. Another type of tolerance was noted by Owens in 1945 when he observed that twin cattle often have two types of erythrocytes— their own and that of their twin. It was subsequently shown that other antigens administered in utero produce tolerance. However, to produce long-lasting tolerance with nonliving antigens, repeated injections of antigen are required. Both of these types of tolerance require the continued presence of antigen and, although the mechanisms are not well understood, they may play important roles in the long-term survival of allogenic grafts.

Still another type of tolerance can be produced by injection of a very small amount of an antigen, followed several days thereafter by another small dose. Subsequent challenge by the same antigen may show a specific nonreactivity. The mechanism for low-dose tolerance is not understood either, but there is undoubtedly a central effect on the lymphoid system. While low-dose tolerance has been induced in a variety of laboratory animals, a very fine line exists between the induction of tolerance and the sensitization of the individual against the antigen. Therefore, induction of low-dose tolerance by administration of antigen before grafting in man is a dangerous undertaking at present. Obviously, much more work is necessary before tolerance mechanism can be clinically applied.

Graft Adaptation. In addition to the decisive role of enhancement and tolerance in extending the life of allogenic organ grafts, the graft itself may change with time. Certain cells (i.e., endothelial cells lining larger blood vessels and hepatic Kupffer cells) are replaced with cells of the host, thus decreasing the opportunity for interaction between sensitized lymphoid cells of the host and donor cells. It has not been demonstrated that parenchymal cells of the donor organ are replaced by host cells.

DISEASES OF AUTOALLERGY

These diseases have previously been called autoimmune diseases, and this term is still widely used. However, since autoimmune is clearly a misnomer (meaning safe from self), we have chosen the more appropriate term, autoallergic (hypersensitive to self).

While clearly different from transplantation, the autoallergic diseases present a similar problem and challenge to the physician, namely, control of an unwanted immunological reaction. Progress in therapy of the autoimmune diseases has been gratifying but, in spite of much accumulating knowledge, many questions remain unanswered. In addition to the primary auto-allergic disorders, there are a number of diseases in which "autoimmunity" plays a secondary role.

Immune reactions against "self" antigens must be considered abnormal. They may arise through a variety of mechanisms, including unmasking of normal antigenic components not normally revealed to the lymphoid apparatus, production of antibody by so-called *forbidden clones* of lymphoid cells, genetic or hereditary factors resulting in appearance of new antigens, neoantigens induced by virus infections or variant forms of bacteria, loss of tolerance to "self" antigens, and immunity against a normal carrier protein initiated by haptene (Fig. 35-4). However, the precise mechanisms responsible for many of the diseases of this extremely diverse group are still obscure. Interestingly, there is a significantly higher incidence of the autoallergic diseases in women and in family members of persons having autoallergic diseases. (A partial listing of diseases in which autoallergic mechanisms are of probable etiological importance is presented in Appendix B of this Section.) Since a detailed analysis of these

diseases is beyond the scope of this presentation, only a few will be discussed in order to conceptualize the pathophysiological mechanisms of immunological injury and autoallergy involved.

Hashimoto's Thyroiditis

Focal collections of lymphocytes may be found in as many as one of four thyroids at autopsy, but the classical goiterous struma lymphomatosa (Hashimoto's disease) is a relatively rare disease. Hashimoto's disease is approximately 30 times more common in women. It is characterized by a diffuse thyroidal infiltration of lymphocytes which replace normal structures and form occasional germinal centers. Besides the fibrosing and hypercellular variants of Hashimoto's disease, other forms of lymphocytic thyroiditis are believed to be the cause of adult primary myxedema and severe atrophic thyroiditis as well as the much more common multifocal thyroiditis. Antithyroid antibodies may be found in a variety of other complex diseases, and there is a significantly higher incidence of antithyroid antibodies in healthy relatives of patients having Hashimoto's disease.

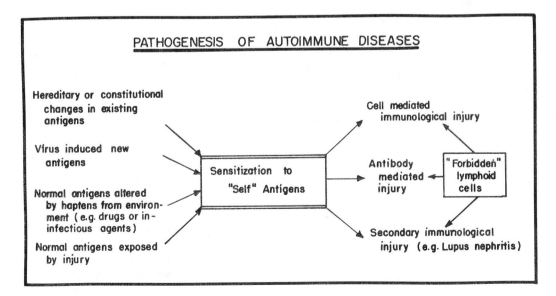

Fig. 35-4. Several possible mechanisms in the pathogenesis of autoallergic diseases.

The mechanisms responsible for the auto-allergic processes are poorly understood in Hashimoto's disease, but 3 well defined thyroid-specific antigens are involved: thyroglobulins, a microsomal antigen and a second colloid antigen. Antibodies against each have been detected. The autoantibodies against the microsomal antigen are organ-specific and may be of particular clinical importance since they are complement-fixing and cytotoxic to thyroid cells in tissue culture. Evidence for a damaging effect of the antibodies against thyroglobulin and the second colloid antigen is less impressive. The relative roles of antibody-mediated damage and cell-mediated damage have not been defined clearly, but many investigators feel both types of damage are important components of the disease. Since the immunological damage in lymphocytic thyroiditis is primarily organ-specific and thyroid failure is easily corrected, immuno-suppressive treatment does not seem to be warranted. Instead, therapy is directed to surgical removal of symptomatic goiters or replacement of thyroid hormones when the gland is hypoactive.

Autoimmune Hemolytic Anemias

The autoimmune hemolytic anemias are among the oldest recognized autoallergic clinical disorders. The antibodies involved have been separated into those which react best at warm temperatures and those which react best at cold temperatures. They are thus designated as warm or cold antibodies. Warm antibodies are the most common cause of the "autoimmune hemolytic anemias;" they are of the IgG type and can be heterogeneous. Most seem to react with the Rh antigens (usually C or E) and are usually not complement-fixing. The lack of complement fixation may result either from the relatively large distance between Rh antigens on the cell surface or because the antigenic constituents of Rh are located on the surface in such a way that there is spatial interference with the fixation of complement. It is well recognized that two

closely situated IgG antibody molecules are necessary to fix complement, in contrast to only one IgM antibody molecule. The presence of IgG on the red cells can be detected by the antiglobulin (Coombs) test. Erythrocytes coated with warm autoantibodies are preferentially eliminated by the spleen rather than by the liver, probably because of the relatively small role of complement in the disease. The majority of the hemolytic anemias caused by warm autoantibodies are unassociated with other diseases, but they can occur with several underlying diseases such as ulcerative colitis, systemic lupus erythematosus and drug reactions.

Those autoantibodies that react best at 0 to 4° C., and become disassociated from the erythrocytes at 37° C., are referred to as the cold autoantibodies. The cold autoantibodies may be either IgG or IgM, but both fix complement. The IgM type of cold antibody is a cold hemagglutinin. Cold agglutinin is usually directed against the erythrocyte antigen I, but examples of cold hemagglutinins to antigen i have also been found, particularly in patients with cirrhosis, malignancies or infectious mononucleosis (Fig. 35-5). The cold agglutinins in the latter disease are almost always directed against antigen i. Hemolytic anemia caused by cold IgM antibody is particularly common during the course of mycoplasma infections, infectious mononucleosis, sarcoidosis and a variety of other diseases. Experimentally, cold agglutinin disease man be produced in rabbits by immunization with *Listeria monocytogenes.* In contrast to the hemolytic anemia due to warm autoantibodies, the affected erythrocytes are removed preferentially by the liver. Since antibody is eluted from the erythrocytes at 37° C., agglutination tests with anticomplement antibody are particularly useful in diagnosis.

Paroxysmal cold hemoglobinuria is another disease caused by cold antibodies, but, in contrast to the preceding condition, IgG is the effector molecule. Relatively

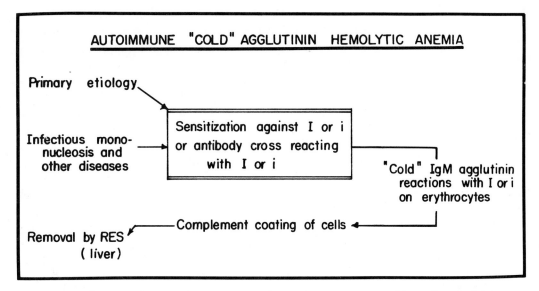

Fig. 35-5. Schema for the pathophysiology of autoimmune cold agglutinin hemolytic anemia.

large amounts of complement can be fixed by the IgG molecules, which are released from the erythrocytes at a slower rate than are the cold agglutinins. These cold hemolysins have anti-P specificity (i.e., they are directed against the erythrocyte antigen designated P). The disease occurs when a body part is cooled below a critical temperature; hemolysis occurs rapidly, inducing hemoglobinemia and hemoglobinuria. Other prominent symptoms, which include fever, chills, hypotension, pain in the back and extremities, and abdominal cramps, are probably explained best as a consequence of activation of the complement system. At the beginning of the century, paroxysmal cold hemoglobinuria was found to be associated frequently with syphilis, but more recently it has been described following viral infections.

Systemic Lupus Erythematosus

Systemic lupus erythematosus (SLE) is a multisystemic disease characterized by a variety of autoallergic phenomena. It is more common in females and occurs with greatest frequency in the second and third decades of life. The etiology is unknown but there appears to be a genetic predispo-

sition. Both viral and bacterial infections as well as nonspecific events such as emotional stress or ultraviolent irradiation have apparently triggered its acute onset, although it seems unlikely that they are etiologically related. Conceivably, viral infections could trigger its onset by causing somatic mutations or inducing an abnormal immune response through a haptene-carrier protein mechanism. In addition, several drugs (i.e., hydralazine, procainamide) can cause a syndrome which is clinically very similar to systemic lupus erythematosus. Fortunately, most of the drug-induced lupus syndromes regress upon discontinuance of the drug. Some authors have felt that a preceding lupus diathesis is important in the pathogenesis of drug-induced lupus.

Regardless of the etiology, the disease SLE is characterized by the development of antibodies against a wide variety of "self" antigens, including nucleoproteins, DNA, histone, nuclear glycoproteins, soluble and membranous intracellular components, IgG, certain coagulation factors, platelets, erythrocytes, leukocytes, thyroglobulin and others. The clinical manifestations are protean and include fever,

742   *Autoallergic Mechanisms and Transplantation*

weight loss, various types of cutaneous eruptions, arthritis, nephropathy, hypertension, serosis, vasculitis, atypical pulmonary infiltrations, hepatomegaly, jaundice, splenomegaly, a variety of gastrointestinal symptoms, and ocular problems. Pathological manifestations include formation of LE bodies (hematoxylin bodies), which are composed of collections of nuclei or nuclear parts, particularly in the peripheral areas of necrosis and in lymph nodes. These LE bodies are the in-vivo counterpart of the in-vitro LE phenomenon, which is the result of phagocytosis by phagocytes of the peripheral blood of nuclear material complexed with antibody. The damaged nuclei are opsonized (made less resistant to phagocytosis) by an IgG antibody in the

serum which cannot penetrate living cells, reaching only nuclei of damaged cells. Complement is utilized in the reaction, thus increasing the opsonizing capability. Fibrinoid deposition is another important pathological characteristic and, most likely, it is the end result of an immunological reaction directed against damaged cells. Fibrinoid, which contains cellular debris, fibrin, immunoglobulin and complement, can be deposited in almost any portion of the body, where it stimulates the deposition of collagen fibers and fibrosis. Immune complexes are frequently deposited in the small vessel endothelium, particularly in the renal glomeruli, resulting in the collection of irregular masses on the basement membranes that are composed of antigen,

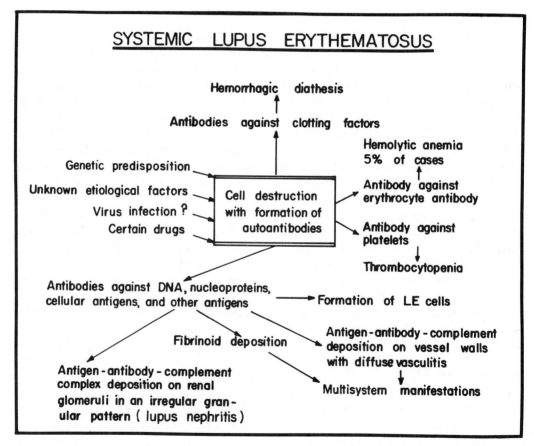

Fɪɢ. 35-6.   Mechanisms of disease in systemic lupus erythematosus.

antibody, complement and fibrinogen. Deposition of immune complexes in the kidney can result in extreme damage, often leading to death as a result of renal failure. (A wide variety of other manifestations are found in SLE, and some of these are depicted in Fig. 35-6.)

For unknown reasons, SLE can be associated with rheumatoid arthritis, Sjögren's disease, hemolytic anemia, pernicious anemia, lymphocytic thyroiditis, and other autoallergic diseases. Occasionally, it may present as another autoallergic disease because of a predominant manifestation such as hemolytic anemia, thrombocytopenic purpura or arthritis. This complex disease exemplifies many of the difficulties encountered by the search for a true understanding of pathophysiological mechanisms.

IMMUNOSUPPRESSIVE THERAPY

In addition to the need for effective immunosuppression in transplantation, treatment of many autoallergic diseases would logically involve attempts to decrease the patient's immunological response against his own autoantigens. Indeed, immunosuppressive therapy has proved to be quite effective in many of them. In general, cell-mediated immunity is easier to control than antibody-mediated immunity, and the IgG type of antibody is much easier to suppress than the IgM type of antibody. Autoimmune diseases involving injury caused by IgG antibody are therefore easier to control than autoimmune diseases associated with damage caused by IgM antibody.

There are numerous ways to produce nonspecific suppression of the immune response, including mechanical removal of lymphoid cells (thoracic duct fistula), destruction of lymphoid cells by ionizing radiation, blockade of antigen binding to cell receptor sites (alpha$_2$ globulin) and administration of a large variety of pharmacological agents.

A review of the pharmacological suppression of immunity is not within the scope of this presentation, but a brief discussion is pertinent. Among the drugs which have been shown to have immunosuppressive activity are the salicylates, benzene, toluene, adrenocortical steroids, purine antagonists, folic acid antagonists, certain antibiotics (such as mitomycin, puromycin, actinomycin, Azaserine, and chloramphenicol), and a variety of other drugs which inhibit protein synthesis, proteolysis or inflammation. The two drugs which are used most often are azathioprine and one of the adrenocorticosteroids, usually prednisone. Each can affect both antigen processing and antibody formation. The adrenal steroids profoundly inhibit the development of the inflammatory lesion. Fortunately, azathioprine and prednisone have different toxic manifestations so that their immunosuppressive effects are cumulative without a similar cumulative increase in the danger of unwanted reactions from either drug during their concurrent administration. Bone marrow suppression is the major clinical toxic complication of azathioprine; gastrointestinal bleeding, neutrophilic dysfunction and delayed wound healing are the major complications of steroid therapy. Unfortunately, both drugs lead to an increased susceptibility to infection.

Heterologous antilymphocyte globulin (ALG) is the most effective immunosuppressive agent known at present. It was first used by Metchnikoff at the turn of the century, but its important effect on immunity was not recognized until recently. In experimental animals, marked prolongation of allograft survival can be obtained when ALG is used in large doses as the only immunosuppressive agent, and its effect is significantly increased when given with prednisone and azathioprine. Many problems still exist in the production of antilymphocyte globulin because it has not been determined which is the best animal in which to produce the antisera and there is as yet no satisfactory method for standardization of the material. In addition, purity and effectiveness of ALG vary considerably because of lack of standard methods of purification and quality control. The mode of action of ALG is multifaceted. Certainly, lymphocyte depletion by direct cytotoxicity of lymphocytes or by the opsonic removal of affected lymphocytes by the reticuloendothelial system is an important mechanism. Lymphocyte inactivation by transformation of the lymphocytes to immunologically immature blast forms may also have an immunosuppressive effect, since it is known that other mitogenic substances such as phytohemagglutinin are immunosuppressive. Antilymphocyte globulin also contains antibodies against histocompatibility antigens and therefore may act as enhancing antibody. Combined therapy is greatly preferable to the use of a single agent because the immunosuppressive effects often become additive, whereas unwanted side effects may be different among the different drugs and thus are not additive. The combination of azathioprine, predisone and ALG has been particularly useful as an immunosuppressive regimen in clinical transplantation.

ANNOTATED REFERENCES

Alexander, J. W., and Good, R. A.: Immunobiology for Surgeons. Philadelphia, W. B. Saunders, 1970. (A concise text containing the important principles and concepts of immunobiology, including chapters on nonspecific immunity, humoral immunity, cellular immunity, immunosuppression and specific negative adaptation, mechanisms of immunological injury, infections, cancer, and transplantation. It provides a useful overview for the student.)

Dacie, J. V.: The Hemolytic Anemias, Congenital and Acquired. Part II. The Autoimmune Hemolytic Anemias. New York, Grune & Stratton, 1962. (A well referenced and complete text. The autoimmune hemolytic anemias are caused by the interaction of antibody with the erythrocyte membrane, usually with the activation of complement and removal of the cells and destruction of the erythrocytes by the reticulodendothelial system.)

Dresser, D. W., and Mitchinson, N. A.: The mechanisms of immunological paralysis. Advances Immunol., 8:128, 1968. (A thorough review of immunological negativity to well defined noncellular antigens with over 250 references)

Dubois, E. L. (ed.): Lupus Erythematosus. New York, McGraw-Hill, 1966. (An excellent review textbook)

Duthie, J. J. R., and Alexander, W. R. M. (eds.): Arthritic Rheumatic Diseases. Baltimore, Williams & Wilkins, 1968. (Contains four excellent chapters on SLE)

Gabrielsen, A. E., and Good, R. A.: Chemical suppression of adaptive immunity. Advances Immunol., 6:91, 1967. (A review of the suppression of immune responses by chemotherapeutic agents with an extensive bibliography)

Glynn, L. E., and Holborow, E. J.: Autoimmunity and Disease. Philadelphia, F. A. Davis, 1965. (A useful review text for the interested student)

Hutchin, P.: Mechanisms and function of immunologic enhancement. Surg. Gynec. Obstet., 126:1331, 1968. (A review relating particularly to organ transplantation; 357 references)

Kaplan, M. H., and Frengley, J. D.: Autoimmunity to the heart in cardiac diseases. Current concepts of the relation of autoimmunity to rheumatic fever, postcardiotomy and postinfarction syndromes and cardiomyopathies. Am. J. Cardiol., 24:459, 1969. (The authors present a current and complete review of the subject with 71 references. Accumulated data are consistent with the hypothesis that rheumatic fever is an autoimmune mechanism caused by the cross reaction of an antigen of the group A streptococcus.)

Levin, J. M., and Boshes, L. D.: Autoimmunity and the central nervous system. Dis. Nerv. Syst., 30:273, 1969. (This brief review relates a variety of autoimmune diseases to their involvement in the central nervous system.)

Miescher, P. A., and Muller-Eberhard, H. J. (eds.): Textbook of Immunopathology. Vol. 2. New York, Grune & Stratton, 1969. (This excellent and well referenced compendium examines immunological dysfunctions and iso- and autoimmune phenomena. It is an excellent reference, highly recommended for further study.)

Paronetto, F.: Immunologic aspects of liver diseases. *In:* Popper, H., and Schaffner, F.: (eds.): Progress in Liver Diseases. p. 229. New York, Grune & Stratton, 1970. (This current review presents a cautious interpretation for the role of immunological injury in the induction and maintenance of liver diseases. The list of 130 references makes it useful to those with further interest in the subject.)

Pirofsky, B.: Autoimmunization and the Autoimmune Hemolytic Anemias. Baltimore, Williams & Wilkins, 1969. (The author has provided an exceptionally complete and well referenced review which will be an excellent source of information for years to come. Familiarity with this text is an absolute necessity for those interested in the hemolytic anemias.)

Rapapport, F. T., and Dausset, J. (eds.): Human Transplantation. New York, Grune & Stratton, 1968. (This comprehensive book presents both the practical and theoretical considerations of transplantation in man. Fortunately, rapid progress in this field will soon outdate the text.)

Rose, N. R., and Witebsky, E.: Thyroid autoantibodies in thyroid disease. *In:* Lavine, R., and Luft, R. (eds.): Advances in Metabolic Disorders. Vol. 3, p. 231. New York, Academic Press, 1968. (Two of the most knowledgeable authorities on thyroid autoimmunity summarize the literature and their own data.)

Russel, P. S., and Monaco, A. P.: The Biology of Tissue Transplantation. Boston, Little, Brown & Co., 1965. (This short book provides an excellent review of the immunology of transplantation through 1964.)

Samter, M., and Alexander, H. L.: Immunological Diseases. Boston, Little, Brown & Co., 1965. (A comprehensive text with 89 contributing authors. This volume represents a compilation of the data available at the time of its publication.)

Schwartz, R. S.: Therapeutic strategy in clinical immunology. New Eng. J. Med., *280:* 367, 1969. (An excellent review of the problem of selective immunosuppression and specific tolerance by one of the founders of the concept of immuno-suppression)

Snell, G. D.: Immunologic enhancement. Surg. Gynec. Obstet., *130:*1109, 1970. (A comprehensive and concise review of the subject by a leading authority, 75 references)

Strauss, A. J. L.: Myasthenia gravis, autoimmunity and the thymus. *In:* Snapper, I., and Stollerman, G. H. (eds.): Advances in Internal Medicine. Vol. 14, p. 24. Chicago, Yearbook Medical Publishers, 1968. (This extensive review presents the clinical and laboratory evidence for myasthenia being a disease of autoimmunity. 260 references)

Unanne, E. R., and Dixon, F. J.: Experimental glomerulonephritis: immunological events and pathogenic mechanisms. Advances Immunol., 6:1, 1967. (A thorough review of the mechanisms of immunological damage in nephritis with 254 references)

Van Rood, J. J.: Leukocyte grouping and organ transplantation. Brit. J. Haematol., *16:*211, 1969. (Unfortunately, histocompatibility typing is not completely perfected at the present time, but its present usefulness and potential future value are both great.)

Walford, R. L.: The Immunologic Theory of Aging. Copenhagen, Munksgaard, 1969. (Interesting reading relating the problem of autoimmunity and aging)

Watson, B. W., and Johnson, A. G.: The clinical use of immunosuppression. Med. Clin. N. Am., *53:*1225, 1969. (A cautious review of the clinical value of immunosuppressive drugs in the autoimmune diseases)

Weigle, W. O.: Natural and Acquired Immunological Unresponsiveness. Cleveland, World Publishers, 1967. (A recent analysis and review)

Appendix A

REFERENCES

The following list of references is provided as a key to the literature in immunology. Several textbooks and review series are cited. (See also original articles cited at the end of each chapter.)

Students are warned not to draw firm conclusions after studying textbooks and reviews alone. In most cases, conclusions reached in such works are generalizations or simplifications which usually explain most, but very rarely all, of the available data.

The clinical application of immunology is in a developmental stage. These references reflect many of the opportunities for further exploitation of the science toward a better understanding of mechanisms of pathophysiology.

General

Abramoff, P., and LaVia, M. (eds.): Biology of the Immune Response. New York, McGraw-Hill, 1970. (An excellent text written by eleven distinguished immunologists)

Dixon, F. J., and Kunkey, H. G. (eds.): Advances in Immunology. New York, Academic Press, 1961-1970. (Twelve volumes of reviews on topics in immunology. All aspects of immunology have been reviewed by scientists most qualified to do so. Each volume contains a list of all the reviews printed to date.)

Good, R. A., and Fisher, D.: Immunobiology. Connecticut, Sinaur Assoc. Inc., 1971. (A very good basic textbook on immunology)

Cell-Mediated Immunity

(Delayed Hypersensitivity)

Lawrence, H. S., and Landy, M. (eds.): Mediators of Cellular Immunity. New York, Academic Press, 1969. (Proceedings of a conference. Thirty-eight distinguished scientists present the theories and facts of cell mediated immunity.)

Transplantation

Moller, G. (ed.): Transplantation Reviews. Baltimore, Williams and Wilkins. (A series of comprehensive and analytical reviews on clinical and experimental transplantation)

Immunology and Cancer

Hellstrom, K. E., and Hellstrom, I.: Immunological enhancement as studied by cell culture techniques. Ann. Rev. Microbiol., 24: 373, 1970. (Comprehensive review of immunological bases of resistance and susceptibility to neoplasms)

Homberger, F. (ed.): Immunological Aspects of Neoplasia, Progress in Experimental Tumor Research. Basel, Switzerland, S. Karger, 1970. (Four reviews concerning immunological bases of resistance to cancer)

Allergy

Lawrence, H. S. (ed.): Cellular and Humoral Aspects of the Hypersensitivity States. New York, Hoeber-Harper, 1959. (A basic reference on the fundamentals of hypersensitivity, current to 1959)

Sherman, W. B.: Hypersensitivity Mechanisms and Management. Philadelphia, W. B. Saunders, 1968. (Text on theoretical and practical basis of the clinical practice of allergy)

Methods

Williams, C. A., and Chase, M. W. (eds.): Methods in Immunology and Immunochemistry. New York, Academic Press, 1967-1971. (Each volume contains descriptions of important laboratory procedures with references to original publications. Each technique is described by authorities who have developed and used the procedure.)

Bloom, B. R., and Glade, P. R. (eds.): In vitro Methods in Cell-Mediated Immunity. New York, Academic Press, 1971. (Proceedings of a combined conference and workshop concerning cell-mediated immunity; includes 335 pages on materials and methods for assessment of cell-mediated immunity)

Campbell, D. H., Garvey, J. S., Cremen, N. E., and Sussdorf, D. H. (eds.): Methods in Immunology. Ed. 2. New York, W. A. Benjamin, 1970. (A laboratory text describing in detail both general and specialized procedures in the immunology laboratory)

Appendix B

The following tables and figures are provided for further reference in the practice of clinical immunology. Although normal and abnormal values could be listed for most of the referred-to assays, it is wiser when analyzing data from these (and many other) assays, to include test material from patient, known normal individuals, and known abnormal individuals. Use the following rule when interpreting tests with these controls:

If the test reconfirms a significant difference between known normal and abnormal test materials, the value for the patient sample is probably reliable. If the expected values for normal and abnormal are not achieved, discard the results! After determining why the tests of knowns were in error, make appropriate adjustments of the test and repeat. This provides the assay with internal control and avoids to a large extent the possibility of misdiagnosis due to laboratory error.

TABLE VIII-2. Characteristics of Human Immunoglobulin

	IgG	IgM	IgA	IgD	IgE
Physicochemical Properties	150,000	890,000	160,000	185,000	200,000
Molecular weight			380,000		
Carbohydrate content (%)	2.9	11.8	7.5	?	10.7
Immunochemical Characteristics					
Heavy chain classes (number of subclasses)	γ (4)*	μ (2)	α (2)	δ (1)	ε (1)
Allotypic markers	Gm	0	Am	0	0
Light chain type	κ & λ	κ & λ	κ & λ	κ & λ	κ & λ
Sedimentation constant	7S	19S	7S, 11S, 18S	7S	8S
Biological Properties					
Serum concentration (mg./100 ml)	1,160 ± 300	100 ± 30	200 ± 60	2.3	0.03
% in intravascular pool	44	70	40	73	50
Half-life (days)	18-23	4-6	4-6	2.8	2.5
Antibody activity	+	+	+	+	+
Complement fixation	+	+	−	−	−
Placental passage	+	−	−	−	− (?)
Skin sensitization					
man	−	−	−	−	+++
guinea pig	+	−	−	−	−

* In normal adults Ig G-1 = 68%, IgG-2 = 20%, IgG-3 = 8% and IgG-4 =4% of total IgG.

TABLE VIII-3.   Interpretation of NBT Test of Blood Neutrophils

PATIENT GROUP	NBT + NEUTROPHILS/cu. mm. NUMBER†	PERCENT
Normal individuals and patients without acute bacterial infection	120-750	3 to 13
Acute infection by bacteria or *C. albicans*	1,115-13,166	12 to 35
Chronic fatal granulomatous disease		< 1

In this test, the number and percentage of circulating neutrophils which will reduce Nitro Blue Tetrazolium (NBT) is determined. The test is very useful in rapid detection of acute bacterial infection and in screening for chronic fatal granulomatous disease. (See Park, B. H., Fikrig, S. M., and Smithwick, E. M.: Lancet, 2:532, 1968.)

† Range

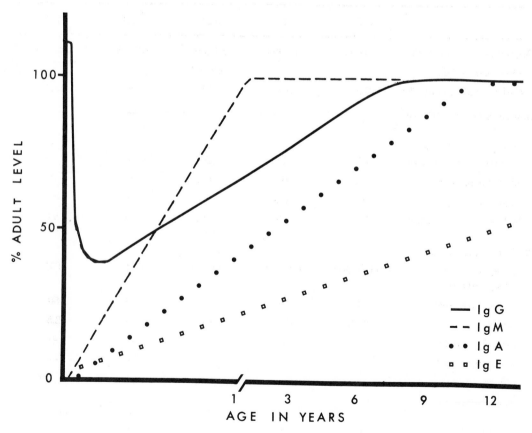

FIG. 8-1.   Quantitation of serum immunoglobulin levels.

TABLE VIII-4. Interpretations of Tests for Adequacy of T-cell Function (See Table 33-10)

1. In-vitro transformation of blood cells in response to phytohemagglutinin (PHA). Large numbers of blastlike cells derived primarily from small lymphocytes are present in phytohemagglutinin-stimulated cultures of human blood cells. Failure of appearance of blast cells within 72 hours of culture incubation is presumptive evidence of T-cell deficiency. Comparison of cultures of patient's cells with those of normal individuals is essential to interpretation of this test.

2. Dose-response analysis of PHA responsiveness in vitro. Blood lymphocytes are cultured with varying concentrations of PHA, and the amount of ^3H thymodine incorporated into DNA at intervals after culture initiation is plotted. Low level responsiveness to PHA is presumptive evidence of T-cell deficiency.

3. In vitro response of blood lymphocytes to allogeneic lymphocytes. T-cells of normal individuals undergo blast cell transformation, and increase synthesis of DNA, RNA, and protein when cultured in the presence of allogeneic lymphocytes. Failure of stimulation indicates T-cell deficiency.

4. Delayed allergy to ubiquitous allergens; SK-SD, mumps, Candida, trichophyton, PPD.

It is considered highly unlikely that an individual will fail to express delayed hypersensitivity to all of the five listed skin test antigens merely because of lack of prior exposure. Typical positive reaction to one or more of these shows that this expression of T-cell function is intact. Lack of reactivity to all five suggests, but is not diagnostic for, T-cell deficiency.

5. Development of contact allergy to 2,4 dinitrochlorobezene (DNCB). This test is most appropriate when individuals fail to demonstrate delayed hypersensitivity (as tested in 4, above). Absence of contact hypersensitivity in individuals sensitized with 2,4 DNCB is strong evidence of T-cell deficiency.

6. Small lymphocyte count.

7. Capacity to reject allograft of skin. This test is somewhat redundant of less risky tests suggested above. Because of the possibilities of transferring infectious agents to the patient and of sensitizing the patient to transplantation antigens, grafting for diagnostic purposes should be done only in most unusual circumstances.

8. Presence of an abundant cell population in deep cervical areas of lymph nodes following antigenic stimulation. Examine H and E and methyl green pyronine stained reactions of axillary or inguinal lymph node. The node should be stimulated by injection of an immunizing dose of toxoid (tetanus or diphtheria) 5 to 6 days prior to biopsy. The antigen should be injected subcutaneously in the arm or leg drained by the node to be biopsied.

9. Vigorous defense against fungi, virus, and facultative intracellular pathogens. Evidence of T-cell deficiency is to be found in the patient's medical history. Recurrent or chronic infection or unusually severe infection by a given agent or host of agents within the groups referred to is presumptive evidence of T-cell deficiency.

TABLE VIII-5. Interpretation of Tests to Evaluate B-cell Function (See Table 33-11).

1. Quantitation of serum immunoglobulin levels (See Table VIII-2 and Figure VIII-1).

2. Evaluation of fractional catabolic and/or synthetic rates of individual immunoglobulin. Derive from steady state level in serum and half life (Table VIII-2).

3. Levels of serum antibodies to widely distributed antigens, e.g., isohemagglutinins, ASO, antidiphtheria toxin (negative Schick test), etc.

Normal antibody levels vary with respect to nature of antigen, age, race, blood type, etc. Absence of detectable antibody is strong presumptive evidence of B-cell deficiency. Levels of detectable antibodies should be compared with values obtained by testing sera from healthy individuals matched for age, race, blood type, etc.

4-6. Quantitation of antibody in sera taken at regular intervals after active artificial immunization. The immune response is quantitated by measuring antibody levels in serum taken before antigen injection and at intervals of 5 to 7 days after antigen injection. The curve generated by plotting antibody level against time after antigen exposure is compared with standard curves.

7. Salivary IgA is quantitated and analyzed for antigenic and physicochemical properties. (See text, Chapter 33.)

8. Quantitation of serum immunoglobulin subclasses (See Table VIII-2).

9. Defense against systemic and respiratory infection by bacterial agents including pneumococci, streptococci, *Haemophilus influenza*, meningococci, and *Pseudomona aeruginosa*. Frequent or recurrent infection by these agents is presumptive evidence of B-cell deficiency.

10. Lymph node morphology. Absence of germinal centers in the cortical region and/or plasma cells in the medulla of a lymph node (prepared as in item 8, Table VIII-4) may be taken as confirming evidence of B-cell deficiency.

TABLE VIII-6. Properties of Normal Human Complement System Proteins. (See Chapter 34)

PROPERTIES		C1q	C1r	C1s	C1INH*	C2	C3	C3INA†	C4	C5	C6	C6INA†	C7	C8	C9
Serum Concentration (µg/ml.)	A	190	–	22	–	20–40	1200	–	430	75	–	–	–	10	10
	B	100–200	–	–	12–30	8–12	900–1500	–	200–600	51–99	–	–	–	–	1–2
Sedimentation Coefficient(s)	A	11.1	7.0	4.0	–	5.5	9.5	–	10.0	8.7	5–7	–	5–6	8.0	4.5
	B	11.0	7.0	4.0	4.0	5.5	9.0	55	10.0	8.7	6.6	6.6	5.6	8.5	4.5
Molecular Weight (approximate)	A	400,000	–	79,000	–	117,000	185,000	–	240,000	–	–	–	–	150,000	79,000
	B	400,000	–	110,000	90,000	120,000	185,000	–	230,000	–	–	–	–	–	75,000
Relative Electrophoresic Mobility	A	$\gamma 2$	β	$\alpha 2$	–	$\beta 2$	$\beta 1$	–	$\beta 1$	$\beta 1$	$\beta 2$	–	$\beta 2$	$\gamma 1$	α
	B	$\gamma 2$	β	$\alpha 2$	2	$\beta 2$	$\beta 1$	β	$\beta 1$	$\beta 1$	$\beta 2$	$\beta 2$	$\beta 2$	$\gamma 1$	α

A, Muller Eberhard, H. J.: Complement. Ann. Rev. Biochem., 38:389, 1969.
B, Ruddy, S., and Austen, K. F.: Inherited abnormalities of the complement system. Prog. Med. Genet., 7:69, 1970.
* INH, Inhibitor
† INA, Inactivator

TABLE VIII-7.

Antigenicity associated with an individual's lymphocyte is characterized by use of antisera which, in the presence of complement, kill lymphocytes bearing a given antigen. This table illustrates the use of this assay for histocompatibility matching within a sibship. The assay has also been used to type and match unrelated donors and recipients, though with less satisfactory results.

TABLE VIII-7. Example of Histocompatibility Typing Using a Cytotoxicity Test (See Chapter 35) Given blood group O parents with HLA genotypes, 1,5/2, 12 (mother) and 1,7/3,8 (father), the possible HLA genotypes of their offspring are: 1,5/1,7; 1,5/3,8; 2,12/1,7; and 2,12/3,8. A typing profile for a prospective recipient in such a family with five children might be as follows:

| | | Antigens in First Sublocus | | | | | | Antigens in Second Sublocus | | | | | Number of Antigenic Incompatibilities | HLA Phenotype |
		1	2	3	9	10	11	5	7	8	12	13		
Recipient		+	−	+	−	−	−	+	−	+	−	−	−	1,3,5,8
Sibling No.	1	+	−	−	−	−	−	+	⊕	−	−	−	1	1,5,7,
	2	+	−	+	−	−	−	+	−	+	−	−	0	1,3,5,8*
	3	+	⊕	−	−	−	−	−	⊕	−	⊕	−	3	1,2,7,12
	4	−	⊕	+	−	−	−	−	−	+	⊕	−	2	2,3,8,12
Mother		+	⊕	−	−	−	−	+	−	−	⊕	−	2	1,2,5,12
Father		+	−	+	−	−	−	−	+	+	−	−	1	1,3,7,8

+, positive reaction of lymphocytes with antisera having the designated specificity
−, negative reaction
⊕, incompatibility of antigen with donor
*, Sibling No. 2 has HLA identity with the recipient, making him an ideal donor.

TABLE VIII-8. Selected Autoallergic Disorders of Man

This table includes several of the diseases of man with clearly demonstrable autoallergic components. Many other diseases (which are not included here) may show some autoallergic phenomena.

TABLE VIII-8.

Organ System and Disease	Probable Antigen	Experimental Model	Comment
Hematological Diseases			
Autoimmune hemolytic anemias	Rh, I, i, P	Rabbit immunized with *Listeria monocytogenes* (cold agglutinin disease), NZB mice, dog	Three distinct types, depending on antigen involved. See text.
Idiopathic thrombocytopenic purpura	Platelet membrane		IgG antibody detected by antiglobulin consumption test
Pernicious anemia	Intrinsic factor parietal cell microsomal antigen	No satisfactory model	Disease may be responsive to immunosuppressive therapy.
Neurological Diseases			
Postinfectious encephalomyelitis	CNS myelin	Experimental allergic encephalomyelitis	Damage probably caused by cell-mediated immunity, but circulating antibody can cause demyelination of cells in tissue culture.

(Continued on p. 754)

Table VIII-8 *(Continued)*

Organ System and Disease	Probable Antigen	Experimental Model	Comment
Multiple sclerosis	Peripheral nerve myelin	Allergic neuritis	Disease not proved to be autoimmune in nature.
Guillain-Barré Syndrome and Landry's paralysis	Myelin	Allergic neuritis	
Peripheral neuritis	Peripheral nerve myelin	Experimental allergic neuritis	
Renal Diseases			
Glomerulonephritis	Glomerular basement membrane (GBM) Possible cross reactions with streptococcal antigens	Production of disease in sheep by immunization with GBM in Freund's adjuvant. Several other models of renal immunopathology such as Masugi nephritis, Dixon model, etc.	Many variants of nephritic disease not well understood
Goodpasture's Disease	Basement membranes of lung and kidney		Cross reacting antibody
Endocrine Diseases			
Lymphocytic thyroiditis	Thyroglobulin micrososomal antigen Second colloid antigen	Thyroiditis in obese chickens Allergic thyroiditis	Antibody demonstrable by a variety of techniques by a variety of techniques.
Thyrotoxicosis	Long acting thyroid stimulator (LATS) antigen on cell surface		LATS is an IgG that stimulates thyroid activity. LATS cross reacts with thyroid tissues from other species. Etiological significance not known.
Gastrointestinal Diseases			
Portal cirrhosis	Nuclei, hepatocytes	No satisfactory model	IgA antibody involved?
Chronic active hepatitis	Ductal cells, smooth muscle, mitochondria, nuclei, hepatocytes	No satisfactory model	IgG antibody. Associated with hypergammaglobulinemia
Biliary cirrhosis	Mitochondria Ductal cells	Damage by rabbit antimitochondrial antibody in the rat	IgM antibody
Ulcerative colitis	Colon mucosal cell	Rabbit and rat models immunized with colon cells or intestinal bacteria	Responds to immunosuppressive therapy. Damage probably cell-mediated. May be triggered by immune reaction to exogenous antigens.

TABLE VIII-8 *(Continued)*

ORGAN SYSTEM AND DISEASE	PROBABLE ANTIGEN	EXPERIMENTAL MODEL	COMMENT
Miscellaneous Diseases			
Sympathetic ophthalmia	Uveal antigen (pigment?)	Homologous uveal tissue in adjuvant induces uveitis in guinea pigs.	Follows perforating wounds of the ciliary body or root of iris. Disease benefited by steroids.
Phacogenic uveitis	Lens material	Allergic uveitis can be produced in many animals.	Follows injury to lens. Not clear how this causes the uveitis.
Postcardiotomy syndrome	Heart muscle	Transient lesions can be produced in rats, rabbits, and other species with difficulty.	Associated with increase in antibody to heart muscle. Role of antibody not well defined.
Pemphigus	Intercellular antigen in stratified squamous epithelium.	Rabbit model	Antibody titers may fluctuate with activity of the disease.
Bullous pemphigoid	Basement membrane area in skin	None	Antibody does not cross react with kidney basement membrane.
Myasthenia gravis	Striated muscle	Rabbits immunized with muscle antigen complexed to dinitrophenol.	Antibody reacts with myoid cell of thymus. Disease improved by thymectomy.
Infertility	Sperm specific	Experimental autoimmune aspermatogenesis. Naturally occurring human counterpart has not been described.	Sperm agglutinins in male can cause infertility.
Systemic lupus erythematosus	Nucleoproteins, DNA, histones, nuclear glycoproteins, membranous organelles, many other antigens (See text.)	NZB mice	Complex multisystem disease (See text.)
Rheumatoid arthritis	IgG	None entirely satisfactory	Rheumatoid factor is an IgM reactive against native IgG. Some have Gm specificity. Pathogenesis of the disease not well understood.
Rheumatic fever	Sarcolemma and subsarcolemmal sarcoplasm in cardiac myofibrils and skeletal muscle, smooth muscle of vessel walls and endocardium.		Exciting antigen is the M protein of group A streptococci. A second antigen may be involved. Certain streptococcal antigens also cross react with antisera against histocompatibility antigens of man.

Index

Page numbers in *italics* indicate figures; "t" indicates tabular matter.

Abscess, formation of, 675
Absorption, and digestion, relationships between, *449*
  decreased, causes of, 450
  systemic effects of, 450
  and digestion, 423
  and intestinal bacterial flora,
    425-426
  and intestinal blood flow, 425
  and intestinal epithelium,
    423-425
  intestinal mucosal contact time
    influencing, 425
  intestinal surface area influ-
    encing, 423
  ion exchange in, 430
  mechanisms of intestinal mem-
    brane transport influenc-
    ing, 426-431
  overload, causes of, 450
  pathway of, 428
  simple diffusion in, 430
  solvent drag in, 430
  transport processes in, 428-429
    active, and facilitated
      diffusion, 429-430
Acetazolamide, in sodium reabsorption inhibition, 224
Acetyl-beta-methacholine, in
  achalasia, 396
Acetylcholine, as neurotransmitter, 73, 654
  formation of, 639
  and gastric peristalsis, 400, 617
  liberation of, 639
    in botulism, 640
    in myasthenia gravis, 639
    in myasthenic syndrome, 640
  and muscle fibers, 631
  and neuroeffector system,
    607-608
Acetyl-CoA, in adipose tissue
  metabolism, 344
  in hepatic metabolism,
    341-342, 342-343
  in muscle metabolism, 339
Achalasia, in Chagas' disease,
  396-397
  of esophagus, 396-397
Achlorhydria, 418
Acetylsalicylic acid, as anti-
  inflammatory agent, 679
Acid(s), gastric mucosal barrier
  destruction by, 416-417

hemorrhage associated with,
  418
properties of, 416, 543
amino, 16
  absorption of, 437
  for gluconeogenesis, 343
  transport of, 436
    congenital defects in, 437
    systems of, 437
bile. *See* Bile, acids
body, buffer pairs in regulation
  of, 251, 252
and buffered solutions, 250
carbonic, in body fluids, 251
  dissociation of, 250
  in metabolic process, 253-254
chenodeoxycholic, 444, *445*,
  447
cholic, 444, *445*, 447
conjugate base of, 249-250
deoxycholic, 446, 447
dissociation of, 249-250
duodenal, neutralization of,
  414
endogenous, production of,
  253-258
  components of, 253-255, *256*
  excretion of, renal, 255-256,
    257
    adrenal cortical hormones
      and, 263
    defective, and metabolic
      acidosis, 260-261
    effects of Na^+, K^+ and Cl^-
      balances on, 262
    measurement of, 276-277
    normal values for, 277
  fixed, production of, metabolic
    processes in, 254-255
folic, absorption of, 438
  deficiency of, causes of, 502
    effects of, 502-503
    evaluation of, 503
  in heme and globin synthesis,
    502, *498*
  sources of, 502
and gastric mucosal barrier,
  414-415
  back diffusion through, 415,
    417-418
gastric secretion of, tests of,
  409-412
hydrogen ion concentration
  of, 249

10-hydroxy stearic, 443-444
lactic, 159
lithocholic, 446
non-volatile, production of,
  metabolic processes in,
  254-255
nucleic, cardiac synthesis of,
  9, 13, 16
  defective, 17-18
organic, excessive, consequences of, 436
in peptic ulceration, 415-416
sialic, in human chorionic
  gonadotropin, 358
titratable, in urine, determination of, 276
  excretion of, 257
types of, examples of, 250
uric, serum, levels of, elevation
  of, 192
Acid-base balance, defective,
  in cardiac disease, 11
  disturbances of, 192
  clinical evaluation of, 263-266
  principles of therapy for,
    265-266
  estimation of, in assessment of
    respiratory disease, 141
  measurement of, in blood,
    274-276
  in urine, 276-277
  tests of, 274-277
Acidemia, buffer pairs counteracting, 252
  systemic, renal response to,
    260
Acidity, duodenal, 413
Acidosis, diabetic, hypokalemia
  in, 230
  inflammatory defect in, 680
  experimental, calcium excretion
    in, 261-262
  hemoglobin affinity for oxygen
    in, 148
  metabolic, 259-263
    acute, characteristics of,
      259-260
  due to decreased acid excre-
    tion, therapy for, 265-266
  due to overproduction of
    acids or losses of base,
    therapy for, 265
  cause(s) of, 259
    carbohydrate malabsorption
      as, 436

757

Acidosis (*continued*)
 chronic, characteristics of,
  260
  therapy for, 266
 defective renal acid excretion
  causing, 260-261
 2,3-diphosphoglycerate
  concentration in, 150
 renal response to, 259-260
 respiratory effects of, on aortic
  and carotid bodies, 116
 ventilation adaptations to,
  119, *118*
 organic, serum electrolytes in
  evaluation of, 264-265
 renal, chronic, calcium carbonate as buffer in, 261
 tubular, distal, 261
  proximal, 261
 respiratory. *See* Hypercapnia
Aciduria, paradoxical, in primary aldosteronism, 329
Acromegaly, 298
ACTH, chemical nature of, 290
 deficiency of, 291
 extra-adrenal effects of, 290
 in growth of adrenal cortex,
  293
 production of, by lungs, 181
  by nonpituitary tissue, 291
  increased, in adrenal insufficiency, 289, 297
 release of, diurnal, 288
  in response to stress, 288
  with hypothalamic lesions,
   288
 secretion of, abnormalities of,
  290-291
  corticosteroid suppression of,
   288-289
  cycle of, 289
  hypothalamic control of,
   287-290
Actin, 8, 9, 17
Addison's disease, effect on
  ascending limb, 327-328
 effect on distal tubule, 328
 effect on proximal tubule, 327
 etiology of, 296, 327
 hyperpigmentation in, 297
 renin production in, increased,
  328
 stress as factor in, 296-297
 therapy in, 296-297
 water metabolism in, abnormal, 322-323
Adenocarcinoma, follicular metastatic, of thyroid, 316
Adenomatosis, peptide-secreting,
  386
 features of, 412-414
 "overflow diarrhea" in, 432
Adenosine triphosphate, 73
 in cardiac failure, 17

 in coupling, 9
 formation of, 337
 storage in mitochondria, 8
Adenosine triphosphatase,
  activity of, 9
 in cardiac failure, 17, 18
 stimulation of, 9
Adenyl cyclase system. *See*
  Cyclic 3'5' AMP.
ADH. *See* Hormone(s), antidiuretic
Adipose tissue, metabolism of,
  343-344, *343*
Adrenal(s), congenital hyperplasia of. *See* Hyperplasia, congenital adrenal
 cortex, aldosterone production
  by, 295
  atrophy of, in corticosteroid
   therapy, 289
  corticosteroidogenesis by,
   292-293
  effects of metyrapone on,
   289-290, *290*
  growth of, and ACTH, 293
  secretion by, 291-297
   hormonal, 287
   normal, 291-293, *292*
  effects of estrogen upon, 293
  effects of metyrapone upon,
   293
  functions of, in normal and
   disease states, 382t
  hormonal secretion of, 287
   regulation of, factors influ-
    encing, 287
  insufficiency of, in corticosteroid therapy, 289
   etiology of, 296
   hyperpigmentation in, 297
   increased ACTH production
    in, 289
   stress as factor in, 296-297
   therapy in, 296-297
  -pituitary axis, 287
   regulatory mechanisms of,
    287-299
  steroids. *See* Steroid(s),
   adrenal
  tumors of, androgen-producing,
   295
   primary, and Cushing's
    syndrome, 294
Adynamia episodica hereditaria,
  642
Afibrinogenemia, 531
 congenital, 533
Agammaglobulinemia, acquired,
  699-701, *700*
 Bruton-type, 694-695, *696*
 Swiss-type, 698, *697*
Age, as factor in urinary concentrating ability, 237-238

 as factor in urinary diluting
  ability, 243
Aging, and immunity, 708
Agranulocytosis, familial,
  benign, 523
 infantile genetic, 523
Air, flow, in airways, pressures
  influencing, 136-137
 rate of, expiratory resistance,
  test of, 137
  maximum mid-expiratory,
   test of, 137
 inspired, distribution of, in
  lungs, 132, *132*
  effect of localized airway
   obstruction on, 132
Airways, air flow in, obstruction
  of, 137
 pressures influencing, 136-137
 obstruction of, tests of,
  137-138
Albumin, biosynthesis of, 208
 catabolism of, 210, 211
  in nephrotic syndrome, 211,
   210t
  during inflammation, 678
 excessive loss of, in urine,
  effects of, 211-213
 excretion of, in exercise,
  202-203
  and posture, 203-204
 in urine, 210-211
 loss of, in urine, in nephrotic
  syndrome, 211
 metabolism of, 207-211
  in nephrotic syndrome, 210,
   210t
  in normal subjects, 208 210t
 plasma level of, low, causes
  of, 211, 211t
  in nephrotic syndrome, 211
 serum, concentrations, 191
  in ascites, 462
 sieving coefficient of, 195-197,
  196
  in massive proteinuria, 206,
   196
  hepatic, 208, 212
  in nephrotic syndrome, 211
 tubular reabsorption of, 199,
  201, 211
 in urine, in glomerular
  disease, 206
Alcohol, and antidiuretic hormone secretion, 320
Aldosterone, action of, levels of,
  297
 as influence on potassium
  secretion, 227-228
 deficiency of, in Addison's
  disease, 296, 327, 328
 excess of, conditions causing,
  295
 in primary aldosteronism, 328

insufficiency of, and abnormal water metabolism, 322
secretion of, excessive, 263
in sodium balance maintenance, 220
and tubular reabsorption of sodium, 324
Aldosteronism, primary, effect on ascending limb, 329
effect on distal tubule, 329
effect on proximal tubule, 328-329
etiology of, 328
and plasma renin activity, 59
plasma volume in, 52, 51t
renin production in, decreased, 329-330
Alkali. *See* Base(s)
Alkalosis, as factor in hypokalemia, 229-230
hemoglobin affinity for oxygen in, 148
hypokalemic, role of chloride in, 230
metabolic, 262
acute, therapy for, 266
chronic, therapy for, 266
2,3-diphosphoglycerate concentration in, 150
respiratory, acute, 258-259
chronic, 259
Allergy(ies), reactions to, in lungs, 180
Allograft(s), renal, rejection of, following "acceptance," 726
immediate, 726
long-term, 726
See also Transplantation
Altitude, high, acclimatization to, regulation of respiration in, 118-119, *118*
Amenorrhea, -galactorrhea syndrome, 313-314
investigation of, 366
primary, in failure of process of pubescence, 367
and Turner's syndrome, 366-367
secondary, 367
Aminoglutethimide, 293-294
Δ-Aminolevulinic acid synthetase, 497
Ammonia, "intoxication," in hepatic encephalopathy, 465-467
levels of, factors increasing, 466
metabolism of, 465
production of, factors decreasing, 466
secretion of, in urine, 257
sources of, 465
in urine, measurement of, 276

Amphenone, 293
Amyloidosis, in nephrogenic diabetes insipidus, 240
Analbuminemia, 315-316
Analysis, as sensory function of nervous system, 580
in sensory and motor mechanisms, 574
Anaphylatoxin, 672
Anaphylaxis, IgE immunoglobulins in, 179
kinins in, 26
vasoactive substances released in, 177
Anderson's disease, 469
Androgen(s), adrenal tumors producing, 295
aromatization of, 356
definition of, 355
fetal activity of, 360-361
metabolism of, effect of thyroid hormones upon, 311t
production of, excess, and amenorrhea, 367
Androstenedione, 356
and estrogen production, 367
Anemia(s), aplastic, stem cell defect in, 496
deficient tissue oxygenation in, 32, 146, 150
diagnosis of, differential, chemical tests in, 563t
laboratory techniques in, 492
of endocrine deficiency, 495
and erythropoietin abnormalities, 494
etiology of, 491
hemolytic, 491
abnormal ion transport in, 508-509
acquired, complement abnormalities in, 724-726
autoimmune, 509-510
cold agglutinin, pathophysiology of, *741*
chemical agents inducing, 510
"cold antibody," 725-726, 740
diagnostic approach in, 511
hereditary nonspherocytic, 508
microangiopathic, 487, 510
and intravascular clotting, 536-537
physical findings in, 510-511
symptoms of, 510
"warm antibody," 725, 740
high output demands of, 14
iron deficiency, diagnosis of, 505-506
symptoms of, 491
microcytic hyperchromic, 679

pernicious, in gastric atrophy, 418, 419
vitamin B_{12} deficiency causing, 502
in renal hypertension, 47
sickle-cell, impaired urinary concentrating ability in, 240-241
symptoms of, 491
vitamin B_{12} deficiency, 491
vitamin C deficiency, 491
Anesthesia, and arterial pressure changes, 72
Angina pectoris, 18
as result of cardiac dilatation, 13
blood flow in, 36
of ischemic heart disease, 18-19
lactate accumulation in, 16
Angioneurotic edema, hereditary, 719
Angiotensin, production of, in syndrome of juxtaglomerular hyperplasia, 330
in renovascular hypertensin, 47
I, conversion to angiotensin II, in lungs, 177
II, in renal pressor system, 55, 58
Anosmia, in hypogonadotropic hypogonadism, 363, 367
Antibody(ies), in immunological injury, 731-733
preformed, and hyperacute graft rejection, 733
secretory, 179
Anticoagulant(s), circulating, 537
and synthesis of vitamin K-dependent factors, 535
Antigen(s), histocompatibility, 736
in immunological adaptation in transplantation, 738
Antihistaminics, as anti-inflammatory agents, 679
α_1-Antitrypsin, deficiency of, and emphysema, 173-174
Anxiety, 25
Aorta, 11
chemoreceptors of, 115-116
coarctation of, 15
hypertension in, 47-48
"compression chamber" of, 43
denervation of, in hypercapnia, 116
in metabolic acidosis, 116
distensibility of, equation defining, 43
factors influencing, 43
stenosis of, 11, 13
stimulation of, results of, 116

Aorta (*continued*)
  valves of, in cardiac arrythmia, 12
  "windkessel effect" of, 43
Apnea, following deprivation of chemoreceptor stimuli, 117
Apneusis, 112
Apraxia, respiratory, 120
Arrhythmia(s), cardiac, as cause of decreased cardiac performance, 11, 15
  electrical conversion of, 97-98
  in ischemic heart disease, 18
  isometric contractions in, 12
  neural structures in, 612
Arteriole(s), autonomous myogenic activity in, 600
  in control of blood flow and filtration pressure, 50-51
  coronary disease of, 19
  dilatation of, in reactive hyperemia, 75-76
  sensory factors in, 73
  disease of, 19
  resistance to blood flow in, 68
Arteriosclerosis, 11, 18, 33
Artery(ies), blood volume in, 49-50
  carotid, 115
    chemoreceptors of, 115
    denervation of, in hypercapnia, 116
    in metabolic acidosis, 116
    and hypercapnia, 116
    local stimulation of, results of, 116
    sinus mechanoreceptor in, 70-72
  chemoreceptors of, 605-606
  coronary, disease of, 18
    inadequate perfusion of, 16
    insufficiency of, 18
    occlusion of, 13
    diastolic pressure of, and plasma volume in hypertension, 52-53, *53*
    insufficiency of, 33
    mechanoreceptors of, 603-604
    myocardial, ischemia of, 13
    pressure of. *See* Blood pressure, arterial
Ascites, albumin concentrations in, 462
  fluid obstruction in, 461
  formation of, "outflow block" theory of, 461
  pathophysiology of, 461-464
  and portal venous pressure, 462
  renal excretion in, 461
Aspergillosis, bronchopulmonary, 180

Aspirin. *See* Acetylsalicylic acid
Asthma, bronchial, release of vasoactive substances in, 177
Ataxia-telangiectasia, 179, 698
  scleral telangiectases characteristic of, *698*
Atherosclerosis, manifestations of, 33
  risk factors in, 33-34
  vessels involved in, 33
Athetosis, 644
Atrioventricular block, complete, 7
  development of, 86
Atrioventricular bundle. *See* His, bundle of
Atrium(a), 5
  conduction in, pathways of, 86
  velocity of, 86
  extrasystoles of, 15
  fibrillation of, 15
  and flutter of, compared, 94
  flutter of, causes of, 91
  and fibrillation of, compared, 94
  myxomatous tumors of, 14
  pacemaker activity in, 83
  septal defect of, 15
Autoallergy, diseases of, 739-742, 753-755t
  immunosuppressive therapy in, 742-744
  possible mechanisms in, 739, *739*
  mechanisms of, and transplantation, 731-745
Autoerythrocyte sensitization, 544
Autoimmunity, diseases of. *See* Autoallergy, diseases of
Automaticity, cardiac, 6, 7, 83
Axis, pituitary-adrenal, regulatory mechanism of, 287-299
  pituitary-thyroid, regulatory mechanisms of, 301-318
Axon reflex, 73
Azathioprine, in immunosuppressive therapy, 743
  and leukocyte migration defect, 680

Bachmann's bundle, pacemaker activity in, 83
Bacterium(a), colonic, 426
  metabolism of selected compounds by, 427t
  intestinal, 425-426
  overgrowth of, tests of, 480
  pathological importance of, 426
Ballismus, 643-644

Barbiturate(s), toxication, 158-159
  ventilation in, 131
Baroreceptor(s). *See* Mechanoreceptor(s)
Base(s), and buffered solutions, 250
  conjugate, of acid, definition of, 250
  hydrogen ion concentration and, 249
Beat(s), cardiac, reciprocal, re-entry as mechanism for, 88, *89*
Beriberi, 19
Beta-adrenergic receptor site, hyperdynamic responsiveness to, in hypertension, 32, 45
3-Beta-ol-dehydrogenase, deficiency of, 295
Bicarbonate, as part of buffer pair, 251-252
  concentration of, in intestinal tract, 432
  in metabolic process, 253-254, 255
  reabsorption of, 256-257, 259
  transport of, mechanisms of, 432
  in urine, increased, and hydroxyl ion loss, 326-327
  measurement of, 276
Bile, acids, abnormalities, in liver disease, patholophysiology of, 467-468
  absorption of, 444-446
  to cholesterol ratio, in bile, 468
  colonic secretion of, 447
  conjugated, absorption of, 444-446, *445*
  deficiency of, 441
  causes of, 441
  consequences of, 441
  intraluminal, 447
  therapy for, 443
  enterohepatic circulation of, 440-441, 446-447, *446*
  in fat digestion, 440, *439*
  free, 447
  glycine-conjugated, in bile acid malabsorption, 448
  malabsorption of, causes of, 447
  diagnosis of, 448
  metabolism of, disturbed, in liver diseases, 468
  pool, in gallstone disease, 468
  measurement of, 440-441
  secondary, formation of, 446

synthesis of, regulation of, 447
components of, 407
and aqueous component of pancreatic juice, compared, 407
duct(s), common, 401
system of, motor disorders of, 401-402
gallbladder, in gallstone disease, 468
human, bile acids in, 444, *445*
salts, in gastric mucosal barrier destruction, 417
Biliary tract, motor disorders of, 401-402
Bilirubin, binding by albumin, 457
conjugation of, 459
early-labeled, 456
and erythrocyte destruction, 512
glucuronide, 457
excretion of, 512
defects in, 458-459
hepatic uptake of, impaired, 457
intracellular conjugation defects of, 457-458
metabolism of, 455-456, *456*
inborn errors of, classification of, 456
and jaundice, 455-459
overproduction of, 456-457
Bleeding time, long, causes of, 539
Block, atrioventricular, complete, 7
development of, 86
cardiac, as cause of re-entry of impulses, 87-88
"entrance," 85
unidirectional, 87
in atrial flutter, experimental, 91-93, *92*
in study of cause of ventricular premature contractions, 96-97
Blood, acid-base measurements in, 274-276
adrenergic substances in, regulating smooth muscle contraction, 25-26
carbon dioxide content of, total, measurement of, 275
carbon dioxide tension of, calculation of, 275
cells. *See* Cell(s), blood
circulating, normal hematological indices of, 563t
clots, abnormal, in pathological fibrinogenolysis, 554

dissolution of, plasminogen activators in, 552-553
fibrin, and degradation products, 550
formation of, 527-528
retraction of, 538
clotting, disorders of, acquired, 534-537
diagnosis of, 530-531
hereditary, 532-534
factors. *See* Factor(s), clotting
factors promoting, 530
inhibition of, 530
intravascular, 535-536
pathway of, extrinsic, 528-529, *528*
disorders in, test of, 531
intrinsic, 529-530, *529*
disorders in, tests of, 530-531
pulmonary factors in, 175, *175*
tests of, normal values for, 564t
flow, autonomic control of, in colon, 628-629
in small intestine, 425, 628
in stomach, 418, 626-628
autoregulation of, 30-31, 47
in coarctation of aorta, 48
in kidney, 51
reactive hyperemia as, 75-76
collateral, 19
control of, intrarenal mechanisms in, 50-51
coronary, 11
neural regulation of, 609
in myocardial failure, 16, 613
maldistribution of, neurogenic factors in, 28-30
measurement of, 24, 42-43
metabolic factors influencing, 31-32
neural regulation of, during exercise, 600-601
mechanisms in, 599-600
pulmonary, distribution of, 132-133, *132*
redistribution of, in posture changes, 76-77
renal, 36-37
decreased, causes of, 225
measurement of, 271
resistance to,
by organs, 24
rheologic and mechanical factors altering, 35-36
structural factors altering, 33-35
vascular, 23-24
factors determining, 27-28

thyroid, 304
vasoconstrictor effects determining, 600
velocity of, 10-11
See also Circulation
-gas equilibrium, pulmonary, assessment of, 140
hemoglobin content of, 491
incompatible, transfusion of, 536
osmolality of, and antidiuretic hormone release, 242
increase in, 68
oxygen content of, 491
pH in, measurement of, 275
plasma. *See* Plasma
platelets. *See* Platelets
sugar, normalization of, in diabetes mellitus, 352
viscosity of, 10, 42, 68
altered, and vascular resistance, 35
volume, cardiac, 10
central, in renovascular hypertension, 46
loss of, 80-81
chemoreceptor activity in, 606
shock associated with, 81
measurement of, 216
whole, clotting time of, test, 531
Blood pressure, arterial, as determinant perfusion, 23
central nervous system in maintenance of, 600
during exercise, 600-601
changes in, 67
posture changes causing, 72, 76-77
decreasing, cardiovascular factors in, 67-70, *69*
environmental stimuli in, 76-77
nervous factors in, 70-76, *71*
pathophysiological states in, 77-81
development of, 23
and hypertension, 41-66
mean, definition of, 67
estimation of, 67
and mechanoreceptor stimulation, 604
neural control of, 60-61
and plasma renin activity, 57
relationship to cardiac output and total peripheral resistance, 42-43
Blood vessels, anomalies of, acquired, 34-35
congenital, 34
autoregulation by, 30-31
changes in, and blood flow, 42-43

Blood vessels (*continued*)
dilatation of, sensory factors in, 73
elasticity of, as influence on arterial pressure, 70
large, role of, in hemodynamics, 43
local depressor responses of, sensory factors in, 73
permeability of, increased, in inflammatory response, 671-672, *673*
physical obstruction of, 36
resistance of, as determinant of perfusion, 23-24
system of, intravascular volume and, 49-50
Blue diaper syndrome, 437
Blushing, 74
Body, composition of, fluid, 215, *216*
defenses of, lymphocyte B-cells as, 694
lymphocyte T-cells as, 693
specialization of, 684-685
Bohr effect, 148
Bone(s), in calcium balance maintenance, 330
and hyperparathyroidism, 333
and hypoparathyroidism, 332
marrow of, aspirates of, in diagnosis of thrombocytopenia, 539
cell content of, measurement of, 521-522
granulocyte reserve in, 515
leukocytes in, 520-522
normal differential proportion of cells in, 563t
production rate of, 521
osteomalacia of, in medullary thyroid carcinoma, 333-334
Botulism, 640
Bowel, functional, syndrome of, 404
small. *See* Intestine(s), small
Bradycardia, 15
Bradykinin, 26-27, 75
in anaphylaxis, 177
in lungs, 177
Brain, organization of, and regulation, 569-582
stem, cellular organization of, 577
reticular formation of, 580
Breath, holding of, anatomical separation of pathways and functional integration in, 111-112
Breathing. *See* Respiration
Breathing capacity, maximal, test, 138

Budd-Chiari syndrome, portal hypertension in, 460
Buffer(s), activity of, 250
body, mechanisms of, 249-268
normal, 251-253
total capacity of, 251-252
capacity, fixed acid load influencing, 255
pairs, 250
H_2CO_3/HCO_3^-, as example of, 251
H_2PO_4/HPO_4^-, 257
Buffering mechanisms, 249-268
Bundle branches, 6
Bundle of His, anatomy of, 6-7
atrioventricular, 6
Bursa, and bursal equivalent sites, 690-691, 691t
of Fabricius, 689-691, *689*
Bysinnosis, histamine in, 178

Calcium, absorption of, 438-439
balance, maintenance of, bone in, 330
gastrointestinal tract in, 330
kidney in, 330
body, 330-334
control of, 330
in cardiac coupling, 9
excretion of, in chronic renal acidosis, 261
in experimental acidosis, 261-262
ion, concentration, in plasma, double feedback loop control of, 331-332, *332*
in hypoparathyroidism, 332
-magnesium competition at gut and tubular membrane, 331
plasma and urine levels of, in normal and disease states, 382t
"pump," 9, 17
renal tubular reabsorption of, factors influencing, 330
and sodium reabsorption, 331
serum concentrations of, lowered, and tetany, 648-649
normal, 648
-sodium competition at tubular membrane, 331
Cancer, gastric, acid secretion in, 410-411
and immune deficiency diseases, 707
Capillary(ies), blood volume in, 49
changes in, shock associated with, 80
filtration, coefficient of, determination of, *626*
control of, 50

function of, in oxygen diffusion, 152-153
hydrostatic pressure of, 625
in maintenance of arterial pressure, 68-70
pulmonary, gas exchange in, 131
stasis, development of, 80
Carbohydrates, absorption of, 434-436
malabsorption of, 434
consequences of, 436
metabolism of, in cirrhosis, 460
disturbances in, in liver disease, 468-469
in lungs, 174-175
Carbon dioxide, abnormal levels of, adaptation to, 119
as influence on hemoglobin affinity for oxygen, 148
as product of metabolism, 253-254
in assessment of pulmonary diffusion capacity, 140-141
chronic retention of, in respiratory insufficiency, 135
excess of, and arterial pressure, 75
excretion of, alteration of, methods for, 265
gaseous, partial pressure of, changes in, renal responses to, 259
determination of, 251
hydration of, and body hydrogen ions, 327
in local regulation of blood flow, 74-75
partial pressure of, 131
production of excess, in cardiopulmonary insufficiency, 131, *130*
retention of, in respiratory disease, 130
sensitivity to, to Cheyne-Stokes respiration, 121
serum total content of, measurement of, 252-253, 275
tension, 115
arterial, changes in, chemoreceptor responses to, 117
serum, calculation of, 275-276
transport and excretion of, 253-254, *254*
urine total content of, measurement of, 276
in water, acid-forming properties of, 115
Carbon monoxide, from heme, 456

intoxication, 158
Carbonic acid, in body fluids, 251
  dissociation of, 250
  in metabolic process, 253-254
Carboxyhemoglobinemia, 32
Carcinoid syndrome, kinins in, 26-27
  and pulmonary metabolism of vasoactive substances, 178
Carcinoma, bronchogenic, 180
  inappropriate antidiuretic hormone secretion in, 226
  "oat-cell," and syndrome of inappropriate secretion of antidiuretic hormone, 321-322
  medullary, of thyroid, 316, 333-334
  thyroid, metabolic derangements associated with, 316-317
Cardiac output, 9-10
  in acute glomerulonephritis, 47
  adequacy of, measurements of, 146-147
  as determinant of arterial pressure, 67-68, 69
  calculation of, 10, 42-43
  in cardiac failure, 14
  characteristics of, in different types of hypertension, 44
  in coarctation of aorta, 48
  in complete heart block, 11-12
  decreased, mechanisms in, 67-68
  and shock, 81
  high, 15-16
  inadequate, blood flow in, 146
  in inlet disorders, 14
  measurement of, 42-43
  oxygen supply to tissues influenced by, 146-147
  and plasma volume, in hypertension, 53
  regulatory factors in, 10
  in renovascular hypertension, 46-47
  in stress and exercise of hypertensive patients, 48
  stroke volume in, 9-10, 11
  Fick, 10
  indicator-dilution, 10
Cardiomyopathy, 6, 20
Cardiospasm, esophageal peristalsis in, 396
Cardiotoxicity, hyperkalemia as cause of, 228, 229
Cardiovascular system, cardiac adaptation in, 5-21

depressor mechanisms in, 67-82
disorders of, and conditioning, pathological model systems of, 656-657
emotions affecting, neural basis for, 601-602
function of, control of, autonomic, 599-614
pressor mechanisms in, 41-66
reflexes of, control of, 603-608
  function of, 603
regulation of, disturbances in, 589-591
levels of, 599
  neural, 589, 590
  higher central, 602-603
  integrated, 599-602
  sympathetic efferent neuroeffector system in, 607-608
tissue perfusion in, 23-39
Cardioversion, in cardiac arrhythmias, 97-98
Catecholamine(s), and adenyl cyclase system, 17
  as adrenergic neuroeffector mechanisms, 61-63
  in blood, effects of, on tissue perfusion, 25-26
  sources of, 25
  and gastric peristalsis, 400
  in hyperglycemia, 350
  in hypertensive patients, 63
  and intestinal motility, 621
  metabolic effects of, 176
  metabolism of, 61, 62t
  in nerves of gut, 391
  and neural mechanisms, 60-65
  plasma concentrations of, measurement of, 63
  synthesis of, 61, 62t
Catechol-O-methyl transferase, 61-63
Causalgia, symptoms of, 583
  therapy for, 583-584
Cell(s), B—, functions of, 694
  tests to evaluate, 694t, 751t
  in peripheral lymphoid system, 693
  system, 690, 690t
  specificity of, 692, 692t
  and T-, interactions of, 694
  B_1, of B-cell system, 690, 690t
  B_2, in immunoglobulin secretion, 690, 690t
  blood, formation of, 488
  development of, differentiation in, 495-496
  red, decrease in, in anemia, 491
  2,3-diphosphoglycerate in, 149-150

evolution of, 496
  glycolytic rate in, factors influencing, 149
  increase in, in erythremia, 491-492
  membrane, in study of red cell alterations, 487
  microscopic examination of, conditions revealed by, 492
  survival time of, 492
  wall rigidity of, and vascular resistance, 35-36
  See also Erythrocyte(s)
cardiac, automaticity of, 83
  depolarization in, 83-84, 84
  neural influence on, 611
of digestive tract, 414
enterochromaffin, 424
erythroid. See Cell(s), blood, red; Erythrocyte(s)
follicle, ovarian, steroid secretion by, 356
granulosa, activity of, 365
  luteinization of, 366
hilus, activity of, 365
intestinal, absorptive, 423-424
  epithelial, 423-424, 424
  activity of, 428
  disturbances of, 424-425
  goblet, 424
  Kulchitsky, 424
Leydig, activity of, factors influencing, 362
  steroid secretion by, 356
  testosterone secretion by, 361
  tumor of, 362
lymphoid, immunoglobulin production by, 689, 688
  in bursa of Fabricius, 689
membranes, 7-8
mononuclear, exudation of, in local inflammatory response, 674-675
myocardial, characteristic features of, 8
  function of, regulation of, 7-8
  and hypoxia, 16
  intercalated discs of, 8, 8, 9
  membrane of, 7
  mitochondria of, 8
  myofibrils of, 7, 8
  nucleus of, 7, 9
  P, 7
  automaticity of, 6, 7
  Purkinje, 7
  sarcotubular system of, 8-9, 8
  transitional, 7
  working, 7
nerve, reticular formation of, 580-581
oxygen delivery by, 146-152
oxygen supply of, 146

Leydig (*continued*)
oxygen uptake by, oxygen
content and partial
pressure in, 152
plasma, human, electron micrograph of, *688*
immunoglobulin secretion by,
689, *688*, 690, 690t
post-thymic, 691
pulmonary, alveolar, brush,
169
large, 167, *169*
macrophage, 170, *170*
small epithelial, 167-169
APUD, 172
argentaffin, 172
argyrophil, 172
clara, 171-172, *171*
corner, 167, *169*
endothelial, 171, *171*
enterochromaffin, 172
Feyrter, 172
granular, type II, 167, *169*
Kulchitsky, 172
mast, 170-171, *170*
stem, defective, diseases
causing, 496
pluripotential, 516
self-regulation by, 496
stromal, activity of, 365
T-, and B-, interactions
of, 694
functions of, 693-694, 693t
tests for adequacy of,
693-694, 693t, 751t
in peripheral lymphoid
system, 693
system, 691, 691t
specificity of, 692
nature of receptor influ-
encing, 692t
T₁, 691
theca interna, activity of, 365
thyroid, follicular, ultrastructure of, *305*
Cerebellum, red nucleus of,
lesions of, 646
Cerebrum, diseases of, inappropriate antidiuretic hor-
mone secretion in, 226
perfusion of, decreased, and
orthostatic hypotension,
28-29
Cervical rib syndrome, 72
Chagas' disease, achalasia of,
396-397
Chalasia, of gastroesophageal
sphincter, 397
Chemoreceptor(s), 70
abnormal function of, breathing patterns associated
with, 124
aortic, 115-116

physiological stimulation of,
response to, 116
arterial, 605-606
carotid, physiological stimulation of, factors causing,
115
response to, 116
central, decreased function of,
breathing patterns asso-
ciated with, 124
intracranial, 116-117
peripheral, 115-116
loss of function of, breathing
patterns associated with,
124
physiological importance of,
116
Chest, pain of, in ischemic
heart disease, 18
wall of, compliance of, 127
Cheyne-Stokes respiration, in
circulatory disorders, 121-122
with central nervous system
lesions, 121
Chloramphenicol, inhibition of
cytochrome oxidase by,
159
Choride, concentration of, in
intestinal tract, 432
in hypokalemic alkalosis, 230
in measurement of extracellular fluid, 49
renal tubular reabsorption of,
in Addison's disease, 327
shift, 253-254
transport of, mechanisms of,
432
Choridorrhea, congenital, 433
Chlorine, balance, effects of, on
renal acid excretion, 262
Chlorothiazide, in sodium reabsorption inhibition, 224
Cholecystitis, 401
Cholecystokinin, 386
in fat digestion, 439
secretory effects of, 409
pancreozymin, in gallbladder
emptying, 401
in intraluminal digestion, 423
in nontropical sprue, 442
Cholelithiasis, 401
Cholestasis, drug-induced, 459
Cholesterol, abnormalities, in
liver disease, pathophysiology of, 467-468
absorption of, 444
to bile acid ratio, in bile, 468
plasma, in nephrotic syndrome,
211-212
in plasma membranes, 7-8
stones, bile salts and lecithin
abnormalities in, 468
Choline, phosphatidyl, absorption of, 444

Cholinesterase, in nerves of gut,
391
Chorea, 644
Christmas disease, 533
Christmas factor, 529
Chylomicron, formation of,
441-442
disturbances of, 442
Chyluria, proteinuria in, 207
Circulation, enterohepatic, 441,
455
hyperkinetic, 14
local, fibrinolytic activity in,
553
lung-to-brain, delay in, in
Cheyne-Stokes respi-
ration, 121-122
nervous control of, 60-61
parasympathetic nervous system influencing, 608
peripheral, in cardiac failure,
613
regional, and essential hypertension, 44-45
respiration influencing, 609
systemic, fibrinolytic activity
in, 553-554
Circulatory state(s), hyperdynamic beta-adrenergic,
32, 45
Circus movement, production
of, in experimental
flutter, 91-94, *92*
Cirrhosis, alcoholic, portal hypertension in, 459
biliary, primary, bile acids
concentrations in, ele-
vated, therapy for,
467-468
lipid and cholesterol abnormalities in, 467
secondary, 467
bleeding syndrome in, 537
carbohydrate and fatty acid
metabolism in, 460
esophageal bleeding in,
460-461
fluid retention in, 460, 461-464
impaired hepatic concentrating ability in, 240
impaired hepatic diluting
ability in, 245
and jaundice, 459
Laennec's, 31-32
"postnecrotic," portal hypertension in, 459
renin production in, 463-464
Coagulation, disseminated intravascular, and accel-
erated local fibrinolysis,
554, *555*
subacute form of, 556
and kallikrein-kinin reactions,
176

in pulmonary disorders, 176
factors of, in nephrotic syndrome, 212
pulmonary factors in, 175-176
tests of, normal values for, 564t
See also Blood, clotting
Coagulopathy, consumption.
    See Coagulation, disseminated intravascular
Cold, as stressful stimulus in hypertension, 48
Colitis, functional, 625
    mucous, 625
Collagen, 173
    diseases, 35
Colon, bacterial flora in, 426
    metabolism of selected compounds by, 427t
    and bile acid secretion, 447
    blood flow in, neural control of, 629
    reflex and central nervous control of, 627
    bypass, in hepatic encephalopathy therapy, 467
    functions of, 622-623
    gastrocolic reflex of, 404, 623
    haustration of, 622-623
    inertia of, 624
    irritable, syndrome of, 404
    mass movement in, 403-404, 623
    motility of, factors influencing, 404
    parasympathetic control of, 623, *617*
    pathophysiological aspects of, 624-625
    reflex and central nervous control of, 623-624
    sympathetic control of, 623, *618*
    motor disorders of, 404-405
    muscle of, 389-390
    slow waves of, 404
    peristalsis in, 404
    propulsion in, 403
    segmentation in, 403
    spastic, 625
    stenosis of, congenital, 404-405
Coma, hyperglycemic nonketotic, syndrome of, 349-350
Complement, abnormalities, acquired, **720-727**
    congenital, 719-720
C1, 713-715
    esterase, inhibitor of, 715
        lack of 719
C1q, 713
    interaction with immunoglobulins, 713, 715

C1r, 713
C1s, 713-715
C2, 715
    deficiency of, **680, 719-720**
C3, 716
    activation of, 715, 176
        alternative modes of, 717-718
    and bacterial endotoxins, 717
    and cobra venom factor, 717-718
    convertase, 715
        deficiency of, two forms of, 720
    fragments of, biological activities of, 716
    immune adherence agglutination activity of, 716
    inactivator, 717
    interaction with C5, 716
    peptidase activity of, 716
    virus neutralization by, 716-717
C4, 711, 715
    deficiency of, 720
$\overline{C42}$ complex, 715
    and C3, 716
C5, 716, 718
    deficiency of, 720
C6 and C7, 718
C8 and C9, 718-719
    cell intermediates, of 712
    in chemotaxis of neutrophils, 674
    functional components of, 711-712
    detection system for, 712
    interactions and resultant biological activities of, *714*
    and hemolytic processes, 725-726
    -kallikrein system, and plasminogen activation, 488
    in phagocytosis, 675-676
    proteins, characteristics of, 712, 752t
    in renal allograft rejection reactions, 726-727
    serum, and red blood cell abnormalities, 509
    system, activation of, 672, 713
        components of, 711
    in disease, anti-glomerular—basement membrane, experimental model of, 728
        immune complex, experimental model of, 727-728
    functions of, 711, 712-719
    and immunoglobulins, 688-689

mechanisms involving, 711-729
    in nephrotoxic nephritis, experimental model of, 728
    status of, evaluation of, three techniques in, 721
Conditioning, and cardiovascular disorders, pathological model systems of, 656-657
    classical, 653
    and instrumental learning, differences between, 658, *658*
    definition of, 656
    and neuroses, experiments in, 657
    in visceral control, 653
Conduction, cardiac, "decremental," 86
    in ischemic heart disease, 18
    neural influences on, 610-612
    one-way, in atrial flutter, experimental, 91-93, *92*
    pathways of, 6-7, 85-86
    block of, 7
    subnormal and supernormal, 87
    velocity of, 86
    drugs influencing, 86
    factors determining, 93
    sectional estimation of, 87
    intracardiac, pathways of, 6-7, *6*
    ventricular, pathways of, 6-7
Constipation, and colonic segmentation, **404**
    rectal, 624
Contraceptives, hormonal, 371-372
    oral, and alterations of physiological state, 371-372
    contents of, 371
    effectiveness of, 371
    and plasma renin activity, 59
    of progestational agents, 372
    side effects of, 371, 458-459, 460
Contractility, changes in, and cardiac performance, 12
Contraction(s), myocardial, 12
    peristaltic, in gastric emptying, 399, 616
    pharyngoesophageal, in swallowing, 394-395
    afferent nerves in, 395
    in small intestine, segmenting, 402
    slow wave control of, 403
    tonic, in gastric emptying, 399
    ventricular premature, 96-97
Convulsive phenomena, 647-648
Coomb's test, 510

Copper, deposition of, in Wilson's disease, 470
Cordotomy, bilateral, respiratory syndrome following, 123
Corpus luteum, activity of, 365
function of, control of, in animals, 366
in women, 366
cerebral, 577
and cardiovascular function, 656
lesions of, and muscular atrophy, 632
and tremors, 646
and respiratory and cardiovascular regulation, 590, 602
stimulation of, effects of, 611
and voluntary movement, 642-643
renal, diluting segment of, sodium transport in, 242
Corticospinal tract, 113
Corticosteroid(s), action of, levels of, 296-297
administration of, and atrophy of adrenal cortex, 289
adrenal, as anti-inflammatory agents, 679
and leukocyte migration defect, 680
plasma concentration of, 295-296
prolonged use of, 297
secretion of, 287
suppression of ACTH secretion by, 288-289
Corticotropin, -releasing factor, 287
in stress, 288
Cortisol, adrenal cortical production of, 291-293, 292
defective, 294
deficiency of, in Addison's disease, 296, 327
insufficiency of, and abnormal water metabolism, 322
plasma concentration of, 295-296
diurnal variation in, 296
therapy, in hydroxylase deficiencies, 294-295
transcortin binding of, 296
Coughing, function of, 120
Coupling, cardiac, digitalis and, 9
mechanism of, 9
Creatine, phosphate, 9, 16
Creatinine, clearance, in measurement of glomerular filtration rate, 270-271
in thyrotoxicosis, 641
urinary excretion of, 189-190

Crigler-Najjar syndrome, phenobarbital therapy for, 459
type I, 458
type II, 458
Crohn's disease, 438
Cryofibrinogens, 550, 556
Cushing's syndrome, ACTH levels in, 291
etiology of, 290, 294
and thyroid carcinoma, 317
Cyanide, intoxication, cardiac failure in, 14
diffusion of oxygen in, 152
toxicity of, 158
Cyanosis, 15
Cyclic 3', 5' AMP, 14t, 17, 18, 174, 330, 608
activities of, 341
Cystinosis, 239-240
Cystinuria, 192, 437
Cytochrome, oxidase, inhibition of, 158
Cytolysis, 712-713
cell-mediated damage and enhancement, relationship between, 732, 737

Deafness, nerve, and congenital goiter, 306
Defecation, 623-624
Defibrination syndrome, acute. See Coagulation, disseminated intravascular
Deglutition, 394-396
Degranulation, in phagocytosis, 676, 677
Dehydration, as result of impaired concentrating ability, 238
and dilutional hyponatremia, 226
and hypernatremia, 227
Dendrites, 580
Denervation, and achalasia, 396-397
cardiac, 11
sensitization of, and epilepsy, 588-589
Denny-Brown, pallidal syndrome of, 644-645
striatal syndrome of, 645
Depolarization, cardiac, 6-7, 6, 9
in cells, 83-84, 84
rate of, neural influence on, 611
diastolic, 83, 84
Depression, 643
Depressor mechanisms, 67-82
Diabetes insipidus, antidiuretic hormone secretion in, 320
impaired urinary concentrating ability in, 241, 320

nephrogenic, amyloidosis in, 240
congenital, 241
quiet standing in, 321
urine concentration in, 321
water reabsorption in, abnormal, 320-321
defective adrenergic reflex in, 28
Diabetes mellitus
circulatory impairment in, 19
esophageal dysfunction in, 398
gastric dysfunction in, 400
hepatic adaptation in, 347-348
hormonal basis of, 350
hyperglycemia in, 348
insulin in, 350-351
ketoacidosis in, 349
metabolic relationships during, 347-348, 349
microvascular complications in, 352
orthostatic hypotension in, 604
osmotic diuresis in, 348-349
and starvation, 347
vascular occlusion in, 643
Diarrhea, in bile acid malabsorption, 448, 446
and body hydrogen ions, 326
in colonic motility disturbance, 625
and colonic segmentation, 404
definition of, 450
infectious, 436
in intestinal motility disturbance, 622
and intestinal secretion of electrolytes and water, 431
osmotic, 436
"overflow," 432, 436
in peptide-secreting adenomatosis, 412, 413
Diet, "acidity" of intake ash of, in hydrogen ion measurement, 326
as influence on production of fixed acids, 254-255
as influence on urinary concentrating ability, 238
Diffusion, facilitated, and active transport, 429-430
definition of, 429
simple, rate of, factors influencing, 430
DiGeorge syndrome, 695-696, 697
Digestion, and absorption, relationships between, 449
fluid and electrolyte movement during, 431-432

intraluminal, 423
luminal, 423
tests of, 478-479
Digitalis, and cardiac coupling, 9
and cardiac impulse conduction, 86
and cardiac impulse generation, 85
response of flutter to, 94
strengthening of cardiac contraction with, 11
Diiodotyrosine, coupling of, defect in, 306-308
Diphenylhydantoin, in atrial fibrillation, 96
2,3-Diphosphoglycerate, 32
concentration, changes in, genetic abnormalities producing, 150
clinical states producing, 150
formation of, 149
in hypoxemia, 149-150
in metabolism of erythrocytes, 507
in placental oxygen exchange, 150
in regulation of oxygen transport, 149
2,3-Diphosphoglyceride deficiency, 19
Disseminated intravascular coagulation. *See* Coagulation
Dissociation curve, oxyhemoglobin, 147-152, *147*
oxymyoglobin, 155, *155*
Diuresis, osmotic, in diabetes mellitus, 348-349
solute, in chronic renal disease, 239
effect of, on urine osmolality during renal concentration, 238, *238*
on urine osmolality during renal dilution, 243, *238*
hydropenia during, and urinary concentrating ability, 235
in postobstructive nephropathy, 240
therapeutic, agents causing, 224
water, 241, 242-243
impaired, consequence of, 243
urinary concentrating ability during, 235
Diuretics, mercurial, in sodium reabsorption inhibition, 224, 244
and renal salt depletion, 221

in sodium reabsorption inhibition, 244
Diverticulosis, 404, 624-625
Diverticulum(a), epiphrenic, 397-398
pulsion, 397-398
Zenker's, 398
Diving, cardiovascular adjustment to, 606-607
Dizziness, decreased cerebral perfusion causing, 28
Dopamine, synthesis of, 61
dP/dt index, of myocardial tension, 12
Dubin-Johnson syndrome, 458
Dumping syndrome, gastric, 400, 620
Duodenum, acidity of, 413
and gastric emptying, 414
neutralization of, 414
bacterial flora of, 426
buffering of, 414
mucosa of, 413, 415, 450
ulcer of, acid secretion in, 410, 411t
Dysautonomia, familial, esophageal dysfunction in, 398
orthostatic hypotension from, 29
Dysfibrinogenemia, congenital, 533
Dyskinesia, biliary, 401-402
Dysphagia, esophageal, 398
oropharyngeal, 398
Dyspnea, contributory mechanisms in, 124-125
definition of, 124
Dysrhythmias, cardiac, electrical conversion of, 97-98
studies of, 83
Dystonia, musculorum deformans, 644
torsion, 644
Dystrophy, muscular, 641
myotonic, esophageal dysfunction in, 398

Eaton-Lambert syndrome, 640
Ebstein's disease, 14
Ecchymoses, 530
"Echoes," as condition for re-entry of impulses, 88
Edema, angioneurotic, hereditary, 680, 719
formation of, pressure gradients in, 222
impaired concentrating ability in, 240
impaired diluting ability in, 245
inflammatory, 680
localized, causes of, 222

in nephrotic syndrome, 212-213
Efficiency, cardiac, equations expressing, 11
Elastin, 173
Electrolyte(s), homeostasis of, maintenance of, 215-231
metabolism of, endocrine control mechanisms of, 319-334
movement of, during digestion, 431-432
during fasting, 432-433, *431*
serum, in evaluation of organic acidosis, 264-265
Electrophysiology, of smooth muscle, 393-394
Eledoisin, in lungs, 177
Embden-Meyerhof pathway, in adipose tissue metabolism, 344
in diabetes mellitus, 348
in hepatic metabolism, 341, 342
in metabolism of erythrocytes, 507
in muscle metabolism, 339
rate-limiting steps in, 507
Embolism, amniotic fluid, 535-536
Emotions, and colonic function, 629
and gastric function, 627
hemodynamic patterns of response to, 74
neural stimulation by, cardiovascular effects of, 601-602
stimulation of thirst by, 238
Emphysema, α_1-antitrypsin deficiency in, 173-174, *174*
chronic obstructive, compliance of lungs in, 128, *128*
oxygen consumption changes in, 130, *130*
ventilation in, 131
patterns of, *139*
Encephalopathy, hepatic, pathophysiology of, 465-467, *466*
Endarterectomy, bilateral carotid, breathing patterns associated with, 124
Endocrine(s), in erythropoiesis regulation, 494-495
functional tests, 380-382t
glands, hypophysis and hypothalamus, interdependence of, 569-570
and nervous system, interdependence of parts of, 569-570

Endocrine(s) (*continued*)
  mechanisms, control, of elec-
    trolyte and water metab-
    olism, 319-334
  of reproduction, 355-373
  -metabolic mechanisms, 283-
    382
  syndromes, ectopic, 180
  lung tumors associated with,
    180-181, 181t
Endotoxin(s), bacterial, 677-
    678
  interaction with complement
    component C3, 717
Energy, cardiac, 8, 9, 11, 13, 15
  conversion of, 336-337
  interrelationships, metabolic
    pathways for food
    through, *336*
  mechanisms of, 335-353
  mitochondrial production of,
    337
  provision of, and oxygen,
    155-159
    metabolic control of, 157-158
    reactions in mitochondrion
      in, 156-157
    raw materials used to produce,
      16
  storage of, 17
    forms of, 335
  stores, utilization of, 17
  transportation of, 335-336
  utilization of, hormonal con-
    trol of, 17-18, *18*
Enhancement, immunological,
    731-733
  as factor in immunological
    adaptation in transplan-
    tation, 737-738
  cell-mediated damage and
    cytolysis, relationship
    between, *732*, 737
  methods of production of,
    737-738
Enteritis, regional, 403
Enterogastrone, 399
Enteropathy, "protein-losing,"
    431
Envenomation, intravascular
    clotting following, 536
Environment, as factor in evo-
    lution of organisms with
    nervous systems, 570-571
Enzyme(s), adenyl cyclase, 17,
    18
  cytochrome C, 19
  fibrinolytic, 173, 547-548
  of glycogen synthesis, in regu-
    lation of metabolic proc-
    esses, 340-341
  induction in liver, by pheno-
    barbital, 459
  lipoprotein lipase, in adipose

  tissue metabolism, 344
  oxidative, 19
    in lungs, 167, *168*, *169*
  pancreatic, 419-420
  proteolytic, in lungs, discharge
    of, factors in, 174
    sources of, 174
    and pulmonary disease,
      173-174
  respiratory, 19
  of respiratory-cytochrome sys-
    tem, 8
  unidirectional, in metabolic
    process, 337
Epilepsy, as symptom complex,
    586-589
  focal, 586
  grand mal, 586-587
  petit mal, 587
  probable pathophysiology of,
    588-589
  psychomotor, 587-588
  seizures of, electroencephalo-
    graphic changes in,
    647-648
    mechanisms of, 647
    origin of, 648
    phases of, 647-648
Epinephrine, in adipose tissue
    metabolism, 344
  and arterial pressure, 75
  in blood, 25
  effect of, on myocardial con-
    tractility, 12
  in exercise, and proteinuria,
    203
  formation of, 61
  plasma concentrations of,
    measurement of, 63
  and tremors, 645
Epithelium, intestinal. *See* In-
    testine, small, epithelium
    of
Erythremia, diagnosis of, labo-
    ratory techniques in, 492
  and inappropriate erythropoi-
    etin activity, 494
  stem cell defect in, 496
  symptoms of, 491
Erythrocyte(s), abnormalities
    of, extracorpuscular, ac-
    quired, associated with
    antibodies, 509-510
    without antibodies, 510
  intracorpuscular, acquired, 50
    inherited, 507-509
  circulating, life cycle of, 489
  concentration of, determination
    of, 492
  destruction of, bilirubin in
    measurement of, 512
  reticuloendothelial system in,
    511-512
  steps in, 511

  function of, 489-491
  mature, 506-512
  metabolism of, 506-507, *508*
  normal senescence of, 507
  production of. *See* Erythro-
    poiesis
  sedimentation rate, in detection
    of inflammation, 678
  spherocytic, 487
  structure and chemical compo-
    nents of, 506
  survival of, 506, 510
Erythron, concept of, 489
  control of, 492-493, *493*, 496
  pathophysiology of, 490t
Erythropoiesis, 489-513
  cellular differentiation in,
    495-496
  and cellular proliferation,
    495-506
  hemoglobin synthesis in,
    496-501
  nutritional factors in, 501-506
  regulation of, endocrine system
    in, 494-495
  erythropoietin in, 492-494
  negative feedback in, 495,
    493
Erythropoietin, abnormalities,
    and anemia, 494
  and erythremia, 494
  actions of, 492-493
    mode of, 493-494
  in mitochondrial oxygen utiliza-
    tion abnormalities, 163
  source of, 493
Esophagus, achalasia of, clinical
    findings in, 396
    incidence of, 396
  body of, 394
  in cirrhosis, 460-461
  motor disorders of, 396-398
  muscle of, 389
    smooth, control of, 395
  sphincter of, lower, 394-395
    chalasia of, 397
    closure of, control of,
      395-396
    hypertension of, 397
Estradiol, 356
  and cytoplasmic binding recep-
    tors, 359-360
  mechanism of action of,
    359-360
  metabolites of, 356-357
Estriol, 356
  determination of, and placental
    insufficiency, 370
  in feto-placental unit, 369-370
Estrogen(s), cytosol receptors
    for, locations of, 359
  definition of, 355
  effects of, on adrenals, 293

metabolism of, effect of thyroid hormones upon, 311t
in oral contraceptives, 371
and plasma corticosteroid levels, 296
production of, 17-hydroxylase deficiency influencing, 295
in ovariectomized and postmenopausal women, 367
synthesis of, 356
ovarian, 365
Estrone, 356
Ethacrynic acid, in sodium reabsorption inhibition, 224, 239, 244
Ethanol, in gastric mucosal barrier destruction, 417
phagocytosis depression by, 178
precipitation assay, 551
properties of, 417
Excitability, in sino-atrial nodes, 6
supernormal, of Purkinje fibers, 87
Exercise, arterial pressure in, 76
neural regulation of, 600-601
as stressful stimulus in hypertension, 48
blood flow in, distribution of, 26
neural regulation of, 600-601
cardiac output in, 10, 76
cardiopulmonary response to, 141-142
gas exchange abnormalities during, 141-142, 141t
hemodynamic responses to, in hypertension, 48
muscle blood flow in, 76
and muscle strength, 638
nervous adjustments during, 74
proteinuria of, 202-203
mechanism of, 203
Exposure, high altitude, hypoxia in, 31
Extrasystole(s), atrial, 15
ventricular, 12, 15
Eye(s), voluntary movement of, control of, impaired, 649-650

Fabricius, bursa of, 689-691, 689
Factor(s), clotting, antihemophilic, 529
functional deficiencies of, 532-533
Christmas, 529
functional deficiency of, 533
deficiencies of, specific localization of, 531
fibrin-stabilizing, 528, 552

deficiency of, functional, 533-534
Hageman. *See* Hageman factor
Stuart, 528
deficiency of, 534
vitamin K-dependent, 534
deficiencies of, 534-535
I, 527
II, 528
deficiency of, 534
V, 528-529
VII, 528
deficiency of, 534
VIII, 529
IX, 529
X, 528
deficiency of, 534
XI, 529
deficiency of, 534
XII. *See* Hageman factor
XIII, 528, 552
deficiency of functional, 533-534
cobra venom, and complement system, 717-718
intrinsic, in gastric atrophy, 419
intrinsic, secretion of, 418-419
Fasciculations, 650
Fasting. *See* Starvation
Fat(s), absorption of, 439
failure of, 420
digestion of, chemical and physical events in, 439-440, 439
disturbances of, 441
lipolytic products of, 439-440
physical and chemical states of, 440
tests of, 479
uptake and chylomicron formation in, 441-442
disturbances in, 442
malabsorption of, combined defects causing, 442-443
maldigestion of, and bile acid malabsorption, 448, 446
metabolism of, disturbances in, in liver disease, 469
in starvation, 345
unabsorbed, fate of, 443-444
Fatty acid(s), 16
free, in adipose tissue metabolism, 344
in hepatic metabolism, 342-343
in lechithin synthesis, 172-173, 172
medium-chain and long-chain, metabolism of, compared, 443
metabolism of, in cirrhosis, 460
oxidation of, 342

synthesis of, 342
Feminization, testicular, syndrome of, 360-361
Ferritin, 505
and iron absorption, 505
Feto-placental unit, steroid metabolism of, 369-371
Fetus, growth of, human placental lactogen in, 358
testis in, activity of, 360-361
Fever, production of, 677-678
Fiber(s), length, end-diastolic, changes in, and cardiac performance, 11-12
Fibrillation, atrial, 15
and atrial flutter, compared, 94
and clinical, compared, 96
production of, 94-96
behavior of nonuniform matrix in, 95
self-sustaining turbulence in, 95-96
persistence of, conditions determining, 96
ventricular, 15
Fibrin, clots, and degradation products, 550
formation of, 527-528
from fibrinogen, and from normal plasma, compared, 528
degradation products of, anticoagulant properties of, 549-550
detection in serum, 550-551
physicochemical properties of, 549
deposition of, 175
factors initiating, 553-554
enzyme splitting of. *See* Fibrinolysis
formation of, in plasma, 527-528, 527
-stabilizing factor, 528
deficiency of, functional, 533-534
test of, 531
function of, 552
Fibrinogen, composition of, 527
conversion to fibrin, 527-528, 527
degradation products of, anticoagulant properties of, 549-550
detection in serum, 550-551
physicochemical properties of, 549
Detroit, 533
disorders of, hereditary, 533
plasma, in nephrotic syndrome, 212
quantitative measurement of, 531

Fibrinogenolysis, pathological, 554-555
Fibrinolysis, 547-557
abnormal states of, 554-557
accelerated local, and disseminated intravascular coagulation, 554, 555-556, *555*
excessive, 555-556
and hypofibrinogenemia, 554, *555*
physiological mechanism for, 551-554
primary disorders of, 554-555
secondary, 555-556
system of activity of, decreased, 556-557
in circulating plasma and in thrombus, compared, *552*
components of, 547-551
microcirculation in, 553
systemic circulation in, 553-554
Fibrinopeptides, in clotting process, 527-528
Fibroblasts, in repair stage of local inflammatory response, 675
Fibrosis, cystic, glycoproteins in, 175
Fick method of cardiac output calculation, 10
Filtered load, determination of, 189, 219
Filtration, glomerular. *See* Glomerular filtration
pressure, 50
control of, intrarenal mechanisms in, 50-51
Flavine adenine dinucleotide, in Krebs cycle, 157
Flocculation test, latex particle, 550
Fluid(s), body, buffers of, 251
composition of, 216-218
daily shifting of, 216
homeostasis of, maintenance of, 215-231
movement of, during digestion, 431-432
normal volumes of, 215, *216*
maintenance of, 218
measurement of, 215-216
sources of, 218
cerebrospinal, carbon dioxide tension of, changes in, 117
composition of, processes determining, 117-118
hydrogen ion, changes in, in altitude changes, 118-119, *118*
regulation of, 117-118
extracellular, components of, 48-49

composition of, 217-218
control of, in hypertension-exaggerated natriuresis, 54-55
distribution of, 48-49
glomerular filtration rate influencing, 50
measurement of, 49, 216
osmotic structure of, 262
volume of, 191
control of, 50-51
expansion of, in chronic renal failure, 260-261
and sodium, 220
in sodium depletion, 221
extravascular, composition of, 217-218
interstitial, 48-49
composition of, 217-218
osmolality of, 217-218
volume of, and plasma volume, relationships of, control of, 50
intracellular, 48
osmolality of, 218
Flutter, atrial, and atrial fibrillation, compared, 94
causes of, 91
electrical conversion in, 98
experimental, frequency of, determinants of, 93
influence of refractory period abbreviation on, 94
influence of refractory period increase on, 93-94
initiation of one-way conduction in, 91-93, *92*
production of, 91
Folate, in erythropoiesis, 502-503
Food(s), digested, water solubility of, 335-336
metabolic pathways for, through energy interrelationships, *336*
storage forms of, 335
Forebrain, lesions of, breathing patterns associated with, 120
Fragment(s), fibrinogen, D, 549
E, 549
X, 549
anticoagulant properties of, 550
detection in serum, 550
Y, 549
anticoagulant properties of, 550
Frank-Starling phenomenon, 11, 612
Fructose, intolerance, hereditary, 469

Furosemide, in sodium reabsorption inhibition, 224, 239, 244

Galactose, absorption of, 435
Galactosemia, 469
Gallbladder, bile in, in gallstone disease, 468
emptying of, 401
filling of, 401
Gallstone disease, cholesterol and bile acid abnormalities in, 468
Gamstorp's disease, 642
Ganglion(a), basal, 577
sympathetic, 18-19
Gas(es), exchange of, in lungs, 131-134
abnormalities of, during exercise, 141-142, 141t
assessment of, 140-141
Gastrectomy, partial, gastric emptying following, 620
Gastrin, in gastric emptying, 393, 399
release of, inhibition of, 408
secretion of, in pancreatic tumor, 413
secretory effects of, 408
Gastrointestinal tract, blood flow in, autonomic control of, 625-629
body fluids in, 216
in calcium balance maintenance, 330
fasting contents of, secretion, absorption, intraluminal flow rate and composition of, *431*
functions of, autonomic control of, 615-630
disturbed, laboratory tests for, 476-481
innervation of, parasympathetic, 616
sympathetic, 616
intramural nerve plexuses of, 615-616
mechanisms of, 383-484
motility of, 389-406
autonomic control of, 616-625
muscles of, 389-390
muscularis propria of, innervation of, 616
nerves of, 390-392
plasma protein secretion into, 210
vasculature of, blood flow in, 625-626
innervation of, 616, 626
Gene(s), in control of hemoglobin synthesis, 498-500
Genitalia, development of, 360, *360*

Gibbs-Donnan phenomenon, 217
Gigantism, etiology of, 298
treatment of, 298
Gilbert's syndrome, 457, 458
Gland(s), responses of, instrumental learning of, 657-659
Glanzmann's disease, 543
Globin, synthesis of, in formation of hemoglobin, 497-498, *498*
Globulin, beta 1C, 716-717
beta 1E, 711, 715
beta 1F, 716, 718
heterologous antilymphocyte, in immunosuppressive therapy, 743-744
thyroxine-binding, binding capacity of, 315
decreased, 315
increased, 315
Glomerular filtration, diffusion theory of, 198
pore theory of, 197-198
rate, and extracellular fluid, 50-51
factors determining, 50, 189
measurement of, 189
creatinine clearance technique in, 270
criteria for substance for, 269
inulin technique in, 269-270, *269*
sieving coefficient in, 195-197
total chromagen technique in, 270-271
and sodium filtration, 324
Glomerulonephritis, acute, 536-537
hypertension of, 47
plasma volume in, 52
post-streptococcal, chronic and systemic lupus erythematosus, complement abnormalities in, summarized, 724
complement abnormalities in, 721-722
and in systemic lupus erythematosus, compared, 722-723
chronic, complement abnormality in, 723-724
experimental model demonstrating, 727-728
in Goodpasture syndrome, 180, 724
inflammatory defect in, 680
systemic lupus erythematosus, 722-723

and acute post-streptococcal, compared, 722-723
Glomerulus(i), as molecular sieve, 195-199
in disease, 205-206
capillary wall of, basement membrane of, 198-199, *198*
in disease states, 199, 206
structure of, 198-199, *198*
changes in, in disease, 206
diseases of, and proteinuria, 205-206
filtration rate of. See Glomerular filtration, rate
ultrafilter of, leakiness of, 205-206
ultrafiltrate of, 189
Glucagon, 17
in adipose tissue metabolism, 344
in diabetes mellitus, 350, 351
in hepatic metabolism, 340, 341
in starvation, 347
in thyrocalcitonin release, 331
Glucocorticoid(s), adrenal, deficiency of, 245
in urinary diluting mechanism, 243
in hyperglycemia, 350
Gluconeogenesis, in adipose tissue metabolism, 344
in fasting man, 345, *346*
in hepatic metabolism, 341, 342-343
and ketogenesis, in diabetes mellitus, 348
in cows with bovine ketosis, 348
Glucose, 16
absorption of, 435, 436
in adipose tissue metabolism, 344
-galactose malabsorption, congenital, 435
in hepatic metabolism, 341
metabolism of, in starvation, 345
in muscle metabolism, 338-339
-6-phosphate, dehydrogenase deficiency, two forms of, 508
in hepatic metabolism, 341
in muscle metabolism, 339
protein-sparing ability of, in starvation, 347
regulation of, disturbances in, in liver disease, 468-469
and sodium, absorption of, 435
pathway of, *433*
Glucuronyl transferase, deficiency of, 458

inhibition of, 457-458
Glyceraldehyde phosphate dehydrogenase, 161
Glycogen, release of, 17
in starvation, 345
synthesis, enzymes of, in regulation of metabolic processes, 340-341
Glycogenesis, type IV, 469
Glycolysis, aerobic, 159, 337
anaerobic, chemical reactions in, 160
definition of, 337
functions of, 159, 337
metabolic control of, long-short-term factors in, 161
and mitochondrial oxygen utilization, interrelationship between, 161-162
pyruvate and lactate formation in, 337
sequential steps of, 160-161, *160*
structural basis of, 160
enzymes in, 19
Glycosides, cardiac, 9
Goiter, congenital, and nerve deafness, 306
incidence of, 301
in iodide organification defect, 306
in iodotyrosine dehalogenase defect, 308
in mono- and diiodototyrosine coupling defect, 306-308
in serum iodopeptide defect, 308
Golgi tendon organ, 634, *633*
Gonad(s), development of, 360, *360*
steroids of, 356-357
Gonadotropin, human chorionic, function of, 358
levels of, in pregnancy, 371, *370*
structure of, 358
synthesis of, 357-358
Goodpasture syndrome, as cytotoxic pulmonary reaction, 180
glomerulonephritis in, 180, 724
Gout, 192
Grand mal seizures, 586-587
Granulocyte(s), blood, half disappearance time of, 518-519, 518t
marrow, production rate of, 521
reserve of, 515, 521-522, 522t
neutrophilic, formation of, 515

Granulocyte(s) (*continued*)
  pool, circulating, size of,
    518, 518t
  marginal, size of, 518, 518t
  total blood, size of, calcula-
    tion of, 518, 518t
  decreased, 520, *519*
  expanded, 520, *519*
  normal, 519-520, *519*
  turnover rate of, calculation
    of, 519
  normal, 518t
Granulocytopoiesis, effective,
  measurement of, 519,
    518t
Granulocytosis, "masked," 520
Granuloma, formation of, 675
Granulomatosis, chronic, in
  females, 681
  inflammatory defect in,
    680-681
  in males, 681
  Wegener's, 180
Graves' disease, 310
  and Plummer's disease,
    compared, 310-311
Gunn rat, glucuronyl transferase
  deficiency in, 458
Gut, and drug receptors, 392
  nerves of, 390-392
  See also Intestine(s), small

Hageman factor, 529
  as fibrinolytic activator, 548
  hereditary absence of, 534
  in inflammatory reactions, 672
  in initiation of clotting and
    kallikrein-kinin reactions,
    176, *175*
Hageman trait, 534
Haptoglobin, urinary excretion
  of, 202
  in exercise, 203
Hartnup's disease, 437
Hashimoto's thyroiditis, characteristics of, 739
  prehypothyroid state in, 313
  thyroid-specific antigens
    involved in, 740
Haustra, familiar, of colon, 403
Heart, adaptation of, 5-21
  arrhthymias of, electrical
    conversion of, 97-98
  studies of, 83
  automaticity of, 6, 7, 83
  autoregulation by, 13
  beats of, reciprocal, re-entry
    as mechanism for, 88, *89*
  block, complete, 7, 11
  and re-entry of impulses of,
    87-97
  blood pumped by, volume of,
    factors determining, 50

central nervous system control
  of, 608-613
conduction by. *See* Conduction, cardiac
dilatation of, 13
disease of, arteriosclerotic, 18
  ischemic, clinical manifesta-
    tions of, 18
  nutrient failure as factor
    in, 19-20
  rheumatic, 14
dysrhythmias of, electrical
  conversion of, 97-98
  studies of, 83
efficiency of, 11
failure of, acute, 16
  mechanisms of, 17
  catecholamines in blood in,
    26
  chronic, 16
  mechanisms of, 17
  compensatory mechanisms for
    maintenance of function
    in, 612
  congestive, 9, 12, 13, 14, 15,
    16, 17-18
  impaired concentrating abil-
    ity in, 240
  impaired diluting ability
    in, 245
  myocardial and peripheral
    circulatory changes in,
    612-613
  oxygen consumption changes
    in, 130, *130*
  sodium retention in, 223
  reversal of, 223
  excessive water reabsorption
    in, 323-324
  intracellular metabolic events
    in, 16-18
  mechanisms of, 14-16
  norepinephrine in, 612
  in prolonged shock, 81
in fibrosis of, 13
function of, central nervous
  system dysfunction influencing, 611-612
  changes in, with hypoxia,
    164-165
hypertrophy of, 5
impulses of. *See* Impulse(s),
  cardiac
inflammation of, 13
mechanoreceptors of, 604-605
morphological features of, 5-9
muscle of. *See* Myocardium
myogenic regulation of, 11
output of. *See* Output,
  cardiac
performance of, measurements
  of, 9-11
  poor, causes of, 15

pumps of, 5
performance of, 9-11
rate, 10, 15
  during exercise, neural
    regulation of, 601
  respiration influencing, 609
reserve mechanism of, 9, 10
residual volume of, 13
rhythmicity of, 83-99
  development of, 83
  See also Rhythm, cardiac
stroke, volume of, 13
valves of, 6
  insufficiency of, 13
  stenosis of, 14-15
work of, in cardiac hypertrophy, 13
  kinetic, 10, 11
  mechanical, 10, 11, 17
  potential, 10, 11
  measurement of, 10, *10*
  pressure-volume diagram in
    measurement of, 10, *10*
Heat, body, cardiac production
  of, 11, 13
  hypothalamic regulation of,
    590, 607
  stroke, 77, 536
Heister, valve of, 401
Hemagglutination, inhibition
  immunoassay, tanned red
  cell, 550-551
Hemangioma, giant cavernomatous, 536
Heme, bilirubin formation from,
  455-456, 457
  biosynthesis of, regulation of,
    497
  steps in, 497
  functions of, 497
Hemochromatosis, 438
Hemodilution, 243
Hemodynamics, basic considerations in, 42-43
  resting, in hypertensive dis-
    eases, 43-45, 44t, *44*
Hemoglobin(s), abnormalities
  of, acquired, as influence
    on hemoglobin affinity
    for oxygen, 150
  and altered oxygen affinity,
    500
  heritable, as influence on
    hemoglobin affinity for
    oxygen, 150-151
  physiological effects of, 499t
  tissue hypoxia with, 32
  affinity for oxygen, 147-152
  abnormal, 19
  acquired abnormal hemo-
    globins influencing, 150
  body temperature influencing,
    148-149

carbon dioxide influencing, 148
decreased, 151
hemoglobins associated with, 151
heritable abnormalities of hemoglobin influencing, 150-151
increased, 151-152
hemoglobins associated with, 151
organic phosphates influencing, 149-150
quantitative definition of, 147-148, *147*
red cell pH influencing, 148
blood content of, 491
in cells, determination of, 492
chain, synthesis of, genetic control of rate of, 499
genetic control of structure of, 498-499
Chesapeake, 500, 499t
component parts of, metabolism of, 512
erythrocytes as transport vehicles for, 489-492
free, in plasma, and proteinuria, 204
Kansas, 500, 499t
M, 500, 499t
molecule, normal, 499-500
sickle, 500
disease associated with, 500-501
stability of, factors influencing, 500
synthesis of, 496-501, *498*
control of steps in, 498-500
heme in, 497
iron in, 503-504
protein in, 497-498
vitamin B₆ in, 503
Zurich, 500, 499t
Hemoglobinuria, paroxysmal, cold, 724-725, 740-741
nocturnal, 509
Hemolysis, in bilirubin overproduction, 457
complement participating in, 725-726
and hyperbilirubinemias, 458
Hemolytic-uremic syndrome, 510
Hemophilia, classic, diagnosis of, 532
inheritance of, 532
symptoms of, 532
therapy for, 532
vascular, 532-533
Hemorrhage, in back diffusion of acid through gastric mucosal barrier, 417-418
in cardiac failure, 14

of newborn, in deficiency of vitamin K-dependent factors, 534
Hemosiderin, definition of, 505
Hemostasis, 527-546
disordered, due to vascular pathology, 543-544
tests of, 564t
See also Blood, clotting
Henderson-Hasselbalch equation, 115
in calculation of carbon dioxide tension in serum, 275
Henderson-Hasselbalch equation, in evaluation of acid-base disturbances, 263-264
in measurement of hydrogen ion activity, 252-253
parameters of, approximate values of, 263-264
interpretation of interrelationships of, 264
Henoch-Schönlein's purpura, 544
Heparin, in disseminated intravascular coagulation syndrome, 556
in hypofibrinogenemia therapy, 536
in lungs, 175
Hepatitis, acute, bleeding syndrome in, 537
"alcoholic," portal hypertension in, 459-460
viral, 459
Hering-Breuer inhibito-inspiratory reflex, 120
Hernia, hiatus, and chalasia of gastroesophageal sphincter, 397
Heroin, intoxication, 14-15
Hexokinase, 161
specific, activity of, 341
Hexose monophosphate shunt, in metabolism of erythrocytes, 507
High-energy stores, utilization of, 17
High-output circulatory states, mechanisms of, 14t
Hirschsprung's disease, 404-405, 624
Hirsutism, as familial trait, 369
in polycystic ovary syndrome, 368
and secondary amenorrhea, 368
His, bundle of anatomy of, 6-7
His-Purkinje system, pacemaker activity in, 83
Histalog, in augmented histamine test, 410
Histamine, in anaphylaxis, 177

and arterial pressure, 75
in peptic ulceration, 415
release of, 672
antigen-antibody reactions in, 672
storage of, 672
test, augmented, of gastric secretion, 409-410
and vascular permeability, 671-672
Histocompatibility testing, in transplantation, 736, 753t
Histoplasma capsulatum, 180
Hodgkin's disease, resistance to pathogens in, 684
Homeostasis, new view of, 660
traditional view of, 660
Hormone(s), adrenal, cortical, antidiuretic hormones and renal function, interrelationships between, 323
excessive, and gastric mucosal barrier, 416
and renal acid excretion, 263
glucocorticoid, deficiency of, 245
in urinary diluting mechanism, 243
metabolism of, 293
effect of thyroid hormones upon, 311t
mineralocorticoid, in maintenance sodium balance, 220
adrenocorticotropic. *See* ACTH
antidiuretic, 27
activity of, evaluation of, 244, *244*
inappropriate, and dilutional hyponatremia, 226
adrenal cortical hormones and renal function, interrelationships between, 323
as factor in permeability of collecting duct or distal tubule to water, 237, 239, 242
circulating, and renal water reabsorption, 319
in diabetes insipidus, 241
and dilutional hyponatremia, 226
in lungs, 177
release of, appropriate, 243-244
factors stimulating, 226
nicotine as stimulus to, 321
inappropriate, 192
in renal tubular reabsorption, 220

Hormone(s) (*continued*)
  secretion of, by lungs, 181
    control of, 320
    inappropriate, in Addison's
      disease, 322-323
    causes of, 320
    syndrome of, 226, 245-246,
      321-322
      diagnosis of, "free
        water" excretion in,
        323-324
      hypothyroidism in, 314
      and "oat-cell" broncho-
        genic carcinoma,
        321-322
      serum hypertonicity and
        hyponatremia in, 321
    inhibition of, alcohol in,
      320
    in diabetes insipidus, 320
    in urinary diluting mecha-
      nism, 242
  in control of energy utilization
    process, 17-18, *18*
  follicle-stimulating, deficiency
    of, and spermatogenesis,
    362
    function of, 357
    in induction of ovulation,
      367-368
    in menstrual cycle, 365
    secretion of, control of, 362
    in treatment of secondary
      hypogonadism, 363
  gastrointestinal, and contrac-
    tion of smooth muscle,
    393
  growth, in hyperglycemia, 350
    metabolic effects of, 297
    secretion of, decreased, 298
    increased, in acromegaly
      and gigantism, 298
    regulation of, 297-298
    stimulators of, 297
  interstitial cell-stimulating,
    deficiency of, 363
    function of, 357
    in menstrual cycle, 365-366
  luteinizing, deficiency of, 363
    function of, 357
    in menstrual cycle, 365-366
  melanocyte stimulating, 290
    in adrenal insufficiency, 297
  natriuretic, 51, 55, 220
  parathyroid, absence of, 332
    in calcium balance mainte-
      nance, 330
    control of, 330-331
      and actions of, *332*
    excess of, in medullary thy-
      roid carcinoma, 333-334
    and hydrogen ion concentra-
      tion, 326-327

    in phosphate reabsorption,
      331
    sustained hypersecretion of,
      in hyperparathyroidism,
      333
    in pregnancy, 369-371, *370*
  protein, 357-358
    blood levels and transport of,
      measurement of, 358-359
    plasma concentrations of,
      359t
    in pubescence, 361
  secretion of, by adrenal
      cortex, 287
    by tumors, 291
  sodium-retaining, autonomous
    excess of, in primary
    aldosteronism, 328-330
    lack of, in Addison's disease,
      327-328
  steroid. *See* Steroid(s)
  in sugar absorption, 436
  thyroid, biosynthesis of,
    304-306
    defects of, 306-308
    circulating, 308
    end-organ refractoriness to,
      314
    excess, effect of, on other
      endocrine function,
      311, 311t
    metabolism of, kinetics of,
      309t
    in regulation of oxygen
      consumption, 159
  thyroid-stimulating, 301
    chemistry and dynamics of,
      302-303
    hypothalamic control of,
      301-302
    and long-acting thyroid
      stimulator, compared, 309
    pituitary secretion of,
      302-304
    releasing factor of, action
      of, 302
      control of, 302
      isolation and chemical
        structure of, 301-302
      in reserve test, 303-304
    reserve test for, 303-304
    secretion of, regulation of,
      303
    in serum, elevated, effect of
      thyroid upon, 313
    increases in, studies of, 303
    and thyroid structure and
      function, 302
  vasoactive, pulmonary han-
    dling of, 176-178, *176*
Huntington's disease, 644
Hyaline membrane disease, 128
Hydrogen, gas, production of,
  436

    reductive, in hepatic
      metabolism, 341
    -sodium exchange, in primary
      aldosteronism, 329
  Hydrogen ion, activity of,
    measurement of, 252, 253
    as influence on potassium
      secretion, 228
    body, 326-330
      balance, 326
      concentrations, factors
        increasing, 326-327
      respiration influencing, 327
      metabolic release of, 326
      tubular secretion of, 327
    concentration, of acids,
      249-250
    in buffered solutions, 250
    chemical processes related
      to, 249
    interconversion between
      [H⁺] and pH in nota-
      tion of, 253t
    maintenance of, 249
    problems of measurement of,
      252
    in urine, 256
    definition of, 249
    gradients, kidney in estab-
      lishment of, 256
    production of, metabolic
      processes in, 254-255
    secretion of, in urine, 256-257
  Hydroxyl ions, loss of, 326-327
  11-Hydroxylase, deficiency of,
    294-295
  17-Hydroxylase, deficiency of,
    295
  21-Hydroxylase, deficiency of,
    294
  Hyperbilirubinemia, conjugated,
    458
    "shunt," 457
    transient familial neonatal,
      457-458
    unconjugated, in Crigler-
      Najjar syndrome, 458
    in Gilbert's syndrome, 458
  Hypercapnia, acute, 258
    characteristics of, 258
    chronic, 258
    in inadequate alveolar
      ventilation, 131
    peripheral chemoreceptors in,
      116
    renal responses in, 259
  Hypercarbia, 74
  Hyperemia, active, 75
    reactive, 75-76
  Hyperglycemia, in diabetes
    mellitus, 348, 350
    hormones in production of,
      350
  Hyperkalemia, 325, 323

in Addison's disease, 328
and cardiotoxicity, 228
in chronic renal failure,
228-229
electrocardiogram of, 228,
229
muscular weakness in, 642
Hyperkinesis, idiopathic, 32, 45
Hyperlipidemia, 33-34
altered viscosity in, 35
in primary biliary cirrhosis,
467, 469
Hyperlipoproteinemia, types
of, 34
Hypermetabolism, in Luft's
syndrome, 159
severe, 14
Hypernatremia, causes of, 227
Hyperparathyroidism, 332-333
effect on bone, 333
effect on gut, 332
effect on tubule, 333
parathyroid hormone in, sustained hypersecretion
of, 333
thyrocalcitonin production in,
333
Hyperpigmentation, 290, 297
Hyperplasia, congenital adrenal,
291, 294
clinical picture of, 294
11-hydroxylase deficiency
causing, 294-295
21-hydroxylase deficiency
causing, 294
gastric, 413
juxtaglomerular, syndrome of,
330
Hyperplasminemia, 554-555
Hypersensitivity, active, natural,
669
cell-mediated, 669
delayed, 669
immunological, development
of, 669
Hypertension, borderline, cardiac output character-
istics in 45, *44*
hemodynamics in, 45, 44t
hyperdynamic beta-adrenergic
circulatory state in, 45
cardiac output characteristics
in different types of, *44*
cardiac work in, 11
cardiovascular reflex responses
in, 63-64
catecholamines in, 26, 63
in coarctation of aorta, 47-48
cardiac output in, 48
common types of, 41t
and defective renal perfusion,
29
diastolic, causes of, 41, 41t

essential, cardiac output
characteristics in, 44, *44*
hemodynamics in, 43-45, 44t
plasma volume in, 52, 51t
regional circulations in, 44-45
and skeletal muscular
system, 44-45
exaggerated natriuresis in,
control of extracellular
fluids in, 54-55
hemodynamic characteristics
in different types of, 44t
hemodynamic responses to
stress and exercise in, 48
in 11-hydroxylase deficiency,
294-295
in 17-hydroxylase deficiency,
295
labile, cardiac output characteristics in, 45, *44*
hemodynamics in, 45, 44t
hyperdynamic beta-adrenergic circulatory state in, 45
"low renin," 59
malignant, 536-537
plasma renin activity in,
58-59
renin-angiotensin system in,
26
neural reflexes in, 63-64
orthostatic, neural reflex
responses in, 64
pathogenesis of, renal blood
flow in, 30-31
physiological relationships of
renin in, 59-60
physiological relationships of
sympathetic vasomotor
activity in, 64-65
plasma renin activity in, 57-59
plasma volume abnormalities
in, 51-52, 51t
plasma volume, and cardiac
output in, 53
and diastolic arterial pressure
in, 52-53, 53
and neural mechanisms in,
53-54
portal, causes of, 459-460
metabolic changes in, 460
vascular changes in, 460
prostaglandins in, 27
pulmonary, severe, 15
renal, 45-47
anephric state in, 45, *46*
renal arterial stenosis in,
45, *46*
renal pressor and renoprival
factors in, 45, *46*
of renal parenchymal disease,
cardiac output in, 47
total peripheral resistance
in, 47

renovascular, cardiac output
in, 46-47
central blood volume in, 46
experimental, 46-47
total peripheral resistance in,
46-47
resting hemodynamics in,
43-45
systemic, 15
Hyperthyroidism, definition of,
310
and erythropoiesis, 494-495
etiology of, 308-310
hyperfunctioning hyperplastic
thyroid in, 310
hyperfunctioning thyroid
nodule(s) in, 310-311
hyperkinetic states associated
with, 32, 45
prehypothyroid state following
treatment of, 313
suppression test for, 310
thyroid hormone in, 159
triiodothyronine, 311
with neoplasia, 311-312
Hypertrophy, cardiac, 5, 13, 19
gastric, 413
infundibular, 15
Hyperventilation, central,
121-122
Hypervolemia, 14
in renal hypertension, 47
Hypocalcemia, in hypoparathyroidism, 332
Hypocapnia, acute, 258-259
characteristics of, 258
chronic, 259
Hypofibrinogenemia, diagnosis
of, 531
and fibrinolysis, 554, 555
intravascular clotting causing,
disease states associated
with, 536
Hypogammaglobulinemia,
transient, of infancy,
701
Hypoglycemia, causes of,
468-469
Hypogonadism, hypogonado-
tropic, and delayed
pubescence, in men,
compared, 363
in women, compared, 367
in panhypopituitarism, 299
primary, 363-364
therapy in, 364
secondary, 362-363
therapy in, 363
Hypokalemia, alkalosis in
development of, 229-230
causes of, 229, 325
in diabetic acidosis, 230
muscular weakness in,
641-642

Hypokalemia (*continued*)
in primary aldosteronism, 329
renal effects of, 230
Hypokinesia, 643
Hypomotility, disease mechanisms producing, 643
Hyponatremia, 191, 192
in Addison's disease, 322, 327-328
definition of, 224
development of, 222
dilutional, 191-192, 224-226, 225
causes of, decreased renal perfusion as, 225
high fluid intake as, 224-225
inappropriate antidiuretic hormone activity as, 226
multiple, 226
renal failure as, 225, 225
clinical picture of, 224
impaired water diuresis causing, 243
in panhypopituitarism, 299
in syndrome of inappropriate secretion of antidiuretic hormone, 322
therapy for, causative factors influencing, 222
Hypoosmolality, impaired water diuresis causing, 243
Hypoparathyroidism, effect on bone, 332
effect on gut, 332
effect on tubule, 332
and pseudohypoparathyroidism, compared, 332
Hypoperfusion syndrome, pulmonary, 128-129
Hypophysis, endocrine glands and hypothalamus interdependence of, 569-570
Hypoproteinemia, sodium retention in, 223
Hypotension, kinins in, 26
orthostatic, arterial mechanoreceptor malfunction causing, 604
arterial pressure in, 76-77
and decreased cerebral perfusion, 28-29
following sympathectomy, 77
idiopathic, arterial pressure in, 72
plasma renin activity in, 57
in studies of neural reflex responses, 64
severe, in adrenal insufficiency, 296
systemic, and mechanoreceptors, 70
reflex, production of, 72
Hypothalamus, anterior, defense reaction of, 627-628

as neural control center, 602-603
composition of, 573
control of ACTH secretion by, 287-290
control of thyroid-stimulating hormone by, 301-302
and heat regulation, 590
hormonal control by, 570
hypophysis and endocrine glands, interdependence of, 569-570
lesions of, and ACTH secretion, 288
location of, 577
substances isolated from, 287-288
in visceral control, 656
Hypothyroidism, abnormal water diuresis in, 245
primary, diagnosis of, 312
etiology of, 312
and hypothyroidism secondary to pituitary failure, compared, 312-313
symptoms of, 312
syndromes associated with, 313-314
secondary to pituitary failure, 312-313
and primary hypothyroidism, compared, 312-313
thyroid hormone in, 159
Hypoventilation, 124
alveolar, 131
arterial blood gas findings in, 134-135, 135t
Hypovolemia, 14
and antidiuretic hormone secretion, 320
severe, orthostatic hypotension from, 28
Hypoxemia, definition of, 145-146
2,3-diphosphoglycerate in, 149-150
and tissue abnormalities, 146
Hypoxia, acute, ventilation in, 116
and arterial pressure, 75
definition of, 145-146
effects of, on central nervous system, 164
on heart, 164-165
on kidney, 165
on liver, 165
on myocardial cell, 16
on pulmonary vascular bed, 165
on skeletal muscle, 165-166
from high altitude exposure, 14, 31
in inadequate alveolar ventilation, 131

insensitivity to, 124
irreversibility of, 124, *125*
local, 74
mitochondrial, detection of, 163
and myocardial failure, 16
phagocytosis depression by, 178
in respiratory insufficiency, 135
sustained, ventilation in, 116

Ileum, distal, bacterial flora in, 426
Ileus, 403
paralytic, 622
Immune response(s), adaptive, 683, 684t
prophylactic and therapeutic usefulness of, 685-686
biological amplification systems used by, 692t
in diagnosis of disease, 686
infectious agents influencing, 685
nonspecific suppression of, 742
primary, 669
secondary, 669
Immunity, active, artificial, 669
natural, 669
adaptive, 667-670, 683-710
host-parasite relationships in, 683-686
adoptive, 670
and aging, 708
antibody-mediated, control of, 742
cell-mediated, 669
control of, 742
development of, 669
humoral, 686-691
and cellular, fundamental relationships in, 686-689
innate, 667
in invertebrates, 703
and malignant adaptation, 707-708
passive, 670
pharmacological suppression of, 742-744
phylogenetic development of, 702-707
Immunization, active, diseases prevented by, 685-686
attenuated virus, viral infections prevented by, 685, 686t
passive, diseases prevented by, 686
Immunobiology, scope of, 668t
Immunodeficiency(ies), 694-702
and cancer, 707

common variable form of, 699-701
family history of, *700*
and persisting virus infection, 701
primary, classification of, 695t
X-linked recessive form of, 694-695, *696*
secondary, 702, 702t
severe dual system, 698, *697*
Immunoelectrophoresis, in detection of fibrin degradation products in serum, 550
Immunoglobulin(s), basic units of, 687, *686*
and complement system, 688-689
of different classes, molecular distinctions of, 687-688, *687*
distribution of, 688
diversity of, evolution of, 705-707
human, characteristics of, 749-750t
IgA, 179, 688, 689
isolated absence of, 701-702
7S, in secretory molecule, 179
IgD, 689
IgE, 179, 688, 689
IgG, 688, *686*
in autoimmune hemolytic anemias, 740-741
in immunological enhancement, 737
in immunological injury, 731
interaction with Clq, 713, 715
IgM, 687, 688-689
in hemolytic anemia, 740
in immunological injury, 731
interaction with Clq, 713, 715
light chains of, in plasma, and proteinuria, 205
N-terminal amino acid sequences in study of, 706
production and secretion of, 689, *688*
serum, levels of, *750*
Immunosuppression, 742-744
Immunosurveillance, 707
Impulse(s), cardiac, block and re-entry of, 87-97
circus movement, conditions causing, 87-88, *89*
experimental production of, 91-93, *92*
generation of, ectopic sites of, 85, *84*
normal site of, 83-85, *84*
propagation of, normal mechanisms in, 85-87

subnormal and supernormal conduction in, 87
turbulence in, 95
re-entry of, conditions causing, 87-88, *89*
experimental production of, 91-93, *92*
reciprocal rhythm caused by, 88-90
in ventricle, as cause of ventricular premature contractions, 96-97
transmission of, 6
Indicator-dilution technique, of cardiac output calculation, 10
in measurement of body fluids, 215-216
Infarction, myocardial, 7, 15, 18
small bowel, 403
Infection(s), recurrent or persistent, specific adaptive barrier to, 683-684
thrombocytopenia during, 542
viral, attenuated virus immunization for, 685, 686t
persisting, and immunodeficiency, 701
Infertility, male, spermatogenic arrest as cause of, 364
Inflammation, abnormalities of, diseases associated with, 680-681
and anti-inflammatory drugs, 679
detection of, erythrocyte sedimentation rate in, 678
kinins in, 26
local, response to, cellular proliferation and repair in, 675
mononuclear cell exudation in, 674-675
neutrophil exudation in, 672-674
phagocytosis and killing in, 675-677
stages of, 671-677
studies of, Rebuck skin window in, 675
vascular permeability in, increased, 671-672, *673*
response to, C3 component of complement in, 716-717
systemic manifestations of, 677-679
Inlet disorders, cardiac, 14-15
Insufficiency, arterial, 33
mesenteric vascular, 33
respiratory. *See* Respiration, insufficiency of
vascular, and peptic ulceration, 628

Insulin, in adipose tissue metabolism, 344
circulating, in muscle metabolism, 338, 339
in diabetes mellitus, juvenile, 350
maturity onset, 350-351
in hepatic metabolism, 340
production of, diminished, in diabetes mellitus, 351
in starvation, 345-347
Interferon(s), 173, 679
production of, in lungs, 179
Intestine(s), absorption by, tests of, 478
contents of, in fat digestion, 440, 442
secretion of electrolytes and water by, increased, 431
small, bacterial flora of, 425-426
overgrowth of, tests of, 480
pathological importance of, 426
blood flow in, 425
neural control of, 628
pathophysiological aspects of, 628
reflex and central nervous control of, 627
disorders of, derangements of motility in, 622
duodenum of. *See* Duodenum
epithelium of, absorptive function of, 425
cells of, 423-424, *424*
activity of, 428
disturbances of, 424-425
in transport processes, 428-429
structure of, 426-428
transport by, mechanisms of, 426-431
flow rate in, 425, 432
function of, disorders of, 403
infarction of, 403
integrating mechanisms in, 402-403
motility of, parasympathetic control of, 621, *617*
pathophysiological aspects of, 622
reflex and central nervous control of, 621-622, *619*
sympathetic control of, 621, *618*
muscle of, 389
slow waves in, 402-403
pendular movements of, 402, 621
peristalsis in, 402, 621

Intestine(s) (*continued*)
proximal, bacterial flora of, 426
relaxation of, 621
reversed segments of, 403
segmenting contractions of, 402, 621
surface area of, 423
Intrinsic factor, in gastric atrophy, 419
secretion of, 418-419
Inulin, clearance, in measurement of glomerular filtration rate, 269-270
in measurement of extracellular fluid, 49
sieving coefficient of, 195, *196*
urinary excretion of, 189-190
Iodide, organification defect, 306
thyroid concentration of, 304
transport defect, 304
Iodine, storage of, 306
Iodopeptides, in serum, abnormal, 308
Iodotyrosine, dehalogenase defect, 308
Iron, absorption of, 438, 504
factors influencing, 504
-binding, 505
deficiency of, 438
anemia associated with, diagnosis of, 505-506
distribution of, kinetics and pattern of, 505
in hemoglobin synthesis, 503-506
loss of, consequences of, 504
"messenger," 504
storage of, 505
depletion of, causes of, 506
Ischemia, myocardial, sites of, 19, *19*
Isosexual sexual precocity syndrome, 314

Jaundice, in cirrhosis of liver, 459
neonatal, 457, 458
obstructive, vitamin K absorption in, 535
"physiological," of newborn, 458
"regurgitation," 457
in viral hepatitis, 459
Juxtaglomerular apparatus, as source of renin, 56
description of, 56, *56*

Kallikrein, activation of, 176, 177
-kinin system, 6

and lungs, 176, *175*
Kallmans syndrome, 363
Kernicterus, in bilirubin overproduction, 457
in glucuronyl transferase deficiency, 458
Ketoacidosis, in diabetes mellitus, 349
Ketogenesis, in diabetes mellitus, 348
in hepatic metabolism, 343
17-Ketosteroids, urinary, 356
analysis of, 361-362
Kidney(s), acid excretion by, 255-256, *257*
adrenal cortical hormones and, 263
defective, and metabolic acidosis, 260-261
effects of Na^+, K^+ and Cl^- balances on, 262
albumin catabolism in, 210, 211
allograft to, rejection of, 726
arterial stenosis of, plasma renin activity and diastolic arterial pressure in, 58, *58*, 59
ascending limb of, in Addison's disease, 327-328
in primary aldosteronism, 329
autoregulation of blood flow by, 30
blood flow in, 36-37
decreased, causes of, 225
measurement of, 271
in calcium balance maintenance, 330
concentrating and diluting capacity of, measures of, gradient, 233-234
rate-dependent, 233-234
concentration of urine in, 235-241
dilution of urine in, 241-246
disease of, arterial, plasma volume in, 52, 51t
chronic, azotemic, acid excretion in, 260
impaired concentrating ability in, 239-240
impaired diluting ability in, 244-245
parenchymatous, acid excretion in, 260
end-stage, hypertension in, 47
hypertensions associated with, 45-47
parenchymal, and hypertension, 47
and erythropoietin, 493
failure of, acute, impaired diluting ability in, 245

chronic, hyperkalemia in, 228-229
circulatory, in severe liver disease, prognosis in, 464-465
symptoms of, 464
in severe liver disease, pathophysiology of, 464-465
sodium retention in, 223
terminal, plasma renin activity in, 59, *46*
fluid balance maintenance by, 218
in fluid and electrolyte homeostasis maintenance, 215-231
function of, 189
adrenal cortical hormones and antidiuretic hormones, interrelationships between, 323
potassium in, 325
tests of, 269-277
normal range of values in, 272t
functional changes in, with hypoxia, 165
hypokalemia affecting, 230
in maintenance of body tonicity, 233-247
mechanisms of, 187-282
medulla(ae) of. *See* Medulla(ae), renal
perfusion of, defective, and hypertension, 29
plasma flow in, measurement of, 271
pressor mechanisms of, 55-60
pressor system of, description of, 55, *55*
measurement of, 55-56
in protein homeostasis maintenance, 195-214
responses to changes in Pa_{co2}, 259
tubules of. *See* Tubule(s), renal
water reabsorption by, 319-322
Kinase(s), actions of, 341
phosphoglycerate, 161
pyruvate, 161
Kinin(s), liberation of, in lungs, 176, *175*
pathophysiological role of, 26-27
physiological role of, 26
production of, 672
and vascular permeability, 671-672
Kininogen, 176, 672
Klinefelter's syndrome, clinical signs of, 364
definition of, 364
testis in, 364

therapy in, 364
Krebs tricarboxylic acid cycle, 156-157, *156*
Kulchitsky, cells of, 424
Kyphoscoliosis, respiratory pattern in, 131

Lactase, activity of, 434
in oligodisaccharidase deficiency, 435
assay, 480
Lactate accumulation of, in cardiac disease states, 16
concentrations of, factors influencing, 162
dehydrogenase, 161
excess, definition of, 163
formation of, 162
in anaerobic glycolysis, 337
pyruvate ratio, measurement of, 162-163
Lactic acid, 159
Lactogen, human placental, function of, 358
secretion of, 370-371
synthesis of, 358
Lactulose, in hepatic encephalopathy therapy, 467
Laennec's cirrhosis, 31-32
Lampreys, immunobiological analysis of, 704
Lead, poisoning, hemolytic anemia in, 510
Learning, definition of, 656
instrumental, and classical conditioning, differences between, 658, *658*
of visceral and glandular responses, 657-659
mechanisms of, 653-661
L-dopa, synthesis of, 61
Lecithin, absorption of, 444
dipalmitoyl, synthesis of, in lung, 172-173, *172*
pancreatic conversion of, 417
Leukapheresis, 521-522
Leukemia(s), cellular disorder in, 523-524
chronic myelocytic, neutrophil production and turnover in, 517
and pluripotential stem cells, 516
classification of, 524
hypofibrinogenemia in, 536
Leukemoid reactions, 524
Leukocyte(s), bone marrow, 520-522
circulating, pool size and distribution of, normal, 517-518
turnover and survival time of, in disease, 519-520

normal, 518-519
"lifespan" of, 515-516
membrane, in phagocytosis, 487
migration of, abnormalities of, 680
production of. *See* Leukopoiesis
reserve of, test of, 521-522
in tissues, 522
Leukocytosis, 678
"shift," 519-520
Leukopoiesis, 515-525
cell division in mitotic compartment during, *520*
control of, 516-517
in disease, 522-524
stem cell in, 516
Lidocaine, in atrial fibrillation, 96
and cardiac impulse conduction, 86
and cardiac impulse generation, 85
Limbic system, 578-579, *579*, 588
and cardiovascular function, 602
in emotional and visceral behavior, 656
seizures of, and neocortical focal seizures, compared, 586
Lipase, deficiency of, 441
therapy for, 443
pancreatic, in fat digestion, 439
Lipid(s), absorption of, 439-448
biosynthesis of, in lungs, 172-173
plasma, in nephrotic syndrome, 212
synthesis of, hepatic, 212
synthesis of, disturbances in, in liver disease, 469
Lipolysis, in adipose tissue metabolism, 344
Lipoprotein(s), 7
in adipose tissue metabolism, 344
plasma, in nephrotic syndrome, 212
β-Lipoproteinemia, 442
Liver, albumin catabolism in, 210, 211
albumin synthesis in, regulation of, 212
anatomical position of, 455
bile acid synthesis by, 447
and carbohydrate metabolism disturbances, 468-469
carcinoma of, bleeding syndrome in, 537

cirrhosis of. *See* Cirrhosis
in diabetes mellitus, 347-348
disease of, cholesterol and bile acid abnormalities in, 467-468
chronic, bleeding syndrome in, 537
edema in, 223
parenchymal, 535
renal failure in, pathological physiology of, 464-465
symptoms of, multiplicity of, 455
syndrome of confusion and neurological changes in, 465-467
fat metabolism by, 469
disturbances in, 469
functions of, categories of, 455
clinically used tests of, 482-484t
disturbances of, 455
with hypoxia, 165
and intestine, recycling of bile acids by, 441, 447
mechanisms of, 455-473
metabolism of, fasting state, 342-343, *340*
fed state, 341-342, *340*
unique features of, 339-341
necrosis of, fulminating, ammonia removal in, 467
and protein metabolism disturbances, 469
secretions of, control of, *408*
sequestration in, 512
in starvation, 345
and trace metals metabolism disturbances, 470
tumor of, neoplastic, and portal hypertension, 460
vascular changes of, and portal hypertension, 459-461
and vitamin metabolism disturbances, 470
Löffler's syndrome, 180
Lucey-Driscoll syndrome, 457-458
Luft's syndrome, 159
Lung(s), allergic reactions in, anaphylactic, 180
arthus type, 180
cell-mediated, 180
cytotoxic, 180
alveolar-capillary block syndrome in, 19
alveolar stability in, maintenance of, 173
alveolar ventilation of, 131-132
arterial occlusion of, surfactant inadequacy in, 173

Lung(s) (*continued*)
arteriovenous fistulae of, oxygen deficiency from, 31
atelectasis of, 128
blood flow in, distribution of, 132-133, *132*
blood-gas equilibrium in, assessment of, 140
carbohydrate metabolism in, 174-175
cells in, alveolar and bronchiolar, 167-172
compliance of, frequency-dependent, 129
  measurement of, 127
  in three different types of lungs, 127-128, *128*
concentrating ability of, impaired, in cirrhosis, 240
defense of, against infection, 178-179
diffusion in, capacity of, assessment of, 140-141
  defect of, 134
diluting ability of, impaired, in cirrhosis, 245
disease of, chronic obstructive, patterns of ventilatory function in, 139, 139t
  diagnosis of, acid-base balance assessment in, 141
  gas exchange asessment in, 140
  endocrine syndromes in, 180-181, 181t
  parenchymal, 14-15
  and proteolysis, 173-174
  edema syndrome of, 14
  surfactant inadequacy in, 173
embolism of hypoperfusion syndrome in, 128-129
  release of vasoactive substances in, 177, 178t
excised, studies of elasticity in, 128, *128*
fibrinolytic activity of, 175, *175*
fibrosis of, compliance of lungs in, 128, *128*, 130, 131
  diffuse, 134
  patterns of ventilatory function in, 140, *139*, 139t
function of, assessment of, 136-140
gas exchange in, 131-134
  abnormalities of, during exercise, 141-142, 141t
  assessment of, 140-141
  and hematologic mechanisms, 175-176

hormone secretion by, potential for, 180-181
immunological responses of, 179-180
inspired air in, distribution of, 132, *132*
  effect of localized airway obstruction on, 132, *132*
  and kallikrein-kinin system, 176, *175*
lecithin synthesis in, 172-173, *172*
lipid and phospholipid biosynthesis in, 172-173
mechanisms of, 109-183
metabolic events in, 167-181
metabolic pathways in, histochemical demonstration of, 167, *168*, *169*
metabolism in, cellular sites of, 167-172
mucociliary action in, 178-179
mushroom-picker's, 180
normal, compliance of, 127-128, *128*
oxygen exchange in, and in tissues, compared, 153, 154-155, *154*
oxygen supply of, 146
oxygen uptake by, 145-146
phagocytosis in, 178
protein synthesis and secretion by, 173-174
reflexes arising from, influencing breathing, 119-120
resistance of, elastic, 127-129
  vital capacity in determination of, 136
  nonelastic, 129
  and elastic interaction between, 129
  forced vital capacity in determination of, 136-138
  normal, 129
  with obstruction, 129
  "shocked," 128-129
  "stiff," 128, 130
surfactant of, function of, 128
thromboembolic obstruction of, 134
tidal volume distribution in, in normal and disease conditions, 129, *129*
total capacity of, 127
  in various states, 136, *136*
true venous admixture in, 133-134
  assessment of, 140
tumors of, endocrine syndromes associated with, 180-181, 181t
vascular bed of, functional changes in, with hypoxia, 165

vasoactive substances in, metabolism of, 177, *176*
  alterations in, and human disease, 178
  release of, experimental and clinical conditions associated with, 177, 178t
  uptake of, 176-177, *176*
ventilation/perfusion relationships of, 133, *132*
  assessment of, 140-141
ventilatory function of, 127-131
  assessment of, 136-139
  impaired, patterns of, obstructive, 139, 139t
  restrictive, 140, 139, 139t
  normal values of, 138t
  volume(s), 127
  dynamic, determination of, 136-139
  residual, 127
  static, determination of, 136
Lupus erythematosus, systemic, 35
  antibody development in, 741-742
  clinical manifestations of, 742
  conditions precipitating, 741
  diseases associated with, 742
  glomerulonephritis in, 722-723
  pathological manifestations of, 742, *743*
Lymph, loss of, in chyluria, 207
Lymphatics, cardiac, 6
  in control of extracellular fluid, 50
  intestinal, in absorption, 425
Lymphocyte(s), B-cells of. See Cell(s), B-
  depletion of, in immunosuppressive therapy, 744
  immunoglobulin secretion by, 689, *688*, 690, 690t
  products of, following T-cell stimulation, 692t
  T-cells of. See Cell(s), T-
  thymus-dependent, in acute graft rejection, 734
  and immunological damage, 731
  two lines of, 691-692
Lymphoid system, malignancy of, immunodeficiencies secondary to, 702, 702t
Lymphokines, currently recognized, 692t
Lysolecithin, absorption of, 444
  in gastric mucosal barrier destruction, 417
Lysosomes, 7, 9

McArdles syndrome, 165-166
α_2Macroglobulin, in nephrotic syndrome, 212
Macroglobulinemia, Waldenström's, altered viscosity in, 35
Macrophage(s), alveolar, 163, 170, *170*
  phagocytic ability of, 178
Magnesium, -calcium competition at gut and tubular membrane, 331
Malabsorption, definition of, 448-450
  mechanisms of, 450
  systemic effects of, 450
Malignancy, and immunity, 707-708
Malnutrition, impaired urinary concentrating ability in, 241
Maltase, activity of, 434
Mannitol, in therapeutic diuresis, 224
Mechanoreceptor(s), arterial, 603-604
  abnormal function of, manifestations of, 604
  cardiac, 604-605
    stimulation of, 605
  carotid sinus, 70-72
  description of, 70
  high pressure, 603-604
  improper function of, results of, 72
  locations of, 70, 72-73
  low pressure, 604-605
    stimulation of, 605
  in reduction of arterial pressure, 70-73
  stimulation of, 70
Mediastinum, tumors of, inappropriate antidiuretic hormone secretion in, 226
Medulla(ae), oblongata, 602
  compression of, respiratory syndrome associated with, 123
  lesions of, breathing patterns associated with, 122-124
  respiratory centers in, 112-113, *113*
  renal, cystic disease of, 239-240
  diluting segment of, sodium transport in, 242
  hyperosmolality of, impaired, 239
  interstitium of, hyperosmotic, 235, 237
    hypertonic, 236-237
  "washout" effect of, 237, 239
Meekrin-Ehlers-Danlos syndrome, 543

Megacolon, congenital aganglionic, 404-405, 624
Megakaryocytes, 175, 537-538
Membrane(s), cellular, 7-8
  external basement, 7, 8
  intestinal. *See* Intestine, small, epithelium of
  plasma, 7, 8
  resistance to oxygen flow by, 153
Memories, storage of, 593-594, *593*
Menstruation, cycle of, normal, hormonal patterns in, 365-366, *365*
6-Mercaptopurine, as anti-inflammatory agent, 679
  and leukocyte migration defect, 680
Mercurials, in sodium reabsorption inhibition, 244
Mesentery, circulation of, 386
  impaired, 386-387
  ischemia of, nonocclusive, pathophysiological mechanisms in, *387*
  vascular diseases of, 386
  vascular insufficiency of, 33
Metabolic pathways, intracellular, *17*
Metabolism, adipose tissue, 343-344, *343*
  altered states and diseases of, 344-352
  anaerobic, adenosine triphosphate produced by, 16, *17*
  general features of, 335-337
  hepatic, 339-343, *340*
  mechanisms of, 335-353
  muscle, 338-339, *338*
Methemoglobinemia, 32
Metyrapone, effects of, on adrenal cortex, *290*, 293
  in investigation of pituitary-adrenal axis abnormalities, 289-290
Micelles, in fat digestion, 440, *439*
Microcirculation, fibrinolytic activity in, 553
β-Microglobulin, in tubular proteinuria, 207
Mitochondrion(a), 7, 8, 17, 19
  in cardiac hypertrophy, 13
  chemical reactions in, 156-157, *156*
  energy production in, 337
  giant, 155-156
  loss of, in cardiac failure, 18
  oxidative metabolism in, 8
  oxygen diffusion pathway in, 156

  oxygen utilization by, abnormalities of, monitoring of, 162-163
  and anaerobic glycolysis, interrelationship between, 161-162
  disorders affecting, 158-159
  long-term regulation of, 158
  short-term regulation of, 158
  structure of, 155
Mole(s) hydatidiform, thyroid hyperfunction with, 311-312
Molecule(s), amphipathic, in fat digestion, 440, *439*
Mönckeberg's sclerosis, 33
Monoidotyrosine, coupling of, defect in, 306-308
Mononeuritis, 638
Monosaccharides, absorption of, 434, 435-436
Motility, autonomic control of, in colon and rectum, 622-625
  in small intestine, 621-622
  in stomach, 616-621
Motor system, analysis and synthesis in, 574
  embryological organization of, 577
  spinal and brain stem, compared, 578, *578*
  synthetic functions of, 580-582
Movement(s), mechanisms of, 642-650
  involuntary, disease mechanisms producing, 643-650
  voluntary, pathways of, 642-643
Mucopolysaccharides, 175
Mucosa, duodenal, 413, 415
  absorptive disturbances of, and decreased absorption, 450
  gastric, barrier of. *See* Stomach, mucosal barrier of
  glandular, oxyntic, 414-415
    pyloric, 415
  uptake by, tests of, 479
Mucoviscidosis, glycoproteins in, 175
Mucus, abnormalities of, 178-179
  and ciliary function, 178
  gastric, 419
Muscle(s), atrophy of, denervation, 631
  disease mechanisms producing, 631-632
  disuse, 632
  in myositis, 631

Muscle(s) (*continued*)
  bulk of, normal, development
      and maintenance of,
      mechanisms in, 631-632
  cardiac. *See* Myocardium
  colonic, 389-390
  contraction of, 634, 635
  deficiency, phosphofructase
      kinase, 165-166
    phosphorylase, 165-166
  dystrophy of, 641
  elongation of, 634
  esophageal, 389
    smooth, control of, 395
    in opossum, 395
  fibers, intrafusal, 632-634
  function of, normal, potassium
      in, 641-642
  gastric, 389
  gastrointestinal, anatomy of,
      389-390
    smooth, electrophysiology
        of, 393-394
      integration of contraction
          in, 392-394
      cell membrane electrical
          properties in, 393
      extrinsic nerves in, 392-393
      hormones in, 393
      local reflexes in, 393
    ultrastructure of, 390
  hypertonia of, disease mech-
      anisms producing, 636-637
  hypertrophy of, disease mech-
      anisms producing, 632
  hypotonia of, disease mecha-
      nisms producing, 635-636
  inhibition of, presynaptic,
      634-635
    recurrent collateral, 635
  intercostal, muscle spindles in,
      as source of input to
      spinal motor neurons, 114
  maximum velocity of shorten-
      ing of, 12
  metabolism of, fasting state,
      339, *338*
    fed state, 338-339, *338*
  respiratory, work of, 129-130
  skeletal, functional changes in,
      with hypoxia, 165-166
    neural control of, 631-652
  system of, and essential
      hypertension, 44-45
  smooth, action potential of,
      393
    electrophysiology of, 393-394
    resting membrane potential
        of, 393
    slow waves of, 394
      in small intestine, 402-403
      in stomach, 399-400
  vascular. *See* Muscle(s),
      vascular smooth

spindles, composition of,
    632-633
  function of, 114
    in dyspnea, 124-125
  and related neural pathways,
      633
  strength of, development and
      maintenance of, mecha-
      nism for, 637-642
  tone of, evaluation of, 636
    normal, development and
        maintenance of, mech-
        anisms in, 632-637
    influences on, 632-635
    vascular smooth, basal tone
        of, 625
    contraction of, humoral fac-
        tors regulating, 24-27
      intrinsic tissue factors
          regulating, 30-32
      neurogenic factors regulat-
          ing, 27-30
    and neural influences, 600
  weakness of, disease mecha-
      nisms producing, 638,
      638t
Muscularis propria, gastroin-
    testinal innervation of,
    616
  parasympathetic control of,
      617-618, *617*, 621
  control of, 618, *618*
Myasthenia gravis, clinical
    findings in, 639
  esophageal dysfunction in, 398
  and thymic disorders, 639
Myasthenic syndrome, 640
Myelocyte(s), production of,
    520-521, *520*
Myelofibrosis, total blood gran-
    ulocyte pool in, 520, *519*
Myeloma, multiple, resistance
    to pathogens in, 684
Myelopoiesis, cellular divisions
    during, 520-521, *520*
Myocardium, adaptability of,
    11-13
  atrophy of, 16
  cells of. *See* Cell(s),
      myocardial
  contractility of, in cardiac
      failure, 612
    measurement of changes in,
        12
  dilatation of, 13
  failure of, 15-16
    mechanisms of, 16-18
  fibrosis of, 16
  force-velocity relation of, 12
  high energy phosphate com-
      pounds in production
      in, 16-17
  hypertrophy of, 16
  infarction of, 7, 15, 18

ischemia of, 18
  sites of, 19, *19*
  macrostructure of, 5-7, *5*
  mass of, increase in, 13
  maximum velocity of shorten-
      ing of, 12
  nerve endings in, morphology
      of, 610
  oxygen consumption by, 12
  pathological changes in, in
      cerebral disease, 611-612
  tension of, measurement of
      level of, 12
  ultrastructure of, 7-9
Myoclonus, 650
Myocytolysis, focal, 611
Myofibrils, 7, 8, 13
Myofils, abnormalities in, 20
Myoglobin, 155
  free, in plasma, and protein-
      uria, 205
Myokymias, 650
Myopathy(ies), 640-641
  thyrotoxic, 640-641
Myosin, 8, 9, 17
Myositis, atrophy in, 631
Myotonia, 637
Myxedema, primary thyroid,
    312
  and hypothyroidism second-
      ary to pituitary failure,
      compared, 312-313
  syndromes associated with,
      313-314

Natriuresis, 219-220
  hypertension-exaggerated, con-
      trol of extracellular fluids
      in, 54-55
  postural, in hydropenia, 238
Nelson's syndrome, and pitui-
    tary tumors, 290-291
Neomycin, in hepatic encepha-
    lopathy therapy, 466
Neoplasia, thyroid hyperfunc-
    tion with, 311-312
Nephritis, nephrotoxic, comple-
    ment in, experimental
    model demonstrating,
    728
Nephrocalcinosis, 240
Nephron, in chronic renal dis-
    ease, 239
  collecting duct of, permeabil-
      ity of, to water, 237
    decreased, 239
    water reabsorption in, re-
        quirements for, 235-236
  diluting segments of, increased
      permeability to water of,
      causes of, 243-244
    sodium transport in, 242
      decreased, 244

distal tubule of, permeability of, to water, decreased, 239
"intact," hypothesis, 190, 244-245
schematic diagram of, *219*
Nephropathy, hypercalcemic, impaired concentrating ability in, 240
hypokalemic, impaired concentrating ability in, 240
postobstructive, impaired concentrating ability in, 240
Nephrosis, impaired diluting ability in, 245
Nephrotic syndrome, albumin metabolism in, 210, 210t, 211
coagulation factors in, 212
edema in, 212-213, 223
reversal of, 223-224
impaired concentrating ability in, 240
plasma fibrinogen in, 212
plasma lipids in, 211-212
plasma protein level changes in, 212, 223
Nerve(s), afferent, of gastrointestinal tract, 615-616
in swallowing, 395
block of, intercostal, 19
cranial, respiratory function of, 117, *113*
deafness, and congenital goiter, 306
efferent, dilator, and skeletal muscle, 74
of gastrointestinal tract, 615
gastrointestinal, extrinsic, 390
and contraction in smooth muscle, 392-393
of gut, 390-392
mechanisms of, and catecholamine, 60-65
and plasma volume, in hypertension, 53-54
parasympathetic vasodilator fibers of, organs supplied by, 74
reflexes of, in hypertensive patients, 63-64
sensory, in dilatation and local depressor responses, 73
endings of, of intrafusal muscle fibers, 634
splanchnic, and biliary system, 401
sympathetic, endings of, in myocardium, morphology of, 610
in norepinephrine release, 61, 608, 610
vasodilator fibers of, activation of, 73, 74

vagus, afferent, in dyspnea, 125
and biliary system, 401
in cardiac control, 609-610
in gastric functioning, 617-618, *617*
and gastrointestinal tract, 390
pneumotaxic center of, 112
Nervous system, autonomic, and blood proteins, interaction of, 488
central aspects of, 655-656
in control of circulation, 60-61, 609
in control of gastrointestinal function, 615-630
peripheral, organization of, 653
parasympathetic division of, 654, *655*
organ functions influenced by, 655
sympathetic division of, 653-654, *654*
organ functions influenced by, 655
regulation of cardiac rate by, 10
central, and ACTH secretion, 288
and ammonia intoxication, 465-467
and arteriolar vasoconstriction, 600
as influence on arterial pressure, 73-74
in blood flow regulation, 599-600
during exercise, 600-601
cardiac control by, 608-613
and cardiovascular reactions to emotions, 601-602
cardiovascular reflex control by, 603-607
chemoreceptors of, 116-117, 605-606
disease of, effects of, on breathing, 120-125
on cardiac function, 611-612
functional changes in, with hypoxia, 164
lesions of, patterns of breathing associated with, 121t
in maintenance of arterial pressure, 600
during exercise, 600-601
and maintenance of thyroid function, 301
neuroeffector function of, and cardiovascular system, 607-608
neurons in, 573-574, *574*

supramedullary regions of, and cardiac rate and conduction, 611
in temperature regulation, 607
and endocrine glands, interdependence of parts of, 569-570
evolution of, regulatory activity of hydra as example of, 571-572, *571*
horizontal and vertical organization of, 575-579, *575*
input and output channels of, 573-575
integration in, by analysis and synthesis, 579-582
mechanisms of, evolution of, 571-572
regulatory, embryological emergence of, 572-573
somatic, origin of, 572
transactional, 570-571
propriospinal path of, 576, 577
regulation, pathophysiological mechanisms affecting, 582-594
in regulation of perceptual processes, 591-592
reticular formations of, 572
reticular scaffolding of, development of, 573
spino-bulbar relay system of, 576-577
storage of information by, 593-594, *593*
sympathetic, adrenergic component of, 27-28
in cardiac failure, 612-613
cholinergic component of, 30
in control of circulation, 61
function of, 27
and plasma renin activity, 57, 59-60
in visceral regulation, 589-591
Neuritis, peripheral, 631
Neuroeffector system, and cardiovascular control, 608
Neuromuscular junction, disorders of, 639-640
normal physiology of, 638-639
poisons, 640
Neuron(s), bulbar, and vasoconstriction, 590
central, 574, *574*
in central nervous system, 573-574, *574*
enteric, 391
motor, alpha, in muscle tone maintenance, 634-635
in spinal motor pool, 114
dysfunction of, and muscular weakness, 638

Neuron(s) (*continued*)
gamma, in muscle tone maintenance, 635
in spinal motor pool, 114
of gastrointestinal tract, 615
and neurotransmitters, 654-655
reticular, embryological emergence of, 572-573
formation of, 580, *581*
sensory, of gastrointestinal tract, 615-616
spinal, 602
motor, function of, in breathing, 114
membrane potential of, systems determining, 113-114
types of, 114
Neuropathy, autonomic, orthostatic hypotension from, 27-28
and muscular weakness, 638
Neurosis(es), experimental, and individual differences, 657
Neurotransmitters, synthesis and release of, 654-655
in visceral regulation, 654
Neutropenia, causes of, 523
clinical findings in, 523
cyclic, inflammatory response in, 680
severe, inflammatory response in, 680
total blood granulocyte pool abnormalities in, 520, *519*
Neutrophil(s), blood, nitro blue tetrazolium test of, 749t
chemotaxis of, 674
control of, 516-517
exudation of, in local inflammatory response, 672-674
maturation of, sequence of, 515, *516*, 521
peripheral blood concentration of, and total blood granulocyte pool, *519*
pool size of, measurement of, 518
production of, at myelocyte stage, 520-521, *520*
Neutrophilia, acute, in acute infection, 523
chronic, disease induced, 522-523
from adrenal corticosteroids, 523
total blood granulocyte pool abnormalities in, 519-520, *519*
Nicotinamide, adenosine dinucleotide, in Kreb's cycle, 156-157
deficiency of, 159

Nicotine, as stimulus to antidiuretic hormone release, 321
Nitroglycerin, 18, 19
Node(s), atrioventricular, 6, 610
function of, neural influence on, 611
sino-atrial, 6, 610
function of, neural influence on, 610-611
sinus, pacemaker activity in, 83
Norepinephrine, in blood, 25
cardiac supply of, 610
depletion of, in cardiac failure, 17, 612
excess of, in thyrotoxicosis, 19-20
in exercise, and proteinuria, 203
function of, in sympathetic efferent neuroeffector system, 608
plasma concentrations of, measurement of, 63
released from nerve endings, fate of, 61-63, 62t, 608
synthesis of, 61
Normoblast, functions of, 489
Nucleic acid, synthesis of, cardiac, 9, 13, 16
defective, 17-18
Nucleus, ambiguus, respiratory cells in, 112, *113*
parabrachialis medialis, 112
retroambiguus, respiratory cells in, 112, 113-114, *113*
Nystagmus, 649-650
labyrinthine, 649
miner's, 649-650
opticokinetic, 650
palatal, 650

Obesity, oxygen consumption changes in, 130, *130*
respiratory pattern in, 131
Oddi, sphincter of, 401
Oligodisaccharidase, deficiency of, 434-435
Oligosaccharidases, 434
Ondine's curse, 123
location of spinal cord incisions producing, *123*
Opisthotonos, 636
Opsonization, 675-676
Osmolality, 217
Osmometry, freezing point depression, in measurement of urine concentration, 233

Osteogenesis imperfecta, 543
Osteomalacia, in medullary thyroid carcinoma, 333-334
Outlet disorders, cardiac, 15
Ovary(ies), compartments of, 364-365
steroid-synthetic capacities of, 365
in menstrual cycle, 365-366, *365*
diseases of, and amenorrhea, 367
interstitium of, activity of, 365
polycystic, syndrome of, 368-369
tumors of, and virilization, 368
virilizing syndromes of, 368-369
Ovulation, 365-366
induction of, clomiphene citrate in, 368
follicle-stimulating hormone in, 367-368
Oxygen, arterial partial pressure of, in disease states, 133, 134, 138, 140
arterial tension of, 115
low, and hypoxemia, 145-146
measurement of, 145, 146
binding to hemoglobin. See Hemoglobin, affinity for oxygen
blood content of, 491
cellular delivery of, 146-152
cellular uptake of, oxygen content and partial pressure in, 152
consumption of, by brain, 164
in breathing, mechanical resistances altering, 130, *130*
cardiac, 164
increased, causes of, 32
myocardial, 12
renal, 165
deficiency of, in tissues, causes of, 31-32
depletion of, profound. See Hypoxia
difference, A-a, 153
V-c, 153-155
anatomical shunting as factor in, 154-155
diffusion limitation in, 153-154
physiological shunting influencing, 154
diffusion of, distances and pathways as factors in, 153
driving pressure as influence on, 152
limitation of, 153-154

mitochondrial pathway of, 156

physicochemical properties of membranes as influence on, 153

surface area as factor in, 152-153

time availability influencing, 153

electron passage from glucose to, in energy production, 157

and energy provision, 155-159

exchange of, placental, 2,3-diphosphoglycerate in, 150

pulmonary and tissue, compared, 153, 154-155, *154*

gradient of, mixed venous-metabolizing cell, 153-155

lack of, and blood pressure, 75

chemoreceptor activity in, 605-606

in local regulation of blood flow, 74-75

metabolism of, measurement for monitoring, 145t

molecular, from substrate, reactions in, 157

movement of, from plasma to intracellular sites of utilization, 152-153

and nonenergy providing processes, 163-164

partial pressure of, 131

pulmonary uptake of, 145-146

supply of, arterial oxygen content influencing, 146

cardiac output influencing, 146-147

hemoglobin affinity for oxygen influencing, 147-152, 500

tension, arterial, low, and hypoxemia, 145-146

mean capillary, factors influencing, 490

in tissues, factors influencing, 489-490

tissue stores of, 155, 490-491

transport of, erythrocytes in, 489-492

measurement for monitoring, 145t

utilization of, cellular, 155-159

mitochondrial. *See* Mitochondrion(a), oxygen utilization by

Oxyhemoglobin dissociation curve, 147-148, *147*

physiological and pathophysiological factors modifying, 148t

shifts to left in, potential pathological effects of, 151

shifts to right in, potential pathological effects of, 151

Oxymyoglobin dissociation curve, 155, *155*

Pacemaker activity, cardiac, 6

abnormal site of, example of, 85

exposure of, 85

drugs influencing, 84-85

ectopic, suppression of, 84

frequency of, 84-85

phases of, 83-84, *84*

in Purkinje system, 85, *84*

sites of, 83

Paget's disease, vascular anomalies in, 34

Pain, as stimulus to antidiuretic hormone secretion, 320

as subjective experience, 582-583

burning sensations with, 583-584

definition of, 582

ischemic, as stressful stimulus in hypertension, 48

perception of, pathophysiological processes involved in, 582-586

reflex disorders and trophic changes associated with, 583, *584*

syndromes, pathophysiology of, 584-586

therapy for, 585

Palate, soft, involuntary movement of, 650

Pancreas, beta cell of, and insulin of diabetes mellitus, 351

deficiency of, 420

endocrine function of, 385-386

enzymes of, 419-420

exocrine function of, 385

clinical assessment of, 386

disordered, causes of, 386

tests of, 478-479

insufficiency of, therapy for, 443

secretion(s) of, components of, 407

control of, 407-409, *408*

dysfunction of, hormonal imbalance in, 409

nervous regulation deviations in, 409

mechanisms of, 407-422

test of, 477

tumor of, course of, "case history" of, 412-413

Pancreozymin, 409

in fat digestion, 439

Panhypopituitarism, characteristics of, 298-299

diagnosis of, 299

etiology of, 298

treatment of, 299

Paraaminohippurate, clearance, in measurement of renal plasma flow, 271

tubular maximum transport capacity of, measurement of, 271-272

Parahemophilia, 534

Paralysis(es), agitans, 646

flaccid, 638

muscular, ventilation in, 131

periodic, hyperkalemic, 642

hypokalemic, 641-642

normokalemic, 642

Parasystole, 87

Parkinson's disease, arteriosclerotic, 646

hypomotility in, 643

idiopathic, 646

postencephalitic, 646

postural disturbance in, 644-645

Pasteur effect, 161-162

Patent ductus arteriosus, 15

Pathogen(s), immunosuppressive actions of, 685

specific, resistance to, 684-685

Pavlov, and conditioned reflex studies, 657

Pellagra, 159

Pendred's syndrome, 306

Pentagastrin, in test of gastric secretion, 411-412

Pepsin, in gastric secretion, 419

Peptide(s), vasoactive intestinal, in lungs, 177

Perception, processes of, regulation of, 591-593

cultural experience influencing, 592

Perfusion, pulmonary, distribution of, 132-133, *132*

-ventilation relationships, 133, *132*

tests of, 133

of tissues. *See* Tissue(s), perfusion of

venous-admixture-like, 133

Pericarditis, constrictive, 14

Peristalsis, antral, gastric slow wave in, 399-400

gastric, 616

pharyngoesophageal, 394-395

in achalasia, 396

secondary, 395

in small intestine, 402, 621

Peritonitis, 622
in phagocytosis, 676-677
Petechiae, 530
Petit mal seizures, 587
pH, arterial, mechanisms altering, 115
in blood, measurement of, 275
changes in, as influence on hemoglobin affinity for oxygen, 148
and [H+], interconversion between, in notation of hydrogen ion concentration, 253t
in urine, measurement of, 276
normal values for, 277
Phagocytosis, as function of inflammatory reaction, 675-677
defects of, 680
leukocyte membrane in, 487
in lungs, 178
process of, 676-677, 676
Pharyngeal pouch syndrome, 695-696, 697
Pharynx, 394-396
muscle of, 389
Phenolsulfonphthalein, clearance, in measurement of renal plasma flow, 271
Phenylethanolamine-N-methyl transferase, and norepinephrine methylation, 61
Pheochromocytoma, 25
catecholamines in, 63
hyperglycemia in, 350
plasma volume in, 52
with thyroid carcinoma, 316
Phosphate, compounds, high energy, myocardial production of, 16-17
levels of, and 2,3-diphosphoglycerate, 150
renal tubular reabsorption of, parathyroid hormone in, 331
Phospholipase A, in gastric mucosal barrier destruction, 417, 419
Phospholipids, 7-8
biosynthesis of, in lungs, 172-173
dipalmitoyl phosphatidyl choline, synthesis of, in lung, 172-173, 172
in nephrotic syndrome, 211-212
Phosphorylation, oxidative, uncoupling of, 19-20, 337
Pinocytosis, 431, 437
Pituitary, -adrenal axis, 287
regulatory mechanism of, 287-299

endocrine glands and hypothalamus, interdependence of, 569-570
failure of, hypothyroidism secondary to, 312-313
regulatory mechanism of, 287-299
secretion of thyroid-stimulating hormone by, 302-304
-thyroid axis, regulatory mechanisms of, 301-318
tumors of, acromegaly and gigantism associated with, 298
and amenorrhea, 367
and Cushing's syndrome, 290-291
in men, 363
Placenta, condition of, human placental lactogen levels revealing, 371
insufficiency of, estriol determinations revealing, 370
premature separation of, intravascular clotting in, 536
Plasma, antiplasmin activity of, 548-549
cell, human, electron micrograph of, 688
immunoglobulin secretion by, 689, 688, 690, 690t
corticosteroid concentration in, 295-296
hypotonicity of, in Addison's disease, 322
normal, composition of, 216-217, 217
osmolality of, 216-217, 233
and antidiuretic hormone secretion, 320
proactivator, 548
progesterone concentration in, 366
proteins of. See Protein(s), plasma
renal, flow of, measurement of, 271
renin activity in, factors decreasing, 57
factors increasing, 57, 64, 463
in hypertension, 57-59
and intravascular volume, in hypertension, 54, 54
steroid and protein hormone concentrations in, 359t
testosterone levels in, 362
thromboplastin antecedent, 529
volume, abnormalities of, in hypertension, 51-52, 51t
in acute glomerulonephritis, 52
and antihypertensive drug therapy, 52

as component of extracellular fluid, 48
and cardiac output, in hypertension, 53
and diastolic arterial pressure, in hypertension, 52-53, 53
in essential hypertension, 52, 51t
and interstitial fluid volume, ratio, 49
relationships of, control of, 50
measurement of, 49, 215
and neural mechanisms, in hypertension, 53-54
in pheochromocytoma, 52
physiological relationships of, in hypertension, 52-55
in primary aldosteronism, 52, 51t
and red cell mass, in normal people and hypertensive patients, 51t
in renal arterial disease, 52, 51t
Plasmin, function of, 547
inhibitors of, 548-549
production of, 547-548
Plasminogen, activation of, and complement-kallikrein system, 488
in lung, 175
activator(s), 548
excessive, in pathological fibrinogenolysis, 554
in microcirculation, 553
regulation of, 551-552
in systemic circulation, 553-554
two-phase concept of, 552
concentration, areas of, 547
conversion to plasmin, 547
proactivator, 548
production of, 547
Platelets, antigen-mediated mechanism of, 672
count, decreased, decreased life span causing, 539-542, 540-541t
dilution causing, 542
reduced production causing, 539, 540t
sequestering in spleen causing, 542, 541t
increased, 542
disorders of, 537-543
function of, 538
tests of, 538-539
and glass, 538
life cycle of, 538
membrane, in clotting, 487-488
qualitative abnormalities of, acquired, 543

inherited, 542-543
role of, in hemostasis, 537-538
thrombin influencing, 538
Plexus(es), of gastrointestinal
   tract, Auerbach, 615
   enteric, 390-391
   Meissner, 615
   mucous, 391
   myenteric, 391, 615
   submucosal, 391, 615
   subserous, 391
Plummers disease, 310-311
   and Graves' disease, compared,
   310-311
Pneumonitis, Ascaris, 180
Pneumonocyte(s), 167, *169*
Poiseuille formula, 42
Poiseuille's law, 430
Poisoning, barbiturate, 158-159
   ventilation in, 131
   lead, hemolytic anemia in, 510
Poliomyelitis, bulbar, esophageal dysfunction in, 398
   respiratory syndrome associated with, 123
   paralytic, 631, 638
Polisteskinin, in lungs, 177
Polyarteritis nodosa, 35, 180
   circulatory impairment in, 19
Polycythemia, 11
   vera, stem cell defect in, 496
   symptoms of, 491
   total blood granulocyte pool
   in, 520, *519*
Polymyositis, 641
Polyneuritis, 638
Polypeptides, vasodilator, 26-27
Polyuria, as result of impaired
   concentrating ability, 238
   isotonic, 191
Polyvinylpyrrolidone, sieving
   coefficient of, 197, *196*
Pons, lesions of, breathing patterns associated with, 122
   respiratory centers in, 112,
   113, 589
Posterohypophysis, antidiuretic
   hormone release by, 242,
   245
Posture, as factor in urinary
   diluting ability, 243
   changes in, and changes in
   arterial pressure, 72,
   76-77
   disturbances of, mechanisms
   of, 644-645
   and proteinuria, 203-204
   upright, hemodynamics of,
   60-61, 63-64
Potassium, balance, effects of,
   on renal acid excretion,
   262
   body, control of, 325
   in cardiac coupling, 9

depletion of, 229-230
   renal effects of, 230
excretion of, 227-228
intoxication, 228
loss of, cellular, 325
in normal muscle functioning,
   641-642
and plasma renin activity, 57
in renal function, 325
secretion of, aldosterone influencing, 227-228
   delivery of sodium influencing, 227
   ratio of potassium to hydrogen ion influencing, 228
   serum, concentration, 191
-sodium exchange, in primary
   aldosteronism, 329
sources of, 227
Prealbumin, thyroxine-binding,
   315
Prednisone, in immunosuppressive therapy, 743
Pregnancy, and amenorrhea,
   367
   pattern of hormones throughout, *370*
   vascular anomalies in, 34-35
Pregnenolone, 356
Prehypothyroid state, 313
Prematurity, surfactant inadequacy in, 173
Pressor mechanisms, 41-66
Pressure relationships, hydrostatic-osmotic, in formation of edema, 222
Pressure-volume diagram, in
   measurement of cardiac
   work, 10, *10*
Proaccelerin, 528-529
   hereditary absence of, 534
Proactivator, plasma, 548
Procainamide, and cardiac impulse conduction, 86
   and cardiac impulse generation, 85
Progesterone, 356
   concentration of, in plasma,
   366
   metabolism of, effect of thyroid hormones upon,
   311t
   in oral contraceptives, 372
   in pregnancy, 370
Progestin(s), definition of, 355
Proinsulin, in diabetes mellitus,
   351
Pronormoblast, 495, 496
Propranolol, in atrial fibrillation, 96
   and cardiac impulse conduction, 86
   and cardiac impulse generation, 85

Prostaglandins, 27, 358
   in anaphylaxis, 177
   as abortifacients, 372
   in lungs, 177
   metabolic effects of, 176
   secretion of, in thyroid
   carcinoma, 317
Prostate, carcinoma of, 536
Prosthesis, cardiovascular,
   536-537
Protein(s), absorption of, 437
   acute phase, as systemic manifestation of inflammation,
   678
   Bence-Jones, in plasma, and
   proteinuria, 205
   binding, in radioligand assay,
   359
   body, maintenance of homeostasis of, 195-214
   breakdown of, in starvation,
   347
   Cl esterase inhibitor, 715
   complement, characteristics
   of, 712, 752t
   contractile, coupling of, 17
   C-reactive, 678
   dietary, deficiency of, and
   erythropoiesis, 503
   digestion of, intraluminal, 437
   surface, 437
   11S, 713
   excessive loss of, in urine,
   effects of, 211-213
   hormones, 357-358
   blood levels and transport of,
   measurement of, 358-359
   plasma concentrations of,
   359t
   in pubescence, 361
   iron-binding, 505
   isotopic labeled, factors affecting studies of, 208t
   in lung, 173
   -M, of streptococci, 675
   metabolism of, disturbances
   in, in liver disease, 469
   in muscle, 339
   plasma, catabolism of, 210
   distribution of, 208
   excretion of, in exercise,
   202-203
   glomerular sieving of,
   195-199
   high levels of, proteinuria
   associated with, 204-205,
   205t
   isotopic labelling of, 207, 208
   kinetics of, 207
   metabolism of, 207-211
   studies of, precautions
   concerning, 207-208
   radioactivity decay curve
   in, *209*

Protein(s) (*continued*)
in nephrotic syndrome, 212
secretion into gastrointestinal
tract, 210
in thrombin formation, 528
transudation of, 431
plasma-binding, alteration in,
315-316
secretion of, pulmonary,
173-174
sieving coefficient for, 195-197
in starvation, 345, 347
synthesis of, cardiac, 9, 13,
16
defective, 17-18
pulmonary, 173-174
in red cell, 497-498, *498*
Tamm-Horsfall, urinary excretion of, 202
tubular absorption and secre-
of, 199-204
tubular reabsorption of, 201
urinary, concentration of,
methods in, 280t
excretion of, measurement of,
272-273
methods in, 279t
normal, 201-202
size of, in disease states, 206,
207
measurement of, methods
in, 282t
specific, measurement of,
methods in, 281t
Y, in bilirubin uptake, 457
Z, in bilirubin uptake, 457
Proteinosis, alveolar, surfactant
inadequacy in, 173
Proteinuria, in chyluria, 207
in disease, types of, 204, 204t
evaluation of, 272-273, 279t,
280t, 281t, 282t
and exercise, 202-203
orthostatic, 203-204
postural, 203-204
mechanism of, 204
prerenal, 204-205, 205t
tubular, 206-207
with abnormal renal tubular
function, 206-207
with high levels of plasma proteins, 204-205, 205t
with increased glomerular permeability, 205-206
Proteolysis, and lung disease,
173-174
Prothrombin, 528
activation of, in lung, 175
activity of, in serum, test of,
531
consumption, test of, 531
-converting principle, 528
deficiency of, 534
time, one-stage, test of, 531

Pseudohermaphroditism, in congenital adrenal hyper-
plasia, 294
Pseudohypoparathyroidism, and
hypoparathyroidism,
compared, 332
Pseudoxanthoma elasticum, 543
PTA, 529
deficiency of, 534
Puberty, delayed, and hypogonadotropic hypogonadism,
in men, compared, 363
in women, compared, 367
precocious, with hypothyroidism, 314
protein hormones in, 361
testis in, 361-364
Pump(s), calcium, 9
cardiac, 5, 16
in cardiac failure, 14
performance of, 9-11
sodium, 9
and cellular potassium, 325
stimulation of, 263
Purkinje fibers, excitability of,
87
system of, 7
Purpura, allergic, 544
anaphylactoid, 544
fulminans, intravascular clotting in, 536
Henoch-Schönlein, 544
idiopathic thrombocytopenic,
539-542
senile, 543-544
simple, 544
thrombopathic, 543
PV/IF ratio, 49
Pyelonephritis, chronic, 239
Pyloroplasty, 400
Pyrogen, endogenous, 677-678
Pyruvate, formation of, in anaerobic glycolysis, 337
in hepatic metabolism,
341-342
lactate ratio, measurement of,
162-163
in muscle metabolism, 339

Quinidine, in atrial fibrillation,
96
and cardiac impulse conduction, 86
and cardiac impulse generation, 85

Radiobromine, in measurement
of extracellular fluid, 49
Rapaport-Luebering shunt, in
metabolism of erythro-
cytes, 507
Raynaud's phenomenon, etiology of, 29

treatment of, 29-30
Rebuck skin window technique,
675
Receptors, alpha-adrenergic,
392
beta-adrenergic, 392
hyperdynamic responsiveness
to, in labile hypertension,
32, 45
cholinergic, muscarinic, 392
nicotinic, 392
cytosol steroid-binding, 359
drug, 391-392
lung irritant, 120
type J, 120
Rectum, defecation reflex of,
623-624
inhibition of, 624
Re-entry of cardiac impulses.
See Impulses(s), cardiac,
re-entry of
Reflex(es), arc, segmental, 634
arterial mechanoreceptor,
603-604
cardiovascular, control of,
603-608
chemoreceptor, 116, 605-606
control of breathing, 119-120
coordinations, interlimb,
576-577
segmental, 575-576
cough, 120
deep tendon, 632, 634
defecation, 623-624
inhibition of, 624
disorders of, in pain, 583, *584*
diving, 606-607
gastric motility controlled by,
618-620
gastrocolic, 393, 404, 623
gastrointestinal, and contraction of smooth muscle,
393
Hering-Breuer inhibito-inspiratory, 120
inflation, 120
intestino-gastric inhibitory,
central nervous control
of, 618-619, *619*
intestino-intestinal inhibitory,
central nervous control
of, 621-622, *619*
inverse myotatic, 634
irritant receptor, 120
muscle spindle, 114
type J receptor, 120
Refractory period, cardiac,
abbreviation of, 86
alterations of, in experimental atrial flutter, 93-94
Renin, activity of, estimation of,
55-56
in plasma. *See* Plasma, renin
activity in

angiotensin system, in tissue
 perfusion, 26
physiological relationships of,
 in hypertensive patients,
 59-60
plasma. *See* Plasma, renin
 activity in
production of, autonomous
 excess, 330
decreased, in primary aldosteronism, 329-330
increased, in Addison's disease, 328
in cirrhosis, 463-464
release of, 56-57
in renal pressor system, 55
source of, 56
Reproduction, endocrine mechanisms of, 355-373
Resistance, pulmonary. *See*
 Lung(s), resistance of
total peripheral, 24
 as determinant of arterial
 pressure, 68, *69*
calculation of, 43
factors influencing, 68
relationship to arterial pressure and cardiac output,
 42-43
vascular, 23-24
definition of, 23
factors determining, 27-28
mechanoreceptor reflex influencing, 604
pulmonary, in respiratory insufficiency, 135
Respiration, "ataxic," 122
centers of, 111
medullary, 112-113, *113*
pneumotaxic, 112, *113*
pontine, 112, *113*
Cheyne-Stokes, in circulatory
 disorders, 121-122
with central nervous system
 disorders, 121
circulatory influence of, 609
control of, 111-126
automatic systems in, 112-119
behavioral, 111
chemical, 115-119
unified concept of, 118-119
disturbances in, 589-591
metabolic, 111
neural, 112-114, 589, 590
and pulmonary hydrogen
 loss, 327
reflex, 119-120
spinal level, 113-114
effect of various lesions on,
 121t
effects of central nervous system disease on, 120-125
forebrain lesions as cause of
 abnormal, 120

function of, assessment of,
 136-142
and heart rate, 609
insufficiency of, arterial blood
 gas findings in, 134-135,
 135
development of, mechanism
 of, 134, *134*
phases of, 135
hypoxia in, 135
pulmonary vascular resistance in, 135
rate of, and mechanical work
 of, 130-131
rhythm of, generation of, 113
rhythmic, 111, 113
work of, 129-131
and rate of, 130-131
Respiratory system, compliance
 of, measurement of, 127
Reticulocyte(s), 489
count, 492
Reticuloendothelial system,
 511-512
Rhythm(s), cardiac, idioventricular, 85
junctional, 85
neural influences on, 610-612
parasystolic, 85
reciprocal, 88-90, *89*
respiratory, generation of, 113
sinoventricular, 86
Rib(s), cervical, syndrome of,
 72, 604
Rigidity, alpha, 636-637
decerebrate, 636
extrapyramidal, 637
from failure of local inhibition
 of motor neuron firing, 637

Salt(s), bile, in gastric mucosal
 barrier destruction, 417
depletion, renal, 221
excretion of, exaggerated, in
 hypertension, 54-55
retention of, mechanisms in,
 191-193
Saluresis, agents causing, 224
Sarcolemma, 7, 8, 9
Sarcomeres, composition of,
 8, *8*
Sarcotubular system, 8-9, *8*, 13,
 17
Scalenus anticus syndrome, 72
Schistosomiasis, and portal
 hypertension, 460
Schwartzman reaction, 36
Scleroderma, 35
esophageal motor disorder in,
 398
Sclerosis, amyotrophic lateral,
 esophageal dysfunction
 in, 398

fasciculations in, 650
Mönckeberg's, 33
multiple, esophageal dysfunction in, 398
systemic, progressive, esophageal motor disorder in,
 398
Scurvy, 491
lack of vascular support in, 543
Secretin, 386
in fat digestion, 439
release of, 414
secretory effects of, 408-409
test, 386
Secretion, gastric. *See* Stomach, secretion(s) of
pancreatic. *See* Pancreas,
 secretion(s) of
Secretory immune system, 179
Seizure(s), epileptic, 647-648
grand mal, 586-587
limbic, 587-588
and neocortical focal seizures,
 compared, 586
neocortical focal, and limbic
 system seizures, com-
 pared, 586
petit mal, 587
Sensory system, analysis and
 synthesis in, 574
analytic functions of, 580
animal, centrifugal controls of,
 591
embryological organization of,
 577
signals of, conditionable control of, 591
individual experience in, 592
spinal and brain stem, compared, 578, *578*
Sepsis, severe, 536
Serotonin, 6
in lungs, 177
and vascular permeability,
 671-672
Serum sickness, as arthus type
 pulmonary reaction, 180
Sham rage, 601
Sheehan's syndrome, 298
Shock, anaphylactic, 78-79
arrhythmias in, 15
blood coagulation in, 36
blood volume decrease causing, 80-81
cardiogenic, 79
endotoxin, 717
venous pooling in, 79
hemorrhagic, 78, 81
hypotensive, 25
and intestinal circulation, 628
and kallikrein-kinin reactions, 176
myocardial, 79
prolonged, 81

Shock (*continued*)
role of veins in, 79-80
septic, 78
spinal, 589
arterial pressure in, 77-78
and systemic hypotension,
78-79
transcapillary exchange in, 80
traumatic, 79
Shunt, cardiac, 10, 13
acquired, 15
congenital, 15
hexose monophosphate, in
metabolism of erythro-
cytes, 507
portacaval, and ascites, 462
Rapaport-Luebering, in metabolism of erythrocytes, 507
Shy-Drager syndrome, orthostatic hypotension in, 29
Sickle-cell disease, impaired
urinary concentrating
ability in, 240-241
Sieving, coefficient, in measurement of glomerular filtration, 195-197
phenomenon of, 197, *196*
diffusion theory of, 198
pore theory of, 197-198
Sjogrens syndrome, 239-240
Sodium, in absorption of amino
acids and sugars, 428-429
absorption of, in intestinal
tract, 432
pathway of, *433*
balance, effects of, on renal
acid excretion, 262
maintenance of, 218-220
body, control of, 324-325
filtration of, 324
-calcium competition at tubular membrane, 331
in cardiac coupling, 9
in countercurrent multiplier
urinary concentrating
mechanism, 236, *236*
cyanide, toxication, cardiac
failure in, 14
delivery, as influence on potassium secretion, 228
depletion of, 221-222
administration of sodium in,
222
amounts required in, 222
administration of water in,
222
causes of, 221, 221t
renal response to, 221-222
water intake in, 222
excess of, in hypernatremia,
227
excretion of, in ascites, 461,
462
factors determining, 219-220

filtered load of, 219
factors influencing, 324, *323*
intake and extrarenal loss of,
218
in intestinal transport process,
428-429
loss of, in Addison's disease,
327-328
in measurement of extracellular fluid, 49
metabolism of, altered, 220-227
normal, 215-220
in monosaccharide absorption,
435
and plasma renin activity, 57,
59
-potassium or -hydrogen exchange in primary aldo-
steronism, 329
"pump," 9
and cellular potassium, 325
stimulation of, 263
renal tubular reabsorption of,
219-220, *219*, 324-325
in Addison's disease, 327-328
agents inhibiting, 223-224,
244
and calcium reabsorption, 331
control of, "third factor" in,
324
impaired, 223-224
urinary concentrating ability
in, 239
urinary diluting ability in,
244
in primary aldosteronism, 329
stimulation of, 263
repletion of, procedure for,
222
retention of, 222-223
transport of, in diluting segments of nephron, 242
decreased, 244
mechanisms of, 432
Solution, buffered, definition
of, 250
Solvent drag, 430
Spasticity, 636
Spermatogenesis, 362-364
Spherocytosis, hereditary, 509
Sphincter, esophageal. *See*
Esophagus, sphincter of
of Oddi, 401
Spinal cord, bulb of, 577
respiratory and cardiovascular centers in, 589
cellular organization of, 577
disorders of, and muscular
weakness, 638
intermediolateral cells of, 602
in respiratory and cardiovascular regulation, 589-590
respiratory control in, 113-114
reticular formation of, 580

transection of, and arterial
pressure, 77-78, 589
Spironolactone, in distal potassium loss, 230
in sodium reabsorption inhibition, 224
Spleen, erythrocyte destruction
by, 511-512
"pitting" function of, 487
sequestration in, 512
structure of, 511-512
Sprue, malabsorption of vitamin B$_{12}$ in, 438
malabsorption of vitamin K in,
535
nontropical, celiac, 424-425,
430
nontropical, fat malabsorption
in, 442
Staphylococcal clumping test,
551
Staphylococcus aureus, in clotting, 530
Starch, components of, 434
Starvation, calories during,
sources of, 345, 347
hepatic adaptation in, 345
hormonal control mechanisms
during, 345-347
metabolic interrelationships
during, 345, *346*
phagocytosis depression by,
178
protein breakdown during,
347
Steatorrhea, 450
in bile acid malabsorption,
448, *446*
following vagotomy, fat malabsorption in, 442-443
in lipase deficiency, 441
in pancreatic deficiency, 420
in peptide-secreting adenomatosis, 412, 413
post-gastrectomy, fat malabsorption in, 442
Stein-Leventhal syndrome,
368-369
Steroid(s), adrenal, cortical,
and neutrophilia, 523
in immunosuppressive therapy, 743
metabolism of, 293
effect of thyroid hormones
upon, 311t
carbohydrate-active, and potassium loss, 325
and sodium filtration, 324
gonadal, 356-357
biosynthesis of, 356, *357*
blood levels and transport of,
measurement of, 358-359
conversion of, into hormones,
356

mechanism of action of, 359-360

plasma concentrations of, 359t

metabolism of, in feto-placental unit, 369-371

sodium-retaining, and tubular reabsorption of sodium, 324-325

Stomach, atrophy of, 418, 419

autoregulatory escape phenomenon of, 627

blood flow in, parasympathetic control of, 626

pathophysiological aspects of, 628

reflex and central nervous control of, 627-628

sympathetic control of, 626-627, 626

contents of, acidity of, 414

contractions of, 616

dilatation of, acute, 620-621

distention of, effects of, 408

divisions of, 398

dumping syndrome of, 400, 620

emptying of, gastric slow wave in, 399-400

net, factors influencing, 399

peristaltic contractions in, 399, 616

rate of, 414

disturbances of, pathological conditions associated with, 620-621

terminal antral contractions in, 616

tonic contractions in, 399

filling of, 399

hyperplasia of, 413

hypertrophy of, 413

motility of, parasympathetic control of, 617-618, 617

pathophysiological aspects of, 620-621

reflex control of, 618-620, 617, 618, 619

sympathetic control of, 618, 618

motor disorders of, 400-401

mucosa of, blood flow of, 418

mucosal barrier of, 414-415

back diffusion of acid through, consequences of, 415, 415

hemorrhage associated with, 417-418

cell desquamation and regeneration in, 416

destruction of, 416-417

permeability of, 415-416

mucus of, 419

muscle of, 389

pepsin in, 419

relaxation of, neural control of, 617, 618

receptive, 619

secretion(s) of, acid, normal variation in, 411, 412

augmented histamine test of, 419-410

and basal gastric secretion, compared, 410t

basal, 409

and Histalog-stimulated acid secretion, in normal and ulcer patients, compared, 410-411, 411t

measurement of, 409

components of, 407

control of, 407-409, 408

dysfunction of, hormonal imbalance in, 409

nervous regulation deviations in, 409

intrinsic factor in, 418-419

and maximum secretory capacity measurement, 409

mechanisms of, 407-422

overnight, in normal patients, 411, 412t

stimulated, 409-410

tests of, 410, 476-477

Stool, fat, test of, 480

weight, test of, 480

Streptococcus(i), M-protein of, 675

Streptokinase, 548

in thrombus dissolution studies, 552-553

Stress, and adrenal insufficiency, 296-297

as factor in release of ACTH, 288

emotional, colonic disorders associated with, 629

intestinal motility disturbance in, 622, 624

hemodynamic responses to, in hypertension, 48

and ischemic heart disease, 18

Stroke volume, 9-10

Struma lymphomatosa, goiterous. See Hashimoto's thyroiditis

Stuart factor, 528

deficiency of, 534

Succus entericus, 425

Sucrase activity of, 434

-isomaltase deficiency, congenital, 435

Sucrose, in measurement of extracellular fluid, 49

Sugar(s), malabsorption of, 436

Sulcus, central of cerebrum, 577

limitans, 577

Sulfate, in therapeutic diuresis, 224

Sulfhemoglobinemia, 32

Surfactant, alveolar, formation of, 167, 173

function of, 173

inadequacy of, situations associated with, 173

main component of, 172, 173

maintenance of, 173

Swallowing, 394-396

buccopharyngeal phase of, 394

afferent nerves in, 395

Sympathectomy, arterial pressure in, 77

Synapse en passant, 61, 608

Syncope, in complete atrioventricular block, 7

decreased cerebral perfusion as cause of, 28

heat exposure as cause of, 77

posture change as cause of, 76-77

vasodepressor, vasodilation in, 29

vasovagal, 30

Sydenham's chorea, 644

Synthesis, as motor function of nervous system, 580-582

in motor and sensory mechanisms, 574

Tabes dorsalis, orthostatic hypotension in, 604

Tachycardia, 15

paroxysmal, atrial, 15

supraventricular, development of, 90

experimental, electrical conversion of, 97-98

intervention in, 90

sinoatrial reciprocation causing, 90-91

reciprocal, development of, 90

intervention in, 90

Tamm-Horsfall protein, urinary excretion of, 202

Tamponade, pericardial, 14

Temperature, body, as influence on hemoglobin affinity for oxygen, 148-149

neural regulation of, 607

environmental, changes in, cardiac output in, 10

vascular responses to, 607

Tension-time index, of myocardial tension, 12

Testicular feminization, syndrome of, 360-361

Testis(es), adult, functions of, 361-362

fetal activity of, 360-361

in pubescence, 361-364

Testosterone, 356
and erythropoiesis stimulation, 494
mechanism of action of, 360
metabolites of, 356
origin of, in normal women, 369
plasma levels of, 362
production of, 361
rates of, in women, 369t
secretion of in ovarian virilizing syndromes, 368
synthesis of, *357*
Tetany, 648-649
Tetrahydrofolate, 502
Tetrazolium, reaction, principle of, in pulmonary histochemical technique, 167, *168*
Thalamus, and reticular neurons, embryological emergence of, 573
Thalassemia(s), 457, 501
Thiamine, deficiency of, in Beriberi, 19
and deficient tissue oxygenation, 31
Thiazides, in sodium reabsorption inhibition, 244
Thiosulphate, in measurement of extracellular fluid, 49
Third factor, 51, 55, 260-261
in maintenance of sodium balance, 220, 324
in proximal reabsorption of sodium in primary aldosteronism, 329
Thirst, in diabetes insipidus, 321
factors influencing, 319
stimulation of, 238, 320
Threshold potential, cardiac, 83-85, *84*
Thrombasthenia, 543
Thrombasthenin, 538
Thrombin, action of, on platelets, 538
in fibrin formation, 527-528
formation of, 175
extrinsic pathway for, 528-529, *528*
intrinsic pathway for, 529-530, *529*
time, test of, 531
Thrombocythemia, 542
Thrombocytopenia, diagnosis of, 539
mechanisms responsible for, 539-542, 540-541t
symptoms of, 539
Thrombocytosis, 542
Thromboembolism, 14-15
in pulmonary disorders, 175
Thrombopathia, 543

Thromboplastin, antecedent, plasma, 529
deficiency of, 534
time, partial, test, 530-531
tissue, 528-529
intravenous injection of, experimental, 535
Thrombus(i), development of, and fibrinolytic mechanism, 556
dissolution of, plasminogen activators in, 552-553
in microcirculation, 554
in systemic circulation, 554
Thymidine, tritiated, in leukocyte reserve studies, 521
Thymocytes, 691
Thymus, disorders of, and myasthenia gravis, 639
lack of, immunodeficiency associated with, 695-696, 697
of primitive vertebrates, aves and normal human, compared, 704, *705*
role of, in cell-mediated immunity, 691-694
Thyrocalcitonin, 301
in calcium balance maintenance, 330
control and actions of, *332*
excess of, in hyperparathyroidism, 333
in medullary carcinoma of thyroid, 333-334
secretion of, control of, 331
Thyroglobulin, biosynthesis of, 305
Thyroid, adenocarcinoma of, metastatic follicular, 316
blood supply of, 304
carcinoma of, and Cushing's syndrome, 317
medullary, 316, 333-334
parathyroid hormone excess in, 333-334
thyrocalcitonin excess in, 333
metabolic derangements associated with, 316-317
pheochromocytoma in, 316
prostaglandin secretion in, 317
effect of, on elevated serum thyroid-stimulating hormones, 313
embryonic development of, 304
follicular cell of, ultrastructure of, *305*
function of, 301
in disease states, 380-381t
level of, in disease states, 307t
tests of, 375-379

hormones. *See* Hormone(s), thyroid
hyperfunctioning hyperplastic, 310
iodide concentration by, 304
iodide transport defect of, 304
iodine storage by, 306
nodule(s) of, hyperfunctioning, 310-311
-pituitary axis, regulatory mechanisms of, 301-318
stimulator, long-acting, activity of, 309
in hyperthyroidism, 309
tumor of, calcitonin-secreting medullary, 316
Thyroidectomy, thyroid-stimulating hormone secretion following, 303
Thyroiditis, Hashimoto. *See* Hashimoto's thyroiditis
Thyrotoxicosis, 19-20, 640-641
Thyrotropin. *See* Hormone(s), thyroid-stimulating
Thyroxine, 301
biosynthesis of, 304-306
congenital defects in, 307t
circulating, 308
excessive, 641
metabolism of, abnormal, 315
kinetics of, 308, 309t
synthesis and secretion of, decreased, 312-314
excess, 308-312
in thyroid-stimulating hormone secretion, 303
Tissue(s), adipose, 343-344
metabolism of, fasting state, 344, *343*
fed state, 343-344, *343*
connective, disorders of, hereditary, 543
leukocytes in, 522
oxygen exchange in, anatomical shunting influencing, 154-155
diffusion limitation influencing, 153-154
and in lungs, compared, 153, 154-155, *154*
physiological shunting in, 154
oxygen storage by, 155
oxygenation deficiency of, causes of, 31-32
perfusion of, 23-39
autoregulation of blood flow in, 30-31
defective, neurogenic factors in, 28-30
factors controlling, 23-24
metabolic factors in, 31-32
vascular smooth muscle contractions determining, 24-32

thromboplastin, 528-529
Torticollis, spasmodic, 645
Transferase, glucuronyl, deficiency of, 458
inhibition of, 457-458
Transferrin, plasma, in iron distribution, 505
in nephrotic syndrome, 212
Transplantation, 733-738
and autoallergic mechanisms, 731-745
histocompatibility testing in, 736, 753t
history of, 733
immunological adaptation in, enhancement influencing, 737-738
graft adaptation as factor in, 738
tolerance as factor in, 738
and immunosuppressive therapy, 742-744
rejection in, acute, 734-736, 735
causative factors in, 734
intensity of, 734-736
symptoms of, 734
chronic, 736, 735
hyperacute, 733-734, 735
causes of, 733-734
renal allograft rejection as example of, 726
structural changes associated with, 735
Transport, active, definition of, 429
and facilitated diffusion, 429-430
processes in, 428-429
Tremor(s), action, 645
ataxic, 646
cerebellar, 646
parkinsonian and essential, compared, 646-647
definition of, 645
essential, 645
familial, 645
intention, 646
parkinsonian, 646
postural, 645
static, 654
Tremorine, 645-646
Triamterene, in sodium reabsorption inhibition, 224
Triglyceride(s), absorption of, 439, *439*
in fat digestion, physical and chemical states of, *440*
medium-chain, 443
plasma, in nephrotic syndrome, 212
resynthesis of, 442
Triiodothyronine, 301
circulating, 308

hyperthyroidism, 311
metabolism of, kinetics of, 308, 309t
synthesis and secretion of, decreased, 312-314
excess, 308-312
in thyroid-stimulating hormone secretion, 303
Tropinin, 9, 17
Trypsin, pancreatic, in clotting, 530
Tryptophan, malabsorption of, 437
Tuberculosis, inappropriate antidiuretic hormone secretion in, 226
mycobacterium, 180
Tubule(s), renal, abnormal function of, and proteinuria, 206-207
acute necrosis of, impaired concentrating ability in, 240
distal, in Addison's disease, 328
in primary aldosteronism, 329
hydrogen ion secretion in, 259
maximum transport capacity of, measurement of, 271-272
protein absorption and secretion by, 199-204
proximal, in Addison's disease, 327
albumin concentration in, 195
in glomerular disease and proteinuria, *200*
hyaline droplets of, 199-201, *200*
in primary aldosteronism, 328-329
reabsorption by, 190
of filtrate, 51
of protein, 201
of sodium, 219-220, *219*, 324-325
of water, 319
Turbulence, in cardiac impulse propagation, 95
Tyrosine, synthesis of, 61

Ulcer(s), duodenal, acid secretion in, 410, 411t
causative agents in, 628
gastric, acid secretion in, 410 411t
causative agents in, 628
development of, 415-416
peptic, causative agents in, 628
development of, 415-416

"stress," 409
Ultrafiltrate, glomerular, 189
Urea, and ammonia, 465
urinary excretion of, 189-190
in urine concentrating process, 237
Uremia, 543
development of, 190
hypertensions in, types of, 47
terminal, plasma renin activity in, 59, *46*
Uric acid, serum, levels of, elevation of, 192
Urine, acid eliminated in, net quantity of, 257
acid-base measurements in, 276-277
acidification, tests of, 276-277
albumin in, 197, 204, 206
ammonia secretion in, 257
ammonium content of, measurement of, 276
bicarbonate in, measurement of, 276
clearance formula for, 234
clearance of metabolic waste products by, 189-190
concentrated, definition of, 233
two portions of, 234
concentration of, in absence of antidiuretic hormone, 321
maximal, factors influencing, 274
normal values in, 274
tests of, 273-274
measure of, 233
mechanism of, countercurrent exchanger, 237
countercurrent multiplier, 236-237, *236*
impaired, consequences of, 238-239
diseases associated with, 239-241
pathophysiology of, 239
normal, 235-237
physiological variables influencing, 237-238
urea entrapment, 237
pathophysiology of, 235-241
dilute, definition of, 233, 241
dilution of, maximal, tests of, 274
mechanism of, impaired, consequence of, 243
diseases associated with, 244-246
pathophysiology of, 243-244
normal, 241-243
physiological variables influencing, 243
pathophysiology of, 241-246

Urine (*continued*)
elaboration of, in kidneys, 189
excretion of, isosmotic concentrating ability during, *235*
regulation of, 189-190
formation of, damaged nephrons in, 190
hydrogen ion concentration in, 256
isosmotic, volume of, determination of, 234
net acid excretion in, measurement of, 276-277
normal values for, 277
normal, proteins in, 201-202
osmolality of, 233
antidiuretic hormone secretion influencing, 319
low, in diabetes insipidus, 320-321
measure of, 233
osmolar clearance of, 234
pH in, measurement of, 276
normal values for, 277
-to-plasma osmotic ratio, 233
proteins in. *See* Protein(s), urinary; Proteinuria
sodium in, 219-220
hypertension-exaggerated, 54-55
postural, 238
specific gravity of, measurement of, 273
specimens of, collection of, 269
titratable acid excretion in, 257
total acidity of, measurement of, 276
total carbon dioxide content of, measurement of, 276
water clearance in, negative, 234, *235*
solute-free, 234, *235*
water reabsorption in, solute-free, 234
Urinometer, use of, 273
Urobilinogen, 512
Urokinase, 548
action of, agents inhibiting, 555
in thrombus dissolution studies, 552-553
Uromucoid, urinary excretion of, 202

Vacuole, phagocytic, 676, *676*
Vagotomy, gastric emptying following, 620
and pyloroplasty, gastric dysfunction associated with, 400

Valsalva maneuver, neural circulatory control in, 61
Valve(s), aortic, stenosis of, 13
cardiac, 6
insufficiency of, 13
of Heister, 401
mitral, stenosis of, 14-15
tricuspid, stenosis of, 14
Vanillylmandelic acid, 61-63
Vasa recta, maintenance of medullary hyperosmolality by, 237
Vasoconstriction, arteriolar, and central nervous system, 600
cerebral systems influencing, 590
hypoxic, 165
Vasodilation, cerebral systems influencing, 590
in orthostatic hypotension, 29
in reduction of arterial pressure, 68, 73
sympathetic nerous system in, 27-30
various systems in, 30
Vasodilator(s), influence of, on blood pressure, 75
origin of ascites formation, 464
Vasomotor activity, sympathetic, physiological relationships of, in hypertensive patients, 64-65
Vasopressin. *See* Hormone(s), antidiuretic
Vein(s), blood return by, 10, 11, 14
blood volume in, 49-50
constriction of, 79
function of, capacitance, 70, 79, 600
resistance, 70
oxygen tension of, measurement of, 147
pooling in, 79
portal, hypertension in, 459-461
pressure in, 14
pulmonary, anomatous insertion of, 15
constriction of, 14-15
in regulation of arterial pressure, 70
shunt, 600
Velocity of shortening, of myocardium, 12
Vena cava, obstruction of, 14
Venom(s), cobra, factor, and complement system, 717-718
snake, clot-promoting, 530

Ventilation, alveolar, 131-132
impaired, 258
dead-space-like, 133
distribution of, 132, *132*
factors influencing, 127-131
maximum voluntary, test, 138
-perfusion relationships, 133, *132*
tests of, 133
Ventricle(s), conduction pathways of, 85-86
dilatation of, 13
end-diastolic volume of, 12-13
13
end-diastolic volume of, 12-13
extrasystoles in, 12, 15
fibrillation of, 15
function of, in cardiac hypertrophy, 13
left, 6-7
inlet disorders of, 14-15
outlet disorders of, 15
premature contractions of, 96-97
right, 6-7, 14
septum of, 6-7
defect of, 15
Virilization, in androgen-producing adrenal tumors, 295
in congenital adrenal hyperplasia, 294
ovarian, syndromes of, 368-369
precocious, atypical, 361-362
and secondary amenorrhea, 368
Virus(es), attenuated, used in immunization, 686t
herpes, parasitism by, 684-685
and host, mutual adaptations of, 684-685
infection, persisting, and immunodeficiency, 701
neutralization of, C3 component of complement in, 716-717
Viscera, control mechanisms of, 577-578, 653
function of, control of, neural, 589-591, 653-655
function of, regulation of, 578, 589-591
responses of, instrumental learning of, 657-659
Viscosity, in reduction of arterial pressure, 68
Vital capacity, in determination of elastic resistance, 136
forced, in determination of nonelastic resistance, 136-138
Vitamin(s), B, absorption of, 419

B-6, as influence on hemoglobin synthesis, 503
B₁₂, absorption of, 438, 501
impaired, 502
deficiency of, 491
and anemia, factors caus-
ing, 501-502
evaluation of, 503
in heme and globin synthesis,
501, *498*
malabsorption of, 438
C, deficiency of, 491
D, in calcium absorption, 330,
438-439
fat-soluble, absorption of, 444
K, impaired absorption of,
diseases associated with,
535
restriction of dietary intake
of, deficiency as result
of, 534-535
in synthesis of plasma clotting factors, 534
metabolism of, disturbances
in, in liver disease, 470
V$_{max}$, as measurement of myocardial contractility, 12
Volume, cardiac, 10
residual, 13
extracellular, fluid. *See*
Fluid(s), extracellular,
volume of
intracellular, potassium in
maintenance of, 191
intravascular, and antidiuretic
hormone secretion, 320
decrease in, renal disorders
associated with, 191
and plasma renin activity, 54,
54, 57, 59
in reduction of arterial pressure, 70
in relation to neurogenic
tone, 53

and vascular system, 49-50
mechanisms, 48-55
pulmonary. *See* Lung(s),
volume(s)
stroke, in cardiac output,
9-10, 11, 15
ventricular end-diastolic,
12-13
Vomiting, of gastric juice, renal
tubular response to, 262
mechanism of, 400-401
Von Willebrand's disease,
532-533
Von Willebrand-Jurgen's syndrome, 543

Waldenström's macroglobulinemia, altered viscosity in,
35
Water, absorption of, and solute absorption, 433, *433*
body, 319-324
balance, control of, 27, 218
"free," excretion of, 323-324
altered, disease states
causing, 323-324
loss of, 218
as result of impaired concentrating ability, 238-239
metabolism of, abnormal, in
Addison's disease,
322-323
altered, 220-227
normal, 215-220
reabsorption of, renal,
319-322
sources of, 218
total, measurement of, 216
volume of, control of, 322-323
normal, 215, *216*
See also Fluid(s), body
chronic drinking of, impaired
urinary concentrating
ability in, 241

excretion of, in ascites, 461,
462
intake of, 319
in intestinal transport process,
428, 430
intoxication, 226, 243, 245,
246
metabolism of, endocrine control mechanisms of, 319-
334
movement of, passive, 433-434
osmotic flow of, movement of
solutes by, 430
in treatment of diabetes insipidus, 321
Waterhouse-Friderichsen syndrome, 536
Wegener's granulomatosis, 180
Weight, body, gain in, 243
Wilson's disease, copper deposition in, 470
Wiskott-Aldrich syndrome,
698-699
herpes simplex lesions in, *699*
Wolff-Parkinson-White syndrome, 90

Xenografts, and hyperacute
rejection, 733
ᴅ-Xylose, in intestinal function
test, 436, 480

Yoga trance, as example of neural influence on cardiovascular activity, 74

Zenker's diverticulum, 398
Zollinger-Ellison syndrome. *See*
Adenomatosis, peptide-
secreting